COMPUTATIONAL TECHNIQUES AND APPLICATIONS: CTAC-83

COMPUTATIONAL TECHNIQUES AND APPLICATIONS: CTAC-83

Proceedings of the 1983 International Conference on
Computational Techniques and Applications
held at the University of Sydney, Australia

Edited by

JOHN NOYE

Associate Dean
Faculty of Mathematical Sciences
The University of Adelaide
South Australia
Australia

and

CLIVE FLETCHER

Senior Lecturer
Department of Mechanical Engineering
The University of Sydney
New South Wales
Australia

1984

NORTH-HOLLAND – AMSTERDAM • NEW YORK • OXFORD

ISBN: 0 444 87527 1

Publishers:
ELSEVIER SCIENCE PUBLISHERS B.V.
P.O. Box 1991
1000 BZ Amsterdam
The Netherlands

Sole distributors for the U.S.A. and Canada:
ELSEVIER SCIENCE PUBLISHING COMPANY, INC.
52 Vanderbilt Avenue
New York, N.Y. 10017
U.S.A.

Library of Congress Cataloging in Publication Data

International Conference on Computational Techniques and Applications (1983 : University of Sydney)
Computational techniques and applications, CTAC-83.

Bibliography: p.
1. Differential equations, Partial--Numerical solutions --Congresses. 2. Finite element method--Congresses. 3. Finite differences--Congresses. 4. Fluid mechanics--Congresses. I. Noye, John, 1930- . II. Fletcher, C. A. J. III. Title.
QA377.I56 1983 515.3'53 84-6135
ISBN 0-444-87527-1 (U.S.)

PRINTED IN THE NETHERLANDS

PREFACE

Following the very successful 1981 Conference on Numerical Solutions of Partial Differential Equations held at Queen's College, University of Melbourne, it was decided to hold a similar meeting in 1983, namely a Computational Techniques and Applications Conference (CTAC-83) at the Faculty of Engineering of the University of Sydney. This volume represents the proceedings of that conference.

There were seven invited speakers, Professor Ron Mitchell of the University of Dundee (who presented a paper on "Recent Developments in the Finite Element Method"), Professor Maurice Holt of the University of California ("The Changing Scene in Computational Fluid Dynamics"), Dr. Carlos Brebbia from the University of Southampton ("The Simulation of Time Dependent Problems using Boundary Elements"), Dr. Ron Davis of the National Bureau of Standards, Washington, D.C. ("Finite Difference Methods for Fluid Flow"), Dr. Robert Anderssen of the CSIRO Division of Mathematics and Statistics, Canberra ("Mathematical Analysis as the Foundation for Successful Computation"), Dr. Andrew Currie of Compumod Pty. Ltd., Sydney ("Computation in Industry: Achievement and Potential"), and Dr. William Bourke of the Australian Numerical Meteorological Research Centre ("Computational Weather Prediction"). The first six of these are included in the section titled "Invited Papers".

In addition, 73 papers were presented in which recent refinements of various numerical techniques used in applications were described, or experiences in applying these methods to solve real world problems were reported. These papers have been categorised into the following groups: finite difference techniques, finite element methods, boundary integral methods, oceanic and atmospheric problems, viscous flow, porous media, thermal problems, turbulence, acoustics and plasmas, dynamics and structures. They were presented by scientists from many parts of the world, including the United Kingdom, the United States of America, Sweden, Italy, Israel, Japan and New Zealand, as well as Australia.

A special mention must be made of the work done by the committee which organised the conference so efficiently. Thanks also go to the various organisations which assisted financially and otherwise — they are listed separately.

Finally, our personal thanks go to Drs. Arjen Sevenster (Mathematics Editor), John Butterfield (Technical Editor) and Sue Kettle (Promotions Department) of Elsevier Science Publishers B.V. in Amsterdam, for their assistance in arranging the publication of this book.

John Noye
The University of Adelaide

Clive Fletcher
The University of Sydney

COMPUTATIONAL TECHNIQUES AND APPLICATIONS CONFERENCE
CTAC-83

CONFERENCE ORGANISING COMMITTEE

Convenor: Dr. JOHN NOYE
Department of Applied Mathematics
The University of Adelaide
Adelaide
South Australia, 5000
Australia

Secretary: Dr. CLIVE FLETCHER
Department of Mechanical Engineering
The University of Sydney
Sydney
New South Wales, 2006
Australia

Committee: Dr. JOHN ATKINSON
Department of Mechanical Engineering
The University of Sydney
Sydney
New South Wales, 2006
Australia

Dr. FRANK BARRINGTON
Department of Mathematics
The University of Melbourne
Parkville
Victoria, 3052
Australia

Dr. ALAN EASTON
Department of Mathematics
Swinburne Institute of Technology
John Street
Hawthorn
Victoria, 3122
Australia

THE COMPUTATIONAL MATHEMATICS GROUP
DIVISION OF APPLIED MATHEMATICS
AUSTRALIAN MATHEMATICAL SOCIETY

The proposal to form a Computational Mathematics Group as a Specialist Group of the Division of Applied Mathematics was initiated by the Committee who organised the Computational Techniques and Applications Conference (CTAC-83) reported in this Proceedings. CTAC-83 was a follow up to the very successful Conference on the Numerical Solution of Partial Differential Equations held at the University of Melbourne in August 1981, which was based on the successful Conference on Numerical Simulation of Fluid Motion held at Monash University in January 1976. Motivation for the proposal to form a Computational Mathematics Group was the desire of the committee to ensure the long term viability of the CTAC concept.

The first meeting of the Computational Mathematics Group of the Division of Applied Mathematics of the Australian Mathematical Society took place on August 30 during CTAC-83. At this meeting the inaugural committee of the computational Mathematics Group was elected. It consists of:

Chairman: Dr. B.J. NOYE, *Department of Applied Mathematics, University of Adelaide, Adelaide, South Australia, Australia.*

Secretary: Dr. C.A.J. FLETCHER, *Department of Mechanical Engineering, University of Sydney, New South Wales, Australia.*

Treasurer: Dr. J.D. ATKINSON, *Department of Mechanical Engineering, University of Sydney, New South Wales, Australia.*

Committee Members:

Dr. F.R. de HOOG, *CSIRO Division of Mathematics and Statistics, Canberra, Australian Capital Territory, Australia.*

Associate-Professor G. de VAHL DAVIS, *School of Mechanical and Industrial Engineering, University of New South Wales, Kensington, New South Wales, Australia.*

Mr. R. KOHOUTEK, *Department of Civil Engineering, University of Melbourne, Parkville, Victoria, Australia.*

Dr. R.L. MAY, *Department of Mathematics, Royal Melbourne Institute of Technology, Melbourne, Victoria, Australia.*

Ex Officio: Chairman, Division of Applied Mathematics, Australian Mathematical Society (presently Dr. R.S. ANDERSSEN, *CSIRO Division of Mathematics and Statistics, Australian Capital Territory, Australia*).

The major responsibility of the Group will be the organisation of a biennial conference on *Computational Techniques and Applications* which is intended to draw into closer and better communication the different disciplines involved with, and the different users of, computational mathematics.

The next such conference (CTAC-85) will be held in Melbourne, Australia, in August 1985. The Conference Director will be Dr. R.L. May, Department of Mathematics, Royal Melbourne Institute of Technology, Melbourne, Victoria 3000, Australia, and all enquiries concerning CTAC-85 should be addressed to him.

ACKNOWLEDGMENTS

The assistance of the following organisations is gratefully acknowledged:

Centre for Mathematical Analysis, Australian National University
The British Council
Perkin-Elmer Data Systems Pty. Ltd.
Finite Element Research Group, Chisholm Institute of Technology
Ansett Airlines of Australia
Applied Mathematics Division, Australian Mathematical Society

CONTENTS

INVITED PAPERS

Computational Techniques & Applications: CTAC-83
J. Noye & C. Fletcher (Editors)

RECENT DEVELOPMENTS IN THE FINITE ELEMENT METHOD

Andrew R. Mitchell

Department of Mathematical Sciences
University of Dundee
Dundee, Scotland

The Finite Element Method (F.E.M.) has now reached a stage of development where it is being used for the solution of non linear problems where interaction takes place between two entirely different physical phenomena. These problems, both steady and time dependent, are solved by the Petrov Galerkin version of the F.E.M. thus allowing maximum freedom of choice for both trial and test functions. The topics covered in the present paper include conduction-convection in fluids, non linear waves in dispersive media, and reaction-diffusion in mathematical biology.

INTRODUCTION

Finite Element methods have come a long way from their origins in structural engineering in the 1950's. They now pervade most areas of science and engineering where the fundamental problem is the solution of a partial differential equation, whether the latter be steady or time dependent. The aim of this paper is to show some recent developments in finite element techniques and to apply these to the solution of selected problems in engineering, physics, and mathematical biology.

By the term finite element method (Mitchell and Wait (1976)) we shall mean the Galerkin formulation or its generalisation the Petrov-Galerkin method, (Anderssen and Mitchell (1979), Mitchell and Griffiths (1980)). The latter is described as follows: consider the partial differential equation

$$P(u(\underline{x},t)) = 0 \ .$$

Discretisation in space is carried out according to

$$(P(\sum_{i=1}^{n} U_i(t)\phi_i(\underline{x})), \psi_j(\underline{x})) = 0 \qquad j = 1,2,\ldots,n$$

where $(\ , \ )$ is the L_2 inner product and

$$u(\underline{x},t) \sim U(\underline{x},t) = \sum_{i=1}^{n} U_i(t)\phi_i(\underline{x}) \ .$$

The finite dimensional spaces $\{\phi_i(\underline{x})\}$ and $\{\psi_j(\underline{x})\}$ involve piecewise polynomial trial and test functions with compact support. After applying Green's theorem (for second and higher order space derivatives) and integrating element by element, a system of ordinary differential equations in time is obtained for $U_i(t)$, $i = 1,2,\ldots,n$. If the problem is stationary, we obtain n equations in the n unknowns U_i, $i = 1,2,\ldots,n$. Identical trial and test spaces give the Galerkin method.

Consider the equation

$$\frac{\partial u}{\partial t} + q_x \frac{\partial u}{\partial x} + q_y \frac{\partial u}{\partial y} = \epsilon \nabla^2 u \tag{1}$$

where ϵ is a parameter. In fluid problems this may, for example, constitute the Navier Stokes equation for incompressible flow, where

$$u = -\nabla^2 \psi \ , \quad q_x = \frac{\partial \psi}{\partial y} \ , \quad q_y = -\frac{\partial \psi}{\partial x}$$

with ψ the stream function, or alternatively it may represent the convection of the temperature u in a fluid where the latter is moving with velocity (q_x, q_y) due to an external field. The Navier Stokes form of (1) is of course time dependent, multidimensional, and non linear, and the question we ask is "how good are finite element methods for the numerical solution of such problems"?

To answer this question, we start with the simplest possible non-trivial model

$$-\epsilon \frac{d^2 u}{dx^2} + q \frac{du}{dx} = 0 \ , \qquad 0 \le x \le 1, \ \epsilon, \ q \ \text{ positive constants} \tag{2}$$

together with the essential boundary conditions

$$u(0) = 1, \quad u(1) = 0 \ . \tag{3}$$

This problem is steady, one dimensional, and linear and has the theoretical solution

$$u(x) = \frac{e^{(qx)/\epsilon} - e^{q/\epsilon}}{1 - e^{q/\epsilon}} \ . \tag{4}$$

The ratio q/ϵ is called the Peclet number (Pe) and we are interested in the finite element solution of (2) and (3) for large Pe. Note that for natural boundary conditions $u(0) = 1$, $\frac{du}{dx}(1) = 0$, the solution is $u(x) = 1$, and so the difficulty in the problem described by (2) and (3) is caused by the combination of a large value of Pe and an unsympathetic downstream boundary condition.

The unit interval is now subdivided into n equal elements with knots at the points $x = ih$ $(i = 0,1,2,\ldots,n)$. The dependent variable u in (2) is approximated on each element by a piecewise polynomial of degree k $(k = 1,2,3)$ using a Lagrangian basis. With elimination of intermediate nodal values where required, the standard Galerkin method leads to a three point difference formula of the form

$$-[1 + d_k L]\delta^2 U_i + L(U_{i+1} - U_{i-1}) = 0 \ , \tag{5}$$

where

$$d_k = \frac{(1 - 1/L)c_k + (1 + 1/L)}{c_k - 1} \ , \quad (k = 1,2,3) \tag{6}$$

and $L = \frac{qh}{2\epsilon}$ is the grid Peclet number. The solution of (5) is

$$U_i = \frac{c_k^i - c_k^n}{1 - c_k^n} \qquad (k = 1,2,3) \tag{7}$$

and the related values of c_k and d_k are

$$c_1 = \frac{1 + L}{1 - L} \qquad\qquad d_1 = 0$$

$$c_2 = \frac{1 + L + \frac{1}{3}L^2}{1 - L + \frac{1}{3}L^2} \qquad\qquad d_2 = \tfrac{1}{3}L$$

$$c_3 = \frac{1 + L + \frac{2}{5}L^2 + \frac{1}{15}L^3}{1 - L + \frac{2}{5}L^2 - \frac{1}{15}L^3} \qquad d_3 = \frac{\frac{1}{3}L}{1 + \frac{1}{15}L^2} .$$

N.B. c_k is the (k,k) Padé approximant to $e^{2L} \equiv e^{qh/\epsilon}$.

Now (7) and (4) are equivalent if

$$d = \coth L - \frac{1}{L} \tag{8}$$

leading to the result that $d \to 1$ as $L \to \infty$. Comparing d_k $(k = 1,2,3)$ with d in (8) for a range of L we get

TABLE 1

L	$\coth L - \frac{1}{L}$	d_1	d_2	d_3
0	0	0	0	0
1	.31	0	.33	.31
2	.53	0	.67	.52
3	.67	0	1.00	.63
5	.80	0	1.67	.63
10	.90	0	3.33	.43

The above results are poor despite the fact that the methods have a local order of accuracy (small h) of $2k$ $(k = 1,2,3)$. More specifically, for $k = 1$, oscillations appear if $L > 1$. (No dissipation.) $k = 2$, results are meaningless for $L > 2.0$. (Too much dissipation.) $k = 3$, oscillations appear if $L > 2.4$. (Too little dissipation.)

Galerkin finite element methods for conduction-convection problems can be saved, however, by the addition of upwinding using Petrov Galerkin methods. For example in the case where the trial functions $\{\phi_i(x)\}$ are piecewise linears, we choose the test functions to be

$$\psi_j(x) = \phi_j(x) + \alpha\sigma\left(\frac{x}{h} - j\right) \qquad \forall\ j$$

where α is a parameter and $\sigma(s)$ is an odd function of s on $[-1,+1]$ which vanishes outside this interval. Examples of $\sigma(s)$ are

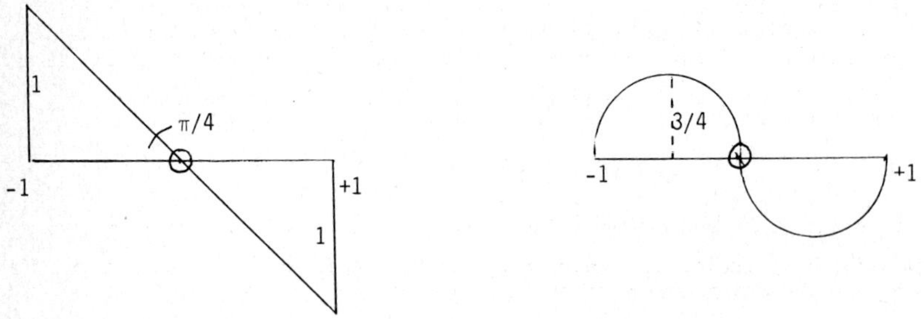

Linear (non-conforming) Quadratic (conforming)

where in both cases $\int_0^1 \sigma(s)ds = -\frac{1}{2}$. This procedure again leads to (5), but this time

$$d_k = \alpha\ .$$

Special cases are

(i) $\alpha = 0$ No upwinding. (1,1) Padé.

(ii) $\alpha = 1$ Full upwinding. (0,1) Padé.

(iii) $\alpha = \coth L - 1/L$ Exact upwinding. $U(x)$ coincides with $u(x)$ at the knots.

Upwinded quadratics and cubics require cubic and quartic perturbations respectively of the test functions. Other ways in which upwinding can be produced are by taking unequal elements or by unsymmetric placement of numerical integration points in equal elements.

From the simple model problem of equation (2) we have identified the principal difficulty in solving conduction-convection problems by the finite element or any other method viz. the amount of diffusion to be added in the direction of the flow. In order to make the model more realistic we consider Burgers' equation

$$\frac{\partial u}{\partial t} + u\,\frac{\partial u}{\partial x} = \epsilon\,\frac{\partial^2 u}{\partial x^2} \tag{9}$$

which introduces time dependence and non-linearity but avoids an increase in the number of space dimensions. Finite element solutions of (9) introduce two new concepts (i) the mass matrix, and (ii) product approximation. These concepts are illustrated by the following three point difference formulae obtained for the solution of (9).

$$L(U_i) \equiv \dot{U}_i + \frac{1}{2h} U_i(U_{i+1}-U_{i-1}) - \frac{\epsilon}{h^2}\delta^2 U_i = 0 \tag{10a}$$

$$G(U_i) \equiv \tfrac{1}{6}(\dot{U}_{i+1}+4\dot{U}_i+\dot{U}_{i-1}) + \frac{1}{6h}(U_{i+1}+U_i+U_{i-1})(U_{i+1}-U_{i-1}) - \frac{\epsilon}{h^2}\delta^2 U_i = 0 \tag{10b}$$

$$\equiv \dot{U}_i + \tfrac{1}{6}\delta^2\dot{U}_i + \frac{1}{2h}(U_i+\tfrac{1}{3}\delta^2 U_i)(U_{i+1}-U_{i-1}) - \frac{\epsilon}{h^2}\delta^2 U_i = 0$$

$$P(U_i) \equiv \tfrac{1}{6}(\dot{U}_{i+1}+4\dot{U}_i+\dot{U}_{i-1}) + \frac{1}{4h}(U_{i+1}+U_{i-1})(U_{i+1}-U_{i-1}) - \frac{\epsilon}{h^2}\delta^2 U_i = 0 \tag{10c}$$

$$= \dot{U}_i + \tfrac{1}{6}\delta^2\dot{U}_i + \frac{1}{2h}(U_i+\tfrac{1}{2}\delta^2 U_i)(U_{i+1}-U_{i-1}) - \frac{\epsilon}{h^2}\delta^2 U_i = 0$$

where (10a) comes from the method of lines, (10b) from the standard Galerkin method with piecewise linear basis functions and (10c) from the linear Galerkin method with the non linear term handled by product approximation which consists of writing $u\,\frac{\partial u}{\partial x}$ in the form $\frac{1}{2}\frac{\partial}{\partial x}(u^2)$ followed by the approximation

$$u^2 = \sum_i U_i^2(t)\phi_i(x)$$

where the basis functions $\{\phi_i(x)\}$ are piecewise linears. One merit of product approximation lies in its increased local accuracy when used to discretise (9) for the case $\epsilon = 0$. Using Taylor expansions, (10c) gives for $\epsilon = 0$,

$$P(U_i) = (1 + \tfrac{1}{6}h^2\frac{d^2}{dx^2})(\dot{U}_i + U_i(\frac{dU}{dx})_i) + O(h^4)$$

and hence the increased local accuracy.

With regard to upwinding by the Petrov-Galerkin method, described earlier for the linear model (2), $G(U_i)$ in (10b), for example, is altered to

$$\bar{G}(U_i) \equiv \tfrac{1}{6}[(1-\tfrac{3}{2}\alpha)\dot{U}_{i+1}+4\dot{U}_i+(1+\tfrac{3}{2}\alpha)\dot{U}_{i-1}]+\frac{1}{6h}(U_{i+1}+U_i+U_{i-1})(U_{i+1}-U_{i-1})-\frac{\alpha}{4h}\delta^2(U_i^2)$$

$$-\frac{\epsilon}{h^2}\delta^2 U_i = 0 \,, \qquad \alpha \text{ is the upwinding parameter.}$$

A final generalisation of the conduction-convection problem takes us to the Navier Stokes equation or Burgers' equation in two space dimensions with the added problem now of not knowing in what direction to add (or subtract) dissipation. Although it is generally accepted that dissipation should not be added crosswind, the overall upwinding problem is far from being resolved and will remain a major research area for some time to come.

Some useful references containing material on finite element methods in conduction-convection problems are Hughes (1979), Mitchell and Griffiths (1979), Christie et al (1981) and Fletcher (1982, 1983).

THE MOVING FINITE ELEMENT METHOD (M.F.E.)

So far in our finite element considerations we have assumed that the elements are equal in shape and size and retain their position in time. This of course is a distinct disadvantage if we are attempting to solve problems with steep fronts, shocks, layers and the like. The ultimate embarrassment is to be unable to solve the simplest of equations

$$\frac{\partial u}{\partial t} + \frac{\partial u}{\partial x} = 0$$

accurately by numerical methods <u>on a fixed grid</u>. As a result we now introduce a relatively new idea for <u>time dependent</u> problems where at each advanced time level the unknowns are the nodal positions as well as the nodal amplitudes. For obvious reasons most progress has been made in one space dimension and so we consider the scalar equation

$$\frac{\partial u}{\partial t} = Au \,, \qquad t \ge 0 \tag{11}$$

where A is an operator involving space derivatives with respect to x only. We introduce a space grid π which is not only non-uniform but is also time dependent, i.e. π is defined as

$$\pi : x_1(t) < x_2(t) < \dots < x_N(t) \,. \tag{12}$$

For fixed boundary problems involving (11), the nodes x_1 and x_N retain their positions with time whereas with pure initial-boundary problems it is more efficient to allow <u>all</u> nodes to change positions.

In the M.F.E. method, we consider a trial solution of the form

$$v(x,t) = v(x\ ;\ a_1(t),\ \dots\ ,\ a_N(t)\ ;\ x_1(t),\ \dots\ ,\ x_N(t)) \tag{13}$$

where $\{a_i(t)\}$ and $\{x_i(t)\}$, the nodal amplitudes and positions respectively, constitute $2N$ unknowns at each time level. Partial differentiation of (13) gives

$$\frac{\partial v}{\partial t}(x,t) = \sum_i (\dot{a}_i(t)\alpha_i(x,t) + \dot{x}_i(t)\beta_i(x,t)) \tag{14}$$

where

$$\alpha_i = \frac{\partial v}{\partial a_i} \,, \qquad \beta_i = \frac{\partial v}{\partial x_i} \,, \tag{15}$$

and a dot denotes the total derivative with respect to the time. In (14), the functions $\{\alpha_i\}$ and $\{\beta_i\}$ are basis functions and for the case of <u>piecewise linear</u>

finite elements, where at a fixed time t,

$$v(x,t) = \begin{cases} a_{i-1} + \dfrac{\Delta a_{i-1}}{\Delta x_{i-1}}(x-x_{i-1}) & x_{i-1} \le x \le x_i \\ a_i + \dfrac{\Delta a_i}{\Delta x_i}(x-x_i) & x_i \le x \le x_{i+1} \\ 0 & \text{otherwise,} \end{cases}$$

we obtain from (15),

$$\alpha_i = \begin{cases} \dfrac{x-x_{i-1}}{\Delta x_{i-1}} & x_{i-1} \le x \le x_i \\ \dfrac{x_{i+1}-x}{\Delta x_i} & x_i \le x \le x_{i+1} \\ 0 & \text{otherwise,} \end{cases} \tag{16}$$

and

$$\beta_i = \begin{cases} -\dfrac{\Delta a_{i-1}}{\Delta x_{i-1}}\alpha_i & x_{i-1} \le x \le x_i \\ -\dfrac{\Delta a_i}{\Delta x_i}\alpha_i & x_i \le x \le x_{i+1} \\ 0 & \text{otherwise,} \end{cases}$$

where Δ denotes the first forward difference. It should be noted that the piecewise linear basis functions $\{\beta_i\}$ are <u>discontinuous</u> at their respective nodes.

We return now to the <u>residual</u>

$$R(v) = \frac{\partial v}{\partial t} - Av \tag{17}$$

which requires to be "minimised" in some sense and at present there are two distinct strategies both resulting in M.F.E. methods.

(1) <u>Least Squares Minimisation</u> (L.S.M.) In mathematical terms this can be expressed as

$$\underset{\alpha_i,\beta_i\ (i=1,2,\ldots,N)}{\text{Min}} \quad \|R\|_{L_2},$$

which leads to the system of $2N$ equations

$$\sum_i (\alpha_j,\alpha_i)\dot{a}_i + \sum_i (\alpha_j,\beta_i)\dot{x}_i = (\alpha_j, Av)$$

$$\sum_i (\beta_j,\alpha_i)\dot{a}_i + \sum_i (\beta_j,\beta_i)\dot{x}_i = (\beta_j, Av)$$

where (,) is the L_2 inner product. A disadvantage of this method is the evaluation of inner products of second derivatives with the discontinuous basis functions $\{\beta_i\}$. However this difficulty can be overcome by a limiting process.

(2) Petrov Galerkin Method (P.G.M.) Here we require $R(v)$ given by (17) to be orthogonal to the Hermite cubic basis functions, i.e.

$$\begin{aligned} (R(v),\ S_j(x)) &= 0 \\ (R(v),\ T_j(x)) &= 0 \end{aligned} \qquad j = 1,2,\ldots,N \tag{18}$$

where

$$S_j(x) = (\alpha_j(x))^2(3 - 2\alpha_j(x)) ,$$

and

$$T_j(x) = (\alpha_j(x))^2(\alpha_j(x) - 1)\left(\frac{d\alpha_j(x)}{dx}\right)^{-1}$$

respectively. This leads to the system of $2N$ equations

$$\sum_i (S_j,\alpha_i)\dot{a}_i + \sum_i (S_j,\beta_i)\dot{x}_i = (S_j,\ Av) \tag{19}$$

$$j = 1,2,\ldots,N$$

$$\sum_i (T_j,\alpha_i)\dot{a}_i + \sum_i (T_j,\beta_i)\dot{x}_i = (T_j,\ Av)\ . \tag{20}$$

The choice of Hermite cubics as test functions is based on three considerations.
(i) There are 2N Hermite cubics readily available which is the exact number of test functions required.
(ii) The Hermite cubics have C' continuity and so evaluation of L_2 inner products involving the discontinuous trial functions $\{\beta_i\}$ present no difficulty even for second order derivatives.
(iii) In problems dominated by convection, upwinding is required to eliminate oscillations and this can be provided by asymmetry in the test functions. The Hermite cubic functions $\{T_i(x)\}$ are asymmetric and these cause the nodal points to space themselves so that the solution may be free of oscillations.

In both of the above methods, the following common features occur.
(i) The test functions at each node consist of one even and one odd function.
(ii) In first order hyperbolic problems, the nodes move along the characteristic curves.
(iii) The nodes move according to an approximate equidistributing principle.
(iv) The solution of $2N$ equations is singular when

$$\frac{\Delta a_{i-1}}{\Delta x_{i-1}} = \frac{\Delta a_i}{\Delta x_i}\ .$$

(v) Neighbouring nodes can coincide or cross over each other. This can be prevented by incorporating a penalty function into the procedure.

Useful references for M.F.E. methods are Gelinas et al (1981), Miller and Miller (1981), Miller (1981), Herbst et al (1982, 1983, 1984), and Watham and Baines (1983).

SOLITON PROBLEMS

In recent years there has been a substantial increase of interest in nonlinear partial differential equations governing nonlinear waves in dispersive media. Examples range from water waves to plasma waves. In these travelling wave problems, the interaction between non linearity and dispersion can result in solitary wave solutions called solitons, the latter maintaining their original shape and speed after collision with other solitons. Examples of soliton bearing equations are

$$
\begin{array}{llll}
\text{(a)} & 0 = \underline{u_t + u_{xxx}} + (uu_x) & \text{Korteweg de Vries} & \\
\text{(b)} & 0 = \underline{iu_t + u_{xx}} + (|u|^2 u) & \text{Non linear Schrödinger} & \\
\text{(c)} & 0 = \underline{u_{tt} - u_{xx}} + (\sin u) & \text{Sine Gordon} & (21) \\
\text{(d)} & 0 = \underline{u_{tt} + u_{xxxx}} - ((u^2)_{xx}) & \text{Good Boussinesq} &
\end{array}
$$

where in each case the linear dispersive component is underlined and the non-linear term is in brackets. Using the elementary solution

$$u = ae^{i(kx-wt)} = ae^{ik(x-ct)}$$

where k is the wave number, w is the frequency, and $c = w/k$ is the speed of the travelling wave, the linear operators in (21) give the dispersion relations $w = -k^3$, $w = k^2$, $w^2 = k^2$ and $w^2 = k^4$ respectively, causing the Fourier components of any initial condition to propagate at different speeds. Equations such as (21) can be solved theoretically by the inverse scattering technique for certain types of initial conditions only and so numerical methods at the moment are the way forward for many initial value problems and for problems with boundary conditions.

The original numerical calculations involving the equations in (21) are based mainly on explicit difference methods although spectral methods are also used. The explicit difference methods in the main impose severe stability restrictions on the time step which is a disadvantage in problems where large time behaviour is often important. As a result the present author and colleagues in some recent papers experimented with Petrov-Galerkin methods on the soliton equations in (21) and a brief summary of the results obtained for 21(a) and 21(b) now follows.

KORTEWEG DE VRIES (K.d.V.) If we approximate u by

$$U(x,t) = \sum_i U_i(t)\phi_i(x) ,$$

the unknown functions $U_i(t)$ are determined from the system of ordinary differential equations

$$(U_t + UU_x + U_{xxx} , \psi_j(x)) = 0 , \quad \forall j .$$

The trial functions $\{\phi_i(x)\}$ are chosen to be piecewise linear "hat" functions located at the nodal points distance h apart, and the test functions $\{\psi_j(x)\}$ to be similar linear functions shifted a distance αh $(0 < \alpha \le \frac{1}{2})$ to the left (or right) of the nodal points i.e. $\psi_j(x) \equiv \phi_{j-\alpha}(x)$. First and second derivatives of linear "hat" functions produce Heaviside and Delta functions respectively, and shifting the test functions ensures that discontinuities in trial and test functions do not occur at the same locations in x. As an example, we evaluate the L_2 inner product $(U_{xxx},\phi_{j-\alpha})$ by integrating by parts twice, neglecting boundary terms, followed by direct evaluation of $(U_x,(\phi_{j-\alpha})_{xx})$ since U_x involves Heaviside functions and $(\phi_{j-\alpha})_{xx}$ a Delta function, the latter being located away from the nodal points. Not surprisingly, optimum convergence occurs when $\alpha = \frac{1}{2}$. If we use quadratic splines (C^1 continuous, 3h support) for test functions, the optimum shift value is $\alpha = \frac{1}{2\sqrt{3}}$ and for cubic splines (C^2 continuous, 4h support) the best shift value is $\alpha = 0$. All of the above methods, with Crank Nicolson used for the time integration of the system of ordinary differential equations, give good numerical results for the K.d.V.. Product approximation is convenient for the non linear term. Further details of the shifted spline technique including numerical experiments can be found in Mitchell and Schoombie (1981) and Schoombie (1982).

NON LINEAR SCHRÖDINGER (N.L.S.) The N.L.S. is given by

$$i \frac{\partial u}{\partial t} + \frac{\partial^2 u}{\partial x^2} + |u|^2 u = 0 , \qquad i = \sqrt{-1} \tag{22}$$

where $u(x,t)$ is a complex field governing the evolution of a weakly nonlinear dispersive almost monochromatic wave. The pure initial value problem involving (22) can be solved exactly using inverse scattering provided the initial condition $u(x,0)$ vanishes for sufficiently large $|x|$. Several finite difference schemes have been proposed for the numerical solution of (22), but in the main these have been unsatisfactory. We proceed as follows, put

$$u(x,t) = v(x,t) + iw(x,t)$$

and so (22) leads to the system

$$\frac{\partial \underline{u}}{\partial t} + A \frac{\partial^2 \underline{u}}{\partial x^2} + \underline{f}(\underline{u}) = 0 , \tag{23}$$

where

$$\underline{u} = (v,w)^T , \quad A = \begin{bmatrix} 0 & 1 \\ -1 & 0 \end{bmatrix}, \text{ and } \underline{f}(\underline{u}) = |\underline{u}|^2 A\underline{u} .$$

For convenience, we consider the initial condition

$$\underline{u}(x,0) = \underline{g}(x) , \qquad x_L \le x \le x_R$$

together with the homogeneous Neumann boundary condition

$$\frac{\partial \underline{u}}{\partial x} = 0 , \quad x = x_L, x_R , \qquad t > 0 .$$

Premultiplication of (23) by $\underline{u}^T$ and integration with respect to x gives

$$\frac{d}{dt} \int_{x_L}^{x_R} \underline{u}^T \underline{u} \, dx = 0 , \tag{24}$$

which follows from $\underline{u}^T \underline{f}(\underline{u}) \equiv 0$ and $\underline{u}^T A\underline{u}_{xx} = (\underline{u}^T A \underline{u}_x)_x$ for a skew symmetric matrix A. Energy in the form of the $[L_2(x_L,x_R)]^2$ norm of the solution is thus conserved.

We now approximate $\underline{u}(x,t)$ by

$$\underline{U}(x,t) = \sum_{i=1}^{N} \underline{\alpha}_i(t)\phi_i(x)$$

where $\underline{\alpha}_i(t) = (\alpha_{i1}(t), \alpha_{i2}(t))^T$, $i = 1,2,\ldots,N$ are time dependent coefficients to be determined, and $\phi_i(x)$, $i = 1,2,\ldots,N$ are piecewise linear "hat" trial functions. Using product approximation for the non linear term $\underline{f}(\underline{u})$ and linear "hat" test functions, we obtain the system of ordinary differential equations

$$M \frac{d\underline{\alpha}}{dt} + \frac{1}{h^2} S \underline{\alpha} + M \underline{F}(\underline{\alpha}) = 0 \tag{25}$$

where

$$M = \frac{1}{6}\begin{bmatrix} 2I & I & & & 0 \\ I & 4I & I & & \\ & \ddots & \ddots & \ddots & \\ 0 & & I & 4I & I \\ & & & I & 2I \end{bmatrix}, \quad S = \begin{bmatrix} -A & A & & & 0 \\ A & -2A & A & & \\ & \ddots & \ddots & \ddots & \\ 0 & & A & -2A & A \\ & & & A & -A \end{bmatrix},$$

and $\underline{F} = (\underline{F}_1, \underline{F}_2, \ldots, \underline{F}_N)^T$ with I the 2×2 unit matrix and $\underline{F}_i = (\underline{\alpha}_i^T \underline{\alpha}_i) A \underline{\alpha}_i$. Note that the matrices M and S are symmetric and skew symmetric respectively.

In order to solve (25) for $\underline{\alpha}(t)$, we discretise in time using the grid points $t = n\tau$, $n = 0,1,2,\ldots$, where τ is the step size. The discretisation involves an intermediate time $t^* = (n+\beta)\tau$, $0 < \beta$, and is given by the Predictor-Corrector (P.C.) pair of equations

$$M\underline{\alpha}^* = (M - \beta r S)\underline{\alpha}^n - \beta\tau M\underline{F}(\underline{\alpha}^n)$$

$$\left(M + \frac{r}{2} S\right)\underline{\alpha}^{n+1} = \left(M - \frac{r}{2} S\right)\underline{\alpha}^n - \tau M\underline{F}\left\{ \frac{\underline{\alpha}^* + \underline{\alpha}^n}{2} \right\}, \tag{26}$$

$n = 0,1,2,\ldots$, where $r = \tau/h^2$.

Returning to energy conservation, this will be the case in the discretised scheme if

$$(\underline{\alpha}^n)^T M\underline{\alpha}^n = (\underline{\alpha}^o)^T M\underline{\alpha}^o . \tag{27}$$

Unfortunately the solution of the P.C. scheme (26) does not satisfy (27) although it comes close if $\beta = 1$ (Griffiths et al (1982)). The principal merit of an energy conserving scheme is that it prevents the solution becoming unbounded i.e. "blowing up".

We now turn our attention to the stability of the numerical scheme (26). Let $B(\underline{u})$ denote the Jacobian of $\underline{f}(\underline{u})$ with respect to $\underline{u}$, i.e.

$$B(\underline{u}) = \begin{vmatrix} 2vw & v^2 + 3w^2 \\ -(3v^2+w^2) & -2vw \end{vmatrix} . \tag{28}$$

The linearisation of (23) in the neighbourhood of a point $(\bar{x},\bar{t})$ leads to the system

$$\frac{\partial \underline{u}}{\partial t} + A\frac{\partial^2 \underline{u}}{\partial x^2} + \bar{B}\underline{u} = 0 , \tag{29}$$

where $\bar{B}$ is the value of $B(\underline{u})$ at $(\bar{x},\bar{t})$. A von Neumann stability analysis of (29) leads to the result that the scheme (26) is stable to small local perturbations if

$$\beta \le 1 + O(\tau) . \tag{30}$$

Thus bearing in mind the combined effects of energy conservation and linearised stability, the optimal choice of β appears to be slightly in excess of $\beta = 1$.

Numerical experiments were carried out for initial conditions taken from the theoretical solutions of (i) a single soliton and (ii) two solitons prior to collision. In the former experiment with $\beta = 0.8$, the solution blows up after about 3,000 time steps. Full details of the numerical experiments are available in Griffiths et al (1982).

MATHEMATICAL BIOLOGY

Biological systems are considerably more complex than those appearing in physics and chemistry. In the analysis of biological models there are two alternatives, brute force computer simulation on a complicated model or drastic reduction of the model in order to make it mathematically viable. Irrespective of ones point of view, the reaction-diffusion equations

$$\frac{\partial \underline{u}}{\partial t} = D\nabla^2 \underline{u} + \sum_{j=1}^{m} M_j(\underline{x},\underline{u}) \frac{\partial \underline{u}}{\partial x_j} + \underline{f}(\underline{u}) , \tag{31}$$

with appropriate initial and boundary conditions, play a major rôle in the modelling of biological systems. Here $\underline{u}(x,t)$ is an $\mathbb{R}^n$-valued function of $x \in \mathbb{R}^m$ and $t \in \mathbb{R}^+$, the diffusion matrix D has non negative constant entries and is diagonal, and $\underline{f}(\underline{u})$ describes the non linear reaction of the system. The convective term $\sum_{j=1}^{m} M_j(\underline{x},\underline{u}) \frac{\partial \underline{u}}{\partial x_j}$ plays little part in the main confrontation between reaction and diffusion and is put to zero. The parabolic nature of the system (31) together with the presence of rest states as $t \to \infty$, makes numerical calculation of (31) straight forward unless of course the non linear coupling due to $\underline{f}(\underline{u})$ is particularly severe. To illustrate the rich variety of solution types which result from reaction diffusion systems, we look at a scalar model of (31) with a cubic non linearity. This is Fisher's equation and is given by

$$\frac{\partial u}{\partial t} = \frac{\partial^2 u}{\partial x^2} + u(1-u)(u-a) \qquad 0 < a \leq \tfrac{1}{2} . \tag{32}$$

The favourite ploy of analysts in order to solve reaction diffusion equations is to assume travelling wave solutions of the form

$$u(x,t) = u(x-ct) = U(\xi) ,$$

where $c(>0)$ is the constant speed of a wave travelling in the direction of ξ increasing. This transforms the partial d.e. (32) into the ordinary d.e.

$$U'' + cU' + U(1-U)(U-a) = 0 \tag{33}$$

where a dash denotes differentiation with respect to ξ and the rest states are $U = 0,1$ (stable) and $U = a$ (unstable). An exact solution of (33) given by Huxley is

$$U(\xi) = \left(1 + \exp \frac{\xi}{\sqrt{2}}\right)^{-1} , \quad c = \sqrt{2}\,(\tfrac{1}{2}-a) .$$

This represents a wave front with $U(-\infty) = 1$, $U(+\infty) = 0$, travelling in the positive ξ direction with speed c.

The assumption of travelling wave solutions can be checked by solving (32) directly by a numerical method. If we approximate $u(x,t)$ by

$$U(x,t) = \sum_i U_i(t)\phi_i(x) ,$$

(32) becomes after integration by parts

$$\left(\frac{\partial U}{\partial t}, \psi_j\right) + \left(\frac{\partial U}{\partial x}, \psi_j'\right) = \left(\sum_i f_i(t)\phi_i(x), \psi_j\right) + \left[\frac{\partial U}{\partial x}\psi_j\right]_B \qquad \forall j , \tag{34}$$

where a dash denotes differentiation with respect to x, $\left[\frac{\partial \psi}{\partial x} \psi_j\right]_B$ is a boundary term, and product approximation has been used to deal with the cubic non linearity. Evaluation of (34) leads to the system of ordinary d.e.s

$$M\dot{\underline{\alpha}} + S\underline{\alpha} = F(\underline{\alpha}) + \underline{b} \tag{35}$$

where $M = ((\phi_i,\psi_j))$, $S = ((\phi_i',\psi_j'))$, $\underline{\alpha} = (U_1,U_2,\ldots,U_N)^T$, $\underline{b} = (b_1,0,0,\ldots,b_N)^T$ and a dot denotes differentiation with respect to time. There are many ways of carrying out the time integration of (35) and we leave the choice open. For basis functions we choose in the main linear hat functions for both trial and test functions, although for increased accuracy linear hat trial functions and cubic spline test functions can be used.

Numerical experiments with linear hat trial and test functions were carried out for the pure initial value problem with a rectangular pulse as initial condition. The following results were obtained for

$$a = \tfrac{1}{4} \quad ; \quad h = k = \tfrac{1}{2} .$$

(1) The rectangular pulse either dies away or grows to unity depending on whether the height of the pulse is less than or greater than $\frac{1}{4}$. In the latter case there are two travelling fronts, one in each direction.

(2) The equilibrium state $(u = 1)$ and the front shape as $t \to \infty$ are correct to four decimal places.

(3) The speed of the front as $t \to \infty$ is 0.3544 compared with the theoretical value of 0.3536.

We now consider (32) over a finite domain and subject to Dirichlet boundary conditions. The solutions are no longer travelling waves and in fact depend critically on the size (and shape in higher dimensions) of the region. To be precise we consider (32) subject to the boundary conditions

$$u(-L,t) = u(L,t) = b \text{ (constant)} , \quad t > 0$$

and the initial condition

$$u(x,0) = u_0(x) \quad , \quad -L \le x \le +L$$

where L is a bifurcation parameter. From analytical considerations the following necessary conditions regarding the solution as $t \to \infty$ have been obtained:

$$\begin{array}{cccc} b = & 0 & a & 1 \\ u(x,t) \to & 0 & a & 1 \\ L < & \dfrac{\pi}{1-a} & \pi & \dfrac{\pi}{a} \end{array} .$$

Numerical experiments more or less confirm these upper limits on L for which the above constant asymptotic $(t \to \infty)$ solutions are obtained. For larger values of L in each case non constant bifurcation solutions are obtained by numerical methods, some of which agree with known asymptotic solutions $(t \to \infty)$ of the particular problem.

The reader interested in more details in the numerical experiments referred to in this section can consult Mitchell and Manoranjan (1982, 1983) and Manoranjan et al (1983).

REFERENCES

[1] Anderssen, R.S. and Mitchell, A.R., Analysis of generalised Galerkin methods in the numerical solution of elliptic equations, Math. Meths in the Appl. Sciences 1 (1979) 1-11.

[2] Christie, I., Griffiths, D.F., Mitchell, A.R. and Sanz Serna, J.M., Product approximation for non linear problems in the finite element method, Inst. Maths. and Applics. Journ. (Numerical Analysis) 1 (1981) 253-266.

[3] Fletcher, C.A.J., The group finite element formulation, Comp. Meths. in Appl. Mechs. and Engng. (1983) (to appear).

[4] Fletcher, C.A.J., A comparison of finite element and finite difference solutions of the one and two dimensional Burgers' equations, J. Comp. Phys.

(1983) (to appear).
[5] Gelinas, R.J., Doss, S.K. and Miller, K., The moving finite element method, J. Comp. Phys., 40 (1981) 202-249.
[6] Griffiths, D.F., Mitchell, A.R. and Morris, J.Ll., A numerical study of the nonlinear Schrödinger equation, Comp. Meths. in Appl. Mechs. and Engng (1983) (to appear).
[7] Herbst, B.M., Moving finite element methods for the solution of evolution equations, Ph.D. Thesis, University of the Orange Free State (1982).
[8] Herbst, B.M., Schoombie, S.W. and Mitchell, A.R., A moving Petrov-Galerkin method for transport equations, Inst. J. Num. Meths. in Engng. 18 (1982) 1321-1336.
[9] Herbst, B.M., Schoombie, S.W., Mitchell, A.R. and Griffiths, D.F., Generalised Petrov-Galerkin methods for the numerical solution of Burgers' equation, Int. J. Num. Meths. in Engng. (1983) (to appear).
[10] Hughes, T.J.R., Finite element methods for convection dominated flows, American Society of Mechanical Engineers, Applied Mechanics Division, Volume 34 (1979).
[11] Manoranjan, V.S. and Mitchell, A.R., A numerical study of the Belousov-Zhabotinskii reaction using Galerkin finite element methods. J. Math. Biology 16 (1983) 151-260.
[12] Manoranjan, V.S., Mitchell, A.R., Sleeman, B.D. and Kuo, P.Y., Bifurcation studies in reaction-diffusion, University of Dundee Report D.E. 83:1 (1983).
[13] Miller, K. and Miller, R., Moving finite elements, Part 1, SIAM J. (N.A.) 18 (1981) 1019-1032.
[14] Miller, K., Moving finite elements, Part 2, SIAM J. (N.A.) 18 (1981) 1033-1057.
[15] Mitchell, A.R. and Wait, R., The Finite Element Method in Partial Differential Equations (John Wiley, New York, 1976).
[16] Mitchell, A.R. and Griffiths, D.F., The Finite Difference Method in Partial Differential Equations (John Wiley, New York, 1980).
[17] Mitchell, A.R. and Griffiths, D.F., Semi-discrete generalised Galerkin methods for time-dependent conduction-convection problems, in: Whiteman, J.R. (ed.), The Mathematics of Finite Elements and Applications III (Academic Press, London, 1979).
[18] Mitchell, A.R. and Manoranjan, V.S., Finite element studies of reaction diffusion, in: Whiteman, J.R. (ed.), The Mathematics of Finite Elements and Applications IV (Academic Press, London, 1982).
[19] Mitchell, A.R. and Schoombie, S.W., Finite element studies of solitons, in: Hinton, E., Bettess, P. and Lewis, R.W. (eds.), Numerical Methods For Coupled Problems (Pineridge Press, Swansea, 1981).
[20] Schoombie, S.W., Spline Petrov-Galerkin methods for the numerical solution of the Korteweg de Vries equation, Inst. Maths. and Applics. Journ. (Numerical Analysis) 2 (1982) 95-110.
[21] Wathan, A.J. and Baines, M.J., On the structure of the moving finite element equations, University of Reading Numerical Analysis Report No. 5 (1983).

Computational Techniques & Applications: CTAC-83
J. Noye & C. Fletcher (Editors)

THE CHANGING SCENE IN COMPUTATIONAL FLUID DYNAMICS

Maurice Holt

Department of Mechanical Engineering
University of California
Berkeley, California
U.S.A.

The evolution of numerical techniques for solving problems in Fluid Dynamics is followed, in outline, from the days when Digital Computers were first available, at the end of the Second World War, to the present time, when the Computer Aerodynamic Simulator is being assembled. In this period the range of numerical methods has been broadened five fold, while the speed and capacity of computers have increased by several orders of magnitude. Two areas close to the author's interests are selected to illustrate these changes. The first concerns the extension of the Method of Integral Relations to apply to laminar and turbulent boundary layer problems, including internal flows, separated flows and turbulent mixing flows. The second area deals with unsteady inviscid compressible flow in one or more dimensions and a discussion is given of the relative merits of Godunov and Glimm techniques.

INTRODUCTION

The need for numerical, as distinct from analytical, methods to solve problems in Fluid Dynamics first arose when problems in Gas Dynamics presented themselves. These are connected with combustion and explosive processes and are essentially non-linear. The equations of motion governing such problems are hyperbolic and, in principle, could be solved numerically by the Method of Characteristics. This has a long history going back to the pioneer paper by Massau (1900). The foundations of the method are laid out in Courant, Friedrichs and Lewy (1928) and developed for application in Courant and Friedrichs (1978).

Two difficulties arise in applying the Method of Characteristics to Gas Dynamic problems. Firstly, shock waves appear frequently in such problems and need to be fitted on an ad hoc basis into the characteristic network. Secondly, this network does not have uniform mesh spacing, is non-orthogonal, curvilinear and frequently highly skewed.

One of the earliest methods avoiding these drawbacks was the Finite Difference method of von Neumann and Richtmyer (1950). This applied to unsteady spherical flow in Lagrangian coordinates, using finite differences in the original orthogonal independent variables, and treated shocks by the introduction of artificial viscosity. This was refined later in the well-known Lax-Wendroff(1964) method. The development of these methods is well described in Richtmyer and Morton (1972).

The search for better numerical techniques to solve Gas Dynamic problems was given further impetus in the Sputnik era, with the need to solve the flow field problem for a space vehicle on re-entry into the earth's atmosphere, the so-called Blunt Body problem.

This led to a spate of new techniques, mostly developed in the USSR. Firstly, Godunov's method (1959) was presented as a novel way to solve the Lagrangian equation in unsteady flow but grew into one for solving the unsteady Eulerian equations in two and three dimensions —yielding the solution to the blunt body problem as the steady flow, asymptotic, limit of the unsteady solution. Godunov used discontinuity breakdown formulae, in place of finite difference formulae, a significant departure from the previous approaches of Lax-Wendroff and others.

A solution to the blunt body problem predating that of Godunov was obtained by Belotserkovskii (1958) who solved the equations of steady motion for symmetrical flow past a circular cylinder or sphere, of mixed elliptic-hyperbolic type, by the first formulation of the Method of Integral Relations due to Dorodnitsyn (1956). In this method moments of the equations of motion are integrated across the flow field, the integrands are represented by suitable interpolation functions and the problem is reduced to the integration of ordinary differential equations in the coefficients of these functions.

The steady flow version of the Blunt Body problem was also solved by Telenin's method (Gilinskii, Telenin and Tinyakov, 1964), in which the unknown is represented by an interpolation function in one of the coordinates, in the original equations of motion. The Method of Lines (Jones and South, 1979) is a similar technique applied over three or five mesh points instead of over the whole flow field.

The last method developed for high speed steady flow problems is due to Babenko and others. The version applied to purely supersonic flow is by Babenko, Voskresenskii, Lyubimov and Rusanov (1964) while the blunt body version was developed by Lyubimov and Rusanov (1970). This is a sophisticated finite difference method applicable to any steady high speed inviscid flow problem, especially in three dimensions.

When viscous effects are important we need to solve the Navier-Stokes equations, either in their original, or approximate, boundary layer form. The full equations are non-linear and elliptic. Several Finite Difference methods have been developed for these but they become prohibitively expensive as the governing Reynolds number is increased. In the low speed range Finite Element methods have been proposed for flow fields limited by boundaries (cavities and steps) while Spectral methods have been used for investigation of flow structures in unlimited regions. A full account of these is given in Peyret-Taylor (1983).

In many applications viscous effects only need to be considered near boundaries and it is sufficient to solve the boundary layer equations in these regions in interaction with inviscid flow in the outer regions. Many finite difference methods have been developed for the boundary layer equations. At the present time, however, these can be replaced by more recently developed techniques such as the Method of Integral Relations, Finite Element methods, Galerkin techniques and Spectral Methods. These are described in full detail by Fletcher (1983).

The remainder of this article deals with two topics connected with the author's own research, Recent Applications of the Method of Integral Relations to Turbulent and Internal Flows and Recent Developments in Methods for Problems in Gas Dynamics and Propagation of Large Amplitude Surface Waves.

RECENT APPLICATIONS OF THE METHOD OF INTEGRAL RELATIONS

The Method of Integral Relations was first formulated for Viscous Incompressible Boundary Layer Problems by Dorodnitsyn (1960). This formulation was extended to compressible flows by Pavlovskii (1963) and to flows with wall injection by Liu (1962). The early applications were all to attached flows but extensions of the method to apply to separated and reversed flows were made by Nielsen, Goodwin and Kuhn (1969) and by Holt (1966,1967) and Holt and Lu (1975).

In the original formulation the basic integral relation is derived by factoring the continuity equation by one of a set of weighting functions $f(u)$ and the streamwise momentum equation by its derivative $f'(u)$ where u is the streamwise velocity component. The results are added and integrated across the boundary layer, using u, rather than η (the transverse coordinate), as variable of integration. The functions $f(u)$ belong to a complete set and taken in fact as integral powers of $(1-u)$ [u is made dimensionless with respect to velocity just outside the boundary layer]. The integrands in the integral relation contain the unknown transverse velocity gradient $\partial u/\partial\eta$, which, in the Dorodnitsyn formulation, is represented as a polynomial in u, factored by a term (1-u) to ensure approach to zero at the boundary layer edge. Successive integral relations in the nth approximation yield n ordinary differential equations for the coefficients in the ($\partial u/\partial\eta$) polynomial. The matrix of these equations (for the first derivatives) is dense and the system becomes progressively more ill-conditioned as the order of approximation is increased.

To eliminate this defect Fletcher and Holt (1975) proposed a modified formulation of the Method of Integral Relations. In this, the weighting functions f(u) belong to a complete orthonormal set in (0,1). Moreover, the unknown ($\partial u/\partial\eta$) is expanded in combinations of the same orthonormal set, factored by (1-u). The resulting ordinary differential equations for the coefficients in the expansion then have a sparse matrix with diagonal elements only. The integrals which occur can all be evaluated by quadratures and the modified version can be applied to attached flows at any level of approximation.

The extension of the Dorodnitsyn MIR to separated flows introduces a great deal of tedious algebra, even in the lowest order approximation and it would be useful to extend the Fletcher-Holt modified version for such flows. The difficulty here is that a factor $(u+\alpha)^{\frac{1}{2}}$ must be included in the representation of ($\partial u/\partial\eta$) to take account of reversed flow between u = 0 (at the wall) and u = -α, with a vertical tangent in the $\partial u/\partial\eta$ - u curve at u = -α.

A possible orthonormal formulation for separated flows is now given, developed as a course term project at the University of California, Berkeley by O. Ozcan and R.-J. Yang. This uses Chebychev polynomials. This is followed by an orthonormal formulation of MIR for turbulent boundary layers, applied to model turbulent flows by Yeung and Yang (1981) and to the turbulent wall jet problem by Yang and Holt (1983).

ORTHONORMAL FORMULATION OF MIR FOR SEPARATING FLOWS

When the original Method of Integral Relations is applied to the incompressible laminar boundary layer equations in Dorodnitsyn variables, the following basic integral relation results

$$\frac{d}{d\xi}\int_A^B uf_k\theta du = \frac{\dot U_1}{U_1}\int_A^B (1-u^2)f_k'(u)\theta du + \frac{f_k'(u)}{\theta}\bigg|_A^B - \int_A^B \frac{f_k''}{\theta}du + uf_k\bigg|_B \frac{d\eta_2}{d\xi} - uf_k\bigg|_A \frac{d\eta_1}{d\xi} - wf_k(u)\bigg|_A^B , \qquad (1)$$

where ξ is the streamwise coordinate, u the velocity component in the ξ direction, θ the reciprocal of ($\partial u/\partial\eta$), η the transverse coordinate and $f_k(u)$ are weighting functions. The points A and B correspond to the u limits of integration over the section of boundary layer in question. The values η_1,η_2 correspond to η at A and B, respectively.

The velocity field is divided into two parts in the attached (but retarded) boundary layer region and into three parts in the separated region.

Attached flow

In the velocity range $0 < u < \varepsilon$, where ε is small ($\varepsilon \sim 0.1$), θ is represented by the second order approximate expression

$$\theta = \frac{c_0}{(u+\alpha)^{\frac{1}{2}}(1-u)} \tag{2}$$

The following two weighting functions are chosen to simplify the calculation of the integrals in Eq. (1)

$$f_k(u) = (\varepsilon-u)(1-u)(u+\alpha)^{k+\frac{1}{2}} \quad , \quad k = 1,2 \quad . \tag{3}$$

Equations (2) and (3) are substituted in Eq. (1) to yield two ordinary differential equations for c_0 and α

$$\frac{d}{d\xi}\,[c_0 E_k(\alpha)] = F_k(c_0,\alpha) \quad , \quad k = 1,2 \quad , \tag{4}$$

where E_k are polynomials in α, and F_k are polynomials in α, linear in c_0, with coefficients containing ε.

In the velocity range $\varepsilon < u < 1$, θ is represented by the following expression

$$\theta = \frac{1}{(u+\alpha)^{\frac{1}{2}}(1-u)} \quad [b_0 + \sum_{i=2}^{N} \sum_{j=1}^{N-1} \frac{b_i\sqrt{u+\alpha}}{u(1-u)^{3/4}} \frac{g_i + g_j}{\sqrt{u-\varepsilon}}] \tag{5}$$

together with weighting functions

$$f_k = (1-u)^{5/4} \; g_k \quad , \tag{6}$$

where g_k are Chebychev polynomials, i and k are even, j is odd.

The orthogonality condition for these polynomials is

$$\int_{-1}^{+1} \frac{g_i(U) g_k(U)}{(1-U^2)^{\frac{1}{2}}} \, dU = \begin{matrix} \pi/2 \text{ when } i = k \\ 0 \quad \text{when } i \neq k \end{matrix} \tag{7}$$

The limits of integration $\varepsilon \to 1$ are changed to $-1 \to 1$ by means of the change in variable

$$U = \frac{1}{\varepsilon-1}\,[\varepsilon + 1 - 2u] \quad . \tag{8}$$

We can show that

$$\underset{\substack{u \to \varepsilon \\ U \to -1}}{\mathrm{Lt}} \; \frac{g_i + g_j}{(u-\varepsilon)^{\frac{1}{2}}} = 0 \quad .$$

Hence, matching expressions (2) and (5) at $u = \varepsilon$,

$$a_0 = b_0$$

Substitution of Eqs. (5) and (6) in Eq. (1) yields

$$\frac{\pi}{2}\,(N-4)\,\frac{db_k}{d\xi} = C'(k) \quad , \quad k = 1,\ldots,N \quad , \tag{9}$$

where $C'(k)$ is a linear combination of the unknowns b_n with coefficients which can be evaluated numerically.

Separated flow

In the separated flow region three representations of $\partial u/\partial \eta$ in terms of u are used. In the reversed flow range $-\alpha < u < 0$, close to the wall, we use

$$\theta = -\frac{a_0}{(u+\alpha)^{\frac{1}{2}}(1+\alpha)}$$

with weighting function (10)

$$f = (u+\alpha)^{\frac{3}{2}} \quad .$$

In the intermediate range $-\alpha < u < \varepsilon$, where ε is small and positive, we use

$$\theta = -\frac{a_0}{(u+\alpha)^{\frac{1}{2}}(1-u)}$$

$$f = (\varepsilon-u)(1-u)(u+\alpha)^{\frac{3}{2}} \qquad (11)$$

Substitution of Eqs. (10) and (11) into Eq. (1) yields two first order ordinary differential equations for a_0 and α.

In the outer range $\varepsilon < u < 1$ we again use representation (5) for θ with weighting functions (6). The coefficients b_n are determined from the same equation (9).

This approach is currently being applied to model separated flow problems.

APPLICATION OF MIR TO TURBULENT BOUNDARY LAYERS

The key to the extension of the Method of Integral Relations to apply to turbulent boundary layer flow is in the representation of the eddy viscosity as a function of mean streamwise velocity component. This was investigated by Abbott and Deiwert (1968) and by Murphy and Rose (1968). Although their models of eddy viscosity were reasonable these were not well adapted to the original formulation of MIR and produced disappointing results. On the other hand, when the modified, orthonormal, version of MIR is applied to turbulent boundary layers, with representations of eddy viscosity similar to those used previously, comparisons of applications to experimental results in model cases prove to be very satisfactory. This advance is described in Yeung and Yang (1981).

The turblent boundary layer equations in two dimensions may be written

$$u\frac{\partial u}{\partial x} + v\frac{\partial u}{\partial y} = -\frac{1}{\rho}\frac{\partial p}{\partial x} + \frac{\partial}{\partial y}\left[(\nu + \frac{\varepsilon}{\rho})\frac{\partial u}{\partial y}\right] \quad , \qquad (12)$$

$$\frac{\partial u}{\partial x} + \frac{\partial v}{\partial y} = 0 \quad , \qquad (13)$$

where u and v are the x and y components of mean velocity, p is the mean pressure, ρ the density, ν the kinematic viscosity. The eddy viscosity ε is given by

$$-\rho\,\overline{u'v'} = \varepsilon\frac{\partial u}{\partial y} \qquad (14)$$

representing the Reynolds stress.

We introduce the dimensionless variables

$$U = \frac{u}{u_e} \; , \; V = \frac{vRe^{1/2}}{u_e} \; , \; \bar{x} = \frac{x}{L} \; , \; \bar{y} = \frac{yRe^{1/2}}{L} \; , \; U_e = \frac{u_e}{u_\infty} \; , \qquad (15)$$

where

$$\mathrm{Re} = u_\infty L/\nu \quad . \tag{16}$$

Bernoulli's equation gives

$$-\frac{1}{\rho}\frac{\partial p}{\partial x} = u_e \frac{du_e}{dx} \tag{17}$$

and Eqs. (12) and (13), in the dimensionless variables, become

$$U\frac{\partial U}{\partial \bar{x}} + V\frac{\partial U}{\partial \bar{y}} = \frac{1}{U_e}\frac{dU_e}{d\bar{x}}(1-U^2) + \frac{1}{U_e}\frac{\partial}{\partial \bar{y}}\left[\left(1+\frac{\varepsilon}{\mu}\right)\frac{\partial U}{\partial \bar{y}}\right] \quad , \tag{18}$$

$$\frac{\partial U}{\partial \bar{x}} + \frac{\partial V}{\partial \bar{y}} = -\frac{U}{U_e}\frac{dU_e}{d\bar{x}} \quad . \tag{19}$$

To apply the Method of Integral Relations we introduce a complete set of orthonormal functions (over (0,1)) $f_i(U)$, multiply Eq. (18) by f_i', add the result and integrate with respect to $\bar{y}$ across the boundary layer. We then change the variable of integration from $\bar{y}$ to U, introducing the reciprocal of the transverse mean velocity gradient

$$Z = \left(\frac{\partial U}{\partial \bar{y}}\right)^{-1} \quad . \tag{20}$$

The basic integral relation is then

$$\frac{\partial}{\partial \bar{x}}\int_0^1 f_i UZdU = \frac{1}{U_e}\frac{dU_e}{d\bar{x}}\int_0^1 [(1-U^2)f_i' - Uf_i]ZdU$$
$$- \frac{1}{U_e} f_i'(0)\frac{1}{Z_o} - \frac{1}{U_e}\int_0^1 \left(1+\frac{\varepsilon}{\mu}\right)\frac{f_i''}{Z}dU \quad . \tag{21}$$

The orthonormal functions $f_i(U)$ are given by

$$f_i(U) = \sum_{k=1}^{i} c_{ik}(1-U)^k \quad , \tag{22}$$

and

$$\int_0^1 f_k f_j \frac{U}{1-U} dU = \delta_{kj} \quad , \tag{23}$$

where δ_{kj} is the Kronecker delta. We represent Z by the factored orthonormal expansion

$$Z = \frac{b_o + \sum_{j=1}^{N-1} b_j f_j(U)}{1-U} \tag{24}$$

The basic integral relation (21) then yields the following set of ordinary differential equations for the unknown coefficients $b_o, b_1, b_2, \ldots$ in Eq. (24):

$$\frac{db_o}{d\bar{x}}\int_0^1 \frac{f_i U}{1-U}dU + \frac{db_i}{d\bar{x}} = \frac{1}{U_e}\frac{dU_e}{d\bar{x}}\int_0^1 [(1-U^2)f_i' - Uf_i]ZdU$$
$$- \frac{1}{U_e}\frac{f_i'(0)}{Z_o} - \frac{1}{U_e}\int_0^1 \left(1+\frac{\varepsilon}{\mu}\right)\frac{f_i''}{Z}dU \quad , \quad i = 1,2,\ldots,N-1 \tag{25a}$$

and

$$\frac{db_o}{d\bar{x}}\int_0^1 \frac{f_N U}{1-U}dU = \frac{1}{U_e}\frac{dU_e}{d\bar{x}}\int_0^1 [(1-U^2)f_N' - Uf_N]ZdU$$
$$- \frac{1}{U_e} f_N'(0)\frac{1}{Z_o} - \frac{1}{U_e}\int_0^1 \left(1+\frac{\varepsilon}{\mu}\right)\frac{1}{Z} f_N'' dU \tag{25b}$$

These can be integrated subject to suitable initial conditions.

The advantages of the modified version of MIR over the original version are now apparent. Firstly, the system of ordinary differential equations [(25a) and (25b)] can easily be reduced to diagonal form while the corresponding equations in the original formulation are highly coupled. Secondly, the representation (24) for Z allows for greater flexibility in reproducing the highly inflected velocity profile characterizing turbulent (as opposed to laminar) boundary layers. Thirdly, in the orthonormal version, the integral

$$\int_0^1 (1 + \frac{\varepsilon}{\mu}) \frac{1}{Z} f_i''(U)dU$$

can be evaluated by quadrature, while in the original version this must be reduced to an algebraic expression. This is a tedious task, increasingly difficult to carry out as the order of approximation is increased, since the term ε/μ is represented by a complicated combination of exponentials and powers in U.

Turbulence modelling

The eddy viscosity ε/μ is represented by two formulae, one based on a model due to Spalding (1961) and Kleinstein (1967) applicable near the wall and the other, applicable in the outer part of the boundary layer, based on the wake model of Clauser (1956). The value of U at the junction of these two representations is denoted by U_m and is determined by conditions that the $\varepsilon/\mu - U$ curve should be continuous with continuous slope at $U = U_m$.

In terms of U and Z the eddy viscosity is represented by:

For $0 \leqslant U < U_m$

$$\frac{\varepsilon}{\mu} = 0.04432[e^{0.4U\sqrt{U_e Z_o Re^{1/2}}} -1 -0.4U\sqrt{U_e Z_o Re^{1/2}} -0.08U^2 U_e Re^{1/2} Z_o] \quad . \tag{26}$$

For $U_m \leqslant U \leqslant 1$

$$\frac{\varepsilon}{\mu} = 0.0168\, U_e Re^{1/2} \int_0^1 (1-U)ZdU \quad . \tag{27}$$

The value of U_m is determined from

$$0.04432[e^{0.4U\sqrt{U_e Z_o Re^{1/2}}} -1 - 0.4U\sqrt{U_e Z_o Re^{1/2}} = 0.08U^2 U_e Z_o Re^{1/2}$$

$$= 0.0168\, U_e Re^{1/2} \int_0^1 (1-U)ZdU \quad . \tag{28}$$

Yeung and Yang (1981) applied this formulation to three model flows based on data presented at the 1968 Stanford Conference on Turbulent Flows (Coles and Hirst (1968)). The first corresponds to a zero pressure gradient (identified as ID 1400 in the Stanford proceedings), the second to adverse pressure gradient (ID 1100) and the third to favorable pressure gradient (ID 1300). In the first case good results were obtained with the third approximation $N = 3$ while agreement with experiment was excellent for $N = 4$, $N = 5$. The CPU time for $N = 4$ was only 8 secs (using a CDC 7600 computer). In the third case sufficient accuracy was obtained with $N = 3$ and higher order approximations were not needed. For adverse pressure gradient flow results were partially satifactory for $N = 5$ but it is evident that the ε/μ model needs to be improved in this area.

The wall jet problem

Yang and Holt (1983) used the orthonormal version of MIR for turbulent boundary layers to investigate the effect of injecting a parallel stream on turbulent flow in a pipe. This has important applications to coal gasification systems, in which certain products (especially sulfides and oxides) can cause serious corrosion at the walls of pipes in the heat exchange section.

To simplify the problem the flow field is divided into three parts. Firstly, just downstream of the station where the wall jet is introduced, turbulent mixing between the uniform flow in the pipe and uniform flow in the parallel jet is investigated. The injected gas and main stream gas contain different species and both turbulent and molecular diffusion are accounted for. Below the mixing region a turbulent boundary layer develops along the wall of parallel jet and its growth is determined by a separate application of MIR. This second part of the flow field is not initially affected by the mixing process. The third part of the flow field extends downstream of the station where the wall boundary layer and lower part of the mixing layer intersect. The velocity profile in this region of interaction initially has two inflexion points but the orthonormal version of MIR is sufficiently flexible to permit a faithful representation of the Z-U behavior. and the growth of this interacting flow until a full turbulent boundary layer is developed and calculated.

The calculation establishes that introduction of a wall jet can protect a pipe wall from corrosive effects for up to 100 jet thicknesses downstream.

NUMERICAL METHODS FOR UNSTEADY FLOW PROBLEMS

The equations of motion of one dimensional unsteady flow are easily reduced to characteristic form and the earliest numerical methods for solving problems governed by such equations were based on the Method of Characteristics. The latter has the disadvantage of being tied to a non-orthogonal coordinate network which has to be built up, step by step, as the numerical procedure is advanced. To overcome this drawback Godunov (1959) proposed a numerical scheme for solving Lagrangian equations of motion which could be executed in the physical x (distance), t (time) plane. He later extended this scheme to apply to the Eulerian equations in one or more dimensions and used this to solve the Blunt Body Problem. These early Godunov schemes are of first order accuracy and are monotonic in character; that is, if the initial form of the unknown is monotonically increasing this property is preserved as the Godunov schemes are applied at successive time intervals. Godunov (1970) subsequently introduced a scheme of second order accuracy, based on a predictor-corrector approach. This is not monotonic in character but has been successfully applied to many problems in Gas Dynamics with plane symmetry and in Shallow Water Theory. In the period between the publication of the two Godunov schemes, Glimm (1965) proposed a modification of the first Godunov scheme which, in principle, has second order accuracy. We now give a brief comparison between Godunov's first scheme and Glimm's scheme, using explanations of the latter by Chorin (1976,1977). The comparison is made with reference to the one dimensional acoustic equations and is a summary of the discussion in Holt (1984).

Solution of the acoustic wave equation

Consider the one dimensional acoustic wave equation

$$\frac{\partial U}{\partial t} + a\,\frac{\partial U}{\partial x} = 0 \qquad (29)$$

where a is a positive constant (the acoustic speed) with initial conditions

$$U = f(x) \quad , \quad t = 0 \quad . \tag{30}$$

The analytical solution to this equation is

$$U = f(x-at) \quad . \tag{31}$$

If the initial wave form is a Heaviside unit function

$$\begin{aligned} f(x) &= 0 \quad , \quad x \leqslant 0 \\ f(x) &= 1 \quad , \quad x > 0 \end{aligned} \tag{32}$$

the general solution is

$$\begin{aligned} U(x,t) &= 0 \quad , \quad x \leqslant at \\ U(x,t) &= 1 \quad , \quad x > at \end{aligned} \tag{33}$$

representing a step function propagated unchanged along the characteristic line $x = at$.

To solve problem (29), (30) by Godunov's first scheme we divide the x axis into a series of small cells (usually of equal spacing) and represent f(x) by a staircase function so that f(x) is constant in each cell. At each cell boundary there is a discontinuity in the initial value of U and, for $t > 0$, we solve the problem of breakdown in discontinuity. In other words, we solve a succession of problems (29), (32). The same procedure is followed at all later time intervals.

Thus, at a general time (after n time steps each of duration k) we solve the initial value problem: Solve (29) with

$$\begin{aligned} U &= 0 \qquad ih < x \leqslant (i+\tfrac{1}{2})h \qquad t = nk \\ U &= 1 \qquad (i+\tfrac{1}{2})h < x < (i+1)h \qquad t = nk \quad . \end{aligned} \tag{34}$$

If we transfer the origin $\{(i+\tfrac{1}{2})h, nk\}$ to (0,0) this problem has the solution

$$\begin{aligned} U(x,t) &= 0 \qquad X \leqslant aT \\ U(x,t) &= 1 \qquad X > aT \quad , \end{aligned} \tag{35}$$

where X,T are coordinates. In the Godunov scheme we always use this solution at the cell boundary itself. To satisfy stability of the Godunov scheme we require

$$ah > k \tag{36}$$

so that the boundary $\{(i+1)h, (n+\tfrac{1}{2})k\}$ is to the right of $X - aT = 0$. Therefore,

$$U^{i+1}_{n+\frac{1}{2}} = 1 \quad . \tag{37}$$

so that the Godunov scheme always moves the step jump 0-1 a distance ½ h to the right. As a consequence it can be shown that, after N whole time steps the Godunov scheme causes the step jump to have moved a distance

$$\{h/k - a\}T \tag{38}$$

beyond its position as given by the analytical solution.

In Glimm's scheme we record the solution to the breakdown problem (29),(34) over a half step range on either side of the cell boundary. We then sample this solution at a randomly chosen point within the range (-h/2,h/2) in X. The stability (Courant-Friedrichs-Lewy) condition ensures that the characteristic line $X = aT$ intersects the line $T = \frac{1}{2}k$ inside (-h/2,h/2). The greater flexibility in the Glimm scheme, as compared with the first Godunov scheme, permits us to sample a certain number of intersection points in $(-\frac{1}{2}h, \frac{1}{2}h)$ on both sides of $X = aT$. If the randomly chosen samples are uniformly distributed over (-h/2,h/2) it can be shown that balance between samples to the left and those to the right of $X = aT$ is that required to ensure that the path of the discontinuity 0-1 stays on course. In this sense, then, the Glimm scheme is more accurate than the first Godunov scheme.

Shallow water wave propagation

Li and Holt (1981) applied Glimm's method to the problem of shallow water waves generated by large, near surface, disturbances. The calculation is an extension of the work by Sod (1977) on spherical, or cylindrical, explosions in gases and the paper by Flores and Holt (1981) on underwater explosions. Li and Holt first tested Glimm's method on the classical Dam Break Problem and then applied it to the Dam Break Problem with cylindrical symmetry. It was then used to calculate large amplitude surface waves generated by near surface explosions.

The shallow water equations, omitting dissipative terms, may be written

$$U_t + \{F(U)\}_r = -W(U) \tag{39}$$

where

$$U = \begin{bmatrix} \eta \\ u \end{bmatrix}, \quad F(U) = \begin{bmatrix} u(\eta + d) \\ u^2/2 + g\eta \end{bmatrix}, \quad W(U) = \begin{bmatrix} i(u/r)(\eta + d) \\ 0 \end{bmatrix} \tag{40}$$

and

$i = 0$ for plane symmetry

$i = 1$ for cylindrical symmetry

In Eqs. (40) d is the undisturbed ocean depth, η is the displacement of the ocean surface from its undisturbed position, r is the space coordinate.

Glimm's method is only applicable to equations of motion in conservation form. Following Sod (1977) we therefore solve Eqs. (39) by a splitting method. We solve the homogeneous equations

$$U_t + \{F(U)\}_r = 0 \tag{41}$$

by the established Glimm technique for plane flow equations and subsequently determine the non-homogeneous term from

$$U_t = -W(U) \quad . \tag{42}$$

To apply Glimm's method to Eqs. (41) we use h as a constant space interval and k as the constant time interval. After time $t = nk$ we represent the solution by a staircase function $\tilde{u}_i^n$ such that

$$\begin{aligned} U(r,nk) &= \tilde{u}_{j+1}^n \qquad r > (i + \tfrac{1}{2})h \quad , \\ U(r,nk) &= \tilde{u}_j^n \qquad r < (i + \tfrac{1}{2})h \end{aligned} \tag{43}$$

Thus U is constant in each cell, with spacing h, along the r axis and jumps discontinuously across each cell boundary. To determine U at $t = (n+\frac{1}{2})k$ we allow each cell boundary to be suddenly removed at $t = nk$ and solve a series of Riemann breakdown problems to determine the solution in $nk \leqslant t \leqslant (n+\frac{1}{2})k$. In Glimm's method we sample this solution at a randomly chosen point in the range $(-\frac{1}{2}h, \frac{1}{2}h)$ on either side of a cell boundary. If θ is an equidistributed random variable in $(-\frac{1}{2}, \frac{1}{2})$ Glimm's method then gives

$$\tilde{u}_{i+\frac{1}{2}}^{n+\frac{1}{2}} = U\{(i+\tfrac{1}{2}+\theta)h,\ (n+\tfrac{1}{2})k\} \tag{44}$$

The grid and the wave fronts used in solving successive Riemann problems are shown in Fig. 1, while the sampling procedure is shown in Fig. 2.

After solving the Riemann problem at $t = (n+\frac{1}{2})k$ the solution (43) is substituted in the right-hand side of Eq. (42) which is solved as a simple difference equation for a corrected $u_{i+\frac{1}{2}}^{n+\frac{1}{2}}$. The corrected values are used as initial data to apply the Glimm method in the next half time interval $(n+\frac{1}{2})k \leqslant t \leqslant nk$. The Glimm solution at $t = nk$ is again corrected from Eq. (42).

The solution of the Riemann problem at each cell boundary is given by algebraic formulae, the details of which depend on the nature of the discontinuities across the boundary. Waves are propagated in both directions, when each discontinuity breaks down, and can be either expansion waves or bores (the shallow water equivalent of shocks). The Riemann problem solutions are given in full in Li and Holt (1981).

The initial conditions for the classical Dam Break Problem are shown in Fig. 3 (Fig. 4, Li and Holt). The dam maintains the drop in water level shown. It is suddenly removed at time $t = 0$. Subsequently a bore is propagated to the left while an expansion is propagated to the left. In the Holt-Li calculations a solid wall is introduced to the left of the dam so that reflection of the bore can be calculated. The solutions before and after reflection are shown in Figs. 4 and 4. These calculations were made by the Glimm method alone, since Eq. (42) was not required in the plane flow case. The Holt-Li method was then applied to the cylindrical dam break problem. In this case a cylindrical bore converged on its center and was reflected there, in a manner similar to a cylindrical implosion in a gas. The wave profiles in this case are shown in Fig. 6.

The final calculation deals with the Upper Critical Depth problem. This concerns the generation of large amplitude surface waves by spherical explosions detonated at different depths below, but near to, the ocean surface. The pressure field from the underwater explosion was calculated previously by Falade and Holt (1978). The data were used to provide initial conditions in the surface wave calculation using Glimm's method. Figure 7 shows the most important result of this calculation namely, that for different depths of explosive charge the maximum amplitude of surface wave generated occurs when the charge depth is equal to one half of the charge radius. This confirms the Upper Critical Depth phenomenon observed in field tests.

Acknowledgment. The author wishes to thank Dr. J. Noye and Dr. C. A. J. Fletcher for their invitation to present this paper. The section dealing with the Method of Integral Relations was partly supported by The U.S. Air Force Office of Scientific Research, Mechanics Branch, Dr. James D. Wilson, Program Manager.

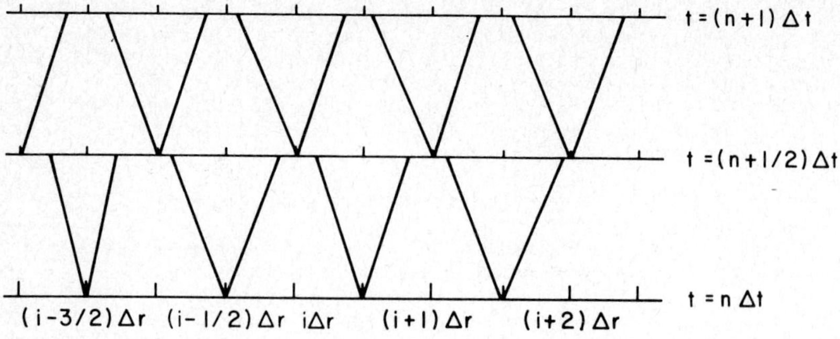

Fig. 1. Sequence of Riemann problems on grid.

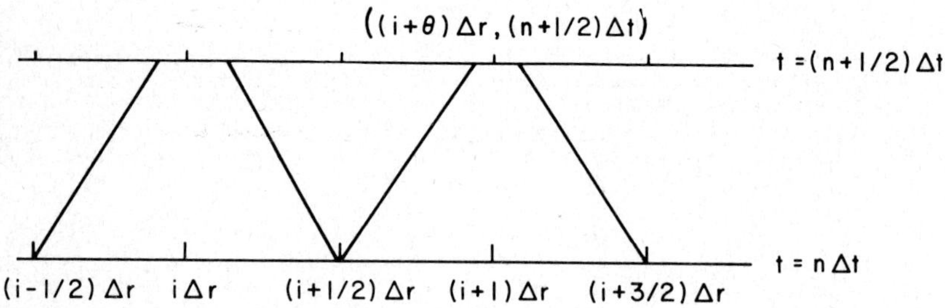

Fig. 2. Sampling procedure for Glimm's scheme.

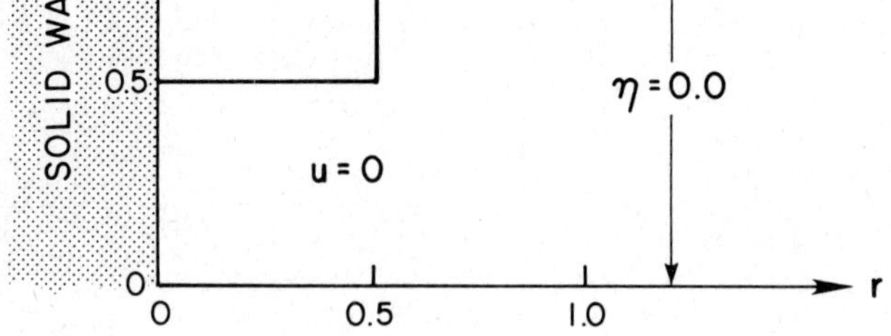

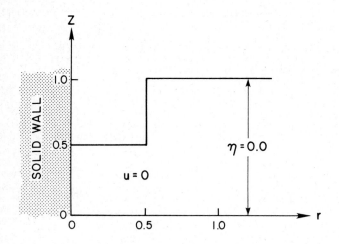

Fig. 3. Initial condition for dam break problem.

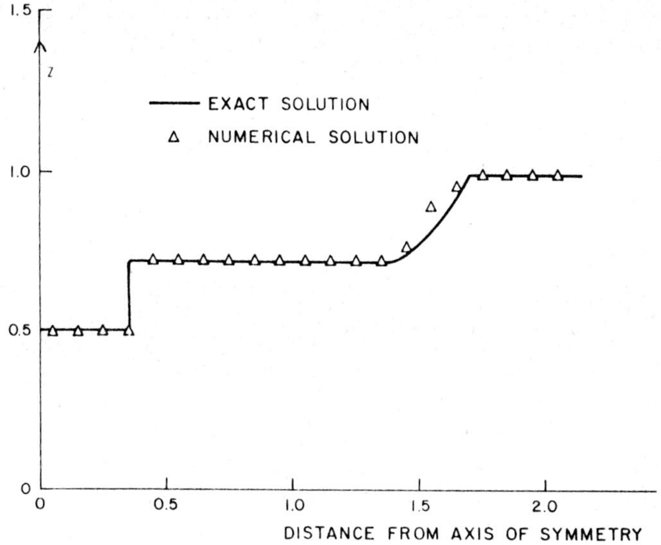

Fig. 4. Dam break problem with plane symmetry.
Time t = 0.69.

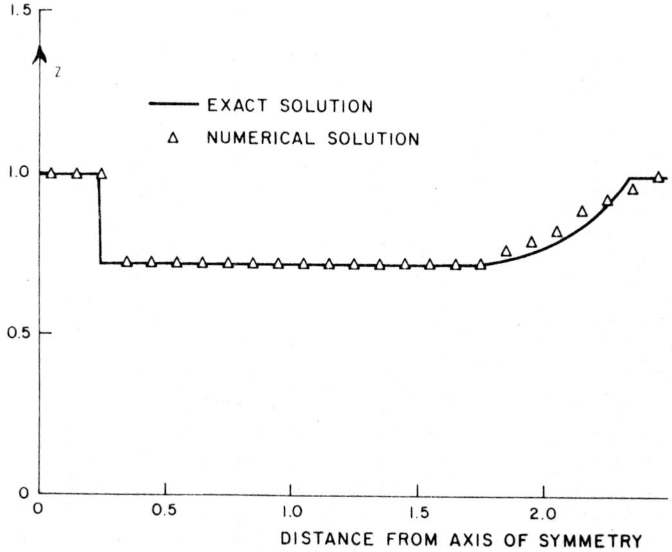

Fig. 5. Dam break problem with plane symmetry.
Time t = 1.36.

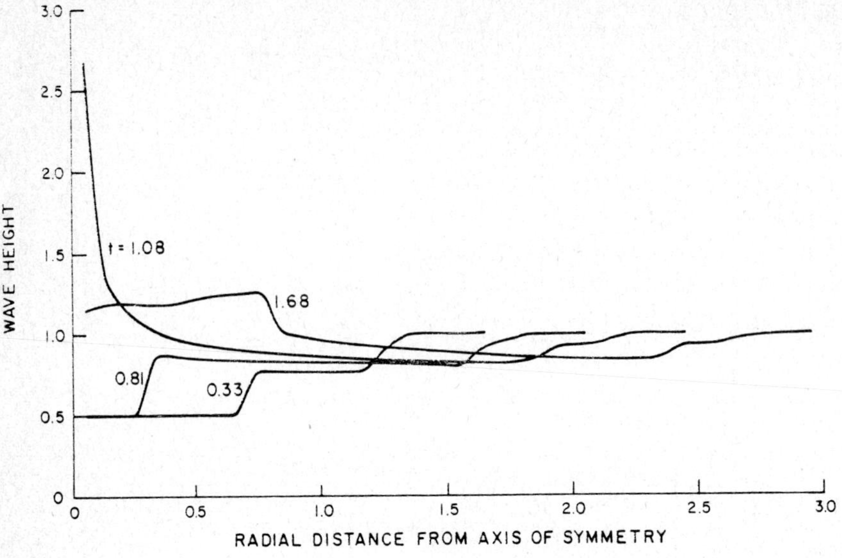

Fig. 6. Dam break problem with cylindrical symmetry.

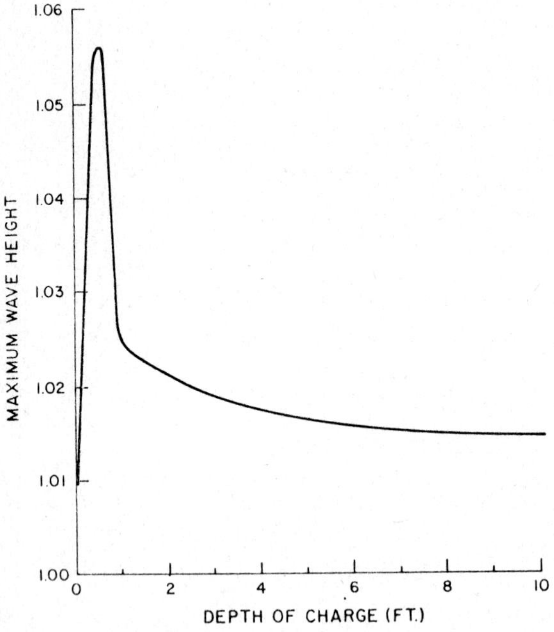

Fig. 7. Variation of maximum wave height with charge depth.

REFERENCES

Abbott, D. E. and Deiwert, G. S., Proceedings, Computation of Turbulent Boundary Layers, 1968 AFOSR-IFP Stanford Conference, Vol. 1, 54-75 (Stanford 1968).

Babenko, K. I., Voskresenskii, G. P., Lyubimov, A. N., Rusanov, V. V., Three-dimensional flow of ideal gases around smooth bodies, NASA TT F-380 (1968) (Russian original published by NAUKA, Moscow 1964).

Belotserkovskii, O. M., Prik. Mat. Mekh. 22 (1958) 206.

Chorin, A. J., J. Comp. Phys. 22 (1976) 517.

Chorin, A. J., J. Comp. Phys. 25 (1977) 253.

Clauser, F. H., The turbulent boundary layer, Advances in Appl. Mech. 4 (1956) 1.

Coles, P. and Hirst, E. (eds.) Proceedings, Computation of Turbulent Boundary Layers, 1968 AFOSR-IFP Stanford Conference, Vol. 2, Compiled Data (Stanford 1968).

Courant, R., Friedrichs, K. O. and Lewy, H., "Uber die partiellen Differenzengleichungen der mathematischen Physik," Math. Ann. 100 (1928) 32.

Courant, R., and Friedrichs, K. O., Supersonic flow and shock waves (Springer-Verlag, Berlin-Heidelberg-New York, 1978).

Dorodnitsyn, A. A., Solution of mathematical and logical problems on high-speed digital computers. Proc. Conf. Develop. Soviet Mach. Machines and Devices, Part 1, 44-52, VINITI Moscow, 1956.

Dorodnitsyn, A. A., Advances in aeronautical sciences, Vol. 3 (Pergamon Press, New York, 1960).

Falade, A. and Holt, M., Phys. Fluids 21 (1978) 1709.

Fletcher, C. A. J., Holt, M., J. Computational Physics 18 (1975) 154.

Fletcher, C. A. J., Computational Galerkin methods (Springer Series in Computational Physics, Springer-Verlag, New York-Heidelberg-Berlin, 1984).

Flores, J. and Holt, M., J. Comp. Phys. 44 (1981) 377.

Gilinskii, S. M., Telenin, G. F., Tinyakov, G. P., Izv. Akad. Nauk, SSSR Mekh. Mash. 4 (1964) 9 (translated as NASA TT F297).

Glimm, J., Comm. Pure Appl. Math. 18 (1965) 497.

Godunov, S. K., Mat. Sborn. 47 (1959) 271.

Godunov, S. K., Alalykin, G. B., Kireeva, I., Pliner, L. H., Solutions of one-dimensional problems in gas dynamics in moving networks, Moscow, NAUKA (1970).

Holt, M., Proc. AGARD Conference on Separated Flows, 69-87 (Rhode St. Genese 1966).

Holt, M., Proc. XVII Int. Astronautical Congress, 383-401 (Polish Scientific Publishers, 1967).

Holt, M., Lu, T. A., Acta Astronautica 2 (1975) 409.

Holt, M., Numerical methods in fluid dynamics (Springer Series in Computational Physics, Springer-Verlag, New York-Heidelberg-Berlin, Second Edition, 1984).

Jones, D. J. and South, J. C. Jr., Application of the method of lines to the solution of elliptic partial differential equations, National Research Council Canada, NRC No. 18021. Aeronautical Report LR 599 (1974).

Kleinstein, G., AIAA Journal 5 (1967) 1402.

Lax, P. D. and Wendroff, B., Difference schemes for hyperbolic equations with high order of accuracy, Comm. Pure Appl. Math. 17 (1964) 381.

Li, K.-M. and Holt, M., Phys. Fluids 24 (1981) 816.

Lyubimov, A. N., Rusanov, V. V., Gas flows past blunt bodies, Part I, NASA TT F-714 (1973) (Russian original published by NAUKA, Moscow 1970).

Massau, J., "Mémoire sur l'intégration graphique des équations aux dérivées partielles. Gand: von Goethem, 1900.

Murphy, J. D. and Rose, W. C., Proceedings, Computation of Turbulent Boundary Layers, 1968 AFOSR-IFP Stanford Conference, Vol. 1, 54-75 (Stanford 1968).

von Neumann, J. and Richtmyer, R. D., A method for numerical calculations of hydrodynamic shocks, J. Appl. Phys. 21 (1950) 232.

Nielsen, J. N., Goodwin, F. K., Kuhn, G. D., Review of the method of integral relations applied to viscous interaction problems including separation, Symp. on Viscous Interaction Phenomena in Supersonic and Hypersonic Flow, Hypersonic Research Lab., Aeronautical Research Labs., Wright Patterson Air Force Base, 1969.

Ozcan, O. and Yang, R.-J., Calculation of supersonic laminar boundary layer separation in an adiabatic concave corner by modified MIR, University of California Course ME 226 term paper, 1982.

Peyret, R. and Taylor, T. D., Computational methods for fluid flow (Springer-Series in Computational Physics, Springer-Verlag, New York-Heidelberg-Berlin, 1983).

Richtmyer, R. D. and Morton, K. W., Difference methods for initial-value problems (Interscience Publishers, New York-London-Sydney, 1967).

Sod, G. A., J. Fluid Mech. 83 (1977) 785.

Spalding, D. B., J. Appl. Mech. 29 (1961) 455.

Yang, R.-J. and Holt, M., to appear in J. Appl. Mech. (1983).

Yeung, W.-S. and Yang, R.-J., J. Appl. Mech. 48 (1981) 701.

Computational Techniques & Applications: CTAC-83
J. Noye & C. Fletcher (Editors)

THE SIMULATION OF TIME DEPENDENT PROBLEMS USING BOUNDARY ELEMENTS

C.A. Brebbia*

University of Southampton
and
Computational Mechanics Institute, England

1. INTRODUCTION

The boundary element method (BEM) has now become an accepted numerical tool for the solution of engineering problems [1,2]. The main advantage of the technique is the reduction in the number of unknowns needed to solve a problem, as the method only requires the definition of elements and nodes on the surface of the domain, rather than everywhere as in finite elements or finite differences. In addition BEM gives more accurate results than domain techniques and it is very well suited to problems extending to infinity.

Finite elements on the contrary requires the definition of internal as well as boundary nodes, plus a series of associated data such as nodal coordinates and connectivity, all of which make the method difficult to use, specially when being implemented on part of computer aided design systems. These difficulties have led engineers to investigate the uses of boundary elements and the technique has gained a great deal of acceptance for solving elliptic type problems in solid as well as fluid mechanics. A substantial body of literature now exists on the applications of boundary elements and the most up to date references can be seen in the last three international conferences on this topic [3,4,5] as well as in [2].

Advances have recently being made in the solution of non-linear and time dependent problems using boundary elements. Some of the resulting papers can be seen in references [3] to [5], more definite work on this topic has been published as chapters in the Progress in Boundary Element Methods Series [6,7,8]. The present paper attempts to review some of the recent advances in the solution of time dependent problems using the BEM, including parabolic and hyperbolic cases. The former case follows the formulation presented in reference [9] while the latter is on the lines of references [10,11].

A new procedure for forming dynamic matrices in boundary elements using only the boundary nodes has been reported in reference [12]. The technique can be applied to solve eigenvalue and transient dynamic analysis in solid and fluid mechanics. It allows for dynamic problems to be treated in a similar manner as finite differences or finite elements, i.e. the problem can be reduced to a set of time dependent differential equations expressed in matrix form. The free oscillations problem can then be reduced to the solution of an algebraic eigenvalue system. The main advantage of the new technique is that the boundary integrals need to be computed only once as they are frequency independent. Hence the procedure is extremely economic for eigenvalue solutions when compared to the ones previously presented.

---

* Presently Visiting Fellow at the Chisholm Institute of Technology, Melbourne, Australia.

2. STEADY STATE POTENTIAL PROBLEMS

In order to define the notation to be used throughout the paper and introduce some basic concepts, let us first consider the case of Poisson's equation, i.e.

$$\nabla^2 u + b = 0 \qquad \text{in } \Omega \tag{1}$$

with the following boundary conditions,

$$\begin{aligned} u &= \bar{u} \qquad \text{on } \Gamma_1 \\ q &= \bar{q} \qquad \text{on } \Gamma_2 \end{aligned} \tag{2}$$

The problem can now be written in weighted residual form by defining a distribution function u* - with derivative q* normal to the boundary - such that [1,2],

$$\int_\Omega (\nabla^2 u + b) u^* d\Omega = \int_{\Gamma_2} (q - \bar{q}) u^* d\Gamma - \int_{\Gamma_1} (u - \bar{u}) q^* d\Gamma \tag{3}$$

Integrating the Laplacian in this expression twice by parts the following relationship is obtained,

$$\int_\Omega (\nabla^2 u^*) u \, d\Omega + \int_\Omega b u^* \, d\Omega = \int_\Gamma u \, q^* \, d\Gamma - \int_\Gamma q \, u^* \, d\Gamma \tag{4}$$

where for simplicity the boundary integrals are written for the whole boundary, i.e. $\Gamma = \Gamma_1 + \Gamma_2$. The necessary boundary conditions - equation (2) - will be imposed later.

In order to eliminate the first domain integral in (4) one can find the fundamental solution for the Laplace's equation, i.e.

$$\nabla^2 u^* + \delta_i = 0 \tag{5}$$

where δ_i is a Dirac delta function zero everywhere but at the point i. The fundamental solutions for Laplace's equation in different dimensions are well known [2].

Substituting (5) into (4) one finds the following expression for a singularity applied at 'i', i.e.

$$u_i + \int_\Gamma u \, q^* \, d\Gamma = \int_\Gamma q \, u^* \, d\Gamma + \int_\Omega b u^* \, d\Omega \tag{6}$$

or in general, [1,2]

$$c_i u_i + \int_\Gamma u\, q^*\, d\Gamma = \int_\Gamma q\, u^*\, d\Gamma + \int_\Omega b u^*\, d\Omega \tag{7}$$

where c_i is a constant depending on the position of the singularity ($c_i = \frac{1}{2}$ on a smooth boundary, $c_i = 1$ for interior points, $c_i = 0$ for points outside Ω and c_i proportional to the solid angle in non-smooth boundaries).

Equation (7) is the starting statement for the BEM. It involves still a domain integration, but this term can be taken to the boundary under the condition that b is harmonic. Then one can define,

$$u^* = \nabla^2 v^* \tag{8}$$

where the v^* function is given in reference [2]. (Similar considerations apply for body forces in elastostatics, see also reference [2].)

After the above integrations have been performed the matrix form of equation (7) can be written,

$$\underset{\sim}{H}\, \underset{\sim}{U} = \underset{\sim}{G}\, \underset{\sim}{Q} + \underset{\sim}{B} \tag{9}$$

One can now apply the boundary conditions (2) and rearrange the columns of (9) to produce the final system of equations, i.e.

$$\underset{\sim}{A}\, \underset{\sim}{X} = \underset{\sim}{F} \tag{10}$$

Solving equation (10) will render the unknown values of u and q on the boundaries. Internal values of potential can be calculated using equation (7) for an internal point 'i' (in this case $c_i = 1$).

3. TIME DEPENDENT DIFFUSION

Time dependent diffusion problems are of parabolic character and can be represented by the following equation

$$\alpha \nabla^2 u + b = \frac{\partial u}{\partial t} \qquad \text{in } \Omega \tag{11}$$

with boundary conditions (2) and initial conditions u_0 known in Ω at the initial time t_0. α is a constant dependent on the material properties.

Some of the first boundary integral solutions for the diffusion equation were obtained by removing the time dependency in equation (11) using the Laplace's transform. The boundary integral equation in the transform space was solved numerically and the inversion was then performed in order to evaluate the time dependent solution. The technique is very elegant but only suitable for well defined problems where the solution varies smoothly and hence it is difficult to use for some general engineering problems. Other authors have proposed using finite differences schemes in time [1,3] applying a step by step procedure to advance the solution in time. This scheme requires the integration of a domain term which represents the effect of the initial conditions. It also has the advantage that no major savings in computer time can be expected when compared against the similar schemes used in domain techniques.

In a series of publications Wrobel and Brebbia [2,9] have investigated the use of boundary elements in time and space. The approach employs time dependent fundamental solutions - for a complete list of these solutions see reference [2].

Consider first the weighted residual expression for equation (11) with the corresponding boundary conditions (2), i.e.

$$\int_{t_0}^{t_F} \int_{\Omega} \left(- \frac{\partial u}{\partial t} + \alpha \nabla^2 u + b\right) u^* \, d\Omega \, dt = \tag{12}$$

$$\int_{t_0}^{t_F} \int_{\Gamma_2} \alpha(q - \bar{q})u^* \, d\Gamma \, dt - \int_{t_0}^{t_F} \int_{\Gamma_1} \alpha(u - \bar{u})q^* \, d\Gamma \, dt$$

Integrating twice by parts the term in ∇^2 and once with respect to time the $\partial u/\partial t$ term, one obtains the following inverse statement after applying the causality principle, i.e.

$$\int_{t_0}^{t_F}\int_{\Omega}\left(\alpha\,\nabla^2 u^* + \frac{\partial u^*}{\partial t}\right) u \, d\Omega \, dt \; + \tag{13}$$

$$\int_{t_0}^{t_F}\int_{\Omega} b u^* \, d\Omega \, dt - \left[\int_{\Omega} u^* \, u \, d\Omega\right]_{t_0}$$

$$= \int_{t_0}^{t_F}\int_{\Gamma} \alpha \, u \, q^* \, d\Omega \, dt - \int_{t_0}^{t_F}\int_{\Gamma} \alpha \, q \, u^* \, d\Gamma \, dt$$

The fundamental solution needed for this problem corresponds to the adjoint of the original equation, i.e.

$$\alpha\,\nabla^2 u^* + \frac{\partial u^*}{\partial t} + \delta(\xi)\,\delta(t_F) = 0 \tag{14}$$

Notice that the Dirac delta correspond to space and time, u* is an exponential type function on both variables - for its explicit formulation, see [2] -. After substituting the fundamental solution of (14) into (13) one obtains,

$$c_i u_i + \int_{t_0}^{t_F}\int_{\Gamma} \alpha \, u \, q^* \, d\Gamma \, dt = \tag{15}$$

$$\int_{t_0}^{t_F}\int_{\Gamma} \alpha \, q \, u^* \, d\Gamma \, dt + \int_{t_0}^{t_F}\int_{\Omega} b u^* \, d\Omega \, dt + \int_{\Omega} u_0 u^* \, d\Omega$$

For the numerical formulation of (15) the boundary Γ is discretized into elements with prescribed space and time interpolation functions. One can also assume that the domain Ω is divided into a number of integration cells, although this is not always necessary as we will see.

Two different numerical procedures can be employed. The first is to consider that we integrate over a sufficiently small time step ($\Delta t = t_F - t_0$) over which the unknown can be assumed to vary on time with a simple function - usually constant, linear or quadratic - this implies that the initial condition integral has to be recalculated at the beginning of each new time step. The second procedure always starts the time integration process at the initial time t_0 and hence there is no need to compute initial value integrals after each step. Furthermore if the initial conditions at the start of the process satisfy Laplace's equation the domain integral in equation (15) can be transformed into an equivalent boundary integral. In this case the integrals need still to be discretized in time as the interpolation functions for the variables are not applicable from time t_0 to t_F. This discretization is simply the decomposition of the total integral with respect to time into a summation of integrals and does not imply computing any domain integrals.

The two processes are completely equivalent from the point of view of accuracy of the results. The computer efficiency however is different. Although the second process does not require domain integrations, thus effectively reducing the dimensionality of the problem, the influence coefficients need to be recalculated for each new time step since they depend, through the fundamental solution on the actual value of time. For the first process, however, they depend on the value of the time step itself and so if Δt is assumed to be constant throughout the analysis these coefficients can all be computed only once and stored (the coefficients include those related to the domain integral). From these arguments one can conclude that in general the first procedure is computationally more efficient than the second, excluding the case of problems with regions extending to infinity. In spite of this the second approach is the one favored by the author for computer coding as it facilitates their use, i.e. it does not require the definition of any internal points.

EXAMPLE 1

This example studies a plane plate, initially at $0^{o}C$, surrounded by a medium at $100^{o}C$. Its cross section is 0.1 x 0.1m and the values of the thermal conductivity, heat capacity and heat transfer coefficients are 18 kcal/(hm ^{o}C), 912 kcal/(m^3 ^{o}C) and 5000 kcal/(h m^2 ^{o}C), respectively.

The surface temperature of the plate is plotted in figure 1. Due to symmetry, one quarter of its cross section needs to be discretized. The finite element analysis employs 5 parabolic or cubic elements in time [6]. Results using a central finite difference scheme are also presented.

The discretization for the BEM analysis consists of only 8 boundary elements (no elements are needed over the symmetry axes). The time steps adopted for this problem is extremely large and because of this, the BEM with linear time variation produced oscillations which are, however, much smaller than the finite difference and finite element ones. No oscillation was detected for the BEM with constant time variation, while the linear time variation introduces an error during the first time interval which damps out very quickly.

EXAMPLE 2

For problems with regions extending to infinity, BEM solutions are much more economical than finite element ones. In order to demonstrate this, the case of a circular opening in an infinite plane region, initially at zero temperature is studied next. The radius of the hole is unity, its ambient temperature equals -10 and the material properties of the medium are also taken to be unity, for simplicity.

The variation of the surface temperature with time is presented in figure 2 for various values of the heat transfer coefficient, and compared with an analytical solution. The agreement between the two solutions is very good. A time step of 0.5 was adopted and the analysis carried out until the surface temperature began to drop significantly. For obvious reasons, only the second numerical approach described in this section was employed in this BEM analysis and, due to symmetry, only one quarter of the interface between hole and medium was divided into 6 boundary elements.

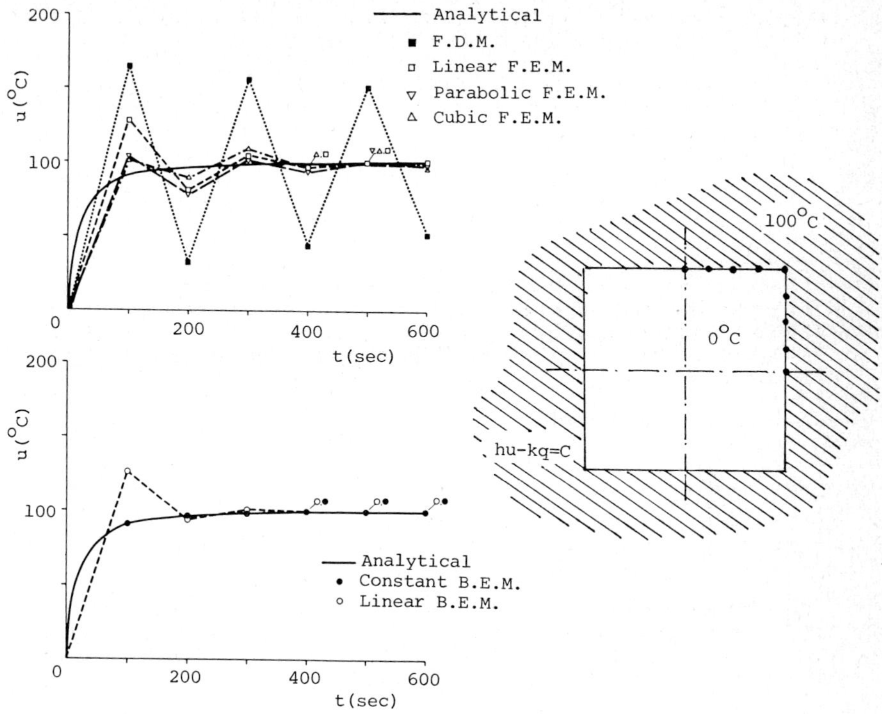

Figure 1 Surface temperature of plane plate

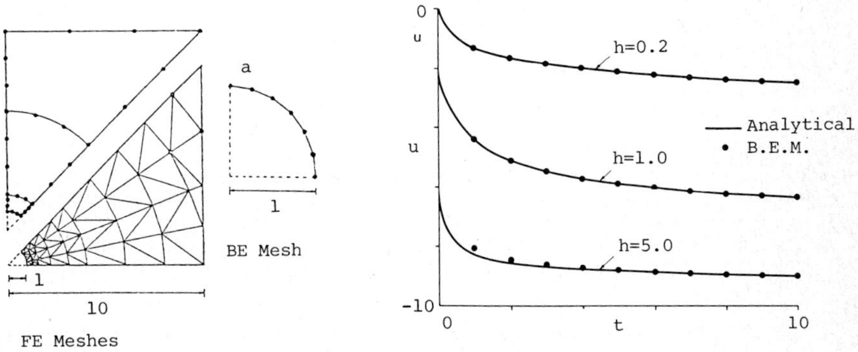

Figure 2 Discretization and surface temperature of cooling hole in an infinite domain

This problem was also studied using finite elements - for references see [6] - but since the finite element method is a domain-type technique, the infinite region had to be limited by a finite, non-conducting boundary. In order to achieve the same level of accuracy, a time step ten times smaller ($\Delta t = 0.05$) was required and the domain discretized into 70 triangular elements or 3 cubic isoparametric elements.

EXAMPLE 3

This example is more practical, an actual turbine disc temperature variation is studied here (figure 3). The initial temperature of the turbine disc is 295.1^oK and the values of the thermal conductivity, density and specific heat of the material are 15W/(m^oK), 8221 kg/m^3 and 550 J/(kg^oK), respectively. There are 18 different zones along the boundary, each of which with a different set of prescribed values for the heat transfer coefficient and the temperature of the surrounding gas. This time variation at one of such boundary zones is shown in figure 4.

The BEM analysis was performed by discretizing the boundary into 90 elements and using no internal points. In order to simplify the computation, a constant temperature was subtracted out so as to make the initial temperature equal to zero, thus avoiding the domain integration. This value was afterwards added to the solution.

Isotherms at various times are plotted in figure 3 and compared to finite element results obtained with 85 parabolic isoparametric elements and 348 nodes. The agreement was, in general, excellent (for further details see chapter 6 of reference [6]).

4. TIME DEPENDENT CONVECTION

Considerable effort has recently being applied to solve the transport equation using boundary elements. The efficient solution of this problem could open up a whole new set of applications for the BEM.

A way of obtaining the necessary integral statement for convection is simply by replacing the distributed source term b in equation (15) by

$$b - \left(v_x \frac{\partial u}{\partial x} + v_y \frac{\partial u}{\partial y} + v_z \frac{\partial u}{\partial z}\right) \tag{16}$$

This gives the following integral relationship,

$$c_i u_i + \int_{t_0}^{t_F} \int_\Gamma \alpha\, u\, q^*\, d\Gamma\, dt = \int_{t_0}^{t_F} \int_\Gamma \alpha\, q\, u^*\, d\Gamma\, dt \tag{17}$$

$$+ \int_\Omega u_o u^*\, d\Omega - \int_{t_0}^{t_F} \int_\Omega \left(v_x \frac{\partial u}{\partial x} + v_y \frac{\partial u}{\partial y} + v_z \frac{\partial u}{\partial z}\right) u^*\, d\Omega\, dt$$

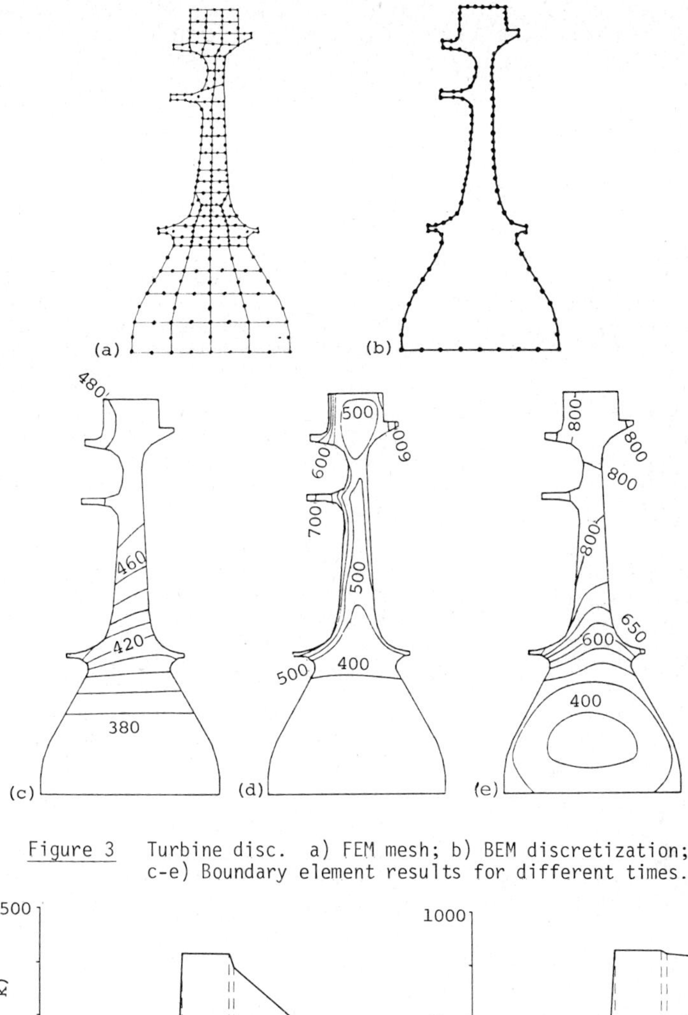

Figure 3 Turbine disc. a) FEM mesh; b) BEM discretization; c-e) Boundary element results for different times.

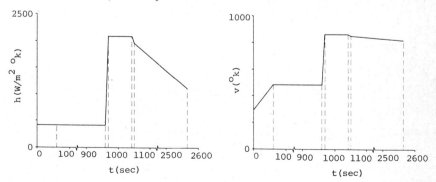

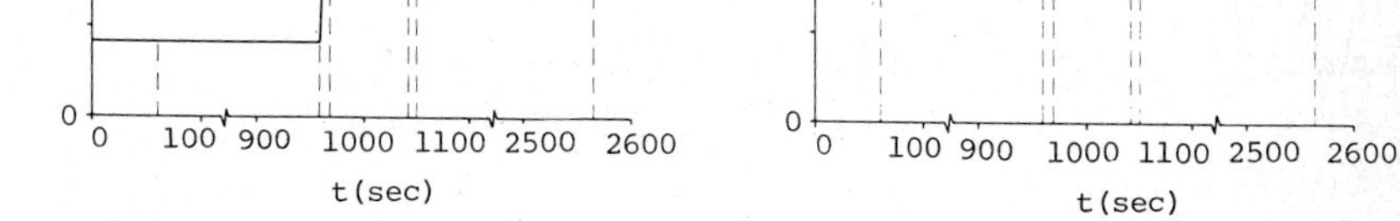

Figure 4 Time variation of heat transfer coefficient and temperature of surrounding for the rod under a Heaviside forcing function.

Usually one prefers to integrate by parts the last term in order to pass the derivatives from the approximate function u to the fundamental solution u*. This gives,

$$c_i u_i + \int_{t_0}^{t_F} \int_\Gamma \alpha\, u\, q^*\, d\Gamma\, dt = \int_{t_0}^{t_F} \int_\Gamma \alpha\, q\, u^*\, d\Gamma\, dt \qquad (18)$$

$$+ \int_\Omega u_0 u^*\, d\Omega - \int_{t_0}^{t_F} \int_\Gamma v_n\, u u^*\, d\Omega\, dt$$

$$+ \int_{t_0}^{t_F} \int_\Omega \left(v_x \frac{\partial u^*}{\partial x} + v_y \frac{\partial u^*}{\partial y} + v_z \frac{\partial u^*}{\partial z} \right) u\, d\Omega\, dt$$

The last two integrals describe now the convection from the boundary and the convective effect in the domain.

The last integral in equation (18) requires the definition of internal nodes in BEM. This makes the procedure computationally not as efficient as the BEM formulation discussed in part 2 and 3. Results obtained using this approach and published in reference [14] indicate that the method gives very accurate results and allows for larger time steps than those required for the FEM solution of similar problems. For a fuller discussion of this problem the reader is referred to reference [14] and [2].

5. THE SCALAR WAVE EQUATION

The scalar wave equation is another type of problem whose solution can now be attempted using boundary elements. The equation to be analysed is hyperbolic and can be written as,

$$\nabla^2 u = \frac{1}{c^2} \frac{\partial^2 u}{\partial t^2} \qquad (19)$$

where c is the wave celerity. The equations requires a set of boundary conditions such as given by equation (2) and initial conditions defined by the function and its velocity at t_0.

The integral statement can be obtained using weighted residuals, i.e.

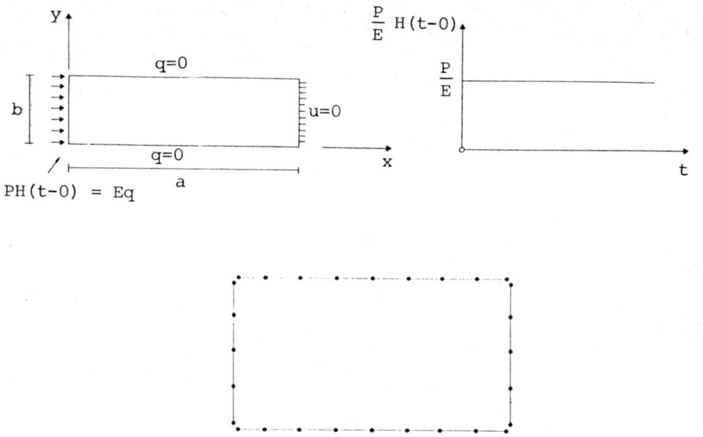

Figure 5a Boundary conditions, geometry definitions and boundary discretization for the rod under a Heaviside forcing function.

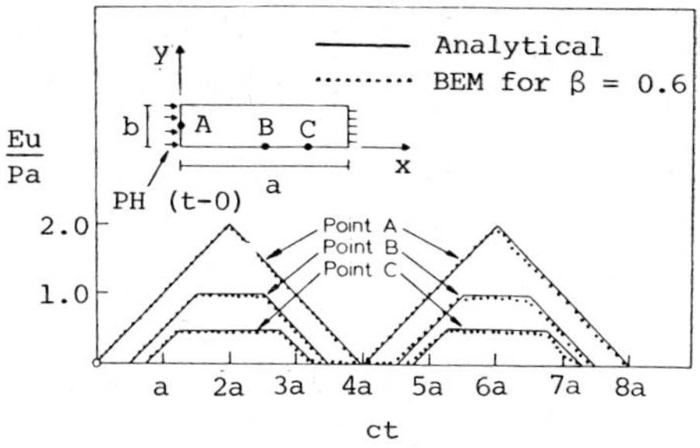

Figure 5b Displacements at boundary points A (0, b/2), B (a/2, 0) and C (3a/4, 0) for one-dimensional rod.

Figure 5c Normal derivative of displacement at point D (a, b/2) for one-dimensional rod.

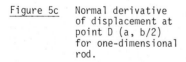

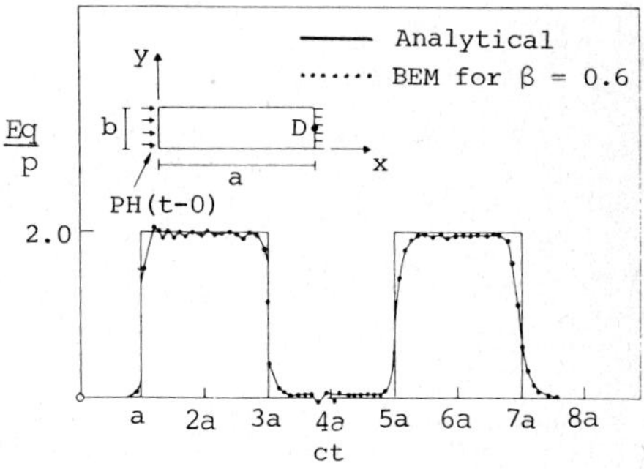

$$\int_{t_0}^{t_F}\int_{\Omega}\left(\nabla^2 u - \frac{1}{c^2}\frac{\partial^2 u}{\partial t^2}\right)u^*\, d\Omega\, dt = \tag{20}$$

$$\int_{t_0}^{t_F}\int_{\Gamma}(q - \bar{q})u^*\, d\Gamma\, dt - \int_{t_0}^{t_F}\int_{\Gamma}(u - \bar{u})q^*\, d\Gamma\, dt$$

Integrating by parts with respect to time and space and applying causality produces the following expression,

$$\int_{t_0}^{t_F}\int_{\Omega}\left(\nabla^2 u^* - \frac{1}{c^2}\frac{\partial^2 u^*}{\partial t^2}\right) u\, d\Omega\, dt + \tag{21}$$

$$\int_{t_0}^{t_F}\int_{\Gamma} u^*\, q\, d\Gamma\, dt - \int_{t_0}^{t_F}\int_{\Gamma} u\, q^*\, d\Gamma\, dt$$

$$\frac{1}{c^2}\int_{\Omega}\left[\frac{\partial u^*}{\partial t}\, u - \frac{\partial u}{\partial t}\, u^*\right]_{t_0} d\Omega = 0$$

The fundamental solution in this case is given by the solution of

$$\nabla^2 u^* - \frac{1}{c^2}\frac{\partial^2 u^*}{\partial t^2} + \delta(\xi)\ \delta(t_F) = 0 \tag{22}$$

which gives

$$c_i u_i + \frac{1}{c^2}\int_{\Omega}\left[\left(\frac{\partial u^*}{\partial t}\right)_o u_o - \left(\frac{\partial u}{\partial t}\right)_o u_o^*\right] d\Omega \tag{23}$$

$$\int_{t_0}^{t_F}\int_{\Gamma}(u^* q - q^* u)\, d\Gamma\, dt$$

It is interesting to point out that the three dimensional fundamental solution of (22) is expressed in terms of a Dirac delta function and gives rise to the so-called Huggens principle or wave retarded potential formulation. A special feature of the three-dimensional analysis is that no time integration is required.

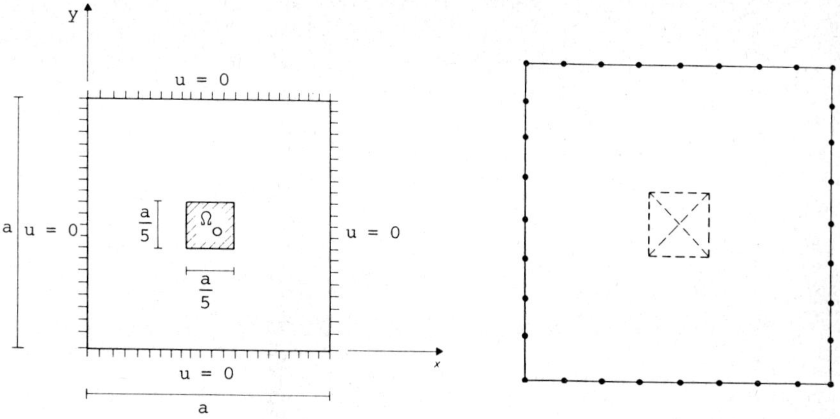

Figure 6a Geometry definition, discretization, boundary and initial conditions for membrane analysis.

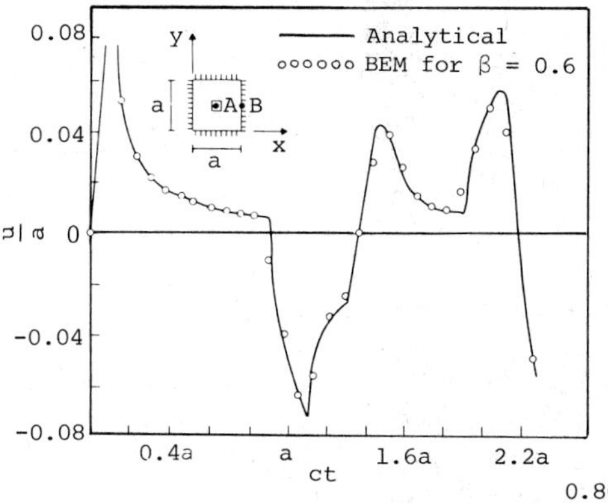

Figure 6b Displacements at point A (a/2, a/2)

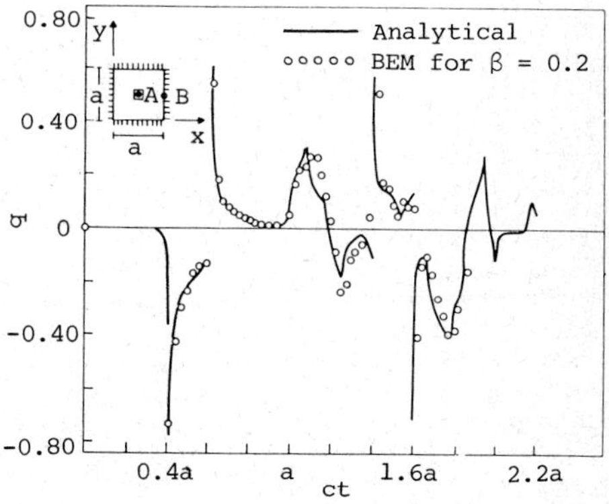

Figure 6c Normal derivative of displacement at point B (a, a/2)

This simplification does not apply to the two dimensional case which because of this has seldom being attempted. The examples in this paper however are for two dimensional case for which the numerical implementation of equation (23) is rather complex. The solution proposed by Mansur and Brebbia [10], [11] follows the same format as the second approach of section 3, i.e. the time origin is always taken at the initial time t_o. This procedure is algebraically more complex but has the advantage, from the user's point of view that it does not require any internal points. In addition, the above authors propose using a function linear in space for the potentials and constant for the fluxes as these variations can represent better some types of shock waves. They recommend the use of linear functions on time for both variables.

EXAMPLE 4

Results obtained using the two-dimensional boundary element program were compared against analytical results for a one-dimensional rod under a Heaviside type forcing function. The boundary element solution considered a rectangular domain with sides of length a and b (b=a/2) as indicated in figure 5a. The u-function represents the displacements which are assumed to be zero at x = a and their normal derivative q is taken as null at y = 0 and y = b for any time. At x = 0 and t = 0 a load Eq is suddenly applied and kept constant until the end of the analysis -E is the Young's modulus of the material. Due to the topology and boundary conditions the problem is actually one-dimensional and its analytical solution can be found easily.

The boundary was discretized into 24 linear elements using double nodes at the corners. The time was subdivided into equal intervals such that,

$$\beta = \frac{c\ \Delta t}{\ell} = \text{constant}$$

where ℓ is the length of an element.

Figure 5c shows the BEM and analytical results at internal and boundary points. The order of accuracy of the BEM results is excellent. In figure 5d the normal derivative of u at point (a,b/2) versus ct is presented. Except for the presence of a comparatively small amount of noise boundary elements and analytical solutions are in good agreement.

As expected it was found that care must be taken on the choice of β in order to avoid noise, which although usually not critical for displacements, can be excessive for its derivatives or tractions. The value of β = 0.6 was found to be the optimum for this problem.

EXAMPLE 5

This example analyses a square membrane with an initial velocity v_o = c prescribed all over the boundary.

The boundary was discretized into 32 elements and Ω_o was divided into four cells. Analytical and BEM results for displacements at point (a/2, a/2) and the normal derivative of displacements at point (a, a/2) more compared.

The values of u and q for β = 0.2 are plotted in figure 6b and 6c respectively. Although the agreement for displacements was reasonable for β = 0.6, q was not so well represented and it was decided to reduce the β until the results coverage to the analytical solution.

6. DYNAMIC MATRIX REPRESENTATION

The recent work by Nardini and Brebbia [12][15] has concentrated on the represtation of dynamic or mass type matrices in function only of the boundary nodes. This new development allows for wave equation problems to be treated in a similar way as in finite differences or finite elements, i.e. the problem is reduced to a set of time dependent differential equations expressed in matrix form. The free oscillations or vibrations problem can then be reduced to the solution of an algebraic eigenvalue problem. The main advantage of the new approach is that the boundary integrals need to be computed only once as they are frequency independent and static fundamental solutions can be employed to generate all matrices. Hence the procedure is extremely economic when compared to the ones previously presented. The technique also allows for the general elastodynamic case to be solved in the time rather than the transform domain.

The approach will be described here for the case of the steady state potential wave equation but once the dynamics or mass matrix is found it could be used to carry out a time integration procedure for instance.

The starting weighted residual statement is

$$\int_\Omega \left(\nabla^2 u + \frac{\omega^2}{C^2} u\right) u^* d\Omega = \int_{\Gamma_2} (q - \bar{q})\, u^* d\,\Gamma - \int_{\Gamma_1} (u - \bar{u})\, u^* d\,\Gamma \tag{24}$$

Integrating the Laplacian by parts one obtains

$$\int_\Omega \left(\nabla^2 u^* + \frac{\omega^2}{c^2} u^*\right) u\, d\Omega = \int_\Gamma (u^* q - u q^*)\, d\Gamma dt \tag{25}$$

Using the statis fundamental solution the following integral statement can be written,

$$c_i u_i = \int_\Gamma u^* q\, d\Gamma - \int_\Gamma q^* u\, d\Gamma + \frac{\omega^2}{c^2} \int_\Gamma u u^* d\Omega \tag{26}$$

It is now proposed to transform the domain integral in the above expression into a boundary integral by using an approximation for the function u, such that,

$$u = \alpha^j f^j \tag{27}$$

One can now consider that the f^j functions are a source distribution and consequently can be associated with a pseudo potential field $\hat{u}$ and its corresponding flux $\hat{q}$. The f^j can be taken to be very simple functions, such as the distance between nodes and hence can be easily integrated to yield the $\hat{u}$ and $\hat{q}$ state). This means that the mass integral can be transformed as follows,

$$\frac{\omega^2}{c^2}\int_\Omega uu^* d\Omega = \frac{\omega^2}{c^2}\alpha^j\left\{-c_i u_i^j + \int_\Gamma u^*\hat{q}^j d\Gamma - \int_\Gamma q^*\hat{u}^j d\Gamma\right\} \tag{28}$$

The new boundary integrals on the right hand side may appear to be difficult but if the same type of interpolation functions are used for the pseudo field $\hat{u}$ and $\hat{q}$ as for the actual field u and q, they produce the same $\underset{\sim}{H}$ and $\underset{\sim}{G}$ matrices as for the boundary terms in equation (26), i.e. the integrals in (28) give the following matrix equation

$$(-\hat{H}\hat{U} + \hat{G}\hat{Q})\{\underset{\sim}{\alpha}\}$$

To find the α's, one can consider the values of the functions f^j at the nodal points and obtain a system of linearly independent equations, i.e.

$$\underset{\sim}{U} = [\underset{\sim}{F}]\,\underset{\sim}{\alpha} \tag{30}$$

$$\therefore\ \underset{\sim}{\alpha} = [\underset{\sim}{E}]\,\underset{\sim}{U}\ ,\quad \text{where } [\underset{\sim}{E}] = [\underset{\sim}{F}]^{-1}$$

Hence the matrix representation of equation (26) can be written using only boundary integrals, i.e.

$$\underset{\sim}{H}\,\underset{\sim}{U} - \underset{\sim}{G}\,\underset{\sim}{Q} = \frac{\omega^2}{c^2}\underset{\sim}{M}\,\underset{\sim}{U}$$

where, (31)

$$\underset{\sim}{M} = (-\underset{\sim}{H}\hat{\underset{\sim}{U}} + \underset{\sim}{G}\hat{\underset{\sim}{Q}})[\underset{\sim}{E}]$$

To solve an algebraic eigenvalue problem one needs to rearrange the system of equations (31) by applying homogeneous boundary conditions on Γ_1 and Γ_2. This finally gives,

$$(\underset{\sim}{H}' - \omega^2\underset{\sim}{M}')\underset{\sim}{U} = \underset{\sim}{0} \tag{32}$$

which can now be solved as usual.

The same procedure applied for time dependent problems for which a dynamic or mass matrix can now be found and the equations written matrix form in terms of the accelerations. A standard time integration scheme can then be used.

Experience with finite elements and finite difference has demonstrated that the representation of mass matrices can be approximated using simple functions than those required for the stiffness part of the system. The following examples validate this experience.

EXAMPLE 6

The following examples illustrate how accurate the results are when using mass matrices formed in accordance with the boundary technique described in this section. Although the technique was explained with reference to the potential wave equation, it is also applicable to elastodynamics following the same procedure [16].

The dynamic properties of a deep cantilever beam (h=6, ℓ=24) was studied first. The structure shown in figure 7, was analysed using several quadratic boundary element measures, and the results compared to the finite element ones. The third and fifth nodes are the longitudinal ones for which the results are very accurate even for the smaller discretization. For the transverse nodes (first, second and fourth) it is to be expected that some elements will be needed to obtain the

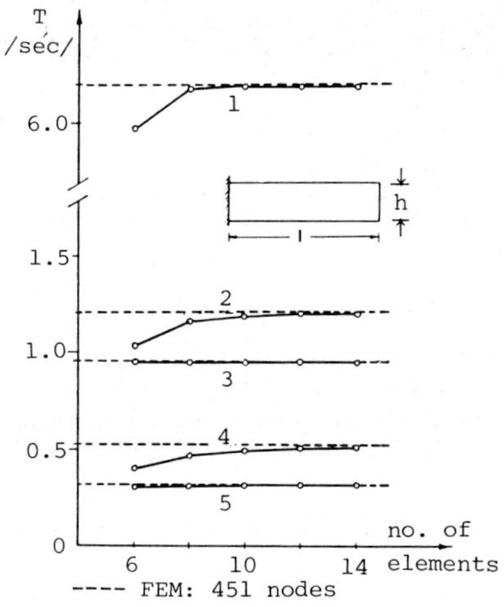

Figure 7 Periods of free vibrations of a deep cantilever beam.

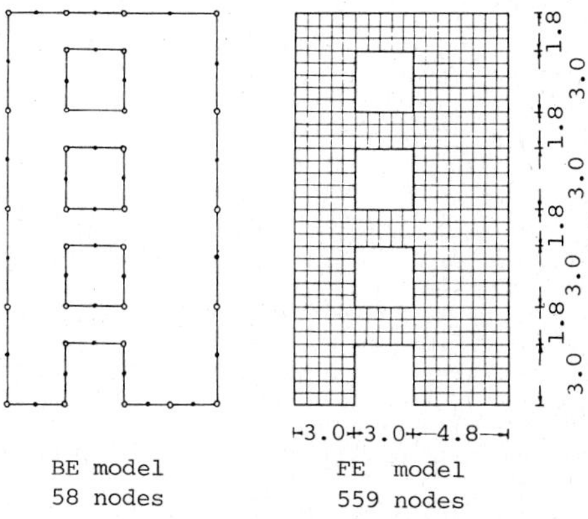

Figure 8 The two numerical models of the shear wall.

same degree of accuracy. In all cases however the solutions show very good agreement with finite element results.

EXAMPLE 7

Figure 8 shows a more realistic application, i.e. a shear wall with four openings. The boundary element model consists of 29 quadratic elements with 58 nodes, as shown in the figure. The finite element mesh comprising of 559 nodes and used for comparison purposes is shown in the same figure. The results for the free vibration periods of the first eight natural nodes are given in table 1. In spite of the complicated geometry and the rather small number of boundary elements used, the agreement of the results is surprisingly good.

Mode	1	2	3	4	5	6	7	8
BEM	3.022	0.875	0.822	0.531	0.394	0.337	0.310	0.276
BEM	3.029	0.885	0.824	0.526	0.409	0.342	0.316	0.283

Table 1 - Periods of free vibrations for the two models

7 CONCLUSIONS

Boundary elements have only recently being applied to solve time dependent problems in a general manner. Previous solution of parabolic and hyperbolic cases tended to concentrate in particular applications. Wave propagation problems for instance were in most cases solved using Laplace's or Fourier's transform and their applications were restricted to a limited range of practical engineering problems.

Deeper understanding of the basic principles and of the approximations involved when solving boundary integral equations in numerical form, was achieved as a result of their interpretation as a weighted residual technique [1][2]. It was then a natural development to interpret parabolic and hyperbolic problems using the concepts of distribution in time and space. This technique, pioneered by Brebbia, Wrobel and Mansur [2][10] produces accurate and stable results. Its acceptance is reflected in the fact that it has been implemented in commercially available computer codes [17].

In spite of the above mentioned advantages, many practical applications do not require using the space and time dependent fundamental solution. One may for instance be only interested in computing the free vibrations frequencies of the system or in carry out - because of non-linearities or other reasons - a full step by step analysis. In these cases one needs to express the problems in function of their mass matrices, which up to now were generally difficult and expensive to formulate.
Nardini and Brebbia [17] in 1982 proposed a very simple and elegant way of using frequency or time independent fundamental solutions to form the mass matrices of the system. The advantages of their approach consisted in that the integrations are carried out only on the boundary considerably reducing the description of a particular problem and the subsequent computations. This technique has opened a new range of applications for the boundary element method in wave propagation problems.

REFERENCES

[1] BREBBIA, C.A. "The Boundary Element Method for Engineers". Pentech Press, London, Halstead Press, N.Y., 1978. Second Edition 1980.

[2] BREBBIA, C.A., TELLES, J. & WROBEL, L. "Boundary Element Methods - Theory and applications in Engineering". Springer-Verlag, Berlin and N.Y., 1983.

[3] "Boundary Element Methods" (Ed. C.A. BREBBIA) Proceedings of the 3rd Int. Conf. on BEM, California, 1981. Springer-Verlag, Berlin and N.Y., 1981.

[4] "Boundary Element Methods in Engineering" (Ed. C.A. BREBBIA) Proceedings of the 4th Int. Conf. on BEM, Southampton, 1982. Springer-Verlag, Berlin and N.Y., 1982.

[5] "Boundary Element Techniques" (Eds. C.A. BREBBIA, T. FUTAGAMI & M. TANAKA). Proceedings of the 5th Int. Conf. on BEM, Hiroshima, 1983. Springer-Verlag, Berlin and N.Y., 1983.

[6] BREBBIA, C.A. (Ed.) "Progress in Boundary Element Methods, Vol. 1" Pentech Press, London, Halstead Press, N.Y., 1981.

[7] BREBBIA, C.A. (Ed.) "Progress in Boundary Element Methods, Vol. 2" Pentech Press, London, Springer-Verlag, N.Y., 1983.

[8] BREBBIA, C.A. (Ed.) "Progress in Boundary Element Methods, Vol. 3" Springer-Verlag, Berlin and N.Y., 1984.

[9] WROBEL, L. & BREBBIA, C.A. "Time dependent Potential Problems" Chapter 6 in Progress in Boundary Element Methods, Vol. 1. Pentech Press, London, Halstead Press, N.Y., 1981.

[10] MANSUR, W.J. & BREBBIA, C.A., "Formulation of the Boundary Element Method for Transient Problems governed by the Scalar Wave Equation". Applied Mathematical Modelling, Vol. 6, August, 1982, pp 307-311.

[11] MANSUR, W.J. & BREBBIA, C.A., "Numerical Implementation of the Boundary Element Method for two Dimensional Transient Scalar Wave Propagation Problems". Applied Mathematical Modelling, Vol. 6, August, 1982, pp 299-306.

[12] BREBBIA, C.A. & NARDINI, D. "Dynamic Analysis in Solid Mechanics by an alternative Boundary Element Procedure". Int. J. of Soil Dynamics and Earthquake Engineering, Vol. 3, 1983.

[13] BREBBIA, C.A. & WALKER, S., "Boundary Element Technique in Engineering". Batterworths, London, 1979.

[14] BREBBIA, C.A. & SKERGET, P. "The Solution of Convective Problems in Laminar Flow" in "Boundary Element Techniques" (Ed. C.A. BREBBIA et.al) Springer-Verlag, Berlin & N.Y., 1983.

[15] NARDINI, D. & BREBBIA, C.A., "Transient Dynamic Analysis by the Boundary Element Method" in "Boundary Element Techniques" (Ed. C.A. BREBBIA et.al). Springer-Verlag, Berlin and N.Y., 1983.

[16] NARDINI, D. & BREBBIA, C.A., " A new Approach to Free Vibrations using Boundary Elements" in "Boundary Element Methods in Engineering" (Ed. C.A. BREBBIA). Springer-Verlag, Berlin and N.Y., 1982.

[17] DANSON, D., BREBBIA, C.A. & ADEY, R.A., "The BEASY System", Advances in Engineering Software, Vol. 4, No. 2, pp 68-74.

Computational Techniques & Applications: CTAC-83
J. Noye & C. Fletcher (Editors)

FINITE DIFFERENCE METHODS FOR FLUID FLOW

R. W. Davis

National Bureau of Standards
Washington, D. C. 20234

This paper examines how finite difference methods can be employed to solve the incompressible Navier-Stokes and continuity equations of fluid flow. The differencing of the various terms in these equations is considered in detail, and a solution procedure is presented which gives reasonable results for two complex flow problems. These problems involve unsteady viscous separated flows in the wake of a rectangular obstacle inside a two-dimensional channel and in an axisymmetric mixing layer. The importance of a priori testing of the numerical methods on appropriate simple model problems is stressed and a useful example is given. Also stressed is the importance of computational flow visualization and data analysis in order to make sense of a flow calculation.

INTRODUCTION

The numerical simulation of fluid flows, i.e., computational fluid dynamics (CFD), first emerged as a distinctly separate branch of fluid dynamics about 20 years ago. At that time finite difference methods composed essentially the whole of CFD. This is no longer the case, of course, as witness the development of finite element, boundary element, vortex-tracking and spectral techniques. Although the oldest of these, finite difference methods have been continually evolving, with the latest techniques representing significant improvements over what was available even five years ago. It is the purpose of this paper to discuss some of the latest finite difference methods for a particularly difficult class of fluid flow problems and to indicate how these techniques are applied in some specific instances.

As is the case for fluid dynamics in general, finite difference methods for fluid flow are composed of numerous diverse areas each with its own specialized techniques. The particular area of concern here will be unsteady, incompressible viscous separated flows in simple geometries. This requires the numerical solution of the complete Navier-Stokes equations. No simplifying assumptions (as per boundary layer theory, for instance) are possible in this type of situation. In addition, when solving the equations in primitive form, a Poisson equation for pressure must be solved. This can, in fact, be the most time-consuming portion of a computation, although the recent use of direct Poisson solvers has been alleviating this problem. The flowfield geometries to be considered can all be handled with simple nonuniform cartesian (or axisymmetric cylindrical) meshes. This avoids the problem of curvilinear grid generation, which is a separate discipline in its own right [1].

Another problem to be avoided here is that of turbulence. As it is the Navier-Stokes (not the Reynolds-averaged) equations which are being considered, no Reynolds stress closure is required. However, in two dimensions the viscous length scales are $O(Re^{-1/2})$, where Re is some appropriate Reynolds number. At

Reynolds numbers of practical interest (at least $O(10^3)$) this means that these viscous length scales are often unresolvable on any reasonable finite difference mesh. Thus the large scales are computed while the small scales are not. Many globally important properties of viscous separated flows can be adequately modeled in this manner. The precise effects of the small length scales on the larger ones, as for instance in the case of the small-scale internal cellular structure seen inside the large coherent structures of a mixing layer [2], are a subject of current debate [3]. Comparisons between computed flows with no small-scale structure and experimental flows with unavoidable small-scale structure should prove interesting and informative.

In the next section the basic equations and concepts will be discussed. The following section will discuss a one-dimensional model problem useful for testing the performance of finite difference schemes. The final two sections will examine specific computations of flow (i) around a rectangle inside a channel and (ii) inside an axisymmetric mixing layer.

BASIC EQUATIONS AND CONCEPTS

The Navier-Stokes and continuity equations for a viscous incompressible fluid are

$$\frac{\partial \tilde{q}}{\partial t} + (\tilde{q} \cdot \nabla)\tilde{q} = -\nabla p + \nu\nabla^2\tilde{q} \tag{1}$$

$$\nabla \cdot \tilde{q} = 0 \quad . \tag{2}$$

Here $\tilde{q}$ is a velocity vector, p is the ratio of pressure to constant density, ν is kinematic viscosity, and t is time. It is eqs. (1) and (2) which must be solved subject, of course, to appropriate initial and boundary conditions. The solution will be accomplished on a staggered mesh whereby pressures are defined at cell centers and normal velocities at cell faces. Thus each mesh cell can be treated as a small control volume, as shown in Fig. 1 for the two-dimensional case ($\tilde{q} = (u,v)$) with mesh indices i and j in the x- and y-directions, respectively. Note that this type of approach leads to "conservative" finite difference schemes which conserve momentum across mesh cell boundaries. Nonconservative schemes can also be employed [4] but will not be discussed here.

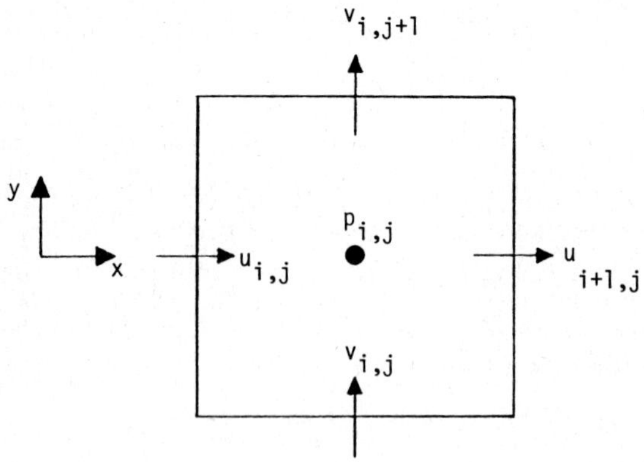

Figure 1
Mesh Cell for Flow Computations

In order to examine the control volume approach more closely, it is useful to consider the following transport equation for the scalar quantity ϕ:

$$\frac{\partial\phi}{\partial t} + (\underset{\sim}{q} \cdot \nabla)\phi = \Gamma\, \nabla^2\phi + s_\phi \quad , \tag{3}$$

where Γ is a diffusion coefficient and s_ϕ is a source term. The quantity ϕ is defined at cell centers, as for p in Fig.1. A typical control volume and its neighboring grid points is shown in Fig. 2 for a uniform mesh with grid spacings Δx and Δy. In order to approximate eq. (3) in the neighborhood of ϕ_P, it is necessary to find the convective and diffusive fluxes of ϕ across the control volume faces surrounding grid point P. This comes about from integrating eq. (3) over the domain of the control volume and using the divergence theorem to obtain the following integral form of eq. (3):

$$\frac{\partial}{\partial t}\int_V \phi \, dA + \int_{\partial V} (\underset{\sim}{q}\phi \cdot \hat{n})d\ell = \Gamma \int_{\partial V} (\nabla\phi) \cdot \hat{n}\, d\ell + \int_V s_\phi dA \ , \tag{4}$$

where V represents the domain of the control volume and ∂V its faces [4]. It is thus necessary to find the values of ϕ along these faces which, since ϕ is defined only at grid points, requires the use of interpolation. Note that since normal velocities are defined at control volume faces, no interpolation is needed to find convection velocities. The form that the interpolation for ϕ takes determines the type of finite difference scheme being employed. It is thus of crucial importance and will be discussed in detail.

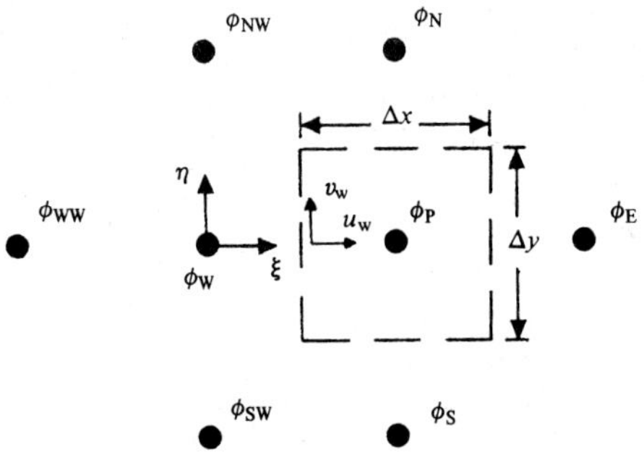

Figure 2
Typical Control Volume

Referring to Fig. 2, it is required, for instance, to obtain the value of $\phi = \phi_w$ on the west face of the control volume. Note that the normal velocity, u_w, is defined here while the tangential velocity, v_w, must be interpolated from the values of v defined on the north and south faces. The simplest apparent method of obtaining ϕ_w would be to let $\phi_w = 1/2(\phi_P + \phi_W)$. If the same type of interpolation is done on the east face, the net horizontal convective flux out of the control volume is

$$\frac{\partial(u\phi)}{\partial x} \approx \frac{u_e}{\Delta x}\left(\frac{\phi_E + \phi_P}{2}\right) - \frac{u_w}{\Delta x}\left(\frac{\phi_P + \phi_W}{2}\right)$$

which, for constant u, gives the centered difference approximation

$$\frac{\partial(u\phi)}{\partial x} = u\,\frac{\partial\phi}{\partial x} \approx u\left(\frac{\phi_E - \phi_W}{2\Delta x}\right) \quad . \tag{5}$$

Similarly, for the horizontal diffusive flux, centered differencing leads to

$$\Gamma \frac{\partial^2\phi}{\partial x^2} \approx \frac{\Gamma}{\Delta x}\left[\frac{(\phi_E - \phi_P)}{\Delta x} - \frac{(\phi_P - \phi_W)}{\Delta x}\right] = \Gamma \left(\frac{\phi_E - 2\phi_P + \phi_W}{\Delta x^2}\right) \quad . \qquad (6)$$

Note that the RHS of eqs. (5) and (6) are only approximations to the indicated derivatives. It can be shown [4] that the leading terms in the truncation errors for both these approximations are $O(\Delta x^2)$. It can also be shown [4] that use of eqs. (5) and (6) together can lead to large bounded spatial oscillations in the solution field, especially in the vicinity of steep gradients, unless $Pe = u\Delta x/\Gamma \leq 2$, where Pe is the cell Peclet number. This is a far too restrictive condition for most convectively-dominated flow computations.

A popular alternative convective differencing scheme is upwind differencing whereby $\phi_w = \phi_W$ for $u_w > 0$ and $\phi_w = \phi_P$ for $u_w < 0$. For positive constant u this gives

$$u \frac{\partial\phi}{\partial x} = u\left(\frac{\phi_P - \phi_W}{\Delta x}\right) + O(\Delta x). \qquad (7)$$

In fact the leading term in the truncation error of eq. (7) is $u\Delta x/2\ \partial^2\phi/\partial x^2$. Thus a false or numerical diffusion is introduced with coefficient $u\Delta x/2$. There are, however, no problems with spatial oscillations. Unfortunately, the false diffusion introduced by this method can be much larger than the physical diffusion, thus leading to erroneous results. This has been the subject of much investigation [4,5,6,7,8]. Generally the severity of the problem increases when the flow is at an angle to the grid lines and/or transients are important [6]. The flow angle problem seems understandable since, from Fig. 2, as the value of (v_w/u_w) increases, the direction along which ϕ is being convected deviates increasingly from the horizontal. This increases the error involved in assuming that ϕ is simply being convected unchanged horizontally over a distance $\Delta x/2$ upstream of the control volume face. A similar type of problem also occurs when transients are important. In order to deal with the flow angle aspect of the problem, various "skew upwind" differencing schemes have been proposed and tested [9,10,11,12]. These trace back from the control volume face along the direction of the net velocity vector in order to find where this vector intersects the surrounding box of eight grid points. For example, if $u_w, v_w > 0$ and $v_w/u_w \leq 2\Delta y/\Delta x$, then this intersection point would be between ϕ_W and ϕ_{SW}. Interpolation is then used to find the value of ϕ here and ϕ_w is set equal to this value. The skew schemes clearly outperform simple upwind differencing but are relatively new and need more testing, especially in situations where the flow is highly time-dependent.

It is also possible to devise "hybrid" schemes which utilize centered differencing when $Pe \leq 2$ and some sort of upwinding when $Pe > 2$, as first proposed by Spalding [13]. When used with simple upwinding, these schemes seldom improve matters much since $Pe > 2$ in most regions of a typical convectively-dominated flow computation. The improvement obtained by "hybridizing" a skew upwind scheme remains an open question, although it seems reasonable to use centered differencing in diffusion-dominated portions of a flowfield. Finally, another sort of hybrid scheme utilizes the same weighted average of centered and simple upwind differencing everywhere in the flowfield regardless of cell Peclet number [14]. However, the results of the flow computation may then turn out to be dependent on the weighting factor, which is not a particularly satisfactory situation [15].

A new development in upwind difference schemes appeared in the late 1970's [16,17]. This is quadratic upwind differencing which, under certain circumstances, leads to a third-order accurate formulation. Referring to Fig. 2, the following quadratic expansion in the vicinity of grid point W is employed:

$$\phi(\xi,\eta) = c_1 + c_2\xi + c_3\xi^2 + c_4\eta + c_5\eta^2 + c_6\xi\eta \ . \qquad (8)$$

For positive u_w and v_w, the six constants in eq. (8) are evaluated from values of ϕ at grid points NW, WW, W, P, SW, and S. Thus ϕ_w, which represents the average value of ϕ along the control volume's west face, is

$$\phi_w = \frac{1}{\Delta y}\int_{-\frac{\Delta y}{2}}^{\frac{\Delta y}{2}} \phi(\tfrac{\Delta x}{2},\eta)d\eta$$

$$= \frac{1}{2}(\phi_P + \phi_W) - \frac{1}{8}(\phi_P - 2\phi_W + \phi_{WW}) + \frac{1}{24}(\phi_{NW} - 2\phi_W + \phi_{SW}) \quad . \tag{9}$$

The absence of ϕ_S from eq. (9) indicates that the sign of v_w does not influence the final result at the west face. Note that eq. (9) is simply the centered difference formula with some local diffusion-like terms added. For positive constant u, quadratic upwinding then gives

$$u\,\frac{\partial\phi}{\partial x} \approx u[\frac{\phi_{WW} - 7\phi_W + 3\phi_P + 3\phi_E}{8\Delta x} + \frac{(\phi_N - 2\phi_P + \phi_S) - (\phi_{NW} - 2\phi_W + \phi_{SW})}{24\Delta x}] \ ,$$

which, for the one-dimensional case, reduces to

$$u\,\frac{\partial\phi}{\partial x} = u[\frac{\phi_{WW} - 7\phi_W + 3\phi_P + 3\phi_E}{8\Delta x}] + \frac{1}{24}\,u\Delta x^2\,\frac{\partial^3\phi}{\partial x^3} + O(\Delta x^3) \quad . \tag{10}$$

Note that the truncation error here indicates that this scheme is second-order accurate, not third-order as has been claimed [16,17]. A more accurate formulation will be discussed subsequently. Spatial oscillations in the vicinity of sharp gradients can still occur with this scheme but in practice often seem to cause little or no problem, thus making this an attractive method [12,15,16,17,18,19,20, 21]. This improved situation results from less dispersion (smaller $O(\Delta x^2)$ term) and more damping ($O(\Delta x^3)$ term) than for centered differencing (no damping).

The time derivative of ϕ (first term in eqs. (3) and (4)) also needs to be differenced in conjunction with the convection and diffusion terms which have just been discussed. The simplest way of doing this is with explicit forward time differencing

$$\frac{\partial\phi}{\partial t} = \frac{\phi_P^{N+1} - \phi_P^N}{\Delta t} + O(\Delta t) \ , \tag{11}$$

where N denotes the time level. The one-dimensional QUICK scheme [17] results when eq. (11) is used in conjunction with eqs. (6) and (10) for diffusion and convection respectively. Unfortunately the stability criterion for QUICK is that $\Delta t \leq 2\Gamma/u^2$ (for the one-dimensional case with constant u), which is the same as if centered differencing, eq. (5), were being used for convection [17]. This is clearly too restrictive for most time-dependent convectively-dominated flow computations. An alternative is to use an explicit Leith-type of temporal differencing [4], thus leading to the QUICKEST scheme [17].

Consider eq. (3) in one dimension with no source term and constant u and Γ:

$$\frac{\partial\phi}{\partial t} + u\,\frac{\partial\phi}{\partial x} = \Gamma\,\frac{\partial^2\phi}{\partial x^2} \quad . \tag{12}$$

Expand ϕ about time level N to obtain

$$\phi^{N+1} = \phi^N + \Delta t\,\frac{\partial\phi}{\partial t}\Big|^N + \frac{1}{2}\,\Delta t^2\,\frac{\partial^2\phi}{\partial t^2}\Big|^N + \frac{1}{6}\,\Delta t^3\,\frac{\partial^3\phi}{\partial t^3}\Big|^N + O(\Delta t^4) \ . \tag{13}$$

Then, from eq. (12),

$$\frac{\partial^2\phi}{\partial t^2} \approx u^2 \frac{\partial^2\phi}{\partial x^2} - 2u\Gamma \frac{\partial^3\phi}{\partial x^3} \quad , \quad \frac{\partial^3\phi}{\partial t^3} \approx - u^3 \frac{\partial^3\phi}{\partial x^3} \quad , \tag{14}$$

where, consistent with the spatial finite difference approximations to be used, fourth- and higher- spatial-derivatives have been dropped. Using eqs. (12) and (14), eq. (13) can be written as

$$\phi^{N+1} - \phi^N = \Delta t(-u \frac{\partial\phi}{\partial x} + \Gamma \frac{\partial^2\phi}{\partial x^2}) \Big|^N + \frac{1}{2} \Delta t^2(u^2 \frac{\partial^2\phi}{\partial x^2} - 2u\Gamma \frac{\partial^3\phi}{\partial x^3}) \Big|^N$$
$$+ \frac{1}{6} \Delta t^3(-u^3 \frac{\partial^3\phi}{\partial x^3}) \Big|^N + O(\Delta t^4) \ . \tag{15}$$

The LHS of eq. (15) represents the difference of the average values of ϕ within the given mesh cell between time levels N+1 and N. If this mesh cell is denoted by i, then these average values are obtained by integrating quadratic fits across grid points i - 1, i, and i + 1 at each time level to obtain

$$\phi^{N+1} - \phi^N = \phi_i^{N+1} - \phi_i^N + \frac{1}{24}[(\phi_{i+1} - 2\phi_i + \phi_{i-1})^{N+1} - (\phi_{i+1} - 2\phi_i + \phi_{i-1})^N] \ . \tag{16}$$

The last two terms in eq. (16) can be interpreted as

$$\frac{\Delta t}{24} \Delta x^2 \frac{\partial^2}{\partial x^2} (\frac{\partial\phi}{\partial t}) \approx - \frac{1}{24} u \Delta t \Delta x^2 \frac{\partial^3\phi}{\partial x^3} \quad , \tag{17}$$

where eq. (12) has been used. Note that the inclusion of this term in eq. (15) exactly cancels the $O(\Delta x^2)$ truncation error of the approximation to $u\ \partial\phi/\partial x$ in eq. (10), thus resulting in third-order spatial accuracy for the approximation to $- \Delta t u\ \partial\phi/\partial x$ in eq. (15). The remaining spatial approximations to the nondiffusion terms in eq. (15) are at least as accurate provided that $u\Delta t/\Delta x \lesssim 1$. When spatial derivatives are approximated as per the QUICK scheme, the following results:

$$\phi_i^{N+1} = \phi_i^N - \frac{1}{2} c(\phi_{i+1}^N - \phi_{i-1}^N) + (\gamma + \frac{1}{2} c^2)(\phi_{i+1}^N - 2\phi_i^N + \phi_{i-1}^N)$$
$$+ c(\frac{1}{6} - \gamma - \frac{1}{6} c^2) (\phi_{i+1}^N - 3\phi_i^N + 3\phi_{i-1}^N - \phi_{i-2}^N) \quad , \tag{18}$$

where $c = u\Delta t/\Delta x$ and $\gamma = \Gamma\Delta t/\Delta x^2$. This is the QUICKEST formulation [17] which is third-order accurate both temporally and spatially as $\Gamma \to 0$ for $c \lesssim 1$. For small γ, it is stable for $c \leq 1$ [17].

The same sort of procedure can be carried out in two dimensions, with the convective terms being written in conservation form $[u \frac{\partial\phi}{\partial x} + v \frac{\partial\phi}{\partial y} \to \frac{\partial(u\phi)}{\partial x} + \frac{\partial(v\phi)}{\partial y}]$ through the use of eq. (2). The final result for ϕ_P^{N+1} is as given in Davis and Moore [15] with the addition of difference forms for the following cross-derivative terms:

$$\frac{\Delta t^2}{2} [\frac{\partial}{\partial x} (uv \frac{\partial\phi}{\partial y}) + \frac{\partial}{\partial y} (uv \frac{\partial\phi}{\partial x})]^N \ .$$

The neglect of various small terms in the two-dimensional formulation reduces the temporal accuracy to $O(\Delta t^2)$ as $\Gamma \to 0$, with the spatial accuracy remaining unchanged. Modifications for a nonuniform mesh $[\Delta x = f(x),\ \Delta y = g(y)]$ add considerable complexity to this scheme but are necessary for most practical flow computations. The source term, s_ϕ, in eqs. (3) and (4) is defined at the cell center, point P.

The preceding discussion has focused on the solution of eq. (1), the Navier-Stokes equations, with ϕ being replaced by u or v. The solution so obtained will probably not satisfy eq. (2), the continuity equation. In fact, it is necessary to find the pressure field at each time step such that eq. (2) is satisfied. One procedure that can be used is as follows [22,23]:

$$\frac{\underset{\sim}{q}^* - \underset{\sim}{q}^N}{\Delta t} + F(\underset{\sim}{q}^N) = \nu\delta^2\underset{\sim}{q}^N \tag{19}$$

$$\frac{\tilde{q}^{N+1} - \tilde{q}^*}{\Delta t} = - \delta p^{N+1} \quad , \tag{20}$$

where $\tilde{q}^*$ is a predicted value of $\tilde{q}^{N+1}$; $F(\tilde{q}^N)$ includes both temporal and convective differencing terms; and δ and δ^2 are discrete central difference analogs of ∇ and ∇^2. Taking the divergence of eq. (20) gives the following discrete Poisson equation for pressure:

$$\delta^2 p^{N+1} = \frac{\delta \cdot \tilde{q}^*}{\Delta t} \quad , \tag{21}$$

since, from eq. (2), $\delta \cdot \tilde{q}^{N+1} = 0$. Thus, the procedure is to solve the momentum equations without pressure gradient terms for $\tilde{q}^*$, solve eq. (21) for pressure, and then determine $\tilde{q}^{N+1}$ from eq. (20). The solution method for eq. (19) has already been described; the procedure for solving eq. (21) will now be discussed.

There are two general classes of methods for solving the Poisson equation for pressure: iterative and direct. The iterative methods, such as successive over-relaxation [4,14,15], are simple and algorithms are easily constructed by non-specialists in elliptic equations. Unfortunately, they converge slowly unless great care is taken in optimizing parameters. Direct methods, on the other hand, require specialized expertise to create but are generally much faster and more accurate than iterative schemes. One good strategy here is to utilize existing direct solvers, such as are found in the FISHPAK package of FORTRAN subprograms for the solution of separable elliptic partial differential equations developed at the National Center for Atmospheric Research [24]. The methods employed here utilize cyclic reduction [25] and can be implemented on irregular domains (e.g., rectangular domains containing obstacles) via the capacitance matrix technique [26]. Note that boundary conditions on pressure are not necessary a priori but can be derived from eqs. (1) and (2) [27]. FISHPAK has been used very successfully in various flow computations carried out at the National Bureau of Standards. Two of these computations will be discussed subsequently. The next section will describe a one-dimensional model problem useful for testing both the transient and the steady-state performance of various numerical schemes.

MODEL PROBLEM

Before attempting to embed a specific numerical scheme in a complex hydrodynamics code, it is useful to test its performance on simple model problems which exhibit some of the features of the flows to be simulated. This can, of course, save much time and effort if the performance of a scheme proves unsatisfactory. A particularly useful model problem which contains some of the essential features of transient shear flows will now be discussed.

The model problem to be considered is an exact solution to the three-dimensional axisymmetric Navier-Stokes equations. It represents one or more moving shear layers of rotating fluid whose thicknesses are functions of Reynolds number and time. Assume radial and axial velocity components, u_r and u_z respectively, representing a time-dependent stagnation-point flow as follows:

$$u_r = - \frac{r}{2} f(t) \quad , \quad u_z = z \, f(t) \quad , \tag{22}$$

where eq. (2) is satisfied for arbitrary f(t). The equation for the swirl, w(r,t) = rv (v = tangential velocity), is

$$\frac{\partial w}{\partial t} - \frac{f(t)}{2} \left[\frac{\partial (rw)}{\partial r} - w\right] = \nu r \frac{\partial}{\partial r} \left(\frac{1}{r} \frac{\partial w}{\partial r}\right) \quad . \tag{23}$$

The initial and boundary conditions for eq. (23) are $w(r,0) = \xi(r)$, $w(0,t) = 0$

and $w(1,t) = g(t)$. Due to the radial influx of swirl (for $f(t) > 0$), the choice of $\xi(r)$ for $r > 1$ determines the boundary condition $g(t)$ at $r = 1$. Thus, the solution to eq. (23) treated simply as an initial value problem can be shown to be

$$w = \int_0^y y_* \, L^{-1}(\bar{\theta}) \, dy_* \quad , \tag{24}$$

where $y = r \exp(\frac{1}{2}\int_0^t f(t)dt)$, L^{-1} is the inverse Laplace transform, and

$$\begin{aligned} \bar{\theta}(p) = {} & K_0(\sqrt{p}\, y) \int_0^y \zeta(y_0) y_0 \, I_0(\sqrt{p}\, y_0) dy_0 \\ & + I_0(\sqrt{p}\, y) \int_y^\infty \zeta(y_0) y_0 \, K_0(\sqrt{p}\, y_0) dy_0 \quad , \end{aligned} \tag{25}$$

where I_0 and K_0 are Bessel functions and $\zeta(y_0) = \zeta(r) = \frac{1}{r}\frac{\partial \xi}{\partial r}$ is the initial vorticity distribution. For the simplified case $f(t) = \text{constant} = \alpha$, $\xi(r) = \{{0,\ r < 1 \atop 1,\ r > 1}\} \Rightarrow g(t) = 1$, the solution to eq. (23) is as follows [18]:

$$w(r,t) = \frac{1}{2}\int_0^r \frac{Re\, \rho e^t}{(e^t - 1)} \exp\left[\frac{-Re(1 + \rho^2 e^t)}{4(e^t - 1)}\right] I_0\left[\frac{Re\, \rho e^{t/2}}{2(e^t - 1)}\right] d\rho \quad , \tag{26}$$

with

$$\lim_{t\to\infty} w(r,t) = 1 - \exp\left[\frac{-Re\, r^2}{4}\right] \quad , \tag{27}$$

where $Re = \alpha/\nu$ for characteristic length 1. This solution, eq. (26), represents a moving shear layer which begins at $r = 1$ and moves toward the origin, its final position being dictated by eq. (27). The thickness of this layer decreases as Re increases. More details on this are given in Baum, Ciment, Davis, and Moore [18].

Figures 3, 4, and 5 show the exact solution, eq. (26), as a function of r and t for Reynolds numbers of 500, 4000, and 30,000. Also shown are numerical solutions with $\Delta r = \Delta t = 0.05$ for simple first-order upwind differencing (UDS), centered differencing (CDS), QUICKEST, and a new cylindrical fundamental solution scheme (CFSS) [18]. The initial condition for the computations is the exact solution shown at $t = t_0 = 0.10$. It can be seen from these figures that: (i) a scheme's performance in modeling the transient phase may differ substantially from its performance in modeling the steady-state (see especially UDS in Fig. 3); (ii) QUICKEST appears to give the best results overall at the two lower Reynolds numbers; (iii) no scheme performs well at Re = 30,000. As the transient shear layer in this model problem is convected and diffused in a manner similar to what one expects to find in a Navier-Stokes computation of the sort being considered in this paper, these results are relevant and useful. The performance of the QUICKEST scheme here leads one to believe that this method will produce reasonable results in a full two-dimensional flow simulation so long as the Reynolds numbers do not become too large for the computational mesh. The next section will discuss the application of this numerical scheme to modeling the flow around a rectangle inside a channel. The following section will describe the application of QUICKEST to an axisymmetric mixing layer computation.

FLOW AROUND A RECTANGLE INSIDE A CHANNEL

The problem being considered here is illustrated in Fig. 6. A rectangle of aspect ratio A/B is situated symmetrically between two walls a distance H apart. The velocity profile a distance x_R upstream of the rectangle is arbitrary, with center-

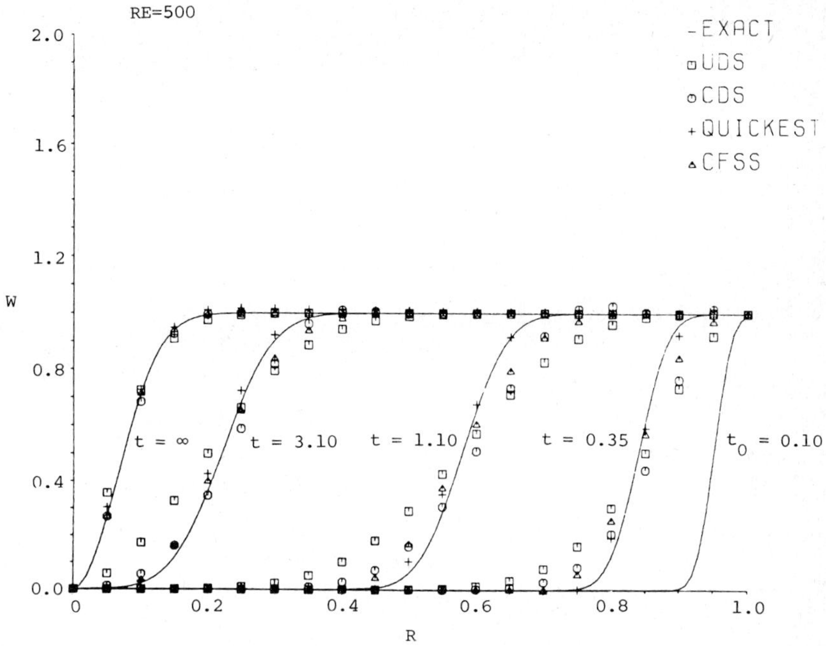

Figure 3
Scheme Comparison for Re = 500

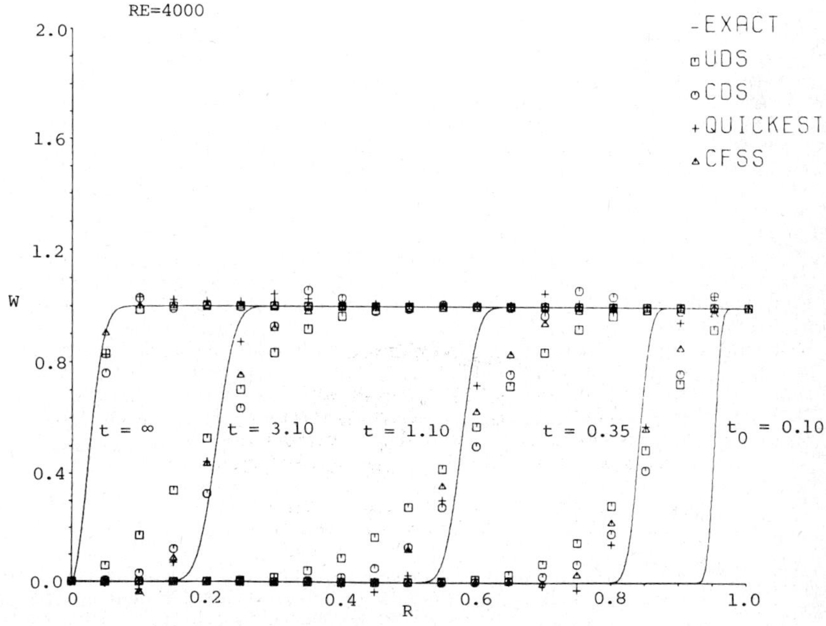

Figure 4
Scheme Comparison for Re = 4000

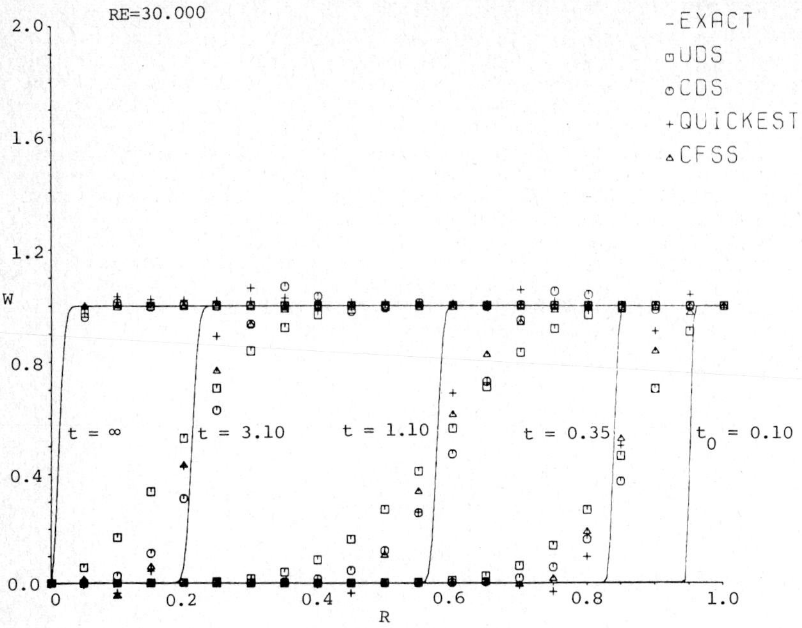

Figure 5
Scheme Comparison for Re = 30,000

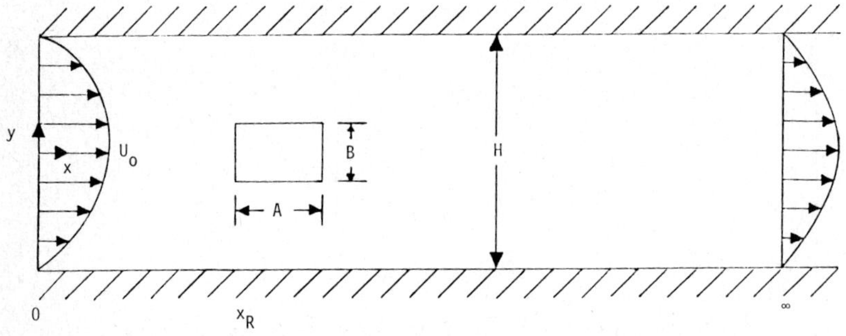

Figure 6
Configuration Definition for Rectangle Inside Channel

line velocity U_o, while the velocity profile at an infinite distance downstream is fully-developed. The objectives are to find the unsteady nature of the wake and of the forces acting upon the rectangle as functions of aspect ratio (A/B), blockage ratio (B/H), Reynolds number (U_oB/ν), and upstream velocity profile shape.

The boundary conditions employed here are that both normal and tangential velocities are zero along the channel walls and on the rectangle. The upstream velocity profile is specified at x = 0, while a fully-developed parabolic profile is assumed as $x \to \infty$. The fully-developed profile at the mesh exit is implemented by means of an infinite-to-finite mapping of the form $Z = K_1 + K_2/x$ (K_1, K_2 are constants) which is applied only near the mesh exit. At the point where the mapping begins, $\partial/\partial Z = \partial/\partial x$ in order to allow for a smooth convective transition into the transformed region. A specialized treatment of the mesh cells in the vicinity of the rectangle's corners is required. The situation near the top front corner is shown in Fig. 7. The control volumes for the two velocities nearest the corners are shown together in Fig. 7a, and separately in Fig. 7b and c. The convective fluxing across the half-faces shown in Fig. 7 must be handled carefully, as described in detail in Davis and Moore [15]. The best method of doing this is by local grid refinement, thus reducing the size of these half-faces. Once the finest grid possible is in place, then the sensitivity of the overall flowfield to changes in the fluxing approximations across these half-faces must be determined. The point here is that the modeling of flow in the vicinity of exterior corners such as this is complicated and must be thoroughly investigated in order to assure confidence in the results of a computation.

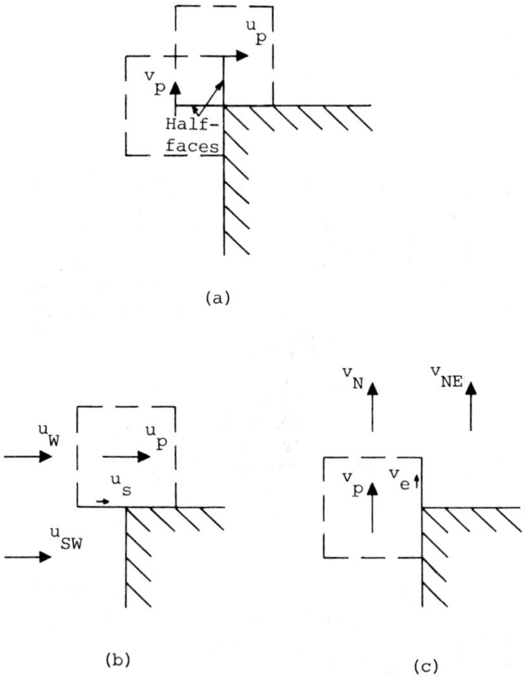

Figure 7
Control Volumes Near Top Front Corner

A typical nonuniform computational mesh for this problem is shown in Fig. 8 for flow around a square. The blockage ratio here is 1/6, and there are 76 mesh cells in the x-direction and 52 in the y-direction. The 76 in the x-direction include 10 not shown which are inside the transformed region beginning at x = 24, i.e., 24 square lengths from the computational inlet. Mesh cells are concentrated in the vicinity of the square and of the walls, since gradients are expected to be largest in these areas. The use of several different meshes to compute a particular configuration such as this is recommended in order to assure that solutions are not mesh dependent. Initial conditions for a computation can be, for example, fully-developed flow everywhere outside the rectangle or the results of a previous computation with slightly different parameters (e.g., a different Reynolds number). No perturbations are necessary to initiate vortex shedding.

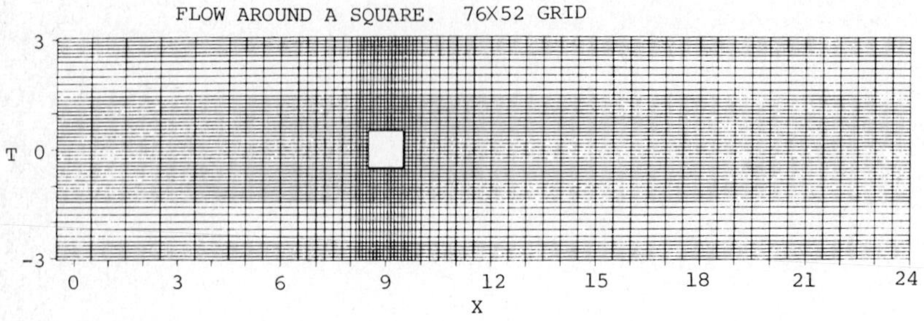

Figure 8
Nonuniform Mesh for Computation of Flow Around
A Square Inside a Channel

A streakline plot for flow around the square in Fig. 8 is shown in Fig. 9 for Re = 1000. This plot is composed of 14 closely-spaced rows of passive marker particles entering the flow ahead of the square. Each entering row in Fig. 9 consists of a different symbol, thus allowing the determination of the entry point of any particle in the domain. Each particle is moved after every time step of the flow computation in accordance with the local fluid velocity in its vicinity. The particles get swept up into the recirculation zone just behind the square and are then carried downstream by the vortices. Thus an excellent picture is obtained of the formation, shedding and subsequent development of these vortices. The thin braids between the vortices mark the boundary between fluid entrained into the wake from each side of the channel centerline. Lift and drag coefficients as functions of time for this configuration are presented in Fig. 10. The irregularities in these plots are functions of Reynolds number, with both lift and drag being simple sine waves for Re $\leq$ 500.

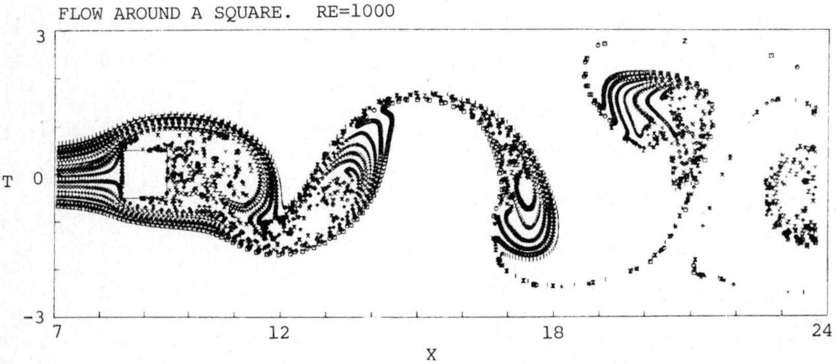

Figure 9
Streakline Plot of Flow Around a Square
Inside a Channel for Re = 1000

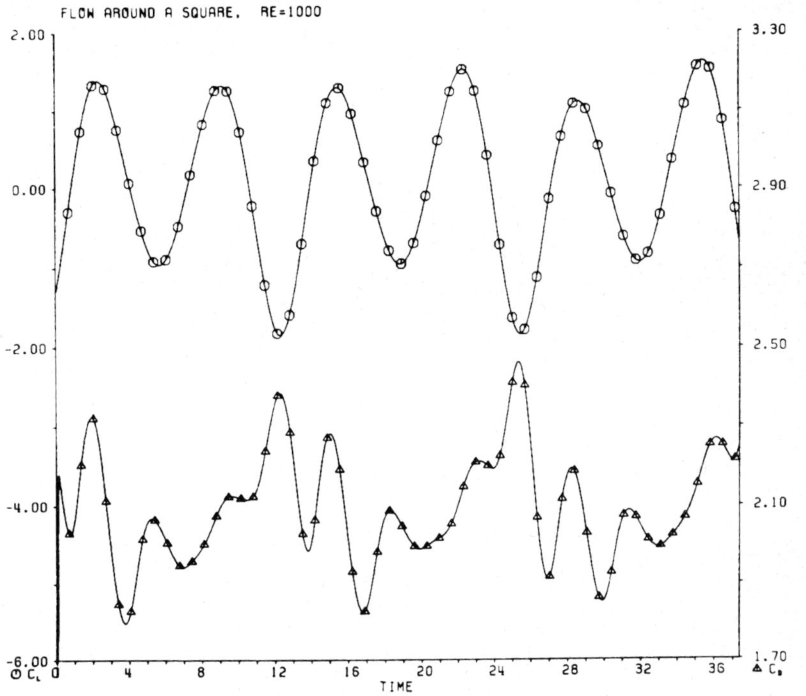

Figure 10
Lift and Drag Coefficients for Square
Inside Channel for Re = 1000
(Symbols Every 50 Time Steps)

More details on this channel computation will be presented in a future paper, including a comparison which shows reasonable agreement for Re $\leq$ 1000 between computed and experimental Strouhal numbers. The primary motive for the previous discussion was to outline how a finite difference computation of a complex flowfield might be performed. The use of high-quality computer graphics and data analysis routines in analyzing the computational results cannot be overemphasized. The mass of numbers that spews forth from a typical time-dependent flow computation is virtually meaningless without appropriate flow visualization and data analysis. This is, of course, the same problem confronting experimentalists in fluid dynamics. Thus, there is much in common between the data analysis and visualization techniques of each group and close collaboration between both types of investigators should prove beneficial.

THE AXISYMMETRIC MIXING LAYER

The problem considered in this section is rather different from that of the previous section in that there are no solid boundaries here. The situation is shown in Fig. 11. A velocity profile characteristic of a coflowing axisymmetric jet is specified as a function of r at Z = 0. The thin shear layer between the inner jet and the outer stream rolls up into vortices for Z > 0. It is the dynamics of the vortices inside this axisymmetric mixing layer that is of interest in this problem. The merging of these vortices is what causes the downstream growth of the mixing layer [28].

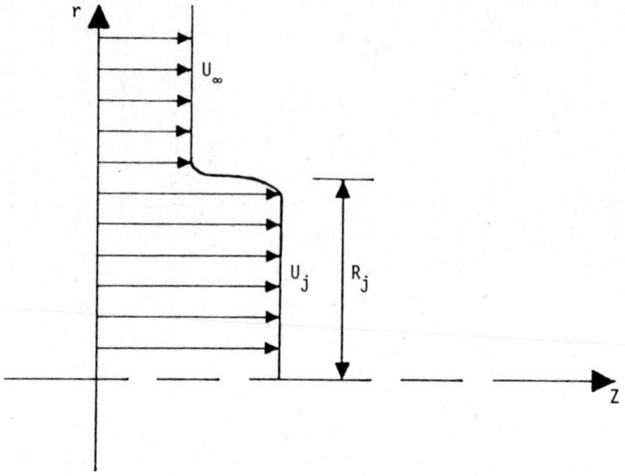

Figure 11
Configuration Definition for Coflowing Jet

The boundary conditions employed in this problem are that the velocity profile at $Z = 0$ is specified; radial derivatives are set to zero along $r = 0$ (i.e., axisymmetry); a simple asymptotic analysis involving small perturbations about the freestream velocity is used for large r; and the freestream velocity is specified as $Z \to \infty$ by utilizing the same type of mapping discussed in the previous section. Typical initial conditions for a computation consist of the initial (at $Z = 0$) velocity profile everywhere outside and the freestream velocity everywhere inside the transformed region.

There is one further, and major, difficulty with this computation. If, in fact, it is carried out as described above, nothing occurs other than the slow diffusive spreading of the initial shear layer. In order to obtain vortex formation, a perturbation must be applied to the shear layer. In any physical experiment involving flow of this type, random background noise will supply the perturbation necessary to trigger roll up. The particular component of the background noise that actually causes the initial vortex formation can be determined from linear inviscid stability theory. What is desired is that frequency component which exhibits the largest spatial growth rate. This can be determined for the coflowing jet as described in Michalke and Hermann [29]. Thus, for the computation, the flowfield is perturbed with the eigenfunctions for any desired combination of the most unstable frequency and its subharmonics as found from the stability analysis. The perturbation is applied for $Z \leq 2R_j$ and has a maximum amplitude of about 1% of the mean velocity. The amplitude and spatial extent of this applied traveling wave disturbance are, of course, somewhat arbitrary. However, the ensuing vortex dynamics has been observed to be independent of any reasonable variation in these parameters. The importance of the presence of subharmonics in the perturbation will be seen from the computational results.

A typical nonuniform computational mesh for this problem is shown in Fig. 12. The 70 mesh points in the axial direction include 14 not shown inside the transformed region beginning at $Z = 15$. The initial shear layer occurs at $r \approx 1$ and thus mesh points are concentrated in this region. The radial extent of the mesh is about 5, which is sufficiently far from the mixing layer so as to negligibly effect it.

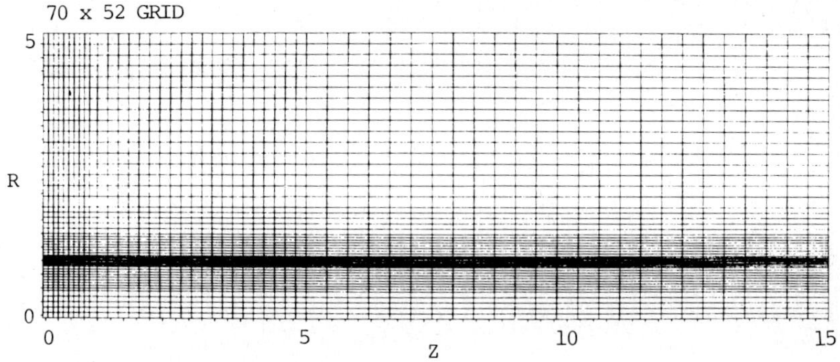

Figure 12
Nonuniform Mesh for Axisymmetric Mixing Layer Computation

Some results from the computations are shown in Figs. 13, 14 and 15 for a jet-to-freestream velocity ratio of 3.33 and a Reynolds number based on jet radius and velocity of 1000. Passive marker particles are used to illuminate the interior of the jet in Fig. 13 and the mixing layer itself in Fig. 14. Isovorticity contours are shown in Fig. 15. Thus, these three figures, each presenting the same results, illustrate three different ways of illuminating the same phenomenon.

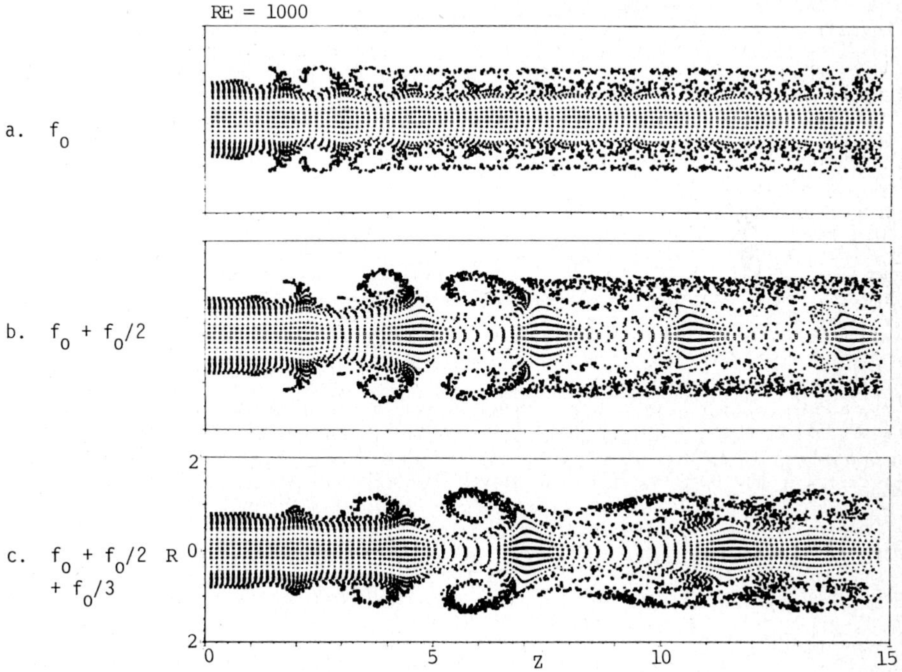

Figure 13
Streakline Plots of Interior of Jet

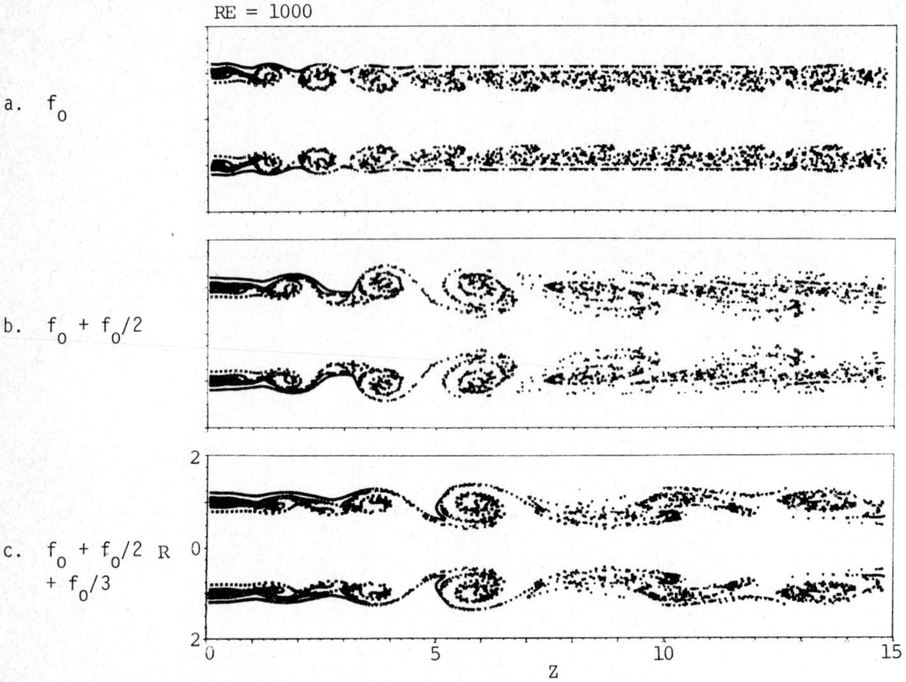

Figure 14
Streakline Plots of Mixing Layer

This phenomenon is the role that the presence of subharmonics plays in the vortex merging inside a mixing layer. The perturbation applied in Figs. 13a, 14a and 15a consists solely of the single most unstable frequency, the fundamental (f_0). Vortices form and shear as they move downstream, but no vortex merging occurs. The perturbation applied in Figs. 13b, 14b and 15b consists of the fundamental (f_0) plus its first subharmonic ($f_0/2$). A single vortex merging is seen to occur around Z = 3 to 4. Figures 13c, 14c and 15c show that when the perturbation consists of the fundamental (f_0) plus its first two subharmonics ($f_0/2$ and $f_0/3$), an additional merging occurs around Z = 9. Thus each subharmonic induces one vortex merging, in agreement with experimental results for the two-dimensional mixing layer [28]. A more complete description of the computational results for the axisymmetric mixing layer will be presented in a future paper. The importance of appropriate computational flow visualization, however, is apparent from the limited results just discussed.

CONCLUSIONS

Some of the fundamental concepts related to the finite differencing of the incompressible Navier-Stokes and continuity equations have been discussed in this paper. For the particular class of flows that was considered here, the techniques and examples presented in the previous sections can serve as a starting point for setting up a complex flowfield computation. Once again it is worth stressing that, prior to actually carrying out this computation, use be made of appropriate model problems to test the basic numerics. The model swirl flow problem presented pre-

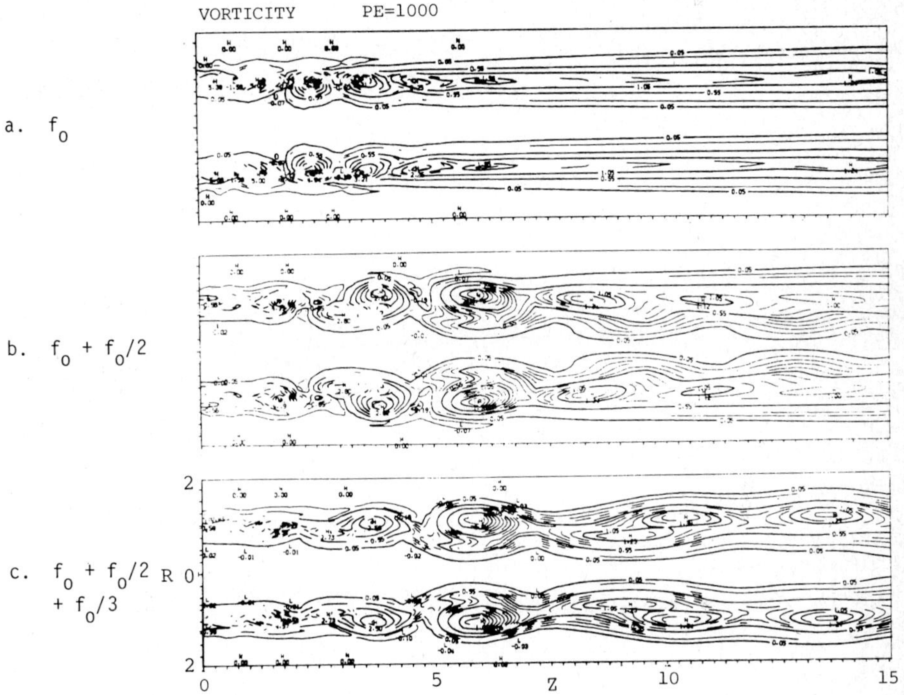

Figure 15
Isovorticity Contours for Coflowing Jet

viously is only one possible example. Other popular model problems include the driven cavity [30] which, unfortunately, has no known exact solution, and the color problem [31], use of which is an excellent means of determining a convective scheme's diffusive and dispersive characteristics [4].

The clear superiority of quadratic upwind differencing over the older first and second order methods is now clear. For any given flow problem, the tradeoff between the improved accuracy due to this method and the increased complexity inherent in its use must be assessed. Certainly in time-dependent computations where Reynolds number is an important parameter, the use of QUICKEST must be seriously considered.

Finally, of course, once the computation has been performed, the output must be analyzed. Thus, the computational fluid dynamicist must not only be familiar with numerical methods for fluid flow but must also be able to use computer graphics packages. Fortunately many of these are easy to use and improved versions are constantly appearing. The use of color graphics seems to offer much future potential in CFD [32], with color movies of time-dependent computations being especially desirable.

ACKNOWLEDGMENTS

The invaluable assistance of E. F. Moore and P. H. Gurewitz in preparing this paper is acknowledged. The axisymmetric mixing layer simulation described herein is being supported by the Air Force Office of Scientific Research.

REFERENCES

[1] Thompson, J. F. (ed.), Numerical Grid Generation (North-Holland, Amsterdam, 1982).

[2] Brown, G. L. and Roshko, A., On density effects and large structure in turbulent mixing layers, J. Fluid Mech. 64 (1974) 775-816.

[3] Cantwell, B., Organized motion in turbulent flow, Ann. Rev. Fluid Mech. 13 (1981) 457-515.

[4] Roache, P. J., Computational Fluid Dynamics (Hermosa, Albuquerque, 1976).

[5] Davis, G. De Vahl and Mallinson, G. D., An evaluation of upwind and central difference approximations by a study of recirculating flow, Computers and Fluids 4 (1976) 29-43.

[6] Raithby, G. D., A critical evaluation of upstream differencing applied to problems involving fluid flow, Comp. Meth. Appl. Mech. and Eng. 9 (1976) 75-103.

[7] Leonard, B. P., Adjusted quadratic upstream algorithms for transient incompressible convection, in: A Collection of Technical Papers: AIAA Computational Fluid Dynamics Conference (AIAA, New York, 1979).

[8] Castro, I. P., Cliffe, K. A. and Norgett, M. J., Numerical predictions of the laminar flow over a normal flat plate, Int. J. for Num. Meth. in Fluids 2 (1982) 61-88.

[9] Raithby, G. D., Skew upstream differencing schemes for problems involving fluid flow, Comp. Meth. Appl. Mech. and Eng. 9 (1976) 153-164.

[10] Leschziner, M. A., Practical evaluation of three finite difference schemes for the computation of steady-state recirculating flows, Comp. Meth. Appl. Mech. and Eng. 23 (1980) 293-312.

[11] Lillington, J. N., A vector upstream differencing scheme for problems in fluid flow involving significant source terms in steady-state linear systems, Int. J. for Num. Meth. in Fluids 1 (1981) 3-16.

[12] Smith, R. M. and Hutton, A. G., The numerical treatment of advection: a performance comparison of current methods, Num. Heat Trans. 5 (1982) 439-461.

[13] Spalding, D. B., A novel finite difference formulation for differential expressions involving both first and second derivatives, Int. J. for Num. Meth. in Eng. 4 (1972) 551-559.

[14] Hirt, C. W., Nichols, B. D. and Romero, N. C., SOLA - A numerical solution algorithm for transient fluid flows, Los Alamos Scientific Laboratory Rep. LA-5852 (1975).

[15] Davis, R. W. and Moore, E. F., A numerical study of vortex shedding from rectangles, J. Fluid Mech. 116 (1982) 475-506.

[16] Leonard, B. P., Leschziner, M. A. and McGuirk, J., Third-order finite-difference method for steady two-dimensional convection, in: Taylor, C., Morgan, K. and Brebbia, C. A. (eds.), Numerical Methods in Laminar and Turbulent Flow (Wiley, New York, 1978).

[17] Leonard, B. P., A stable and accurate convective modelling procedure based on quadratic upstream interpolation, Comp. Meth. Appl. Mech. and Eng. 19 (1979) 59-98.

[18] Baum, H. R., Ciment, M., Davis, R. W. and Moore, E. F., Numerical solutions for a moving shear layer in a swirling axisymmetric flow, in: Reynolds, W. C. and MacCormack, R. W. (eds.), Proc. 7th Int. Conf. on Num. Meth. in Fluid Dyn. (Springer, Berlin, 1981).

[19] Han, T., Humphrey, J. A. C. and Launder, B. E., A comparison of hybrid and quadratic-upstream differencing in high Reynolds number elliptic flows, Comp. Meth. Appl. Mech. and Eng. 29 (1981) 81-95.

[20] LeQuere, P., Humphrey, J. A. C. and Sherman, F. S., Numerical calculation of thermally driven two-dimensional unsteady laminar flow in cavities of rectangular cross section, Num. Heat Trans. 4 (1981) 249-283.

[21] Leschziner, M. A. and Rodi, W., Calculation of annular and twin parallel jets using various discretization schemes and turbulence-model variations, J. Fluid Eng. 103 (1981) 352-360.

[22] Chorin, A. J., Numerical solution of the Navier-Stokes equations, Math. Comp. 22 (1968) 745-762.

[23] Goda, K., A multistep technique with implicit difference schemes for calculating two- or three-dimensional cavity flows, J. Comp. Phy. 30 (1979) 76-95.

[24] Swarztrauber, P. and Sweet, R., Efficient FORTRAN subprograms for the solution of elliptic equations, National Center for Atmospheric Research Tech. Note IA-109 (1975).

[25] Swarztrauber, P. N., A direct method for the discrete solution of separable elliptic equations, SIAM J. Num. Anal. 11 (1974) 1136-1150.

[26] Buzbee, B. L., Dorr, F. W., George, J. A. and Golub, G. H., The direct solution of the discrete Poisson equation on irregular regions, SIAM J. Num. Anal. 8 (1971) 722-736.

[27] Cebeci, T., Hirsh, R. S., Keller, H. B. and Williams, P. G., Studies of numerical methods for the plane Navier-Stokes equations, Comp. Meth. Appl. Mech. and Eng. 27 (1981) 13-44.

[28] Ho, C. M. and Huang, L. S., Subharmonics and vortex merging in mixing layers, J. Fluid Mech. 119 (1982) 443-473.

[29] Michalke, A. and Hermann, G., On the inviscid instability of a circular jet with external flow, J. Fluid Mech. 114 (1982) 343-359.

[30] Olson, M. D. and Tuann, S. Y., Computing methods for recirculating flow in a cavity, in: Taylor, C., Morgan, K. and Brebbia, C. A. (eds.), Numerical Methods in Laminar and Turbulent Flow (Wiley, New York, 1978).

[31] Liu, C. Y., Goodin, W. R. and Lam, C. M., Numerical problems in the advection of pollutants, Comp. Meth. Appl. Mech. and Eng. 9 (1976) 281-299.

[32] Belie, R. G. and Rapagnani, N. L., Color computer graphics, Astro. and Aero. 19 (1981) 41-48.

Computational Techniques & Applications: CTAC-83
J. Noye & C. Fletcher (Editors)

MATHEMATICAL ANALYSIS AS THE FOUNDATION FOR SUCCESSFUL COMPUTATION

R.S. Anderssen

CSIRO Division of Mathematics and Statistics
Box 1965
GPO Canberra ACT 2601
Australia

It is initially argued that the strength of mathematics cannot in general be harvested without exploiting its depth. After discussing the essential nature of computational mathematics and then introducing and illustrating with simple examples reasons why depth in mathematics plays such a crucial role in computational mathematics, the remainder of the paper discusses some specific examples which illustrate how depth in mathematics has been used to advance in non-trivial ways the theory and practice of computational mathematics. These examples include: adaptivity strategies for the numerical solution of ordinary and partial differential equations; estimation of the stabilization parameters in regularization; and uniformly valid approximation methods for differential eigenvalues.

1 INTRODUCTION

In giving science, industry and the community the ability to experiment mathematically, the computer has caused a fundamental change in the way mathematically and statistically based research is now being pursued. This has occurred because computers allow, in no other way possible, for the problem under investigation to remain of central importance. Results about a problem can now be derived without forcing it to be modelled mathematically in a form which fits some classical framework or methodology.

By making more explicit the crucial and underpinning role played by the results of (pure) mathematics in all aspects and all levels of problem solving, this change is forcing a reassessment of the pessimistic attitude that

> "... the great bulk of mathematics is useless."
>
> p.135, Hardy (1967),

and

> "... the trivial mathematics is, on the whole, useful, and that the real mathematics, on the whole, is not ..."
>
> p.139, Hardy (1967).

The reason for this reassessment is the contribution that mathematics gives to problem solving and that is its *depth*. Even though he does not explicitly introduce a concept of depth, Hardy (1967) himself argues that the real power behind mathematics and an explanation of what real mathematics is all about is, along with its generality, its "depth":

"It seems that mathematical ideas are arranged somehow in strata, the ideas in each stratum being linked by a complex of relations both among themselves and with those above and below. The lower the stratum, the deeper (and in general more difficult) the idea. Thus, the idea of an "irrational" is deeper than that of an integer; and Pythagoras's theorem is, for that reason, deeper than Euclid's."

p.110, Hardy (1967).

From a problem solving point of view, the strength of mathematics is in part its ability to

(i) either prove the obvious or identify the counter-intuitive; and

(ii) allow the differences between methods for the same problem to be quantified.

It is the need to cope with situations such as are typified by (i) and (ii) that forces problem solving methodologies to be examined within an appropriate mathematical framework. The counter-intuitive can only be assumed not to hold if it is known that the intuitive does. Numerical procedures for the same problem can only be conclusively compared when their differences with respect to given criteria have been explicitly characterized.

However, this strength of mathematics cannot in general be harvested without exploiting its depth. In many ways, it is this fact more than anything else which makes efforts to examine the relative merits of pure and applied mathematics a more or less meaningless exercise. Except for their motivations and aims, there is no difference between pure and applied mathematics. Both exploit the depth of mathematics though often by necessity at different levels.

Remark. Implicit in the above statement that "this strength of mathematics cannot in general be harvested without exploiting its depth" is the assumption that mathematics itself has a separate reality in the sense explained by Hardy (1967), p.123:

"... mathematical reality lies outside us, that our function is to discover or observe it, and that the theorems which we prove, ..., are simply our notes of our observations."

The aim of this paper is to illustrate the contributions depth in mathematics has played in advancing the theory and practice of computational mathematics. More specifically, attention is limited to contributions made by mathematical analysis.

The essential nature of computational mathematics is first discussed in §2. Reasons why depth in mathematics plays such a crucial role in computational mathematics are then introduced and illustrated with simple examples in §3. Some specific examples are discussed in §4 which illustrate how depth in mathematics has been used to advance in non-trivial ways the theory and practice of computational mathematics.

We conclude this introduction by noting that, for a given problem formulation, there is a price which must be paid with respect to the depth of the stratum of the mathematical ideas which are used to "solve" this formulation. It is not only a matter of working with more difficult concepts. It is also necessary to either apply appropriate regularity conditions to the problem formulation (exclude certain possibilities from further consideration) or to reinterpret the meaning of solution to a much weaker form than that which is really required. The relevance of these comments will be examined in §3 when discussing the exemplifications given there.

In fact, there often appears to be a trade-off between utility and the sophistication of the mathematical concepts used. It could be argued that it was a lack of awareness of the existence of such trade-offs that led Hardy (1967) to take the pessimistic attitude mentioned above. In this paper, the depth of the stratum of the mathematics actually discussed in the examples and illustrations is at the level of that used by professional mathematicians in the normal course of research. It is certainly not at the level of much of the mathematics published by Hardy which examined extremely deep questions in number theory and analysis.

§2 THE NATURE OF COMPUTATIONAL MATHEMATICS

In addressing this topic, we follow quite closely the point of view developed by Anderssen and de Hoog (1983).

In any examination of the use of computers as a problem solving tool, the inescapable facts are that, as sophisticated as their development has been:

(a) the basic operations of most computers are restricted to the elementary arithmetic operations of addition, subtraction, multiplication, division, and logical comparison;

(b) most computers do not work with all the real numbers, but with a finite subset (usually, a floating-point number system), because of the finite resources they have available for representing numbers (normally, they cannot manipulate numbers abstractly).

From (a) it follows that a problem cannot be solved computationally until it is reduced to or approximated by a finite sequence of such elementary operations. In many applications, such as arise in commercial data processing, the problem itself is defined in terms of such operations. In others, such as arise in data acquisition and management as well as combinatorial processing, the problem can be reduced to an equivalent formulation in terms of such operations. In most scientific and industrial applications, however, it is necessary to *approximate* the given problem by a finite sequence of such operations.

For a given problem, the task of constructing such a finite sequence is not a trivial one. As a strategy for action, it reduces to constructing:

1. *The problem formulation* which aims to model the problem of interest mathematically. There is a need to distinguish between formulations which lead to simple rather than involved computational procedures.

2. *The computational formulation* which aims to mimic approximately the appropriate problem formulation in terms of the building blocks of numerical

analysis (viz. matrices, iterations, summations, etc.) which have representations in terms of a finite sequence of the elementary arithmetic operations.

3. *The algorithm* which defines the finite sequence of elementary arithmetic operations which will be used by a computer to carry out the computational formulation.

The distinction between the computational formulation and the algorithm is crucial. The construction of the computational formulation must be based on the nature of the mathematical formulation, whereas the algorithm is just one of a number of possible ways for evaluating the computational formulation. Usually, it is chosen so as to optimize some characteristic which is essential to the efficient use of the computer such as storage, growth of rounding error, or speed. For example, if the computational formulation reduces to evaluating the quartic polynomial

$$p_4(x) = ax^4 + bx^3 + cx^2 + dx + e$$

then the algorithm defines how this polynomial will be evaluated. Thus, if the number of multiplications is to be minimized, nested multiplication must be used:

$$p_4(x) = (((ax+b)x+c)x+d)x + e \ .$$

The need for making this distinction between the computational formulation and the algorithm appears as an essential feature of any discussion of the stability of numerical processes.

Because this finite sequence of elementary arithmetical operations, which the algorithm generates, is the actual representation of the problem used by the computer, it is given the special name of *numerical* (or *computational*) *process*.

Though the choice of algorithm is governed directly by the choice of computational formulation, the crucial nature of the second step is a direct consequence of the fact that, no matter how good the algorithm is, it will only yield reliable results if an appropriate choice has been made for the computational formulation. Thus, a detailed understanding of the ways computational formulations can be constructed and of their interrelationship with the original problem formulation and the resulting algorithms is needed before the performance of numerical processes can be successfully assessed. An introduction to such material, which explores the most important factors which influence the available choices, and which discusses the key results which characterize why numerical analysis and computation mathematics must be viewed as a discipline in its own right, is contained in Anderssen and de Hoog (1983).

When constructing algorithms, it is not only necessary to approximate a given mathematical formulation by a finite sequence of elementary operations (i.e. construct a computational process to approximate the given mathematical formulation), but also to construct such approximations so that their computational performance is acceptable. Without checking that it is, there is no guarantee that a given numerical process will yield an adequate approximation to the solution of the given mathematical problem. It is the need to guarantee that it does which makes numerical analysis and computational mathematics far more involved than might appear at first sight.

In any examination of the behaviour of numerical processes, the three key factors are the following.

1.1 *Discretization of the mathematical formulation.* The mathematical formulation must be replaced by a finite-dimensional approximation. However, the solution of the resulting finite-dimensional formulation may not yield a good approximation to the solution of the original problem. It is necessary to construct conditions under which good approximations can be guaranteed.

1.2 *The floating-point number system.* Because of their finite resources, computers only work with a finite subset of the real numbers. This is usually a floating-point number system. As each elementary operation is performed, the resulting numbers are approximated by adjacent floating-point numbers. The resulting error is called *rounding error*. The behaviour and growth of this error can exhibit quite unexpected patterns.

1.3 *Stability.* As a consequence of 1.1 and 1.2, one of the potential limitations of any numerical process is its inability to cope with the effects of discretization and rounding error. When such situations occur, the underlying computational process is said to be *unstable*. Much of numerical analysis is concerned with identifying conditions which guarantee that a given numerical process will be 'stable'.

From a problem solving point of view, the process "mathematical formulation $\rightarrow$ computational formulation $\rightarrow$ algorithm" must be embedded in the larger framework of applied mathematics of "formulate (the mathematical model for the given problem) $\rightarrow$ solve $\rightarrow$ interpret", and thereby corresponds to the solve step. On the other hand, from a computational mathematics point of view, the process "mathematical formulation $\rightarrow$ computational formulation $\rightarrow$ algorithm" must be embedded within appropriate mathematical frameworks which allow relevant questions connected with consistency, convergence and stability as well as computational complexity and efficiency to be answered.

It is the latter aspect of the process "mathematical formulation $\rightarrow$ computational formulation $\rightarrow$ algorithm" that is pursued in the remainder of this paper. Clearly, the framework chosen will depend heavily on the question being asked; and, in general, there exists no unique framework in which all possible questions can be answered.

§3. REASONS WHY DEPTH IN MATHEMATICS IS NEEDED IN COMPUTATIONAL MATHEMATICS

In examining how depth in mathematics has been used to advance in non-trivial ways the theory and practice of computational mathematics, it is not simply a matter of listing some examples. In order to give meaning and motivation to these ideas, it is first necessary to list and illustrate the reasons why depth in mathematics plays such a crucial role in computational mathematics.

We examine the following reasons.

3.1 *The need to move into a more appropriate framework before the formalism under consideration can be manipulated to yield the required results.*

The need to define an appropriate framework, before any results can be derived, arises every time it is necessary to analyse a finite set of observations d_i, $i = 1,2,\ldots,n$. Except for the obvious, no progress can be made with their interpretation until an appropriate framework, consistent with the problem context, is proposed. For example, statistical methodology can be invoked if the framwork of random variable modelling is used. The observations are thereby interpreted as realizations of some specific random variable structure such as a parametric signal $s(x;\underset{\sim}{\alpha})$ with additive noise ε_i with known distribution properties:

$$d_i = s(x_i;\underset{\sim}{\alpha}) + \varepsilon_i \ , \qquad i = 1,2,\ldots,n \ , \tag{1}$$

where $\underset{\sim}{\alpha}$ denotes the vector of unknown parameters.

About the simplest example in numerical analysis, of where an appropriate mathematical analysis framework must be invoked before convergence criteria can be derived, is the analysis of one-point iteration: *for a given function* g *and starting value* x_0 , *evaluate*

$$x_n = g(x_{n-1}) \ , \quad n = 1,2,\ldots \ . \tag{2}$$

It is an automatic consequence of (2) that

$$x_{n+1} - x_n = g(x_n) - g(x_{n-1}) \ , \quad n = 1,2,\ldots \ . \tag{3}$$

From a convergence point of view, the left hand side of (3) has the desired form, and therefore it is necessary to manipulate its right hand side in some appropriate manner before convergence results can be derived. The usual procedure is to restrict g to the class of functions with continuous first derivatives defined on some suitably large compact interval $[a,b]$ of the real line which contains x_0 , and then invoke the mean value theorem

$$g(x_n) - g(x_{n-1}) = g'(\xi_n)(x_n - x_{n-1}) \ , \quad x_n < \xi_n < x_{n-1} \ . \tag{4}$$

Together, (3) and (4) yield

$$x_{n+1} - x_n = g'(\xi_n)(x_n - x_{n-1}) \ , \quad n = 1,2,\ldots \ ,$$

and hence,

$$x_{n+1} - x_n = \left\{ \prod_{i=1}^{n} g'(\xi_i) \right\} (x_1 - x_0) \ , \quad n = 1,2,\ldots \ . \tag{5}$$

Thus, within the framework of the regularity conditions imposed above on $g(x)$, convergence criteria for the one-point iteration (1) is reduced to an examination of conditions which guarantee that the iterates x_n, $n = 1,2,\ldots,$ form a Cauchy sequence; viz.

$$|x_{n+1} - x_n| \to 0 \quad \text{as} \quad n \to \infty \ . \tag{6}$$

Clearly, this is achieved if the above framework is further reduced by invoking the additional condition that

$$|g'(\xi)| < 1 \ , \quad \xi \in [a,b] \ . \tag{7}$$

The compactness of $[a,b]$ guarantees that the Cauchy sequence of iterates has a limit point which corresponds to a fixed point of $x = g(x)$ located in $[a,b]$.

3.2 *The strategy of taking the idea that solves a simple problem and constructing a new framework in which the same idea solves more complex problems.*

One of the most powerful examples of this is the extension of the ideas used above, to derive convergence criteria for one-point iteration, to the various *contraction mapping principles* used in the study of theoretical and numerical methodologies for the solution of ordinary and partial differential equations.

For example, consider the first-order ordinary differential equation

$$\dot{x} = f(t,x) \ , \tag{8}$$

with initial condition

$$x(t_0) = x_0 \ . \tag{9}$$

Using the indefinite integral representation for a first derivative (i.e. integrating (8) from t_0 to t and applying (9)), the following equivalent integral representation for (8) and (9) can be constructed

$$x(t) = x_0 + \int_{t_0}^{t} f(\tau, x(\tau))\,d\tau \ . \tag{10}$$

Because the right hand side of (10) can be reinterpreted as a mapping from given functions x to corresponding functions x^* defined by

$$x^* = x_0 + \int_{t_0}^{t} f(\tau, x(\tau))\,d\tau \ ,$$

it can be reinterpreted as an *operator* $\underset{\sim}{A}$ defining this correspondence between x and x^*. In this way, (10) can be rewritten as the operator equation

$$x = \underset{\sim}{A}\, x \ . \tag{11}$$

The one-point iteration (2) generalizes naturally to define *the method of successive approximations* for the solution of (11) : *for a given operator* $\underset{\sim}{A}$ *and starting value* $\phi_0(t)$, *evaluate*

$$\phi_n = \underset{\sim}{A}\,\phi_{n-1}\ , \qquad n = 1,2,\ldots\ . \tag{12}$$

It is now an automatic consequence of (12) that

$$\phi_{n+1} - \phi_n = \underset{\sim}{A}\,\phi_n - \underset{\sim}{A}\,\phi_{n-1} = \int_{t_0}^{t} \left\{ f(\tau,\phi_n(\tau)) - f(\tau,\phi_{n-1}(\tau)) \right\} d\tau\ . \tag{13}$$

The argument sketched above for proving that the iterates x_n , $n = 1,2,\ldots,$ define a Cauchy sequence generalizes naturally to this situation. On invoking appropriate regularity conditions on f and the family of functions Φ to which the ϕ_n , $n = 0,1,2,\ldots$, belong, and introducing the maximum norm

$$\|\phi\|_\infty = \max_{t_0 \le t \le T} |\phi(t)|\ , \tag{14}$$

it can be shown that, for suitably small T , the sequence (12) satisfies the inequality

$$\|\phi_{n+1}-\phi_n\|_\infty \le k\|\phi_n-\phi_{n-1}\|_\infty\ , \qquad n = 1,2,\ldots\ , \tag{15}$$

with $0 < k < 1$. A careful discussion of the details can be found in Pontryagin (1962), Chapter 4.

It then follows from (15) that

$$\|\phi_{n+1}-\phi_n\|_\infty \le k^n\|\phi_1-\phi_0\|_\infty\ , \quad n = 1,2,\ldots\ , \tag{16}$$

which proves that, with respect to the norm (14), the iterates ϕ_n , $n = 0,1,2,\ldots$, form a Cauchy sequence and therefore converge (uniformly) to a fixed point of (11) in Φ .

This proof, that the iterates ϕ_n , $n = 0,1,2,\ldots$, form a Cauchy sequence and therefore converge to a fixed point of (11) in Φ , can be derived independently if $\underset{\sim}{A}$ can be shown to define a contraction mapping on Φ . In fact, with respect to a given $\underset{\sim}{A}$, it is only necessary to show that the functions Φ are such that

(a) $\underset{\sim}{A}\,\phi \in \Phi$ for all $\phi \in \Phi$; and

(b) $\|\underset{\sim}{A}\psi-\underset{\sim}{A}\chi\|_\infty \le k\|\psi-\chi\|_\infty$ with $0 < k < 1$ for all $\psi , \chi \in \Phi$,

for $\underset{\sim}{A}$ to define a *contraction mapping* on Φ .

It is clear that this concept of contraction mapping can be used to generalize the simple argument, invoked above for one-point iteration, to the analysis of more complex situations (cf. Gilbarg and Trudinger (1977), Chapter 5; Simmons (1963)).

3.3 *The need to move into a larger framework where the problem can be shown to have a solution before being able to show that its solutions have required or desired properties.*

As well as being the motivation for the use of concepts such as weak solutions and Sobolev spaces in the analysis of variational and finite element methods for the numerical solution of ordinary and partial differential equations (cf. Mikhlin (1971), Oden and Reddy (1976)), this strategy also plays a key role in the study of mathematical analysis itself (cf. Mikhlin (1970), Gilbarg and Trudinger (1977)).

To illustrate, we examine the significance of the weak form of the Rayleigh quotient definition for differential eigenvalues. The aim is not only to discuss the importance mathematically and computationally of this weak formulation, but also to stress that the weak formulation is not necessarily some purely mathematical construct without any physical significance. In particular, we examine the Sturm-Liouville eigenvalue problem defined by the differential equation

$$-(pu')' + qu = \lambda ru, \quad u = u(x) \;, \quad u' = du/dx \;, \quad p = p(x) \;, \quad q = q(x) \;, \quad 0 \le x \le 1 \;, \tag{17}$$

and the boundary conditions

$$u(0) = u(1) = 0 \;. \tag{18}$$

To solve (17) and (18) as they stand, it is first necessary to make appropriate assumptions about the form of p , q and r (e.g. $p(x)$ continuously differentiable; $p > 0$, $x \in [0,1]$; q and r continuous and positive for $x \in [0,1]$.) Then it is necessary to find a scalar λ and a corresponding function u (which is twice differentiable) which satisfy (17) and (18) identically. The difficulty with (17) is that, even though λ is a scalar, (17) itself is non-scalar. This difficulty is removed, if (17) and (18) are transformed to a more appropriate framework where they can be replaced by a scalar representation. Though there exist numerous ways in which this could be done, the aim is to find a framework in which the resulting scalar structure can be exploited mathematically and computationally.

Because it can be given a Hilbert space interpretation and it also leads naturally to the Rayleigh quotient definition for eigenvalues, the scalar conversion usually adopted is to multiply both sides of (17) by u and then integrate the resulting equation over the interval $[0,1]$ to obtain

$$\int_0^1 \{-(pu')'u + qu^2\}dx = \lambda \int_0^1 ru^2 dx \;. \tag{19}$$

This integral form is chosen rather than some other as it is the simplest under which integration by parts can be applied to the left hand side of (19) in conjunction with the boundary conditions (20) to reduce the regularity conditions on both $p(x)$ and $u(x)$ by yielding

$$\int_0^1 \{p(u')^2 + qu^2\}dx = \lambda \int_0^1 ru^2 dx \ . \tag{20}$$

The advantages of (20) both mathematically and computationally are obvious: it allows the eigenvalue problem (17) and (18) to be given a more general interpretation; for given p and q, the class of solutions on which (20) is defined is larger than that on which (17) and (18) is defined (the essence of 3.3); simpler basis functions can be used to construct approximate methods for (20) than for the corresponding (17) and (18) (practical consequence of 3.3).

In addition, the Rayleigh quotient definition for eigenvalues follows automatically from (20); viz.

$$\lambda = \int_0^1 \{p(u')^2 + q\, u^2\}dx \Big/ \int_0^1 ru^2 dx \ . \tag{21}$$

This fact has important consequences for the weak formulation (20). On the one hand, it yields a physical interpretation for the weak formulation, since (20) represents a generalization of the equivalencing of average potential energy and average kinetic energy of a vibrating system which Rayleigh himself used to define the eigenfrequencies of such a system. On the other, (21) (following Rayleigh) can be used as the basis for a quite simple, but very effective strategy for estimating λ approximately, which exploits the facts that

(i) if a first order approximation for u is substituted in the Rayleigh quotient (21), then the resulting eigenvalue approximation is second order accurate;

(ii) the eigenvalues defined by (21) have a variational formulation in terms of the Rayleigh quotient (cf. Weinstein and Stenger (1972)); e.g., for the fundamental eigenvalue λ_0,

$$\lambda_0 = \min_{\substack{u\in U \\ u\neq 0}} \left[\int_0^1 \{p(u')^2 + q\, u^2\}dx \Big/ \int_0^1 ru^2 dx\right] , \tag{22}$$

where U denotes the set of functions u on which (21) is defined.

In fact, for the approximation of λ_0, one form of this procedure becomes: with respect to an appropriate parametric representation $u(x;\underset{\sim}{\alpha})$ for u, where $\underset{\sim}{\alpha}$ denotes the vector of parameters, estimate λ_0 as

$$\hat{\lambda}_0 = \min_{\underset{\sim}{\alpha}} \left[\int_0^1 \{p(u'(x;\underset{\sim}{\alpha}))^2 + q(u(x,\underset{\sim}{\alpha}))^2\}dx \Big/ \int_0^1 r(u(x;\underset{\sim}{\alpha}))^2 dx\right] . \tag{23}$$

A detailed discussion of these and related aspects associated with the use and interpretation of Rayleigh's quotient can be found in Lapwood and Anderssen (1984).

The importance of the physical interpretation of weak solutions is not limited to the Rayleigh quotient interpretation given above. For example, it plays a crucial role in the interpretation and determination of the evolving thermal history in solidification (cf. Atthey (1974))

3.4 *The strategy of modifying the mathematical formalism to answer the specific question under examination, since the form of the mathematics used depends heavily on the question which must be answered.*

This strategy has been employed successfully to avoid the need to solve numerically the improperly posed formalism which defines the given problem (cf. Anderssen (1980) and Anderssen and Jackett (1983)). To illustrate, consider the situation where the problem formulation has an operator equation representation

$$\underset{\sim}{A}\, u = s \ , \quad \underset{\sim}{A} : \underline{\underline{H}} \rightarrow \underline{\underline{H}} \ , \tag{24}$$

where $\underline{\underline{H}}$ denotes a Hilbert space with inner product $(\cdot,\cdot)$. In many applications, it is not the solution of (1) {which is often improperly posed (cf. de Hoog (1980))} which is required for inference purposes, but a linear functional

$$\gamma = (\theta, u) \tag{25}$$

defined on the solution u with θ known.

If it is assumed that $\theta \in \underline{\underline{R}}(\underset{\sim}{A}^*)$ (i.e. the range of the adjoint operator $\underset{\sim}{A}^*$), then it is a simple matter to show (cf. Golberg (1979)) that

$$\gamma = (\phi, s) \tag{26}$$

with

$$\underset{\sim}{A}^*\phi = \theta \ . \tag{27}$$

In fact,

$$(\theta,u) = (\underset{\sim}{A}^*\phi,u) = (\phi,Au) = (\phi,s) \ .$$

From a numerical point of view, the appeal of this result is the ease with which it copes with the common practical situation where the only information available about s is observational data of the form

$$d_i = s(x_i) + \varepsilon_i \ , \quad i = 1,2,\ldots,n \ , \tag{28}$$

where the ε_i are assumed to be random errors. The estimation of γ is reduced to evaluating (ϕ,s) using only the d_i , $i = 1,2,\ldots,n$.

The solution of $\underset{\sim}{A}\, u = s$ is replaced by the need to solve $\underset{\sim}{A}^*\phi = \theta$. But, where $\underset{\sim}{A}^*$ and θ are such that ϕ can be determined analytically (e.g. the Abel integral equation case (Anderssen (1980)), the overall numerical procedure is greatly simplified. Even in situations where it is in fact necessary to solve $\underset{\sim}{A}^*\phi = \theta$ numerically to determine an approximation to ϕ , the position is far superior to that defined by (24) and (28), since θ is known analytically.

3.5 *The need to examine and possibly exploit the problem formulation mathematically in order to optimize any resulting computational procedure.*

Computationally, the most important illustration of this point is the need to examine whether the given problem formulation is properly posed. On the one hand, it is necessary to ensure that, computationally, a unique solution is being sought. On the other, it may be necessary to stabilize it in some appropriate manner if its solution fails to depend continuously on the data.

Numerous other illustrations exist. For example, the linear functional example given to illustrate 3.4 could be reinterpreted as an exploitation mathematically of the problem formulation.

In part, the importance of 3.5 is the implication that, given a problem formulation, it is incorrect to assume that the direct application of numerical techniques will necessarily yield the best results computationally. For example, consider the situation where the regularization formulation

$$\min_{u} \left\{ \sum_{i=1}^{n} (d_i - u(x_i))^2 + \alpha \int_{x_1}^{x_n} (u'')^2 dx \right\}, \tag{29}$$

(assuming $x_1 < x_2 < \ldots < x_n$ and writing u'' for d^2u/dx^2) is used to smooth the observational data (1). For a given α, discretization could be applied directly to (29) to generate an appropriate algorithm. This is not the optimal strategy in the present situation since it is known that the unique minimizer of (29) is a cubic smoothing spline (with a knot at every data point). This fact can be exploited numerically in various ways (cf. de Boor (1978) and Craven and Wahba (1979)) including the construction of procedures to determine an optimal estimate of the value of α which should be used in (29) (cf. Lukas (1980)).

§4 ADVANCES IN THE THEORY AND PRACTICE OF COMPUTATIONAL MATHEMATICS

In this section, we discuss briefly some examples of how depth in mathematics (and in particular, mathematical analysis) has been used to make major advances with the theory and practice of computational mathematics. Because the book in which this paper is published, reports the proceedings of an Australian conference, examples are limited to advances to which Australians have made a contribution.

Adaptivity and Superconvergence

In the analysis of the numerical performance of different approximate methods for the solution of ordinary and partial differential equations as well as integral equations, a key concept is the *error estimate*. On the one hand, it can be used to derive *a priori* theoretical information about the order of convergence of methods which can be used as a basis for their comparison. On the other, it can be used to derive *a posteriori* practical information about the particular method being applied to some given problem which can be used to modify the method in appropriate ways to improve its performance.

It is this latter aspect which is exploited for the *adaptive refinement* of numerical methods as they are being applied. To illustrate the nature of the mathematical problems involved, consider the situation where it is necessary to approximate the solution u of a given problem defined on some region D. Let $u_h(x;G_h)$ define the current approximate (e.g. finite element) solution on a non-even mesh G_h constructed on D, with representative mesh spacing h. With respect to some error measure (e.g. mean square error or maximum modulus error), the aim is to refine G_h to obtain $\hat{G}_h$ so that, compared with $u_h(x;G_h)$, the new approximation $u_h(x;\hat{G}_h)$ has smaller error.

For this, an estimate of the error

$$|u(x) - u_h(x;G_h)| \,, \quad x \in D \,, \tag{30}$$

is required. It is then simply a matter of adding points to G_h to form $\hat{G}_h$ so that

$$|u(x) - u_h(x;\hat{G}_h)| \leq |u(x) - u_h(x;G_h)| \tag{31}$$

for x at appropriate places in D (e.g. $\hat{G}_h$).

But, this is impossible as it stands, since $u(x)$ is unknown. What is needed is a better estimate $u_h^*(x;G_h)$ of u than $u_h(x;G_h)$ which can be used instead of $u(x)$ in (30). Clearly, it cannot be $u_h(x;\hat{G}_h)$. The optimal strategy is to determine $u_h^*(x;Gh)$ using only the information obtained on G_h for the construction of $u_h(x;G_h)$. This can only be achieved if some appropriate mathematical aspect of the problem context is exploited.

Though it is conceptually simple to motivate the need for the estimate $u_h^*(x;G_h)$, something special must be done mathematically before an appropriate candidate for $u_h^*(x;G_h)$ can be constructed.

Two such procedures for ordinary differential equations are deferred correction (cf. Fox (1957)) and defect correction (cf. Ueberhuber (1979)). For partial differential equations, considerable work on *a posteriori* error estimates has been done by Babuska and Rheinboldt (1981). The paper by Miller (1983), which is published in these proceedings, reviews the different approaches being developed for the finite element solution of elliptic partial differential equations.

For the second kind integral equation,

$$u + \underset{\sim}{K} u = u(x) + \int_0^1 k(x,y)u(y)dy = f(x) \ , \quad 0 \le x \le 1 \ , \tag{32}$$

Sloan (1976) proposed the use of the iteration

$$u_h^*(x) = f(x) - \int_0^1 k(x,y)u_h(y)dy \tag{33}$$

to improve the Galerkin approximation $u_h(x)$ constructed for the approximation solution of (32) using polynomial splines of degree r . In fact, it is well known that

$$\|u-u_h\| = O(h^{r+1}) \ , \tag{34}$$

where $\|\cdot\|$ corresponds to the standard L_2-norm. The utility of the iterate $u_h^*(x)$ is easily motivated by observing that, in constructing u_h^* , the oscillatory nature of the error in $u_h(x)$ is smoothed if the kernel $k(x,y)$ is itself appropriately smooth.

Subsequently, Chandler (1980) has shown that, for appropriately smooth $k(x,y)$ the approximations u_h^* are globally superconvergent; viz.

$$\|u-u_h^*\| = O(h^{2r+2}) \ . \tag{35}$$

This, u_h^* represent an obvious candidate to use for adaptivity purposes.

Clearly, such results have important ramifications for the use of boundary integral methods for the numerical solution of partial differential equations. More recently, work has been done on extending Chandler's result to more general kernels and more general contexts (cf. Chandler (1983), Graham (1983) and Sloan (1982)). In the present context, it is not the actual iteration (33) which is deep mathematically, but the tools needed to establish the range of validity of results like (35) in terms of regularity conditions on $k(x,y)$, $f(x)$ and the method used to generate u_h . Clearly, u_h^* cannot be exploited for adaptive purposes until the form of $\|u-u_h^*\|$ has been suitably characterized.

The key to a superconvergence result such as (35) is the observations that (cf. Chandler (1980))

(i) the Galerkin approximation $u_h \in S_h$ (the polynomial spline subspace) is defined by

$$(I-\underset{\sim}{P}_h\underset{\sim}{K})u_h = \underset{\sim}{P}_h f$$

where $\underset{\sim}{P}_h : L_2([0,1]) \to S_h$ denotes the orthogonal projection of $L_2([0,1])$ onto S_h ;

(ii) $$(u_h^*-u)(x) = (k_x, u_h-u) \ , \quad k_x = k(x,y) \ ,$$

where $(\cdot,\cdot)$ denotes the standard L_2-inner product;

(iii) for suitably smooth $\Phi = (I-\underset{\sim}{K}^*)\Psi$, $\underset{\sim}{K}^*$ the adjoint of $\underset{\sim}{K}$,

$$(\Phi, u_h - u) = (\Psi, (I-\underset{\sim}{K})(u_h - u))$$
$$= (\Psi - \phi, (I-\underset{\sim}{K})(u_h - u)) \; , \quad \phi \in S_h \; ,$$

because, from (1), $(\phi, (I-\underset{\sim}{K})(u_h - u)) = 0$.

More recently, Sloan and Thomée (1984) have shown how such superconvergence results can be obtained using functional analytic arguments.

Estimation of the Stabilization Parameter in Regularization

For the approximate solution of first kind Fredholm integral equations

$$\int_0^1 k(x,y)u(y)dy = s(x) \; , \qquad 0 \le x \le 1 \; , \tag{36}$$

a major difficulty arises computationally when the kernel $k(x,y)$ is smooth (e.g. contains no singularities). There is then no guarantee that the solution $u(y)$ depends continuously on its right hand side $s(x)$. For this reason, first kind Fredholm integral equations with smooth kernels are said to be *improperly posed.* There are various ways in which this concept can be quantified mathematically. For example, the rate at which the eigenvalues of the operator $\underset{\sim}{K}^*\underset{\sim}{K}$, with

$$\underset{\sim}{K}\, u = \int_0^1 k(x,y)u(y)dy \; , \tag{37}$$

decay to zero (cf. Smithies (1958) and de Hoog (1980)). The faster the decay, the more acute is the improperly posedness and hence the computational difficulties; especially if the only information about s is the observational data (28). This point has already arisen when discussing exemplification for 3.3.

Computationally, the way around this difficulty is to replace the solution of (36) by some properly posed formulation which can be made to approximate (36) arbitrarily closely. The usual approach is to use some form of regularization, which reduces the solution of (36) to a variational formulation, such as

$$\underset{u \in \underline{\underline{W}}}{\text{minimize}} \; \{\|\underset{\sim}{K}u - s\|^2 + \alpha\|\underset{\sim}{T}u\|\} \; , \tag{38}$$

where $\underline{\underline{W}}$ denotes some space of smooth functions, $\alpha > 0$ is the regularization constant which determines the closeness of the solutions of (38) to (36), $\underset{\sim}{T}$ is an *a priori* chosen operator and $\|\cdot\|$ corresponds to the standard L_2-norm.

Once $\underset{\sim}{T}$ and α have been specified, the solution of (38) presents no difficulties. The Euler-Lagrange equations for (38) reduce it to solving

$$(\underset{\sim}{K}^*\underset{\sim}{K} + \alpha\underset{\sim}{T}^*\underset{\sim}{T})u = \underset{\sim}{K}^*s \; . \tag{39}$$

The difficulty mathematically is characterizing and comparing the various choices for $\underset{\sim}{T}$ and the different procedures proposed for determining α .

Such examinations involve non-trivial aspects of functional analysis and delicate estimates concerning the behaviour of the eigenvalues of various operators. A detailed discussion and analysis of these and related aspects of regularization can be found in Lukas (1980, 1981) and Wahba (1983).

A principal motivation for such work is the unification it often yields for various classes of methods. For example, Anderssen and Bloomfield (1974) have shown how an explicit equivalence between regularization and Wiener filtering can be established for numerical differentiation and numerical linear filtering. The fact that regularization methods have a Wiener filtering interpretation has been used by Anderssen and Prenter (1981) to construct an objective basis of comparison for the different methods proposed for the solution of first kind integral equations.

Uniformly Valid Approximate Eigenvalues.

In many applications, the underlying computational problem reduces to finding the eigenvalues of specific Sturm-Liouville problems

$$-(pu')' + qu = \lambda ru \ , \quad u = u(x) \ , \quad u' = du/dx \ , \quad x \in (0,\pi) \ , \tag{40}$$

$$a_1 u(a) + a_2 u'(a) = 0 \ , \tag{41}$$

$$b_1 u(b) + b_2 u'(b) = 0 \ , \tag{42}$$

$$u, \ pu' \in C[0,\pi] \quad \text{(continuous functions on the interval } [0,\pi])$$

where p , q and r are functions of x , and a_1, a_2, b_1 and b_2 are constants. In situations where only the fundamental and first few harmonics are required, standard algebraic and variational (e.g. minimize the Rayleigh quotient) procedures yield excellent approximations. However, when accurate approximations to the first 10 or more eigenvalues are required, then the standard procedures are not applicable since the approximations they generate do not yield uniformly valid estimates of the corresponding Sturm-Liouville eigenvalues (cf. Paine and de Hoog (1980) and Paine and Anderssen (1980)).

It is therefore necessary to find alternative procedures which do in fact generate uniformly valid estimates.

All methods which can be applied to (40) - (42) involve two basic steps: (i) replace the exact problem by an approximate problem which is (theoretically) soluble; (ii) solve the approximate problem numerically. Clearly, the total error will be the sum of the errors associated with these two steps. Although the importance of the error introduced at the second step should not be underestimated, the most important component is that arising from the first, since the error thereby generated usually dominates all other considerations. Thus, the choice of the approximate problem in step (i) should be tailored to the nature of the eigenvalues to be approximated.

In fact, the standard algebraic and variational procedures fail to yield uniformly valid approximations because the first step in their application corresponds to approximating the infinite set of Sturm-Liouville eigenvalues by a finite set of algebraic or variational eigenvalues. This is easily illustrated.

Consider the eigenvalue problem

$$-y'' = \mu y\ ,\ y = y(x)\ ,\ y'' = d^2y/dx^2\ ,\ 0 \le x \le \pi\ ,\ y(0) = y(\pi) = 0\ . \tag{43}$$

We know its eigenvalues exactly: $\mu_j = j^2$. If, on the grid

$$\underline{\underline{G}} = \{x_i ; x_i = ih\ ,\ i=0,1,2,\ldots,n+1,\ h = \pi/(n+1)\} \tag{44}$$

y'' is replaced by its central difference approximation, the following algebraic eigenvalue problem is obtained

$$A\,\underset{\sim}{y} = \mu^{(n)}\underset{\sim}{y} \tag{45}$$

with

$$A = h^{-2}\begin{bmatrix} 2 & -1 & & & \\ -1 & 2 & -1 & & \\ & \ddots & \ddots & \ddots & \\ & & -1 & 2 & -1 \\ & & & -1 & 2 \end{bmatrix}$$

a matrix of order n . In fact, for this simple algebraic problem, we know that

$$\mu_j^{(n)} = 4\sin^2(jh/2)/h^2\ ,\quad j = 1,2,\ldots,n\ ; \tag{46}$$

and hence the error $\varepsilon_j^{(n)} = \mu_j - \mu_j^{(n)}$ takes the form

$$\varepsilon_j^{(n)} = j^2 - 4\sin^2(jh/2)/h^2 = O(j^4h^2)\ . \tag{47}$$

This clearly illustrates the significance of step (i) relative to step (ii) (which has been performed exactly here) and the rapid growth of $\varepsilon_j^{(n)}$ as a function of j .

It is therefore natural to conclude that, for the construction of methods which yield uniformly valid estimates, step (i) should yield a problem which corresponds to approximating the infinite set of Sturm-Liouville eigenvalues by an infinite set of approximations. The obvious strategy is to replace (40) - (42) by a Sturm-Liouville problem with coefficients $\bar{p}$, $\bar{q}$ and $\bar{r}$, which are simple (e.g. piecewise constant) approximations to p , q and r , and the same boundary conditions. In this way, the required Sturm-Liouville eigenvalues are approximated by eigenvalues of a simpler but similar eigenvalue problem.

The utility of this strategy has been examined in some detail by Paine and de Hoog (1980) and Andrew (1982).

The interesting facet of the above argument is its failure to allow for the possibility that the error $\lambda_j - \lambda_j^{(n)}$ can be estimated theoretically with sufficient accuracy to allow the $\lambda_j^{(n)}$ to be corrected to yield uniformly valid estimates of the corresponding λ_j. To illustrate, consider the following canonical Liouville normal form

$$-y'' + q\,y = \lambda y\ , \quad y = y(x)\ , \quad q = q(x)\ , \quad 0 \le x \le \pi\ , \tag{48}$$

$$y(0) = y(\pi) = 0\ . \tag{49}$$

If, on the grid $\underset{\approx}{G}$ of (44), y'' is replaced by its central difference approximation, then (48) and (49) are replaced by the algebraic eigenvalue problem

$$(A+D)\,\underset{\sim}{y} = \lambda^{(n)}\underset{\sim}{y} \tag{50}$$

with $D = \text{diag}(q(x_1),\ldots,q(x_n))$ and A defined in (45).

Paine, de Hoog and Anderssen (1981) have shown, using the $\varepsilon_j^{(n)}$ above, that

$$\lambda_j - \lambda_j^{(n)} = \varepsilon_j^{(n)} + O(kh^2)\ , \quad 1 \le j \le \alpha n\ , \quad \alpha < 1\ ;$$

and hence, that the following correction procedure

$$\tilde{\lambda}_j^{(n)} = \lambda_j^{(n)} + \varepsilon_j^{(n)} = \lambda_j + O(kh^2)\ , \quad 1 \le j \le \alpha n\ , \quad \alpha < 1\ ,$$

yields uniformly valid approximations. This result has been extended to the canonical Sturm-Liouville equation (48) in conjunction with the general boundary conditions (41) and (42) by Anderssen and de Hoog (1982).

§5. CONCLUSION

Obviously, the list of advances in §4 is not comprehensive, and does not pursue mathematical aspects in great detail or rigour. The aim has been to illustrate with broad brush strokes the crucial role that mathematical analysis plays and will continue to play in the theory and practice of computational mathematics.

However, in making this point, it is important to stress that it is incorrect to extrapolate to any conclusion which implies that future advances in the theory and practice of computational mathematics can only be made with the assistance of mathematical analysis. This is not true for the examples cited. Though mathematical analysis has played a crucial role, the ability to change or adapt the interpretation of the problem under examination has been equally important.

In conclusion, it is important to note that, in the above discussion, it is tacitly assumed that one starts with some continuously defined mathematical model and proceeds to discretize it to construct computational procedures. This ignores the situation where one starts with the problem context and derives computational procedures directly, from the explicit use of the laws that define that context. Such a strategy underlies much of the engineering input to the development of the finite element method. It is also the foundation behind the use of stochastic processes to model real world phenomenon such as sedimentation (cf. Gani and Todorovic (1983)).

The use of mathematical analysis to formalize and interpret such strategies (e.g. Strang and Fix (1973) for the finite element method) has played a major role in the recent developments of computational mathematics and therefore should continue to be seen as such a source for future developments.

Acknowledgement

The author wishes to thank Frank de Hoog for valuable comments and useful discussions during the preparation of this paper.

REFERENCES:

[1] Anderssen, R.S., On the use of linear functionals for Abel-type integral equations in applications, in: The Application and Numerical Solution of Integral Equations (eds.) R.S. Anderssen, F.R. de Hoog and M.A. Lukas, (Sijthoff and Noordhoff International Publishers, 1980).

[2] Anderssen, R.S. and Bloomfield, P., Numerical differentiation procedures for non-exact data, Numer. Math. 22 (1974) 157-182.

[3] Anderssen, R.S. and de Hoog F.R., On the correction of finite difference eigenvalue approximations for Sturm-Liouville problems with general boundary conditions, CMA Research Report, RO5-82, Centre for Mathematical Analysis, Australian National University, (1982).

[4] Anderssen, R.S. and de Hoog, F.R., The nature of numerical processes, The Math. Scientist (in press), (1983).

[5] Anderssen, R.S. and Jackett, D.R., Linear functionals of foliage angle density, J. Aust. Math. Soc., Series B (in press) (1983).

[6] Anderssen, R.S. and Prenter, P.M., A formal comparison of methods proposed for the numerical solution of first kind integral equations, J. Aust. Math. Soc. (Series B) 22 (1981) 488-500.

[7] Andrew, A.L., Computation of higher Sturm-Liouville eigenvalues, Congressus Numerantium, 34 (1982) 3-16.

[8] Atthey, D.R., A finite difference scheme for melting problems, J. Inst. Math. Appl. 13 (1974) 353-366.

[9] Babuška, I. and Rheinboldt, W.C., A posteriori error analysis of finite element solutions for one-dimensional problems, SIAM J. Numer. Anal. 18 (1981) 565-589.

[10] Chandler, G.A., Superconvergence for second kind integral equations, in: The Application and Numerical Solution of Integral Equations (eds.) R.S. Anderssen, F.R. de Hoog and M.A. Lukas, (Sijthoff and Noordhoff International Publishers,

[11] Chandler, G.A. Superconvergent approximations to the solution of a boundary integral equation on polygonial domains, CMA Research Report, R15-83, Centre for Mathematical Analysis, Australian National University, (1983).

[12] Craven, P. and Wahba, G., Smoothing noisy data with spline functions: estimating the correct degree of smoothing by the method of generalized cross-validation, Numer. Math. 31 (1979) 377-403.

[13] de Boor, C., A Practical Guide to Splines (Springer-Verlag, New York, 1978).

[14] de Hoog, F.R., Review of Fredholm equations of the first kind, in: The Application and Numerical Solution of Integral Equations (eds.) R.S. Anderssen, F.R. de Hoog and M.A. Lukas, (Sijthoff and Noordhoff International Publishers, 1980).

[15] Fox, L., Numerical Solution of Two-Point Boundary Problems (Clarendon Press, Oxford, 1957).

[16] Gani, J. and Todorovic, P., A model for the transport of solid particles in a fluid flow, Stochastic Processes and their Appl. 14 (1983) 1-17.

[17] Gilbarg, D. and Trudinger, N.S., Elliptic Partial Differential Equations of Second Order (Springer-Verlag, Berlin, 1977).

[18] Golberg, M.A., A method of adjoints for solving some ill-posed equations of the first kind, J. Appl. Math. and Comp. 5 (1979) 123-130.

[19] Graham, I.G., Galerkin methods for second kind integral equations with singularities, Math. Comp. (to appear) (1983).

(20) Hardy, G.H., A Mathematician's Apology (Cambridge University Press, Cambridge, 1967).

[21] Lapwood, E.R. and Anderssen, R.S., Rayleigh's Principle: Theory and Practice (in preparation) (1984).

[22] Lukas, M.A., Regularization, in: The Application and Numerical Solution of Integral Equations (eds.) R.S. Anderssen, F.R. de Hoog and M.A. Lukas, (Sijthoff and Noordhoff International Publishers, 1980).

[23] Lukas, M.A., Regularization of Linear Operator Equations, Ph.D. Thesis, Australian National University, (January, 1981).

[24] Mikhlin, S.G., Mathematical Physics, An Advanced Course (North-Holland Publishing Company, Amsterdam, 1970).

[25] Mikhlin, S.G., The Numerical Performance of Variational Methods (Wolters-Noordhoff Publishing, The Netherlands, 1971).

[26] Miller, A.D., Adaptive techniques in finite element analysis, Proceedings Comp. Techniques and Appl. Conf., Sydney, (August 29-31, 1983).

[27] Oden, J.T., and Reddy, J.N., An Introduction to the Mathematical Theory of Finite Elements, (A Wiley-Interscience Publication, John Wiley and Sons, New York, 1976).

[28] Paine, J.W. and Anderssen, R.S., Uniformly valid approximation of eigenvalues of Sturm-Liouville problems in geophysics, Geophys. J.R. astr. Soc. 63 (1980) 441-465.

[29] Paine, J.W. and de Hoog F.R., Uniform estimation of the eigenvalues of Sturm-Liouville problems, J. Aust. Math. Soc. (Series B) 21 (1980) 365-383.

[30] Paine, J.W., de Hoog, F.R. and Anderssen, R.S., On the correction of finite difference eigenvalue approximations for Sturm-Liouville problems, Computing 26 (1981) 123-139.

[31] Pontryagin, L.S., Ordinary Differential Equations (Pergamon Press, London, 1962).

[32] Simmons, G.F., Introduction to Topology and Modern Analysis (McGraw-Hill, Kogakusha, Tokyo, 1963).

[33] Sloan, I.H., Improvement by iteration for compact operator equations, Math. Comp. 30 (1976) 758-764.

[34] Sloan, I.H., Superconvergence and the Galerkin method for integral equations of the second kind, Durham Symposium on the Numerical Treatment of Integral Equations (July, 1982).

[35] Sloan, I.H. and Thomée, V., Superconvergence of the Galerkin iterates for integral equations of the second kind, submitted (1984).

[36] Smithies, F., Integral Equations (Cambridge University Press, London, 1958).

[37] Strang, G. and Fix, G.J., An Analysis of the Finite Element Method (Prentice-Hall, New Jersey, 1973).

[38] Ueberhuber, C.W., Implementation of defect correction methods for stiff differential equations, Computing 23 (1979) 205-232.

[39] Wahba, G., Practical approximate solutions to linear operator equations when the data are noisy, SIAM J. Numer. Anal. 14 (1977) 651-667.

[40] Wahba, G., Bayesian "confidence intervals" for the cross-validation smoothing spline, J.R. Statist. Soc. B 45 (1983) 133-150.

[41] Weinstein, A. and Stenger, W., Methods of Intermediate Problems for Eigenvalues (Academic Press, New York, 1972).

Computational Techniques & Applications: CTAC-83
J. Noye & C. Fletcher (Editors)

COMPUTATION IN INDUSTRY: ACHIEVEMENTS AND POTENTIAL

Dr Andrew O Currie

Compumod Pty Ltd
Sydney Australia

The classical basis for computation in industry has evolved from the work of mathematicians, scientists and engineers over the last few centuries. The advent of the digital computer and numerical algorithms capable of solving these classical formulations for practical applications in industry has given today's engineers very powerful simulation tools. This paper investigates some of the achievements made to date and looks at the potential for greater use. Particular emphasis is given to applications of the finite element method in structural analysis.

INTRODUCTION

Computation as a tool in the civilised world has grown from a means of trying to explain physical events to be part of today's pool of information technology which aids daily decision making. Computation has, of course, influenced many fields of science and engineering and this paper describes some areas but concentrates on the subject once known as 'strength of materials'. Timoshenko (1) gives a colourful account of developments in this field, from Leonardo da Vinci's attempts to describe tension tests of wire and Galileo's early model for beam bending, to the development and use of the theory of elasticity. While workers in mathematics and variational calculus, such as the Bernoulli family and Euler, were based in academic institutions it is interesting to note that some key figures were involved with a major industry of the day, military advancement. For example Lagrange worked at the Royal Artillery School in Turin and Coulomb entered the military corps of engineers and gave papers on compass construction as well as his currently used friction theory. Military establishments are still a major source of funding and justification for scientific research in engineering and many computational tools used by general industry had military origins.

The fields of transportation and civil engineering services for urban communities have also provided major needs for computation to improve the design and performance of systems such as arch and suspension bridges, plate structures in concrete floor systems and the many mechanical systems developed during the industrial revolution. The ages of railway and automotive transportation were major driving forces as is space transportation today.

Computation did not, of course, proceed or even support many designs that were successfully built. For example, in the early 1900's some reinforced concrete slab design methods were based on bending moments which did not obey static equilibrium. The correct model of Nichols (2) was not accepted until rigorous plate theory solutions were given by Westergard and Slater (3) and compared with test results. With hindsight we can see why the slabs as built were satisfactory since cracking of the concrete results in a redistribution of bending moments. This example shows the importance of the basis for a computational model. If these assumptions do not agree with the physical occurrence then the computational model will have an uphill battle.

Models based on proven classical theory are best and the theory of elasticity is a good example. Here differential equations were developed by people like Poisson, Navier, Kirchhoff, Cauchy and Lame for plates and shells and general 3D solids. The solution of these equations for practical structures, however, had to wait for the computational engine provided by the digital computer and the necessary modelling software. With this power and extensions of theory a much wider range of problems are being solved. This paper addresses some of the areas which have gained acceptance in day to day engineering design and looks at the potential for future uses.

COMPUTATION AND INDUSTRY

Industry has different meanings for different people but generally involves three primary motivators for deciding a course of action; namely time, money and risk. Since the financial captains of industry equate time with money and risk as probability of loss or gain of money they can put it all into a financial model for computation.

Economic viability is certainly the long term key for industry but actual delivery of a successful finished product involves a complex mix of engineering, production, marketing and management skills. A number of areas where computation can be of assistance are investigated here.

Computation in its purest form is usually restricted to research and development departments where high risk alternatives can be investigated in the hope of long term gain. The need to spread the risk over several projects means that this approach can only be afforded by large organisations or small 'high technology' companies (hopefully lower risk) which are funded by venture capital. Computation in this form is to support the design and manufacture of a new product or technique. Examples of these R and D groups can be found wtihin major automotive and oil companies and the electronics and computer hardware industries.

A major area where computation is being used and has the potential for even greater use in industry is for production design. Here an organisation's product is simulated and refined within the computer prior to actual manufacture. The main benefits are savings in the time taken to complete a successful design and the reduced risk of failure of the design when production starts. To provide these benefits the computational tool (usually computer software) and the modelling approach must have the following features:

- reliability
- capability
- speed

- accuracy
- comprehension of model and results

For commercial applications these features are provided by tried and proven programs from suppliers who have a commitment to ongoing support and development. The use of interactive graphics is a key to timely model building and verification and also for interpretation and communication of results.

Another area where computation can have great benefit, as long as the modelling is performed fast enough, is in failure investigation or troubleshooting. Obviously a fix must be made that will work and a computational model can help decide on the best fix. An example of such use is for structural failures in earthmoving or mining equipment where a finite element model is kept 'on the shelf' to allow quick modification and investigation.

The field of process control is also a major user of computation where algorithms are used, in real time, to change the process operation as a result of measurements such as flow, temperature or pressure. This application is very much process dependent and not discussed further.

Summarising the use of computational models in industry, the writer can see an enormous potential for cost effective use as long as the analyst, or engineer has a sound appreciation of the physical phenomena he is modelling and is skilled in using a reliable modelling program.

SOME ACHIEVEMENTS

A partial list of computational achievements in engineering is given below together with comments on the success, or limitation, of the modelling capabilities.

1. The Finite Element Method

   This modelling technique has proved to be very successful in solving a wide range of problems governed by classical differential equations and is finding wide use in industry. Some examples are given here while structural applications are examined further in a following section of the paper.

   - Structural Analysis - Linear elastic systems in static or dynamic domains provide the major use of FE methods. Non-linearities such as large deformations, contact problems, plasticity and creep are also being modelled successfully. Aniostropic material failure and metal forming or excessive yield, contact and deformation problems are still restricted to R and D departments.

   - Heat Transfer problems with conduction and radiation can be readily solved. Convection can be successfully modelled where equivalent conduction boundary layer properties are known. General convection energy transport problems fall into the research area of fluid mechanics.

   - Fluid/Structure Interaction problems can be successfully modelled for a range of dynamic cases by the addition of fluid elements or by coupling for aeroelastic modelling.

- Field problems such as electro magnetism or diffusion can be solved, via anology with heat transfer, when expressed in the Laplace equation form of differential equation.

- Geomechanics problems are a form of structural analysis with cohesive and/or frictional materials plus the additional time dependent influence of seepage and consolidation. Realistic modelling of geologic structures such as joints and fissures present extra modelling problems.

- Fluid Flow modelling is still very much a research field and appears to need a breakthrough in handling viscous behaviour before reliable computational tools are available for use by industry.

2. Mechanical CAD/CAM

The use of computer models for mechanical components can provide many benefits for designers and some of these are given below:

- Geometric Models can form the basis of a sound mechanical CAD/CAM system if they provide a unique and consistent representation. Unfortunately conventional 2D drawings and even wire frame or surface models can be ambiguous. Solid or volume based models provide many benefits and software with computationally reliable and efficient algorithms for visualisation and Boolean operations is now being used by parts of industry.

- Use of 3D Space by a mechanical designer requires a tool which automatically checks for interference or clash and allows simple simulation of alternative geometric arrangements. A solid modeller can provide this design tool.

- Visual simulation is very important during design of a mechanical component as well as a means of communication of manufacturing details. Conventional drawings are not ideal for communication but if augmented by exploded isometric assembly diagrams they convey information which is much easier to understand. The computational power and unique data structure of a solid modeller can also provide these visually accurate isometric and assembly details.

- Computer Aided Manufacture can take the form of automated generation of bills of material or the production of numerical control tool paths directly from the geometric model. For 3D tool path movements considerable computation can be required.

3. Electronics

Computer models are used extensively in electronics design from the simulation of proposed digital or analog circuits to layout of printed circuit boards or layers of integrated circuits. The generation of wiring routing or schematic diagrams is also a major use of computer aided electronic design.

4. Chemical Engineering

Process modelling of components such as pumps, distillation

columns, flash drums, compressors and heat exchangers is very common in gas and petrochemical plants. The thermodynamic properties of a library of hydrocarbons are usually stored in the program and an iterative technique is used to find the equilibrium pressures, temperatures, flows and chemical compositions in the process.

5. Civil Engineering, Mining & Surveying

In addition to the structural areas covered by finite element methods, computational tools are used for terrain or surface modelling, road and bridge layouts and mine planning. While similar to mechanical geometric models in some ways they are usually defined in terms of geological and surveying data rather than simple geometric entities. Computational tools are available and used for cut and fill calculations, survey reduction, mapping, prediction of geologic structures from drill hole or seismic data and the calculation of drainage requirements and flood simulation models.

COMPUTATIONAL TOOLS

Computer modelling relies on a blend of adequate processing power, capable software based on efficient algorithms and sound data structures all combined in a system the user can readily utilize. While hardware is important it is only the delivery vehicle and software is the real key to successful modelling. Good quality software is continually improved, updated and fully supported on a range of processor and graphic devices. This ensures long term customer usage and program life. This software can be thought of as 'living' since the next upgrade is the lastest model. Typical features of living software which continue to provide the user with powerful and flexible modelling tools are:

- internal user language
- range of modelling capabilities
- modular programming
- sound data structure
- development and support track record
- data processing efficiency

Many of these features are concerned with computer science aspects rather than the modelling discipline itself. This is a major difference between research programs and commercial or industrial programs since the efficient implementation of proven modelling techniques is the major task facing commercial program developers rather than the undertaking of original research.

FINITE ELEMENT MODELS FOR STRUCTURAL ANALYSIS - AN EXAMPLE OF COMPUTATION IN INDUSTRY

Finite element methods in structural analysis have been very successful due to their ability to fill an arbitrary geometric shape of a structure with elements. The subdivision of the structure into standard element types allows the complex mathematics of the theory of elasticity (or even plasticity) to be completely hidden from the user. MacNeal (4) gives some 1982 measures of this success:

* about 500 FE structures programs worldwide

* approximately 20,000 users around the world

* about $500 million spent annually on FE structural analysis.

The rapid increase in capability of FE methods since computers were first introduced is shown in figure 1 by the growth in problem size

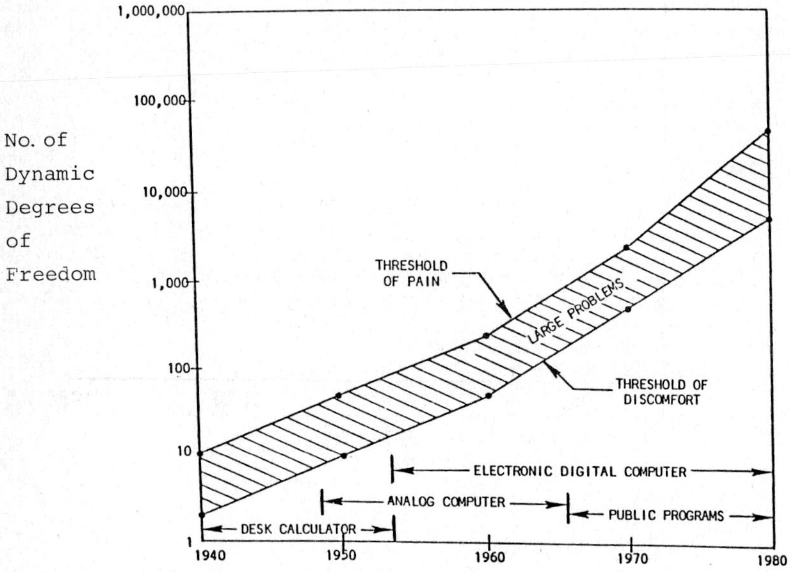

Figure 1
Growth in Problem Size for
Dynamic Analysis - from MacNeal (4)

modelled in dynamic analysis. To get some feeling for the 'threshold of pain' in this figure, consider a 1000 dynamic degree of freedom problem analysed on a CDC 6600 (the fastest machine by far in the late 1960's) to extract all natural frequencies by the Givens method. This eigenvalue solution would have required about 5 hours of dedicated cp time so that the elapsed time for a complete dynamic response analysis might have been several painful days of trying to keep other users away from the 6600! The relative cost increase for these large problems over the years is shown in fig 2 and

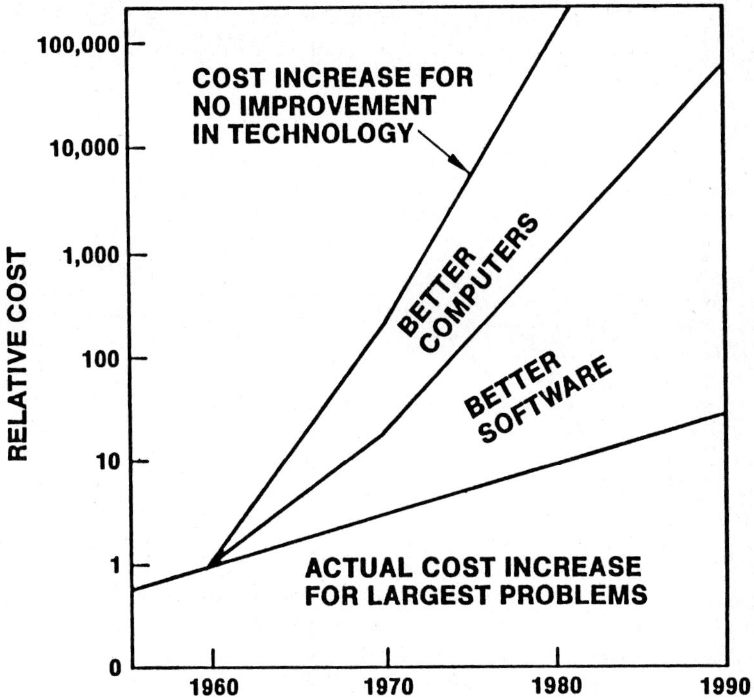

Figure 2
Cost For Large Dynamic Problems
of Figure 1

improvements in hardware price/performance are only part of the reason why larger problems have become economical. Improvements in software have also been great and these include sparse matrix solution techniques such as those of McCormick (5) and algorithms for dynamic reduction such as subspace iteration and generalised dynamic reduction (Joseph (6)). Another area where great improvements have been made, with very little mention, is that of the element capabilities themselves. This improvement is illustrated by figure 3 which shows element performance for a simple shell roof

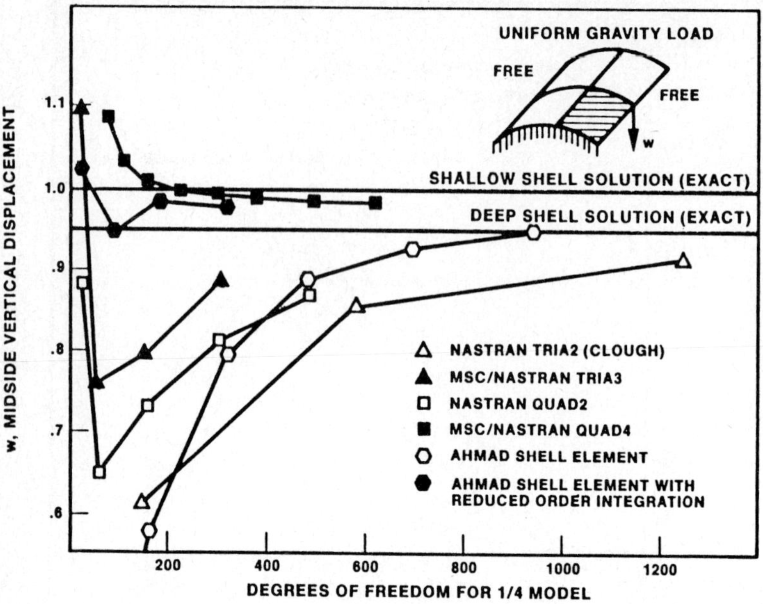

Figure 3
Performance of Pre-1970 Elements (△, □, ○)
Versus Post-1970 Elements (▲, ■, ●) - from MacNeal (4)

structure with known exact solution. Element stiffness matrix generation techniques such as reduced integration and selective integration to avoid excessive shear stiffness in rectangular elements have allowed greater accuracy or, alternatively, less elements for the same accuracy. These results are for low order elements (with the exception of the 8-noded Ahmad shell element) and it appears that the low order elements will survive the claims of superiority made by the high order isoparametric elements. This is particularly true for 3D problems and nonlinear iterations where element stiffness matrix formation and data recovery calculations form a major part of the total processing load. This is illustrated by the fact that generation of the stiffness matrix for a MSC/NASTRAN QUAD8 element (Ahmad shell with reduced integration) takes about 9 times longer than that for a QUAD4 element.

New modelling techniques have also resulted in cost improvements and use of cyclic symmetry is a good example. This combines conventional finite element solutions with a Fourier approach which uses a conventional model for one geometrically symmetric segment to determine the non-symmetric response for the complete structure. Figure 4 shows a segment of a tank cluster which was modelled by

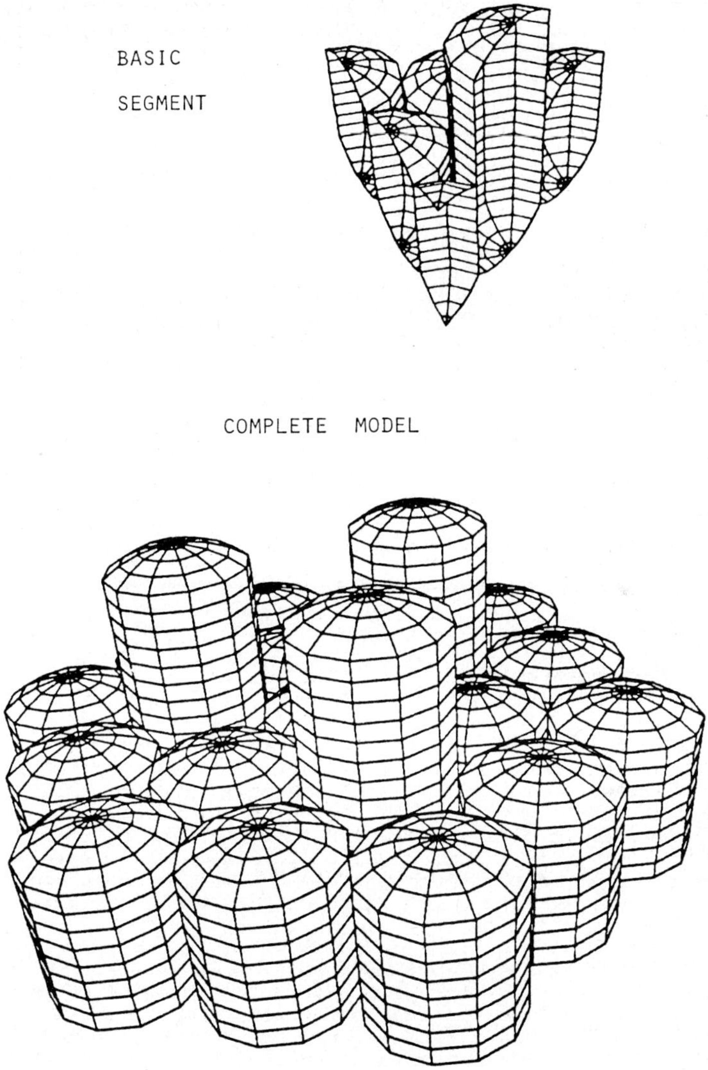

Figure 4
Basic Segment and Visualisation of Complete Model for Cyclic Symmetry Analysis

plate elements and also shows the visualisation of the complete structure being modelled. This computer graphics visualisation was achieved by one mirror and two 120 degree rotation operations. The user of cyclic symmetry simply has to define the boundary points on the cyclic segment and to specify the load (in a static solution) which acts on each segment in the complete structure. The Fourier solution and treatment of common segment boundary conditions is handled automatically by the program although the user can select computation of only the necessary harmonics in the solution if he knows the form of response (for example, rotational symmetry of response is given by the zeroth harmonic solution alone).

Further details on the theory behind cyclic symmetry are given by Joseph (6) while Currie and Wilson (7) describe use of the technique for an automotive wheel.

Finite element structural methods have also been applied to a wide range of topics and recent dynamic analysis examples include the investigation by Johnson and Kienholz (8) of alternative methods for predicting damping due to viscoelastic materials, and use of penalty function and predictor-corrector methods in dynamic redesign by Anderson et al (9). A novel application of radiation heat transfer capabilities was used by Genberg et al (10) to solve diffuse illumination problems that occur in copiers and microfilm units. This analogy is exact for ideal diffuse surfaces since they accept all incoming light regardless of direction and re-radiate it in the same fashion as a radiating black body. Materials such as proprietary reflectance coatings closely approximate an ideal diffuser.

Another application of heat transfer analogy is for magnetic field analysis and Root et al (11) used this analogy to determine fields around a 3D magnetic recording head.

FUTURE POTENTIAL FOR MODELLING

As the information revolution expands it is inevitable that more and more complex simulations or models will be investigated as part of the decision making process. In addition to increasing complexity (such as coupled solutions of thermo-structural problems) the availability of cheaper computing means that more people will have access to powerful modelling. Figure 5 shows how the introduction of 32-bit mini and micro computers have lowered the cost for purchase of an in house computer system suitable for general purpose finite element analysis. The rapid growth in the micro computer market indicates that prices will continue to fall and this means that nearly every design engineer could have processing power on his desk. When combined with high speed interactive graphics and good software this will provide designers with a very powerful tool. To realise the full potential, of course, will still depend heavily on the skills and aptitude of the people building and driving these models. While micros on every desk are important, the linking of computing facilities by a high speed network is also a requirement for the future. This will allow the sharing of expensive hardware such as large disk drives, tape units, printers, plotters and access to larger computers. The network will also allow transfer of data to other disciplines in an organisation (such as production, documentation or numerical control machines) and the off loading of computationally intensive analysis to faster processors.

While integration of computer modelling and manufacturing or construction is a worthwhile goal, the concept of a centralized massive data base, proposed by some people, appears to be a myth. The data structures required by different modelling disciplines

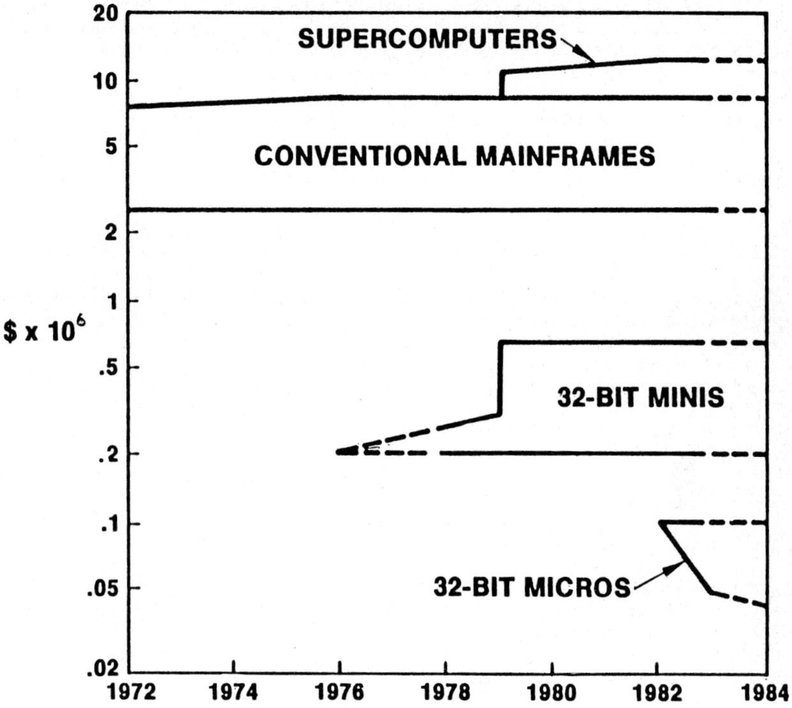

Figure 5
Initial Cost for a New Computer System
Suitable for Major Finite Element Programs
- from MacNeal (4)

vary greatly and organisational structures correspond to the people skills and various departmental functions. The best solution appears to be a set of strong links which can transfer the required information through an organisation. For example, a design and manufacture operation requires a sound definition of geometry to which are added attributes such as material and surface data, construction details and parts list summaries. Different disciplines need to access this data, use it in their models and certainly modify it during the design process. The better the definition and data structure for geometry the greater the potential for productivity. For example, a Boolean solid modeller with general boundary representation of shape can define geometry as a combination of many parts in an assembly. If one part is changed and the linking of the parts preserved then the modeller can propagate the new assembly automatically. This can be computationally intensive for large models but offers tremendous benefits for the future.

Returning to finite element analysis for some specific examples of the potential for future computational models, there are two trends. The first is the addition of new and more general capabilities for structural analysis and the second is the demand (probably insatiable) for easier use of analysis systems.

Recent capabilities which are starting to be used more regularly in industry include plasticity, creep, gap opening or closure and large deformations such as snap through or buckling. These capabilities can be used with either static or transient loads. The main limitations to use of nonlinear analysis are the availability of people who can use these modelling tools efficiently and the availability of material properties. Analysis of nonlinear problems requires careful modelling and usage of computer resources for efficiency (see Currie (12)). Methods such as substructuring (or superelements) to contain the nonlinear iterations to small matrices are valuable tools. Capabilities which are still in the research field and not generally available for industry include large strain and friction/sliding problems that occur in metal forming and the modelling of aniostropic material failure. Development and acceptance of adequate material and kinematic models should proceed general use of modelling in these fields.

Design optimization is another area where the potential for modelling is good. Lahey (13) describes new capabilities in finite element analysis where numerical derivatives of design constraints are calculated with respect to design variables. Design constraints can include stress, force or displacement limits in static analysis and natural frequencies or buckling load factor limits in normal modes or critical buckling modes analysis. Design variables include material and element properties such as plate thickness or beam depth and these variables can be defined so that all elements in a region would change properties together or could individually change. The derivatives of design constraint are used (perhaps via linear programming) to select the next design iteration. As long as these derivatives have not changed markedly for the change in design variables then convergence towards the 'optimum' solution would occur.

In the area of ease of use or 'user friendliness' the problem is to provide interactive tools which aid the user but do not remove flexibility or capability. For example, there are many finite element pre-processors which generate X,Y,Z coordinates and element connectivities for a model but not many allow use of alternative coordinate systems to enforce, say, cylindrical boundary conditions. Certainly the use of interactive graphics to display component geometry and to map an element mesh onto this geometry is a prime requirement for complex models. Casale and Crain (14) describe state of the art techniques for utilizing graphics hardware and geometric definition. However, for simple geometry (planar, cylindrical or spherical regions) a batch generator as part of the analysis program can be more productive when combined with interactive viewing capabilities. This is due to the fact that the complete model, including loads, boundary conditions and special modelling features is built in one single operation.

For results checking and interpretation, graphics is also very important to display deformed shapes, stress or temperature contours and XY graphs of results. Colour and animation can be effectively used as mediums for interpreting and understanding results. Once again it is important that the graphics program have a strong link to the analysis program to ensure that conventions for displacement or stress quantities or directions are correctly translated. For interactive inquiry of a results data base this link is even more important.

Summarizing the potential for easier access to modelling we can expect to see powerful interactive software as the main means of communication with the analysis engine. However, this will not substitute for the engineers skill or experience in effectively using the modelling capabilities

CONCLUSIONS

Computational modelling as a tool in engineering provides industry with the ability to refine and simulate new or changed designs prior to expensive prototype manufacture and test. Modelling can be successfully used to reduce the risk of failure of the first prototype and to streamline manufacture. A number of different modelling disciplines have been described in this paper and emphasis has been given to the finite element method in structural design. This technique already has a good track record in industry and has great potential for even better use. Advances in modelling capabilities and computer science, combined with improved price/performance of computer processors and graphics devices ensure a healthy future for computer modelling. The greatest challenge is for design engineers to efficiently use these tools in the evolution of their organisation's products.

REFERENCES

(1) Timoshenko, S.P., History of Strength of Materials (McGraw-Hill, New York, 1953)

(2) Nichols, J.R., Statical Limitations upon the Steel Requirements in Reinforced Concrete Flat Slab Floors, Trans. ASCE, Vol 77 (1914), pp 1670-1736.

(3) Westergard, H.M. and Slater, W.A., Moments and Stresses in Slabs, Proc. ACT, Vol. 17 (1921), pp 415-538.

(4) MacNeal, R.H., Trends in Finite Element Structural Analysis, AIAA/ASME/ASCE 23rd Structures, Structural Dynamics and Materials Conference, New Orleans, May, 1982.

(5) McCormick, C.W., Data Structure and Computational Techniques - MSC/NASTRAN, ACADS Symposium - CAD/CAM in the 80's and Integration through Databases, Melbourne, Oct., 1980.

(6) Joseph, J.A., Ed., MSC/NASTRAN Application Manual, MacNeal-Schwendler Corp., Los Angeles, 1983.

(7) Currie, A.O. and Wilson, B., Finite Element Analysis of an Automotive Wheel - A Case History, Inst. Engineers Aust Symposium on Stress Analysis for Mechanical Design, Sydney, August, 1981.

(8) Johnson, C.D. and Kienholz, D.A., Prediction of Damping in Structures with Viscoelastic Materials using MSC/NASTRAN, Proc. MSC/NASTRAN User's Conf., Los Angeles, March, 1983.

(9) Anderson, W.J., Kim, K.O., Zhi, B., Bernitsas, M.M., Hoff, C. and Cho, K.N., Nonlinear Perturbation Methods in Dynamic

Redesign, Proc. MSC/NASTRAN User's Conf., Los Angeles, March 1983.

(10) Genberg, V., Oinen, D. and Fronheiser, S., Diffuse Illumination with MSC/NASTRAN, Proc. MSC/NASTRAN Users Conf., Los Angeles, March 1983.

(11) Root, R.R., Tranchitta, C.J., Anderson, R.B., Fernandez, R.B. and McDaniel, T.W., On the Solution of Large Three-Dimensional Magnetostatic Problems Utilizing MSC/NASTRAN, in:

Conaway, J.H. (ed.), First Chautauqua on Finite Element Modelling, Schaeffer Analysis, Harwichport, Sept., 1980.

(12) Currie, A.O., An Evaluation of Commercial Finite Element Programs for NonLinear Analysis, in: Hoadley, P.J. and Stevens, L.K. (eds.), Finite Element Methods in Engineering, University of Melbourne, Aug., 1982.

(13) Lahey, R.S., Design Sensitivity in MSC/NASTRAN, Proc. MSC/NASTRAN Users Conf., Los Angeles, March, 1983.

(14) Casale, M.S. and Crain, L.M., Quantel Methods for the Display of Geometry and Finite Element Models, in: Schaeffer, H. (ed.), Second Chautauqua on Productivity in Engineering and Design, Kiawah Island, Schaeffer Analysis, Nov., 1982.

FINITE DIFFERENCE METHODS

Computational Techniques & Applications: CTAC-83
J. Noye & C. Fletcher (Editors)

THIRD-ORDER UPWINDING AS A RATIONAL BASIS FOR COMPUTATIONAL FLUID DYNAMICS

B.P. Leonard

City University of New York
Staten Island, N.Y.
U.S.A.

Classical numerical methods applied to problems in diffusion, wave-motion, and elasticity, have been highly successful -- due to inherent feedback stability of central differencing in modelling even-order spatial derivatives. However, central methods lack this stability when applied to odd-order derivatives -- the major cause of problems in modelling the convection in fluid flow. Irrational "remedies" have compounded the difficulties; but rational design of stable and accurate algorithms shows that third-order upwinding provides a clean and robust foundation for computational fluid dynamics.

INTRODUCTION

Computational fluid dynamics is distinguished from the other branches of computational physics by the importance of the convection term -- a first-order spatial derivative. "Classical" numerical techniques (essentially second-order central differencing) evolved in response to problems dominated by diffusion, wave-motion, or elasticity -- all involving even-order spatial derivatives. When applied to the first-derivative convection term of fluid dynamics, central-difference methods generate a variety of "difficulties", ranging from bizarre and clearly unphysical oscillations to catastrophic numerical divergence. Historically, various characteristics of the convection term have been blamed; e.g., the problem is sometimes known as the "nonlinear instability" or put down to the variable coefficient (convecting velocity). But the primary cause of the difficulty is much more basic, stemming from an habitual use of the wrong old tools for the "new" problem: it occurs whenever central-difference methods (of any order of accuracy) are used to model odd-order spatial derivatives.

Some insight can be gained by studying the "feedback sensitivity" of difference operators; i.e., the sensitivity to perturbations of the central node value. If this is negative, stable negative feedback occurs and numerical noise is damped out. If positive, of course, divergence will follow quickly. In the case of the classical central difference models of even-order derivatives, the feedback sensitivity happens to be strictly negative. This is the primary (and somewhat fortuitous) reason for central difference methods' notable success for problems dominated by even-order derivatives. For these problems, the classical techniques are accurate and stable; or, in popular jargon: clean and robust. In hindsight, one could say that, for successful computation it is quite sensible to require that, in general:

> The inherent feedback sensitivity of finite-difference operators should be strictly negative.

However, for odd-order derivatives, central differencing results in essentially neutral sensitivity, and undamped parasitic oscillations may pervade the computations. In a control-volume formulation of convection, the feedback sensitivity may actually be positive (in decelerating regions); and in coupled equations, a computed diffusion coefficient may take on an unphysical negative value. In either case, computational disaster usually follows rapidly!

One of the central aims of this paper is to drive home the fact that, because of a lack of inherent feedback sensitivity:

> Central difference methods (of any order of accuracy) are not appropriate tools for modelling odd-order spatial derivatives, in particular the first-order convection term.

Non-centered schemes offer a possible improvement; e.g., over the past decade or so, first-order upwinding techniques have become very popular. But there is now a widening realization that, by using these methods, stability is obtained at too high a cost in accuracy. The difficulty is best portrayed in terms of the artificial numerical diffusion introduced by the truncation error in the usual form of first-order upwinding. This is most pronounced in problems involving (what should be) thin shear layers at an oblique angle to the grid mesh, and in problems involving sources or unsteady terms. The problem is that, in the convection-diffusion equation, modelling high convection is equivalent to correctly simulating very small (or negligible) physical diffusion. The second-derivative truncation error inherent in first-order upwinding totally corrupts this intention. Again in hindsight (and on making reasonable smoothness assumptions), it seems clear that one should require that:

> The order of the spatial derivatives introduced by the truncation error in finite-difference operators should be higher than that of any modelled physical term.

This simple and rational criterion (quite properly) excludes first-order upwinding as a technique for modelling the convection-diffusion equation. The same conclusion has been reached in several studies concerned with directly measuring the effects of the artificial diffusion introduced by first-order upwinding [1-14].

A number of techniques have been developed in an attempt to model high convection, which could be called Lagrangian in nature [15]. For one-dimensional pure convection of a scalar ϕ at constant velocity, the equation

$$\phi(x, t+\Delta t) = \phi(x-u\Delta t, t) \qquad (1)$$

is, of course, exact in principle. Given discrete spatial values of ϕ at time t, the numerical problem becomes one of interpolation in estimating the right-hand side of Equation (1). Various interpolation schemes will be briefly described. It will be found, for example, that centered polynomial interpolation schemes of even degree are much more oscillatory than odd-degree upwind-biassed methods. Once again, this distinction can be related to feedback sensitivity and spatial truncation error. The following summarizes the behaviour of Lagrangian methods of this kind:

> Methods involving odd-order spatial derivatives in the leading truncation error term are much more oscillatory (dispersive) than those involving even-order derivatives.

This is true even in the case of upwinded even-degree polynomial methods; e.g., second-order upwinding involves a dispersive third-derivative in the truncation error. In the test problem of a uniformly convected step profile, for example, one sees this distinction in dispersive characteristics quite clearly. In addition, central methods become much more dispersive as the time step is decreased, whereas the dispersion of upwinded odd-degree methods is practically insensitive to Δt.

When considering the attributes of higher-order methods, of course, one must be concerned with the computational costs in comparison with the benefits gained. For a given method, accuracy improves in principle with spatial grid refinement;

for a given grid, accuracy may or may not improve with an increase in the formal order of the algorithm (e.g., for pure convection, Nth-order central methods are less accurate than (N-1)th-order upwinding). However, for methods in which fixed-grid accuracy improves with order, the following is true:

> For a prescribed global accuracy, higher-order methods are computationally more economical.

The cost per space-time grid point certainly increases with order (roughly arithmetically); but, to achieve the same global accuracy, a lower-order method will need the number of spatial grid points to be increased by a factor in each coordinate direction, and the number of time (or iteration) steps must be increased by the same factor. The cost of this geometric increase in the number of space-time grid points of the lower-order method far outweighs the arithmetic increase in local cost of the higher-order algorithm. In terms of machine running costs, there is thus a strong motivation toward higher-order methods. However, one must also consider developmental costs: above fourth-order, algorithmic complexity begins to escalate rapidly, and soon outstrips the normal powers of human comprehension and associated debugging skills.

Since central-difference methods of any order are not appropriate for convection-dominated flows, and since first-order upwinding must be excluded on the basis of its artificial diffusion, and since second- (and fourth-) order upwind methods are too dispersive, and since fifth- and higher-order upwind methods are excessively complex, one is drawn to the logical conclusion that:

> THIRD-ORDER UPWINDING FORMS A RATIONAL BASIS FOR THE DEVELOPMENT OF CLEAN AND ROBUST ALGORITHMS FOR COMPUTATIONAL FLUID DYNAMICS.

This is not to say that third-order upwinding has absolutely no problems of its own. For example, under pure convection conditions, third-order upwinding may generate small (~5%) overshoots in modelling a step, or a few wiggles near a downstream boundary [7]; but these problems can be overcome by using alternate interpolation techniques locally. By monitoring local curvature of the convected variables, an algorithm can easily be designed which changes over automatically from third-order upwinding in the "smooth" regions (i.e., most of the flow domain) to a more appropriate interpolation (e.g., exponential upwinding) in local regions of rapidly varying gradients. In this way, sudden jumps can be negotiated monotonically. The EULER-QUICK shock-capturing compressible Euler code, for example, produces shocks at the correct location, with the proper strength, and with a numerical shock structure only two grid points wider than that obtainable with exact solutions [16]. In isentropic regions, third-order upwinding generates results which are graphically indistinguishable from theoretically exact; in particular:

> Third-order upwinding conserves stagnation pressure in isentropic regions.

This inherent property of conserving stagnation pressure in inviscid regions is a strong indication of the algorithm's non-dissipative nature, and has been confirmed for incompressible flows, as well [5]. Static pressure behaviour is correspondingly good [4, 16].

Finally, some comments on computational grids are in order. For one-dimensional flow with non-reversing convecting velocities at control-volume cell faces, second-order central differencing requires three points for updating the central node value due to convection and diffusion. By contrast, first-order upwinding requires only two for convection (physical diffusion usually being ignored), whereas third-order upwinding requires four. These are shown in Figure 1a. When reversing velocities are allowed for, the situation is as shown in Figure 1b. Note that, in this case, first- and second-order methods both require three points, whereas third-order upwinding needs four (as does fourth-order central). In two dimensions, allowing for velocity reversals, first- and second-order methods involve the well-known five-point "star" (as does the Laplacian operator for diffusion).

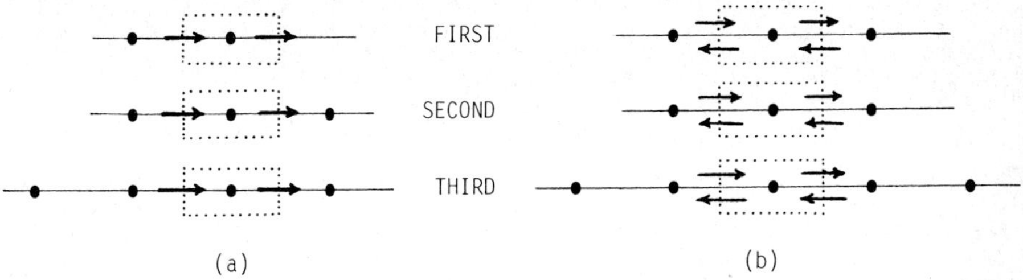

Figure 1. One-dimensional spatial grids.

Third-order upwinding requires an additional point in the normal direction for each face (i.e., four more, making a total of nine); this is enough to guarantee the essential convective stability of third-order upwinding, and from a practical point of view generates results which are essentially third-order accurate. However, for strict formal third-order accuracy, four additional corner points are required, to take account of transverse curvature effects. The complete star is shown in Figure 2 (with the latter corner points shown by hollow circles).

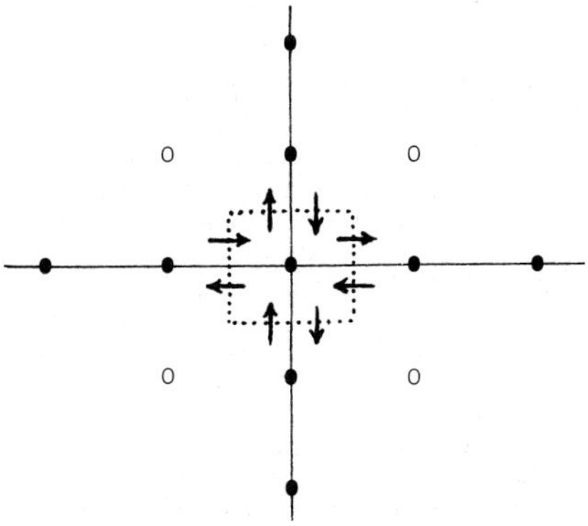

Figure 2. Computational star for third-order upwinding in two dimensions.

By an obvious extension of these ideas, it can be seen that in three dimensions, first- and second-order require a seven-point configuration; quasi-third-order-upwinding requires one more for each of the six faces (i.e., thirteen in all), whereas a formally fully third-order algorithm would require a total of twenty-five. The effects of including or deleting transverse curvature terms have not yet been fully tested; however, if one makes a worst-case assumption (i.e., only second-order accurate if deleted), the inherent stability of the retained normal curvature terms is enough to justify the additional normal-direction grid points.

FEEDBACK SENSITIVITY AND TRUNCATION ERROR

Consider the unsteady convection and diffusion of a scalar in the presence of source terms:

$$\frac{\partial \phi}{\partial t} = - \nabla \cdot (\underset{\sim}{u} \phi) + \nabla \cdot (\Gamma \nabla \phi) + S \tag{2}$$

Assume that a computational algorithm for a node value ϕ_i can be written

$$\frac{\partial \phi_i}{\partial t} = \text{RHS} \tag{3}$$

where RHS represents the modelled terms on the right-hand side. In general, RHS will involve some dependence on ϕ_i ; thus the evolution of numerical perturbations in ϕ_i can be studied by taking the variation of Equation (3) with respect to ϕ_i , giving

$$\frac{\partial \delta \phi_i}{\partial t} = \frac{\partial \text{RHS}}{\partial \phi_i} \delta \phi_i \tag{4}$$

which has a formal solution

$$\delta \phi_i = \exp[\ \int \Sigma\ \partial t\ ] \tag{5}$$

where the feedback sensitivity, Σ , is given by

$$\Sigma = \frac{\partial \text{RHS}}{\partial \phi_i} \tag{6}$$

Clearly, a positive Σ would lead to explosive growth of numerical perturbations -- a highly undesirable occurrence. If $\Sigma = 0$, there is no inherent feedback and certain types of perturbations (e.g., wiggles) can be superimposed on the solution without affecting the RHS, thus generating no automatic corrective action. A negative Σ assures that the algorithm will provide inherent damping of random fluctuations -- a highly desirable property.

In a model one-dimensional version of Equation (2) involving constant u and Γ, the "classical" (second-order) finite-difference model of diffusion is

$$\Gamma \left[\frac{\phi_{i+1} - 2\phi_i + \phi_{i-1}}{\Delta x^2} \right] = \Gamma \left[\left(\frac{\partial^2 \phi}{\partial x^2} \right)_i + \frac{1}{12} \phi_i^{(iv)} \Delta x^2 + \ldots \right] \tag{7}$$

Note that this operator is exact for a cubic polynomial interpolated through (i-1), i, (i+1), and either (i-2) or (i+2). The feedback sensitivity is $-2\Gamma/\Delta x^2$; i.e., strictly negative for finite (physical) diffusion. The fourth-order model for diffusion has a similar structure (as does the second-order model of the biharmonic operator).

Second-order central differencing of the model convection term leads to

$$- u \left[\frac{\phi_{i+1} - \phi_{i-1}}{2\Delta x} \right] = - u \left[\left(\frac{\partial \phi}{\partial x} \right)_i + \frac{1}{6} \phi_i''' \Delta x^2 + \ldots \right] \tag{8}$$

which has the same formal order of accuracy as Equation (7), although it is exact only for quadratic interpolation. However, a major problem exists for this operator: $\Sigma \equiv 0$. Thus, in a second-order central-difference model of the convection-diffusion equation, feedback stabilization depends on the physical diffusion term. Sudden jumps in the streamwise direction lead to upstream-penetrating oscillations (the penetration distance is linearly proportional to the cell Reynolds or Peclet number [10]).

Going to higher-order central methods does not correct this inherent neutral sensitivity in modelling odd-order derivatives. For example, the fourth-order central-difference model for convection is

$$- u \left[\frac{- \phi_{i+2} + 8\phi_{i+1} - 8\phi_{i-1} + \phi_{i-2}}{12\,\Delta x} \right]$$

$$= - u\left[\left(\frac{\partial \phi}{\partial x} \right)_i - \frac{1}{4} \phi_i^{(v)} \Delta x^4 + \ldots \right] \qquad (9)$$

and the corresponding second-order version of the third derivative is

$$\frac{\phi_{i+2} - 2\phi_{i+1} + 2\phi_{i-1} - \phi_{i-2}}{2\,\Delta x^3} = \left(\frac{\partial^3 \phi}{\partial x^3} \right)_i + \frac{1}{4} \phi_i^{(v)} \Delta x^2 + \ldots \qquad (10)$$

both of which have neutral sensitivity. In practical flow situations (variable u), the convective feedback sensitivity becomes positive in decelerating regions; this is very often the cause of numerical divergence.

First-order upwinding of convection results in

$$- u \left[\frac{\phi_i - \phi_{i-1}}{\Delta x} \right] = - u\left[\left(\frac{\partial \phi}{\partial x} \right)_i - \frac{1}{2} \phi_i'' \Delta x + \ldots \right] \qquad u > 0 \qquad (11)$$

and

$$- u \left[\frac{\phi_{i+1} - \phi_i}{\Delta x} \right] = - u\left[\left(\frac{\partial \phi}{\partial x} \right)_i + \frac{1}{2} \phi_i'' \Delta x + \ldots \right] \qquad u < 0 \qquad (12)$$

The feedback sensitivity is strictly negative: $\Sigma = -|u|/\Delta x$. But note that the leading truncation error term is equivalent to a numerical diffusion term with an artificial diffusion coefficient

$$\Gamma_{num} = \frac{|u|\,\Delta x}{2} \qquad (13)$$

which, of course, corrupts (or dominates) the physical diffusion unless the grid is fine enough so that the cell peclet number, $P_\Delta = |u|\,\Delta x/\Gamma$ is very much smaller than 2.

The basic algorithm of the well-known TEACH code [17] uses central differencing for both convection and diffusion if $P_\Delta \leq 2$, and first-order upwinding (with physical diffusion neglected) if $P_\Delta > 2$. In this hybrid strategy, the effective cell Peclet number is

$$\left. \begin{aligned} P^*_\Delta &= P_\Delta \quad \text{for } P_\Delta \leq 2 \\ P^*_\Delta &\equiv 2 \quad \text{for } P_\Delta > 2 \end{aligned} \right\} \qquad (14)$$

A slight modification of this procedure is advocated by the "optimal upwinding" fallacy [18]:

$$P^*_\Delta = 2 \tanh(P_\Delta/2) \qquad (15)$$

A recent text-book on numerical heat transfer and fluid flow [19] has introduced a further red herring by going to considerable lengths to approximate Equation (15) by (less expensive) quintic polynomials. The futility of such exercises should, by now, be obvious. For high convection ($P_\Delta > 2$ for hybrid or $P_\Delta \gtrsim 5$ for optimal), these schemes consist of pure first-order upwinding for convection with all physical diffusion terms totally neglected. Very often, these methods are used to test quite sophisticated turbulence models under highly convective conditions. One goes to the expense of solving (at least) two additional transport equations for turbulence properties and several auxiliary variables, and then proceeds to totally ignore them in modelling the momentum and thermal or species transport equations! The unacceptable effects of the artificial diffusion associated with first-order methods are now well documented in the fundamental research literature. Unfortunately, because of natural time lags, these methods are currently becoming even more firmly established as standard "handbook" techniques in the heat-transfer industry and related areas.

Second-order upwinding can be written, for $u > 0$,

$$- u \left[\frac{3\phi_i - 4\phi_{i-1} + \phi_{i-2}}{2\Delta x}\right] = - u \left[\left(\frac{\partial \phi}{\partial x}\right)_i - \frac{1}{3}\phi_i''' \Delta x^2 + \frac{1}{4}\phi_i^{(iv)} \Delta x^3 + \ldots\right] \qquad (16)$$

with a similar expression for $u < 0$. Regardless of velocity direction, the feedback sensitivity is $\Sigma = - 3|u|/2\Delta x$, which is encouraging. However, the truncation error, being dominated by the third-derivative term, is responsible for oscillatory behaviour (although much less than that of central methods).

The troublesome third-derivative term can be entirely eliminated by going to third-order upwinding, which can be written in symmetrical form (for $u \gtrless 0$) as:

$$- u \left[\frac{-\phi_{i+2} + 8\phi_{i+1} - 8\phi_{i-1} + \phi_{i-2}}{12\Delta x}\right] - |u| \left[\frac{\phi_{i+2} - 4\phi_{i+1} + 6\phi_i - 4\phi_{i-1} + \phi_{i-2}}{12\Delta x}\right]$$

$$= - u\left(\frac{\partial\phi}{\partial x}\right)_i - \frac{|u|}{12}\phi_i^{(iv)}\Delta x^3 + \frac{u}{4}\phi_i^{(v)}\Delta x^4 + \ldots \qquad (17)$$

in which the left-hand side is recognized as the fourth-order operator of Equation (9), modified by a fourth-difference term. The particular combination is consistent with interpolating a cubic polynomial through (i-1), i, (i+1), and (i-2) if $u > 0$, or (i+2) if $u < 0$. As noted in relation to Equation (7), this is also consistent with the second-order operator for diffusion. The corresponding combined model convection-diffusion equation can be written

$$\frac{\partial \phi_i}{\partial t} = S - u\left(\frac{\partial \phi}{\partial x}\right)_i + \Gamma\left(\frac{\partial^2\phi}{\partial x^2}\right)_i - \frac{|u|}{12}\left(1 - \frac{1}{|P_\Delta|}\right)\phi_i^{(iv)}\Delta x^3 + \ldots \qquad (18)$$

and the combined feedback sensitivity is

$$\Sigma = -\frac{|u|}{2\Delta x} - \frac{2\Gamma}{\Delta x^2} = -\frac{|u|}{2\Delta x}\left(1 + \frac{4}{|P_\Delta|}\right) \qquad (19)$$

which, of course, is always stabilizing.

LAGRANGIAN CONVECTION SCHEMES

Figure 3 shows the basis of several Lagrangian schemes for pure convection based on Equation (1).

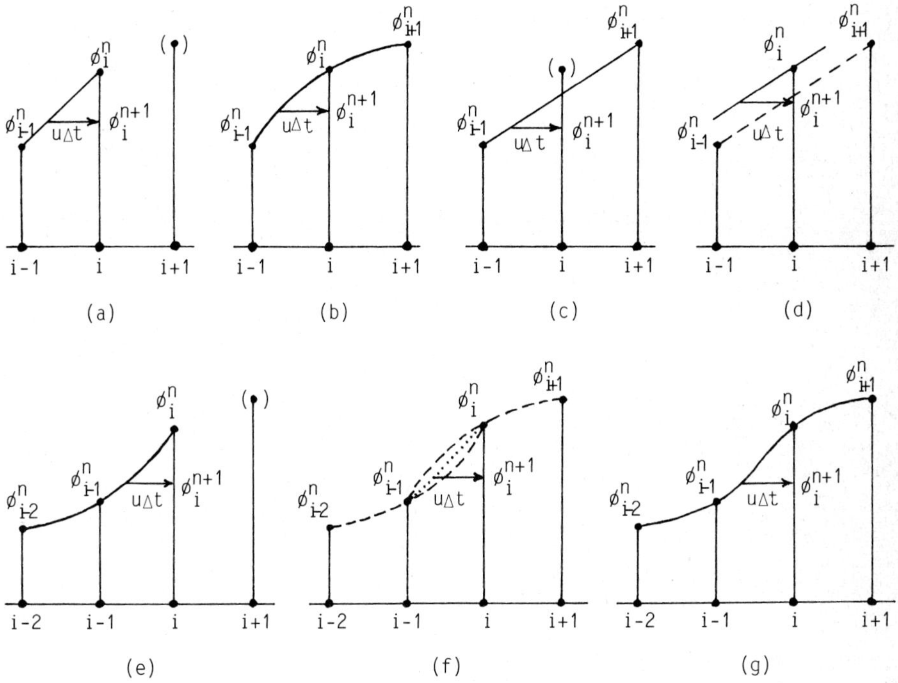

Figure 3. Lagrangian schemes for pure convection at constant velocity. (a) First-order upwinding; (b) second-order central (Leith's method); (c) the Lax scheme; (d) forward-time-central-space (FTCS); (e) second-order upwinding; (f) Fromm's method; (g) third-order upwinding (QUICKEST in inviscid limit).

In the case of upwind linear interpolation, for example, it is clear that the update algorithm is

$$\phi_i^{n+1} = \phi_i^n - u\Delta t\left(\frac{\phi_i^n - \phi_{i-1}^n}{\Delta x}\right) \tag{20}$$

which, of course, is the well-known first-order upwind method (with second-derivative truncation error).

All methods of this type can be written in conservative control-volume form as:

$$\frac{\phi_i^{n+1} - \phi_i^n}{\Delta t} = \frac{u}{\Delta x}(\phi_\ell - \phi_r) \tag{21}$$

where $\phi_\ell(i) = \phi_r(i-1)$. The notation refers to the left and right control-volume face values (half-way between nodes). For constant u and Δx, a measure of feedback sensitivity in this case is

$$\sigma = \frac{\partial}{\partial \phi_i} (\phi_\ell - \phi_r) \tag{22}$$

For example, for first-order upwinding ($\phi_\ell = \phi_{i-1}^n$, $\phi_r = \phi_i^n$),

$$\sigma_1 = -1 \tag{23}$$

Note, in particular, that this is independent of Δt.

The centered parabolic interpolation scheme, shown in Figure 3(b), results in

$$\phi_i^{n+1} = \phi_i^n - \frac{c}{2}(\phi_{i+1}^n - \phi_{i-1}^n) + \frac{c^2}{2}(\phi_{i+1}^n - 2\phi_i^n + \phi_{i-1}^n) \tag{24}$$

from which, using Equation (21), one can identify the left face convective flux value as

$$\phi_\ell = \frac{1}{2}(\phi_i^n + \phi_{i-1}^n) - \frac{c}{2}(\phi_i^n - \phi_{i-1}^n) \tag{25}$$

introducing the Courant number, $c = u\Delta t/\Delta x$. Leading truncation error is proportional to $\phi_i'''\Delta x^2$, implying second-order accuracy and (as seen later) oscillatory behaviour. The corresponding feedback sensitivity is

$$\sigma_2 = -c \tag{26}$$

Thus, as Δt (i.e., c) is reduced, the feedback sensitivity of the second-order method decreases in linear proportion. This scheme is also known as Leith's method [20] or (in the variable velocity form) the Lax-Wendroff method [21].

Another method due to Lax [22] is shown in Figure 3(c). In this case, the update algorithm is

$$\phi_i^{n+1} = \phi_i^n - \frac{c}{2}(\phi_{i+1}^n - \phi_{i-1}^n) + \frac{1}{2}(\phi_{i+1}^n - 2\phi_i^n + \phi_{i-1}^n) \tag{27}$$

which is Leith's method together with an added artificial diffusion equivalent to $\Gamma_{LAX} = (1-c^2)u\Delta x/2c$ (!).

The so-called forward-time-central-space (FTCS) method is portrayed in Figure 3(d); in this case, instead of interpolation, one approximates ϕ in the vicinity of node i by a first-order Taylor series about i, using central-differencing to approximate the first derivative. This gives

$$\phi_\ell = \frac{1}{2}(\phi_i^n + \phi_{i-1}^n) \tag{28}$$

and, in particular,

$$\sigma_{FTCS} = 0 \tag{29}$$

[As is well known, this algorithm is unconditionally unstable.]

In the case of second-order upwinding,

$$\phi_\ell = \frac{1}{2}(\phi_i^n + \phi_{i-1}^n) - \frac{c}{2}(\phi_i^n - \phi_{i-1}^n) - \frac{(1-c)}{2}(\phi_i^n - 2\phi_{i-1}^n + \phi_{i-2}^n) \tag{30}$$

and

$$\sigma_{2UP} = -c - \frac{3}{2}(1-c) = -\frac{3}{2} + \frac{c}{2} \tag{31}$$

Truncation error again involves ϕ_i''' (an oscillatory term).

Fromm's so-called second-order method of zero-average-phase-error [23] averages two quadratics (the central and upwind) to give

$$\phi_\ell = \frac{1}{2}(\phi_i^n + \phi_{i-1}^n) - \frac{c}{2}(\phi_i^n - \phi_{i-1}^n) - \frac{(1-c)}{4}(\phi_i^n - 2\phi_{i-1}^n + \phi_{i-2}^n) \tag{32}$$

and

$$\sigma_{FROMM} = -c - \frac{3}{4}(1-c) = -\frac{3}{4} - \frac{c}{4} \tag{33}$$

approaching $-3/4$ as $c \to 0$.

Finally, Figure 3(g) shows the pure-convection form of the consistent third-order upwind algorithm known as QUICKEST -- Quadratic Upstream Interpolation for Convective Kinematics with Estimated Streaming Terms -- [7]. The quadratic upstream interpolation is used in the control-volume formulation; for the Lagrangian form, the appropriate interpolation is an upwind-biassed cubic through the four points shown:

$$\phi_\ell = \frac{1}{2}(\phi_i^n + \phi_{i-1}^n) - \frac{c}{2}(\phi_i^n - \phi_{i-1}^n) - \frac{(1-c^2)}{6}(\phi_i^n - 2\phi_{i-1}^n + \phi_{i-2}^n) \tag{34}$$

gives the convected left face value, and the feedback sensitivity is

$$\sigma_3 = -c - \frac{(1-c^2)}{2} \tag{35}$$

which, of course, approaches $-1/2$ as $c \to 0$. Leading truncation error involves, in this case, $\phi_i^{(iv)}\Delta x^3$ (a non-oscillatory term).

Fourier von Neumann Analysis of Lagrangian Schemes

By studying the evolution of Fourier components of the form

$$\phi = A(t)\, e^{ikx} \tag{36}$$

for a range of the wave number k, one can determine the complex amplitude ratio (or "amplification factor") corresponding to a time increment Δt, i.e.,

$$G = \frac{\phi(x, t+\Delta t)}{\phi(x,t)} = \frac{A(t+\Delta t)}{A(t)} \tag{37}$$

For the exact (constant u) result, in the absence of diffusion, $A = e^{-iukt}$, and

$$G_{EXACT} = e^{-ic\theta} = \cos(c\theta) - i\sin(c\theta) \tag{38}$$

where $\theta = k\Delta x$. By computing G for the various polynomial convection schemes, one can get a measure of the accuracy of a given algorithm. If the expansion of the particular method's G in Cartesian form agrees with that of Equation (38) through θ^N terms, the algorithm is correctly denoted as being Nth-order accurate in both space and time (in terms of modelling unsteady, purely convective flow). Perhaps not surprisingly, Nth-degree polynomial Lagrangian methods generate Nth-order accurate G's. For example, it is not difficult to derive the following for first-order upwinding:

$$G_1(\theta) = 1 + c(\cos\theta - 1) - i\,c\sin\theta \tag{39}$$

$$= 1 - \frac{c}{2}\theta^2 + O(\theta^4) - i\,[\,c\,\theta - \frac{c}{6}\theta^3 + O(\theta^5)\,] \tag{40}$$

which, of course agrees with Equation (38) only through the first-order term in θ.

Leith's method has

$$G_2(\theta) = 1 + c^2(\cos\theta - 1) - i\,c\sin\theta \tag{41}$$

$$= 1 - \frac{1}{2}(c\,\theta)^2 + O(\theta^4) - i\,[\,c\theta - \frac{c}{6}\theta^3 + O(\theta^5)\,] \tag{42}$$

which has the correct second-order term, but is deficient at third-order. It is this third-order deficiency in the imaginary part which is reflected in poor phase behaviour of Leith's method.

For a particular algorithm, the solution for each wave component can be written

$$\phi = \phi_0 e^{i\,k(x - u_{ph}t)} \tag{43}$$

where the phase velocity is given by

$$u_{ph} = -\frac{ph(G)}{c\,\theta}\,u \tag{44}$$

[the exact phase velocity being the convecting velocity u, of course]. The fact that u_{ph} in general depends on θ means that in the computational simulation, different wavelengths move at different speeds, instead of the correct constant speed, u. This is the phenomenon known as dispersion. Centrally interpolated methods tend to have far greater dispersion than upwinded methods, the effect being worse for smaller c values (i.e., smaller time-step).

Third-order upwinding has a complex amplitude ratio given by

$$G_3(\theta) = 1 + c^2(\cos\theta - 1) - \frac{c}{6}(1-c^2)(\cos 2\theta - 4\cos\theta + 3) - i\,[c\sin\theta - \frac{c}{6}(1-c^2)(\sin 2\theta - 2\sin\theta)] \tag{45}$$

$$= 1 - \frac{1}{2}(c\,\theta)^2 + \frac{c}{24}(2c^2+c-2)\,\theta^4 + O(\theta^6) - i\,[c\theta - \frac{1}{6}(c\theta)^3 + \frac{c}{120}(5c^2-4)\,\theta^5 + O(\theta^7)] \tag{46}$$

which agrees with the expansion of G in Equation (38) through the critical third-order term. It might be noted at this point that the full QUICKEST [7] algorithm (including both convection and diffusion) has

$$G_{QUICKEST} = 1 - (2\alpha + c^2)(1 - \cos\theta) - 2c\,[\tfrac{1}{6}(1-c^2) - \alpha]\,(1 - \cos\theta)^2$$
$$- i\,\{c\sin\theta + 2c\sin\theta\,[\tfrac{1}{6}(1-c^2) - \alpha]\,(1 - \cos\theta)\} \quad (47)$$

where α is the diffusion parameter $\Gamma\,\Delta t/\Delta x^2$. This agrees with the expansion of the exact complex amplitude ratio for the model convection-diffusion equation -- the Gaussian spiral [24]:

$$G_{EX} = e^{-\alpha\theta^2} e^{-ic\theta} \quad (48)$$

through θ^3 terms involving both c and α ! In this interpretation, the QUICKEST algorithm is the "canonical explicit third-order" method for fluid transport for any combination of convection and diffusion (within the appropriate von Neumann stability boundary [7], of course).

The process of interpolating higher-degree polynomials can clearly be continued indefinitely, leading to correspondingly higher difference terms in the expressions for convected face values. Even-degree, centrally distributed schemes always result in highly dispersive algorithms (the leading truncation-error derivative is of odd order and the feedback sensitivity goes to zero linearly with c). Upwinded odd-degree methods involve even-order leading truncation-error derivatives and the feedback sensitivity is much less dependent on c (and approaches a constant as c goes to zero). Figure 4 shows the results of simulating the pure convection (at constant velocity) of an initial step profile in a convected passive scalar, for polynomial Lagrangian schemes from first through fifth degree; the Courant number is 0.5 in each case (for smaller c values, the shape of the third and higher odd-order profiles remains virtually insensitive to c, whereas the extent of the even methods' trailing oscillations increases rapidly).

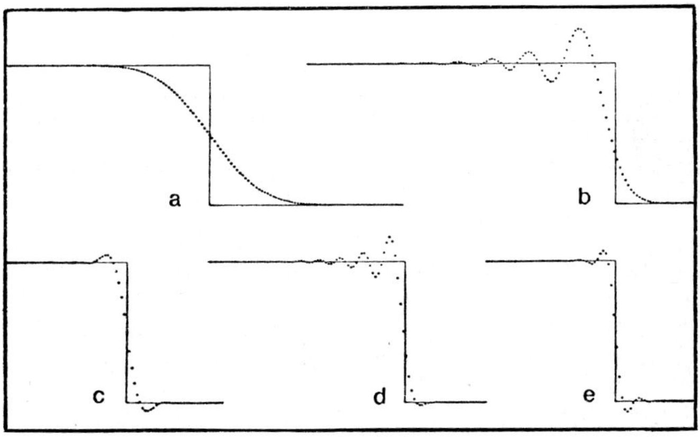

Figure 4. Details of model step convection at a nondimensional time of $t = 100$. (a) first-order upwinding; (b) Leith's method; (c) third-order upwinding; (d) fourth-order; (e) fifth-order.

Note the gross artificial diffusion of the first-order scheme giving typical error-function profiles. The trailing oscillations of the even order methods reflect the poor phase behaviour of these schemes: shorter wavelengths have sharply lagging phase velocities and not very much damping. [Upwind even-order methods (not shown) generate leading oscillations.] By contrast, upwind odd-order methods generally have excellent phase behaviour and, except for first-order (in which all wavelengths are strongly damped), they have good amplitude characteristics, as well.

From consideration of Figure 4 alone, it can be seen that third-order upwinding is clearly preferable to first-, second-, and fourth-order methods. Although sharper profiles can be obtained with fifth- and higher odd-order schemes, the complexity of the algorithms effectively eliminates such methods from serious consideration as practical simulation codes for multidimensional problems. Again, third-order upwinding emerges as the most rational basis for further development of computational techniques for unsteady highly convective flow. For example, the 5% overshoots in the third-order profile in Figure 4(c) can be entirely eliminated by using alternate interpolation (such as exponential upwinding) in the strong-curvature region. A similar strategy is used in the Exponential-Upwinding or Linear-Extrapolation Refinement of QUICK for solving the compressible Euler equations [16].

SUMMARY

Space does not permit a comprehensive survey outlining details of all the advantages of third-order upwinding over other techniques. The brief one-dimensional considerations discussed above indicate the essential features. Other attributes are apparent from the results of comparative tests and practical applications reported elsewhere. For example, an unsteady two-dimensional QUICKEST scheme has been applied to the simulation of vortex shedding from bluff bodies [25], showing excellent resolution of the vortex dynamics and agreement with experiments on shedding frequency; the same study showed that lower-order hybrid schemes generate anomolous results.

Several workers have successfully QUICKened 2D and 3D versions of TEACH-type codes. This is a relatively straight-forward procedure for any code based on a control-volume flux formulation [4, 5, 11, 12]. There have now been a number of carefully chosen numerical tests of QUICK against hybrid and other (e.g., skew-differencing) first-order schemes, involving linear problems (e.g., oblique convection of a scalar to test for cross-grid false diffusion) and nonlinear (laminar and turbulent) transport (such as driven-cavity flows, back steps, and impinging jets). Although there are clearly many problems which will need further development, it is now quite firmly established that third-order upwinding provides a firm foundation for building further refinements in specific applications. To quote from one of the most recent reports [5]:

> "QUICK emerged, overall, as decisively the most successful of the schemes. ... In our view it is the best of the simple interpolation schemes currently available and is well suited for incorporation in a robust, general-purpose solver for laminar or turbulent flows."

REFERENCES

[1] Brooks, A. and Hughes, T.J.R., Streamline-Upwind/Petrov-Galerkin Fomulation for Convection-Dominated Flows with Particular Emphasis on the Incompressible Navier-Stokes Equations, Computer Methods in Applied Mechanics and Engineering 32 (1982) 199.

[2] deVahl Davis, G. and Mallinson, G.D., An Evaluation of Upwind and Central Difference Approximations by a Study of Recirculating Flow, Computers and Fluids 4 (1976) 29.

[3] Hackman, L.P., Raithby, G.D., and Strong, A.B., Numerical Predictions of Flows Over Backward-Facing Steps, International Journal for Numerical Methods in Fluids (to appear).

[4] Han, T., Humphrey, J.A.C., and Launder, B.E., A Comparison of Hybrid and Quadratic-Upstream Differencing in High Reynolds Number Elliptic Flows, Computer Methods in Applied Mechanics and Engineering 29 (1981) 81.

[5] Huang, P.G., Launder, B.E., and Leschziner, M.A., Discretization of Nonlinear Convection Processes: A Broad Range Comparison of Four Schemes, Mechanical Engineering Department Report TDF/83/1, University of Manchester Institute of Technology (1983).

[6] Leonard, B.P., A Consistency Check for Estimating Truncation Error due to Upstream Differencing, Applied Mathematical Modelling 2 (1978) 239.

[7] Leonard, B.P., A Stable and Accurate Convective Modelling Procedure Based on Quadratic Upstream Interpolation, Computer Methods in Applied Mechanics and Engineering 19 (1979) 59.

[8] Leonard, B.P., The QUICK Finite Difference Method for the Convection-Diffusion Equation, in: Vichnevetsky, R. and Stepleman, R.S. (eds.), Advances in Computer Methods for Partial Differential Equations--III (IMACS, 1979) 292.

[9] Leonard, B.P., A Survey of Finite Differences of Opinion on Numerical Muddling of the Incomprehensible Defective Confusion Equation, in: Hughes, T.J.R. (ed.), Finite Element Methods for Convection Dominated Flows (AMD-34, ASME, New York, 1979) 1.

[10] Leonard, B.P., A Survey of Finite Differences with Upwinding for Numerical Modelling of the Incompressible Convective Diffusion Equation, in: Taylor, C. (ed.), Recent Advances in Numerical Methods in Fluids--Volume 2 (Pineridge Press, 1981).

[11] Leschziner, M.A., Practical Evaluation of Three Finite Difference Schemes for the Computation of Steady-State Recirculating Flows, Computer Methods in Applied Mechanics and Engineering 23 (1980) 293.

[12] Leschziner, M.A. and Rodi, W., Calculation of Annular and Twin Parallel Jets Using Various Discretization Schemes and Turbulence Model Variants, ASME Journal of Fluids Engineering 103 (1981) 352.

[13] Raithby, G.D., A Critical Evaluation of Upstream Differencing Applied to Problems Involving Fluid Flow, Computer Methods in Applied Mechanics and Engineering 9 (1976) 75.

[14] Raithby, G.D., Skew-Upstream Differencing for Problems Involving Fluid Flow, Computer Methods in Applied Mechanics and Engineering 9 (1976) 151.

[15] Forester, C.K., Higher Order Monotonic Convective Differencing Schemes, Journal of Computational Physics 23 (1977) 1.

[16] Leonard, B.P., The EULER-QUICK Code, in: Taylor, C. (ed.), Numerical Methods in Laminar and Turbulent Flow--III (Seattle, 1983).

[17] Gosman, A.D., Pun, W.M., Runchal, A.K., and Wolfshtein, M., Heat and Mass Transfer in Recirculating Flows (Academic Press, 1969).

[18] Christie, I., Griffiths, D.F., and Mitchell, A.R., Finite Element Methods for Second Order Differential Equations with Significant First Derivatives, International Journal for Numerical Methods in Engineering 10 (1976) 1389.

[19] Patankar, S.V., Numerical Heat Transfer and Fluid Flow (McGraw-Hill, 1980).

[20] Leith, C.E., Numerical Simulation of the Earth's Atmosphere, Methods in Computational Physics 4 (1965) 1.

[21] Lax, P.D. and Wendroff, B., Difference Schemes with High Order of Accuracy for Solving Hyperbolic Equations, Communications of Pure and Applied Mathematics 17 (1964) 381.

[22] Lax, P.D., Weak Solutions of Nonlinear Hyperbolic Equations and Their Numerical Computation, Communications on Pure and Applied Mathematics 7 (1954) 159.

[23] Fromm, J.E., A Method for Reducing Dispersion in Convective Difference Schemes, Journal of Computational Physics 3 (1968) 176.

[24] Leonard, B.P., Note on the von Neumann Stability of the Explicit FTCS Convective Diffusion Equation, Applied Mathematical Modelling 4 (1980) 401.

[25] Davis, R.W. and Moore, E.F., A Numerical Study of Vortex Shedding from Rectangles, Journal of Fluid Mechanics 116 (1982) 475.

Computational Techniques & Applications: CTAC-83
J. Noye & C. Fletcher (Editors)

GENERALIZED VARIABLE GRID SIZE METHODS WITH APPLICATION TO THE DIFFUSION EQUATION

Roger Braddock and John Noye

Australian Environmental Studies,
Griffith University,
Nathan, Queensland,
Australia

Department of Applied Mathematics,
University of Adelaide,
Adelaide, South Australia,
Australia

The solutions to partial differential equations frequently include regions in which a physical variable is rapidly varying. In order to improve accuracy, a finer grid is often used in the region of steep fronts. This paper investigates some of the mathematical and computational problems associated with the use of variable-spaced grids. Three methods of constructing variably-spaced grids are discussed in detail. Using the variable-spaced grid thus constructed, finite-difference approximations of the governing equations are developed. Explicit variable-grid schemes are given for three and five-point representations of the spatial derivatives, and the truncation errors are derived. These are applied to the solution of the one-dimensional linear diffusion equations, and stability criteria are developed in detail. Finally, some results of applying some of these techniques to the nonlinear diffusion equation are discussed. The results indicate that the grids need to be constructed with a great deal of care in order to avoid instabilities.

INTRODUCTION

In the numerical solution of partial differential equations, the spacing of a finite-difference grid is normally selected so that the resulting numerical solution agrees as closely as possible with the solution of the continuous system being studied. The commonly-used grid has uniform spacing and this method can give accurate approximations (Ames, 1977). In some circumstances, the solution to the physical system changes rapidly in some regions and detail is lost with such a grid (Parlange et al., 1982). In such circumstances, the usual advice is to "insert a finer grid in the area of interest".

A generalized technique for obtaining finite-difference approximations to spatial derivatives on a variable grid is described in this paper. Operational formulae for the representation of first and second derivatives are derived and expressions obtained for the higher-order terms of the local truncation error. In particular, both three and five-point formulae for first and second order derivatives are considered.

The choice of grid structure depends on the nature of the solution being sought, as well as on the truncation error involved in the finite-difference method being used. Three methods of constructing a variable grid are considered, then tested by application to the solution of the nonlinear diffusion equation. The first method is based on minimisation of the truncation error (the Kappa-method, Noye, 1978), the second uses the slope of the front as a criterion, and the third is based on co-ordinate transformation. Precise stability criteria are derived for the three-point explicit schemes, and sufficient bounds for stability of the five-point forms are obtained.

FINITE DIFFERENCING ON A VARIABLE-SPACED GRID

Consider the function, $f = f(x)$, defined on a suitable region of the x-axis, such that f is sufficiently differentiable for Taylor series to exist. Define the grid spacings by

$$\Delta x_i = x_{i+1} - x_i \,, \qquad (2.1)$$

and denote $f_i = f(x_i)$ for all relevant i. For a uniform grid, we will denote

$$\Delta x_i = h \,, \qquad (2.2)$$

where h is the constant step size.

Consider the linear combination,

$$q_i = \sum_{j=-\ell}^{k} \alpha_{i+j} f_{i+j} \,, \qquad (2.3)$$

where the α_{i+j} are coefficients which are to be determined. In the variable-grid case, the α_{i+j} are functions of position i in the grid, while on a uniform grid, they are generally independent of i. With these assumptions, each f_{i+j} can be expanded as a Taylor series about the point x_i to any desired order. Thus,

$$f_{i+j} = f(x_i + \delta_{i+j}) = f_i + \delta_{i+j} \left(\frac{df}{dx}\right)_i + \frac{\delta_{i+j}^2}{2!} \left(\frac{d^2f}{dx^2}\right)_i + \ldots \,, \qquad (2.4)$$

where $\delta_{i+j} = \Delta x_i + \Delta x_{i+1} + \ldots + \Delta x_{i+j-1}$, for $j > 0$, $\delta_i = 0$,

and $\delta_{i+j} = - \Delta x_{i-1} - \Delta x_{i-2} \ldots - \Delta x_{i+j}$, for $j < 0$.

Substituting for f_{i+j} in Equation (2.3) yields

$$q_i = f_i \left(\sum_{j=-\ell}^{k} \alpha_{i+j}\right) + \left(\frac{df}{dx}\right)_i \left(\sum_{j=-\ell}^{k} \alpha_{i+j} \delta_{i+j}\right) + \frac{1}{2!} \left(\frac{d^2f}{dx^2}\right)_i \left(\sum_{j=-\ell}^{k} \alpha_{i+j} \delta_{i+j}^2\right) + \ldots$$

$$= \beta_0 f_i + \beta_1 \left(\frac{df}{dx}\right)_i + \frac{1}{2!} \beta_2 \left(\frac{d^2f}{dx^2}\right)_i + \ldots \,, \qquad (2.5)$$

where $(df/dx)_i$ is the value of f' at x_i, with obvious extensions to higher derivatives. The extent of the summation is determined by the number of α's which are to be determined. The β's are defined by (2.5), and are

$$\beta_n = \sum_{j=-\ell}^{k} \alpha_{i+j} \delta_{i+j}^n \,. \qquad (2.6)$$

Now q_i, as defined in Equation (2.3), is a linear combination of f_{i+j}, $j=-\ell, 1-\ell, 2-\ell, \ldots, k$, and the corresponding values of x_{i+j} form a basis set. The coefficients β_n depend on the α's and on the grid spacing Δx_i and they can be chosen to obtain difference representations for derivatives of f(x) of any order. Thus, to obtain a finite-difference representation of $(df/dx)_i$ on this generalized variable grid, we set

$$\beta_0 = 0, \ \beta_2 = 0, \ \beta_3 = 0, \ \ldots \,. \qquad (2.7)$$

to any prescribed order, say m. We call the condition $\beta_0 = 0$, the consistency relation. Equation (2.7) is a linear system in the unknown coefficients α_{i+j}, expressed in terms of the spacings Δx_{i+j}. The extent of the basis set limits the order of the linear set (2.7). If the range is small or m is large, the system may be overdetermined and no solution may exist. On the other hand, if the range is large, an infinite number of solutions results and the extra degrees of freedom can be used to satisfy other requirements.

For example, consider approximations to df/dx, at $x = x_i$, using a three-point basis. These may be "centred", where x_i is the middle point, or skewed, where x_i is at the right or left end of the basis. For the "centred" approximation, we select $\ell = 1$ and $k = 1$, so that the β's are linear combinations of α_{i-1}, α_i, α_{i+1}. We can set at most two of the β's to zero and calculate two of the α's in terms of Δx_i and Δx_{i-1}.

Set $\beta_0 = 0$ (consistency relation) and $\beta_2 = 0$, so $\alpha_i = -(\alpha_{i+1} + \alpha_{i-1})$, and $\alpha_{i+1}\,\delta_{i+1}^2 + \alpha_{i-1}\,\delta_{i-1}^2 = 0$, or $\alpha_{i+1}\,(\Delta x_i)^2 + \alpha_{i-1}\,(\Delta x_{i-1})^2 = 0$.

Selecting α_{i+1} as the arbitrary parameter and then dividing Equations (2.3) and (2.5) by α_{i+1} is equivalent to setting $\alpha_{i+1} = 1$, which yields

$$\alpha_{i-1} = -\nu_{i-1}^2,\quad \alpha_i = \nu_{i-1}^2 - 1\ , \tag{2.8}$$

where $\nu_{i-1} = \Delta x_i/\Delta x_{i-1}$. Furthermore,

$$\beta_1 = \alpha_{i+1}\,\delta_{i+1} + \alpha_{i-1}\,\delta_{i-1} = \Delta x_i\,(1 + \nu_{i-1})\ ,$$

$$\beta_3 = \alpha_{i+1}\,\delta_{i+1}^3 + \alpha_{i-1}\,\delta_{i-1}^3 = \Delta x_{i-1}\,(\Delta x_i)^2\,(\nu_{i-1} + 1)\ ,$$

$$\beta_4 = \alpha_{i+1}\,\delta_{i+1}^4 + \alpha_{i-1}\,\delta_{i-1}^4 = (\Delta x_i)^2\,(\Delta x_{i-1})^2\,(\nu_{i-1}^2 - 1)\ . \tag{2.9}$$

Equating (2.3) and (2.5) gives

$$-\nu_{i-1}^2\,f_{i-1} + (\nu_{i-1}^2 - 1)\,f_i + f_{i+1} = \beta_1\left(\frac{df}{dx}\right)_i + \frac{1}{6}\beta_3\left(\frac{d^3f}{dx^3}\right)_i + \frac{1}{24}\beta_4\left(\frac{d^4f}{dx^4}\right)_i + \dots\ .$$

Rearrangement gives

$$\left(\frac{df}{dx}\right)_i = \frac{-\nu_{i-1}^2\,f_{i-1} + (\nu_{i-1}^2 - 1)\,f_i + f_{i+1}}{\Delta x_i\,(\nu_{i-1} + 1)} - \frac{\beta_3}{6\beta_1}\left(\frac{\partial^3 f}{\partial x^3}\right) + \dots\ . \tag{2.10}$$

This is the three-point "centred" difference form for the derivative df/dx at $x = x_i$. The coefficients depend on the local grid spacings Δx_i and Δx_{i-1}, while $T_i = -\beta_3\,(d^3f/dx^3)_i/6\beta_1 + \dots$ is the local truncation error. For a uniform grid, $\Delta x_i = \Delta x_{i-1} = h$ (const), so that $\nu_{i-1} = 1$, and Equation (2.10) reduces to the standard "centred" difference form (Noye, 1983, p. 109)

$$\left(\frac{df}{dx}\right)_i = \frac{-f_{i-1} + f_{i+1}}{2h} + T_i,\ \text{with } T_i = -\frac{h^2}{6}\left(\frac{d^3f}{dx^3}\right)_i + \dots\ .$$

Note that $\beta_4 = 0$ for $\nu_{i-1} = 1$. In fact, all even β's are zero.

Skewed representations of $(df/dx)_i$ are readily obtained by applying the above process to the appropriate bases. For the basis x_i, x_{i+1}, x_{i+2}, select $\ell = 0$ and $k = 2$ in Equations (2.3) and (2.5), then set $\beta_0 = \beta_2 = 0$. This yields

$$\alpha_{i+2} = 1,\quad \alpha_{i+1} = -(1 + \nu_i)^2,\quad \alpha_i = \nu_i\,(2 + \nu_i)\ . \tag{2.11}$$

Thus,

$$\beta_1 = -\Delta x_i\,\nu_i\,(1 + \nu_i),\quad \beta_3 = (\Delta x_i)^3\,\nu_i\,(1 + \nu_i)^2\ , \tag{2.12}$$

and

$$\left(\frac{df}{dx}\right)_i = \frac{-\nu_i(2 + \nu_i)\,f_i + (1 + \nu_i)^2\,f_{i+1} - f_{i+2}}{\Delta x_i\,\nu_i\,(1 + \nu_i)} + \frac{1}{6}\,(\Delta x_i)^2\,(1 + \nu_i)\left(\frac{d^3f}{dx^3}\right)_i + \dots\ .$$

Since the δ_{i+j} have all the same sign in this case (see Equation (2.4)), no cancellation occurs among the β_i's, $i \geq 4$. Similar expressions are found for the skewed basis x_{i-2}, x_{i-1}, and x_i.

The above method is readily extended to obtain expressions for higher derivatives, and also to the use of more extensive bases. To obtain expressions for

$(d^2f/dx^2)_i$ using a three-point basis, we select $\beta_0 = 0$ (consistency condition), and $\beta_1 = 0$ (Equation (2.7)). For the centred three-point basis, we obtain

$$\alpha_{i+1} = 1,\ \alpha_i = -(1+\nu_{i+1}),\ \alpha_{i-1} = \nu_{i-1}\ , \tag{2.13}$$

$$\left.\begin{aligned}\beta_2 &= (\Delta x_{i-1})^2\ \nu_{i-1}\ (\nu_{i-1}+1),\ \beta_3 = (\Delta x_{i-1})^3\ \nu_{i-1}\ (\nu_{i-1}^2 - 1)\ ,\\ \beta_4 &= (\Delta x_{i-1})^4\ \nu_{i-1}\ (\nu_{i-1}^3 + 1)\ .\end{aligned}\right\} \tag{2.14}$$

The obvious pattern from the β-values arises from the symmetry and sign changes of the δ's in Equation (2.4). Thus,

$$\begin{aligned}\left(\frac{d^2f}{dx^2}\right)_i &= \frac{\alpha_{i-1}\ f_{i-1} + \alpha_i\ f_i + \alpha_{i+1}\ f_{i+1}}{(\beta_2/2)} - \frac{\beta_1}{3\beta_2}\left(\frac{d^3f}{dx^3}\right)_i - \frac{\beta_4}{12\beta_2}\left(\frac{d^4f}{dx^4}\right)_i + \ldots\ .\\ &= \frac{\nu_{i-1}\ f_{i-1} - (1+\nu_{i-1})\ f_i + f_{i+1}}{(\Delta x_{i+1})^2\ \nu_{i-1}\ (\nu_{i-1}+1)} + T_i\ ,\end{aligned} \tag{2.15}$$

where T_i is the truncation error. For a variable-grid spacing, the truncation error is given by

$$T_i = -\frac{\Delta x_{i-1}\ (\nu_{i-1} - 1)}{3}\left(\frac{d^3f}{dx^3}\right)_i + \ldots\ .$$

Setting $\nu_{i-1} = 1$ gives the standard finite-difference form for a uniform grid see Noye, 1983, p. 109)

$$\left(\frac{d^2f}{dx^2}\right)_i = \frac{f_{i-1} - 2f_i + f_{i+1}}{2h^2} - \frac{h^2}{3}\left(\frac{d^4f}{dx^4}\right)_i + \ldots\ .$$

Note that $\beta_3 = 0$ for $\nu_{i-1} = 1$, and the order of the truncation error decreases to the next term β_4. Here, all β_i's with odd i are zero. The set of β_i, i even or i odd, which are zero in the uniform case, depends on the evenness (or oddness) of the order of the δ derivative being represented.

More extensive bases can be used to represent the derivatives, and consider the "centred" five-point representation for $(df/dx)_i$. With $\ell = k = 2$, it can be shown that

$$\alpha_{i+2} = (\Delta x_{i-1} + \Delta x_{i-2})^2\ \Delta x_{i-2}\ (\Delta x_{i-2} + \Delta x_{i-1} + \Delta x_i)$$

$$\alpha_{i+1} = \frac{-\Delta x_{i-2}(\Delta x_i + \Delta x_{i+1})^2(\Delta x_{i-1} + \Delta x_{i-2})^2(\Delta x_{i-2} + \Delta x_{i-1} + \Delta x_i + \Delta x_{i+1})(\Delta x_i + \Delta x_{i+1} + \Delta x_{i-1})}{(\Delta x_i)^2\ (\Delta x_i + \Delta x_{i-1})}$$

$$\alpha_{i-1} = \frac{\Delta x_{i+1}(\Delta x_i + \Delta x_{i+1})^2(\Delta x_{i-1} + \Delta x_{i-2})^2(\Delta x_{i-1} + \Delta x_{i-2} + \Delta x_i)(\Delta x_{i-2} + \Delta x_{i-1} + \Delta x_i + \Delta x_{i+1})}{(\Delta x_{i-1})^2\ (\Delta x_{i-1} + \Delta x_i)}$$

$$\alpha_{i-2} = -\ (\Delta x_i + \Delta x_{i+1})^2\ \Delta x_{i+1}\ (\Delta x_{i-1} + \Delta x_i + \Delta x_{i+1})\ ,$$

$$\alpha_i = -\ (\alpha_{i+2} + \alpha_{i+1} + \alpha_{i-1} + \alpha_{i-2})\ , \tag{2.16}$$

These results are obtained by setting $\beta_0 = \beta_2 = \beta_3 = \beta_4 = 0$, while β_1 and β_5 are obtained from Equation (2.6),

$$\left.\begin{aligned}\beta_1 &= \alpha_{i+2}\ \delta_{i+2} + \alpha_{i+1}\ \delta_{i+1} + \alpha_{i-1}\ \delta_{i-1} + \alpha_{i-2}\ \delta_{i-2}\ ,\\ \beta_5 &= \alpha_{i+2}\ \delta_{i+2}^5 + \alpha_{i+1}\ \delta_{i+1}^5 + \alpha_{i-1}\ \delta_{i-1}^5 + \alpha_{i-2}\ \delta_{i-2}^5\ .\end{aligned}\right\} \tag{2.17}$$

For a grid of uniform spacing h, $\alpha_{i+2} = 12h^4$, $\alpha_{i+1} = -96h^4$, $\alpha_{i-1} = 96h^4$, $\alpha_{i-2} = -12h^4$, and $\beta_1 = -144h^5$, $\beta_5 = -1536h^9$.

Note the symmetry of the α-values for the uniform grid. Again, the order of the truncation error is not altered by the choice of a uniform spacing; this is evident from the equivalence of sign between δ_{i+j} and δ^5_{i+j}, $j = \pm 1, \pm 2$ in Equation (2.13).

In similar fashion, we obtain a "centred" five-point expression for $(d^2f/dx^2)_i$. Here, we set $\beta_0 = \beta_1 = \beta_3 = \beta_4 = 0$, and obtain

$$\alpha_{i+2} = (\Delta x_{i-1} + \Delta x_{i-2})[(\Delta x_i^3 + \Delta x_{i-1}^3)\{(\Delta x_{i-1} + \Delta x_{i-2})^2 - \Delta x_i^2\} - (\Delta x_{i-1}^2 - \Delta x_i^2).$$
$$\{(\Delta x_i)^3 + (\Delta x_{i-1} + \Delta x_{i-2})^3\}] ,$$
$$\alpha_{i+1} = (\Delta x_i + \Delta x_{i+1})(\Delta x_{i-1} + \Delta x_{i-2})\, \Delta x_i^{-1}\, [\{\Delta x_{i-1}^3 - (\Delta x_{i-1} + \Delta x_{i-2})^3\}.$$
$$\{(\Delta x_i + \Delta x_{i+1})^2 - \Delta x_{i-1}^2\} - \{(\Delta x_{i-1} + \Delta x_{i-2})^2 - \Delta x_{i-1}^2\}.$$
$$\{\Delta x_{i-1}^3 + (\Delta x_i + \Delta x_{i+1})^3\}] , \tag{2.18}$$
$$\alpha_{i-1} = -\Delta x_{i-1}^{-1}\,(\Delta x_i + \Delta x_{i+1})(\Delta x_{i-1} + \Delta x_{i-2})[\{(\Delta x_i + \Delta x_{i+1})^2 - \Delta x_i^2\}.$$
$$\{(\Delta x_i^3 + (\Delta x_{i-1} + \Delta x_{i-2})^3\} - \{(\Delta x_i^3 - (\Delta x_i + \Delta x_{i-1})^3\}\{(\Delta x_{i-1} + \Delta x_{i-2})^2$$
$$- \Delta x_i^2\}] ,$$
$$\alpha_{i-2} = (\Delta x_{i+1} + \Delta x_i)[\{\Delta x_i^3 + \Delta x_{i-1}^3\}\{(\Delta x_i + \Delta x_{i+1})^2 - \Delta x_i^2\} - \{\Delta x_{i-1}^2 - \Delta x_i^2\}.$$
$$\{(\Delta x_i^3 - (\Delta x_i + \Delta x_{i+1})^3\}] ,$$
$$\alpha_i = -(\alpha_{i+2} + \alpha_{i+1} + \alpha_{i-1} + \alpha_{i-2}) .$$

We also have

$$\beta_2 = \alpha_{i+2}\,\delta_{i+2}^2 + \alpha_{i+1}\,\delta_{i+1}^2 + \alpha_{i-1}\,\delta_{i-1}^2 + \alpha_{i-2}\,\delta_{i-2}^2 , \tag{2.19}$$
$$\beta_5 = \alpha_{i+2}\,\delta_{i+2}^5 + \alpha_{i+1}\,\delta_{i+1}^5 + \alpha_{i-1}\,\delta_{i-1}^5 + \alpha_{i-2}\,\delta_{i-2}^5 .$$

For a uniform spacing h, $\alpha_{i+2} = 12h^6$; $\alpha_{i+1} = -192h^6$; $\alpha_{i-1} = -192h^6$, $\alpha_{i-2} = 12h^6$, $\alpha_i = 360h^6$, and

$$\beta_2 = -288h^8,\ \beta_5 = 0,\ \beta_6 = 1152h^{12} . \tag{2.20}$$

Here, the signs of δ_{i+j}^2 and δ_{i+j}^5 are opposite, cancellation occurs and $\beta_5 = 0$, so that the truncation error is reduced in order. The five-point "centred" expression for $(d^2f/dx^2)_i$ on a variably-spaced grid has non-symmetric α's (see Equation (2.18)). For the uniform grid,

$$\alpha_{i+2} = \alpha_{i-2},\ \alpha_{i+1} = \alpha_{i-1} ,$$

and the α's are symmetric about the central point x_i. This holds for the "centred" three-point representation of (d^2f/dx^2).

AN EXPLICIT THREE-POINT VARIABLE-GRID SCHEME FOR THE DIFFUSION EQUATION

Consider the one-dimensional linear diffusion equation

$$\frac{\partial\theta}{\partial t} = D\,\frac{\partial^2\theta}{\partial x^2} , \tag{3.1}$$

where D is a constant, and $\theta = \theta(x,t)$ (Ames, 1977). Consider boundary conditions of the form

$$\theta(0,t) = \Theta_0,\ \text{all } t,\ x = 0,\ \theta(1,t) = \Theta_1,\ \text{all } t,\ x = 1 ,$$

where Θ_0 and Θ_1 are constants, and

$$\theta(x,0) = g(x),\ \text{all } x \in [0,1],\ t = 0 .$$

A finite-difference grid is constructed in the solution domain, so that a numerical solution to Equation (3.1) can be computed. Time lines are defined by

$$t_{n+1} = t_n + \Delta t_n,\ t_0 = 0,\ n = 0, 1, 2, \ldots , \tag{3.2}$$

where the spacings Δt_n are not necessarily equal. Each time line is discretised by a variable spatial grid defined by

$$x_{i+1,n} = x_{i,n} + \Delta x_{i,n},\ i = 1, 2, \ldots I-1,\ x_{0,n} = 0,\ x_{I,n} = 1 . \tag{3.3}$$

The procedures for selecting a suitable variable spatial grid, not necessarily the same on each time line, is discussed in the next Section. In this Section, we are concerned with the stability of the numerical process as the solution is advanced from time line n to n+1. For this single step, it is convenient to drop the subscript n in Equations (3.2), (3.3). For convenience, denote $\theta(x_i, t_n)$ by $\theta_{i,n}$.

Equation (3.1) may be discretised in an explicit or an implicit manner (Ames, 1977), and the spatial derivative can be specified using a variety of bases. Here we concentrate on two explicit cases. Firstly, using a "centred" three-point spatial and a forward time discretisation, gives, (Equation (2.15))

$$\frac{\theta_{i,n+1} - \theta_{i,n}}{\Delta t} = D\left(\frac{\alpha_{i+1}\,\theta_{i+1,n} + \alpha_i\,\theta_{i,n} + \alpha_{i-1}\,\theta_{i-1,n}}{\beta_3/2}\right) \tag{3.4}$$

for $i = 1, 2, \ldots I-1$; the boundary conditions are given by $\theta_{0,n} = \Theta_0$, $\theta_{I,n} = \Theta_1$, for all n. The local truncation error is given by

$$\tau_{i,n} = -(\beta_3/3\beta_2)\left(\frac{\partial^3\theta}{\partial x^3}\right)_{i,n} - (\beta_4/12\beta_2)\left(\frac{\partial^4\theta}{\partial x^4}\right)_{i,n} + \frac{\Delta t}{2}\left(\frac{\partial^2\theta}{\partial t^2}\right)_{i,n} + \cdots . \tag{3.5}$$

By differentiation of Equation (3.1), the last term in the truncation error can be replaced by $D\Delta t(\partial^4\theta/\partial x^4)_{i,n}/2$. On rearranging Equation (3.4), the following three-point explicit scheme results:

$$\theta_{i,n+1} = \theta_{i,n} + 2D\Delta t\ \beta_2^{-1}\ (\alpha_{i+1}\,\theta_{i+1,n} + \alpha_i\,\theta_{i,n} + \alpha_{i-1}\,\theta_{i-1,n}) , \tag{3.6}$$

where the α_{i+1}, α_i, α_{i-1} and β_2^{-1} are defined in Equations (2.13) and (2.14).

The stability of finite-difference schemes has been discussed by many authors (Ames, 1977, and Noye, 1983, pp. 124-134). The widely-used von Neumann, or Fourier series, method is used here. Local applications of this method to the finite-difference equations used at the boundaries also yield the influence of the boundary conditions on numerical stability (see Trapp and Ramshaw, 1976) on numerical stability. Following Noye (1983), assume a solution to Equation (3.6) in the form

$$\theta_{i,n} = G^n e^{i\phi i} , \tag{3.7}$$

where $i = \sqrt{-1}$, G is an amplification factor, and ϕ is a phase angle. Substituting Equation (3.7) into (3.6) then yields,

$$G = 1 + \alpha_i\,\Gamma + \Gamma(\alpha_{i+1} + \alpha_{i-1})\cos\phi + i\Gamma(\alpha_{i+1} - \alpha_{i-1})\sin\phi , \tag{3.8}$$

where $\Gamma = 2D\Delta t/\beta_2$. Note that G is complex in general, but is real if $\alpha_{i+1} = \alpha_{i-1}$. This last case arises when $\Delta x_i = h$ (constant) for all i, that is, for the uniform grid. The necessary condition for stability is $|G| \leqslant 1$, or $|G|^2 - 1 \leqslant 0$, for all values of the phase angle ϕ. From (3.8), we obtain

$$|G|^2 = 1 + 2\Gamma(\alpha_{i+1} + \alpha_{i-1})(\cos\phi - 1) + 2\Gamma^2(\alpha_{i+1} + \alpha_{i-1})^2(1 - \cos\phi)$$
$$- 4\Gamma^2\,(\alpha_{i+1} + \alpha_{i-1})\sin^2\phi .$$

Thus, the condition $|G|^2 - 1 \leqslant 0$, leads to

$$(\cos\phi - 1)[\Gamma(\alpha_{i+1} + \alpha_{i-1}) - \Gamma^2(\alpha_{i+1} + \alpha_{i-1})^2 + 2\Gamma^2(\cos\phi + 1)\,\alpha_{i+1}\,\alpha_{i-1}] \leqslant 0\ ,$$

which must hold for all ϕ. Since $\cos\phi - 1 \leqslant 0$, for all ϕ, this is equivalent to $2(1 + \cos\phi)\,\Gamma^2\,\alpha_{i+1}\,\alpha_{i-1} \geqslant \Gamma(\alpha_{i+1} + \alpha_{i-1})(\Gamma(\alpha_{i+1} + \alpha_{i-1}) - 1)$, for all ϕ. From Equation (2.13), $\alpha_{i+1} = 1 \geqslant 0$, and $\alpha_{i-1} = \nu_{i-1} > 0$, so this becomes

$$\Gamma(1 + \nu_{i-1})(\Gamma(1 + \nu_{i-1}) - 1) \leqslant 0\ , \tag{3.9}$$

or

$$0 \leqslant \frac{\Delta t D}{\frac{1}{2}(\Delta x_i\ \Delta x_{i-1})} \leqslant 1\ . \tag{3.10}$$

For the scheme to be stable at all points in the variable grid, Equation (3.10) must hold for all points i where Equation (3.6) is used. Thus, it is equivalent to the stability condition

$$0 \leqslant D\Delta t\ [\operatorname*{Min}_i\ (\tfrac{1}{2}\Delta x_i\ \Delta x_{i-1})]^{-1} \leqslant 1\ . \tag{3.11}$$

For the uniform grid spacing h, this reduces to (Noye, 1983, p. 133)

$$0 \leqslant \frac{\Delta t D}{h^2} \leqslant \tfrac{1}{2}\ . \tag{3.12}$$

There is a problem which must be resolved in the use of variable-grid spacings. On the one hand, the stability of the scheme and hence the selection of the length of the time step, is governed by the smallest spatial grid spacing; on the other hand, the local truncation error is bounded by the largest spacings multiplied by the appropriate derivatives. The latter can be resolved by use of large spacings only where the appropriate derivatives are small. When the grid spacings are uniform, the truncation error decreases by one order. Thus, in using a variable-grid technique, we are faced with the dual problems of relatively large truncation errors and relatively small time steps which cause computation count to rise rapidly.

Finally, consider the nonlinear diffusion equation in the form

$$\frac{\partial\theta}{\partial t} = \frac{\partial}{\partial x}\,(D(\theta)\ \theta_x)\ , \tag{3.13}$$

where $D(\theta)$ is a function of dependent variable θ. Equations such as this arise in the problem of percolation of ground water (Braddock et al., 1981). By writing Equation (3.13) in the form

$$\frac{\partial\theta}{\partial t} = D(\theta)\,\frac{\partial^2\theta}{\partial x^2} + \left(\frac{dD}{d\theta}\right)\left(\frac{\partial\theta}{\partial x}\right)^2\ ,$$

the derivatives may be approximated with finite-difference forms. However, the resulting finite-difference equation is highly nonlinear and the stability for the finite-difference representation of the fully nonlinear form has not been determined. However, noting that locally $D(\theta)$ is approximately constant and that an approximate stability condition is obtained by applying Equation (3.10) locally, we obtain the approximate stability condition for Equation (3.13), namely

$$0 \leqslant \Delta t\ \operatorname*{Max}_i\ \left[\frac{D(\theta_{i,n})}{\frac{1}{2}\Delta x_i\ \Delta x_{i-1}}\right] \leqslant 1\ . \tag{3.14}$$

Clearly, as $\theta_{i,n}$ depends on the particular time line involved, the time step which satisfies (3.14) must be established at each time line.

SELECTION OF A VARIABLE GRID

It has been shown that a stable explicit numerical scheme can be constructed for a grid with variable spacing. Some workers have used them to handle particular physical problems (Anthes, 1970, Sundquist and Veronis, 1970, Denny and Landis, 1972, and Chong, 1978), but the method of selecting the grid has been largely ad hoc. The most commonly used variable grid is one in which there is a sudden change from a large to a small spacing, as in Crowder and Dalton (1971). Usually the ratio of coarse to fine mesh size is an integer to simplify the matching of the two regions (Greenberg, 1975).

Less work has been done on application of grids with gradual changes of spacing. Dane and Mathis (1981) studied the soil-water equation (Equation (3.13)) and used a cubic spline analysis to locate new spatial points x_i at each time step. The values $\theta_{i,n}$ on the time line n, were used to locate the knots of the cubic spline which gave the best approximation to the function $\theta(x,t_n)$. This method consumes much computer time in calculating the positions of the knots along each time line. Further, no attention has been paid to the stability and accuracy of the resulting difference schemes.

We consider two methods for generating a variable grid, the first based on minimizing the order of the truncation error in Equation (3.5), the second based on ideas used in the numerical solution of ordinary differential equations. From Equation (2.14), the local truncation error can be written in the form

$$T_{i,n} = -(\Delta x_i - \Delta x_{i-1})\left(\frac{\partial^3\theta}{\partial x^3}\right)_{i,n} - \Delta x_{i-1}^2\,(\nu_{i-1}^2 - \nu_{i-1} + 1)\left(\frac{\partial^4\theta}{\partial x^4}\right)_{i,n} + D\frac{\Delta t}{2}\left(\frac{\partial^4\theta}{\partial x^4}\right)_{i,n} + \cdots .$$

If $\nu_{i-1} = O(1)$, the leading term of the local error $T_{i,n}$ can be minimized by an appropriate choice of the grid spacings and time steps. If

$$\Delta x_i = \Delta x_{i-1}\,(1 + K_1\,\Delta x_{i-1}) + K_2\,\Delta t\ , \tag{4.1}$$

where

$$K_1 = -(\nu_{i-1}^2 - \nu_{i-1} + 1)\left(\frac{\partial^4\theta}{\partial x^4}\right)_{i,n}\left[\left(\frac{\partial^3\theta}{\partial x^3}\right)_{i,n}\right]^{-1}\ , \quad \text{and} \quad K_2 = D\left[2\left(\frac{\partial^3\theta}{\partial x^3}\right)_{i,n}\right]^{-1}\ ,$$

then the leading error terms cancel. Because K_1 and K_2 are dependent both on the time line and on the spatial position, then estimating K_1 and K_2 requires approximate values for the derivatives $\partial^3\theta/\partial x^3$ and $\partial^4\theta/\partial x^4$, which can only be obtained from the numerical solutions produced using Equation (3.4).

In the following, we consider K_1 and K_2 constant and note that the stability condition equation (3.11) requires selection of $D\Delta t \leq \text{Min}\ \frac{1}{2}(\Delta x_i\ \Delta_{i-1})$. Assuming $D\Delta t = O(\Delta x_{i-1}^2)$, we absorb this term into K_1. Equation (4.1) is thus replaced by

$$y_i = F(y_{i-1}) = y_{i-1}\,(1 + K_1\,y_{i-1})\ , \tag{4.2}$$

where $y_{i-1} = \Delta x_{i-1}$, and we need to select $y_0 > 0$ to generate grid spacings. The limit points, ℓ, of this sequence are obtained from the condition $\ell = F(\ell)$, that is, $\ell^2 K_1 = 0$. The case $K_1 = 0$ corresponds to the uniform grid. The other possibility is $\ell = 0$, where $\partial F/\partial y = 1 + 2K_1 y_1$, so that $\partial F/\partial y|_{y=\ell} = 1$, and the limit point is neutrally stable. For y_i small and positive, when $K_1 < 0$, $(\partial F/\partial y)\,y_i < 1$ and the sequence diverges monotonically near the limit point; for $K_1 > 0$, $(\partial F/\partial y)\,y_i > 1$ and the sequence converges monotonically near the limit. Equation (4.2) thus provides us with a one-parameter scheme of generating an expanding or contracting variable grid.

When constructing the variable grid, a prescribed number of points x_i, $i = 0, 1, 2, .., I$, must be fitted into some predetermined interval, say $x \in [0, 1]$, for a chosen K_1. K_1 is determined by the nature of the numerical solution and the truncation error. Given x_0 and an initial choice Δx_0, a trial grid for the given K_1 is generated using Equation (4.2). In general, the final grid-point $x_I \neq 1$; if $x_I > 1$, a smaller estimate of Δx_0 is required, and if $x_I < 1$, a larger estimate is needed. Linear interpolation of Δx_0 based on the value of x_I, iteratively generates the value Δx_0 which gives x_I as close to 1 as we desire.

Based on the nature of the solution at the nth time line, regions of uniform, contracting and expanding grids may be required on different parts of the spatial interval. Provided that the break points can be specified, this is not difficult since Equation (4.2) can be applied, with appropriate values for K_1 between any two such points.

The selection of the break points may be made by considering the various strategies used when solving ordinary differential equations. Variable-step methods such as the Runge-Kutta techniques change the step size according to the nature of the local solution. One rule of thumb is: where $|y'(x)| \leqslant 1$, then equal-spaced steps in x are used; where $|y'(x)| \geqslant 1$, the differential equation can be inverted and solutions to $x = x(y)$ are found numerically using equi-spaced y steps. The choice of the step length in either x or y may be further dictated by the slope of the appropriate solution. This concept can be applied to the time line t_n, since the derivative $(\partial\theta/\partial x)_{t_n}$ can be calculated with reasonable accuracy.

In the following, break points are chosen to occur where $|\partial\theta/\partial x| = 1$, a uniform spatial grid is used in regions with $|\partial\theta/\partial x| \leqslant 1$, and appropriate expanding or contracting grids used where $|\partial\theta/\partial x| \geqslant 1$.

The use of the gradient $\partial\theta/\partial x$ suggests an alternative method of generating the variable grids. This technique is based on concentrating calculations in the regions of a rapidly varying solution. Knowing the solution on the time line t_n, $(\partial\theta/\partial x)_{t_n}$ can be estimated along with the break points where $|\partial\theta/\partial x| = 1$. Based on a computed estimate of $\partial\theta/\partial x$, the grid is constructed so that

a) for $|\partial\theta/\partial x| \leqslant 1$, uniform steps $\Delta x = h$ are used;

b) for $|\partial\theta/\partial x| \geqslant 1$, the grid is generated by holding $\Delta\theta$ constant, i.e. $\Delta\theta = \pm h$ (the sign depends on the sign of $\partial\theta/\partial x$), the values of Δx being computed from $\Delta x = \Delta\theta/(\partial\theta/\partial x) = h/|\partial\theta/\partial x|$.

This technique concentrates grid points in the region of rapid changes in the solution.

Finally, a co-ordinate-stretching method may be used (Noye, 1983, p. 306). Consider the linear diffusion equation, Equation (3.1). We seek to rescale the space variable as $x = H(z)$ so that an equi-spaced grid in z leads to non-uniform spacings in x. The function $H(z)$ is selected such that $H'(z)$ is small in regions of the x-axis where the solution is changing rapidly. Equation (3.1) transforms to

$$\frac{\partial\theta}{\partial t} = - \frac{\alpha H''(z)}{(H'(z))^3}\frac{\partial\theta}{\partial z} + \frac{\alpha}{(H'(z))^2}\frac{\partial^2\theta}{\partial z^2}, \qquad (4.3)$$

where $H'(z) = dH/dz$. Equation (4.3) is a transport-like equation with a variable diffusivity and a variable convective velocity.

A special case arises where the convective velocity $H''/(H')^3 = a$, is independent of z. In this case,

$$H(z) = c - (b - 2az)^{\frac{1}{2}}/a, \qquad (4.4)$$

where b and c are integration constants. Equation (4.3) is of the form,

$$\frac{\partial\theta}{\partial t} = - g_1(z)\frac{\partial\theta}{\partial z} + g_2(z)\frac{\partial^2\theta}{\partial z^2} \qquad (4.5)$$

and may now be solved by discretising on a uniformly spaced grid. Such a scheme will yield a truncation error containing a higher order time derivative such as $\partial^2\theta/\partial t^2$, $\partial^3\theta/\partial t^3$, depending on the nature of the discretisation applied to $d\theta/dt$ in Equation (4.5). Now

$$\frac{\partial^2\theta}{\partial t^2} = (- g_1\frac{\partial}{\partial z} + g_2\frac{\partial^2}{\partial z^2})^2\,\theta = \frac{\partial\theta}{\partial z}(g_1\frac{dg_1}{dz} - g_2\frac{d^2g_1}{dz^2}) + \frac{\partial^2\theta}{\partial z^2}(g_1^2 - g_1\frac{dg_2}{dz}$$

$$- 2g_2\frac{dg_1}{dz} + g_2\frac{d^2g_2}{dz^2}) + \ldots$$

and the expansions for $\partial^3\theta/\partial t^3$, etc. have the same general structure. The term $\partial\theta/\partial z\,[g_1\,(dg_1/dz) - g_2\,(d^2g_1/dz^2)]$ is a convection-like term, whilst the term containing $\partial^2\theta/\partial z^2$ is a diffusion-like term; both introduced by the numerical method. Thus, false convection and false diffusion are present for all but a few special forms of $H(z)$. The numerically induced convection is zero if

$$g_1 \frac{dg_1}{dz} - g_2 \frac{d^2 g_1}{dz^2} = \frac{dg_1}{dz} [- \frac{\alpha H''}{(H')^3} - \frac{\alpha}{(H')^2} \frac{d}{dz} (- \frac{\alpha H''}{(H')^3})] .$$

Thus, either g_1 = constant or

$$H'''H' - 3(H'')^2 - H''(H')^3 = 0 . \qquad (4.6)$$

One possible solution of this equation is $H'' = 0$, so that $H(z)$ is linear in z; this leads to uniform grid-spacing on the original x-axis. Apart from this case, Equation (4.6) appears to be intractable. A similar problem arises when setting the false diffusion at zero in order to determine an optimal form for $H(z)$. In summary, such transform methods introduce false diffusion and convection in the explicit finite-difference forms except for a very limited number of special functions $H(z)$.

AN APPLICATION OF THE THREE-POINT SCHEME

As a test problem, the nonlinear diffusion equation (3.13) was considered, with the auxiliary conditions,

$$\theta = 0, \text{ at } x = 0, \text{ all } t , \quad \theta = \theta^*, \text{ at } x = 1, \text{ all } t$$

and

$$\theta = \theta^* x, \text{ at } t = 0 ,$$

where $D(\theta) = \exp(p\theta)$, and p and θ^* are parameters. Equation (3.13) arises in soil physics (Braddock et al., 1981) and typical values of p lie in the range $p = 4$ to 10. An analytic solution of Equation (3.13) is not yet available except for $p = 0$, when D = constant. However, the steady-state solution in analytic form is

$$\theta_A (x, t \to \infty) = p^{-1} \ln [1 + (\exp(p\theta^*) - 1) x] . \qquad (5.1)$$

The parameter θ^* can be used to vary the steepness of the curve $\theta_A (x, t \to \infty)$, while increasing p increases the steepness of the steady-state solution near $x = 0$. This solution may be used to test the numerical techniques under investigation.

Expanding the derivative in Equation (3.13) gives

$$\frac{\partial \theta}{\partial t} = D(\theta) \frac{\partial^2 \theta}{\partial x^2} + \frac{dD}{d\theta} (\frac{\partial \theta}{\partial x})^2 , \qquad (5.2)$$

which is similar to Equation (4.5). Thus, numerical diffusion and convection will be present in any computed solutions. The numerical convection term is absent if $\partial D/\partial x = 0$, i.e. if D is a function of t only. Equation (5.2) formed the basis of the numerical solution procedures, with the time derivative replaced by the forward-difference form and the spatial derivatives, using three-point "centred" formulae, discretised using the results of Section 2.

Three types of grid construction were used; a uniform grid, a variable grid constructed using the difference equation (4.2) (the Kappa-technique), and a variable grid based on the condition $|\partial\theta/\partial x| \gtrless 1$.

The time step Δt used in all the calculations was chosen to be 0.65 times the value obtained using Equation (3.14). This choice was based on the following reasoning. Equation (5.2) can be considered as a form of the transport equation with a variable convective speed. For a constant convective speed u, and constant diffusivity D_0, the stability criterion for the explicit three-point space-centred scheme for the transport equation, is

$$\Delta t \leqslant \text{Min } (2D_0/u^2, (\Delta x)^2/2D_0) , \qquad (5.3)$$

if the grid is uniform (Noye, 1983, p. 216). Local application of (5.3) to the case of variable convection speeds and variable diffusivities, yields

$$\Delta t \leqslant \text{Min } \{2/(p^2 \exp(p\theta)), (\Delta x)^2/(2 \exp(p\theta))\} . \qquad (5.4)$$

For the range of values of p and Δx which are involved here, the stability limit on Δt is always obtained from the second term. The reduction factor of 0.65 was obtained on the basis of experience in various numerical experiments; values larger than 0.65 correspond to time steps nearer to the stability limit and produced unstable methods, particularly for larger values of p and θ^*.

Errors in the calculations were estimated by the measure

$$E = \int_0^1 (\theta_A(x, t \to \infty) - \theta(x,t))^2 \, dx \ , \tag{5.5}$$

where $\theta(x,t)$ was the numerical approximation. The integral was evaluated using a trapezoidal rule for which the truncation error is of the order of the step length cubed. This compares more than favourably with the local truncation errors obtained in Section 2, for the space derivatives, which are at best the order of the step length squared. The trapezoidal rule also has the advantage of being easily adapted for a variable-step length.

The solution using a uniform grid was found using J = 10, 20, 30 and 40 grid spaces in the interval $x \in [0,1]$, for the cases p = 1(1)7, and θ^* = 1(1)7. Some results are shown in Table 1.

TABLE 1

Parameter values	Number of intervals on x-axis	Smallest error E	Time of occurrence of smallest error	Number of time steps to reach smallest error
p = 1, θ^* = 1	10	0.7×10^{-9}	0.361	360
	20	0.56×10^{-10}	0.435	1451
	30	0.49×10^{-10}	0.495	3152
	40	0.29×10^{-11}	0.520	5202
p = 1, θ^* = 7	10	1.203	0.00096	
	20	0.0187	0.00132	
	30	0.024	0.00136	
	40	0.0245	0.00139	
p = 7, θ^* = 1	10	0.0248	0.00096	320
	20	0.266×10^{-3}	0.00131	1755
	30	0.492×10^{-3}	0.00136	4297

For all parameters used, the error estimated by Equation (5.5) fell to a minimum as time stepping proceeded and then increased again as rounding errors built up. In the decreasing stage, the error was approximately a quadratic function of the number of time steps taken, while in the increasing stage, it was initially a linear function of this number. For the larger values of p and θ^*, the method developed instabilities which first became evident near x = 1, the point where the diffusivity and its gradient were largest. For smaller values of p and θ^*, the linear growth due to the accumulated rounding error continued for some time as the error built up. In such cases, the solution process finally became unstable.

The results in Table 1 show that, for small p and θ^*, the error E was small, and decreased as the number of grid points increased. However, this involved an increasing computational burden since the number of time steps required to reach the steady state increased rapidly, and each time step involved more computations due to the increasing number of grid points. Thus, for 10 grid points on the x-axis, the minimum error of 0.7×10^{-9} was achieved with just 3,600 applications of the difference equation. For 40 grid points on the axis, about 200,000 applications were required. The results for p = 1, θ^* = 7, and p = 7, θ^* = 1 were very poor. The error remained large, and its minimum was reached quickly. This occurred because of the extreme steepness of the solution being sought. Thus, the derivative terms in the local truncation error are very large, the local truncation error is large, and these errors rapidly accumulate in the numerical solution, effectively preventing the true steady state from being reached. These effects may possibly be controlled by using a variable-grid scheme.

The Kappa-technique (Noye, 1983, p. 301) was also used to generate variable grids for the numerical solution of Equation (3.13) and (5.2). The parameter K in Equation (4.2) was varied systematically to investigate how its value affected accuracy. As before, grids were constructed in the interval $x \in [0,1]$, using J = 10, 20, 30 and 40. The grid was then held fixed whilst the finite-difference equation was solved on the computer and estimates made of the error E. Some results are given in Table 2. These represent the limit of applicability of this method.

TABLE 2

Parameter values	Value of K	Minimum Error	Time of occurrence of minimum error	Number of iterations to minimum error
J = 10, p = 1, θ* = 1	2	0.22×10^{-2}	0.0954	280
	4	0.32×10^{-2}	0.0141	100
	6	0.20×10^{-2}	0.0114	150
	8	0.12×10^{-2}	0.0109	230
	10	0.76×10^{-3}	0.0108	330
	20	0.315×10^{-4}	0.0105	1100
	22	0.134×10^{-4}	0.0105	1300
	24	0.802×10^{-5}	0.0104	1500
	26	0.940×10^{-5}	0.0104	1750
	28	0.162×10^{-4}	0.0104	2000
	30	0.227×10^{-4}	0.0102	2240
	40	0.687×10^{-4}	0.0100	3700
	50	0.977×10^{-4}	0.0096	5400
J = 40, p = 1, θ* = 1	5	0.416×10^{-2}	0.00294	710
	10	0.102×10^{-2}	0.00369	3300
	20	0.253×10^{-3}	0.00438	5000
	25	0.590×10^{-3}	0.00438	22500
	30	0.935×10^{-3}	0.00425	32250
J = 10, p = 1, θ* = 7	5	6.82	0.35×10^{-4}	140
	10	5.39	0.796×10^{-4}	985
J = 40, p = 1, θ* = 7	5	7.51	0.304×10^{-5}	300
	10	6.45	0.144×10^{-4}	5200
J = 10, p = 6, θ* = 1	5	0.1176	0.818×10^{-4}	120
	10	0.0921	0.164×10^{-3}	750
	20	0.0603	0.356×10^{-3}	5550

The results obtained using this approach were disappointing. J = 10, p = θ* = 1 was the simplest case considered. For K = 1, Equation (4.2) generates a uniform grid. For the non-uniform grid generated by K = 2, the error is smaller and is attained more quickly on the computer, although at the expense of more computational time. As K increases beyond 4, the minimum error drops to 0.8×10^{-5}, for K = 24, then increases again. This is comparable with the results in Table 1, where an error of 0.7×10^{-9} was obtained using 10 equal grid spacings. In the variable-grid case, the local truncation error is one order of magnitude larger than for the uniform grid, which leads to an increasing rounding or accumulating numerical error. This limits the number of stable calculations which can take place. In all cases, the instability was first noticed near x = 1. Similar comparisons can be made for other values of J, p and θ^*; in all cases, the variable-grid method fared badly.

The best result was obtained for the large value K = 24. This is because the bulk of the grid points are then located near x = 0 where the solution is steepest. Much larger intervals, such as $\Delta x_{10} \simeq 0.447$ are located near x = 1 and this increases the size of the truncation error near x = 1. It was noticed that the K-method must produce a grid which fits very accurately into the range [0,1]. Any small errors arising near x = 1, because the method of fitting the grid had not proceeded through enough iterations, produced instability there very rapidly. Finally, the variable-slope method of calculating the grid spacings was tested. In this case, the spacings were adjusted as the solution developed and the numbers of time steps taken between computing new sets of grid spacings varied. Some results are given in Table 3.

TABLE 3

Parameter values	Number of time steps between changes of grid	Minimum error obtained	Time of minimum error	Total number of iterations to the minimum
J = 10, p = 1, θ^* = 1	5	0.450×10^{-3}	0.68×10^{-1}	110
	10	0.404×10^{-3}	0.48×10^{-1}	60
	15	0.862×10^{-4}	0.17	255
	20	0.868×10^{-4}	0.63	940
	30	0.50×10^{-3}	went unstable quickly	
	40	0.30×10^{-3}	went unstable quickly	
J = 20, p = 1, θ^* = 1	5	0.556×10^{-3}	0.456	3005
	10	0.50×10^{-2}	0.8×10^{-4}	20
	15	0.198×10^{-3}	0.54×10^{-1}	270
	20	0.811×10^{-4}	0.717×10^{-1}	360
	25	0.808×10^{-4}	0.15×10^{-3}	900
	30	0.808×10^{-4}	0.15×10^{-3}	1020

These are much more haphazard in their nature, as the calculations proved to be very sensitive to any small errors in fitting the grid to the interval [0,1]. The fit-at-the-end points, as well as the fit-at-the-break points, had to be very accurate or very poor results were obtained. The result for J = 20, p = θ^* = 1, using 10 time steps between recomputing the grid spacings, illustrates this point. The minimum error was large, was attained quickly, and then the calculation became unstable due to poor grid fitting.

THE FIVE-POINT EXPLICIT SCHEME

Equation (3.1) can also be discretised using a five-point centred form for the second-order derivative on a spatially varying grid. This yields

$$\frac{\theta_{i,n+1} - \theta_{i,n}}{\Delta t} = \frac{2D}{\beta_2} (\alpha_{i+2}\,\theta_{i+2,n} + \alpha_{i+1}\,\theta_{i+1,n} + \alpha_i\,\theta_{i,n} + \alpha_{i-1}\,\theta_{i-1,n} + \alpha_{i-2}\,\theta_{i-2,n}) + \frac{\Delta t}{2}\left(\frac{\partial^2\theta}{\partial t^2}\right)_{i,n} - \frac{\beta_5}{60\beta_2}\left(\frac{\partial^5\theta}{\partial x^5}\right)_{i,n} + \dots, \tag{6.1}$$

where the α's are determined by Equation (2.18) and the β's by Equation (2.19). Equation (6.1) holds for i = 2, 3, ..., I-2; the boundary conditions provide values for $\theta_{0,n}$ and $\theta_{I,n}$. To compute $\theta_{1,n}$ and $\theta_{I-1,n}$, skewed bases and appropriate representations of the second derivative (see Section 2) must be used. Using the basis x_0, x_1, x_2, x_3, x_4, a skewed representation of $\partial^2\theta/\partial x^2$ can be developed and used in the explicit scheme to calculate $\theta_{1,n}$. The five-point scheme, Equation (6.1), is readily converted to the three-point scheme by setting $\alpha_{i+2} = \alpha_{i-2} = 0$, altering the definitions of the α's and β's and modifying the truncation error. The stability analysis of Equation (6.1) closely follows that for the three-point scheme. Rearrangement of Equation (6.1) yields

$$\theta_{i,n+1} = \theta_{i,n} + \Gamma\,(\alpha_{i+2}\,\theta_{i+2,n} + \alpha_{i+1}\,\theta_{i+1,n} + \alpha_i\,\theta_{i,n} + \alpha_{i-1}\,\theta_{i-1,n} + \alpha_{i-2}\,\theta_{i-2,n})\,, \qquad (6.2)$$

where the truncation error has been omitted, $\Gamma = 2\Delta t\, D/\beta_2$, the α's are determined from Equation (2.18), and β_2 is determined from Equation (2.19). Substitution of Equation (3.7) in (6.2) yields the von Neumann amplification factor

$$G = 1 + \alpha_i\,\Gamma + \Gamma\,(\alpha_{i+2} + \alpha_{i-2})\cos(2\phi) + \Gamma\,(\alpha_{i+1} + \alpha_{i-1})\cos(\phi) + i\Gamma\,((\alpha_{i+2} + \alpha_{i-2})\sin(2\phi) + (\alpha_{i+1} - \alpha_{i-1})\sin(\phi))\,. \qquad (6.3)$$

Note that the gain G is complex in general, but is real if $\alpha_{i+2} = \alpha_{i-2}$ and $\alpha_{i+1} = \alpha_{i-1}$, which applies when the grid is uniform.
The condition for stability, namely $|G|^2 - 1 \leqslant 0$, must hold for all values of the phase angle ϕ. From Equation (6.3), we then have

$$\begin{aligned}|G|^2 = {} & 1 + 2\alpha_i\,\Gamma + \alpha_i^2\,\Gamma^2 + 2\Gamma\,(\alpha_{i+2} + \alpha_{i-2})\cos(2\phi) + 2\Gamma\,(\alpha_{i+1} + \alpha_{i-1})\cos(\phi) \\ & + 2\alpha_i\,\Gamma^2\,(\alpha_{i+2} + \alpha_{i-2})\cos(2\phi) + 2\alpha_i\,\Gamma\,(\alpha_{i+1} + \alpha_{i-1})\cos(\phi) \\ & + \Gamma^2\,(\alpha_{i+2} + \alpha_{i-2})^2\cos^2(2\phi) + 2\Gamma^2\,(\alpha_{i+2} + \alpha_{i-2})(\alpha_{i+1} + \alpha_{i-1}). \\ & \cos(2\phi)\cos(\phi) + \Gamma^2\,(\alpha_{i+1} + \alpha_{i-1})^2\cos^2(\phi) + \Gamma^2\,(\alpha_{i+2} - \alpha_{i-2})^2. \\ & \sin^2(2\phi) + 2\Gamma^2\,(\alpha_{i+2} - \alpha_{i-2})(\alpha_{i+1} - \alpha_{i-1})\sin(\phi)\sin(2\phi) \\ & + \Gamma^2\,(\alpha_{i+1} - \alpha_{i-1})^2\sin^2(\phi)\,. \end{aligned} \qquad (6.4)$$

Using the relations

$$(\alpha_{i+2} - \alpha_{i-2})^2 = (\alpha_{i+2} + \alpha_{i-2})^2 - 4\alpha_{i+2}\,\alpha_{i-2}\,,$$

$$(\alpha_{i+1} - \alpha_{i-1})^2 = (\alpha_{i+1} + \alpha_{i-1})^2 - 4\alpha_{i+1}\,\alpha_{i-1}\,,$$

$$(\alpha_{i+1} - \alpha_{i-1})(\alpha_{i+2} - \alpha_{i-2}) = (\alpha_{i+1} + \alpha_{i-1})(\alpha_{i+2} + \alpha_{i-2}) - 2\alpha_{i+2}\,\alpha_{i-1} - 2\alpha_{i+1}\,\alpha_{i-2}\,,$$

$$\cos(\phi) = \cos(\phi)\cos(2\phi) + \sin(\phi)\sin(2\phi)\,,$$

this expression for $|G|^2$ reduces to

$$\begin{aligned}|G|^2 = {} & 1 + 2\alpha_i\,\Gamma + \alpha_i^2\,\Gamma^2 + 2\Gamma\,(\alpha_{i+2} + \alpha_{i-2})\cos(2\phi) + 2\Gamma\,(\alpha_{i+1} + \alpha_{i-1})\cos(\phi) \\ & + 2\alpha_i\,\Gamma^2\,(\alpha_{i+2} + \alpha_{i-2})\cos(2\phi) + 2\alpha_i\,\Gamma^2\,(\alpha_{i+1} + \alpha_{i-1})\cos(\phi) \\ & + \Gamma^2\,(\alpha_{i+2} + \alpha_{i-2})^2 + \Gamma^2\,(\alpha_{i+1} + \alpha_{i-1})^2 + 2\Gamma^2\cos(\phi)(\alpha_{i+2} + \alpha_{i-2}). \\ & (\alpha_{i+1} + \alpha_{i-1}) - 4\Gamma^2\,\alpha_{i+2}\,\alpha_{i-2}\sin^2(2\phi) - 4\Gamma^2\,\alpha_{i+1}\,\alpha_{i-1}\sin^2(\phi) \\ & - 4\Gamma^2\sin(\phi)\sin(2\phi)\,\alpha_{i-1}\,\alpha_{i+2} - 4\Gamma^2\sin(\phi)\sin(2\phi)\,\alpha_{i+1}\,\alpha_{i-2}\,. \end{aligned} \qquad (6.5)$$

However, since the α's satisfy the consistency relation (Equation (2.16)), then

$$\alpha_i^2 = (\alpha_{i+2} + \alpha_{i-2})^2 + (\alpha_{i+1} + \alpha_{i-1})^2 + 2\,(\alpha_{i+2} + \alpha_{i-2})(\alpha_{i+1} + \alpha_{i-1})\,.$$

On substitution into Equation (6.5), we obtain

$$|G|^2 = 1 + H\,(\alpha_{i+2}, \alpha_{i+1}, \alpha_{i-1}, \alpha_{i-2}, \Gamma, \phi)\,, \qquad (6.6)$$

where

$$H = 2\Gamma(\alpha_{i+1} + \alpha_{i-1})(\cos(\phi) - 1) + 2\Gamma(\alpha_{i+2} + \alpha_{i-2})(\cos(2\phi) - 1)$$
$$+ 2\Gamma^2(\alpha_{i+1} + \alpha_{i-1})^2(1 - \cos(\phi)) + 2\Gamma^2(\alpha_{i+2} + \alpha_{i-2})^2(1 - \cos(2\phi))$$
$$+ 2\Gamma^2(\alpha_{i+2} + \alpha_{i-2})(\alpha_{i+1} + \alpha_{i-1})(1 - \cos(2\phi)) - 4\Gamma^2\alpha_{i+2}\alpha_{i-2}\sin^2(2\phi)$$
$$- 4\Gamma^2\alpha_{i+1}\alpha_{i-1}\sin^2(\phi) - 4\Gamma^2\sin(\phi)\sin(2\phi)(\alpha_{i-1}\alpha_{i+2}$$
$$+ \alpha_{i+1}\alpha_{i-2}) \, . \qquad (6.7)$$

For stability, we seek conditions on α_{i+2}, α_{i+1}, α_{i-1}, α_{i-2} and Γ, such that $H \leqslant 0$, for all $\phi \in [0,2\pi]$. Using obvious trigonometric identities, then it can be shown that

$$H = (\cos\phi - 1)[2\Gamma(\alpha_{i+1} + \alpha_{i-1}) - 2\Gamma^2(\alpha_{i+1} + \alpha_{i-1})^2 + 2(\cos\phi + 1)\{2\Gamma(\alpha_{i+2}$$
$$+ \alpha_{i-2}) - 2\Gamma^2(\alpha_{i+2} + \alpha_{i-2})^2 - 2\Gamma^2(\alpha_{i+1} + \alpha_{i-1})(\alpha_{i+2} + \alpha_{i-2})\}$$
$$+ 16\Gamma^2\alpha_{i+2}\alpha_{i-2}\cos^2\phi(1 + \cos\phi) + 4\Gamma^2\alpha_{i+1}\alpha_{i-1}(1 + \cos\phi)$$
$$+ 8\Gamma^2(\alpha_{i+2}\alpha_{i-1} + \alpha_{i-2}\alpha_{i+1})\cos\phi(1 + \cos\phi)] \, .$$

Since $\cos\phi - 1 \leqslant 0$ for all ϕ, the condition $H \leqslant 0$ can be replaced by

$$(1 + \cos\phi)[-4\Gamma^2(\alpha_{i+2} + \alpha_{i-2})(\alpha_{i+2} + \alpha_{i+1} + \alpha_{i-1} + \alpha_{i-2}) + 4\Gamma^2\alpha_{i+1}\alpha_{i-1}$$
$$+ 16\Gamma^2\alpha_{i+2}\alpha_{i-2}\cos^2\phi] \geqslant 2\Gamma(\alpha_{i+1} + \alpha_{i-1})(\Gamma(\alpha_{i+1} + \alpha_{i-1}) - 1)$$
$$- 4\Gamma(\alpha_{i+2} + \alpha_{i-2})(1 + \cos\phi) - 8\Gamma^2(\alpha_{i-2}\alpha_{i+1} + \alpha_{i+2}\alpha_{i-1}) \cdot$$
$$\cos\phi(1 + \cos\phi) \, , \qquad (6.8)$$

which must hold for all ϕ. Equation (6.8) can be written in the form

$$0 \geqslant \Gamma^2\psi - \Gamma\chi \, , \qquad (6.9)$$

where

$$\psi = 2(\alpha_{i+1} + \alpha_{i-1})^2 + 4(1 + \cos\phi)(\alpha_{i+2} + \alpha_{i-2})^2 + 4(1 + \cos\phi) \cdot$$
$$(\alpha_{i+1} + \alpha_{i-1})(\alpha_{i+2} + \alpha_{i-2}) - 16\cos^2\phi(1 + \cos\phi)\alpha_{i+2}\alpha_{i-2}$$
$$- 4(1 + \cos\phi)\alpha_{i+1}\alpha_{i-1} - 8(1 + \cos\phi)\cos\phi(\alpha_{i-2}\alpha_{i+1} + \alpha_{i+2}\alpha_{i-1}) \, ,$$

$$\chi = 2(\alpha_{i+1} + \alpha_{i-1}) + 4(1 + \cos\phi)(\alpha_{i+2} + \alpha_{i-2}) \, .$$

The zeros of the right side of Equation (6.9) are $\Gamma = 0$, $\Gamma = \Gamma_1 = \chi/\psi$, and we note that the signs of the α's, β_2 and Γ are dependent on the nature of the grid spacing selected, and on position in the grid. Further, for a uniform grid, the α's and β_2 given by Equation (2.20) are not all of the same sign, and lead to a negative value of Γ. Thus, the four independent quantities Δx_{i-2}, Δx_{i-1}, Δx_i and Δx_{i+1} govern the values of the α's and hence the quantities χ and ψ. For the appropriate grid spacings, Equation (6.9) indicates either (a) no stable region, (b) a stable region to the left or right of the origin for Γ (the choice of sign will depend on the sign of β_2), or (c) a stable region between some Γ^* and positive or negative infinity, depending on the sign of β_2.

These sign changes do not occur for the three-point formulae (see Equations (2.13) and (2.14)), so the analysis of the five-point form is more difficult. In this case, we must check the signs of χ and ψ for all values of ϕ, to see if a stable region exists, and also determine $\mathrm{Min}_\phi|\chi/\psi|$, where the signs of the various coefficients depend, in a complicated manner, on the independent grid spacings Δx_i, $i = 0, 1, \ldots, I-1$.

The stability properties have not been completely established, but a procedure has been developed which provides a lower bound on the stable region, if it exists. This method will be described for the sign structure in the uniform grid, i.e. in Equation (2.20). Since Equations (2.18) are continuous in the Δx_i, there will be variable grids with this sign structure. For the uniform grid, Equations (2.20) show that α_{i+2}, α_i and α_{i-2} are positive, α_{i+1}, α_{i-1} and β_2 are negative, and hence, that the corresponding Γ is also negative.

Thus, the left-hand side of the inequality (6.8) is non-negative for all ϕ, and the stability condition reduces to

$$2\Gamma(\alpha_{i+1}+\alpha_{i-1})(\Gamma(\alpha_{i+1}+\alpha_{i-1})-1)+4(-\Gamma)(\alpha_{i+2}+\alpha_{i-2})(1+\cos\phi)$$
$$+8\Gamma^2(\alpha_{i-2}(-\alpha_{i+1})+\alpha_{i+2}(-\alpha_{i-1}))\cos\phi\,(1+\cos\phi)\leqslant 0\ . \qquad (6.11)$$

Using obvious trigonometric identities, we obtain from Equation (6.11) the conditions

$$\Gamma^2(2(\alpha_{i+1}+\alpha_{i-1})^2+16(\alpha_{i+2}|\alpha_{i-1}|+\alpha_{i-2}|\alpha_{i+1}|))-\Gamma(2(\alpha_{i+1}+\alpha_{i-1})$$
$$+8(\alpha_{i+2}+\alpha_{i-2}))\leqslant 0\ . \qquad (6.12)$$

Since the coefficient of Γ^2 is positive in (6.12), we require

$$\alpha_{i+1}+\alpha_{i-1}<-4(\alpha_{i+2}+\alpha_{i-2})\ , \qquad (6.13)$$

to obtain a range of values of Γ for stability. Thus, the stability condition is

$$0\geqslant\Gamma\geqslant\Gamma_1=\frac{\alpha_{i+1}+\alpha_{i-1}+4(\alpha_{i+2}+\alpha_{i-2})}{(\alpha_{i+1}+\alpha_{i-1})^2+8(\alpha_{i+2}|\alpha_{i-1}|+\alpha_{i-2}|\alpha_{i+1}|)}\ , \qquad (6.14)$$

subject to Equation (6.13). Using the equally-spaced grid and the α's of Equation (2.20) yields $\Gamma_1 = -0.0015625h^{-6}$. Therefore we obtain the condition

$$0\leqslant\frac{D\Delta t}{h^2}\leqslant 0.225\ . \qquad (6.15)$$

The constant 0.225 in Equation (6.15) is a lower bound on the stable region for the uniform grid; this arises from some of the approximations used. The precise limit for the uniform grid is readily obtained from Equation (6.3) using the observation that G is real in the case of a uniform grid. Then the stability condition becomes

$$-1\leqslant G\leqslant 1\ . \qquad (6.16)$$

Using the consistency relation, the upper limit yields, since $\Gamma < 0$,

$$2(\alpha_{i+2}+\alpha_{i-2})(\cos\phi+1)+(\alpha_{i+1}+\alpha_{i-1})\leqslant 0\ ,$$

which must hold for all ϕ. Hence,

$$\alpha_{i+1}+\alpha_{i-1}\leqslant -4(\alpha_{i+2}+\alpha_{i-2})\ .$$

The lower limit in Equation (6.16) yields

$$(\cos 2\phi-1)(\alpha_{i+2}+\alpha_{i-2})+(\cos\phi-1)(\alpha_{i+1}+\alpha_{i-1})\leqslant -2/\Gamma\ . \qquad (6.17)$$

The left side of Equation (6.17) is a function of ϕ and the zeros of its derivative occur at $\phi = 0, \pi, \ldots$, and at the roots of

$$\cos\phi=-4\left(\frac{\alpha_{i+1}+\alpha_{i-1}}{\alpha_{i+2}+\alpha_{i-2}}\right)\ .$$

The condition in Equation (6.13) precludes a real root of this equation, and the left-hand side of Equation (6.17) attains its maximum at $\phi = \pi, 3\pi, \ldots$. Thus,

$$-2(\alpha_{i+1}+\alpha_{i-1})\geqslant -2/\Gamma\ ,$$

which yields the condition

$$0 < \frac{D\Delta t}{h^2} \leqslant 0.375 \ .$$

The analysis of the general case for a variable grid follows the procedure described above. The property used to handle Equation (6.8) is the nature of the signs of the terms in Equation (6.7), i.e. the terms with positive coefficients of $(1 + \cos \phi)$ are grouped on the left-hand side in Equation (6.8). Equation (6.7) then contains only linear or quadratic terms in Γ. Thus, for any given distribution of Δx_i values, the α's and β's are obtained from Equation (2.20). These values are then used to determine the signs of the terms containing $(1 + \cos \phi)$ in Equation (6.7), and the positive terms are collected on the left side. This yields a quadratic inequality for Γ which is readily solved to obtain the local stability criteria.

DISCUSSION

The use of a variable grid for the numerical solution of partial differential equations adds a new degree of freedom which can be utilised in the treatment of physical problems. Thus we are free to locate the grid points to satisfy some solution strategy, such as increasing the density of grid points in regions of interest. Various techniques can be used to locate grid points, such as considering the solution as a spline and locating the knots to provide the best spline representation (Haverkamp et al., 1981). The Kappa-method and the gradient method offer alternative techniques for the construction of the variable grid. However, difficulties arise from the need to carefully fit variable grids smoothly to the domain of the problem. It has been found that slight imperfections in this regard produce inaccurate results through induced errors, and instability results usually at the point of greatest diffusivity.

The penalties of using a variable-grid spacing include the change in the order of the truncation error for the second derivative, and the need to use the smallest grid spacing in the appropriate stability criteria. In a consideration of the bounds on the local truncation error, the effects are two-fold; the order of the truncation error is reduced, and its magnitude along the grid is affected by the changing grid spacing, as well as a higher-order spatial-derivative term. It may be possible to balance these effects as a strategy for selecting the grid spacings. Non-uniform grid spacings are generally used when the solution, and its derivatives, are changing rapidly, which may require grid spacings to change by several orders of magnitude. Since the relevant stability criterion has to hold for all points on the variable grid, including those of smallest spacing, very small time steps may be needed to retain numerical stability, resulting in large increases in computational time. These drawbacks are apparent in results obtained by solving the nonlinear diffusion equation using the methods described. The results are uniformly poor in the explicit scheme with an equi-spaced grid, but they are better than those obtained using a variable-grid scheme. This lack of accuracy is caused by the form of the local truncation error, and any errors in calculating the new grids.

The stability analysis for the variable grid is complicated by the nature of the amplification factor G, which is real for a uniform grid and complex for a variable grid. In the three-point schemes, precise stability limits have been obtained for both the explicit schemes described. In the corresponding five-point schemes, the analysis depends on the signs of the various parameters in the difference representations, and ultimately on the grid spacings selected. The analysis is not complete, but conditions sufficient to guarantee stability have been obtained. These are complicated and require a point-by-point check along the entire grid at each time level.

REFERENCES

[1] Ames, W.E., Numerical solution of partial differential equations (1977).

[2] Anthes, R.A., Numerical experiments with a two dimensional horizontal variable grid, Monthly Weather Review, Vol. 98, No. 11 (1970), pp. 810.

[3] Braddock, R.D., Parlange, J.-Y. and Lisle, I.G., Properties of the sorptivity for exponential diffusivity, Soil Sci. Soc. Am. J. 45 (1981), pp. 705.

[4] Chong, T.H., A variable mesh finite difference method for solving a class of parabolic differential equations in one space variables, Siam, J. Num. Anal. 15 (1978), pp. 835.

[5] Crowder, H.J. and Dalton, C., Errors in the use of non-uniform mesh systems, J. Compt. Phys. 7 (1971), pp. 32-45.

[6] Dane, J.H. and Mathis, F.H., An adaptive finite difference scheme for the one dimensional water flow equation, Soil Sci. Soc. Am. J. 45 (1981), pp. 1048.

[7] Denny, V.E. and Landis, R.B., A new method for solving two point boundary value problems using optimal distributions, J. Compt. Phys. 9 (1972), pp. 120.

[8] Greenberg, D.A., Mathematical Studies of Tidal Behaviour in the Bay of Fundy, Ph.D. Thesis, University of Liverpool (1975).

[9] Noye, B.J., An introduction to finite difference techniques, in: Noye, J. (ed.), Numerical Simulation of Fluid Motion (North-Holland, Amsterdam, 1978), pp. 2-113.

[10] Noye, B.J., Finite difference techniques for partial differential equations, in: Noye, J. (ed.), Computational Techniques for Differential Equations (North-Holland, Amsterdam, 1983), pp. 95-354.

[11] Parlange, J.-Y., Lockington, D.A. and Braddock, R.D., Nonlinear diffusion in a finite layer, Bull. Aust. Math. Soc. 26 (1982), pp. 249-262.

[12] Sundquist, H. and Veronis, G., A simple finite difference grid with non constant intervals, Tellus, Vol. 22 (1970), pp. 26.

[13] Trapp, J.A. and Ramshaw, J.D., A simple heuristic method for analysing the effect of boundary conditions on numerical stability, J. Compt. Phys. 20 (1976), pp. 238-242.

Computational Techniques & Applications: CTAC-83
J. Noye & C. Fletcher (Editors)

THE NUMERICAL SOLUTION OF PARTIAL DIFFERENTIAL EQUATIONS USING THE METHOD OF LINES

N.G. Barton

C.S.I.R.O. Division of Mathematics and Statistics,
P.O. Box 218, Lindfield, Australia 2070

In a recent review of software for the numerical solution of partial differential equations (PDEs), Machura & Sweet (1980) assert that the method of lines is the most popular technique for solving systems of time-dependent PDEs. A number of robust and user-friendly software interfaces are now available to implement the method, and this note describes the application of one such interface, PDEONE (Sincovec & Madsen, 1975) to two consulting problems. Overall, the results support Sincovec & Madsen's claim that PDEONE is useful and reliable, and eliminates much of the expensive and time consuming human effort involved in the numerical solution of PDEs.

INTRODUCTION

The numerical solution of partial differential equations (PDEs) is a bread-and-butter issue in applied mathematics and most branches of the physical and engineering sciences. In the author's experience, about half the PDEs from consulting applications which require a numerical solution are (possibly systems of) parabolic PDEs for which the method of lines is ideally suited. The utility of this method derives from highly developed and robust software for the numerical solution of ordinary differential equations (ODEs). A number of software packages (as described in Section 12 of the survey by Machura & Sweet, 1980, on software for PDEs) have been developed since the mid 1970s to act as an interface between the PDE requiring solution and the ODE integrators.

The application of PDEONE, one of the earliest of these software interfaces, to two consulting problems is described in the present contribution. The first problem is a nonlinear heat generation and conduction problem arising from transmission line theory, and is posed by one quasi-linear parabolic PDE under a non-linear boundary condition at a cylindrical surface. The other describes the transport of two species of ions through reactive soil. Computationally, it is a good test of a numerical method as it involves coupled parabolic quasi-linear PDEs with strong convective and weak diffusive terms. In an Appendix, PDEONE is also shown to work reasonably well for a model hyperbolic PDE.

In both consulting problems, numerical solutions computed using PDEONE are presented and the particular claims of PDEONE - namely portability, robustness and ease of programming - are confirmed. The author's belief is that the combination of PDEONE and GEARB (a standard ODE integrator modified for systems with a banded Jacobian) is highly successful and deserving a place in every applied mathematician's tool-bag.

THE METHOD OF LINES

For the purposes of exposition, consider the application of the method of lines to the single quasi-linear parabolic PDE

$$u_t = f(x,t,u,u_x,u_{xx}), \quad 0 < x < 1, \quad t > 0 \tag{1a}$$

under boundary conditions

$$\alpha u + \beta u_x = \gamma \quad \text{at} \quad x = 0,\ 1 \tag{1b}$$

and the initial condition

$$u(x,0) = g(x). \tag{1c}$$

This problem may be discretized on the mesh

$$x_i = ih, \quad i = 0,1,2,\ldots,N+1, \quad h = 1/(N+1)$$

so that if $v_i(t)$ is an approximation to $u(x_i,t)$ and the spatial derivatives in (1a) are replaced by second order central difference approximations, the $v_i(t)$ satisfy the system of ODEs

$$\frac{dv_i}{dt} = f\left(x_i,t,v_i,\frac{v_{i+1}-v_{i-1}}{2h},\frac{v_{i-1}-2v_i+v_{i+1}}{h^2}\right) \tag{2a}$$

$$= F_i(t,v_{i-1},v_i,v_{i+1})$$

for i = 1,2,...,N. ODEs for $v_0(t)$ and $v_{N+1}(t)$ may be obtained by using (1b) and, if necessary, one-sided difference approximations for u_x, u_{xx} at the boundaries x=0,1. This gives the additional ODEs

$$\frac{dv_0}{dt} = F_0(t,v_0,v_1,v_2), \qquad \frac{dv_{N+1}}{dt} = F_{N+1}(t,v_{N-1},v_N,v_{N+1}). \tag{2b}$$

The system of ODEs (2a,2b) has a banded structure and can be efficiently solved subject to the initial condition

$$v_i(0) = g(x_i) \tag{2c}$$

using a standard ODE integrator modified for banded systems. The usual ODE package for this task appears to be GEARB (Hindmarsh, 1977).

There are, of course, many embellishments on the above theme. For example, higher order or upwind differencing could be used. Also, the spatial variation could be described by writing

$$u(x,t) = \sum_{i=0}^{N+1} C_i(t)b_i(x) \tag{3}$$

where the $\{b_i(x)\}_{i=0}^{N+1}$ are a suitable basis (e.g. Fourier series, orthogonal polynomials or B-splines). ODEs for the $\{C_i(t)\}_{i=0}^{N+1}$ are then obtained by arithmetizing the spatial variation using some form of the method of weighted residuals (e.g. collocation, Galerkin, least squares, etc.).

A survey of nine software interfaces written since 1975 for the method of lines is given by Machura & Sweet (1980, section 12). The present work is concerned with one of the earliest and possibly the best known of these - PDEONE (Sincovec & Madsen, 1975). PDEONE follows the basic structure for discretization outlined in this section as applied to the system of NPDE coupled parabolic PDEs

$$\frac{\partial u_k}{\partial t} = f_k(t,x,\underset{\sim}{u},\frac{\partial \underset{\sim}{u}}{\partial x},x^{-c}\frac{\partial}{\partial x}(x^c\underset{\sim}{D}_k.\frac{\partial \underset{\sim}{u}}{\partial x})) \tag{4a}$$

$$(a < x < b, \quad t > t_0, \quad k = 1,2, \ldots,\text{NPDE}),$$

$$\underset{\sim}{u} = [u_1 \; u_2 \; \ldots \; u_{\text{NPDE}}]^T,$$

subject to boundary conditions

$$\alpha_k u_k + \beta_k \frac{\partial u_k}{\partial x} = \gamma_k \quad \text{at} \quad x = a,b, \quad t > t_0, \quad k = 1,2,\ldots,\text{NPDE}, \tag{4b}$$

and initial condition

$$\underset{\sim}{u}(x,0) = \underset{\sim}{g}(x), \; a \leqslant x \leqslant b. \tag{4c}$$

The constant c in (4a) is 0,1 or 2 depending on whether the problem is posed in Cartesian, cylindrical or spherical co-ordinates respectively. A complete algorithm for PDEONE consisting of only 153 lines of code is freely available (Sincovec & Madsen, 1975). The user merely has to write a driver program and three subroutines defining the boundary conditions (4b), the diffusion coefficients $\underset{\sim}{D}_k$ and the function $\underset{\sim}{f}$ in (4a). PDEONE then performs the discretization - on a variable mesh if desired - and GEARB solves the resulting ODE initial value problem with automatic timestep control to achieve a specified accuracy. Both PDEONE and GEARB are written in FORTRAN and the computations described subsequently in this note were performed in single precision on a VAX 11/750.

TRANSMISSION LINE PROBLEM

The first problem discussed is the quasi-linear parabolic PDE

$$\gamma(\theta)c(\theta)\frac{\partial \theta}{\partial t} = \frac{1}{r}\frac{\partial}{\partial r}(\lambda(\theta)r\frac{\partial \theta}{\partial r}) + Q(r,\theta) \tag{5a}$$

for the temperature $\theta(r,t)$ in a solid metal conductor carrying an electric current. Here the temperature dependent coefficients are given by

$$\lambda(\theta) = \lambda_0(1 + \kappa\theta) \quad \text{(thermal conductivity, W m}^{-1}\text{ K}^{-1}\text{)},$$

$$\gamma(\theta) = \gamma_0(1 + \delta\theta) \quad \text{(density, kg m}^{-3}\text{)}, \tag{5b}$$

$$c(\theta) = c_0(1 + \beta\theta) \quad \text{(specific heat, J kg}^{-1}\text{ K)},$$

and the problem is to be solved subject to the initial and boundary conditions

$$\theta(r,0) = 0, \tag{5c}$$

$$\theta_r(0,t) = 0, \qquad \theta_r(a,t) = -h\{\theta(a,t)\}^q. \tag{5d}$$

The heating function $Q(r,\theta)$ is defined by (Dwight, 1945, p.159)

$$Q(r,\theta) = |J(r,\theta)|^2\rho(\theta) \tag{5e}$$

where

$$J(r,\theta) = \frac{Im}{2\pi a}\frac{\text{ber } mr + i \text{ bei } mr}{\text{bei' } ma - i \text{ ber' } ma} \tag{5f}$$

and I is the total current (A), a is the outer radius (m), $m = \{2\pi f\mu\mu_0/\rho(\theta)\}^{1/2}$ (m^{-1}), f is the frequency (Hz), $\rho(\theta) = \rho_0(1 + \alpha\theta)$ is the temperature dependent resistivity (Ωm), μ is the relative permeability of the conductor (dimensionless), μ_0 is the permeability of free space (Ω sec m^{-1}), and {ber, bei} are the Kelvin functions (Abramowitz & Stegun, 1965, section 9.9).

The problem (5) required a numerical solution for a number of different conductors for various values of the parameters h and q in the boundary condition (5d). (This boundary condition is an empirical model of heat dissipation by convection under various ambient wind conditions as in Morgan, 1967). A separation of variables solution of problem (5) is possible if the temperature dependence is neglected in the coefficients (5b) and the heating (5c) <u>and</u> if the index q in (5d) is unity. This analytic solution served as one check on the program. A further check on the complete program with all temperature dependent terms retained was that the heat input should balance the surface dissipation in the ultimate steady state. In all applications, this balance was accurate to within 0.2%.

Numerical solutions of problem (5) were originally found using an explicit forward time centre space (FTCS) method. The major difficulties were coding the heating function $Q(r,\theta)$, and the slow execution time of the program due to the stability requirement which typically takes the form (see e.g. Roache, 1976, p.39).

$$\Delta t \leqslant (\Delta r)^2/[2\lambda/\gamma c].$$

The FTCS solution served as a check on the subsequent solution using the method of lines.

Once the heating function $Q(\theta,t)$ was understood and coded, it took less than a day to design and implement a numerical solution of problem (5) using PDEONE. As expected, the execution speed of the program depended on the method chosen in the ODE integrator GEARB : with the best setting for the method flag (MF), the method of lines solution computed the initial temperature rise 40 times faster than the FTCS solution of comparable accuracy; whilst the worst setting for MF gave an execution speed about 4 times slower than the FTCS solution.

The execution speeds are summarized in Table 1, and typical values of the surface temperature are plotted in Figure 1 for various values of the index q in the boundary condition (5d). It should be noted that the execution speed of the method of lines solution for a given accuracy depends on the automatic timestep control available in GEARB, and the timesteps become bigger once the initial heating is over. This feature is not available in the FTCS solution.

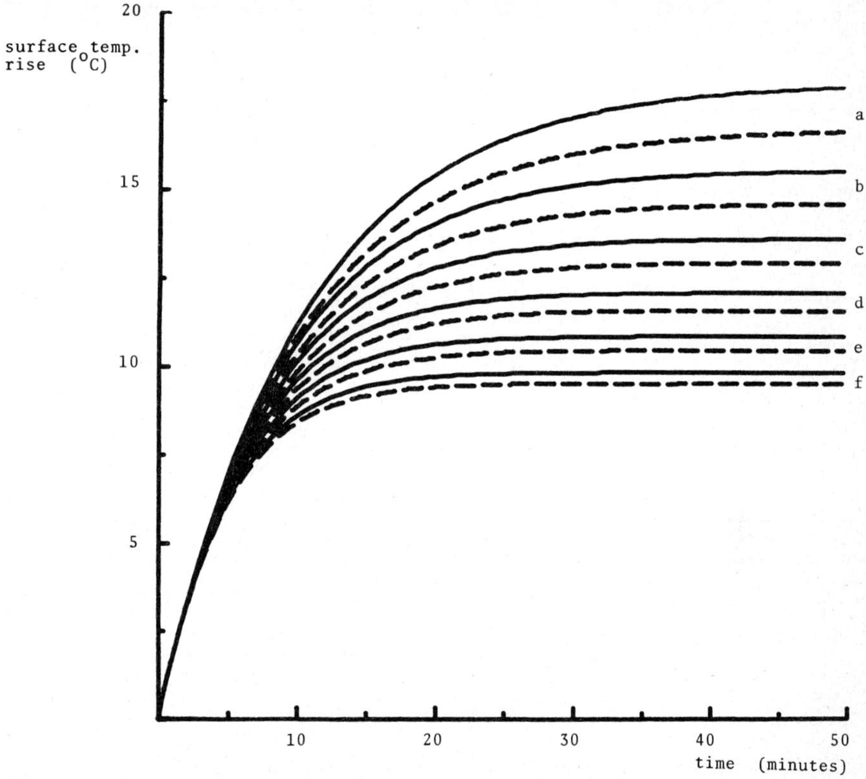

Figure 1. The surface temperature as a function of time in an aluminium conductor of radius 0.01m carrying a current of 500 A; number of radial points = 7; the constants (mksA units) are $\lambda_0 = 236$, $\kappa = 0.17 \times 10^{-3}$, $\gamma_0 = 2.7 \times 10^{3}$, $\delta = 0.72 \times 10^{-4}$, $c_0 = 860$, $\beta = 0.38 \times 10^{-3}$, $\alpha = 0.445 \times 10^{-2}$, $\rho_0 = 0.263 \times 10^{-7}$, $h = 20$; the dashed lines show the results for temperature independent coefficients ($\kappa = \delta = \beta = \alpha = 0$); (a) $q = 1.00$, (b) $q = 1.05$, (c) $q = 1.10$, (d) $q = 1.15$, (e) $q = 1.20$ (f) $q = 1.25$.

Table 1. Comparison of c.p.u. time for the numerical solution of problem (5) for the first 60 seconds heating in an aluminium conductor. Properties and constants as in the legend for Figure 1; ϵ is the error control setting on GEARB, MF denotes the method flag in GEARB

method	c.p.u. time (s)	timestep	ϵ	comments
FTCS	126.5	0.01		timestep for stability
MF=10	510.2	0.006–0.032	10^{-1}	less accurate than FTCS
MF=12	12.0	0.287–3.20	10^{-6}	superior accuracy to FTCS
MF=12	3.0	3.98–16.9	10^{-4}	superior accuracy to FTCS
MF=13	53.4	0.114–1.01	10^{-5}	comparable accuracy to FTCS
MF=20	502.0	0.005–0.034	10^{-1}	less accurate than FTCS
MF=22	3.5	3.99–9.83	10^{-4}	superior accuracy to FTCS
MF=23	24.4	0.037–4.32	10^{-5}	comparable accuracy to FTCS

SOLUTE TRANSPORT PROBLEM

The second problem discussed is posed by the coupled dimensionless quasi-linear parabolic PDEs

$$(\theta + \rho\frac{\partial s_1}{\partial c_1})\frac{\partial c_1}{\partial t} + \rho\frac{\partial s_1}{\partial c_2}\frac{\partial c_2}{\partial t} = \frac{\partial^2 c_1}{\partial x^2} - \frac{\partial c_1}{\partial x}, \tag{6a}$$

$$\rho\frac{\partial s_2}{\partial c_1}\frac{\partial c_1}{\partial t} + (\theta + \rho\frac{\partial s_2}{\partial c_2})\frac{\partial c_2}{\partial t} = D_2\frac{\partial^2 c_2}{\partial x^2} - \frac{\partial c_2}{\partial x}, \tag{6b}$$

which describe the concentration of two species of ions in a column of reactive soil through which water is flowing at a steady rate. In the equations, θ and D_2 are constants characteristic of the process, and $\rho s_1(c_1,c_2)$ and $\rho s_2(c_1,c_2)$ are instantaneous adsorption isotherms describing the amount of the ions adsorbed onto soil particles as a function of concentration. A description of the process is given by Charbeneau (1981) and Barnes & Aylmore (1983) reduce the general problem to the form given by (6a,b).

In the present application, the adsorption isotherms are taken to be

$$\rho s_1 = 0.1\, c_1\{1 + c_2 - \frac{1.25}{c_1+c_2}\}, \tag{7a}$$

$$\rho s_2 = -0.1\, c_2\{1 - c_1 - \frac{1.25}{c_1+c_2}\}, \tag{7b}$$

the boundary conditions are set to be

$$c_1(0,t) = 0.25, \quad c_2(0,t) = 0.75, \tag{7c}$$

$$c_{1_x}(x_\infty, t) = 0, \qquad c_{2_x}(x_\infty, t) = 0 \tag{7d}$$

(where the subscript x denotes a partial derivative), and the PDEs are to be solved subject to the initial conditions

$$c_1(x,0) = 0.25, \qquad c_2(x,0) = 0.25. \tag{7e}$$

Equations (6 a,b) need to be cast in the form of (4a) prior to using PDEONE, and this is achieved by forming suitable linear combinations of (6 a,b). Thus, the PDEs to be solved are found to be

$$\frac{\partial c_1}{\partial t} = \left[e_{11} + g_{11}\frac{\partial c_1}{\partial x} + g_{12}\frac{\partial c_2}{\partial x}\right]\frac{\partial c_1}{\partial x} + \left[e_{12} + g_{12}\frac{\partial c_1}{\partial_x} + g_{13}\frac{\partial c_2}{\partial x}\right]\frac{\partial c_2}{\partial x}$$

$$+ \frac{\partial}{\partial x}\left\{d_{11}\frac{\partial c_1}{\partial x} + d_{12}\frac{\partial c_2}{\partial x}\right\}, \tag{7f}$$

$$\frac{\partial c_2}{\partial t} = \left[e_{21} + g_{21}\frac{\partial c_1}{\partial x} + g_{22}\frac{\partial c_2}{\partial x}\right]\frac{\partial c_1}{\partial x} + \left[e_{22} + g_{22}\frac{\partial c_1}{\partial x} + g_{23}\frac{\partial c_2}{\partial x}\right]\frac{\partial c_2}{\partial x}$$

$$+ \frac{\partial}{\partial x}\left\{d_{21}\frac{\partial c_1}{\partial x} + d_{22}\frac{\partial c_2}{\partial x}\right\} \tag{7g}$$

where the coefficient functions $\underset{\sim}{e}$, $\underset{\sim}{g}$, $\underset{\sim}{d}$ involve lengthy but straightforward derivatives of S_1 and S_2 and the details are omitted.

The problem posed by equations (7) was solved numerically using PDEONE and GEARB. It took about two days to perform the preliminary analysis to arrive at equations (7f,g), and then another couple of days to write, test and debug the program. A test on the program was afforded by considering the associated problem for $c_0 = c_1 + c_2$ under the definitions (7a,b) for the adsorption isotherms: if D_2 is unity, c_0 satisfies

$$\alpha\frac{\partial c_0}{\partial t} = \frac{\partial^2 c_0}{\partial x^2} - \frac{\partial c_0}{\partial x} \qquad (\alpha \text{ constant}, \ 0 < x < x_\infty, \ t > 0), \tag{8a}$$

with

$$c_0(0,t) = 1.0, \qquad c_{0_x}(x_\infty, t) = 0, \tag{8b}$$

$$c_0(x,0) = 0.5. \tag{8c}$$

This problem for C_0 possesses an analytic solution (van Genuchten & Alves, 1981) which was to check the numerical method.

Experimentally, measurements of C_1 and C_2 would be made at various times at fixed $x = L \leqslant x_\infty$ in the belief that the results are not sensitive to the choice of x_∞ in (7d). These measurements are called breakthrough curves. The computer program is required to simulate the experiments, in particular to examine the effects of varying the adsorption isotherms and the boundary and initial conditions. In practice, it was found the results were quite insensitive to the value chosen for x_∞ and breakthrough curves for $x_\infty = 1.6L$ are shown in Figure 2. The execution speed of the program again depended on the choice of the method flag in GEARB as shown in Table 2.

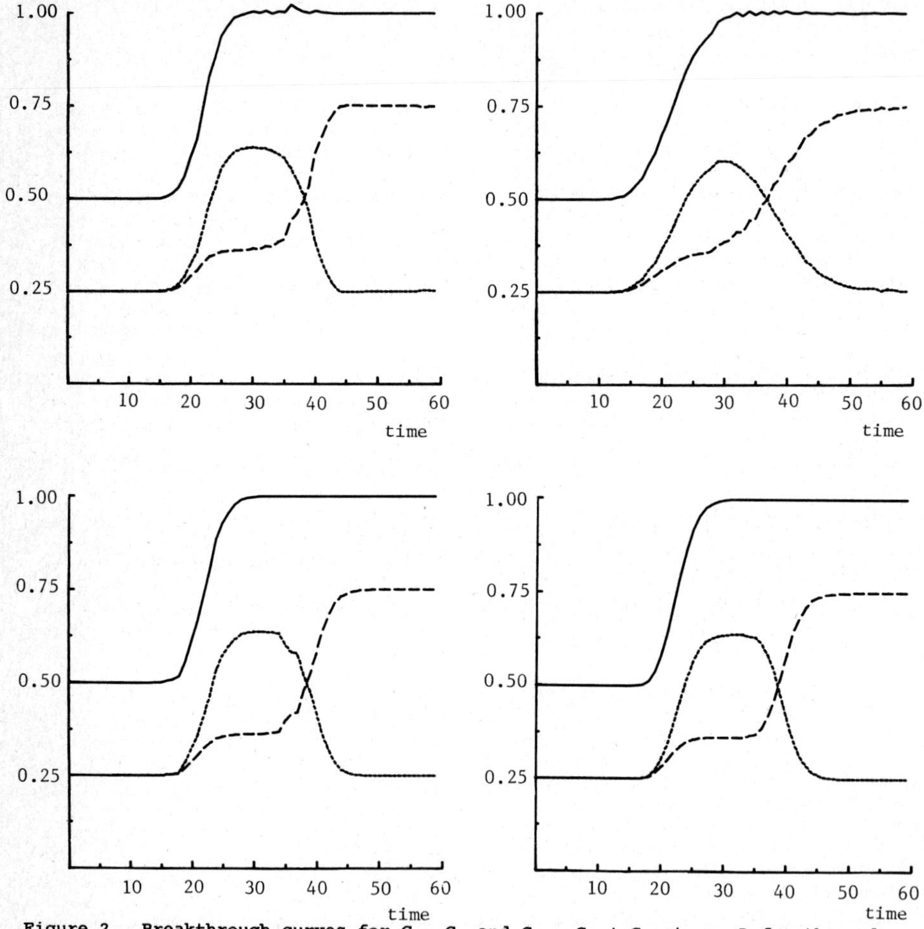

Figure 2. Breakthrough curves for C_1, C_2 and $C_0 = C_1 + C_2$ at $x = L$ for the solutions of problem (6) subject to conditions (7a–e), $\epsilon = 10^{-6}$, $\theta = 0.25$, $D_2 = 1.0$, $L = 150$, $x_\infty = 240$: ...... C_1, – – – – C_2, ——— C_0. Top left, NPTS=61, c.p.u. time 179 sec; top right, NPTS=61, first derivative in PDEONE calculated using upwind differences, c.p.u. time 206 sec; bottom left, NPTS=121, c.p.u. time 565 sec; bottom right, NPTS=241, c.p.u. time 1623 sec.

Table 2. Comparison of c.p.u. time taken for the first 5 seconds of a test numerical solution of problem (6) using 16 points and with $x_\infty = L = 7.5$, ϵ (error tolerance in GEARB) is 10^{-3}, $\theta = 0.25$, $D_2 = 1$, MF denotes the method flag in GEARB.

	non-stiff options			stiff options	
MF	c.p.u. time (sec)	timestep	MF	c.p.u. time (sec)	timestep
10	112.6	0.006–0.009	20	117.9	0.006–0.055
12	15.8	0.110–0.323	22	21.5	0.101–0.368
13	76.9	0.005–0.129	23	124.7	0.012–0.041

DISCUSSION

Sincovec & Madsen (1975) begin their general comments on PDEONE with "We have been using the numerical method of lines approach for solving PDEs for some time and in general have found the method to be quite powerful, reasonably efficient from a computer time point of view, and extremely versatile and easy to implement for most problems. ... When our interface is used with one of the recently developed stiff ODE integrators, we feel that one has a reasonably robust piece of software for obtaining numerical solutions for fairly broad classes of problems." My experiences with PDEONE supports these claims. The method is probably best suited to parabolic problems - which can include convection terms - although the following Appendix shows that it works adequately for hyperbolic problems as well.

The number of packages available for the method of lines has increased greatly since 1975 when Sincovec & Madsen wrote "Measuring the efficiency and effectiveness of general purpose software is extremely difficult if not impossible... Comparable software against which comparisons could be made is not readily (if at all) available". Since then, a number of packages based on the method of lines have been introduced. These have been summarized by Machura & Sweet (1980) who rank PDEONE as the second lowest in complexity and power of the nine mentioned. Other packages include PDETWO (Melgaard & Sincovec, 1976) for problems with two space dimensions, PDECOL (Madsen & Sincovec, 1976) where the spatial approximation is by B-splines, and FORSIM VI (Carver et al, 1978) with options for more sophisticated (e.g. upwind) differencing. These packages are designed to solve a different class of problems from PDEONE and a detailed comparison of their relative efficiency has not been attempted.

For computing efficiency, it is essential to exploit the structure of the system of ODEs by using an ODE integrator designed for banded systems. For robustness, it is also essential that the ODE integrator has a stiff option (even though the fastest execution speeds for the problems described above were obtained using non-stiff methods). On the efficiency of PDEONE, Sincovec & Madsen remark that "General purpose software such as PDEONE is typically aimed at reducing the 'human time' spent rather than computer time. Obviously, for any specific problem, one could write a more efficient program for its solution." Based on the author's experience with the transmission line problem, numerical solutions using PDEONE/GEARB are to be strongly preferred over explicit FTCS solutions. Moreover, in the transmission line problem, as with many problems, it was impractical to develop an implicit method of solution which would permit longer timestops. For such cases, it is hardly worthwhile trying to develop a better method of solution than PDEONE/GEARB.

In conclusion, the method of lines in general and PDEONE in particular have a place in every applied mathematician's inventory of methods for numerically solving PDEs. The method probably deserves wider publicity in Australia, despite the fact that local specialists have been using variations of the method - e.g. spectral methods - for years.

APPENDIX SOLUTION OF A HYPERBOLIC PDE

Hyperbolic PDEs are reckoned to be the most difficult of all to solve numerically and it was therefore decided to test PDEONE on the problem

$$\frac{\partial u}{\partial t} + \frac{\partial u}{\partial x} = 0, \qquad 0 < x < 10, \qquad t > 0 \tag{9a}$$

$$u(0,t) = 0, \qquad u(10,t) = 0, \tag{9b}$$

$$u(x,0) = \begin{cases} x & 0 < x \leqslant 2 \\ 2-x & 1 < x \leqslant 2 \\ 0 & x > 2 \end{cases} \tag{9c}$$

Here the condition at $x = 10$ in (9b) is a fictitous condition. Morrow (1982) has surveyed numerical methods for this problem and concluded that flux-corrected transport algorithms are the most accurate, albeit slightly troublesome to program. The true solution has the initial triangular shape propagating in the x direction without change, and some results obtained using PDEONE/GEARB are shown in Figure 3. It can be seen that a reasonable solution to this problem can be obtained using PDEONE, although there is some numerical diffusion evidenced at the peak and leading edge of the pulse and there is a number of spurious oscillations trailing the pulse. These features can be reduced but not eliminated by using more points. There was no need to introduce any artificial diffusion as recommended by Sincovec & Madsen (1975, section 6B) for a hyperbolic system possessing shock solutions.

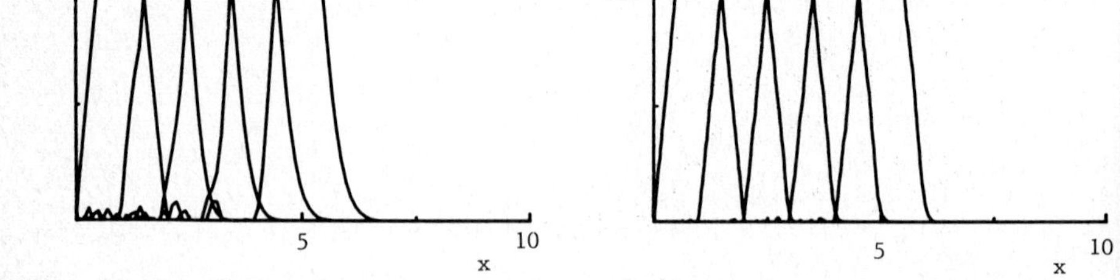

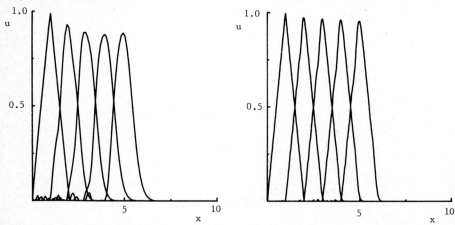

Figure 3. Numerical solution of the hyperbolic problem (9) using PDEONE/GEARB. Left: NPTS=100, $\epsilon = 10^{-4}$. Right: NPTS=500, $\epsilon = 10^{-6}$. The solutions shown are $u(x,t)$ for $t = 0.0(1.0)4.0$.

The results displayed in Figure 3 could presumably be improved using upwind differencing (as available in FORSIM VI, Carver et al, 1978), but it is expected that a flux-corrected algorithm would still be the most accurate for the problem. It is reassuring, however, that PDEONE still works adequately for problem (9) at very little expense of development and programming time.

The author would like to thank Dr J.H. Knight for his generous assistance with many aspects of this work.

REFERENCES

[1] Abramowitz, M. and Stegun, I.A., Handbook of mathematical functions (Dover, New York, 1965).

[2] Barnes, C.J. and Aylmore, L.A.G., A theoretical treatment of the effects of ionic and non-ionic competitive adsorption during solute transport in soils, Aust.J.Soil.Res. to appear.

[3] Carver, M. et al., The FORSIM VI simulation package for the automated solution of arbitrarily defined partial differential and/or ordinary differential equation systems. Rep. AECL 5821, Chalk River Nuclear Laboratories, Ontario, Canada, 1978.

[4] Charbeneau, R.J., Groundwater contaminant transport with adsorpton and ion exchange chemistry: method of characteristics for the case without dispersion, Water Resources Res. 17 (1981) 705-713.

[5] Dwight, H.B., Electrical coils and conductors (McGraw-Hill, New York, 1945).

[6] Hindmarsh, A.C., GEARB : solution of ordinary differential equations having banded Jacobian, Rep. UCID-30059, Lawrence Livermore Lab., Livermore, Calif., 1977.

[7] Machura, M. and Sweet, R.A., A survey of software for partial differential equations, ACM Trans. Math. Software 6 (1980) 461-488.

[8] Madsen, N.K. and Sincovec, R.F., PDECOL, general software for partial differential equations, ACM Trans Math Software 5 (1979) 326-351.

[9] Melgaard, D.K. and Sincovec, R.F., General software for two-dimensional nonlinear partial differential equations, ACM Trans Math Software 7 (1981) 106-125.

[10] Morgan, V.T., Rating of bare overhead conductors for continuous currents, Proc. IEE 114 (1967) 1473-1482.

[11] Morrow, R., A review of the methods of solution of hyperbolic equations for election and plasma motion in uniform and non-uniform electric fields, in Numerical solution of partial differential equations, J. Noye (ed.) (North Holland, 1982).

[12] Roache, P.J., Computational fluid mechanics (Hermosa, Albuquerque, 1976).

[13] Sincovec, R.F. and Madsen, N.K., Software for nonlinear partial differential equations, ACM Trans Math. Software 3 (1975) 232-260; PDEONE, solutions of systems of partial differential equations, ACM Trans Math. Software 3 (1975) 261-263.

[14] van Genuchten, M. Th. and Alves, W.J., Analytical solutions of the one-dimensional convective-dispersive solute transport equation, Tech. Bull. 1661, U.S. Dept. Agric. Res. Service, 1981.

Computational Techniques & Applications: CTAC-83
J. Noye & C. Fletcher (Editors)

ADE METHODS FOR THE SOLUTION OF TWO-DIMENSIONAL PROBLEMS IN COMPUTATIONAL SOLIDIFICATION

R.S. Anderssen and B.A. Ellem

CSIRO Division of Mathematics and Statistics,
GPO Box 1965, Canberra ACT, Australia.

In 1974, Fox examined in some detail "What are the best numerical methods?" for solving solidification problems. He advanced a number of independent reasons to support his conclusion that numerical methods should be constructed from the enthalpy (i.e. energy) formulation for solidification. Independently, a number of authors (e.g. Larkin (1964)) have explored the advantages of alternating direction explicit (ADE) methods for the solution of diffusion equations. In this paper, we review and examine the use of ADE methods for the solution of two-dimensional problems in computational solidification.

1. Introduction

Even with the latest instrumentation, it is technically difficult to monitor the thermal history of even a small simple casting, and economically unrealistic for large complex structures which are only ever manufactured in small numbers, such as large ships propellers and big bells. Even when experimental techniques are used the process is time-consuming and expensive. For example, a popular method for monitoring the progress of solidification in a casting is to destructively stop the solidification at regular intervals by breaking open the moulds, pouring out the remaining liquid and inspecting the current state of the solidification (cf Kotschi and Plutshack (1982)). What is required are models of the solidification process which can be solved computationally with the results being displayed graphically and interactively.

Physically, solidification is modelled as a process of heat conduction in an idealised material with assumed thermal properties in which a phase change (solidification) occurs at a specified temperature. A moving surface (solidification front) is assumed to exist with material in its solid phase on one side and liquid on the other. Even though the temperature is continuous throughout, the temperature gradient has a jump discontinuity across the solidification front due to the liberation of the latent heat of fusion. The essential features of the mathematical and computational problem are discussed in **2** for one-dimensional solidification. The alternating direction explicit (ADE) scheme is introduced in **3** and is applied to one-dimensional solidification. The mathematical model and algorithm used for solidification in a two dimensional rectilinear moulding configuration are given in 4. A discussion of numerical results, along with some conclusions, is contained in 5.

2. One-dimensional Solidification

2.1 The Heat Conduction Model

In a one-dimensional problem in which the material is solid for $x < X(t)$ and liquid for $x > X(t)$ the position of the solidification front is defined by

$$- k \left. \frac{\partial \theta}{\partial x} \right|_{\text{solid}}^{\text{liquid}} = \rho L \frac{dX}{dt} , \tag{1}$$

where θ is the temperature, k is the thermal conductivity, ρ is the density and L is the latent heat of fusion per unit mass. The liberation of the latent heat of fusion is the source of the non-linearity in a solidification problem. For such problems, the heat conduction equation takes the form

$$\rho C \frac{\partial \theta}{\partial t} = \frac{\partial}{\partial x} \left(k \frac{\partial \theta}{\partial x}\right) , \tag{2}$$

where C is the heat capacity. The condition (1) can be studied via the following one-dimensional problem.
A bar of length ℓ is heated to melting and is then allowed to cool at one end. The progress of the solidification front is shown in Figure 1.

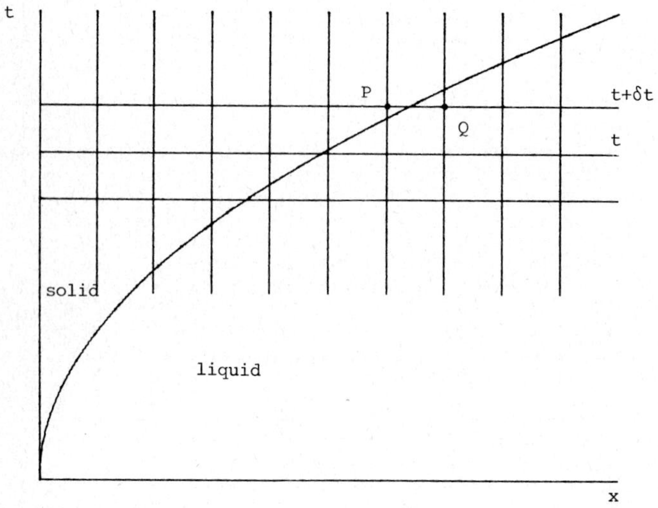

Figure 1. Position of the solidification front in one-dimensional solidification.

The obvious method that suggests itself is to solve the coupled linear system

$$\rho_1 c_1 \frac{\partial \theta_1}{\partial t} = \frac{\partial}{\partial x}\left(k_1 \frac{\partial \theta_1}{\partial x}\right), \quad \rho_2 c_2 \frac{\partial \theta_2}{\partial t} = \frac{\partial}{\partial x}\left(k_2 \frac{\partial \theta_2}{\partial x}\right), \tag{3}$$

where 1=solid and 2=liquid, which, in conjunction with (1), the interface condition

$$k_2 \frac{\partial \theta}{\partial x} - k_1 \frac{\partial \theta}{\partial x} = \rho_1 L \frac{ds(t)}{dt}, \qquad \theta_1 = \theta_2 \text{ at } x = s(t), \tag{4}$$

and appropriate boundary conditions, defines the solidification process in the bar. While these equations can be solved independently in the main part of the solid and liquid regions, problems arise at the solidification front in coupling them through the condition (4). At each time step the condition (4) must be used to advance the position of the solidification front. But this poses difficulties because the x-coordinate of the new position of the solidification front at the time $t+\delta t$ is not known in advance. Thus, it is not known in advance whether points like P and Q in Figure 1 will lie in the solid or the liquid region.

This shows that the construction of computational schemes which track the solidification front accurately across a finite difference mesh is difficult and awkward.

2.2 The Enthalpy (Energy) formulation

If h denotes the heat content per unit mass, the heat conduction equation (2) can be written as

$$\rho \frac{\partial h}{\partial t} = \frac{\partial}{\partial x}\left(k \frac{\partial \theta}{\partial x}\right), \quad 0 \leqslant x \leqslant \ell . \tag{5}$$

Because (5) corresponds to the energy form of the heat conduction equation, it will be valid throughout the whole of the region under investigation (solid and liquid). The non-linearity associated with the liberation of the latent head of fusion is now modelled directly as the temperature-enthalpy relation

$$\theta = F(h,x) . \tag{6}$$

A typical form of this relation is shown graphically in Figure 2. The advantage of this formulation is that the solidification front now corresponds to the "zero" temperature contour (isotherm) on the surface $\theta(x,t)$. This reformulation obviates the need to track the solidification front. Another advantage of the enthalpy formulation is that finite difference methods constructed to solve it yield approximations which converge to the weak solution of the underlying heat conduction problem. A discussion of the details and obvious advantages is contained in Atthey (1974). As will be shown below, the inherent non-linearity associated with (6) does not pose major difficulties when constructing such finite difference schemes. A more detailed discussion of other numerical methods can be found in Fox (1974).

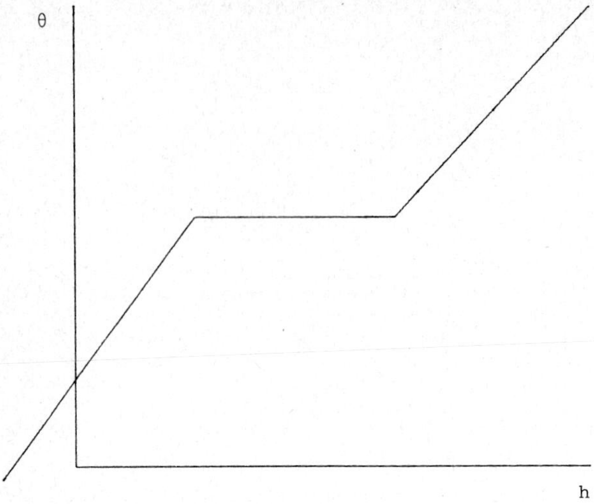

Figure 2. A typical form of a temperature-enthalpy relationship $\theta = F(h)$.

3. The Alternating Direction Explicit (ADE) Scheme.

The various explicit finite difference schemes, proposed for the approximate solution of the heat conduction equation (cf. Mitchell and Griffiths (1980)), can be easily adapted for the approximate solution of the enthalpy equation (5) in conjunction with the temperature-enthalpy relationship (6) and appropriate initial and boundary conditions. Because accurate approximations for complex casting geometries and moulding configurations will require the use of fine meshes in two and three dimensions, the possible use of explicit schemes has been under examination. In particular, attention has concentrated on the use of alternating direction explicit (ADE) methods. If the one-dimensional enthalpy equation (5) is discretized on the even grid.

$$G = \{x_i;\ x_i = i\,\delta x,\ i = 0,1,2,\ldots,n+1,\ \delta x = 1/(n+1)\}\ , \tag{7}$$

then, as well as the standard explicit and implicit methods, the following two finite difference schemes.

$$\rho_i\{h_i^{(m+1)} - h_i^{(m)}\} = r[k_{i+1}^{(m)}\{\theta_{i+1}^{(m)} - \theta_i^{(m)}\} - k_i^{(m)}\{\theta_i^{(m+1)} - \theta_{i-1}^{(m+1)}\}] \tag{8}$$

and

$$\rho_i\{h_i^{(m+2)} - h_i^{(m+1)}\} = r[k_{i+1}^{(m+1)}\{\theta_{i+1}^{(m+2)} - \theta_i^{(m+2)}\} - k_i^{(m+1)}\{\theta_i^{(m+1)} - \theta_{i-1}^{(m+1)}\}] \tag{9}$$

can be constructed with $\rho_i = \rho(x_i)$, $h_i^{(m)} = h(i\,\delta x,\ m\,\delta t)$,

$\theta_i^{(m)} = \theta(i\,\delta x,\ m\,\delta t)$, $k_i^{(m)} = k(i\,\delta x,\ m\,\delta t)$ and $r = \delta t/(\delta x)^2$,

where δt denotes the chosen time step. For reasons which will become clear below, we have written these two schemes for successive time steps; viz. $m\,\delta t$ to $(m+1)\,\delta t$, and $(m+1)\,\delta t$ to $(m+2)\,\delta t$.

The interesting fact about these two schemes is that the first is explicit if it moves across the grid G from left to right, while the second is explicit if it moves across G from right to left. In particular, using the temperature-enthalpy relationship (6), the finite difference scheme (8) and (9) can be rewritten as

$$\rho_i h_i^{(m+1)} + rk_i^{(m)}F(h_i^{(m+1)}) = rk_i^{(m)}F(h_{i-1}^{(m+1)}) + \rho_i h_i^{(m)} + rk_{i+1}^{(m)}\{F(h_{i+1}^{(m)}) - F(h_i^{(m)})\}\ , \tag{10}$$

and

$$\rho_i h_i^{(m+2)} + rk_{i+1}^{(m+1)}F(h_i^{(m+2)}) = rk_{i+1}^{(m+1)}F(h_{i+1}^{(m+2)}) + \rho_i h_i^{(m+1)} + rk_i^{(m+1)}\{F(h_{i-1}^{(m+1)}) - F(h_i^{(m+1)})\}\ , \tag{11}$$

Individually, because they correspond to explicit schemes, the standard stability analysis shows that they are conditionally stable for suitably small δt. However, if used jointly by applying them alternatively (as indicated by the way they have been presented in (10) and (11), they yield unconditionally stable schemes for the solution of (5). Each compensates for the errors introduced by the other so that rounding error growth is controlled. Algebraic verification of this point can be found in Larkin (1964).

It is an immediate consequence of the form of both (10) and (11) that the application of the iteration reduces at each step to solving the non-linear equation

$$h_i^{(p)} + \{rk_i^{(p-1)}/\rho_i\}F(h_i^{(p)}) = \Phi/\rho_i \tag{12}$$

after Φ has been evaluated using the current right hand side of (10) or (11), and the temperature-enthalpy relationship (6) under investigation. Even when (6) contains a jump discontinuity corresponding to the liberation of latent heat of fusion, and therefore does not always have an inverse, the non-linear equation

$$h + cF(h) = \phi$$

can always be solved for h for given values of c and ϕ. This is illustrated in Figure 3. As a direct consequence, the iterations (10) and (11) are independent of any discontinuities in (6) and pose no difficulties numerically. In fact, when $\theta = F(h)$ has a piecewise linear structure, as shown in Figure 2, then the solution of (12) reduces at each step to, one of a series of explicit linear inversions.

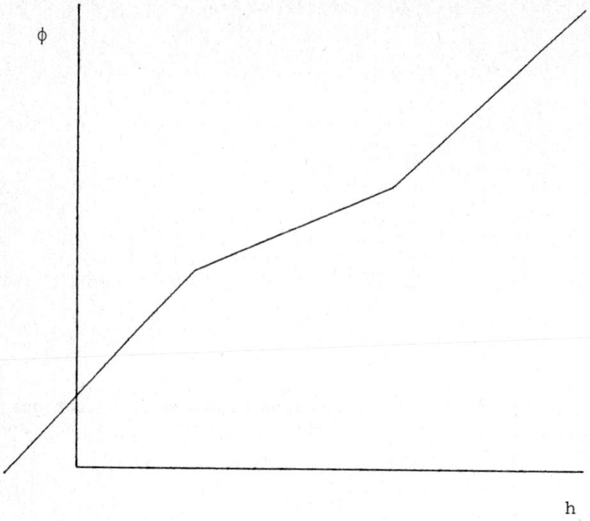

Figure 3. A typical form of the augmented temperature - enthalpy relationship $h+cF(h) = \phi$.

It is clear from this discussion that, even though the enthalpy formulation for solidification is non-linear, the use of the ADE method in the manner described above yields a simple, efficient, unconditionally stable and economical procedure for its approximate solution. The only difficulty posed for its implementation is the treatment of the jump discontinuities in thermal conductivity which occur at the various interfaces in the mounding configuration (sand-chills, sand-casting, chills-casting, etc). Numerical experimentation on symmetric problems indicated that symmetry could only be maintained in the approximations if, when crossing an interface, the schemes (10) and (11) were modified so that they take the form

$$\rho_i h_i^{(m+1)} + rk_i^{(m)} F(h_i^{(m+1)}) = rk_i^{(m)} F(h_{i-1}^{(m+1)}) + \rho_i h_i^{(m)} + rk_i^{(m)} \{F(h_{i+1}^{(m)}) - F(h_i^{(m)})\} , \tag{13}$$

and

$$\rho_i h_i^{(m+2)} + rk_i^{(m+1)} F(h_i^{(m+2)}) = rk_i^{(m+1)} F(h_{i+1}^{(m+2)}) + \rho_i h_i^{(m+1)} + rk_i^{(m+1)} \{F(h_{i-1}^{(m+1)}) - F(h_i^{(m+1)})\} , \tag{14}$$

where $k_i^{(m)}$ denotes the thermal conductivity of the region to the left of the interface, while $k_{i+1}^{(m+1)}$ denotes the thermal conductivity of the region to the right of the interface.

It appears that the ADE scheme fails if, at an interface, there is too large a difference between the values of the thermal conductivity which enter the iteration being used to step across the interface. The use of (13) and (14) at interfaces, which completely removes this difference in thermal conductivity values, circumvents the difficulty. In fact, it was found, that excellent approximations were generated if (13) and (14) were used exclusively.

4. Two-Dimensional Rectilinear Geometries

For two-dimensional rectilinear moulding configurations which include chills and pads as well as sand and casting, the problem formulation used consists of the following parabolic partial differential equation

$$\rho \frac{\partial h}{\partial t} = \frac{\partial}{\partial x}\left(k\frac{\partial \theta}{\partial x}\right) + \frac{\partial}{\partial y}\left(k\frac{\partial \theta}{\partial y}\right) , \tag{15}$$

$$\theta = \theta(x,y,t),\ h = h(x,y,t),\ -a \leqslant x \leqslant a ,\quad -b \leqslant y \leqslant b ,\quad t \geqslant 0 ,$$

with the nonlinear temperature-enthalpy relationship

$$\theta = F_1(h) , \tag{16}$$

the constitutive relationships

$$k = G_2(\theta,x,y) = G_2(F_1(h),x,y) = F_2(h,x,y) , \tag{17}$$

$$\rho = G_3(\theta,x,y) = G_3(F_1(h),x,y) = F_3(h,x,y), \tag{18}$$

the initial conditions

$$h(x,y,0) = f(x,y), \tag{19}$$

the boundary conditions

$$\theta(\pm a,y,t) = g_a^{\pm}(y,t),\quad \theta(x,\pm b,t) = g_b^{\pm}(x,t), \tag{20}$$

and interface conditions of continuity of temperature and flux across interfaces between sand and casting, sand and chills, and sand and pads.

When applying the relationships (16), (17) and (18), the form of $F_1(h)$, $F_2(h,x,y)$ and $F_3(h,x,y)$ used will depend on whether the point (x,y) under consideration lies in the sand, casting, chills or pads. Though it leads to additional complexity in the algorithm used by forcing the need to check for the material in which the given point (x,y) lies and then to choose the temperature-enthalpy relationship and the constitutive relationships accordingly, it is not necessary to introduce such complexity into the present discussion.

The rectilinear discretization applied below to (15)-(20) exploits numerically one of the principal practical motivations behind constructing computational procedures for modelling the solidification process. Any computational procedure is satisfactory if it allows comparative studies to be made of the effect, on the thermal history of the casting, of changing the relative positions of the casting, chills and pads in the mould. In fact, it follows that, if a reasonably fine two-dimensional grid is used for the region $-a \leq x \leq a$, $-b \leq y \leq b$, then the boundaries of the castings, chills and pads can be placed along the grid lines. The algorithmic advantages are two-fold. On the one hand, a simple procedure is easily constructed to define moulding configuration. On the other, it avoids the messy complications which arise if the interfaces between sand and casting, sand and chills, and sand pads are not along grid lines. For the grid G in (x,y,t)-space, we take

$$G = \{(x_i, y_j, t_k); x_i = i\,\delta x,\ i=0,1,\ldots,n,\ \delta x = 2a/n\ ,\quad y_j = j\,\delta y,\ j=0,1,\ldots,m,\ \delta y = 2b/m,\quad t_y = k\,\delta t,\ k = 0,1,\ldots\} \tag{21}$$

The ADE schemes derived above for (5) generalises naturally to define corresponding schemes for the solution of (15) on G. In fact, the following ADE scheme for (15) on G is explicit if it moves across G from west to east in the x-direction and from south to north in the y-direction:

$$\rho_{ij}^{(2k-1)} h_{ij}^{(2k)} + \{r_x k_{ij}^{(2k-1)} + r_y k_{ij}^{(2k-1)}\} F_{ij}^{(2k)} = \phi_{ij}^{(2k-1)} \tag{22}$$

with

$$\phi_{ij}^{(2k-1)} = k_{ij}^{(2k-1)} \{r_x F_{i-1,j}^{(2k)} + r_y F_{i,j-1}^{(2k)}\} + \rho_{ij}^{(2k-1)} h_{ij}^{(2k-1)} + r_x k_{i+1,j}^{(2k-1)} F_{i+1,j}^{(2k-1)} + r_y k_{i,j+1}^{(2k-1)} F_{i,j+1}^{(2k-1)} - \{r_x k_{i+1,j}^{(2k-1)} + r_y k_{i,j+1}^{(2k-1)}\} F_{ij}^{(2k-1)} \tag{23}$$

where i proceeds through the values $1,2,\ldots,n-1$, as j successively steps through the values $1,2,\ldots m-1$; $r_x = \delta t/(\delta x)^2$ and $r_y = \delta t/(\delta y)^2$.

The corresponding ADE scheme, which is explicit as it moves across G from east to west in the x-direction and from north to south in the y-direction, is given by

$$\rho_{ij}^{(2k)} h_{ij}^{(2k+1)} + \{r_x k_{i+1,j}^{(2k)} + r_y k_{i,j+1}^{(2k)}\} F_{ij}^{(2k+1)} = \psi_{ij}^{(2k)} \tag{24}$$

with

$$\psi_{ij}^{(2k)} = r_x k_{i+1,j}^{(2k)} F_{i+1,j}^{(2k+1)} + r_y k_{i,j+1}^{(2k)} F_{i,j+1}^{(2k+1)} + \rho_{ij}^{(2k)} h_{ij}^{(2k)} + k_{ij}^{(2k)} \{ -(r_x + r_y) F_{ij}^{(2k)} + r_x F_{i-1,j}^{(2k)} + r_y F_{i,j-1}^{(2k)} \} \tag{25}$$

where i proceeds through the values $n-1, n-2, \ldots, 1$, as j steps successively through the values $m-1, m-2, \ldots, 1$. Different algorithms arise from the way in which the interfaces are treated. If the aim is to treat all interfaces equivalently, then it is necesary to construct an implementation for the ADE method which cycles to cover all 4 combinations of east-west and south-north movements across G. Recalling that (as explained in 3) the value of a property at an interface is taken to be that of the material from which the ADE scheme is coming (spatially) (not that into which it is moving), the following procedure based on (22) and (24) cycles over all possible movements across the grid G:

For each $k=1,3,5,\ldots,$

(i) evaluate $\phi_{ij}^{(2k-1)}$

and solve (22) for $\{(i=1,2,\ldots,n-1)\ ,\ j=1,2,\ldots,m-1\}$;

(ii) evaluate $\psi_{ij}^{(2k)}$

and solve (24) for $\{(i=n-1,n-2,\ldots,1)\ ,\ j=m-1,m-2,\ldots,1\}$;

(iii) repeat (i) with $k=k+1$ for $\{(j=1,2,\ldots,m-1),\ i=1,2,\ldots,n-1\}$.

(iv) repeat (ii) with $k=k+1$ for $\{j=m-1,m-2,\ldots,1),\ i=n-1,n-2,\ldots,1\}$.

5. Results and Conclusions

In order to examine the effectiveness of the ADE method and the use of two-dimensional rectilinear models for moulding configurations, a number of numerical experiments were performed. Initially, the method was applied to a symmetric two-dimensional casting (without chills and pads) to test the degree to which it could retain the given symmetry during the period over which solidification took place. The results indicated that this is easily achieved as long as the size of the time step was not chosen too large (in the sense of advancing the solidification process too quickly).

Subsequently, the method was applied to a moulding configuration of a swiss-cross casting between two vertical chills (with initial temperature higher than the sand but much less than the casting) as shown in Figures 4 and 5. The aim was to test the potential of the method for making comparative studies of the effect, on the thermal history of the casting, of changing the relative positions of the casting, chills and pads in the mould.

For this reason, the only difference between the moulding configurations in Figures 4 and 5 are the lengths of the chills. Together, both illustrate the following facts about the use of chills in a moulding configuration:

(i) The speed with which the initial solidification takes place. As Figures 4 and 5 show, close to half the volume of the casting has solidified 0.2 seconds after pouring.

(ii) The slowness with which the sand is heated. This is clear from Figures 4 and 5 which show that the lowest temperature profile has only moved outside the casting and the chills when solidification is close to completion.

(iii) To graphical accuracy, the solidification front retains symmetry even though the surrounding temperature contours do not.

(iv) The ability of the ADE method to maintain the symmetries of the given moulding configuration.

A comparison of Figures 4 and 5 yields the following conclusions:

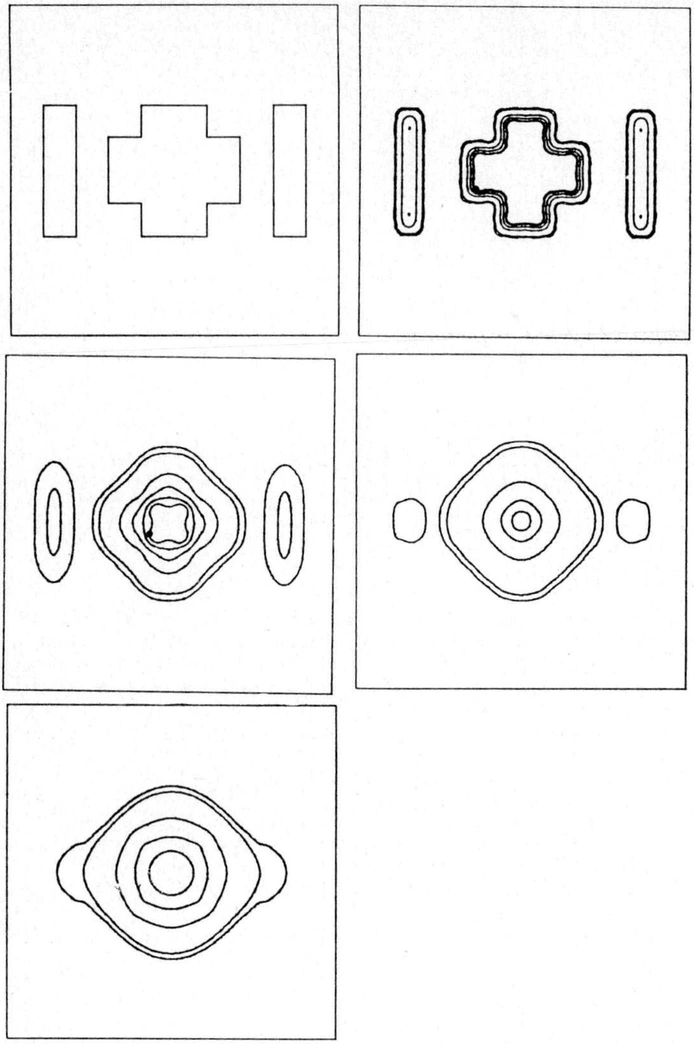

Figure 4. The sequence of plots illustrates the use of the ADE method in tracing the thermal evolution of solidification in a moulding configuration. The geometry of the cross-shaped casting, the rectangular chills and the sand mould is shown in Figure 4(a). The temperature profiles throughout the mould at 0.2, 8.2, 16.2 and 32.2 seconds after pouring are shown in Figures 4(b)-4(e).

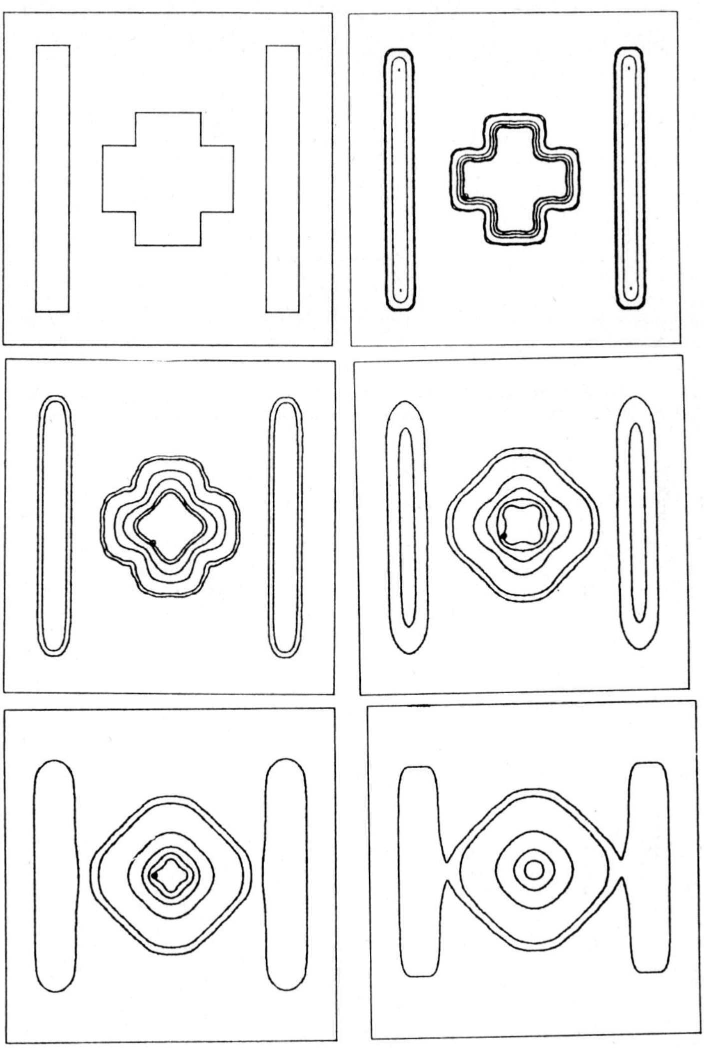

Figure 5. The sequence of plot illustrates the same situation as in Figure 4, except that the length of the chills has been increased, with the temperature profiles at 0.2, 4.2, 8.2, 12.2 and 16.2 seconds after pouring shown in Figures 5(b)-5(f).

(i) Solidification occurs faster if larger rather than smaller chills (of the same shape, placed in the same positions relative to the casting) are used; but this does not initially effect greatly the geometry of the temperature evolution in the casting and can have little effect on the final geometry of the solidification front.

(ii) It is clear from the final diagrams in both Figures 4 and 5 that the chills eventually redistribute heat back to the casting and thereby slow its long term cooling rate (a desirable feature).

(iii) In both Figures 4 and 5, the vertical temperature gradient across the centre of the casting is steeper than the corresponding horizontal temperature gradient. This is exactly how the presence of chills in the mould is expected to influence its thermal evolution.

(iv) A comparison of Figures 4 and 5 shows that, relative to the horizontal gradient, the vertical gradient in Figure 5 is steeper than that in Figure 4. Thus, making the chills longer has a greater effect on redistributing heat away from the closest sections of the casting rather than from more distant sections of the casting.

The fact that it is possible to draw such conclusions from simple situations as represented by Figures 4 and 5 demonstrates the effectiveness, for comparative purposes, of the rectilinear modelling of moulding configurations.

Acknowledgement

The authors acknowledge with thanks the helpful advice and comments received from Frank de Hoog about the use of the ADE method for solidification modelling.

References

D.R. Atthey (1974) A finite difference scheme for melting problems, J.Inst.Maths. Applics. 13 (1974), 353-366.

L. Fox (1974) What are the best numerical methods?, in Moving Boundary Problems in Heat Flow and Diffusion (Ed. J.R. Ockendon and W.R. Hodgkins), Oxford University Press, Oxford, 1974.

R.K. Kotschi and L.A. Plutshack (1982) An easy and inexpensive technique to study the solidification of castings in three dimensions, AFS Int. Cast Metals J. 7 (1982), 29-37.

B.K. Larkin (1964) Some stable explicit difference approximations to the diffusion equation, Math. Comp. 18 (1964), 196-202.

A.R. Mitchell and D.F. Griffiths (1980) The Finite Difference Method in Partial Differential Equations, J. Wiley and Sons, Chichester, 1980.

Computational Techniques & Applications: CTAC-83
J. Noye & C. Fletcher (Editors)

SOME ASPECTS OF THE NUMERICAL SOLUTION OF A STEADY STATE CONVECTION-DIFFUSION EQUATION

Frank de Hoog and David Jackett

CSIRO Division of Mathematics and Statistics
Box 1965 GPO, Canberra ACT 2601.

The steady state convection diffusion equation that governs the isotope separation in a high speed centrifuge has a number of features that make its numerical solution somewhat difficult. Firstly, a coefficient in the equation changes very rapidly and secondly convection dominates in the boundary layers. Nevertheless, as is shown in this paper, it is possible to get good numerical approximations using very simple finite difference schemes.

INTRODUCTION

An effective device to separate a mixture of two gases with different molecular weights is a gas centrifuge. This technique was pioneered in Germany in 1895 by Bredig and was extended in the USA in 1934 by Beams to separate isotopes. During the second world war, attempts were made to use gas centrifuges to separate isotopes of uranium for the atomic weapons project but because the technology for constructing high speed centrifuges was inadequate at that time, this project was abandoned in 1943. However, the subsequent development of gas centrifuges has led to a renewed interest in using these devices to efficiently and economically separate isotopes of uranium.

In addition, the advent of high speed computers and the associated developments in the construction and analysis of numerical schemes for partial differential equations has made it possible to obtain new insight into the various mechanisms taking place in these devices.

The aim of this paper is to describe some of the features of a numerical scheme that is used to obtain an approximation to the solution of the equation governing the isotope separation. However before this can be done it is necessary to briefly describe the underlying mathematical model. Such a description is given in section 2.

It turns out that the convection term in the isotope diffusion equation cannot in general be determined analytically and must be calculated numerically. Since the structure of this term and the numerical scheme used for its solution plays an important role in the numerical solution of the isotope diffusion equation, relevant features of the scheme used for its calculation are also presented in section 2. Finally numerical schemes for the isotope diffusion equation are discussed in section 3.

SOME ASPECTS OF THE MATHEMATICAL MODEL AND THE CALCULATION OF FLOWS

Consider gas in a cylinder of height h and radius a rotating with angular velocity Ω as shown in figure 1. The natural coordinate system for this geometry is the cylindrical polar coordinate system and we denote by r, Θ and z the radial, azimuthal and vertical coordinates. In addition we denote by u, v and w the velocities of the case in the radial, azimuthal and vertical directions.

Figure 1

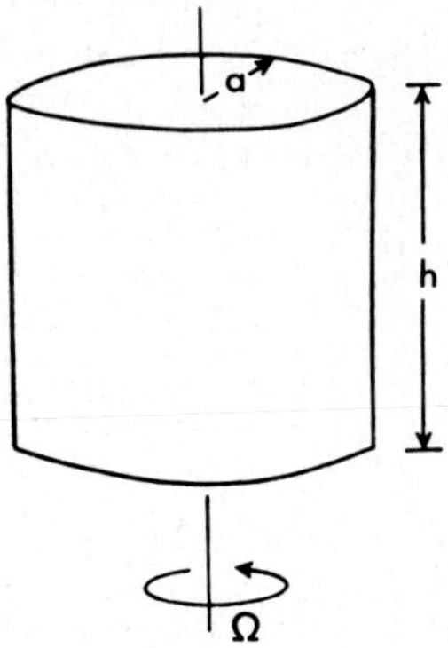

When the temperature T on the wall is a constant (T_o say) and the gas is ideal, the steady state flow is 'rigid body' rotation (see for example Soubbaramayer[7]) with velocities

$$u = u_o = 0; \quad v = v_o = \Omega r, \quad w = w_o = 0$$

and thermodynamic variables (density ρ, pressure p and temperature T),

$$\rho = \rho_o = \rho_o(0)\exp\left(\frac{Gr^2}{a^2}\right), \quad T = T_o, \quad p = p_o = R\rho_o T_o/M$$

where R is the perfect gas constant, M is the molar mass and $G = M\Omega^2 a^2/2RT_o$. Furthermore, if two gases with molecular weights M_1 and M_2 ($M_1 < M_2$) are involved then (Soubbaramayer[7]) the mole fraction f of the light gas satisfies

$$\frac{f}{1-f} = K \exp\left(\frac{-\Delta G}{a^2} r^2\right)$$

where $\Delta G = (M_2 - M_1)\Omega^2 a^2/2RT_o$ and K is a constant that can be calculated when the mole fraction of light gas for the centrifuge as a whole is given. Thus, when f is small, the elementary separation factor is approximately

$$S = f(0)/f(a) = \exp(\Delta G).$$

It should be clear from the above discussion that the case of 'solid body' rotation is relatively straightforward. In practice however it has been found that substantially higher separation factors can be achieved if small counter current flows (small relative to the solid body rotation) are induced. This is possible by prescribing a non constant temperature profile on the boundary for example. Other methods used to induce counter current flows are discussed in Soubbaramayer[7] and Orlander[5].

The equations governing the steady state flow of gas in a centrifuge under the assumption of axial symmetry are given in Ratz[6]. They consist of the Navier Stokes equations (three momentum equations - one for each dimension), the continuity equation, an energy equation and an equation of state. Since the counter current flow is expected to be small compared to the solid body rotation, these equations are usually simplified by linearization about the solid body flow. Details of such linearization and associated scaling may be found in Soubbaramayer[7] and Dickinson and Jones[1].

We shall see in the sequel that the velocities u and w occur in the 'convection' term of the isotope separation equation. It is therefore necessary to make some remarks about the velocities and the numerical scheme used for their calculation. Perhaps the most significant feature of the flow is the occurrence of boundary layers near the walls of the cylinder. These layers have been the subject of a number of investigations (Soubbaramayer[7], Ratz[6], Landahl[3], Louvet[4]) and the method of matched asymptotic expansions can be used to obtain approximate solutions. For our purposes it is sufficient to note that the velocities very significantly near the boundary, as is shown in figures 2 and 3, which show some normalized velocities for a typical centrifuge that have been calculated using a finite difference scheme. The counter current in this case was induced by a linear temperature gradient on the wall. As a consequence of these boundary layers, numerical schemes such as finite element and finite difference schemes must be based on a non-uniform discretization of the domain. In our case, we have discretized the domain $\mathcal{D} = \{(r,z);\ 0 \leqslant r \leqslant a,\ 0 \leqslant z \leqslant h\}$ by the grid $G=\{(r_i,z_j);\ i = 0,\dots,2M+1,\ j = 0,\dots,2N+2\}$ and a finite difference scheme similar to those proposed by Kai[2] and Dickinson and Jones[1] has been used to obtain a numerical solution. The variation in the grid spacing is illustrated for a typical case in figures 4 and 5. As is customary in finite difference schemes for fluid flow problems, the numerical approximations are obtained on a staggered grid with

u approximated by u_{ij} at (r_{2i},z_{2j})
v approximated by v_{ij} at (r_{2i},z_{2j})
w approximated by w_{ij} at $(r_{2i+1},\ z_{2j+1})$
ρ approximated by ρ_{ij} at $(r_{2i+1},\ z_{2j})$
T approximated by T_{ij} at $(r_{2i+1},\ z_{2j})$
p approximated by p_{ij} at $(r_{2i+1},\ z_{2j})$

Our final remark on the calculation of the flow concerns the enormous variation of the density in the centrifuge. From the solid body rotation we see that

$$\rho_o(r) = \rho_o(0)\exp\left(\frac{Gr^2}{a^2}\right)$$

and typically G varies from 10 to 35. The linearized continuity equation is

$$\frac{1}{r}\frac{\partial}{\partial r}\left(r\,\rho_o(r)u\right) + \frac{\partial}{\partial z}\left(\rho_o(r)w\right) = 0$$

and the simplest discretization of this is

$$(r_{2i+2}\rho_o(r_{2i+2})u_{i+1,j} - r_{2i}\rho_o(r_{2i})u_{i,j})/(r_{2i+2}-r_{2i})$$

$$+\ r_{2i+1}\rho_o(r_{2i+1})(w_{i,j+1} - w_{i,j})/(z_{2j+2} - z_{2j}) = 0.$$

Figure 2

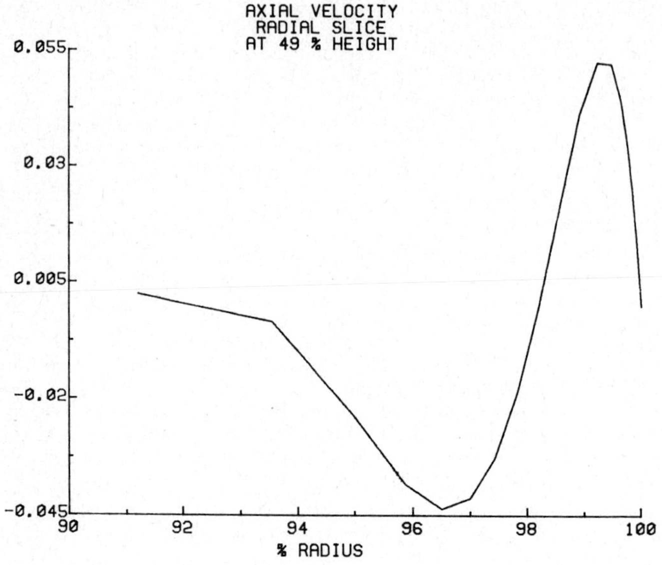

Figure 3

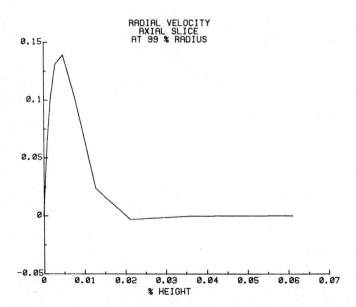

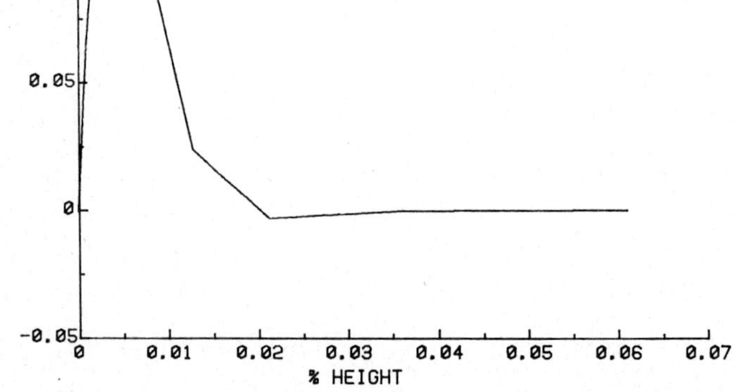

Figure 4

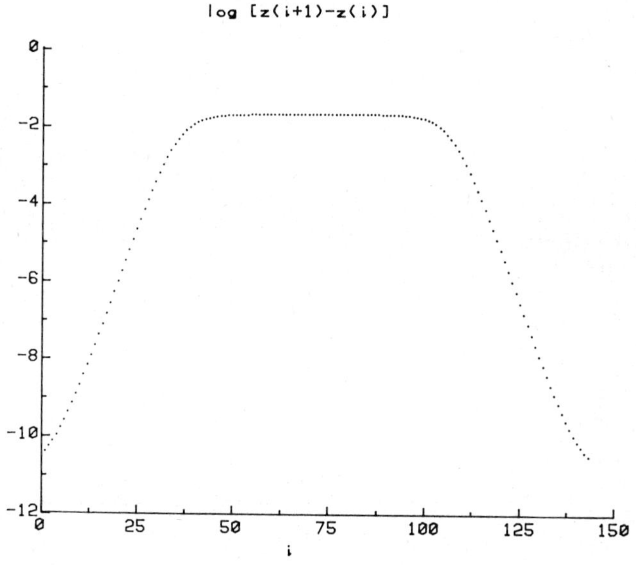

Figure 5

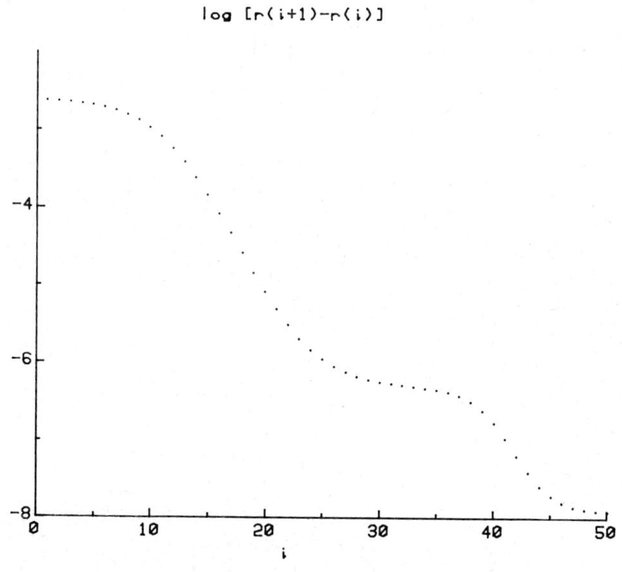

However, because of the large radial variation in $\rho_o(r)$ it was found that more satisfactory results were obtained when $r_{2i+1}\rho_o(r_{2i+1})$ was replaced with its 'average' value. This gives the discretization

$$(r_{2i+2}\rho_o(r_{2i+2})u_{i+1,j} - r_{2i}\rho_o(r_{2i})u_{i,j})$$
$$+ \int_{r_{2i}}^{r_{2i+2}} r\rho_o(r)dr \quad (w_{i,j+1} - w_{i,j})/(z_{2j+2} - z_{2j}) = 0$$

When f, the molar fraction of light isotope is small, the linearized steady state equation is given by (Soubbaramayer[7], Orlander[5])

$$\frac{1}{r}\frac{\partial}{\partial r}\{r\,D\left(\frac{\partial f}{\partial r} + 2\frac{\Delta G}{a^2} r\,f\right) - \rho_o r\,u\,f\}$$
$$+ \frac{\partial}{\partial z}\{D\frac{\partial f}{\partial z} - \rho_o\,w\,f\} = 0$$

where D is a constant. The boundary conditions for the case of a closed centrifuge are

$$\frac{\partial f}{\partial r} + 2\frac{\Delta G}{a^2} r\,f = 0 \ , \ r = 0 \ , \ a$$

$$\frac{\partial f}{\partial z} = 0 \ , \ z = 0, \ h$$

As we have remarked previously, the density ρ varies significantly with radius and becomes very small away from the boundary. It is therefore not too surprising that for a typical centrifuge the 'diffusion' term dominates away from the walls. However, near the boundary the 'convection' term tends to dominate.

THE NUMERICAL SOLUTION OF THE SEPARATION EQUATION

Recall that the velocities u and w are approximated by u_{ij} and w_{ij} at (r_{2i},z_{2j}) and (r_{2i+1}, z_{2j+1}) respectively. It was found that schemes based on interpolating the velocities did not work well and it is therefore necessary to use the same discretization of the domain for the separation equation as for the flow calculation. A cursory examination of the grid shows that it is most natural to approximate f at the points (r_{2i+1}, z_{2j}). We denote the resulting approximation by f_{ij}.

First, we show that not all difference schemes yield satisfactory results. To demonstrate this, consider a scheme based on expanding the separation equation and approximating each term separately. On expanding the separation equation and subtracting the equation of continuity we obtain

$$D\frac{\partial^2 f}{\partial r^2} + D\frac{\partial^2 f}{\partial z^2} + \left\{D\left(\frac{1}{r} + 2\frac{G}{a^2}\right) - \rho_o(r)u\right\}\frac{\partial f}{\partial r}$$
$$- \rho_o(r)w\frac{\partial f}{\partial z} + 4\frac{\Delta G}{a^2} f = 0$$

subject to

$$\frac{\partial f}{\partial r} + 2\frac{\Delta G}{a^2} r\,f = 0 \ , \ r = 0, \ a$$

Now consider the discretization obtained by approximating the following quantities at (r_{2i+1}, z_{2j})

$$\frac{\partial^2 f}{\partial r^2} \approx 2\left\{\frac{(f_{i+1,j}-f_{i,j})/(r_{2i+3}-r_{2i+1}) - (f_{i,j}-f_{i-1,j})/(r_{2i+1}-r_{2i-1})}{(r_{2i+3} - r_{2i-1})}\right\}$$

$$\frac{\partial^2 f}{\partial z^2} \approx 2\left\{\frac{(f_{i,j+1}-f_{i,j})/(z_{2j+2}-z_{2j}) - (f_{i,j}-f_{i,j-1})/(z_{2j}-z_{2j-2})}{(z_{2j+2} - z_{2j-2})}\right\}$$

$$\frac{\partial f}{\partial r} \approx \frac{(f_{i+1,j} - f_{i-1,j})}{(r_{2i+3} - r_{2i-1})}, \quad \text{and} \quad \frac{\partial f}{\partial z} \approx \frac{(f_{i,j+1} - f_{i,j-1})}{z_{2j+2} - z_{2j-2}}$$

The solution of the separation equation is arbitrary up to a constant multiplier but we can eliminate this arbitrariness by specifying f at some point on the boundary (say $f(a,h/2) = 1$). This is achieved in the numerical scheme by adding the term $\lambda(f_{M,N/2} - 1)$ to the equation for the discretization at the point (r_{2M+1}, z_N).

The finite difference scheme outlined above was applied to the case of solid body rotation (i.e. $u = w = 0$) with the grids shown in figures 4 and 5 and $\lambda = 10$. Clearly, the analytical solution is

$$f = \exp\left(\frac{\Delta G}{a^2}\ (a^2 - r^2)\right)$$

and $f_{M,j}$ plotted in figure 6 is obviously a rather poor approximation to 1.

Figure 6

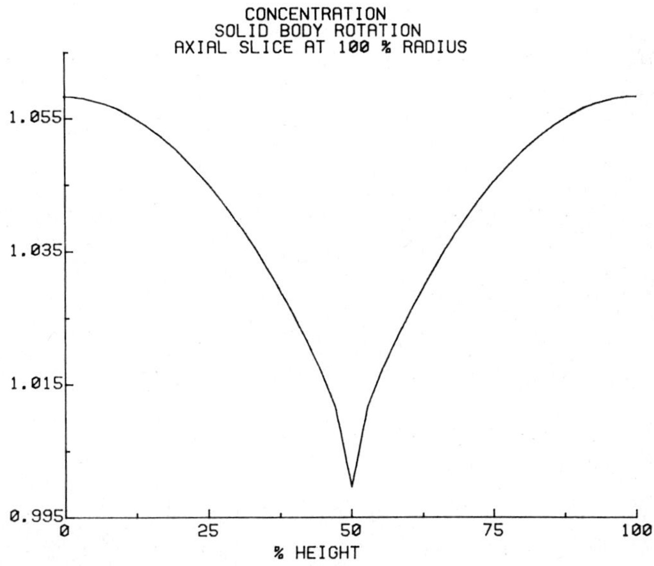

The reason for the behaviour is that although the differential equation (which is homogeneous) has a nontrivial solution the corresponding difference equation has only a trivial solution. Thus the extra condition $f(a,h/2) = 1$ is not compatible with the difference equation.

The above difficulty can be avoided if we use a scheme that is conservative. We can rewrite the spearation equation as

$$\frac{1}{r}\frac{\partial}{\partial r}(r\,Q) + \frac{\partial P}{\partial z} = 0$$

subject to

$$Q = 0\ ,\ r = 0,\ a$$

$$P = 0\ ,\ z = 0,\ h$$

where

$$Q = D\left(\frac{\partial f}{\partial r} + \frac{2\Delta G}{a^2}\, r\, f\right) - \rho_o\, u\, f$$

and

$$P = D\frac{\partial f}{\partial z} - \rho_o\, w\, f$$

Given approximations $Q_{i,j}$ and $P_{i,j}$ to Q and P at (r_{2i+2}, z_{2j}) and (r_{2i+1}, z_{2j+1}) respectively, we use the discretization

$$(r_{2i+2}\,Q_{i,j} - r_{2i}\,Q_{i-1,j})/(r_{2i+2} - r_{2i})$$
$$+\ r_{2i+1}(P_{i,j+1} - P_{i,j})/(z_{2j+1} - z_{2j-1}) = 0.$$

It now remains to define $Q_{i,j}$ and $P_{i,j}$. We have used

$$Q_{i,j} = D\{\ (f_{i+1,j} - f_{i,j})/(r_{2i+3} - r_{2i+1})$$
$$+\ \Delta G\, r_{2i+2}\,(f_{i+1,j} + f_{i,j})/a^2\ \}$$
$$-\ \rho_o(r_{2i+2})\, u_{i+1,j}\,(f_{i+1,j} + f_{i,j})/2$$
$$P_{i,j} = D(f_{i,j+1} - f_{i,j})/(z_{2j+2} - z_{2j})$$
$$-\rho_o(r_{2i+1}) w_{i,j}\,(f_{i,j+1} + f_{i,j})/2.$$

For the scheme outlined above, the case of flow induced by a linear temperature profile was implemented with two different grid spacings. The profiles plotted in figure 7 however are not satisfactory as they show a large variation between the numerical solutions.

In section 2, it was mentioned that the radial variation in density ρ was considerable and that the continuity equation in the flow calculation had been modified because of this fact. It is therefore also reasonable to modify the separation equation. This was achieved by replacing $r_{2i+1}\rho_o(r_{2i+1})$ in P_{ij} by $\int_{r_{2i}}^{r_{2i+2}} \rho_o(r) r\, dr/(r_{2i+2} - r_{2i})$. The profiles corresponding to those in figure 7 are plotted in figure 8 for this scheme. The scheme now shows very little variability with respect to changes in the grid spacings. Further numerical experimentation has shown that the scheme performs quite well generally.

Figure 7

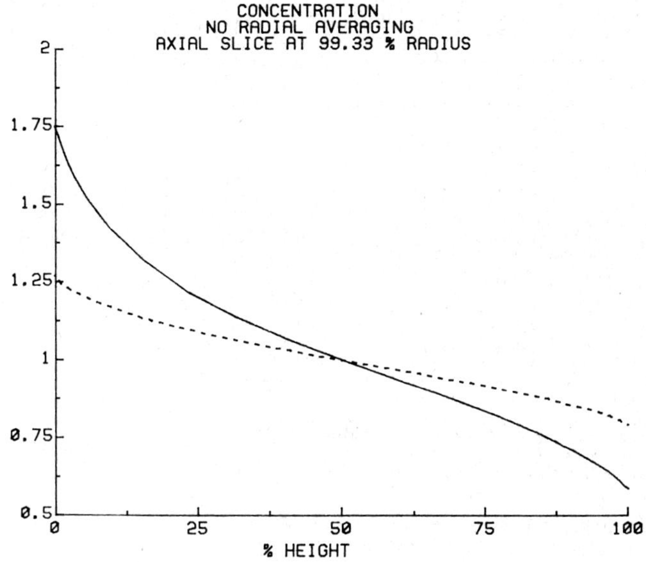

Figure 8

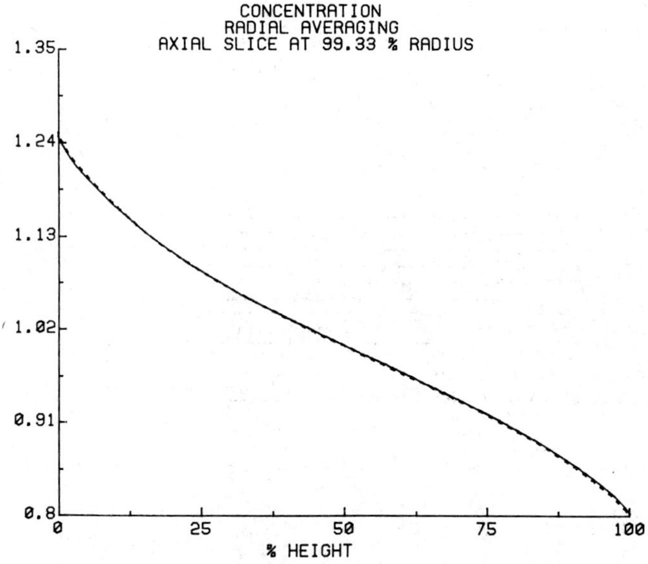

CONCLUDING REMARKS

The problem of calculating isotope separation in a high speed centrifuge involves equations with rapidly changing coefficients which lead to solutions with boundary layers. Because of the fine grids necessary to resolve the flow in the layers, no instabilities associated with the fact that convection dominates near the walls was detected for practical operating parameters. In addition when a small change has been made to take account of the large radial variation in the density, a simple conservative finite difference scheme appears to give good numerical approximations.

REFERENCES

[1] Dickinson, G.J. and Jones, I.P. Numerical solutions for the compressible flow in a rapidly rotating cylinder. J.Fluid Mech. 107, pp89-107 (1981)

[2] Kai, T. Basic characteristics of centrifuges, III Analysis of fluid flow in centrifuges. J.Nuc.Sci.Tech. 14, pp267-281 (1977)

[3] Landahl, M.T. Boundary layers and shear layers in a rapidly rotating gas. Proceedings of the 2nd Workshop on Gases in Strong Rotation ed. by Soubbaramayer, C.E.A., France (1977)

[4] Louvet, P. Two and three dimensional rotating flows. Proceedings of the 2nd Workshop on Gases in Strong Rotation ed. by Soubbaramayer, C.E.A., France (1977)

[5] Orlander, D.R. The theory of uranium enrichment by the gas centrifuge. Prog. Nuc.Energy, 8, pp1-33 (1981)

[6] Ratz, E. Uranium isotope separation in the gas centrifuge. Aerodynamic Separation of Gases and Isotopes, Lecture series 1978-8, (Von Karman Institute for Fluid Dynamics. Belgium 1978)

[7] Soubbaramayer, Centrifugation. Topics in Applied Physics, 35, pp183-243 (1979)

Computational Techniques & Applications: CTAC-83
J. Noye & C. Fletcher (Editors)

A NEW METHOD OF DERIVING FINITE DIFFERENCE FORMULAS FOR ARBITRARY MESHES

S. K. KWOK

Department of Civil Engineering
University of Western Australia
Nedlands, W.A., 6009
AUSTRALIA

An improved method of computer generation of finite difference approximations for partial derivatives in arbitrary meshes is described. The proposed method aims at improving the numerical accuracy of the original Curvilinear Finite Difference (CFD) Method (1)[1] when highly irregular grids are dealt with. The concept of using a local curvilinear coordinate system at each node is retained. However, Taylor's Series Expansion Theorem is employed in the current method to derive the required transformation matrix that correlates the local and global partial derivatives.

INTRODUCTION

Over the recent years, the application of the Finite Difference Method to irregular meshes has drawn much attention. Chu (3) used a so-called machine transformation to transform points in a global x-y plane onto an 'equipotential' plane, on which finite difference approximations were generated with reference to a local triangular mesh. Jensen (4) employed a six-point star in his finite difference evaluation on an arbitrary grid using the two-dimensional Taylor's Series Expansion Theorem. However, the 5x5 matrix that correlates the nodal function values and the partial derivatives, so obtained may be singular or ill-conditioned. Much effort was then devoted to a careful selection of local mesh to avoid singularity and to improve the accuracy of the so derived finite difference approximations. This problem of avoiding singularity was pursued further by Perrone & Kao (5). They also obtained better finite difference approximations by averaging four sets of finite difference coefficients, obtained by applying Jensen's method to four carefully chosen six-point stars selected from a local nine-point mesh. However, four 5x5 matrix inversions are required in such a process for each node.

Later on, a new line of thought was introduced by the development of the Curvilinear Finite Difference (or abbreviated as CFD) method. Although the name CFD method was firstly attached to Lau (1), a different account of the same method was presented by Frey (6) around the same time under the name of 'flexible finite difference stencils'. In the CFD method, the concept of a coordinate transformation between the global Cartesian and a local curvilinear coordinate system was employed. Within every defined local coordinate system, a complete polynomial surface fit of the unknown function, expressed in terms of its nodal values, is sought to approximate its true value. Partial derivatives in the local curvilinear coordinate system can therefore be obtained by successive differentiation of the approximate complete polynomial surface fit with respect to the curvilinear coordinates. These local partial derivatives are then transformed back to the global Cartesian coordinate system according to the Chain Rule of Partial Differentiation to give the required finite difference approximations of the global partial derivatives. However, the use of the Chain Rule of Partial Differentiation implies that the transformation of the finite difference approximations of, say the nth-order partial derivatives, is independent of those finite difference approximations of any partial derivatives with order higher than n. This implication may have a significant effect on the numerical accuracy of the method, especially when the global grid coordinates cannot be mapped conformally onto the local curvilinear coordinate system. This is the reason why Liszka and Orkisz (2) reported that their least square fit method performs better than the CFD method on irregular grids.

In the present paper, a new finite difference method for arbitrary mesh systems is proposed. The concept of a coordinate transformation between the global Cartesian and a local curvilinear coordinate system employed in the CFD method is retained. However, in the proposed method, instead of using the Chain Rule of Partial Differentiation, the transformation matrix that correlates the local and global partial derivatives will be derived from the Taylor's Series Expansion Theorem. By doing so, finite difference approximations for all the partial derivatives will be ensured to be derived from a same-order truncated Taylor's series expansion of the unknown function.

In the next section, the algorithm of the proposed method will be described. A numerical example involving a single 2nd-order linear mixed-type partial differential equation will also be worked out to illustrate the application of the method. Since extensive use is made of matrix algebra, the current method is highly systematic and can be easily extended to derive higher-order finite difference approximations. To demonstrate this, the numerical example solved earlier by 2nd-order finite difference approximations will be solved again using 4th-order finite difference approximations.

FINITE DIFFERENCE APPROXIMATIONS FOR LOCAL PARTIAL DERIVATIVES

Let $\phi(x,y)$ be an unknown function defined in some region R of the global Cartesian x-y plane (see Fig. 1), and suppose that its behaviour in R is governed by the following 2nd-order linear partial differential equation.

$$A_i \, \phi,_i + B_{ij} \, \phi,_{ij} = f \tag{1}$$

Furthermore, Eq.(1) is subject to the following boundary condition along the boundary S.

$$C_i \, \phi,_i + D\phi = g \tag{2}$$

where in Eq.(1) & (2), f, g and all the coefficients are functions of x and y defined in R or constants.

For convenience, the familiar indicial notation used in tensor calculus will be employed throughout this paper to represent partial derivatives, i.e., the comma appears in Eq.(1) & (2) denotes partial differentiation and the subscripts i, j refer to the global x,y coordinates. The subscripts i, j will take the range of 1 and 2 in this case. Like the Finite Element Method, the numerical computation starts with a discretization of the domain R to form a computational mesh. At each node of the arbitrary mesh formed, a local α-β curvilinear coordinate system is defined as it is shown in Fig. 1.

Within the local curvilinear mesh, the function value of $\phi(\alpha,\beta)$ at an arbitrary point is now approximated by the following 2nd-order complete polynomial.

$$\phi = a_1 + a_2\alpha + a_3\beta + a_4\alpha^2 + a_5\beta^2 + a_6\alpha\beta + a_7\alpha^2\beta + a_8\alpha\beta^2 + a_9\alpha^2\beta^2 \tag{3}$$

Eq.(3) is the lowest order complete polynomial that one can use if the function $\phi(\alpha,\beta)$ is to be ensured to be continuous up to its 2nd-order partial derivatives. In order to express the nine unknown coefficients, i.e. $a_1, \ldots, a_9$ in terms of

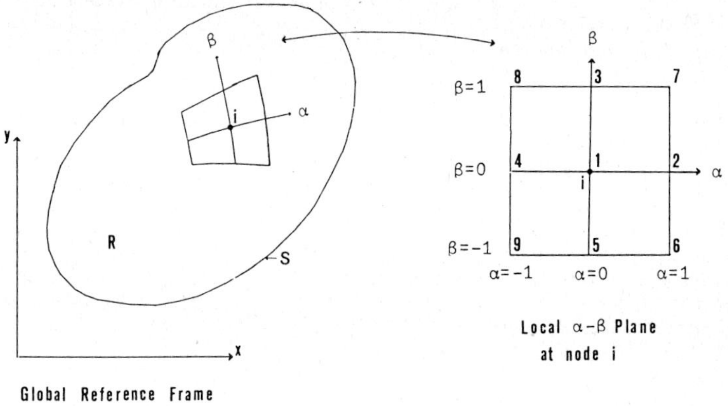

FIG. 1 The definition of the local curvilinear 9-point mesh

the nodal function values, a minimum number of nine nodes will be required to define a local curvilinear mesh. This is the case shown in Fig. 1. By applying Eq.(3) to each node of the local mesh, the following matrix equation can be arrived.

$$\{\phi_N\} = \{A\}\{a_i\} \tag{4}$$

where $\{\phi_N\}^T = \{\phi_1, \phi_2, \dots, \phi_9\}$

$\{a_i\}^T = \{a_1, a_2, \dots . a_9\}$

and

$$\{A\} = \begin{bmatrix} 1 & \alpha_1 & \beta_1 & \cdots & \alpha_1^2\beta_1^2 \\ 1 & \alpha_2 & \beta_2 & \cdots & \alpha_2^2\beta_2^2 \\ \vdots & \vdots & & \vdots & \vdots \\ 1 & \alpha_9 & \beta_9 & \cdots & \alpha_9^2\beta_9^2 \end{bmatrix}$$

By multiplying $\{A\}^{-1}$ to each side of Eq.(4), $\{a_i\}$ can be expressed in terms of $\{\phi_N\}$, i.e.,

$$\{a_i\} = \{A\}^{-1}\{\phi_N\} \tag{5}$$

Hence, within the local curvilinear mesh, ϕ can be obtined by the following interpolation function

$$\phi = \{\psi\}\{\phi_N\} \tag{6}$$

where $\{\psi\} = \{\, 1\ \alpha\ \beta\ \alpha^2\ \alpha\beta\ \beta^2\ \alpha^2\beta\ \alpha\beta^2\ \alpha^2\beta^2\}\{A\}^{-1}$

By successive partial differentiating Eq.(6) with respect to the curvilinear coordinates α and β, the finite difference approximations of the five partial derivatives in the local curvilinear coordinate system can be obtained. This is best summarized by the following matrix equation.

$$\{LD\} = \{DC\}\{\phi_N\} \tag{7}$$

where $\{LD\}^T = \{\ \phi,_\alpha\ \phi,_\beta\ \phi,_{\alpha\alpha}\ \phi,_{\alpha\beta}\ \phi,_{\beta\beta}\}$

and {DC} is a 5x9 matrix. Subsequently, by substituting the appropriate α, β

values at a point within the local mesh into those expressions of the {DC} matrix, finite difference approximations of the local partial derivatives at that particular point can be easily obtained.

TRANSFORMATION OF THE LOCAL PARTIAL DERIVATIVES TO THE GLOBAL X-Y PLANE

Instead of using the Chain Rule of Partial Differential to transform those derived finite difference approximations of the local partial derivatives from the local curvilinear coordinate system to the global x-y plane, the two-dimensional Taylor's Series Expansion Theorem is used. For the purpose of illustration, the central differencing case is considered. The function $\phi(x,y)$ at each node of the local curvilinear mesh will be approximated by the following 2nd-order truncated Taylor's series expansion about the center node.

$$\phi_i \simeq \phi_1 + h_i\phi_{1,x} + k_i\phi_{1,y} + \frac{h_i^2}{2}\phi_{1,xx} + k_ih_i\phi_{1,xy} + \frac{k_i^2}{2}\phi_{1,yy} \quad (8)$$

where $\phi_i = \phi(x_i,y_i)$, $h_i = x_i - x_1$, $k_i = y_i - y_1$

By writing Eq.(8) at each node of the local curvilinear mesh, the following matrix equation is obtained.

$$\{\phi_N\} = \{D\}\{GD\} \quad (9)$$

where $\{GD\}^T = \{\phi_{1,x} \quad \phi_{1,y} \quad \phi_{1,xx} \quad \phi_{1,xy} \quad \phi_{1,yy}\}$

and

$$\underset{9\times5}{\{D\}} = \begin{bmatrix} h_1 & k_1 & h_1^2/2 & h_1k_1 & k_1^2/2 \\ h_2 & k_2 & \cdots & \cdots & \cdots \\ \vdots & \vdots & \vdots & \vdots & \vdots \\ h_9 & k_9 & h_9^2/2 & h_9k_9 & k_9^2/2 \end{bmatrix}$$

Now, $\{\phi_N\}$ in Eq.(9) can be substituted into Eq.(7) to obtain

$$\{LD\} = \{DC\}\{D\}\{GD\} \quad (10)$$

By inverting Eq.(10), one gets

$$\{GD\} = \left(\{DC\}\{D\}\right)^{-1}\{LD\} \quad (11)$$

Further substitution of Eq.(7) into Eq.(11) yields the finial desirable equation, i.e.,

$$\{GD\} = \left(\{DC\}\{D\}\right)^{-1}\{DC\}\{\phi_N\} \quad (12)$$

It can be observed from Eq.(12) that the current method is highly systematic. One needs to form the {D} and the {DC} matrices at each node, and the finite difference approximations for the global partial derivatives will be generated automatically by subsequent numerical inversion and multiplication of matrices.

No special geometric constraints have to be specified for the formation of the local curvilinear mesh in the proposed method, except that the determinant of the Jacobian of the transformation matrix, i.e.,

$$|J| = \begin{vmatrix} x_{,\alpha} & y_{,\alpha} \\ x_{,\beta} & y_{,\beta} \end{vmatrix}$$

cannot be zero. This condition merely states the fact that in the local curvilinear coordinate system, the angle between the α and β coordinate axes cannot be zero. In other words, those nodes which define a local mesh cannot lie on a curve or a straight line at the same time. If this condition is met, the transformation matrix defined in Eq.(10) will not be singular. Physically, this means that in the local mesh, if the nodal points can be spreaded around to form a surface, one can always find a complete surface polynomial to fit the unknown function values at these points.

NUMERICAL EXAMPLE

Consider the following nonhomogenous mixed-type partial differential equation,

$$y\,\phi,_{xx} + \phi,_{yy} = (y+1)\,e^{x+y} \tag{13}$$

defined in the inverted teardrop domain shown in Fig.2. The domain is bounded by three curves, namely, a smooth curve S_3 in the elliptic subregion and two characteristic arcs S_1 and S_2 in the hyperbolic subregion.

Furthermore, Eq.(13) is subject to the following boundary conditions. Firstly,

$$\phi = e^{x+y} \qquad \text{along } S_2 \text{ and } S_3$$

and secondly, no data will be prescribed on the free characteristic boundary S_1. This problem was solved before by Cheung (7) using higher order finite elements and the exact solution is

$$\phi = e^{x+y} \tag{14}$$

In this problem, the irregular boundary geometry of the domain highlights the necessity to use irregular mesh in the numerical computation. Fig.2 shows a possible way of discretizing the domain automatically by computer. The method used here for the automatic mesh generation was the one described by Amsden and Hirt (8). Because of the presence of the free characteristic boundary S_1, backward differencing technique has to be used for those nodes lying along S_1 to provide one algebraic finite difference equation at each of these nodes. Although a 143-point mesh is shown in Fig.2, a 90-point mesh was actually used in the analysis, and the results obtained are presented in Table 1 for some selected nodes in the mesh, and compared with the analytic solutions. Numerical accuracy will be assessed in terms of the discrepancy E_{rr}, which is defined by the following ratio expressed in percentages.

$$E_{rr} = (\text{Numerical result} - \text{Analytical result})/\text{Analytical result} \tag{15}$$

x	y	E^*_{rr} ϕ	$\phi,_x$	$\phi,_y$
-0.237	-1.095	4.01	14.93	8.40
-0.601	-0.710	2.58	8.41	10.49
0.195	-0.862	1.88	0.78	6.95
-0.636	-0.668	2.45	9.76	10.12
0.000	-0.640	2.32	2.73	2.88
0.190	-0.630	2.05	1.08	2.89
0.367	-0.613	1.65	0.11	1.77
-0.748	-0.523	2.41	14.87	9.35
-0.885	-0.310	2.73	16.36	11.69
-0.642	-0.234	0.65	1.07	0.81

Table 1: Comparsion of results for Eq.(13)

*Discrepance defined in Eq.(15) in percentages.

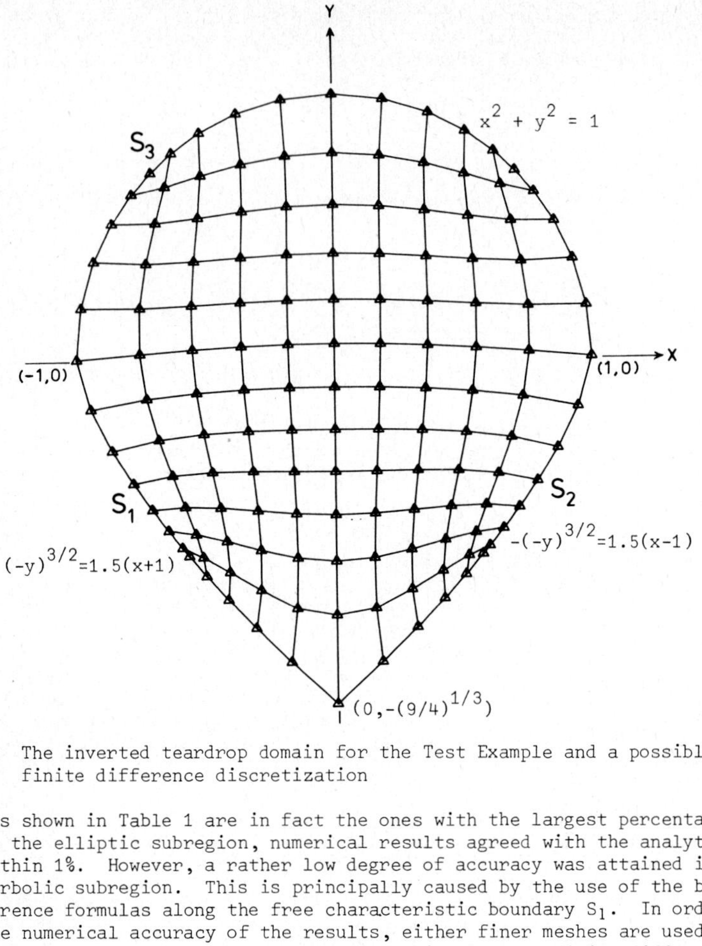

Fig. 2 The inverted teardrop domain for the Test Example and a possible finite difference discretization

The results shown in Table 1 are in fact the ones with the largest percentages of error. In the elliptic subregion, numerical results agreed with the analytical ones to within 1%. However, a rather low degree of accuracy was attained in the lower hyperbolic subregion. This is principally caused by the use of the backward difference formulas along the free characteristic boundary S_1. In order to improve the numerical accuracy of the results, either finer meshes are used, or as it will be demonstrated in the next section, higher-order finite difference approximations for the global partial derivatives must be employed.

EXTENSION TO HIGHER-ORDER FINITE DIFFERENCE APPROXIMATIONS

In this section, the current method will be extended to derive those higher-order finite difference approximations for arbitrary meshes.

Employing the same local curvilinear 25-point mesh as used by Lau (9) (see Fig.3), a function ϕ is now approximated by the following 4th-order complete polynomial within the local region.

$$\phi = (a_1 + a_2\alpha + a_3\alpha^2 + a_4\alpha^3 + a_5\alpha^4) + \beta(a_6 + a_7\alpha + a_8\alpha^2 + a_9\alpha^3 + a_{10}\alpha^4) + \beta^2(a_{11} + a_{12}\alpha + a_{13}\alpha^2 + a_{14}\alpha^3 + a_{15}\alpha^4) + \beta^3(a_{16} + a_{17}\alpha + a_{18}\alpha^2 + a_{19}\alpha^3 + a_{20}\alpha^4) + \beta^4(a_{21} + a_{22}\alpha + a_{23}\alpha^2 + a_{24}\alpha^3 + a_{25}\alpha^4) \quad (16)$$

where $a_1, a_2, \ldots, a_{25}$ are constant coefficients.

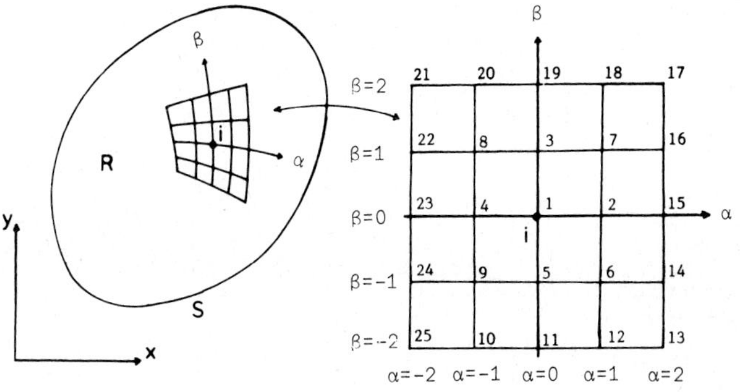

Fig. 3 A 25-point local curvilinear coordinate system

Having gone through similar computations as it is shown in the 2nd-order case, Eq.(7) can be obtained, in which {LD} becomes $\{ \phi,_{\alpha}\ \phi,_{\beta}\ \phi,_{\alpha\alpha}\ \phi,_{\alpha\beta}\ \phi,_{\beta\beta}\ \phi,_{\alpha\alpha\alpha}\ \phi,_{\alpha\alpha\beta}\ \phi,_{\alpha\beta\beta}\ \phi,_{\beta\beta\beta}\ \phi,_{\alpha\alpha\alpha\alpha}\ \phi,_{\alpha\alpha\alpha\beta}\ \phi,_{\alpha\alpha\beta\beta}\ \phi,_{\alpha\beta\beta\beta}\ \phi,_{\beta\beta\beta\beta}\}$ and {DC} is now a 14x25 matrix. Furthermore, $\{\phi_N\}^T$ is the transpose of the nodal vector, i.e., $\{\phi_1, \phi_2, \ldots, \phi_{25}\}$

Now, with the help of Eq.(12) and by means of a truncated 4th-order Taylor's series expression, this finite difference approximations for the local partial derivatives can easily be transformed to the global x-y plane to give the desired finite difference approximations for the global partial derivatives. The vector {GD} in Eq.(12) becomes $\{ \phi,_x\ \phi,_y\ \phi,_{xx}\ \phi,_{xy}\ \phi,_{yy}\ \phi,_{xxx}\ \phi,_{xxy} \ldots\ldots \phi,_{yyyy}\}$ and the matrix {D} is now a 25x14 matrix defined as

$$\underset{25\times14}{\{D\}} = \begin{bmatrix} h_1 & k_1 & \frac{h_1^2}{2} & h_1k_1 & \frac{k_1^2}{2} & \frac{h_1^3}{6} & \frac{h_1^2k_1}{2} & \frac{h_1k_1^2}{2} & \frac{k_1^3}{6} & \frac{h_1^4}{24} & \frac{h_1^3k_1}{4} & \frac{h_1^2k_1^2}{6} & \frac{h_1k_1^3}{4} & \frac{k_1^4}{24} \\ h_2 & k_2 & & \cdots & & \cdots & & \cdots & & & \cdots & & & \cdots \\ \vdots & \vdots & & \vdots & & \vdots & & \vdots & & & \vdots & & & \vdots \\ h_i & k_i & & \cdots & & \cdots & & \cdots & & & \cdots & & & \cdots \\ \vdots & \vdots & & \vdots & & \vdots & & \vdots & & & \vdots & & & \vdots \\ h_{25} & k_{25} & \frac{h_{25}}{2} & & & \cdots & & \cdots & & & \cdots & & & \cdots \end{bmatrix}$$

Once the {D} and {DC} matrices are formed, it becomes relatively simple to compute the finite difference expressions for the global partial derivatives. It should be pointed out that the method involves considerable computational effort, as a 14x14 matrix inversion is required for each node.

The numerical example solved earlier was re-solved again using the 4th-order finite difference approximations and the results are presented in Table 2.

x	y	Analytical ϕ	Numerical ϕ	E^*_{rr} ϕ	$\phi_{,x}$	$\phi_{,y}$
-0.234	-1.095	0.26370	0.26419	0.18	0.59	0.13
-0.601	-0.710	0.26994	0.26952	0.15	1.48	0.02
0.195	-0.862	0.51259	0.51366	0.21	0.17	0.54
-0.636	-0.668	0.27199	0.27144	0.20	0.01	1.09
0.000	-0.640	0.52635	0.52771	0.26	0.43	0.30
0.190	-0.630	0.64245	0.64397	0.24	0.11	0.15
0.367	-0.613	0.78067	0.78192	0.16	0.18	0.10
-0.748	-0.523	0.28137	0.28058	0.28	1.08	0.78
-0.885	-0.310	0.30303	0.30273	0.10	2.05	0.31
-0.642	-0.234	0.41762	0.41665	0.23	0.08	0.22

Table 2: Summary of solution to the Test Example obtained by the current method using 4th-order finite difference approximations.

*Discrepancy defined in Eq.(15) in percentages.

CONCLUDING REMARKS

A new method is proposed to generate finite difference approximations for partial derivatives in arbitrary meshes. The main difference between the proposed method and the CFD method developed by Lau (1) is the use of truncated Taylor's series expansions to derive the global-local partial derivative transformation matrix in the proposed method, while in the CFD method the Chain Rule of Partial Differentiation is used. The adoption of such a computational procedure will greatly simplify the amount of work involved in the computer implementation of the method and puts the Finite Difference Method in a highly systematic and flexible fashion. Finite difference approximations can now be easily generated for partial derivatives at any point within a local curvilinear mesh without any cumbersome complication of the computer implementation. The numerical example worked out in this paper indicated that the current method is not only successful in solving 2nd-order partial differential equations, but it can also be extended easily to generate higher-order finite difference approximations. Further extension of the method to derive finite difference approximations for partial derivatives in three-dimensional problems should be a straightforward matter.

REFERENCES

(1) Lau, P.C.M., Numerical solution of Poisson's equation using curvilinear finite differences, Appl. Math. Modelling, 1 (1977) 349-350.
(2) Liszka, T. and Orkisz, J., The finite difference method at arbitrary irregular grids and its applications in applied mechanics, Computers & Structures, 11 (1980) 83-95.
(3) Chu, W.H., Development of a general finite difference approximation for a general domain, J. Comp. Phys., 8 (1971) 392-408.
(4) Jensen, P.S., Finite difference techniques for variable grids, Computers & Structures, 2 (1972) 17-29.
(5) Perrone, N. and Kao, R., A general finite difference method for arbitrary meshes, Computers & Structures, 5 (1975), 45-58.
(6) Frey, W.H., Flexible finite-difference stencils from isoparametric finite elements, Int. J. Num. Methods Eng., 11 (1977) 1653-1665.
(7) Carey, G.F., Cheung, Y.K. and Lau, S.L., Mixed operator problems using least square finite element collocation, Computer Methods in Appl. Mech. and Eng., 22 (1980) 121-130.

(8) Amsden, A.A. and Hirt, C.W., A simple scheme for generating general curvilinear grids, J. Comp. Phys., 11 (1973) 348-359.
(9) Lau, P.C.M., Curvilinear finite difference method for biharmonic equation, Int. J. Num. Methods Eng., 14 (1979) 791-812.

[1] Numerals in parentheses refer to corresponding items in References.

Computational Techniques & Applications: CTAC-83
J. Noye & C. Fletcher (Editors)

A NUMERICAL MODEL OF DROPLET FORMATION

L.E. Cram

CSIRO Division of Applied Physics, Sydney, Australia 2070

The demand for efficient, high-quality arc welding has stimulated attempts to develop simple yet realistic models of the consumable-electrode (MIG or GMA) welding process. We present here a one-dimensional model of droplet formation which will be used to study the physics of droplet growth, detachment and transfer in this form of welding. The model consists of two coupled partial differential equations describing the conservation of mass and axial momentum. We first provide an analytic description of the linear stability of the system, and then present a numerical solution describing its non-linear evolution.

INTRODUCTION

With the increasing need for efficient, high-quality production methods stimulated by the introduction of automatic welding equipment, there has emerged a need for simple yet realistic models of the consumable-electrode welding process. A holistic model must treat the simultaneous action of many phenomena, including the heating and melting of the feed wire, the formation, detachment and transfer of droplets, the arc as a source of heat and momentum, and the hydrodynamic behaviour of the weld pool. This paper describes a model for droplet formation and detachment which will be part of a holistic description of consumable-electrode welding.

Only a few studies of the problem have been undertaken in the past. Amson (1962) assumed that droplet detachment occurs when the "increasing forces acting on the drop have reached limiting equilibrium". He computed this limiting equilibrium by considering the static balance of forces acting on a droplet whose shape was specified by a set of intersecting spheres and tori, without a self-consistent treatment of the form of the free surface of the drop. Greene (1960) and later Voropai and Kolesnichenko (1979) described methods for calculating self-consistently the shapes of pendant droplets, and showed how the point of marginal stability could be identified by a shooting method. Models based on equilibrium configurations have two drawbacks: they cannot describe the temporal development of the system, and they ignore the influence of inertia and impulse on the droplet behaviour.

Lancaster (1980) provided an alternative description of droplet formation and detachment based on a theory of the instability of a liquid cylinder conducting an electric current. The theory was developed by Murty (1960) and verified by experiments on falling streams of mercury. Murty's analysis is based on an assumed sinusoidal perturbation of the surface of the cylinder. To draw an analogy between this configuration and the problem of droplet formation in welding, Lancaster argued that droplet detachment would occur suddenly when the droplet radius reaches a certain critical value which could be determined from measurements of drop radii made during the welding process.

Measured values of drop lengths and velocities are consistent with predictions based on Lancaster's model. The model does include the inertia and impulse effects missing from other models, and it does provide an estimate of the time-scale of the process. However, it is not clear that the initial state of Lancaster's model, namely a liquid cylinder of infinite length, is appropriate in the welding context and, moreover, a perturbation analysis cannot describe accurately the evolution of droplets to the point of detachment.

A non-linear, time-dependent model of droplet formation and detachment would evidently provide valuable insight into an important aspect of metal transfer in consumable-electrode welding. A model of this kind, introduced by Lee (1974) to describe the behaviour of ink-jets in printers, is presented in this paper.

A MODEL OF DROPLET FORMATION

We consider an axisymmetric stream as shown in Figure 1. If the axial component of velocity, v, and the hydrostatic pressure, p, are independent of radial position in the flow, the equation describing conservation of axial momentum is (Lee (1974))

$$\frac{\partial v}{\partial t} + v\frac{\partial v}{\partial z} = -\frac{1}{\rho}\frac{\partial p}{\partial z} + \nu\frac{\partial^2 v}{\partial z^2} + \frac{f}{\rho} \tag{1}$$

Here ρ is the density, ν the kinematic viscosity, and f is the body force, also assumed to be independent of radial position.

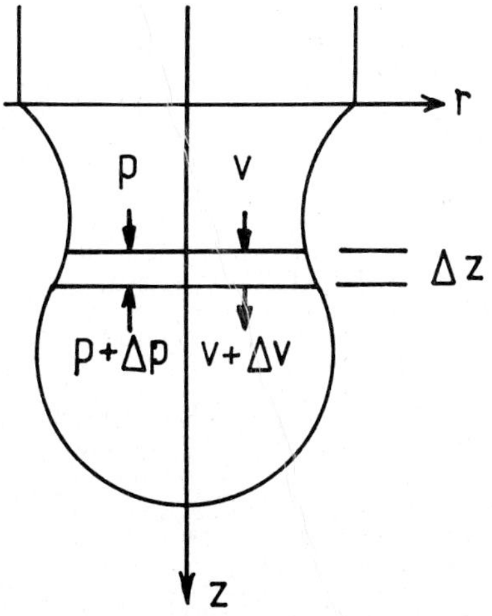

Figure 1
Coordinate system and forces acting on a volume element of a pendant drop

The continuity equation is

$$\frac{\partial r^2}{\partial t} + \frac{\partial v r^2}{\partial z} = 0 \tag{2}$$

where r(z,t) is the radius of the surface of the stream.

The hydrostatic pressure contains two components, one due to the external gas pressure and the other to surface tension. If the gas pressure is uniform, the only contribution to the hydrostatic pressure gradient is due to the surface tension,

$$p_s = T\left(\frac{1}{R_1} + \frac{1}{R_2}\right) \tag{3}$$

Here T is the surface tension and R_1, R_2 are the two principal radii of curvature of the surface of the stream. Using standard results for the radii of curvature of an axisymmetric body (Weatherburn (1939)) we find that the pressure gradient can be written

$$\frac{\partial p}{\partial z} = -\frac{r'r''p_s}{(1 + r'^2)} + \frac{T}{\sqrt{(1 + r'^2)}}\left| -\frac{r'}{r^2} - \frac{r'''}{(1 + r'^2)} + \frac{2r'r''^2}{(1 + r'^2)^2}\right| \tag{4}$$

where $r' = \partial r/\partial z$, etc.

In the process of droplet formation in welding, several forces act in addition to the gradient of surface tension. These include gravity, electromagnetic forces, reaction due to evaporation, and drag from the arc plasma jet. Only gravity is included in the present discussion.

PERTURBATION ANALYSIS

To verify the utility of the one-dimensional model and to explore the stability of the system described above, we compare the results of a perturbation analysis of equations (1) and (2) with exact solutions of two-dimensional models (Rayleigh (1899)). We begin with a uniform, stationary liquid cylinder of radius a, and let v', r' be small-amplitude perturbations in axial velocity and radius, respectively. Equations (1), (2) and (4) then give, to first-order accuracy, the coupled system

$$\frac{\partial r'}{\partial t} + \frac{a}{2}\frac{\partial v'}{\partial z} = 0 \tag{5}$$

$$\frac{\partial v'}{\partial t} - \frac{T}{\rho a^2}\left[\frac{\partial r'}{\partial z} + a^2\frac{\partial^3 r'}{\partial z^3}\right] = 0. \tag{6}$$

The velocity perturbation v' may now be eliminated by differentiating (5) with respect to time and (6) with respect to position, to yield the wave equation

$$\frac{\partial^2 r'}{\partial t^2} + \frac{T}{\rho a}\left[\frac{\partial^2 r'}{\partial z^2} + a^2\frac{\partial^4 r'}{\partial z^4}\right] = 0. \tag{7}$$

If the perturbation in radius, r', has the form

$$r'(z,t) = A\exp[\gamma t + ikz] \tag{8}$$

where γ is the growth rate and ik the wavenumber, the dispersion equation is

$$\gamma^2 = \Gamma^2 a^2k^2(1 - a^2k^2) \tag{9}$$

where $\Gamma^2 = T/2\rho a^3$. The most dangerous mode (maximum growth rate) occurs when

γ^2 is maximum. By differentiating (9) we find that the wavelength λ_d and e-folding time t_d of the most dangerous mode are given by

$$\lambda_d = \sqrt{2}\,\pi d \qquad \text{and} \qquad t_d = \sqrt{d\rho/T} \tag{10}$$

where d = 2a is the diameter of the cylinder. The value of λ_d found using the one-dimensional model (4.443d) is close to the value (4.508d) found by Rayleigh (1899) for the two-dimensional case. It should be noted that the liquid cylinder is unstable for all modes with $0 < k < 1/a$: only modes with wavelengths satisfying $\lambda < \pi d$ are stable.

NUMERICAL ALGORITHM

Two factors complicate the numerical resolution of the nonlinear, coupled partial differential equations (1) and (2). Firstly, as shown above the system is unstable to a wide range of perturbations, so that care must be taken to introduce appropriate damping of unwanted solutions. Second, the surface tension gradient represented by equation (4) contains high-order derivatives which must be computed accurately to avoid catastrophic cancellation errors. The numerical scheme described below appears to overcome these problems.

The differential equations are represented by finite difference equations on discrete meshes $(x_i \mid i = 1,2,\ldots N_x)$ and $(t^j \mid j = 1,2\ldots)$. Time derivatives of the form $\partial f/\partial t$ are expressed by the explicit relation due to Lax (see Roache (1976), p.242)

$$f_i^{j+1} = \frac{1}{2}(f_{i-1}^j + f_{i+1}^j) - \Delta t(\mathrm{RHS})_i^j \tag{11}$$

where $(\mathrm{RHS})_i^j$ contains only spatial derivatives. The application of Lax's method introduces numerical damping which suppresses the development of instabilities.

The time step Δt in equation (11) cannot exceed the value given by the Courant condition, $\Delta t^j < \min(\Delta x/v_i^j)$, without leading to numerical instabilities. However, it is found that the solution of (1) and (2) evolves rapidly, so that time steps are required that are much smaller than those demanded by the Courant condition. A value of Δt given approximately by 0.01 t_d (equation 10) ensures accurate solutions.

Advective spatial derivatives of the form $\partial uf/\partial z$ are represented by "second" upwind differencing (Roache 1976, p.73). We first define donor cell velocities by

$${}_+u_i^j = \frac{1}{2}(u_i^j + u_{i+1}^j); \qquad {}_-u_i^j = \frac{1}{2}(u_i^j + u_{i-1}^j). \tag{12}$$

Advective derivatives are then approximated by

$$\left.\frac{\partial(fu)}{\partial z}\right|_i^j = \frac{{}_+f_i^j\,{}_+u_i^j + {}_-f_i^j\,{}_-u_i^j}{\Delta x},$$

where

$${}_+f_i^j = f_i^j \quad \text{if} \quad {}_+u_i^j > 0; \quad {}_+f_i^j = f_{i+1}^j \quad \text{otherwise},$$

and $\quad {}_{-}f_i^j = f_{i-1}^j \quad$ if $\quad {}_{-}u_i^j > 0; \quad {}_{-}f_i^j = f_i^j \quad$ otherwise.

This formula accurately conserves the advected scalar: after 2000 time steps, the mass conservation of the system of difference equations is accurate to better than 0.5%. Other conservative systems could of course be used, but this algorithm is probably the simplest to implement while still preserving the accuracy and stability required for the problem at hand.

The evaluation of the pressure gradient given by equation (4) is difficult on account of the high order derivatives which appear. Strong cancellation of terms in this equation occurs at points which have a hyperbolic character (i.e. where the Gaussian curvature is negative -- see Weatherburn (1939), p.75), and at points where the second curvature is zero [points of inflection in r(z)]. For this reason the second and third derivatives must be evaluated with a relative accuracy approaching 0.01. It has been found that 5-point central difference formulae are sufficiently accurate, provided that the spatial resolution is sufficiently fine. The relevant formulae can be found in the handbook of Abramowitz and Stegun ((1964), Table 25.2). By experimenting with hemispherical surfaces whose curvature is known exactly it was found that the spatial mesh should satisfy $\Delta x < 0.1a$ to ensure satisfactory accuracy.

EXAMPLE

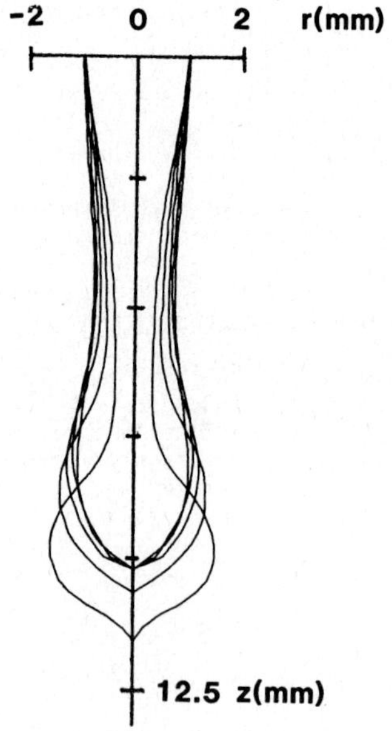

Figure 2
The time development of a liquid drop evolving under the influence of gravitational forces. The shape of the drop is shown at five instants separated by equal time steps of 11 msec.

Figure 2 illustrates the evolution of a droplet under the simultaneous action of gravity and surface tension. The material properties adopted for this calculation correspond to those of molten iron: $T = 1500$ dyne cm^{-1}, $\nu = 0.0028$ stoke, and $\rho = 7.86$ gm cm^{-3}. It is assumed that the droplet hangs from the planar face of a rigid cylinder of radius 0.1 cm. The droplet contains initially 0.3 g of liquid, and this total mass is not changed during the calculation. The boundary condition imposed at the three-phase intersection of the droplet, cylinder and surrounding atmosphere ensures that the liquid surface is perpendicular to the planar face of the cylinder (more general boundary conditions can be specified in terms of the surface energies). A moving boundary occurs at the end of the drop: this is represented by a spherical "cap" with a radius of curvature equal to that of the liquid at the terminal mesh point.

The droplet evolves through a sequence of complex shapes which do not correspond to a sequence of equilibrium configurations: the inertia term in the momentum equation is of the same order as the pressure gradient term, especially in the bulbous, almost spherical, incipient droplet. Characteristic features of the flow include the acceleration of the tip of the droplet, and the formation of a neck which extends over a length corresponding to many times the diameter of the flow. Work is in progress to explore the influence of boundary conditions and material constants on the numerical model.

DISCUSSION

The one-dimensional model presented in this paper appears to describe the formation of droplets with sufficient accuracy to be used in studies of consumable-electrode welding. By using a one-dimensional approximation, rather than the full system of two- or three-dimensional, time-dependent Navier-Stokes equations it will be possible to incorporate a numerical model of droplet formation into a holistic model of the welding process. As an indication of the computer time required by the one-dimensional model we may note that the results illustrated in Figure 2 were obtained in about 1 minute on a VAX 750 computer.

It might be of some interest to note that the problem of computing droplet shapes was responsible for some of the earliest developments of numerical techniques to solve differential equations, including the classic methods due to Bashforth and Adams (1883). Since it took the present author considerable time and effort to develop computing techniques of sufficient accuracy to correctly evaluate the effects of surface tension, he has a developed a considerable respect for these workers. It seems that it took Bashforth about two years to calculate the theoretical forms of droplets using Adams' methods, and the cost was 50 pounds (in 1856 currency).

REFERENCES

[1] Amson, J.C., Brit. Weld. J. 9 (1962) 232-248.

[2] Greene, W.J., Trans. Amer. Inst. Elec. Eng. 2 (1960) 194-202.

[3] Voropai, N.M. and Kolesnichenko, A.F., Avt. Svarka. 9 (1979) 27-32.

[4] Lancaster, J.F., "Arc Physics and Weld Pool Behaviour", ed. W. Lucas (The Welding Institute, Cambridge, 1980), p.135-146.

[5] Murty, G.S., Arkiv f. Fysik 18 (1960) 241-250.

[6] Lee, H.L., IBM J. Res. Develop. (1974) 364-369.

[7] Rayleigh, Lord, Scientific Papers, I (1899) 377.

[8] Roache, P.J., "Computational Fluid Dynamics" (Hermosa, Alberquerque, 1976).

[9] Weatherburn, C.E., "Differential Geometry" (Cambridge University Press, 1939).

[10] Abramowitz, M. and Stegun, I.A., "Handbook of Mathematical Functions" (Dover, New York, 1964).

[11] Bashforth, F. and Adams, J.C., "An attempt to test the theories of capillary action" (Cambridge University Press, 1883).

FINITE ELEMENT METHODS

Computational Techniques & Applications: CTAC-83
J. Noye & C. Fletcher (Editors)

UPPER BOUND A-POSTERIORI ERROR ESTIMATES FOR THE FINITE ELEMENT METHOD APPLIED TO LINEAR ELASTICITY

D.W. Kelly and J. Donovan

University of New South Wales

A-posteriori estimates of the error in energy, displacement and stress are obtained for finite element solutions to problems in linear elasticity. A complementary analysis of the error is used to try to ensure the error is overestimated. The analysis is implemented in a post-processing mode. It is carried out independent of the parent finite element analysis in much the same way as graphical processing of the results. Use of the error measures to guide local refinement of the mesh if the error is unacceptably high is discussed.

INTRODUCTION

A-posteriori estimates of the energy of the error in a finite element solution have been developed by Babuška and Rheinboldt (e.g. 1979) to guide local mesh refinement in self-adaptive finite element codes. A great deal of emphasis has been given to the effectiveness of these estimates but it has only been possible to show that the estimate becomes accurate in the limit of the adaptive refinement as the element size tends to zero and the mesh becomes "optimal". Here the word optimal is used in the sense of removing the detrimental influence of singularities on the rate of convergence.

Recently it has been shown (Kelly [1982] and [1983]) that a-posteriori error estimates can be based on a complementary approach and provide an upper bound on the error for arbitrary meshes. In this paper an attempt is made to expand the basis of that work to include linear elasticity and to describe a computer program which is being developed to implement the theory in a post-processing mode. It is intended that this program should interface with commercial finite element systems in much the same way as a graphics post-processor.

We will first generalise the fundamental theory by making reference to the Principle of Virtual Work. Application of the finite element method to two- and three-dimensional problems in elasticity and fluid mechanics, including in elasticity the analysis of plates and shells, can be based on integral relations expressing this principle (see, for example, Zienkiewicz [1977]). The equivalence between these integrals and weighted residual approximations to the equilibrium equations is shown in that reference. In the finite element method the weighting functions have a local base. Satisfaction of equilibrium in a weighted integral sense is therefore enforced on the same local base, usually centred on a node and extending only over neighbouring elements.

Satisfaction of the equilibrium equations in a pointwise sense does not occur without the application of a residual force system. These residuals include a distributed body force on the element and tractions applied along the interface between elements. For a compatible finite element formulation the error in the solution can be interpreted as the response to these forces.

For an a-posteriori analysis of the error it is advantageous to be able to try to determine this response separately on each element and to over-estimate rather than under-estimate the error. In Kelly [1983] it was shown that the residual forces, forced to equilibrate locally in the weighted integral sense identified above, can be further partitioned into self-equilibrating systems on each element. A complementary analysis of the error can then be performed element by element and the energy of the error on each element summed to provide an upper bound on the energy of the error in the solution. Upper estimates of the error in stress and displacement follow by the application of theory which closely parallels the classical unit load method.

In the following sections we only deal with the compatible displacement formulation of the finite element method and the four node bilinear element. We concentrate on demonstrating the existence of self-equilibrating partitions of the residual forces on each element. We then discuss the error estimates and the computer program which is being developed to implement the error analysis in a post-processing mode. The essential features are then demonstrated on three examples. In the short term we seek not only to identify errors caused by discretization in the parent analysis (that is, the division of the domain into elements on which assumptions are made about the variation of displacement), but also the error caused by geometric non-linearity and round-off error. Discussion of these applications is confined to the concluding section.

THE PRINCIPLE OF VIRTUAL WORK AND WEIGHTED RESIDUAL APPROXIMATIONS

Consider a body in equilibrium loaded by body forces $\underset{\sim}{b}$ and applied tractions $\bar{\underset{\sim}{t}}$. The virtual work statement requires that for arbitrary displacements $\delta\underset{\sim}{u}^T = [\delta u, \delta v, \delta w]$

$$\int_V \delta\underset{\sim}{\varepsilon}^T \underset{\sim}{\sigma}\, dV - \int_V \delta\underset{\sim}{u}^T \underset{\sim}{b}\, dV - \int_\Gamma \delta\underset{\sim}{u}^T \bar{\underset{\sim}{t}}\, d\Gamma = 0 \tag{1}$$

where $\underset{\sim}{\sigma} = [\sigma_{xx}, \sigma_{xy} \ldots]$ is the stress state in the body in equilibrium, and $\delta\underset{\sim}{\varepsilon}$ are the strains caused by the virtual displacements. Integrating the first term using Gauss's Theorem

$$\int_V \delta\underset{\sim}{\varepsilon}^T \underset{\sim}{\sigma}\, dV = -\int_V \left[\left(\frac{\partial\sigma_{xx}}{\partial x} + \frac{\partial\sigma_{xy}}{\partial y} + \frac{\partial\sigma_{xz}}{\partial z}\right)\delta u + (\quad)\,\delta v + (\quad)\,\delta w\right] dV$$

$$+ \oint \left[(\sigma_{xx} l + \sigma_{xy} m + \sigma_{xz} n)\delta u + (\quad)\delta v + (\quad)\delta w\right] d\Gamma \tag{2}$$

where l, m and n are the usual direction cosines of the normal to the surface. It is standard in finite elements to take

$$\delta\underset{\sim}{u} = \underset{\sim}{N}(x,y,z)\ \delta\hat{u}_i \tag{3}$$

where the $\underset{\sim}{N}$ are shape functions defined on a local base (see for example Figure 1) and $\hat{u}_i$ is a nodal displacement. The virtual work statement (1) is therefore enforcing satisfaction of the following equilibrium equations in a weighted integral sense on a local base;

$$\left.\begin{array}{l}\dfrac{\partial\sigma_{xx}}{\partial x} + \dfrac{\partial\sigma_{xy}}{\partial y} + \dfrac{\partial\sigma_{xz}}{\partial z} + b_x = 0 \\[2ex] \dfrac{\partial\sigma_{yx}}{\partial x} + \dfrac{\partial\sigma_{yy}}{\partial y} + \dfrac{\partial\sigma_{yz}}{\partial z} + b_y = 0 \\[2ex] \dfrac{\partial\sigma_{zx}}{\partial x} + \dfrac{\partial\sigma_{yz}}{\partial y} + \dfrac{\partial\sigma_{zz}}{\partial z} + b_z = 0\end{array}\right\} \quad \text{on } V \qquad (4)$$

and

$$\underset{\sim}{\sigma}^T \underset{\sim}{n} = \bar{\underset{\sim}{t}} \qquad (5)$$

on Γ_t. The equations (4) express force equilibrium on a differential volume, and (5) expresses equilibrium on the surface Γ.

RESIDUALS IN THE FINITE ELEMENT SOLUTION

The equilibrium equations (4) and (5) are not satisfied pointwise so that, for example,

$$\frac{\partial\sigma_{xx}}{\partial x} + \frac{\partial\sigma_{xy}}{\partial y} + \frac{\partial\sigma_{xz}}{\partial z} + b_x = R_x \qquad (6)$$

with similar expressions in the y and z directions, and

$$\underset{\sim}{\sigma}_1^T \underset{\sim}{n} - \underset{\sim}{\sigma}_2^T \underset{\sim}{n} = \rho \qquad (7)$$

on any boundary (including Γ_t but also element interfaces) on which there is a discontinuity in traction between sides 1 and 2. It is possible to identify ρ as a singular part of the residual in (6) caused by discontinuities in the stresses.

We can now combine (3) and (1). The finite element solution ensures that

$$\int_V N_i R dV + \int_\Gamma N_i \rho d\Gamma = 0 \qquad (8)$$

where a typical shape function N_i is shown in Figure 1. Obviously this applies wherever the virtual work expression can be identified as the basis of the finite element approximation. It includes the most common formulations of linear elasticity. Here we will concentrate on plane stress, and the special applications in plates and shells.

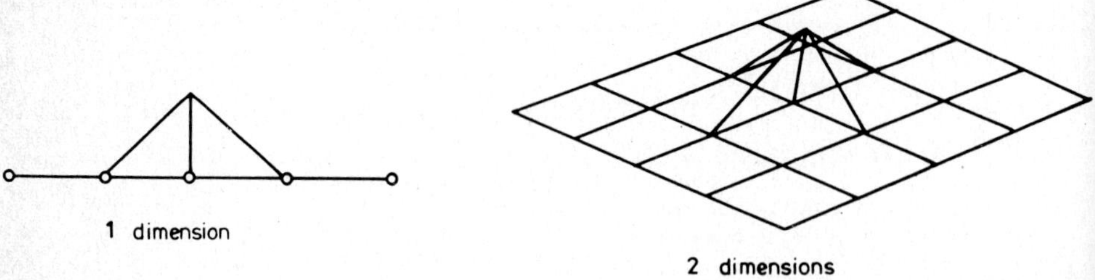

Figure 1. Bases for finite element shape functions

EXAMPLE: RESIDUALS FOR THE BILINEAR ELEMENT AND PLANE STRESS.

For small displacements and plane stress

$$\varepsilon_{xx} = \frac{\partial u}{\partial x},\ \varepsilon_{yy} = \frac{\partial v}{\partial y},\ \varepsilon_{xy} = \frac{\partial u}{\partial y} + \frac{\partial v}{\partial x} \quad \text{and } \sigma_{zz} = \sigma_{xz} = \sigma_{yz} = 0. \tag{9}$$

where u and v are the in-plane displacement components. Defining the material constants E and ν as Young's Modulus and Poisson's ratio respectively

$$\sigma_{xx} = \frac{E}{1-\nu^2}\left(\frac{\partial u}{\partial x} - \nu\frac{\partial v}{\partial y}\right)$$

$$\sigma_{yy} = \frac{E}{1-\nu^2}\left(\frac{\partial v}{\partial y} - \nu\frac{\partial u}{\partial x}\right) \tag{10}$$

and

$$\sigma_{xy} = \sigma_{yx} = \frac{E}{2(1+\nu)}\left(\frac{\partial u}{\partial y} + \frac{\partial v}{\partial x}\right)$$

Substituting in (4) gives the equilibrium equations in terms of displacement components as

$$\frac{E}{1-\nu^2}\left[\frac{\partial^2 u}{\partial x^2} + \left(\frac{1-\nu}{2}\right)\frac{\partial^2 u}{\partial y^2}\right] + \frac{E(1-2\nu)}{2(1-\nu^2)}\frac{\partial^2 v}{\partial x \partial y} + b_x = 0$$

$$\frac{E}{1-\nu^2}\left[\frac{\partial^2 v}{\partial y^2} + \left(\frac{1-\nu}{2}\right)\frac{\partial^2 v}{\partial x^2}\right] + \frac{E(1-2\nu)}{2(1-\nu^2)}\frac{\partial^2 u}{\partial x \partial y} + b_y = 0 \tag{11}$$

Now consider the four node bilinear element in Figure 2. On this element displacements are approximated as

$$\hat{u} = a_1 + b_1 x + c_1 y + d_1 xy$$

and

$$\hat{v} = a_2 + b_2 x + c_2 y + d_2 xy \tag{12}$$

so that the only non-zero derivatives appearing in (11) are

$$\frac{\partial^2 \hat{u}}{\partial x \partial y} = d_1 \quad \text{and} \quad \frac{\partial^2 \hat{v}}{\partial x \partial y} = d_2.$$

Substituting in (11) gives

$$\frac{E(1-2\nu)}{2(1-\nu^2)} d_2 + b_x = R_x$$

and

$$\frac{E(1-2\nu)}{2(1-\nu^2)} d_1 + b_y = R_y \tag{13}$$

where d_1 and d_2 are defined using the four nodal displacements on the element obtained from the finite element solution.

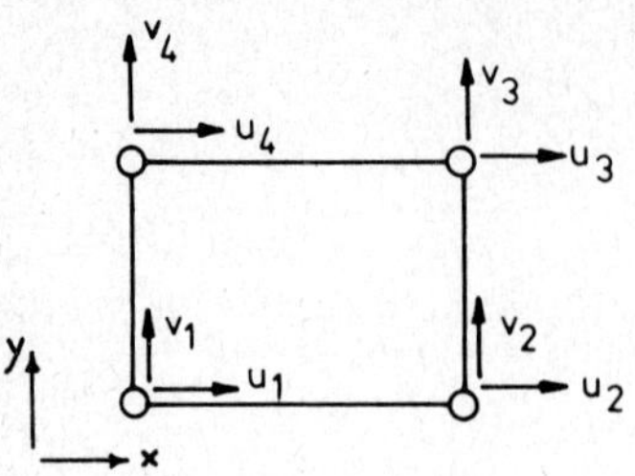

Figure 2: The four node bilinear element.

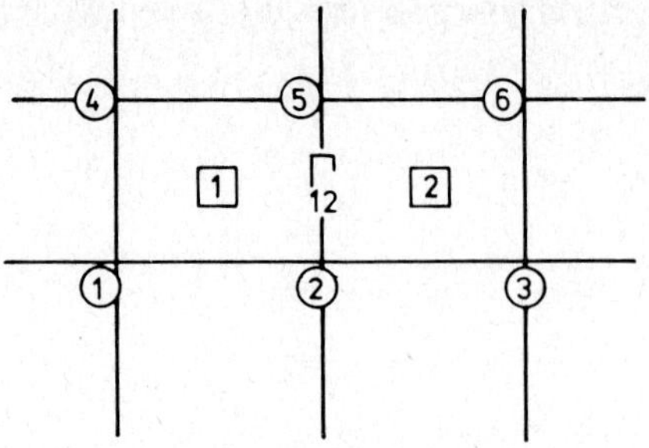

Figure 3: Element configuration for interface residual calculation.

On the element interfaces stresses are not continuous. Referring to the two elements in Figure 3, it is obvious that

$$\text{on element } 1 \qquad \hat{\sigma}^1_{xx} = \frac{E}{1-\nu^2}\left(\frac{\partial \hat{u}}{\partial x} - \nu \frac{\partial \hat{v}}{\partial y}\right) = f(u_1, u_2, u_4, u_5) \tag{14}$$

and

$$\text{on element } 2 \qquad \hat{\sigma}^2_{xx} = f(u_2, u_3, u_5, u_6).$$

where the superscript denotes the element number. The normal and interface tractions (t_n and t_t respectively) on the interface Γ_{12} are

$$t_n{}^1 = \sigma_{xx}{}^1\Delta_1, \quad t_n{}^2 = \sigma_{xx}{}^2\Delta_2$$

and

$$t_t{}^1 = \sigma_{xy}\Delta_1, \quad t_t{}^2 = \sigma_{xy}{}^2\Delta_2 \tag{15}$$

where Δ_i is the thickness of the ith element and the stresses σ_{xx} and σ_{xy} are evaluated at the interface. Equilibrium across the interface requires

$$t_n{}^1 + t_n{}^2 = 0$$

and

$$t_t{}^1 + t_t{}^2 = 0 \tag{16}$$

In general $\sigma_{xx}{}^1 \neq \sigma_{xx}{}^2$ and $\sigma_{xy}{}^1 \neq \sigma_{xy}{}^2$ even if the thickness of the two elements is the same. Therefore external tractions

$$\left(\rho_{12}\right)_n = \left[t_n{}^2 - t_n{}^1\right]_{\Gamma_{12}}$$

$$\left(\rho_{12}\right)_t = \left[t_t{}^2 - t_t{}^1\right]_{\Gamma_{12}} \tag{17}$$

must be applied to ensure pointwise satisfaction of equilibrium on the interface.

EQUILIBRATION OF RESIDUALS IN THE FINITE ELEMENT METHOD

The weighted residual process ensures equilibration of the residuals in the sense of (8)

i.e. $$\int_V N_i R \, dV + \int_\Gamma N_i \rho \, ds = 0$$

where the N_i are usually defined on a local base as indicated in Figure 1.

To facilitate the computations in the post-processing mode we wish to isolate elements with self-equilibrating residual force systems. In one dimension the equilibrating residual force system can be easily identified. Referring to the linear finite elements in Figure 4 we define.

$$\bar{\rho}_2^1 = -\int_{\Omega_1} N_2 \, r \, dx \tag{18a}$$

and

$$\bar{\rho}_2^2 = -\int_{\Omega_2} N_2 \, r \, dx \tag{18b}$$

where ρ_i^j is the residual ρ at node i associated with element j.

From (8) the finite element solution satisfies

$$\int_{\Omega_1 + \Omega_2} N_2 \, r \, dx + \rho_2 = 0 \tag{19}$$

Obviously adding (18a) and (18b) and comparing with (19)

$$\bar{\rho}_2^1 + \bar{\rho}_2^2 = \rho_2 \tag{20}$$

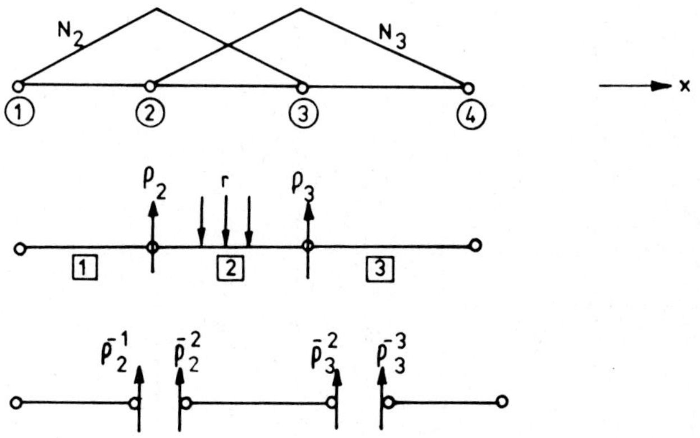

Figure 4. Linear finite elements and their residuals in one dimension

Similarly at node 3 we define

$$\bar{\rho}_3^2 = - \int_{\Omega_2} N_3 \ r \ dx \tag{21a}$$

and

$$\bar{\rho}_3^3 = - \int_{\Omega_3} N_3 \ r \ dx \tag{21b}$$

Summing forces on element 2 and substituting from (20) and (21a)

$$\bar{\rho}_2^2 + \int_{\Omega_2} r \ dx + \bar{\rho}_3^2$$

$$= - \int_{\Omega_2} N_2 \ r \ dx - \int_{\Omega_2} N_3 \ r \ dx + \int_{\Omega_2} r \ dx$$

$$= - \int_{\Omega_2} (N_2 + N_3) \ r \ dx + \int_{\Omega_2} r \ dx$$

$$= \ 0 \tag{22}$$

since at any point on element 2 $N_2 + N_3 = 1$. The result in (22) ensures that the partitioning of ρ_2 and ρ_3 into the components in (18a) and (18b),(21a) and (21b) and associating $\bar{\rho}_2^2$ and $\bar{\rho}_3^2$ with element 2 gives a self-equilibrating residual force system on the element. We note here the fundamental role played by the weighted residual process in equation (8) in this proof.

To extend this analysis to two dimensions, consider the addition of a new element to the two dimensional mesh in Figure 5. The new element is identified by a square. Existing elements in the mesh to which it is added are identified in the figure by stars. The following process has already been applied to the elements marked with stars.

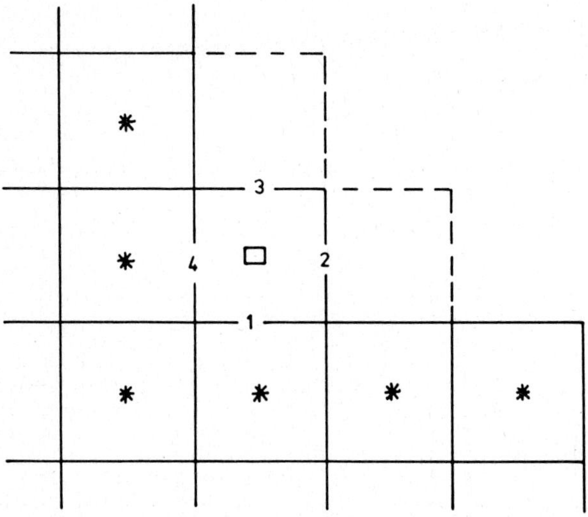

Figure 5. Assembling a two-dimensional mesh

On the element sides 1 and 4 $\bar{\rho}_i^{\square}$ i = 1 and 4 for the element $\square$ are defined from

$$\bar{\rho}_i^{\square} = \rho_i - \bar{\rho}_i^{*} \qquad i = 1 \text{ and } 4 \tag{23}$$

where ρ_i is the residual on the interface i and $\bar{\rho}_i^{*}$ the component preassigned to the neighbouring element *. We can now arbitrarily set $\bar{\rho}_2^{\square}$ and $\bar{\rho}_3^{\square}$ to ensure equilibration on the element $\square$. That is, to ensure

$$\int_{\square} r \, dx + \int_{\Gamma_{\square}} \bar{\rho}^{\square} \, ds = 0 \tag{24}$$

the values of $\bar{\rho}_2^{\square}$ and $\bar{\rho}_3^{\square}$ are obviously not unique.

The interface residuals ρ_2 and ρ_3 will be preserved when the new elements are added into the dotted positions on Figure 5 which will have the values on the interfaces 2 and 3 defined using (23).

We see that in general there will be an infinity of ways to partition the residuals to form self-equilibrating systems on each element and procedures could be defined (see Kelly [1983]) to identify exactly a self-equilibrating system. In that paper it is noted that for a mesh of square bilinear elements the partitioning factors for the residuals ρ to form the self-equilibrating systems on each element were equal to ½ when the error was quadratic. In this application the residuals were constant throughout the mesh. The one-dimensional proof indicates that partitioning factors of ½ should be reasonable on meshes in which neither the residuals nor the elements exhibit rapid variations.

Finally we point out that the search for equilibrating residual systems on each element is only for the convenience of the post-processing computation. We could revert to the weighted systems identified in (8) if the element partitioning proves unreliable, and hence retain the theoretical guarantee of exact self-equilibrating systems.

THE ERROR ESTIMATES

In the standard complementary approach in linear elasticity stresses rather than displacements are discretized. The stresses are expressed as

$$\underset{\sim}{\sigma} = \underset{\sim}{\sigma}_b + \underset{\sim}{\sigma}_a \tag{25}$$

where both $\underset{\sim}{\sigma}_b$ and $\underset{\sim}{\sigma}_a$ satisfy the equilibrium equations. The basic distribution $\underset{\sim}{\sigma}_b$ satisfies the force boundary conditions on Γ_t. The additional distribution $\underset{\sim}{\sigma}_a$ are self-equilibrating systems satisfying

$$\underset{\sim}{\sigma}_a = 0 \text{ on } \Gamma_t$$

whose amplitude can be chosen to minimise the complementary energy in the solution.

The essential features are that a feasible complementary solution is obtained by the superposition of self-equilibrating systems, and that the complementary energy functional exhibits a positive minimum when all the prescribed displacements are to be set to zero. Any feasible complementary solution will therefore provide an upper bound on the energy.

The residual forces in the finite element solution have been partitioned into self-equilibrating systems on each element. The a-posteriori error analysis can therefore proceed one element at a time and the energy of the error in the total solution obtained as a summation of the energy of the error on each element. If exact equilibration is maintained there is no approximation and the error estimate will be a guaranteed upper bound.

Determination of upper bounds for errors in stress and displacement requires the following extension of the theory. We write the governing differential equation (11) in the form

$$A(u) = b$$

For linear elastic analysis an estimate of the error in the vicinity of any point x can be found by considering the auxiliary problem

$$A(v) = p(x) \tag{26}$$

where the role of the function p(x) associated with the point x will be indicated below. The condition

$$\int_\Omega vb\,d\Omega = \int_\Omega vA(u)\,d\Omega = \int_\Omega A(v)u\,d\Omega = \int_\Omega p(x)u\,d\Omega = \bar{u}(x) \tag{27}$$

can be used to obtain a measure of the error in u(x). Let the approximate finite element solution be $\hat{u}$. Then

$$A\hat{u} = \hat{b} \quad \text{and } A\hat{v} = \hat{p} \tag{28}$$

and

$$\int_\Omega (b-\hat{b})(v-\hat{v})\,d\Omega = \int_\Omega bv\,d\Omega - \int_\Omega \hat{b}v\,d\Omega - \int_\Omega (b-\hat{b})\hat{v}\,d\Omega$$

$$= \bar{u}(x) - \bar{\hat{u}}(x) - \int_\Omega A(u-\hat{u})\hat{v}\,d\Omega\,. \tag{29}$$

If either the Galerkin method or Rayleigh Ritz is used the integral

$$\int_\Omega A(u-\hat{u})\hat{v}\,d\Omega \tag{30}$$

is automatically zero. Then the error in $\bar{u}(x)$

$$\varepsilon = \bar{u}(x) - \bar{\hat{u}}(x) = \int_\Omega (b-\hat{b})(v-\hat{v})\,d\Omega\,. \tag{31}$$

$$\text{Now} \quad \varepsilon^2 = \left[\int (b-\hat{b})(V-\hat{v})d\Omega\right]\left[\int (b-\hat{b})(v-\hat{v})d\Omega\right]$$

$$= \left[\int A(u-\hat{u})(v-\hat{v})d\Omega\right]\left[\int A(u-\hat{u})(v-\hat{v})d\Omega\right]$$

$$\leq \left|\int A(u-\hat{u})(u-\hat{u})d\Omega\right|\left|\int A(v-\hat{v})(v-\hat{v})d\Omega\right|$$

$$= \left|\left|e(u)\right|\right|_E^2 \quad \left|\left|e(v)\right|\right|_E^2 \quad . \tag{32}$$

where $e(u) = u-\hat{u}$ and $e(v) = v-\hat{v}$ are the errors in both solutions. Hence the error in u at the point x is given by

$$\varepsilon \leq \left|\left|e(u)\right|\right|_E \quad \left|\left|e(v)\right|\right|_E \tag{33}$$

where the sense of the error ε is defined by the last integral in (27)

$$\text{i.e.} \quad \varepsilon = \int p(x)(u(x) - \hat{u}(x))d\Omega \tag{34}$$

Here we note that $\left|\left|e(u)\right|\right|_E^2$ is the energy of the error in u which is, of course, unknown. However the complementary estimate of the error provides an upper bound for this energy, and for the energy of the error in v. Both can be substituted into (33) without violating the upper bound indicated in that equation.

Errors in stresses follow by direct implementation of (34) at two neighbouring points to define a "strain". The success of these pointwise estimates will obviously depend on the choice of p(x) since a finite element solution to (26) is required (perhaps as an additional load case in the original finite element solution).

IMPLEMENTATION OF THE THEORY

The program which has been developed to implement this theory has two fundamental features. First it is intended to operate in a post-processing mode requiring as data only the results sent to file by a finite element program for graphics post-processing. Second, to achieve generality in applications to plates and shells and to arbitrary distortion of the elements, the complementary solution is at present, performed numerically. We will refer to the parent mesh and the submesh. The parent mesh is the mesh of the original finite element solution for which the error is sought. The submesh comprises a simple mesh of 4 or 9 elements into which each parent element is subdivided to perform the complementary analysis. This analysis considers separately the response to the self-equilibrating set of residual forces on the domain of each parent element.

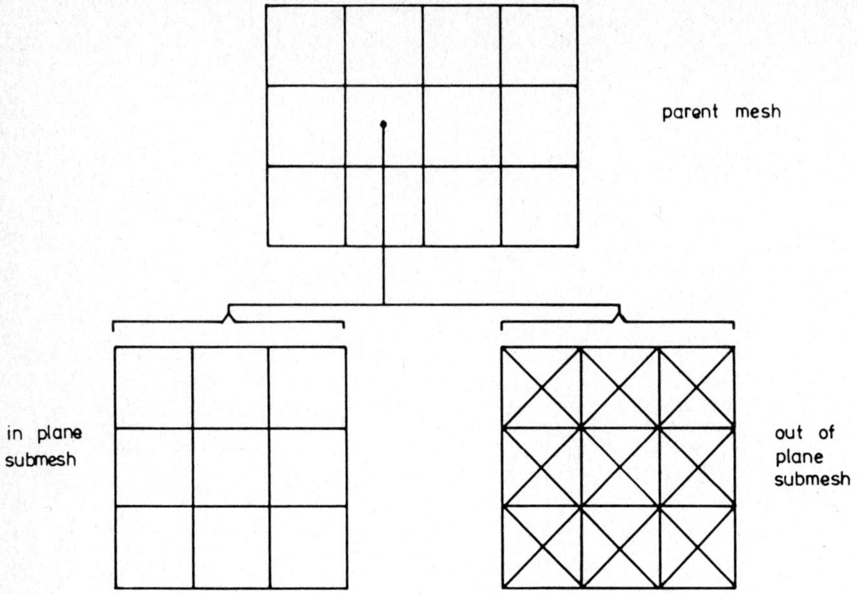

Figure 6. Submesh configuration

A parent mesh and the associated submesh for one element is shown in Figure 6. For applications in thin plates the submesh analysis combines plane stress and out -of-plane bending. The traction discontinuities on the boundary of the parent element include in-plane normal and shear tractions, and out-of-plane shear and bending moments. These tractions can vary on the element side (at most linearly for the bilinear element) but for simplicity here we take only the mean value and assume it is constant on the element side. Also for simplicity we partition the interface residuals between elements using simple splitting factors of ½.

The membrane element for the submesh is a four-node bilinear finite element whose stiffness matrix is based on reduced integration. The fact that this element will provide upper bounds on the response to the residual forces is discussed in Kelly [1982] and proved in special circumstances. The out-of-plane response is provided by the constant moment triangle developed by Morley [1971] who showed that it provides a complementary solution with an upper bound on the energy.

The program is intended to operate in a post-processing mode. At present the data required includes only nodal co-ordinates, element nodal connectivity and nodal displacements. Stresses, shear forces and moments are calculated independently in the error analysis because they are required on element boundaries which may not be a standard option in the parent finite element program.

Finally it is to be noted that it should be possible to greatly simplify the submesh analysis. Applications in Kelly [1983] consider only the Laplace equation and square elements and the complementary solution was obtained analytically with negligible computational expense. In the present work emphasis has so far been placed on demonstrating generality, feasibility as a post-processor and the accuracy of the error analysis on a wider range of problems. Attention will soon have to be given to the economics of the process.

APPLICATIONS

In the following applications the parent finite element program is a version of SAPIV (Bathe et al [1973]) maintained and updated at the University of New South Wales. The first two examples consider only plane stress and the standard Type 3 bilinear finite element is used. The third example uses the Type 6 plate/shell element in the program. It is a four node quadrilateral element comprising four triangles from which degrees of freedom, except at the four nodes, have either been condensed or constrained.

Example 1. In-Plane Bending (plane stress).

A linear variation of displacement was prescribed on uniform meshes of 16 and 64 square elements as shown in Figure 7a. The residual forces defined by (13) and (17) on each element in the 16 element mesh are shown in Figure 7b. Here a factor of ½ is used to partition the boundary residuals to neighbouring elements.

Summing the residuals we note

$$\Sigma F_x = 0$$

$$\Sigma F_y = 2 \times 80 \times 2.5 + 2 \times 93 \times 2.5 - 277 \times 2.5 \times 2.5 \times .5$$

$$= 0$$

Also shown in Figure 7c is a physical interpretation of the role played by the residual forces. They distort the finite element from the exact shape to the shape allowed by the assumed displacement variation on the element.

The energy for the exact solution is 1.300. The energy of the finite element solutions are given in Table 1 together with the energy of the error predicted by the complementary post-processor. Valid upper bounds are obtained. Also we noted that the energy of the error is the same on each element. The work by Babuška (e.g. 1979) indicates that this mesh is "optimal" and we note that halving the mesh size has led to a reduction of the error by a factor of 4 which is the optimal rate of convergence for energy using the bilinear element.

No. Elements	Exact Strain Energy	Strain Energy in F.E. sol.	Predicted energy of error	Exact Energy of error
16	1.300	1.338	$\leqslant .057$	0.038
64	1.300	1.309	$\leqslant .014$	0.009

TABLE I

Example 2. Shear Load on Cantilever (plane stress).

In this second example we concentrate on the error in stress and displacement. The cantilevered beam shown in Figure 8 has a tip shear load applied parabolically. The beam has a length to depth ratio of 10 so shear deformation will be small. The "exact" vertical and horizontal tip deflections and horizontal stress at A given in Table II have therefore been obtained using simple bending theory.

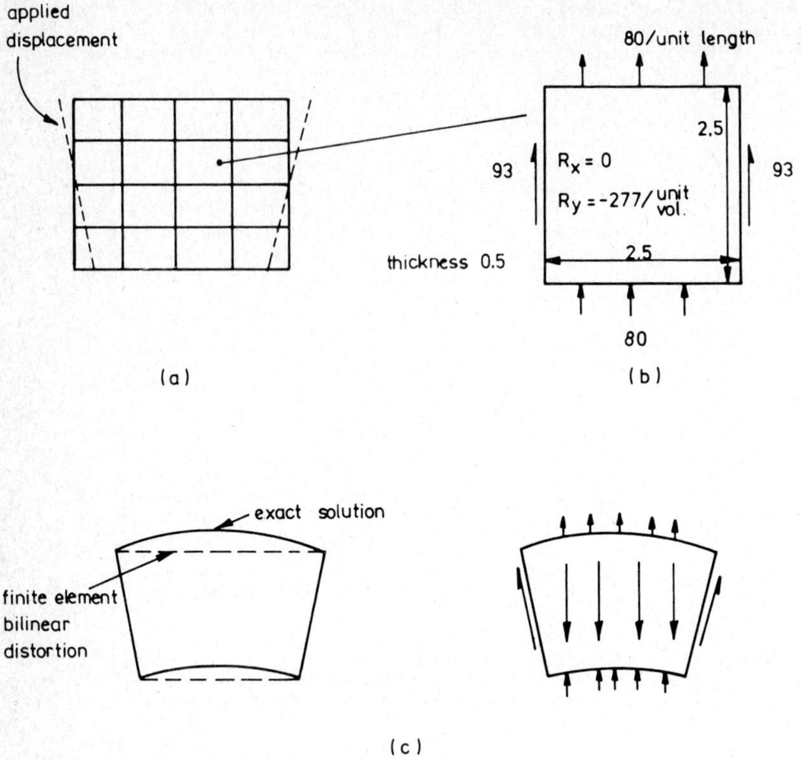

Figure 7. In plane bending example

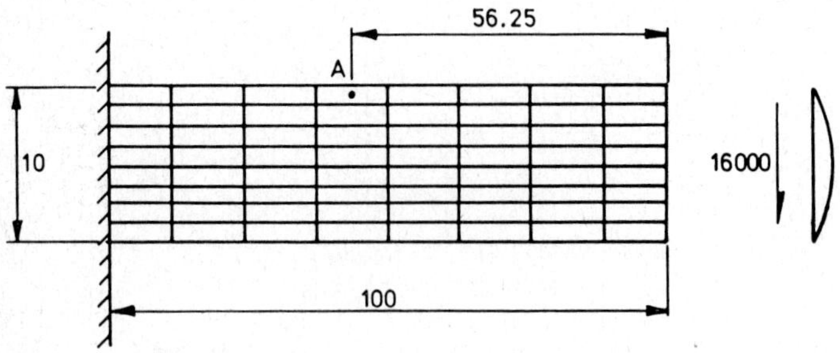

Figure 8. Shear Load on Cantilever

	"exact" sol.	f.e. solution	$\left\lVert e(u) \right\rVert_E^2$ equation (32).	$\left\lVert e(v) \right\rVert_E^2$	Predicted bound on error	"exact" error
energy	84715	53268	69560		69560	31447
vert. tip defl.	5.2947	3.3293	69560	2.715×10^{-4}	4.3475	1.9654
horz.tip defl.	0	10^{-12}	69560	10^{-24}	3×10^{-10}	0
Stress at A	94500	58969	69560		62372	35531

Table II

The 4 node bilinear element is unsuited to this problem and a substantial error is detected. The error in vertical tip deflection was assessed using equation (32) with the applied load associated with the auxiliary problem equal to the distributed shear load scaled to give a unit vertical load. Therefore from (32)

$$\text{error in vertical tip displacement} \leqslant (69560)^{\frac{1}{2}} \left(\frac{69560}{16000^2}\right)^{\frac{1}{2}}$$

$$= 4.3475.$$

For the error in horizontal displacement the loading was a uniform horizontal traction applied to the end of the beam and scaled again to give a total load of 1.

To predict the error in stress at A uniform tractions giving a total load of 1 were applied separately to either side of the element to predict the error in displacements at the midsides of the element. The error in strain was then obtained and converted to an error in stress.

We note with satisfaction that the bounds are roughly the same distance from the exact solution as the parent finite element solution. The bound and the finite element solution therefore form a symmetric envelope on the exact solution.

Example 3. Asymmetric stiffener.

The stiffened plate shown in Figure 9 is subject to uniform extension of the stiffener and base plate. The boundary conditions applied on ends A and B simulate stiff bulkheads connecting to both the stiffener and plate. Stiff beams at C and D suppress the vertical deflection of the plate. Horizontal deflections applied at A and B induced the loads caused by bending on the tension side of a box beam.

Application of the finite element program SAPIV required use of the plate/shell element because the analysis is attempting to determine the out-of-plane twisting of the stiffener and the out-of-plane movement of the base plate caused by the geometric asymmetry provided by the cutout in the stiffener. Analysis using the mesh shown in Figure 9b gave the deformed shape shown in Figure 9c. The energy of the finite element solution was 424.1 and the post-processor predicted a bound on the error in energy of 11.02.

The finite element mesh was then subdivided so that every element became four

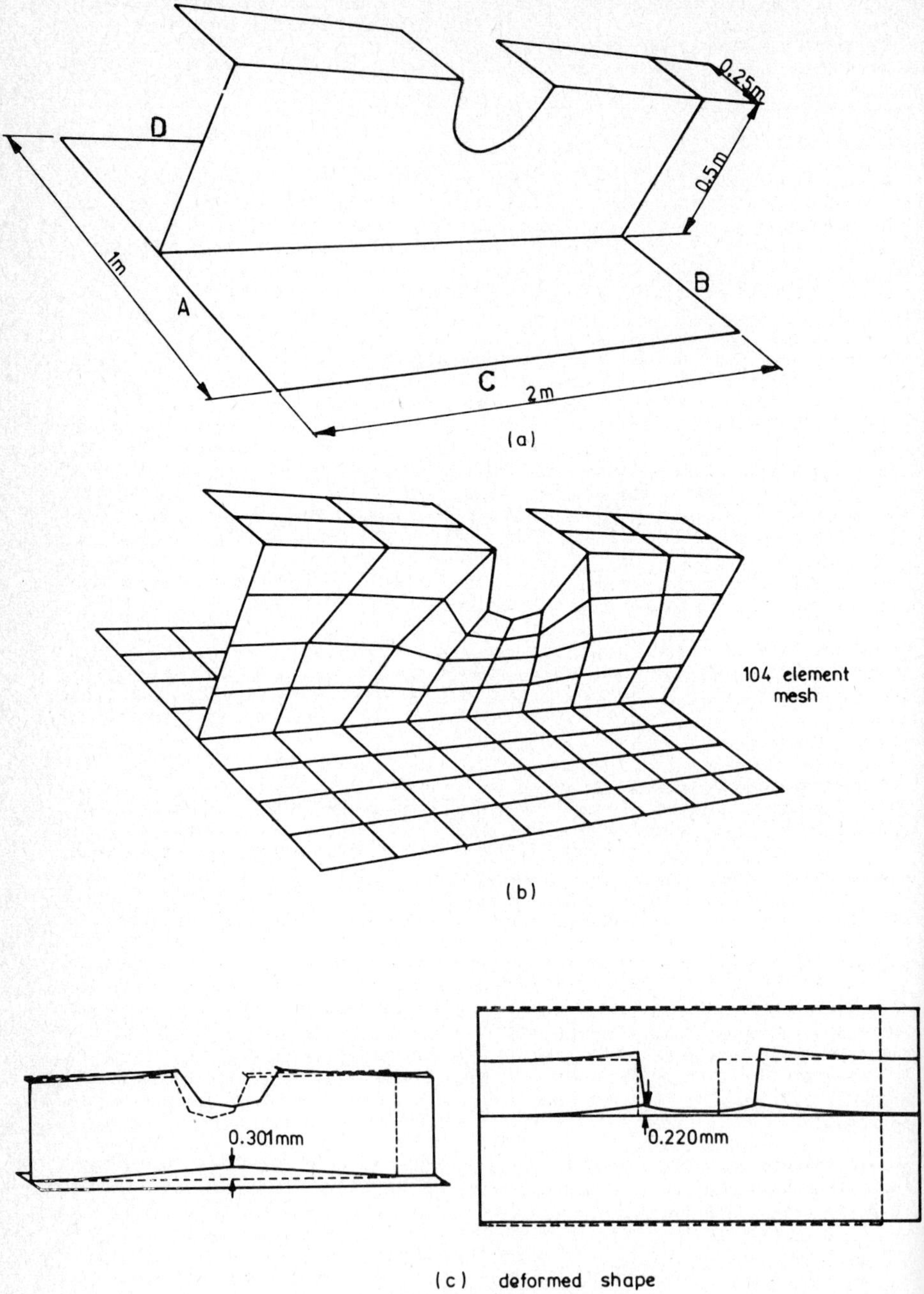

Figure 9. Asymmetric Stiffener

and a re-analysis was performed. Only minor changes in the deformed shape were noted and the energy of the new solution was 419.5. We again anticipate a reduction in the energy of the error for the second solution of approximately four which indicates that a bound on the energy of the error has been obtained by the a-posteriori error analysis.

CONCLUDING DISCUSSION

The work reported in this paper demonstrates the feasibility of developing a post-processing algorithm to determine the accuracy of the finite element solution obtained using a given mesh. We seek not only to identify errors caused by the discretization in the parent analysis (ie. the division into elements on which assumptions are made about the variation of displacements) but also to detect when the error associated with geometric non-linearity and round-off in the computer calculations is signif ant. For the former we note that linear elastic analysis usually assumes deformations are small. Compatability is maintained so that, if the co-ordinates of the structure are up-dated by the deflections the error caused by large deformations will be sensed (though perhaps not bounded) by the a-posteriori error analysis which is based on the lack of satisfaction of the equilibrium equations. Similar arguments apply to round-off error. We force compatability so destruction of the solution by round-off error will again be sensed as a lack of satisfaction of equilibrium.

It should be noted that we are attempting to bound the error to ensure that an inaccurate solution will not pass unchecked. It is therefore encouraging that the predicted errors are all approximately a factor of 2 higher than the actual error. This indicates that a plot would show the parent finite element solution and the bound form a symmetric envelope on the exact solution.

We acknowledge that considerable effort must now be applied to reduce the expense of the a-posteriori error analysis. Here there are many possibilities and the character of the complementary analysis could be radically changed. In particular, for elements of simple shape, it should be possible to provide analytical expressions for the complementary analysis.

The error analysis does proceed element by element and therefore is suited to the minicomputers which are being given the task of the graphical post-processing. Prediction of pointwise errors however requires the solution of additional load cases which can be considered at the time of the parent analysis. This feature also appears in the recent work of Babuška and Miller [1983].

Finally we note that the estimates of the energy of the error on each element provide the basis for the adaptive refinement algorithms described by Gago et al [1983]. In the "optimal" mesh the energy of the error should be the same on each element so obviously we must refine those with higher error. The error estimates developed here therefore indicate where local refinement is required to obtain an improved solution if the error predicted is unacceptably high. In this paper we have resorted to a complete division of all elements into four in the first and last examples only as a basis for assessing the accuracy of our error measures.

ACKNOWLEDGMENT.

We are grateful for the financial assistance provided for this project by the Australian Research Grants Scheme.

REFERENCES

[1] Babuška, I. and Miller, A., The post-processing approach in the finite element method. Parts 1, 2 and 3. To appear: International Journal for Numerical Methods in Engineering (1983).

[2] Babuška, I. and Rheinboldt, W.C., Adaptive approaches and reliability estimations in finite element analysis. Computer Methods in Applied Mechanics and Engineering 17/18 (1979) 519-540.

[3] Bathe,K., Wilson, E.L. and Peterson, F.E., SAPIV a structural analysis program for static and dynamic response of linear systems. University of California, Berkeley, Report No. EERC 73-11 (1973).

[4] Gago, J., Kelly, D.W., Zienkiewicz, O.C. and Babuška, I. A-posteriori error analysis and adaptive processes in the finite element method, parts I and II. To appear: International Journal for Numerical Methods in Engineering. (1983).

[5] Kelly, D.W., A bound theorem for reduced integrated finite elements and error analysis, in: Whiteman, J.R. (ed), The Mathematics of Finite Elements and Applications IV, (Academic Press, 1982).

[6] Kelly, D.W., The self-equilibration of residuals and complementary a-posteriori error estimates. To appear: International Journal for Numerical Methods in Engineering (1983).

[7] Morley, L.S.D., On the constant moment plate bending element, J. Strain Analysis, 6, (1971), 20-24.

[8] Zienkiewicz, O.C. The finite element method, third edition, McGraw-Hill (1977).

Computational Techniques & Applications: CTAC-83
J. Noye & C. Fletcher (Editors)

THE APPLICATION OF A FINITE ELEMENT PACKAGE TO ROAD VEHICLE AERODYNAMICS

Peter D. Cenek[1], Gary F. Fitz-Gerald[2] and Jeffrey W. Saunders[2]

[1] Central Laboratories, Ministry of Works and Development, Wellington, New Zealand.
[2] Royal Melbourne Institute of Technology, Melbourne, Australia.

This paper demonstrates how a general purpose finite element package can be used for simulating automotive wind tunnel testing. Mathematical formulations which enable the inclusion of engine cooling, passenger ventilation, flow incidence and circulation around the body are outlined, as is the procedure for defining the body shape. The method is assessed by comparing computed centre-line pressure distributions for the Alfa Romeo Alfasud, Porsche 924 and GMH's Torana Hatchback with experimental results. Use of the method as a design aid for achieving desired aerodynamic performance is also illustrated.

INTRODUCTION

With any motor car the aerodynamic shaping of the body exerts a considerable influence on fuel consumption and safety through stability and wind noise. It is therefore desirable to have an efficient procedure for accurately predicting the aerodynamic performance of an intended design, either by calculating the air pressure distribution over the surface of the car or the aerodynamic forces and moments.

The conventional motor car has a bluff, low fineness ratio body characterized by boundary-layer separation, a turbulent wake behind the body and low base pressure. The relatively low ground clearance, rough underbody and boundary layers (which build up on the road and underbody), lead to a partial blockage of the flow underneath the car with an increase in pressure on the forward underbody. This in turn tends to direct the air over the top of the body and to an extent laterally. Photographs of experimentally obtained streamlines indicate a stagnation point positioned low on the front of the motor car, typically at bumper mid-height. Over the front of the car the flow is accelerated which produces a reduced pressure and with the increased pressure on the underbody creates nose lift. Any separation of flow on this part of the surface is a result of body detailing, such as sharp corners. If separated regions do occur, they are usually of limited extent although the vortices formed can trail back to influence the remainder of the flow field. The flow at the rear of a vehicle is invariably separated and the near-wake can contain vortices of any orientation. In the near-wake the flow velocities are small and the pressure, which is less than the pressure in the undisturbed flow, is nearly constant. Beyond the near-wake a turbulent far-wake is formed in which the pressure increases from that in the near-wake to the pressure in the undisturbed flow. In addition flow is also developed by the rotating wheels and from inside the wheel arches as well as through the engine compartment.

At present there appears to be no method available to calculate a flow field of the complexity outlined above (see the review by Bearman [1] of three-dimensional bluff flows applicable to road vehicle aerodynamics and the problems created by the presence of a ground plane). Nevertheless, outside the wake and the attached boundary layer the flow is irrotational and a two-dimensional potential flow model can be used to represent the flow in this region. Although such a model is limited to describing the mean flow, it is hoped that it can serve as a guide in the early development of a vehicle design.

The simulation of the irrotational flow in the region described above may proceed using either:

(a) finite differences,

(b) surface singularity methods (such as source-sink panel method, vortex-lattice method etc.), or

(c) finite elements.

To date activity appears to have been concentrated on surface singularity methods to calculate the potential flow over actual road vehicles (Stafford [2], Chometon et al. [3], Ahmed and Hucho [4] and Berta et al. [5]), principally because these methods have the ability to represent complex vehicle shapes accurately and also the domain of numerical solution is limited to only the body surface. However, although there is a considerable library of software available for source-panel methods applied to aircraft aerodynamics, the difficulties related to simulation of the ground plane and the vehicle's wake means that panel method programs have to be specifically written for road vehicle aerodynamics and these are apparently not commercially available.

The finite element method appears to have received only a token recognition from land vehicle aerodynamicists. A possible explanation is the need for a flow domain of finite dimensions which requires correction for the proximity of lateral boundaries and some uncertainty with regard to appropriate placement of inflow and outflow boundaries. Nevertheless, as finite element software is already extensively used by automobile manufacturers in the stress analysis of the body structure and the forming of sheet metal pressings and glass windscreens, it seems appropriate to extend its application to vehicle aerodynamics. Accordingly, the suitability of utilizing a general purpose finite element package for generating flow patterns and pressure distributions on motor cars has been investigated by Cenek [6]. This paper introduces the resulting computational procedure and demonstrates the capabilities and limitations of this approach.

Nomenclature

c = total length of vehicle

C_D, C_L, C_{MP} = drag, lift, and pitching moment coefficients

C_P = pressure coefficient = $1 - (q/U_\infty)^2$

C_{P_b} = base pressure coefficient

C_{P_f} = equivalent free stream pressure coefficient

F_{AB} = prescribed boundary condition on arc AB

h = tunnel height

L = distance in x direction from front stagnation point

ε = solid correction term

m = maximum height of vehicle above the ground plane

n = unit outward normal vector

q = fluid velocity vector, with magnitude $\left(\left(\frac{\partial\psi}{\partial x}\right)^2 + \left(\frac{\partial\psi}{\partial y}\right)^2\right)^{1/2}$

s = variable of the parametric equations used in defining the region under analysis ($0 \leq s \leq 1$)

u,v = components of q in x,y directions

U_∞ = free stream velocity

χ = velocity correction factor

λ = body shape function

Γ = fluid circulation

ψ = stream function

Abbreviations

CPU = Central Processor Unit

FEM = Finite Element Method

GMH = General Motors Holden

IMSL = International Mathematical and Statistical Libraries,Inc.

RMIT = Royal Melbourne Institute of Technology

DESCRIPTION OF THE FINITE ELEMENT PACKAGE

The second edition of the IMSL finite element package TWODEPEP has been applied to the above idealized problem in bluff body aerodynamics.

TWODEPEP is a small, easy to use package that purports to solve general two-dimensional linear and non-linear elliptic, parabolic and eigenvalue problems. The finite element used is the standard six node triangle with quadratic basis function with one edge curved when adjacent to a curved boundary according to the isoparametric method. The user inputs an initial mesh with only enough triangles to define the region and supplies a function which guides the refinement of this mesh.

TWODEPEP also contains a preprocessor program which allows the user to supply the problem description in a greatly simplified form and controls the dimension sizes so that the program uses only as much storage as required for that particular problem.

A complete description of TWODEPEP is given by Sewell [7], and Fitz-Gerald [8] outlines its application to problems in fluid dynamics.

TWODEPEP was considered more suitable than other finite element packages available at RMIT because of its facility of user controlled automatic mesh refinement and flexibility in output format.

DEFINITION OF THE PROBLEM

The infinite domains associated with flows over isolated bodies present a problem as to the size of the approximating finite domain required in a finite element solution. Typically the domain boundaries are chosen "far enough" from the body to satisfy the assumption of unbounded, uniform flow. Such a vaguely defined condition obviously affects the accuracy of the solution, as well as the computational costs, since it is advantageous to make the finite element domain as small as possible. Most information concerning road vehicle aerodynamics has however been obtained from wind tunnel testing. Hence the resulting flow condition about a motor car immersed in a wind tunnel is readily simulated by the finite element method since the tunnel boundaries define the region under analysis. This nevertheless means that problems associated with wind tunnel testing, predominately blockage and wall interference effects, will also occur in the numerical simulation.

Since the aerodynamic pressures acting on the various surfaces of a vehicle make up the aerodynamic forces and moments on the vehicle, a comprehensive centre-line correlation study allows a good general assessment to be made of the quality of the numerical simulation. Therefore to enable direct comparisons between the measured and calculated centre-line pressure distribution, the output format for the TWODEPEP analysis has been arranged to give the velocity components at points corresponding to the pressure tappings on the actual vehicle. Upon calculating the resultant velocity, the pressure coefficient at each pressure tapping follows from

$$C_{P_i} = 1 - \left(\frac{q_i}{U_\infty}\right)^2 \tag{1}$$

where subscript i refers to the number assigned to a pressure tapping.

The mathematical formulation is defined for a two-dimensional "wind tunnel". The fluid in the tunnel is considered to be irrotational, incompressible and inviscid. By assuming that the front stagnation point is at the middle of the front bumper, the flow regime can be divided into two parts. In this way, analysis of the flow over the underbody and upper surface of the motor car can be carried out separately, with the car body (defined by its centre-line profile) forming part of the boundary of the resulting mathematical formulation (refer Figure 1).

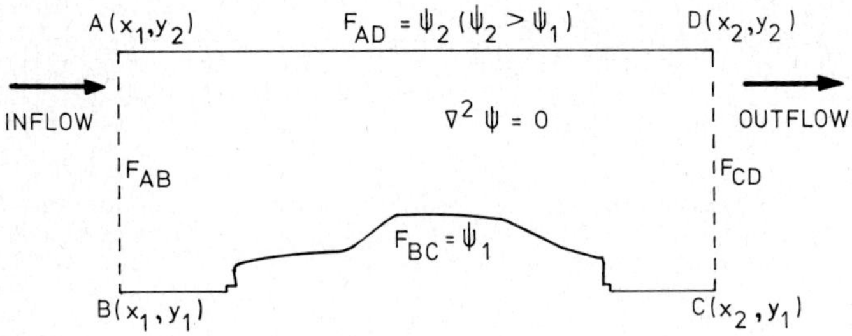

Where: $F_{AB} = (\psi_2 - \psi_1)(1-s) + \psi_1$

$F_{CD} = (\psi_2 - \psi_1).s + \psi_1$

Fig. 1(a) Boundary Conditions for Top Surface Analysis

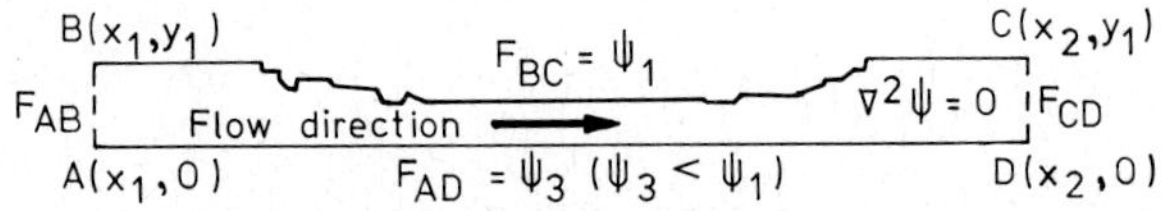

Where: $F_{AB} = (\psi_3 - \psi_1).s + \psi_1$

$F_{CD} = (\psi_3 - \psi_1)(1-s) + \psi_1$

Fig. 1(b) Boundary Conditions for Underbody Analysis

Figure 1
Dirichlet Boundary Conditions for Wind Tunnel Simulation

To permit a systematic comparison with the computed results, extensive pressure and force measurements were conducted with a quarter scale model in RMIT's 3×2 metre industrial wind tunnel [6]. The dimensions of the numerically modelled wind tunnel were hence chosen to duplicate the working section of the RMIT wind tunnel.

MATHEMATICAL FORMULATIONS

Equations of Potential Flow. For two-dimensional incompressible, irrotational flow, the governing equations are:

$$\nabla^2\psi = 0 \tag{2}$$

$$\text{with} \quad u = \frac{\partial\psi}{\partial y} \quad v = -\frac{\partial\psi}{\partial x} \quad \text{(for a left to right flow)} \tag{3}$$

Here, ψ is the stream function, and u and v are the x and y components of the velocity vector q.

The boundary condition at a solid surface is that the fluid and the surface have the same normal velocities, i.e.

$$\frac{d\psi}{ds} = 0 \tag{4}$$

The flow rate between any pair of streamlines in two-dimensional flow is numerically equal to the difference in their ψ-values. Thus, with reference to Figure 1(a), if the streamline describing the boundary BC is $\psi = 0$, then the incident fluid stream velocity is given by

$$U_\infty = \frac{\psi_{max}}{(y_2 - y_1)} \tag{5}$$

Unlike actual automotive wind tunnel testing, which is Reynolds number dependent, the magnitude of U_∞ has no effect on the potential flow derived pressure distribution. To simplify the C_p calculations, U_∞ was chosen to be 10 m/s.

For a left to right flow, the Dirichlet boundary conditions on the vertical faces CD and AB are

$$F_{CD} = \psi_{max}.s \tag{6}$$

$$\text{and} \quad F_{AB} = \psi_{max}(1-s) \tag{7}$$

where $0 \leqslant s \leqslant 1$ and the sense of s (i.e. the path direction) is positive in a counterclockwise direction.

Note that if the value of ψ defining either boundary AD or BC is not constant, then fluid is either drawn in or injected out over that part of the boundary that ψ varies. This property was successfully used to model engine cooling and passenger ventilation of a vehicle [6].

Flow Incidence. The boundary conditions on the inflow and outflow boundaries can alternatively be described in Neumann form.

For example, on inflow boundary AB

$$\left(\frac{\partial\psi}{\partial n}\right)_{AB} = \left(\frac{\partial\psi}{\partial x}, \frac{\partial\psi}{\partial y}\right).(-1,0) = -\frac{\partial\psi}{\partial x} = V \tag{8}$$

An incoming flow at an angle θ to the tunnel floor can therefore be specified using

$$V = U_\infty \sin \theta \tag{9}$$

This same concept can be used in simulating the effect of grille slats when engine cooling and passenger ventilation are considered (see Cenek [6]).

Circulation Modelling. To enable the TWODEPEP package to be used in solving the flow field around a body totally immersed in the finite element tunnel, the problem needs to be formulated in such a way that a singly-connected region results (see Figure 2).

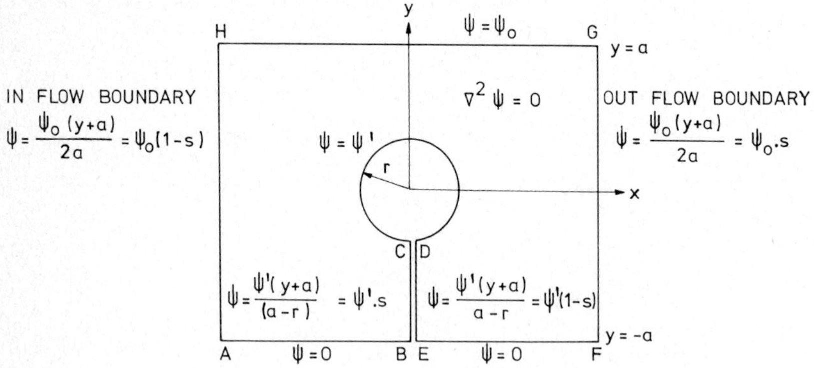

Figure 2
Boundary Conditions for Circulation Simulation

By making the slit underneath the body extremely narrow, the body can be treated in the computation as completely enclosed. That is, since the stream function ψ changes values on boundary BC from 0 to ψ', fluid is ejected across the boundary, whereas the stream function change from ψ' to 0 on boundary DE injects exactly the same amount of fluid back into the tunnel. Note that $\psi' > 0$ for the fluid to flow in a left to right direction.

With reference to Figure 2 the velocity at points C and D is given by

$$q = \frac{\psi'}{(a-r)} \tag{10}$$

The larger the value of ψ', i.e. fluid flux, the more the approaching flow is diverted under the object. Physically this has the same effect as increasing the circulation around the object (Fitz-Gerald [8]) and the location of the stagnation points and the pressure distribution will be altered.

Inflow Boundary Placement. By considering the inflow boundary analogous to the contraction exit in the RMIT industrial wind tunnel, it was assumed that a minimum upstream fetch of approximately half of a body length was sufficient to allow for the full effect of the bluff body on the incoming flow. The accuracy of this assumption was numerically tested by moving the inflow boundary further away from the body to establish that the solution remained unchanged. This investigation resulted in the finite element tunnel being 2 metres long with the 1:4 scaled car body placed some 0.5 metres from the inflow boundary.

Pressure Coefficient Corrections. To account for the effect of the tunnel roof compressing the streamlines around the body and thereby effectively increasing the flow velocity past the body, the calculated pressure coefficient (C_P) is corrected to the equivalent free stream pressure coefficient (C_{P_f}) by the following relation:

$$C_{P_f} = C_P(1-2\varepsilon) + 2\varepsilon \qquad (11)$$

Note that equation 11 is not valid for the underbody analysis.

If the top surface of the car body is approximated to half an ellipse, the expression for the correction term (ε), relevant to a two-dimensional situation, is given by Maskell [9] as

$$\varepsilon = \frac{\pi^2}{12}\left(\frac{c}{h}\right)\lambda \qquad (12)$$

where

$$\lambda = \frac{1}{2}\left(1 + \left(\frac{m}{c}\right)\right) . \qquad (13)$$

For the three motor cars analysed, ε was taken to be approximately equal to 0.03.

Equation 11 is particularly suited to this application since the position of the front stagnation point remains unaltered.

Results obtained in Cenek and Fitz-Gerald [10] suggest that the numerically generated pressures are too high downstream from the windscreen due probably to the transverse loss of airflow. Because this was felt to be most important downstream from the base of the windscreen, the velocity was multiplied by the factor

$$\chi = 1.0 - \alpha H(L-L_0) \qquad (14)$$

where $L=L_0$ is the location of the base of the windscreen, α is a convenient empirical factor used to account for this reduction in fluid flux and H is the Heaviside step function. After comparing the theoretical and experimental results for the 1976 LX Torana Hatchback, the average value of α was found to be 0.18 [6]. This same multiplicative factor was used on all the subsequent body shapes analysed.

Hence

$$C_{P_f} = \left(1 - \left(\frac{\chi . q}{U_\infty}\right)^2\right)(1-2\varepsilon) + 2\varepsilon \qquad (15)$$

Car Shape Description. The procedure adopted for determining the mathematical equations that describe the centre-line contour of the car body are:

(a) segmentation of the contour into a series of arcs (governed by the position of sharp changes in curvature),

(b) manual acquisition (via 2-D digitiser) of cartesian co-ordinates of points that lie on each arc,

(c) curve fitting by least squares, and

(d) computer plot of the mathematical equations to highlight discrepancies between the engineering layout drawing and the mathematical representation.

The method used to ensure continuity between the various contour segments is as follows. Suppose that after performing a least squares analysis on any arc, the curve that best fits the data is a polynomial of degree N-1. If the analysis is repeated, but with the data reduced to only N coordinates, then the resulting polynomial will be an exact fit, i.e. the sum of the squares of errors is zero. Continuity between arcs is therefore preserved, provided the co-ordinates of the arc limits are included in the reduced data set.

Each arc used in defining the vehicle profile, in general, forms a side of a triangle element and so the initial mesh of the finite element tunnel is primarily determined by the segmentation of the car body.

A total of 43 arcs (18 top surface, 25 underbody) was required for the centre-line profile of the Torana Hatchback given in Figure 3. Note that the only simplifications necessary to the complete configuration were at the front bumper and grille.

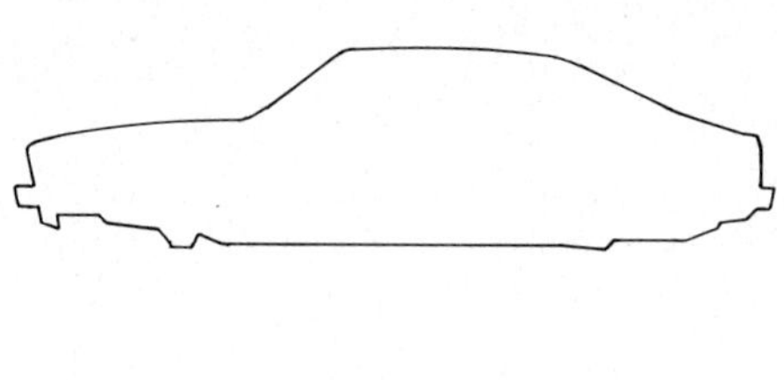

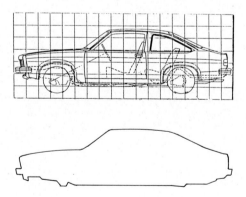

Figure 3
Comparison of Engineering Drawing with
Mathematical Representation

RESULTS OF THE NUMERICAL INVESTIGATIONS

Centre-line Surface Pressure Distribution. Figure 4 shows the theoretically predicted pressure distribution over the 1976 LX Torana Hatchback for the standard formulation (i.e. the upper surface of the vehicle analysed separately with no simulation of internal air flow) in comparison with that measured in the RMIT wind tunnel. It is clear that the general form of the experimentally derived pressure distribution is well reproduced by the numerical simulation, especially the positions of minimum and maximum pressure. Application of the velocity correction factor (χ) results in the modified Figure 4(b). Flow visualisation studies suggested that the discrepancy over the boot-lid area is due to flow separation.

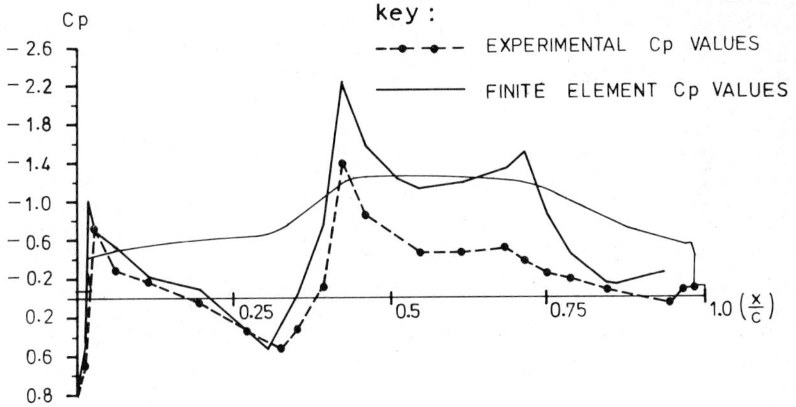

Fig. 4(a) Standard Formulation

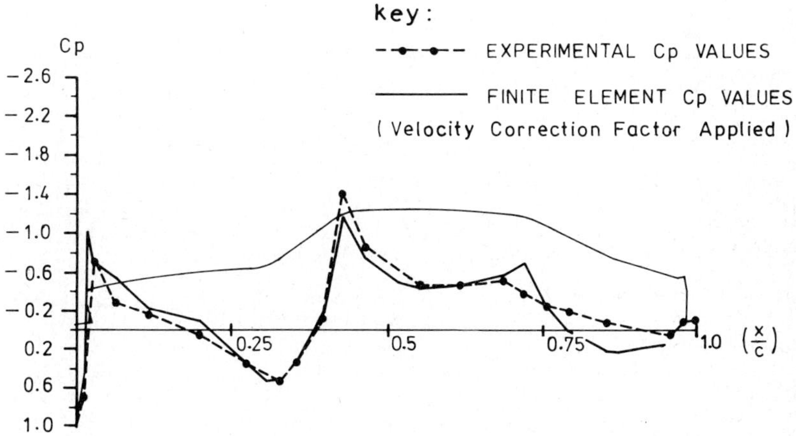

Fig. 4(b) Application of Velocity Correction Factor χ

Figure 4
Torana Centre-line Surface Pressure Distribution: Experiment versus FEM Simulation

Circulation Modelling. Although it is expedient to divide the flow field into an upper lower region by a horizontal plane for separate analysis of the top surface and underbody, the location and shape of both the front and rear stagnation streamlines are consequently fixed. The circulation around the car's body is thereby set to zero. The formulation given in Figure 2 was thus used to assess the need for correctly specifying circulation.

From the resulting numerical investigation it appears that the flow over the upper surface of the vehicle is substantially less sensitive to circulation than the underbody flow. For example, when the amount of incoming fluid flux diverted under the car is increased from 2.7% to 10%, the pressure coefficient on the underbody at point A changed from -0.2 to -15.7 whereas on the surface at point B, the pressure coefficient changed from -2.7 to -1.7 (refer Figures 5(b) and 6(b)). This result duplicates the findings of Tuck [11]. Note also that the streamline contour plots vividly demonstrate the effect of circulation on the incoming flow and the position of the stagnation points, i.e. the presence of circulation in such a direction so as to slow down the under-vehicle flow shifts both the front and rear stagnation points downwards whereas circulation in the opposite direction shifts the stagnation upwards.

Our comparisons between the experimental and theoretical front stagnation points suggest that zero or near zero longitudinal circulation occurs in practice with motor cars (see also Tuck [11]). In addition a close correlation was obtained when the pressure distribution plot for the case of zero circulation was compared with the results obtained when the flow regime was divided into two parts.

VERIFICATION OF THE NUMERICAL PROCEDURE

Regrettably few wind tunnel measurements on motor cars have appeared in open literature. However the centre-line pressure distributions for the Alfa Romeo Alfasud and Porsche 924, two cars currently in production, were published in Style Auto 34/35 [12] and 37 [13] respectively. To investigate the relevance of equation 14, in particular the 0.18 value for α, the method was applied to these two significantly different fastback cars. Results of the computations are as follows:

Alfa Romeo Alfasud - The resulting numerical prediction of the pressure distribution over the top surface of the Alfasud is shown in Figure 7. Although the pressure survey is twice as comprehensive as for the Torana (i.e. 43 locations compared to 22 for the Torana), the results of this analysis demonstrated that the position of minimum and maximum pressure and features of the overall pressure distribution correlated exceptionally well with the published results.

When compared to the Torana, the fastback angle of the Alfasud is less steep and it will be noted that the agreement of the inviscid flow model with the experimental results is much better in this region. This result suggests that the flow is attached right to the rear edge of the boot-lid for the Alfasud.

The CPU time for the solution using a final mesh of 475 triangles on the CYBER 73 computer was 275 seconds, and storage amounting to 173K words was required.

Porsche 924 - Like the Alfasud, the pressure survey over the Porsche 924 was very comprehensive with the experimental pressures known at 41 locations. The wind tunnel tests were performed in the Volkswagen wind tunnel on a full sized prototype (refer Style Auto No 37 and Burst and Srock [14]).

Referring to Figure 8, the closeness between theoretical and experimental pressures for the Porsche 924, a car with a drag coefficient of only 0.36, is extremely encouraging, the only significant discrepancy occurring at the leading edge of the

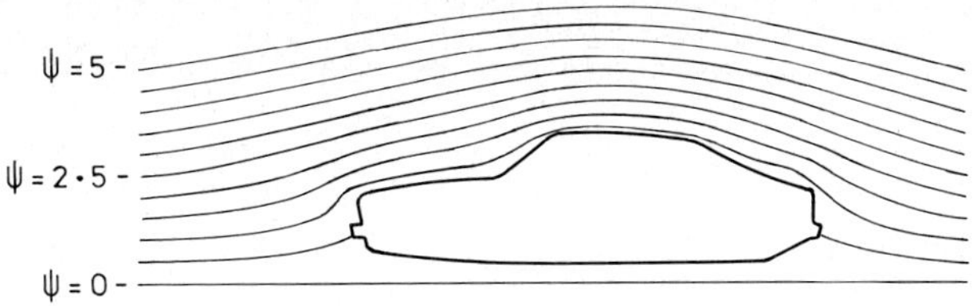

Fig. 5(a) Computed Streamline Contour Plot

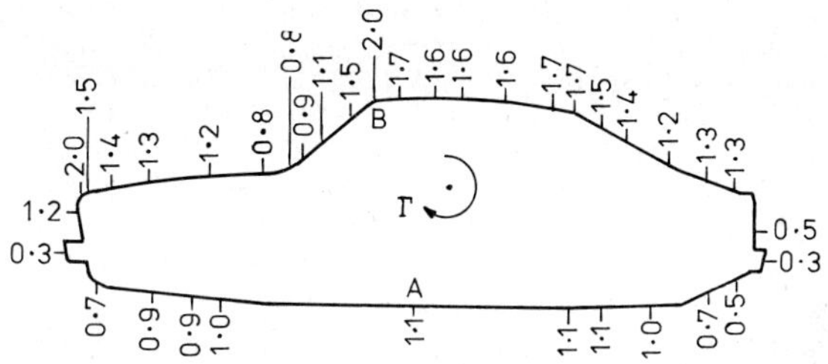

Fig. 5(b) Surface Velocity Distribution (Numbers outside contour are for resultant velocity per unit stream velocity)

Figure 5
Results for Clockwise Circulation

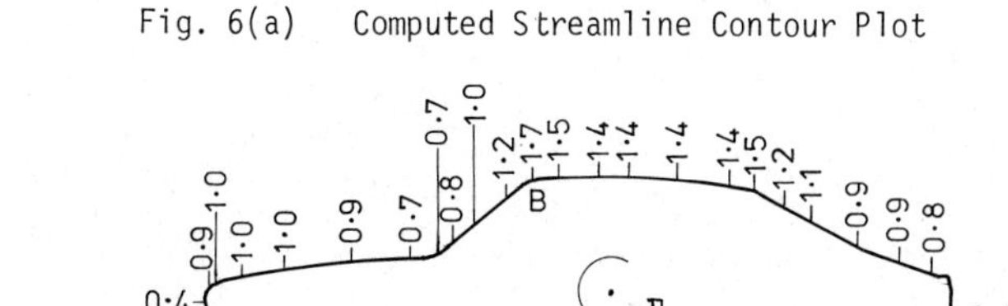

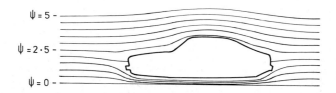

Fig. 6(a) Computed Streamline Contour Plot

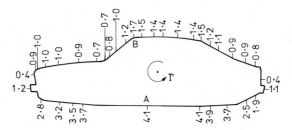

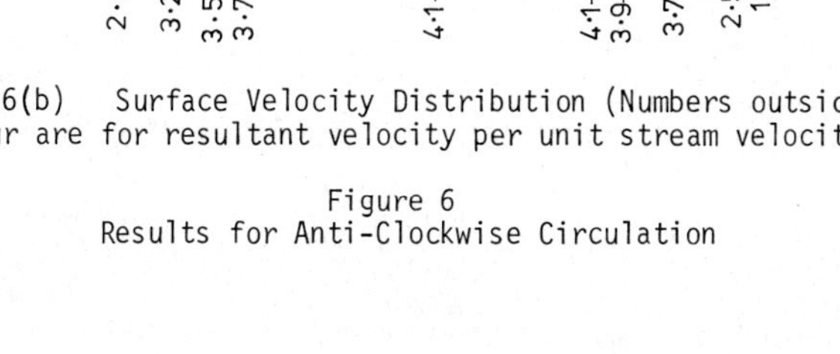

Fig. 6(b) Surface Velocity Distribution (Numbers outside contour are for resultant velocity per unit stream velocity)

Figure 6
Results for Anti-Clockwise Circulation

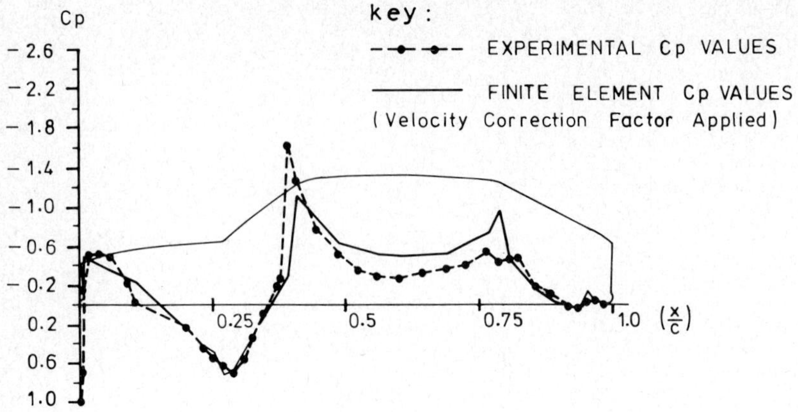

Figure 7
Alfa Romeo Alfasud Centre-line Pressure Distribution

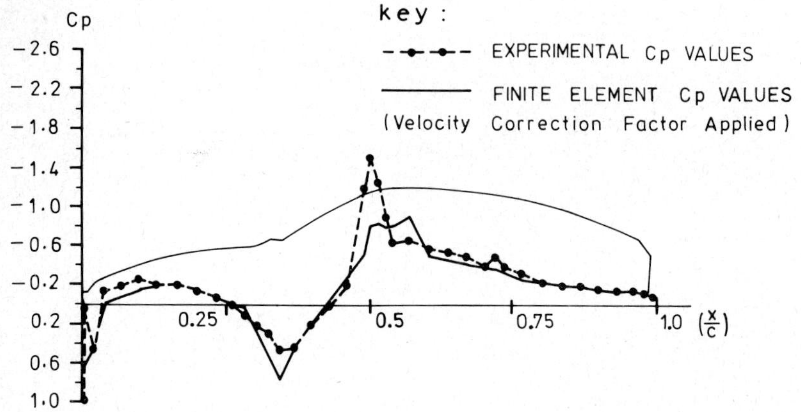

Figure 8
Porsche 924 Centre-line Pressure Distribution

roof. A possible explanation for this discrepancy is that the windscreen of the 1977 Porsche 924 is recessed in relation to the roof, thereby forming a 15mm step at the junction of the windscreen and roof thus causing the flow to trip. In contrast, the windscreens for both the Torana and Alfasud are set in a rubber surround which constitutes a ridge of about 5 to 8mm around the windscreen. This small ridge appears to have very little effect on the flow as witnessed by the potential flow results.

A final mesh of only 460 triangles was required to obtain the results given in Figure 8, corresponding to a CPU time of 270 seconds.

NUMERICAL AERODYNAMIC TUNING OF CAR BODIES

The possible use of the finite element method as an economical means for the aerodynamic tuning of car bodies is demonstrated by investigating the effect of modifying the Torana Hatchback profile to incorporate a variation of the currently fashionable "shovel nose" front.

Referring to Figure 9, the results of the numerical computation indicates that this configuration change reduces the likelihood of separation at the leading edge of the bonnet since the pressure gradients are now not as severe in this region when compared with the standard configuration. Moreover a reduction in front axle lift is expected since the extent of low pressure over the bonnet has been reduced.

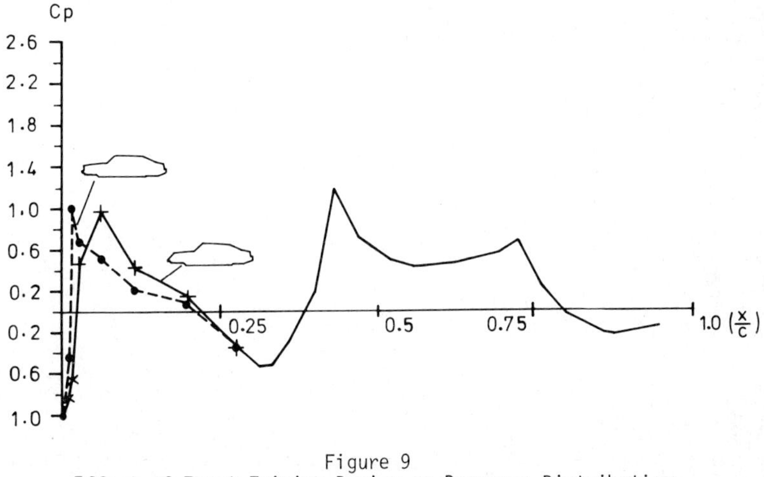

Figure 9
Effect of Front Fairing Design on Pressure Distribution

To quantify these changes in the aerodynamic performance an attempt was made to estimate the drag and lift coefficients together with the pitching moment by integrating the centre-line pressure distributions given in Figure 9. The corresponding results are tabulated in Figure 10. As can be seen from the numerically derived force coefficients this relatively minor profile change, caused by altering the slope of the front fairing, results in a decrease in both drag and lift coefficients of respectively 11% and 15%. The stability of the car is also improved since the change in sign of the pitching moment indicates that the loading on the front wheels has increased.

Centre Section

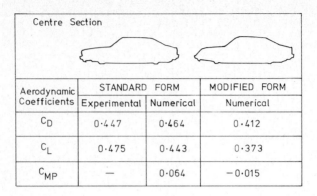

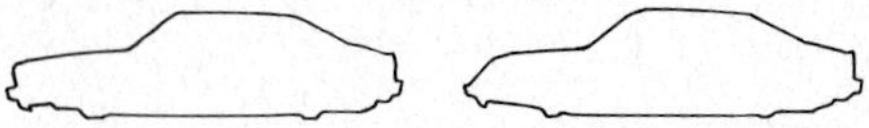

Aerodynamic Coefficients	STANDARD FORM		MODIFIED FORM
	Experimental	Numerical	Numerical
C_D	0·447	0·464	0·412
C_L	0·475	0·443	0·373
C_{MP}	—	0·064	−0·015

Figure 10
Theoretically Derived Aerodynamic Coefficients

In estimating the aerodynamic force coefficients the following steps were taken in an attempt to compensate for the shortcomings of the two-dimensional inviscid and incompressible flow regime.

1. The underbody experimental centre-line pressure distribution was used.

2. The base pressure must be assumed. Generally the base pressure for fastback/hatchback cars is in the range $0 \geq C_{P_b} \geq -0.1$. To obtain a conservative estimate a C_{P_b} of -0.1 was used for the results given in Figure 10 although reference [6] shows that the results would not have changed significantly for an arbitrary specification of base pressure within the above limits.

3. Morelli [15] has theoretically shown that the ratio of the average lift coefficient to the lift coefficient calculated at the vehicle's longitudinal symmetry plane is 2/3. It could be expected that this factor may be also used to convert the numerically derived centre-line form drag and pitching moment to total force and moment coefficients.

4. The numerical drag coefficient was calculated from the following relation

$$C_{D_{Total}} = C_{D_{Form}} + C_{D_{Induced}} \tag{16}$$

where

$$C_{D_{Induced}} = 0.36 \times (C_L)^2 \tag{17}$$

The analysis and derivation of the formula for induced drag can be found in Scibor-Rylski [16].

With reference to Figure 10, note that the numerically derived aerodynamic force coefficients are within 6% of the wind tunnel results.

CONCLUSION

This work suggests that a general purpose finite element package can be used to predict with reasonable accuracy the centre-line pressure distributions of cars having a fastback/hatchback configuration.

The main features of the computational model are seen to be:

(a) the relative ease and economy of application;

(b) the facility to model (i) internal air flows,
(ii) flow incidence, and
(iii) flow circulation;

(c) the apparent ability to handle changes in the geometric detailing of the body; and

(d) the facility to generate flow streamlines which in turn can be used in the design of wind tunnel working sections for testing at high blockage ratios as presented in Stafford [17].

ACKNOWLEDGEMENTS

The authors gratefully acknowledge the assistance of both RMIT and Ministry of Works and Development staff during the course of the research and in the preparation of this paper.

REFERENCES

[1] Bearman, P.W., Bluff body flows applicable to vehicle aerodynamics, in: Morel T. (ed), Aerodynamics of Transportation (A.S.M.E., 1979).

[2] Stafford, L.G., An improved numerical method for the calculation of the flow field around a motor vehicle, in: Kramer C. and Gerhardt H.J. (ed), Proc. Colloquium on Industrial Aerodynamics (Aachen, 1974).

[3] Chometon, F., Fontanet, P., Migeot, J.C., Contribution à l'étude theorique et experimentale de l'écoulement autour d'un profil bidimensionnel de véhicule automobile, Paper 6.3, 16th International Automobile Technical Conf. (F.I.S.I.T.A., 1976).

[4] Ahmed, S.R. and Hucho, W.-H., The calculation of the flow field past a van with the aid of a panel method, SAE Paper No 770390, 1977.

[5] Berta, C., Tacca, T. and Zucchelli, A., Aerodynamic study on vehicle shape with the 'Panel Method' - An effort to calculate the influence of shape characteristics on aerodynamic performance, SAE Paper No 801401, 1980.

[6] Cenek, P.D., Application of the Finite Element Method to Automotive Aerodynamics, ME Thesis, Dept. of Mech. and Prod. Engineering, RMIT, (August, 1982).

[7] Sewell, G., A finite element program with automatic, user controlled, mesh grading, in: Proc. of 3rd I.M.A.C.S. International Symposium on Computer Methods for P.D.E.'s, Rutgers Univ. (New Brunswick, 1979).

[8] Fitz-Gerald, G.F., The application of a finite element package to problems in fluid dynamics, in: Proc. of 2nd Applied Physics Conf. of the Australian Inst. of Physics, RMIT (December 1981).

[9] Maskell, E.C., A theory of the blockage effects on bluff bodies and stalled wings in a closed wind tunnel (ARC R and M 3400, 1965).

[10] Cenek, P.D. and Fitz-Gerald, G.F., Finite element prediction of pressure distribution on motor cars, in: Proc. of 2nd Applied Physics Conf. of the Australian Inst. of Physics, RMIT (December 1981).

[11] Tuck, E.O., Irrotational flow past bodies close to a plane surface, Jrnl. of Fluid Mechanics, 50 (1971) 481-491.

[12] Style Auto, Architettura della Carrozzeria 34/35 (Dinarch, Torino, 1974).

[13] Style Auto, Architettura della Carrozzeria 37 (Torino, 1978).

[14] Burst, H.E. and Srock, R., The Porsche 924 body - main development objectives, SAE Paper No 770311, 1977.

[15] Morelli, A., Theoretical method for determining the lift distribution on a vehicle, Paper No. 2.2, 10th International Automobile Technical Conf. (F.I.S.I.T.A., 1964).

[16] Scibor-Rylski, A.J., Road Vehicle Aerodynamics (Pentech Press, London, 1979).

[17] Stafford, L.G., A streamline wind-tunnel working section for testing at high blockage ratios, Jrnl. of Wind Eng. and Indust. Aerodynamics, Vol. 9 (Nov. 1981) 23-31.

Computational Techniques & Applications: CTAC-83
J. Noye & C. Fletcher (Editors)

GENERATION OF FINITE ELEMENT MESHES FROM NODAL ARRAYS

M.B. McGirr, D.J.H. Corderoy, P.C. Easterbrook
and A.K. Hellier

School of Metallurgy
University of New South Wales
Kensington 2033
AUSTRALIA

Two methods for automatically generating meshes of triangular elements from arbitrary arrays of nodes are discussed. Both techniques involve the construction of the Wigner-Seitz cell around each node. One technique, which has been reported previously, leads to a large proportion of inefficient, high aspect-ratio triangles. An alternative approach is proposed which yields a mesh of near-equilateral triangles, even in those regions where the density of the nodal array is varying rapidly.

INTRODUCTION

In most schemes for finite element mesh generation the elements are obtained by virtue of the intersection of families of curves which form a local system of coordinates. The elements so formed are usually quadrilateral in type with triangular elements used as "fillers", particularly on boundaries. However, the nodal density, and hence the element density, may be varied in a more general manner if the array of nodes is constructed first (McGirr, Corderoy, Easterbrook and Hellier (1982), Suhara and Fukuda (1972)).

The generation of a mesh starting from an arbitrary array of nodes necessitates the use of triangular elements, and a reliable procedure for forming well-proportioned elements is required. While several such procedures have been reported (Frederick, Wong and Edge (1970), Suhara and Fukuda (1972), Cavendish (1974), Lewis and Robinson (1978), Nelson (1978)), they are ad hoc in nature and the form of the resulting mesh is generally neither predictable nor unique.

This paper describes two methods for forming a mesh of triangles from an arbitrary array of nodes. Both methods are based on the construction of the Wigner-Seitz cell around each node, and the meshes obtained are unique. One technique, which has been briefly reported previously (McGirr, Corderoy, Easterbrook and Hellier (1982)), is difficult to program and produces a mesh containing a number of high aspect-ratio triangles which must later be removed. The second technique involves fewer programming difficulties and produces arrays of surprisingly well-proportioned triangles, even in situations where the nodal density varies rapidly.

WIGNER-SEITZ CELL

This device has been used in the field of quantum mechanics to apportion space between the atoms in either regular or irregular atomic arrays (Wigner and Seitz (1933), Raimes (1961). It is formed by considering each atom in turn as the central atom, and drawing the perpendicular bisectors of the lines which join the central atom to its neighbours. When sufficient near neighbours have been considered a closed polyhedron is formed, which is not intersected when the bisectors of the lines joining the central atom to more distant neighbours are constructed. The resulting polyhedron is termed the Wigner-Seitz (WS) cell, and the space contained within each cell may be regarded as belonging to the central atom of that cell.

The procedure described above may also be applied to a two-dimensional array, in which case the WS cell becomes a polygon. Figure 1 illustrates the outcome of this procedure for one particular atom (node) of a planar array, and its near neighbours. The polygonal framework formed by constructing the WS cell around each node is both space-filling and unique.

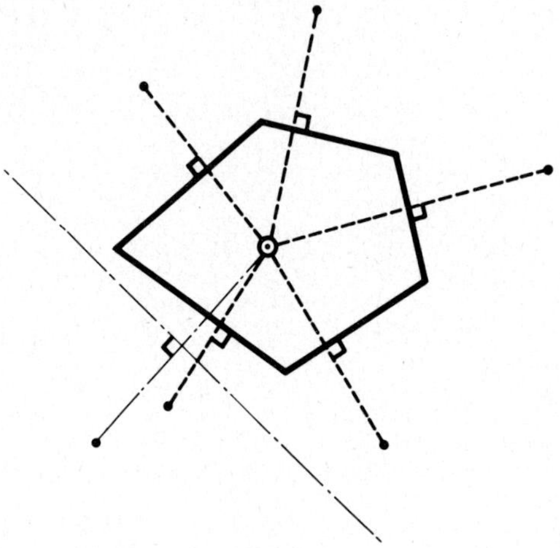

Figure 1
Construction of Wigner-Seitz Cell around a Node

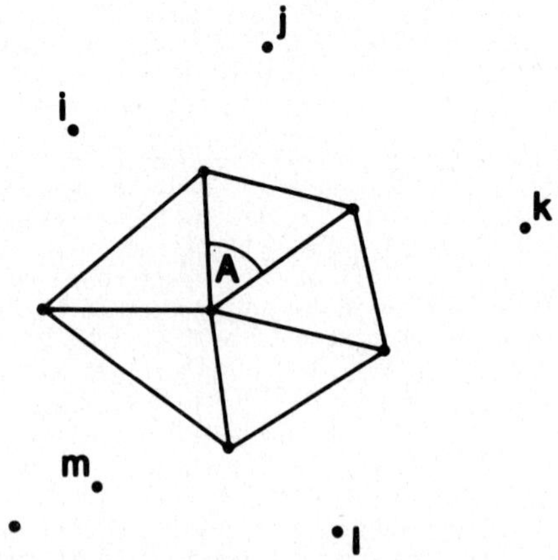

Figure 2
Element Formation by Procedure A

PROCEDURE A

In this first procedure a triangular mesh is constructed from the polygonal framework by joining each central node to all the vertices of the WS cell which surrounds that node (Figure 2). As a result, the vertices are added to the array as new nodes and the total number of nodes is increased by a factor of approximately three; the overall nodal density, however, still varies in the same manner.

There are a number of programming difficulties associated with this technique. The program for constructing the WS cell is quite complex and its execution will consume considerable computer time if all nodes are considered as possible near neighbours of a given node. A substantial increase in speed may be gained by setting a "window" (whose size is inversely proportional to the local nodal density) around each node, and omitting the bisector construction for nodes which lie outside this window. Such a procedure will lead to error if the nodal array is entirely arbitrary, but causes little problem if the nodal density varies in a continuous manner.

Lines joining central nodes to neighbours must, of course, be checked to ensure that they do not cross re-entrant boundaries, except to join to an adjacent node on the same boundary. However, the major programming problem arises when the WS cell surrounding a node projects outside the boundaries of the array. This situation must be detected and a protocol adopted for truncating the WS polygon.

Despite the major problems mentioned above, and a host of lesser difficulties, this procedure has been programmed and runs quite reliably. An occasional difficulty is encountered when the exclusion "window" referred to above is too small to cope with an extremely rapid variation in nodal density, in which case cross-linking of elements occurs. If this possibility is suspected then mesh generation may be followed by a program which checks the integrity of the mesh.

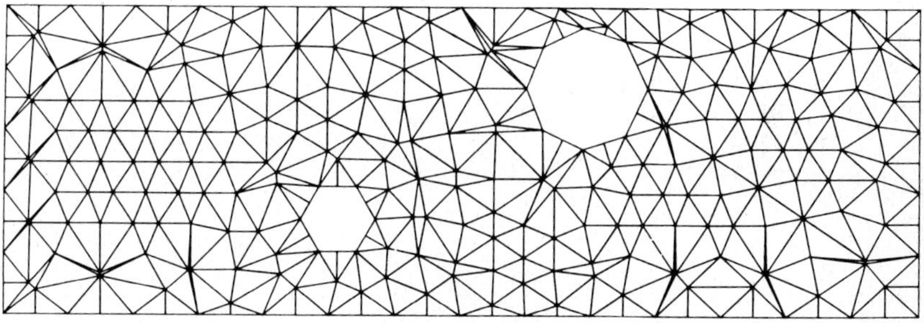

Figure 3
Constant Density Mesh Generated using Procedure A

Figure 3 shows a mesh generated from a nodal array with essentially constant density using procedure A. There are a number of high aspect-ratio triangles present; these are removed by a program which reduces angles of the type A in Figure 2 to zero if they are less than twenty degrees. This process is not completely effective, but it eliminates the majority of ill-proportioned triangles.

PROCEDURE B

The process of forming the WS cell not only associates each node with part of the space occupied by the array, but also associates the node with that group of neighbouring nodes used in the construction of the cell. Thus the central node of Figure 2 is uniquely associated with the surrounding nodes i, j, k, l, m because bisectors to the lines from the central node to these nodes form WS cell boundaries. After the WS cell is constructed, a unique set of triangles may be formed by joining the central node to the nodes which were involved in formation of the WS cell, as shown in Figure 4. In this procedure the construction of the WS cell serves only to identify the neighbours involved.

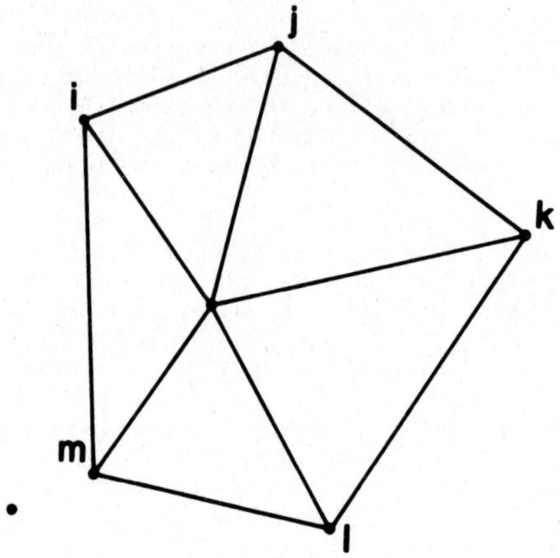

Figure 4
Element Formation by Procedure B

One flaw in this unique assignment of neighbours occurs when a WS cell corner is formed by the intersection of three or more bisectors. Under these conditions there is some ambiguity concerning which neighbouring nodes actually form a given WS cell. This situation is highly unlikely when the array of nodes is arbitrary, but can occur in regions where the density is constant and the node generator tends to produce a regular array, such as the square array of Figure 5.

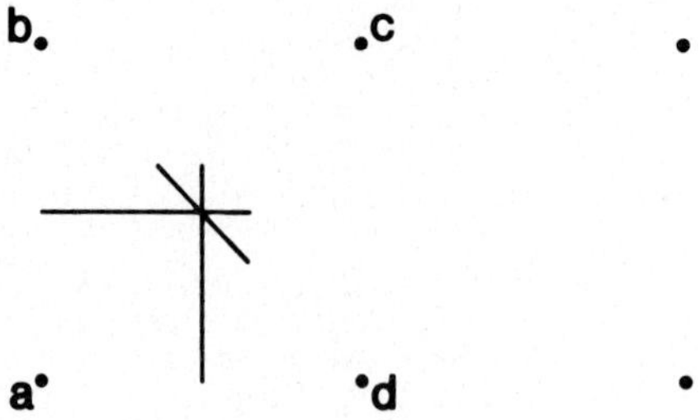

Figure 5
Square Array of Nodes showing Triple Intersection during Formation of WS Cell around Node a

In this case, if b, c and d are regarded as neighbours of a, and a, b and c are regarded as neighbours of d, the triangles formed by the procedure shown in Figure 4 are cross-linked. Thus a rule must be adopted for rejecting nodes when three or more contribute to a WS cell corner. In the case of regular arrays it appears that rejection of the more remote neighbours leads to a mesh without cross-linking; however, there is no guarantee that this procedure will be effective in more complex situations. In the work reported in this paper, mesh generation was followed by a program which checks mesh integrity. Consequently, it has been easier to circumvent the cross-linking problem by adding a small, random number to the coordinates of each internal node after generation; this technique drastically reduces the probability of cross-linking, while the mesh-checking program removes any cross-linking that does occur.

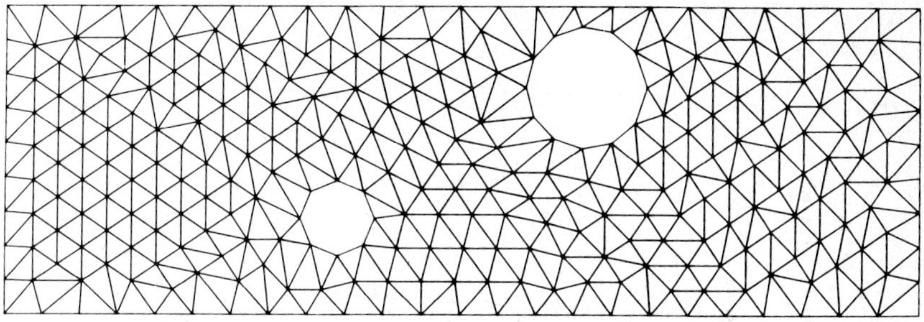

Figure 6
Constant Density Mesh Generated using Procedure B

Figure 6 shows a constant density mesh generated by procedure B. This mesh shows a much more even distribution of well-proportioned triangles than that produced by procedure A (Figure 3). A further mesh whose nodal density varies considerably, but in a continuous manner, was generated using procedure B and is shown in Figure 7. It can be seen that all the triangular elements are of near-equilateral shape.

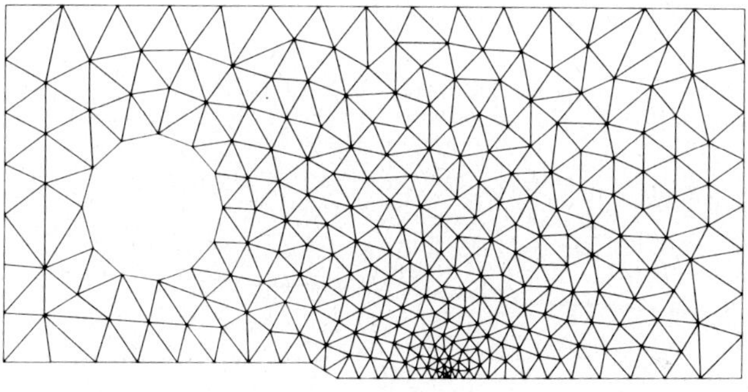

Figure 7
Mesh of Compact Tension Specimen with Varying Nodal Density Generated using Procedure B

FUTURE DEVELOPMENT

The performance of procedure B has been found to be superior to that of procedure A in producing a well-proportioned triangular mesh from an arbitrary array of nodes, even under circumstances where the nodal density varies rapidly. Procedure B also has the advantage of not increasing the number of nodes beyond that of the original array.

In procedure B the constructed WS cell is not required; it merely serves to identify a unique set of near neighbours. An algorithm for deciding directly which neighbours of a central node are involved in the WS cell around that node is currently under test. It is hoped that the use of this algorithm will both reduce the programming complexity of mesh generator B, especially when the nodes concerned are near boundaries, and substantially reduce the time involved in the mesh generation process.

The WS cell has been used in the present work in its two-dimensional form. However, it is equally applicable to the generation of tetrahedral elements from an arbitrary three-dimensional nodal array. In three dimensions the programming problems involved are much more severe, especially near the boundaries of the array; it is felt that the simplicity of procedure B as well as its tendency to produce well-proportioned elements, will be of great value in this regard.

CONCLUSION

Both techniques (A and B) produce unique triangular meshes from arbitrary nodal arrays. Technique B is simpler to program, does not increase the number of nodes and requires less conditioning of the mesh after construction. Technique B also appears to have greater potential for development, particularly as a tetrahedral element generator for three-dimensional finite element analysis.

REFERENCES

[1] McGirr, M.B., Corderoy, D.J.H., Easterbrook, P.C. and Hellier, A.K., A new approach to automatic mesh generation in the continuum, Proc. 4th Int. Conf. in Australia on Finite Element Methods, Hoadley, P.J. and Stevens, L.K. (eds.), Univ. of Melbourne (1982) 36-40.

[2] Suhara, J. and Fukuda, J., Automatic mesh generation for finite element analysis, in: Oden, J.T., Clough, R.W. and Yamamoto, Y. (eds.), Advances in Computational Methods in Structural Mechanics and Design (UAH Press, Alabama, 1972) 607-624.

[3] Frederick, C.O., Wong, Y.C. and Edge, F.W., Two-dimensional automatic mesh generation for structural analysis, Int. Jrnl. for Num. Meth. in Engng. 2 (1970) 133-144.

[4] Cavendish, J.C., Automatic triangulation of arbitrary planar domains for the finite element method, Int. Jrnl. for Num. Meth. in Engng. 8 (1974) 679-696.

[5] Lewis, B.A. and Robinson, J.S., Triangulation of planar regions with applications, Computer Jrnl. 21 (1978) 324-332.

[6] Nelson, J.M., A triangulation algorithm for arbitrary planar domains, Appl. Math. Modelling, 2 (1978) 151-159.

[7] Wigner, E. and Seitz, F., On the constitution of metallic sodium, Phys. Rev. 43 (1933) 804-810.

[8] Raimes, S., The Wave Mechanics of Electrons in Metals (North-Holland, Amsterdam, 1961) 247-248.

Computational Techniques & Applications: CTAC-83
J. Noye & C. Fletcher (Editors)

THE AUTOMATIC GENERATION OF ARRAYS OF NODES WITH VARYING DENSITY

M.B. McGirr and P. Krauklis

School of Metallurgy
University of New South Wales
Kensington 2033
AUSTRALIA

A self-adaptive technique is discussed in which a constant number of nodes is repositioned after successive finite element analyses. Selection of a family of functions that can satisfactorily control the nodal density, both in the presence and absence of stress singularities, is difficult. It is shown that the nodal density control parameter (e.g. strain energy density) obtained from each solution iteration may be used in tabular form to control the nodal density. This procedure removes any bias associated with the choice of specific functions, and is easily adapted to the use of nodal density control parameters other than strain energy density.

INTRODUCTION

Finite element schemes in which arrays of nodes are generated prior to the creation of the element mesh must contain a procedure for controlling the density of nodes. This procedure may be as simple as the manual positioning of nodes using a magnetic pen (Frederick, Wong and Edge (1970)), or the assignment of differing, but constant, densities to various regions of the problem domain (Suhara and Fukuda (1972)). These forms of density control require manual intervention, and either rely on the experience of the analyst in choosing a suitable node distribution, or are unwieldy if the finite element procedure is to be used adaptively. Alternatively an attempt may be made to allow a continuous variation in density, in which case a density function is required (McGirr, Corderoy, Easterbrook and Hellier (1982)).

The role of this density function ($\rho(\underset{\sim}{r})$) is best illustrated by reference to a simple node generation technique.

NODE GENERATION TECHNIQUE

(1) The function $\rho(\underset{\sim}{r})$ is defined everywhere within the finite element solution domain; in the case of a self-adaptive finite element procedure $\rho(\underset{\sim}{r})$ may initially be constant. The integral of $\rho(\underset{\sim}{r})$ over the solution domain is equal to the total number of nodes required.

(2) A rectangular region is defined which includes the solution domain, and this region is subdivided by a fine grid.

(3) In the interior of the solution domain, points on the fine grid are examined successively, and accepted as nodes if no other node lies within a circle of radius inversely proportional to $\sqrt{\rho(\underset{\sim}{r})}$, as shown in Figure 1. The boundaries of the domain are usually treated first, with the nodes being selected in a similar manner.

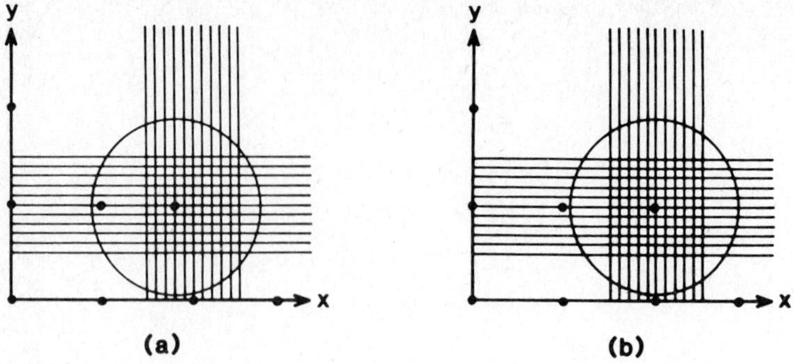

Figure 1
Simple Node Generation Technique:
(a) Grid Point Rejected; (b) Grid Point Accepted as a Node

This method of node generation is extremely slow, particularly when the required nodal density varies considerably throughout the problem domain, entailing the use of a very fine reference grid. There are a number of obvious techniques for increasing the speed of the procedure, but in the simple form described above the role of $\rho(\underset{\sim}{r})$ is most clearly seen.

DENSITY CONTROL FUNCTIONS

The type of density control function used depends to a large extent on the objective of the finite element analysis. If a detailed analysis of the stresses near some feature is required then $\rho(\underset{\sim}{r})$ may simply be a function, imposed by the user, which gives a high nodal density in the vicinity of the feature concerned. Alternatively, if it is intended that the finite element analysis be made self-adaptive, $\rho(\underset{\sim}{r})$ must be in a form that may be easily modified after each solution. In this latter case the most obvious choice for $\rho(\underset{\sim}{r})$ is a functional form whose coefficients can be changed on each iteration. The solution domain may contain regions whose nodal density requirements are incompatible, e.g. regions of essentially constant density combined with a rapid variation in nodal density near a crack. Under these circumstances the domain may be divided into regions, and a different type of density function used in each region. Although piecewise definition of $\rho(\underset{\sim}{r})$ has been deprecated in the introduction to this paper, in this case the density in each region need not be constant and self-adaption is still possible.

Two functional forms, which accommodate non-singular and singular stress intensification respectively, have been studied previously (Corderoy, McGirr and Easterbrook (1981), McGirr, Corderoy, Easterbrook and Hellier (1982)), and are described briefly below.

$$\rho(\underset{\sim}{r}) = \frac{\Sigma \dfrac{\rho_i w_i}{|\underset{\sim}{r}-\underset{\sim}{r}_i|}}{\Sigma \dfrac{w_i}{|\underset{\sim}{r}-\underset{\sim}{r}_i|}} \tag{1}$$

where $\underset{\sim}{r}_i$ are the position vectors of nominated density control points and ρ_i, w_i are density and weighting coefficients which are modified after each solution iteration. This function has the properties of being equal to the density ρ_i at each point $\underset{\sim}{r}_i$ and of varying smoothly in between.

$$\rho(\underset{\sim}{r}) = \frac{1}{|\underset{\sim}{r}|}\ (\text{polynomial in } x,\ y) \tag{2}$$

This second function is well suited to the representation of the singularity produced at the tip of a sharp crack. Its performance for a specific geometry (Figure 2) using a small number of nodes and constant strain triangles is shown in Table I (McGirr, Corderoy, Easterbrook and Hellier (1982)).

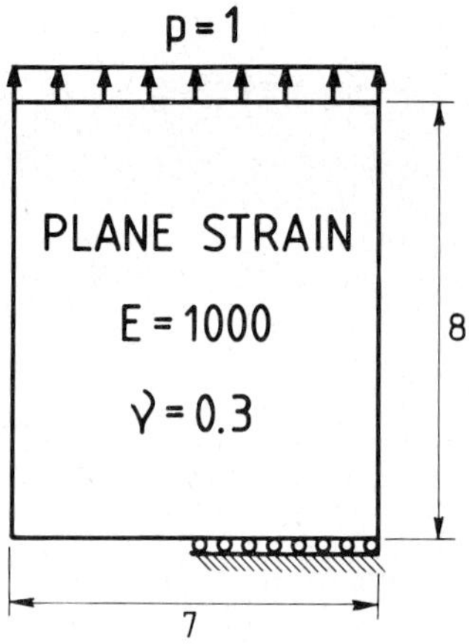

Figure 2
Edge-Cracked Plate Used as Test Problem

TABLE I

COMPARISON OF RESULTS USING VARIOUS DENSITY FUNCTIONS AND MORE EXACT CALCULATIONS

Density function or procedure used	No.of Nodes	Total strain energy (unit thickness)	Stress intensity factor
$\rho(\underset{\sim}{r}) = \alpha$, constant	106	.05105	7.14
$\rho(\underset{\sim}{r}) = \frac{\alpha}{\lvert\underset{\sim}{r}\rvert}$	99	.05413	7.91
$\rho(\underset{\sim}{r}) = \frac{1}{\lvert\underset{\sim}{r}\rvert}$ (quadratic polynomial in x, y)	144	.05415	7.92
$\rho(\underset{\sim}{r}) = \frac{1}{\lvert\underset{\sim}{r}\rvert}$ (cubic polynomial in x, y)	176	.05552	8.37
Contour integral	-	-	9.01
Hybrid element	377	-	9.16
Boundary collocation	-	-	9.3

While combinations of the above functions are adequate for geometries involving sharp cracks and regions of otherwise slowly varying nodal density, they are unsuitable for investigations into the process of crack blunting, and the effect of alloy microstructure on this process. Their use, and also the use of special crack tip elements, imposes a functional form on the stress near the yielded region around a crack tip which cannot be achieved in real materials. While this discrepancy may be ignored in fracture mechanics studies concerned with the relatively far field effects of a crack, it is vital when the processes at the crack tip are being considered.

Consequently, a representation of $\rho(\underset{\sim}{r})$ is sought which imposes the least possible constraint on the manner in which the nodal density may vary, even though its performance, in the case of a sharp crack, may be inferior to that of nodal configurations specifically designed for this situation.

SELF-ADAPTION

Any suitable solution parameter may be used to modify the density function. In this study, and those reported in Table I, $\rho(\underset{\sim}{r})$ was made proportional to the strain energy density (SED) obtained after each solution. Consequently, the self-adaptive process is one in which the elements are being changed in size to have constant strain energy. It has been argued that the definition of near-optimum finite element meshes depends on the variation in SED (Shephard, Gallagher and Abel (1980)); however, when a singularity is present the integrals of the derivatives of SED are not finite. The total strain energy is finite, and moreover, SED is considered to be an important factor in determining both fracture and yield behaviour (Sih (1982)).

TABULAR FUNCTIONS

The direct use of the SED values obtained after each solution avoids the imposition of possibly incorrect functional forms on the variation of SED throughout the problem domain. Following this procedure, $\rho(\underset{\sim}{r})$ would be described by the table of SED values obtained at the centroid of each constant strain triangle, or by a table of values which describe the variation of SED within each triangle for analyses involving higher order elements.

This technique has so far only been tested using constant strain triangles and in this case $\rho(\underset{\sim}{r})$, as defined above, would vary in a stepwise fashion. In order to provide a smoothly varying function the nodal density $\rho'(\underset{\sim}{r})$ described below was used.

$$\rho'(\underset{\sim}{r}) = \frac{\Sigma \dfrac{\rho_i}{|\underset{\sim}{r}-\underset{\sim}{r}_i|}}{\Sigma \dfrac{1}{|\underset{\sim}{r}-\underset{\sim}{r}_i|}} \qquad (3)$$

where the ρ_i are proportional to the SED at the centroid ($\underset{\sim}{r}_i$) of each element.

The test problem was that used for the investigations reported in Table I: a plate in uniaxial tension with a sharp edge crack perpendicular to the tension axis. The node control technique under examination is not intended for the investigation of sharp cracks, which are better treated using other techniques. However, the unsuitability of the geometry makes it useful as a test, and the solutions obtained using more appropriate methods provide a performance target. In addition, the stress intensity factor for the crack provides a single, extremely sensitive, parameter by which the performance of a given mesh may be assessed. In all cases the stress intensity factor was determined by a displacement-based extrapolation method.

The nodal array was generated by a technique whose details are shortly to be published elsewhere (McGirr, Corderoy, Hellier and Easterbrook), and the element generator used was a modification of the Wigner-Seitz cell procedure (McGirr,

Corderoy, Easterbrook and Hellier (1983). Figure 3a shows the initial mesh produced by setting $\rho'(\underset{\sim}{r})$ constant. After three iterations the mesh developed was that shown in Figure 3b, with little change being observed after any further iterations.

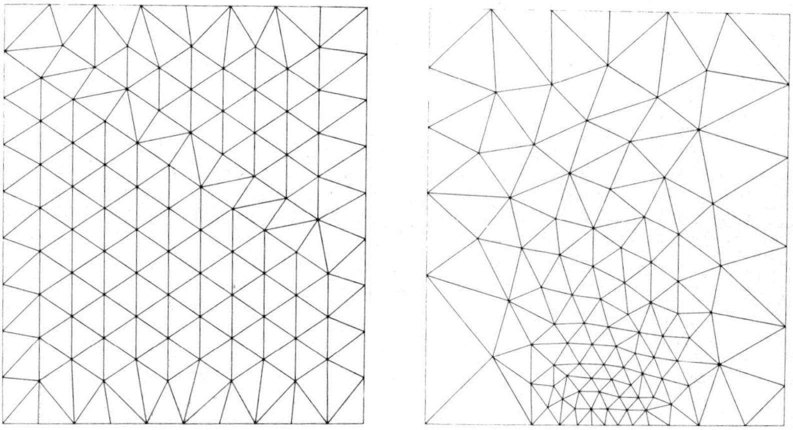

(a) (b)

Figure 3
Meshes for Test Problem (~100 nodes):
(a) Constant Density; (b) Third Iteration Using Tabular SED Values

TABLE II

PERFORMANCE OF TABULAR FUNCTION WITH ~100 NODES

Mesh	No. of nodes	Total strain energy (unit thickness)	Stress intensity factor
Constant density	116	.05118	7.20
1st iteration	101	.05332	7.96
2nd iteration	108	.05410	8.12
3rd iteration	108	.05393	8.14

Table II shows the values of total strain energy and stress intensity factor achieved. It is apparent from this table, and from the observed meshes, that the mesh had stabilised after three iterations. Theoretically, this should occur when the strain energy of each element is the same. Table III shows the frequency of occurrence of element strain energy for the initial mesh and the mesh obtained after three iterations. While it appears that the variation in the strain energy of elements has decreased dramatically, it is still far from constant. The node generator and element generator used are known to be capable of coping with much larger variations in nodal density, and are unlikely to be the source of this early mesh stabilisation; it is felt that use of the averaging process described in equation (3) is the most likely cause. If so, this problem should not be present in later tests using linear strain triangles. The number of nodes and the polynomial order of the elements has been kept low in order to improve the sensitivity of the test. The effect of number of nodes is illustrated by the improved performance of a 150 node mesh, which gave a stress intensity factor of 8.54 on the third iteration.

TABLE III

FREQUENCY OF OCCURRENCE OF ELEMENT STRAIN ENERGY

Strain energy range	Frequency: Initial mesh	Frequency: 3rd iteration
10^{-7} - $3x10^{-7}$	1	0
$3x10^{-7}$ - 10^{-6}	1	0
10^{-6} - $3x10^{-6}$	4	0
$3x10^{-6}$ - 10^{-5}	1	2
10^{-5} - $3x10^{-5}$	7	2
$3x10^{-5}$ - 10^{-4}	23	13
10^{-4} - $3x10^{-4}$	121	90
$3x10^{-4}$ - 10^{-3}	30	69
10^{-3} - $3x10^{-3}$	2	4
$3x10^{-3}$ - 10^{-2}	3	0
	193	180

CONCLUSION

It appears that automatic mesh generation controlled by a table of SED values, using constant strain triangles and a low number of nodes, can result in a performance approaching that of techniques more appropriate to the test problem. Stabilisation of the mesh occurs at a stage where element strain energies still vary considerably, and it would be expected that further adaption is possible. Neither the node generator nor the element generator seem to be the likely source of this early stabilisation; it may be due to the SED averaging procedure, which will be unnecessary when further tests with linear strain triangles are carried out.

REFERENCES

[1] Frederick, C.O., Wong, Y.C. and Edge, F.W., Two-dimensional automatic mesh generation for structural analysis, Int. Jrnl. for Num. Meth. in Engng. 2 (1970) 133-144.

[2] Suhara, J. and Fukuda, J., Automatic mesh generation for finite element analysis, in: Oden, J.T., Clough, R.W. and Yamamoto, Y. (eds.), Advances in Computational Methods in Structural Mechanics and Design (UAH Press, Alabama, 1972) 607-624.

[3] McGirr, M.B., Corderoy, D.J.H., Easterbrook, P.C. and Hellier, A.K., A new approach to automatic mesh generation in the continuum, Proc. 4th Int. Conf. in Australia on Finite Element Methods, Hoadley, P.J. and Stevens, L.K. (eds.) Univ. of Melbourne (1982) 36-40.

[4] Corderoy, D.J.H., McGirr, M.B. and Easterbrook, P.C., Finite element analysis of microscopic stresses produced by inclusions in steel using variable node density coordinate independent mesh generation, Proc. Symp. on Stress Analysis for Mechanical Design 1981, The Institution of Engineers, Australia, Sydney (1981) 33-36.

[5] Shephard, M.S., Gallagher, R.H. and Abel, J.F., The synthesis of near-optimum finite element meshes with interactive computer graphics, Int.Jrnl. for Num. Meth. in Engng. 15 (1980) 1021-1039.

[6] Sih, G.C., Prediction of crack growth characteristics, Proc. of Int. Symp. on Absorbed Specific Energy and/or Strain Energy Density Criterion, Sih, G.C., Czoboly, E. and Gillemot, F. (eds.), Sijthoff and Noordhoff, The Netherlands (1982) 3-16.

[7] McGirr, M.B., Corderoy, D.J.H., Hellier, A.K. and Easterbrook, P.C., An automatic self-adaptive procedure for two-dimensional mesh generation, in preparation for submission to Int. Jrnl. for Num. Meth. in Engng.

[8] McGirr, M.B., Corderoy, D.J.H., Easterbrook, P.C. and Hellier, A.K., Generation of finite element meshes from nodal arrays, Computational Techniques and Applications Conf., Univ. of Sydney (1983).

Computational Techniques & Applications: CTAC-83
J. Noye & C. Fletcher (Editors)

THE PERFORMANCE OF AN AUTOMATIC, SELF-ADAPTIVE FINITE ELEMENT TECHNIQUE

M.B. McGirr, D.J.H. Corderoy, A.K. Hellier
and P.C. Easterbrook

School of Metallurgy
University of New South Wales
Kensington 2033
AUSTRALIA

Progress in the development of an automatic, self-adaptive, two-dimensional finite element technique, whose basic philosophy has been reported previously, is presented. The procedure involves the following steps: (1) generation of an array of nodes according to a nodal density function; (2) construction of a triangular mesh from the nodal array, and (3) modification of the nodal density function in light of the finite element solution. The various algorithms that have been used to execute these steps are discussed, and the performance of the technique on a trial problem containing a stress singularity is reported.

INTRODUCTION

A number of powerful commercial finite element packages exist which provide varying degrees of automatic mesh generation. A fully automatic package would require only unavoidable input such as the problem geometry, boundary conditions and the number of nodes to be used in the analysis. While existing packages are not automatic in this strict sense they contain, in general, sufficient aids to mesh generation to come close to satisfying the requirements of the design engineer.

There is, however, no package available which satisfies the requirements of the metallurgical scientist who, operating at the scale of an alloy microstructure, must determine the reasons why these microscopically inhomogeneous materials have the physical properties observed by engineers. Specifically, in the field of fracture mechanics it is necessary to follow the blunting of cracks, their propagation and the effect on these phenomena of the microstructure through which a crack is passing.

In these situations stress distributions are unpredictable, making the selection of specific element generators for each region of the specimen unreliable, and special crack-tip elements of little use. Moreover, the variety of possible geometries involving combinations of crack, phase and defect means that any investigation will involve a large number of analyses, and user invervention must be reduced to a minimum. The requirements of a finite element analysis package designed for metallurgical research are, therefore, that it should be self-adaptive, so that initial estimates of the stress variation are not needed, and fully automatic so that a minimum of user time is involved.

Such a package has been under development at the School of Metallurgy (UNSW) for the past three years. The purpose of this paper is to report the structure of the package, some details of the procedures used for mesh generation and the stage of development that the package has reached.

PACKAGE STRUCTURE

The programs are written in BASIC and were developed on a DEC PDP 11/10, although modifications are currently being made to enable transfer to a VAX 11/780 system.

Among the operations performed by the package, the following group contain features that are sufficiently unusual to warrant detailed discussion.

(1) Entry of shape, loads and fixities.

(2) Node generation under the control of a density function.

(3) Element generation.

(4) Self-adaption by modification of the density function.

ENTRY OF SHAPE, LOADS AND FIXITIES

While the boundary curves of the shape are being entered, program lines are simultaneously written to two files which are later automatically overlaid as subroutines in other modules. One file contains test conditions that determine whether or not a nominated point lies within the shape; the other file is a routine which accepts a value of x and returns the corresponding y values of points on boundaries. Information relating to loads and fixities is retained and applied at the stage when mesh generation is complete.

NODE GENERATION

This module generates nodes at a density determined by a nodal density function, $\rho(\underset{\sim}{r})$, whose integral over the solution domain is approximately equal to the number of nodes required. Several node generators have been used in conjunction with a number of techniques for controlling the nodal density.

In order to properly outline the limits of the shape, nodes are first placed on the boundaries in all cases. This is done by performing a numerical integration of $\rho^{\frac{1}{2}}(\underset{\sim}{r})$ along a boundary, resulting in the value I; this value is then rounded to the next highest integer I'. An integration of $(I'/I)\rho^{\frac{1}{2}}(\underset{\sim}{r})$ is now carried out along the same boundary, with a node being generated each time the accumulated integral passes an integer value. The end points of each boundary curve are automatically added to the list of nodes.

All internal node generators that have been used rely on the definition of a rectangular "outline" which completely encloses the shape. In all cases a point $\underset{\sim}{r}$ is accepted as a node if no other node, including boundary nodes, lies inside a test circle of radius inversely proportional to $\rho^{\frac{1}{2}}(\underset{\sim}{r})$. Within this general framework the generators tested have differed mainly in the manner in which test points are selected for consideration as possible nodes. The only procedure that is totally unbiased by the external coordinate system is random selection. Apart from the extreme slowness of this process it has been found that it leads to badly proportioned triangles when mesh generation is carried out. Indeed, the meshes generated following random selection of nodes prompt the conclusion that some degree of bias towards a regular array of nodes in regions of constant density is necessary.

More realistically, considerable work has been done using modifications of the technique of selecting nodes from a very fine, rectangular grid of points defined within the "outline" of the shape; the fineness of this grid sets the maximum density of nodes. Such techniques, a simple example of which is reported elsewhere in these proceedings (McGirr and Krauklis (1983)), are extremely slow unless some method is used to avoid examination of all point on the grid. Specifically, a coarse grid may be used initially, and each grid element subdivided further only if examination of the density function in its vicinity suggests that subdivision will result in the placement of more nodes. This procedure is difficult to program and tends to result in a stepwise variation of nodal density. The use of superimposed grid search nodal generators has been abandoned in favour of the active search generator described below.

In order to simplify this brief description it will be assumed that the lower left-hand corner of the shape lies at the origin of the "outline". By incrementally changing x and y a point is found, the test circle around which approximately touches two existing nodes. This point is then added to the list of nodes, the x value is incremented to a position outside the test circle and the process is repeated until the boundary of the shape is reached. The search point is then returned to the origin, y is incremented and the process repeated. This technique clearly involves many tests for the position of a point with respect to other nodes and boundaries, as well as procedures for varying x and y increments depending on the previous search history. Despite this complexity the search operates quickly, since each sequence of increments results in the placement of a node. Moreover, this active search is capable of producing nodal arrays with extreme variations in density; a mesh of a compact tension test specimen generated using this algorithm exhibited densities varying by a factor of 1000 or more.

NODAL DENSITY FUNCTIONS

Several approaches to the problem of supplying a suitable nodal density function have been used, and are described in detail in another paper in this conference (McGirr and Krauklis (1983)). In the earliest investigations a family of functions was assigned to each region of the specimen, and the coefficients of these functions varied after the finite element solution so that the functions were best fitted to the spatial variation of the selected adaptive parameter. More recently it has been found that the table of values of the adaptive parameter, unfitted to any functional form, is quite effective as a controller of nodal density. In view of the fact that such a table is unbiased by any preconception concerning the stress distribution, most current work is done using this procedure.

ELEMENT GENERATION

Two element generators have been tested, both of which are based on the construction of the Wigner-Seitz cell around each node. These generators are discussed fully in a companion paper (McGirr, Corderoy, Easterbrook and Hellier (1983)).

ADAPTIVE PARAMETER

Using the tabular technique mentioned above, any parameter whose spatial variation can be defined by a table of coefficients for each element, may be used to control the nodal density. In all the work carried out so far strain energy density has been used as the adaptive parameter. Strain energy density is a scalar and may readily be determined at any point of the mesh for which stresses are calculated; in the type of investigation for which the package is designed the strain energy density is a good indicator of regions of interest in the material.

PERFORMANCE

The package has been tested in the elastic regime only, using small numbers (100-200) of nodes and constant strain triangles. These latter restrictions are useful for highlighting performance differences between the techniques tested. Particular emphasis has been placed on the examination of sharp cracks. Although it is not possible for a stress singularity to occur in a real alloy, there will exist regions of rapid variation in stress; moreover the stress intensity factor for a crack provides a single, sensitive parameter by which the performance of a given mesh may be assessed. The test problem used for the current work consists of a single edge-cracked plate in uniform tension, as shown in Figure 1a. A constant density mesh of approximately 150 nodes was generated and iterated self-adaptively using tabular strain energy density values. The mesh obtained on the third iteration is shown in Figure 1b.

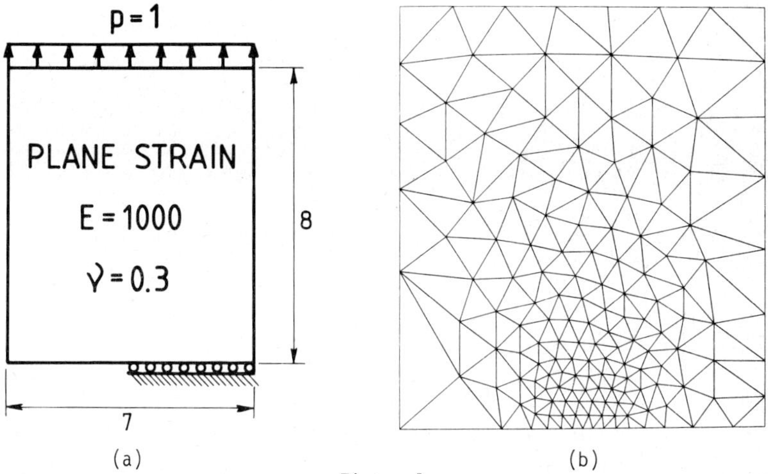

Figure 1
Test Problem
(a) Geometry of Edge-cracked Plate; (b) Mesh on Third Iteration (~150 nodes)

TABLE I

PERFORMANCE OF TABULAR FUNCTION WITH ~150 NODES

Mesh	No. of nodes	Total strain energy (unit thickness)	Stress intensity factor
Constant density	161	.05094	7.12
1st iteration	152	.05499	8.45
2nd iteration	154	.05534	8.51
3rd iteration	157	.05521	8.54

Table I contains values of total strain energy and stress intensity factor obtained from the solution of each mesh. During the iterations the stress intensity factor, calculated using the displacment method (Chan, Tuba and Wilson (1970)), increased from 7.12 to 8.54; a crack-tip element method and a boundary collocation technique used on this problem resulted in stress intensity factors of 9.16 and 9.3 respectively (Becker, Dunham and Stern (1974)).

The distribution of element strain energies is shown in Table II, both for the original, uniform mesh and the third iteration of this mesh.

The adaptive scheme used should cause the strain energy per element to become constant. It can be seen that the range of strain energy values has decreased markedly, the remaining spread presumably being a consequence of the relatively low number of nodes used for this test.

TABLE II

DISTRIBUTION OF ELEMENT STRAIN ENERGIES

	Frequency	
Strain energy range	Initial mesh	3rd iteration
10^{-7} - 3×10^{-7}	1	0
3×10^{-7} - 10^{-6}	0	0
10^{-6} - 3×10^{-6}	7	0
3×10^{-6} - 10^{-5}	6	1
10^{-5} - 3×10^{-5}	11	11
3×10^{-5} - 10^{-4}	117	29
10^{-4} - 3×10^{-4}	106	217
3×10^{-4} - 10^{-3}	26	17
10^{-3} - 3×10^{-3}	4	2
	278	277

CONCLUSION

A fully automatic, self-adaptive, two-dimensional finite element package has been designed, whose performance on a problem involving a stress singularity is comparable to that of techniques specific to this type of problem. The active search node generator has performed well in the relatively simple test geometry and will be tested on more complex geometries in the future. The element generator is believed to be near optimal, and the use of a table of solution parameters to control the nodal density is a totally unbiased procedure which works satisfactorily. Apart from an overall effort to improve the efficiency of individual operations, the package is essentially complete as regards its development for use in linear elasticity. Future efforts will be concentrated on the provision of a plastic capability.

REFERENCES

[1] McGirr, M.B. and Krauklis, P., The automatic generation of arrays of nodes with varying density, Computational Techniques and Applications Conf., University of Sydney (1983).

[2] McGirr, M.B., Corderoy, D.J.H., Easterbrook, P.C. and Hellier, A.K., Generation of finite element meshes from nodal arrays, Computational Techniques and Applications Conf., University of Sydney (1983).

[3] Chan, S.K., Tuba, I.S. and Wilson, W.K., On the finite element method in linear fracture mechanics, Engng. Fract. Mech. 2(1970) 1-17.

[4] Becker, E.B., Dunham, R.S. and Stern, M., Some stress intensity calculations using finite elements, Proc. Int. Conf. on Finite Element Methods in Engineering, University of New South Wales (1974) 117-138.

Computational Techniques & Applications: CTAC-83
J. Noye & C. Fletcher (Editors)

ADAPTIVE TECHNIQUES IN FINITE ELEMENT ANALYSIS

A.D. Miller

Centre for Mathematical Analysis
The Australian National University
GPO Box 4 Canberra ACT 2601
Australia

Finite element analysis has become one of the major tools in contemporary computational mechanics. However, the practical success or failure of many finite element computations often depends critically on certain user decisions - for instance, choices for the finite element mesh and the degree of the elements to be employed in the finite element model. Usually, such decisions are made a priori on the basis of the user's training and experience. Recently however, there has been considerable interest in finite element algorithms which make some of these decisions automatically. In this paper we review a class of algorithms with this kind of automatic character that has been suggested in recent years. Particular attention will be paid to the FEARS program.

§1 INTRODUCTION

Finite element analysis has become one of the major tools in contemporary computational mechanics. However, the practical success or failure of many finite element computations often depends critically on the user's choices of finite element mesh and element type. As a simple illustration of this, consider the elliptic boundary value problems that arise in classical plane linear elasticity. For such problems it is well known that in the neighbourhood of certain critical boundary points (e.g. angular boundary points, or points where the boundary conditions change from specified tractions to specified displacements) the stresses (i.e. the first derivatives of the solution) exhibit some form of singular behaviour. Unless such critical points are handled properly in the finite element model (for instance, by a suitable mesh refinement etc.), the resulting finite element solution may have disappointing accuracy. Such poor accuracy is usually a global phenomenon, not only local to the immediate vicinity of the critical points themselves. So, even if the user's primary interest is in features of the solution in regions distant from any critical boundary point, the accuracy of the results may still be adversely affected. (This is the so-called "pollution effect" of boundary singularities.)

Usually the decisions concerning choices of mesh and element type are made a priori, and somewhat arbitrarily, based upon the user's training and previous experience with like problems. Obviously, it would be an advantage if these choices could be made automatically, thus freeing the user from what is often a not inconsiderable amount of work. Moreover, if these choices could be made according to some "intelligent" criteria which takes account of the ultimate goal of the computation, the advantage would be even greater. In this paper we shall discuss a technique that has gained popularity in recent years for making at least some of these choices automatically. This technique is adaptive in nature; that is, for some initial finite element model the finite element solution is found, and based upon some information obtained from this approximation, the model is modified in some way to yield a "better" model. This process can then be iterated, using the last model as the new initial model. The process is terminated when some stopping criterion is satisfied. The permitted model modification at each step can be of various kinds, for instance, mesh refinement but with the degree of the elements remaining fixed (h-refinement), or an increase in the degree of the local approximating functions but with the mesh fixed (p-refinement).

The ideal adaptive algorithm would be one which, roughly speaking, minimized in some "optimal" fashion a specified measure $E(u,\tilde{u})$, say, of the "difference" between the exact solution u and the approximate solution $\tilde{u}$. Typically, for the structural mechanics problems we are considering here $E(u,\tilde{u})$ would be the strain energy of the difference $(u-\tilde{u})$. The word "optimal" here is meant to convey some sense of the "minimal" computational work to achieve a prescribed accuracy in $\tilde{u}$ as measured by $E(u,\tilde{u})$. This is obviously asking a lot, not the least because, in general, one has no way of knowing $E(u,\tilde{u})$. Nonetheless, this ideal serves to indicate three important features that any practical adaptive algorithm should try to possess:

(i) Each model modification step should aim to minimize some error measure. This error measure should be related to the ultimate goal of the computation. It is not unreasonable to expect that adaptive techniques based on different error measures lead to finite element models of quite different character.

(ii) The adaptive process needs a stopping criterion. This criteria is obviously not arbitrary, but should be related to the ultimate goal of the computation, usually through the error measure in (i).

(iii) The entire adaptive process should seek to attain its goal (as defined by the stopping criterion in (ii)) in some "optimal" fashion. Again, it is hard to be precise about what "optimal" should mean here.

As one does not know $E(u,\tilde{u})$ exactly in general, the most natural practical approach would be to rely upon some "estimate" for E. Since the "estimates" that we shall discuss are not necessarily estimates in the strict mathematical sense of the word, we shall use the term estimator to convey a looser, and less precise concept. (Of course, a major mathematical concern is to try to clarify in what sense an estimator is an estimate.) The adaptive technique that we shall discuss here is based upon some estimator $e(\tilde{u})$ for $E(u-\tilde{u})$ which is of the form

$$e(\tilde{u}) = \sum_{\substack{\text{all elements} \\ \Delta}} \eta(\tilde{u};\Delta) \ . \tag{1.1}$$

Here $\eta(\tilde{u};\Delta)$ is termed the error indicator associated with the element Δ of the finite element mesh. It is a non-negative quantity, and it is calculated only using information about the finite element solution $\tilde{u}$ on Δ itself and on the immediately adjacent elements. Given such error indicators $\eta(\tilde{u};\Delta)$ for some finite element model with corresponding solution $\tilde{u}$, the next model modification step, if necessary, entails a refinement of those elements Δ whose indicators $\eta(\tilde{u};\Delta) \geq \tau$ for some cut off value τ. Various methods for calculating the error indicators and cut off value τ for each modification step have been proposed.

In §2 we illustrate some of the kinds of error estimators and indicators that have been proposed for the case where the error measure E is the "strain energy" of the error. Our discussion here will be in the context of a simple model problem. Section 3 contains a brief outline of a particular example of an adaptive algorithm of the kind we are discussing. In §4 we show how estimators and indicators can be derived in cases where the error measure E is now the error in the value of some functional. Finally in §5 we briefly mention some open mathematical problems relating to the algorithms we have been discussing.

§2 CALCULATION OF ERROR ESTIMATORS $e(\tilde{u})$

§2.1 The Model Problem

To illustrate the basic ideas behind the various error estimators that have been proposed we shall limit our discussion here to the simple model problem:

$$\begin{aligned} \nabla^2 u &= f \quad \text{in} \quad \Omega = (-1,1)^2 \\ u &= 0 \quad \text{on the boundary } \partial\Omega \text{ of } \Omega \ . \end{aligned} \tag{2.1}$$

Let $H^1(\Omega)$ be the usual Sobolev space of all functions which together with their generalized first derivatives are square integrable on Ω, and let $H_0^1(\Omega) = \{w \in H^1(\Omega) : w = 0 \text{ on } \partial\Omega\}$ For any $w,z \in H^1(\Omega)$ define the bilinear form

$$B(w,z) = \int_\Omega \nabla w \cdot \nabla z \, dx \ .$$

The problem (2.1) has a Galerkin formulation: Find $u \in H_0^1(\Omega)$ such that

$$B(u,v) = \int_\Omega f\, v\, dx \qquad \forall v \in H_0^1(\Omega) \ . \tag{2.2}$$

If we assume that the load f is "smooth" then u is also quite smooth with only a mild singularity at the corner points of $\partial\Omega$. Of course, the full power of an adaptive algorithm is not needed here as, a priori, we know that a uniform mesh should perform adequately. Nonetheless, the example suffices to illustrate the basic flavour of the various error estimates without the need to be concerned with extra technicalities introduced by non-uniform meshes etc. For details of how to proceed in the case of more general meshes than considered here see the quoted references.

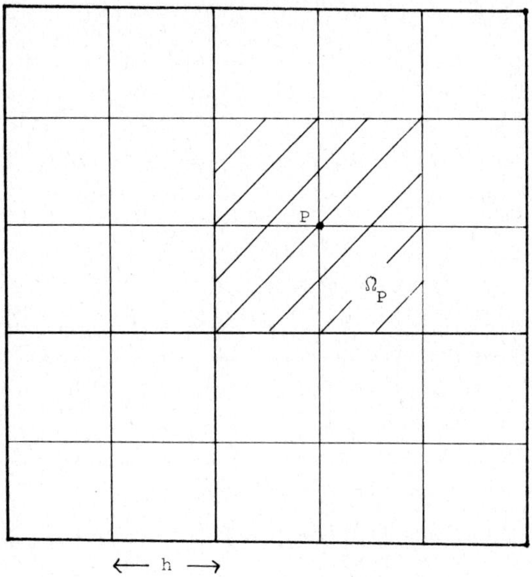

Fig. 1 The region for problem (2.1).

Consider a uniform square mesh on Ω (see Fig.1) of mesh size h. Let $\tilde{S} \subset H_0^1(\Omega)$ be the corresponding finite element subspace associated with C^0 bilinear elements. The corresponding finite element solution $\tilde{u}$ is defined by

$$B(\tilde{u},v) = \int_\Omega f\, v \qquad \forall v \in \tilde{S} \ . \tag{2.3}$$

We shall be interested in trying to "estimate" the strain energy of the error $(u-\tilde{u})$,

$$E(u,\tilde{u}) = B(u-\tilde{u},u-\tilde{u}) = \int_\Omega |\nabla(u-\tilde{u})|^2 dx \ .$$

§2.2 The "star" Estimator

This estimate was first proposed in [1]. In the present context their basic result becomes:

Theorem 1 For any interior node P of the mesh on Ω let Ω_P be the square ("star") formed by the four elements around P (see Fig.1). Let $z_P \in H_0^1(\Omega_P)$ be the (unique) solution of the local subproblem

$$\int_{\Omega_P} \nabla z_P \cdot \nabla v = \int_{\Omega_P} (fv - \nabla\tilde{u} \cdot \nabla v) \qquad \forall v \in H_0^1(\Omega_P) .$$

Then, there exist constants $C_1, C_2 > 0$, independent of h, such that

$$C_1 \sum_P |z_P|^2_{1,\Omega_P} \leq E(u,\tilde{u}) \leq C_2 \sum_P |z_P|^2_{1,\Omega_P}$$

where $\sum_P$ denotes summation over all interior nodes of the mesh and $|z_P|_{1,\Omega_P} = \left(\int_{\Omega_P} |\nabla z_P|^2 dx\right)^{\frac{1}{2}}$.

For a proof see [1]. Of course, the exact solutions of the local subproblems (2.4) are not in general known. However it is possible to estimate $|z_P|^2_{1,\Omega_P}$ and so obtain the following computable version of Theorem 1:

Theorem 2 There are constants C_1, C_2 and h_0 such that for all $0 < h < h_0$

$$C_1 \sum_\Gamma h \int_\Gamma |J_\Gamma(\nabla\tilde{u} \cdot \hat{n})|^2 ds \leq E(u,\tilde{u}) \leq C_2 \sum_\Gamma h \int_\Gamma |J_\Gamma(\nabla\tilde{u} \cdot \hat{n})|^2 ds$$

where $\sum_\Gamma$ denotes the sum over all inter-element boundaries and $J_\Gamma(\nabla\tilde{u} \cdot \hat{n})$ is the difference between the outward pointing normal derivatives of $\tilde{u}$ from the two sides of Γ. For a proof see [2]. The theorem suggests that we use as an estimator

$$e(\tilde{u}) = \sum_\Gamma h \int_\Gamma |J_\Gamma(\nabla\tilde{u} \cdot \hat{n})|^2 ds .$$

(This can be put in the form (1.1) by simply assigning $\frac{h}{2} \int_\Gamma |J_\Gamma(\nabla\tilde{u} \cdot \hat{n})|^2 ds$ to each of the elements on either side of Γ.) Theorem 2 says that, at least for h sufficiently small,

$$0 < C_1 \leq \frac{e(\tilde{u})}{E(u,\tilde{u})} \leq C_2 . \qquad (2.4)$$

Roughly speaking then, $e(\tilde{u})$ cannot depart too outrageously from $E(u,\tilde{u})$ in the sense that their ratio can neither approach zero or infinity.

Estimators of this type have also been discussed by [3].

§2.3 Complementary-type Estimators

For convenience let us write $z = u-\tilde{u}$. Then we have after an integration by parts,

$$\begin{aligned} B(z,v) &= B(u-\tilde{u},v) \\ &= \int_{\Omega} f\ v\ dx - B(\tilde{u},v) \\ &= \int_{\Omega} f\ v\ dx - \sum_{\Gamma} \int_{\Gamma} J_{\Gamma}(\nabla\tilde{u}\cdot\hat{n})v\ ds \qquad \forall v \in H_0^1(\Omega) . \end{aligned}$$

So in mechanical terms, the error z is the "displacement" corresponding to a "body load" of density f and "line loads" along each Γ of density $-J_{\Gamma}(\nabla\tilde{u}\cdot\hat{n})$. The complementary variational principle in this setting would say that if $q = (q_1,q_2)$ is any stress field on Ω that satisfies the conditions,

$$\begin{aligned} \nabla \cdot q &= f \quad \text{inside each element } \Delta \\ J_{\Gamma}(q\cdot\hat{n}) &= -J_{\Gamma}(\nabla\tilde{u}\cdot\hat{n}) \quad \text{on each } \Gamma , \end{aligned} \tag{2.5}$$

then

$$E(u,\tilde{u}) = B(z,z) \le \int_{\Omega} |q|^2 dx , \tag{2.6}$$

with equality if and only if $q = \nabla z$. The relation (2.6) provides a strict upper bound on the error. The difficulty, of course, is to find a stress field satisfying (2.5). A natural first approach is to try to handle each element Δ independently. However, for this to be possible, we need to equilibrate the loading data on each non-boundary element individually (since the necessary condition for a problem of the form

$$\begin{aligned} \nabla \cdot q &= f \quad \text{on } \Delta \\ q \cdot \hat{n} &= g \quad \text{on the boundary } \partial\Delta \end{aligned} \tag{2.7}$$

to have a solution is that $\int_{\Delta} f\ dx = \int_{\partial\Delta} g\ ds$). The most natural way to proceed is to try to distribute the global load $-J_{\Gamma}(\nabla\tilde{u}\cdot\hat{n})$ between the two elements on either side of Γ so that for each local subproblem (2.7) the equilibration condition is satisfied. It is not obvious how this is to be done. The matter has been investigated by [4] who proposes a penalty algorithm to achieve this, at least approximately.

§2.4 The FEARS Asymptotically Exact Estimate

This estimate is based upon a comparison of $\tilde{u}$ with a superconvergent approximation constructed by differencing $\tilde{u}$. The basic results is:

Theorem 3 Let ∂_h^j be the difference operator defined by

$$(\partial_h^j w)(x) = \frac{1}{2h}\,(w(x+he_j) - w(x-he_j)) ,$$

whenever x is further than h from $\partial\Omega$ in the j coordinate direction (e_j is the unit vector in the j direction); otherwise $(\partial_h^j w)(x)$ is whichever of the difference quotients $\frac{1}{\pm h}\,(w(x\pm he_j) - w(x))$ that is defined. Then,

$$\left\|\frac{\partial u}{\partial x_j} - \partial_h^j \tilde{u}\right\|_{L_2(\Omega)} = o(h) . \tag{2.8}$$

The proof is implicit in [5] and is presented in [2]. The underlying idea is the interior superconvergence results of [6]. (The definition of ∂_h^j when x is within h of $\partial\Omega$ is somewhat arbitrary, as in one part of the proof it is shown that this small region near $\partial\Omega$ may, in some sense, be neglected.) The significance of (2.8) is that if we use the usual derivative of $\tilde{u}$, we only have

$$\left\|\frac{\partial u}{\partial x_j} - \frac{\partial \tilde{u}}{\partial x_j}\right\|_{L_2(\Omega)} = O(h) .$$

Thus $\partial_h^j \tilde{u}$ is an asymptotically better approximation to $\frac{\partial u}{\partial x_j}$ than $\frac{\partial \tilde{u}}{\partial x_j}$.

Consider an error estimator given by

$$e(\tilde{u}) = \sum_{j=1}^{2} \left\|\partial_h^j \tilde{u} - \frac{\partial \tilde{u}}{\partial x_j}\right\|_{L_2(\Omega)} . \tag{2.9}$$

This can readily be written in the form (1.1). Theorem 3 implies that

$$\frac{e(\tilde{u})}{E(u,\tilde{u})} \to 1 \quad \text{as} \quad h \to 0 .$$

Any estimator for which this asymptotic relation holds has been called an asymptotically exact estimator for E . Such an estimator not only provides an error measure for the model modification process, but also should be a reliable stopping criterion for the entire adaptive process (at least asymptotically).

The list of estimators mentioned above is far from complete. Some additional estimators are described for instance in [7], [8] and [9].

§3 THE FEARS PROGRAM

Many of the ideas discussed in §1 have been implemented in the FEARS program. FEARS is an experimental adaptive finite element code developed over the past few years at the University of Maryland. For a detailed description of the operation and mathematical background of the program see [10] and [2]. For the purposes of this paper, the following few remarks will suffice. FEARS is able to handle symmetric elliptic systems of two equations on regions which are unions of curvilinear quadrilaterals. Within the program each of these quadrilaterals is mapped onto a unit square. The actual finite element modelling is carried out on these mapped squares using conforming square bilinear elements. The permitted model modification at each adaptive step is local mesh refinement. The model modification process proceeds as outlined in §1. The particular choice of cut off value for the error indicators $\eta(\tilde{u},\Delta)$ at each step is determined from some prediction of the expected decrease in $e(\tilde{u})$ if Δ were to be refined. This prediction is based upon the past history of the local refinement process.

In FEARS the finite element equations are solved by a direct method. Usually, after each refinement step a full new solution is calculated and a new set of error indicators is found. However, calculating a new solution each time is quite expensive, and as it is just an intermediate step, only being used to compute the new error indicators for the next refinement step, one would like to avoid it, if possible. FEARS has a number of "economy" modes which do this to varying degrees. In these a full solution is computed only after a specified increase in the total number of elements has occurred since the last full solution. After any refinement step between two such full solution phases, the new error indicators are only approximated on the basis of the past history of the local refinement process. This "economy" mode permits a multi-level refinement to take place between two full solutions. This possibility is particularly desirable for efficient operation for problems with severe singularities. Provided the number of such "short passes" between full solution steps is not too great, the resulting refinement pattern does not differ too much from that obtained by the all full solution method, but, of course, with a considerable saving in computational cost.

As an illustration of some of the features of the FEARS program consider the boundary value problem

$$\begin{aligned} \nabla^2 u &= 0 \quad \text{in } \Omega \\ u &= 0 \quad \text{on } \Gamma_1 \\ \frac{\partial u}{\partial n} &= 1 \quad \text{on } \Gamma_2 \\ \frac{\partial u}{\partial n} &= 0 \quad \text{on } \Gamma_3 \end{aligned} \tag{3.1}$$

where Ω is the unit circle slit along the positive X_1 axis, and Γ_1, Γ_2 and Γ_3 are as shown in Fig. 2. This can be thought of as modelling a slit membrane, one face of the slit being fixed, the other uniformly loaded. It can be shown that $u(x) = O(|x|^{\frac{1}{4}})$ and $\nabla u(x) = O(|x|^{-3/4})$. So the solution u has a rather severe singularity at the slit tip (0,0), and we would expect that a quite strong refinement would be called for about this point.

Table 1 lists some properties of a sequence of five adaptively refined meshes produced by FEARS for (3.1). Ω was considered as being made up of twelve

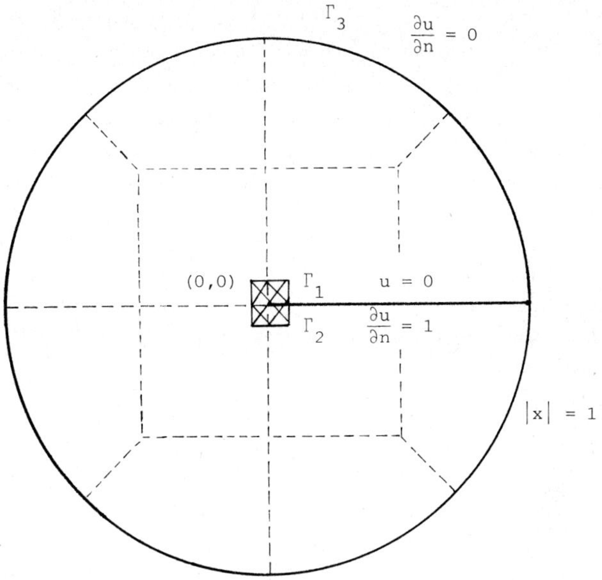

Fig. 2 The region for problem (3.1)

Table 1 Numerical results for the model problem (3.1)

Mesh Label (degrees-of-freedom)	$E(u,\tilde{u})^{\frac{1}{2}}$	$\frac{E(u,\tilde{u})^{\frac{1}{2}}}{(\int_\Omega \lvert\nabla u\rvert^2 dx)^{\frac{1}{2}}}$	$e(\tilde{u})^{\frac{1}{2}}$	$\frac{e(\tilde{u})^{\frac{1}{2}}}{E(u,\tilde{u})^{\frac{1}{2}}}$
I (56)	.6577	30.9%	.3747	.57
II (89)	.5156	24.2%	.3544	.69
III (118)	.3875	18.2%	.3108	.80
IV (171)	.2622	12.3%	.2434	.93
V (391)	.1618	7.6%	.1638	1.01

Notes: (i) The error measure $E(u,\tilde{u}) = \int_\Omega |\nabla(u-\tilde{u})|^2 \, dx$

(ii) The estimator $e(\tilde{u})$ is an extended version of (2.9).

curvilinear quadrilaterals as shown in Fig. 2. The mesh refinement process was directed by an extended version of the estimator (2.9). The initial mesh I was uniform on each mapped square. Subsequent meshes exhibit, as we expected, quite severe refinement around the slit tip. (Fig. 3 shows details of the refinement on the cross-hatched region shown in Fig. 2 for Mesh IV.) For this problem we are able to find an exact solution in closed form by the method of separation of variables, and so we are able to calculate an exact value for the error measure $E(u,\tilde{u})$.

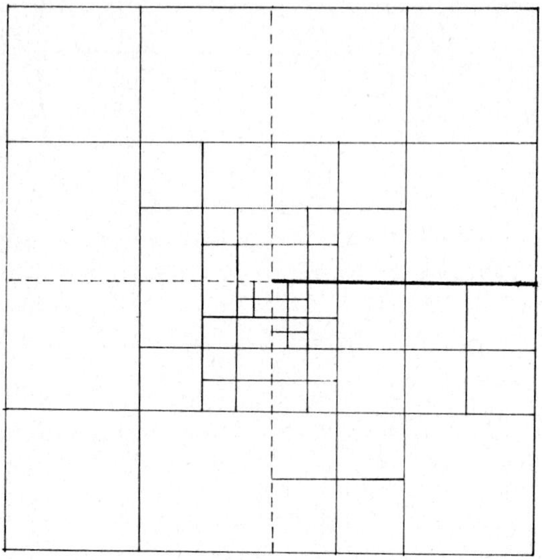

Fig. 3 Detail of refinement on cross-hatched region of Fig. 2 for Mesh IV (This region contains 32% of the total number of elements.)

There are two points worth noting about the results listed in Table 1:

(i) Despite the presence of the $|x|^{\frac{1}{4}}$-type singularity at the slit tip, the sequence of adaptively created meshes gives an apparent rate of convergence for the energy norm $E(u,\tilde{u})^{\frac{1}{2}}$ of the error that is at least as good as the theoretical "optimal refinement" rate of $N^{-\frac{1}{2}}$ (N is the number of degrees-of-freedom). Had only uniform meshes been used on the mapped squares, theory would have predicted an $N^{-1/8}$ rate. In that case, to obtain an accuracy comparable to that of our present mesh V would have required 4×10^6 degrees-of-freedom.

(ii) The ratio $\left(\frac{e(\tilde{u})}{E(u,\tilde{u})}\right)^{\frac{1}{2}}$ appears to converge to 1, showing that the estimator $e(\tilde{u})$ is effectively asymptotically exact.

For this problem FEARS was run in an "economy" mode, so permitting multi-level refinements between full solutions. The meshes I - V referred to here are a selection of meshes created by the full solution steps. The efficiency of the economy mode can be gauged by the fact that the total solution time for the entire process up to and including mesh V was about $2\frac{1}{2}$ times the solution time for mesh V alone.

§4 APPROXIMATE VALUES FOR FUNCTIONALS

So far we have only explicitly dealt with cases where the error measure E was the strain energy of the error $(u-\tilde{u})$. However, in practice, the goal of a computation is usually not to obtain high accuracy in the strain energy as such, but rather to find sufficiently accurate approximations for values of certain functionals (for instance, stress or displacement at a point, stress intensity factor at a crack tip, average flow rate through a surface). The most natural choice of error measure E^* , say, in this case is just the difference between the exact and calculated values of the functional. In this section we shall illustrate how for a certain class of post-processing calculations error estimators for such error measures can be constructed. We shall only treat a simple model problem. For a more complete discussion see [11], [12] and [13].

Consider the stress function formulation of the classical torsion problem for a square bar. The governing boundary value problem can be taken to be

$$\begin{aligned} \nabla^2 u &= -1 \quad \text{in} \quad \Omega = (-1,1)^2 \\ u &= 0 \quad \text{on} \quad \text{the boundary } \partial\Omega \text{ of } \Omega . \end{aligned} \tag{4.1}$$

We shall be interested in finding the (maximum) stress component

$$\sigma = \frac{\partial u}{\partial x_1}(1,0) .$$

Given a finite element approximation $\tilde{u}$, the most common approach would be to use $\frac{\partial \tilde{u}}{\partial x_1}(1,0)$ as the approximation to σ . However we shall employ a post-processing approach which gives better accuracy than this pointwise evaluation approach, and for which error estimators can be readily constructed.

It can be shown that for the problem (4.1)

$$\sigma = \int_\Omega (u\xi+\phi)\,dx \tag{4.2}$$

where

$$\xi = -\nabla^2\phi_0 ,$$

and

$$\phi = \frac{1}{\pi}\,\frac{x_1-1}{(x_1-1)^2+x_2^2} - \phi_0 ,$$

with ϕ_0 being any smooth function which ensures that $\phi = 0$ on $\partial\Omega - (1,0)$. It is not difficult to construct such a ϕ_0 . The identity (4.2) can be readily verified by integration by parts and a simple limiting argument (see [12]). Obviously (4.2) suggests that we use

$$\tilde{\sigma} = \int_\Omega (\tilde{u}\xi+\phi)\,dx$$

as an approximation for σ. This is the approximation we shall consider here.

Let us now turn to an analysis of the accuracy of $\tilde{\sigma}$. Clearly

$$E^*(u,\tilde{u}) = \sigma - \tilde{\sigma} = \int_\Omega (u-\tilde{u})\xi\, dA ,$$

and proceeding with a standard duality argument we introduce the auxiliary problem

$$\begin{aligned} \nabla^2\psi &= -\xi \quad \text{in } \Omega \\ \psi &= 0 \quad \text{on } \partial\Omega \end{aligned} \tag{4.3}$$

and obtain

$$E^* = \sigma - \tilde{\sigma} = \int_\Omega \nabla(u-\tilde{u}) \cdot \nabla(\psi-\tilde{\psi})\,dx$$

where $\tilde{\psi}$ is the finite element solution of (4.3) (with the same mesh etc as used for $\tilde{u}$). We see clearly that the accuracy of $\tilde{\sigma}$ depends not only on the accuracy of $\tilde{u}$, but also on the error $(\psi-\tilde{\psi})$ arising from the auxiliary problem. Simple algebraic manipulation allows us to rewrite the above expression for E^* as

$$E^* = \sigma - \tilde{\sigma} = \frac{1}{4}(\|u+\psi-(\tilde{u}+\tilde{\psi})\|^2 - \|u-\psi-(\tilde{u}-\tilde{\psi})^2\|) . \tag{4.4}$$

On the other hand, applying Schwarz's inequality leads to

$$|E^*| = |\sigma-\tilde{\sigma}| \leq \|u-\tilde{u}\|\ \|\psi-\tilde{\psi}\| \tag{4.5a}$$

$$\leq \frac{1}{2\alpha}(\|u-\tilde{u}\|^2 + \alpha^2\|\psi-\tilde{\psi}\|^2) \quad (\alpha>0) . \tag{4.5b}$$

Here $\|\cdot\|$ denotes the appropriate strain energy norm $(\int_\Omega|\nabla\cdot|^2dx)^{\frac{1}{2}}$.

Notice that (4.4) and (4.5) are expressed solely in terms of the strain energies of the finite element errors $(u+\psi) \pm (\tilde{u}+\tilde{\psi})$, $u-\tilde{u}$ and $\psi-\tilde{\psi}$. So if we have estimators $e(\tilde{u}\pm\tilde{\psi})$, $e(\tilde{u})$ and $e(\tilde{\psi})$ respectively for these we could obtain in a natural way estimators

$$e^{(1)}(\tilde{u},\tilde{\psi}) = \frac{1}{4}|e(\tilde{u}+\tilde{\psi}) - e(\tilde{u}-\tilde{\psi})| \tag{4.6a}$$

$$e^{(2)}(\tilde{u},\tilde{\psi}) = e(\tilde{u})^{\frac{1}{2}}e(\tilde{\psi})^{\frac{1}{2}} \tag{4.6b}$$

$$e^{(3,\alpha)}(\tilde{u},\tilde{\psi}) = \frac{1}{2\alpha}(e(\tilde{u}) + \alpha^2 e(\tilde{\psi})) \quad (\alpha>0) . \tag{4.6c}$$

for the error measure $|E^*| = |\sigma-\tilde{\sigma}|$. We remark that the presence of $\tilde{\psi}$ in these estimators is no real computational drawback. Comparing (4.1) and (4.3) we see that these problems can be considered a multiple loading pair. The linear equations to be solved for $\tilde{u}$ and $\tilde{\psi}$ thus only differ in their right hand sides, so only one stiffness matrix decomposition need be performed for their solution. Of the estimators (4.6) only $e^{(3,\alpha)}$ can be expressed in the form (1.1), and so is suitable for use in the model modification process. Nonetheless, $e^{(1)}$ and $e^{(2)}$ can still be used in the role of stopping criteria. If e is an asymptotically exact then we could normally expect $e^{(1)}$ also to be asymptotically exact, though in general $e^{(2)}$ and $e^{(3,\alpha)}$ will, at least asymptotically, only be upper bounds for $|\sigma - \tilde{\sigma}|$.

Table 2   Numerical results for the example of §4

Number of elements (Uniform mesh, h = element size)	16 (h=0.5)	64 (h=0.25)	256 (h=0.125)
Strain energy error measure: $(E(u,\tilde{u}) / \int_{\Omega} \|\nabla u\|^2 dx)^{\frac{1}{2}}$	30.1%	15.2%	7.62%
Efficiency of estimator: $(e(\tilde{u}) / E(u,\tilde{u}))^{\frac{1}{2}}$	0.94	0.96	0.98
Relative error in pointwise approximation: $\left\| \sigma - \frac{\partial \tilde{u}}{\partial x_1} (1,0) \right\| / \|\sigma\|$	29%	16%	8.7%
Relative error in $\tilde{\sigma}$: $E^\star / \|\sigma\|$ $\quad$ $(E^\star(u,\tilde{u}) = \sigma - \tilde{\sigma})$	1.5%	0.37%	0.089%
Efficiency of estimators for $E^\star$: $e^{(1)}(\tilde{u},\tilde{\psi}) / E^\star$ $\qquad$ (see (4.6))	.99	1.02	1.01
$e^{(2)}(\tilde{u},\tilde{\psi}) / E^\star$	1.24	1.28	1.29

Some numerical results for the model problem (4.1) were obtained using FEARS. These results are summarized in Table 2 (again, we have an exact solution in closed form available for comparison). In this case the mesh refinement was directed by the estimator $e^{(3,1)}(\tilde{u},\tilde{\psi}) = \frac{1}{2}(e(\tilde{u}) + e(\tilde{\psi}))$ where e is the same estimate used in the example §3. Since both u and ψ are rather smooth it is not surprising that, at least initially, all meshes produced were uniform. Notice the following points about the results in Table 2.

(i) $E(u,\tilde{u})^{\frac{1}{2}}$ displays the usual $O(h)$ rate of convergence expected of bilinear elements for a smooth exact solution u on a uniform mesh.

(ii) The estimator $e(\tilde{u})$ and exact error measure $E(u,\tilde{u})$ are very close.

(iii) The post-processed value $\tilde{\sigma}$ is markedly superior to the pointwise evaluated value $\frac{\partial \tilde{u}}{\partial x_1}(1,0)$. The pointwise value converges as $O(h)$, whereas the convergence rate of the post-processed value is $O(h^2)$. This high accuracy can be explained by (4.5a). Indeed, for this problem, since ψ is smooth we expect $E(\psi-\tilde{\psi})^{\frac{1}{2}} = O(h)$; so by what was said in (i), (4.5a) gives $|\sigma-\tilde{\sigma}| = O(h^2)$.

(iv) The estimator $e^{(1)}(\tilde{u},\tilde{\psi})$ is quite close to E^* , while $e^{(2)}(\tilde{u},\tilde{\psi})$ is a slight overestimate of E^* .

§5 OPEN MATHEMATICAL QUESTIONS

Mathematically, much of what is done in adaptive finite element algorithms cannot at present be strictly justified. Heuristic considerations and experimentation play a major part in the design of such algorithms. The questions perhaps most in need of some form of mathematical analysis are:

(i) In what sense does an estimator "estimate" the exact error measure?

(ii) In what sense is the overall adaptive process "optimal"?

As an indication of the type of answers one could expect to these questions, recall the example of §3. The numerical results there seem to indicate that

(i) the estimator is asymptotically exact. (This can in fact be proved under some restricted conditions. See [2].)

(ii) The adaptive sequence of solutions $\tilde{u}$ has the optimal rate of convergence with respect to the number of degrees of freedom of the finite element model.

REFERENCES

[1] Babuška, I. and Rheinboldt, W.C., Error Estimates for Adaptive Finite Element Computations, SIAM J. Numer. Anal., 15 (1978), pp736-754.

[2] Babuška, I. and Miller, A., A Posteriori Error Estimates and Adaptive Techniques for the Finite Element Method, provisional title, in preparation.

[3] Weiser, A., Local Mesh, Local Order, Adaptive Finite Element Methods with A Posteriori Error Estimators for for Elliptic Partial Differential Equations, Technical Report No. 213, December 1981, Dept. of Computer Science, Yale University.

[4] Kelly, D.W., The Self-Equilibration of Residuals and Complementary A Posteriori Error Estimates in the Finite Element Method, submitted to Int. J. Num. Meth. Engng.

[5] Miller, A., On Some A Posteriori Error Estimates in the Finite Element Method, Ph.D. thesis, Dept. of Mathematics, Univ. of Qld (May 1981).

[6] Nitsche, J.A. and Schatz, A.H., Interior Estimates for Ritz-Galerkin Methods, Math. Comp., 28 (1974) pp937-958.

[7] Zienkiewicz, O.C. and Craig, A.W., Adaptive Mesh Refinement and A Posteriori Error Estimation for the p-version of the Finite Element Method, in Proc. U.S. Army Research Office Workshop on Adaptive Methods for Partial Differential Equations, Feb 1983, to appear.

[8] Zienkiewicz, O.C., Kelly, D.W., Gago, J. and Babuska, I., Hierarchical Finite Element Approaches, Adaptive Refinement and Error Estimates, Proc. MAFELAP 1981, ed. J.R. Whiteman (Academic Press).

[9] Shephard, M.S. and Yunus, S.M., A Posteriori Error Estimates for Linear Triangles using h- and p- Hierarchic Modes", submitted to Int. J. Num. Meth. Engng.

[10] Mesztenyi, C. and Szymczak, W., FEARS User's Manual for Univac 1100, Institute for Physical Science and Technology, University of Maryland, Tech. Note No. BN-991 (Oct. 1982).

[11] Babuška, I. and Miller, A., The Post-Processing Approach in the Finite Element Method. Part I: Calculation of Displacements, Stresses and other Higher Derivatives of the Displacements, to appear in Int. J. Num. Meth. Engng.

[12] Babuška, I. and Miller, A., The Post-Processing Approach in the Finite Element Method. Part II: The Calculation of Stress Intensity Factors, to appear in Int. J. Num. Meth. Engng.

[13] Babuška, I. and Miller, A., The Post-Processing Approach in the Finite Element Method. Part III: A-Posteriori Error Estimates and Adaptive Mesh Selection, to appear in Int. J. Num. Meth. Engng.

Computational Techniques & Applications: CTAC-83
J. Noye & C. Fletcher (Editors)

INCOMPRESSIBLE FLOW IN NOZZLES USING THE GALERKIN FINITE ELEMENT METHOD

T. Doan and R.D. Archer

School of Mechanical and Industrial Engineering
University of New South Wales
Kensington, N.S.W., 2033
Australia

The Galerkin Finite Element Method has been used to calculate inviscid, incompressible, fluid flow in arbitrarily shaped axisymmetric nozzles including entrance length. The present formulation uses the continuity and irrotationality equations which are linear when written in terms of velocity components. The flow field was discretized into triangular elements with linear shape functions approximating the velocity components.

NOMENCLATURE

A	cross sectional area; or two-dimensional domain of solution co-ordinates
ℓ	co-ordinate index (= 0 for Cartesian, =1 for cylindrical co-ordinates)
$\dot{m}$	mass flow rate
N_i, n	shape function, number of nodal points
$\vec{n}$, $\vec{t}$	unit vectors normal, tangential to surface
p	pressure
r, z	radial and axial directions (Fig. 1), (cylindrical co-ordinates)
R_*	throat radius of curvature (Fig. 1)
S	element boundary
u, w, V	radial, axial velocity components, and total velocity
ρ	fluid density

Subscripts:

i, j	node number
in, out	inlet, outlet faces AB, CD (Fig. 1)
o	stagnation condition

INTRODUCTION

There are a number of well known numerical techniques for solving fluid flow problems including Time Dependent (Midgal, et al, 1969), Finite Difference (Murman and Cole, 1971), Method of Lines (South, et al, 1972), Integral Relations (Liddle and Archer, 1973) and Finite Volume (Caughey and Jameson, 1977), although many of these have long computation time, numerical singularities, or difficulty in handling arbitrarily shaped geometrical boundaries. Interest has turned recently to the FEM, so successful elsewhere. Some of these solutions (Labujère and Van der Vooren, 1974; Periaux, 1975 and Chan, et al, 1979) have adopted the velocity potential (ϕ) or the stream function (ψ) as dependent variables in the governing equations of motion. In solving for flows over cylinders and aerofoils, Fletcher (1979) used groups of (primitive) variables, namely (ρu, ρw, ρuw).

In this paper, the incompressible flow of an inviscid fluid in an arbitrary nozzle will be investigated by the FEM in conjunction with the Galerkin criterion, Fletcher (1978). The equations of motion are written in terms of the variables (u,w) and are used together with the irrotationality condition. Results will be presented for air.

BASIC EQUATIONS AND ASSUMPTIONS

The shape and corresponding geometrical parameters of the nozzle follow the simple design of Back, et al (1965) and are shown in Figure 1. It is made up of a constant diameter inlet length and simple convergent-divergent cones, interconnected by sections whose wall shapes are circular arcs.

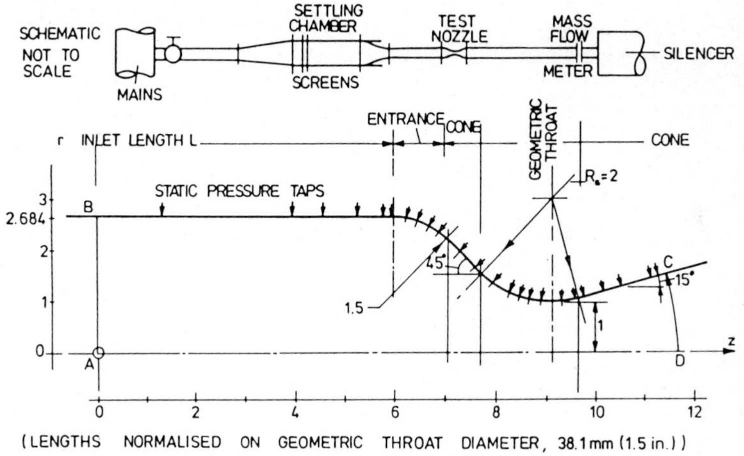

Figure 1
Co-ordinate System and Nozzle Profile

Using simple cylindrical (r,z) co-ordinates, the continuity equation under steady state conditions is:

$$\nabla \cdot \vec{V} r^{\ell} = \frac{\partial}{\partial r}(ur^{\ell}) + \frac{\partial}{\partial z}(wr^{\ell}) = 0 \tag{1}$$

Assuming that boundary layer effects can be neglected, we have the irrotationality condition:

$$\nabla \times \vec{V} = 0 \quad , \quad \text{i.e.} \quad \frac{\partial u}{\partial z} - \frac{\partial w}{\partial r} = 0 \tag{2}$$

Refer velocities to the maximum velocity $V_{max} = \sqrt{2 \times (p_o/\rho)}$, p to the stagnation value p_o, and all lengths to the radius at the geometric throat. Adopt the same symbols for normalized variables. Integrated Euler equation becomes Bernoulli's equation:

$$p = (1 - u^2 - w^2) \tag{3}$$

Equations (1) and (2) will now be solved for the unknowns u and w. Applying the Galerkin criterion to these equations would result in a system of linear equations.

Chain differentiate (1), to give:

$$r^{\ell}\frac{\partial u}{\partial r} + u\ell r^{\ell-1} + r^{\ell}\frac{\partial w}{\partial z} = 0 \tag{4}$$

Approximate (u,w) by linear shape functions N_i between nodal values u_i, w_i:

$$\begin{bmatrix} u \\ \\ w \end{bmatrix} = \sum_i N_i(r,z) \begin{bmatrix} u_i \\ \\ w_i \end{bmatrix} \qquad (5)$$

Substitute Equation (5) into (2) and (4). The following residuals are produced:

$$R_1 = \sum_i u_i \left[\frac{\partial N_i}{\partial r} r^{\ell} + \ell r^{\ell-1} N_i\right] + \sum_i w_i \left[\frac{\partial N_i}{\partial z} r^{\ell}\right] \qquad (6)$$

$$R_2 = \sum_i u_i \frac{\partial N_i}{\partial r} - \sum_i w_i \frac{\partial N_i}{\partial r} \qquad (7)$$

Apply the Galerkin criterion to Equations (6) and (7):

$$\left.\begin{aligned} \iint_A N_j R_1 \, drdz &= 0 \\ \iint_A N_j R_2 \, drdz &= 0 \end{aligned}\right\} \quad j = 1,n \qquad (8)$$

Applying Green's theorem to Equations (8) gives:

$$\left.\begin{aligned} \sum_{i=1}^{n} a_{ji} u_i + \sum_{i=1}^{n} b_{ji} w_i &= 0 \\ \sum_{i=1}^{n} c_{ji} u_i + \sum_{i=1}^{n} d_{ji} w_i &= 0 \end{aligned}\right\} \quad j = 1,\ldots,n \qquad (9)$$

where:

$$a_{ji} = \int_S r^{\ell} N_i N_j dz - \int_A r^{\ell} N_i \frac{\partial N_j}{\partial r} drdz \qquad (10)$$

$$b_{ji} = \int_S r^{\ell} N_i N_j dr - \int_A r^{\ell} N_i \frac{\partial N_j}{\partial z} drdz$$

$$c_{ji} = \int_S N_i N_j dr - \int_A N_i \frac{\partial N_j}{\partial z} drdz$$

$$d_{ji} = -\int_S N_i N_j dz + \int_A N_i \frac{\partial N_j}{\partial r} drdz$$

Solving (9) we obtain u_i and w_i.

BOUNDARY CONDITIONS

The disturbance caused by the inlet contraction curvature R_{in} can nominally be transmitted up stream indefinitely. However, because the strength of this disturbance dies down over a short distance, it can be assumed that the flow at the inlet face AB (Figure 1) is uniform if the length L of parallel inlet duct exceeds a (small) finite value, at which upstream location and beyond:

$$\left.\begin{aligned} u_{in} &= 0 \\ w_{in} &= \text{constant} = \frac{\dot{m}}{\rho_{in} A_{in}} \end{aligned}\right\} \qquad (11)$$

Along the centreline and nozzle wall the impermeability condition applies:

$$\vec{V} \cdot \vec{n} = 0 \qquad (12)$$

If the outlet face CD (Figure 1) is part of a spherical surface having origin at the apex of, and bounded by, the divergent cone then this surface can be considered as a potential surface of a conical source flow, for a sufficiently long diverging cone. Thus:

$$\vec{V} \cdot \vec{t} = 0 \qquad (13a)$$

or:

$$\left.\begin{aligned} u_{out,i} &= V_{out} \sin\alpha_i \\ w_{out,i} &= V_{out} \cos\alpha_i \end{aligned}\right\} \qquad (13b)$$

where $V_{out} = (\dot{m}/(\rho_{out} A_{out}))$ and α_i is the inclination to the nozzle centreline, of V_{out} at node i on CD.

EXPERIMENT

As well as the need for measuring nozzle wall static pressure distribution to compare with numerical results, it was particularly desirable to seek evidence of the adverse wall pressure gradient predicted by the FEM near the inlet contraction. This is the region where the parallel upstream duct is connected to the circular arc of the convergence section, ahead of the converging cone.

The nozzle as tested had a nominal throat diameter of 38 mm (1.5 in) and inlet diameter of 102 mm (4 in). Air at ambient temperature and ~ 4.5 atmospheres abs. pressure was used as working fluid. The schematic in Figure 1 shows the layout of the test system. A manually controlled 102 mm (4 in) gate valve regulated stagnation pressure. A conical diffuser inlet led to the stagnation chamber containing two screens and settling length which exhausted through a 4 to 1 contraction to the primary test components. These were a 408 mm (16 in) length of 102 mm (4 in) internal diameter pipe followed by the test nozzle itself. A 3.4 m (11 ft) length of 102 mm (4 in) pipe connected the nozzle exit to mass flow rate metering station. A short diffuser led from this nozzle to a silencer. 'O' rings, gaskets and sealants were used throughout.

Ambient air compressed to 9 atmospheres abs. (130 psia, 900 kPaa) by two Holman screw-type air compressors with a rated capacity of 0.34 kgs^{-1} (0.75 lb/sec) each, is stored in five 5.7 m^3 (200 ft^3) reservoirs. Alumina dessicant is used to dry the compressed air to -45°C (-50°F) dew point

Six wall static pressure holes were located in the run up, five in the entrance and the remainder along the nozzle. Taps were 0.8 mm (0.032 in) in diameter and drilled normal to the nozzle profile. The axial position of the taps were known to ±0.1 mm and were also spaced circumferentially. A liquid manometer with a range of 0-1000 mm water with graduations in 1 mm was used for measuring small pressure differences. The low pressure differences in the parallel and inlet contraction sections were measured using the same manometer inclined at 15° to the horizontal, with reading error of ±0.26 mm water.

Other pressures were measured by a Druck pressure transducer of ±350 kPag (±50 psig) range, with digital readout accurate to 0.1 kPag (0.0145 psig). The tests were run near the full rated pressure of the transducer.

The mass flow rate was measured by a standard VDI nozzle of 20 mm (0.8 in) internal diameter. The Mach number at the throat of the test nozzle was just under 0.2. Mass flow rate measurements were accurate to ±1%. The Reynolds numbers of the flow in the parallel pipe section is 1.2×10^5.

RESULTS AND DISCUSSION

Numerical results for the nozzle defined in Figure 1 are presented for the triangulated flow field of Figure 2. Linear shape functions, Equation (5), were used to represent variations of u and w within elements. Three meshes having successively smaller sized elements were generated. Mesh I is the coarsest, Figure 2, with 68 nodes and 96 triangles. Meshes II and III have element sizes ½ and ¼ of those of Mesh I. The numerical solution is obtained directly without iteration.

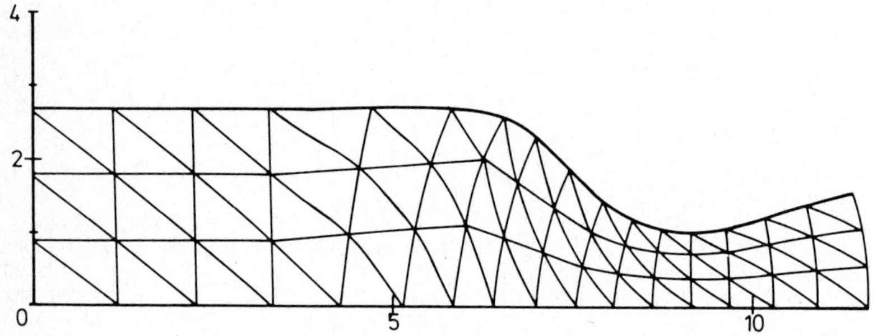

Figure 2
Finite Element Mesh I

The numerical method was in fact generalized to include compressible flow, Doan (1981), and so carried extra calculations involving density. Consequently, it is not as efficient as it could be for incompressible flow, the solution for which was obtained by setting the density field at the "stagnation" value.

Figure 3 shows the streamline and airspeed contours generated by the 3 meshes. The effect of the inlet, convergent, throat and divergent sections is well predicted. The smoothest contours are those from Mesh III which are, however, little different from those of Mesh II. This trend can be confirmed by comparing the wall pressure distributions in Figure 4. It is seen that the numerical results are very close to the experimental values, even those of Mesh I. The maximum error between pressures computed by Mesh I and experimental results occurs at the geometric throat and is only 0.12%. The extent of the precursor effect involving a length of positive unfavourable wall pressure gradient approaching the inlet contraction is well predicted by the FEM. Even though many experimental data points are consistently higher than the computed values, the maximum difference is only about 5×10^{-5}. Experimental error is 3×10^{-4}. We observe that essentially uniform inlet conditions, Equation (11), prevail upstream from no further than one inlet diameter ahead of the inlet. The correlation in the divergent section breaks down after the flow separates at some point downstream of the throat and thus violates our assumption of inviscid flow.

Another measure of convergence of the FEM solution is obtained by comparing the ratios of the computed mass flow rates to the value predicted by simple one dimensional theory.

	Mesh I	Mesh II	Mesh III	Experiment ±1%
$\frac{\dot{m}}{\dot{m}_{1-D}}$ (%)	89.04	89.43	89.59	90.5

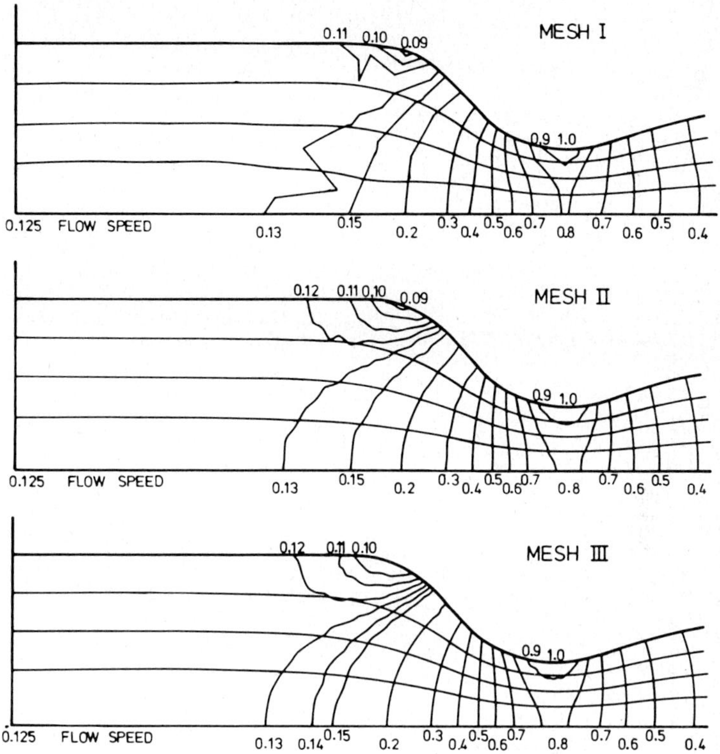

Figure 3
Predicted Incompressible Field

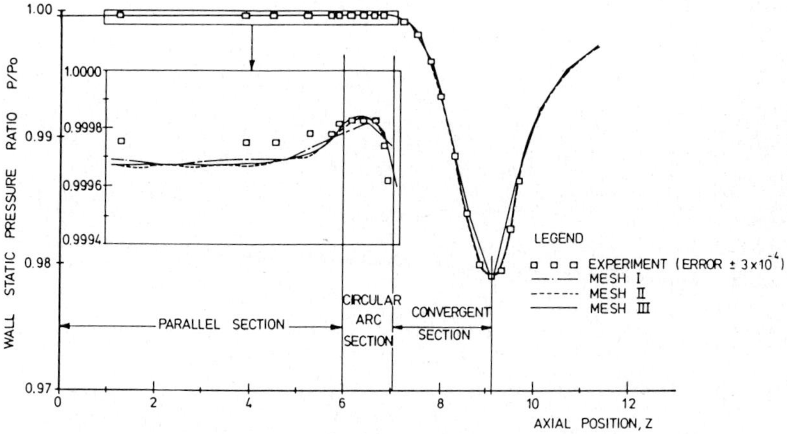

Figure 4
Wall Static Pressure - Incompressible Flow

CONCLUSIONS

The FEM in conjunction with the Galerkin criterion has been applied to the equations of motion of an inviscid, incompressible fluid in an axisymmetric nozzle of arbitrary wall shape, and its associated entrance length of constant diameter pipe. The equations are written in conservation form using simple cylindrical coordinates and primitive variables. The flow field was discretized into triangular elements with linear shape functions approximating the velocity components. A coarse mesh ($\simeq 32 \times 6$ elements, Mesh II) was found to be sufficient to obtain a satisfactory flow solution, in one pass.

The correlation between computed and experimental wall pressures is good, even to the prediction of a small adverse pressure gradient just upstream of the nozzle inlet contraction. Confirmation of uniform pipe flow conditions upstream of one pipe diameter was obtained. The maximum difference between the computed (Meshes II and III) and the experimental results was found to be within the experimental error.

Although this paper describes results for a nozzle with throat radius of curvature of 2, the method has been applied by Doan (1981) with similar results down to a radius of 0.625.

REFERENCES

[1] Midgal, D., Klein, K. and Moretti, G., Time-dependent calculations for Transonic nozzle flow, AIAA J., V. 7, No. 2 (Feb. 1969), 372-374.

[2] Murman, E.M. and Cole, J.D., Calculation of plane steady transonic flows, AIAA J., V. 9 (January 1971), 114-120.

[3] South, J.C., Flunker, E.Q. and Jones, D.J., On the numerical solution of elliptic partial differential equations by the method of lines, J. Comp. Phys., V. 9 (1972), 496-527.

[4] Liddle, S.G. and Archer, R.D., Transonic flow in nozzles using the method of integral relations, J. Spacecraft, 8 (July 1973), 722-728.

[5] Caughey, D.A. and Jameson, A., A finite volume method for transonic potential flow calculations, Proc. AIAA 3rd Comp. Fluid Dynamics Conf., Albuquerque (1977), 35-53.

[6] Labujère, Th.E. and Van der Vooren, J., Finite element calculations of axisymmetric subcritical compressible flow, NLR-TR 74162 U (1974).

[7] Periaux, J., 3-D analysis of compressible potential flow with the finite element method, Int. J. Num. Meth. in Eng., 9 (April 1975), 775-832.

[8] Chan, S.T.K., Brashers, M.R. and Young, V.C.Y., Finite element analysis of transonic flow by method of weighted residuals, AIAA Paper (1979), 75-79.

[9] Fletcher, C.A.J., The primitive variables finite element formulation for inviscid compressible flow, J. Comp. Phys., 33 (1979),301-312.

[10] Fletcher, C.A.J., The Galerkin method: An introduction, Numerical Simulation of Fluid Motion, ed. Noye, J., North Holland (1978).

[11] Back, L.H., Massier, P.F. and Gier, H.L., Comparison of measured and predicted flows through conical supersonic nozzles with emphasis on the transonic region, AIAA J., 3 (September 1965), 1606-1614.

[12] Doan, T., Finite element analysis of fluid flow in axisymmetric nozzles, Ph.D. Thesis (1981), University of New South Wales.

BOUNDARY INTEGRAL METHODS

Computational Techniques & Applications: CTAC-83
J. Noye & C. Fletcher (Editors)

BOUNDARY ELEMENTS IN FLUID MECHANICS

M.B. Bush

Department of Mechanical Engineering,
University of Sydney,
Sydney. N.S.W. 2006
Australia.

This paper reports progress in the application of a direct boundary integral formulation to viscous fluid mechanics problems. The formulation is written in terms of velocities and surface traction forces, together with 'pseudo-body forces' and extra stresses in the domain when required. The method has been applied to a range of linear and non-linear problems: Stokes flow, Oseen flow, flow governed by the Navier-Stokes equations and the flow of non-Newtonian fluids. Some representative examples are included in this report.

INTRODUCTION

Since the first serious applications of boundary integral equation methods to mechanics problems (Rizzo, 1967; Hess and Smith, 1967) the technique has rapidly evolved into a powerful tool. Examples of the state of the art can be found in recent conference proceedings (Brebbia, 1981, 1982). The evolution has not been as rapid in the realm of fluid mechanics as it has been in solid mechanics; with most interest focussed on inviscid flows. Relatively few reports deal with viscous flows. However, some authors (Wu and Thompson, 1973; Youngren and Acrivos, 1975; Coleman, 1981; Bézine and Bonneau, 1981; Bush and Tanner, 1983) have illustrated the advantages of applying boundary integral methods to viscous fluid mechanics problems. These advantages include the relatively small system of algebraic equations generated and features such as the automatic treatment of flow conditions at infinity in external linear flows.

In the present paper we report some investigations into the application of a direct boundary element formulation (see Brebbia, 1980) to a range of viscous fluid mechanics problems. This research has been concerned with the solution of the Navier-Stokes equations, Oseen's equations and the equations governing some types of non-Newtonian flow. We illustrate the method by presenting results for the expansion of an axisymmetric jet of Newtonian fluid at non-zero Reynolds number, the circular Couette flow of a power-law fluid and the drag force on a circular cylinder in Oseen flow.

THE GOVERNING EQUATIONS

We will employ Cartesian tensor notation throughout. In the absence of prescribed body forces, the governing differential equations for the steady flow of an incompressible fluid are

$$\frac{\partial \sigma_{jk}}{\partial x_k} - \rho\, a_j = 0 \qquad \text{(momentum)} \tag{1}$$

and

$$d_{kk} = 0 \qquad \text{(continuity)} \tag{2}$$

where σ_{jk} is the total stress tensor, ρ is the fluid density, a_j represents acceleration terms and d_{jk} is the rate-of-strain tensor. In terms of the velocity, u_j, the acceleration terms are given by

$$a_j = u_k \frac{\partial u_j}{\partial x_k} \tag{3}$$

and the rate-of-strain tensor is written as

$$d_{jk} = \frac{1}{2}\left\{\frac{\partial u_j}{\partial x_k} + \frac{\partial u_k}{\partial x_j}\right\} \quad . \tag{4}$$

We will also need to include constitutive equations relating the stress tensor to other flow variables. These will be considered shortly. Finally, to these equations suitable boundary conditions will be added, thereby completing the description of the problem.

The problem of a rigid body moving with a constant speed U in an infinite expanse of fluid at low Reynolds number is often treated using Oseen's approximation. If the motion is in the x_1 direction, we write

$$a_j = -\,U \frac{\partial u_j}{\partial x_1} \tag{5}$$

Oseen's approximation makes the governing equations linear (given that the fluid is Newtonian) and more amenable to solution. Together with the boundary condition $u_j = U\delta_{j1}$ on the surface of the body, we also require that $u_j \to 0$ and pressure $\to 0$ at infinity.

In general we will assume the total stress tensor is made up of an isotropic pressure term, a Newtonian deviatoric component and an 'extra' deviatoric component:

$$\sigma_{jk} = -p\,\delta_{jk} + 2\eta_0\, d_{jk} + \varepsilon_{jk} \quad . \tag{6}$$

In equation (6), p represents the pressure, δ_{jk} is the Kronecker delta, η_0 is a constant viscosity and ε_{jk} is the 'extra' stress component. The extra stress is zero if the fluid is Newtonian. Otherwise, the expression for ε_{jk} depends on the particular non-Newtonian model chosen. In the present report we will consider the 'generalized Newtonian fluid' model. This allows for a dependence of viscosity on shear rate but is incapable of predicting viscoelastic effects. The viscosity will be represented by $\eta(\dot{\gamma})$, where $\dot{\gamma}$ is the shear rate. The shear rate is simply the magnitude of the rate-of-strain tensor and is given by the expression:

$$\dot{\gamma}^2 = 2d_{jk}\, d_{jk} \quad . \tag{7}$$

There are many expressions relating the viscosity to the shear rate; see Bird et al (1977). The simplest model is the 'power-law' model written as

$$\eta(\dot{\gamma}) = m\, \dot{\gamma}^{n-1} \quad . \tag{8}$$

When n = 1, the power-law fluid becomes a Newtonian fluid with viscosity m. The stress tensor in the generalized Newtonian fluid can be written in the form of equation (6) if we take $\varepsilon_{jk} = 2\{\eta(\dot{\gamma}) - \eta_o\}d_{jk}$. The constant η_o is arbitrary.

BOUNDARY INTEGRAL FORMULATIONS

Boundary integral equation formulations of Newtonian fluid mechanics problems, in terms of velocities and surface tractions, have been considered by Ladyzhenskaya (1963), Youngren and Acrivos (1975) and Bush and Tanner (1983). In the last report the effects of fluid inertia were treated as 'pseudo-body forces'. The formulation can also be extended to include non-Newtonian effects. In this we initially treat the nonlinear terms as 'pseudo-body forces' given by the spatial derivatives of ε_{jk}. However, we find that it is more convenient to recast the formulation so that we deal directly with the stress ε_{jk}. We will not repeat the derivation here since it follows directly from the Newtonian formulations. The final integral equation is

$$c_{ij}(P)\, u_j(P) = \int_\Gamma u^*_{ij}(P,Q)t_j(Q)d\Gamma - \int_\Gamma t^*_{ij}(P,Q)\, u_j(Q)d\Gamma - \int_\Omega \left\{\rho a_j(Q)\, u^*_{ij}(P,Q) + \varepsilon_{jk}(Q)\, \frac{\partial u^*_{ij}}{\partial x_k}\right\} d\Omega. \tag{9}$$

In this, Ω and Γ represent the domain and boundary respectively, P and Q are points in Ω or on Γ and $t_j(Q)$ represents the boundary traction ($\sigma_{jk}n_k$, where n_k is the unit outward normal vector) at Q. Likewise, the terms $u_j(Q)$, $a_j(Q)$ and $\varepsilon_{jk}(Q)$ represent the velocity, acceleration terms and extra stresses at Q. The starred weighting function $u^*_{ij}(P,Q)$ indicates the velocity field at Q due to a point force directed in the x_i direction at P; the 'Stokeslet'. The Stokeslet is taken to act in an infinite expanse of Newtonian fluid with viscosity η_o. The function $t^*_{ij}(P,Q)$ is the corresponding traction field at a boundary point Q. The Stokeslet is well known and can be found in the previously cited references. Finally, the term $c_{ij}(P)$ is included in recognition that the integral containing the singular function $t^*_{ij}(P,Q)$ must be interpreted in the sense of its 'Cauchy principal value' when P lies on the boundary (see for example Mikhlin, 1957). In particular, if the surface at P possesses a normal vector and tangent plane it is easily shown that $c_{ij}(P) = \frac{1}{2}\delta_{ij}$. If P lies in Ω then $c_{ij}(P) = \delta_{ij}$. Coupled with suitable discretization and iterative schemes, equation (9) can be used to solve nonlinear fluid mechanics problems. If inertia can be neglected and the fluid is Newtonian, the equation reduces to a simple boundary integral representation of Stokes flow.

In the case of flow caused by a body moving in an infinite expanse of Newtonian fluid governed by Oseen's equations, we find that a simple boundary integral representation is also possible. On substituting (5) into (9), integrating by parts and taking account of the boundary conditions for this problem, it can be shown (see Bush, 1983) that

$$u_i(P) = -\int_\Gamma u^o_{ij}(P,Q)\, t_j(Q)\, d\Gamma \qquad (10)$$

where Γ represents the boundary of the body and $u^o_{ij}(P,Q)$ is the velocity field at Q due to an 'Oseenlet' directed in the negative x_i direction at P (see Rosenhead, 1963). Following discretization this can be used to find the traction distribution on the body; hence the net drag and lift forces can be evaluated.

A range of viscous fluid mechanics problems can therefore be treated using boundary integral formulations. The equations presented above also apply to axisymmetric problems (formulated in terms of cylindrical coordinates) if we make use of axisymmetric fundamental solutions. In this case, however, there is an additional term in the domain integral due to the 'extra' hoop stress, $\varepsilon_{\phi\phi}$.

NUMERICAL FORMULATION

Further discussion will be restricted to plane and axisymmetric problems. Equations (9) and (10) were discretized by representing the boundary with a series of geometrically linear 'boundary elements' and dividing the domain into an array of triangular 'domain cells'. In the case of Oseen or Stokes flows the domain discretization is obviously not required. The velocity and traction were taken to be constant on each boundary element and were defined at a centrally located node. The velocity in Ω was defined at the vertices of each domain cell, allowing a linear variation across each cell. The velocity gradients in Ω were obtained by differentiation of the piecewise linear velocity field. These were then smoothed by assigning to each cell vertex the value of gradient obtained by algebraically averaging the values contributed by all cells sharing the vertex. This was then used to determine the inertia terms or viscosity at the vertices.

The integrals over individual boundary elements or domain cells were evaluated using ordinary Gaussian quadrature rules, except for the case where the singular point P was positioned on the boundary element or domain cell in question. In this case the leading singular terms were extracted and integrated exactly while the remainder of the integrand (if any) was integrated numerically.

For a given distribution of $a_j(Q)$ and $\varepsilon_{jk}(Q)$ (these will be zero for Newtonian Stokes flows or Oseen flows) the prescribed boundary conditions were incorporated in the discretized version of (9) or (10) and a system of linear algebraic equations for the unknown boundary data generated. After solution the discretized equations were reapplied to find the velocity field in Ω if required. In the nonlinear problems the updated nonlinear terms so obtained were integrated and the cycle repeated. This continued until little change in the solution was observed.

NUMERICAL RESULTS

An axisymmetric free jet

In this example we consider the extrusion of a Newtonian fluid from a circular die at low Reynolds number. This illustrates the modelling of inertia effects and extends the creeping solution previously presented (Bush et al, 1982).

The effects of fluid inertia on the shape of an axisymmetric free jet of Newtonian fluid have been investigated by a number of researchers. The most noteworthy examples are the experimental measurements of Goren and Wronski (1966) and the numerical finite element results of Reddy and Tanner (1978). In both of these reports the range of Reynolds number treated was Re = 0 to Re = 50, where the Reynolds number is given by $2\bar{u}_0 R_0\ \rho/\eta_0$ ($\bar{u}_0$ represents the mean velocity far upstream and R_0 is the nozzle radius). Descriptions of the behaviour of the jet profiles throughout this range are given, although the experimental and numerical jet profiles do not agree. More recently, Gear et al (1983) reported new experimental and numerical finite element results for the same problem. By comparing their results with those of Goren and Wronski they show that the latter appear to be plotted incorrectly and should be scaled by a factor of 2 in the axial direction. When this is done, good agreement between all results is obtained.

Boundary element solutions of this problem were obtained for Reynolds numbers up to Re = 4.2, allowing a comparison to be made with the other published results. Solutions at higher Reynolds numbers could be obtained. However, this was not investigated since the solution at Re = 4.2 is sufficient to illustrate the method. The domain discretization scheme used is shown in Figure 1. The number of domain cells is 276. The boundary discretization consists of 41 boundary elements. The boundary conditions are also shown in the figure. The starting profile was $R(z) = R_0$ throughout, as in Figure 1, and the free surface was adjusted during each iteration to ensure that the updated surface was a material boundary. At Re = 4.2 the number of iterations required before little change in the jet profile occurred was 5.

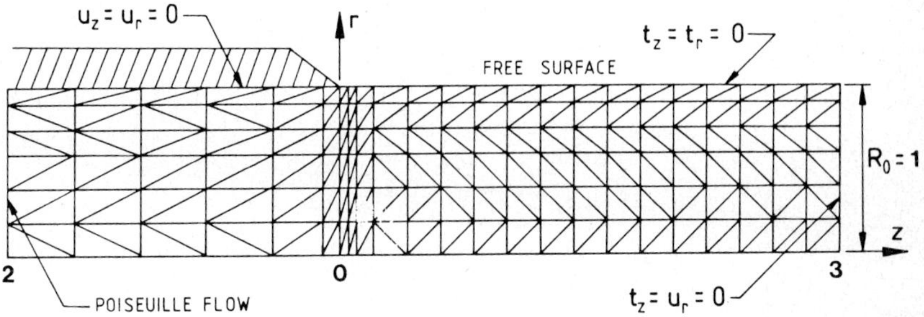

Figure 1
Grid Pattern and Boundary Conditions
for the axisymmetric free jet problem

The solutions at Re = 0, 2.0 and 4.2 are shown in Figure 2. The solution at Re = 4.2 agrees very well with the other results. It should be noted that the results by Gear et al (1983) are at Re = 4.1 while the measurements of Goren and Wronski (1966) and the present boundary element results are at Re = 4.2. However, this small discrepancy will have little effect. Of more importance is the influence of surface tension. The experimental test fluids (glycerol-water) exhibit a small amount of surface tension. The boundary element results do not include this factor. Such effects have been studied numerically by Reddy and Tanner (1978). They show that surface tension tends to reduce the final expansion,with the greatest influence occurring at low values of the Reynolds number. On the basis of these predictions and the value of 0.3 reported by Goren and Wronski (1966) for the surface tension parameter, one would expect the final expansion ratio to be reduced by approximately 0.01. The boundary element results then agree very well with the experimental measurements, much more closely than the finite element results which have taken surface tension into account. Surface tension effects could be included in the boundary element formulation in the manner of Reddy and Tanner (1978) if required. Unfortunately, large surface tension effects cause very slow convergence rates or even divergence with finite elements and this is also expected to be true with boundary elements.

Circular Couette flow of a power-law fluid

We will now illustrate the modelling of non-Newtonian effects by solving the problem of circular flow of a power-law fluid between two coaxial cylinders, in

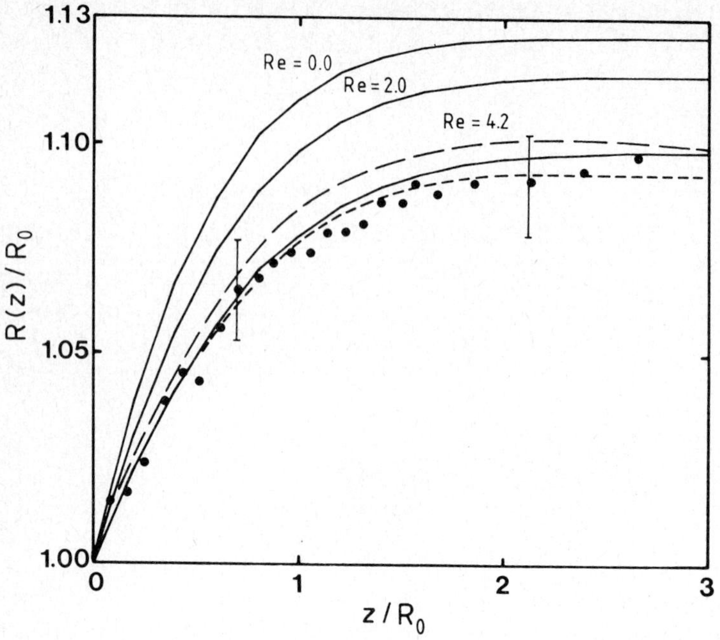

Figure 2
The axisymmetric jet profiles: ———, present results; — —, finite element results (Gear, et al, 1983); •, experimental measurements (Gear, et al, 1983) with typical error bars; -----, experimental measurements (Goren and Wronski, 1966).

the absence of inertia effects. We consider the case of the outer cylinder, of radius s_o, being fixed while the inner cylinder, of radius s_i, rotates with tangential velocity U.

Two grid patterns were used to solve this problem. These are shown in Figure 3. The first grid, GRID A, consists of 36 boundary elements and 160 domain cells. The second grid, GRID B, includes a finer discretization in the vicinity of the inner cylinder where large velocity gradients are encountered; 42 boundary elements and 220 domain cells. The boundary conditions enforced were no-slip velocity conditions on the cylinder surfaces, together with zero radial velocity and zero circumferential traction on the surfaces AB and CD. The problem was solved with $U = 1.0$, $s_i = 1.0$, $s_o = 2.0$ and $m = 1.0$. The power-law index was given the values $n = 0.5$ and $n = 0.3$.

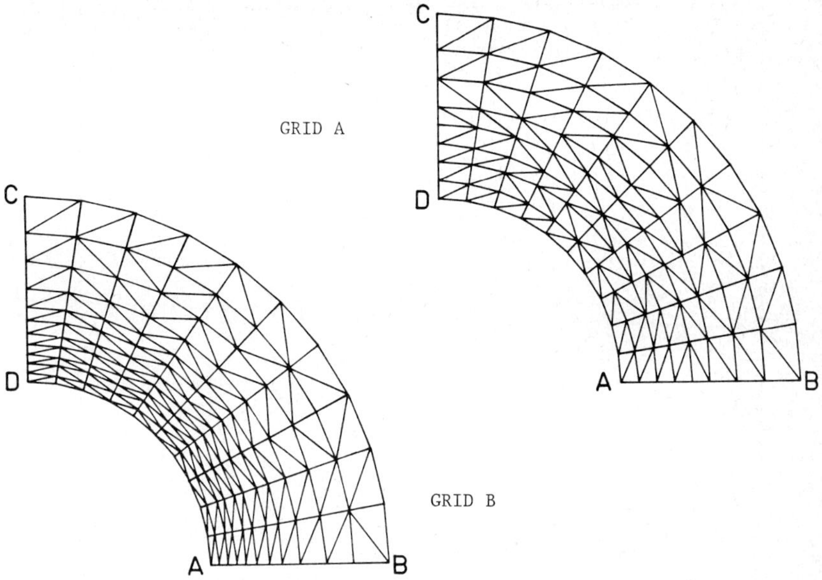

Figure 3
Grid Patterns used for the Circular Couette Flow Problem

The value of η_0 chosen was the minimum value of viscosity at each cycle of iteration. Thus, the size of the nonlinear terms $\{\eta(\dot{\gamma}) - \eta_0\}\, d_{jk}$ was controlled in this way.

The final velocity field solutions are shown in Figure 4. The results from GRID B are an improvement over those from GRID A due to the refined discretization in the region of rapid velocity variation.

The viscosity variation η/η_i, where η_i represents the true viscosity at the inner cylinder, can be seen in Figure 5. Since the shear rate is nowhere zero, the viscosity is bounded throughout the region considered.

Finally, the torque experienced by the cylinders can be evaluated by integrating the computed traction forces over the boundary elements. The computed torque on the two cylinders differed by a small amount due to numerical error; the average value was adopted as the final estimate. The results are tabulated in Table 1. We see good agreement with the exact solution, together with the expected improvement with grid refinement. We can conclude that this problem has been successfully solved.

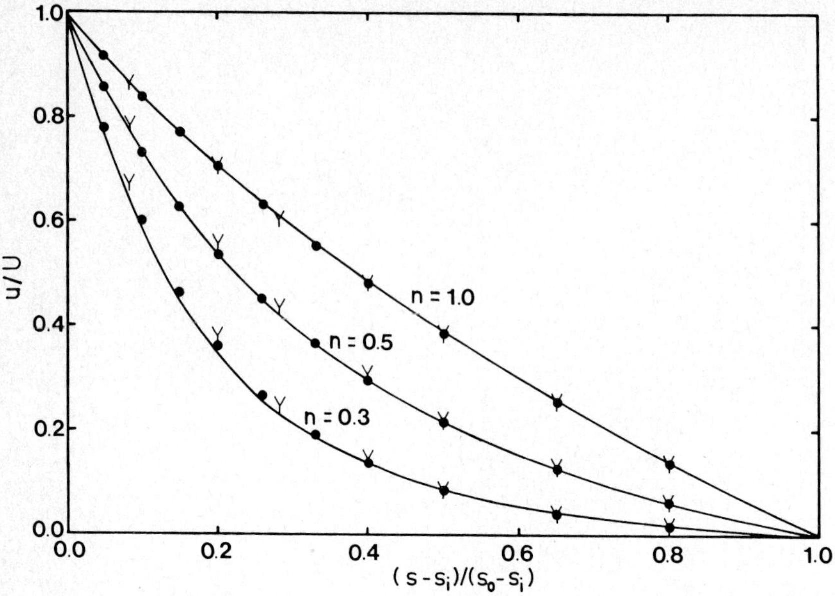

Figure 4
The velocity profile in circular Couette flow as a function of the distance, s, from the cylinder axis: Y, GRID A; •, GRID B.

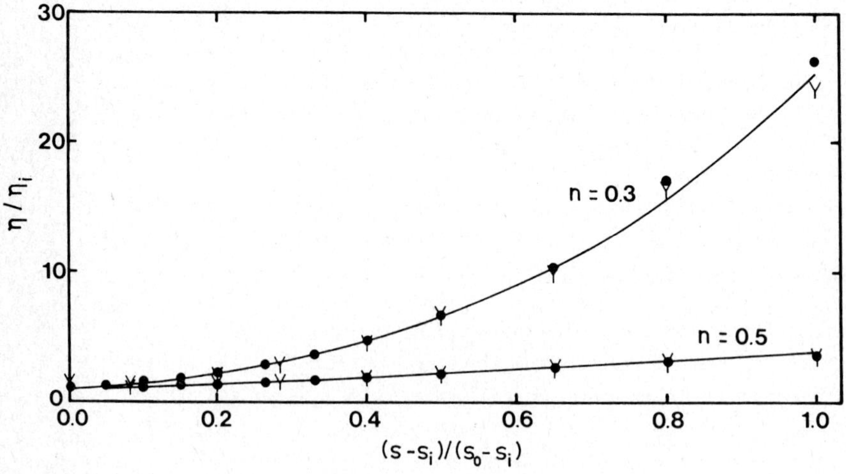

Figure 5
The Viscosity Profile in Circular Couette Flow of a Power-law Fluid: Y, GRID A; •, GRID B.

n	GRID	$\frac{T_i}{T_{exact}}$	$\frac{T_o}{T_{exact}}$	$\frac{T_{average}}{T_{exact}}$
1.0	A	1.000	1.008	1.004
	B	0.997	1.007	1.002
0.5	A	1.015	1.022	1.018
	B	1.001	1.012	1.007
0.3	A	1.016	1.024	1.020
	B	1.008	1.017	1.013

TABLE 1
The computed torque per unit length on the inner cylinder, T_i, and the outer cylinder, T_o.

The drag on a single circular cylinder in Oseen flow

Finally, we will illustrate the solution of Oseen's equations by considering the drag force on a single circular cylinder in cross-flow. An expansion in terms of Reynolds number for the drag coefficient on a circular cylinder (radius a) in uniform flow of speed U (density ρ, viscosity η_0) was reported by Tomotika and Aoi (1951). The Reynolds number is given by $Re = 2Ua\rho/\eta_0$ and the drag coefficient by $C_D = DRAG/(\rho U^2 a)$. Tomotika and Aoi (1950) also obtained a numerical solution of Oseen's equations for the drag on a circular cylinder, valid at higher Reynolds numbers. A comparison between these results and the expansion shows that the expansion significantly over-estimates the drag for $Re > 3$. However, below $Re = 2$ it provides a reliable expression for the drag coefficient.

The boundary element solution of this problem was obtained using N elements of equal length. To test the convergence of the solution as N increases, the problem was solved with N = 20, 30, 40 and 50. The drag coefficient computed with N = 20 is probably of sufficient accuracy for most applications. In fact, the relative difference between the solution with N = 20 and N = 50 was no more than 0.55 per cent. This indicates the efficiency and accuracy with which the problem can be solved using boundary element methods. The results for N = 20 are shown graphically in Figure 6 together with the reliable experimental measurements of Tritton (1959) and Kaplun's matched asymptotic expansion (Van Dyke, 1964) for the Navier-Stokes equations. In comparison with Tritton's data it is clear that the Oseen drag provides a reasonable first approximation of the true drag coefficient over the whole range of Reynolds number. Kaplun's expansion of course agrees most closely with the experimental data at low Reynolds number ($Re < 1$) but is invalid at higher values. Other Oseen flow examples can be found in Bush (1983).

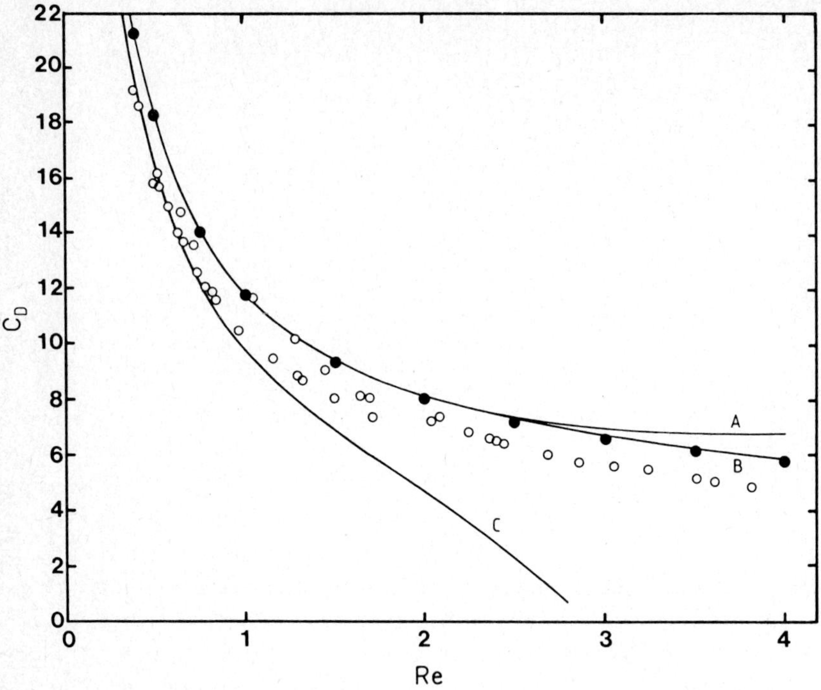

Figure 6
The drag force on a circular cylinder: A, Tomotika and Aoi (1951); B, Tomotika and Aoi (1950); ●, present results; o, Tritton (1959); C, Kaplun (see Van Dyke, 1964).

CONCLUSIONS

In this paper we have illustrated the application of the boundary element method to a range of fluid mechanics problems. Although the simple iterative scheme employed here limits the degree of nonlinearity that can be modelled, the method appears to be a useful alternative to conventional domain discretization methods for this class of problems. Its main advantage appears to be the smaller number of equations to be solved simultaneously, since this depends solely on the boundary discretization. Furthermore, the number of degrees of freedom at internal nodes is reduced by one in comparison with conventional velocity-pressure finite element formulations, since pressure is not treated. The pressure can be evaluated in an auxiliary calculation if required. We can expect the boundary element method to become as important in the field of fluid mechanics as it has already become in the field of solid mechanics.

ACKNOWLEDGEMENT

This work forms part of a Ph.D. thesis. I wish to thank Professor R.I. Tanner for supervising the research project.

REFERENCES

Bézine, G. and Bonneau, D. (1981) Integral equation method for the study of two dimensional Stokes flow, Acta Mechanica, 41, 197-209.

Bird, R.B., Armstrong, R.C. and Hassager, O. (1977) Dynamics of Polymeric Fluids VI (J. Wiley and Sons, U.S.A.)

Brebbia, C.A. (1980) The Boundary Element Method for Engineers (Pentech Press, Great Britain).

Brebbia, C.A. (ed.) (1981) Boundary Element Methods, Proc. 3rd Int. Seminar on Boundary Element Methods, Irvine, 1981 (CML Pub., Germany).

Brebbia, C.A. (ed.) (1982) Boundary Element Methods in Engineering, Proc. 4th Int. Seminar on Boundary Element Methods, Southampton, 1982 (CML Pub., Great Britain).

Bush, M.B. (1983) Modelling two-dimensional flow past arbitrary cylindrical bodies using boundary element formulations, Appl. Math. Mod. (to appear).

Bush, M.B. and Tanner, R.I. (1983) Numerical solution of viscous flows using integral equation methods, Int. J. Num. Meth. in Fluids, 3, 71-92.

Bush, M.B., Tanner, R.I. and Milthorpe, J.F. (1982) Application of boundary element methods to fluid mechanics problems, Proc. 4th Int. Conf. in Australia on Finite Element Methods, Melbourne, 74-78.

Coleman, C.J. (1981) A contour integral formulation of plane creeping Newtonian flow, Q.J. Mech. and Appl. Math. XXXIV, 453-464.

Gear, R.L., Keentok, M., Milthorpe, J.F. and Tanner, R.I. (1983) On the shape of low Reynolds number jets, Phys. Fluids, 26, 7-9.

Goren, S.L. and Wronski, S. (1966) The shape of low-speed capillary jets of Newtonian liquids, J. Fluid Mech., 25, 185-198.

Hess, J.L. and Smith, A.M.O. (1967) Calculation of potential flow about arbitrary bodies, in: Küchemann, D. (ed.), Progress in Aeronautical Sciences 8 (Pergamon Press, Great Britain, 1967).

Ladyzhenskaya, O.A. (1963) The Mathematical Theory of Viscous Incompressible Flow (Gordon and Breach, U.S.A.)

Mikhlin, S.G. (1957) Integral Equations (Pergamon, Poland).

Reddy, K.R. and Tanner, R.I. (1978) Finite element solution of viscous jet flows with surface tension, Comp. Fluids, 6, 83-91.

Rizzo, F.J. (1967) An integral equation approach to boundary value problems in classical elastostatics, Q. Appl. Math., 25, 83-95.

Rosenhead, L. (ed.) (1963) Laminar Boundary Layers (Oxford University Press, Great Britain).

Tomotika, S. and Aoi. T. (1950) The steady flow of a viscous fluid past a sphere and a circular cylinder at small Reynolds numbers, Q. J. Mech. and Appl. Math., 3, 140-161.

Tomotika, S. and Aoi, T. (1951) An expansion formula for the drag on a circular cylinder moving through a viscous fluid at small Reynolds numbers, Q. J. Mech. and Appl. Math. IV, 401-406.

Tritton, D.J. (1959) Experiments on the flow past a circular cylinder at low Reynolds numbers, J. Fluid Mech., 6, 547-567.

Van Dyke, M. (1964) Perturbation Methods in Fluid Mechanics (Academic Press, Great Britain).

Wu, J.C. and Thompson, J.F. (1973) Numerical solutions of time dependent incompressible Navier-Stokes equations using an integro-differential formulation, Comp. Fluids, 1, 197-215.

Youngren, G.K. and Acrivos, A. (1975) Stokes flow past a particle of arbitrary shape: a numerical method of solution, J. Fluid Mech., 69, 377-403.

Computational Techniques & Applications: CTAC-83
J. Noye & C. Fletcher (Editors)

THE EFFECT OF APPROXIMATING THE GEOMETRY FOR A SIMPLE BOUNDARY INTEGRAL EQUATION

W. McLean

Department of Mathematics,
Faculty of Science,
Australian National University,
CANBERRA A.C.T. 2601 AUSTRALIA

Sometimes, a linear boundary value problem for a domain Ω can be reformulated as an integral equation on the boundary $\Gamma = \partial\Omega$ of Ω. When solving this integral equation numerically, it is a common practice to approximate Γ by a curve (or surface) which admits a simple and cheaply computable parameterization. Using a simple example, we investigate the errors which arise.

1. GENERAL PRELIMINARIES

Throughout the paper we employ standard notation from the theory of P.D.E.'s (see eg [3]) to describe the smoothness of functions and to formulate error estimates.

Let Ω be a bounded, connected, open set in the plane $\mathbb{R}^2$. Denote the boundary of Ω by Γ, and consider the classical Dirichlet problem : given $g \in C^0(\Gamma)$, find $W \in C^0(\bar{\Omega}) \cap C^2(\Omega)$ satisfying

(1.1) $$\Delta W = 0 \quad \text{in} \quad \Omega \qquad \text{and} \qquad W = g \quad \text{on} \quad \Gamma .$$

A traditional way of solving (1.1) is to look for a representation of W as the potential of a double layer of sources with density u on Γ :

(1.2) $$W(y) = \int_\Gamma k(y,x')\, u(x')\, dH_1(x') , \quad y \in \Omega .$$

Here,

$$k(y,x') := \frac{\partial}{\partial \nu(x')} \log |y-x'| = \frac{\nu(x')\cdot(x'-y)}{|x'-y|^2}$$

is the potential at y due to a dipole at x' (with axis $\nu(x')$ and strength 2π), $\nu(x')$ is the unit outer normal to Γ at x', and H_1 is the one-dimensional Hausdorff measure (ie. the arc length along Γ).

If Γ is a Jordan curve of class C^m, $m \geq 2$, then (1.1) certainly has a unique solution which admits a representation of the form (1.2) (see eg [9]). In fact, the identities

$$\int_\Gamma k(y,x')\, dH_1(x') = 2\pi \quad , \quad y \in \Omega$$

(1.3)

$$\int_\Gamma k(x,x')\, dH_1(x') = \pi \quad , \quad x \in \Gamma ,$$

which are consequences of the divergence theorem, imply

(1.4) $$W(y) = 2\pi\, u(x) + \int_\Gamma k(y,x')\, [u(x')-u(x)]\, dH_1(x') \quad , \quad y \in \Omega$$

and

(1.5) $$g(x) = \lim_{\substack{y\to x\\ y\in\Omega}} W(y) = \pi\, u(x) + \int_\Gamma k(x,x')\, u(x')\, dH_1(x') \quad , \; x \in \Gamma ,$$

so the density u satisfies a second kind integral equation on Γ. In order to analyse (1.5), let us parameterize Γ by

$$x = \gamma(t) \quad , \quad x \in \Gamma \quad , \quad 0 \le t \le 1 ,$$

where γ is a C^m function $\mathbb{R} \to \mathbb{R}^2$ satisfying

$$\gamma(t+1) = \gamma(t) \quad \text{for all} \quad t \in \mathbb{R} ,$$

$$\gamma \text{ is one-one on } [0,1) \quad ,$$

(1.6) $$|\dot\gamma(t)| \neq 0 \quad \text{for all} \quad t \in \mathbb{R} \quad ,$$

γ gives Γ a positive (ie. anticlockwise) orientation.

Via γ , (1.5) is converted into a periodic integral equation on the real line

$$\pi\, u(t) + \int_0^1 k(t,t')\, u(t')\, |\dot\gamma(t')|\, dt' = g(t) \quad , \quad t \in \mathbb{R} .$$

Notice that here, as in what follows, we adopt the obvious convention of writing $u(t)$, $k(t,t')$ etc. instead of $u[\gamma(t)]$, $k[\gamma(t),\gamma(t')]$ etc.

It will prove convenient to introduce

$$\tilde\nu(t) := |\dot\gamma(t)|\, \nu(t) \quad \text{and} \quad \tilde k(t,t') := k(t,t')\, |\dot\gamma(t)| ;$$

then simple calculations show

(1.7) $$\tilde\nu(t) = (\dot\gamma^2(t), -\dot\gamma^1(t))$$

(1.8) $$\tilde k(t,t') = \begin{cases} \dfrac{\tilde\nu(t')\cdot[\gamma(t')-\gamma(t)]}{|\gamma(t')-\gamma(t)|^2} & \text{if} \quad \gamma(t') \neq \gamma(t) \\[2ex] \dfrac{1}{2}\, \dfrac{\dot\gamma^1(t)\ddot\gamma^2(t)-\dot\gamma^2(t)\ddot\gamma^1(t)}{|\dot\gamma(t)|^2} & \text{if} \quad \gamma(t') = \gamma(t) \end{cases}$$

and we can write

$$\pi\, u(t) + \int_0^1 \tilde{k}(t,t')\, u(t')dt' = g(t) \quad , \quad t \in \mathbb{R} . \tag{1.9}$$

To obtain an approximate solution to (1.9), define a uniform grid

$$t_i := ih \quad , \quad h := \frac{1}{N} \quad , \quad i \in \{0,1,2,\ldots,N\} , \tag{1.10}$$

on the unit interval, let

$$u_i \simeq u(t_i) \quad , \quad g_i := g(t_i) ,$$

and then use the Nyström method (see [1]) with the composite trapezoidal rule (remembering that $u_N = u_0$) :

$$\pi\, u_i + h \sum_{j=1}^{N} \tilde{k}(t_i,t_j)\, u_j = g_i \quad , \quad i \in \{1,2,\ldots,N\} . \tag{1.11}$$

If it is inconvenient to compute $\tilde{k}(t_i,t_i)$, which by (1.8) involves the second derivative of γ , one can first use (1.3) to modify (1.9) along the lines of (1.4), and then discretize to obtain

$$2\pi\, u_i + h \sum_{\substack{j=1 \\ j\neq i}}^{N} \tilde{k}(t_i,t_j)(u_j-u_i) = g_i \quad , \quad i \in \{1,2,\ldots,N\} . \tag{1.12}$$

The values u_i can now be used to compute an approximation to the potential W . Simply use the trapezoidal rule on the integral in (1.2) or (1.4), substituting u_i for $u(t_i)$. The choice of (1.2) is suitable provided y is not too close to Γ . For y near Γ , $k(y,x')$ will become large so it would be better to use (1.4) with a suitable choice of x near y .

From (1.8), we can see that $\tilde{k}$ is C^{m-2} , and so the integral operator in (1.9) is compact $L^1(\Gamma) \to C^r(\Gamma)$ for $r \in \{0,1,2,\ldots,m-2\}$. Since it can be shown that $-\pi$ is not an eigenvalue of this operator, u is unique and satisfies the a priori estimate

$$\|u\|_{W^{r,\infty}} \le c\, \|g\|_{W^{r,\infty}} \quad , \quad r \in \{0,1,2,\ldots,m-2\} . \tag{1.13}$$

Moreover, it is not difficult to show that the two schemes (1.11) and (1.12) are each stable and consistent (see eg [1]), and that if $\dot{u}$ is Lipschitz and $m \ge 4$, then in both cases the asymptotic error estimate

$$\max_{i\in\{1,2,\ldots,N\}} |u_i-u(t_i)| \le ch^2\, \|u\|_{W^{2,\infty}}$$

holds for $h \to 0+$ (ie. $N \to \infty$). In fact, because we are using the trapezoidal rule in (1.11) and (1.12) to integrate a periodic function over one whole period the Euler-MacLaurin summation formula yields the improved estimate

$$\max |u_i - u(t_i)| \le ch^{2r+1} \|u\|_{W^{2r+1,\infty}} , \quad 1 \le 2r + 1 \le m - 2 , \; r \text{ integral},$$

provided, of course, that u is sufficiently smooth, ie. $u \in W^{2r+1,\infty}(\Gamma)$. Notice that (1.13) allows us to infer the smoothness of u from that of g . Also, it is easily seen that the method explained in the last paragraph yields an order h^{2r+1} approximation to $W(y)$.

The preceding remarks show why this boundary integral method gives such accurate results provided Γ and g are smooth, and why in this case more elaborate schemes than (1.11) or (1.12) could be inappropriate. Unfortunately, in most applications, one wants to solve (1.1) for a domain Ω which does not have a smooth boundary, but which has corners. In this case, the kernel k is unbounded, the associated integral operator is not compact, and we cannot expect the density u to be piecewise smooth. The theory of double-layer potentials has been extended to cope with these difficulties (see [2] and [6]), and some of the problems arising in the numerical solution of the resulting integral equations have also been investigated (see the paper of G.A. Chandler in these proceedings). Obviously, in such cases the Euler-MacLaurin formula cannot be applied, and higher order schemes are of interest. In this context, if we want to develop software which does not rely on explicit analytical expressions for γ , or if γ is expensive to compute exactly, we may decide to replace Γ by a "simpler" curve Γ_n , parameterized by

$$(1.14) \qquad x = \gamma_n(t) \; , \quad x \in \Gamma_n \; , \quad 0 \le t \le 1 \, .$$

(Here, we suppose that the sequence of approximations $(\gamma_n)_{n=0}^{\infty}$ converge to γ in a suitable sense.) It might also happen that we do not know Γ exactly, but only experimently determined approximations Γ_n . In either case, we would like to estimate the error incurred by using Γ_n instead of Γ .

From now on, we make the restrictive assumption that Γ is smooth, with the parameterization γ satisfying all the conditions specified above. This means, in view of what was said in the last paragraph, that the results of our analysis will, in themselves, be of rather limited applicability. Nevertheless, it is to be hoped that some insight will be gained for dealing with the more interesting and difficult case of when Ω has corners.

We are indebted to the work of J.C. Nedelec and J. Giroire in [4] and [7], dealing with related singular integral equations of the first kind on smooth surfaces in $\mathbb{R}^3$. In particular, the idea of using the simple identity (3.5) is taken from [7]. Finally, we mention that the usual practice is adopted of denoting by c (and sometimes C) a generic (strictly) positive constant, which may depend on γ but not on t , g , u and n .

2. THE CURVES Γ , Γ_n

The functions $\gamma_n : \mathbb{R} \to \mathbb{R}^2$ appearing in (1.14) will be assumed to satisfy

$$\gamma_n(t+1) = \gamma_n(t) \qquad \text{for all} \quad t \in \mathbb{R} ,$$

(2.1) $\quad \|\gamma_n\|_{W^{2,\infty}} \le c$, (so the second derivatives are uniformly bounded),

(2.2) $\quad \varepsilon_n := \|\gamma_n - \gamma\|_{W^{1,\infty}} \to 0$ (so the first derivatives converge in $L^\infty(0,1)$) .

Defining the "distance modulo 1"

$$|t',t| := \min_{p \in \mathbb{Z}} |t'-t+p| , \qquad \text{all} \quad t',t \in \mathbb{R} ,$$

we can state

2.1 Lemma (i) $|(\gamma_n-\gamma)(t') - (\gamma_n-\gamma)(t)| \le \varepsilon_n |t',t|$

(ii) $c|t',t| \le |\gamma(t') - \gamma(t)| \le C|t',t|$

(iii) $c|t',t| \le |\gamma_n(t') - \gamma_n(t)| \le C|t',t|$ for n sufficiently large.

Proof. (i) Choose t'' such that $\gamma(t'') = \gamma(t')$ and $|t''-t| = |t',t|$,

$$|(\gamma_n-\gamma)(t')-(\gamma_n-\gamma)(t)| = \left|\int_t^{t''} (\dot\gamma_n-\dot\gamma)\right| \le \|\dot\gamma_n-\dot\gamma\|_{L^\infty} |t''-t| \le \varepsilon_n |t',t| .$$

(ii) The right hand inequality is simply the statement that γ is Lipschitz, ie. $\gamma \in W^{1,\infty}$. For the left hand inequality, one first proves the analogous estimate for an arc-length parameterization. This is known as a chord-arc condition, and is valid for compact Liptschitz boundaries. (One simply proves the estimate locally by considering the graph of a Lipschitz function, and then uses compactness.) The result follows from the chord-arc condition because we have assumed (1.6).

(iii) The right hand inequality is a consequence of the uniform bound $\|\gamma_n\|_{W^{1,\infty}} \le c$, guaranteed by (2.2). The left hand inequality follows from parts (i) and (ii) :

$$\begin{aligned}|\gamma_n(t')-\gamma_n(t)| &= |[\gamma(t')-\gamma(t)] - [(\gamma-\gamma_n)(t')-(\gamma-\gamma_n)(t)]| \\ &\ge \big|\,|\gamma(t')-\gamma(t)| - |(\gamma-\gamma_n)(t')-(\gamma-\gamma_n)(t)|\,\big| \\ &\ge (c-\varepsilon_n)|t',t| . \qquad \square\end{aligned}$$

2.2 Example We could construct $\gamma_n = (\gamma_n^1, \gamma_n^2)$ by periodic, cubic spline interpolation of the two components of γ. In other words, for $p = 1,2$ let γ_n^p be the unique periodic C^2 function $\mathbb{R} \to \mathbb{R}$ which satisfies $\gamma_n^p(t_i) = \gamma^p(t_i)$ and which coincides with a cubic polynomial on each subinterval $[t_{i-1}, t_i]$ (here t_i is as in (1.10) with $N = n$). If Γ is C^4, then (see [8])

$$\|\gamma_n^{(r)} - \gamma^{(r)}\|_{L^\infty} \le 5\, h^{4-r} \|\gamma^{(4)}\|_{L^\infty} \quad , \qquad r \in \{0,1,2,3\}$$

where $\gamma^{(r)}$ denotes the rth derivative of γ. In particular, by taking $r = 1$, we see that in this example

$$\varepsilon_n = O(h^3) = O(\frac{1}{n^3}) \ .$$

3. THE INTEGRAL EQUATIONS

Using γ_n, the integral equation (1.9) is approximated by

$$\pi\, u_n(t) + \int_0^1 \tilde{k}_n(t,t')\, u_n(t')\, dt' = g_n(t) \quad , \qquad t \in \mathbb{R} \ , \tag{3.1}$$

where g_n is some approximation to g, and (cf (1.7) and (1.8))

$$\nu_n(t) := |\dot{\gamma}_n(t)|\ \nu_n(t) = (\dot{\gamma}_n^2(t), -\dot{\gamma}_n^1(t)) \tag{3.2}$$

$$\tilde{k}_n(t,t') := k_n(t,t')|\dot{\gamma}_n(t')| = \frac{\tilde{\nu}_n(t')\cdot[\gamma_n(t')-\gamma_n(t)]}{|\gamma_n(t')-\gamma_n(t)|^2} \ . \tag{3.3}$$

We emphasize that u_n is not a numerical solution, it is the exact solution to the perturbed integral equation (3.1).

If we write

$$Tu(t) := \frac{1}{\pi}\int_0^1 \tilde{k}(t,t')\, u(t')\, dt' \quad ,$$

$$T_n u(t) := \frac{1}{\pi}\int_0^1 \tilde{k}_n(t,t')\, u(t')\, dt' \ ,$$

then (1.9) and (3.1) can be written in operator notation as

$$(I+T)u = g/\pi \qquad \text{and} \qquad (I+T_n)u_n = g_n/\pi \ . \tag{3.4}$$

It has already been remarked that T is compact $L^1(\Gamma) \to C^r(\Gamma)$ for $r \in \{0,1,2,\ldots,m-2\}$. The assumption (2.1) implies $k_n(t,t')$ is continuous for $t' \neq t$, and $|k_n(t,t')| \le C$, so that T_n is bounded (and even compact) $L^\infty(\Gamma) \to C^0(\Gamma)$. Our aim is to estimate $T_n - T$, for which we need

3.1 Lemma $|\tilde{\nu}_n(t) - \tilde{\nu}(t)| \le \varepsilon_n$.

Proof. Recalling (3.2) and (1.7),

$$|\tilde{\nu}_n-\tilde{\nu}| = |(\dot{\gamma}_n^2,-\dot{\gamma}_n^1)-(\dot{\gamma}^2,-\dot{\gamma}^1)| = |(\dot{\gamma}_n^2-\dot{\gamma}^2,\dot{\gamma}^1-\dot{\gamma}_n^1)| = |\dot{\gamma}_n-\dot{\gamma}| \le \varepsilon_n . \qquad \Box$$

3.2 Theorem (i) For n sufficiently large and for $|t',t| \neq 0$,

$$|\tilde{k}_n(t,t')-\tilde{k}(t,t')| \le \frac{c\varepsilon_n}{|t',t|} .$$

(ii) $\lim_{n\to\infty} \|T_n-T\|_{L(C^0,C^0)} = 0$.

Proof. (i) For any two vectors $A,B \in \mathbb{R}^2$,

$$|A|^2 - |B|^2 = (A-B)\cdot(A-B) + 2(A-B)\cdot B . \tag{3.5}$$

Put $A = \gamma_n(t') - \gamma_n(t)$ and $B = \gamma(t') - \gamma(t)$, then by Lemma 2.1,

$$|A-B| = |(\gamma_n-\gamma)(t')-(\gamma_n-\gamma)(t)| \le \varepsilon_n|t',t| ,$$

$$c|t',t| \le |A| \le C|t',t| \quad\text{and}\quad c|t',t| \le |B| \le C|t',t| .$$

Using (3.5), we obtain

$$\left|\frac{1}{|B|^2} - \frac{1}{|A|^2}\right| = \left|\frac{|A|^2-|B|^2}{|B|^2\,|A|^2}\right| \le \frac{|A-B|^2+2|A-B|\cdot|B|}{|B|^2\,|A|^2} \le \frac{c\varepsilon_n}{|t',t|^2}$$

and hence, noting Lemma 3.1,

$$|\tilde{k}_n(t,t')-\tilde{k}(t,t')| = \left|\frac{\tilde{\nu}_n(t')\cdot A}{|A|^2} - \frac{\tilde{\nu}(t')\cdot B}{|B|^2}\right|$$

$$= \left|\frac{(\tilde{\nu}_n-\tilde{\nu})\cdot A+\tilde{\nu}\cdot(A-B)}{|A|^2} - \tilde{\nu}\cdot B\left\{\frac{1}{|B|^2} - \frac{1}{|A|^2}\right\}\right|$$

$$\le \frac{c\varepsilon_n|t',t|}{|t',t|^2} + c|t',t|\frac{\varepsilon_n}{|t',t|^2} \le \frac{c\varepsilon_n}{|t',t|} .$$

(ii) Since $|\tilde{k}_n(t,t')| \le c$ and $|\tilde{k}(t,t')| \le c$, part (i) implies

$$\pi|(T_n-T)u(t)| = \left|\int_0^1 [\tilde{k}_n(t,t')-\tilde{k}(t,t')]u(t)dt\right|$$

$$\le \|u\|_{C^0}\int_0^1 |\tilde{k}_n(t,t')-\tilde{k}(t,t')|\,dt'$$

$$= \|u\|_{C^0}\left|\int_{|t',t|<\varepsilon_n} + \int_{|t',t|\ge\varepsilon_n}\right|$$

$$\le \|u\|_{C^0}\,[c\varepsilon_n+c\varepsilon_n \log(\frac{1}{\varepsilon_n})]$$

and hence

$$\|T_n - T\|_{L(C^0,C^0)} \leq c\varepsilon_n \log\left(\frac{1}{\varepsilon_n}\right) .$$

3.3 Corollary $\|(I+T_n)^{-1}\|_{L(C^0,C^0)} \leq c$.

4. THE POTENTIAL W_n

Let Ω_n be the bounded component of $\mathbb{R}^2 \setminus \Gamma_n$, ie. Ω_n is the perturbed domain having boundary Γ_n , and introduce the corresponding potential

$$W_n(y) := \int_{\Gamma_n} k_n(y,x')\, u_n(x')\, dH_1(x') = \int_0^1 \tilde{k}_n(y,t')\, u_n(t')\, dt' , \quad y \in \Omega_n .$$

Notice that W_n satisfies (cf. (1.1))

$$\Delta W_n = 0 \text{ in } \Omega_n \quad \text{and} \quad W_n = g_n \text{ on } \Gamma_n .$$

Defining, for $y \in \mathbb{R}^2$,

$$d(y) := \operatorname{dist}(y,\Gamma) \quad \text{and} \quad d_n(y) := \operatorname{dist}(y,\Gamma_n)$$

where, as usual, $\operatorname{dist}(y,S) := \inf\{|s-y| : s \in S\}$ for a nonempty subset $S \subset \mathbb{R}^2$, we can state

4.1 Lemma (i) $|d_n(y)-d(y)| \leq \|\gamma_n-\gamma\|_{L^\infty}$

(ii) If $y \in \Omega$ and $\|\gamma_n-\gamma\|_{L^\infty} < d(y)$, then $y \in \Omega_n$.

Proof. (i) For all t ,

$$|\gamma(t)-y| = |[\gamma_n(t)-y]-[\gamma_n(t)-\gamma(t)]| \geq |\gamma_n(t)-y| - |(\gamma_n-\gamma)(t)| \geq d_n(y) - \|\gamma_n-\gamma\|_{L^\infty}$$

so $d(y) \geq d_n(y) - \|\gamma_n-\gamma\|_{L^\infty}$, ie. $d_n(y) - d(y) \leq \|\gamma_n-\gamma\|_{L^\infty}$. Interchanging the roles of γ and γ_n , we get $d(y) - d_n(y) \leq \|\gamma_n-\gamma\|_{L^\infty}$, and hence the result.

(ii) $d_n(y) = d(y) - [d(y)-d_n(y)] > \|\gamma_n-\gamma\|_{L^\infty} - \|\gamma_n-\gamma\|_{L^\infty} = 0$. □

To estimate $W_n(y) - W(y)$ in section 5, we shall need (cf. Theorem 3.2):

4.2 Theorem If $\|\gamma_n-\gamma\|_{L^\infty} \leq \frac{1}{2} d(y)$, then

(i) $\frac{1}{2}|\gamma(t)-y| \leq |\gamma_n(t)-y| \leq \frac{3}{2}|\gamma(t)-y|$

(ii) $|\tilde{k}_n(y,t)-\tilde{k}(y,t)| \leq \dfrac{c\varepsilon_n}{|\gamma(t)-y|^2}$, where c is independent of y .

Proof. (i) Using part (i) of Lemma 4.1,

$$d_n(y) = d(y) - [d(y)-d_n(y)] \geq 2\|\gamma_n-\gamma\|_{L^\infty} - \|\gamma_n-\gamma\|_{L^\infty} = \|\gamma_n-\gamma\|_{L^\infty}$$

so

$$|\gamma(t)-y| \leq |\gamma(t)-\gamma_n(t)| + |\gamma_n(t)-y| \leq d_n(y) + |\gamma_n(t)-y| \leq 2|\gamma_n(t)-y|$$

and

$$|\gamma_n(t)-y| \leq |\gamma_n(t)-\gamma(t)| + |\gamma(t)-y| \leq \tfrac{1}{2}d(y) + |\gamma(t)-y| \leq \tfrac{3}{2}|\gamma(t)-y| .$$

(ii) For any two vectors $A,B \in \mathbb{R}^2$

$$\left| |A|^2-|B|^2 \right| = |A\cdot(A-B)+(A-B)\cdot B| \leq |A-B|(|A|+|B|) .$$

Put $A = \gamma_n(t) - y$, $B = \gamma(t) - y$ then $|A-B| \leq \|\gamma_n-\gamma\|_{L^\infty} \leq \varepsilon_n$ and, by part (i), $\frac{1}{2}|B| \leq |A| \leq \frac{3}{2}|B|$, so

$$|\tilde{k}_n(y,t)-\tilde{k}(y,t)| = \left| \frac{\tilde{\nu}_n(t)\cdot A}{|A|^2} - \frac{\tilde{\nu}(t)\cdot B}{|B|^2} \right|$$

$$= \left| \frac{[\tilde{\nu}_n(t)-\tilde{\nu}(t)]\cdot A+\tilde{\nu}(t)\cdot(A-B)}{|A|^2} - \tilde{\nu}(t) \cdot B \left\{\frac{1}{|B|^2} - \frac{1}{|A|^2}\right\} \right|$$

$$\leq \frac{c\varepsilon_n}{|A|^2} + \frac{c|B|\varepsilon_n(|A|+|B|)}{|A|^2\,|B|^2} \leq \frac{c\varepsilon_n}{|B|^2} ,$$

where we have used Lemma 3.1 to estimate $\tilde{\nu}_n(t) - \tilde{\nu}(t)$. □

5. SOME ERROR ESTIMATES

Having established Theorem 3.2, it is now easy to prove

5.1 Theorem (i) $\|u_n-u\|_{C^0} \leq c(\|g_n-g\|_{C^0} + \|(T_n-T)u\|_{C^0})$

(ii) If u is Lipschitz, then

$$\|(T_n-T)u\|_{C^0} \leq c\varepsilon_n \|u\|_{W^{1,\infty}} .$$

Proof. (i) Remembering (3.4), we have

$$(I+T_n)(u_n-u) = \frac{g_n}{\pi} - (\frac{g}{\pi} - Tu) - T_n u$$

$$= \frac{1}{\pi}(g_n-g) - (T_n-T)u$$

from which the result follows by Corollary 3.3.

(ii) Using (0.3) (which also holds for the curve Γ_n) ,

$$Tu(t) = u(t) + \frac{1}{\pi}\int_0^1 \tilde{k}(t,t')[u(t')-u(t)]\,dt'$$

$$T_n u(t) = u(t) + \frac{1}{\pi}\int_0^1 \tilde{k}_n(t,t')[u(t')-u(t)]\,dt'$$

so by Theorem 3.2 (i),

$$\begin{aligned}|(T_n-T)u(t)| &= \frac{1}{\pi}\left|\int_0^1 [\tilde{k}_n(t,t')-\tilde{k}(t,t')][u(t')-u(t)]dt'\right|\\ &\le c\int_0^1 \frac{\varepsilon_n}{|t',t|}\,\|u\|_{W^{1,\infty}}\,|t',t|\,dt'\\ &= c\varepsilon_n\,\|u\|_{W^{1,\infty}}\,. \quad \square\end{aligned}$$

5.2 Corollary If g_n satisfies

$$\|g_n-g\|_{C^0} \le c\varepsilon_n\,\|g\|_{W^{r,\infty}} \tag{5.1}$$

for some $r \ge 1$, then

$$\|u_n-u\|_{C^0} \le c\varepsilon_n(\|g\|_{W^{r,\infty}}+\|u\|_{W^{1,\infty}}) \le c\varepsilon_n\,\|g\|_{W^{r,\infty}}\,. \tag{5.2}$$

We now use the estimate of the error in the density to bound the error in the potential.

5.3 Theorem If g_n satisfies (5.1) and $\|\gamma_n-\gamma\|_{L^\infty} < \frac{1}{2}d(y)$, then $y \in \Omega_n$ and

$$|W_n(y)-W(y)| \le \frac{c\varepsilon_n}{d(y)}\,\|g\|_{W^{r,\infty}}\,. \tag{5.3}$$

Proof. Part (ii) of Lemma 4.1 implies that $y \in \Omega_n$. Write

$$\begin{aligned}W_n(y) - W(y) &= \int_0^1 \tilde{k}_n(y,t')\,u_n(t')\,dt' - \int_0^1 \tilde{k}(y,t')\,u(t')\,dt'\\ &= \int_0^1 \tilde{k}_n(y,t')[u_n(t')-u(t')]\,dt' + \int_0^1 [\tilde{k}_n(y,t')-\tilde{k}(y,t')]\,u(t')\,dt'\,.\end{aligned}$$

Since $|\tilde{k}_n(y,t')| \le c/d(y)$, we can use Corollary 5.2 to bound the first term by $c\varepsilon_n\|g\|_{W^{r,\infty}}/d(y)$ as required. Choose t such that $d(y) = |\gamma(t)-y|$, then since, by (1.3)

$$\int_0^1 [\tilde{k}_n(y,t')-\tilde{k}(y,t')]\,dt' = 2\pi - 2\pi = 0\,,$$

the second term equals

$$\int_0^1 [\tilde{k}_n(y,t')-\tilde{k}(y,t')][u(t')-u(t)]\,dt'\,.$$

Using theorem 4.1, this is bounded by

$$\int_0^1 \frac{c\varepsilon_n}{|\gamma(t')-y|^2} \|u\|_{W^{1,\infty}} |t',t| \, dt' \le \frac{c\varepsilon_n \|g\|_{W^{1,\infty}}}{d(y)} \cdot \int_0^1 \frac{|t',t|\,dt'}{|\gamma(t')-y|}$$

and the result follows since

$$c|t',t| \le |\gamma(t')-\gamma(t)| \le |\gamma(t')-y| + |y-\gamma(t)| = |\gamma(t')-y|+d(y) \le 2|\gamma(t')-y| \ . \ \square$$

6. SOME REMARKS

The assumption (2.1) means that our analysis does not include the case where Γ_n is piecewise linear (ie. Γ_n is polygonal). This assumption was only needed to guarantee the uniform bound $|k_n(t,t')| \le c$, used in the proof of Theorem 3.2 and it should be possible, by employing the Stieltjes integral formulation set out in [2], to weaken (2.1) sufficiently to allow for polygonal Γ_n. Since for piecewise linear interpolation

$$\|\gamma_n-\gamma\|_{L^\infty} \le ch^2 \|\gamma\|_{W^{2,\infty}} \quad \text{and} \quad \varepsilon_n := \|\gamma_n-\gamma\|_{W^{1,\infty}} \le ch \|\gamma\|_{W^{2,\infty}}$$

we would then have the estimate

$$|W_n(y)-W(y)| \le \frac{ch}{d(y)} \|g\|_{W^{1,\infty}} = \frac{1}{d(y)} O(h) \ . \tag{6.1}$$

Unfortunately, this estimate cannot be optimal, at least for some cases, as the following simple analysis shows. If Ω is $C^{2,\alpha}$, and g coincides with the restriction to Γ of a function in $C^{2,\alpha}(\bar{\Omega})$, then the solution W of (1.1) has a bounded gradient (in fact, it is shown in chapter 6 of [3] that $W \in C^{2,\alpha}(\bar{\Omega})$). We shall suppose for simplicity that Ω is convex, so that $\Omega_n \subset \Omega$ for all n. The maximum principle implies

$$\max_{\Omega_n} |W_n-W| = \max_{\Gamma_n} |W_n-W| = \max_{\Gamma_n} |g_n-W| ,$$

and the mean value theorem implies

$$W[\gamma_n(t)] - g(t) = W[\gamma_n(t)] - W[\gamma(t)] = [\gamma_n(t)-\gamma(t)] \cdot \text{grad } W(z)$$

for some z on the line segment between $\gamma_n(t)$ and $\gamma(t)$. Thus, for $y \in \Omega_n$,

$$\begin{aligned} |W_n(y)-W(y)| &\le \max_t |g_n(t) - W[\gamma_n(t)]| \\ &\le \max_t (|g_n(t)-g(t)| + |g(t)-W[\gamma_n(t)]|) \\ &\le \|g_n-g\|_{L^\infty} + \|\gamma_n-\gamma\|_{L^\infty} \max_{\bar{\Omega}} |\text{grad } W| \\ &= O(h^2) , \end{aligned}$$

assuming we construct g_n from g by piecewise linear interpolation. Not only is this estimate one order higher than (6.1), it is also independent of $d(y)$.

An obvious question to ask at this point is : if we can estimate the error in the potential W directly using P.D.E. theory, do we really need to estimate the error in the density u ? One answer to this is to look at how one would estimate errors arising in the numerical solution of (3.1). The difficulty here is that the derivatives of $\tilde{k}_n$ are unbounded at every point where Γ_n is not smooth. If we know how $\tilde{k}_n$ differs from $\tilde{k}$, we can hope to exploit the smoothness of $\tilde{k}$. But the difference $\tilde{k}_n - \tilde{k}$ was just what was required for estimating $u_n - u$.

Finally, it is interesting to compare the results here with those of J.L. Hess in [5], who found that for a consistent approximation of an integral similar to our $W(y)$, it was necessary to use a higher order discretization of Γ than of u .

Acknowledgement

I would like to thank Bob Anderssen and Frank de Hoog for many helpful criticisms and suggestions, and Anna Zalucki for her patience and care in preparing the typescript.

References

[1] Atkinson, K.E., A Survey of Numerical Methods for the Solution of Fredholm Integral Equations of the Second Kind (SIAM, Philadelphia, 1976).

[2] Cryer, C.W., The solution of the Dirichlet problem for Laplace's equation when the boundary data is discontinuous and the domain has a boundary which is of bounded rotation by means of the Lebesgue-Stieltjes integral equation for the double layer potential, Technical Report #99, Computer Sciences Department, University of Wisconsin (1970).

[3] Gilbarg, D. and Trudinger, N.S., Elliptic Partial Differential Equations of Second Order (Springer-Verlag, Berlin, Heidelberg, New York, 1977).

[4] Giroire, J. and Nedelec, J.C., Numerical Solution of an Exterior Neumann Problem Using a Double Layer Potential, Math. of Comp., 32 (1978) 973-990.

[5] Hess, J.L., Higher Order Solution of the Integral Equation for the Two-Dimensional Neumann Problem, Comp. Meth. in Appl. Mech. and Eng., 2 (1973) 1-15.

[6] Kral, J., Integral Operators in Potential Theory, Lecture Notes in Mathematics 823 (Springer-Verlag, Berlin, Heidelberg, New York, 1980).

[7] Nedelec, J.C., Curved Finite Element Methods for the Solution of Singular Integral Equations on Surfaces in $\mathbb{R}^3$, Comp. Meth. in Appl. Mech. and Eng., 8 (1976) 61-80.

[8] Sharma, A. and Meir, A., Degree of Approximation of Spline Interpolation, Jour. of Math. and Mech., 15 (1966) 759-767.

[9] Smirnov, V.I., A Course of Higher Mathematics, Vol. IV, 568ff (Pergamon Press, Oxford, 1964).

Computational Techniques & Applications: CTAC-83
J. Noye & C. Fletcher (Editors)

MESH GRADING FOR BOUNDARY INTEGRAL EQUATIONS

G.A. Chandler

Department of Mathematics
University of Queensland
Brisbane, Qld. 4067.

When an indirect boundary element method is used to solve a two dimensional potential problem corners introduce singularities which degrade the rates of convergence. These may be restored by grading the mesh. A careful error analysis shows how fully superconvergent approximations to the double layer potential may be recovered.

INTRODUCTION

The simplest problem leading to a boundary integral equation is the interior Dirichlet problem for a smoothly bounded region in $\mathbb{R}^2$. Thus we are given a domain Ω with a smooth boundary Γ and seek the unknown function U satisfying

$$\Delta U(x) = 0 \quad x \in \Omega$$
$$U(x) = f(x) \quad x \in \Gamma ,$$

where f is the prescribed boundary data. This may be solved by direct or indirect methods. The direct formulation computes approximations to the normal derivative of U around the boundary, and this approach is more commonly used in elastostatic problems (See Hartmann (1981)). However first kind integral equations need to be solved. While they give no problems in practice, the mathematics required to cope with their slightly ill-posed nature is more complex. We refer to Wendland (1982) and the references given there for a description.

The indirect formulation uses the fact that U is the double layer potential of a double layer distribution. That is

$$U(x) = \int_\Gamma G(x,\xi)^1 u(\xi) d\xi \quad x \in \mathrm{Int}(\Omega) ,$$

where

$$G(x,\xi) = \frac{1}{2\pi} \ln \frac{1}{|x-\xi|} , \quad G(x,\xi)^1 = \frac{\partial}{\partial \nu(\xi)} G(x,\xi)$$

and $\nu(\xi)$ is the outward normal at ξ. Then u can be found as the solution of a second kind integral equation

(1) $$(I - T)u = -2f ,$$

with $$(Tu)(x) = 2\int_\Gamma G(x,\xi)^1 u(\xi) d\xi \quad x \in \Gamma .$$

This is a classical problem, and there is a well developed theory for the corresponding numerical methods (Atkinson (1976)). Using the fact that T is a compact operator the convergence of numerical methods can be proved. Since the kernel of T is smooth rates of convergence of various methods can be established to give a qualitative guide to the choice of method.

It is more difficult if corners are permitted in the domain. Practically the corners cause slight difficulties, but ways to overcome them have been discovered. The effect on the mathematical theory is more damaging. The integral operator is no longer compact and it becomes more difficult to prove convergence. Indeed proofs have not been found for the most successful methods. This paper aims to describe the difficulties and outline some partial results. We refer to Wendland (1982), Atkinson and de Hoog (1983a, b), Costabel and Stephan (1981) and the references given there for additional information, including some results for the direct formulation. Detailed statements of most of the theorems described here can be found in Chandler (1983a, b).

CONVERGENCE OF GALERKIN'S METHOD

The essential difficulties created by the corners are best illustrated by dealing with one in isolation. Consider the vertex P_1OP_2 with the exterior angle $\chi\pi$ shown in fig. 1, and the distances $|OP_1| = |OP_2| = 1$.

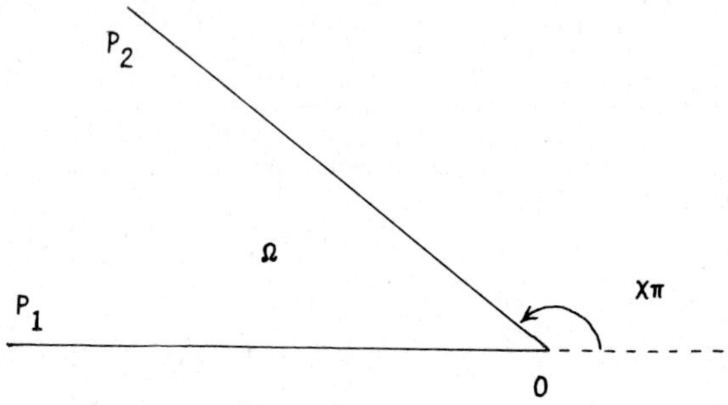

Fig. 1

If $\chi < 0$, the corner is reentrant. We are unable to deal with cracks, and assume $-1 < \chi < 1$.

Parameterize the boundary by letting s denote the distance along P_1OP_2, with $s = -1, 0, 1$ corresponding to P_1, O, P_2 respectively. Let $u(s)$ denote the double layer potential at s. To avoid dealing with a 2×2 system of integral equations write for $s \geq 0$,

$$v_0(s) = \tfrac{1}{2}(u(s) + u(-s)) \ , \ v_1(s) = \tfrac{1}{2}(u(s) - u(-s))$$

$$g_0(s) = -(f(s) + f(-s)) \ , \ g_1(s) = -(f(s) - f(-s)) \ ,$$

for the even and odd components of $u(s)$ and $-2f(s)$ respectively. The boundary integral equation (1) then becomes

$$(2.0) \qquad (I + R)v_0 = g_0$$

$$(2.1) \qquad (I - R)v_1 = g_1 \ ,$$

where R is the integral operator defined by

$$(3) \qquad (Rv)(s) = \int_0^1 r_\chi(\sigma)v(\sigma)d\sigma = \frac{\sin\chi\pi}{\pi}\int_0^1 \frac{s}{s^2+2s\sigma\cos\chi\pi+\sigma^2}\, v(\sigma)d\sigma$$

When the corner O is included in the full polygon, the integral equation (2) must be augmented to $(I + R)v_0 + K_0\tilde{u} = f_0$ for example. The term K_0 being the potential at O due to the strength of the double layer, $\tilde{u}$, on the boundary away from O. As K_0 is compact it is simpler to deal with, and may be neglected.

To solve (2) numerically the unit interval is divided by n grid points $0=x_0<x_1<\ldots x_n=1$. The set of basis functions S_n is then the set of piecewise polynomials of degree r having ν continuous derivatives at the grid points $(-1 \le \nu \le r$, with $\nu = -1$ corresponding to discontinuous functions.) Numerical methods seek a member of S_n which approximates the exact solution v_0. (We suppose we are solving (2.0); (2.1) is the same). It does this by selecting a $v_n \in S_n$ which "approximately" satisfied (2.0), with different senses of approximately leading to different methods and solutions (Galerkin, collocation etc). If the method is successful the method will be

(i) *Stable*. That is v_n is uniquely determined, and

(ii) *Convergent*. As the grid is refined (i.e. $n \to \infty$) v_n becomes an increasingly accurate approximation to v.

A theoretical numerical analyst is also obliged to prove these facts: practical people are satisfied if they just happen to be true.

The different numerical methods are described by introducing a projection operator P_n. If v is any function $P_n v \in S_n$ is a basis function approximating v. Two common examples of P_n are:

(i) $\phi \in S_n : \int_0^1 (P_n w)(\sigma)\phi(\sigma)d\sigma = \int_0^1 w(\sigma)\phi(\sigma)d\sigma$

(ii) For $v = -1$ select $0 < \xi_0 < \ldots < \xi_r < 1$

and define the collocation points

$$\xi_{ij} = x_{i-1} + \xi_j(x_i - x_{i-1}) \ .$$

Then $P_n w \in S_n$ is defined by $(P_n w)(\xi_{ij}) = w(\xi_{ij})$

(In (ii) we may allow $\nu = 0$ if $\xi_0 = 0$ and $\xi_r = 1$). For either (i) or (ii) v_n is defined by

(4) $$v_n + P_n R v_n = P_n g_0 \ .$$

With (i) v_n is the Galerkin solution and (4) becomes the Galerkin equations

(5) $$((I + R)v_n , \phi) = (g_0, \phi) \quad \phi \in S_n \ .$$

With (ii) v_n is the collocation solution with (4) reducing to

(6) $$v_n(\xi_{ij}) + (Rv_n)(\xi_{ij}) = g_0(\xi_{ij})$$

In both instances (5) and (6) are a system of linear equations $((I + P_n R)$ defines a linear operator on the finite dimensional vector space S_n). Stability requires us to show that the corresponding matrix is non-singular. This is the mathematical difficulty, and only the Galerkin method has been shown completely.

To describe the results define

$$||v|| = \left(\int_0^1 |u(s)|^2 ds\right)^{\frac{1}{2}}$$

and

$$|v| = \max\{|u(s)| : 0 \le s \le 1\}$$

These represent ways of measuring the error in an approximation. For example $|v_n - v_0|$ represents the maximum deviation between the computed and exact solutions.

A simple observation about R is

$$(Rv)(s) = \frac{\sin \chi\pi}{\pi}\int_0^{1/s} \frac{1}{1+2\sigma\cos\chi\pi + \sigma^2} v(\sigma s)d\sigma$$

and hence

$$|Rv| \le |v| \left(\frac{\sin \chi\pi}{\pi} \int_0^\infty \frac{d\sigma}{1+2\sigma \cos\chi\pi + \sigma^2} \right) = |\chi| |v| .$$

As $\chi < 1$, R is a contraction (i.e. Rv is smaller than v). Using Mellen Transforms (Chandler (1983a)).

$$||Rv|| \le \sin(\tfrac{1}{2}\chi\pi) ||v|| ,$$

and R is a contraction with respect to $||.||$. Consequently the series

$$g_0 - Rg_0 + R^2 g_0 - R^3 g_0 + \ldots\ldots$$

converges to a well defined function, which is the solution to (2.0). (And so we have proved that (2.0) and (2.1) have unique solutions). Similarly if

$$|P_n Rv| < |v| \text{ or } ||P_n Rv|| < ||v|| \text{ , then}$$

$$P_n g_0 - P_n R P_n g_0 + (P_n R)^2 P_n g_0 - \ldots$$

will be the unique solution of (4).

Since $||P_n u|| \le ||u||$, and hence $||P_n Ru|| \le ||Ru|| < ||u||$; the Galerkin equations are non-singular and for any $\phi \in S_n$

(7)
$$\begin{aligned} ||(I + P_n R)^{-1} \phi|| &= ||\sum_k (P_n R)^k \phi|| \\ &\le \left(\sum_k || P_n R \phi||^k \right) \\ &\le \frac{1}{1 - \sin\frac{1}{2}\chi\pi} || \phi|| . \end{aligned}$$

This stability inequality is basic to the subsequent theory.

There is no such easy theorem for the collocation solution. In general $||P_n u|| \ge ||u||$ and $|P_n u| > |u|$. Other techniques are partially successful. For example if either $r = 0$ or $r = 1$ and ξ_{ij} are the Gaus points inequalities similar to (7) are known (Atkinson and de Hoog (1983a, b), and also Arnold and Wendland (1982)). However we will subsequently deal only with the Galerkin method. Furthermore we assume $\nu = -1$, and avoid more complicated approximation theory.

Once stability is known convergence is simple to prove. From (2.0), $P_n v_0 + P_n R v_0 = P_n g_0$, and hence $P_n v_0 + P_n R P_n v_0 = P_n g_0 - P_n R P u' v_0$, where $P_n' = I - P_n$. Subtracting from (4) gives

$$(I + P_n R)(v_n - P_n v_0) = P_n R P_n' v_0 .$$

In the Galerkin case we can now apply the stability inequality to give

(8)
$$||v_n - P_n v_0|| = ||(I + P_n R)^{-1} (P_n R P_n' v_0)|| \le \frac{1}{1 - \sin\frac{1}{2}\chi\pi} ||P_n' v_0||.$$

One can now examine the behaviour of $P_n' v_0$, and deduce the behaviour of $v_n - P_n v_0$ and then $v_n - v_0$. Since $P_n' v_0$ depends on the unknown solution v_0, it is unlikely that useful statements can be made about the absolute size of $P_n' v_0$. However one can use knowledge of the behaviour of v_0 near $s = 0$ to find the rate at which $P_n' v_0$ and hence $v_n - v_0$ converges. If a method increases the rate at which $P_n' v_0$ converges to zero, it is likely to improve the convergence behaviour of $||v_n - v_0||$. It is this sort of qualitative guidance we seek from an abstract theory.

The disadvantage of (8) is that $||v_n - v_0||$ may be small, but the error can still be large at specific points. This should not be allowed in practice, and so results with $|v_n - v_0|$ are preferred. These may be obtained, but the proofs are difficult. We thus quote a result of Chandler (1983b) which shows

(9)
$$|v_n - P_n v_0| \le C \ln^\gamma n \; |RP_n' v_0| .$$

C and γ are constants independent of n and v_0, and depend on the grid and on r. (The factor $\ln^{\gamma} n$ is probably the product of inefficient proofs.) Once (9) is accepted it is easier to examine $|RP_n{}'v_0|$ than $\|RP_n{}'v_0\|$.

If v is smooth, it can be shown that a uniform grid may be used ($x_i = i/n$) and

$$|P_n v - v| = |P_n{}'v| \le 0(1/n^{r+1})$$

However, taking the more difficult case of a reentrant corner ($\chi < 0$), the solution v_0 is not smooth and

$$v_0(s) = as^{\alpha} + \text{"smoother terms"} ,$$

where $\alpha = 1/1 -$ and a is some constant determined by v (Grisvard (1976)). Thus D v_0 is infinite at 0. If a uniform grid is still used.

$$|P_n v_0 - v_0| = 0(1/n^{\alpha}) .$$

Now when n is doubled for instance the error in the approximation $P_n v_0$ decreases only by a factor of 1/2 , $\frac{1}{2} < \chi < 1$, and the advantages of using higher order polynomials is lost. This slow convergence would also appear in the solution of the boundary integral equation; for the simple reason that the basis functions are unable to approximate the solution well near s = 0. The approximation can be improved by adding a singular function s^{α} to the basis. Alternatively the grid can be graded, and we consider

$$x_i = (i/n)^q$$

($q \ge 1$ is the grading exponent). As q increases more grid points cluster towards 0. Provided $q \ge (r + 1)/\alpha$ an elementary Taylor series argument shows

$$|P_n v_0 - v_0| \le 0(1/n^{r+1}) . \tag{10}$$

Notice that the grading depends on both the corner and the degree of the basis functions.

Using this graded mesh $|RP_n{}'v_0| \le C|P_n{}'v_0| \le 0(1/n^{r+1})$ and so

$$\begin{aligned} |v_n - v_0| &\le |v_n - P_n v_0| + |P_n v_0 - v_0| \\ &\le C\ln^{\gamma} n\, |P_n v_0 - v_0| \\ &\le 0(\ln^{\gamma} n/n^{r+1}) \end{aligned} \tag{11}$$

using (9) and (10). With the exception of the very small $\ln^{\gamma} n$ term, this is the same qualitative error estimate that can be proved for a smooth domain. In this sense a graded mest overcomes the effect of the corners.

The same result is true for collocation methods. The above proof holds, provided a stability result is known.

SUPERCONVERGENCE OF GALERKIN'S METHOD

We now study very carefully the structure of the error $v_n - v_0$. It is highly oscillatory, being alternatively positive and negative. This oscillation is quite systematic and we can find ways of counteracting it and recovering improved approximations from v_n.

Observe firstly that for any functions u and v and $\phi \in S_n$

$$\int_0^1 \phi(s)(P_n{}'v)(s)ds = 0$$

(by (i)) and hence

$$\int_0^1 u(s)(P_n{}'v)(s)ds = \int_0^1 (u(s)-(P_n u)(s))(P_n{}'v)(s)ds = \int_0^1 (P_n{}'u)(s)(P_n{}'v)(s)ds$$

Thus from the definition of $r_s(\sigma)$ as the kernel function of R (equation (3))

$$(RP_n{}'v_0)(s) = \int_0^1 r_s(\sigma)(P_n{}'v)(\sigma)d\sigma = \int_0^1 (P_n{}'r_s)(\sigma)(P_n{}'v)(\sigma)d\sigma .$$

Thus provided s is not too near zero

(12) $$|(RP_n{'}v_0)(s)| \le |P_n{'}r_s||P_n{'}v_0| \le O(1/n^{2r+2})$$

by (10). This is exactly twice the rate of convergence for $|RP_n{'}v_0|$ used previously. As s approaches 0 more technicalities intervene (because r_s becomes increasingly singular). Taking the most difficult case of s = 0 we note that for any v

$$(Rv)(0) = \lim_{s\to 0}(Rv)(s) = \chi v(0) .$$

Hence $(RP_n{'}v_0)(0) = (P_n{'}v_0)(0)$. If $|RP_n{'}v_0|$ is to be $O(1/n^{2r+2})$ we must therefore have

(13) $$|P_n{'}v_0(0)| \le O(1/n^{2r+2})$$

But on $[0,x_1]$ the error $v_0 - P_nv_0$ is $O(x_1\alpha)$, and so (13) will be true only if $x_1^{\alpha} = (1/n)^{q\alpha} = O(1/n^{2r+2})$, i.e. $q \ge (2r+2)/\alpha$. Thus with the stronger grading $q \ge (2r+2)/\alpha$

$$|RP_n{'}v_0| \le O(\ln n/n^{2r+2}) ,$$

and hence from (9)

(14) $$|v_n - P_nv_0| \le O(\ln n/n^{2r+2})$$

This increased order of convergence is known as *superconvergence*. Thus comparing with (10) we see that v_n is extremely close to P_nv_0, while P_nv_0 is only an inaccurate approximation to v_0. Equivalently, the error in v_n closely follows the same pattern as the error in P_nv_0. A method of recovering a better approximation to v_0 from P_nv_0 should also improve v_n. There are two such devices: the *natural iteration* (Sloan (1976, 1982), Lin Qun (1979), Graham (1980)) and *averaging* (Chandler (1980, 1983b)).

The natural iteration follows by rearranging (2.1) to observe $v_0 = g \;\; - Rv_0$. If we take P_nv and computer $(P_nv)^*$ defined by $(P_nv_0)^* = g_0 - RP_nv_0$, then

$$(P_nv_0)^* - v_0 = -R(P_nv_0 - v_0) = RP_n{'}v_0 ,$$

and so $|(P_nv_0)^* - v_0| \le O(1/n^{2r+2})$ by previous arguments. Since this works for P_nv_0 it should work for v_n. Thus if

$$v_n^* = g_0 - Rv_n ,$$

$$|v_n^* - v_0| \le |(v_n - P_nv_0)^*| + |(P_nv_0)^* - v_0| \le O(\ln^{\gamma} n/n^{2r+2})$$

Notice that v_n^* requires O(n) operations each time it is evaluated, and involves the value of v_n at remote points.

The averaging procedure comes by considering P_nv_0 on two successive sub-intervals $[x_{i-1},x_i]$ and $[x_i,x_{i+1}]$ say. There exists a unique polynomial p of degree 2r + 1 such that on $[x_{i-1},x_{i+1}]$ $P_np = P_nv_0$. To within $O(1/n^{2r+2})$ this is simply the Taylor polynomial of degree 2r + 1 at x_i. To illustrate, suppose r = 1 and the grid is uniform with $x_{i+1} - x_i = x_i - x_{i-1} = h$. If

$$\begin{aligned}(P_nv_0)(s) &= a_1 + b_1(s-x_i) \qquad x > x_i\\ &= a_{-1} + b_{-1}(s-x_i) \quad x < x_i\end{aligned}$$

then $$p(x) = A_0 + A_1(s-x_i) + A_2(s-x_i)^2 + A_3(s-x_i)^3$$

where $$A_2 = (b_1 - b_{-1})/2h , \; A_3 = -5(a_1 - a_{-1})/2h^3$$

$$A_0 = \tfrac{1}{2}(a_1 + a_{-1}) + \frac{1}{12}h^2A_2 , \; A_1 = \tfrac{1}{2}(b_1 + b_{-1}) - \frac{9}{20}h^2A_3 .$$

As $(p - v_0)(s) = O(1/n^4)$, $s \in [x_{i-1}, x_{i+1}]$
we may define

$$p_i = p(x_i) = \tfrac{1}{2}(a_1 + a_{-1}) + \frac{1}{24}(b_1 - b_{-1})h$$

$$p_i^{(1)} = p^{(1)}(x_i) = \tfrac{1}{2}(b_1 + b_{-1}) + \frac{9}{8}(a_i - a_{-1})h^{-1}$$

and prove the estimates

(15) $$p_i = v_0(x_i) + O(h^4) \; , \; p_i^{(1)} = v_0^{(1)}(x_i) + O(h^3) \; .$$

With a non-uniform mesh the formulae are more complicated, but (15) remain true with $h = \max\{x_{i+1}-x_i,\ x_i-x_{i-1}\}$. A slight modification at the end points produces approximations to $v_0(x_i)$ and $v_0^{(1)}(x_i)$, $i = 0,n$. Now let $(P_nv_0)^{**}$ denote the cubic Hermite spline defined by

$$(P_nv_0)^{**}(x_i) = p_i \; , \; D(P_nv_0) = p_i^{(1)} \; , \; i = 0, \ldots, n \; .$$

The estimates (13) can then be used to show

$$(P_nv_0)^{**} - v_0 = O(1/n^4)$$

provided $q \ge 4/\alpha$. This grading (which is coincidently the same needed for (14)) is required before any piecewise cubic can approximate v_0 to with the optimal order $O(1/n^4)$.

The calculations which produced $(P_nv)^{**}$ from P_nv_0 can also be applied to v_n to produce v_n^{**}. Again since v_n matches P_nv_0 we expect v_n^{**} to be an improved approximation. This follows by

$$\begin{aligned} |v_n^{**} - v_0| &\le |(v_n - P_nv_0)^{**}| + |(P_nv_0)^{**} - v_0| \\ &\le C_1|v_n - P_nv_0| + |(P_nv_0)^{**} - v_0| \\ &\le O(\ln^{\gamma} n/n^4) \end{aligned}$$

(The penultimate step follows as $|\phi^{**}| \le C_1|\phi|$ for all $\phi \in S_n$.) Thus v_n^{**} has twice the rate of convergence of v_n. Provided n the number of elements is large enough for the qualitative estimates to be good guides in practice the averaging should be beneficial. For general r v_n^{**} will be a Hermite spline of degree $2r+1$, and approximations p_i, $p_i^{(1)}, \ldots, p_i^{(r)}$ are needed at each grid point. However only $O(n)$ operations are required to evaluate v_n^{**} at every mesh point, compared with $O(n^2)$ for v_n^*. Furthermore to calculate $v_n^{**}(x)$, only the coefficients of v_n on the two or three elements near x are involved.

Averaging should be applied only to Galerkin solutions. For consider the case of piecewise linear functions and collocation at the Gauss points. Averaging is unnecessary for v_n is already superconvergent at the collocation points $(|(v_n-v_0)(\xi_{ij})| \le O(1/n^4))$ provided $q > (2r + 2)/\alpha$. But this collocation method may also be viewed as a Galerkin method in which the inner products required to set up the Galerkin equations have been replaced by the two-point Gaussian integration rule. This reasonable strategy does not disturb the convergence of v_n (they are both $O(1/n^2)$) but it does alter the fine error structure (by $O(1/n^3)$). These differences becomes crucial if we are attempting to recover all the information about v_0 contained in v_n.

REFERENCES

[1] Arnold, D.N. and Wendland, W.L., On the asymptotic convergence of collocation methods, Math. Comp. (to appear), (Reprint Nr. 665, Dept. Math., Technical Univ. Darmstadt)(1982)

[2] Atkinson, K.E., A Survey of Numerical Methods for the Solution of Fredholm Integral Equations of the Second Kind (SIAM)(1976).

[3] Atkinson, K.E. and de Hoog, F.R., Collocation methods for a boundary integral equation on a wedge, in: Baker, C.T.H. and Miller, G.F. (eds.), Treatment of Integral Equations by Numerical Methods (Academic Press)(1983a).

[4] Atkinson, K.E. and de Hoog, F.R., The numerical solution of Laplace's equation on a wedge, IMA J.Num.Anal. (to appear), (preprint)(1983b).

[5] Chandler, G.A., Superconvergence for second kind integral equations, in: R.S. Anderssen et al (eds), The Application and Numerical Solution of Integral Equations (Sijthoff & Noordhoff)(1980).

[6] Chandler, G.A., Galerkin's method for boundary integral equations on polygonal domains, Numerical Analysis Report 83/1, Math. Dept., University of Queensland (1983a).

[7] Chandler, G.A., Superconvergent approximations to the solution of a boundary integral equation on polygonal domains, Research Report CMA-R15-83, Centre for Mathematical Analysis, Australian National Univ., (1983b).

[8] Costabel, M. and Stephan, E., The boundary integral method for a mixed boundary value problem in a polygonal domain, Advances in Computer Methods for Partial Differential Equations IV, IMACS (1981).

[9] Graham, I.G., Galerkin methods for second kind integral equations with singularities, Math. Comp. (to appear), (preprint 1980).

[10] Grisvard, P., Behaviour of the solutions of an elliptic boundary value problem in a polygonal or polyhedral domain, in: Hubbard B. (ed.), Numerical Solution of Partial Differential Equations - III, Synspade 1975, Academic Press (1976).

[11] Hartmann, F., Elastostatics, chapter 5 in: Brebbia, C.A. (Ed.), Progress in Boundary Element Methods, Vol 1, (Pentech Press)(1981).

[12] Lin Qun, Some problems about the approximate solution operator equations, Acta. Math. Sinica 22, 219-230, (1979).

[13] Sloan, I.H., Improvement by iteration for compact operator equations, Math. Comp. 30, 758-764, (1976).

[14] Sloan, I.H., Superconvergence and the Galerkin method for integral equations of the second kind, in: Baker, C.T.H. and Miller, G.F. (Eds.), Treatment of Integral Equations by Numerical Methods (Academic Press)(1983).

[15] Wendland, W., Boundary element methods and their asymptotic convergence, in: Filippi, P. (Ed.), Lecture Notes of the CISM Summer School on Theoretical Acoustics and Numerical Techniques, Lecture Notes in Physics (Springer-Verlag) 1982.

Computational Techniques & Applications: CTAC-83
J. Noye & C. Fletcher (Editors)

NUMERICAL INTEGRATION OF A SINGULAR INTEGRAL EQUATION

B.L. Karihaloo

Department of Civil Engineering and Surveying
The University of Newcastle, N.S.W. 2308 Australia

The paper describes an efficient numerical technique for integrating a singular integral equation. This equation arises in the study of elastic-plastic fracture behaviour of solids containing a distribution of cracks when the displacement discontinuity across the cracks is replaced by an appropriate distribution of dislocations. The technique relies on the decomposition of the kernel into singular and non-singular parts and the expansion of the latter in a series of orthogonal polynomials. Several illustrative examples are given.

INTRODUCTION

All engineering materials contain inhomogeneities of various types, shapes and sizes, although for mathematical simplicity they are generally regarded as homogeneous and isotropic. It is recognised that the inhomogeneities - voids, cracks, inclusions - strongly influence the elastic behaviour and fracture characteristics of the material. However, in order to study this behaviour mathematically several restrictions are imposed on the orientation and scale of the inhomogeneities. Thus, the randomness of the orientation and scale of inhomogeneities is disregarded and one- or two-dimensional periodic arrays of inhomogeneities of like shape and size are considered. In this paper attention is restricted to inhomogeneities in the form of slitlike cracks. Dislocation formalism is employed to make the problem amenable to mathematical treatment. The equivalence of slitlike cracks (displacement discontinuities) and suitable distributions of straight dislocations was established by Bilby, Cottrell and Swinden (1963).

Using the dislocation formalism and satisfying the traction free requirement across the faces of the cracks leads to the following singular integral equation for each of the cracks in the periodic array (Karihaloo (1977), (1978), (1979)).

$$\int_{-1}^{1} \frac{f(x')\,G(x',x)}{(x-x')}\,dx' + P(x) = 0 \quad . \qquad (1)$$

$f(x')$ is the distribution of straight dislocations (displacement discontinuity across the crack) appropriate to the particular stress state, $G(x',x)$ is the kernel that arises due to the interaction of cracks in the array and depends on the type of array, and $P(x)$ is a piece-wise constant function representing either the externally applied stress or the yield stress of the material in the plastic zone ahead of the crack tips. The singular integral equation (1) is written in a non-dimensional form. The linear dimension is non-dimensionalised by one half the length of the crack, a, the stress function $P(x)$ by the yield stress of the material, σ_y, and the dislocation distribution function by σ_y/A where A is a material constant depending on the Burgers vector, b, of the dislocation, and on the shear modulus, μ, and Poisson's ratio, ν, of the material. It is the purpose of this paper to describe an efficient method for calculating f(x') with a view to determining the relative displacement of crack faces

$$\Delta(x) = b \int_0^x f(x')\, dx' \quad , \tag{2}$$

The relative displacement of crack faces or the crack opening displacement is a measure of the fracture toughness of the material.

FORMULATION OF THE PROBLEM

To make the presentation self-contained, the derivation of the singular integral equation (1) is briefly outlined below.

Consider an infinite, isotropically elastic solid containing a doubly-periodic (rectangular or diamond-shaped) array of equal cracks. In the rectangular configuration (Fig. 1(a)) the traction-free cracks occupy the positions $(md_1 - c) \le x \le (md_1 + c)$; $y = \pm nh_1$ $(n, m = 0, \pm1, \pm2, \ldots\ldots)$, while the coplanar plastic regions ahead of the crack tips are located at $(md_1 - a) \le x \le (md_1 - c)$ and $(md_1 + c) \le x \le (md_1 + a)$; $y = \pm nh_1$. It should be noted that, provided $d_1/a > 2$, the adjacent plastic regions in any row of cracks do not coalesce. Similarly, in the diamond-shape configuration of cracks (Fig. 1(b),(c)) the freely-slipping parts of the cracks occupy the positions $(md_1/2 - c) \le x \le (md_1/2 + c)$; $y = \pm nh_1/2$ $(n = 0, \pm2, \pm4, \ldots\ldots)$ or $(md_1 + d_1/2 - c) \le x \le (md_1 + d_1/2 + c)$; $y = \pm nh_1/2$ $(n = \pm1, \pm3, \ldots\ldots)$, and the coplanar plastic regions are located at either $(md_1 - a) \le x \le (md_1 - c)$ and $(md_1 + c) \le x \le (md_1 + a)$; $y = \pm nh_1/2$ $(n = 0, \pm2, \pm4, \ldots\ldots)$ or $(md_1 + d_1/2 - a) \le x \le (md_1/2 - c)$ and $(md_1 + d_1/2 + c) \le x \le (md_1 + d_1/2 + a)$; $y = \pm nh_1/2$ $(n = \pm1, \pm3, \ldots\ldots)$.

Replacing each crack in the array by an appropriate distribution of dislocations, an expression for the interactive stresses is readily obtained by considering the influence of a vertical array of relevant dislocations situated along a plane $x = x'$ passing through the point x along one of the planes $y = \pm nh_1$. It should be mentioned that each crack and its associated coplanar plastic region is represented by the same continuous distribution of infinitesimal dislocations. Moreover, because of symmetry, the interactions of all but one component of the stress tensor (σ_{yy} in opening Mode I, σ_{xy} in plane shear Mode II or σ_{yz} in anti-plane shear Mode III) cancel one another, and only one singular integral equation in each mode remains to ensure traction-free state of crack faces. The horizontal and vertical crack spacing are non-dimensionalised as follows: $h = h_1/\pi c$, $d = d_1/a$. The piece-wise constant function P(x) takes the values σ/σ_y and $(\sigma/\sigma_y - 1)$ in the freely slipping parts of the cracks $|x| \le \alpha$ and the plastic regions $\alpha < |x| \le 1$, respectively, where $\alpha = c/a$.

It is convenient to decompose the kernel, G(x',x) into singular and non-singular parts. This decomposition is dictated by the general numerical technique used to integrate the singular integral equation (see below). Here, it is sufficient to point out that the decomposition of the kernel permits recovery of the solution to the problem of an isolated relaxed crack. This may not sound to be a great advantage if it is not realised that the singularity in the solution of equation (1) is of the same type as that for an isolated crack and is associated with the sudden jump in P(x) at $|x| = \alpha$. With the indicated decomposition, equation (1) may be rewritten as

$$\int_{-1}^{1} f(x') \left\{\frac{1}{x - x'} + K(x',x)\right\} dx' + P(x) = 0 \tag{3}$$

The non-singular part of the kernel reflects the crack configuration and external stress state. In other words, it is a function of the horizontal and vertical

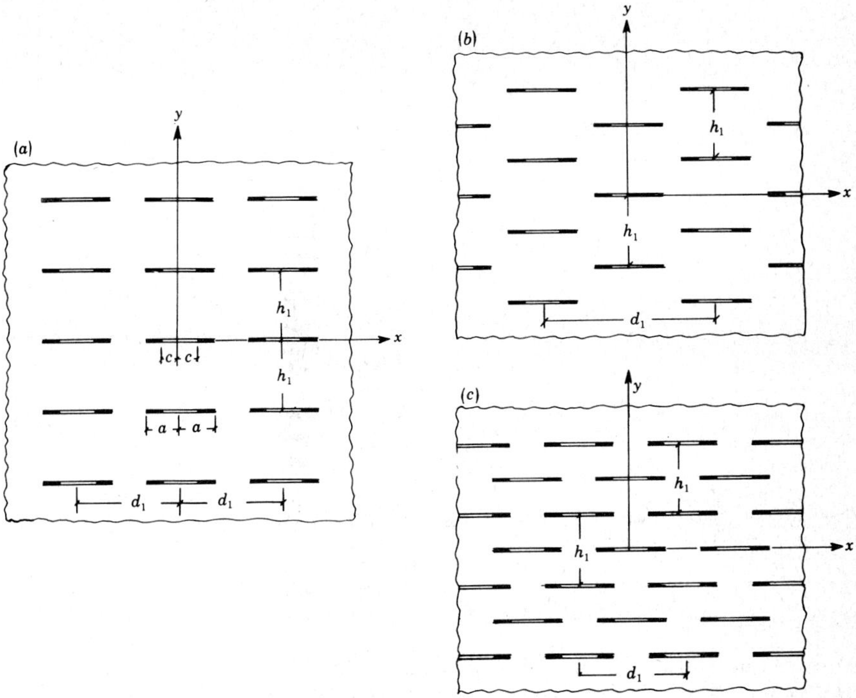

FIGURE 1
An elastic body subject to a stress σ at infinity and containing (a) a rectangular array of cracks; (b) a diamond-shaped array of cracks without overlap $d_1 > 4a$; (c) a diamond-shaped array of cracks with overlapping plastic regions in adjacent rows, $d_1 < 4a$.

crack spacings and the non-zero external stress (σ_{yy} in Mode I, σ_{xy} in Mode II or σ_{yz} in Mode III). Expressions for $K(x',x)$ for the two crack configurations shown in Fig. 1 are:

Doubly-periodic rectangular array

$$K(x',x) = \begin{cases} -\dfrac{1}{x - x'} + \sum\limits_{m=-\infty}^{\infty} K^*_I(x',x) & \text{, Mode III} \\ -\dfrac{1}{x - x'} + \sum\limits_{m=-\infty}^{\infty} K^*_{II}(x',x) & \text{, Mode II} \\ -\dfrac{1}{x - x'} + \sum\limits_{m=-\infty}^{\infty} \{2K^*_I(x',x) - K^*_{II}(x',x)\} & \text{, Mode I} \end{cases} \tag{4}$$

where

$$\begin{aligned} K^*_I(x',x) &= \frac{1}{\alpha h} \coth\{(x - x' + md)/\alpha h\} \\ K^*_{II}(x',x) &= \frac{1}{(\alpha h)^2}(x - x' + md)\operatorname{csch}^2\{(x - x' + md)/\alpha h\}. \end{aligned} \tag{5}$$

Doubly-periodic diamond-shaped array

$$K(x',x) = \begin{cases} -\dfrac{1}{x - x'} + \sum\limits_{m=-\infty}^{\infty} \{K^*_I(x',x) + K^{**}_I(x',x)\} & \text{, Mode III} \\ -\dfrac{1}{x - x'} + \sum\limits_{m=-\infty}^{\infty} \{K^*_{II}(x',x) - K^{**}_{II}(x',x)\} & \text{, Mode II} \\ -\dfrac{1}{x - x'} + \sum\limits_{m=-\infty}^{\infty} \{2K^*_I(x',x) + 2K^{**}_I(x',x) & \\ \qquad -K^*_{II}(x',x) + K^{**}_{II}(x',x)\} & \text{, Mode I} \end{cases} \tag{6}$$

where

$$K^{**}_I(x',x) = \frac{1}{\alpha h} \tanh\{(x - x' + md + md/2)/\alpha h\},$$
$$K^{**}_{II}(x',x) = \frac{1}{(\alpha h)^2} (x - x' + md + d/2) \operatorname{sech}^2\{(x - x' + md + d/2)/\alpha h\} \tag{7}$$

It should be pointed out that despite the appearance of individually singular terms in the expressions for $K(x',x)$, each of the expressions is non-singular. It may be verified that as $x' \to x$, $K(x',x) \to 0$. Moreover, as $h \to \infty$ and $m = 0$, $K(x',x) \to 0$, and the singular integral equation for an isolated relaxed crack is recovered.

METHOD OF SOLUTION

A closed-form solution of the singular integral equation (3) appears to be feasible only for a stack of cracks under Mode III conditions (Smith (1964)), and for a row of collinear cracks under plane strain conditions (Bilby et al (1964)). A perturbation solution for the same crack configuration under plane strain conditions was given by Karihaloo (1977) for widely spaced cracks. For all other crack configurations and loading conditions the equation has to be integrated numerically. Several methods have been proposed for the case of unrelaxed cracks, i.e. when the singular points coincide with the limits of the region of integration. (Erdogan & Gupta (1972), Krenk (1975), Ioakimidis & Theocaris (1979), Ioakimidis (1983)). All the methods are based on an approximation of the non-singular part of the equation, i.e. $f(x')\,K(x',x)$ by a series of orthogonal polynomials of degree $(2p - 1)$ in x' where p is the number of interpolation points.

The method used in the present paper was first proposed by Erdogan (1969). In its original form it was applicable only to unrelaxed cracks when the singularity in $f(x')$ occurs at the limits of the interval. This method was later adapted by Karihaloo (1977) to the solution of relaxed crack problems. It relies on an expansion of the non-singular part of the kernel $K(x',x)$ in a series of orthogonal polynomials and offers two advantages over other methods. Firstly, by making it necessary to extract explicitly the singularity in $f(x)$ it allows the solution for an isolated crack to be recovered. Indeed, as will be clear subsequently, the singular behaviour of $f(x')$ is similar to that of an isolated relaxed crack. The second, and perhaps more important, advantage from a mathematical viewpoint is that one of the interpolation points coincides with the sharp discontinuity in $P(x)$ at $|x| = \alpha$. In fact, it is because of this discontinuity that $f(x)$ suffers a singularity at the transition from the freely slipping crack tips to the coplanar plastic regions.

For prescribed values of crack spacings, h and d, and loading mode the non-singular part of the kernel $K(x',x)$ is known. It should be mentioned that the external stress σ is related to α through the necessary condition for existence of a solution to equation (3) (see below). $K(x',x)$ may thus be expanded in a series of orthogonal polynomials in the spatial variable x, the coefficients of the series being functions of the other spatial variable x':

$$K(x',x) = \sum_{i=0}^{\infty} A_i(x')\, T_i(x) \quad , \qquad (8)$$

where $T_i(x)$ is the i-th Chebyshev polynomial of the first kind, defined by

$$T_i(x) = \cos(i \cos^{-1} x) \quad , \qquad (9)$$

with $T_0(x) = 1$, $T_1(x) = x$ and $T_{i+1}(x) = 2\,x\,T_i(x) - T_{i-1}(x)$, $i = 1, 2, 3, \ldots$. The choice of the Chebyshev polynomials of the first kind rather than the second, as in the solution of unrelaxed cracks (Delameter et al. (1975)), is dictated by the expected form of the solution for an isolated relaxed crack (Bilby et al. (1963)). Chebyshev polynomials are chosen because their orthogonality properties considerably simplify the resulting expressions. For instance, the coefficients $A_i(x')$ are easily determined from the following orthogonality properties of $T_i(x)$:

$$\int_{-1}^{1} \frac{T_i^2(x)}{\sqrt{(1-x^2)}}\, dx = \begin{cases} \pi & ; \quad i = 0 \\ \pi/2 & ; \quad i = 1, 2, 3, \ldots\ldots , \end{cases} \qquad (10)$$

whence it follows that

$$\begin{aligned} A_0(x') &= \frac{1}{\pi} \int_{-1}^{1} \frac{K(x',x)}{\sqrt{(1-x^2)}}\, dx \\ A_i(x') &= \frac{2}{\pi} \int_{-1}^{1} \frac{K(x',x)\, T_i(x)}{\sqrt{(1-x^2)}}\, dx \ ; \quad i = 1, 2, 3, \ldots , \end{aligned} \qquad (11)$$

Substituting (8) into the singular integral equation (3) gives

$$\int_{-1}^{1} \frac{f(x')}{x-x'}\, dx' + \sum_{i=0}^{\infty} B_i\, T_i(x) + P(x) = 0 \quad , \qquad (12)$$

where

$$B_i = \int_{-1}^{1} f(x')\, A_i(x')\, dx' \ . \qquad (13)$$

The condition for the existence of a solution to (12) requires that (Muskhelishvili (1953))

$$\int_{-1}^{1} \frac{f(x')}{(x-x')\sqrt{(1-x^2)}}\, dx = 0 \qquad (14)$$

Physically, this condition ensures a smooth closure of the plastic zone tips and thus specifies the distance to which the yield zones can spread under a given applied stress σ. In fact, substituting (12) into (14) and using the identities

$$\int_{-1}^{1} \frac{T_0(x)\, T_i(x)}{\sqrt{(1 - x^2)}}\, dx = \begin{cases} \pi & ; \quad i = 0 \\ 0 & ; \quad i = 1, 2, 3, \ldots\ldots , \end{cases} \qquad (15)$$

yields

$$\pi B_o + \int_{-1}^{1} \frac{P(x)}{\sqrt{(1-x^2)}} dx = 0. \tag{16}$$

The function P(x) takes the value σ/σ_y and $(\sigma/\sigma_y - 1)$ in the intervals $|x| \leq \alpha$ and $\alpha < |x| \leq 1$, whereupon (16) reduces to

$$\alpha = \cos \frac{\pi}{2} \left(\frac{\sigma}{\sigma_y} + B_o\right) . \tag{17}$$

The expression for the extent of plastic yield (17) reduces to the corresponding expression for an isolated crack (Bilby et al. (1963)) if $B_o = 0$. If the expression for an isolated crack is identified by the subscript "iso", it is easy to compare the extent of plastic deformation with that for an isolated crack

$$\alpha/\alpha_{iso} = \cos \frac{\pi}{2} B_o \left(1 - \tan \frac{\pi}{2} \frac{\sigma}{\sigma_y} \tan \frac{\pi}{2} B_o\right) . \tag{18}$$

When the existence condition (17) is satisfied, the solution of the singular integral equation (12) is (Muskhelishvili (1953))

$$f(x) = - \frac{\sqrt{(1-x^2)}}{\pi^2} \left[\sum_{i=o}^{\infty} B_i \int_{-1}^{1} \frac{T_i(x')}{(x'-x)\sqrt{(1-x'^2)}} dx' + \int_{-1}^{1} \frac{P(x')}{(x'-x)\sqrt{(1-x'^2)}} dx' \right] , \tag{19}$$

or

$$f(x) = \eta(x) - \sum_{i=o}^{\infty} B_i \psi_i(x) , \tag{20}$$

where, for brevity,

$$\psi_i(x) = \frac{\sqrt{(1-x^2)}}{\pi^2} \int_{-1}^{1} \frac{T_i(x')}{\sqrt{(1-x'^2)}(x'-x)} dx' , \tag{21}$$

$$\eta(x) = - \frac{\sqrt{(1-x^2)}}{\pi^2} \int_{-1}^{1} \frac{P(x')}{(x'-x)\sqrt{(1-x'^2)}} dx'$$

$$= \frac{1}{\pi^2} \left\{\cosh^{-1} \left|\frac{1 - \alpha x}{\alpha - x}\right| - \cosh^{-1} \left|\frac{1 + \alpha x}{\alpha + x}\right|\right\} . \tag{22}$$

In fact, $\eta(x)$ is identical in form to the solution for an isolated relaxed crack and reduces to it if α is replaced by α_{iso}. One of the abovementioned reasons for choosing the present numerical integration technique thus becomes apparent. $\eta(x)$ has an integrable singularity at $|x| = \alpha$, i.e. at the crack tips. The coefficients B_i in (20) which are still unknown, cannot be determined until the function itself has been evaluated. However, before proceeding to determine f(x) and hence B_i, it is convenient to simplify the functions ψ_i. It is easily shown that

$$\psi_o(x) = 0 ,$$
$$\psi_{i+1}(x) = \frac{\sqrt{(1-x^2)}}{\pi} U_i(x) , \quad i = 0, 1, 2, 3, ... \tag{23}$$

The second set of relations is a consequence of the identity

$$\int_{-1}^{1} \frac{T_{i+1}(x')}{(x'-x)\sqrt{(1-x'^2)}}\, dx' = \pi\, U_i(x) \quad , \tag{24}$$

where $U_i(x)$ is the i-th Chebyshev polynomial of the second kind, defined by

$$U_i(x) = \frac{\sin\{(i+1)\cos^{-1}(x)\}}{\sin(\cos^{-1} x)} \quad ,$$

with $U_0(x) = 1$, $U_1(x) = 2x$, $U_{i+1}(x) = 2x\, U_i(x) - U_{i-1}(x)$, $i = 1, 2, 3, \ldots\ldots$

The symmetry of the problem suggests that the dislocation distribution function $f(x)$ is an odd function. Moreover, since $\psi_0(x) = 0$ and $\eta(x)$ is an odd function, it is possible to express $f(x)$ through odd Chebyshev polynomials of the second kind

$$f(x) = \eta(x) - \frac{\sqrt{(1-x^2)}}{\pi} \sum_{i=1}^{\infty} C_i\, U_{2i-1}(x) \quad , \tag{25}$$

where

$$C_i = \int_{-1}^{1} f(x')\, A_{2i}(x')\, dx' \ . \tag{26}$$

The coefficients C_i and hence the coefficient B_0 appearing in the expression (17) are determined from an infinite system of linear algebraic equations obtained after substituting (25) into (26) and rearranging the terms

$$\sum_{m=1}^{\infty} E_{km}\, C_m = D_k \quad ; \quad k = 1, 2, 3, \ldots\ldots \tag{27}$$

where

$$\begin{aligned} D_k &= \int_{-1}^{1} \eta(x)\, A_{2k}(x)\, dx \ , \\ E_{km} &= \delta_{km} + \int_{-1}^{1} \frac{\sqrt{(1-x^2)}}{\pi}\, A_{2k}(x)\, U_{2m-1}(x)\, dx, \end{aligned} \tag{28}$$

δ_{km} being the Kronecker delta.

To investigate the fracture behaviour of a solid containing a given crack distribution it is necessary to calculate the displacement discontinuity $\Delta(x)$ across each of the cracks. In particular, the most important quantity is the crack tip opening displacement $\Delta(\alpha)$, which would accommodate the plastic zone spread predicted by (17). This displacement is proportional to the number of dislocations between the crack tip, $x = \alpha$, and the end of the plastic zone, $x = 1$. It is easily shown that

$$\Delta^*(\alpha) = \ln(1/\alpha) - \frac{\pi}{2\alpha} \sum_{i=1}^{\infty} C_i \int_{\alpha}^{1} \sqrt{(1-x'^2)}\; U_{2i-1}(x')\, dx' \ , \tag{29}$$

where $\Delta^*(\alpha) = \Delta(\alpha)\, \pi\mu/4c\, \kappa\, \sigma_y$ and κ takes the values 1 or $(1 - \nu)$ under anti-plane and plane strain conditions, respectively. The expression for an isolated relaxed crack is recovered by ignoring the second term on the right-hand side of (29) and replacing α by α_{iso}.

RESULTS AND DISCUSSION

For prescribed values of the non-dimensional parameters α, h and d the coefficients C_i were determined by solving the system of linear algebraic equations (27). The infinite system of equations was truncated at k = m = 11. This assured sufficient numerical accuracy. The accuracy of the numerical procedure was judged against

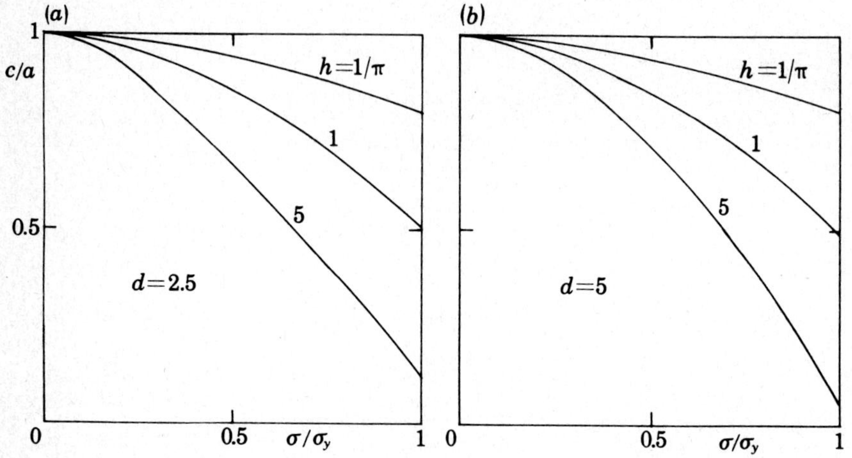

FIGURE 2
The extent of spread of plasticity, a, from a rectangular array of cracks in an infinite solid as a function of the applied stress $\sigma_{yy} = \sigma$ and the non-dimensional distance h for two selected values of distance, d. (a) d = 2.5; (b) d = 5.

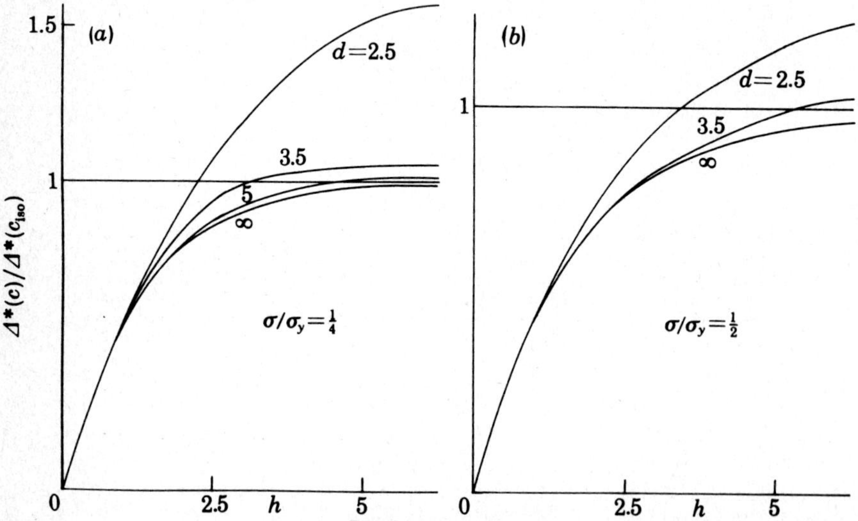

FIGURE 3
The ratio $\Delta^*(c)/\Delta^*(c_{iso})$ as a function of non-dimensional distances h and d for two selected values of $\sigma_{yy} = \sigma$ acting on an infinite solid containing a rectangular array of cracks. (a) $\sigma/\sigma_y = \frac{1}{4}$; (b) $\sigma/\sigma_y = \frac{1}{2}$, d = ∞ corresponds to a stack of relaxed cracks.

known closed-form solutions for the cases $h \to \infty$ (a row of collinear relaxed cracks in all deformation modes Bilby et al (1964)) and $d \to \infty$ ($m = 0$) in Mode III (a stack of relaxed cracks in anti-plane shear Smith (1964)). The absolute value of the maximum difference was within 5% for values of α close to zero;

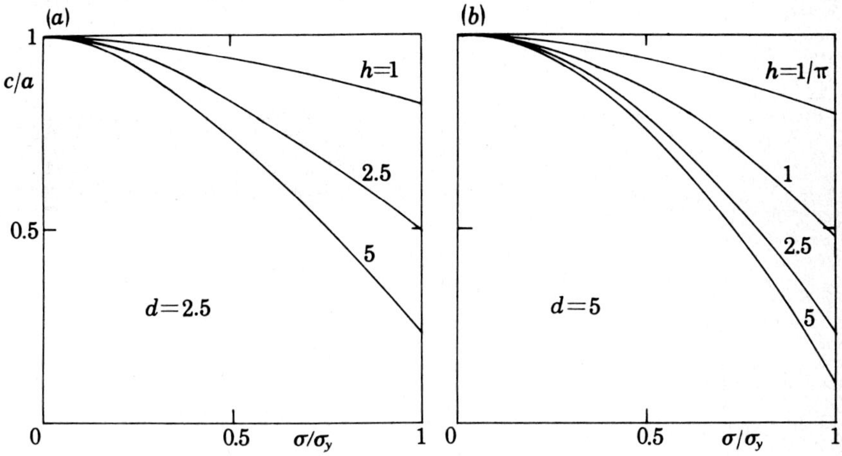

FIGURE 4
The extent of spread of plasticity, a, from a diamond-shaped array of cracks in an inifinite solid as a function of the applied stress $\sigma_{yy} = \sigma$ and the non-dimensional distance h for two selected values of d. (a) d = 2.5 (overlapping plastic regions in adjacent rows); (b) d = 5 (no overlapping).

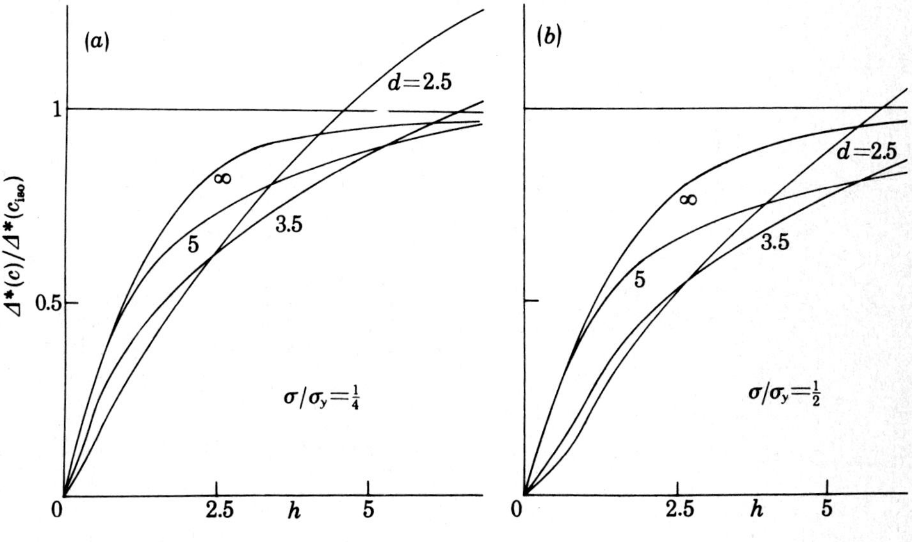

FIGURE 5
The ratio $\Delta^*(c)/\Delta^*(c_{iso})$ as a function of non-dimensional distances h and d for two selected values of $\sigma_{yy} = \sigma$ acting on an infinite solid containing a diamond-shaped array of cracks (a) $\sigma/\sigma_y = \frac{1}{4}$; (b) $\sigma/\sigma_y = \frac{1}{2}$.

the accuracy increased with increasing α, i.e. with decreasing extent of the spread of plasticity.

All integrals were evaluated by Simpson's rule after making a change of variables, where necessary, to render the integrands non-singular. The value of σ/σ_y was found in an inverse manner, whereby a value of α was assumed and σ/σ_y determined from (17) at the end of the interactions, i.e. after having computed the distribution function f(x) and hence the coefficient B_0 appearing in (17).

Typical results for the extent of spread of plasticity α (= $\frac{c}{a}$), and for the crack tip opening displacement $\Delta^*(c)$ as a proportion of the displacement for an isolated crack $\Delta^*(c_{iso})$ are shown in Fig. 2 - 5.

REFERENCES

[1] Bilby, B.A., Cottrell, A.H. and Swinden, K.H., *Proc. Roy. Soc. London, A272,* (1963), 304.

[2] Karihaloo, B.L., *Int. J. Solids Structures,* 13, (1977), 367.

[3] Karihaloo, B.L., *Proc. Roy. Soc. London,* A360, (1978), 373.

[4] Karihaloo, B.L., *J. Engng. Frac. Mech.,* 12, (1979), 49.

[5] Smith, E., *Proc. Roy. Soc. London.,* A282, (1964), 422.

[6] Bilby, B.A., Cottrell, A.H., Smith, E. and Swinden, K.H., *Proc. Roy. Soc. London.,* A279, (1964), 1.

[7] Erdogan, F. and Gupta, G.D., *Quart. Appl. Math.,* 30, (1972), 525.

[8] Krenk, S., *Quart. Appl. Math.,* 32, (1975), 479.

[9] Ioakimidis, N.I. and Theocaris, P.S., *Int. J. Fracture,* 15, (1979), 299,

[10] Ioakimidis, N.I., *Int. J. Fracture,* 21, (1983), 115.

[11] Erdogan, F., *SIAM J. Appl. Math.,* 17, (1969), 1041.

[12] Delameter, W.R., Herrmann, G. and Barnett, D.M., *J. Appl. Mech., ASME,* 42, (1975), 74.

[13] Muskhelishvili, N.I., *Singular Integral Equations,* (P. Noordhoff, Groningen, 1953).

Computational Techniques & Applications: CTAC-83
J. Noye & C. Fletcher (Editors)

A HYBRID DISTINCT ELEMENT - BOUNDARY ELEMENT METHOD FOR SEMI-INFINITE AND INFINITE BODY PROBLEMS

B.H.G. Brady, M.A. Coulthard
CSIRO, Division of Geomechanics
P.O. Box 54, Mount Waverley
Victoria 3149, Australia

J.V. Lemos
Department of Civil and Mineral Engineering
University of Minnesota
Minneapolis, MN 55455, U.S.A.

When an excavation is created in a highly jointed rock mass, rock blocks near the excavation periphery may undergo slip and separation. Outside this zone, the rock tends to behave as an elastic continuum. The computational technique described in this paper uses the Distinct Element Method to model the near-field, discontinuous material, and the Boundary Element Method to represent the far-field continuum. The performance of the program is illustrated by some verification and demonstration problems. The potential for development and application of such a hybrid scheme is indicated.

INTRODUCTION

Engineering activity in geologic media such as soil and rock masses takes place in a system subject to a state of initial stress. The design of structures in and on such media requires a capacity to predict induced displacements and total stresses resulting from the applied or induced loads. Numerical methods of analysis are now firmly established as the preferred method of undertaking this aspect of geomechanics practice.

Computational procedures used for design analysis in geomechanics are identical in principle to those used in other fields of engineering mechanics. Either the Finite Element, Boundary Element or Distinct Element Methods for analysis of stress and displacement may be used when key aspects of the design problem can be represented in the constitutive equations implemented in the computational scheme. Each method of analysis has clear conceptual and operational advantages and disadvantages when applied in engineering design.

The differential methods of analysis (Finite Elements, Distinct Elements) require discretization of the complete problem domain, so that discretization errors are distributed throughout the problem volume. An arbitrary external boundary must be assumed for the problem domain, and the conditions to be imposed on this surface are not always obvious. Also, the size of the numerical problem to be solved is related to the volume of the problem domain. Thus, as the physical size of the problem domain increases, the size of the numerical problem can become prohibitively large.

The clear advantage of the differential methods of analysis is their capacity to model complex constitutive behaviour of the geologic system. Slip and separation of blocky systems, and plastic flow and creep in continuous media, can be represented with various degrees of accuracy and efficiency (Cundall et al. (1978), Owen and Hinton (1980)).

In geomechanics, the host medium for the prospective engineering activity may be conceptualized as an infinite or semi-infinite body. Various formulations of the Boundary Element Method have been developed to resolve problems in bodies with these limiting geometries. In this method, a problem is defined and solved, for the complete domain, in terms of the conditions imposed on the surfaces of openings , or over loaded segments of any surface. A property of the method is that the size of the numerical problem increases in proportion with the size of the surface problem. The volume of the problem domain is not considered explicitly in the analysis.

Efficiency in a Boundary Element formulation is achieved by assuming simple, linear constitutive behaviour for the medium. For geologic media, linear elastic behaviour is often assumed. Many versions of the Boundary Element Method have been presented, by, for example, Lachat and Watson (1976), Brady and Bray (1978) and Watson (1979), for plane strain, complete plane strain and three-dimensional problem geometries.

A HYBRID DISTINCT ELEMENT - BOUNDARY ELEMENT SCHEME

The complementary advantages of differential methods and boundary element methods may be realized when a differential domain is embedded in a boundary element domain, as illustrated in Figure 1. These include, firstly, elimination of uncertainties about the conditions to be applied at the outer boundary of the differential domain. Secondly, far-field behaviour can be represented in a computationally economical and mechanically appropriate way as an elastic continuum. Finally, zones of complex constitutive behaviour in a geologic structure are usually small and localized, so that only these zones require the analytical power associated with a differential formulation. The implied reduction in the differential domain again favours computational efficiency.

The coupling of the Distinct Element Method (Cundall et al. (1978)) and a Boundary Element Method are described here. Such a hybrid method is of interest in excavation design problems in jointed and fractured rock masses. In these cases, rigid body translations and rotations of rock blocks may occur in the medium, in the periphery of an excavation.

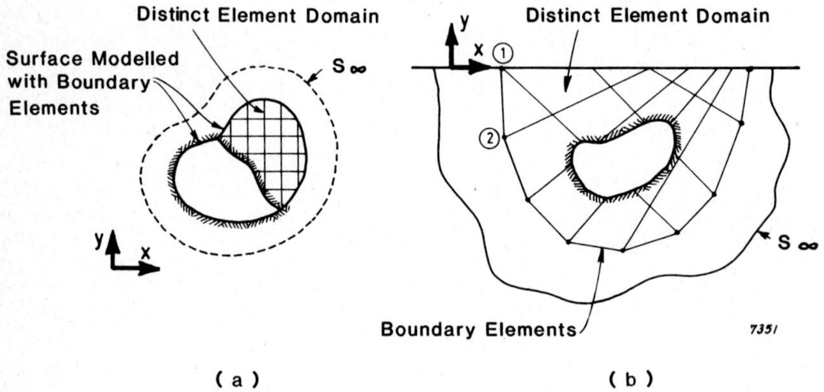

Figure 1
Hybrid distinct element - boundary element systems.
(a) excavation in infinite elastic region;
(b) excavation near surface of half-plane.

The formulation of a hybrid D.E.-B.E. scheme follows that proposed by Brady and Wassyng (1981) for coupling Finite Element and Boundary Element routines. The requirement is to satisfy the conditions for continuity of traction and displacement across the interface between the two domains, as illustrated in Figure 1. The respective governing equations are satisfied in each separate domain, in the separate computational formulations.

BOUNDARY ELEMENT FORMULATION

The B.E. Method exploits fundamental results of elasticity theory, to establish relationships between values of traction and displacement over a surface defining a problem domain. The direct formulation of the method, used in the current approach, is based on the integral equation (Watson (1979)):

$$c_{ij}(x)\, u_j(x) + \int_S T_{ij}(x,y)\, u_j(y)\, dS_y = \int_S U_{ij}(x,y)\, t_j(y)\, dS_y \qquad (1)$$

where point y lies on the boundary S of the domain, and the kernels $U_{ij}(x,y)$ and $T_{ij}(x,y)$ represent, respectively, displacements and tractions at point y, due to a unit load at x. When point x lies within the body (in which case $c_{ij}(x) = \delta_{ij}$), equation (1) becomes the Somigliana identity, and allows the determination of the displacements of interior points, once boundary unknowns are determined. Differentiation of this equation with respect to the coordinates of the load point x produces a similar expression for calculating the strain field within the body, from which stresses can be determined. When point x lies on the boundary S, equation (1) relates only boundary values of traction and displacement.

The numerical treatment of the boundary integral equation involves the discretization of the boundary into segments, within which some type of interpolation of unknowns is assumed. In the present model, a linear variation of both displacements and tractions is used. Displacements within an element can therefore be expressed in terms of the nodal values as

$$u_i(\xi) = N_i^k(\xi)\, u_i^k \qquad (2)$$

where ξ is an intrinsic coordinate, $N_i^k(\xi)$ the well-known linear shape functions, and u_i^k the displacement vector at node k. Tractions are interpolated in a similar way.

Introducing these expressions in equation (1), the nodal unknowns can be isolated from the integrals, and a linear algebraic equation between surface tractions and displacements is established. Performing the collocation of this discretized equation at the nodal points produces a system of equations that can be represented in matrix form as

$$[T]\,\{u\} = [U]\,\{t\} \qquad (3)$$

where nodal displacements have been assembled into a vector {u}, and tractions into {t}. The coefficients of the [T] and [U] matrices consist of integrals of kernel-shape function products and can be evaluated numerically. As either displacement or traction must be prescribed at a node, the number of equations equals the number of unknowns, so system (3) can be rearranged into a standard system of linear equations.

For deep-sited problems, the two-dimensional Kelvin solution for a unit load in an infinite plane is used for the kernels in equation (1). For near-surface problems, it is convenient to use a half-plane solution that automatically includes the traction-free condition at the surface of the half-plane (Telles and Brebbia (1981)). This avoids the need to discretize the portion of the free surface where the traction-free condition is actually met, as in the case depicted in Figure 1(b).

As the half-plane solution can be expressed as the sum of the infinite plane problem and a complementary part, it is easy to include in the code the possibility of specifying the type of geometric problem (i.e. infinite body or half-plane) to be solved.

STIFFNESS MATRIX

The coupling of boundary elements and distinct elements is accomplished by means of a stiffness matrix relating nodal loads and displacements along the interface between the respective solution domains.

Nodal loads can be defined by integration of tractions on the adjacent elements and are given, for a node p, by Brady and Wassyng (1981):

$$q_i^p = a^p \, t_i^p \quad \text{(p no sum)} \tag{4}$$

where the scalar a^p is a function of the local geometry. Equation (3) can then be rewritten as

$$[T]\,\{u\} = [U']\,\{q\} \tag{5}$$

where $[U']$ is obtained from $[U]$ by dividing each column by the appropriate a^p factor. By inversion,

$$\{q\} = [U']^{-1}\,[T]\,\{u\} = [K]\,\{u\} \tag{6}$$

and $[K]$ is the required stiffness matrix of the boundary element domain.

DISTINCT ELEMENT FORMULATION

The Distinct Element Method models a rock mass as an assembly of discrete blocks. Interaction forces between blocks are obtained as a function of the displacements of the block corners relative to the edges of the surrounding blocks that they contact, by application of the assumed constitutive equations for the joints.

Initially, the method used rigid blocks, but new versions allow the blocks to deform and to crack and split. The present hybrid model includes a simple formulation for block deformability, which corresponds to the assumption of constant stress within a block (Cundall et al. (1978)). A more elaborate formulation exists, where blocks are discretized by means of a finite-difference mesh, thus allowing a rather general deformability.

The algorithm of the Distinct Element Method is based on the explicit integration in time of the equations of motion. At each time step, the accelerations of the block centroids are obtained, as in the rigid block model, from the balance of forces and moments, given the mass and moment of inertia of each block. The introduction of block deformability is achieved by defining deformation accelerations

$$\ddot{\varepsilon}_{ij} = \frac{\sigma_{ij}^A - \sigma_{ij}^I}{m^e_{(ij)}} \tag{7}$$

where σ_{ij}^A are average stresses equivalent to the forces applied to the block,

σ_{ij}^{I} are internal stresses obtained from the strains, and $m^{e}_{(ij)}$ are effective masses, evaluated in such a way that, for rigid joints, the block assembly displays the correct wave propagation velocities. Integrating equation (7) produces strain increments. Then, use of the constitutive equations for the block material gives the increments of internal stresses. Also, new corner and edge locations can be found, and contact forces between blocks can be updated.

The method can be used to obtain static solutions, by inclusion of damping coefficients, as a dynamic relaxation technique. The procedure outlined enables the method to include, without significant computational effort, large displacements and general nonlinear material behaviour, both for the intact rock and the discontinuities.

THE COUPLING PROCEDURE

As seen in Figure 1, the boundary element nodes along the interface are made to correspond with block corners. Kinematic continuity is approximately ensured by identifying each boundary element nodal displacement with the average of the displacements of the corners of blocks that contact that node. The essential assumption behind the hybrid model is that the nonlinear zone must be confined within the block domain. Therefore, contacts between blocks along the interface should remain elastic and this averaging procedure is a reasonable assumption, as the results given subsequently confirm.

At each time-step, the blocks will define the interface displacements. Using the stiffness matrix of equation (6), the interface reaction forces can be calculated and applied to the block corners along the interface. In this way, the blocks are constrained by the stiffness of the surrounding elastic region. After calculation of the behaviour of the system of blocks, the final displacements of the boundary nodal points can be input to the B.E. part of the program. The resulting displacements and stresses at specified points within the elastic region may then be evaluated.

VERIFICATION AND DEMONSTRATION PROBLEMS

The validity of the hybrid code has been demonstrated by Lemos and Brady (1983) by means of an analysis of a strip load on an elastic half-plane. The region in the vicinity of the load was modelled with a system of triangular distinct elements which were coupled to a semi-infinite region using boundary elements. The interfaces between distinct elements were given high stiffness and neither slip nor separation was allowed, so that the block region also approximated an elastic continuum. Despite the coarse block and boundary element discretisation that was employed, the hybrid model yielded quite good agreement with analytical expressions for elastic displacements and stresses under the load. Results of another analysis, with rectangular blocks in the distinct element domain (Lemos, 1983), differed markedly from the analytic solution. This reflects limitations of the simply-deformable block model used in the distinct element part of the code - an assembly of pairs of triangular blocks has more degrees of freedom, and hence can respond more accurately to applied loads, than an equivalent assembly of rectangular blocks.

An underground excavation near the surface of a half-plane has been analysed in several ways to further check the operation of the hybrid code and demonstrate its potential. The boundary element part of the code was tested first, using the mesh in Figure 2(a). The elastic properties and pre-mining stress field used in the analysis are given in the Figure. Some results from this analysis are compared with corresponding results from an analysis with a different boundary element program, BITEMJ (Crotty (1983)), in Table 1. Program BITEMJ also uses a direct formulation, with linear boundary elements, but does not include a half-plane

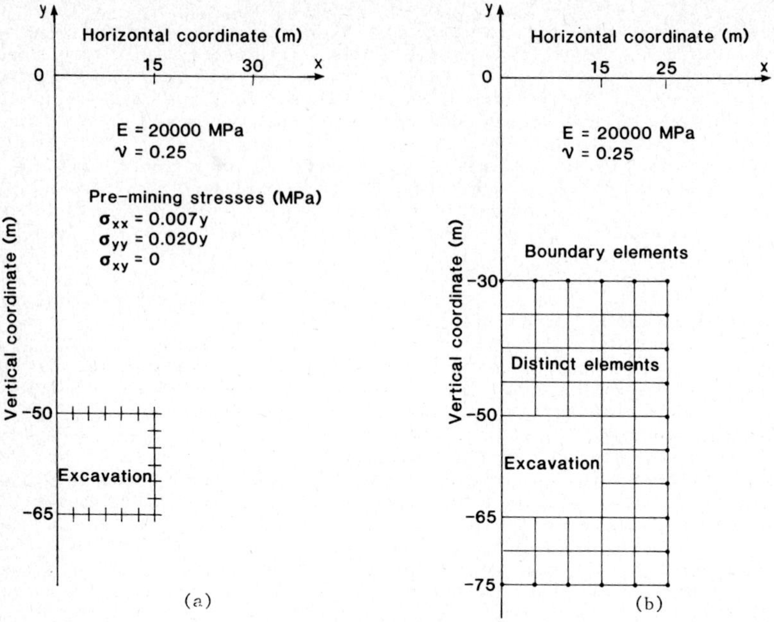

Figure 2
Computational grids for analysis of simple near-surface excavation.
(a) boundary element analysis;
(b) hybrid D.E.-B.E. analysis.
In each case, y=0 is a free surface, and the problem is symmetrical about the y-axis.

Table 1

Comparison of Boundary Element Results for Near-surface Excavation (Figure 2(a))

		Hybrid code (this work)	BITEMJ (Crotty (1983))
Vertical displacement (mm)			
centre top (0,-50)		-1.39	-1.24
centre bottom (0,-65)		2.39	2.45
Stresses (MPa)			
at (10,-40):	σ_{xx}	-0.240	-0.259
	σ_{yy}	-0.386	-0.389
	σ_{xy}	0.401	0.403
at (20,-75):	σ_{xx}	-0.330	-0.334
	σ_{yy}	-1.695	-1.707
	σ_{xy}	-0.286	-0.276

solution kernel or a fixed, distant, point to avoid an indeterminacy in the solution of half-plane problems. Consequently, the free surface of the half-plane and a distant bounding region must also be discretized when it is used for this type of problem. Further, with BITEMJ, double nodes may be used at corners of boundaries to account for the discontinuities in normal and tangential tractions between neighbouring elements. Given these differences between the programs, the two sets of results are in good agreement and serve to confirm the validity of the boundary element part of the hybrid code.

Similar analyses have been performed with the excavation being made in a region of distinct elements composed entirely of square blocks (Figure 2(b)). Ideally, forces equivalent to the assumed pre-excavation stress field should first be applied to the outside of the (non-excavated) block region. After equilibrium is reached, coupling with the boundary elements would be completed, the excavation created by deletion of blocks, and the system allowed to relax. The incremental displacements in the second stage would reflect ground movements induced by the excavation. However, the present version of the code does not allow for deletion of blocks, so the analysis has been performed in a single stage. This procedure ignores the pre-mining stress field and considers the movement of the illustrated block system under gravity, as constrained by the surrounding elastic region.

To enable comparison with the elastic solutions above, the joints between blocks were initially ascribed very high stiffnesses and strength. Calculated deformations of the block system were of the same order as the boundary element results in Table 1, but otherwise showed no correlation with them. This is only to be expected, given the different initial conditions for the two analyses.

In the next analysis, the horizontal and vertical joint planes defining the blocks were all assumed to be of low strength. Massive movement of the roof material resulted, with shear failure propagating upwards from the upper corners of the excavation, as shown in Figure 3(a). The uppermost row of blocks in the roof was constrained by connection, via boundary elements, to the elastic half-space.

Finally, the shear strength of the vertical joints in the roof was increased considerably, but with their tensile strength remaining low. None of the other joints was altered. In this case, each layer of blocks in the roof yielded as a hinged beam, with a tensile crack opening from the base of the centre of the span, and similar cracks opening at the top of the "beam", above the corners of the roof. This behaviour is illustrated in Figure 3(b).

The two modes of behaviour shown in Figure 3 are precisely those usually considered in simple models of the failure of roofs of excavations in bedded rock (see, for example, Beer and Meek (1982)). The first mode follows from shear failure of the roof rock and the second, which corresponds to a 'voussoir' beam model (Evans (1941)), is initiated by tension cracking.

Whilst these mechanisms of behaviour have previously been demonstrated with a simple blocky model (Voegele, 1979), the hybrid code has the added capacity to include the far-field elastic response of the rock mass and so, potentially, be able to model real systems more accurately. In this context, it is noted that the elastic confinement across the upper surface of the system of blocks was found to be essential in allowing the 'voussoir' beam mechanism to develop - if the vertical joints continued to the surface, the shear mode of failure occurred even when those joints were given greatly increased strength.

One of the advantages of a hybrid model such as this over the more common continuum methods for analysing faulted rock is illustrated by one further calculation. Program BITEMJ (Crotty (1983)) can also model major faults in rock by means of explicit joint elements embedded in an elastic continuum. Results from analysis of a system akin to that modelled with the hybrid code are shown in Figure 4. The "beams" in the roof deform elastically and separate as shear takes

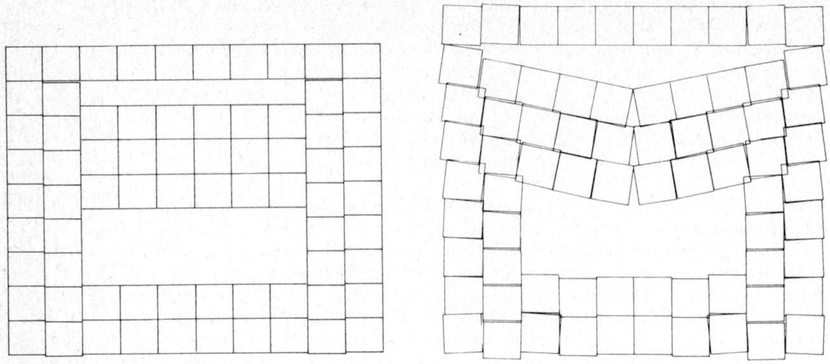

(a) (b)

Figure 3
Modes of deformation of jointed roof rock of near-surface excavation, as predicted by hybrid model.
(a) Horizontal and vertical joints weak - shear failure;
(b) Increased shear strength in vertical joints above excavation - 'voussoir' beam failure.
Displacements magnified 40 and 750 times respectively.

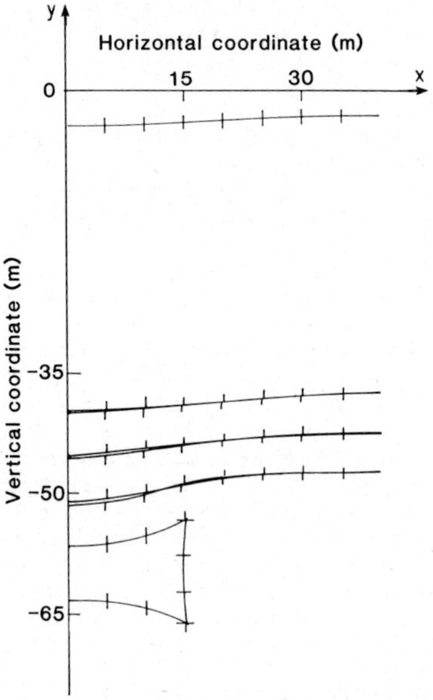

Figure 4
Deformation of roof containing three discrete faults, as calculated with boundary element program BITEMJ.
Displacements magnified 2000 times.

place on the joint planes over the abutments. However, the development of ultimate failure of these beams could only be modelled in this program, or in a similar finite element program (e.g. Beer et al. (1983)), by inclusion of a complex nonlinear material model for the rock. The distinct elements in the hybrid code can treat the full failure mechanism automatically and in a completely natural way.

CONCLUSIONS

The theoretical background to the development of a hybrid boundary element - distinct element code has been oultined. Some of its capabilities have been demonstrated in simple analyses of the behaviour of the roof of a near-surface excavation in highly jointed rock. The present version of the computer program is only experimental, and will need to be extended to allow analysis of larger systems, pre-mining stresses and more complex block deformations before it can be used for reliable predictive analyses of proposed excavations in rock. However, it is apparent that this type of hybrid model has the potential to become an important tool for assisting engineers to design larger and safer excavations in highly jointed rock masses.

ACKNOWLEDGEMENTS

Bruce Perkins converted the original version of the code (Lemos, 1983) for operation on a Hewlett-Packard 1000/65 minicomputer at CSIRO, and he and Russell MacKinnon modifed the code to allow semi-automatic generation of distinct elements and plotting of displaced systems of blocks. Their assistance is greatly appreciated.

REFERENCES

Beer, G. and Meek, J.L., Design curves for roofs and hanging-walls in bedded rock based on 'voussoir' beam and plate solutions, Trans. Instn Min. Metall. (Sect. A: Min. industry) 91 (1982) A18-A22.

Beer, G., Meek, J.L. and Cowling, R., Prediction of the behaviour of shale hanging walls in deep underground excavations, Proc. 5th Int. Congr. on Rock Mech., D45-D51 (Melbourne, 1983).

Brady, B.H.G. and Bray, J.W., The boundary element method for determining stresses and displacements around long openings in a triaxial stress field, Int. J. Rock Mech. Min. Sci. & Geomech. Abstr. 15 (1978) 21-28.

Brady, B.H.G. and Wassyng, A., A coupled finite element - boundary element method of stress analysis, Int. J. Rock Mech. Min. Sci. & Geomech. Abstr. 18 (1981) 475-485.

Crotty, J.M., User's manual for program BITEMJ - Two-dimensional stress analysis for piecewise homogeneous solids with structural discontinuities, Geomechanics Computer Program No. 5, CSIRO, Division of Applied Geomechanics (1983).

Cundall, P.A., Marti, J., Beresford, P.J., Last, N.C. and Asgian, M.I., Computer modelling of jointed rock masses, Technical Report N-78-4, U.S. Army Engineer Waterways Experimental Station, Vicksburg, Miss. (1978).

Evans, W.H., The strength of undermined strata, Trans. Instn Min. Metall., 50 (1941) 475-532.

Lachat, J.C. and Watson, J.O., Effective numerical treatment of boundary integral equations: A formulation for three-dimensional elastostatics, Int. J. Numer. Meth. Engng. 10 (1976) 991-1005.

Lemos, J.V., A hybrid distinct element - boundary element computational model for the half-plane, M.S. Thesis, Dept. of Civil and Mineral Engng., Univ. of Minnesota (1983).

Lemos, J.V. and Brady, B.H.G., Stress distribution in a jointed and fractured medium, Proc. 24th U.S. Symp. on Rock Mech., 53-59 (1983).

Owen, D.R.J. and Hinton, E., Finite elements in plasticity: theory and practice (Pineridge Press, Swansea, 1980).

Telles, J.C.F. and Brebbia, C.A., Boundary element solutions for half-plane problems, Int. J. Solids Struct. 17 (1981) 1149-1158.

Voegele, M.P., Rational design of tunnel supports: an interactive graphics based analysis of the support requirements of excavations in jointed rock masses, Ph.D. Thesis, Dept. of Civil and Mineral Engng., Univ. of Minnesota (1979).

Watson, J.O., Advanced implementation of the boundary integral equation method for two- and three-dimensional elastostatics, in: Banerjee, P.K. and Butterfield, R. (eds.), Developments in Boundary Element Methods, (Applied Science Publishers, London, 1979).

Computational Techniques & Applications: CTAC-83
J. Noye & C. Fletcher (Editors)

THE GROWTH OF FINGERS BETWEEN TWO IMMISCIBLE FLUIDS IN A TWO-DIMENSIONAL POROUS MEDIUM OR HELE-SHAW CELL

Malcolm R. Davidson

CSIRO Division of Mineral Physics
Lucas Heights Research Laboratories
Private Mail Bag 7, Sutherland, NSW.
Australia

A stepwise numerical procedure is described for calculating the motion of an unstable periodic interface (from some initial configuration) between two immiscible fluids flowing in a two-dimensional porous medium or Hele-Shaw cell when one fluid displaces the other, interfacial velocity at each time step being derived as a numerical solution of a boundary integral equation. The growth of numerical errors is discussed and it is shown how the usable part of the calculation can be extended. Numerical results are compared with an exact solution.

INTRODUCTION

Unstable displacement of one fluid by another in a porous medium is characterised by the development of long fingers of displacing fluid which penetrate the displaced fluid region. Fingering can occur, for example, when flow is directed from the less viscous to the more viscous fluid, provided the velocity is large enough. This phenomenon is of particular interest in the oil industry where it reduces the efficiency of oil recovery (e.g. during water or gas drive). It may also be encountered in the fields of groundwater hydrology (saltwater-freshwater interface in coastal aquifers) and soil sciences (infiltration flows).

Saffman and Taylor (1958) and Chuoke et al. (1959) derived conditions governing the onset of instability of an initially plane interface between two immiscible fluids flowing in a two-dimensional porous channel or Hele-Shaw cell. They used a first order perturbation analysis which is valid for asymptotically small amplitudes to show that, for a small displacement,

$$y = \varepsilon \exp(inx + \sigma t)$$

of the undisturbed interface $y = 0$,

$$(\mu_1+\mu_2)\frac{\sigma}{n} = (\mu_1-\mu_2)V + kg(\rho_1-\rho_2)-n^2\gamma k \quad . \tag{1}$$

Hence, this disturbance is unstable ($\sigma > 0$) when its wavelength ($\lambda = \frac{2\pi}{n}$) is greater than the critical value λ_c, where

$$\lambda_c^2 = 4\pi^2\gamma/(V(\mu_1-\mu_2)/k + g(\rho_1-\rho_2)) \tag{2}$$

whenever the denominator is positive. Motion is upward with uniform velocity V, and (x,y) are rectangular coordinates, with y the upward vertical. Suffix 1 refers to the upper fluid and suffix 2 to the lower, and μ, ρ, γ denote fluid viscosity, density and interfacial tension, respectively. The permeability of the medium is taken to be the same (k) for both fluids and g is the acceleration due to gravity.

Few exact solutions of developing fingers exist (Saffman, 1959; Jacquard and Seguier, 1962) and it would be useful to integrate finger solutions numerically to finite amplitudes.

In a recent paper, Davidson (1983a) derived an integral equation for the normal velocity of the interface in terms of the physical parameters, a Green's function and the interface itself. For stable flows, in which small perturbations of the interface decay with time, a stepwise numerical procedure could, in principle, be used to determine completely the evolution of the interface, from some initial configuration, by progressively solving the integral equation for the normal velocity at each point on the interface at each time step. In this paper, the above numerical method is applied to the more physically interesting case of unstable flow. In that case, numerical perturbations of the interface must grow, along with the desired solution, regardless of the computational method. The objective is to minimise the error growth rate and thus extend the usable part of the calculation.

THE PROBLEM

Consider an infinitely long, two-dimensional vertical porous channel of width L (Figure 1) in which one fluid is being displaced upwards by another. Darcy's Law for each fluid (i = 1,2) is

$$u_i = -\frac{k}{\mu_i}\frac{\partial p_i}{\partial x} \quad , \quad v_i = -\frac{k}{\mu_i}\left(\frac{\partial p_i}{\partial y} + \rho_i g\right) \qquad (3)$$

and hence, when the densities are constants, the continuity equation gives

$$\nabla^2 p_i = 0 \qquad (4)$$

where p_i is the pressure and u_i, v_i are the velocity components in the x,y directions, respectively. In the case of a Hele-Shaw cell, the thickness-averaged flow equations are mathematically analogous to Darcy's Law.

The macroscopic interface (denoted by C) separating the two fluids is assumed to be sharp and periodic, where the channel width L defines a half period (i.e. x = 0 and x = L are lines of symmetry). Across this interface the normal velocity (U_n) is continuous, so that

$$U_n = -\frac{k}{\mu_1}\left(\frac{\partial p_1}{\partial n} + \rho_1 g\cos\theta\right) = -\frac{k}{\mu_2}\left(\frac{\partial p_2}{\partial n} + \rho_2 g\cos\theta\right) \qquad (5)$$

and there is a pressure jump chosen with the form

$$p_2 - p_1 = \gamma H(x,y) + P_c \quad . \qquad (6)$$

Here, the normal n is directed from the lower to the upper fluid, θ denotes the angle between the normal and the positive y axis, H is the macroscopic curvature of the interface and P_c is a constant "capillary" pressure associated with the microscopic interfaces underlying C.

At the sides of the channel, we require $u_i = 0$, hence

$$\frac{\partial p_i}{\partial x} = 0 \quad \text{for} \quad x = 0 \text{ and } x = L \quad , \qquad (7)$$

and the motion at large distances upstream and downstream of the interface is taken to be uniform with velocity V so that

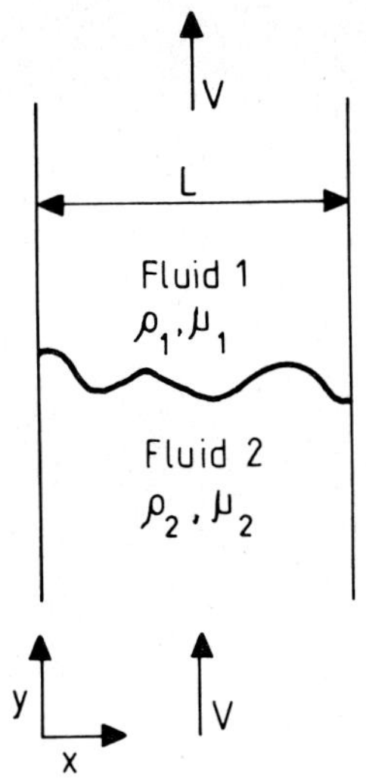

Figure 1
Schematic representation of the vertical displacement of fluid 1 (density ρ_1 and viscosity μ_1) by fluid 2 (density ρ_2 and viscosity μ_2) in a two-dimensional porous medium or Hele-Shaw cell of width L, where the interface is assumed to be periodic and V is the velocity at infinity. (Davidson, 1983a)

$$\frac{\partial p_1}{\partial y} = -\left(\frac{\mu_1 V}{k} + \rho_1 g\right) \quad , \quad y \to \infty$$

$$\frac{\partial p_2}{\partial y} = -\left(\frac{\mu_2 V}{k} + \rho_2 g\right) \quad , \quad y \to -\infty \quad .$$

The flow is specified by equation (4) subject to conditions (5)-(8).

Davidson (1983a) used the Green's function

$$G(x,y;\zeta,\eta) = \frac{1}{4\pi} \log(\cosh \tfrac{\pi}{L}(y-\eta)-\cos\tfrac{\pi}{L}(x+\zeta))(\cosh\tfrac{\pi}{L}(y-\eta)-\cos\tfrac{\pi}{L}(x-\zeta)) \quad (9)$$

which satisfies condition (7), to derive the following integral equation for the normal velocity of the interface:

$$U_n(X) = V\cos\theta(X) + 2\int_C F(\xi)\,\frac{\partial G(X;\xi)}{\partial n(X)}\,ds(\xi)$$

$$\frac{-2\gamma k}{\mu_1+\mu_2}\left(\int_C M(X;\xi)ds(\xi) - \frac{dH(X)}{ds}\left(G(X;\xi)\right)_{\zeta=0}^{L}\right) \quad , \quad (10)$$

with $F(\xi) = \left(\frac{\mu_1-\mu_2}{\mu_1+\mu_2}\right)U_n(\xi) + kg\left(\frac{\rho_1-\rho_2}{\mu_1+\mu_2}\right)\cos\theta(\xi)$

and $M(X;\xi) = (H(\xi)-H(X))\dfrac{\partial^2 G}{\partial n(X)\partial n(\xi)} + \dfrac{dH(X)}{ds}\dfrac{\partial G}{\partial s(\xi)}$

where points $X = (x,y)$, $\xi = (\zeta,\eta)$ and s denotes arc length. By symmetry about the channel walls, $\theta = dH/ds = 0$ at the end points of C.

An alternative form of equation (10) may be derived (Davidson, 1983b) as

$$U_n(X) = V\cos\theta(X) + 2\int_C \left(F(\xi)\frac{\partial G}{\partial n(X)} - F(X)\frac{\partial G}{\partial n(\xi)}\right)ds(\xi)$$

$$+ \frac{2\gamma k}{\mu_1+\mu_2}\int_C Q(X;\xi)ds(\xi)$$

$$\text{where } Q(X;\xi) = \frac{dH(X)/ds}{\cos\theta(X)}\left(\cos\theta(\xi)\frac{\partial G}{\partial s(X)} - (\eta-y)\frac{\partial^2 G}{\partial s(X)\partial n(\xi)}\right)$$

$$- (H(\xi)-H(X))\frac{\partial^2 G}{\partial n(X)\partial n(\xi)} \tag{11}$$

When $\xi = X$,

$$\frac{\partial G}{\partial n(X)} = \frac{\partial G}{\partial n(\xi)} = \frac{H}{4\pi} - \frac{\sin\theta}{4L}\cot\frac{\pi x}{L}$$

and $$Q(X;X) = \frac{1}{4\pi}(H_{xx}\cos^2\theta - H H_x \sin\theta) + \frac{1}{4L} H_x \cot\frac{\pi x}{L}\cos^2\theta$$

If the interface is described by a single valued function $y(x,t)$, then its motion may be expressed as

$$\partial y/\partial t = U_n/\cos\theta \tag{12}$$

and followed along lines of constant x. Given an initial displacement of the interface, the objective is to use equations (11) and (12) to follow numerically the position of the interface with time for as long as the growth of numerical errors permits. For reasons to be discussed in the next section, it is important to use equation (11) rather than equation (10) during time stepping. A similar treatment should be possible in the more general case when C is expressed parametrically, by following the motion of points $(x(s,t),y(s,t))$ on C.

THE NUMERICAL METHOD

We solve equations (11) and (12) along equally spaced lines $x = x_j = j\Delta x$ $(j = 0,1,\ldots,N)$. On each line $x = x_j$, we use the Adams-Bashforth-Moulton scheme (see Hamming, 1962), applied to equation (12), to advance through time with local errors of order $(\Delta t)^5$ where Δt is the time step. Then

$$y_{1p} = y_0 + \frac{\Delta t}{24}(55\, y'_0 - 59\, y'_{-1} + 37\, y'_{-2} - 9\, y'_{-3})$$

and $$y_{1c} = y_0 + \frac{\Delta t}{24}(9\, y'_{1p} + 19 y'_0 - 5\, y'_{-1} + y'_{-2})$$

where ' denotes $\partial/\partial t$, y_k denotes $y(x_j, t+k\Delta t)$ and y_{1p}, y_{1c} denote predicted and corrected values of y_1. A fourth order Runge-Kutta formula is used to start the process.

At each time step, quadrature of the integral in equation (11) (in terms of variable x) using Simpson's rule followed by collocation at $x = x_j$, gives a set of simultaneous linear algebraic equations, having errors of order $(\Delta x)^4$, for the values of U_n. When $\gamma = 0$, the form of the first integral in equation (11) avoids the need to determine H and also numerically conserves mass, following quadrature of the integral. When $\gamma \neq 0$, both the slope and the curvature, together with its first two derivatives, are required to be determined.

Davidson (1983a) solved equation (10) for U_n corresponding to example interfaces which were known analytically. However, during time stepping, the interface is only known approximately, and the term $\cot\pi x/L$ in $M(X;X)$ and $\partial G/\partial n(X)$ and the logarithmic singularity in $(G(X;\xi))^L_{\zeta=0}$ can lead to the catastrophic growth of errors near the side walls if equation (10) is used. Equation (11) eliminates this difficulty in the first and last terms of equation (10); however, the term $\cot\pi x/L$ in $Q(X;X)$ remains. This suggests a truncated Fourier series approximation for H_x, so that, if

$$H_x = \frac{-\pi}{L} \sum_{k=1}^{K} k\, a_k \sin \frac{k\pi x}{L} \tag{13}$$

where
$$a_k = \frac{2}{L} \int_0^L H(x) \cos \frac{k\pi x}{L}\, dx , \tag{14}$$

then the product $H_x \cot\pi x/L$ in $Q(X;X)$ becomes

$$-\frac{\pi}{L} \cos \frac{\pi x}{L} \sum_{k=1}^{K} k\, a_k\, f_k(x) , \tag{15}$$

where the functions

$$f_k(x) = \sin(k\pi x/L)/\sin(\pi x/L)$$

can be evaluated in several forms which avoid any computational difficulty near $x = 0$ or L (see e.g. Gradshteyn and Ryzhik, 1980; p.27). One such form is

$$f_k(x) = \sum_{i=1}^{I} (-1)^{i-1} \binom{k-i}{i-1} \left(2 \cos \frac{\pi x}{L}\right)^{k-2i+1} \tag{16}$$

where
$$I = \tfrac{1}{2}(k+1) , \quad k \text{ odd}$$
$$= \tfrac{1}{2}k , \quad k \text{ even} .$$

We choose here to calculate the slope and the curvature of the interface at each time step using cubic splines. The recommended procedure (Ahlberg et al., 1967) is to first derive the slope by spline-fitting $y(x_j)$ and then to spline fit the slopes themselves and use the resulting derivatives as the required $y_{xx}(x_j)$ for determining the curvature $H(x_j)$.

Having obtained the curvature, we integrate equation (14) numerically (Simpson's rule) to give the Fourier coefficients a_k. Derivatives of curvature H_x and H_{xx} are now estimated from equation (13), and it is also important to re-evaluate $H(x_j)$ according to

$$H(x) = \tfrac{1}{2}a_0 + \sum_{k=1}^{K} a_k \cos \frac{k\pi x}{L} \tag{17}$$

so that the values for H and its derivatives used in the evaluation of

$Q(X;\xi)$ are consistent. Failure to do this results in the rapid growth of errors arising from the cancellation of the singularities near $\xi = X$ in the terms involving $\partial G/\partial s(X)$ and $\partial^2 G/\partial n(X)\partial n(\xi)$. Other methods of estimating H and its derivatives, involving the repeated use of cubic and quintic splines, proved unsucessful.

During unstable flows, the rapid growth of numerical errors results in a calculated solution which developes a characteristic saw-toothed appearance. Here we draw on the numerical experiences of Longuet-Higgins and Cokelet (1976) who calculated the motion of surface waves on water. These authors encountered a weak instability in their solution which they removed by smoothing; the exact curve was presumed to lie midway between two smooth curves passing through alternate points. In this paper we approximate the bounding upper and lower smooth curves separately by cubic splines, and carry out the smoothing at every time step.

An attempt to combine the elimination of high frequency error modes of y with a computationally suitable form for $H_x \cot \pi x/L$ by approximating y with a truncated Fourier series met with some success, but the method described here is more accurate and integrates the solution further.

RESULTS AND DISCUSSION

Saffman (1959) considered the motion of a viscous fluid which is driven by a fluid of negligible viscosity ($\mu_2 = 0$) in the absence of gravity, and derived a family of exact solutions for the interface which are approximately sinusoidal at small times and are asymptotically equal to the Saffman-Taylor profile at large times. These solutions may be used to gauge directly the accuracy of our numerical procedure when $\gamma = 0$. Unfortunately, the author is unaware of any similar exact solution corresponding to $\gamma \neq 0$.

In Figure 2, numerical solutions for $\gamma = 0$ are compared with the corresponding exact Saffman profiles showing the development of the instabilty in the absence of smoothing and the subsequent improvement in the solution when smoothing is applied. In the latter case, numerical errors still grow (as they must) but they do so much more slowly than before so that the solution may be integrated further. Similar behaviour occurs when $\gamma \neq 0$.

The difference $|y_{smooth}-y|_{max}$ is an index of the error growth in the numerical solution and, of course, it increases with time for unstable flows. Alternatively, when $\gamma \neq 0$ we could monitor the error in the calculated mass balance which follows a similar pattern with time. However, when $\gamma = 0$, mass is conserved numerically. The above comparison between the numerical and exact Saffman solutions for $\gamma = 0$ indicates that they are almost indistinguishable when

$$|y_{smooth}-y|_{max} < 0.4 \times 10^{-4}$$

We now apply this criterion to define a valid calculation when $\gamma \neq 0$.

Figure 3 illustrates the effect of changes in Δx and Δt for an initial cosine interface. In the absence of smoothing, halving Δx from 0.05 to 0.025 accelerates the onset of the instability. However, when smoothing is applied, Figure 3 shows that such a reduction in Δx reduces the error as does halving Δt from 0.003125 to 0.0015625. It also illustrates the error reduction which can be achieved by increasing the number of terms K (from 5 to 10) in the Fourier approximation of curvature and its derivatives (equations (13),(15),(17)) when $\gamma \neq 0$. (Note that we must have $\Delta x < L/K$

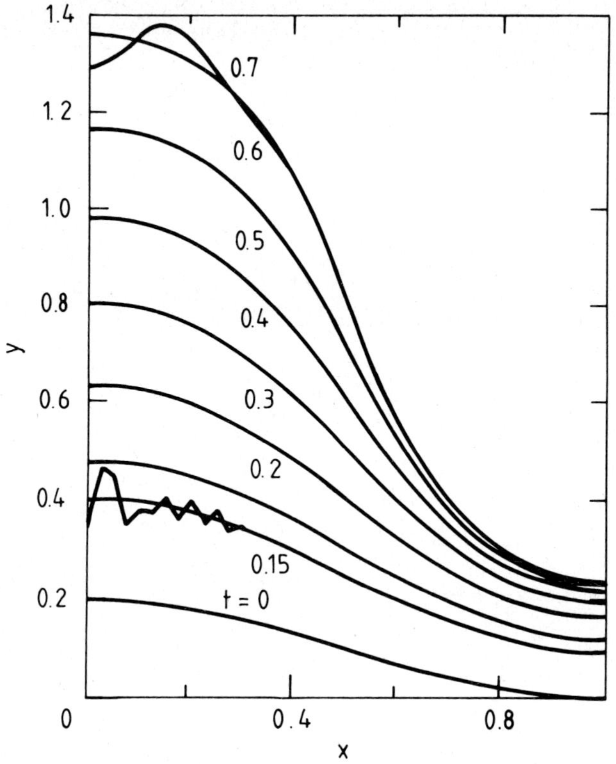

Figure 2
Comparison of exact and numerical solution (with and without smoothing) for an initial Saffman interface with amplitude 0.1, where V = L = 1, $\mu_2 = 0$, gravity is ignored and $\gamma = 0$. (Davidson, 1983b).

for the computation to resolve the Kth Fourier component).

In Figure 3, the physical parameters are in the unstable range (specifically, $\lambda_c/\lambda = \pi(0.02)^{\frac{1}{2}} < 1$). However with increasing λ_c/λ we found that, when K = 10, smaller values of Δt were required to achieve error growth rates similar to those in Figure 3. Furthermore, for physically stable wavelengths given by $\lambda_c/\lambda > 1$, there appears to be a restriction on Δt (perhaps analogous to a Courant condition) which ensures that the error decreases with time (note that, when $\lambda_c/\lambda > 1$, smoothing at each time step remains necessary, indicating that the basic formulation is still computationally unstable; however, in stable regimes which occur when the denominator of equation (2) is negative, the calculation is also stable). The (unknown) bound on Δt presumably depends on Δx, K and the physical parameters. One can speculate that this same condition on Δt, together with a formulation

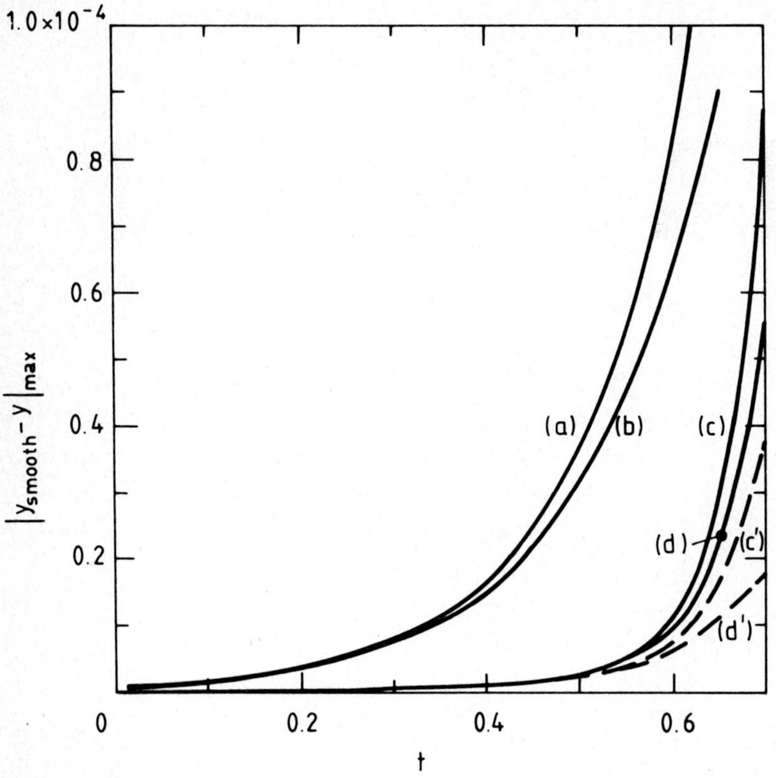

Figure 3
Error index vs time when V = L = 1, $\mu_2 = 0$, gravity is ignored and $\gamma k/\mu_1 = 0.02$ for an initial interface y = 0.1 cosπx. Solid curves (a)-(d) correspond to a choice of K = 5 Fourier components describing the curvature and to the following choices of Δx and Δt : (a) $\Delta x = 0.05$, $\Delta t = 0.003125$, (b) $\Delta x = 0.05$, $\Delta t = 0.0015625$, (c) $\Delta x = 0.025$, $\Delta t = 0.003125$, (d) $\Delta x = 0.025$, $\Delta t = 0.0015625$. The dashed curves (c') and (d') correspond to curves (c) and (d), respectively, with K = 10. (Davidson, 1983b).

(if such exists) which is computationally stable when $\lambda_c/\lambda > 1$, would also ensure "minimum" error growth rates of physically unstable wavelengths.

In Figure 4, the numerical solution for an initial displacement y = εcosπx/L is compared with the corresponding first order perturbation solution which has an error of only $O(\varepsilon)^3$ rather than the expected $O(\varepsilon)^2$ (Outmans, 1962). The two solutions are almost identical when t = 0.2 but differences occur at larger times when a perturbation analysis becomes inapplicable. In each case, the amplitude growth rate is less than that predicted by the linear theory (a result also shown by Outmans (1962) based on the inclusion of higher order terms) and

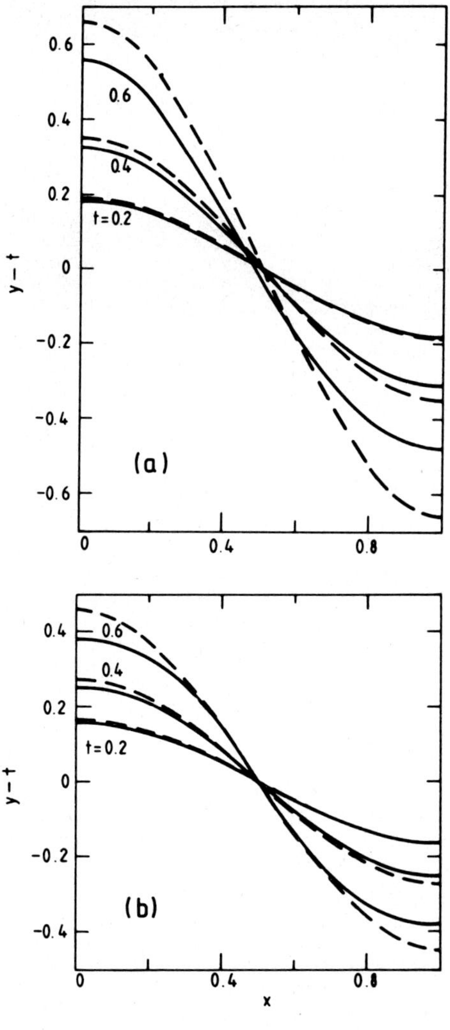

Figure 4
Comparison of numerical solutions (solid curves) and first order perturbation solutions (dashed curves) for an initial interface $y = 0.1 \cos \pi x$ when $V = L = 1$, $\mu_2 = 0$ and gravity is ignored. Results are plotted relative to the mean flow position. In (a) $\gamma = 0$ and in (b) $\gamma k/\mu_1 = 0.02$. (Davidson, 1983b).

the rate in Figure 4b ($\gamma \neq 0$) is less that that in Figure 4a ($\gamma = 0$) as is expected.

REFERENCES

[1] Ahlberg, J.H., Nilson, E.N. and Walsh, J.L. (1967). The Theory of Splines and Their Applications. Academic, New York.
[2] Chuoke, R.L., van Meurs, P. and van der Poel, C. (1959). The instability of slow, immiscible, viscous liquid-liquid displacement in permeable media. Trans. AIME 216, 188-194.
[3] Davidson, M.R. (1983a). An integral equation for immiscible fluid displacement in a two-dimensional porous medium or Hele-Shaw cell. J. Austral. Math. Soc. Ser. B, in press.
[4] Davidson, M.R. (1983b). Numerical simulation of unstable immiscible fluid displacement in a two-dimensional porous medium or Hele-Shaw Cell. Submitted to J. Austral. Math. Soc. Ser. B.
[5] Gradshteyn, I.S. and Ryzhik, I.M. (1980). Table of Integrals, Series and Products (enlarged edition). Academic, New York.
[6] Hamming, R.W. (1962). Numerical Methods for Scientists and Engineers. McGraw-Hill, New York.
[7] Jacquard, P. and Seguier P. (1962) Mouvement de deux fluides en contact dans un milieu poreux. J. de Mecanique 1, 367-394.
[8] Longuet-Higgins, M.S. and Cokelet, E.D. (1976). The deformation of steep surface waves on water. I. A numerical method of computation. Proc. Roy. Soc. A350, 1 - 26.
[9] Outmans, H.D. (1962). Nonlinear theory for frontal stability and viscous fingering in porous media. Soc. Pet. Eng. J. 2, 165 - 176.
[10] Saffman, P.G. and Sir Geoffrey Taylor (1958). The penetration of a fluid into a porous medium or Hele-Shaw cell containing a more viscous fluid. Proc. Roy. Soc. A245, 312 - 329.
[11] Saffman, P.G. (1959). Exact solutions for the growth of fingers from a flat interface between two fluids in a porous medium or Hele-Shaw cell. Quart. J. Mech. Appl. Math. 12, 146 - 150.

Computational Techniques & Applications: CTAC-83
J. Noye & C. Fletcher (Editors)

A BOUNDARY ELEMENT FORMULATION FOR DIFFUSER AUGMENTED WIND TURBINES

E. Zapletal and C.A.J. Fletcher

University of Sydney
Sydney, N.S.W. 2006
Australia

A boundary element formulation is presented for the flow through a diffuser augmented wind turbine. For the representative flow about a prolate spheroid it is shown that higher-order elements on a curved surface are computationally more efficient than constant strength elements on a flat surface. Preliminary results indicate that the turbine simulation is producing velocity fields that are qualitatively correct.

1 INTRODUCTION

Recent studies (Fletcher et al, 1979, 1983) have shown that the generation of electricity from jet-stream winds over Australia would be economically competitive with conventional power stations. The jet-stream winds are produced by a global focusing effect of solar energy. They are confined to two bands, one each in the northern and southern hemispheres, that circle the globe in a west to east direction. Over Australia at altitudes of 9-13 kms the average yearly wind speed is 35 m/s with peak speeds above 100 m/s. As an illustration of the magnitude of this energy resource it is estimated (Fletcher et al, 1979) that the average annual energy carried by the jet-stream above Australia is the equivalent of all of Australia's proven coal reserves (some 50 billion tonnes).

Jet stream energy extraction schemes envisage the use of tethered aerodynamic platforms carrying compact, efficient wind turbines. By enclosing a conventional wind turbine with a static diffuser the power extracted per unit area can be increased by 3 to 4 times. The diffuser achieves this by acting as a "ring-wing" and inducing a greater mass flow through the turbine than would otherwise be the case. Since small diameter turbines are inherently more reliable than large turbines the diffuser augmented wind turbine (DAWT) with its small turbine diameter for given power output is seen to be well suited to jet-stream energy extraction schemes.

Several experimental (Igra, 1977; Gilbert et al, 1978, 1979) and theoretical (Fletcher, 1981) analyses of DAWTs have been undertaken. The experimental studies consisted of wind tunnel tests on many different small scale models of DAWTs followed by more detailed larger scale tests on selected configurations. The theoretical studies have been essentially one-dimensional analyses similar to conventional turbine analysis methods combined with empirical parameters relevant to the particular diffuser.

The work presented in this paper is an attempt to increase the accuracy and speed of analysis of DAWT configurations by using the numerical boundary element (BE) method. The method used is similar to and based on the "panel" methods that have been used by the aerospace industry for the past two decades. It is for similar reasons, of ease of model redesign and speed of data reduction, that the numerical approach was adopted rather than experimental methods.

The method has been written and implemented on the University of Sydney's CDC Cyber 720 computer. The general theory behind the method is explained in section 2. A more detailed explanation of the workings of the basic method, together with some comparative results are presented in section 3. Refinements of the method used to model the turbine and boundary layers are presented in section 4.

2. THEORY

The strength of the boundary element formulation lies in its generality to fluid flows of arbitrary geometry. The only initial assumption of the flow field one makes is that it can be modelled by potential flow. Non-potential effects such as viscous boundary layers or turbulent wakes can be modelled by programs that alternately iterate with the pure potential flows.

The general problem being solved is the Laplace equation which governs potential flow. The Laplace equation can be used with two or three independent variables (x,y and/or z) and the boundary conditions specified can be of Dirichlet, Neumann or mixed types. Green's theorem then gives an integral equation formulation which is approximated at discrete points on the surfaces and solved numerically. The physical abstraction implied by Green's theorem is that the surfaces consist of varying source and doublet sheets.

The present work most closely follows the work of Hess (1975). It uses an axi-symmetric implementation for the surfaces with Neumann boundary conditions. It may be noted that although the formulation is three-dimensional, it is only the surface elements that must be axisymmetric. Onset flows perpendicular to the geometric axis can be calculated, and since solutions to Laplaces equation can be linearly superposed, it is possible to calculate onset flows at any angle of attack.

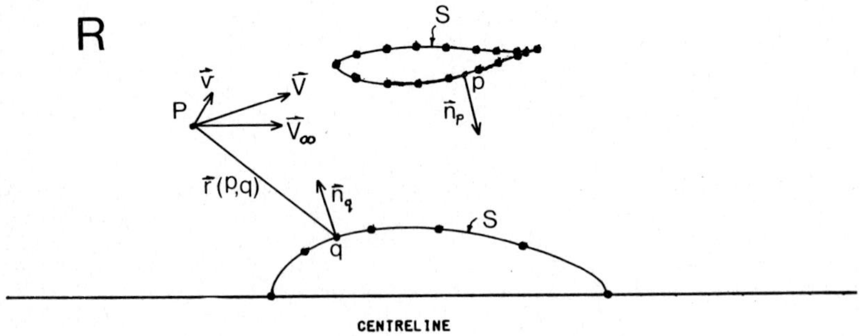

Figure 1
Potential Flow Representation

Fig. 1 shows a typical half-section of a configuration that has been studied. The ring wing and centrebody are discrete bodies each formed of several elements with a control point on each element. The region of interest is designated R. The fluid velocity at any point P can be expressed as

$$V = V_\infty + v \qquad (1)$$

where V_∞ is the onset flow (usually a uniform parallel stream) and v is the disturbance velocity field due to the surface S. Since the flow is considered incompressible V_∞ and v have zero divergence. It is assumed that v is irrotational so we may express v in terms of a potential function ϕ as,

$$v = -\operatorname{grad} \phi \qquad (2)$$

which then yields the Laplace equation,

$$\nabla^2 \phi = 0 \qquad (3)$$

in region R. A Neumann boundary condition on the surface S is defined as

$$\frac{\partial \phi}{\partial n} = \operatorname{grad} \phi . n = f(S), \qquad (4)$$

where f(S) is usually set equal to $V_\infty.n$. In the usual exterior problem the regularity condition

$$|\operatorname{grad} \phi| \to 0 \qquad (5)$$

applies at infinity.

For simply connected regions R, equations (3) to (5) comprise a well posed problem for the potential ϕ. Considering regions that are exterior to the surfaces S a unique solution can be obtained by representing the disturbance potential by a surface distribution of sources, alone, in the form

$$\phi(P) = \iint_S \frac{1}{r(P,q)} \sigma(q) \, dS \; , \qquad (6)$$

where $\sigma(q)$ denotes the source strength over S. Combining (6) with (4) and evaluating the velocity on the exterior of the surfaces yields the integral equation for $\sigma(q)$,

$$2\pi\sigma(p) - \iint_S \frac{\partial}{\partial n_p} \frac{1}{r(p,q)} \sigma(q) \, dS = - n_p . V_\infty \; . \qquad (7)$$

Once σ has been determined the perturbation velocity is obtained from

$$v(p) = - \iint_S \operatorname{grad}\left(\frac{1}{r(p,q)}\right) \sigma(q) \, dS \; . \qquad (8)$$

In regions R that are not simply connected (such as the ring wing cases here) equations (3) to (5) do not define a unique velocity field due to the possibility of arbitrary values of circulation in paths enclosing the surfaces. The non-zero circulations are associated with "lifting flows" and are dealt with here by setting the arbitrary value of circulation around any surface equal to zero. The circulation required to produce the lift is introduced as on onset flow. The disturbance velocity potential is thus rendered unique and can be combined linearly with solutions to uniform and other circulatory onset flows. The relative strengths of the circulatory onset flows are determined by the Kutta condition. There is one circulatory onset flow per body with a sharp trailing edge and the strengths of the onset flows are adjusted so as to yield velocities of equal magnitude on each side of a trailing edge.

Equation (7) is solved by representing the surface S by a total of N elements. Boundary conditions are set at N control points, one control point being associated with each element. In the usual case the boundary conditions imply that the flow is tangential to the surface at the control point.

We define V_{ij} as the velocity at the ith control point due to a unit source strength on the jth element. Letting n_i be the unit normal vector at the ith control point we define the normal component of V_{ij} as

$$A_{ij} = n_i.V_{ij} \quad . \tag{9}$$

The integral equation (7) now reduces to the set of linear equations

$$\sum_{j=1}^{N} A_{ij}\sigma_j = - n_i.V_\infty \ (i = 1,\ldots,N) \quad , \tag{10}$$

which is solved for the σ_j. The perturbation velocities at the control points are calculated by the numerical approximation of equation (8),

$$v_i = \sum_{j=1}^{N} V_{ij}\sigma_j \quad . \tag{11}$$

Once the σ_j values are calculated velocities at any off-body points can be found by a similar process to equation (11). The solution set of σ_j values thus fully defines the perturbation flow field.

3. GEOMETRY AND NUMERICAL INTEGRATION

In the simplest implementation of the method the boundary elements are flat (conical frustrums in this axisymmetric case) and the source density on each element is constant. The velocities of equation (11) can then be calculated to any desired accuracy and can be thought of as exact for the flat element, idealised model. The numerical model can be made to approach the integral equation formulation by increasing the number of elements. However since the solution time for the system of equations (10) is proportional to N^3 a more efficient procedure is to use a smaller number of elements and to allow the elements to be curved with varying source density, thus more accurately representing the particular geometry.

To implement the method used here only the end points of the elements together with some control flags are input to the program. The program then interpolates between element end points giving a polygonal profile that approximates a cubic curve. The vertices of the profile (8 per element) are stored for later use. Storing actual points on the curved elements rather than parameters defining curved element segments was chosen as it facilitates later programming and plotting routines.

When calculating the elements of the V_{ij} matrix it is assumed that each surface element has associated with it an unknown "average" value of source strength. The element source distribution can then be set to vary linearly or quadratically about the average value. The linear or quadratic variations across each element are functions of element geometry and neighbouring source strengths only and add no further unknowns to the problem.

Calculation time for the V_{ij} matrix is proportional to N^2 and hence a significant proportion of total running time. The components of the V_{ij} matrix are calculated by first subdividing the jth surface element into a number of sub-elements. If the ith control point is relatively close to the surface element than the number of sub-elements is made large (to a maximum of sixteen). When the ith control point is distant from the element then the whole element is considered to be the sub-element. The sub-element is then approximated by three source rings and a three-point Gaussian integration is performed. The effects of the individual source rings are calculated using elliptic integrals.

For the exterior flow case the resulting V_{ij} matrix is, in general, well-conditioned, full and non-symmetric. It has a large diagonal component, but is not strictly diagonally dominant.

There are therefore no problems with a direct solution of the system of linear equations (10) by Gaussian triangularization. The program uses a standard library subroutine available under the Cyber operating system to solve the system of linear equations.

Since the V_{ij} matrix is dependent only on the surface geometry intermediate solutions to the system of linear equations can be saved and multiple right hand sides calculated. This is useful when there are several onset flows such as perpendicular uniform flows and circulatory flows due to "lifting" surfaces. The resulting total calculation time is then considerably less than if each onset flow was calculated separately.

As a check on the relative efficiencies of the "flat" element and higher-order methods several analyses of ellipsoids were carried out and compared with exact analytic results. Figure 2 shows the results for two such cases. It is seen that the higher-order method with approximately half the number of elements is considerably more accurate than the basic method. The largest errors occur at the ends of the ellipsoid as is expected due to the large gradients of curvature and source density. Total program running time for the higher-order method was approximately half that of the basic method. Greater savings in running time would be expected for cases with larger numbers of elements due to the N^3 proportional component of computation required for the linear equation solutions.

4. BOUNDARY LAYER AND WAKE MODELLING

The method, as described so far, gives the potential velocity field due to surfaces such as the centrebody and diffuser sections. The turbine, however, has a considerable influence on the total flow field and its effect must be included to give an adequate analysis of the DAWT. Boundary layer effects are included at the same time to further refine the resulting velocity field.

The turbine and boundary layer effects are included by first analysing the flow around the centrebody and diffuser sections alone. The turbine blade geometry

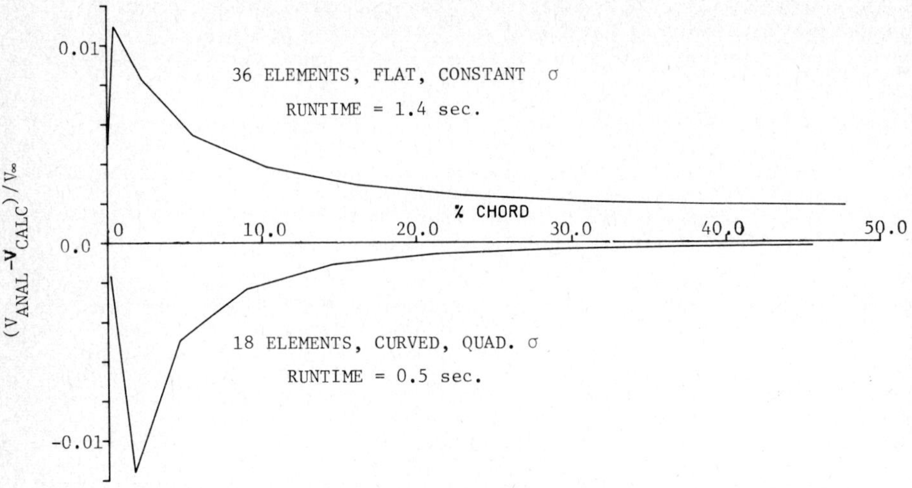

Figure 2
Element Accuracy Comparison for a Prolate Spheroid

(blade section, twist, angular velocity, etc.) is combined with this flow field to predict the strength and geometry of the turbine wake. Likewise the growth of the boundary layers on the centrebody and diffuser sections are predicted from the surface velocity profiles. The turbine wake is then approximated by a distribution of vortices that model pressure gradient effects on the turbine disc, and sources that model the momentum loss of the flow due to the extraction of power. The boundary layers are approximated by allowing a small outwards normal flow at the body surfaces. This results in a small net source value for each body which is equivalent to a small momentum loss in the flow.

The method of modelling momentum loss in the flow by introducing sources has the correct effects upstream and to the sides of the introduced source, but not directly downstream. The introduced source produces a semi-infinite body that extends downstream from the source point. The velocity field downstream of the source must be adjusted by eliminating the velocities within the semi-infinite body and then averaging the velocities on either side of the body to produce a smooth velocity profile.

The models of the turbine wake and boundary layers are combined with the original potential flow to produce a new resultant velocity field. The new velocity field leads to slight variations in the turbine and boundary layer wake models so the procedure must be iterated until convergence occurs. With cases that have been examined to date the velocities have stabilised after only two or three iterations.

Figure 3 shows the flow field around a DAWT with and without the turbine effect. As can be seen from the streamline pattern the turbine effectively blocks some of the flow through the diffuser. Since momentum is removed from the flow by the turbine and boundary layers the flow velocity downstream of the DAWT is less than the upstream freestream velocity, V_∞. This is apparent from the increased downstream diameter of the streamtube that encloses the DAWT.

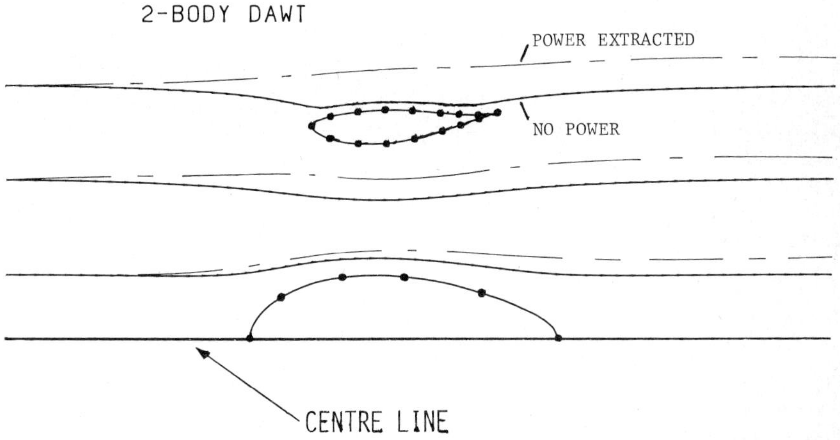

Figure 3
Turbine Influence on Velocity Distribution

5. CONCLUSION

A boundary element formulation of the flow through a diffuser augmented wind turbine (DAWT) has been constructed to provide rapid and comprehensive data generation for different DAWT configurations. Application of the present formulation to the flow about a prolate spheroid indicates that higher-order boundary elements on curved surfaces are computationally more efficient than constant strength elements on flat surfaces. Preliminary results for the simulation of a power-extracting turbine indicate that the correct qualitative trends of reduced flow through the turbine, compared with the non-power situation, do occur.

REFERENCES

(1) Fletcher, C.A.J. and Roberts, B.W., Electricity generation from jet-stream winds, J. Energy, 3 (1979), 241-249

(2) Fletcher, C.A.J., Honan, A.J. and Sapuppo, J.S., Aerodynamic platform comparison for jet-stream electricity generation, J. Energy, 7 (1983,) 17-23

(3) Igra, O., Compact shrouds for wind turbines, Energy Conversion, 16 (1977) 149-157

(4) Gilbert, B.L., Oman, R.A. and Foreman, K.M., Fluid dynamics of diffuser

augmented wind turbine concept, J. Energy, 2 (1978) 368-374

(5) Gilbert, B.L. and Foreman, K.M., Experimental investigation of the diffuser augmented wind turbine concept, J. Energy, 3 (1979) 235-240

(6) Fletcher, C.A.J., Computational analysis of diffuser-augmented wind turbines, Energy Conversion, 21 (1981) 175-183

(7) Hess, J.L., Review of integral equation techniques for solving potential flow problems with emphasis on the surface-source method, Comp. Meth. Appl. Mech. Eng., 5 (1975) 145-196

Computational Techniques & Applications: CTAC-83
J. Noye & C. Fletcher (Editors)

NUMERICAL METHODS FOR MULTIDIMENSIONAL INTEGRAL EQUATIONS

Ivan G. Graham

Department of Mathematics
University of Melbourne
Parkville, Victoria 3052
Australia

We consider the numerical solution of multidimensional Fredholm integral equations of the second kind, using the Galerkin and collocation methods and their iterative variants. We prove convergence, and examine superconvergence phenomena for the iterated solution. In particular $O(h^2)$ convergence is proved for methods based on simple piecewise constant functions. Numerical results are presented for the two dimensional singular equation arising from the skin effect current distribution problem. Iterative refinement methods are used to minimise the computational cost of the numerical scheme.

1. INTRODUCTION

In this paper we discuss the numerical solution of the Fredholm integral equation of the second kind:

$$y(t) = f(t) + \lambda \int_{\bar{\Omega}} k(t,s)y(s)ds \ , \qquad t \in \bar{\Omega} \ . \tag{1.1}$$

Here, $\bar{\Omega}$ is the closure of a bounded domain Ω in $\mathbf{R}^m$, λ is a given scalar, k and f are given functions, and y is the solution to be determined. We can rewrite (1.1) in the abbreviated form:

$$y = f + \lambda Ky \ , \tag{1.1}$$

where K is the linear integral operator given by

$$Ky(t) = \int_{\bar{\Omega}} k(t,s)y(s)ds \ .$$

Equations of this form arise in a variety of applications. For the univariate case m = 1 examples of applications include polymer physics [38], transport theory [30], [34], and more generally, the solution of a wide range of mixed boundary value problems (see the many examples in [48]). In the multivariate case, m > 1, applications (usually for m = 2 or 3) again arise generally in the reformulation of boundary value problems (see, for example, [36], [33], [53]). More concrete examples occur in the papers [3] (on electromagnetic wave scattering), [42] (on magnetic field problems) and [31], [32] (on crack problems in fracture mechanics).

In this paper we emphasise one particular application which arises in electrical engineering - that of modelling the well known "skin effect" current distribution problem. This problem arises when an alternating current flows in a conducting bar and the magnetic field produced sets up eddy currents which cause the current to be displaced towards the surface of the conductor. The current distribution y (which may be defined as a complex valued function defined over the cross-section $\bar{\Omega}$, say, of the conductor) then satisfies the equation

$$y(t,t') = C + i\lambda \int_{\bar{\Omega}} \ln\left[\frac{\sqrt{(t-s)^2 + (t'-s')^2}}{r}\right] y(s,s')dsds' , \tag{1.2}$$

where C and r are constants which may be chosen arbitrarily, and $\lambda \in \mathbf{R}$ depends on the conductivity of the material and the frequency of the alternating current [41], [21]. This is an equation of type (1.1) where the kernel function k has a weak (logarithmic) singularity. The singularity in the kernel induces singularities in the solution y as (t,t') approaches the boundary of Ω, in line with the expected physical "skin effect". The solution of this problem is not known in closed form except for the case when the conducting bar has circular cross-section. However, the solution for more general geometries is of relevance to the optimal design of conductors for carrying heavy current loads, a problem which has received quite a lot of attention in the literature [19], [37], [41], [42]. Thus there is a demand for an accurate and efficient numerical procedure for approximating the solution of (1.2) or more generally, the solution of (1.1).

The aim of this paper is to describe some numerical methods which are generally applicable to equation (1.1), to prove convergence (and also certain super-convergence phenomena), and to apply one of these methods to the skin effect problem discussed above. Our analysis will make the following assumptions on (1.1).

(i) The homogeneous equation

$$y(t) = \int_{\bar{\Omega}} k(t,s)y(s)ds$$

has only the trivial solution $y \equiv 0$;

(ii) k satisfies the mild integrability conditions

$$\int_{\bar{\Omega}} |k(\tau,s)|ds < \infty , \quad \lim_{t\to\tau} \int_{\bar{\Omega}} |k(t,s) - k(\tau,s)|ds = 0 , \quad \tau \in \bar{\Omega} ; \tag{1.3}$$

(iii) f is continuous on $\bar{\Omega}$. (1.4)

It can then be shown that there exists a unique continuous solution y of (1.1). (A proof may be obtained from the Fredholm alternative - see, for example, [21].) For the skin effect equation (1.2) direct verification of (i), (ii), (iii) is straightforward: (i) follows on physical grounds, (iii) since C is constant and (ii) may be shown using the tests in [27].

We remark that, since $\bar{\Omega}$ is the closure of a bounded domain, the equation (1.1) is not of the usual "boundary integral equation" type. In this latter type of equation, the region of integration is usually the <u>boundary</u> of some bounded domain in $\mathbf{R}^m$ (e.g. a closed contour in $\mathbf{R}^2$ or a closed surface in $\mathbf{R}^3$). Nevertheless the numerical methods and analysis discussed here can also be adapted to the boundary integral equation case. In this context we take the opportunity of mentioning [4], [6], [8], [14], [15], [29], [47], and the other papers in this session of the conference as important recent contributions to the boundary integral method development.

2. NUMERICAL METHODS

For each $n \in \mathbf{N}$, we introduce a mesh (partition) Π_n of $\bar{\Omega}$ by choosing N (= N(n)) open simply-connected pairwise-disjoint subsets of $\bar{\Omega}$, $\{\Omega_i : i = 1,\ldots,N\}$, with the property that each Ω_i contains its centroid and

$$\bar{\Omega} = \bigcup_{i=1}^{N} \bar{\Omega}_i .$$

We define the mesh diameter h by

$$h = h(n) = \max_{i=1,\dots,N} \operatorname{diam}(\Omega_i) ,$$

where $\operatorname{diam}(\Omega_i)$ is the diameter of Ω_i with respect to the uniform norm on $\mathbf{R}^m$. We make the natural assumption that

$$h \to 0 , \quad \text{as} \quad n \to \infty .$$

Thus we have a sequence of meshes on $\bar{\Omega}$ which is very general indeed. On these meshes we then take the very simple basis set comprising the functions on $\bar{\Omega}$ which are constant over each of the subregions Ω_i. Denoting this set by S_n, we then note that S_n is spanned by the characteristic functions

$$u_i(t) = \begin{cases} 1 , & t \in \Omega_i \\ 0 , & t \in \bar{\Omega}\backslash\Omega_i \end{cases} ,$$

for $i = 1,\dots,N$.

The theory presented below will be developed rigorously just for the simple basis set S_n. However, the results obtained can often be extended to the case of higher degree piecewise polynomial basis functions, and we shall highlight these possible extensions as we proceed. Nevertheless, the choice of S_n as the main subject of our analysis has a stronger motivation than just simplicity of exposition. In fact we shall show that quite respectable orders of convergence can be obtained using this basis set, for a relatively small amount of numerical work. In addition, we shall show in Section 6 that for equations with singularities (such as (1.2)), this simple basis set often yields orders of convergence which are just as good as those obtained by analogous piecewise polynomial basis sets of higher degree. Finally, the piecewise constant functions have the attraction that the basic convergence theory goes through for the very general meshes discussed above. Higher degree basis functions, by contrast, usually require some more regularity in the meshes for even the basic analysis to go through.

We shall describe two possible methods for obtaining a first approximation to y out of S_n. Denoting our first approximation by y_n^I we can write

$$y_n^I = \sum_{i=1}^{N} a_i u_i , \tag{2.1}$$

where the scalars $\{a_i\}_{i=1}^N$ are determined according to one or other of the following schemes.

(a) Galerkin scheme

In this case we define a Galerkin solution y_n^I by requiring that

$$(y_n^I, u_j) = (f, u_j) + (\lambda K y_n^I, u_j) , \quad j = 1,\dots,N , \tag{2.2}$$

where $(\cdot,\cdot)$ denotes the usual inner product $(f,g) = \int_{\bar{\Omega}} f\bar{g}$. Substituting (2.1) into (2.2) and rearranging we obtain the linear system

$$\sum_{i=1}^{N} a_i \left[D_{ij} - \lambda(Ku_i, u_j)\right] = (f, u_j) , \quad j = 1,\dots,N , \tag{2.3}$$

where $D_{ij} = \delta_{ij} \int_{\Omega_i} ds$, and δ_{ij} is Kronecker's delta.

(b) Collocation scheme

In this case we choose a set of N collocation points $\{t_j\}_{j=1}^N$ with $t_j \in \Omega_j$ for each j and we choose a collocation solution y_n^I by requiring that

$$y_n^I(t_j) = f(t_j) + \lambda K y_n^I(t_j) \ , \quad j = 1,\dots,N \ . \tag{2.4}$$

Substitution of (2.1) into (2.4) yields

$$\sum_{i=1}^{N} a_i [\delta_{ij} - \lambda K u_i(t_j)] = f(t_j) \ . \tag{2.5}$$

Thus the implementation of scheme (a) (respectively (b)) requires the setting up of the linear system (2.3) (respectively (2.5)) and its solution for the coefficients $\{a_i\}_{i=1}^N$. For the scheme (a) we must thus calculate the integrals

$$(Ku_i,u_j) = \int_{\Omega_j}\int_{\Omega_i} k(t,s)dsdt \ , \quad i,j = 1,\dots,N \tag{2.6}$$

and

$$(f,u_j) = \int_{\Omega_j} f(s)ds \ , \quad i,j = 1,\dots,N \ . \tag{2.7}$$

These usually will have to be done by quadrature unless k or f is fairly simple. The first integral above, being 2m-fold, can be particularly arduous. For scheme (b), on the other hand, the only integrals which are required are

$$Ku_i(t_j) = \int_{\Omega_i} k(t_j,s)ds \ , \quad i,j = 1,\dots,N \ , \tag{2.8}$$

the other requirement being the point evaluations $f(t_j)$. The integrals (2.8) must also be done by quadrature (in general) but it is clear that in overall terms scheme (b) is much simpler to set up on the machine than scheme (a).

Once y_n^I has been chosen we may generate a second approximation y_n^{II} to y by the natural iteration

$$y_n^{II} = f + \lambda K y_n^I \ , \tag{2.9}$$

which implies

$$y_n^{II} = f + \lambda \sum_{i=1}^{N} a_i K u_i \ . \tag{2.10}$$

If y_n^I is the Galerkin (respectively collocation) solution, then y_n^{II} is the iterated Galerkin (respectively iterated collocation) solution. Now note that K, being an integral operator is, in general, a smoothing operator. In fact under condition (1.3), K maps a piecewise continuous function into a continuous function [27]. Thus in view of (1.4), y_n^{II} is continuous whereas y_n^I is only piecewise continuous. By iterating then, we are using (1.1) as a natural smoothing process to generate a smooth second approximation y_n^{II} from a piecewise smooth initial approximation y_n^I.

Let us suppose that we have calculated y_n^I so that we know the coefficients $\{a_i\}_{i=1}^N$. What is the cost of iterating to obtain y_n^{II}? By (2.10), the calculation of y_n^{II} at arbitrary $t \in \bar{\Omega}$ requires the extra integrals

$$Ku_i(t) = \int_{\Omega_i} k(t,s)ds \ , \quad i = 1,\dots,N \ , \tag{2.11}$$

but does not require any further solution of linear equations. In fact sometimes the integrals (2.11) will be available to the user as a side result of the calculation of y_n^I, in which case their storage until y_n^{II} is actually calculated is the only extra work that iterating would require. For example, consider the Galerkin scheme, and note that finding y_n^I entailed the calculation of the repeated integrals (2.6). If these are found by genuine repeated integration, then at the end of the first stage of integration, (2.11) would be available. Now consider the collocation scheme and recall that finding y_n^I required the calculation of (2.8). It is often (but not always, of course) the case that the technology for the required extra integrals (2.11) can be obtained by simple adaptation of the technology (e.g. a quadrature routine or a piece of analysis) which has already been set up for (2.8). (After all (2.8) are just a specialised version of (2.11).) Indeed it is easy to see from (2.4) and (2.9) that

$$y_n^{II}(t_j) = y_n^I(t_j) \quad \text{for} \quad j = 1,\dots,N \ , \tag{2.12}$$

and so no extra work is involved in iterating if we restrict ourselves to evaluation at collocation points only.

3. PROJECTION METHOD ANALYSIS

On examination of (2.2) and (2.4) we see that in both of the schemes described in Section 2, y_n^I is chosen by requiring that

$$y_n^I = P_n f + \lambda P_n K y_n^I \ , \tag{3.1}$$

where, in the case of the Galerkin scheme, P_n is the orthogonal projection:

$$P_n \phi = \sum_{i=1}^{N} \frac{(\phi,u_i)}{(u_i,u_i)} u_i \ ,$$

and, in the case of the collocation scheme, P_n is the interpolatory projection:

$$P_n \phi = \sum_{i=1}^{N} \phi(t_i) u_i \ .$$

Note that in both cases

$$P_n x_n = x_n \ , \tag{3.2}$$

for all $x_n \in S_n$. Applying P_n to each side of (2.9) and comparing with (3.1) we obtain

$$y_n^I = P_n y_n^{II} \ ,$$

and hence y_n^{II} satisfies the approximate operator equation

$$y_n^{II} = f + \lambda K P_n y_n^{II} \ . \tag{3.3}$$

Let C denote the space of all bounded uniformly continuous functions on Ω, equipped with the norm

$$\|\phi\|_\infty = \sup_{x\in\Omega} |\phi(x)| \ .$$

Let L_∞ denote the space of all essentially bounded measurable functions on Ω equipped with the norm

$$\|\phi\|_\infty = \operatorname{ess\,sup}_{x\in\Omega} |\phi(x)| \ .$$

Note that $S_n \subset L_\infty$ but $S_n \not\subset C$. It is then easy to show that P_n is a bounded linear operator from C (and from S_n) to L_∞ and that

$$\|P_n\| = 1 \ . \tag{3.4}$$

Standard arguments (e.g. [7]) can then be used to obtain the following result:

Theorem 1. For sufficiently large n y_n^I exists in S_n, y_n^{II} exists in C and

$$\|y - y_n^I\|_\infty \leq C_1 \|y - P_n y\|_\infty , \tag{3.5}$$

and

$$\|y - y_n^{II}\|_\infty \leq C_2 \|K(y - P_n y)\|_\infty , \tag{3.6}$$

where C_1 and C_2 are constants.

Using (3.2), (3.4), (3.5), it is easily seen that

$$\begin{aligned} \|y - y_n^I\|_\infty &\leq C_1 \|(I - P_n)(y - x_n)\|_\infty \\ &\leq 2C_1 \|y - x_n\|_\infty , \end{aligned} \tag{3.7}$$

for any $x_n \in S_n$. Since $y_n^I \in S_n$ we then have

$$\inf_{x_n \in S_n} \|y - x_n\|_\infty \leq \|y - y_n^I\|_\infty \leq 2C_1 \inf_{x_n \in S_n} \|y - x_n\|_\infty ,$$

and so the convergence to y of y_n^I is asymptotically as fast as the convergence to y of its best approximation from the space S_n. Such convergence is called "quasi-optimal".

Clearly (3.6) and the boundedness of K imply that the convergence of y_n^{II} is also at least quasi-optimal. However, since K is an integral operator and since $y - P_n y$ is oscillatory, the quantity $\|K(y - P_n y)\|_\infty$ is often of higher order than $\|y - P_n y\|_\infty$. Hence the convergence of y_n^{II} to y is often faster than the quasi-optimal order discussed above. This phenomenon, observed in the integral equations context, is called "superconvergence". Thus in this paper, by the term "superconvergence" we mean the construction of smooth approximate solutions (usually from initial non smooth approximate solutions) which have global convergence rate faster than quasi-optimal. Such constructions have been studied, for example, in [11], [12], [13], [16], [17], [22], [24], [29], [43]-[47]. On the other hand the term is also often used (especially in the differential equations context) when it is observed that the initial approximate solution (which converges globally with quasi-optimal order) exhibits higher order local convergence at some distinguished set of points. (For a very extensive survey of this latter phenomenon in the integral equations context see [10].)

4. SUPERCONVERGENCE

By the result of Theorem 1, we see that a bound on $\|y - y_n^{II}\|_\infty$ may be obtained by estimating $\|K(y - P_n y)\|_\infty$. Note that

$$\begin{aligned} |K(y - P_n y)(t)| &= |\int_{\bar{\Omega}} k(t,s)(y(s) - P_n y(s))ds| \\ &= |(y - P_n y, \bar{k}_t)| , \end{aligned} \tag{4.1}$$

where $k_t(s) = k(t,s)$, for $t,s \in \bar{\Omega}$. Careful exploitation of the inner product structure of (4.1) yields superconvergence results for y_n^{II}.

(a) Galerkin scheme

In this case the orthogonality of P_n and the Hölder inequality yield

$$\begin{aligned} |K(y - P_n y)(t)| &= |(y - P_n y, \bar{k}_t - s_n)| \\ &\leq \|\bar{k}_t - s_n\|_1 \|y - P_n y\|_\infty , \end{aligned}$$

for any $s_n \in S_n$ with $\|\cdot\|_1$ denoting norm in $L_1(\Omega)$. Recalling (3.7) there obtains

$$|K(y - P_n y)| \leqslant C\beta_n(t) \inf_{x_n \in S_n} \|y - x_n\|_\infty$$

where

$$\beta_n(t) = \inf_{s_n \in S_n} \|\bar{k}_t - s_n\|_1 .$$

If k_t is smooth enough the quantity $\beta_n(t)$ can be shown to converge to zero uniformly in t, yielding a superconvergence estimate for y_n^{II}. The arguments used here are due to Graeme Chandler [12], [13]. More general proofs of superconvergence for y_n^{II} are found in [44], [45]. A recent analysis of Galerkin methods in the multidimensional context is given in [43], [47].

(b) Collocation scheme

To obtain superconvergence for y_n^{II} in this case we choose, for each $i = 1,\ldots,N$, the collocation point t_i to be the centroid of Ω_i. In this case P_n is not orthogonal so the analysis of (a) above does not apply. Nevertheless, we can write

$$|K(y - P_n y)(t)| \leqslant |(y - P_n y,\bar{k}_t - s_n)| + |(y - P_n y,s_n)| , \tag{4.2}$$

for $s_n \in S_n$. Now, since

$$s_n = \sum_{i=1}^{N} \alpha_i u_i ,$$

for some constants $\{\alpha_i\}_{i=1}^N$, we have

$$\begin{aligned}(y - P_n y,s_n) &= \sum_{i=1}^{N} \bar{\alpha}_i \int_{\Omega_i} (y - P_n y)(s)ds \\ &= \sum_{i=1}^{N} \bar{\alpha}_i \int_{\Omega_i} (I - P_n)(y - \xi_n)(s)ds ,\end{aligned}$$

where $\xi_n \in S_n^{(1)}$, and $S_n^{(1)}$ is the set of all piecewise linear functions (with respect to the mesh Π_n). This last equality follows since the approximate integration rule

$$\int_{\Omega_i} \phi(s)ds \simeq \phi(t_i) \int_{\Omega_i} ds$$

integrates exactly any ϕ which is linear on Ω_i. (This rule is the m dimensional analogue of the simple one dimensional midpoint rule.) Thus,

$$|(y - P_n y,s_n)| = |((I - P_n)(y - \xi_n),s_n)|$$

for any $\xi_n \in S_n^{(1)}$, and so, using (4.2) and (3.4), we have

$$|K(y - P_n y)(t)| \leqslant C'\|\bar{k}_t - s_n\|_1 \inf_{x_n \in S_n} \|y - x_n\|_\infty + C\|s_n\|_1 \inf_{\xi_n \in S_n^{(1)}} \|y - \xi_n\|_\infty .$$

It then follows that

$$|K(y - P_n y)(t)| \leqslant C'\beta_n(t) \inf_{x_n \in S_n} \|y - x_n\|_\infty + C[\|\bar{k}_t\|_1 + \beta_n(t)] \inf_{\xi_n \in S_n^{(1)}} \|y - \xi_n\|_\infty .$$

Now if k_t, y are smooth enough it can be shown that both terms on the right hand side converge uniformly in t with a rate faster than $\inf_{x_n \in S_n} \|y - x_n\|_\infty$. In the case of the second term this is because approximation by piecewise linears is more accurate than approximation by piecewise constants, provided the function to be approximated is smooth enough. The arguments described above were first given for the two dimensional case in [22]. In the one dimensional case superconvergence for iterated collocation is examined in [11], [17], [46]. We note finally that global superconvergence for y_n^{II} implies local superconvergence for y_n^I at the collocation points because of (2.12).

5. RATES OF CONVERGENCE

In order to quantify the rates of convergence attained by the methods described above we introduce some Sobolev spaces. For $\beta = (\beta_1, \beta_2, \ldots, \beta_m) \in \mathbf{R}_+^m$, we let D^β denote the differential operator

$$\left(\frac{\partial}{\partial x_1}\right)^{\beta_1} \left(\frac{\partial}{\partial x_2}\right)^{\beta_2} \cdots \left(\frac{\partial}{\partial x_m}\right)^{\beta_m}$$

of order $|\beta| = \sum_{i=1}^{m} \beta_i$. Then, for $k \in \mathbf{N}$, C^k denotes the space of all functions $\phi \in C$ which have $D^\beta\phi \in C$ for $|\beta| \leq k$, and W_1^k denotes the space of functions $\phi \in L_1$ which have $D^\beta\phi \in L_1$ for $|\beta| \leq k$. (L_1 denotes the usual space of Lebesgue integrable functions.) C^k and W_1^k can be made complete normed spaces in the usual way [1]. It is easy to show that if $y \in C^1$, then

$$\inf_{x_n \in S_n} \|y - x_n\|_\infty = O(h) , \tag{5.1}$$

and hence we have, recalling (3.7), the result:

<u>Theorem 2</u>. If $y \in C^1$, then

$$\|y - y_n^I\|_\infty = O(h) .$$

To obtain superconvergence estimates for y_n^{II}, however, it is clear from Section 4 that we need more approximation theoretic results, both for $S_n^{(1)}$ with respect to the uniform norm and for S_n with respect to the L_1-norm. In the first case the estimate

$$\inf_{\xi_n \in S_n^{(1)}} \|y - \xi_n\|_\infty = O(h^2) \tag{5.2}$$

for $y \in C^2$ is easy to obtain by choosing ξ_n to be an appropriate multivariate Taylor polynomial of degree 1 for y. To obtain results in the second case we assume that the meshes Π_n have the property that each Ω_i contains an (m dimensional) sphere of diameter τ diam(Ω_i) where τ is a fixed constant independent of i and n. Such meshes are often called <u>regular</u> or <u>non-degenerate</u> and occur naturally in the analysis of the Finite Element Method [18], [49]. We also assume that the boundary $\partial\Omega$ of Ω is at least piecewise smooth. More precisely, we assume that each point $x \in \partial\Omega$ should have a neighbourhood U_x such that $\partial\Omega \cap U_x$ is the graph of a Lipschitz continuous function. Under these conditions it may be shown (see [26]) that if $k_t \in W_1^1$ with $\|k_t\|$ (in W_1^1) bounded independently of t, then

$$\inf_{s_n \in S_n} \|\bar{k}_t - s_n\|_1 = O(h) . \tag{5.3}$$

Thus with regular meshes and a piecewise smooth boundary as described above, we can obtain the following quantitative order of convergence estimates for y_n^{II}.

Theorem 3.

(a) In the case of the Galerkin scheme, if $y \in C^1$ and $k_t \in W_1^1$ with $\|k_t\|$ (in W_1^1) bounded independently of t, then

$$\|y - y_n^{II}\|_\infty = O(h^2) .$$

(b) In the case of the collocation scheme, if $y \in C^2$ and $k_t \in W_1^1$ with $\|k_t\|$ (in W_1^1) bounded independently of t, then

$$\|y - y_n^{II}\|_\infty = O(h^2) .$$

The proofs follow readily from (3.6), the discussion in Section 4 and estimates (5.1), (5.2), and (5.3).

Remark 1. The above results for the Galerkin scheme can easily be generalised to the case of higher degree piecewise polynomial basis functions. As an example consider the case when the meshes Π_n can be chosen to be composed of regular rectangular regions Ω_i in $\mathbf{R}^m$ (i.e. each Ω_i is a product of open intervals of $\mathbf{R}^1$). Then, although (3.4) is no longer true, it can be shown [20] that P_n is uniformly bounded, and this allows us to prove the basic convergence result, Theorem 1. If $S_n^{(r)}$ denotes the set of piecewise polynomials of degree r on Π_n then it can be shown (see, for example [9]), that

$$\text{(i)} \qquad \inf_{x_n \in S_n^{(r)}} \|y - x_n\|_\infty = O(h^r)$$

when $y \in C^r$, and

$$\text{(ii)} \qquad \inf_{s_n \in S_n^{(r)}} \|\bar{k}_t - s_n\|_1 = O(h^r) ,$$

when $k_t \in W_1^r$ with $\|k_t\|$ bounded independently of t. Under the above conditions, then, we have

$$\|y - y_n^I\|_\infty = O(h^r) ,$$

$$\|y - y_n^{II}\|_\infty = O(h^{2r}) .$$

Analogous results are given in [43].

Remark 2. In Theorem 3 more smoothness assumptions (on y) are required to prove superconvergence of y_n^{II} for the collocation scheme than for the Galerkin scheme. This observation has led us in [25] to consider the question: Are such extra smoothness requirements on y really necessary? Looking at one dimensional equations only, we demonstrate by example that the answer to this question is, in general, yes. This observation gives the Galerkin scheme a rather subtle advantage over the collocation scheme, which goes some way towards compensating for its extra computational cost (see Section 2).

However, the sort of smoothness requirements discussed above are very general. If we consider more specific equations, we see that there are important classes of problems for which the iterated Galerkin and collocation schemes have the same convergence properties. Such a class is considered in the next section.

6. WEAKLY SINGULAR EQUATIONS

In many practical cases, including the current distribution problem (1.2), the kernel k(t,s) of the integral equation has a weak singularity along the diagonal t = s. Such equations often take the form

$$y(t) = f(t) + \lambda \int_{\bar{\Omega}} k_\alpha(|t - s|)h(t,s)y(s)ds \ , \quad t \in \bar{\Omega} \ , \tag{6.1}$$

for some $0 < \alpha \leqslant m$, where

$$k_\alpha(|t - s|) = |t - s|^{\alpha-m} \qquad 0 < \alpha < m \ ,$$
$$k_m(|t - s|) = \ell n|t - s| \ ,$$

and h is a smooth factor. (Here $|\cdot|$ denotes the Euclidean norm.) In the case $m = 1$, the regularity properties and numerical solution of such equations has been well studied [13], [16], [23], [24], [39], [40], [50], [51]. In higher dimensions a very nice regularity study has been made by Pitkäranta [35], and (for the case $m = 2$) the convergence of numerical methods has been analysed ([22], [26], [43], [47]).

Consider the general case $m \geqslant 1$. Suppose for convenience $\alpha \notin \mathbf{N}$, and let ℓ be the integer part of α. Then the results of [35] show that $y \in C^\ell$ and, for any $|\beta| = \ell$, $D^\beta y$ satisfies a Lipschitz condition of order $\alpha - \ell$. This result requires sufficient smoothness conditions on f and h, and that the boundary of $\partial\Omega$ of Ω be piecewise smooth (see [35] for details). Moreover, y can be made arbitrarily smooth in the interior of Ω, by increasing the smoothness of f and h. In such a case, then, the singular behaviour of y(t) occurs (as one would expect from the current distribution problem), as t approaches $\partial\Omega$.

When $0 < \alpha < 1$, $y \in C$, and y satisfies a Lipschitz condition of order α, but in general is no smoother. For $\alpha > 1$ on the other hand, y is at least C^1, and we would assume that approximation of y by S_n would work well. In fact it can be shown that

$$\inf_{x_n \in S_n} \|y - x_n\|_\infty = O(h^\gamma)$$

$$\inf_{s_n \in S_n} \|\bar{k}_t - s_n\|_1 = O(h^\gamma)$$

for all $\alpha \notin \mathbf{N}$, where $\gamma = \min\{\alpha,1\}$, $k_t(s) = k_\alpha(|t - s|)h(t,s)$, and the meshes Π_n are assumed regular. By Sections 3 and 4 we then have, for the Galerkin scheme:

$$\left.\begin{aligned} \|y - y_n^I\|_\infty &= O(h^\gamma) \\ \|y - y_n^{II}\|_\infty &= O(h^{2\gamma}) \end{aligned}\right\} \tag{6.2}$$

where $\gamma = \min\{\alpha,1\}$.

For the collocation scheme (with the collocation points chosen as centroids) we can show, at least for the case when $\bar{\Omega}$ is a rectangle in $\mathbf{R}^2$, and when our regular meshes Π_n are obtained as tensor products of one dimensional meshes, that the estimates (6.2) remain true. The proof of this result is rather tricky, involving a sharper analysis than that given in Section 4. Details will be given in [26].

Now consider the current distribution equation (1.2). If we use approximation by S_n, with regular tensor-product meshes then our estimates are (analogously to (6.2), since $\alpha = 2 = m$)

$$\|y - y_n^I\|_\infty = O(h)$$

$$\|y - y_n^{II}\|_\infty = O(h^2)$$

for either the Galerkin or collocation scheme, when $\bar{\Omega}$ is a rectangle.

Thus, despite the fact that the kernel is singular, a full superconvergence rate is attained by y_n^{II}. We illustrate this result numerically for the collocation scheme in Section 7.

I would like to finish this section by pointing out that, in some ways, the piecewise constant basis functions are the best choice for the kind of singular equations discussed here. Note that they often work optimally even in the presence of singularities. Return to (6.2) and note that for the Galerkin scheme on S_n, $\|y - y_n^{II}\|_\infty$ attains the order $O(h^2)$ for any α just marginally bigger than 1. However, using piecewise linear bases in such a case the best estimate for $\|y - y_n^{II}\|_\infty$ (using arbitrary regular meshes) can be shown to be $O(h^{2\alpha})$ and the impovement over the piecewise constant case is negligible as $\alpha \to 1$. Substantially more work is needed to compute the piecewise linear approximant. Thus if the equation we are solving is known to have singularities in its kernel or solution, good advice would be to use piecewise constant (rather than higher degree piecewise polynomial) bases. That way, one should be able to afford a fine mesh (as piecewise constants are cheapest to compute with), while at the same time one would minimise the risk of unrewarded investment in computing time.

There is one qualification to the above rule of thumb: If one knows precisely the form of the singularity in the solution then the meshes may be graded appropriately so that the optimal orders are returned for higher order piecewise polynomial approximation [14], [15], [24], [40], [51]. However, finding the precise form of the singularity (if it is not already known) can be a lengthy process and numerical results are often required urgently!

7. NUMERICAL RESULTS

We solve the current distribution equation (1.2) for the case when $\bar{\Omega}$ has a rectangular cross-section of length 0.1 and breadth 0.05. The parameter λ is equal to $\mu g\omega/2\pi$ where g is the conductivity of the conductor, μ is the permeability of free space and ω is the frequency of the current. These quantities were given the numerical values $\mu = 4\pi \times 10^{-7}$, $g = 10^8/2.83$, $\omega = 60 \times 2\pi$, corresponding to the physical situation of a copper bar and an alternating current of frequency 60. The arbitrary parameter r was chosen to take the value 0.1 (= length of conductor) according to a strategy for scaling the equation [21]. We use a sequence of uniform tensor product meshes Π_n, each of which divides $\bar{\Omega}$ into $N = n^2$ congruent subrectangles of length 0.1/n and breadth 0.05/n. We determine $y_n^I \in S_n$ by collocation at the centroids of subregions and y_n^{II} by iteration as described in Section 2. We fix the arbitrary parameter C in (1.2) so that the average of the values of the piecewise constant approximation y_5^I over the whole domain $\bar{\Omega}$ is unity.

By Section 6 we expect y_n^{II} to be fully superconvergent (i.e. $O(h^2)$) despite the mildly singular kernel and solution. Therefore we make no attempt to grade the mesh. The integrals (2.8) and (2.11) which are needed for collocation and iterated collocation can in this case be found analytically. Additional computational savings can be made by observing that the collocation matrix (i.e. the matrix whose (i,j)th element is $\delta_{ij} - \lambda K u_j(t_i)$) is symmetric. Some previous numerical results were reported in [21], [22], but these were rather unsatisfactory in that only very coarse meshes were used, yielding rather poor accuracy. To get a better idea of how well the method works, we performed the calculations for the cases n = 14, 16, 18, 20, 22, 24, the number of basis functions being n^2 in each case. Then the theory implies that

$$\|y - y_n^{II}\|_\infty = O\left(\frac{1}{n^2}\right) . \tag{7.1}$$

Experimental verification of this is tricky; since the solution y is unknown we cannot calculate the error norms $\|y - y_n^{II}\|_\infty$. However, we do know that at each point t, $y_n^{II}(t)$ is a sequence of complex numbers converging to $y(t)$. Thus it is reasonable to make the conjectures:

$$\left.\begin{aligned} \mathrm{Re}\, y(t) - \mathrm{Re}\, y_n^{II}(t) &= C_1(t)n^{-\mu_1} , \\ \mathrm{Im}\, y(t) - \mathrm{Im}\, y_n^{II}(t) &= C_2(t)n^{-\mu_2} , \\ \mathrm{Arg}\, y(t) - \mathrm{Arg}\, y_n^{II}(t) &= C_3(t)n^{-\mu_3} , \\ |y(t)| - |y_n^{II}(t)| &= C_4(t)n^{-\mu_4} . \end{aligned}\right\} \qquad (7.2)$$

In each case μ_i can be calculated by extrapolation using data for three different values of n. We would hope to obtain $\mu_i \geqslant 2$ for each i, and indeed such results are obtained for the points $t = (0.05,0.025)$, $(0.05,0.0)$, $(0.0,0.025)$, $(0.0,0.0)$. Results for these four points in the order listed are given in Table 1(a), (b), (c), (d) respectively. For these four points the conjecture $\mu_i \geqslant 2$ is verified consistently, with rather high orders of convergence obtained at the point $(0.05,0.025)$, which is situated at the centre of the conductor, where the solution is very smooth. At $(0.0,0.0)$ (where the solution is least smooth), convergence is still consistently $O(1/n^2)$. However, the four points chosen to test convergence are special in that they all bear a constant relationship to the meshes as they vary (for example $(0.05,0.025)$ is always at a corner of four mesh subdivisions). As a result of this consistent values of μ_i are obtained at those points. If we choose an arbitrary t and test the hypotheses (7.2), consistent values of μ_i are often not obtained. In fact the convergence of $\mathrm{Re}\, y_n^{II}(t)$, $\mathrm{Im}\, y_n^{II}(t)$, $\mathrm{Arg}\, y_n^{II}(t)$ and $|y_n^{II}(t)|$ is often not even monotonic, a possibility excluded by (7.2). Thus it is difficult to devise a sensible procedure for extrapolation of our $y_n^{II}(t)$ values at a general point t. (However, see [52] for a possible strategy.)

The alternative to extrapolation is to increase n until agreement in the required number of digits is obtained. However, at $n = 24$ there are 576 basis functions and so we must solve a 576×576 linear system with complex coefficients. Moreover, although the matrix is symmetric, it is also full and thus is rather costly to solve.

To avoid such problems, we applied an iterative procedure of Atkinson [5]. Recalling (3.3), we may write

$$y_n = f + \lambda K_n y_n ,$$

where y_n is just y_n^{II} and K_n is KP_n. Then for $m > n$ we define an iterative method by:

$$y_m^{(0)} = y_n$$

$$(I - \lambda K_n)y_m^{(\nu+1)} = f + \lambda(K_m - K_n)y_m^{(\nu)} .$$

Defining

$$r^{(\nu)} = f - (I - \lambda K_m)y_m^{(\nu)} ,$$

then our iteration is simply

$$y_m^{(\nu+1)} = y_m^{(\nu)} + (I - \lambda K_n)^{-1} r^{(\nu)} . \qquad (7.3)$$

n	y_n^{II}	Arg y_n^{II}	$\lvert y_n^{II} \rvert$	μ_1	μ_2	μ_3	μ_4	
14	-0.0797 - 0.3432i	-1.799	0.3524	2.6	3.8	2.6	3.1	(a)
16	-0.0787 - 0.3431i	-1.796	0.3520	2.7	3.0	2.7	2.8	
18	-0.0781 - 0.3430i	-1.795	0.3518	2.2	3.2	2.2	2.6	
20	-0.0777 - 0.3430i	-1.794	0.3516	2.6	4.8	2.5	3.3	
22	-0.0774 - 0.3429i	-1.793	0.3515					
24	-0.0772 - 0.3429i	-1.792	0.3515					
14	1.248 + 1.100i	0.7224	1.663	2.1	2.0	2.0	2.4	(b)
16	1.252 + 1.096i	0.7190	1.664	2.0	2.0	2.0	1.9	
18	1.255 + 1.093i	0.7166	1.664	2.1	1.9	2.0	3.1	
20	1.257 + 1.091i	0.7150	1.664	1.9	2.2	2.1	-	
22	1.258 + 1.089i	0.7137	1.664					
24	1.259 + 1.088i	0.7128	1.664					
14	2.014 + 1.483i	0.6348	2.501	1.9	2.1	2.0	1.4	(c)
16	2.036 + 1.466i	0.6240	2.508	1.9	2.1	2.0	1.5	
18	2.050 + 1.453i	0.6166	2.513	1.9	2.0	2.0	1.6	
20	2.061 + 1.445i	0.6113	2.517	1.9	2.0	2.0	1.7	
22	2.069 + 1.438i	0.6074	2.520					
24	2.076 + 1.433i	0.6044	2.523					
14	2.467 + 3.829i	0.9985	4.555	1.9	2.1	2.0	CONVERGENCE EFFECTIVELY COMPLETE	(d)
16	2.484 + 3.819i	0.9941	4.556	1.9	2.1	2.0		
18	2.496 + 3.812i	0.9911	4.556	2.0	2.2	2.0		
20	2.504 + 3.807i	0.9890	4.557	2.0	2.0	2.0		
22	2.510 + 3.803i	0.9874	4.557					
24	2.515 + 3.801i	0.9862	4.557					

Each value of μ_i obtained by extrapolation using y_n^{II}, y_{n+2}^{II}, y_{n+4}^{II}.

Table 1.

Computationally this method has the advantage that the linear collocation system need only be solved for some relatively small n, and the approximations $y_m^{(\nu)}$ to y_m (the solution of the larger system) are then computed by iterative refinement. Full details including an idea of how to program the method and a proof of convergence are given in [5]. Related methods have been studied in [2], [16], [28], [52]. It is important to emphasise that we are talking here about iteratively refining our y_m (= y_m^{II}) values. This is a different process than "natural iteration" described in Section 2 by which we obtain y_n^{II} from y_n^I.

To get an idea of how the method (7.3) performs, we first took m = 16 and found the time taken to calculate y_m at the collocation points and at the four special points used in Table 1, by the standard iterated collocation method. Extrapolations based on Table 1 allowed us to estimate the error in $|y_m|$ at the four special points. With n = 6 we iterated using (7.3) until $|y_m^{(\nu)}|$ agreed with $|y_m|$ at each of the four special points to within the estimated error in $|y_m|$. The experiment was repeated using n = 8. There was no substantial saving in computation time by the iterative scheme compared with the standard iterated collocation method. The situation was quite different for m = 24. Again y_m was found at the collocation points and the special points using standard iterated collocation. Errors in $|y_m|$ at the special points were estimated by extrapolation. Then iterations (7.3) were performed with various starting values of n until $|y_m^{(\nu)}|$ agreed with $|y_m|$ to within the estimated error in $|y_m|$. Results are given in Table 2 (all times are in milliseconds of CPU time).

n	No. of iterations required	Time taken
6	5	1425670
8	4	1311270
10	3	1085480
12	4	1672190

Time taken for standard iterated collocation (m = 24) = 2719960.

Table 2.

In order to obtain a highly accurate solution we set m = 28, n = 12 and iterated until three decimal place accuracy was obtained between successive iterates at the four special points. This process was repeated with m = 30, n = 12. Results are in Table 3.

	No. of iterations required	Time taken
m = 28 n = 12	3	2067560
m = 30 n = 12	3	2620170

Time taken for standard iterated collocation (m = 28) = 8608980.
Insufficient virtual memory for standard iterated collocation when m = 30.

Table 3.

To see that three decimal place accuracy has almost been obtained, we give the values of $|y_{28}|$ and $|y_{30}|$ on a 6 × 6 grid over bottom left hand quarter of conductor in Table 4(a), (b) respectively. The skin effect is clearly visible.

0.025	2.526	1.312	0.852	0.575	0.406	0.351
0.02	2.602	1.376	0.873	0.569	0.396	0.343
0.015	2.835	1.567	0.954	0.596	0.426	0.385
0.01	3.231	1.885	1.148	0.760	0.622	0.599
0.005	3.801	2.349	1.525	1.149	1.035	1.016
0.00	4.558	3.039	2.192	1.819	1.692	1.664
	0.00	0.01	0.02	0.03	0.04	0.05

(a) $|y_{28}|$

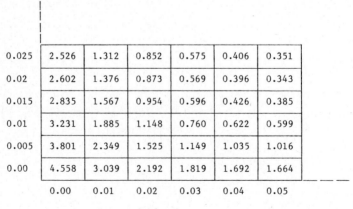

0.025	2.527	1.311	0.852	0.574	0.406	0.351
0.02	2.604	1.375	0.872	0.569	0.396	0.343
0.015	2.836	1.566	0.954	0.596	0.426	0.385
0.01	3.232	1.884	1.148	0.760	0.623	0.599
0.005	3.802	2.348	1.525	1.149	1.035	1.016
0.00	4.558	3.039	2.192	1.819	1.692	1.664
	0.00	0.01	0.02	0.03	0.04	0.05

(b) $|y_{30}|$

Table 4.

REFERENCES

[1] Adams, R.A., Sobolev Spaces (Academic Press, New York, 1975).
[2] Ahués, M., d'Almeida, F., Chatelin, F. and Telias, M., Iterative refinement techniques for the eigenvalue problem of compact integral operators, in: Baker, C.T.H. and Miller, G.F. (eds.), Treatment of Integral Equations by Numerical Methods (Academic Press, London, 1982).
[3] Andreo, R.H. and Krill, J.A., Stochastic variational formulations of electromagnetic wave scattering, in: Varadan, V.K. and Varadan, V.V. (eds.), Acoustic, Electromagnetic and Elastic Wave Scattering - Focus on the T-Matrix Approach (Pergamon Press, New York, 1980).
[4] Arnold, D.N. and Wendland, W.L., Collocation versus Galerkin procedures for boundary integral methods, Technische Hochschule Darmstadt Preprint No. 671 (1982).
[5] Atkinson, K.E., Iterative variants of the Nyström method for the numerical solution of integral equations, Numer. Math. 22 (1973) 17-31.
[6] Atkinson, K.E., Piecewise polynomial collocation for integral equations on surfaces in three dimensions, preprint.
[7] Atkinson, K.E., Graham, I. and Sloan, I., Piecewise continuous collocation for integral equations, SIAM J. Numer. Anal. 20 (1983) 172-186.
[8] Atkinson, K.E. and de Hoog, F., The numerical solution of Laplace's equation on a wedge, preprint.
[9] Brudnyi, Ju.A., Piecewise polynomial approximation, embedding theorem and rational approximation, in: Schaback, R. and Scherer, K. (eds.), Approximation Theory, Bonn 1976 (Lecture Notes No. 556, Springer-Verlag, Heidelberg, 1976).
[10] Brunner, H., The application of the variation of constants formula in the numerical analysis of integral and integro-differential equations, Utilitas Math. 19 (1981) 255-290.
[11] Brunner, H., Iterated collocation methods and their discretizations for Volterra integral equations, submitted for publication.
[12] Chandler, G.A., Superconvergence for second kind integral equations, in: Anderssen, R.S., de Hoog, F.R. and Lukas, M.A. (eds.), The Application and Numerical Solution of Integral Equations (Sijthoff and Noordhoff, Alphen aan den Rijn, 1980).

[13] Chandler, G.A., Superconvergence of numerical solutions to second kind integral equations, Ph.D. Thesis, Australian National University, 1979.
[14] Chandler, G.A., Galerkin's method for boundary integral equations on polygonal domains, Numerical Analysis Report No. 83/1, University of Queensland, 1983.
[15] Chandler, G.A., Superconvergent approximations to the solution of a boundary integral equation on polygonal domains, Research Report No. CMA-R15-83, Australian National University, 1983.
[16] Chatelin, F., Spectral Approximation of Linear Operators (Academic Press, New York, 1983).
[17] Chatelin, F. and Lebbar, R., The iterated projection solution for the Fredholm integral equation of second kind, J. Austral. Math. Soc. Ser. B. 22 (1981) 439-451.
[18] Ciarlet, P.G., The Finite Element Method for Elliptic Problems (North-Holland, Amsterdam, 1980).
[19] Deeley, E.M. and Okon, E.E., A numerical method of solving Fredholm integral equations applied to current distribution and inductance in parallel conductors, Internat. J. Numer. Methods Engrg. 11 (1977) 447-467.
[20] Douglas, J., Dupont, T. and Wahlbin, L., The stability in L^q of the L^2-projection into finite element function spaces, Numer. Math. 23 (1975) 193-197.
[21] Graham, I.G., Some application areas for Fredholm integral equations of the second kind, in: Anderssen, R.S., de Hoog, F.R. and Lukas, M.A. (eds.), The Application and Numerical Solution of Integral Equations (Sijthoff and Noordhoff, Alphen aan den Rijn, 1980).
[22] Graham, I.G., Collocation methods for two dimensional weakly singular integral equations, J. Austral. Math. Soc. Ser. B. 22 (1981) 456 -473.
[23] Graham, I.G., Singularity expansions for the solutions of second kind Fredholm integral equations with weakly singular convolution kernels, J. Integral Equations 4 (1982) 1-30.
[24] Graham, I.G., Galerkin methods for second kind integral equations with singularities, Math. Comp. 39 (1982) 519-533.
[25] Graham, I.G., Joe, S. and Sloan, I.H., Galerkin versus collocation for integral equations of the second kind, in preparation.
[26] Graham, I.G. and Schneider, C., Product integration for two dimensional weakly singular integral equations, in preparation.
[27] Graham, I.G. and Sloan, I.H., On the compactness of certain integral operators, J. Math. Anal. Appl. 68 (1979) 580-594.
[28] Hemker, P.W. and Schippers, H., Multiple grid methods for the solution of Fredholm integral equations of the second kind, Math. Comp. 36 (1981) 215-232.
[29] Hsiao, G.C. and Wendland, W.L., The Aubin-Nitsche lemma for integral equations, J. Integral Eqns. 3 (1981) 299-315.
[30] Kaper, H.G. and Kellogg, R.B., Asymptotic behaviour of the solution of the integral transport equation in slab geometry, SIAM J. Appl. Math. 32 (1977) 191-200.
[31] Martin, P.A., Diffraction of elastic waves by a penny-shaped crack, Proc. R. Soc. Lond. A378 (1981) 263-285.
[32] Martin, P.A. and Wickham, G.R., Diffraction of elastic waves by a penny-shaped crack: analytical and numerical results, preprint.
[33] Mikhlin, S.G., Integral Equations (Pergamon Press, London, 1957).
[34] Pitkäranta, J., On the differential properties of solutions to Fredholm equations with weakly singular kernels, J. Inst. Maths. Applics. 24 (1979) 109-119.
[35] Pitkäranta, J., Estimates for the derivatives of solutions to weakly singular Fredholm integral equations, SIAM J. Math. Anal. 11 (1980) 952-968.
[36] Pogorzelski, W., Integral Equations and their Applications (Pergamon Press, London, 1966).

[37] Schaffer, G. and Banderet, P., Skin effect in heavy current conductor bars, Brown Boveri Review 52 (1965) 623-628.
[38] Schlitt, D.W., Numerical solution of a singular integral equation encountered in polymer physics, J. Math. Phys. 9 (1968) 436-439.
[39] Schneider, C., Regularity of the solution to a class of weakly singular Fredholm integral equations of the second kind, Integral Equations Operator Theory 2 (1979) 62-68.
[40] Schneider, C., Product integration for weakly singular integral equations, Math. Comp. 36 (1981) 207-213.
[41] Silvester, P., Modern Electromagnetic Fields (Prentice-Hall: Englewood Cliffs, 1968).
[42] Simpkin, J. and Trowbridge, C.W., On the use of the total scalar potential in the numerical solution of field problems in electromagnetics, Internat. J. Numer. Methods Engrg. 14 (1979) 423-440.
[43] Sloan, I.H., Superconvergence and the Galerkin method for integral equations of the second kind, in: Baker, C.T.H. and Miller, G.F. (eds.), Treatment of Integral Equations by Numerical Methods (Academic Press, London, 1982).
[44] Sloan, I.H., Error analysis for a class of degenerate-kernel methods, Numer. Math. 25 (1976) 231-238.
[45] Sloan, I.H., Burn, B.J. and Datyner, N., A new approach to the numerical solution of integral equations, J. Comput. Phys. 18 (1975) 92-105.
[46] Sloan, I.H., Noussair, E. and Burn, B.J., Projection methods for equations of the second kind, J. Math. Anal. Appl. 69 (1979) 84-103.
[47] Sloan, I.H. and Thomée, V., Superconvergence of the Galerkin iterates for integral equations of the second kind, in preparation.
[48] Sneddon, I.N., Mixed Boundary Value Problems in Potential Theory (North-Holland, Amsterdam, 1966).
[49] Strang, G. and Fix, G.J., An Analysis of the Finite Element Method (Prentice-Hall, Englewood Cliffs, 1973).
[50] Vainikko, G. and Pedas, A., The properties of solutions of weakly singular integral equations, J. Austral. Math. Soc. Ser. B. 22 (1981) 419-430.
[51] Vainikko, G. and Uba, P., A piecewise polynomial approximation to the solution of an integral equation with weakly singular kernel, J. Austral. Math. Soc. Ser. B. 22 (1981) 431-438.
[52] Volk, W., An automatic iteration scheme and its application to nonlinear operator equations, in: Baker, C.T.H. and Miller, G.F. (eds.), Treatment of Integral Equations by Numerical Methods (Academic Press, London, 1982).
[53] Zabreyko, P.P., Koshelev, A.I., Krasnosel'skii, M.A., Mikhlin, S.G., Rakovshchik, L.S. and Stet'senko, V.Ya., Integral Equations - A Reference Text (Noordhoff, Gröningen, 1975).

Computational Techniques & Applications: CTAC-83
J. Noye & C. Fletcher (Editors)

FAST CONVERGENCE OF THE ITERATED GALERKIN METHOD FOR INTEGRAL EQUATIONS

Ian H. Sloan

Department of Applied Mathematics
University of New South Wales
Sydney, Australia.

For integral equations of the form $y = f + Ky$, the iterated variant of the Galerkin method is known, under very general conditions, to converge faster than the Galerkin method itself. In recent joint work with V. Thomée, precise estimates of rates of convergence have been obtained for some higher-dimensional integral equations that arise in practice, under the assumption that the approximating spaces are of finite-element character. Some non-trivial examples are discussed in this paper.

INTRODUCTION

Many of the integral equations that arise in practice are integral equations of the second kind, i.e. they are of the form

$$y(t) = f(t) + \int_{\Omega} k(t,s)y(s)d\sigma(s) , \qquad (1)$$

where Ω is either a bounded region in 1, 2 or 3 dimensions, or the boundary of a smooth bounded region in 2 or 3 dimensional space. Here $d\sigma(s)$ denotes an element of length, surface area or volume, as appropriate. In many problems of this type the kernel $k(t,s)$ has an integrable singularity at $s = t$, and kernels of that type are covered in the present discussion. The two examples that follow are typical of equations that arise in practice, and are used in this paper for purposes of illustration.

Example 1

$$y(t_1,t_2) = 1 + i\lambda \int_{\Omega} \log(|t - s|)y(s_1,s_2)ds_1ds_2 ,$$

where

$$|t - s| = [(t_1 - s_1)^2 + (t_2 - s_2)^2]^{\frac{1}{2}} ,$$

Ω is a bounded region in the plane (for example a rectangle), and λ is a real number. This example describes the distribution of alternating electric current over the cross section of a conducting bar (see, for example, [8]). In that application $y(t_1,t_2)$ is the relative current density at the point $(t_1,t_2) = t$, and λ is a number that depends on the physical parameters.

Example 2

$$y(t) = f(t) + \frac{1}{2\pi} \int_{\Omega} \frac{\partial}{\partial n_s} \left(\frac{1}{|t - s|}\right) y(s)d\sigma(s) ,$$

where Ω is the smooth boundary of a bounded region of three-dimensional space (for example a sphere). This, a typical example of a second-kind boundary integral equation, arises in connection with the Laplace equation for the region enclosed by Ω for the case of Dirichlet boundary conditions, if the solution is expressed in terms of a 'double layer' over the boundary.

A popular method for the solution of such integral equations is the Galerkin method, in which y is approximated by a function y_h belonging to some finite-dimensional space S_h. (For a short description of the Galerkin method see the Appendix.) In this paper S_h is assumed to be a finite-element space, in which case h denotes the maximum diameter of the sub-regions $\Omega_1, \ldots, \Omega_N$ into which Ω is partitioned. For the sake of definiteness, the functions in S_h will be assumed to be of piecewise-linear character. The precise details of the construction of the functions in S_h are not important, since our only essential requirement is that S_h has the approximation properties characteristic of approximation by piecewise-linear functions: most importantly, that the error of best approximation (in any of the usual senses of approximation) of a smooth function by the functions in S_h is of order $O(h^2)$ as $h \to 0$.

Since y_h belongs to S_h, the most we can hope for from the Galerkin approximation is that it achieves the same order of convergence as the *best* approximation of y by functions in S_h. Thus the order of convergence of the Galerkin approximation is at best $O(h^2)$, unless y actually belongs to S_h. It is well known that under certain standard conditions the Galerkin approximation does achieve the same order of approximation as the best approximation to y, and we shall assume that this is the case in the examples discussed below.

In this paper the main topic is not the Galerkin method itself, but rather its iterative variant, obtained by substituting the Galerkin approximation into the right-hand side of (1). The resulting approximation, which we shall denote by $y_h^{(1)}$, turns out to have an interesting property: under very general conditions it converges faster than y_h. This phenomenon, often called 'superconvergence', has been much studied by numerical analysts [1-7, 9-14], but so far does not seem to have had much impact on practical calculations. The early results [9, 10] gave little idea of the rates of convergence, but by now much is known in this direction. It turns out that the gain achieved by the iteration can be very striking. For example Chandler [1-3] has shown, for the case of one-dimensional equations with sufficiently smooth inhomogeneous term f and kernel k, that

$$\max_t |y_h^{(1)}(t) - y(t)| = O(h^4), \tag{2}$$

which is to be compared with $O(h^2)$ for the Galerkin error. The same result also holds (see [12]) for integral equations in higher dimensions, if f and k are sufficiently smooth. (It is sufficient, for example, that the second-order partial derivatives of f and k be bounded and continuous.)

Recently Vidar Thomée and I [12] have sought precise orders of convergence for some of the difficult integral equations that arise in practice, such as Examples 1 and 2. In the remainder of this paper I shall describe some of the results of that work as applied to Examples 1 and 2, without giving any of the theory. It is worth saying that the setting in [12] is more general than that considered here: for example piecewise-polynomial spaces of arbitrary degree are considered, and more general error norms are allowed. The reader is referred to that paper for further details, and also for the precise conditions placed on Ω and S_h. A preliminary report of the results in [12], but presented from a somewhat different point of view, has appeared in [11].

ERROR ESTIMATES

Consider first Example 1, the case of the two-dimensional logarithm. We first consider the mean-square error, defined by

$$\|y_h - y\|_2 = [\int_\Omega |y_h(s) - y(s)|^2 \, d\sigma(s)]^{\frac{1}{2}} . \tag{3}$$

It is shown in [12] that with this definition of the error the Galerkin method achieves the optimal order of convergence for our case of piecewise-linear approximating functions, namely

$$\|y_h - y\|_2 = O(h^2) ,$$

whereas the iterated Galerkin method achieves the maximum possible order of superconvergence,

$$\|y_h^{(1)} - y\|_2 = O(h^4) .$$

At this stage I want to introduce another attractive feature of the iterated Galerkin method, namely that it can yield *derivatives* of very high quality. Intuitively this may perhaps be understood by observing, from (1), that the principal analytic features of the exact solution $y(t)$ are either present in the driving term $f(t)$ or arise from the integral term in (1). Thus $y_h^{(1)}(t)$, which has the same structure, of inhomogeneous term plus integral, might be expected to have the same analytic features as $y(t)$. However, even after this observation one might be surprised at the rapid convergence of the derivatives.

To be specific, for the case of Example 1 it is shown in [12] that the gradient of $y_h^{(1)}$ satisfies

$$\|\nabla y_h^{(1)} - \nabla y\|_2 = O(h^3) .$$

Thus even the gradient $\nabla y_h^{(1)}$ converges faster than the $O(h^2)$ rate that is characteristic of best approximation by piecewise-linear functions.

In practice one often prefers to deal with the maximum error

$$\|y_h - y\|_\infty = \max_t |y_h(t) - y(t)| , \tag{4}$$

rather than with the mean-square error defined by (3). For the case of Example 1 the maximum error estimates obtained in [12] converge slightly more slowly than the corresponding mean-square estimates: it is shown in [12] that

$$\|y_h - y\|_\infty = O(h^2 \log \frac{1}{h}) ,$$

$$\|y_h^{(1)} - y\|_\infty = O(h^4 \log^2 \frac{1}{h}) ,$$

$$\|\nabla y_h^{(1)} - \nabla y\|_\infty = O(h^3 \log^2 \frac{1}{h}) ,$$

whose only difference from the corresponding mean-square estimates above is that now one or more factors of $\log(1/h)$ occur. Roughly, the logarithmic factor in the first result, that for the Galerkin error $\|y_h - y\|_\infty$, arises because the partial derivatives of second order for the exact solution y are not quite bounded: they have singularities on the boundary of Ω, which can be traced to the logarithmic singularity in the kernel of Example 1. The logarithmic factors in the other estimates enter for related reasons.

Analogous results are also given in [12] for Example 2, the integral equation from potential theory, though there are a number of differences. Let us assume for simplicity that $f(t)$, the inhomogeneous term, is smooth. Then the first difference from the previous example is that $y(t)$ also is smooth. (In the previous example the essential difference is that Ω has a boundary. Here, however, Ω has no boundary, so that there is no reason for the solution to be singular at any point.) Thus the Galerkin error satisfies

$$\|y_h - y\|_\infty = O(h^2) ,$$

for our case of piecewise-linear trial functions. The first iterate $y_h^{(1)}$, on the other hand, is shown in [12] to satisfy

$$\|y_h^{(1)} - y\|_\infty = O(h^3 \log \frac{1}{h}) .$$

This is (roughly) one power of h better than the corresponding estimate for y_h, compared to two in the previous example.

The reason for the reduced effectiveness of the iteration in Example 2, compared to Example 1, is simply that the integral operator in Example 2 is markedly more singular than that in Example 1. For in essence superconvergence is produced by the smoothing properties of the integral operator in (1) – or more precisely, by the smoothing properties of the adjoint integral operator, obtained by replacing $k(t,s)$ by $\overline{k(s,t)}$ on the right-hand side of (1). The greater the smoothing produced by the adjoint integral operator, the faster is the superconvergence. Thus the reduced smoothing produced by the adjoint of the integral operator in Example 2, compared to that in Example 1, leads to reduced superconvergence of $y_h^{(1)}$

On the other hand Example 2 has a property not shared by Example 1, namely that a second iteration of the approximation will further improve the rate of convergence: with $y_h^{(2)}$ denoting the approximation obtained by substituting $y_h^{(1)}$ into the right-hand side of (1), it is shown in [12] that

$$\|y_h^{(2)} - y\|_\infty = O(h^4 \log^2 \frac{1}{h}) .$$

It is also shown in [12] that

$$\|\nabla y_h^{(2)} - \nabla y\|_\infty = O(h^3 \log^2 \frac{1}{h}) ,$$

where ∇y is the gradient of y with respect to suitable local coordinates. The improved rate of convergence of $y_h^{(2)}$ in this case rests on the fact that the adjoint of the integral operator in Example 2 has a successive smoothing property: repeated application of the operator yields a progressively smoother result. Such a property generally does not hold for a region with a boundary, as in Example 1.

DISCUSSION

It is clear that iteration of the Galerkin approximation can often significantly improve the theoretical rate of convergence, and can also yield derivative information of high quality. However, there are some problems:

(i) The above estimates assume that all the integrals that are needed in the Galerkin method (see the Appendix for details) are evaluated exactly. In practice approximate integrations are almost always necessary, and these need to be sufficiently accurate if the full order of convergence of $y_h^{(1)}$

is to be preserved. As yet little seems to be known about the accuracy that is needed, except for the case of a smooth kernel in one dimension (see [1, 3, 13]).

(ii) Similarly, additional approximations may be needed to handle curved boundaries, or curved regions Ω , but little work has yet been done on the effect of such approximations.

(iii) The error estimates quoted above give the order of convergence in terms of powers of h , but do not give the associated constants. It is well known that estimates of this kind are sometimes misleading, because the constants turn out to be large.

(iv) Very few numerical calculations have yet been carried out. (On the other hand the calculations that have been carried out are encouraging. For example, Dewar [4], using piecewise-constant trial functions in Example 1, found the approximation $y_h^{(1)}$ to be both qualitatively reliable and quantitatively accurate, even for relatively crude partitions of Ω .)

Much therefore remains to be done. However, perhaps the available results are already encouraging enough to persuade some users of the Galerkin method for integral equations of the second kind to try out the iterative variant. In principle the first iterate $y_h^{(1)}$ is easy enough to compute, because the information required, namely the integral operator in (1) applied to a basis in the finite-element space S_h , is already required for the computation of the Galerkin approximation itself. (To see this, compare equations (A6) and (A4) in the Appendix.)

APPENDIX

In this appendix we define precisely the Galerkin method y_h and its iterates $y_h^{(1)}, y_h^{(2)}, \ldots$.

The integral equation (1) may conveniently be written as

$$y = f + Ky \ . \qquad \text{(A1)}$$

Thus K is the integral operator on the right-hand side of equation (1), and $k(t,s)$ is its kernel.

Let $u_1, \ldots, u_n$ be a basis for the finite-element space S_h . Then the Galerkin approximation can be expressed as

$$y_h = \sum_{i=1}^{n} a_i u_i \ , \qquad \text{(A2)}$$

where the coefficients $a_1, \ldots, a_n$ have yet to be determined. The coefficients are determined by requiring that

$$(y_h - Ky_h, u_j) = (f, u_j) \ , \qquad j = 1, \ldots, n \ , \qquad \text{(A3)}$$

where the inner product is the natural one for the integral equation (1), namely

$$(v,w) = \int_\Omega v(s)\, \overline{w(s)} d\sigma(s) \ .$$

On substituting (A2) into (A3) we obtain a set of linear equations for the coefficients,

$$\sum_{i=1}^{n} [(u_i,u_j) - (Ku_i,u_j)]a_i = (f,u_j) , \quad j = 1, \ldots, n . \tag{A4}$$

The maximum-norm error estimates given in the text assume that the Galerkin method converges 'optimally' in the maximum norm as $h \to \infty$, i.e. that it converges as fast as the best uniform approximation of y by an element of S_h. This property often holds, at least for partitions that are 'quasi-uniform'. Precise conditions for this to hold are given in [12].

The first iterate of the Galerkin approximation is defined by

$$y_h^{(1)} = f + Ky_h . \tag{A5}$$

This can also be written, using (A2), as

$$y_h^{(1)} = f + \sum_{i=1}^{n} a_i Ku_i . \tag{A6}$$

Higher iterates are defined by

$$y_h^{(2)} = f + Ky_h^{(1)} , \tag{A7}$$

and so on.

REFERENCES

[1] Chandler, G.A., Global Superconvergence of Iterated Galerkin Solutions for Second Kind Integral Equations, Technical Report, Australian National University (1978).

[2] Chandler, G.A., Superconvergence for second kind integral equations, in: Anderssen, R.S., de Hoog, F.R., Lukas, M.A. (eds.), The Application and Numerical Solution of Integral Equations (Sijthoff and Noordhoff, Alphen aan den Rijn, 1980) 103-117.

[3] Chandler, G.A., Superconvergence of Numerical Solutions to Second Kind Integral Equations, Ph.D. Thesis, Australian National University (1979).

[4] Dewar, F.A., Solution of Fredholm Integral Equations by the Galerkin and Iterated Galerkin Methods, M.Sc. Thesis, University of New South Wales (1980).

[5] Graham, I.G., The Numerical Solution of Fredholm Integral Equations of the Second Kind, Ph.D. Thesis, University of New South Wales (1980).

[6] Graham, I.G., Galerkin methods for second kind integral equations with singularities, Math. Comp. 39 (1982) 519-533.

[7] Hsiao, G.C. and Wendland, W.L., The Aubin-Nitsche lemma for integral equations, J. Integral Equations 3 (1981) 299-315.

[8] Silvester, P., Modern Electromagnetic Fields (Prentice Hall, Englewood Cliffs NJ, 1968).

[9] Sloan, I.H., Error analysis for a class of degenerate-kernel methods, Numer. Math. 25 (1976) 231-238.

[10] Sloan, I.H., Improvement by iteration for compact operator equations, Math. Comp. 30 (1976) 758-764.

[11] Sloan, I.H., Superconvergence and the Galerkin method for integral equations of the second kind, in: Baker, C.T.H. and Miller, G.F. (eds.), Treatment of Integral Equations by Numerical Methods (Academic Press, London, 1982) 197-207.

[12] Sloan, I.H. and Thomée, V., Superconvergence of the Galerkin Iterates for Integral Equations of the Second Kind, Technical Report, University of New South Wales (August 1983).

[13] Spence, A. and Thomas, K.S., On Superconvergence Properties of Galerkin's Method for Compact Operator Equations, Technical Report, University of Southampton (1981).

[14] Wendland, W., Boundary element methods and their asymptotic convergence, to appear in: Lecture Notes of the CISM Summer School on Theoretical Acoustics and Numerical Techniques, Udine, Italy (Springer-Verlag Lecture Notes in Physics, Berlin).

OCEANIC AND ATMOSPHERIC PROBLEMS

Computational Techniques & Applications: CTAC-83
J. Noye & C. Fletcher (Editors)

WAVE PROPAGATION CHARACTERISTICS OF A NUMERICAL MODEL OF TIDAL MOTION

John Noye
University of Adelaide
South Australia

Errors in modelling tide propagation have two main sources. Firstly, the partial differential equations and associated boundary conditions may not describe correctly the physics of the flows involved. Secondly, the numerical technique used to solve these equations may not be accurate. Only if the physics is modelled correctly and the equations solved accurately can the tidal model be calibrated and used with confidence for predictive purposes. This article examines one way of assessing the accuracy of a particular numerical solution of the long wave equations used for modelling tidal propagation, namely the study of its wave propagation characteristics.

1. INTRODUCTION

Numerical methods of finding the response of coastal seas to tidal forcing have been developed progressively since the time of Hansen (1956). The importance of including terms such as sea-bed friction, Coriolis and convective accelerations have been studied by Welander (1964), Bretschneider (1967), and others. Omission of these terms results in an incomplete representation of the physical processes involved in the sea motions.

Jelesnianski (1965), Leendertse (1967), Reid and Bodine (1968), Heaps (1969), and others, have realistically represented both the geometrical and dynamical propertie of coastal seas by means of the long wave equations. Because of their non-linear nature and the complexity of the boundary conditions which must be satisfied, this set of multi-dimensional equations is not amenable to analytical solution techniques, and so must be solved using computational methods on a digital computer

Finite difference methods have been used almost exclusively in such solutions, although finite element methods may be used (see, for example, Taylor and Davis, 1975). Various finite difference procedures were used by the workers referred to above: another such method is described in Noye and Flather (1984).

When assessing a particular finite difference technique, stability has been the main focus of attention in the past (see, for example, Weare, 1976). However, "stability without accuracy has little to commend it " (Fox, 1962). Accuracy may be estimated by several methods, the usual ones being comparison of numerical solutions with exact solutions to simplified problems or a study of wave propagatio characteristics. In this paper, the method in Noye and Flather is assessed by mean of the latter criterion.

2. THE GOVERNING EQUATIONS

The motion of a homogenous coastal sea with constant density ρ, under the influence of tidal forcing at its junction with the open ocean, is described by the continuity equation and three components of the Navier-Stokes equation for a homogeneous, incompressible fluid. Direct numerical solutions of these equations are not yet feasible: instead, they may be Reynold's (time) averaged and then solved in three dimensions using one of the various turbulence closure schemes. Thus Noye et. al. (1981, 1982) and Bills and Noye (1984) define a realistic vertical eddy viscosity function to close their system of equations, while Arnold and Noye (1984) use a turbulent kinetic energy closure scheme.

The full three-dimensional solution only appears necessary where fluid density differs over depth or the vertical flow structure is highly variable, which is the case in very shallow water. In most other cases, an adequate description of tidal propagation may be obtained from a two dimensional vertically integrated form of the Reynolds equations, the so-called long wave equations (see Fortak, 1962, or Welander, 1964). These equations represent conservation of mass and momentum in the x and y spatial directions, as follow:

$$\frac{\partial\hat{\zeta}}{\partial t} + \frac{\partial\hat{U}}{\partial x} + \frac{\partial\hat{V}}{\partial y} = 0 \quad , \tag{2.1}$$

$$\frac{\partial\hat{U}}{\partial t} + \frac{\partial}{\partial x}\left(\frac{\hat{U}^2}{\hat{H}}\right) + \frac{\partial}{\partial y}\left(\frac{\hat{U}\hat{V}}{\hat{H}}\right) - f\hat{V} + g\hat{H}\frac{\partial\hat{\zeta}}{\partial x} + \frac{\tau_{bx}}{\rho} = 0 \quad , \tag{2.2}$$

$$\frac{\partial\hat{V}}{\partial t} + \frac{\partial}{\partial x}\left(\frac{\hat{U}\hat{V}}{\hat{H}}\right) + \frac{\partial}{\partial y}\left(\frac{\hat{V}^2}{\hat{H}}\right) + f\hat{U} + g\hat{H}\frac{\partial\hat{\zeta}}{\partial y} + \frac{\tau_{by}}{\rho} = 0 \ . \tag{2.3}$$

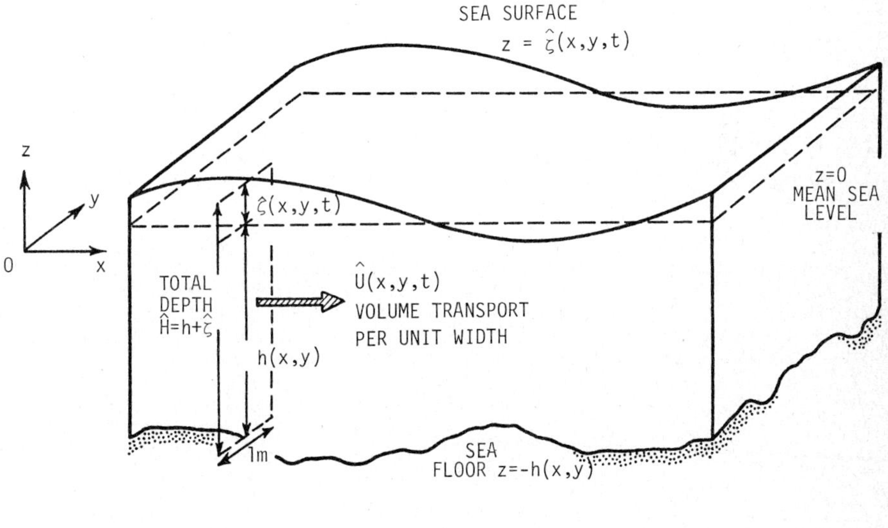

Figure 2.1: Definition of symbols used in the mass and momentum conservation equations.

The (x,y) datum plane (Figure 2.1) is located at the mean sea level with the z-axis directed vertically upwards. The water surface elevation above datum is $\hat{\zeta}(x,y,t)$, the sea floor is $h(x,y)$ below datum, and $\hat{U}$ and $\hat{V}$ are the components of total volume transport over depth, namely $\hat{Q} = (\hat{U},\hat{V})$ per unit width. The effect of bottom stress is represented through the shear stress vector $\underset{\sim}{\tau_b}(x,y,t)$, which has components τ_{bx} and τ_{by} . The Coriolis parameter, f, is given by

$$f = 2\Omega \sin \phi , \tag{2.4}$$

where Ω is the speed of rotation of Earth and ϕ is the latitude in degrees North. ρ is the density of the sea water (assumed constant) and $\hat{H}$ the total depth of water, namely,

$$\hat{H}(x,y,t) = h(x,y) + \hat{\zeta}(x,y,t) \quad . \tag{2.5}$$

At this point it is worth noticing the main assumptions and approximations involved in deriving equations (2.1 - 3) from the Navier-Stokes description of fluid motion. The flow field is assumed nearly horizontal with little vertical motion, the variation of velocity with depth is assumed small, the water mass is assumed homogeneous and incompressible, the horizontal component of the Coriolis force and the spherical shape of the earth are neglected, and frictional effects are confined to vertical shear only. All of these approximations can be justified, but they do introduce small errors in the mathematical representation of the physics involved.

Closure of the system of equations (2.1 - 3) requires a description of the bottom stress τ in terms of the other variables present. Firstly, it is assumed that the sea is sufficiently shallow for the flow to have the same direction over depth; this is not true in very deep water due to the Ekman-spiral effect (see Welander, 1964, Vreugdenhil and Voogt, 1975). The bottom stress vector may then be expressed as a quadratic function of the total transport $\hat{Q}$ and the water depth $\hat{H}$. One such form, based on the Darcy-Weisbach friction factor equation for open channel flow (ASCE, 1963) is

$$\underset{\sim}{\tau} = \frac{K\rho}{8H^2} |\underset{\sim}{\hat{Q}}| \underset{\sim}{\hat{Q}} \quad , \tag{2.6}$$

where K is a friction factor. This may be calculated using the asymptotic aproximation

$$K = \left\{ 2 \log_{10}\left(\frac{14.8\ H}{k}\right)\right\}^{-2} , \tag{2.7}$$

where k is the "roughness-height" for the type of sea floor, normally about 0.025m

The partial differential equations to be solved are therefore (2.1) together with

$$\frac{\partial \hat{U}}{\partial t} + \frac{\partial}{\partial x}\left(\frac{\hat{U}^2}{\hat{H}}\right) + \frac{\hat{U}}{\hat{H}}\frac{\partial \hat{V}}{\partial y} + \hat{V}\frac{\partial}{\partial y}\left(\frac{\hat{U}}{\hat{H}}\right) - f\hat{V} + g\hat{H}\frac{\partial \hat{\zeta}}{\partial x} + \frac{K\hat{Q}\hat{U}}{8\hat{H}^2} = 0 \quad , \tag{2.8}$$

$$\frac{\partial \hat{V}}{\partial t} + \frac{\hat{V}}{\hat{H}}\frac{\partial \hat{U}}{\partial x} + \hat{U}\frac{\partial}{\partial x}\left(\frac{\hat{V}}{\hat{H}}\right) + \frac{\partial}{\partial y}\left(\frac{\hat{V}^2}{\hat{H}}\right) + f\hat{U} + g\hat{H}\frac{\partial \hat{\zeta}}{\partial y} + \frac{K\hat{Q}\hat{V}}{8\hat{H}^2} = 0 \quad , \tag{2.9}$$

within the coastal sea, subject to associated boundary conditions of no flow normal to the coastline and the prescription of either the sea surface variation or normal flow across the connection with the open ocean. The expansion of the third term in Equation (2.2) to give the third and fourth terms of (2.8), and the expansion of the second term in Equation (2.3) to give the second and third terms in (2.9), were carried out by Noye and Flather (1984) to permit easier discretisation of these terms when (2.8) and (2.9) are applied at the lateral boundaries.

3. A NUMERICAL SOLUTION

In order to apply the finite difference method to solve Equations (2.1) and (2.8-9), the region of flow, the coastline, the sea bathymetry and the junction with the open ocean, are represented on a rectangular grid as in Figure 3.1. At each time level $t_n = n\Delta t$, discrete values of the variables are specified in space staggered form as in Figure 3.2, with grid spacings Δx, Δy in the x and y directions respectively. The computer model described in Noye and Flather (1984) was specifically designed for use on computers with small storage capacity. To do this, storage was minimised by arranging the ζ, U, V variables in one-dimensional arrays in such a way that no variables are assigned to land regions. However, for convenience in the following analysis, a double subscripted spatial array is used. Thus the position of a variable is denoted in terms of grid-points labelled (j,k,n) which represent points $(j\Delta x, k\Delta y, n\Delta t)$. The elevation $\hat{\zeta}$ and the depth of water h below mean sea level are then located at grid-points (j,k,n), the volumetric flow $\hat{U}$ at the grid-points $(j+\frac{1}{2}, k, n)$ and the volumetric flow $\hat{V}$ at $(j, k+\frac{1}{2}, n)$ grid-points.

The long-wave equations may be solved by various numerical methods (see Hinwood and Wallis, 1975 a,b). Thus Leendertse (1967) uses an Alternating Direction Implicit Method, and Harper, Sobey and Stark (1978) use a Leapfrog Scheme. Noye and Flather (1984) use central differences in space with forward or backward differences in time. Further space saving in the computer memory was achieved by always using the latest available variable values when calculating spatial derivatives, thereby reducing the number of variables to be stored. It also resulted in updated values of the variable being calculated in an explicit manner.

In this system, the finite difference representation of the continuity equation (2.1) is

$$\frac{\zeta^{n+1}_{j,k} - \zeta^{n}_{j,k}}{\Delta t} + \frac{U^n_{j+\frac{1}{2},k} - U^n_{j-\frac{1}{2},k}}{\Delta x} + \frac{V^n_{j,k+\frac{1}{2}} - V^n_{j,k-\frac{1}{2}}}{\Delta y} = 0 , \qquad (3.1)$$

where $\zeta^n_{j,k}$ is an approximation to the value of $\hat{\zeta}(j\Delta x, k\Delta y, n\Delta t)$,

$U^n_{j+\frac{1}{2},k}$ is an approximation to $\hat{U}((j+\frac{1}{2})\Delta x, k\Delta y, n\Delta t)$,

$V^n_{j,k+\frac{1}{2}}$ is an approximation to $\hat{V}(j\Delta x,(k+\frac{1}{2})\Delta y, n\Delta t)$.

Thus, knowing the approximations ζ,U,V at the n-th time level, the approximation ζ at the (n+1)th time level is computed using

$$\zeta^{n+1}_{j,k} = \zeta^n_{j,k} - r_x(U^n_{j+\frac{1}{2},k} - U^n_{j-\frac{1}{2},k}) - r_y(V^n_{j,k+\frac{1}{2}} - V^n_{j,k-\frac{1}{2}}) , \qquad (3.2)$$

where $r_x = \Delta t/\Delta x$, $r_y = \Delta t/\Delta y$.

The x-momentum equation (2.8) is discretised in a similar way; values of U are then computed at the (n+1)th time level from values of ζ at the same time level, and values of U, V at the n-th time level. Finally, the y-momentum equation (2.9) is discretised which leads to values of V being computed at the (n+1)th time level from values of ζ, U at the same time level, and values of V at the n-th time level. Complete details of the discretisation of Equations (2.1), (2.8 - 9) for all grid-points, including the modified forms required near boundaries, are given in Noye and Flather (1984).

The fact that the solution technique demands that the derivatives in the long wave equations be approximated by finite difference forms introduces a further source of error in addition to approximations made in deriving the long wave

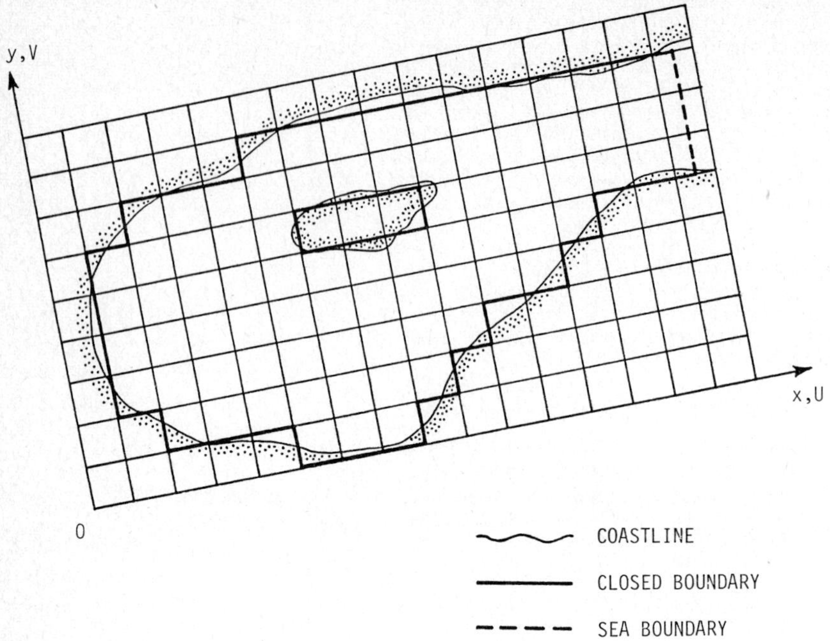

Figure 3.1: Example of coastal sea with grid super-imposed.

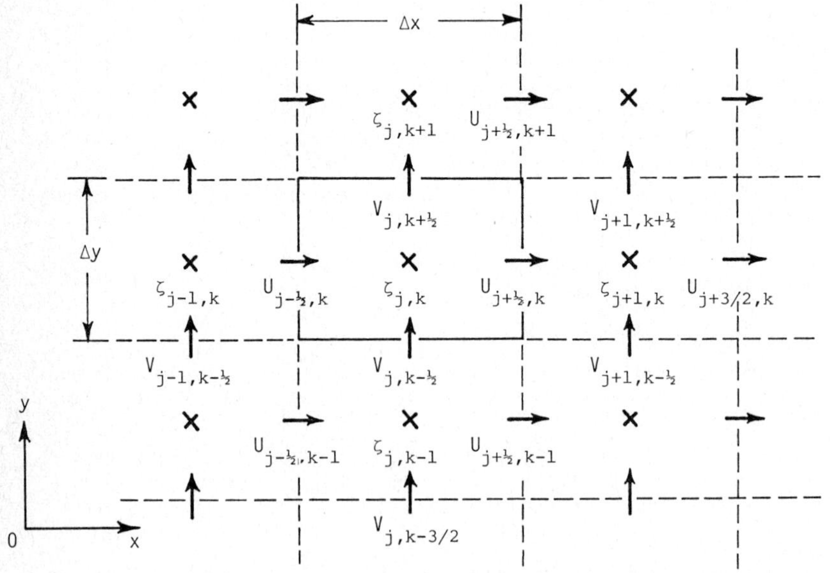

Figure 3.2: Position of the variables in a two-dimensonal array. The lines in this figure correspond to the grid-lines marked in Figure 3.1

equations. The accuracy of any numerical method may be estimated in a number of ways, including comparison with analytic solutions to problems which may have been simplified by linearisation and/or simplification of the sea geometry (see Lynch and Gray, 1978). This approach was used in testing the three-dimensional model described in Bills and Noye (1984). Another method is to determine the wave propagation characteristics of the numerical technique. This involves application of the same finite difference technique to a quasi-linearised form of Equations (2.1), (2.8 - 9), and comparison of the resulting numerical solution with the corresponding analytic solution for an initial condition consisting of an infinitely long wave.

4. WAVE PROPAGATION CHARACTERISTICS

This approach has been used to estimate the accuracy of finite difference solutions of the transport and long-wave equations by Leendertse (1967), Sobey (1970), Abbott (1979), Tong (1980) and Noye (1983).

The locally linearised forms of the governing equations are

$$\frac{\partial\hat{\zeta}}{\partial t} + \frac{\partial\hat{U}}{\partial x} + \frac{\partial\hat{V}}{\partial y} = 0 \quad , \tag{4.1}$$

$$\frac{\partial\hat{U}}{\partial t} + \frac{U_a}{H_a}\frac{\partial\hat{U}}{\partial x} + \frac{U_a}{H_a}\frac{\partial\hat{V}}{\partial y} + \frac{V_a}{H_a}\frac{\partial\hat{U}}{\partial y} - f\hat{V} + gH_a\frac{\partial\hat{\zeta}}{\partial x} + \frac{KQ_a}{8H_a^2}\hat{U} = 0 \quad , \tag{4.2}$$

$$\frac{\partial\hat{V}}{\partial t} + \frac{V_a}{H_a}\frac{\partial\hat{U}}{\partial x} + \frac{U_a}{H_a}\frac{\partial\hat{V}}{\partial x} + \frac{V_a}{H_a}\frac{\partial\hat{V}}{\partial y} + f\hat{U} + gH_a\frac{\partial\hat{\zeta}}{\partial y} + \frac{KQ_a}{8H_a^2}\hat{V} = 0 \quad , \tag{4.3}$$

where U_a, V_a, H_a, Q_a, f_a are constant and representative of the magnitude of $\hat{U}$, $\hat{V}$, $\hat{H}$, $\hat{Q}$, and f in the neighbourhood of $(j\Delta x, k\Delta y)$. Using the same method of discretisation as Noye and Flather (1984), the corresponding finite-difference forms to (4.1 - 3) are:

$$\zeta^{n+1}_{j,k} = \zeta^n_{j,k} - r_x(U^n_{j+\frac12,k} - U^n_{j-\frac12,k}) - r_y(V^n_{j,k+\frac12} - V^n_{j,k-\frac12}) \quad , \tag{4.4}$$

$$\begin{aligned} U^{n+1}_{j+\frac12,k} &= (1-K')U^n_{j+\frac12,k} - U'_x(U^n_{j+\frac32,k} - U^n_{j-\frac12,k}) - 2U'_y(\overline{V}^n_{j+\frac12,k+\frac12} - \overline{V}^n_{j+\frac12,k-\frac12}) \\ &\quad - V'_y(U^n_{j+\frac12,k+1} - U^n_{j+\frac12,k-1}) + f'\,\overline{V}^n_{j+\frac12,k} - r_xS^2(\zeta^{n+1}_{j+1,k} - \zeta^{n+1}_{j,k}) \quad , \end{aligned} \tag{4.5}$$

$$\begin{aligned} V^{n+1}_{j,k-\frac12} &= (1-K')V^n_{j,k-\frac12} - 2V'_x(\overline{U}^{n+1}_{j+\frac12,k-\frac12} - \overline{U}^{n+1}_{j-\frac12,k-\frac12}) - U'_x(V^n_{j+1,k-\frac12} - V^n_{j-1,k-\frac12}) \\ &\quad - V'_y(V^n_{j,k+\frac12} - V^n_{j,k-\frac32}) - f'\,\overline{U}^{n+1}_{j,k-\frac12} - r_yS^2(\zeta^{n+1}_{j,k} - \zeta^{n+1}_{j,k-1}) \quad , \end{aligned} \tag{4.6}$$

where

$$U'_x = r_xU_a/2H_a \quad , \qquad V'_x = r_xV_a/2H_a \quad ,$$

$$U'_y = r_yU_a/2H_a \quad , \qquad V'_y = r_yV_a/2H_a \quad ,$$

$$f' = \Delta t f_a \quad , \qquad K' = \Delta t\, KQ_a/8H_a^2, \qquad S^2 = gH_a \quad ,$$

$$\overline{U}^{n+1}_{j+\frac12,k-\frac12} = \frac{1}{2}(U^{n+1}_{j+\frac12,k-1} + U^{n+1}_{j+\frac12,k}) \quad ,$$

$$\overline{V}^{n}_{j+\frac{1}{2},k+\frac{1}{2}} = \frac{1}{2}(V^{n}_{j,k+\frac{1}{2}} + V^{n}_{j+1,k+\frac{1}{2}}) \quad ,$$

$$\overline{\overline{U}}^{n+1}_{j,k-\frac{1}{2}} = \frac{1}{4}(U^{n+1}_{j+\frac{1}{2},k} + U^{n+1}_{j-\frac{1}{2},k} + U^{n+1}_{j+\frac{1}{2},k-1} + U^{n+1}_{j-\frac{1}{2},k-1}) \quad ,$$

$$\overline{\overline{V}}^{n}_{j+\frac{1}{2},k} = \frac{1}{4}(V^{n}_{j+1,k+\frac{1}{2}} + V^{n}_{j,k+\frac{1}{2}} + V^{n}_{j+1,k-\frac{1}{2}} + V^{n}_{j,k-\frac{1}{2}}) \quad .$$

We now consider the behaviour of the system of equations (4.4 - 6) in the large, that is with the computational element assumed far away from the disturbing influence of any boundary. In fact, we will seek a solution to the initial value problem

$$\underset{\sim}{\zeta}^{\circ} = \sum_{p} \underset{\sim}{\zeta}^{*}_{p} \exp\{im_{p}s\} \quad , \quad -\infty < s < \infty \quad , \tag{4.7}$$

where m_p is the wave number of the p-th Fourier component,

i is the operator $\sqrt{-1}$,

s is the coordinate in the direction of propagation of the plane waves,

$\underset{\sim}{\zeta}^{n} = (\zeta^{n}, U^{n}, V^{n})^{T}$ is the solution of (4.4 -6) at time $n\Delta t$,

$\underset{\sim}{\zeta}^{*}_{p} = (\zeta^{*}_{p}, U^{*}_{p}, V^{*}_{p})^{T}$ is the amplitude of the p-th Fourier component of $\underset{\sim}{\zeta}^{\circ}$.

If the direction of wave propagation is at the angle θ_p to the positive direction of the x-axis, we define wave number components

$$m_{px} = m_p \cos\theta_p \quad , \quad m_{py} = m_p \sin\theta_p \quad , \tag{4.8}$$

so that

$$\underset{\sim}{\zeta}^{\circ} = \sum_{p} \underset{\sim}{\zeta}^{*}_{p} \exp\{i(m_{px}x + m_{py}y)\} \quad . \tag{4.9}$$

Since the system of equations (4.4 -6) is linear, superposition is valid and only one component of the Fourier series need be considered, namely

$$\underset{\sim}{\zeta}^{\circ} = \underset{\sim}{\zeta}^{*} \exp\{i(m_{x}x + m_{y}y)\} \quad , -\infty < x < \infty \quad , -\infty < y < \infty \quad , \tag{4.10}$$

in which

$$m^2 = (m_x)^2 + (m_y)^2 \quad . \tag{4.11}$$

We seek a solution of the form

$$\underset{\sim}{\zeta}^{n}_{j,k} = \underset{\sim}{\zeta}^{*} \exp\{i(\gamma n\Delta t + m_{x}j\Delta x + m_{y}k\Delta y)\} \quad . \tag{4.12}$$

Direct substitution of (4.12) into Equations (4.4 -6) with $\overline{\overline{U}} = \overline{U} = U$, $\overline{\overline{V}} = \overline{V} = V$, yields the matrix equation

$$\begin{bmatrix} (\xi - 1) & i\{2r_x\sin(\beta_x/2)\} & i\{2r_y\sin(\beta_y/2)\} \\ i\{2r_xS^2\sin(\beta_x/2)\}\xi & \xi-1+K'+i\{2U'_x\sin\beta_x+2V'_y\sin\beta_y\} & i\{4U'_y\sin(\beta_y/2)\} - f' \\ i\{2r_yS^2\sin(\beta_y/2)\}\xi & [i\{4V'_x\sin(\beta_x/2)\}+f']\xi & \begin{matrix}\xi - 1 + K' \\ +i\{2U'_x\sin\beta_x+2V'_y\sin\beta_y\}\end{matrix} \end{bmatrix} \begin{bmatrix} \zeta^* \\ U^* \\ V^* \end{bmatrix} = \underset{\sim}{0}$$

where $\xi = \exp(i\gamma\Delta t)$, $\beta_x = m_x\Delta x$ and $\beta_y = m_y\Delta y$. A solution of the form (4.12) exists if and only if the determinant of the coefficient matrix is zero. When $\Delta x = \Delta y = \Delta s$, so $r_x = r_y = r$, $U'_x = U'_y = U'$ and $V'_x = V'_y = V'$, this leads to the cubic equation for ξ ,

$$\xi^3 + E\xi^2 + F\xi + G = 0 \quad , \tag{4.13}$$

where $E = -1 + 2A - BD + 4Cr^2(L-PD)$,

$F = -2A + BD + A^2 + 4Cr^2(AL-BP)$,

$G = -A^2$,

where

$A = i\{2U' \sin\beta_x + 2V' \sin\beta_y\} + K' - 1$,

$B = i\{4U' \sin(\beta_y/2)\} - f'$,

$Cr = rS$,

$D = i\{4V' \sin(\beta_x/2)\} + f'$,

$L = \sin^2(\beta_x/2) + \sin^2(\beta_y/2)$,

$P = \sin(\beta_x/2)\sin(\beta_y/2)$.

Clearly Equations (4.4 -6) will remain stable so long as $|\xi| \leqslant 1$. However, this does not imply that the solution of these equations is a good approximation to the solution of Equations (4.1 - 3) with initial condition (4.10). In order to find this we seek a solution of the form

$$\hat{\underset{\sim}{\zeta}}(x,y,t) = \underset{\sim}{\zeta}^* \exp\{i(\hat{\gamma}t + m_x x + m_y y)\} \quad , \tag{4.14}$$

which will be referred to as the true wave.

Substitution of (4.14) into Equations (4.1 - 3) yields the matrix equation

$$\begin{bmatrix} \hat{\gamma} & m_x & m_y \\ S^2 m_x & \hat{\gamma}+\delta & R \\ S^2 m_y & W & \hat{\gamma}+\delta \end{bmatrix} \begin{bmatrix} \zeta^* \\ U^* \\ V^* \end{bmatrix} = \underset{\sim}{0} \quad , \tag{4.15}$$

in which $R = m_y U_a/H_a + if$, $W = m_x V_a/H_a - if$, and

$$\delta = m_x \frac{U_a}{H_a} + m_y \frac{V_a}{H_a} - i\frac{KQ_a}{8H_a^2} \quad . \tag{4.16}$$

For Equation (4.15) to have a non-trivial solution $[\zeta^*, U^*, V^*]^T$, the determinant of the coefficient matrix must be zero. This yields the cubic equation

$$\hat{\gamma}^3 + 2\delta\hat{\gamma}^2 - [S^2m^2+RW-\delta^2]\hat{\gamma} + S^2[m_x m_y(R+W) - m^2\delta] = 0 \quad , \tag{4.17}$$

from which the appropriate value of the complex frequency $\hat{\gamma}$ of the true wave may be found. This can then be compared with the complex frequency γ of the computed wave, which is given by

$$\gamma = \frac{-i}{\Delta t}\,\ell n\,\xi \tag{4.18}$$

where ξ is found by solving the cubic equation (4.13).

Two parameters which describe the differences between the computed wave and the true wave are the ratios of the amplitudes and the speeds of these two waves. These may be derived from a knowledge of γ and $\hat{\gamma}$ in the following way.

If

$$\hat{\sigma} = - \text{Re}\,\{\hat{\gamma}\}\ , \quad \hat{\eta} = \text{Im}\,\{\hat{\gamma}\}$$

then

$$\hat{\gamma} = - \hat{\sigma} + i\hat{\eta}\ . \tag{4.19}$$

Also, if

$$\zeta^* = a \exp\,\{i\varepsilon\}\ , \quad a \text{ and } \varepsilon \text{ real and positive,}$$

then we may write Equation (4.14) as

$$\hat{\zeta} = a \exp\{-\hat{\eta}t\}.\exp\{i(ms - \hat{\sigma}t + \varepsilon)\}\ ,$$

so that

$$\text{Re}\{\hat{\zeta}\} = ae^{-\hat{\eta}t} \cos(ms-\hat{\sigma}t+\varepsilon)\ . \tag{4.20}$$

The true wave is therefore travelling with speed

$$u_T = \hat{\sigma}/m \tag{4.21}$$

in the positive s-direction, and after one wave period of $t=2\pi/|\hat{\sigma}|$ the amplitude is multiplied by the factor

$$\alpha_T = \exp(-2\pi\hat{\eta}/|\hat{\sigma}|)\ . \tag{4.22}$$

Similar expressions, u_N and α_N, hold for the computed wave.

Therefore, after one wave period the ratio of the amplitudes of the computed and true waves is

$$\chi = \alpha_N/\alpha_T = \exp\{-2\pi(\eta/|\sigma| - \hat{\eta}/|\hat{\sigma}|)\}. \tag{4.23}$$

This may be called the amplitude response/wave period of the method, namely

$$\chi = \exp\{2\pi(\text{Im}\{\hat{\gamma}\}/\text{Re}\{\hat{\gamma}\} - \text{Im}\{\gamma\}/\text{Re}\{\gamma\})\}, \quad \text{Re}\{\hat{\gamma}\} < 0, \quad \text{Re}\{\gamma\} < 0. \tag{4.24}$$

The relative wave speed is

$$\mu = u_N/u_T = \sigma/\hat{\sigma} = \text{Re}\{\gamma\}/\text{Re}\{\hat{\gamma}\}\ . \tag{4.25}$$

Values of χ and μ can be computed for the expected range of values of U', V', K', f' and expressed as functions of N_λ, the number of grid-spacings per wave length λ,

$$N_\lambda = \lambda/\Delta s = 2\pi/m\Delta s\ , \tag{4.26}$$

and the Courant number Cr. Ideally, we wish χ and μ to have the value 1 for all N_λ and Cr.

5. THE GULF OF CARPENTARIA MODEL

The updated finite difference method of solving the long wave equations has been applied to the Gulf of Carpentaria by Mitchell et al (1984). Typical values of the parameters in this model are Δt = 90s, Δs = 25×10^3m, $U_a = V_a = 0$ (that is, the mean current is zero), H_a = 50m, Q_a = $0.7m^2/s$, $f_a = -3.2\times10^{-5}s^{-1}$. The corresponding values of the non-dimensional parameters are

$$U' = 0, \quad V' = 0, \quad K' = 3\times10^{-5}, \quad f' = -2.8\times10^{-3}. \tag{5.1}$$

The amplitude response per wave period and the relative wave speed for plane waves travelling parallel to the x-axis ($\theta_p = 0^0$), obliquely ($\theta_p = 30^0$) and diagonally ($\theta_p = 45^0$) have been computed for these parameters.

In all cases, the value of the amplitude response per wave period, χ, was one to three decimal places, and the quasi-linearised version of the method was stable whenever the Courant number, Cr, did not exceed .78. In fact, for $\theta_p = 0^0$, the linearised version of the method was stable for Cr $\leq$ .999 for $\theta_p = 30^0$ it was stable for Cr $\leq$ .828 and for $\theta_p \leq 45^0$ it was stable for Cr $\leq$ .789. The graphs of the relative wave speed, μ, are shown in Figure 5.1 for $\theta_p = 0^0$, in Figure 5.2 for $\theta_p = 30^0$, and in Figure 5.3 for $\theta_p = 45^0$. For small values of N_λ, the number of gridspacings per wavelength, the numerical wave speed differs noticeably from the true wave speed. However for $N_\lambda > 20$ the error is less than 0.4%. For waves travelling along the axes the numerical wave speeds are always too small (see Figure 5.1), whereas, for waves propagating obliquely to the axes modelling with a value of Cr near one produces numerical wave speeds which are greater than the true wave speed (see Figures 5.2-3).

For the Gulf of Carpentaria model, the parameters given above yield Cr $\simeq$ 0.1 and $N_\lambda \simeq 40$, when the error in the wave speed is less than 0.1% and in the amplitude it is less than 0.001%. Based on wave propagation characteristics as a measure of accuracy, the numerical methods described is very accurate so long as there is sufficient spatial resolution of the waves being modelled.

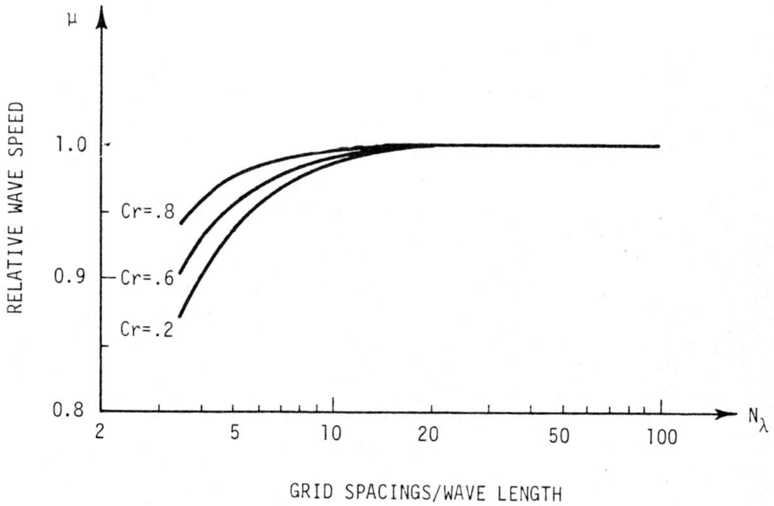

Figure 5.1: The relative speed, μ, for long waves propagating along the x-axis ($\theta_p = 0^0$).

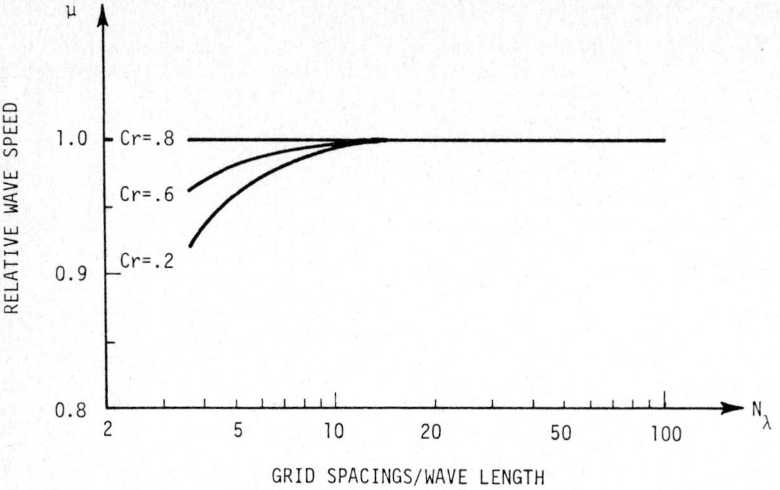

Figure 5.2: The relative wave speed, μ, for waves propagating obliquely ($\theta_p = 30^o$) to the x-axis.

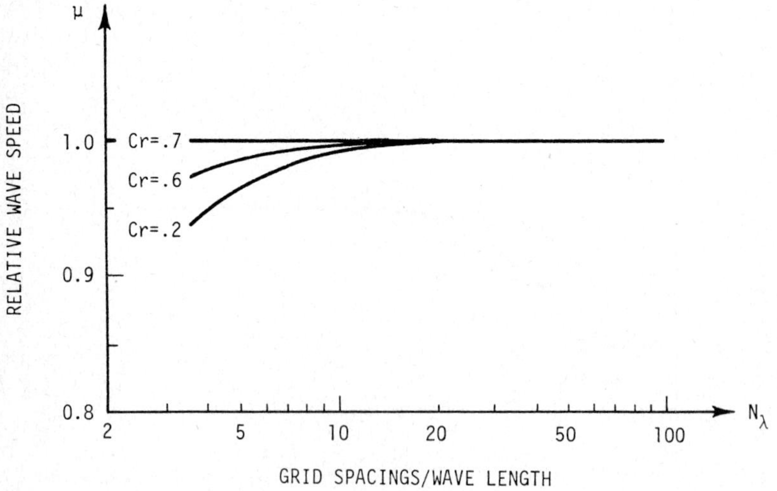

Figure 5.3: The relative wave speed, μ, for waves propagating diagonally across the grid ($\theta_p = 45^o$)

6. MODIFIED DISCRETISATION OF THE BOTTOM STRESS TERM

In order to improve the stability of finite difference analogues of the long-wave equations, the bottom stress terms in the x and y momentum equations are sometimes discretised with the velocity components evaluated at the new time level. That is, the $\hat{U}$ and $\hat{V}$ values in the last terms on the left side of Equations (2.8) and (2.9) are written at the new level $(n+1)$, instead of at the old time level n. We now examine the effect of this on the wave propagation characteristics of the numerical model.

Discretisation of the quasi-linear forms of these momentum equations, namely (4.2) and (4.3), gives the explicit finite difference equations

$$U^{n+1}_{j+\frac{1}{2},k} = K_\star \left\{ \begin{array}{l} U^n_{j+\frac{1}{2},k} - U'_x(U^n_{j+\frac{3}{2},k} - U^n_{j-\frac{1}{2},k}) \\ -2U'_y(\overline{V}^n_{j+\frac{1}{2},k+\frac{1}{2}} - \overline{V}^n_{j+\frac{1}{2},k-\frac{1}{2}}) - V'_y(U^n_{j+\frac{1}{2},k+1} - U^n_{j+\frac{1}{2},k-1}) \\ +f'\overline{\overline{V}}^n_{j+\frac{1}{2},k} - r_x S^2(\zeta^{n+1}_{j+1,k} - \zeta^{n+1}_{j,k}) \end{array} \right\} \quad (6.1)$$

$$V^{n+1}_{j,k-\frac{1}{2}} = K_\star \left\{ \begin{array}{l} V^n_{j,k-\frac{1}{2}} - 2V'_x(\overline{U}^{n+1}_{j+\frac{1}{2},k-\frac{1}{2}} - \overline{U}^{n+1}_{j-\frac{1}{2},k-\frac{1}{2}}) \\ -U'_x(V^n_{j+1,k-\frac{1}{2}} - V^n_{j-1,k-\frac{1}{2}}) - V'_y(V^n_{j,k+\frac{1}{2}} - V^n_{j,k-\frac{3}{2}}) \\ -f'\overline{\overline{U}}^{n+1}_{j,k-\frac{1}{2}} - r_y S^2(\zeta^{n+1}_{j,k} - \zeta^{n+1}_{j,k-1}) \end{array} \right\} \quad (6.2)$$

where $K_\star = (1+K')^{-1}$. Substitution of (4.12) into Equations (4.4), (6.1) and (6.2), with $\overline{\overline{U}} = \overline{U} = U$, $\overline{\overline{V}} = \overline{V} = V$, and $\Delta x = \Delta y = \Delta s$, gives the matrix equation

$$\begin{bmatrix} (\xi - 1) & i\{2r\sin(\beta_x/2)\} & i\{2r\sin(\beta_y/2)\} \\ i\{2K_\star rS^2\sin(\beta_x/2)\}\xi & \xi + M & K_\star B \\ i\{2K_\star rS^2\sin(\beta_y/2)\}\xi & K_\star D\xi & \xi + M \end{bmatrix} \begin{bmatrix} \zeta^* \\ U^* \\ V^* \end{bmatrix} = \underset{\sim}{0} ,$$

where $M = K_\star\{i(2U'\sin\beta_x + 2V'\sin\beta_y) - 1\}$. Setting the determinant of the coefficient matrix to zero gives the cubic equation satisfied by ξ, namely

$$\xi^3 + E_\star\xi^2 + F_\star\xi + G_\star = 0 , \quad (6.3)$$

where

$$E_\star = 2M - BD(K_\star)^2 - 1 + 4K_\star LCr^2 - 4(K_\star)^2 DPCr^2$$

$$F_\star = M^2 - 2M + BD(K_\star)^2 + 4K_\star LMCr^2 - 4(K_\star)^2 BPCr^2$$

$$G_\star = -M^2$$

With the values of U', V', K', f' given in (5.1) for the Gulf of Carpentaria, the amplitude response per wave period and the relative wave speed for this modified method were calculated for $\theta_p = 0^0$ and 45^0. The results were the same to three significant figures as those obtained by the original method, and the stability range was the same.

7. CONCLUSION

The wave-propagation characteristics of the updated explicit finite difference method of solving the long-wave equations which govern tidal motion, indicates that it can be very accurate when applied to the modelling of tides in coastal seas such as the Gulf of Carpentaria. In particular, the amplitude response per wave period is practically one for all values of the Courant number, Cr, and all numbers of grid-spacings per wave length, N_λ. However, the relative wave speed may differ significantly from one if the value of N_λ is less than 20. This implies that all waves of importance which are to be modelled, must be resolved so that their wavelengths are represented spatially by more than 20 grid spacings.

The modified method, in which the friction term is discretised at the new time-level instead of the old, appears to offer no improvement over the original method, according to this analysis of the quasi-linearised system using typical parameters for the Gulf of Carpentaria. Both the original and the modified methods are stable for Courant numbers less than one, and are very accurate so long as the number of gridspacings in a wavelength exceeds twenty.

ACKNOWLEDGEMENT

The computer programs from which Figures 5.1 - 3 were prepared, were written by D. Beard and R. Arnold.

REFERENCES

Abbott, M.B. (1979), *Computational Hydraulics,* Pitman, London.

Arnold, R.J., and Noye, B.J. (1984), "On the performance of turbulent energy closure schemes for wind driven flows in shallow seas," *Computational Techniques and Applications,* editor J. Noye, North-Holland Publishing Company, Amsterdam. To be published.

ASCE Task Force On Friction Factors In Open Channels (1963), "Friction factors in open channels", *Journal of the Hydraulics Division, Proceedings of ASCE,* Vol. 89, pp. 97-143.

Bills, P.J., and Noye, B.J. (1984), "Verification of a three-dimensional tidal model for coastal seas". *Computational Techniques and Applications,* editor John Noye, North-Holland Publishing Company, Amsterdam. To be published.

Bretschneider, C.L. (1967), "Storm surges", *Advances in Hydroscience,* Vol. 4, pp. 321-419.

Fortak, H.G. (1962), *Hydrodynamic Equations with Respect to Basic Storm Surge Equations,* National Hurricane Research Project, Report No. 21.

Fox, L. (1962), *Numerical Solution of Ordinary and Partial Differential Equations,* Addison-Wesley Publishing Company, Inc., Reading, Massachusetts.

Hansen, W. (1956), "Theorie zur Errechnung des Wasserstandes und der Stromungen in Randmeeren nebst Anwendungen", *Tellus,* Vol. 8, pp. 287-300.

Heaps, N.S. (1969), "A two-dimensional numerical sea model", *Proceedings of the Royal Society of London, Series A,* Vol. 265, pp. 93-137.

Hinwood, J.B., and Wallis, I.G. (1975a), "Classification of models of tidal waters", *Journal of the Hydraulics Division, Proceedings of ASCE,* Vol. 101, pp. 1315-1331.

Hinwood, J.B., and Wallis, I.G. (1975b), "Review of models of tidal waters", *Journal of the Hydraulics Division, Proceedings of ASCE,* Vol. 101, pp. 1405-1421. (Also discussions by Abraham, G. and Karelse, N., 1976, Vol. 102, pp. 808-811, and Abbott, M.B., 1976, Vol. 102, pp. 1145-1148.)

Jelesnianski, C.P. (1965), "A numerical calculation of storm tides induced by a tropical storm impinging on a continental shelf", *Monthly Weather Review,* Vol. 93, pp. 343-358.

Leendertse, J.J. (1967), *Aspects of a Computational Model for Long Period Water-Wave Propagation,* Memorandum of Rand Corporation, Santa Monica, RM-5294-PR.

Lynch, D.R., and Gray, W.G. (1978), "Analytic solutions for computer flow model testing", *Journal of the Hydraulics Division, Proceedings of ASCE,* Vol. 104, pp. 1409-1428.

Mitchell, W.M., Beard, D.A., Bills, P.J., and Noye, B.J. (1984), "An application of a three-dimensional tidal model to the Gulf of Carpentaria". *Computational Techniques and Applications,* editor John Noye, North-Holland Publishing Company, Amsterdam. To be published.

Noye, B.J. (1983), "Finite difference techniques for partial differential equations", *Computational Techniques for Differential Equations,* editor John Noye, North-Holland Publishing Company, Amsterdam, pp. 95-354.

Noye, B.J., and Flather, R.A. (1984), "Hydrodynamical-numerical modelling of tides and storm surges". *Tides and Surges,* editor John Noye, Monograph of the American Geophysical Union. To be published.

Noye, B.J., May, R.L., and Teubner, M.D. (1981), "Three-dimensional numerical model of tides in Spencer Gulf", *Ocean Management,* Vol. 6, pp. 137-148.

Noye, B.J., May, R.L., and Teubner, M.D. (1982), "A three-dimensional tidal model for a shallow gulf", *Numerical Solutions of Partial Differential Equations,* editor John Noye, North-Holland Publishing Company, Amsterdam, pp. 417-436.

Reid, R.O., and Bodine, B.R. (1968), "Numerical model for storm surges in Galveston Bay", *Journal of the Waterways and Harbors Division, Proceedings of ASCE,* Vol. 94, pp. 33-57.

Sobey, R.J. (1970), *Finite Difference Schemes Compared for Wave Deformation Characteristics in Mathematical Modelling of Two-Dimensional Long Wave Propagation,* U.S. Army Corps of Engineers, Coastal Engineering Research Center, Technical Memorandum, No. 32.

Taylor, C., and Davis, J. (1975), "Tidal and long wave propagation - A finite element approach", *Computers and Fluids,* Vol. 3, pp. 125-148.

Tong, G.D. (1980), "Environmental hydrodynamic modelling problem identification and a strategy to model assessment", *Industrialised Embayments and their Environmental Problems*, edited by Collins et. al., Pergamon Press, Oxford, pp. 383-392.

Vreugdenhil, C.B., and Voogt, J. (1975), "Hydrodynamic transport phenomena in estuaries and coastal waters: Scope of mathematical models", *Proceedings of the Symposium on Modelling Techniques, ASCE,* San Francisco, pp. 690-708.

Weare, T.J. (1976), "Instability of tidal flow computational schemes", *Journal of the Hydraulics Division, Proceedings of ASCE,* Vol. 102, pp. 569-580. (Also discussion by Abbott, M.B., 1976, Vol. 102, pp. 1787-1790, and Leendertse, J.J., 1977, Vol. 103, pp. 206-207.)

Welander, P. (1964), "Numerical prediction of storm surges", *Advances in Geophysics,* Vol. 8, pp. 315-379.

Computational Techniques & Applications: CTAC-83
J. Noye & C. Fletcher (Editors)

AN IMPROVED THREE-DIMENSIONAL TIDAL MODEL FOR A SHALLOW GULF

Malcolm Stevens & John Noye

Department of Applied Mathematics
The University of Adelaide
Adelaide, South Australia
Australia

Most numerical models that have been developed to predict circulation in coastal seas due to external tidal forcing apply only to waters where the tidal range is small compared to the total depth of water. The method described here follows the approach of Noye et al (1981,1982) in that at each instant of time the irregular body of water being considered is mapped onto a unit cube. The resulting transformed set of equations are then solved on an improved space staggered grid using a more accurate quasi time-centred finite difference technique than the earlier approach. This method applies equally well to shallow waters, in which the tidal variations are large compared to the total depth of water, as to deep waters. Also, the new spatial grid permits inclusion of islands and peninsulas in the model which was not possible in the earlier version.

1. INTRODUCTION

Previously the usual method for modelling the dynamics of well mixed coastal seas which are forced by the tidal motion in connecting oceans, was to numerically solve the depth averaged equations of momentum and mass conservation (e.g. Heaps, 1969). There are two disadvantages with this approach: firstly, these models only work well in deep water as then the tidal variations are relatively small compared to the total depth of water and the horizontal velocity is nearly constant over depth, and secondly, they can only be used when information about the vertical structure of the resulting flow is not required, as this information is lost during the depth averaging process.

Few methods have been developed for predicting sea level variations and currents at all depths in coastal seas. The main ones are those developed for tidal and wind forcing in relatively deep water by Heaps (1972) and Leendertse et al (1973,1975). A different numerical technique for computing tidal currents at all depths in very shallow coastal seas was described in Noye et al (1981,1982). It was based on transforming the region of interest, the body of water in a gulf for example, on to a unit cube and solving the appropriately transformed equations. Besides the advantage this gives when simulating irregular boundary conditions if finite difference methods are used, tidal currents, which tend to flow parallel to the coast in gulfs are aligned along grid lines.

The numerical model described in this paper is an improvement over the forward time differencing used in Noye et al (1982). The transformed equations are differenced on the unit cube using an improved space-staggered grid which permits the easy exclusion of regions representing islands or peninsulas, and a more accurate quasi time-centred differencing.

It has a variable vertical grid spacing and incorporates an implicit method in the vertical direction so that a larger time step can be used than with explicit methods, without introducing instability. This also ensures greater accuracy where it is most needed, that is, in the differencing in the vertical direction.

Tests have been carried out in which the model is applied to a rectangular bay of constant depth and width. This test model uses an artificial vertical eddy viscosity function which, even though unrealistic, is useful because the result can be compared to the exact solution obtained by Johns (1966). Tests of the model against a truncated series solution to a simplified set of tidal motion equations obtained by Rienecker and Teubner (1980) have also been carried out. The results from these two comparisons indicate that the finite differencing is much more stable and accurate than the previous method used.

2. TRANSFORMATION OF THE THREE-DIMENSIONAL TIDAL EQUATIONS

The Reynolds averaged equations of mass and momentum conservation, with density assumed constant, are described by Nihoul (1975). Replacing the vertical component of the momentum conservation equation by the hydrostatic approximation permits replacement of the pressure gradients by surface displacement gradients in the two horizontal components of the momentum equations. Closure of the resulting system of equations is achieved by assuming constant values of the horizontal eddy viscosity coefficients, while the vertical eddy viscosity is taken to be a function of the depth-averaged horizontal velocity and the depth.

The resulting three-dimensional tidal equations may be used to compute the surface elevation, ζ, and the current components, u, v, w, once the boundary conditions are prescribed.

2.1 The Geometry of the Three-Dimensional Region

The plan view of a typical gulf is shown in Figure 2.1. The x- and y-axes are taken to be in the plane of mean sea level and aligned so that two of the boundaries are defined by $x = 0$, $x = L$, where L is the length of the gulf. The other two boundaries are defined by $y = b(x)$ and $y = b(x) + B(x)$, where $b(x)$ is the distance from the closer boundary to an arbitrary x-axis and $B(x)$ is the breadth of the gulf across a section parallel to the y-axis. A small straight boundary along the line $x = 0$ is used to avoid a singularity being produced by the transformation.

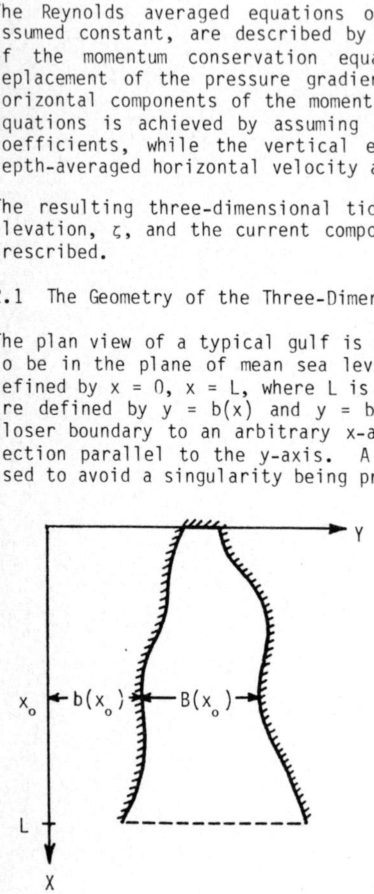

Figure 2.1: Plan view of gulf ($z = 0$ plane).

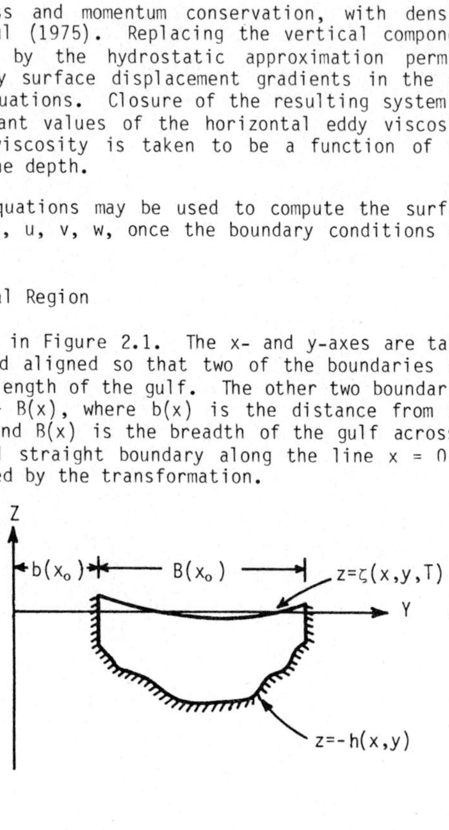

Figure 2.2: The cross-section of the gulf (with the vertical scale enlarged) at position x_0 along the x-axis.

A vertical cross-section of the typical gulf is shown in Figure 2.2 with the positive z-direction upwards. Again the basin is bounded by short vertical sides similar to that on the boundary $x = 0$ or any other closed boundary.

2.2 The Transformation

The region in which the governing equations are to be solved is in general quite irregular in shape (see Figures 2.1 and 2.2) and, because the surface displacement changes with time, the transformation must be time dependent. To enable the tidal equations to be solved more easily using finite difference techniques, the variables are transformed in such a way that the original region being considered is mapped onto a unit cube.

Consider the transformation

$$\underset{\text{original system}}{(x,y,z,T)} \rightarrow \underset{\text{transformed system}}{(\chi,\lambda,\eta,t)}$$

in which the relationship between the two coordinate systems is defined by

$$\chi(x) = x/L\,, \qquad 0 \leqslant \chi \leqslant 1\,, \tag{2.1}$$

$$\lambda(x,y) = (y - b)/B\,, \qquad 0 \leqslant \lambda \leqslant 1\,, \tag{2.2}$$

$$\eta(x,y,z,T) = (\zeta - z)/H\,, \qquad 0 \leqslant \eta \leqslant 1\,, \tag{2.3}$$

$$t(T) = T\,, \qquad t \geqslant 0\,, \tag{2.4}$$

where L is the length of the gulf in the x direction,
b,B are defined as in Figure 2.1, and depend only on x,
h is the depth of the sea floor below mean sea level,
and H is the total depth of water, i.e. $H = h + \zeta$.

Using the chain rule and equations (2.1) to (2.4) the differential operators can be expressed in terms of the new coordinates.

In addition, we define transformed velocity components $\mu = D\chi/DT$, $\nu = D\lambda/DT$ and $\omega = D\eta/DT$, where D/DT is the derivative following the motion. Expanding the operator D/DT and using equations (2.1) to (2.4) the original velocity components can be expressed in terms of the new coordinates, namely,

$$u = L\mu\,, \qquad v = B\nu + \mu\frac{\partial\beta}{\partial\chi}\,, \qquad w = -H\omega + (1-\eta)\frac{\partial\zeta}{\partial t} + \mu\frac{\partial\gamma}{\partial\chi} + \nu\frac{\partial\gamma}{\partial\lambda} \tag{2.5-7}$$

where $\beta(\chi,\lambda) = b(\chi) + \lambda B(\chi)$ and $\gamma(\chi,\lambda,\eta,t) = \zeta(\chi,\lambda,t) - \eta H(\chi,\lambda,t)$.

3. THE TRANSFORMED EQUATIONS AND BOUNDARY CONDITIONS

3.1 The Transformed Equations

Using the transformed differential operators and the velocities given in equations (2.5-7) the tidal equations transform to

$$\frac{\partial\zeta}{\partial t} + \frac{1}{B}\frac{\partial(BH\bar{\mu})}{\partial\chi} + \frac{\partial(H\bar{\nu})}{\partial\lambda} = 0, \tag{3.1}$$

where $(\bar{\mu},\bar{\nu}) = \int_0^1 (\mu,\nu)\,d\eta$,

$$\frac{\partial(H\mu)}{\partial t} + \frac{1}{B}\frac{\partial(BH\mu^2)}{\partial\chi} + \frac{\partial(H\mu\nu)}{\partial\lambda} + H\frac{\partial(\mu\omega)}{\partial\eta}$$

$$= -\frac{gH}{L^2}\left(\frac{\partial\zeta}{\partial\chi} - \frac{1}{B}\frac{\partial\zeta}{\partial\lambda}\frac{\partial\beta}{\partial\chi}\right) + \frac{HN_x}{L^2}\frac{\partial^2\mu}{\partial\chi^2} + \frac{HN_y}{B^2}\frac{\partial^2\mu}{\partial\lambda^2}$$

$$+ \frac{1}{H}\frac{\partial}{\partial\eta}\left(N_\eta\frac{\partial\mu}{\partial\eta}\right) + \frac{Hf\nu}{L}\,, \tag{3.2}$$

$$\frac{\partial(Hv)}{\partial t} + \frac{1}{B}\frac{\partial(BH\mu v)}{\partial X} + \frac{\partial(H\nu v)}{\partial \lambda} + H\frac{\partial(\omega v)}{\partial \eta}$$

$$= -\frac{gH}{B}\frac{\partial\zeta}{\partial\lambda} + \frac{HN_x}{L^2}\frac{\partial^2 v}{\partial X^2} + \frac{HN_y}{B^2}\frac{\partial^2 v}{\partial\lambda^2}$$

$$+ \frac{1}{H}\frac{\partial}{\partial\eta}\left(N_\eta \frac{\partial v}{\partial\eta} \right) - LHf\mu \ , \qquad (3.3)$$

$$\omega = \frac{1}{H}\left\{ \frac{1}{B}\frac{\partial}{\partial X}\left(BH(\bar{\mu}\eta - \int_0^\eta \mu\, d\eta)\right) + \frac{\partial}{\partial\lambda}\left(H(\bar{\nu}\eta - \int_0^\eta \nu\, d\eta)\right) \right\}, \qquad (3.4)$$

where g is gravitational acceleration and f is the Coriolis parameter.

The five unknowns ζ, μ, ν, v and ω may be found by solving equations (3.1-4) together with

$$\nu = \frac{1}{B}\left(v - \mu\frac{\partial\beta}{\partial X} \right) . \qquad (3.5)$$

The horizontal eddy viscosity coefficients N_x and N_y are taken to be constant, while the transformed vertical eddy viscosity parameter is

$$N_\eta = \alpha + \beta F(\eta)\{(L\mu)^2 + v^2\}^{1/2} \ , \qquad (3.7)$$

where $F(\eta) = (0.5 + \eta)(1 - \eta)$ and α and β are constants.

The above equations are to be solved on the unit cube $0 \leq X \leq 1$, $0 \leq \lambda \leq 1$ and $0 \leq \eta \leq 1$, with the surface displacement being prescribed on the open boundary which may occur along all or any part of the lines $X = 0$ or 1, $\lambda = 0$ or 1.

3.2 The Boundary Conditions

The boundary conditions which apply to the transformed equations are:

a "no-slip" condition on the sea floor, so that

$$v = \mu = \nu = \omega = 0 \quad \text{at} \quad \eta = 1 \ ; \qquad (3.8)$$

a "slip" condition with no flow across any of the closed boundaries, so that

$$\omega = 0 \quad \text{at} \quad \eta = 0 \ , \qquad (3.9)$$

$$\nu = 0 \quad \text{at any closed boundary where } \lambda = \text{constant}, \qquad (3.10)$$

$$\mu = 0 \quad \text{at any closed boundary where } X = \text{constant}; \qquad (3.11)$$

and no wind stress acting on the surface, which implies that

$$\frac{\partial\mu}{\partial\eta} = \frac{\partial\nu}{\partial\eta} = 0 \quad \text{at} \quad \eta = 0 \ . \qquad (3.12)$$

4. THE FINITE DIFFERENCE GRID

4.1 The Grid-Scheme

The grid-scheme is the basis of any finite difference model. The plan view of the transformed region, which is a unit square, is divided into rectangular blocks or elements of length $2\Delta X$ and breadth $2\Delta\lambda$. Within each of these elements, at each time step, a value for the surface displacement ζ is calculated as well as values for the transformed velocity components μ, ν and ω, at each depth level.

1	2	3	4	5
6	7	8	9	10
11	12	13	14	15
16	17	18	19	20
21	22	23	24	25
26	27	28	29	30
31	32	33	34	35
36	37	38	39	40
41	42	43	44	45

Figure 4.1: The plan view of the index numbers of the elements in a rectangular gulf.

Elements are labelled with an index number for location purposes (see Figure 4.1), so that either the surface displacement or the water velocity, from any particular point in the gulf can be obtained as long as the index number of that element is known. In the finite difference scheme, when calculating the approximation at the next time level for any variable, information is needed only from the adjacent elements. Thus the convention of labelling as shown in Figure 4.2 has been adopted, where for element i the relative positions of the variables are indicated by the subscripts in the following way:

i1: indicates the element immediately above the ith element, so that corresponding variables are located a distance of $2\Delta X$ in the negative X-direction from i.

i2: in a similar way to i1, indicates the element immediately below the ith element, so that corresponding variables are located a distance of $2\Delta X$ in the positive X-direction from i.

(e±1): indicates the next element to the right (left) of the element e, so that corresponding variables are located a distance of $2\Delta\lambda$ in the positive (negative) λ-direction from the element e, where e may be either i, i1 or i2.

m: the subscript used on the b and B variables which only depend on X and thus are the same for each element in a particular row. In Figure 4.2 the mth row is the row which contains the elements indexed by i-1, i and i+1.

i1-1	i1	i1+1	← Row m-1
i-1	i	i+1	← Row m
i2-1	i2	i2+1	← Row m+1

Figure 4.2: The general method of indexing with respect to the element i.

Therefore, if the element being considered is indexed by the number 17 in Figure 4.1, then the value of i1 is 12, the value of i2 is 22 with the values of i1±1, i±1 and i2±1 being obvious. The notation i1 and i2 is used since there may not a constant number of elements in a each row, such as when there is an island in the gulf. The program determines values for i1 and i2 for each particular element i.

The three velocity components and the surface displacement are calculated at the positions within each element as indicated in Figure 4.3. At each of the velocity points in an element, a number of depth levels are defined and at each of these levels the three velocity components are calculated. The depth levels are determined by the vertical grid spacing which can be arbitrarily chosen to suit the situation. If there are K depth levels, the vertical grid is defined by assigning values to η_k ranging from $\eta_0 = 0$ at the surface to $\eta_K = 1$ at the sea floor. Thus the subscript k on the velocity components indicates that these velocities are measured a distance of η_k from the surface in the transformed system.

In the plan view of an element, such as the one enclosed by the dashed line of Figure 4.3, the vertical velocity component ω is calculated at the same point as the surface displacement ζ is calculated. The two velocity components μ and ν are calculated at points a distance of ΔX and $\Delta\lambda$, in the X and λ directions respectively, from the point used for the calculation of ω.

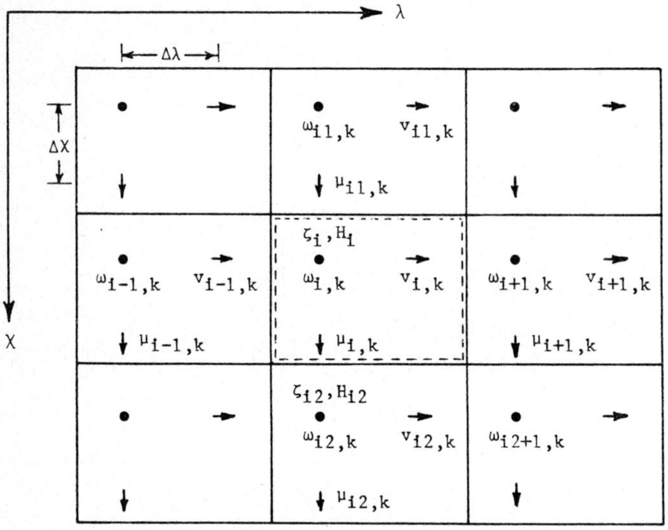

Figure 4.3: The location within the ith and adjacent elements and at the kth depth level of the velocity components and surface displacements in their relative positions. Note that not all variables are shown.

Since the transformed equations described in Section 3 omit terms multiplied by $(\partial\beta/\partial X)^2$, it is required that $\partial\beta/\partial X$ be relatively small so that these terms can be neglected. This means that the boundary lines defined by b(x) and B(x) must be nearly parallel. This is not the case for a gulf with a shape similar to that shown in Figure 4.4. In this case the lines b(x) and B(x) as shown, which do not follow the real sea-land boundaries, must be used. When this area is mapped onto the plan of the unit cube the region corresponding to the water area consists of the unit cube, with a rectangular "chunk" missing (see Figure 4.5). The equations used to numerically solve the tidal problem for this shape are the same as those used on the unit cube, once the program is modified so that it can distinguish which elements are land and which are water.

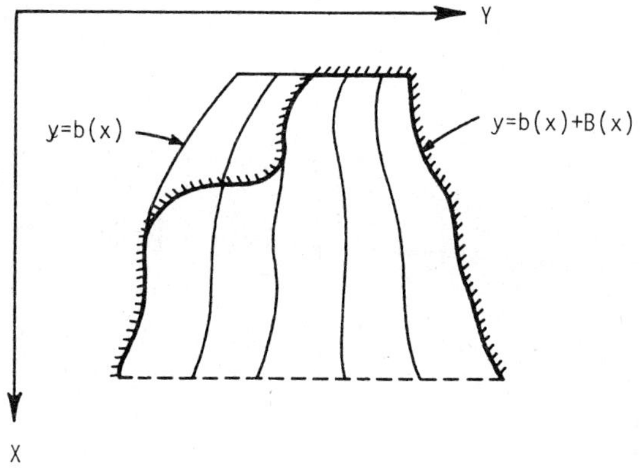

Figure 4.4: The plan of a gulf with b(x) and B(x) redefined so that $(\partial \beta/\partial \chi)^2$ is negligible.

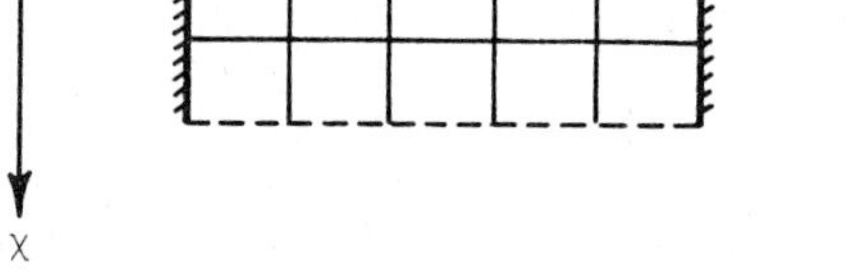

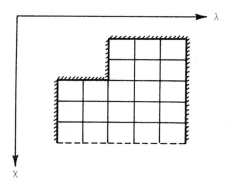

Figure 4.5: The shape of the transformed gulf of Figure 4.4.

The closed boundaries of the gulf are aligned so that the horizontal velocity components of each element are perpendicular to the closed boundaries in order to satisfy the boundary conditions (3.10) and (3.11).

For an open boundary along the line $\chi = 1$, corresponding to $x = L$, the boundary must pass through the points where the surface displacements are computed, since on the open boundary the tide height is specified as input data to the system.

4.2 The Identification Numbers for each Element

In the computer program an element is considered "active" if the point where the surface displacement ζ is calculated (i.e. the • point) lies within the physical boundaries which define the gulf. Thus in Figure 4.6 there are seven active elements.

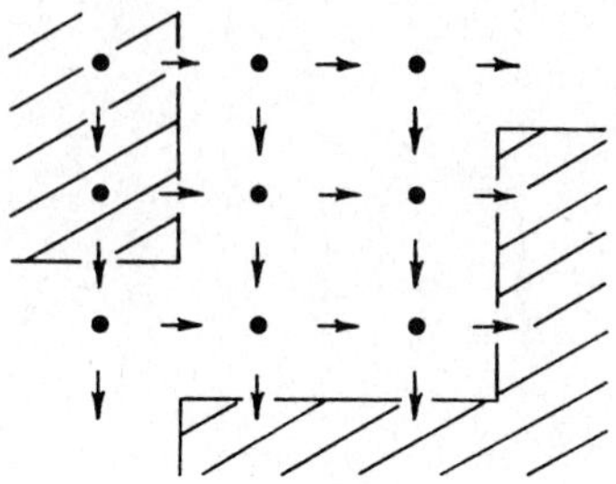

Figure 4.6

Each active element is given an identification number according to which of the elements surrounding it are active or non-active. Elements surrounding an active element have different weights which are incorporated into a single identification number. These weights are powers of two so that they can be added together to form a single number which uniquely determines the nature of the surrounding elements.

There are eight elements surrounding any particular active element and each of these is given a weight of 2^i, $i = 0$ to 7. In determining the identification number for an active element, the weight of each of the surrounding elements is included in the identification number only if that element is non-active, that is if there is land there. A single identification number for an element, obtained in this manner, contains the information necessary to indicate if any of the surrounding elements represent land and, if so, where it is located. For example, if the identification number is 56 then this number can be decomposed into the powers of two given by 8, 16 and 32. Consequently the elements surrounding one with this identification number must be as in Figure 4.7.

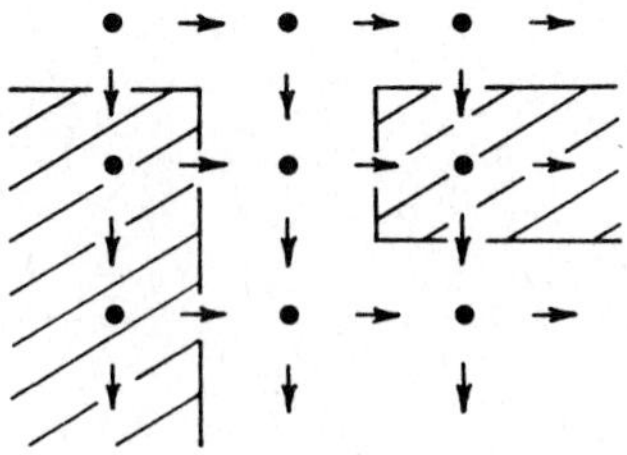

Figure 4.7

4.3 The Vertical Grid Spacing

In the finite difference method it is desirable to have an arbitrary grid spacing in the vertical direction. This is achieved by evaluating the transformed velocity components at the K+1 depth levels defined by

$$\eta = \eta_k \ , \quad k = 0(1)K \ , \tag{4.1}$$

where the sea surface is represented by $k = 0$ and sea floor by $k = K$. The grid spacing $\Delta\eta_k$ is therefore

$$\Delta\eta_k = \eta_k - \eta_{k-1} \ , \quad k = 1(1)K \ . \tag{4.2}$$

A variable vertical grid spacing permits use of small spacings near the sea floor where velocity gradients are large, and large spacings elsewhere. A fine grid spacing over the entire depth is not practical because the storage requirements are then beyond the capacities of most computers.

The variable grid spacing is defined by the κ-method described in Noye (1983, p.301) by means of the equation

$$\Delta\eta_k = \Delta\eta_{k-1}(1 - \kappa\Delta\eta_{k-1}) \ , \quad k = 1(1)K \ , \tag{4.3}$$

where κ is a constant of $O\{1\}$.

5. THE FINITE DIFFERENCE EQUATIONS

The indexing system described in Section 4.1 is used in the following difference schemes. It should be noted that the value of the variable β is measured at the same point in each element as the v velocity component. Thus

$$\beta_i = b_m + \lambda_i B_m$$

where λ_i is the distance from $\lambda = 0$ to the position of the v velocity component in the ith element and m is the row number.

In the difference schemes the symbols used are defined as follows: a ~ above a variable indicates an averaging between adjacent elements in the mth row and the $(m+1)^{th}$ row and an → above a variable indicates an averaging between two adjacent elements in the same row. Thus

$$\tilde{H}^n_{i+\phi} = (H^n_{i+\phi} + H^n_{i2+\phi})/2 \ , \qquad \tilde{H}^n_{i1+\phi} = (H^n_{i1+\phi} + H^n_{i+\phi})/2 \ ,$$

$$\vec{H}^n_e = (H^n_e + H^n_{e+1})/2 \ , \qquad \vec{H}^n_{e-1} = (H^n_{e-1} + H^n_e)/2 \ ,$$

$$\tilde{B}_m = (B_m + B_{m+1})/2 \ , \qquad \tilde{B}_{m-1} = (B_{m-1} + B_m)/2 \ ,$$

$$\vec{\beta}_i = (\beta_i + \beta_{i+1})/2 = b_m + \lambda_{i+1/2} B_m \ , \qquad \vec{\beta}_{i-1} = b_m + \lambda_{i-1/2} B_m \ ,$$

$$\vec{\beta}_{i1} = b_{m-1} + \lambda_{i1+1/2} B_{m-1} \ , \qquad \vec{\beta}_{i2-1} = b_{m+1} + \lambda_{i2-1/2} B_{m+1} \ ,$$

$$\tilde{f}_i = (f_i + f_{i2})/2 \ , \text{ where } f_i \text{ is the value of the Coriolis parameter for the } i^{th} \text{ element,}$$

$$\vec{f}_i = (f_i + f_{i+1})/2 \ .$$

5.1 The Transformed Depth Integrated Continuity Equation

Equation (3.1) evaluated at the nth time level and at the ζ point in the ith element in the gulf is

$$\left.\frac{\partial\zeta}{\partial t}\right|_i^n + \frac{1}{B}\frac{\partial}{\partial X}(BH\bar{\mu})\Big|_i^n + \frac{\partial}{\partial\lambda}(H\bar{\nu})\Big|_i^n = 0 \ . \tag{5.1}$$

Using forward differencing in time and central differencing in space, equation (5.1) becomes

$$\zeta_i^{n+1} \simeq \zeta_i^n - \frac{\Delta t}{2B_m\Delta X}\{\tilde{B}_m\tilde{H}_i^n\bar{\mu}_i^n - \tilde{B}_{m-1}\tilde{H}_{i1}^n\bar{\mu}_{i1}^n\}$$
$$- \frac{\Delta t}{2\Delta\lambda}\{\vec{H}_i^n\bar{\nu}_i^n - \vec{H}_{i-1}^n\bar{\nu}_{i-1}^n\} , \tag{5.2}$$

where ν is calculated using equation (3.5) and $\bar{\nu}$ is the depth averaged value of ν:

$$\nu_i^n = \frac{1}{B_m}\{v_i^n - 1/4\,(\mu_{i1}^n + \mu_{i1+1}^n + \mu_i^n + \mu_{i+1}^n)(\frac{\beta_{i2} - \beta_{i1}}{4\Delta X})\} , \tag{5.3}$$

$$\bar{\nu}_i^n = \frac{1}{B_m}\{\bar{v}_i^n - 1/4\,(\bar{\mu}_{i1}^n + \bar{\mu}_{i1+1}^n + \bar{\mu}_i^n + \bar{\mu}_{i+1}^n)(\frac{\beta_{i2} - \beta_{i1}}{4\Delta X})\} , \tag{5.4}$$

where the approximations for $\bar{\mu}$ and $\bar{v}$ are obtained by using Simpson's rule for integration over an uneven grid spacing.

Equation (5.2) is used to obtain a value for ζ at the new time level. This value is then used to calculate μ, v and ω at the new time level. Once the new values for the three velocity components have been calculated a more accurate value for ζ can be obtained using equation (5.5), which is centrally differenced in time and is therefore more accurate than equation (5.2).

$$\zeta_i^{n+1} \simeq \zeta_i^n - \frac{\Delta t}{4B_m\Delta X}\{\tilde{B}_m\tilde{H}_i^{n+1}\bar{\mu}_i^{n+1} - \tilde{B}_{m-1}\tilde{H}_{i1}^{n+1}\bar{\mu}_{i1}^{n+1}\}$$
$$- \frac{\Delta t}{4\Delta\lambda}\{\vec{H}_i^{n+1}\bar{\nu}_i^{n+1} - \vec{H}_{i-1}^{n+1}\bar{\nu}_{i-1}^{n+1}\} - \frac{\Delta t}{4B_m\Delta X}\{\tilde{B}_m\tilde{H}_i^n\bar{\mu}_i^n - \tilde{B}_{m-1}\tilde{H}_{i1}^n\bar{\mu}_{i1}^n\}$$
$$- \frac{\Delta t}{4\Delta\lambda}\{\vec{H}_i^n\bar{\nu}_i^n - \vec{H}_{i-1}^n\bar{\nu}_{i-1}^n\} , \tag{5.5}$$

5.2 The Transformed x- and y-momentum Equations

Equation (3.2) evaluated at the $(n+1/2)$ time level and at the μ point of the ith element in the gulf, for the variables at the kth depth level below the surface, is

$$\frac{\partial}{\partial t}(H\mu)\Big|_{i,k}^{n+1/2} + \frac{1}{B}\frac{\partial}{\partial X}(BH\mu^2)\Big|_{i,k}^{n+1/2} + \frac{\partial}{\partial\lambda}(H\mu\nu)\Big|_{i,k}^{n+1/2} + \frac{\partial}{\partial\eta}(H\mu\omega)\Big|_{i,k}^{n+1/2}$$
$$= -\frac{gH}{L^2}(\frac{\partial\zeta}{\partial X} - \frac{1}{B}\frac{\partial\zeta}{\partial\lambda}\frac{\partial\beta}{\partial X})\Big|_{i,k}^{n+1/2} + \frac{HN_x}{L^2}\frac{\partial^2\mu}{\partial X^2}\Big|_{i,k}^{n+1/2} + \frac{HN_y}{B^2}\frac{\partial^2\mu}{\partial\lambda^2}\Big|_{i,k}^{n+1/2}$$
$$+ \frac{1}{H}\frac{\partial}{\partial\eta}(N_\eta\frac{\partial\mu}{\partial\eta})\Big|_{i,k}^{n+1/2} + \frac{Hfv}{L}\Big|_{i,k}^{n+1/2} \ . \tag{5.6}$$

The finite difference approximations for each term of equation (5.6) are as follows using the values of ζ and H which have already been calculated at the new time level:

$$\text{(a)}\quad \frac{\partial}{\partial t}(H\mu)\Big|_{i,k}^{n+1/2} \simeq \left\{ \frac{\tilde{H}_i^{n+1}\mu_{i,k}^{n+1} - \tilde{H}_i^n\mu_{i,k}^n}{\Delta t} \right\} = HU1.\mu_{i,k}^{n+1} + HU2 \tag{5.7}$$

where $HU1 = \dfrac{\tilde{H}_i^{n+1}}{\Delta t}$ and $HU2 = \dfrac{-\tilde{H}_i^n\mu_{i,k}^n}{\Delta t}$.

$$\text{(b)}\quad \frac{1}{B}\frac{\partial}{\partial X}(BH\mu^2)\Big|_{i,k}^{n+1/2} \simeq \frac{1}{\tilde{B}_m}\ \{ B_{m+1}H_{i2}^n[(\mu_{i2,k}^n + \mu_{i,k}^n)/2]^2 - B_m H_i^n[(\mu_{i,k}^n + \mu_{i1,k}^n)/2]^2 \} / 2\Delta X = HUU. \tag{5.8}$$

$$\text{(c)}\quad \frac{\partial}{\partial\lambda}(H\mu\nu)\Big|_{i,k}^{n+1/2} \simeq \frac{1/2(\tilde{H}_i^n + \tilde{H}_{i+1}^n)}{2\Delta\lambda\tilde{B}_m}\{1/4(\mu_{i,k}^n + \mu_{i+1,k}^n)(v_{i,k}^n + v_{i2,k}^n) - [(\mu_{i,k}^n + \mu_{i+1,k}^n)/2]^2(\frac{\beta_{i2} - \beta_i}{2\Delta X})\}$$
$$- \frac{1/2(\tilde{H}_{i-1}^n + \tilde{H}_i^n)}{2\Delta\lambda\tilde{B}_m}\{1/4(\mu_{i-1,k}^n + \mu_{i,k}^n)(v_{i-1,k}^n + v_{i2-1,k}^n) - [(\mu_{i-1,k}^n + \mu_{i,k}^n)/2]^2(\frac{\beta_{i2-1} - \beta_{i-1}}{2\Delta X})\} = HUV. \tag{5.9}$$

$$\text{(d)}\quad \frac{\partial}{\partial\eta}(H\mu\omega)\Big|_{i,k}^{n+1/2} \simeq HUW1.\mu_{i,k-1}^{n+1} + HUW2.\mu_{i,k}^{n+1} + HUW3.\mu_{i,k+1}^{n+1} + HUW4 \tag{5.10}$$

where
$$HUW1 = -\frac{(\tilde{H}_i^n + \tilde{H}_i^{n+1})}{8\Delta\eta_k(r_k+1)}r_k^2(\omega_{i2,k-1}^n + \omega_{i,k-1}^n),$$

$$HUW2 = \frac{(\tilde{H}_i^n + \tilde{H}_i^{n+1})}{8\Delta\eta_k(r_k+1)}(r_k^2 - 1)(\omega_{i2,k}^n + \omega_{i,k}^n),$$

$$HUW3 = \frac{(\tilde{H}_i^n + \tilde{H}_i^{n+1})}{8\Delta\eta_k(r_k+1)}(\omega_{i2,k+1}^n + \omega_{i,k+1}^n),$$

$$HUW4 = \{ HUW1.\mu_{i,k-1}^n + HUW2.\mu_{i,k}^n + HUW3.\mu_{i,k+1}^n \},$$

and $r_k = \Delta\eta_k/\Delta\eta_{k-1}$.

$$\text{(e)}\quad -\frac{gH}{L^2}\frac{\partial\zeta}{\partial X}\Big|_{i,k}^{n+1/2} \simeq -\frac{g}{2L^2}\{ \tilde{H}_i^n[\frac{\zeta_{i2}^n - \zeta_i^n}{2\Delta X}] + \tilde{H}_i^{n+1}[\frac{\zeta_{i2}^{n+1} - \zeta_i^{n+1}}{2\Delta X}] \} = GHDZDX. \tag{5.11}$$

$$\text{(f)}\quad \frac{gH}{L^2B}\frac{\partial\zeta}{\partial\lambda}\frac{\partial\beta}{\partial X}\Big|_{i,k}^{n+1/2} \simeq \frac{g}{2\tilde{B}_m L^2}\{[1/2\tilde{H}_i^n(\frac{\zeta_{i+1}^n - \zeta_{i-1}^n}{4\Delta\lambda} + \frac{\zeta_{i2+1}^n - \zeta_{i2-1}^n}{4\Delta\lambda})\beta_i^*]$$
$$+ [1/2\tilde{H}_i^{n+1}(\frac{\zeta_{i+1}^{n+1} - \zeta_{i-1}^{n+1}}{4\Delta\lambda} + \frac{\zeta_{i2+1}^{n+1} - \zeta_{i2-1}^{n+1}}{4\Delta\lambda})\beta_i^*]\} = GHLDZDY, \tag{5.12}$$

where $\beta_i^* = \{\dfrac{\vec{\beta}_{i2-1} - \vec{\beta}_{i-1}}{2\Delta X}\}$.

$$\text{(g)}\quad \frac{HN_X}{L^2}\frac{\partial^2\mu}{\partial X^2}\Big|_{i,k}^{n+1/2} \simeq \frac{N_X}{4L^2(\Delta X)^2}\tilde{H}_i^n(\mu_{i2,k}^n - 2\mu_{i,k}^n + \mu_{i1,k}^n) = NXHDUX. \tag{5.13}$$

(h) $$\frac{HN_y}{B^2}\frac{\partial^2\mu}{\partial\lambda^2}\bigg|_{i,k}^{n+1/2} \simeq \frac{N_y}{4\tilde{B}_m^2(\Delta\lambda)^2}\tilde{H}_i^n(\mu_{i+1,k}^n - 2\mu_{i,k}^n + \mu_{i-1,k}^n) = \text{NYHDUY}. \quad (5.14)$$

(i) $$\frac{1}{H}\frac{\partial}{\partial\eta}\left(N_\eta\frac{\partial\mu}{\partial\eta}\right)\bigg|_{i,k}^{n+1/2} \simeq \text{HNIU1}.\mu_{i,k-1}^{n+1} + \text{HNIU2}.\mu_{i,k}^{n+1} + \text{HNIU3}.\mu_{i,k+1}^{n+1} + \text{HNIU4} \quad (5.15)$$

where $\text{HNIU1} = \frac{r_k^2}{\text{DENOM}}\{(2-r_k)\tilde{N}_{i,k}^n + r_k\tilde{N}_{i,k-1}^n\}$,

$\text{HNIU2} = \frac{-1}{\text{DENOM}}\{\tilde{N}_{i,k+1}^n - (r_k+1)(r_k^2-3r_k+1)\tilde{N}_{i,k}^n + r_k^3\tilde{N}_{i,k-1}^n\}$,

$\text{HNIU3} = \frac{1}{\text{DENOM}}\{\tilde{N}_{i,k+1}^n + (2r_k-1)\tilde{N}_{i,k}^n\}$,

$\text{HNIU4} = \{\text{HNIU1}.\mu_{i,k-1}^n + \text{HNIU2}.\mu_{i,k}^n + \text{HNIU3}.\mu_{i,k+1}^n\}$,

$\text{DENOM} = (\Delta\eta_k)^2(r_k+1)(\tilde{H}_i^{n+1} + \tilde{H}_i^n)$,

$\tilde{N}_{i,k}^n = \alpha + \beta F(\eta_k)\{(L\bar{u}_i^n)^2 + (\tfrac{1}{4}[\bar{v}_{i-1}^n + \bar{v}_i^n + \bar{v}_{i2-1}^n + \bar{v}_{i2}^n])^2\}^{1/2}$,

$F(\eta_k) = (0.5 + \eta_k)(1 - \eta_k)$,

and $r_k = \Delta\eta_k/\Delta\eta_{k-1}$.

The forms given for the vertical eddy viscosity coefficient $\tilde{N}_{i,k}^n$ and the associated depth function $F(\eta_k)$ are derived from their definitions in equation (3.7).

(j) $$\frac{Hfv}{L}\bigg|_{i,k}^{n+1/2} \simeq \tfrac{1}{4}\tilde{H}_i^n\tilde{f}_i\{v_{i-1,k}^n + v_{i,k}^n + v_{i2-1,k}^n + v_{i2,k}^n\} = \text{HFV}. \quad (5.16)$$

From the difference forms of each term of equation (5.6) a set of equations is obtained for each element of the gulf, namely

$$P_{i,k}^n\mu_{i,k-1}^{n+1} + Q_{i,k}^n\mu_{i,k}^{n+1} + R_{i,k}^n\mu_{i,k+1}^{n+1} = S_{i,k}^n , \quad (5.17)$$

where $P_{i,k}^n = \text{HUW1} - \text{HNIU1}$,

$Q_{i,k}^n = \text{HU1} + \text{HUW2} - \text{HNIU2}$,

$R_{i,k}^n = \text{HUW3} - \text{HNIU3}$,

and $S_{i,k}^n = -\text{HU2} - \text{HUU} - \text{HUV} - \text{HUW4} + \text{GHDZDX} + \text{GHLDZDY}$

$+ \text{NXHDUX} + \text{NYHDUY} + \text{HNIU4} + \text{HFV}$,

This set of equations holds for $k = 1(1)K-1$, so it is implicit in μ in the η-direction. For $k = 0$, at the surface, it is assumed that

$$\frac{\partial\mu}{\partial\eta}\bigg|_{i,0}^{n+1/2} = 0 , \quad (5.18)$$

which implies that there is no shear stress acting on the surface of the water.

Using this equation the following expressions can be derived for P, Q, R and S whe k = 0 in equation (5.17),

$P_{i,0}^n = 0$,

$$Q^n_{i,0} = P^n_{i,1} - r_1(r_1+2)R^n_{i,1} , \quad \text{where } r_1 = \Delta\eta_1/\Delta\eta_0 ,$$
$$R^n_{i,0} = Q^n_{i,1} + (r_1+1)^2 R^n_{i,1} \quad \text{and} \quad S^n_{i,0} = S^n_{i,1} .$$

At the bottom of the cube, where k = K, the boundary condition implies that $\mu = 0$. This simplifies equation (5.17) when k = K-1, since

$$\mu^{n+1}_{i,K} = 0 ,$$

giving the following equation

$$P^n_{i,K-1}\mu^{n+1}_{i,K-2} + Q^n_{i,K-1}\mu^{n+1}_{i,K-1} = S^n_{i,K-1} . \tag{5.19}$$

Thus we have K equations with the K unknowns $\mu_{i,k}$, k = 0(1)K-1, at the new time level, forming a tri-diagonal set of implicit equations. This can be easily and efficiently solved using the algorithm developed by Thomas (1949).

The y-momentum equation, which is evaluated at the (n+½) time level and at the ν point in the ith element in the gulf, is differenced in a similar way to the x-momentum equation.

5.3 The Transformed Continuity Equation used to Compute ω

At this stage all variables except ω are known at the (n+1)th time level. At this time level, equation (3.4) is evaluated at the ζ point of the ith element in the gulf and at the kth depth level, giving

$$\omega\Big|^{n+1}_{i,k} = \{\frac{1}{HB}\frac{\partial}{\partial\chi}(BH(\eta_k\bar{\mu} - \int_0^{\eta_k}\mu\, d\eta)) + \frac{1}{H}\frac{\partial}{\partial\lambda}(H(\eta_k\bar{\nu} - \int_0^{\eta_k}\nu\, d\eta))\}\Big|^{n+1}_{i,k} . \tag{5.20}$$

The finite difference form of equation (5.20), obtained using centred finite difference approximations, is

$$\begin{aligned}\omega^{n+1}_{i,k} = \frac{1}{2H^{n+1}_i}\{ &\frac{1}{B_m\Delta\chi}[\tilde{B}_m\tilde{H}^{n+1}_i(\eta_k\bar{\mu}^{n+1}_i - \int_0^{\eta_k}\mu^{n+1}_i d\eta) \\ &\quad - \tilde{B}_{m-1}\tilde{H}^{n+1}_{i1}(\eta_k\bar{\mu}^{n+1}_{i1} - \int_0^{\eta_k}\mu^{n+1}_{i1} d\eta)] \\ &+ \frac{1}{\Delta\lambda}[\vec{H}^{n+1}_i(\eta_k\bar{\nu}^{n+1}_i - \int_0^{\eta_k}\nu^{n+1}_i d\eta) \\ &\quad - \vec{H}^{n+1}_{i-1}(\eta_k\bar{\nu}^{n+1}_{i-1} - \int_0^{\eta_k}\nu^{n+1}_{i-1} d\eta)]\} , \quad k = 0(1)K-1.\end{aligned} \tag{5.21}$$

5.4 The Boundary Conditions

The finite difference equations given in Section 5.1 to 5.3, are those used for calculating ζ, μ, ν and ω for the element i, which is completely surrounded by water (i.e. with identification number of 0). When one or more of the surrounding elements are land elements, some of the finite difference expressions for the terms in equations (5.1), (5.6) and (5.20) have to be modified or omitted depending on which of the surrounding elements are land. The third term of equation (5.6) is used as an illustration to show how this works. The finite difference form for this term is

$$\begin{aligned}\frac{\partial}{\partial\lambda}(H\mu\nu)\Big|^{n+1/2}_{i,k} \simeq\ & \frac{\tfrac{1}{2}(\tilde{H}^n_i + \tilde{H}^n_{i+1})}{2\Delta\lambda}\{\tfrac{1}{4}(\mu^n_{i,k} + \mu^n_{i+1,k})(\nu^n_{i,k} + \nu^n_{i2,k})\} \\ & - \frac{\tfrac{1}{2}(\tilde{H}^n_{i-1} + \tilde{H}^n_i)}{2\Delta\lambda}\{\tfrac{1}{4}(\mu^n_{i-1,k} + \mu^n_{i,k})(\nu^n_{i-1,k} + \nu^n_{i2-1,k})\} ,\end{aligned} \tag{5.22}$$

which is the same as equation (5.9) except that the v values in this equation have been replaced using the expression given in equation (3.5). In this example only four of the surrounding elements are of importance, namely those represented by i±1 and i2±1. The element represented by i2 will always be water, otherwise $\mu_i = 0$ from the boundary condition of equation (3.11). It is not important if any of the elements represented by i1 or i1±1 are land or sea elements, as no information is required from these elements in calculating the term given in equation (5.22).

If one or more of the elements represented by i±1 and i2±1 are land elements then the finite difference approximation given in equation (5.22) has to be modified. Consider the case in Figure 5.1 when the elements i+1 and i2+1 are land elements.

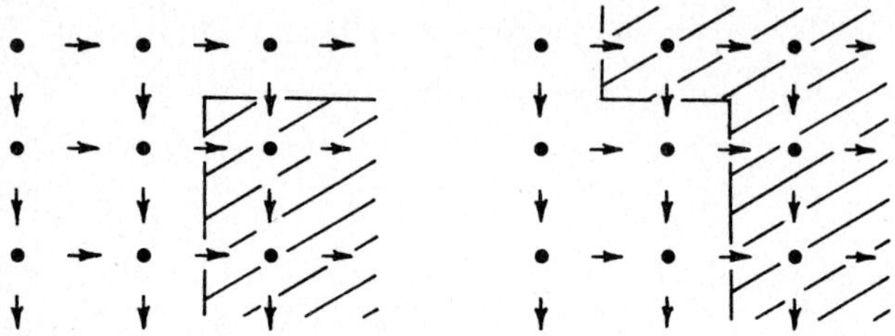

Figure 5.1: Two situations where, with respect to the central element i, the elements i+1 and i2+1 (see Figure 4.3) are both land elements.

When this occurs both ν_i and ν_{i2} are zero due to the boundary condition in equation (3.10) and therefore need not be calculated using equation (5.3). Equation (5.22) can therefore be reduced to

$$\frac{\partial}{\partial\lambda}(H\mu\nu)\bigg|_{i,k}^{n+1/2} \simeq -\frac{(\tilde{H}_{i-1}^n + \tilde{H}_i^n)}{16\Delta\lambda}\{(\mu_{i-1,k}^n + \mu_{i,k}^n)(\nu_{i-1,k}^n + \nu_{i2-1,k}^n)\} . \qquad (5.23)$$

This process is carried out for every term in equation (5.1), (5.6) and (5.20) and for each different combination of surrounding land elements.

5.5 The Organization of the Computer Program

When all the input data has been read into the program and all the required initialisation performed (which includes initialisation of the variables μ, v, ω and ζ at time t = 0), the program proceeds in iterations consisting of time steps of fixed size which are determined by the input data.

Each iteration consists of computing the values of ζ at the new time level using equation (5.1) for all the active elements in the gulf, then the values of μ and v at each depth level of every active element are calculated using equations (5.17) and the corresponding equation for the y-momentum equation. Once these have been calculated, ω at the new time level is obtained using equation (5.21). With all the velocity components available at the new time level, a more accurate approximation for ζ at the new time level can be obtained using equation (5.5). The program continues iterating until a desired time level has been reached.

6. COMPARISONS WITH EXACT SOLUTIONS FOR SIMPLIFIED PROBLEMS

The numerical method of solving the governing equations has been tested by comparing the results with the analytic solutions to two simplified problems. The first of these is the case of a rectangular bay of constant depth and width in which the disturbance of the surface is small compared with the depth of water, the effect of the Coriolis force is neglected and the vertical eddy viscosity is a function of depth only. The second case is that of a rectangular bay of constant depth in which the effect of the Coriolis force is included and the surface disturbance is again relatively small compared to the depth of water. The horizontal velocities are considered almost unchanged over depth so that the depth integrated equations are solved to obtain an analytic solution in the form of an infinite series.

6.1 Johns' Exact Solution

The program has been adapted to model the situation described by Johns (1966). For a rectangular gulf of constant depth in the absence of Coriolis force, Johns developed an exact solution for the predicted surface displacements ζ and the u velocity component if the amplitude of the tidal input is small compared with the total depth of water. The rectangular basin is closed on three of the vertical sides and open on the remaining side along which the surface displacement is the input. The details of the solution are given in Bills and Noye (1984).

The numerical model was forced by prescribing the analytic solution for the elevation at the open boundary, commencing with all elevation and velocity values initialised to zero. The dimension and model parameters used in the comparison were:

Δt = 600 secs (10 mins), Δx = 7436 m (space step), No. of depth levels = 15,
L = 290000 m (length 290 km), h = 20 m, a = 0.5 m,
$\sigma = 2\pi/(12 \times 3600)$ (i.e. period of 12 hrs), γ = 9, ν = 0.001 m^2/sec.

See Bills and Noye (1984) for a detailed description of the meaning of the parameters a, γ and ν.

After running the model for five cycles to allow starting transients to damp out, the numerical solutions were compared with the analytic solutions. Errors of 0.001 m in elevations which averaged about 0.25 m were typical, while errors of 0.0002 m/sec in horizontal velocities which averaged 0.02 m/sec were usual.

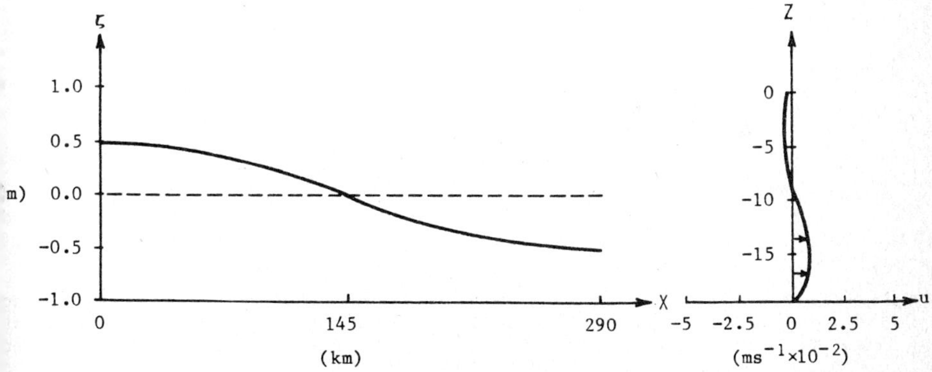

Figure 6.1: The surface displacement ζ and the corresponding u velocity profile, evaluated at the gridpoint where x = 145 km. These results were obtained after 5 tide cycles with a 12 hour period.

In Figure 6.1 the exact solutions for ζ and u are indistinguishable from the values obtained from the numerical model for the given scales on the diagrams. Clearly, the finite difference solution was a good approximation to the solution of the given set of differential equations which model the tidal movements for this situation. When the non-linearity of the vertical transformation was removed from the numerical

Figure 6.2

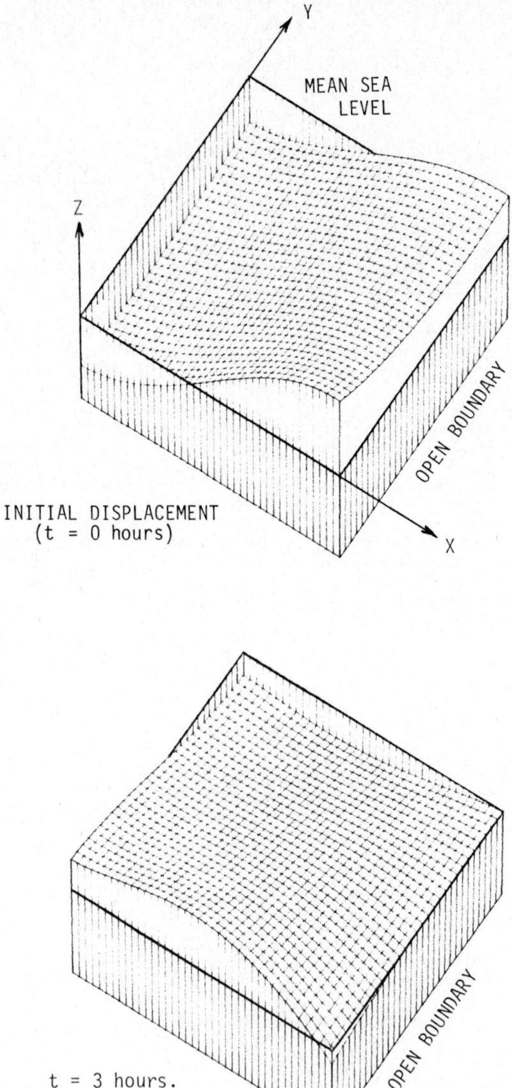
Y
MEAN SEA
LEVEL
Z
OPEN BOUNDARY
INITIAL DISPLACEMENT
(t = 0 hours)
X
OPEN BOUNDARY
t = 3 hours.

Figure 6.2 (cont.)

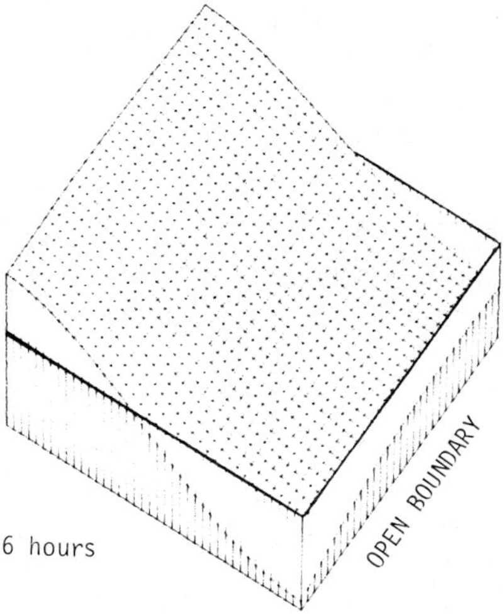

t = 6 hours

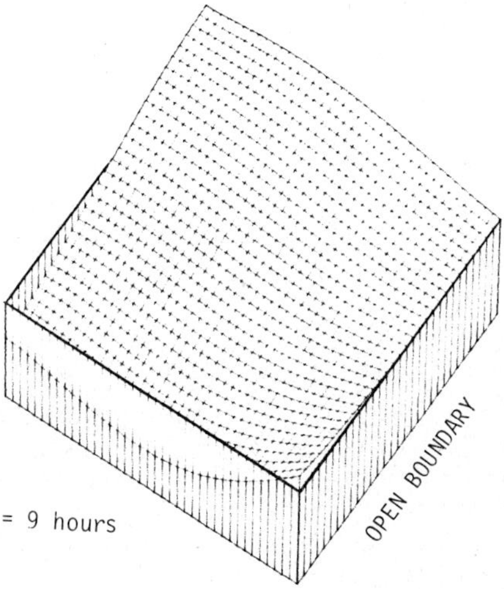

t = 9 hours

model, the results were accurate to three figures.

6.2 A Truncated Series Solution

This section describes a comparison between results from the model, adapted to simulate the idealised conditions described by Rienecker and Teubner (1980), and the solution they gave to the depth integrated system of tidal motion equations.

The depth integrated equations with constant Coriolis parameter, linearized friction terms, surface displacement considered negligible compared to the sea depth and zero transverse velocities across the open boundaries, have been solved for a rectangular bay of constant depth which is open at one end. Analytic solutions for ζ, $\bar{u}$ and $\bar{v}$ given as an infinite series in Rienecker and Teubner (1980), are obtainable to any degree of accuracy by evaluating sufficient terms in the series. The analytic solutions were used to supply open boundary surface displacements at each time step of the numerical model, and to give exact solutions for comparison with output from the suitably modified tidal model.

The dimension of the bay and the model parameters used were:

Δt = 300 secs (5 mins),
L = 290,000 m (length 290 km),
Δx = 7436 m (x space step),
h = 20 m (constant depth),
a = 1/18 m (incident amplitude),
K = 0.001 m/sec (linear friction coefficient)
W = 200,000 m (width 200 km),
Δy = 6667 m (y space step),
f = -0.000126 (i.e. latitude 60°S).

The open boundary forcing in this case produced a Kelvin wave with a maximum elevation amplitude of approximately 0.23 m, which rotated in a clockwise direction around the bay. Figure 6.2 shows the surface elevation at time intervals of one quarter of the semi-diurnal period.

Comparison of the analytic solution with the model results indicate an average numerical error of 0.0005 m in a maximum elevation value of 0.23 m, the maximum error being 0.001 m. The maximum numerical error in the depth averaged x-component of velocity was 0.001 m/sec in a maximum value of 0.109 m/sec, and a maximum error in the depth averaged y-component was 0.001 m/sec in a maximum value of 0.026 m/sec.

7. CONCLUSIONS

A three-dimensional numerical model of tidal propagation has been developed which checks favourably with Johns' (1966) exact solution. It has also been compared with truncated series solution obtained by Rienecker and Teubner (1980) under similar idealised conditions. In both of these modified situations the model produced solutions which are very accurate when compared to the analytic solutions. This both verifies the coding of the linearized terms in the program and the validity and accuracy of the associated finite difference approximations.

The next stages of the development of this model include:

1) Incorporating wind stresses and atmospheric pressure gradients which act on the surface of the water. This permits modelling of storm surges superimposed on tides.

2) Including variable grid spacing for differencing of the transformed equations in the horizontal directions.

3) Allowing the exposing and covering of mud flats and sand banks due to tidal motion.

4) Incorporating equations for the transport and dissipation of turbulent kinetic energy, which enables the vertical eddy viscosity coefficient to be calculated as a function of space and time, instead of being defined "a priori".

CKNOWLEDGMENT

uring the preparation of this material the first mentioned author was partly upported on a project to develop a three-dimensional tidal model, under the uspices of the Australian Research Grants Scheme.

EFERENCES

ills, P.J., and Noye, B.J. (1984), "Verification of a three-dimensional tidal model for coastal seas". *Computational Techniques and Applications,* editor John Noye, North-Holland Publishing Company, Amsterdam. This Volume.

ooper, C.K., and Pearce, B.K. (1977) *A Three-Dimensional Numerical Model to Calculate Currents in Coastal Waters Utilizing a Depth Varying Vertical Eddy Viscosity,* R.M. Parsons Laboratory for Water Resources and Hydrodynamics, Report No. 266, 147 pp.

eeman, N.G., (1970), *The Application of Sigma Coordinates to the Numerical Modelling of Great Lakes' Hydrodynamics,* M.Sc. Thesis, University of Waterloo, 144 pp.

aps, N.S. (1969), "A two-dimensional numerical sea model", *Proceedings of the Royal Society of London,* Series A, Volume 265, pp. 93-137.

aps, N.S., (1972) "On the numerical solution of the three-dimensional hydrodynamical equations for tides and storm surges", *Memoires Societe Royale des Sciences de Liege,* 6 serie, tome 11, pp. 143-180.

hns, B., (1966), "On the vertical structure of tidal flow in river estuaries", *Geophysical Journal of the Royal Astronomical Society,* Volume 12, pp. 103-110.

endertse, J.J., Alexander, R.C. and Liu, S.K. (1973), *A Three-Dimensional Model for Estuaries and Coastal Seas, Volume 1, Principles of Computation,* Memorandum of Rand Corporation, New York, R-1417-OWRR, 57 pp.

endertse, J.J., and Liu, S., (1975), *A Three-Dimensional Model for Estuaries and Coastal Seas, Volume 2, Aspects of Computation,* Memorandum of Rand Corporation, R-1764-OWRT, 123 pp.

houl, J.C.J., (1975), "Marine systems analysis", *Modelling of Marine Systems,* edited by J.C.J. Nihoul, Elsevier Oceanography Series, Volume 10, Elsevier, Amsterdam, pp. 3-40.

ye, B.J. (1983), "Finite difference techniques for partial differential equations", *Computational Techniques for Differential Equations,* editor John Noye, North-Holland Publishing Company, Amsterdam, pp. 95-354.

ye, B.J., May, R.L., and Teubner, M.D. (1981), "Three-dimensional numerical model of tides in Spencer Gulf", *Ocean Management*, Volume 6, pp.137-148.

ye, B.J., May, R.L., and Teubner, M.D. (1982), "A three-dimensional tidal model for a shallow gulf", *Numerical Solutions of Partial Differential Equations*, editor John Noye, North-Holland Publishing Coy., pp. 417-436.

eneker, M.M. and Teubner, M.D., (1980), "A note on frictional effects in Taylor's problem", *Journal of Marine Research,* Volume 38, no. 2, pp. 183-191.

omas, L.H. (1949), *Elliptic Problems in Linear Difference Equations over a Network*, Watson Scientific Computing Laboratory, Columbia University, New York.

Computational Techniques & Applications: CTAC-83
J. Noye & C. Fletcher (Editors)

VERIFICATION OF A THREE-DIMENSIONAL TIDAL MODEL FOR COASTAL SEAS

P.J. Bills & B.J. Noye

Department of Applied Mathematics
The University of Adelaide
Adelaide, South Australia
Australia

A brief description of a computer model for the solution of the three-dimensional non-linear tidal equations with constant density and prescribed vertical eddy viscosity function is presented. Coriolis, atmospheric pressure and wind stress terms are included. The tidal equations are depth transformed and solved on a three-dimensional spatial grid using finite difference techniques to advance the solution in time.

The discretization used in this model is an improvement over that used for a set of transformed equations described in Noye et al (1981). The present scheme uses explicit differencing in the horizontal and implicit differencing in the vertical, whereas Noye et al (1981) used an updated explicit scheme in the horizontal with semi-implicit differencing in the vertical. The new scheme is essentially of the leap-frog type, with an interconnected elevation calculation at each half time step. Boundary elements are handled in a flexible manner, allowing the model to be readily applied to arbitrarily different boundary configurations. The method described in Noye et al (1981) only applies to gulfs with slowly varying widths. All velocity points within the model region are solved using the full set of equations, but on open boundaries they are obtained by extrapolation from interior values.

The performance and accuracy of the model has been checked by comparing numerical and analytical results for a channel and a rectangular bay. The equivalence of the computer program code to varying orientation has been checked by successively rotating a test bay through 45° and verifying that the numerical results are identical for each orientation, for a given value of the Coriolis parameter.

1. INTRODUCTION

The development of time dependent numerical models capable of solving the equations governing tidal flow using finite difference techniques on a three-dimensional spatial grid has been investigated only relatively recently; for example, by Heaps (1981), Heaps and Jones (1981) and Noye et al (1981).

Some of these earlier models are based on finite difference techniques in the horizontal and spectral methods in the vertical - firstly, predictions of depth-averaged character are obtained, followed by computation of depth variation of velocity from a prescribed vertical eddy viscosity. For the present model, full three-dimensional information about velocity is computed each time step. Since access to very fast computers such as the CRAY is only available to a fortunate few, coastal engineers and environmental consultants must make use of machines usually no bigger than the University of Adelaide's DEC VAX 11. Thus there remains a need to continue to develop ways of optimizing calculation time and storage requirements. Development of two- and three-dimensional models on moderately fast computers has met with difficulty in the inclusion of open and closed boundary elements. Often, model elements have been labelled according to the different configuration of their neighbouring elements. In such cases, an elaborate table of 20 or more different types of elements has been developed for each of which an appropriate form of the discretized equations is identified.

In the present work, only three classifications of element type are required, and they are defined according to the nature of the elevation point of the element, and without regard to the element's neighbourhood. Thus, rather than constructing different forms of the discretized equations according to the type of neighbourhood an element has, each equation is discretized once and written as a set of expressions contributing to the central element from each of the eight surrounding elements. Each contribution is locally modified according to its element type - the information required is either available or, if it is not, an approximation is constructed. Thus each discretized equation is written in general partitioned form and the appropriate differencing for any element with its particular neighbourhood is automatically selected by the program code by making use of the three types of element classification referred to above.

Consequently, the method of solution is generally applicable to any bay or ocean for which the governing equations are an adequate description of the tidal flow.

The accuracy of the method of differencing has been thoroughly checked by comparison with analytical solutions to linear forms of the governing equations. As well, the equivalence of the numerical differencing for different orientations of model regions has been tested. For this, a test rectangular bay was successively rotated by 45° increments and for each orientation, the results were compared both with each other and with the analytical solution to ensure that they remained in agreement.

A complete and detailed description of the model appears in Bills and Noye (1983).

2. THE MATHEMATICAL MODEL

The tidal equations used are non-linear and include terms for Coriolis and atmospheric pressure effects. Wind effects are included by incorporating a surface stress term in the approximation used to obtain the surface velocity components. Since coastal seas are generally well mixed, density is assumed constant. The hydrostatic approximation is used to replace the vertical component of the momentum conservation equation and simultaneously eliminate the pressure term in the two horizontal components. The horizontal eddy viscosity coefficients are assumed constant, but the vertical eddy viscosity function has a variable form which is dependent on the depth-averaged horizontal velocity and is a quadratic function over depth.

The equations are written in the rectangular cartesian (x,y,z) coordinate system with the z axis directed vertically upward from the x-y plane located at mean sea level. Velocity components (u,v,w) are defined in the usual way. A transformation of the depth coordinate, similar to that used by Noye et al (1982), is then introduced. It has the advantage of mapping the time variable sea surface defined by $z=\zeta(x,y,t)$ on to the flat surface $\eta=0$, and the sea bed $z=-h(x,y)$ to the flat surface $\eta=1$. To emphasize the fact that a transformation has taken place, the x-coordinate is written as X, the y-coordinate as Y and the time coordinate as T. The transformation from the system (x,y,z,t) to the system (X,Y,η,T) can therefore be defined by:

$$X = x\ ,\quad Y = y\ ,\quad \eta = (\zeta - z)/(\zeta + h)\quad \text{and}\quad T = t\ .$$

Under this transformation, the u and v velocity components are not changed, but the vertical velocity component is modified and its modified form is denoted ω.

The equations modelled then, are -

Mass-conservation equation:

$$\frac{\partial \zeta}{\partial T} = -\frac{\partial}{\partial X}(H\overline{U}) - \frac{\partial}{\partial Y}(H\overline{V}) \tag{2.1}$$

where the depth-averaged velocity components are defined by

$$\overline{U}(\eta) = \int_0^\eta u\, d\eta' \ , \ \overline{V}(\eta) = \int_0^\eta v\, d\eta' \quad \text{and for brevity} \quad \overline{U} = \overline{U}(1),\ \overline{V} = \overline{V}(1) \ ,$$

and the total sea depth is

$$H = h + \zeta.$$

Momentum-conservation equations:

$$\frac{\partial Hu}{\partial T} + \frac{\partial Hu^2}{\partial X} + \frac{\partial Huv}{\partial Y} + H\frac{\partial u\omega}{\partial \eta} - Hfv$$
$$= -Hg\frac{\partial \zeta}{\partial X} - H\frac{1}{\rho}\frac{\partial P_a}{\partial X} + HN_X\frac{\partial^2 u}{\partial X^2} + HN_Y\frac{\partial^2 u}{\partial Y^2} + \frac{1}{H}\frac{\partial}{\partial \eta}\left(N_\eta\frac{\partial u}{\partial \eta}\right) \tag{2.2}$$

$$\frac{\partial Hv}{\partial T} + \frac{\partial Hvu}{\partial X} + \frac{\partial Hv^2}{\partial Y} + H\frac{\partial v\omega}{\partial \eta} + Hfu$$
$$= -Hg\frac{\partial \zeta}{\partial Y} - H\frac{1}{\rho}\frac{\partial P_a}{\partial Y} + HN_X\frac{\partial^2 v}{\partial X^2} + HN_Y\frac{\partial^2 v}{\partial Y^2} + \frac{1}{H}\frac{\partial}{\partial \eta}\left(N_\eta\frac{\partial v}{\partial \eta}\right) \tag{2.3}$$

where

p_a = atmospheric pressure at the sea surface,
f = Coriolis parameter,
g = gravitational acceleration (9.81 m/sec^2),
ρ = sea water density (1,027 kg/m^3),
N_X, N_Y are coefficients of horizontal eddy viscosity (typically 5 m^2/sec),
$N_\eta = \alpha + \beta HF(\eta)\sqrt{\overline{U}^2 + \overline{V}^2}$
is the vertical eddy viscosity formulation with suitable values chosen for the parameters α and β, and
F = the prescribed quadratic variation of vertical eddy viscosity over depth.

Equation to evaluate ω, the transformed vertical velocity component:

$$\omega = -\frac{1}{H}\frac{\partial}{\partial X}\{H[\overline{U}(\eta) - \eta\overline{U}]\} - \frac{1}{H}\frac{\partial}{\partial Y}\{H[\overline{V}(\eta) - \eta\overline{V}]\} \ , \tag{2.4}$$

and the equation used to deduce the cartesian vertical velocity component w is:

$$w = -H\omega + (1-\eta)\left[\frac{\partial \zeta}{\partial T} + u\frac{\partial \zeta}{\partial X} + v\frac{\partial \zeta}{\partial Y}\right] - \eta\left[u\frac{\partial h}{\partial X} + v\frac{\partial h}{\partial Y}\right] . \tag{2.5}$$

For the application of the model to the Gulf of Carpentaria region, described by Mitchell et al (1983), the following values are used in the formulation for vertical eddy viscosity:

α = 0.0001 (m^2/sec),
β = 0.0033,
$F(\eta) = (0.5 + \eta)(1 - \eta)$.

The boundary conditions employed are:

(1) A condition of no flow normal to the coastline leads to the conditions that if the coastline runs parallel to

(i) the X-axis, then $v = 0$ on the coastline,

(ii) the Y-axis, then $u = 0$ on the coastline.

(2) A no-slip condition is used at the sea bed so that

$$u = v = \omega = w = 0 \text{ on } \eta = 1.$$

(3) By applying the free surface condition $w = \frac{\partial \zeta}{\partial T} + u\frac{\partial \zeta}{\partial X} + v\frac{\partial \zeta}{\partial Y}$ at $\eta = 0$ in equation (2.5), it follows that

$$\omega = 0 \text{ on } \eta = 0 \text{ for all } T.$$

(4) At the sea surface, the wind stress components S_X, S_Y are taken to be proportional to the η-gradient of the current in the following way:

$$S_X = -\rho\frac{1}{H}\left[N_\eta \frac{\partial u}{\partial \eta}\right]_{\eta=0} ,$$

$$S_Y = -\rho\frac{1}{H}\left[N_\eta \frac{\partial v}{\partial \eta}\right]_{\eta=0} ,$$

where S_X and S_Y may be expressed in terms of the wind velocity field as follows:

$$S_X = c_W \rho_a w_X |\underline{w}| ,$$

$$S_Y = c_W \rho_a w_Y |\underline{w}| ,$$

and

$\underline{w} = (w_X, w_Y)$ is the horizontal wind velocity field at some specified reference height,

c_W is the wind stress coefficient and is dependent on the reference height and wind velocity magnitude according to the drag laws of Wu (1969), and

ρ_a is air density.

The wind velocity field is specified at all elements of the model for all time.

(5) At each element of the model, the atmospheric pressure is also specified for all time.

(6) At the open boundaries of the model, the elevation is specified for all time by means of the amplitude and phase of the four dominant tidal components.

3. NUMERICAL DIFFERENCING PROCEDURE

The finite difference technique is used to discretize the transformed tidal equations. The equations are differenced on a rectangular staggered grid with a view to maximizing the accuracy of the approximation while minimizing the number of grid elements involved in the differencing. This confines the difference expressions for each equation to nine elements: the computational element and its eight immediate neighbours.

Each grid element consists of three grid points, one for each of the two horizontal velocity components, and the third for elevation and the vertical velocity component and all other quantities. This is illustrated in Fig. 1.

In the η-direction, a smoothly non-uniform grid spacing defined by the Kappa-method described in Noye (1983) is used, with a fine grid near both the transformed sea floor and surface, as illustrated in Fig. 2.

The conservation of mass equation is differenced about the elevation point of the general computational element, the X-component of the conservation of momentum equation is differenced about the u velocity point of the element at the appropriate depth level and the Y-component about the v velocity point. The ω and w equations are evaluated at the appropriate depth level below the elevation point.

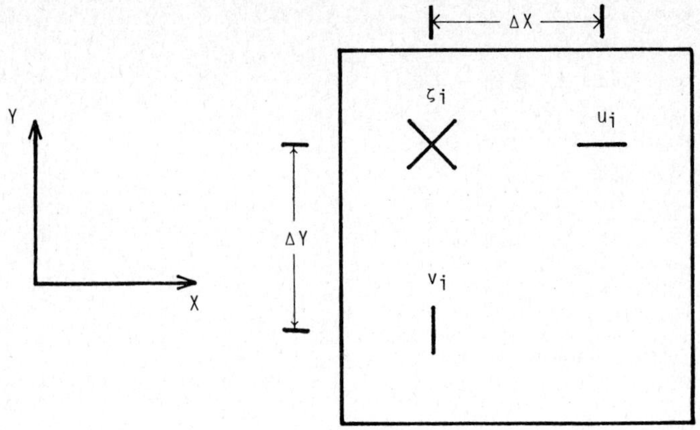

Figure 1
Plan view of the ith grid element, with the ζ_i, u_i and v_i grid points indicated, and the horizontal grid step sizes.

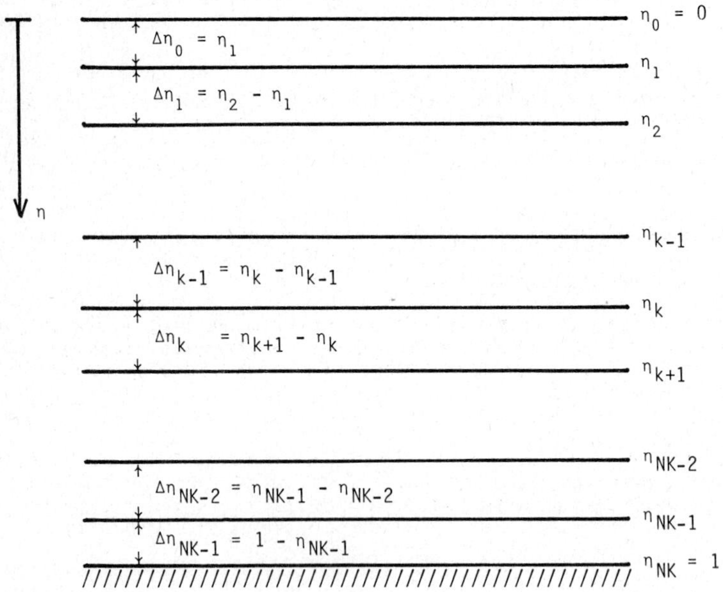

Figure 2
A section of the transformed region illustrating the unequally spaced depth levels and their spacings. There are (NK+1) distinct levels.

Before the advent of the fast vector processing computer, modellers had to be careful to minimize both the number of calculations involved in their computations and the demand on computer storage. Even now, computers such as the CRAY are not readily accessible and so there remains a need for algorithms that discern between model and non-model elements and do their calculations only on the former. Our own computer facility is the moderately fast DEC VAX 11.

In previous approaches to the application of the difference equations to model elements using computer facilities of conventional type, in both depth-averaged and three-dimensional tidal modelling, each element was classified according to its position with respect to neighbouring coastal and open sea boundaries, and an accompanying formulation of the equations was derived (eg. see Flather, 1972). Then, when solving the equations at a particular element, its classification code was used to select the appropriate form of the equations to be solved. These approaches required much preparatory work in programming the various forms of each difference equation and, unless all types of elements were taken into account, resulted in a computer program of limited usefulness since it could often only be applied to a different region after considerable modification by hand.

In the present approach, each element is assigned one of just three codes according to the position of its elevation grid point. That is, the proposed new element coding does not depend on the neighbourhood of the elements. An element is coded

"0" if its elevation point lies outside the model boundaries,
"1" if its elevation point lies within the model boundaries, or
"2" if its elevation point lies on any open sea boundary of the model.

Computations are carried out only at type 1 and type 2 elements.

Since the differencing of each equation involves just nine grid elements, a way of applying the equations to a model element with any neighbourhood configuration has been developed. Each neighbour can be considered to contribute information to the calculation of some quantity at the central element. The contribution from a particular element can be written in such a form as to make use of its element code to either

(1) exclude its contribution, if it is not required,
(2) include its contribution, if it is required, or
(3) include an approximation to its contribution when the actual value of its contribution is not known.

Thus for example, if the code of the nth neighbour is NC, and the quantity being calculated at the computational element is q, an expression can be written down for that neighbour's contribution toward the calculation of q by way of its local value q_n. That expression is :

$$NC * (2-NC) * q_n + (NC/2) * q_n(approx) , \qquad (3.1)$$

where q_n(approx) is a second order accurate approximation in the absence of a value being known for q_n. Substituting the possible values that NC may take and making use of integer computer arithmetic, it can be seen that the contribution is

(1) 0 , if NC = 0,
(2) q_n , if NC = 1, or
(3) q_n(approx), if NC = 2.

Using expressions like (3.1), each of the difference equations can be written in partitioned form so that any allowable boundary configuration of the nine-element block can be handled within a single difference equation for the quantity being calculated.

Understandably, the partitioned forms of the difference equations are sometimes long, but some computational brevity is achieved by testing the neighbourhood and determining whether it

(1) is a mixture of type 0 and type 1 elements and thus occurs in the vicinity of a closed boundary, in which case the form of (3.1) that applies is NC^*q_n,
(2) consists only of type 1 elements and is therefore surrounded by water, in which case (3.1) reduces to q_n, or
(3) is a mixture of all three types of element and occurs in the vicinity of an open sea boundary and possibly a land boundary as well, whence (3.1) is used in its entirety.

For each difference equation, three sets of code have therefore been written to accommodate these three broad classifications of neighbourhood. Option (2) immediately above involves fewest computations; option (1) involves more, and option (3) is the completely general partitioned form of the equations. A simple algebraic test using the neighbourhood codes has been written to guide the program to the correct option, thus saving unnecessary operations and computer time.

This approach requires a minimum of preparatory work before applying the equations to a different model region since all that is required is to correctly code the elements of the new region as being of types 0,1 or 2.

4. NUMERICAL SOLUTION PROCEDURE

The finite difference scheme used is of the leap-frog type with a two-level elevation calculation, as indicated in Fig. 3.

At the beginning of each time step, a space-centred explicit scheme is used to calculate the ω field at each depth level from a knowledge of the $\overline{U},\overline{V},\zeta$ fields at that time level. The required numerical integration is carried out using the Trapezoidal rule for unequally spaced intervals. Then a time- and space-centred scheme is used to calculate the time derivative $\partial\zeta/\partial T$ which is needed for the calculation of the vertical velocity w. In the course of this, the elevation field is updated for the next half time step.

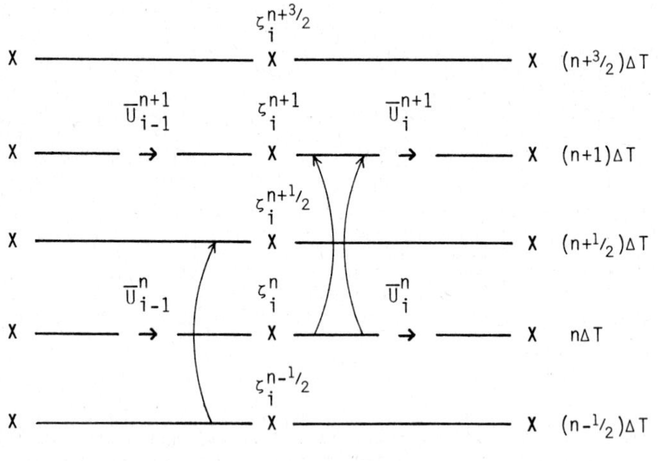

Figure 3
The basic leap-frog nature of the numerical scheme with the additional calculation of elevation after the velocity leap.

Central space schemes, centred at the half time step, are then used to calculate the horizontal velocity components u and v at each element inside the model at the next time level, using an implicit solution procedure over depth. From these horizontal velocity fields the depth-averaged fields $\overline{U}$ and $\overline{V}$ are computed using Simpson's Rule for numerical integration over unequally spaced intervals.

Finally, using the latest depth-averaged velocity fields, the tidal elevation ζ is calculated for the end point of the time step, using a time- and space-centred explicit scheme.

The execution of the computer program therefore follows the sequence:

(1) Input model data such as element type and undisturbed sea depth, and set initial conditions.
(2) Update time to T.
Calculate ω and also the time derivative $\partial\zeta/\partial T$ - this leads to the calculation of elevation for time $T+\Delta T/2$.
If required, calculate w from ω.
(3) Update time to $T+\Delta T/2$.
Read in open boundary elevations.
Calculate total sea depth.
Read in atmospheric pressure field.
(4) Update time to $T+\Delta T$.
Read in wind field values.
(5) Calculate horizontal velocity components, for each velocity point within the model boundaries.
(6) Calculate depth-averaged horizontal velocity field.
(7) Calculate elevation.
Calculate total sea depth.
(8) Transfer depth-averaged velocity values from the computational arrays into the holding arrays, in readiness for the next time loop.
(9) Repeat steps 2-8 until the required results are obtained.

Since the elevation is updated each half time step, this scheme is not truly leap-frog, though it is of leap-frog type. The inclusion of the calculation of elevation at each half time step and the high degree of coupling between the elevation calculations and velocity calculation appears to have reduced the effect of computational modes which occur with the usual form of the leap-frog scheme.

5. VERIFYING THE MODEL FOR DEPTH DEPENDENT CHANNEL FLOW

There exists an analytical solution for the linearized problem of depth dependent tidal flow along a channel of constant depth and constant width.

Omitting Y dependence, non-linear transport terms, Coriolis effects, wind effects, atmospheric pressure and X-diffusion terms, the linearized forms of the tidal equations for flow in a channel of constant depth D are:

Mass-conservation equation:

$$\frac{\partial \zeta_L}{\partial T} = - D\frac{\partial \overline{U}_L}{\partial X} , \tag{5.1}$$

Momentum-conservation equation:

$$D\frac{\partial u_L}{\partial T} = - Dg\frac{\partial \zeta_L}{\partial X} + \frac{1}{D}\frac{\partial}{\partial \eta}\left(N_\eta \frac{\partial u_L}{\partial \eta}\right) , \tag{5.2}$$

Equation for the evaluation of transformed vertical velocity:

$$\omega_L = - \frac{\partial}{\partial X}\left(\overline{U}_L(\eta) - \eta \overline{U}_L\right) , \tag{5.3}$$

Equation for the evaluation of vertical velocity:

$$w_L = - D\omega_L + (1 - \eta)\left[\frac{\partial \zeta_L}{\partial T} + u_L\frac{\partial \zeta_L}{\partial X}\right] , \tag{5.4}$$

where

$$\overline{U}_L(\eta) = \int_0^\eta u_L \, d\eta' \quad \text{and} \quad \overline{U}_L = \overline{U}_L(1) \; . \qquad (5.5)$$

The boundary conditions that apply are:

(1) At the closed end of the channel, $u_L = 0$.

(2) At the sea bed,

$$u_L = \omega_L = w_L = 0 \text{ on } \eta = 1.$$

(3) At the sea surface, given there is no applied stress, $\partial u_L/\partial\eta = 0$ and, applying the free surface condition to (5.4), $\omega_L = 0$ on $\eta = 0$.

(4) At the open boundary, ζ_L is specified for all time.

A solution for equations (5.1) and (5.2) has been obtained by Johns and Odd (1966) for a particular formulation of N_η and a particular frequency σ:

$$u_L(X,\eta,T) = \text{Re}\{\frac{aigm}{\sigma} U_J(\eta).\sin(mX).\exp(-i\sigma T)\} \; , \qquad (5.6)$$

$$\zeta_L(X,T) = \text{Re}\{a\cos(mX).\exp(-i\sigma T)\} \; , \qquad (5.7)$$

where, $i = \sqrt{-1}$. With the following specified form of N_η:

$$N_\eta(\eta) = \nu\{1+\gamma(1-\eta)\}^2 \; ,$$

the solution for $U_J(\eta)$ is then

$$U_J(\eta) = 1 + C_1\{1+\gamma(1-\eta)\}^{n_1} + C_2\{1+\gamma(1-\eta)\}^{n_2} \; , \qquad (5.8)$$

where

$$n_1, n_2 = -\frac{1}{2} \pm \sqrt{\frac{1}{4} - \frac{iD^2\sigma}{\gamma^2\nu}} \; ,$$

and C_1, C_2 are found by solving the simultaneous equations

$$C_1 + C_2 = -1 \; ,$$

$$n_1(1+\gamma)^{n_1}C_1 + n_2(1+\gamma)^{n_2}C_2 = 0 \; ,$$

which arise upon application of the boundary conditions for u_L.

In the above,

a is the amplitude,
m is the wave number,
σ is the circular frequency,
γ is a number used to specify the surface value of N_η, and
ν is a reference bed viscosity.

From equations (5.5-8), the solution for ω_L can be derived (Bills and Noye, 1983):

$$\omega_L(X,\eta,T) = \text{Re}\{\frac{-i\sigma\zeta_L}{D}\,[\frac{m^2}{\sigma^2}Dg(\eta - \frac{1}{\gamma n_1(n_1+1)}(C_1 n_1\{1+\gamma(1-\eta)\}^{n_1+1} + C_2 n_2\{1+\gamma(1-\eta)\}^{n_2+1})) - \eta]\} \; .$$

This solution, along with expressions (5.6) and (5.7) can then be used to calculate w_L by applying

$$w_L(X,\eta,T) = \text{Re}\{-D\omega_L + (1-\eta)[\,-i\sigma\zeta_L - u_L am.\sin(mX).\exp(-i\sigma T)]\} \; .$$

The computer program has been written to solve the complete set of non-linear equations (2.1-5). Therefore, when comparing the accuracy of the numerical solution with an analytical solution, it is essential that the program solution be obtained from precisely the same set of equations as are solved to obtain the analytical solution. In the present case, the program was modified to solve the linear set (5.1-4) by omitting all terms not involved in those equations, and the subroutine updating the total depth was not invoked - instead, the total depth array H was set to the constant value D.

The numerical model is then forced by prescribing the analytical solution for elevation at the open boundary. These values are prescribed every half time step. Apart from this, the elevation and velocity fields are initialized at zero value.

When the model is started, there is a certain length of time, often several model tidal periods in duration, when the transient solutions generated by the explicit difference scheme in the early stages of the run, and by the governing differential equations, damp out. These transients are a response to a type of shock transmitted through the model due to the sudden imposition of motion at the open boundary. During and after this, there is a further period of internal adjustment within the scheme as it settles down to accommodate the cyclic forcing at the boundary.

The model parameters and dimensions used for this comparison are as follows:

ΔT = 720 secs (12 mins), ΔX = 5000 m (5 km), No. of depth levels = 21,
L = 95000 m (ie. length is 95 kms), D = 20 m, a = 0.5 m,
$\sigma = 2\pi/(12*3600)$ (period of 12 hours), γ = 9, ν = 0.001 m^2/sec.

The wavelength of the linear wave is approximately 580,000 m (580 kms).

The following results for elevation, horizontal velocity, vertical velocity in the transformed system and vertical velocity in the cartesian system were obtained after the numerical results had settled down which, for a "cold" start, required 12 model tidal cycles. At this stage, the values generated began to repeat themselves from one cycle to the next with three figure accuracy in the elevation values. It should be noted that for this run, the prescribed open boundary elevation was specified to be at maximum displacement at the initial time. The ability of the program to handle to abrupt starts is thereby demonstrated.

In Table 1, analytical and computed elevations (metres) are presented for a point six minutes before the end of the tidal cycle. The computed values were obtained using the first elevation step (listed as step (2) in section 4). In Table 2, analytical and computed elevations are presented for the end of the tidal cycle. These elevations were obtained using the second elevation step (step (7)).

TABLE 1. Analytical and numerical results using the first elevation step.

Distance from closed end (km) ...	5	15	25	35	45	55	65	75	85	95
Anal.	-0.485	-0.480	-0.470	-0.455	-0.434	-0.410	-0.380	-0.347	-0.310	-0.269
Num.	-0.487	-0.481	-0.471	-0.456	-0.435	-0.410	-0.381	-0.347	-0.310	-0.269

TABLE 2. Analytical and numerical results using the second elevation step.

Distance from closed end (km) ...	5	15	25	35	45	55	65	75	85	95
Anal.	-0.491	-0.486	-0.475	-0.459	-0.439	-0.413	-0.383	-0.349	-0.311	-0.269
Num.	-0.491	-0.486	-0.475	-0.460	-0.439	-0.413	-0.383	-0.349	-0.311	-0.269

Tables 3,4,5 compare results obtained at the end of the tidal cycle. Table 3 compares results for horizontal velocity (m/sec) along the channel. Results are given for only five of the 21 depth levels for which calculations were carried out. Table 4 compares results for transformed vertical velocity (sec^{-1}) at seven of the calculated levels, and Table 5 compares results for cartesian vertical velocity (m/sec) for five of the calculated levels.

TABLE 3. Analytical and numerical solutions for horizontal velocity.

ANALYTICAL SOLUTION

Depth level	Distance from closed end (km.) ... 10	20	30	40	50	60	70	80	90
0	0.0084	0.0166	0.0245	0.0317	0.0382	0.0437	0.0482	0.0514	0.0533
5	0.0084	0.0165	0.0243	0.0315	0.0380	0.0435	0.0479	0.0511	0.0530
10	0.0077	0.0153	0.0225	0.0291	0.0350	0.0400	0.0440	0.0468	0.0484
15	0.0045	0.0088	0.0128	0.0165	0.0197	0.0222	0.0241	0.0251	0.0252
20	0.0000	0.0000	0.0000	0.0000	0.0000	0.0000	0.0000	0.0000	0.0000

NUMERICAL SOLUTION

Depth level	Distance from closed end (km.) ... 10	20	30	40	50	60	70	80	90
0	0.0085	0.0167	0.0246	0.0319	0.0384	0.0440	0.0484	0.0517	0.0536
5	0.0084	0.0166	0.0245	0.0317	0.0382	0.0437	0.0481	0.0514	0.0533
10	0.0078	0.0154	0.0226	0.0293	0.0352	0.0403	0.0443	0.0472	0.0488
15	0.0045	0.0089	0.0131	0.0168	0.0201	0.0227	0.0245	0.0256	0.0258
20	0.0000	0.0000	0.0000	0.0000	0.0000	0.0000	0.0000	0.0000	0.0000

TABLE 4. Analytical and numerical solutions for transformed vertical velocity.

ANALYTICAL SOLUTION (x 10^{-7})

Depth level	Distance from closed end (km.) ... 5	15	25	35	45	55	65	75	85
0	0.000	0.000	0.000	0.000	0.000	0.000	0.000	0.000	0.000
1	-0.052	-0.051	-0.049	-0.046	-0.043	-0.038	-0.033	-0.027	-0.021
5	-0.309	-0.303	-0.292	-0.276	-0.254	-0.228	-0.198	-0.164	-0.126
10	-0.806	-0.791	-0.762	-0.719	-0.663	-0.594	-0.513	-0.423	-0.323
15	-0.777	-0.762	-0.732	-0.687	-0.629	-0.557	-0.474	-0.381	-0.278
19	-0.187	-0.183	-0.174	-0.162	-0.145	-0.125	-0.102	-0.754	-0.047
20	0.000	0.000	0.000	0.000	0.000	0.000	0.000	0.000	0.000

NUMERICAL SOLUTION (x 10^{-7})

Depth level	Distance from closed end (km.) ... 5	15	25	35	45	55	65	75	85
0	0.000	0.000	0.000	0.000	0.000	0.000	0.000	0.000	0.000
1	-0.052	-0.051	-0.049	-0.046	-0.043	-0.038	-0.033	-0.027	-0.021
5	-0.311	-0.305	-0.294	-0.277	-0.256	-0.229	-0.199	-0.164	-0.126
10	-0.812	-0.797	-0.767	-0.723	-0.666	-0.596	-0.515	-0.423	-0.322
15	-0.790	-0.775	-0.744	-0.698	-0.638	-0.565	-0.480	-0.385	-0.280
19	-0.205	-0.201	-0.191	-0.178	-0.160	-0.138	-0.113	-0.084	-0.053
20	0.000	0.000	0.000	0.000	0.000	0.000	0.000	0.000	0.000

TABLE 5. Analytical and numerical solutions for cartesian vertical velocity.

ANALYTICAL SOLUTION (x 10^{-4})

Depth level	Distance from closed end (km.) ... 5	15	25	35	45	55	65	75	85
0	-0.133	-0.129	-0.122	-0.111	-0.097	-0.081	-0.061	-0.040	-0.017
5	-0.103	-0.101	-0.095	-0.086	-0.075	-0.062	-0.047	-0.030	-0.011
10	-0.050	-0.049	-0.046	-0.041	-0.036	-0.029	-0.021	-0.012	-0.002
15	-0.007	-0.007	-0.006	-0.006	-0.004	-0.003	-0.001	0.000	0.002
20	0.000	0.000	0.000	0.000	0.000	0.000	0.000	0.000	0.000

NUMERICAL SOLUTION (x 10^{-4})

Depth level	Distance from closed end (km.) ... 5	15	25	35	45	55	65	75	85
0	-0.133	-0.130	-0.123	-0.113	-0.099	-0.083	-0.065	-0.044	-0.022
5	-0.104	-0.101	-0.096	-0.088	-0.077	-0.064	-0.050	-0.033	-0.015
10	-0.050	-0.049	-0.046	-0.042	-0.036	-0.030	-0.022	-0.014	-0.004
15	-0.007	-0.007	-0.006	-0.006	-0.005	-0.003	-0.002	0.000	0.002
20	0.000	0.000	0.000	0.000	0.000	0.000	0.000	0.000	0.000

The numerical results obtained for the present example are in very good agreement with the corresponding analytical results.

This indicates the accuracy of the linear form of the difference scheme, and in particular, provides a check on the differencing in the vertical.

6. VERIFYING THE MODEL IN PLAN

Omitting non-linear transport terms, atmospheric pressure and horizontal diffusion terms, ignoring wind effects and using a linear friction approximation, the linear tidal equations for depth-averaged flow are:

Mass-conservation equation:

$$\frac{\partial \zeta_L}{\partial T} = - D\frac{\partial \overline{U}_L}{\partial X} - D\frac{\partial \overline{V}_L}{\partial Y} \tag{6.1}$$

Momentum-conservation equations:

$$D\frac{\partial \overline{U}_L}{\partial T} - Df\overline{V}_L = - Dg\frac{\partial \zeta_L}{\partial X} - K_B\overline{U}_L \tag{6.2}$$

$$D\frac{\partial \overline{V}_L}{\partial T} + Df\overline{U}_L = - Dg\frac{\partial \zeta_L}{\partial Y} - K_B\overline{V}_L \tag{6.3}$$

where K_B is the linear bottom drag coefficient.

Rienecker and Teubner (1980) provide a set of truncated series solutions for equations (6.1-3) for a rectangular bay. The boundary conditions applied to the equations for this region are:

$$\overline{U}_L(0,Y,T) = 0 \ , \text{ for } 0 \leqslant Y \leqslant W \text{ and } T \geqslant 0 \ ,$$

$$\overline{V}_L(X,0,T) = \overline{V}_L(X,W,T) = 0 \ , \text{ for } 0 \leqslant X \leqslant L \text{ and } T \geqslant 0 \ ,$$

with the special condition across the open boundary at $X = L$,

$$\overline{V}_L(L,Y,T) = 0 \ , \text{ for } 0 \leqslant Y \leqslant W \text{ and } T \geqslant 0 \ .$$

Because of their length, the series solutions are not reproduced here.

Again, the numerical algorithm was appropriately modified so as to solve the set of equations for which the series solution is valid. This was achieved by the following:

(1) Delete the call to the depth updating subroutine.
(2) Delete the call to the depth-averaging subroutine for velocity.
(3) Use the boundary condition $\overline{V}_L = 0$ on the open boundary which, for the linear equations, only appears in the $\overline{U}$-equation Coriolis term, for an open boundary to the east.
(4) Modify the vertical diffusion terms of the velocity subroutines so as to run the linear bottom friction form. Specifically, set $K_B = 0.001$.
(5) Delete the advective, atmospheric pressure and horizontal diffusion terms, and set the wind stress field to zero.

The model parameters and dimensions used for this comparison are as follows:

ΔT = 300 secs (5 mins), ΔX = 5000 m (5 kms), ΔY = 5000 m (5 kms),
L = 95000 m (ie. length is 95 kms), W = 90000 m (ie. width is 90 kms),
$f = -8.16 \times 10^{-5}$ (ie. latitude is 34° South) K_B = 0.001 m/sec, D = 20 m.

The period of the forcing tidal oscillation at the open boundary was 12 hours and the incident amplitude 0.5 m.

The following results for elevation and depth-averaged horizontal velocity components were taken from the tenth cycle after a "cold" start - at this stage, the tidal regime was established. These results settle down faster than the results for depth dependent tidal flow partly because the time step of 12 mins. used in that case was at the limit of the model's range of stable operation whereas the above step of 5 mins. was well within its stable range. Table 6 gives comparative results for depth-averaged velocity (m/sec) in the X-direction at the three hour point of the tide cycle. Tables 7 and 8 compare results for the depth-averaged velocity (m/sec) in the Y-direction and for elevation (metres), respectively, at the end of the cycle.

The numerical results obtained in each case are in excellent agreement with the corresponding analytical values. The results essentially verify the accuracy of the differenced formulation for the Coriolis term in both velocity equations.

TABLE 6. Comparative results for $\overline{U}$, the depth-averaged u velocity.

Distance across bay (km)		Distance from closed end (km.) ... 10	20	30	40	50	60	70	80	90
5	Anal.	0.074	0.155	0.238	0.321	0.402	0.479	0.552	0.619	0.679
	Num.	0.074	0.155	0.238	0.320	0.400	0.476	0.547	0.611	0.666
25	Anal.	0.087	0.173	0.259	0.344	0.425	0.502	0.574	0.640	0.698
	Num.	0.086	0.173	0.258	0.342	0.423	0.499	0.570	0.635	0.693
45	Anal.	0.094	0.188	0.279	0.366	0.450	0.528	0.601	0.667	0.725
	Num.	0.094	0.187	0.277	0.365	0.448	0.526	0.598	0.664	0.723
65	Anal.	0.104	0.205	0.302	0.394	0.481	0.561	0.635	0.702	0.761
	Num.	0.103	0.204	0.300	0.392	0.479	0.560	0.634	0.702	0.761
85	Anal.	0.122	0.233	0.336	0.432	0.522	0.604	0.680	0.747	0.806
	Num.	0.121	0.231	0.334	0.430	0.520	0.603	0.680	0.749	0.812

TABLE 7. Comparative results for $\overline{V}$, the depth-averaged v velocity.

Distance across bay (km)		Distance from closed end (km.) ... 5	15	25	35	45	55	65	75	85
10	Anal.	0.066	0.042	0.028	0.020	0.014	0.010	0.007	0.005	0.004
	Num.	0.066	0.043	0.029	0.020	0.014	0.010	0.007	0.005	0.004
30	Anal.	0.131	0.095	0.068	0.048	0.035	0.025	0.018	0.013	0.009
	Num.	0.132	0.095	0.068	0.049	0.035	0.025	0.018	0.012	0.008
50	Anal.	0.142	0.104	0.076	0.055	0.039	0.028	0.020	0.014	0.010
	Num.	0.142	0.105	0.076	0.055	0.039	0.028	0.020	0.014	0.009
70	Anal.	0.105	0.073	0.051	0.036	0.026	0.018	0.013	0.009	0.007
	Num.	0.106	0.074	0.052	0.037	0.026	0.018	0.013	0.008	0.005

TABLE 8. Comparative results for elevation.

Distance across bay (km)		Distance from closed end (km.) ... 5	15	25	35	45	55	65	75	85
5	Anal.	1.32	1.31	1.28	1.23	1.18	1.11	1.03	0.94	0.84
	Num.	1.31	1.30	1.27	1.23	1.17	1.10	1.02	0.94	0.84
25	Anal.	1.32	1.30	1.26	1.22	1.16	1.09	1.01	0.92	0.82
	Num.	1.31	1.29	1.26	1.21	1.15	1.08	1.00	0.91	0.82
45	Anal.	1.32	1.29	1.26	1.21	1.15	1.08	1.00	0.91	0.80
	Num.	1.31	1.29	1.25	1.20	1.15	1.08	0.99	0.90	0.80
65	Anal.	1.32	1.29	1.26	1.21	1.15	1.08	1.00	0.90	0.80
	Num.	1.31	1.29	1.25	1.21	1.15	1.08	0.99	0.90	0.80
85	Anal.	1.32	1.30	1.27	1.22	1.16	1.09	1.01	0.91	0.80
	Num.	1.31	1.30	1.26	1.22	1.16	1.09	1.00	0.91	0.80

7. IMPROVING MODEL ACCURACY

The value of checking simplified forms of the numerical model with analytical solutions, even though the latter may be solutions for linear forms of the tidal equations, is that it will indicate the usefulness of, for example, using one method of numerical integration in the place of another, or of using one finite difference formulation in the place of another, and so on. In this way, model accuracy can be improved and the speed of computer solution can be increased by choosing more efficient algorithms to perform a certain function.

Because, generally speaking, the linear part of the equations accounts for the bulk of the flow, the results of such experimentation usually apply to the full non-linear set of discretized equations.

Certain cases tested to find the best method of numerical quadrature, differencing, and so on, follow.

Optimizing the number of depth levels

The accuracy of solution was found to increase as the number of depth levels was increased. However beyond a certain number of levels, improvement was almost negligible, while the cost in terms of computer storage and computational time continued to rise.

Since the gulfs to be studied, such as the Gulf of Carpentaria and Spencer Gulf, are not deep, the required number of depth levels needed to determine the vertical current structure accurately does not need to be very large. Each extra depth level adds another three components of velocity over the entire field and must also be taken into account when performing depth-averaging. While the computer available, a VAX system, has a virtual memory facility, there is a limit to the size of this facility and as well, time is lost if there is the need to continually exchange information into and out of core memory during computation.

On the other hand, as will be seen in the next section, a suitable spacing of the available depth levels can improve the accuracy of solution. When all these factors are taken into account, 10 depth levels spaced in the optimal fashion suffice to optimize accuracy, use least computer storage and computer time and give optimal representation of vertical profiles of velocity, etc.

Optimizing the arrangement of depth levels

Near the bed, the velocity profiles are steep because of the requirement of the no-slip condition at the bed. A grid that is fine near the sea floor and coarse above it, is better suited to convey the required information from this region to the upper level of the water body than a uniform grid of the same number of depth levels. Similarly, a fine grid near the surface is better suited to convey information about an applied surface wind stress into the lower levels of the water.

It is found to be best to have a fine grid near the surface, graded to coarse at the middle levels, back to fine near the bottom, even in the absence of a surface wind stress.

The actual variation of the grid spacing is systematically chosen, based on a second-order smoothing relationship between consecutive grid levels governed by the parameter κ as indicated in Noye (1983), namely:

$$\Delta\eta_k = \Delta\eta_{k-1}(1+\kappa\Delta\eta_{k-1}).$$

Some numerical tests were carried out to determine the optimum κ for the depth dependent channel flow problem. These indicated that a value of $\kappa = 3$ gave best results for a fixed number of depth levels with a symmetrical vertical grid-spacing about mid-depth.

Surface velocity approximation

In early tests, a first order accurate approximation to the surface derivative boundary condition was used. An alternative to this is to use a second-order approximation. However, such an approximation involves many more calculations, and may possibly reduce stability. When the two approximations were checked with an analytical result, it is found that the second-order approximation provided more accuracy, more consistently over the total depth field. In fact, the degree of accuracy of the surface velocity calculation was now the same as the calculation of the rest of the field, while stability was retained for quite large time steps. Therefore, the second-order surface velocity approximation was used.

Integration techniques for depth-averaging the horizontal velocity components

Theoretically, Simpson's Rule is a more accurate numerical integration technique than the Trapezoidal rule. Simpson's Rule for unequally spaced intervals is third-order accurate, and the Trapezoidal rule for unequally spaced intervals is second-order accurate. Because the finite difference equations are at best second-order it might be expected that a second-order numerical integration scheme should provide sufficient accuracy.

In tests, it was found that the two methods of integration gave answers that were essentially equivalent, but Simpson's Rule always produced slightly more accurate results. Therefore, from the point of view of overall accuracy for models run for at least 58 semi-diurnal tidal cycles to enable proper tidal analysis, Simpson's Rule was used. It should be noted that to calculate both the ω and w fields the

Trapezoidal rule is used rather than Simpson's Rule because of the need to compute these fields at every depth level.

8. TESTS WITH DIFFERENT BAY ORIENTATIONS

The above comparisons with analytical solutions for depth dependent channel flow and depth-averaged flow were initially carried out for rectangular channels and bays with an east-west aspect, that is an open boundary to the east, a closed boundary to the west.

The overall aim of the present modelling is to produce a program code that is independent of the orientation of the region, and so it is necessary to rotate the channel and bay through various angles to verify its generality. In doing this, errors present in the program code, particularly for the open boundary, can be located because identical results should be obtained with each orientation, provided the Coriolis force is held constant over the whole model.

It was discovered, for orientations which require energy to enter the model from the north and west, that it is necessary to calculate the velocity components at each depth level for each interior velocity point of the model including those next to an open boundary in type 2 elements. Depth-averaged approximations for the depth dependent velocity components near northern and western open boundaries, with the type of elements shown in Figure 1, have been shown in our tests to be poor approximations because they are an attempt to impose a constant vertical velocity profile at grid points which naturally have a depth varying profile with a sharp gradient near the sea floor.

Initially, the basic orientations with open boundaries facing toward the north, south, east and west, were tested. Later, the oblique (45°) orientations were also tested to enable errors in the rest of the code to be detected. For a spatially constant Coriolis force, identical results were obtained for each orientation.

In the process of carrying out these various tests, confidence in the accuracy and robustness of the model was established. It is not possible to find analytical solutions to directly check every discretized term of the equations, but given that results from eight basic orientations of the bay can be shown to be identical, in the fully non-linear case, and verified for accuracy in the case of the linear terms of the equations, assurance that the model works correctly seems well grounded.

9. CONCLUSION

This model has been successfully applied to the Gulf of Carpentaria by Mitchell et al (1984). The model time step used in that case was 10 mins. for the fully non-linear scheme, with $\Delta X = 9.4$ km. and $\Delta Y = 16.45$ km.

The model is robust, and has proven to be very stable in the fully non-linear case. Where the model has been compared with analytical solutions for the linear case, it has proven to be accurate.

It is intended to carry out further tests with analytical solutions now available for the non-linear problem: this will test the differencing of the non-linear terms.

The model will then be applied to Spencer Gulf in South Australia, with particular emphasis on the northern region, near Stony Point.

ACKNOWLEDGMENT

During the preparation of this material the first mentioned author was supported by a South Australian Transport Department Scholarship.

REFERENCES

Bills, P.J. and Noye, B.J. (1983). "A three-dimensional tidal model for coastal seas." To be published.

Flather, R.A. (1972). "Analytical and numerical studies in the theory of tides and storm surges". Ph D thesis, University of Liverpool.

Heaps, N.S. (1981). "Three-dimensional model for tides and surges with vertical eddy viscosity prescribed in two layers - I. Mathematical formulation". Geophys. J. R. astr. Soc., 64, 291-302.

Heaps, N.S. and Jones, J.E. (1981). "Three-dimensional model for tides and surges with vertical eddy viscosity prescribed in two layers - II. Irish Sea with bed friction layer". Geophys. J. R. astr. Soc., 64, 303-320.

Johns, B. and Odd, N. (1966). "On the vertical structure of Tidal Flow in River Estuaries". Geophys. J. R. astr. Soc., 12, 103-110.

Mitchell, W.M., Beard, D.A., Bills, P.J. and Noye, B.J. (1984). "An application of a three-dimensional tidal model to the Gulf of Carpentaria", in this volume: 'Computational Techniques and Applications', editor J. Noye, North-Holland Publishing Coy.

Noye, B.J., May R.L. and Teubner, M.D. (1981). "Three-dimensional Numerical Model of Tides in Spencer Gulf". Ocean Management, 6,137-148.

Noye, B.J., May R.L. and Teubner, M.D. (1982). "A Three-dimensional Model for a Shallow Gulf". 'Numerical Solutions of Partial Differential Equations', editor J. Noye, North-Holland Publishing Coy, 417-436.

Noye, B.J. (1983). "Finite Difference Techniques for Partial Differential Equations" in 'Computational Techniques for Differential Equations', editor J. Noye, North-Holland Publishing Coy, 95-354.

Reinecker, M.M. and Teubner, M.D. (1980). "A note on frictional effects in Taylor's Problem". J. Mar. Res. 38, No. 2, 183-191.

Wu, J. (1969). "Wind stress and surface roughness at air-sea interface." J. Geophys. Res., 74, 444-55.

Computational Techniques & Applications: CTAC-83
J. Noye & C. Fletcher (Editors)

AN APPLICATION OF A THREE-DIMENSIONAL TIDAL MODEL TO THE GULF OF CARPENTARIA

W.M. Mitchell, D.A. Beard, P.J. Bills and B.J. Noye

The Department of Applied Mathematics
The University of Adelaide
Adelaide, South Australia
Australia.

The tidal dynamics of the Gulf of Carpentaria are investigated by utilizing the three-dimensional model described by Bills and Noye (1984). Open boundary information is generated by a depth-integrated model of a larger region which includes the Gulf. Results are presented for the major tidal components of sea surface elevation generated by both models. Comparisons between observations and computations are made at interior points of the three-dimensional model. It is verified that the diurnal tides consist of a clockwise rotating Kelvin wave, while the semidiurnal tidal energy is mostly confined to the northern half of the Gulf.

1. INTRODUCTION

The three-dimensional model described by Bills and Noye (1984) is applied to the Gulf of Carpentaria (Figure 1). The model requires specification of open boundary sea surface elevation which proves difficult for the Gulf. The information required is obtained from the depth-averaged model described by Noye and Flather (1984). This is applied to an area which has sufficient open boundary information and includes the Gulf.

The models are calibrated by comparing computed and observed sea surface elevations at interior coastal locations. Although the depth-averaged model uses a quadratic drag law to specify frictional effects at the lower boundary, while the three-dimensional model uses a spatially varying vertical eddy viscosity, the final results for both models are in substantial agreement. The influence of the interior open boundary on the three-dimensional model results is found to be negligible.

Comparisons between observed and computed currents are made at various depths and locations and demonstrate agreement, although the amount of such data is small and its geographical distribution is poor.

A Kelvin wave dominates the propagation of the major diurnal components of the tide within the Gulf. However, the major semidiurnal constituents show a progressive wave entering the north of the Gulf from the Arafura Sea which excites a standing wave across the Gulf in the east west direction. This effectively blocks the penetration of tidal energy at the semidiurnal periods into the southern half of the Gulf.

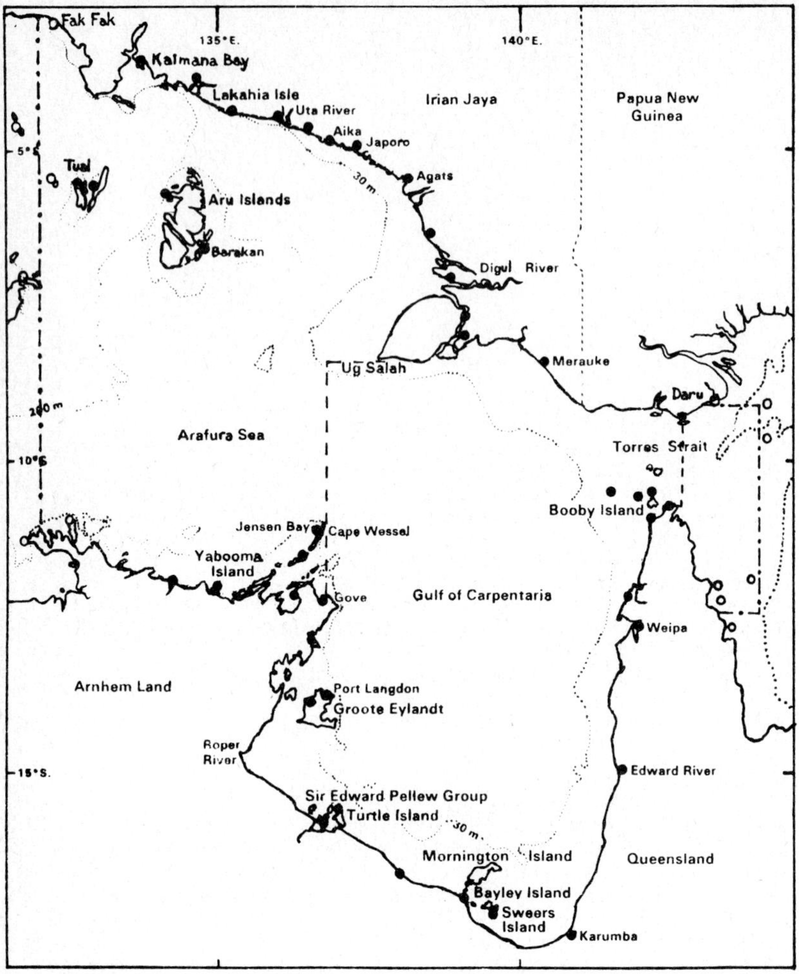

Figure 1: Map of the study area. o Tidal stations used to determine the boundary sea surface elevations for the depth-averaged model. ● Other tidal stations. •—•— Open boundary for the depth-averaged model. – – – Open boundary for the three-dimensional model.

2. THE OPEN BOUNDARY ELEVATIONS

The main area of interest includes all the Gulf of Carpentaria north to Irian Jaya and from Torres Strait in the east to a line joining Cape Wessel and Ug Salah in the west (Figure 1). Along the western and eastern open boundaries of this region there are problems with the specification of elevations of the sea surface. The eastern boundary across Torres Strait is almost choked by islands and submerged reefs so that the tides differ remarkably between the Gulf and the Coral Sea. The difference is not in the diurnal components, since these are relatively uniform in range and occurrence over the whole area, but in the semidiurnal components. When it is spring tide on one side of the Strait, it is neap tide on the other, while at some phases of the moon it is high water on one side and low water on the other. Previously, hydrodynamic models of the region have closed Torres Strait (Williams, 1972; Buchwald and Williams, 1975; Webb, 1981; Rienecker and Noye, 1982) or calculated the tidal flow through the major channels (Church and Forbes, 1981). Rienecker (1980) considers an open boundary at Torres Strait but finds only a small effect on the phases and subsequently makes it a closed boundary (Rienecker and Noye, 1982).

On the western boundary there is also a lack of adequate tidal elevation information. The surface elevations around Jensen Bay on the north coast of Australia are well determined, but there is no useful information near Ug Salah. Consequently, the specification of tides along these two boundaries is difficult and normally involves a great deal of intuition.

In this study the problem is approached differently. Figure 1 shows the major ports in the region listed in the Admiralty Tide Tables (1982). On the extreme western edge of the figure there is a line joining Fak Fak on the southern coast of Irian Jaya to the northern coast of Australia. Close to this line are five evenly distributed, well exposed tidal stations. Along the eastern edge of the figure there are seven stations that enable the elevations there to be well defined. Although the area of interest is relatively small in comparison to the area contained between these boundaries it is clearly better to use the larger area boundary information.

This enlarged area is therefore modelled by a depth-averaged tidal model and the resultant elevations at the interior boundaries are utilized as open boundary information for a three-dimensional model. The results presented are from two uncoupled models, the first providing accurate information for the second at points on open boundaries where this did not exist before. Alternatively, the former can be viewed as an interpolator which calculates tide heights according to the simplified depth-averaged equations of motion.

3. THE DEPTH-AVERAGED MODEL

This model solves the depth-averaged equations for mass conservation and momentum conservation in an orthogonal Cartesian coordinate system and is described in Noye and Flather(1984). The derivative terms of the equations are approximated by forward time, central space, finite difference schemes based on a space staggered grid and the discretized equations are solved using an updated explicit method. At all land-water boundaries the normal velocities to the coastline are set to zero and at open boundaries the surface elevations are specified as a function of time.

Running the depth-averaged model for 35 days of simulation time with a time step of ninety seconds takes about 20 hours of computation time on a VAX 11/780. The grid is aligned in the north to south direction and hence the Coriolis parameter is varied from row to row only. The region is digitized onto a 52x70 grid using horizontal element sizes of 18.8 km x 32.9 km. Of the total number of grid elements available, only 1736 elements are active and the method only requires storage for these in the calculations. The value chosen after calibration for the bottom drag coefficient of the quadratic friction term is 0.0012. Figure 2 shows the values of the tidal elevations (amplitudes and phases) in use along the open boundaries.

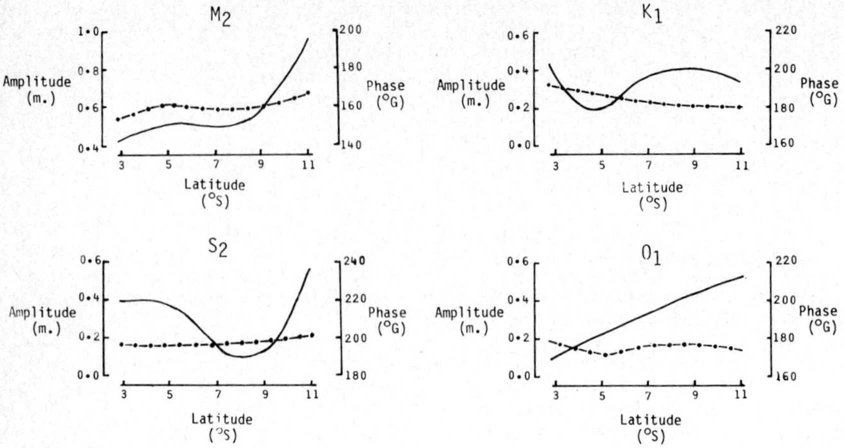

Figure 2(a): Amplitudes (m) and phases (°G) for the open boundary sea surface elevations along the western boundary of the depth-averaged model.

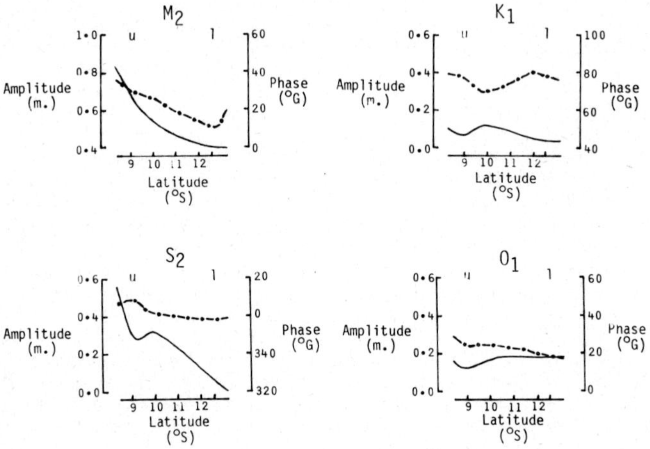

Figure 2(b): Amplitudes (m) and phases (°G) for the open boundary sea surface elevations along the eastern boundary of the depth-averaged model. Points to the left and right of u and l are at the same latitude as u and l respectively, and sequence is clockwise from the most north westerly point of the boundary.

These elevations are based on the observed tides at the twelve tidal stations very close to the boundaries. The penetration of the tide from the Pacific Ocean through Torres Strait is very small so the amplitudes are not critical here. On the western boundary however, the amplitudes are adjusted in areas where little or no information exists in order to improve the resultant signals at the interior tidal stations. The input along the open boundaries is the time series consisting of all four main tidal components, since the model is nonlinear.

Results and Discussion

The amplitudes and phases of the diurnal constituents O_1 and K_1 are given in Figure 3. These are calculated by a tidal analysis program based on the matrix inversion method (Easton, 1977). Hourly sea surface elevations for the last 29 days of the 35 computed are correlated and corrected for long term effects so that the constituent amplitudes and phases could be related to the tidal forcing at Greenwich at the start of the epoch (1/1/1900).

The O_1 tidal amplitudes indicate the presence of a progressive wave along the south coast of Irian Jaya and a large clockwise amphidrome centred in the Gulf of Carpentaria. There is also a gradual increase in amplitude along the eastern coast of the gulf towards Karumba in the south east corner and from the Aru Islands to Digul River in the Arafura Sea. Larger amplitudes are found for the K_1 constituent near Digul River, Torres Strait and near Karumba, while the features of the phases are similar to those of the O_1 phases.

The semidiurnal constituents M_2 and S_2 however, show a markedly different character to the diurnal tides (see Figure 4). The M_2 amplitudes are largest at Yabooma Island and Merauke, and smallest in the Arafura Sea and the south eastern quarter of the gulf. There is a saddle point in amplitudes lying between Ug Salah and Cape Wessel. The Arafura Sea region has an anticlockwise rotating amphidrome and indicates a resonance close to the period of the M_2 tide (12.42 hours). There is also a clockwise amphidrome centred on Groote Eylandt in the west of the Gulf and a clockwise amphidrome near Mornington Island in the south of the Gulf. Similar features are shown by the S_2 constituent, but with far less amplitude variation.

4. THE THREE-DIMENSIONAL MODEL

This model solves the complete three-dimensional equations of mass and momentum conservation in a depth transformed orthogonal Cartesian co-ordinate system. The discretized equations use time-centred and space-centred approximations, where possible, for the derivative terms and are solved explicitly in the horizontal and implicitly in the vertical at each time step (Bills and Noye, 1984). The overall solution scheme in time is of leapfrog type. Running the three-dimensional model for 32 days simulation time at time steps of 10 minutes takes about 20 hours of computation time on a VAX 11/780. A much larger time step of ten minutes is used due to the more stable nature of the leapfrog process, but the additional calculation of vertical information means that the computation time is about the same as for the depth-averaged model.

The grid size is once again 18.8 km x 32.9 km, but since a smaller region is considered there are only 871 active elements used in the computation. Added to this are ten depth levels, distributed finely near the surface and bottom, and coarsely at the middle depths. The form of the eddy viscosity is found by calibration of the model and is described in Bills and Noye (1984).

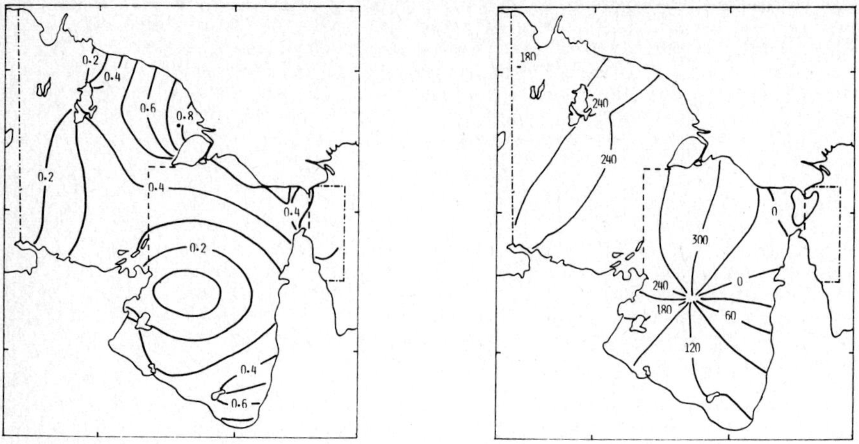

Figure 3(a): The computed amplitude (m) and phase (°G) contours for the O_1 tidal constituent from the depth-averaged model.

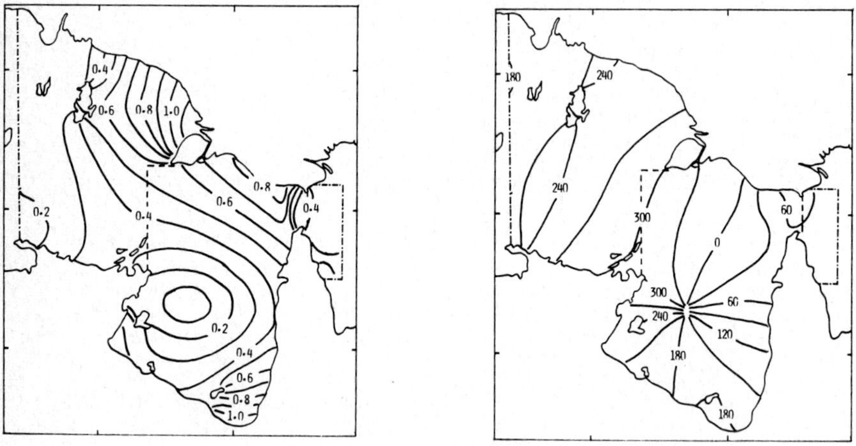

Figure 3(b): The computed amplitude (m) and phase (°G) contours for the K_1 tidal constituent from the depth-averaged model.

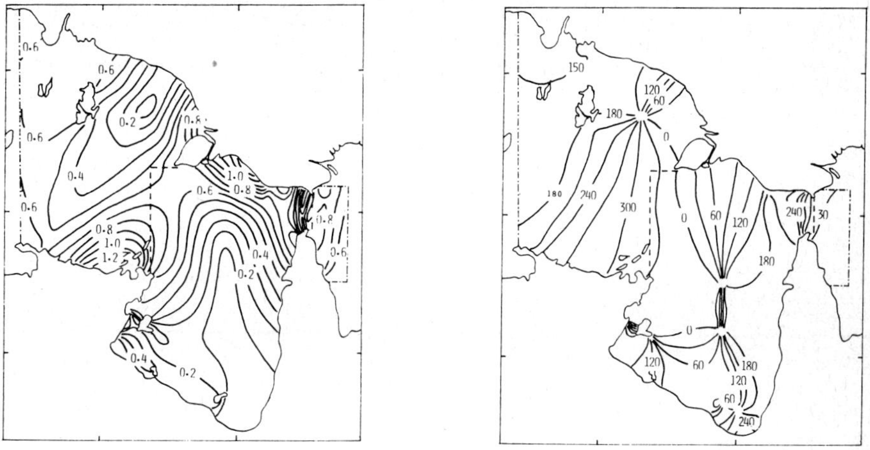

Figure 4(a): The computed amplitude (m) and phase (°G) contours for the M_2 tidal constituent from the depth-averaged model.

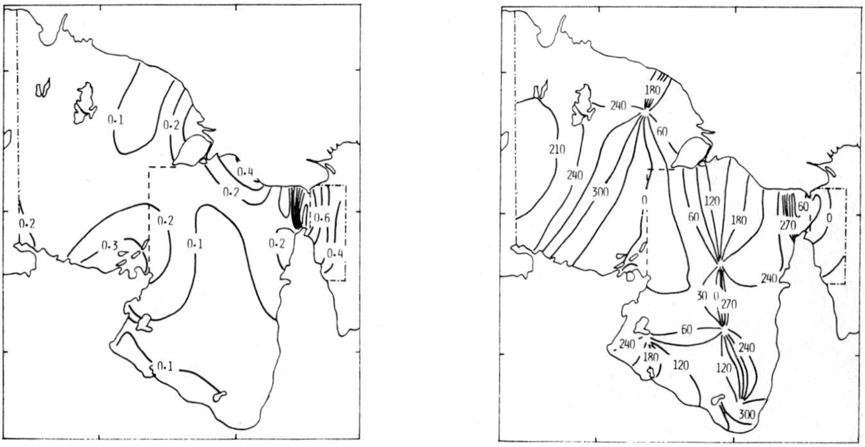

Figure 4(b): The computed amplitude (m) and phase (°G) contours for the S_2 tidal constituent from the depth-averaged model.

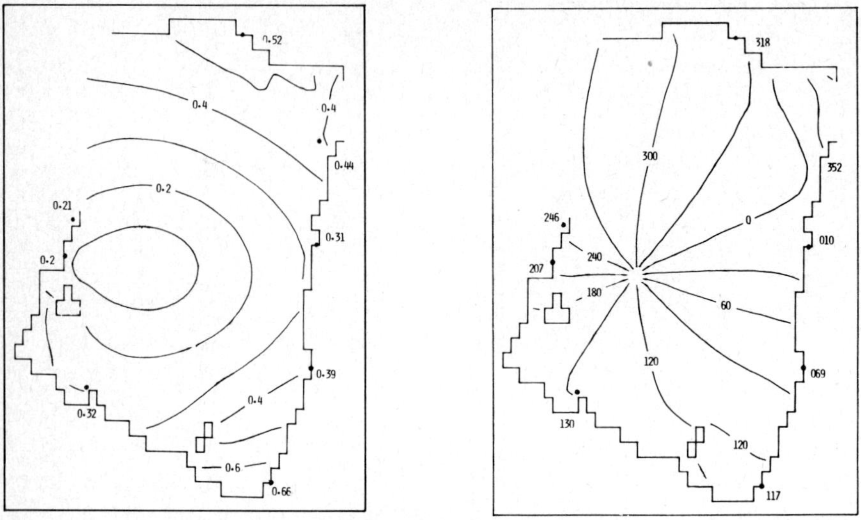

Figure 5(a): The computed amplitude (m) and phase (°G) contours for the O_1 tidal constituent from the three-dimensional model. Numbers on land are the observed values as given by the Admiralty Tide Tables (1982).

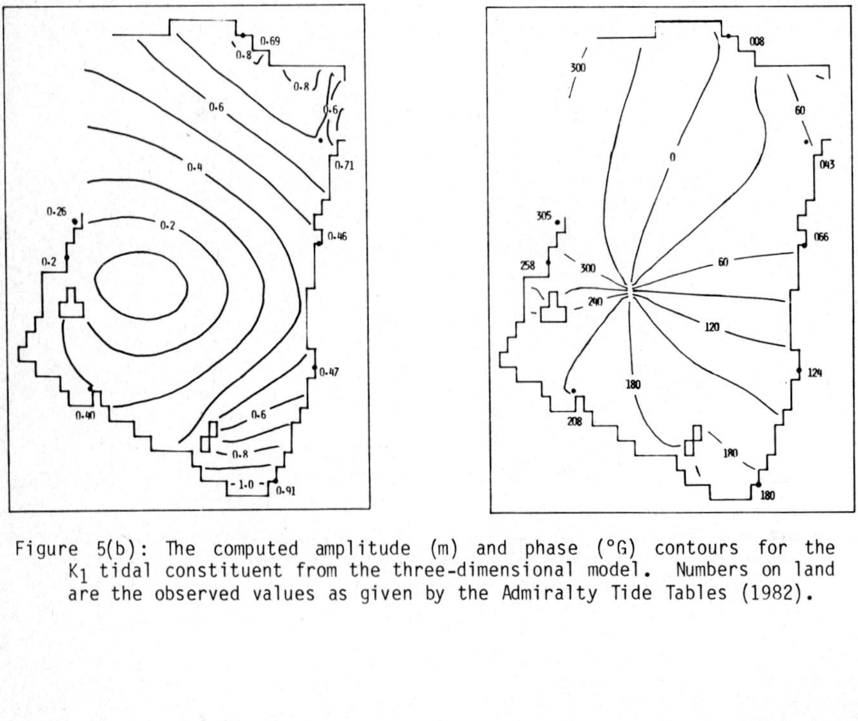

Figure 5(b): The computed amplitude (m) and phase (°G) contours for the K_1 tidal constituent from the three-dimensional model. Numbers on land are the observed values as given by the Admiralty Tide Tables (1982).

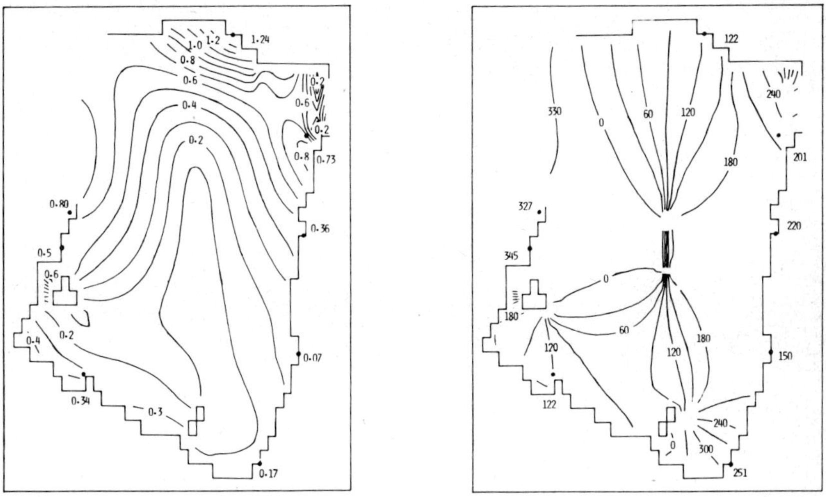

Figure 6(a): The computed amplitude (m) and phase (°G) contours for the M_2 tidal constituent from the three-dimensional model. Numbers on land are the observed values as given by the Admiralty Tide Tables (1982).

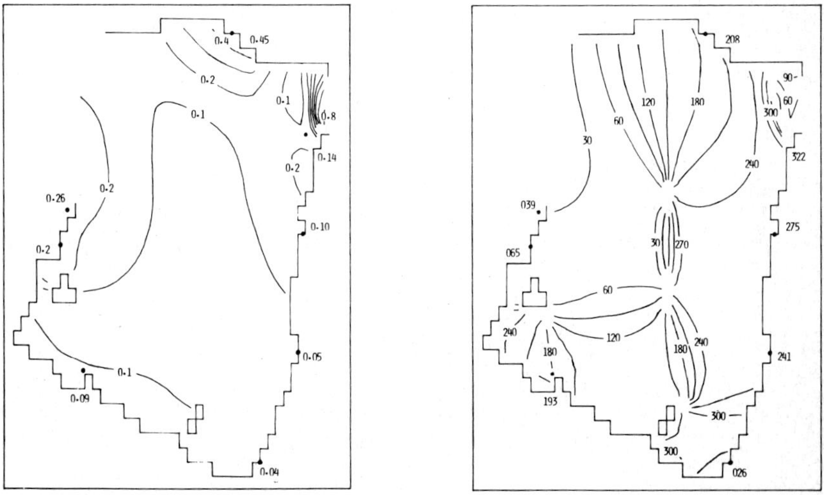

Figure 6(b): The computed amplitude (m) and phase (°G) contours for the S_2 tidal constituent from the three-dimensional model. Numbers on land are the observed values as given by the Admiralty Tide Tables (1982).

Results and Discussion

The amplitude and phase characteristics of the vertical tide are given in Figures 5 and 6. There are no substantial differences in the overall features of the results of both models. The observed and computed values at various coastal locations are given for each of the four major constituents in Table 1. The root mean square (r.m.s.) deviations between the model results and observations are given in Table 4.

Location		O_1		K_1		M_2		S_2	
		O	C	O	C	O	C	O	C
Merauke	Amplitude	0.52	0.59	0.69	0.85	1.24	1.40	0.45	0.42
	Phase	318	324	008	006	122	120	208	192
Booby I	Amplitude	0.44	0.46	0.71	0.69	0.73	0.70	0.14	0.12
	Phase	352	010	043	051	201	212	322	310
Weipa	Amplitude	0.31	0.33	0.46	0.48	0.36	0.43	0.10	0.15
	Phase	010	014	066	055	220	200	275	255
Caledon B.	Amplitude	0.2	0.11	0.2	0.15	0.5	0.54	0.2	0.22
	Phase	207	213	258	287	345	349	065	044
Edward R.	Amplitude	0.39	0.40	0.47	0.52	0.07	0.21	0.05	0.08
	Phase	069	083	124	125	150	193	241	263
Turtle I.	Amplitude	0.32	0.32	0.40	0.34	0.34	0.31	0.09	0.12
	Phase	130	155	208	210	122	120	193	180
Karumba	Amplitude	0.66	0.68	0.91	0.99	0.17	0.17	0.04	0.01
	Phase	117	135	180	184	251	275	026	341

Table 1: The observed (O) and computed (C) values of the four major tidal constituents for the sea surface elevations. Observed values are from the Admiralty Tide Tables (1982). Computed values are from the three-dimensional tidal program. Phases are relative to the Greenwich Meridian (G°) and amplitudes are in metres. The amplitudes and phases for Caledon Bay are inferred and amplitudes are therefore only given to one decimal place.

A large contribution to the discrepancy between observed and computed diurnal tide amplitudes came from the values at Karumba and Merauke. These regions are where the largest amplitudes occur. The large gradient of the M_2 amplitude in the Merauke region also contributes a large amount to the r.m.s. value for that constituent.

The semidiurnal constituents show a nodal line running from the south east corner of the Gulf to a point midway between Gove and Weipa. Along this line are three amphidromes; the northern and southern ones rotate clockwise while the central one rotates anticlockwise. There is also a near nodal line connecting these to Groote Eylandt in the west, which is the site of another clockwise amphidrome. The difference in the semidiurnal phases between the east coast and the west coast is about 150° which implies that the motion is essentially a standing wave, oriented in an east west direction and excited by the progressive wave in the northern half of the Gulf.

Of note in Figure 6 is the nodal line across Torres Strait combined with very large gradients of amplitude and phase. This demonstrates the problems that would be encountered by interpolating the open boundary elevations in this region from observations alone. The diurnal constituents in Figures 3 and 5 however, vary quite uniformly through the area in accordance with observations.

It is possible to see that a much closer agreement in the semidiurnal phases along the east coast can be obtained if the 210° and 180° contours of the M_2 phases and the 270° and 240° contours of the S_2 phases were to meet the coast above and below Weipa and Edward River. This indicates that, perhaps, the eastern amphidromes for these constituents are rotating a little more slowly than predicted by the model. Overall, however, the results are quite reasonable.

Location			O_1		K_1		M_2		S_2	
			O	C	O	C	O	C	O	C
Mooring 1										
	Northerly Component	Amplitude	0.20	0.20	0.32	0.33	0.03	0.05	0.02	0.02
		Phase	204	202	265	256	331	028	140	084
	Easterly Component	Amplitude	0.03	0.01	0.03	0.02	0.02	0.02	0.01	0.01
		Phase	018	015	045	057	117	107	160	194
Mooring 2										
	Northerly Component	Amplitude	0.19	0.24	0.32	0.40	0.05	0.08	0.02	0.01
		Phase	216	226	270	279	328	021	150	181
	Easterly Component	Amplitude	0.04	0.01	0.07	0.02	0.01	0.03	0.01	0.00
		Phase	252	265	310	316	054	089	098	277
Mooring 3										
	Northerly Component	Amplitude	0.10	0.03	0.14	0.04	0.04	0.05	0.01	0.03
		Phase	186	217	262	271	215	180	234	234
	Easterly Component	Amplitude	0.06	0.05	0.06	0.09	0.06	0.05	0.02	0.02
		Phase	343	356	049	056	167	168	238	205
Weipa										
	Northerly Component	Amplitude	0.15	0.12	0.23	0.17	0.17	0.08	0.04	0.02
		Phase	207	206	253	257	093	100	149	133
Arnhem										
	Northerly Component	Amplitude	0.31	0.14	0.39	0.13	0.28	0.13	0.11	0.04
		Phase	292	192	350	238	218	200	269	228

Table 2: The observed (O) and computed (C) values of the four major tidal constituents (where possible), for the tidal currents. Observed values are from Church and Forbes (1981), and Cresswell (1971). Computed values are from the three-dimensional tidal program. Phases are relative to the local time zone (-10 hours) and in degrees (g°) while amplitudes are in metres per second.

Location		O_1 O	O_1 C	K_1 O	K_1 C	M_2 O	M_2 C	S_2 O	S_2 C
Mooring 1	Upper Level								
Northerly	Amplitude	0.20	0.23	0.33	0.37	0.03	0.06	0.03	0.02
Component	Phase	343	340	033	045	261	320	080	030
Easterly	Amplitude	0.03	0.01	0.04	0.02	0.02	0.02	0.01	0.03
Component	Phase	157	150	159	206	047	042	100	120
	Lower Level								
Northerly	Amplitude	0.16	0.21	0.28	0.34	0.02	0.06	0.02	0.02
Component	Phase	348	339	061	044	278	318	104	026
Easterly	Amplitude	0.02	0.01	0.03	0.02	0.02	0.02	0.01	0.01
Component	Phase	159	146	191	201	061	040	109	118
Mooring 2	Upper Level								
Northerly	Amplitude	0.10	0.29	0.32	0.46	0.06	0.10	0.02	0.07
Component	Phase	355	003	060	068	258	322	090	072
Easterly	Amplitude	0.04	0.01	0.07	0.18	0.01	0.03	0.01	0.02
Component	Phase	031	042	100	098	344	030	038	150
	Lower Level								
Northerly	Amplitude	0.17	0.27	0.27	0.44	0.05	0.09	0.02	0.07
Component	Phase	349	003	062	068	273	322	085	070
Easterly	Amplitude	0.03	0.01	0.06	0.02	0.02	0.03	0.01	0.02
Component	Phase	077	041	134	099	037	028	109	148
Mooring 3	Upper Level								
Northerly	Amplitude	0.11	0.03	0.15	0.04	0.04	0.05	0.01	0.03
Component	Phase	325	356	052	063	145	111	175	166
Easterly	Amplitude	0.06	0.06	0.07	0.11	0.07	0.05	0.02	0.02
Component	Phase	122	137	199	207	097	100	178	124
	Lower Level								
Northerly	Amplitude	0.08	0.03	0.11	0.04	0.04	0.05	0.01	0.02
Component	Phase	329	354	044	060	145	106	194	160
Easterly	Amplitude	0.05	0.06	0.08	0.09	0.06	0.05	0.02	0.01
Component	Phase	108	133	179	204	102	096	178	118

Table 3: The observed (O) and computed (C) values of the four major tidal constituents for the tidal currents at different depths. Observed values are from Church and Forbes (1982). Computed values are from the three-dimensional tidal program. Phases are relative to the local time zone (-10 hours) (g°) and amplitudes are in metres per second.

The comparisons between depth-averaged current observations and computed values are given in Table 2. The current measurements are taken from Church and Forbes (1981) and Cresswell (1971). Cresswell's data (Weipa and Arnhem) are included only for interest since there are likely to be errors associated with their rather short lengths, and hence they are not used in the calculation of r.m.s. errors in Table 4. A large contribution to the S_2 phase error is the computed value of the eastern component at Mooring 2, where the calculated amplitude is less than 0.01 m per sec. If the value for phase is omitted due to the very small contribution this constituent makes to the signal, the r.m.s. error in S_2 phase drops from 80° to 36°.

Finally, the r.m.s. errors associated with the variation of currents over depth as given in Table 3 are given in Table 4. Again, there are large errors associated with the phase of the easterly component of Mooring 2 for the S_2 constituent and if these are omitted from the calculation, the r.m.s error drops from 51° to 41° for this constituent. However, the overall results are pleasing and demonstrate the success of the eddy viscosity formulation and the three-dimensional model.

Variable		O_1	K_1	M_2	S_2
Elevations	Amplitude	0.05	0.08	0.09	0.03
	Phase	15	12	21	24
				(14)	(18)
Depth Averaged Currents	Amplitude	0.04	0.06	0.02	0.01
	Phase	12	9	38	80
					(36)
Depth Dependent Currents	Amplitude	0.07	0.09	0.02	0.02
	Phase	19	21	38	51
					(41)

Table 4: The root mean square (r.m.s.) of the difference between observations and computed values for various quantities. Observed values for sea surface elevations are from the Admiralty Tide Tables (1982). Observed values for currents are from Church and Forbes (1981,82). Computed values are from the three-dimensional tidal model. Figures in brackets are the r.m.s. differences in phase when the largest difference is omitted from the calculation. All dimensional quantities are in MKS units.

5. CONCLUSION

The method of generating open boundary information for the three-dimensional model by a depth-averaged model proves to be successful for the Gulf of Carpentaria. There is substantial agreement between the two models in the region common to both.

The results from the three-dimensional model agree well with the observations of currents at various depths and locations, and sea surface elevation at several coastal locations.

A Kelvin wave with a clockwise rotating amphidrome just west of the centre of the Gulf is the dominant feature of the diurnal components of the tide. The semidiurnal components exhibit a line of minimum amplitude with associated amphidromes located centrally in the Gulf. This feature prevents a large amount of the potential energy at the latter periods penetrating the southern half of the region.

ACKNOWLEDGMENT

Part of the work reported in this paper has been supported by various agencies, W. Mitchell by a Marine Sciences and Technology grant from the Australian Department of Science, D. Beard by the Australian Research Grants Scheme, and P. Bills by a South Australian Transport Department Scholarship.

REFERENCES

Admiralty Tide Tables (1982). *Vol. 3, Pacific Ocean*, N.P.203-82, The Hydrographer of the Navy.

Buchwald, V.T., and N.V. Williams (1975). *Rectangular resonators on infinite and semi-infinite channels*, J. Fluid Mech., 67, 497-511.

Bills, P.J. and B.J. Noye (1984). *Verification of a three-dimensional tidal model for coastal seas*, In this volume: 'Computational Techniques and Applications', editor J. Noye, North-Holland Publishing Company, Amsterdam.

Church, J.A. and A.M.G. Forbes (1981). *Nonlinear model of the tides in the Gulf of Carpentaria*, Aust. J. Mar. Freshwater Res., 32, 685-697.

Church, J.A. and A.M.G. Forbes (1982). *Circulation in the Gulf of Carpentaria. Part I. Direct observations of currents in the south east corner of the Gulf of Carpentaria*, Aust. J. Mar. Freshwater Res., 34, (in press).

Cresswell, G.R. (1971). *Current Measurements in the Gulf of Carpentaria*, CSIRO Aust. Div. Fish. Oceanogr., Rep. No. 50.

Easton, A.K. (1977). *Selected programs for tidal analysis and prediction*, Flinders Institute for Atmospheric and Marine Sciences, Computing Report No. 9, 78 pages.

Melville, W.K. and V.T. Buchwald (1976). *Oscillations of the Gulf of Carpentaria*, J. Phys. Oceanogr., 6, 394-398.

Noye, B.J. and R.A. Flather (1984). *Hydrodynamical numerical modelling of tides and storm surges*, Tides and Surges, editor J. Noye, Monograph of the American Geophysical Union, to be published.

Rienecker, M.M. (1980). *Tidal propagation in the Gulf of Carpentaria*, Ph.D. thesis, University of Adelaide.

Rienecker, M.M. and B.J. Noye (1982). *Numerical simulation of tides in the Gulf of Carpentaria*, Numerical Solutions to Partial Differential Equations, editor J. Noye, North-Holland Publishing Company, Amsterdam, 399-416.

Webb, D.J. (1981). *Numerical model of the tides in the Gulf of Carpentaria and the Arafura Sea*, Aust. J. Mar. Freshwater Res., 32, 31-44.

Williams, N.V. (1972). *The application of resonators to problems in oceanography*, Ph.D. thesis, University of New South Wales.

Computational Techniques & Applications: CTAC-83
J. Noye & C. Fletcher (Editors)

ON THE PERFORMANCE OF TURBULENT ENERGY CLOSURE SCHEMES FOR WIND DRIVEN FLOWS IN SHALLOW SEAS

Robert Arnold & John Noye

Department of Applied Mathematics
The University of Adelaide
Adelaide, South Australia.

The validity of using turbulent energy closure schemes in modelling wind induced flows in shallow seas is examined. A two-dimensional (x-z) numerical model has been developed and used to simulate several experiments on the laboratory scale. All experiments used for comparison recorded either the surface set-up or the depth variation of the horizontal velocity component for wind driven flows in closed channels or small basins acted upon by unidirectional winds.

INTRODUCTION

Most hydrodynamic-numeric models of tidal and wind affected sea flows are depth-averaged in nature, as in Noye and Tronson (1978). Such models have been found to accurately predict surface elevations but they give no information about the depth variation of the velocity components. In fact, it is not well known that a basic assumption made in the mathematical development of the governing equations used in the depth-averaged models, is that there is little velocity variation over depth. While this approximation is reasonable in relatively well mixed waters in which the thickness of the bottom boundary layer is negligible compared with the total depth of water, it is not true in the shallow waters of coastal seas, gulfs and inlets. For very shallow seas, the boundary layer may be of the same order of magnitude as the total depth of water.

Three-dimensional hydrodynamic-numeric models give more accurate results than depth-averaged models as well as giving information about the depth variation of current velocity. A few such models have been developed, for example, by Noye et al (1982). All these models achieve closure of the governing equations of conservation of mass and momentum using an eddy viscosity approach. However, the form of the vertical eddy viscosity coefficient must then be specified. A wide range of formulations, varying from absolute constants to quadratic and exponential variations of eddy viscosity over depth, have been used. While the vertical eddy viscosity is often assumed to be independent of horizontal position, in more realistic forms it is taken to be equal to a linear function of the depth-averaged velocity.

In recent years, however, work has been carried out which enables the vertical eddy viscosity coefficient to be calculated in addition to the other flow properties. This approach is superior to the above-mentioned approach in which it must be specified 'a priori', since the eddy viscosity is a property of the flow and not of the fluid. Equations describing the turbulent kinetic energy and the dissipation of such energy may be used to provide estimates for the vertical eddy viscosity coefficient. Such an approach will now be described.

GOVERNING EQUATIONS

The three-dimensional Reynolds-averaged equations for the conservation of mass and momentum are well documented and will not be presented here. Because the experiments to be simulated later are all two-dimensional in nature, equations involving only a vertical and a horizontal component of velocity will be used. The set of equations governing the motion is, therefore,

$$\frac{\partial u}{\partial t} + \frac{u\partial u}{\partial x} + \frac{w\partial u}{\partial \eta} = -\frac{1}{\rho}\frac{\partial p}{\partial x} + \frac{1}{\rho(\zeta+h)^2}\frac{\partial}{\partial \eta}\left(\frac{\rho N\partial u}{\partial \eta}\right), \qquad (1a)$$

$$\frac{\partial \zeta}{\partial t} + \frac{\partial}{\partial x}\left[(\zeta+h)\int_0^1 u\partial\eta\right] = 0 , \qquad (1b)$$

$$w = \frac{1}{(\zeta+h)}\left\{\eta\frac{\partial}{\partial \eta}\left[(h+\zeta)\int_0^1 u\partial\eta\right] - \frac{\partial}{\partial x}\left[(h+\zeta)\int_0^\eta u\partial\eta\right]\right\}, \qquad (1c)$$

$$0 = \frac{1}{(\zeta+h)}\frac{\partial p}{\partial \eta} + \rho g \qquad (1d)$$

where

$\eta = (z+h)/(\zeta+h)$	is a transformed vertical coordinate,
x,z	are the usual cartesian coordinates with the z-axis directed vertically upwards from the earth's surface and the x-axis in the plane of mean water level (M.W.L.),
$\zeta(x,t)$	is the surface elevation,
$h(x)$	is the depth of the channel floor below M.W.L.
$u(x,\eta,t)$, $w(x,\eta,t)$	are the horizontal and transformed vertical components of velocity,
$p(x,\eta,t)$	is the pressure at the space point (x,η) at time t.
$\rho(x,\eta,t)$	is the density,
g	is the acceleration due to gravity, and
$N(x,\eta,t)$	is the vertical eddy viscosity coefficient.

It is assumed that the horizontal eddy viscosity is negligible. The vertical coordinate z is transformed to η as indicated to facilitate numerical modelling since in the new coordinate system the surface of the body of water is at $\eta = 1$ and the bottom is at $\eta = 0$.

The above equations describe wind driven or tidal flows or a combination of both of these. The boundary conditions prescribed determine which flow is being modelled. The laboratory experiments which will be simulated are all concerned with wind forced motions for which the appropriate boundary conditions are:

$$u = 0 \quad \text{at} \quad x = 0,L \qquad (2a)$$
$$u = w = 0 \quad \text{at} \quad \eta = 0 \qquad (2b)$$
$$w = 0 \quad \text{at} \quad \eta = 1 \qquad (2c)$$
$$N\partial u/\partial\eta = (\zeta+h)\tau_0/\rho \quad \text{at} \quad \eta = 1 \qquad (2d)$$
$$p = p_a \quad \text{at} \quad \eta = 1 \qquad (2e)$$

where

τ_0	is the surface stress,
L	is the length of the channel being considered, and
p_a	is the atmospheric pressure at the surface.

Note also that the explicit equation for the transformed vertical velocity, w, given by Equation (1c) automatically, satisfies the boundary conditions $w = 0$ at $\eta = 0, 1$.

The above set of equations and boundary conditions form a closed system if the function $N(x,\eta,t)$ is specified. However, N may be also calculated as part of the solution procedure by using the relationship (see, for example, Launder and Spalding (1972))

$$N(x,\eta,t) = cE^2(x,\eta,t)/\varepsilon(x,\eta,t) \quad , \tag{3}$$

in which,

$c = 0.08$,
$E(x,\eta,t)$ is the turbulent kinetic energy, and
$\varepsilon(x,\eta,t)$ is the dissipation of turbulent kinetic energy.

Additional equations must now be solved to calculate the functions E and ε. The first of these is

$$\frac{\partial E}{\partial t} + u\frac{\partial E}{\partial x} + w\frac{\partial E}{\partial \eta} = \frac{N}{(\zeta+h)^2}\left(\frac{\partial u}{\partial \eta}\right)^2 + \frac{1}{(\zeta+h)^2}\frac{\partial}{\partial \eta}\left(\frac{N}{\sigma_E}\frac{\partial E}{\partial \eta}\right) - \varepsilon \tag{4}$$

where $\sigma_E = 1.0$.

The dissipation of energy, ε , may be calculated using a mixing length hypothesis (see, for example, Johns (1977, 1978), Vager and Kagan (1969 (a), 1969 (b), 1971)). This approach yields

$$\varepsilon = c^{3/4}\, E^{3/2}\, l^{-1} \tag{5}$$

where the mixing length, l , may be calculated by solving (see Vager and Kagan (1971))

$$l = -\frac{khE^{1/2}\, l^{-1}}{\partial(E^{1/2}\, l^{-1})/\partial\eta} \quad , \tag{6}$$

where k is Von Karman's constant which has the value 0.41.

However, instead of using Equations (5) and (6) to calculate ε , an equation similar to Equation (4) may be used. This approach has been found to provide better results.
For channel flow this equation is

$$\frac{\partial \varepsilon}{\partial t} + u\frac{\partial \varepsilon}{\partial x} + w\frac{\partial \varepsilon}{\partial \eta} = \frac{c_1 N\varepsilon}{(\zeta+h)^2 E}\left(\frac{\partial u}{\partial \eta}\right)^2 + \frac{1}{(\zeta+h)^2}\frac{\partial}{\partial \eta}\left(\frac{N}{\sigma_\varepsilon}\frac{\partial \varepsilon}{\partial \eta}\right) - c_2\frac{\varepsilon^2}{E} \quad , \tag{7}$$

where $c_1 = 1.44$, $c_2 = 1.92$ and $\sigma_\varepsilon = 1.3$.

The appropriate boundary conditions which will enable Equations (4) and (7) to be solved are:

$$E = \varepsilon = 0 \qquad \text{at} \quad \eta = 0 \ , \tag{8a}$$

$$\partial E/\partial\eta = \partial\varepsilon/\partial\eta = 0 \quad \text{at} \quad \eta = 1 \ . \tag{8b}$$

The set of equations given by (1) and (2) now form a closed system if solved in conjunction with Equations (3), (4), (7) and (8). A similar system of equations has been used by Mellor and Yamada (1974), Marchuk et al (1977), Blumberg and Mellor (1978) and Svenson (1979). However, few comparisons with controlled experiments have been undertaken to test the performance of this set of equations. After a brief report on how the equations were modelled numerically, their performance will be examined.

NUMERICAL MODEL

A finite difference scheme is used to solve the above mentioned set of partial differential equations and associated boundary conditions. A smoothly varying grid spacing is used in the vertical direction (see Figure 1) since this permits more accurate differencing of velocities which change rapidly over depth, in particular this occurs in the boundary layers near the sea floor and the sea surface. Constant grid spacings using a staggered grid system as shown in Figure 1 are used in the horizontal plane.

A computational element of the horizontal difference scheme consists of two points labelled 0 and X respectively, situated a distance Δx apart. Corresponding points in each element are separated by a distance $2\Delta x$. At the 0-points the horizontal component of velocity is calculated. All other variables are calculated at X-points. The notation $u|_{i,k}^{n}$ is used to represent the approximation for $u(x,\eta,t)$ computed at the 0-point of the ith element at the kth depth level and at the time $n\Delta t$, where Δt is the constant time step. The $(k-1)$th and kth layers are separated by a distance $\Delta\eta_k$, that is,

$$\Delta\eta_k = \eta_k - \eta_{k-1} \quad , \qquad k = 1(1)ND \ , \tag{9}$$

where η_k defines the height of the kth layer above the bottom of the channel.

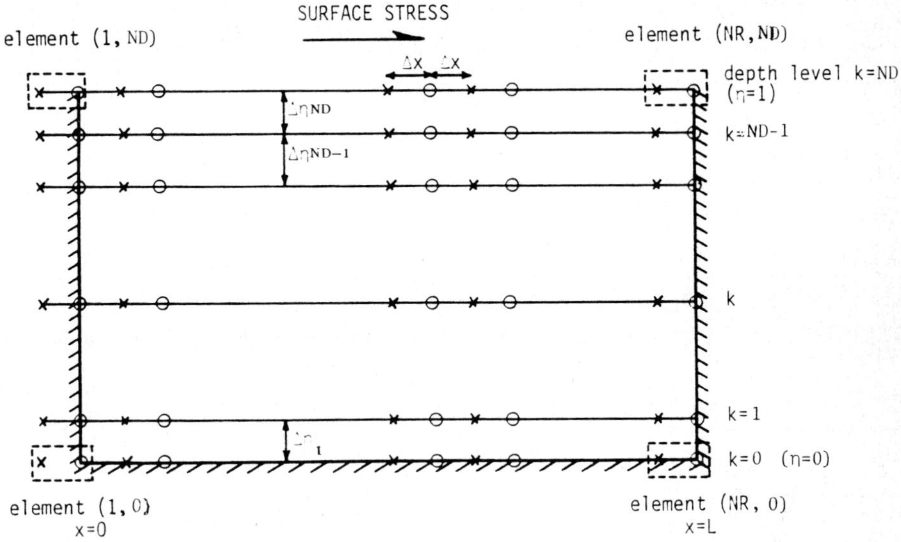

Figure 1
Channel divided into NR horizontal elements with ND depth intervals.

The governing equations (1), (3), (4) and (7) are discretised in the form

$$\left.\frac{\partial\zeta}{\partial t}\right|_{i}^{n} + \frac{\partial}{\partial x}\left[(\zeta+h)\int_0^1 u\,d\eta\right]\Bigg|_{i}^{n} = 0 \quad , \tag{10a}$$

$$\left.\frac{\partial u}{\partial t}\right|_{i,k}^{n} + u\left.\frac{\partial u}{\partial x}\right|_{i,k}^{n} + w\left.\frac{\partial u}{\partial \eta}\right|_{i,k}^{n} = -\left.\frac{1}{\rho}\frac{\partial p}{\partial x}\right|_{i,k}^{n+1} + \frac{1}{\rho(\zeta+h)^2}\frac{\partial}{\partial\eta}\left(\rho N\frac{\partial u}{\partial\eta}\right)\Bigg|_{i,k}^{n+1} , \tag{10b}$$

$$\left.\frac{\partial E}{\partial t}\right|_{i,k}^{n} + u\left.\frac{\partial E}{\partial x}\right|_{i,k}^{n} + w\left.\frac{\partial E}{\partial \eta}\right|_{i,k}^{n} = \frac{N}{(\zeta+h)^2}\left(\frac{\partial u}{\partial\eta}\right)^2\Bigg|_{i,k}^{n} + \frac{1}{(\zeta+h)^2}\frac{\partial}{\partial\eta}\left(\frac{N}{\sigma_E}\frac{\partial E}{\partial\eta}\right)\Bigg|_{i,k}^{n+1} - \varepsilon\Big|_{i,k}^{n+1} , \tag{10c}$$

$$\left.\frac{\partial \varepsilon}{\partial t}\right|_{i,k}^{n} + u\left.\frac{\partial \varepsilon}{\partial x}\right|_{i,k}^{n} + w\left.\frac{\partial \varepsilon}{\partial \eta}\right|_{i,k}^{n} = \frac{c_1 N \varepsilon}{(\zeta+h)^2 E}\left(\frac{\partial u}{\partial\eta}\right)^2\Bigg|_{i,k}^{n} + \frac{1}{(\zeta+h)^2}\frac{\partial}{\partial\eta}\left(\frac{N}{\sigma_E}\frac{\partial \varepsilon}{\partial\eta}\right)\Bigg|_{i,k}^{n+1}$$

$$- \left.\frac{c_2\varepsilon^2}{E}\right|_{i,k}^{n} , \tag{10d}$$

$$w\Big|_{i,k}^{n} = \frac{1}{(\zeta+h)}\left\{\eta\frac{\partial}{\partial x}\left[(\zeta+h)\int_0^1 u d\eta\right] - \frac{\partial}{\partial x}\left[(h+\zeta)\int_0^\eta u d\eta\right]\right\}\Big|_{i,k}^{n}, \tag{10e}$$

$$0 = \frac{1}{(\zeta+h)}\frac{\partial p}{\partial \eta}\Big|_{i,k}^{n} + \rho g\Big|_{i,k}^{n} \tag{10f}$$

$$N\Big|_{i,k}^{n} = c^2E^2\Big|_{i,k}^{n} /\varepsilon\Big|_{i,k}^{n} \tag{10g}$$

Equations (10a - 10d) are used together with values of ζ, u, E and ε at the nth time level to calculate the value of these variables at the new time level (n+1). The remaining equations are all explicit and use these values to calculate N, p and w at the new time level. Note that the dissipation, ε, which appears on the right hand side of the turbulent energy equation is calculated at the (n+1)th time level. It was found that this was essential for stability of the numerical method. It is therefore, necessary that Equation (10d) be solved before (10c). Note that the second order derivatives with respect to η which appear in Equation (10a -10d) are also calculated at time level (n+1). The implicit nature of this scheme also contributes to the overall stability of the method.

The discretisation of the individual terms in the governing equations is now discussed.

All the time derivatives are modelled using the forward difference form,

$$\frac{\partial \zeta}{\partial t}\Big|_{i}^{n} = \frac{1}{\Delta t}\{\zeta_i^{n+1} - \zeta_i^n\} + 0\{\Delta t\} . \tag{11}$$

Centred differencing is used to approximate all the spatial derivatives. For example,

$$\frac{\partial u}{\partial x}\Big|_{i,k}^{n} = \frac{1}{4\Delta x}\{u_{i+1,k}^n - u_{i-1,k}^n\} + 0(\Delta x)^2\} . \tag{12}$$

Because variable grid spacing is used in the vertical, to keep the differencing first order accurate the following formula must be used for derivatives with respect to η :

$$\frac{\partial u}{\partial \eta}\Big|_{i,k}^{n} = \frac{1}{r_k\Delta\eta_k\Delta\eta_{k+1}}\{\Delta\eta_k u_{i,k+1}^n - \Delta\eta_{k+1}(r_k-1)u_{i,k-1}^n + r_k(r_k-2)\Delta\eta_k u_{i,k}^n\}$$

$$+ 0\{\Delta\eta_k\Delta\eta_{k+1}\} \tag{13}$$

where $r_k = \Delta\eta_{k+1}/\Delta\eta_k + 1$ (see Noye, 1983, p.300).

The corresponding formula for the second order η derivatives which appear in the equations is derived by first using Equation (13) with u replaced by $\rho N \partial u/\partial\eta$ for example, and the three depth levels are taken to be the kth, the level halfway between the kth and (k-1)th, denoted (k-1/2) and the level (k+1/2) which is halfway between the kth and (k+1)th levels. This yields an equation containing derivatives at levels (k-1/2), (k+1/2) and k. Centred differences are then used to represent the first order derivatives at the first two levels. For example,

$$\left.\frac{\partial u}{\partial \eta}\right|^{n}_{i,k+1/2} = \frac{(u^n_{i,k+1} - u^n_{i,k})}{\Delta\eta_{k+1}} + 0\,\{(\Delta\eta_{k+1})^2\} \tag{14}$$

At the kth level, Equation (14) is applied. This yields the formula

$$\begin{aligned}\left.\frac{\partial}{\partial\eta}\left(\rho N\frac{\partial u}{\partial\eta}\right)\right|^{n}_{i,k} &= \frac{u^n_{i,k-1}}{r_k(\Delta\eta_k)^2}\Big[(r_k-1)(\overline{\rho N}^n_{i,k-1} + \overline{\rho N}^n_{i+1,k-1}) - (r_k-2).\\ &\quad \left(\rho N^n_{i,k} + \rho N^n_{i+1,k}\right)\Big] + \frac{u^n_{i,k}}{r\,(\Delta\eta_k)^2(\Delta\eta_{k+1})^2}\Big[-\left(\Delta\eta_k\right)^2\left(\rho N^n_{i,k}+\rho N^n_{i+1,k}\right)\\ &\quad - \left(r_k-1\right)\left(\Delta\eta_{k+1}\right)^2\left(\overline{\rho N}^n_{i,k-1} + \overline{\rho N}^n_{i+1,k-1}\right) + r_k\left(r_k-2\right)^2\Delta\eta_k^2\\ &\quad \left(\rho N^n_{i,k} + \rho N^n_{i+1,k}\right)\Big] + \frac{u^n_{i,k+1}}{r_k(\Delta\eta_{k+1})^2}\Big[\left(\overline{\rho N}^n_{i,k} + \overline{\rho N}^n_{i+1,k}\right) +\\ &\quad \left(r_k-2\right)\left(\rho N^n_{i,k} + \rho N^n_{i+1,k}\right)\end{aligned} \tag{15}$$

where $\overline{\rho N}^n_{i,k} = \left(\rho N^n_{i,k+1} + \rho N^n_{i,k}\right)/2$.

The integrals which appear in the equations are computed using the trapezoidal rule, namely

$$\begin{aligned}\left.\int_0^1 u d\eta\right|^n_i &= \frac{1}{2}\Big[\Delta\eta_1\left(u^n_{i,0}+u^n_{i,1}\right)+\Delta\eta_2\left(u^n_{i,1}+u^n_{i,2}\right)+\ldots\ldots\\ &\quad \ldots\ldots+\Delta\eta_{ND}\left(u^n_{i,ND-1}+u^n_{i,ND}\right)\Big]\ ,\end{aligned} \tag{16}$$

if there are (ND+1) depth levels of variable spacing.

All the boundary conditions for which boundary values are known are easily handled. These values are used in the finite difference equations applied at adjacent grid points.

Derivative conditions are somewhat more difficult to incorporate. When approximating the derivatives with respect to η at the boundaries, first order accurate differencing has been used. For example, at the surface $\eta=1$, the following scheme was used:

$$\left.\frac{\partial u}{\partial \eta}\right|_{i,ND}^{n} = \frac{1}{\Delta\eta_{ND}}\left\{\frac{(r_1+1)}{r_1}\,u^n_{i,ND} - \frac{r_1}{(r_1-1)}\,u^n_{i,ND-1} + \frac{1}{(r_1(r_1-1)}\,u^n_{i,ND-2}\right\}, \quad (17)$$

where $r_1 = \Delta\eta_{ND-1}/\Delta\eta_{ND}+1$. A similar method as was used to derive Equation (15) was then used to construct the surface double derivative approximation, giving,

$$\begin{aligned}\left.\frac{\partial}{\partial \eta}\left(\rho N\frac{\partial u}{\partial \eta}\right)\right|_{i,ND}^{n} &= \frac{3\tau_0}{\Delta\eta_{ND}}\left(\zeta^n_{i+1} + h_{i+1} + h_i + \zeta^n_i\right)\\
&+ \frac{u^{n+1}_{i,ND}}{(\Delta\eta_{ND})^2}\left\{-4\,\frac{\left(\overline{\rho N}\,\big|^n_{i+1,ND} + \overline{\rho N}\,\big|^n_{i,ND-1}\right)}{2} + \frac{\Delta\eta_{ND-1}(\rho N)\,\big|^n_{i,ND-1}}{(\Delta\eta_{ND} + \Delta\eta_{ND-1})}\right\}\\
&+ \frac{u^{n+1}_{i,ND-1}}{(\Delta\eta_{ND})^2}\left\{4\,\frac{\left(\overline{\rho N}\,\big|^n_{i+1,ND} + \overline{\rho N}\,\big|^n_{i,ND-1}\right)}{2}\right.\\
&\left.- \frac{\left(\Delta\eta_{ND-1} - \Delta\eta_{ND}\right)}{\Delta\eta_{ND-1}}\,(\rho N)\Big|^n_{i,ND-1}\right\} + u^{n+1}_{i,ND-2}\left\{\frac{-(\rho N)\,\big|^n_{i,ND-1}}{(\Delta\eta_{ND} + \Delta\eta_{ND-1})\Delta\eta_{ND-1}}\right\}.\end{aligned} \quad (18)$$

At the bottom, $\eta = 0$, a similar expression may be derived.

Modelling the equations as shown results in a tridiagonal system of linear algebraic equations which must be solved. For example, the value of $u^{n+1}_{i,k}$ for $k = 1(1)ND$ may be found from a system like

$$A^{n+1}_{i,k}\,u^{n+1}_{i,k-1} + B^{n+1}_{i,k}\,u^{n+1}_{i,k} + C^{n+1}_{i,k}\,u^{n+1}_{i,k+1} = D^n_{i,k}, \quad k = 1(1)ND-1 \quad (19)$$

in which $A^{n+1}_{i,k}$, $B^{n+1}_{i,k}$, $C^{n+1}_{i,k}$ and $D^n_{i,k}$ are known and $u^{n+1}_{i,0} = 0$. Systems of this form may be solved by the very efficient Thomas algorithm for tri-diagonal systems of equations described by Noye (1983, p331).

Note that the formulation used for the derivatives in the vertical direction at the bottom and surface need to be specially incorporated into this system before the Thomas algorithm can be applied. For instance, consider the system of equations (19). Using Equation (18) results in an expression connecting terms $u^{n+1}_{i,ND}$, $u^{n+1}_{i,ND-1}$ and $u^{n+1}_{i,ND-2}$. However, Equation (19) with k=ND-1 also contains terms in $u^{n+1}_{i,ND}$, $u^{n+1}_{i,ND-1}$ and $u^{n+1}_{i,ND-2}$ which may be used to eliminate $u^{n+1}_{i,ND-2}$

from Equation (18) thus yielding an expression involving only the terms in $u^{n+1}_{i,ND}$ and $u^{n+1}_{i,ND-1}$ in the set of equations obtained at the surface. This gives the additional equation

$$A^{n+1}_{i,k}\, u^{n+1}_{i,k-1} + B^{n+1}_{i,k}\, u^{n+1}_{i,k} = D^{n}_{i,k} \quad \text{for } k = ND. \tag{20}$$

The system formed by Equations (19) and (20) is now in the required form for solution by the Thomas algorithm. A similar treatment can be applied to these equations involving a derivative boundary condition at the bottom so as to yield a tri-diagonal system of (ND+1) equations which can be solved using the same algorithm.

COMPARISON WITH OBSERVATION

The set-up caused by a wind blowing over the surface of a rectangular pond of dimensions 240m x 60m and of depth 2m was measured by Van Dorn (1953) and his results are shown in Figure 2. The wind velocities were measured at a height of 10m above the surface and the set-up produced by only those winds which were within 20° of the longitudinal axis of the pond are included in the figure. Hence, the performance of the present two-dimensional numerical model can be tested using this data. The stress at the surface was calculated using

$$\tau_s = c_D \rho_a W^2 \tag{21}$$

where

ρ_a is the density of the air,
W is the longitudinal component of the wind speed (at 10m), and
c_D is the appropriate drag coefficient.

For consistency with Van Dorn's experiment, a value of $c_D = 1.1 \times 10^{-3}$ was used. In this experiment, as in the others which will be simulated, the density is assumed to be constant throughout the fluid.

When various wind speeds were used as input to the numerical model, the results obtained (see Figure 2) indicated that the mathematical model accurately predicted the set-up.

In another experiment carried out by Francis (1953), the vertical profile of horizontal velocity was measured. The dimensions of his basin were considerably smaller, namely 6.1m x 0.127m x 0.44m and a surface stress was simulated by forcing a jet of water over the fluid surface. Although not a wind driven flow, the resulting motion represents turbulent flow in a closed basin produced by a surface stress. The velocity profiles obtained experimentally may therefore be tested against those predicted by this numerical model. This comparison is made in Figure 3. Clearly, the computed velocity profile closely resembles the shape of the experimentally observed profile.

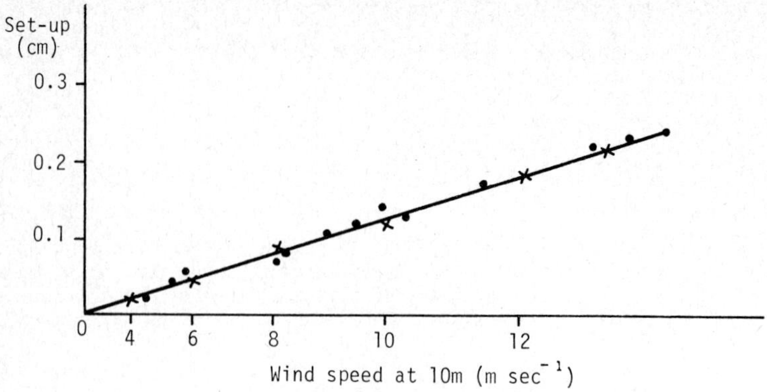

FIGURE 2: Plot of numerical prediction of the wind set-up (X) and the experimental results of Van Dorn (·). The smooth curve drawn is the best curve of fit passing through the numerical results.

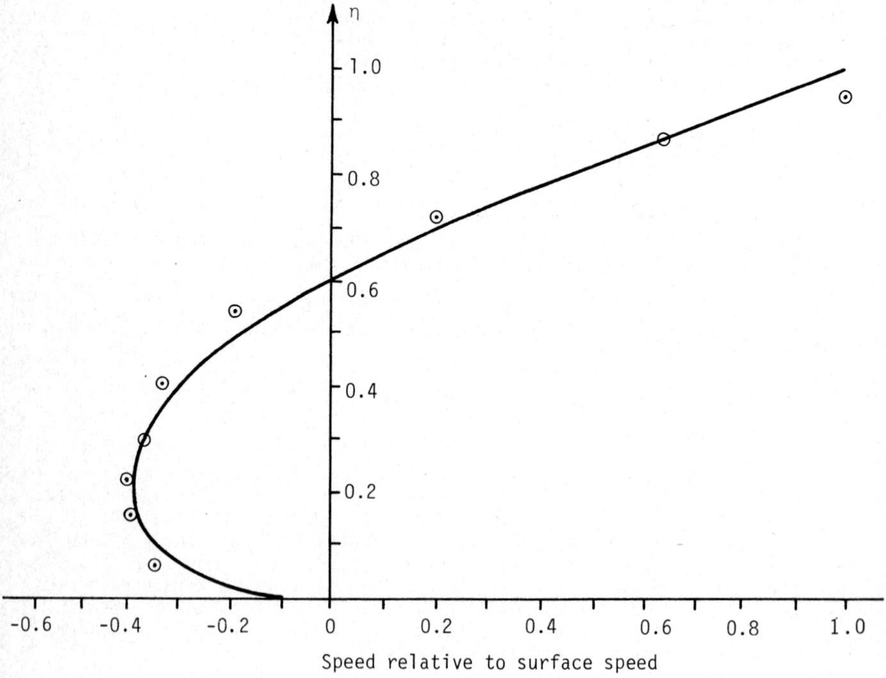

Figure 3: Comparison of velocity profile obtained using the numerical model (solid line) and the results of the experiments of Francis (circled dots).

Several other experiments have been carried out to measure the set-up and velocity profiles of wind driven flows. Fitzgerald (1963) describes one experiment in which the set-up of a body of water of length 1.83m, width 0.15m and depth 0.15m was recorded when it was placed in a wind tunnel. When modelled numerically, experimental results were once again accurately predicted. Similar agreement as shown in Figure 1 was achieved and so the results are not presented here. In Fitzgerald and Mansfield (1965), measurements of the velocity profile were made using the same experimental apparatus as used by Fitzgerald (1963). Baines and Knapp (1965) carried out a similar experiment on a body of water of depth 0.3m and length 9.14m. The Reynolds numbers of the flows in each experiment were of comparable magnitude and both flows were turbulent in nature. A comparison of the two measured velocity profiles is made in Figure 4.

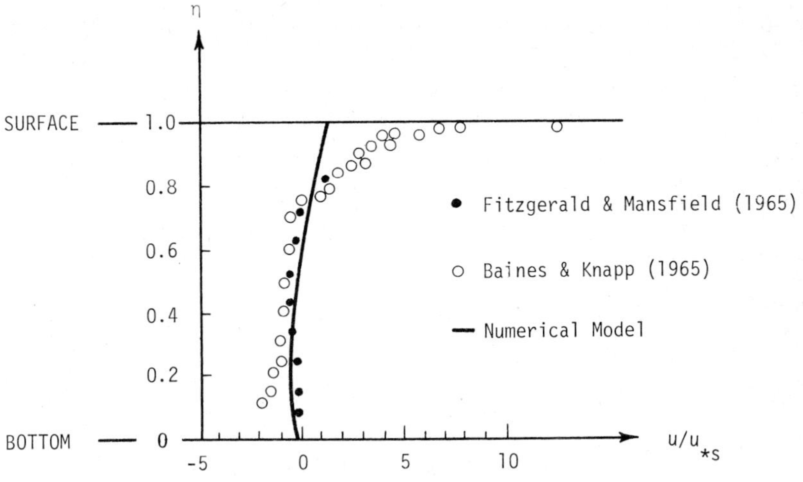

FIGURE 4: Comparison of velocity profiles (u/u*s) for wind driven flows.

Clearly, there is a substantial difference between the two observed profiles. It is uncertain why this should be so, but the reason may be associated with the difficulties encountered in taking the measurements. For example, Fitzgerald and Mansfield state that their measurements could be in error by ± 0.3 cm/sec. Both experiments were simulated using the numerical model and the computed velocity profiles are shown in Figure 4. Agreement is not as close as for the previous cases described. However, because of the discrepancey between the two sets of experimentally observed profiles, it is not possible to properly assess the difference between the observed and numerical results. As well, it must be remembered that the Reynolds number of tidal or wind driven flow in coastal seas will be different than the Reynolds number of the wind driven flows created in laboratories as described above.

CONCLUSIONS

The results obtained from the present numerical solution of the turbulent energy closure scheme have been very encouraging. When controlled laboratory experiments were simulated, the set-up predicted by the model closely matched the set-up which was observed. The predicted depth variation of the horizontal velocity component did not agree so well with experimental measurements but the large variation in the experimental results obtained by different workers made this part of the comparison more difficult. It would be advantageous for such comparisons if the measurements of the velocity profile were made in a basin of a size similar to that used by Van Dorn (1953) since such a depth is more comparable to that of shallow coastal seas.

Because of the difficulty in prescibing the eddy viscosity, a great deal of time and cost must be spent using observed values of set-up and current to "tune" conventional models which define a formulation for the vertical eddy viscosity 'a priori'. Using a turbulent energy closure scheme avoids this problem. Hence, such a scheme lends itself to use in a more general model - one which is applicable for tidal and wind driven flows in any coastal region. A three dimensional numerical model, using a turbulent energy closure scheme, is being developed and will be applied to Bass Strait.

ACKNOWLEDGMENT

During the preparation of this material the first mentioned author was supported by a Marine Sciences and Technology Grant supplied by the Commonwealth Government of Australia.

REFERENCES

Baines, W.D. and Knapp, D.J. (1965) "Wind driven water currents," Journal of Hydraulics Division, ASCE, 91, pp 205-221.

Blumberg, A.F. and Mellor, G.L. (1978) "A coastal ocean numerical model," Proc. Int. Symp. on Mathematical Modelling of Estuarine Physics, Ed. J. Sundermann and K.-P. Holz, pub. by Springer - Verlag, pp 203-219.

Fitzgerald, L.M. and Mansfield, W.W. (1965) "The response of closed channels to wind stresses," Aust. Journal of Physics, 18, pp. 219-226.

Francis, J.R.D. (1953) "A note on the velocity distribution and bottom stress in a wind-driven water current system," Journal of Marine Research, 12(1), pp. 93-98.

Johns, B. (1977) "Residual flow and boundary shear stress in the turbulent bottom layer beneath waves," Journal of Physical Oceanography, 8, pp. 1042-1049.

Johns, B. (1978) "The modelling of tidal flow in a channel using a turbulent energy closure scheme," Journal of Physical Oceanography, 8 pp. 1042-1049.

Launder, B.E. and Spalding, D.B. (1972) "Lectures in mathematical models of turbulence," Academic Press, 169 pp.

Marchuk, G.I., Kochergin, V.P., Klimok, V.Z. and Sukhorukov, V.S. (1977), "On the dynamics of the ocean surface mixed layer," Journal of Physical Oceanography, 7, pp. 865-875.

Mellor, G.L. and Yamada T. (1974) "A hierarchy of turbulence closure models for planetary boundary layers," Journal Atmos. Sci., 31, pp. 1791-1806.

Noye, B.J. (1983) "Finite difference techniques for partial differential equations," In: Computational Techniques for Differential Equations, ed. J. Noye (North-Holland) pp. 95-354.

Noye, B.J., May, R.L. and Teubner, M.D. (1982) "A three dimensional tidal model for a shallow gulf," in Numerical Solutions of Partial Differential Equations, ed. J. Noye (North-Holland Publishing Coy, Amsterdam) pp. 417-436.

Noye, B.J. and Tronson, K.C. (1978) "Finite difference techniquew applied to the simulation of tides and currents in gulfs," in Numerical Simulation of Fluid Motion, ed. J. Noye (North-Holland Publishing Coy, Amsterdam) pp. 285-356.

Svesson, V., (1974) "The structure of the turbulent Ekman layer," Tellus, 31 PP. 340-350.

Vager, B.G. and Kagan, B.A. (1969(a)) "The dynamics of the turbulent boundary layer in a tidal current," Izv., Atmospheric and Oceanic Physics, 5(2), pp. 168-179.

Vager, B.G. and Kagan B.A. (1969(b)) "The determination of the turbulence characteristics and velocity distribution of a tidal flow in a deep ocean," Izv., Atmospheric and Oceanic Physics, 5(8), pp. 836-845.

Vager, B.G. and Kagan B.A. (1971) "Vertical structure and turbulence in a stratified boundary layer of a tidal flow," Izv., Atmospheric and Oceanic Physics, 7(7), pp. 766-777.

Van Dorn, W.G. (1953) "Wind stress on an artificial pond," Journal of Marine Research, 12, pp. 249-276.

Computational Techniques & Applications: CTAC-83
J. Noye & C. Fletcher (Editors)

WAVE HINDCAST MODELLING FOR BASS STRAIT

D.R. Blackman & A.D. McCowan
Department of Mechanical Engineering
Monash University, Victoria
Australia

SUMMARY This paper describes the application of a spectral hindcast model to the generation and propagation of wind waves and swell in the Bass Strait. The model is based on the numerical solution of the spectral energy balance equation. Numerical results are compared with field measurements at the western entrance to Bass Strait.

1 INTRODUCTION

1.1 Physical Situation

The results presented here are part of a project which involves the collection of wave data, particularly swell, and the modelling numerically of the generation, propagation and decay of waves in arbitrary bathymetry. The work has arisen from research in coastal and ocean engineering at Monash University which has been planned to address topics of specific relevance to Australia. Bass Strait washes the most densely populated part of the continent and is therefore of economic significance, but rather little systematic work has been done until recently, and wave data are sparse and largely subjective.

The Southern Ocean is a source of strong swell to which the whole southern half of the continent is exposed. The depth of the Strait means that waves with periods of 10 seconds or more will be affected by interaction with the bottom. This limits the development of local wind-sea and causes it to steepen, a quite notorious feature of the area. The impinging swell feels this effect strongly, and suffers in consequence rather severe dissipation and refraction.

2 MODEL DEVELOPMENT

Most of the early numerical wave hindcast models (e.g. Pierson et al, 1966; Barnett, 1968; Ewing, 1971; Karlsson, 1972) were discrete spectral type models. These models are based on the numerical solution of the spectral energy balance equation, carried out over a range of discrete frequency and direction bands. They include empirically derived source terms for describing various mechanisms associated with wave growth and decay.

From the results of the JONSWAP study, Hasselmann et al (1973) concluded that non-linear wave-wave interactions were the dominant processes in the development of the wind-sea spectrum. Of the early discrete spectral models only those of Barnett (1968) and Ewing (1971) attempted approximations to these processes. However, as discussed by Resio (1979) and by Young and Sobey (1981), the approximations both of Barnett and of Ewing could lead to significant errors for "non-standard" spectral shapes. Problems associated with determining realistic approximations to the wave-wave interaction terms have, until recently, remained as the main limitations to further developments with this type of model.

As the inclusion of rigorous wave-wave interaction terms is impossible with current computer speeds (Cardone and Ross, 1978), Hasselmann et al (1976) proposed a simplified parametric type wave prediction model. With this type of model, the spectral energy balance equation is projected on a set of prognostic equations in the main parameters of the JONSWAP spectrum; in this way the effects of the non-linear wave-wave interaction terms are included implicitly. However, owing to the way in which they are formulated, parametric models model situations with rapid variations in wind speed or direction, or swell-dominated seas less easily.

Weare and Worthington (1978) developed a hybrid for their NORSWAM model. In this model, parametric methods are used for describing the wind sea, while a discrete spectral approach is used for modelling swell. A copy of the NORSWAM model was set up at Monash, and applied to Bass Strait and adjacent parts of the Southern Ocean. It was found that the discrete part of the model was too coarse to describe adequately swell in the Strait, and could not be easily modified to include shallow water effects such as shoaling, bed friction and refraction. It was therefore decided to develop a discrete spectral model for application to Bass Strait, using existing empirical growth terms. In this way, swell could be more accurately resolved, shallow water effects could be easily included, and the source terms could be later modified to take advantage of more accurate approximations to the wave-wave interaction terms (e.g. Resio, 1979) as they are developed.

2.1 The Spectral Energy Balance Equation

For the approaches to Bass Strait, only the deep water version of the spectral energy balance equation need be considered. For modelling the Strait itself, shallow water effects such as shoaling, refraction and bed friction need to be included. The shallow water form of the spectral energy balance equation used for this purpose can be given as

$$\frac{\partial}{\partial t}(ECC_g) + C_g \cos\theta \frac{\partial}{\partial x}(ECC_g) + C_g \sin\theta \frac{\partial}{\partial y}(ECC_g)$$

$$+ \frac{C_g}{C}\left(\sin\theta \frac{\partial C}{\partial x} - \cos\theta \frac{\partial C}{\partial y}\right) \frac{\partial}{\partial \theta}(ECC_g) = S \qquad (1)$$

where E is the spectral wave energy density for frequency f, direction θ, position x,y and time t.

C is the phase celerity for frequency f.

C_g is the group celerity for frequency f.

S represents source/sink terms.

The first line of the left-hand side of (1) is simply a two-dimensional transport equation describing the transport of wave action density (ECC_g) from one point to another. The remainder of the left hand side of (1) then describes the effects of refraction, while the right hand side comprises various source terms used to describe wave generation and decay mechanisms. The treatment of these three separate aspects of (1) is dealt with in more detail below.

The formulation of (1) in terms of wave action density (ECC_g), instead of the more normal wave energy density (E), has the advantage that shoaling effects are included implicitly and do not need to be included separately.

2.2 Transport Modelling

Throughout much of the Bass Strait region, swell propagating from the Southern Ocean can make a significant contribution to the local wave action. It is therefore essential that the left-hand side of (1) be modelled accurately to ensure that model predictions of swell will not be compromised by numerical propagation errors. However, high accuracy transport schemes are generally more complex and require more computer time than simpler less accurate methods. To achieve a satisfactory balance between accuracy and computational cost, several possible finite difference methods were evaluated.

Overall it was concluded that the third order upstream method (Leonard, 1979) provided a satisfactory compromise between the conflicting requirements of accuracy and computational effort (McCowan & Blackman, 1983).

In one dimensional form, the third order upstream difference scheme describing the transport of one component $E(f,\theta)$ of the spectrum can be given as

$$E_j^{n+1} = E_j^n - Cr\left(aE_{j+1} + bE_j + cE_{j-1} + dE_{j-2}\right)^n \tag{2}$$

where
- j denotes grid point $j\Delta x$
- n denotes time step $n\Delta x$
- $Cr = C_g\Delta t/\Delta x$ is the Courant number
- $a = 1/3 - Cr/2 + Cr^2/6$
- $b = 1/2 + Cr - Cr^2/2$
- $c = -1 - Cr/2 + Cr^2/2$
- $d = 1/6 - Cr^2/6$.

At upstream boundaries the sea state is assumed to be duration limited at ocean boundaries (i.e. no net transfer of wave energy density across the boundary) and fetch limited at land boundaries (i.e. no transfer of wave energy density across the boundary). For both land and ocean downstream boundaries, a radiation condition is applied using simple first order differences.

2.3 Refraction

In the shallow water formulation of the model the transport part of the computation is carried out separately. The effects of refraction are then included as an additional source term on the right hand side of (1). For this purpose the change of direction due to refraction for each spectral component is determined from (3) and (4):

$$\frac{d\theta}{dt} = \frac{C_g}{C}\left[\sin\theta\,\frac{\partial C}{\partial x} - \cos\theta\,\frac{\partial C}{\partial y}\right] \tag{3}$$

$$\Delta\theta \approx \frac{d\theta}{dt}\,.\,\Delta t \tag{4}$$

A corresponding amount of spectral energy density is then transferred to or from adjacent directional bands. Following Golding (1978), the maximum change in direction for any component is limited to the width of one directional band.

2.4 Source Terms

Following Hasselmann et al (1973), the source term in (1) can be expressed as a sum of terms, each of which represents a class of physical processes. This is given schematically as:

$$S = S_{in} + S_{nl} + S_{ds} + S_{sw} \tag{5}$$

where S_{in} is the energy input from the atmosphere, S_{nl} the non-linear transfer of energy into this frequency by conservative wave-wave interactions,

S_{ds} dissipative processes, and S_{sw} represents additional shallow water effects. It is now generally accepted that S_{nl} may be of the same order as S_{in}. Some of the main phenomena attributed to S_{nl} include the sharply-peaked spectra of developing seas, the rapid growth rates on the steep-forward face of the spectrum, and the tendency of growing spectral components to overshoot their equilibrium values. In the present form of the model relatively simple empirical growth terms are used in which the effects of S_{nl} are included implicitly.

2.4.1 Wave Growth

The growth term takes a form similar to that proposed by Pierson et al (1966),

$$S_{in} = (A + B.E) \times \left[1 - \left(\frac{E}{E_\infty}\right)^2\right] \tag{6}$$

where A is a linear growth term representing the turbulent resonance mechanism of Phillips (1957).

B is an exponential growth term representing the shear flow mechanism of Miles (1957).

E_∞ is an upper limit to wave growth.

The actual A, B and E_∞ values used in the determination of S_{in} follow from those used by Karlsson (1972), as developed by Brink-Kjaer (1981).

The linear growth term A is based on that given by Barnett (1968), who set a proportional to the sixth power of the wind speed, U^6. As an alternative, Dexter (1974) modified Barnett's original term to become proportional to the fourth power of the wind speed, U^4. Following Brink-Kjaer (1981), Barnett's original formulation is used unchanged for wind speeds less than 15 m/s, and with U^6 replaced by $225\ U^4$ for higher wind speeds.

The linear growth term A is responsible for initiating wave growth; the exponential growth term, B.E, is included to represent the rapid growth rates observed in nature. In the present model the exponential B.E term of Inoue (1966) is used for this purpose.

The Pierson-Moskowitz spectrum E_∞ spectrum for a fully developed sea is given as the upper limit to wave growth. This takes into account the fact that wave breaking ultimately limits any further growth of waves. At present, both the exponential growth term, B.E and the limiting spectral energy density, E_∞ are assumed to have a cosine-squared distribution for directions within ± 90° of the wind direction.

The validity of the chosen growth terms was checked by comparing numerically determined growth rates against those obtained from the SMB nomograms (C.E.R.C., 1977) and the JONSWAP results (Carter, 1982). Overall it was found that the numerical results compared quite favourably.

2.4.2 Wave Dissipation

The use of the limiting spectrum in eqn (6) empirically includes dissipation mechanisms, such as white capping, for growing and fully developed seas. However it does not consider dissipation effects once the energy source is removed. To take this into account, Karlsson (1972), following the work of Gelci and Devillaz (1970), proposed a turbulent dissipation term of the form

$$S_{ds} = \alpha H_s^2 (2\pi f)^4 \tag{7}$$

where α is a dissipation coefficient

H_s is the significant wave height.

This term is highly frequency dependent, and leads to rapid dissipation of high frequency wave energy while low frequency wave energy remains relatively unaffected. This is, qualitatively at least, what is observed in nature. For the present application it was found that best results were obtained with a damping coefficient $\alpha = 10.0 \times 10^{-6}$.

2.4.3 Shallow Water Effects

In the present version of the model, depth limitation to growth and bed resistance effects are included as shallow water source terms.

Following Sand, Brink-Kjaer and Nielsen (1981) the effects of depth limitation to wave growth are introduced by modifying the E_∞ upper limit to growth (E_∞) used in (6) such that the total energy is given by

$$\frac{gH_s}{U^2} = 0.283 \tanh\left(0.578 \frac{gd}{U^2}\right) \tag{8}$$

where
- H_s is the significant wave height
- U is the local wind speed
- d is the water depth

The effects of bed resistance are included using the approach of Hasselman and Collins (1968) as described by Collins (1972) and Young and Sobey (1981). In this way, the effects of bed friction on each spectral component can be given by

$$S_{bf} = \frac{C_f\, g}{f^2 C^2 \cosh\left(\frac{2\pi d}{L}\right)} E\left[\langle u\rangle + \cos^2(\theta-\gamma)\left\langle\frac{u_1^2}{u}\right\rangle + \sin^2(\theta-\gamma)\left\langle\frac{u_2^2}{u}\right\rangle\right] \tag{9}$$

where
- C_f is a bed friction coefficient, typically 0.01
- $\langle u\rangle$ is the ensemble averaged bed velocity
- $\langle u_1\rangle$ is the component of $\langle u\rangle$ in the principal direction γ
- $\langle u_2\rangle$ is the component of $\langle u\rangle$ in the normal to the principal direction γ

3 WIND FIELDS

The weather in the region is characterized by prevailing westerly winds, modulated by depressions recurring with a period of four or five days. During summer the tracks of these depressions generally fall to the south of Tasmania but during winter move north, some of them crossing the Strait itself. These depressions operating in the Southern Ocean give rise to a considerable swell of quite long period (often in excess of 20 seconds); when they traverse the Strait they also generate a local wind-sea.

Unlike the European situation where there is a high density of permanent land observing stations, permanent weather ships, and many casual ships reporting at sea, there is a great paucity of observational data for the region. The permanent stations on the Australian mainland lie wholly on the northern boundary of the Southern Ocean model area, and are frequently very widely spaced. Shipping is sparse and predominantly coastal; almost no vessels now ply the open spaces of the Southern Ocean. Conventional reporting is, therefore, largely absent. The standard weather satellites do not provide quantitative data on wind fields (less still of sea state). We have had recourse, therefore, to very indirect estimates of the wind field.

The traditional method is to estimate the geostrophic wind from the isobaric charts and apply some factor to correct for height and stability to arrive at sea level wind speed and direction. Even when the data are available the method is very tedious and requires considerable skill if it is to give reliable estimates (Harding and Binding, 1978). We have, for the purpose of comparison, performed some analyses of this type. We have no stability data available, which limits accuracy at the outset. We have corrected the geostrophic wind derived from the pressure field to sea level by applying a factor of 0.7 to the magnitude and 15° to the direction.

For the Southern Ocean we have been using the 1000 mb surface winds taken from the analysis model run by the Bureau of Meteorology (Seaman et al, 1977). This model has as input both quantitative measurements and qualitative satellite data which are linked using a variational form of the equations of motion for the atmosphere. The wind field from this model has been made available to us in machine readable form at 12-hour intervals on a 254 km grid. This is rather coarse for our purpose, not only in space but also in time. It does not give wind at the surface, and the data from which this can be recovered cannot be obtained from the windfields. Nevertheless, the wind fields have the virtue of self-consistency and comparisons with observations suggest that the errors may be no worse than those from individual weather stations. These last errors are, in fact, considerable. The wind speeds of the regional analysis wind model are corrected by a factor of 0.8 to take boundary layer effects into account.

There are some 20 coastal stations around the Strait regularly reporting winds. In addition, on any day, there may be about 15 ship reports. However, the reliability of these data is variable, and not especially good. Only about a third of the ships reporting, for example, are equipped with anemometers; most ship reports are subjective estimates. The exposure of land anemometers, and even their physical condition, varies considerably. We attempted to test the reliability of these observations by correlating them with values of wind interpolated from the 1000mb analysis model. This procedure is, of course, suspect since these same observations are incorporated in the model, but there are other independent inputs to the model which make the process somewhat less risky. The correlations are based on the period Nov/Dec 1981 and July/August 1982, a total of about 3 months. The results are summarized in the following table:

Correlation	Stations in this range
1.0 -0.9	nil
0.9 - 0.8	Cape Schanck, Cape Otway
0.8 - 0.7	Wilson's Promontory, Tarwin, Pt. Lonsdale, Murrumah
0.7 - 0.6	Cape Northumberland
< 0.5	average of ship reports, Mt. Gambier, Gabo Island, Point Hicks, Lakes Entrance, Lorne Pier, Bernie, Low Head, Eddystone Point.

These stations are marked on Figure 1 which also shows the skeleton of the grid used for computations in the Strait. The wind field used to run the model in the Strait is computed by interpolating these data on the computational grid. It will be seen that there are no observations at all on about half the boundary, which leaves most curve-fitting schemes in confusion. To maintain the stability of the interpolating scheme the values from the 1000mb model are included (at the nodes in Figure 1) with a weighting of 1.0 . The separate observations are included with a weighting equal to the correlation coefficient described above. Given all these imponderable inaccuracies an involved interpolation scheme did not seem warranted; a third degree polynomial in the two coordinates has therefore been used, which permits an approximate description of frontal systems crossing the Strait if necessary. The 10 unknown coefficients in the polynomial are fitted to the data at 12 hour intervals by least-squares.

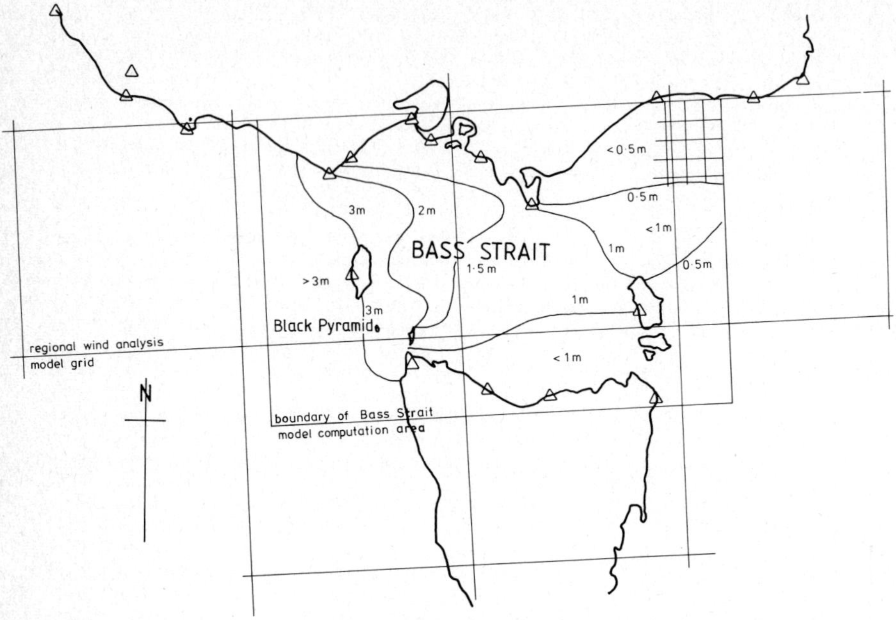

Figure 1
Computational grid in Bass Strait. Wind observations are available at the points Δ . Also shown are predicted contours of equal significant waveheight.

4 FIELD MEASUREMENTS OF SWELL

Three periods of field measurements have been completed at Black Pyramid, a site between King Island and Tasmania. The instrument used is a bottom-resident pressure recorder (Blackman and Evans, 1983). This is capable of conditional sampling, and records only when prescribed conditions are met at the regular awakening which occured every four hours in these experiments. The instrument responds poorly to local wind-sea and principally measures swell. The recorders will take a 40-minute sample when prescribed conditions are met. Even when they are not the recorder attempts to evaluate the sea state by noting the difference between the maximum and minimum surface elevations encountered in a sampling period of 10 minutes. Since this quantity can be approximated by the model these measurements permit a comparison with the model to be made over extended periods.

5 MODEL RESULTS

The sensitivity of the model to input parameters is discussed in McCowan & Blackman (1983). For comparisons of model results against the field measurements from Black Pyramid, a deep water Southern Ocean model, as shown in Figure 2, was used.

For the Southern Ocean the model uses a 127 km square grid, half that of the regional analysis wind model, and a time step of 1 hour. The direction-frequency spectrum was divided into 16 equally spaced direction bands and 12 frequency bands covering a range of wave periods from 4 seconds to 26 seconds, with a uniform spacing of 2 seconds. Wind speeds for each time step are obtained by linear interpolation between the 12-hourly frames provided.

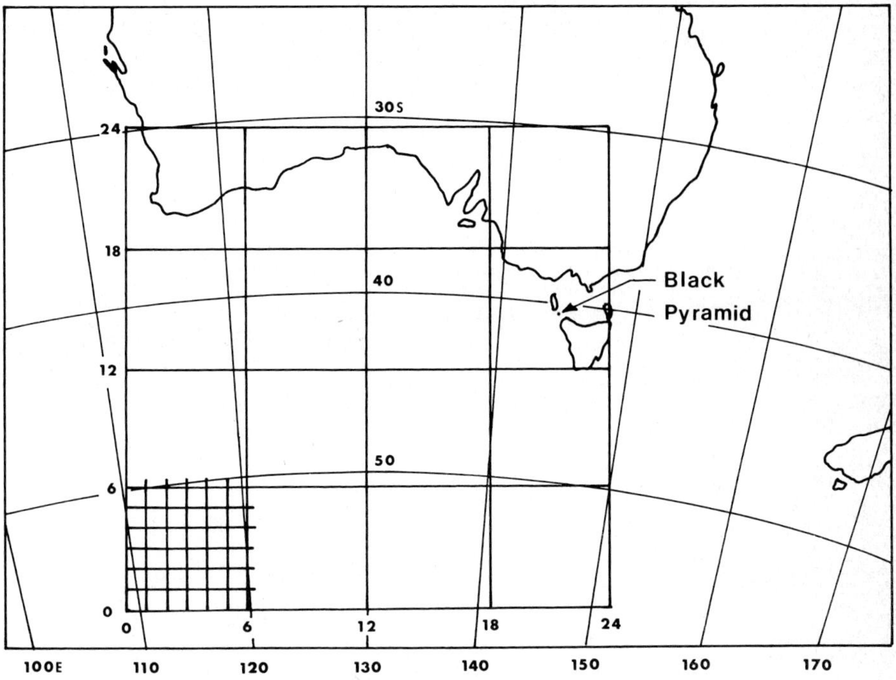

Figure 2
Computational grid for the Southern Ocean

Results from the two field measurement periods at Black Pyramid showed that for much of the time the sea state parameter (the difference between the extrema detected in a 5 minute period) was generally in the range of 0.5 m to 1.5 m, with values of 2.0 m rarely exceeded. However, much higher sea states, up to 4.1 m, were recorded from 0600 hours to 1800 hours on 25 July 1982.

Examination of pressure charts showed that during the period 1982 July 20-26, a series of quite strong depressions, as low as 956 mb, crossed the Southern Ocean at latitudes of 50° to 60° South. These depressions gave rise to strong westerly winds of 20 ms^{-1} (and in places 25 ms^{-1} or more) blowing over the central and southern parts of the model, but for much of the period 1982 July 20-26 winds over Bass Strait were less than 12 ms^{-1}. Thus the high waves which were measured were due predominantly to swell arriving from the Southern Ocean.

Figure 3 gives a comparison of measured and computed sea states. Peak measured and computed surface elevation spectra are compared in Figure 4. With high energy densities at low frequencies, $T_p \approx 18$ seconds, these show that the wave energy was indeed predominantly swell.

For Bass Strait the full shallow water form of the model was applied to the area shown in Figure 1. The model used a grid size of 25.2 km and a time step of 0.25 hour. Boundary conditions at the open west and southern boundaries were supplied from the results of the Southern Ocean model. Figure 5 shows predicted contours of significant wave height throughout Bass Strait at a time near the peak of the calibration storm.

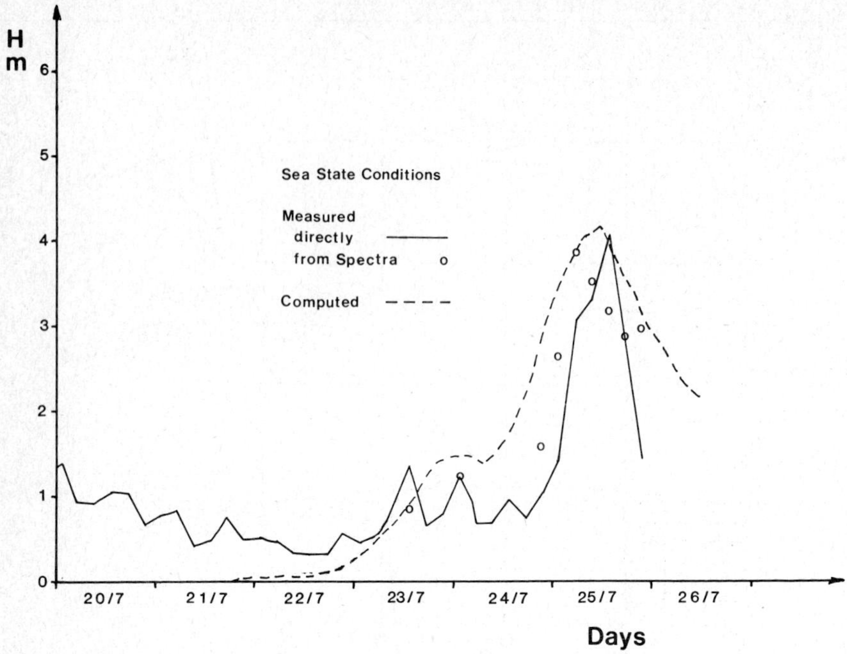

Figure 3
Sea-state comparisons in the calibration period

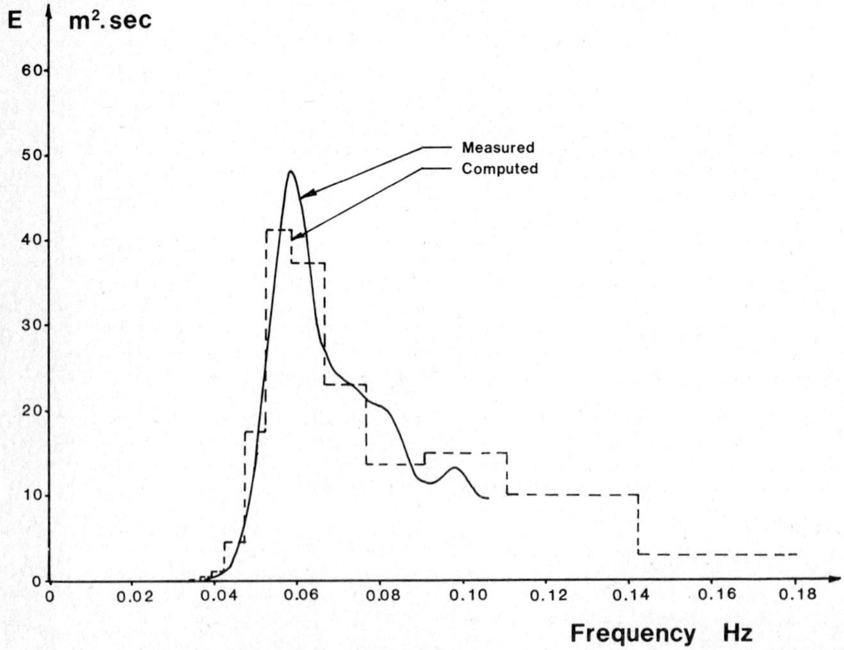

Figure 4
Comparison of spectra in the calibration period

6 ACKNOWLEDGEMENTS

The numerical model was developed by A.D. McCowan whose research fellowship has been supported by AMSTAC Grant 80/0040. Mr. McCowan wishes to acknowledge the assistance of Dr. O. Brink-Kjaer of the Danish Hydraulic Institute in the selection of the source terms used. The cooperation of the Bureau of Meteorology in supplying analysis model wind fields is gratefully acknowledged.

7 REFERENCES

BARNETT (1968). On the generation, dissipation and prediction of ocean wind waves. J. Geophys. Res., Vol. 73, pp 513-29.

BLACKMAN, D.R. & EVANS, R.J. (1983) An intelligent tide and wave recorder, Proc. 6th Australian Conf. on Coastal and Ocean Engineering.

BRINK-KJAER, O. (1981). Personal communication.

CARDONE, V.J. and ROSS, D.B. (1978). State of the art wave prediction methods and data requirements, in Ocean Wave Climate, M.D. Earle and A. Malahoff editors, Plenum, New York.

CARTER, D.S.T. (1982). Prediction of wave height and period for a constant wind velocity using the JONSWAP results. Ocean Engineering, Vol. 9, No. 1, pp 17-33.

COASTAL ENGINEERING RESEARCH CENTRE, (C.E.R.C.) (1977). Shore Protection Manual.

COLLINS, J.J. (1972). Prediction of shallow water spectral, J. Geophys. Res. Vol. 77, No. 15, pp 2693-2707.

DEXTER, P.E. (1974). Tests on some programmed numerical wave forecast models. J. Phys. Oceanogr., Vol. 4, pp 635-44.

EWING, J.A. (1971). A numerical wave-prediction method for the North Atlantic Ocean. Deutsche Hydrographische Zeitschrift, Vol. 24, No. 6, pp 241-61.

GELCI, R. and DEVILLAZ, E. (1970). Le calcul numérique de l'état de la mer. La Houille Blanche, Vol. 25.

GOLDING, B.W. (1978). A depth dependent wave model for operational forecasting. Proc. NATO Symp. on Turbulent Fluxes through the Sea Surface - Wave Dynamics and Prediction. A. Favre and K. Hasselmann editors, Plenum New York.

HARDING, J. and BINDING, A.A. (1978). The specification of wind and pressure fields in the North Sea and some areas of the North Atlantic during 42 gales from the period 1966 to 1976. Institute of Oceanographic Science Report 55.

HASSELMANN, K. and COLLINS, J.I. (1968). Spectral dissipation of finite-depth gravity waves due to turbulent bottom friction. J. Marine Res., Vol. 26, pp 1-12.

HASSELMANN, K. et al (1973). Measurements of wind-wave growth and swell decay during the Joint North Sea Wave Project (JONSWAP). Deutsche Hydrographische Zeitschrift, Supplement A8, 12.

HASSELMANN, K., ROSS, D.B., MÜLLER, P. and SELL, W. (1976). A parametric wave prediction model. J. Phys. Oceanogr., Vol. 6, pp 200-28.

HOLLY, F.M. and PRIESSMANN, A. (1977). Accurate calculation of transport in two dimensions. Proc. ASCE, J. Hyd. Divn., Vol. 103, No. HY11, pp 1259-77.

INOUE, T. (1967). On the growth of the spectrum of a wind generated sea according to a modified Miles-Phillips mechanism and its application to wave forecasting. Ph.D. thesis, School of Arts and Science, New York University.

KARLSSON, T. (1972). An Ocean Wave Forecasting Scheme for Iceland. Report to Science Inst., Div. of Earth Sciences, University of Iceland.

LEONARD, B.P. (1979). A stable and accurate convective modelling procedure based on quadratic upstream interpolation. Computer Methods in App. Mech. and Engineering, Vol. 19.

McCOWAN, A.D. and BLACKMAN, D.R. (1983). A wave hindcast model for the Southern Ocean. Proc. 6th Australian Conf. on Coastal and Ocean Engineering.

MILES, J.W. (1957). On the generation of surface waves by shear flows. J. Fluid Mech., No. 3, pp 185-204.

PHILLIPS, O.M. (1957). On the generation of waves by turbulent wind. J. Fluid Mech., No. 2, pp 417-45.

PIERSON, W.J., TICK, L.J. and BAER, L. (1966). Computer based procedures for preparing global wave forecasts and wind field analyses capable of using wave data obtained by a spacecraft. 6th Naval Hydrodynamics Symp., Office of Naval Research, Washington, D.C., pp 499-532.

RESIO, D.T. (1979). The estimation of wind-wave generation in a discrete spectral model. J. Phys. Oceanogr., Vol. 11, pp 510-25.

RESIO, D.T. and VINCENT, C.L. (1979). A comparison of various numerical wave prediction techniques. 11th Offshore Technology Conf., Houston, pp 2471-81.

SAND, S.E., BRINK-KJAER, O. and NIELSEN, J.B. (1981). Directional numerical models as design basis, in "International Conference on Wind & Wave Directionality", TECHNIP, Paris.

SEAMAN, R.S., FALCONER, R.L. and BROWN, J. (1977). Application of a variational blending technique to numerical analysis in the Australian region. Aust. Meteorological Magazine, Vol. 25, pp 3-23.

WEARE, J. and WORTHINGTON, B.A. (1978). A numerical model hindcast of severe wave conditions for the North Sea. Proc. NATO Symp. on Turbulent Fluxes Through the Sea Surface - Wave Dynamics and Prediction, A. Favre and K. Hasselmann editors, Plenum, New York.

YOUNG, I.R. and SOBEY, R.J. (1981). The Numerical Prediction of Tropical Cyclone Wind-Waves. Res. Bull. No. CS20, Dept. of Civil and Systems Engineering, James Cook Uni. of North Queensland.

Computational Techniques & Applications: CTAC-83
J. Noye & C. Fletcher (Editors)

MODELLING SEICHES

A.K. Easton

Mathematics Department
Swinburne Institute of Technology
Hawthorn, Victoria, 3122

Seiches are initiated by meteorological and seismic events but their period depends on the local topography. In this paper, their motion is represented by the two dimensional shallow water equations. The problem is reduced to finding the eigenvalues and eigenvectors of the corresponding Helmholtz equation. Solutions for an ideal model lake are obtained by finite difference and finite element methods and compared with an analytic solution.

1. INTRODUCTION

Seiches are the natural oscillations of harbours and lakes. They may be caused by meteorological or seismic events and may persist for more than a day. Figure 1.1 (Fryer and Easton (1981)) illustrates a seiche observed in Lake Wellington, Victoria, which was initiated by a change in wind stress due to the passage of a front. The water levels at the eastern and western ends of the lake are shown together with the wind measurements at a nearby meteorological station.

These waves are common at many Australian locations and particularly in those harbours and lakes which are regular in shape. The period of the waves is determined by the local topography and is usually within the range of a few minutes up to several hours. Table 1.1 lists the periods of seiches recorded at selected Australian locations.

Table 1.1

Seiche periods in selected Australian locations.

Location	Period (min)
Coffs Harbour, NSW	15
Newcastle Harbour, NSW	22
Sydney Harbour, NSW	25
Macquarie Harbour, TAS	60
Lake Wellington, VIC	110
Lake George, NSW	131

These waves have caused some inconvenience to shipping in Australian ports and temporary flooding near lakes but, since they are rarely catastrophic, the interest in them is largely scientific. The historical development of the theory of seiches has been summarised by Wilson (1972).

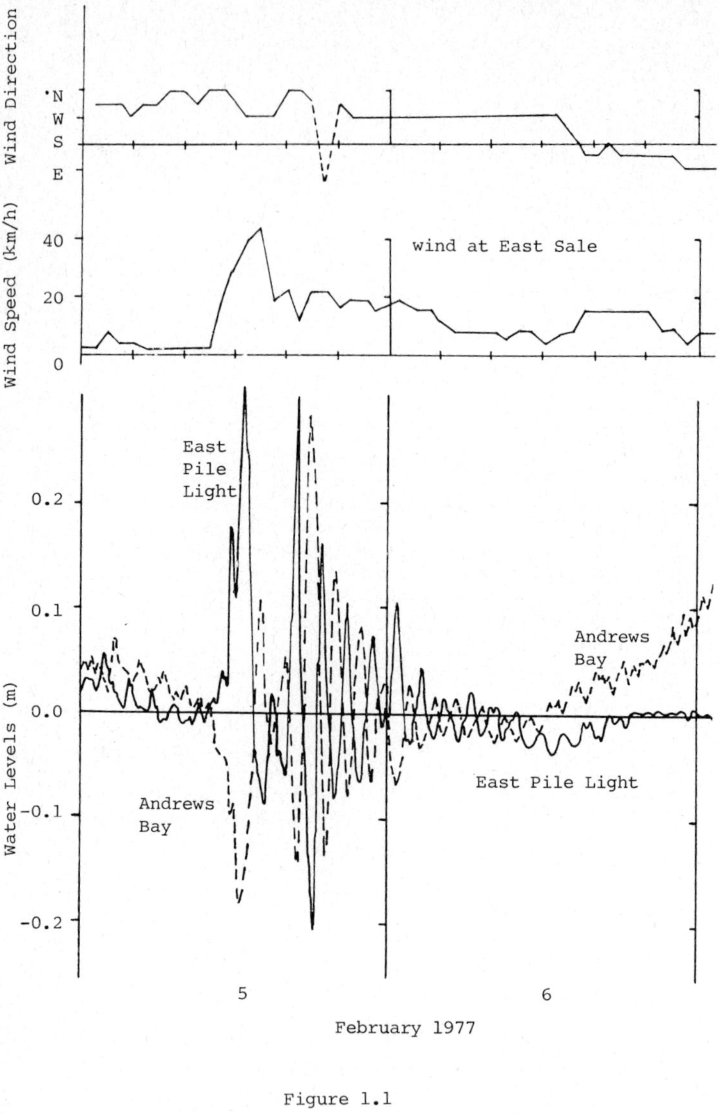

Figure 1.1
Seiching in Lake Wellington.

2. MATHEMATICAL EQUATIONS

Seiche motion may be represented by the 2D linear shallow water equations of conservation of mass and momentum

$$\frac{\partial u}{\partial t} = - g \frac{\partial \eta}{\partial x} \quad , \qquad (2.1)$$

$$\frac{\partial v}{\partial t} = - g \frac{\partial \eta}{\partial y} \quad , \qquad (2.2)$$

$$\frac{\partial \eta}{\partial t} = -(\frac{\partial hu}{\partial x} + \frac{\partial hv}{\partial y}) \quad , \qquad (2.3)$$

where $\eta(x,yt)$ is the surface elevation above mean water level,

$u(x,y,t)$ and $v(x,y,t)$ are the particle velocities in the x and y directions and

$h(x,y)$ is the water depth below mean level.

The boundary value problem is completed by specifying that the displacement $\eta = 0$ on any open water boundary and that the normal velocity is zero on any closed land boundary.

3. ANALYTIC SOLUTION

For convenience, we will consider a rectangular 'lake' with the surface in the x,y plane as $0 \leq x \leq \ell$ and $0 \leq y \leq b$ and with depth h. This will allow a comparison of results for the finite difference and finite element numerical schemes with those for an analytic solution using mathematics which is within the grasp of undergraduate students. Indeed the analysis can be carried out by hand and confirmed using computer programs or by experiment.

When comparing the results, we will use ℓ = 2m, b = 1m and h = 1m and we will list the surface displacement at the corners and the mid points of the longer sides. These points are numbered 1 through 6 (Figure 3.1).

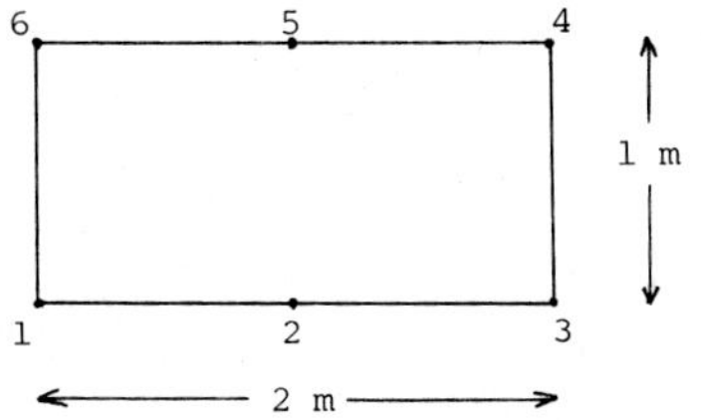

Figure 3.1

Surface nodes of the ideal lake.

Differentiating (2.3) and substituting from (2.1) and (2.2), the surface displacement is shown to be a solution of the wave equation

$$\frac{\partial^2 \eta}{\partial t^2} = gh\, \nabla^2 \eta \,. \qquad (3.1)$$

This form assumes that the depth h is constant. The velocities u and v also satisfy this equation.

Knowing that the motion is harmonic in time, we may assume that

$$\eta(x,y,t) = z(x,y)\cos\sigma t$$

and (3.1) reduces to the Helmholtz equation

$$\nabla^2 z + \lambda^2 z = 0 \,, \qquad (3.2)$$

where $\lambda^2 = \dfrac{\sigma^2}{gh}$.

The appropriate boundary conditions are that the normal velocities are zero

$$\frac{\partial z}{\partial x} = 0 \quad \text{on} \quad y = 0 \text{ and } y = b,$$

$$\frac{\partial z}{\partial y} = 0 \quad \text{on} \quad x = 0 \text{ and } x = \ell.$$

Now the problem can be solved by the method of separation of variables to show that each mode of oscillation has the form

$$\eta(x,y,t) = A \cos \frac{m\pi x}{\ell} \cos \frac{n\pi y}{b} \cos\sigma t,$$

where the resonant periods are given by

$$T = \frac{2\pi}{\sigma} = \frac{2\ell}{\sqrt{gh}} \{m^2 + (\ell n/b)^2\}^{-\frac{1}{2}}$$

and m and n are integers.

The periods and relative nodal displacements for the first six modes are given in Table 3.1.

Table 3.1

Principal modes of vibration for the ideal lake using the analytic soltion.

Eigenvalue		Period	Eigenvector					
m	n	Tsec	z_1	z_2	z_3	z_4	z_5	z_6
1	0	1.28	1	0	-1	-1	0	1
2	0	0.64	1	-1	1	1	-1	1
0	1	0.64	1	1	1	-1	-1	-1
1	1	0.57	1	0	-1	1	0	-1
2	1	0.45	1	-1	1	-1	1	-1
3	0	0.43	1	0	-1	-1	0	1

4. FINITE DIFFERENCE SOLUTION

The method of finite differences has been used to solve such equations for a number of years (a tenth order system is given by Stoker (1957)).

Using the notation that

$$z(x_i, x_j) = z_{ij}$$

with step lengths

$$x_{i+1} - x_i = d = y_{i+1} - y_i \quad ,$$

the derivatives can be approximated by central differences

$$\frac{\partial^2 z}{\partial x^2}(x_i, x_j) \cong \frac{z_{i+1\,j} - 2z_{ij} + z_{i-1\,j}}{d^2} \quad ,$$

$$\frac{\partial^2 z}{\partial y^2}(x_i, x_j) \cong \frac{z_{i\,j+1} - 2z_{ij} + z_{i\,j-1}}{d^2} \quad , \qquad (4.1)$$

and (3.2) becomes

$$z_{i+1\,j} + z_{i-1\,j} + z_{i\,j+1} + z_{ij-1} - 4z_{ij} + \mu z_{ij} = 0,$$

where $\mu = \lambda^2 d^2$.

One of the simplest schemes is to use the nodes numbered 1 through 6 and this numbering avoids the necessity of double subscripts. The boundary conditions can be satisfied by reflecting points in the boundary as shown in Figure 4.1.

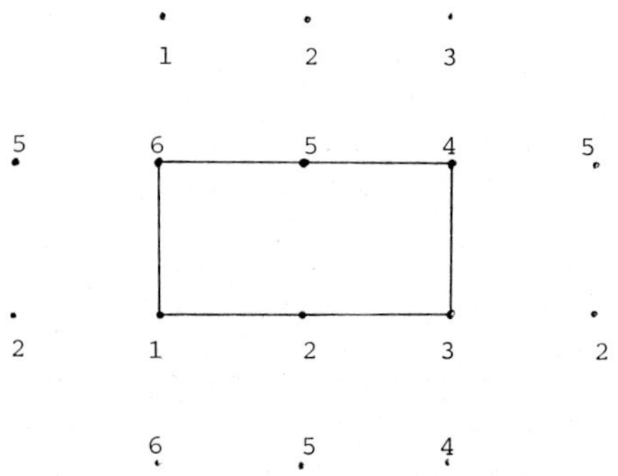

Figure 4.1

Basic Finite Difference Grid.

For this six node approximation, a sixth order system is obtained

$$\begin{pmatrix} 4 & -2 & 0 & 0 & 0 & -2 \\ -1 & 4 & -1 & 0 & -2 & 0 \\ 0 & -2 & 4 & -2 & 0 & 0 \\ 0 & 0 & -2 & 4 & -2 & 0 \\ 0 & -2 & 0 & -1 & 4 & -1 \\ -2 & 0 & 0 & 0 & -2 & 4 \end{pmatrix} \begin{pmatrix} z_1 \\ z_2 \\ z_3 \\ z_4 \\ z_5 \\ z_6 \end{pmatrix} = \mu \begin{pmatrix} z_1 \\ z_2 \\ z_3 \\ z_4 \\ z_5 \\ z_6 \end{pmatrix} , \qquad (4.2)$$

where $\mu = \dfrac{\sigma^2 d^2}{gh}$.

This is the classical eigenvalue problem

$$[A]\underset{\sim}{z} = \mu \underset{\sim}{z}$$

which can be solved for the eigenvalues μ and the eigenvectors $\underset{\sim}{z}$.

Note that (1) the matrix [A] is singular and has a zero eigenvalue
and that (2) this is a coarse grid and does not allow vibrations corresponding to the higher modes.

Table 4.1 lists the eigenvalues and eigenvectors for this six node finite difference representation and can be readily compared with the values for the analytic solution of Table 3.1.

Table 4.1

Principal modes of vibration for the ideal lake using the finite difference solution.

Eigenvalue	Period	Eigenvector					
μ	Tsec	z_1	z_2	z_3	z_4	z_5	z_6
0	∞	1	1	1	1	1	1
2	1.42	1	0	-1	-1	0	1
4	1.00	0	1	0	-1	0	-1
6	0.82	1	0	-1	1	0	-1
8	0.71	1	-1	1	-1	1	-1

Of course, the solutions can be refined by choosing mcre nodes and a finer grid. For a larger system it would be preferable to find the first few eigenvalues and eigenvectors only.

5. FINITE ELEMENT SOLUTION

A simple finite element mesh can be formed by dividing the rectangular basin into triangular elements using the same nodes as before (Figure 5.1). Each triangle must be numbered and each node in each triangle given a local node number (in an anticlockwise order) for that element.

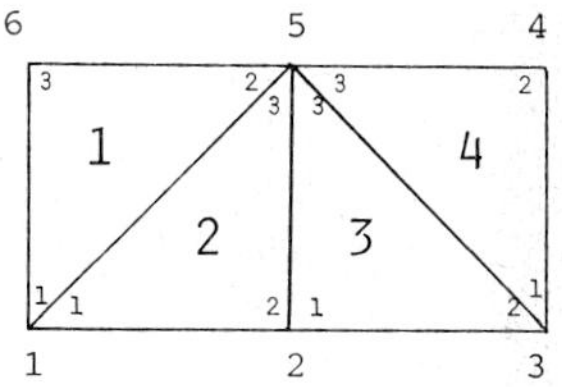

Figure 5.1

Finite element grid with local node and element numbers within each triangle.

On each element, z can be expressed as a linear function of x and y

$$z(x,y) = A + Bx + Cy,$$

or alternatively as

$$z(x,y) = \phi_1(x,y)z_1 + \phi_2(x,y)z_2 + \phi_3(x,y)z_3,$$

where z_i , i = 1, 2, 3 are the nodal values and

$$\phi_i(x,y) = a_i + b_i x + c_i y$$

are the so called shape functions or area coordinates.

This expression can be written in vector form

$$z(x,y) = \underset{\sim}{\phi}^T \underset{\sim}{z} , \qquad (5.1)$$

where $\underset{\sim}{\phi}^T = [\phi_1 \quad \phi_2 \quad \phi_3]$ and $\underset{\sim}{z} = \begin{pmatrix} z_1 \\ z_2 \\ z_3 \end{pmatrix}$.

It is a relatively straight forward procedure to show the equivalence of these forms and to derive the formula

$$\phi_1(x,y) = \frac{1}{2\Delta} \begin{vmatrix} 1 & x & y \\ 1 & x_2 & y_2 \\ 1 & x_3 & y_3 \end{vmatrix} , \qquad (5.2)$$

where $2\Delta = \begin{vmatrix} 1 & x_1 & y_1 \\ 1 & x_2 & y_2 \\ 1 & x_3 & y_3 \end{vmatrix}$

and Δ is the area of the triangle formed by those vertices. The formulas for ϕ_2 and ϕ_3 can be found by rotation of the subscripts.

Using (5.2) or noting that these shape functions are linear with value unity at one node and zero at the other two in the triangle, the shape functions are derived to be as follows:

$$\text{element 1} \qquad \underset{\sim}{\phi}_1(x,y) = \begin{pmatrix} 1 - y \\ x \\ y - x \end{pmatrix} ,$$

$$\text{element 2} \qquad \underset{\sim}{\phi}_2(x,y) = \begin{pmatrix} 1 - x \\ x - y \\ y \end{pmatrix} ,$$

$$\text{element 3} \qquad \underset{\sim}{\phi}_3(x,y) = \begin{pmatrix} 2 - x - y \\ x - 1 \\ y \end{pmatrix} ,$$

$$\text{element 4} \qquad \underset{\sim}{\phi}_4(x,y) = \begin{pmatrix} 1 - y \\ x + y - 2 \\ 2 - x \end{pmatrix} .$$

The Helmholtz boundary value problem (3.2) is equivalent to minimising the functional

$$I = \iint_R \tfrac{1}{2}[(\nabla z)^2 - \lambda^2 z^2]dx\, dy , \qquad (5.3)$$

where the region of integration is the surface of the lake. This variational approach to the finite element method has been used by Taylor et. al. (1969) in an investigation of harbour resonance.

Since z is represented from (5.1) as a piecewise linear function over the total region R, the functional I can be considered as the sum of the values in each element

$$I = \sum_e I_e .$$

This is to be minimised with respect to each of the coordinate values z_i and

$$\frac{\partial I}{\partial z_i} = \sum_e \frac{\partial I_e}{\partial z_i} = 0$$

or in vector form

$$\frac{\partial I}{\partial \underset{\sim}{z}} = \sum_e \frac{\partial I_e}{\partial \underset{\sim}{z}} = 0$$

so that the contributions can be found for each element (local coordinates) and added (global coordinates).

Substituting (5.1) into (5.3) yields

$$I = \iint_R \tfrac{1}{2}\{\underset{\sim}{z}^T[\nabla\phi][\nabla\phi]^T\underset{\sim}{z} - \lambda^2\underset{\sim}{z}^T\underset{\sim}{\phi}\underset{\sim}{\phi}^T\underset{\sim}{z}\}dx\,dy = 0.$$

Differentiating for each element

$$\frac{\partial I_e}{\partial \underset{\sim}{z}} = \begin{pmatrix} \frac{\partial I}{\partial z_1} \\ \frac{\partial I}{\partial z_2} \\ \frac{\partial I}{\partial z_3} \end{pmatrix} = \iint_e \{[\nabla\phi][\nabla\phi]^T\underset{\sim}{z} - \lambda^2\underset{\sim}{\phi}\underset{\sim}{\phi}^T\underset{\sim}{z}\}dx\,dy$$

$$= [K_e]\underset{\sim}{z} - \lambda^2[M_e]\underset{\sim}{z}$$

where
$$[K_e] = \iint_e [\nabla\phi][\nabla\phi]^T\,dx\,dy\,, \tag{5.4}$$

$$[M_e] = \iint_e \underset{\sim}{\phi}\underset{\sim}{\phi}^T\,dx\,dy\,, \tag{5.5}$$

and combining for the global system

$$[K]\underset{\sim}{z} = \lambda^2[M]\underset{\sim}{z}. \tag{5.6}$$

It is worth noting that this system of equations can be generated using the Galerkin weighted residual method. This approach begins with the Helmholtz equation in the form

$$-\nabla^2 z - \lambda^2 z = 0.$$

With the approximate solution

$$z = \underset{\sim}{\phi}^T\underset{\sim}{z}\,,$$

the weighted residual is

$$\iint_R \underset{\sim}{\phi}\{-[\nabla^2\phi]^T\underset{\sim}{z} - \lambda^2\underset{\sim}{\phi}^T\underset{\sim}{z}\}\,dx\,dy = 0.$$

The first of these terms can be 'integrated by parts' using Greens theorem to give as before

$$\iint_R \{[\nabla\phi][\nabla\phi]^T\underset{\sim}{z} - \lambda^2\underset{\sim}{\phi}\,\underset{\sim}{\phi}^T\underset{\sim}{z}\}\,dx\,dy = 0,$$

since the normal derivative of z is zero on the closed boundaries. This is identical with (5.6).

Since (5.4) requires that first derivatives of the shape functions must exist, linear functions are the simplest possible form. We will proceed briefly for element 1 for which we found earlier that the linear shape functions are

$$\underset{\sim}{\phi}_1 = \begin{pmatrix} 1 - y \\ x \\ y - x \end{pmatrix}.$$

Then

$$[\nabla\phi_1] = \begin{pmatrix} \frac{\partial\phi_1}{\partial x} & \frac{\partial\phi_1}{\partial y} \\ \frac{\partial\phi_2}{\partial x} & \frac{\partial\phi_2}{\partial y} \\ \frac{\partial\phi_3}{\partial x} & \frac{\partial\phi_3}{\partial y} \end{pmatrix} = \begin{pmatrix} 0 & -1 \\ 1 & 0 \\ -1 & 1 \end{pmatrix}$$

Since this is constant, (5.4) gives

$$[K_e] = [\nabla\phi][\nabla\phi]^T \iint_e dx\,dy$$

$$= \frac{1}{2}\begin{pmatrix} 1 & 0 & -1 \\ 0 & 1 & -1 \\ -1 & -1 & 2 \end{pmatrix}.$$

The global [K] matrix is formed by summing the similar results for each element

$$[K] = \sum_e [K_e]$$

while noting that some nodes occur in different elements to give

$$[K]\underset{\sim}{z} = \frac{1}{2}\begin{pmatrix} 2 & -1 & 0 & 0 & 0 & -1 \\ -1 & 4 & -1 & 0 & -2 & 0 \\ 0 & -1 & 2 & -1 & 0 & 0 \\ 0 & 0 & -1 & 2 & -1 & 0 \\ 0 & -2 & 0 & -1 & 4 & -1 \\ -1 & 0 & 0 & 0 & -1 & 2 \end{pmatrix}\begin{pmatrix} z_1 \\ z_2 \\ z_3 \\ z_4 \\ z_5 \\ z_6 \end{pmatrix} \qquad (5.7)$$

The element matrix $[M_e]$ can be calculated by (5.5) using double integration of the linear shape functions to give

$$[M_e] = \frac{1}{24}\begin{pmatrix} 2 & 1 & 1 \\ 1 & 2 & 1 \\ 1 & 1 & 2 \end{pmatrix}.$$

It is somewhat easier to use the formula for each part of the matrix

$$\iint_e \phi_p^\alpha \phi_q^\beta \, dx\,dy = \frac{2\Delta\alpha!\beta!}{(\alpha + \beta + 2)!}$$

where p and q = 1, 2 and 3 (Davies (1980)). This can be used as a check on ones ability to perform double integrations.

Combining in a global matrix by summing the results over all elements,

$$[M] = \sum_e [M_e] \ ,$$

and
$$[M]\underset{\sim}{z} = \frac{1}{24}\begin{pmatrix} 4 & 1 & 0 & 0 & 2 & 1 \\ 1 & 4 & 1 & 0 & 2 & 0 \\ 0 & 1 & 4 & 1 & 2 & 0 \\ 0 & 0 & 1 & 2 & 1 & 0 \\ 2 & 2 & 2 & 1 & 8 & 1 \\ 1 & 0 & 0 & 0 & 1 & 2 \end{pmatrix}\begin{pmatrix} z_1 \\ z_2 \\ z_3 \\ z_4 \\ z_5 \\ z_6 \end{pmatrix} \qquad (5.8)$$

The matrices given by (5.7) and (5.8) can now be used to find the eigenvalues and eigenvectors from (5.6)

$$[K]\underset{\sim}{z} = \lambda^2 [M]\underset{\sim}{z}$$

where $\lambda^2 = \dfrac{\sigma^2}{gh}$.

Some computer packages can use this form of the matrix eigenvalue problem but if preferred it can be rewritten in the standard form

$$[A]\underset{\sim}{z} = \lambda^2 \underset{\sim}{z} \ , \qquad (5.10)$$

where $[A] = [M]^{-1}[K]$.

It can readily be verified that for the example being considered here that

$$[M]^{-1} = \frac{24}{224}\begin{pmatrix} 74 & -14 & 10 & 2 & -14 & -30 \\ -14 & 70 & -14 & 14 & -14 & 14 \\ 10 & -14 & 74 & -30 & -14 & 2 \\ 2 & 14 & -30 & 134 & -14 & 6 \\ -14 & -14 & -14 & -14 & 42 & -14 \\ -30 & 14 & 2 & 6 & -14 & 134 \end{pmatrix}$$

and hence that

$$[A] = \frac{24}{112}\begin{pmatrix} 48 & -28 & 8 & 2 & 0 & -30 \\ -28 & 84 & -28 & 14 & -56 & 14 \\ 8 & -28 & 48 & -30 & 0 & 2 \\ -4 & 28 & -52 & 78 & -56 & 6 \\ 0 & -28 & 0 & -14 & 56 & -14 \\ -52 & 28 & -4 & 6 & -56 & 78 \end{pmatrix}$$

Note that (1) although [K] and [M] are symmetric, [A] is not,

(2) [K] is singular and hence [A] is singular,

(3) [A] has real eigenvalues, one of which is zero.

The eigenvalue problem can be solved using the standard packages EISPAK, IMSL or NAG to give the values listed in Table 5.1.

Table 5.1

Principal modes of vibration for the ideal lake using the finite element solution.

Eigenvalue	Period	Eigenvectors					
λ^2	Tsec	z_1	z_2	z_3	z_4	z_5	z_6
0	∞	1	1	1	1	1	1
2.9	1.18	1	0	-1	-1	0	1
12.0	0.58	1	0	1	0	-1	0
12.0	0.58	0	1	0	-1	0	-1
21.1	0.44	1	0	-1	2	0	-2
36.0	0.33	1	-2	1	-2	1	-2

CONCLUDING REMARKS

This paper presents a comparison of methods and results of finite difference and finite element procedures for a commonly observed physical phenomenon. Special emphasis is given to the development of the finite element procedure. This work can by applied to real situations of harbours and lakes and could also be used to investigate the motion in a fish tank.

REFERENCES

[1] Davies, A.J. (1980) *The Finite Element Method: A First Approach*, Oxford.

[2] Fryer, J.J. and Easton, A.K., (1981) *Hydrodynamics of the Gippsland Lakes, Gen. Engg. Trans.*, I.E. Aust., GE 5, 109-114.

[3] Stoker, J.J., (1957) *Water Waves*, Interscience.

[4] Taylor, C., Patil, B.S. and Zienkiewics, O.C., (1969) *Harbour Oscillation: A numerical treatment for undamped modes*, Proc. Inst. Civil Engineers, 43, 141-155.

[5] Wilson, B.W., (1972) *Seiches*, Advan. Hydrosci., 8, 1-94.

Computational Techniques & Applications: CTAC-83
J. Noye & C. Fletcher (Editors)

THE DEVELOPMENT OF A FINITE ELEMENT COMPUTER PACKAGE FOR ATMOSPHERIC DYNAMICS

Ken J. Mann

Mathematics Department
Chisholm Institute of Technology
Caulfield, Victoria
Australia

Finite element modelling of structural mechanics is well supported with a range of computer packages appropriate for one or more of research, consultancy and education. In many of these packages, there are associated computer graphics routines which are readily implemented on line printer, VDU or other peripheral devices. In the field of finite element modelling of fluid dynamics there is a scarcity of such support. With the rapid increase in the general availability of computer graphics hardware, it is desirable to incorporate computer graphics software routines within the package, or have them accessible to it. This paper describes the author's development of a self-contained computer package for the finite element modelling of atmospheric dynamics and the incorporation of a suite of computer graphics within this package. The package is not limited to large computers. There is a discussion on the current achievements and overall aim of the research and the envisaged immediate next steps in this development.

INTRODUCTION

This research is the first stage in the development of a finite element method (FEM) and computer package for thermal convection in the atmospheric boundary layer. The model is gradually developed to provide stability, accuracy and economy in the FEM modelling of absolutely unstable atmospheric convection through to moist atmospheric convection to the inclusion of anisotropic effects. Many models are used and comparisons made to select the "best" model. Many checks are employed to analyse, as far as possible, the accuracy of the algorithms as well as their stability. Wherever possible, evaluations are carried out analytically. This considerably increases the manual work particularly in the use of the least squares criterion. However, it is felt desirable to limit any likely sources of instabilities such as approximating techniques in numerical quadrature. Research into the FEM in convective processes in general has shown numerical stabilities to be a problem: thermal convective processes could expect to present at least as significant a problem in this area.

MATHEMATICAL MODEL

The non-dimensionalized system of equations describing two-dimensional, unsteady convective motion above a heated strip, with the Boussinesq approximation, are presented as follows

$$\frac{\partial \eta}{\partial t} = J(\eta,\psi) - \frac{R}{\sigma}\frac{\partial T}{\partial x} + \nabla^2 \eta \qquad (1)$$

$$\frac{\partial T}{\partial t} = J(T,\psi) + s\,\frac{\partial \psi}{\partial x} + \frac{1}{\sigma}\,\nabla^2 T \tag{2}$$

$$\eta = -\nabla^2 \psi \tag{3}$$

with ψ (streamfunction), η (vorticity), T (temperature), J the Jacobian operator and R, σ the Rayleigh number and Prandtl number respectively. The "S" may be the latent heat factor for moist convection or, for dry convection, it may represent the linear temperature profile leaving T to consist only of the nonlinear part of the temperature (Mann [1979]).

FINITE ELEMENT LEAST SQUARES CRITERION

In line with the approach of Lynn and Arya [1973] and Steven [1976], the order of the derivatives in the Laplacian operator in the Least Squares formulation is reduced by the introduction of "auxiliary functions" α, β, γ, δ, ε, ζ which are defined as follows

$$\alpha = \frac{\partial \psi}{\partial x} \qquad \beta = \frac{\partial \psi}{\partial z}$$
$$\gamma = \frac{\partial \eta}{\partial x} \qquad \delta = \frac{\partial \eta}{\partial z} \tag{4}$$
$$\varepsilon = \frac{\partial T}{\partial x} \qquad \zeta = \frac{\partial T}{\partial z}$$

The substitution of these into (1) - (3), the application of a linear backward-difference analogue for the time derivative and the inclusion of the nonlinear terms in the load matrix produces a coupled system describing the motion at the jth time step, which is represented in matrix form by,

$$\begin{bmatrix} 1 & 0 & 0 & 0 & 0 & 0 & -\frac{\partial}{\partial x} & 0 & 0 \\ 0 & 1 & 0 & 0 & 0 & 0 & -\frac{\partial}{\partial z} & 0 & 0 \\ 0 & 0 & 1 & 0 & 0 & 0 & 0 & -\frac{\partial}{\partial x} & 0 \\ 0 & 0 & 0 & 1 & 0 & 0 & 0 & -\frac{\partial}{\partial z} & 0 \\ 0 & 0 & 0 & 0 & 1 & 0 & 0 & 0 & -\frac{\partial}{\partial x} \\ 0 & 0 & 0 & 0 & 0 & 1 & 0 & 0 & -\frac{\partial}{\partial z} \\ \frac{\partial}{\partial x} & \frac{\partial}{\partial z} & 0 & 0 & 0 & 0 & 0 & 1 & 0 \\ 0 & 0 & \frac{\partial}{\partial x} & \frac{\partial}{\partial z} & -\frac{R}{\sigma} & 0 & 0 & -\frac{1}{\Delta t} & 0 \\ S & 0 & 0 & 0 & \frac{1}{\sigma}\frac{\partial}{\partial x} & \frac{1}{\sigma}\frac{\partial}{\partial z} & 0 & 0 & -\frac{1}{\Delta t} \end{bmatrix} \begin{bmatrix} \alpha \\ \beta \\ \gamma \\ \delta \\ \varepsilon \\ \zeta \\ \psi \\ \eta \\ T \end{bmatrix}^{j} = \begin{bmatrix} 0 \\ 0 \\ 0 \\ 0 \\ 0 \\ 0 \\ 0 \\ -\beta\gamma + \alpha\delta - \eta/\Delta t \\ -\beta\varepsilon + \alpha\zeta - T/\Delta t \end{bmatrix}^{j-1}$$

$$\text{i.e.}\quad [L]\;\{\chi\} = \{f\} \tag{5}$$

where
- L : linear matrix differential operator
- $\{\chi\}$: a vector of dependent variables
- $\{f\}$: a vector which is a function of the solution from the previous iteration.

To satisfy compatibility, completeness and error consistency as detailed in Lynn and Arya [1973], the base functions (ψ,η,T) are represented by quadratic shape functions based upon the vertices and mid-points of the sides of a triangle, and the auxiliary functions are represented by linear shape functions based on the vertices only. The shape functions, N_i, are taken as expressions in the area coordinates, L_i, to facilitate the analytic evaluation of the integrals that arise during the formulation of the stiffness and load matrices.

With

$$[L]\,\{\chi^*\}^e - \{f^*\}^e = r^e \tag{6}$$

where

$\{\chi^*\}^e$: element approximations to $\{\chi\}$

$\{f^*\}^e$: element approximations to $\{f\}$

r^e: element residual

and

$$\{\chi^*\}^e = [N]\,\{\bar{\chi}^*\}^e \tag{7}$$

where

$\{\bar{\chi}^*\}^e$: element nodal approximations to $\{\chi\}$

$[N]$: the 9 x 36 matrix of shape functions

then

$$[L]\,[N]\,\{\bar{\chi}^*\}^e - \{f^*\}^e = r^e \tag{8}$$

i.e.

$$[K]\,\{\bar{\chi}^*\}^e - \{f^*\}^e = r^e \tag{9}$$

and, after applying the Least Squares criterion (Finlayson [1972]), the stiffness and load matrices for an element are written

$$\iint_{A^{(e)}} [K]^T\,[K]\,\{\bar{\chi}^*\}^e\,dA^{(e)} = \iint_{A^{(e)}} [K]^T\,\{f^*\}.dA^{(e)} \tag{10}$$

where the stiffness matrix is 36 x 36, symmetric and positive-definite. The system is assembled, and solved by Gaussian elimination, using a "frontal" solver which has been based on the program of Irons [1970]. This involves large usage of virtual storage rendering the use effective on minicomputers. Storage of the upper-triangular portion is all that is required of the stiffness matrices and this storage is on disk, only the assembled portion of the upper-triangular stiffness matrix being in-core. To cope with the inclusion, or otherwise, of the latent heat factor S at the various nodes of the triangle, S is assigned to be a matrix

$$S = \text{transpose}\,[S_i\ S_j\ S_k] \tag{11}$$

where i, j, k are the vertex nodes of the triangle.

A flexible computer program and associated graphics has been constructed. Comparisons were made with the Galerkin formulation resulting in the Galerkin being superior in accuracy and stability, with the result that development of the least squares model has been suspended until a better formulation of the element matrix equation is obtained. Some comparisons were presented in Mann [1979]. There are many possibilities to try including the weighting of the equations as detailed in Lynn and Arya [1974]. With the large element stiffness matrix involved in this system, future development of this model may involve numerical integration.

FINITE ELEMENT GALERKIN CRITERION

The application of the finite element Galerkin criterion and Green's theorem to the system of equations (1) - (3) produces the following element equations

$$\iint_{A^{(e)}} [N^T N\{\dot{\eta}\} - N^T N_x\{\eta\} N_z\{\psi\} + N^T N_x\{\psi\} N_z\{\eta\} + \frac{R}{\sigma} N^T N_x\{T\} + N_x^T N_x\{\eta\} + N_z^T N_z\{\eta\}]\, dxdz - \int_\ell N^T N_n\{\eta\} d\ell = 0 \qquad (12)$$

$$\iint_{A^{(e)}} [N^T N\{\dot{T}\} - N^T N_x\{T\} N_z\{\psi\} + N^T N_x\{\psi\} N_z\{T\} - S\, N^T N_x\{\psi\} + \frac{1}{\sigma}[N_x^T N_x\{T\} + N_z^T N_z\{T\}]]\, dxdz - \frac{1}{\sigma}\int_\ell N^T N_n\{T\} d\ell = 0 \qquad (13)$$

$$\iint_{A^{(e)}} - [N^T N\{\eta\} - N_x^T N_x\{\psi\} - N_z^T N_z\{\psi\}]\, dxdz - \int_\ell N^T N_n\{\psi\} d\ell = 0 \qquad (14)$$

where $N = [N_1\ N_2\ \ldots\ N_m]$

$\{\psi\} = [\psi_1\ \psi_2\ \ldots\ \psi_m]^T$

$\psi = \sum_{i=1}^{m} N_i(x,z)\psi_i(t) = N\{\psi\}$ (similarly for η, T)

superscript dot refers to time derivative
subscripts x, z refer to partial derivatives in those directions
subscript n refers to the derivative normal to the contour
ℓ : contour of boundary of domain
m : number of nodes per element
N_i : shape functions
ψ_i : nodal values of function ψ

These equations are then summed over all elements. Solution is effected through various nonlinear and quasi-linear models using a fully-coupled system and a frontal-technique for solving the large sparse systems of equations.

Using triangular elements with linear interpolation functions and expressing all matrices in terms of the area coordinates, the integrals of (4) - (6) are evaluated analytically.

In the vertical plane, a rectangular region is analysed with the vertical $z : 0 \rightarrow 1$ divided into NZ equal divisions and the horizontal $x : 0 \rightarrow D$ divided into NX equal divisions, the x and z extents being non-dimensional. This results in there being (NZ x NX) nodes and 2(NZ - 1)(NX - 1) 3-node triangles. Because in atmospheric cellular convection the location of the gradients of greatest value is unknown - mainly because it is unknown how many cells should occur - different sized elements are not employed. The numbering of the elements proceeds along the vertical strips from the z = 0 axis to minimize the "front", for solution purposes.

There have been 8 models used. For 3 of the models, the Jacobian has been placed in the load matrix after employing for the time derivative the backward-difference, Crank-Nicolson and Dufort-Frankel analogues. A further model was developed in which, in conjunction with the backward-difference time analogue, the ψ values of the Jacobian are maintained from the previous iteration, thus enabling the Jacobian contribution to be included in the stiffness matrix. These four quasi-linear models are placed in subroutines BTIME, CRANK, DUFORT and QUASI respectively. Four nonlinear models are used. The Newton-Raphson algorithm is used in conjunction with the Crank-Nicolson and backward-difference time analogues to form subroutines NRCN and NRBT respectively. A Taylor approximation of the Jacobian is also used in conjunction with the Crank-Nicolson and backward-difference time analogues to form subroutines TCN and TBT respectively. Details of the extensive comparisons of these will be given another time. At this stage, it is noted that for accuracy and economy the NRCN and TCN models produced very good results.

The self-contained program consists of twenty-two segments in the form of a main segment and twenty-one subroutines. A schematic representation of the interaction of the segments of the program is given in Figure 1. The main segment, MAIN calls, in turn, subroutines MESH, PREUNS, BNDRY, START and PRINT, each one returning directly to the main segment. MAIN the calls UNSYMM, which calls ELFORM once for each element. ELFORM calls one of BTIME, CRANK, QUASI, DUFORT, NRCN, NRBT, TBT or TCN. The returning sequence proceeds to UNSYMM/RESOL which then calls BACSUB and the returning sequence to MAIN is executed. MAIN then calls in turn subroutines VERTW, FLUX, PRINT and WHEN. If convergence has not been achieved, the sequence recommences at the call to UNSYMM/RESOL.

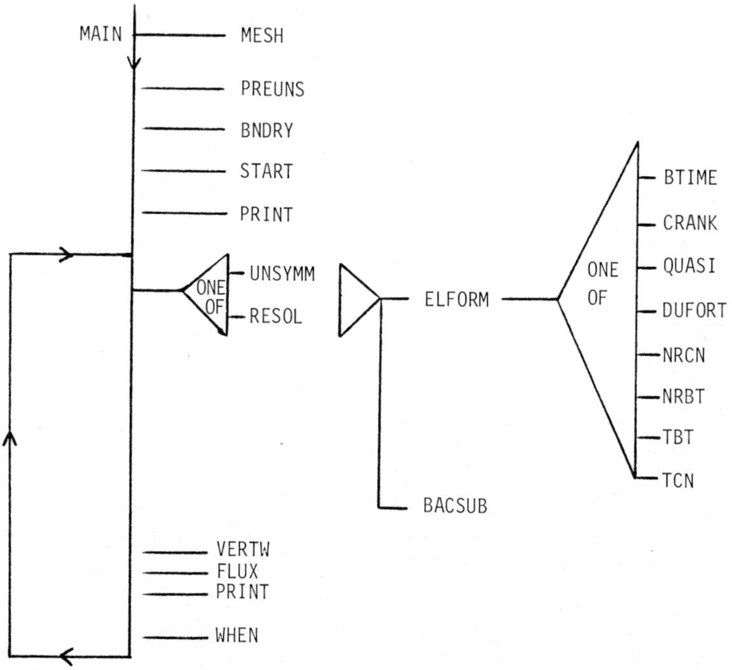

FIGURE 1 A schematic representation of the interaction of segments of the finite element Galerkin program.

The parameters for the particular job are input, using free format, in the main segment. The order of this input data, and its description, is listed in Table 1. This data includes control over the horizontal extent of the convective layer, the density of the mesh, the use of more dense mesh on the lateral extremities of the domain, the time increment, the s-matrix, whether the flow is to be moist and/or anisotropic, the Rayleigh and Prandtl numbers, the boundary conditions, the initial conditions, the grid formation, the minimum number of completed equations before Gaussian elimination takes place and the particular numerical model to be used. Allowance is also made for many jobs to be run simultaneously with separate disk output. At each iteration, unless indicated otherwise, the output data is given as an unformatted record to disc file for use by the associated computer graphics program or for use in a continuation of the current job. The descriptions in Table 1, in general explain the variety of the development of the program at this stage.

Table 1 Listing of the input data to the finite element Galerkin program.

Variable or Array	Description
NX	the number of nodes in the x-direction in each row
NZ	the number of nodes in the z-direction in each column
D	the range in the x-direction
DELT	the time increment
SIE,SJE,SKE	the S-matrix
R	the Rayleigh number
PR	the Prandtl number
NBN	the number of nodes in each element
KUREL	the total number of degrees of freedom in each element
INIT	a coded value to indicate whether this is a new job (zero) or a continuation of a previous job (unity)
KREAD	the number of iterations to be read from a disc file if this is a continuation of a previous job
NAVER	the number of iterations to elapse before averaging starts
NSM	the number of columns of small elements near both lateral boundaries
MOIST	a coded value to indicate whether this is dry (zero) or moist convection (unity)
ND2	a coded value indicating the particular numerical model and, if it's a new job, the new number of its disc output file
ND4	a coded value indicating the particular numerical model and, for a continuation of a previous job, the number of the disc output file
ISTART	the number of elapsed iterations before output to disc file
IFIN	the last iteration to be output to disc file
NSIDER	a coded value indicating no extra emphasis (zero) or extra emphasis (unity) on lateral temperature gradient
MINSOL	the minimum number of completed equations before elimination
IDRYS	a coded value indicating T is temperature (zero) or only nonlinear part of temperature (unity): the S-model
INDIC(12)	a coded array indicating whether there is not (zero) or there is (unity) a function boundary conditions. The order refers to $\psi(x,0)$, $\psi(x,1)$, $\eta(x,0)$, $\eta(x,1)$, $T(x,0)$, $T(x,1)$, $\psi(0,z)$, $\psi(D,z)$, $\eta(0,z)$, $\eta(D,z)$, $T(0,z)$, $T(D,z)$
NMAX	a number that is to exceed the largest frontwidth plus the next element equations to be assembled

Table 1 (Cont.)

Variable or Array	Description
NCRIT	a number (<NMAX) to allow enough equations that within this lies sufficient pivotal choice; increases automatically from this value
NLARG	a value that NCRIT may not exceed as it accumulates
KTEMP	a coded value indicating the particular initialization scheme used, whether it be random initial temperature~10^{-3} (zero), uniform initial temperature as at top boundary (unity), linear temperature profile (2), uniform temperature as at base (3) or sinusoidal streamfunction (4) or sinusoidal temperature (5)
KCELL	the number of complete initial half cells over the range
KDERIV	a coded value indicating normal derivatives are zero (zero) or non-zero (unity) on boundaries
KMESH	a coded value indicating program's mesh generator (zero) or mesh data input from disc file (unity)
KGRID	a coded value indicating the direction of the diagonals of the mesh, whether it be towards the corners of the domain (zero), 45° to x-axis (unity), 135° to x-axis (2), across the corners of the domain (3)
KSTAG	a coded value indicating no stagger (zero) or stagger (unity) of regular mesh
NITER	the number of iterations within each time step for Newton-Raphson programs
KPRINT	the number indicating output to line printer at iterations that are multiples of this
ER	the degree of anisotropy
KKND2	an alternative output disc file for ND2
KKND4	an alternative output disc file for ND4
HA(66)	an array indicating the name of the particular model
NCR	the card reader input channel
NLP	the line printer output channel

Subroutine MESH assigns numbers to the nodes, coordinate values to each node in arrays X and Z, assigns a number to each element and stores, with it, the element node numbers in array NOP, assigns the degrees-of-freedom of each node in array MDF and the first system degree-of-freedom at each node in array NOPP. This segment also deals with staggered grids, smaller elements on lateral boundaries and grids with diagonals in various orientations.

Subroutine PREUNS contains the preliminary process, which indicates the last occurrence of each system node number in element by assigning a negative to that node in that element, proceeding in counting order of elements. This is later used to indicate when equations are fully summed. The process also checks for repeated nodes and forms arrays NTOTEL, to indicate the total number of elements contiguous with each node, and NSUREL, which contains the numbers of the elements contiguous with each node. It also checks for unassigned nodes. Hood [1977] provided the basis, of this preliminary process.

Subroutine BNDRY assigns the prescribed function boundary conditions in array BC and assigns unity to those corresponding entries of array NCOD to indicate a function boundary condition has been applied. Subroutine START invokes the chosen initialization scheme as indicated in Table 1.

Subroutine PRINT outputs, to the line printer, at every iteration, unless directed otherwise, the solution for ψ, η and T at each node together with the other relevant values special to a particular job including the number of the iteration and the particular numerical model. Also printed out are the vertical velocity, the horizontally-averaged absolute vertical velocity at the middle-level of the convective layer and the Nusselt number.

The main segment then commences a sequence that is repeated, to a large extent, at each iteration. Subroutine UNSYMM is called. This assembles the element matrix equations formulated by other subroutines to form the system matrix equation, applies the boundary conditions and places on disc file the pivotal equations using Gaussian elimination. UNSYMM calls subroutine ELFORM once for each element. Subroutine ELFORM calculates the element characteristics and sets, if necessary, the S-matrix for moisture for that element before calling one of BTIME, CRANK, QUASI, DUFORT, NRCN, NRBT, TBT and TCN, which formulates the relevant element matrix equation for its particular numerical model.

Subroutine VERTW evaluates the vertical velocity at each node, using data from the iterative solution. It also evaluates the horizontally-averaged absolute vertical velocity at the middle level of the convective layer. Subroutine Flux evaluates the Nusselt number.

Then in the MAIN the output is placed on the appropriate output disc file at each iteration as unformatted records. Subroutine WHEN evaluates the cpu and I/O time taken for an iteration. This is done for 3 consecutive iterations to give a comparison of the various methods.

MAIN calls either UNSYMM or RESOL as indicated in Figure 7.1. If the system "stiffness" matrix is invariant then RESOL is called. Otherwise UNSYMM is called and the sequence, as earlier described, is repeated. Subroutine RESOL is used on subsequent iterations when the values in the stiffness matrix at each iteration are unaltered. This subroutine calls ELFORM which calls subroutines as above (see Figure 7.1) to form and accumulate the new element load matrices only : the subroutines have controls over this. BACSUB is called for the back-substitution.

SOME EXAMPLES

In determining a preferred mode of convection, in an absolutely unstable layer without latent heat release due to condensation of water vapour, a region with D = 20 has been chosen. With no flow over the boundaries and with non-dimensional values of T = 0 on lower boundary and T = -1 on upper boundary, and no heat flux across the lateral boundaries, the boundary conditions may be tabulated for the Galerkin model as in Table 2.

Table 2 Boundary values for atmospheric cellular convection using the finite element Galerkin criterion

Boundary	ψ	η	T	$\frac{\partial T}{\partial x}$
z = 0	0	0	0	
z = 1	0	0	-1	
x = 0	0	0		0
x = D	0	0		0

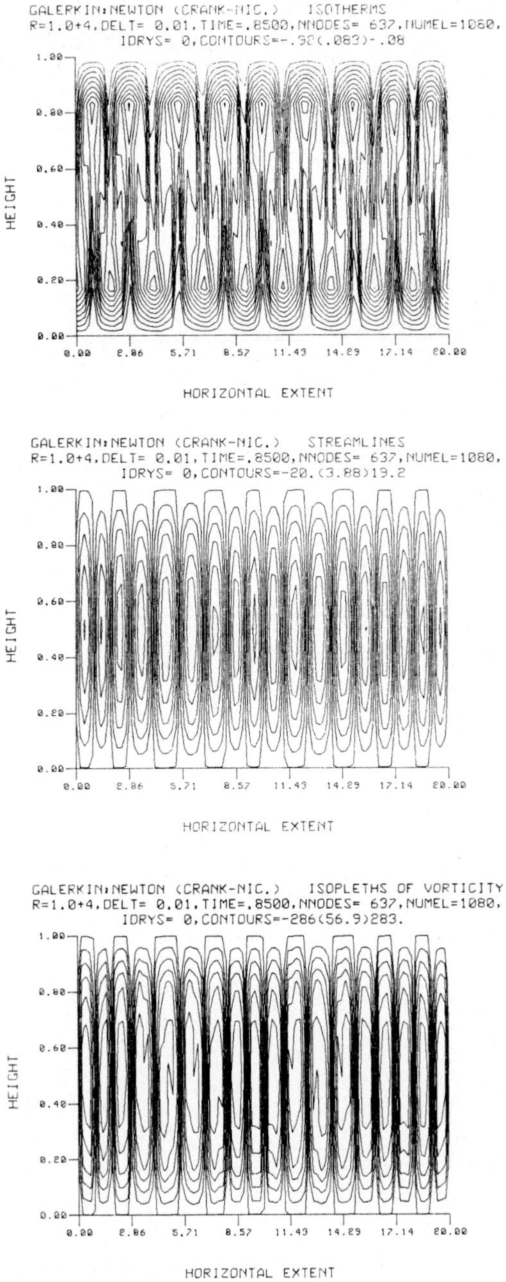

Figure 2 Isotherms, streamlines and vorticity isopleths of atmospheric cellular convection

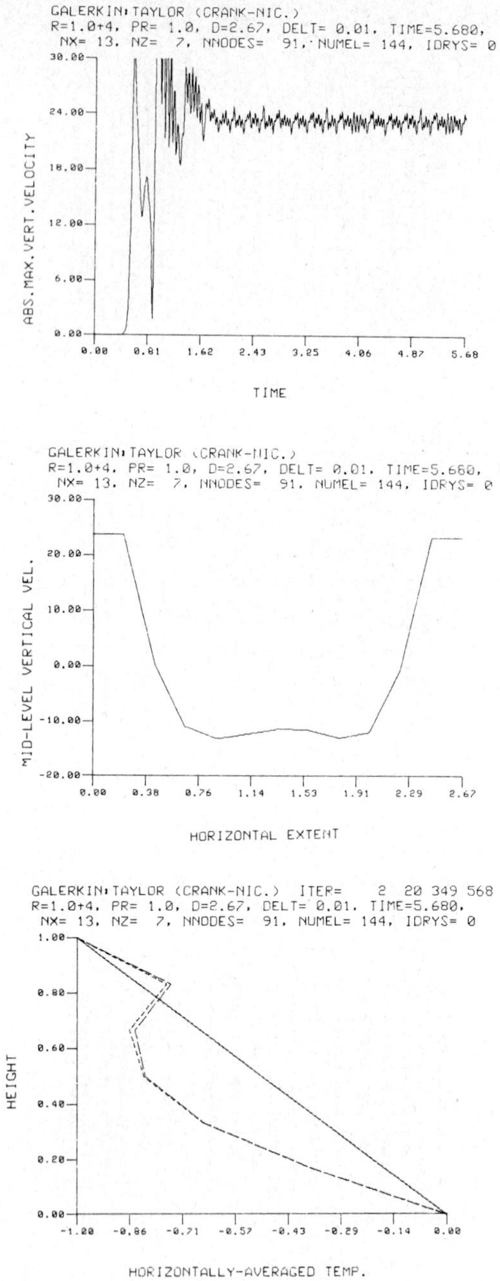

Figure 3 Velocity and temperature profiles of atmospheric cellular convection.

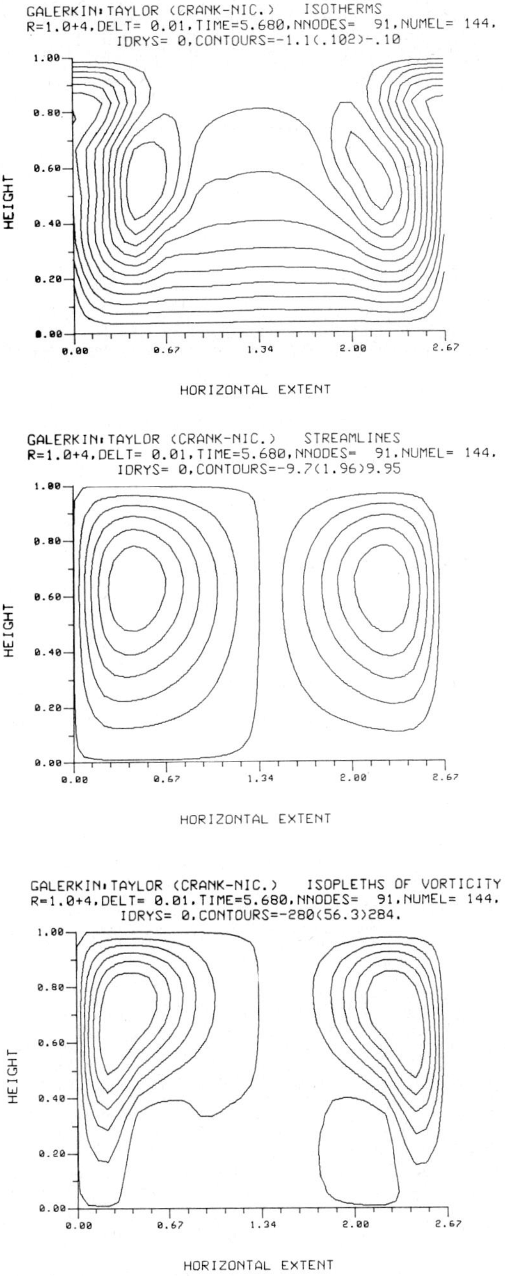

Figure 4 Isotherms, streamlines and vorticity isopleths of cellular convection in a moist atmosphere

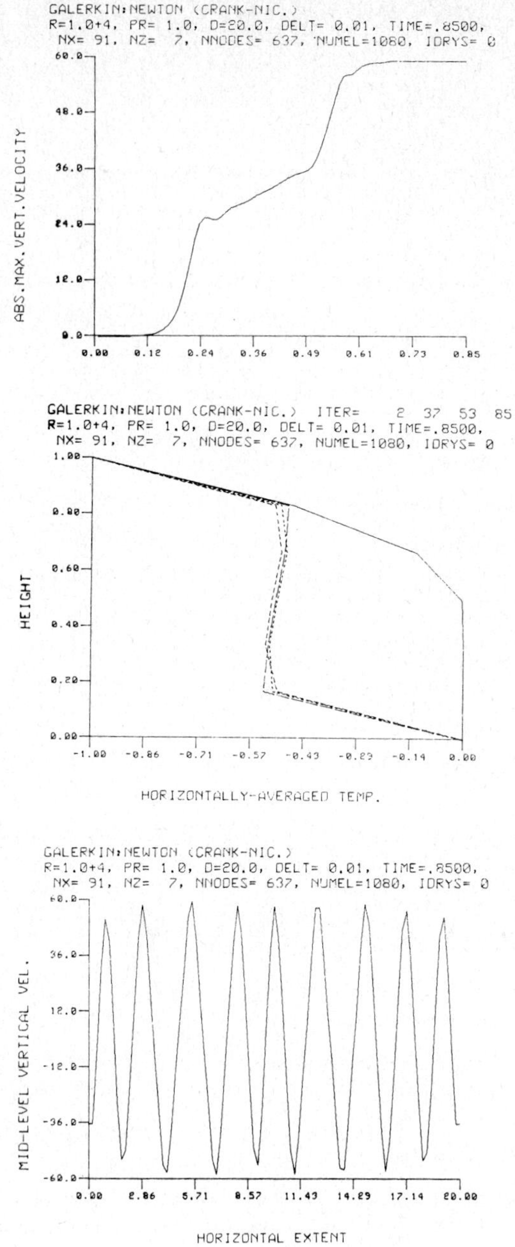

Figure 5 Vorticity and temperature profiles of cellular convection in a moist atmosphere

The imposition of zero heat flux across the lateral boundaries fixes the region to contain multiples of half cells. With no heat flux across the lateral boundaries, one would expect the isotherms to be horizontal at these boundaries as obtained in Figure 2, 3. For this output there was an initialization of $T \approx 0$ across the layer, simulating a cooling from above and producing closed cells with an aspect ratio (diameter:depth) of each cell as 2.22, which agrees with the theory of Busse [1967] and others. The variation in vertical velocity profiles indicates the changing of the number of cells during the transient growth, there being 11 cells at one stage, before the steady state regime of 9 cells was reached. The temperature profile indicates the boundary layers.

To analyse the effects of moist convection, using similar boundary conditions to that of Table 3 and allowing for compensating downdraft as expressed by Asai and Kasahara [1967], with $D = 2.67$, an output as in Figure 4, 5 was obtained. This models the narrow updraft of greater velocity and the wider downdraft of slower speed. The temperature profile indicates the updraft due to the latent heat effects increasing the buoyancy drive. The convergence of such a model is slow. The small oscillations in the vertical velocity would no doubt be due to the coarseness of the mesh. The latent heat adds to the buoyancy when the vertical velocity is upward : the accuracy of the computation of the vertical velocity is dependent on the mesh density. As outlined in Mann [1979] a value of $s = 0.5$ has been used.

CONCLUSION

The finite element Galerkin has been preferred to the finite element least squares on the basis of accuracy and stability. The Galerkin model with Lagrangian elements has been found to model dry, moist and anisotropic convection situations very well. An isoparametric model is now being developed. At this stage it has not produced any advantage over the Lagrangian model and has been found to model moist convection inaccurately : its modelling of anisotropic convection is good. This program uses the "quadratic" isoparametric element and all integrals are computed numerically using Gauss-Legendre integration. Again, the program is fully self-contained.

REFERENCES

Asai, T. and Kasahara, A. (1967). A Theoretical Study of the Compensating Downward Motions Associated with Cumulus Clouds. J. Atmos. Sci., 24, 487-496.

Busse, F. H. (1967). On the Stability of Two-dimensional Convection in a Layer Heated from Below. J. Maths. & Phys., 46, 140-150.

Finlayson, B. A. (1972). The Method of Weighted Residuals and Variational Principles. Academic Press, N.Y..

Hood, P. (1977). Frontal Solution Program for Unsymmetric Matrices. Int. J. Num. Meth. Eng., 10, 379-399.

Irons, B. M. (1970). A Frontal Solution Program for Finite Element Analysis. Int. J. Num. Meth. Eng., 2, 5-32.

Lynn, P. P. and Arya, S. K. (1974). Finite Elements Formulated by the Weighted Discrete Least Squares Method. Int. J. Num. Meth. Eng., 8, 71-90.

Lynn, P. P. and Arya, S. K. (1973). Use of the Least Squares Criterion in Finite Element Formulation. Int. J. Num. Meth. Eng., 6, 75-89.

Mann, K. J. (1979). Finite Element Modelling of Atmospheric Convection. Proc. 3rd. Int. Conf. in Australia on Finite Element Methods, Sydney, July.

Steven, G. P. (1976). Dynamics of a Fluid Subject to Thermal and Gravity Diffusion. Proc. Int. Conf. Fin. Elem. Meth. Eng., Adelaide.

VISCOUS FLOW

Computational Techniques & Applications: CTAC-83
J. Noye & C. Fletcher (Editors)

A GENERALISED COORDINATE TIME-SPLIT FINITE ELEMENT METHOD FOR COMPRESSIBLE VISCOUS FLOW

K. Srinivas and C.A.J. Fletcher

Department of Mechanical Engineering,
University of Sydney,
Sydney. N.S.W. 2006
AUSTRALIA.

A generalised time-split finite element method is developed to solve Navier-Stokes equations for compressible flow. The method handles the non-rectangular elements in the physical domain by transforming the governing equations to the *generalised space* (ξ,η). Such an approach is found to be superior to the use of an isoparametric formulation. The method makes use of the *group* formulation which offers considerable advantages over a formal finite element approach. The generalised time-split finite element method is applied to laminar flow past symmetric and asymmetric trailing edges. Imposition of *non-reflecting boundary conditions* is another feature of the present work. A procedure called the "Variable Sweep" cycle is described and is found to contribute significantly towards saving in execution time.

1. INTRODUCTION

To have uniform rectangular elements in the flow domain is highly desirable from a computational point of view. Rectangular elements produce a simple form for the governing equations while a uniform mesh contributes significantly to the accuracy of the solution (Fletcher and Srinivas, 1984). There are a few flows like the flow past a flat plate or the flow past a backward-facing step, which do allow the prescription of a rectangular mesh. But there are many flows which require a non-rectangular mesh. The flow about a complete aerofoil or just the trailing edge of an aerofoil can be cited as examples. In these cases the mesh close to the body is far from being rectangular.

In many problems, there are certain regions where the gradients of the dependent variables are large and a fine mesh is required to achieve a good resolution of the flow. The near wake, the separated region and shock waves are some common examples. In these situations it is not economical to have a uniform fine mesh throughout the domain. It is a common practice to have a fine mesh in the regions of interest and a coarse mesh in regions where a uniform flow or an almost uniform flow (small gradients) occurs. In such a situation a non-uniform mesh cannot be avoided.

There are methods available in the literature to handle distorted and non-uniform meshes. The isoparametric formulation is frequently used in conjunction with finite element methods for this purpose. In the isoparametric formulation the evaluation of the algebraic coefficients is carried out in the transform plane. Although this idea is conceptually convenient it is computationally expensive and is inaccurate if the elements become too distorted (Strang and Fix, 1973).

In this study, we present a different approach to the problem. We consider the compressible flow past an aerofoil trailing edge which is governed by the Navier-Stokes equations. Instead of first applying the Galerkin formulation in the physical domain with a distorted mesh, we transform the governing equations to the *generalized coordinate space* (ξ,η). Then we apply the Galerkin formulation in the (ξ,η) space, where the mesh can be uniform and rectangular. Information about the physical domain is carried by the transformation Jacobian.

To solve the Navier-Stokes equations in generalized coordinates (ξ,η) we introduce a time-split finite element method which is outlined in a companion paper (Fletcher, 1983a). The method makes use of the *group* formulation (Fletcher, 1983b) where trial solutions are assumed for groups of dependent variables like ρuv. The advantages, particularly in relation to the economy in employing a group formulation, are detailed by Fletcher (1983b, 1983c). After discretisation the resulting set of ordinary differential equations are solved by a three-level time-split scheme.

The time-split method is employed here to compute steady, laminar flow past an aerofoil trailing edge which is a problem of practical interest. This work is a part of our continuing study of complex viscous flows (Srinivas and Fletcher, 1983).

Imposition of *non-reflecting* boundary conditions (Rudy and Strikwerda, 1981) at the boundaries is another feature of the present study. The application of this type of boundary condition (to be described later) absorbs the spurious pressure signals that reach the boundary thus producing faster convergence of the solution.

The example under consideration requires a large number of time steps (of the order of 1000) to give the steady-state solution; this leads to a large execution time. We present, here, a method in which the part of the flow domain where the flow variables undergo large changes is swept more often than the rest of the flow domain. This technique gives a significant gain in economy depending upon the nature of the flow field considered.

The rest of the paper is as follows. The governing equations in the Cartesian coordinates and their transformation to (ξ,η) space are described in Section 2. The time-split finite element method is outlined in Section 3. The application of the method to compute trailing-edge flows, the non-reflecting boundary conditions and a method to accelerate convergence of the solution are discussed in Section 4.

2. GOVERNING EQUATIONS

The flow to be investigated is governed by the steady, compressible Navier-Stokes equations. But we employ an *unsteady* approach to solve the problem, i.e. the unsteady, Navier-Stokes equations are integrated numerically until a steady state is reached. Time plays the role of an iteration parameter. In Cartesian coordinates the governing equations are:

$$\frac{\partial\bar{q}}{\partial t} + \frac{\partial\bar{F}}{\partial x} + \frac{\partial\bar{G}}{\partial y} = \frac{\partial^2\bar{R}}{\partial x^2} + \frac{\partial^2\bar{S}}{\partial x\partial y} + \frac{\partial^2\bar{T}}{\partial y^2} \tag{2.1}$$

where

$$\bar{q} = \{\rho\ ,\ \rho u,\ \rho v\}$$
$$\bar{F} = \{\rho u,\ p + \rho u^2,\ \rho uv\}$$
$$\bar{G} = \{\rho v,\ \rho uv,\ p + \rho v^2\} \tag{2.2}$$
$$\bar{R} = \{\theta^d_\rho\ \rho,\ \frac{4}{3}\ \mu u,\ \mu v\}$$
$$\bar{S} = \left\{0,\ \frac{\mu}{3}\ v,\ \frac{\mu}{3}\ u\right\}$$
$$\bar{T} = \left\{\theta^d_\rho\ \rho,\ \mu u,\ \frac{4}{3}\ \mu v\right\}$$

ρ = density, u,v = velocity components in x and y directions, p = pressure, μ = molecular viscosity.

We assume that the temperature changes in the solution domain are negligible which eliminates the energy equation from consideration. Further, the coefficient of viscosity, μ, is assumed to be constant. Thus pressure can be calculated from the relation

$$p = \rho\left\{RT_o - \frac{\gamma-1}{2\gamma}\ (u^2 + v^2)\right\} \tag{2.3}$$

The governing equations are non-dimensionalized with respect to the free stream velocity U_∞, characteristic length L and free stream density ρ_∞. The non-dimensional equations retain the same form except that 1/Re replaces μ. Re is the Reynolds number and is given by $\rho_\infty U_\infty L/\mu$. The non-dimensional pressure equation assumes the following form:

$$1 + \gamma M_\infty^2\ p = \rho\left\{1 + \frac{\gamma-1}{2}\ M_\infty^2\ (1 - u^2 - v^2)\right\} \tag{2.4}$$

2.1 Transformation of the Governing Equations

The isoparametric formulation effectively applies the finite element method on distorted elements in the physical plane even though evaluation of the algebraic coefficients, by numerical integration, is carried out on uniform elements in the transform plane. But transformation of the governing equations to the (ξ,η) space allows the finite element method to be applied on a uniform grid in the transform plane. As a consequence, greater accuracy is achieved and the computationally expensive numerical integration is avoided.

We introduce a general transformation -

$$\xi = \xi(x,y)\quad ,\quad \eta = \eta(x,y) \tag{2.5}$$

under which

$$u_x = \xi_x\ u_\xi + \eta_x\ u_\eta,\quad \text{etc.} \tag{2.6}$$

when Eqns.(2.1) are transformed to the (ξ,η) space one obtains

$$\bar{q}^*_t + \bar{F}^*_\xi + \bar{G}^*_\eta - \bar{R}^*_{\xi\xi} - \bar{S}^*_{\xi\eta} - \bar{T}^*_{\eta\eta} = 0 \tag{2.7}$$

where

$$\bar{q}^* = \frac{q}{J}\quad ,$$

$$\bar{F}^* = \frac{1}{J}\begin{Bmatrix} \rho U_c \\ \xi_x p + \rho u U_c + \left(\frac{4\mu}{3}\xi_{xx} + \mu\xi_{yy}\right)u + \left(\frac{\mu}{3}\xi_{xy}\right)v \\ \xi_y p + \rho v U_c + \left(\mu\,\xi_{xx} + \frac{4\mu}{3}\xi_{yy}\right)v + \left(\frac{\mu}{3}\xi_{xy}\right)v \end{Bmatrix}$$

$$\bar{G}^* = \frac{1}{J}\begin{Bmatrix} \rho V_c \\ \eta_x p + \rho u V_c + \left(\frac{4\mu}{3}\eta_{xx} + \mu\eta_{yy}\right)u + \left(\frac{\mu}{3}\eta_{xy}\right)v \\ \eta_y p + \rho v V_c + \left(\mu\eta_{xx} + \frac{4\mu}{3}\eta_{yy}\right)v + \left(\frac{\mu}{3}\eta_{xy}\right)u \end{Bmatrix}$$

$$\bar{R}^* = \frac{1}{J}\begin{Bmatrix} \rho \\ \left(\frac{4\mu}{3}\xi_x^{\,2} + \mu\xi_y^{\,2}\right)u + \frac{\mu}{3}\xi_x\xi_y\,v \\ \left(\mu\xi_x^{\,2} + \frac{4}{3}\mu\xi_y^{\,2}\right)v + \frac{\mu}{3}\xi_x\xi_y\,u \end{Bmatrix} \qquad (2.8)$$

$$\bar{S}^* = \frac{1}{J}\begin{Bmatrix} 0 \\ 2\left(\frac{4\mu}{3}\xi_x\eta_x + \mu\xi_y\eta_y\right)u + \frac{\mu}{3}(\xi_x\eta_y + \xi_y\eta_x)v \\ 2\left(\mu\xi_x\eta_x + \frac{4\mu}{3}\xi_y\eta_y\right)v + \frac{\mu}{3}(\xi_y\eta_x + \eta_y\xi_x)u \end{Bmatrix}$$

$$\bar{T}^* = \frac{1}{J}\begin{Bmatrix} \rho \\ \left(\frac{4\mu}{3}\eta^2{}_x + \mu\eta^2{}_y\right)u + \left(\frac{\mu}{3}\eta_x\eta_y\right)v \\ \left(\mu\eta^2{}_x + \frac{4}{3}\mu\eta^2{}_y\right)v + \left(\frac{\mu}{3}\eta_x\eta_y\right)u \end{Bmatrix}$$

In the above equations J is the Jacobian of the transformation. U_c and V_c are the contravariant velocities along the ξ and η directions and are given by

$$U_c = \xi_x u + \xi_y v \text{ and } V_c = \eta_x u + \eta_y v \qquad (2.9)$$

The structural similarity of Eqns. (2.7) in the transform plane and Eqns. (2.1) in the physical plane is noteworthy and is exploited by the group formulation (Section 3).

3. TIME-SPLIT FINITE ELEMENT METHOD

An application of the group Galerkin finite element method (Fletcher, 1983b) with linear Lagrange elements gives the following set of ordinary differential equations:

$$M_\xi \otimes M_\eta \bar{q}^*_t + M_\eta \otimes L_\xi \bar{F}^* + M_\xi \otimes L_\eta \bar{G}^* - M_\eta \otimes L_{\xi\xi} \bar{R}^* - L_\xi \otimes L_\eta \bar{S}^* - M_\xi \otimes L_{\eta\eta} \bar{T}^* = 0 \tag{3.1}$$

In Eqn. (3.1) the components of $\bar{F}^*$, $\bar{G}^*$ etc. are given by Eqn. (2.8). The directional mass and difference operators are defined as follows:

$$M_\xi \equiv \left\{\frac{1}{6}, \frac{1+r_\xi}{3}, \frac{r_\xi}{6}\right\}, \quad M^t_\eta \equiv \left\{\frac{r_\eta}{6}, \frac{1+r_\eta}{3}, \frac{1}{6}\right\} \tag{3.2}$$

$$L_\xi \equiv \left\{-1,0,1\right\}/2\Delta\xi, \quad L^t_\eta \equiv \left\{1,0,-1\right\}/2\Delta\eta$$

$$L_{\xi\xi} \equiv \left\{1, -(1+\frac{1}{r_\xi}), \frac{1}{r_\xi}\right\}/\Delta\xi^2, \quad L^t_{\eta\eta} \equiv \left\{\frac{1}{r_\eta}, -(1+\frac{1}{r_\eta}), 1\right\}/\Delta\eta^2$$

These operators are appropriate to a non-uniform grid in the (ξ,η) plane (Fig. 1). The relevance of these operators is discussed by Fletcher (1983a).

A time-split algorithm to solve Eqn. (3.1) is developed as follows. On introducing a general three-level evaluation of the time derivative, Eqn. (3.1) takes the form:

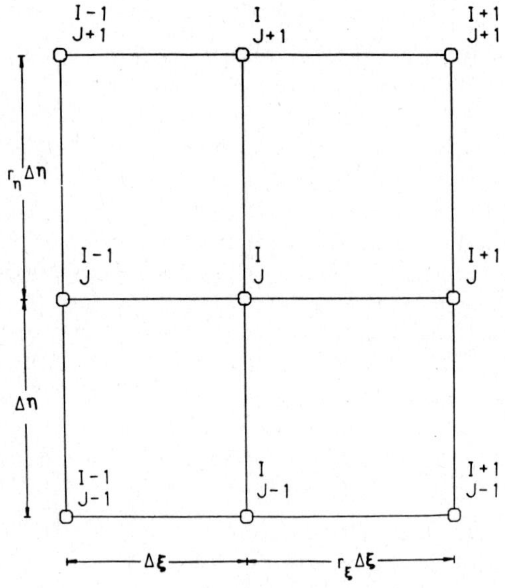

Figure 1
Nodal Geometry

$$M_\xi \otimes M_\eta \left[\alpha \frac{\Delta\bar{q}^{*n+1}}{\Delta t} + (1-\alpha)\frac{\Delta\bar{q}^{*n}}{\Delta t}\right] = \beta\, RHS^{n+1} + (1-\beta)\, RHS^{n}$$

where

$$RHS = M_\eta \otimes L_{\xi\xi}\bar{R}^* + L_\xi \otimes L_\eta \bar{S}^* + M_\xi \otimes L_{\eta\eta}\bar{T}^* - M_\eta \otimes L_\xi \bar{F}^* - M_\xi \otimes L_\eta \bar{G}^* \qquad (3.3)$$

In Eqn. (3.3) α and β are free parameters. The choice $\alpha = 1$ and $\beta = \frac{1}{2}$ gives the Crank-Nicolson scheme while the choice $\alpha = 3/2$ and $\beta = 1$ gives the three-level inplicit scheme used in the present study. Fletcher (1983a) gives a comparison of the two algorithms for a closely related problem.

Equation (3.3) for $\Delta\bar{q}^*$, i.e. $\{\frac{\Delta\rho}{J}, \frac{\Delta u}{J}, \frac{\Delta v}{J}\}$, is nonlinear and hence a Taylor expansion of the equation about the nth time level is introduced. Thus,

$$(RHS)^{n+1} = (RHS)^n + \frac{\partial}{\partial\bar{q}^*}(RHS)\,\Delta\bar{q}^{*n+1} \qquad (3.4)$$

where

$$\Delta\bar{q}^{*n+1} = \bar{q}^{*n+1} - \bar{q}^{*n} \qquad (3.5)$$

Equation (3.3) is now written as

$$M_\xi \otimes M_\eta\left[\alpha\,\Delta\bar{q}^{*n+1}\right] - \beta\Delta t\left[M_\eta \otimes L_{\xi\xi}\frac{\partial\bar{R}^*}{\partial\bar{q}^*} + M_\xi \otimes L_{\eta\eta}\frac{\partial\bar{T}^*}{\partial\bar{q}^*} - M_\eta \otimes L_\xi\frac{\partial\bar{F}^*}{\partial\bar{q}^*} - M_\xi \otimes L_\eta\frac{\partial\bar{G}^*}{\partial\bar{q}^*}\right]\Delta\bar{q}^{n+1}$$
$$= \Delta t\,(RHS)^n + \beta\,\Delta t\, L_\xi \otimes L_\eta\frac{\partial\bar{S}^*}{\partial\bar{q}^*}\,.\,\Delta\bar{q}^{*n+1} - (1-\alpha)M_\xi \otimes M_\eta\,\Delta\bar{q}^{*n} \qquad (3.6)$$

The term, $\beta\,\Delta t.L_\xi \otimes L_\eta \frac{\partial\bar{S}^*}{\partial\bar{q}^*}\;\Delta\bar{q}^{*n+1}$ in eq. (3.6) is due to cross derivatives and it does not have any contribution from the implicit grid lines (i,J). Hence the term is calculated explicitly by replacing $\Delta\bar{q}^{*n+1}$ by $\Delta\bar{q}^{*n}$.

Further, the addition of the following term of $O(\Delta t^2)$,

$$\frac{\beta^2\Delta t^2}{\alpha}\left(L_{\xi\xi}\frac{\partial\bar{R}^*}{\partial\bar{q}^*} - L_\xi\frac{\partial\bar{F}^*}{\partial\bar{q}^*}\right)\otimes\left(L_{\eta\eta}\frac{\partial\bar{T}^*}{\partial\bar{q}^*} - L_\eta\frac{\partial\bar{G}^*}{\partial\bar{q}^*}\right), \qquad (3.7)$$

to the left hand side of eqn.(3.6) enables the time-split algorithm to be constructed. Such an algorithm is implemented as

$$\left[M_\xi - \frac{\beta}{\alpha}\Delta t\left(L_{\xi\xi}\frac{\partial\bar{R}^*}{\partial\bar{q}^*} - L_\xi\frac{\partial\bar{F}^*}{\partial\bar{q}^*}\right)\right]\Delta\bar{q}^{*i} = \frac{\Delta t}{\alpha}(RHS)^A - \left(\frac{1-\alpha}{\alpha}\right)M_\xi \otimes M_\eta\Delta\bar{q}^{*n}$$

$$\left[M_\eta - \frac{\beta}{\alpha}\Delta t\left(L_{\eta\eta}\frac{\partial\bar{T}^*}{\partial\bar{q}^*} - L_\eta\frac{\partial\bar{G}^*}{\partial\bar{q}^*}\right)\right]\Delta\bar{q}^{*n+1} = \Delta\bar{q}^{*i} \ , \tag{3.8}$$

where

$$(RHS)^A = (RHS)^n + \beta.L_\xi \otimes L_\eta \frac{\partial\bar{S}^*}{\partial\bar{q}^*}\Delta\bar{q}^{*n} \tag{3.9}$$

In eqns.(3.8) contributions to $\bar{R}^*$ etc., arising from the transformation parameters ξ_x, η_x etc, are evaluated by using the relationships -

$$\xi_x = Jy_\eta, \ \eta_x = -Jy_\xi, \ \xi_y = -Jx_\eta \text{ and } \eta_y = Jx_\xi \tag{3.10}$$

where J is the Jacobian of the transformation and is given by

$$J = 1/(x_\xi y_\eta - x_\eta y_\xi) \ . \tag{3.11}$$

The following expressions have been used to evaluate x_ξ etc.

$$x_\xi = (x_{\substack{i+1\\j}} - x_{\substack{i-1\\j}})/(1 + r_\xi)\,\Delta\xi, \ x_\eta = (x_{\substack{i\\j+1}} - x_{\substack{i\\j-1}})/(1 + r_\eta)\Delta\eta$$

$$x_{\xi\xi} = \frac{2}{(1+r_\xi)\Delta\xi^2}(x_{\substack{i-1\\j}} - (1 + \frac{1}{r_\xi})\,x_{\substack{i\\j}} + x_{\substack{i+1\\j}}/r_\xi)$$

$$x_{\eta\eta} = \frac{2}{(1+r_\eta)\Delta\eta^2}(x_{\substack{i\\j-1}} - (1 + \frac{1}{r_\eta})x_{\substack{i\\j}} + x_{\substack{i\\j+1}}/r_\eta) \tag{3.12}$$

$$\text{and } x_{\xi\eta} = \frac{1}{(1+r_\xi)(1+r_\eta)\Delta\xi\,\Delta\eta}(x_{\substack{i+1\\j+1}} - x_{\substack{i-1\\j+1}} - x_{\substack{i+1\\j-1}} + x_{\substack{i-1\\j-1}})$$

Equivalent expressions can be written for y_ξ etc.

It may be noted that eqns.(3.8) are a decoupled implicit local system of equations associated with each grid line in the ξ and η directions. The block tridiagonal system is solved in O(N) operations and there is no need for global factorisation at each time step.

The term $(RHS)^A$ in eqn.(3.8) approaches zero as the steady state is approached and it provides a measure of the 'closeness' of the solution to the steady state. In the present study the steady state was assumed to have been reached when $(RHS)^A$ was reduced to less than 10^{-6}.

4. RESULTS AND DISCUSSION

Here we apply the time-split finite element method described in previous sections to laminar, compressible flow past a trailing edge. Both symmetric and asymmetric cases are considered. (See Fig. 2). All computations are carried out at a Reynolds number of 100 and a free stream Mach number of 0.4.

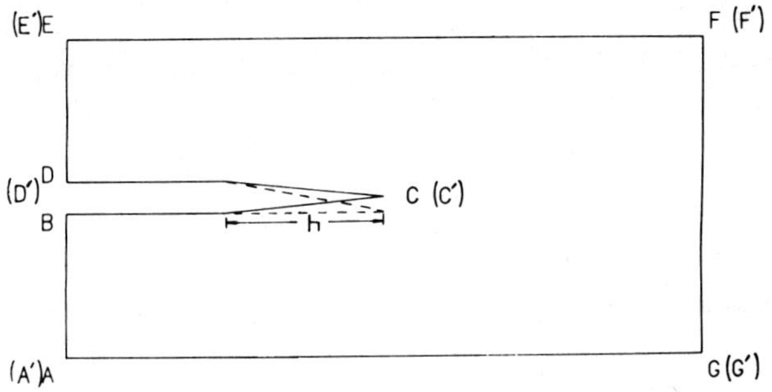

Figure 2
Geometry of the trailing edge. (— symmetric, --- asymmetric)

The rectangle AGFE in the physical domain is transformed to another rectangle A'G'F'E' in the transform plane. The wedge BCD is transformed to a straight line D'C'. Region AGFE is divided into 34 x 42 linear elements. The line DC occupies 14 elements. Mesh spacing is uniform in the y-direction and the mesh width is 10% of the boundary layer thickness at D. The mesh in the x-direction is non-uniform, the smallest being at C and of width 10% of the boundary layer thickness at D. The mesh width is increased in both the upstream and downstream directions from C in a geometric progression by a factor of 1.1. Such a mesh is necessary to achieve a good resolution of the near-wake flow i.e. region close to C. A non-uniform, rectangular mesh is used in the ξ and η directions in the transform plane.

4.1 Boundary Conditions:

The boundary conditions applied are as follows (Fig. 2)

On BCD, $$u = v = 0 \tag{4.1}$$

On AB and DE, velocity components u and v are specified. Pressure is calculated from the extrapolation of the outgoing characteristic variable (Rudy and Strikwerda, 1981)

$$p_1^{n+1} = p_2^{n+1} - \rho_1^n C_1^n (u_2^{n+1} - u_1^{n+1}) \tag{4.2}$$

where C is the speed of sound.

On the freestream boundaries EF and AG, values of ρ and u are prescribed at their non-dimensional freestream values, i.e. 1.0.

On FG, the outflow boundary the following conditions on u and v are applied,

$$\frac{\partial^2 u}{\partial x^2} = \frac{\partial^2 v}{\partial x^2} = 0 \tag{4.3}$$

Pressure is calculated from the nonreflecting condition,

$$p_N^{n+1} = [\, p_N^n + \alpha \Delta t p_\infty + p_N^n c_N^n (u_N^{n+1} - u_N^n)]\,\frac{1}{1+\beta\Delta t}$$

This nonreflecting boundary condition is discussed in detail by Rudy and Strikwerda (1981). One of the properties of this condition is that it reduces reflections at the outflow boundary of the spurious disturbances generated by the transient solution. As a result a faster convergence is achieved. β is set equal to 0.3 after Rudy and Strikwerda (1981).

4.2 Symmetric Trailing Edge:

The u-velocity profiles for various x-stations are given in Fig. 3. A gradual development of the wake profile (downstream of the trailing edge) from the boundary

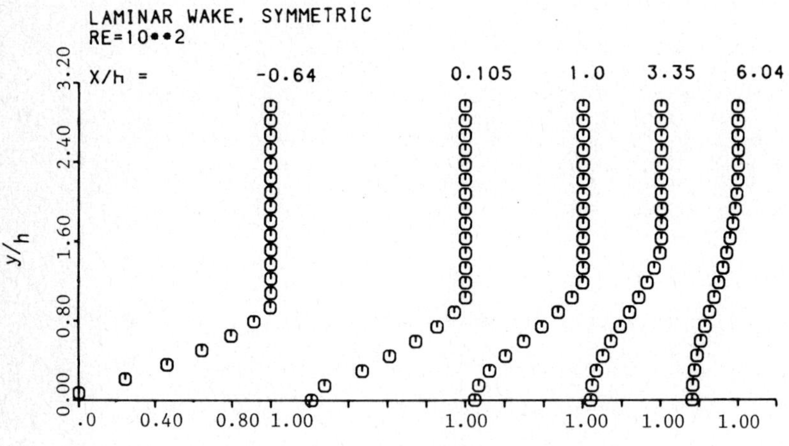

Figure 3
u-velocity profiles for the Symmetric Trailing Edge

layer profile (upstream of the trailing edge) is noticed. In the near wake changes in velocity are concentrated closer to the centreline whereas in the far wake changes are observed all along the profile.

4.3. Asymmetric Trailing Edge:

Figure 4 shows the velocity profiles for the asymmetric trailing edge. Any effect of asymmetry of the trailing edge geometry is strongly felt in the near wake. As the far wake is approached the velocity profile assumes more of a symmetric character.

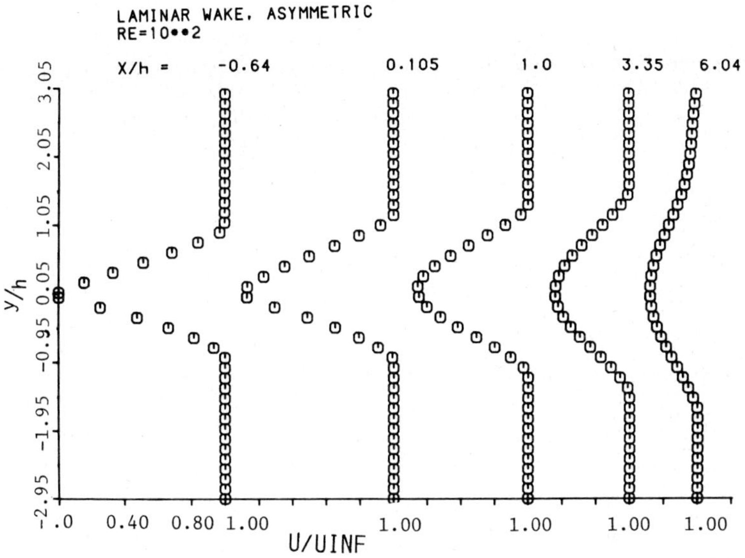

Figure 4
u-velocity profiles for the Asymmetric Trailing Edge

4.4 Acceleration of Convergence:

In the flow examples considered the trailing edge effectively controls the flow behaviour. The disturbances originating at the trailing edge are subsequently communicated to the rest of the flow field. A study of $(RHS)^A$ (see eq. (3.8)) for each of the mesh points, during convergence, indicates that it is a maximum near the trailing edge and decreases towards the freestream and downstream boundaries. Therefore, it appears that the flow field close to the trailing edge has to undergo substantial changes before it can reach a steady state. The flow field near the freestream boundaries requires relatively minor changes.

Computationally, this implies that the region near the trailing edge requires more iterative sweeps than those near the freestream boundary. A procedure wherein the region close to the trailing edge is swept more often than regions near the freestream suggests itself. Such a procedure could contribute significantly to a reduction in the execution time.

In order to test such a procedure and its effect on the convergence properties, the transformation plane E'F'G'A'(Fig. 5) was divided into three parts - an inner

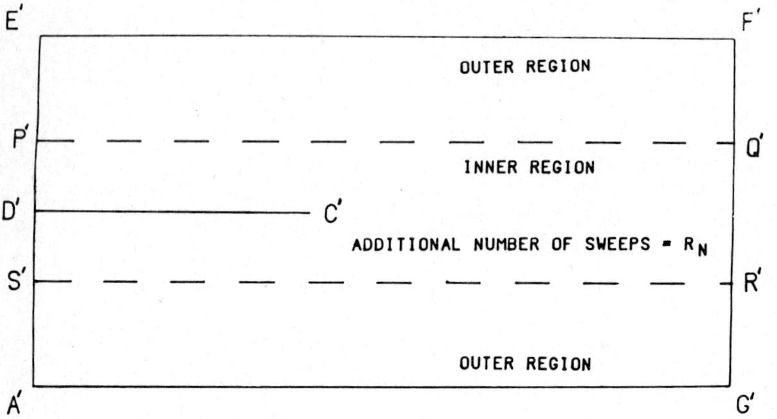

Figure 5
Inner and outer regions for 'variable sweep' cycle

region P'Q'R'S' of height two boundary layer thicknesses (P'D' = D'S' = one boundary layer thickness) and two outer regions, E'F'Q'P' and S'R'G'A', adjacent to the freestream boundaries. Starting with a sweep in the entire domain E'F'G'A' in the first iteration, the next two iterations were limited to the inner region, P'Q'R'S'. This cycle of one full sweep and two inner sweeps was continued until convergence was achieved. The rms value of $(RHS)^A$ is plotted against the number of time steps in Fig. 6. We observe that the convergence is as good and sometimes even better than that for the case where the entire domain E'F'G'A' is swept at every time step. Only in the first few cycles, the convergence seems to be slower. But after about 400 time steps the cycle of "variable sweeps" demonstrates better convergence properties. Figure 6 also shows the result obtained when the inner region is swept four times for every sweep in the entire domain E'F'G'A'. Although the solution converges, the rate of convergence, measured in terms of number of time steps, is less than that for the two sweep cycle. But it should be noted that the execution time is reduced considerably in a four sweep cycle.

A measure of the saving in execution times achieved by using the "variable sweep" is given by

$$\frac{R_N(1 - R_S)}{1 + R_N} \qquad (4.5)$$

where R_S is the ratio of the number of elements in P'Q'R'S' to that in E'F'G'A'. R_N is the number of additional sweeps carried out in the inner layer i.e. P'Q'R'S'. When $R_N = 2$, $R_S = 1/2$ the saving in execution time is 1/3 and when $R_N = 4$ and $R_S = 1/2$ saving is 2/5. The "variable sweep" cycle procedure thus offers a considerable saving in execution time. But its applicability depends strongly on the particular flow problem under consideration. More work is needed before a general prescription of the procedure can be made.

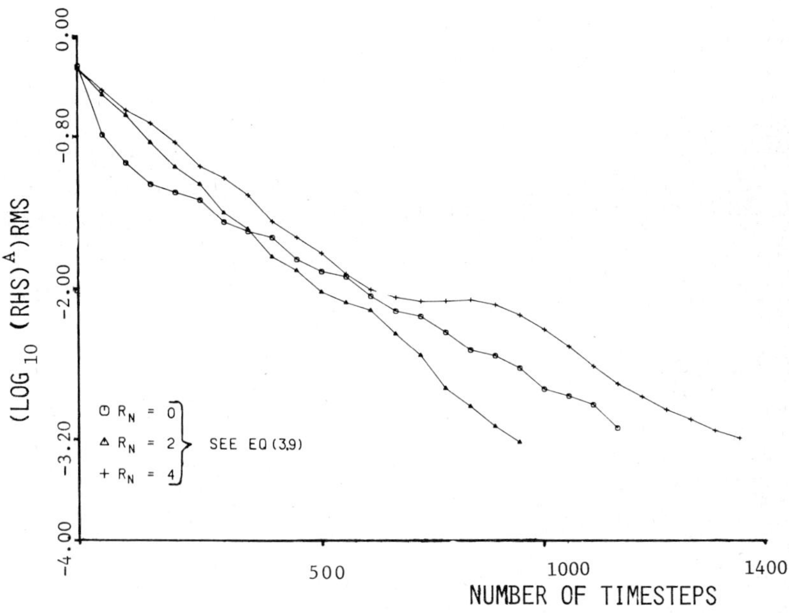

Figure 6
Convergence Pattern for the 'variable sweep' cycles

5. CONCLUSION

The generalised coordinate, time-split group finite element method described in Section 3 is found to be effective in handling flows in physical domains which require an irregular mesh. It is possible to accelerate the convergence of the solution by employing the 'variable sweep' cycle. The time-split, group finite element method is now being applied to turbulent flow past trailing edges and the results obtained are encouraging.

6. ACKNOWLEDGEMENT

The authors are grateful to the Australian Research Grants Scheme for their continued financial support.

7. REFERENCES

Fletcher, C.A.J. and Srinivas, K. (1984), "Stream Function Vorticity Revisited," *Comp. Meth. Appl. Mech. Engn.*, (to appear).

Fletcher, C.A.J. (1983a), "Time-splitting and the Group Finite Element Formulation," Proc. CTAC-83 (ed. J. Noye) North-Holland, Amsterdam.

Fletcher, C.A.J., (1983b), "The Group Finite Element Formulation," *Comp. Meth. Appl. Mech. Engn.*, 37, 225-243.

Fletcher, C.A.J., (1983c), "*Computational Galerkin Methods*", Springer-Verlag, Heidelberg.

Rudy, D.H. and Strikwerda, (1981), "Boundary Conditions for Subsonic, Compressible, Navier-Stokes Calculations," *Computers and Fluids*, 9, 327-338.

Srinivas, K. and Fletcher, C.A.J. (1983), "Finite Element Solutions for Laminar and Turbulent Compressible Flow," *International Journal for Numerical Methods in Fluids* (to appear).

Strang, G. and Fix, G.J. (1973), "*An Analysis of the Finite Element Method*," Prentice-Hall, N.J.

Computational Techniques & Applications: CTAC-83
J. Noye & C. Fletcher (Editors)

FINITE ELEMENT ANALYSIS OF VISCOUS FLOWS

G. A. Mohr,
Department of Theoretical and Applied Mechanics,
University of Auckland,
Auckland, New Zealand.

A finite element model for two dimensional viscous flows is developed using the virtual work method. The model is in part based on a finite element shell model which uses the same reduced integration of quadratic interpolations for all variables [1]. Differences from preceding formulations are that integration by parts is applied to the continuity equation, yielding different loading terms which are more easily defined in some problems, and the convective inertia terms are distributed to both sides of the nonlinear recurrence relation. In the case of compressible flow, for which comparatively few formulations have been proposed to date, the thermal energy equation is used to form a two stage iterative solution.

INTRODUCTION

Finite element formulations for the analysis of viscous fluid flows employ three basic approaches [2]

(a) stream functions [3,4,5]

(b) stream and vorticity functions [6]

(c) formulations with velocity and pressure freedoms [7-10]

The disadvantage of the first two approaches is that boundary conditions for stream functions [11] and vorticity freedoms [2] involve some complications. The physically natural freedoms of the third approach simplify specification of boundary conditions and this approach has thus proved the most popular [12]. It can be derived by either the Galerkin [12] or virtual work methods [13], these methods being equivalent [9].

Remarkably the six node quadratic triangle was chosen in all the first published uvp solutions for viscous flow [7-10] but doubts existed on whether linear or quadratic interpolation should be used for the pressure. The problem here is that quadratic pressure interpolation leads to six incompressibility constraint equations and as these involve first derivatives of the velocities the six velocity sampling points are insufficient to provide six independent equations. As a consequence one sometimes obtains a spurious 'checkerboard' mode in the pressure solutions, i.e. these exhibit sometimes large oscillations about the correct solution which are slow to converge out, if at all [14, 17-21]. Taylor and Hood [8], Kawahara et al. [9] and Yamada et al. [10] use 'mixed interpolation' to overcome this difficulty (i.e. quadratic interpolation for the velocities and linear interpolation for the pressures).

Another approach is to use manipulation of the momentum equations to relegate the pressure freedoms to a separate Poisson equation. Then once the velocity solution has been obtained the pressures are able to be calculated at the same sampling points. Oden and Carter-Wellford [7] used this approach successfully with the quadratic triangle and consistent interpolation but introduced an additional Neumann boundary condition $\partial p/\partial n = 0$ on stationary walls. Note, however, that the problems they dealt with (those of Couette flow, Blasius flow and hydrodynamic lubrication) are all pseudo one dimensional [15,16] and thus do not provide conclusive evidence regarding the checkerboard effect. Olson and Tuann [17], however, find such a "segregated" solution successful without the use of Neumann boundary conditions for the pressures. In their work an Hermitian cubic element (with consistent interpolation) is used (with which the $\partial p/\partial n = 0$ boundary condition could easily have been included if required) and a square cavity problem in which checkerboard pressures are observed (unless mixed interpolation is used) is studied.

Segregated formulations can also be obtained by the use of penalty factors [18-21] and mixedinterpolation [18,19] or selectively reduced integration [20] of the penalty contributions can be used in these. Most of these formulations still fail to remove the checkerboard effect [18-20] but Engelman et al. [21] obtain penalty factor formulations with consistent integration which remove the spurious pressure modes.

Another approach is to apply "smoothing" and "filtering" techniques to extract satisfactory pressures [14,22] from the polluted (by the checkerboard effect) numerical results obtained with consistent interpolation and Sani et al. [22] find that penalty methods themselves have an "automatic" filtering effect.

A similar dilemma to that experienced with the pressure freedoms in uvp flow formulations was posed by the use of a quadratic element for the analysis of thick plates [23]. In this instance, though, reduced interpolation would have to be applied to two of the three variables (i.e. the slopes) rather than one and as a consequence of this, quadratic interpolation of all variables was found to give clearly better results.

This precedent is followed at the outset of the present work, using a quadratic integration at three internal points [24] which has proved useful in shell analysis with the quadratic triangle because it yields a good approximation to cubic and quartic terms, whereas integration at the midside nodes, thought to be the "best" integration points [13], yields poor approximations for these terms [24].

STOKESIAN FLOWS

Assuming incompressible and isothermal conditions (and thus constant viscosity μ) the Navier-Stokes equations for two dimensional viscous flows may be expressed as [15]

$$Du/Dt = X - \partial p/\partial x + \mu(\partial^2 u/\partial x^2 + \partial^2 u/\partial y^2) \quad (1)$$

$$Dv/Dt = Y - \partial p/\partial y + \mu(\partial^2 v/\partial x^2 + \partial^2 v/\partial y^2) \quad (2)$$

where X,Y are the body forces and the material or total derivatives Du/Dt, Dv/Dt may be expanded as

$$Du/Dt = \partial u/\partial t + (\partial u/\partial x)(\partial x/\partial t) + (\partial u/\partial y)(\partial y/\partial t)$$

$$= \partial u/\partial t + (\partial u/\partial x)u + (\partial u/\partial y)v \quad (3)$$

$$Dv/Dt = \partial v/\partial t + (\partial v/\partial x)u + (\partial v/\partial y)v \quad (4)$$

For steady state slow creeping Stokesian flows both the time dependent terms and the convective inertia terms (i.e. $u\partial u/\partial x$ etc) may be neglected, so that the left hand sides of Eqns (1) and (2) vanish. In addition one requires for conservation of mass

$$\partial\rho/\partial t + \nabla\cdot(\rho V) = \partial\rho/\partial t + \partial(\rho u)/\partial x + \partial(\rho v)/\partial y = 0 \tag{5}$$

where ρ is the density and $V = u\hat{i} + v\hat{j}$, which reduces to

$$\partial u/\partial x + \partial v/\partial y = 0 \tag{6}$$

for steady state incompressible flow.

Now allowing an arbitrary variation δu one multiplies Eqn (1) by this and integrates over the element volume to obtain the corresponding virtual work. Now integrating the second derivates by parts e.g.

$$\mu\int\int\delta u(\partial^2 u/\partial x^2)dxdy = \mu\int\delta u(\partial u/\partial x)dy - \mu\int\int(\partial\delta u/\partial x)(\partial u/\partial x)dxdy \tag{7}$$

and using $\delta u = \{f\}^t\{\delta u\}$, $\partial u/\partial x = \{\partial f/\partial x\}^t\{u\} = \{f_x\}^t\{u\}$

$$\partial\delta u/\partial x = \{f_x\}^t\{\delta u\} \tag{8}$$

(where $\{f\}$ is the vector of interpolation functions) one obtains

$$\{\delta u\}^t\int\int[\mu(\{f_x\}\{f_x\}^t + \{f_y\}\{f_y\}^t \{u\} + \{f\}\{f_x\}^t\{p\}]dxdy$$
$$= \{\delta u\}^t\int\int\{f\}Xdxdy + \mu\{\delta u\}^t\int\{f\} (\partial u/\partial x)dy + \mu\{\delta u\}^t\int\{f\}(\partial u/\partial y)dx \tag{9}$$

and the arbitrary common factor $\{\delta u\}^t$ is deleted.

Note that only integration by parts, not the Green-Gauss theorem, has been used. This seems a more fundamental approach and leads to a clearer interpretation of the boundary integrals. To this end, note that

$$\partial x/\partial n = c_x,\quad \partial y/\partial n = c_y,\quad \partial x/\partial s = c_y,\quad \partial y/\partial s = c_x \tag{10}$$

where c_x, c_y are the direction cosines of the normal to the boundary relative to the x and y axes respectively. Thus the last two terms of Eqn (9) may be written (after deletion of $\{\delta u\}$)

$$\mu\int\{f\}(\partial u/\partial x)dy + \mu\int\{f\}(\partial u/\partial y)dx = \mu\int\{f\}(\partial u/\partial x)c_x dS + \mu\int\{f\}(\partial u/\partial y)c_y ds \tag{11}$$

where $\partial u/\partial n = (\partial u/\partial x)(\partial x/\partial n) + (\partial u/\partial y)(\partial y/\partial n) = (\partial u/\partial x)c_x + (\partial u/\partial Y)c_Y$ (12)

so that Eqn (9) finally reduces to

$$\int\int[\mu(\{f_x\}\{f_x\}^t + \{f_y\}\{f_y\}^t)\{u\} + \{f\}\{f_x\}^t\{p\}]dxdy = \int\int\{f\}Xdxdy$$
$$+ \mu\int\{f\}(\partial u/\partial n)dS \tag{13}$$

and it is often more convenient to return to Eqn (11) to interpret the last term of Eqn (13).

Note that, having deleted the arbitrary variations δu, Eqn (13) is exactly that obtained by the Galerkin method, i.e. virtual work with $\delta u = \{\delta u\}^t\{f\}$ is equivalent to the Galerkin method with residuals weighted by $\{f\}$.

Allowing a variation δv and applying this to Eqn (2) in the same fashion as above one obtains

$$\iint[\mu(\{f_x\}\{f_x\}^t + \{f_y\}\{f_y\}^t \{v\} + \{f\}\{f_y\}^t\{p\}]dxdy = \iint\{f\}Ydxdy + \mu\int\{f\}(\partial v/\partial n)dS \qquad (14)$$

Allowing also a variation $\delta p = \{\delta p\}^t\{f\}$ and applying this to the continuity equation (Eqn (6)) and integrating each term by parts one obtains after deleting $\{\delta p\}$

$$\iint\{f\}(\partial u/\partial x)dxdy = \int\{f\}udy - \iint\{f_x\}\{f\}^t\{u\}dxdy \qquad (15)$$

$$\iint\{f\}(\partial v/\partial y)dxdy=\int\{f\}vdx - \iint\{f_y\}\{f\}^t\{v\}dxdy \qquad (16)$$

Combining Eqns (15) and (16) and applying Eqns (10) one obtains

$$-\iint[\{f_x\}\{f\}^t\{u\} + \{f_y\}\{f\}^t\{v\}]dxdy = -\int\{f\}V_n dS \qquad (17)$$

where V_n is the velocity perpendicular to the boundary and is positive if directed outwards. Writing Eqns (13), (14) and (17) in matrix form one obtains the final formulation for Stokes flow as

$$\iint \begin{bmatrix} C & 0 & -A \\ 0 & C & -B \\ -A^t & -B^t & 0 \end{bmatrix} dxdy \begin{Bmatrix} \{u\} \\ \{v\} \\ \{-p\} \end{Bmatrix} = \int \begin{Bmatrix} \{f\}X \\ \{f\}Y \\ 0 \end{Bmatrix} dxdy + \int \begin{Bmatrix} \mu\{f\}(\partial u/\partial n \\ \mu\{f\}(\partial v/\partial n) \\ \{f\}v_n \end{Bmatrix} ds \qquad (18)$$

where $A = \{f\}\{f_x\}^t$, $B = \{f\}\{f_y\}^t$

$$C = \mu[\{f_x\}\{f_x\}^t + \{f_y\}\{f_y\}^t]$$

using $\{-p\}$ as freedoms to obtain a symmetric result. Note in passing that the viscosity matrix of Eqn (18) takes the form associated with the use of Lagrange multipliers [25], the pressure freedoms acting as such to enforce the continuity condition.

Eqn (18) differs from the results given in Refs [10,12,13] where pressure terms in Eqns (1) and (2) are integrated by parts (i.e. the pressure column entries in Eqn (18) are transposed - this should lead to pressure loading terms which in fact do not appear in these works). Refs [2,7] on the other hand, obtain the same results for these pressure terms (i.e. A and B in Eqn (18)) but do not obtain $\int\{f\}V_n dS$ loading terms and do not require negation of the pressure freedoms. A difference from preceding works is the velocity loading terms in Eqn (18) involving v_n. These are no inconvenience as when the velocities are known at entry or exit (usually the former) the values set as boundary conditions on the left hand side are also used to evaluate the $\int\{f\}V_n dS$ forcing term on the right.

CONVECTIVE AND TRANSIENT FLOWS

The usual method of incorporating the convective terms of Eqns (3) and (4) into Eqn (18) is to directly integrate them and add the nonlinear term

$$\rho[\{f\}u_n\{f_x\}^t + \{f\}v_n\{f_y\}^t] \tag{19}$$

to C in Eqn (18). [2,12] Here u_n, v_n are scalar integration point values determined at the last cycle of an iterative solution. The result is an unsymmetric matrix, doubling the computation requirements.

Alternatively one can use the derivatives of u, v as the interpolated scalar multipliers, i.e. the virtual work of the first convective term of Eqn (3) is written as

$$\{\delta u\}^t \int\{f\}\rho u(\partial u/\partial x) = \{\delta u\}^t \rho\int\{f\}\{f\}^t\{u\}(\partial u/\partial x)_n dxdy \tag{20}$$

so that the convection matrix

$$\begin{bmatrix} D(\partial u/\partial x)_n & D(\partial u/\partial y)_n & 0 \\ D(\partial v/\partial x)_n & D(\partial v/\partial y)_n & 0 \\ 0 & 0 & 0 \end{bmatrix} \tag{21}$$

where $D = \iint\rho\{f\}\{f\}^t dxdy$ can be added to the left hand side of Eqn (18).

The result is still unsymmetric but this can be remedied by shifting the offending terms to the right hand side, i.e. the nonlinear correction added to Eqn (18) is then

$$\begin{bmatrix} D(\partial u/\partial x)_n & 0 & 0 \\ 0 & D(\partial v/\partial y)_n & 0 \\ 0 & 0 & 0 \end{bmatrix} \begin{Bmatrix} \{u\} \\ \{v\} \\ \{-p\} \end{Bmatrix} = \begin{Bmatrix} -(\partial u/\partial y)_n D\{u\}_n \\ -(\partial v/\partial x)_n D\{v\}_n \\ 0 \end{Bmatrix} \tag{22}$$

This has precedent in the Crank-Nicolson method for first order time stepping which also distributes damping matrix contributions to both sides of the recurrence relation and one hopes to achieve similar stability advantages in the present application [26]. There is also another precedent for Eqn (22) in the analysis of structures with large strains [13] and curvatures [27] where nonlinear stiffnesses matrix and residual load effects are included on both sides of the recurrence relation, these being calculated using derivatives $(\partial v/\partial x)_n$ and $(\partial^2 v/\partial x^2)_n$ as scalar multipliers in forming the nonlinear terms.

Eqn (22) has an interesting physical interpretation. The left hand side has the form of a mass damping matrix, the damping effect depending upon the velocity gradients, whilst the right hand side could be viewed as a vorticity loading, i.e. when $\partial u/\partial y = \partial v/\partial x$, or the vorticity is zero, the forcing effect is the same upon both the horizontal and vertical velocities.

For transient flows the accelerations $\partial u/\partial t$, $\partial v/\partial t$ of Eqns (3) and (4) must also be included, yielding the mass damping matrix

$$k^* \begin{Bmatrix} \{\dot{u}\} \\ \{\dot{v}\} \\ \{p\} \end{Bmatrix} = \begin{bmatrix} D & 0 & 0 \\ 0 & D & 0 \\ 0 & 0 & 0 \end{bmatrix} \begin{Bmatrix} \{\partial u/\partial t\} \\ \{\partial v/\partial t\} \\ \{p\} \end{Bmatrix} \tag{23}$$

and a recurrence relation is obtained by using the finite difference approximations $\partial u/\partial t = (u_{n+1} - u_n)/\delta t$, $\partial v/\partial t = (v_{n+1} - v_n)/\delta t$ to form a recurrence relation, e.g. if one uses forward differences with $u = u_n$ (the backward point) as the point of reference one obtains

$$k^*\{u, v, p\}_{n+1} = [k^* - (\delta t)k]\{u, v, p\}_n + (\delta t)\{q\} \tag{24}$$

where δt is the time step length and k and {q} are given by Eqns (18) and (22). Though this involves a truncation error of $O(\delta t)$, rather than $O((\delta t)^2)$ for the Crank-Nicolson method, it has the advantage of providing a fully explicit solution if Eqn (23) is diagonalized by mass lumping but as in k one must also include small diagonal entries in k* (see Eqn (35)) in the diagonal locations corresponding to {p}.

COMPRESSIBLE FLOWS

For compressible flows one requires, in addition to Eqns (1), (2) and (5), satisfaction of the thermal energy equation which is, when direct heating from chemical reaction and radiation are absent and c_p and Φ are constant [15]

$$\rho u c_p \partial T/\partial x + \rho v c_p \partial T/\partial y - \kappa\nabla^2 T - \mu\Phi - u\partial p/\partial x - v\partial p/\partial y = 0 \tag{25}$$

where κ is the isotropic thermal conductivity and the viscous dissipation function Φ is given by

$$\Phi = (\partial u/\partial y)^2 + (\partial v/\partial x)^2 + 2(\partial u/\partial y)(\partial v/\partial x) + (4/3)[(\partial u/\partial x)^2 - (\partial u/\partial x)(\partial v/\partial y) + (\partial v/\partial y)^2] \tag{26}$$

Allowing a variation δT and using integration by parts to reduce the second order derivatives the element energy equations are obtained as

$$k_T\{T\} + k_p\{p\} = \{q_T\} \tag{27}$$

where
$$k_T = \int \rho c_p[u_n\{f\}\{f_x\}^t + v_n\{f\}\{f_y\}^t]dV + \int \kappa[\{f_x\}\{f_x\}^t + \{f_y\}\{f_y\}^t]dV \tag{28}$$

$$k_p = -\int[u_n\{f\}\{f_x\}^t + v_n\{f\}\{f_y\}^t]dV \tag{29}$$

$$\{q_T\} = \int\{f\}\Phi dV + \int\kappa\{f\}(\partial T/\partial n)dS = \int\{f\}\Phi dV + \int\kappa\{f\}[(\partial T/\partial x)c_x + (\partial T/\partial y)c_y]dS \tag{30}$$

the viscous dissipation function being used as a loading term. The nonlinear terms in both k_T and k_p render them unsymmetric but are otherwise dealt with in the same manner as Eqn (22).

The complete formulation now involves velocity, pressure and temperature freedoms and μ, ρ and κ also vary from element to element but may be recalculated during the iterative solution using

$$(\mu/\mu_0) = (\kappa/\kappa_0) = (T/T_0)^m \tag{31}$$

and
$$p = \rho RT \tag{32}$$

where μ_0 is the viscosity at the reference temperature T_0. One can also include nodal density freedoms [28] but in view of Eqn (32) these will be uncoupled from the other freedoms and it is more economical to treat ρ as a secondary variable (i.e. in the same way as the elastic moduli in nonlinear analysis of solids).

Eqn (27) can be superposed over Eqns (18) + (22) but the two stage approach outlined by Heubner [12] in connection with compressible lubrication problems seems more economical, i.e. Eqns (18) + (22) are first solved for the velocities and pressures and then Eqn (27) is solved to determine the temperatures, from which μ, κ and ρ are updated in each element and the two solution steps repeated. Further, as the pressures are already determined in the first step, k_p and the nonlinear terms of k_T can be shifted to the right hand side of Eqn (27), relying upon Eqns (18) + (22) to adjust the pressures. Thus the two matrix solution steps involve only symmetric matrices, the second involving only one degree of freedom whilst the first step is used to greater advantage, incorporating temperature corrections as well as the nonlinear velocity corrections.

FINITE ELEMENT IMPLEMENTATION

Finite element implementation of the foregoing formulation is carried out using the quadratic six node triangle as a basis. At the outset the same quadratic areal coordinate interpolation is used for velocities and pressures, i.e.

$$\{f\} = \{L_1(2L_1 - 1),\ L_2(2L_2 - 1),\ L_3(2L_3 - 1),\ 4L_1L_2,\ 4L_2L_3,\ 4L_3L_1\} \tag{33}$$

and using straight sided elements

$$\begin{Bmatrix} \{f_x\}^t \\ \{f_y\}^t \end{Bmatrix} = \frac{1}{2\Delta} \begin{bmatrix} -y_{32} & -y_{13} & -y_{21} \\ x_{32} & x_{13} & x_{21} \end{bmatrix} \begin{Bmatrix} \partial\{f\}^t/\partial L_1 \\ \partial\{f\}^t/\partial L_2 \\ \partial\{f\}^t/\partial L_3 \end{Bmatrix} \tag{34}$$

Three point quadratic integration is used, generally at (L_1,L_2,L_3) = (4/6,1/6,1/6), (1/6,4/6,1/6), (1/6,1/6,4/6), and this reduced integration exactly integrates quadratic terms and provides a good approximation for cubic and quartic terms in Eqns (18), (22), (23) and (27) [24, 29].

Observing Eqn (18) there will in most cases be a need for pivoting because of the absence of diagonal entries in the continuity constraint rows. Alternatively one can use small diagonal entries which were here calculated as

$$(\Sigma_i \Sigma_j\ C_{ij}^{\ 2})^{1/2}/\beta, \quad \beta = 10^5 \tag{35}$$

where C_{ij} are the matrix entries calculated from Eqn (18), or Eqn (22) in the case of convective flows (in which $\rho >> \mu$). In relation to the factor 10^5 used in Eqn (35) it should be noted that single precision 6.3 decimal digit computation is used in this work.

Horizontal velocity, u			Pressure, p		
x,y	FEM	Exact	x,y	FEM	Exact
2,0	0.0000	0.0000	0,2	4.0000	4.0000
2,1	1.1875	1.1875	1,2	3.5003	3.5000
2,2	1.7500	1.7500	2,2	3.0025	3.0000
2,3	1.6875	1.6875	3,2	2.5004	2.5000
2,4	1.0000	1.0000	4,2	2.0000	2.0000

Table 1: Finite element solutions for the analysis of Figure 1.

NUMERICAL APPLICATIONS

Fig. 1 shows the element applied to the Couette flow problem. Pressure boundary conditions are set at inlet and outlet as one seeks to determine the velocity distribution here and elsewhere. Table 1 shows the results for the velocity gradient at $x = 2$, $y = 0 \rightarrow 4$ and the pressure gradient at $y = 2$, $x = 0 \rightarrow 4$. The results throughout the flow field were of equal accuracy to these, which are in excellent agreement with the exact solution [15]. For the special case where $u = 0$ for the upper boundary (plane Poiseille flow), of course, one obtains equally good results and the formulation is readily generalized to deal with axisymmetric flows by in the numerical integration for the viscosity and other matrices replacing $\partial(\ )/\partial y$ by $\partial(\ )/\partial r$ and $\omega_i\Delta$ by $2\pi r\omega_i\Delta$ (here $\omega_i = 1/3$, $i = 1,2,3$ are the integration point weights).

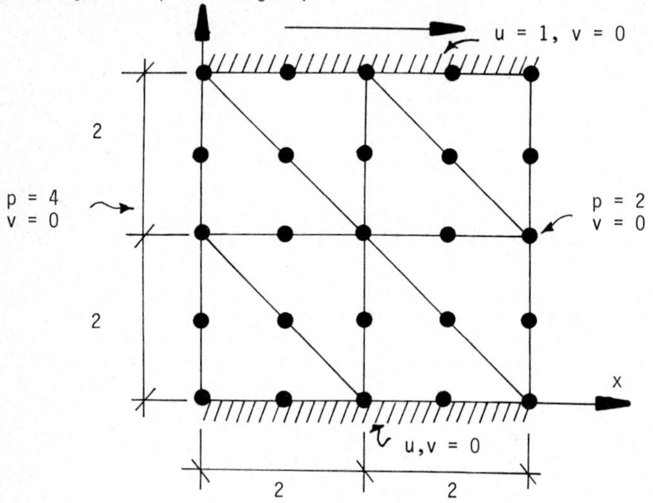

Fig.1. Finite element model of incompressible Couette flow.

Fig. 2 shows a more demanding problem - that of entry flow between parallel plates. Here integration at the midsides was used to obtain the results shown in Fig. 3 and the velocity loading terms of Eqn (18) required at the inlet were evaluated using

$$\{q_p\} = \int\{f\}\{f\}^t\{u\}dy = (a/4)\{U,\ 4U,\ U\}/6$$

for the nodes along the vertical element edges at the inlet.

The developing velocity profiles are in good agreement with those obtained by Atkinson et al. [30] and others [9] with integration at three internal points the unexpected maximum in the first developing profile is not obtained and the solution then corresponds to Schlicting's original solution to the problem [31] which is that applicable to low Reynold's numbers.

The pressure profiles (see ref. 32) were in good agreement with those observed by Yamada et al. [10] though they show a little uneveness at the midside nodes at which pressure freedoms are not allowed in their work. Nevertheless the present formulation gives better results with quadratic interpolation for all freedoms for this problem.

It is also worth noting that generally one must specify either velocity or pressure at inlet and outlet to force the flow, but not both. In the problem of Fig. 2, unlike that of Fig. 1, velocity and pressure are respectively specified at inlet and outlet.

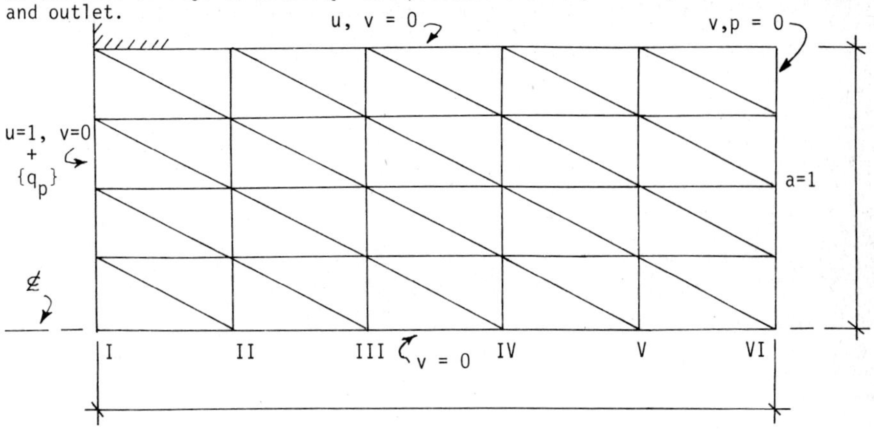

Fig. 2. Finite element model of entry flow between parallel plates (μ = 1).

Fig. 4 shows a flat plate in an infinite medium initially at rest and suddenly given a constant horizontal velocity U. The pressure is constant and the streamlines will be parallel to the plate so that v = 0. Thus the boundary conditions for the vertical boundaries are v = 0, p = constant., u = U, v = 0 at y = 0 and u = 0 , v = 0 at y = ∞ and to deal with the latter condition one can solve the problem using (a) u = 0, (b) u unspecified for y = 2 and average the two solutions so obtained. This is crude but effective and is based on observation of results in elasticity problems in infinite media where averaging of results with rigid and free boundaries yields results of comparable accuracy to those obtained using more precise elastic boundary conditions based upon the known 'far field' behaviour [33].

Here lumping was used in the damping matrix on the left side of Eqn (23), i.e. $2\rho\Delta/9$ at the midside nodes and $\rho\Delta/9$ at the corner nodes, yielding an explicit solution at each time step. The time step lengths δt where 0.1 and the results shown in Fig. 4 are expressed in terms of the similarity parameter

$$\eta = y/(2\sqrt{\nu t}) \tag{36}$$

where $\nu = \mu/\rho$ is the kinematic viscosity and the thickness of the laminar boundary layer (in which the viscous stresses are negligible) can be estimated by the limits [15]

$$0.564 < \eta < 1.253 \tag{37}$$

and the lower limit defines a region in which η/u is approximately constant (i.e. the laminar boundary layer) and the range in Eqn (37) can be interpreted as a transition zone.

To test the formulation for compressible flow the problem of compressible Couette flow under constant pressure between parallel plates with relative velocity U was analysed using the same mesh as in Fig. 1 but setting the inlet and outlet pressures to zero. As the flowis nonconvective Eqn (22) is not required and Eqn (27) simplifies to

$$[\int\kappa(\{f_x\}\{f_x\}^t + \{f_y\}\{f_y\}^t)dxdy]\{T\} = \int\{f\}\Phi dxdy \qquad (38)$$

The analysis assumes $PrM^2 = 2$, where $Pr = \mu c_p/\kappa$ is the Prandtl number and $M = U/c$ is the Mach number of the moving plate. Then as the velocity of sound at the plate is given by $c^2 = \gamma(p/\rho) = \gamma RT$ where the Gas constant is given by $R = c_p - c_v = c_p(\gamma - 1)/\gamma$ we have

$$\mu U^2/\kappa = 2c^2/c_p = 2\gamma RT/c_p = 2\gamma c_p(\gamma - 1)T/(c_p\gamma) = 2(\gamma - 1)T \qquad (39)$$

so that with the specific heat ratio $\gamma = 1.405$ (the value for air) and setting $\mu = 1$, $\kappa = 1$ and $T = 1$ at the plate one also specifies $U = \sqrt{0.81}$.

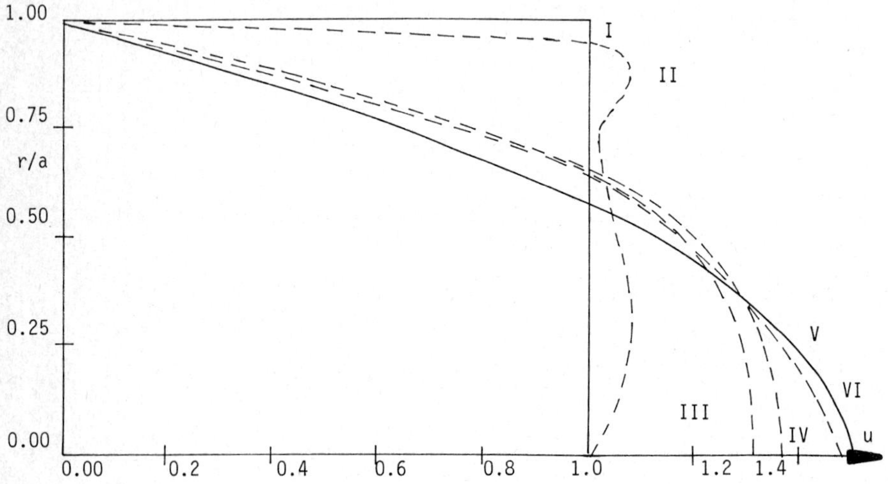

Fig. 3. Developing velocity profiles in the entry flow of Fig. 2.

The analysis commences by assuming the values of μ, κ, T for the plate to apply throughout the medium and solving Eqn (18) with p,v = 0 at inlet and exit, u,v = 0 at y = 0 and u = U at y = h. This first trial solution is that for incompressible flow, giving a linear velocity field between the plates. Next Eqn (38) is solved using these velocities, assuming $\kappa = 1$ throughout and setting $T_p = 1$ as the boundary condition at the moving plate.

The resulting temperature distribution is then used to calculate average temperatures in each element and Eqn (31) with m = 1 is used to adjust the element viscosities from the initially constant values. These are then used in Eqn (18) to obtain a velocity solution, which in turn is used to solve Eqn(38), adjusting the thermal conductivities using Eqn (31). After this second iteration of the two stage solution process the solution is already of sufficient accuracy and the results obtained at x = 2, y = 0 → 4 in Fig. 1 are shown in Fig. 5 and agree well with the exact solution [15]. Indeed one obtains for the temperature at the fixed plate (the recovery temperature) T = 1.4050 whereas the exact solution is given by [15]

$$T/T_p = 1 + PrM^2(\gamma - 1) = 1.405 \qquad (40)$$

exactly the same as the finite element result. Observing the velocity distributions in Fig. 5, however, the compressibility effect upon these is evidently very small and larger numbers of iterations are generally needed to obtain the same accuracy.

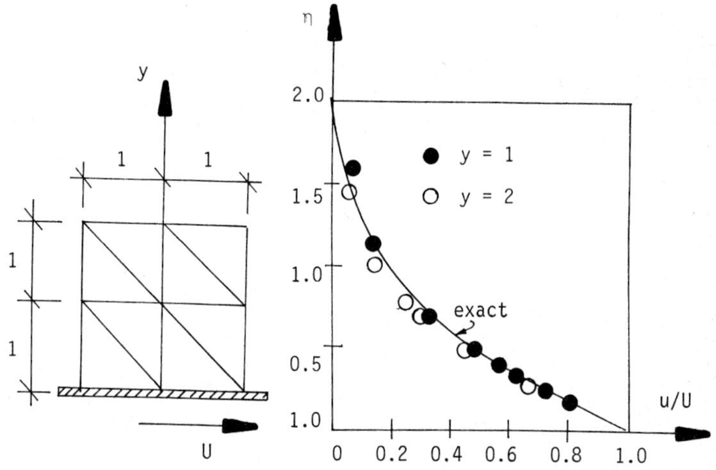

Fig. 4. Finite element model of flow induced by infinitely long plate instantaneously given velocity U in an infinite medium.

The use of a two-stage solution procedure for compressible flows bears comparison with the use of segregated formulations (inlcuding those with penalty factors) for Stokes' flows and follows naturally from the use of the thermal energy equation as a supplementary equation to the momentum and continuity equations. This raises the question of how to extend the segregated Stokes flow formulations to include compressibility: presumably a third solution stage (corresponding to the second in the present work) would need to be added and this might lead to a relatively unwieldy overall formulation with slower convergence than the two stage approach recommended here.

Furthermore, the wide applicability of viscous compressible flow formulations such as that used in the present work is worth remark. For example, inviscid flows can be dealt with by the use of a small artifical viscosity whilst incompressible flow problems can be dealt with by allowing an artifical compressibility.

As a final example the cavity flow problem shown in Fig.6 is solved using a regular mesh of 32 elements (9 x 9 nodes), this being relatively coarse in view of the singularities at the upper corners. The velocity boundary conditions shown are those of a sliding lid with the pressure set to zero at the central basal node to establish a datum. It was found that integration at the midside nodes or three internal points gave much the same results for the velocities. Reduction to linear pressure interpolation, though, yielded slightly better velocity solutions. This is illustrated in Table 2, where the velocity profiles at the vertical centreline for Re = 0 obtained using both quadratic and linear pressure interpolation are compared to Burggraf's finite difference results [34]. Also included is the profile obtained using the continuation method for Re = 100, again indicating reasonable agreement with Burggraf's results (the agreement is as good as obtained by Hughes et al. [35] for Re = 400).

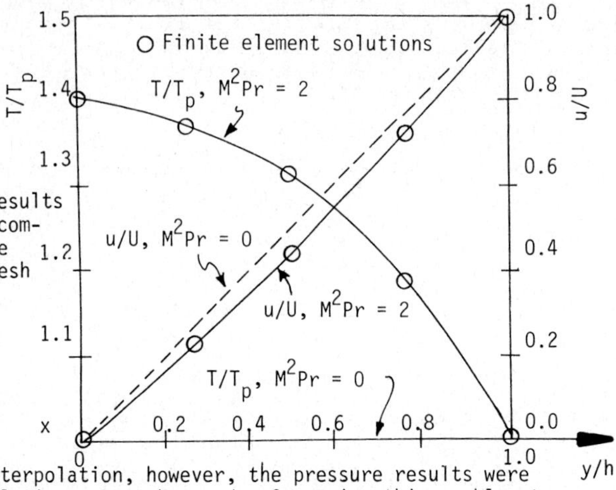

Fig. 5. Finite element results for analysis of compressible Couette flow using the mesh of Fig. 1.

With quadratic pressure interpolation, however, the pressure results were unstable, i.e. large oscillations were observed. Comparing this problem to that of Fig. 2 it seems that this may be owing to the relative scarcity of pressure boundary conditions. The only remedy found to this point is to use linear pressure interpolation, when satisfactory pressure results are obtained. This is illustrated by comparison of the minimum nodal pressure obtained. This is illustrated by comparison of the minimum nodal pressure obtained for Re = 100 (indicating the vortex centre) with values obtained by Olson and Tuann [17] and Burggraf [34] in Table 3. The complete pressure contours were also in satisfactory agreement with those obtained by these authors.

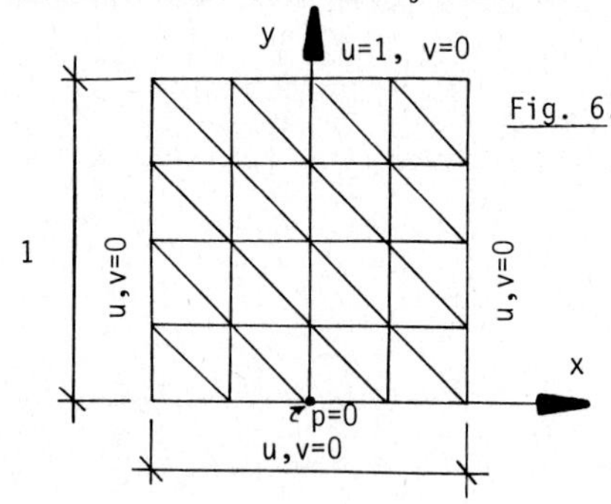

Fig. 6. Finite element grid for square cavity problem.

CONCLUSIONS

1. The quadratic triangle formulation with consistent interpolation and reduced integration proves effective in the first two problems studied. Indeed the results for the entry flow problem are the first satisfactory ones reported without mixed interpolation, segregation or penalty factors. In problems with very few pressure boundary conditions, such as the cavity flow problem studied here, reduction to linear interpolation for the pressures is necessary to obtain satisfactory results for these.

2. The velocity loading terms resulting from integration by parts of the continuity equation, rather than the pressure loading terms which occur in other formulations, prove no inconvenience. In the entry flow problem, for example, the specified velocity boundary conditions at inlet are also used to calculate the velocity loadings. Moreover these loadings provide a more accurate means of specifying fluid fluxes through narrow inlets (and hence few elements) and success without mixed interpolation in the entry flow problem is perhaps evidence of some computational advantage, poor results being obtained without their inclusion.

3. The numerical integration at three internal points, which proved an advantageous alternative to integration at the midside nodes in a shell element [24], did not in fact yield as good agreement with previous numerical results as the latter integration in the entry flow problem (elsewhere there was little difference).

4. The new approach to the convective inertia terms proves economical, informative and effective and seems worth further study with higher Reynold's numbers.

5. The explicit time stepping scheme used has only been tested with relatively few steps in a simple problem. In general an implicit scheme, at least for the pressures, must be recommended at this stage [36].

6. The use of a two stage solution procedure for compressible flow follows naturally from the use of the thermal energy equation as a supplementary equation to the momentum and continuity equations. Whilst this again has been tested only upon a simple problem it can be proposed with more confidence as with k shifted to the right in Eqn (27), as is permitted by the two stage solution, checkerboard problems and the like do not arise so long as they have been prevented in the first stage. The question of how best to deal with Eqn (28), however, requires careful consideration as removal of the unsymmetric terms to the right will increase the number of iterations required. In Fig. 5 the compressibility effect is evidently very small (at least upon the velocities) and further study of other problems is needed.

	R = 0			R = 100	
y	quadratic p	linear p	[34]	linear p	[34]
1.000	1.00	1.00	1.00	1.00	1.00
0.875	0.35	0.37	0.35	0.27	0.28
0.750	0.01	-0.02	-0.04	-0.07	0.01
0.625	-0.12	-0.16	-0.17	-0.25	-0.11
0.500	-0.14	-0.20	-0.20	-0.21	-0.20
0.375	-0.15	-0.18	-0.16	-0.13	-0.19
0.250	-0.07	-0.13	-0.12	-0.06	-0.12
0.125	-0.05	-0.07	-0.05	-0.03	-0.05
0.000	0.00	0.00	0.00	0.00	0.00

Table 2: Numerical results for horizontal velocities at vertical centreline of cavity problem of Fig. 6.

7. The checkerboard effect problem has received a great deal of attention in recent years and this is continuing. In the present work the 'velocity loadings' appear to have a smoothing effect in the entry flow problem but resort is had to mixed interpolation to deal with cavity flows and further investigation is warranted to avoid this. One possibility is raised by quadratic thick plate and shell elements [1,23] where there is a tendency towards a checkerboard effect in some of the stress solutions: here the problem is avoided by appropriate integration point stress sampling [24]. Likewise, in consistent interpolation unsegregated viscous flow formulations pressure sampling at integration points not coincident with the nodes might prove helpful.

Method	Location		Pressure
	x	y	
u,p integrated [17]	-0.13	0.73	-9.557
u,p segregated [17]	-0.13	0.75	-7.028
stream function [17]	-0.11	0.72	-9.447
present method	0.00	0.75	-8.440
finite difference [34]	-0.13	0.74	-9.1

Table 3: Numerical results for pressure at the vortex centre in cavity problem of Figure 6.

ACKNOWLEDGEMENT

Carol Everard and Sylvia Parker for sterling efforts in typing a difficult manuscript.

REFERENCES

[1] Mohr, G.A., Application of penalty factors to a doubly curved quadratic shell element, Comput. Structures 14 (1981) 15-19.

[2] Connor, J.J. and Brebbia, C.A., Finite Element Techniques for Fluid Flow (Butterworths, London, 1977).

[3] Tong, P., The finite element method for fluid flow, in: Gallagher, R.H. et al. (eds), Recent Advances in Structural Analysis and Design (UAH Press, Amsterdam, 1971).

[4] Olson, M.D., Formulation of a variational principle - the finite element method for viscous flows, in: Brebbia, C.A. and Tottenham, H., Variational Methods in Engineering 1 (SU Press, Southampton, 1973).

[5] Lieber, P., Wen, K.S. and Attia, A.V., Finite element methods as an aspect of the principle of maximum uniformity. Proc. Symp. Finite Elements Methods in Flow Problems. UC Wales, Swansea (1974).

[6] Baker, A.J., Finite element solution algorithm for viscous incompressible fluid dynamics. Int. Jour. Num. Meth. Engrg 6, no. 1 (1973).

[7] Oden, J.T. and Carter Wellford, L., Analysis of flow of viscous fluids by the finite element method. Jour. Am. Inst. Aeron. Astron. 10, 1590-99 (1972).

[8] Hood, P. and Taylor, C., Navier-Stokes equations using mixed interpolation. Proc. Symp. Finite Element Methods in Flow Problems. UC Wales, Swansea (1974).

[9] Kawahara, M., Yoshimura, N., Nakagawa, K., and Ohsaka, H. Steady and unsteady finite element analysis of incompressible viscous fluid. Int. Journ. Num. Meth. Engrg 10, 437-456 (1976).

[10] Yamada, Y., Ito, K., Yokouchi, Y., Tamano, T. and Ohtsubo, T. Finite element analysis of steady fluid and metal flow. Finite Element Methods in Flow Problems, eds J.T. Oden et al. (UAH Press, Huntsville 1974).

[11] Argyris, J.H. and Dunne, P.C. The finite element method applied to fluid mechanics. Proc. Conf. Comp. Meth. and Problems in Aeron. Fluid Mech., Univ. of Manchester. (Academic Press, London 1976).

[12] Heubner, K.H., The Finite Element Method for Engineers. (McGraw-Hill, New York 1975).

[13] Zienkiewicz, O.C., The Finite Element Method, 3rd ed. (McGraw-Hill, London 1977).

[14] Sani, R.L., Gresho, P.M., Lee, R.L. and Griffiths, D.F., The cause and cure (?) of the spurious pressures generated by certain FEM solutions of the incompressible Navier-Stokes equations: part 1. Int. Jour. Num. Meth. Fluids 1, 17-43 (1981).

[15] Yuan, S.K., Foundations of Fluid Mechanics. (Prentice-Hall, Englewood-Cliffs NJ 1967).

[16] Mohr, G.A. and Broom, N.D., A finite element lubrication model. Proc. 4th Aust. Int. Conf. Finite Element Methods, Univ. of Melbourne (1982).

[17] Olson, M.D. and Tuann, S.Y., Primitive variables versus stream function finite element solutions of the Navier-Stokes equations. Finite Elements in Fluids Vol. 3, eds R.H. Gallagher et al. (Wiley, New York 1978).

[18] Carey, G.F. and Krishnan, R., Penalty approximation of Stokes flow. Comp. Meth. Appl. Mech. Engrg 35, 169-206 (1982).

[19] Reddy, J.N., On penalty function methods in the finite element analysis of flow problems. Int. Journ. Num. Meth. Fluids 2, 151-171 (1982).

[20] Oden, J.T., Kikuchi, N. and Song, Y.J. Penalty-finite element methods for the analysis of Stokesian flows. Comp. Meth. Appl. Mech. Engrg 31, 297-329 (1982).

[21] Engelman, M.S., Sani, R.L., Gresho, P.M. and Bercovier, M. Consistent vs reduced integration penalty methods for incompressible media using several old and new elements. Int. Jour. Num. Meth. Fluids 2, 25-42 (1982).

[22] Sani, R.L., Gresho, P.M., Lee, R.L., Griffiths, D.F. and Engleman, M. The cause and cure (!) of the spurious pressures generated by certain FEM solutions of the incompressible Navier-Stokes equations: part 2. Int. Jour. Num. Meth. Fluids 1, 171-204 (1981).

[23] Mohr, G.A., A triangular finite element for thick slabs. Comput. Structures 9, 595-598 (1978).

[24] Mohr, G.A., On displacement finite elements for the analysis of shells. Proc. 4th Aust. Int. Conf. Finite Element Methods, Univ. of Melbourne (1982).

[25] Harvey, J.W. and Kelsey, S., Triangular plate bending elements with enforced compatibility. Jour. Am. Inst. Aeron. Astron. 9, 1023-1028 (1971).

[26] Irons, B.M. and Ahmad, S., Techniques of Finite Elements. Ellis-Horwood, Chichester (1980).

[27] Mohr, G.A. and Milner, H.R., Finite element analysis of large displacements in flexural systems. Comput. Struct. 13, 533-536 (1981).

[28] Chung, T.J., Finite Element Analysis in Fluid Dynamics. (McGraw-Hill, New York, 1978).

[29] Mohr, G.A., Quadratic finite elements for shells. Mathematics and Models in Engineering Science, ed. A. McNabb, R.A. Wooding and M.S. Rosser, DSIR, Wellington, New Zealand (1982).

[30] Atkinson, B., Brocklebank, M.P., Card, C.C.H. and Smith, J.M., Low Reynolds number developing flows. Jour. Am. Inst. Ch. Engrg. 15, 548-553 (1969).

[31] Schlicting, H., Boundary-Layer Theory, 7th ed. (McGraw-Hill, New York, 1979).

[32] Mohr, G.A., Finite element analysis of viscous fluid flows. Comput. Fluids (submitted for publication).

[33] Mohr, G.A. and Power, A.S. Elastic boundary conditions for finite elements of infinite and semi-infinite media. Proc. ICE part II 65, 675-685 (1978).

[34] Burggraf, O.C., Analytical and numerical studies of the structure of steady separated flows. Jour. Fluid Mech. 24, part I, 113-145 (1966).

[35] Hughes, T.J.R., Taylor, R.L. and Levy, J.F., High Reynolds number, steady, incompressible flows by a finite element method. Finite Elements in Fluids Vol. 3, ed R.H. Gallagher et al. (Wiley, New York, 1978).

[36] Gresho, P.M., Lee, R.L. and Sani, R.L., On the time-dependent solution of the incompressible Navier-Stokes equations in two and three dimensions. Recent Advances in Numerical Methods in Fluids Vol. 1, ed. C. Taylor and K. Morgan. (Pineridge Press, Swansea 1980).

Computational Techniques & Applications: CTAC-83
J. Noye & C. Fletcher (Editors)

A UNIFICATION OF FINITE ELEMENT APPROACHES FOR PRIMITIVE VARIABLE SOLUTIONS TO VISCOUS INCOMPRESSIBLE FLOWS

A.N.F. Mack

Department of Mechanical Engineering
University of Sydney
New South Wales

In connection with solutions to viscous incompressible flows, an idea which has received scant attention is the solenoidal approach. So as to redress the situation, the present effort is devoted to an examination of this particular approach. Also included is a comparison with the other more established alternatives. All these are shown to be different manifestations of the imposition of the solenoidal constraint.

INTRODUCTION

In spite of an early inclination towards the auxiliary variables, the preferable course is to procure solutions based on the primitive variables. With this, the task is complicated by the imposition of the solenoidal constraint. Apart from direct minimisation of the square of the error, various approaches have evolved. The references cited here, however, are representative samples and, as such, do not suggest respective popularities. The solenoidal approach satisfies the constraint at an element level [1-2]. Another approach incorporates the constraint with the assistance of a penalty function [3-4]. Another approach incorporates the constraint with the assistance of a Lagrange multiplier [5-6]. Nevertheless, it seems that some misconceptions have arisen about these approaches. Since such notions can lead to incorrect formulations, the present paper aims to promote the true position.

GOVERNING EQUATIONS

Suppose a region is defined by the domain A. Suppose its perimeter is defined by the boundary b. Any point in the domain is specified by the cartesian coordinates x,y. Any normal to the boundary is specified by the direction cosines α,β. The constant properties are the viscosity μ, the density ρ. As previously indicated, the solution variables are the pressure p, the velocity components u,v.

Subject to appropriate boundary conditions, the differential equations under consideration are

$$u_x + v_y = 0 \tag{1}$$

declaring zero net mass change,

$$\begin{aligned} \mu\nabla^2 u - p_x &= \rho(uu_x + vu_y) \\ \mu\nabla^2 v - p_y &= \rho(uv_x + vv_y) \end{aligned} \tag{2}$$

denoting balances between the forces of diffusion, pressure and advection.

An integral formulation is derived from an inner product, in which all expressions are assumed to exist. That is, for arbitrary variations $\delta p, \delta u, \delta v$

$$\int_A \delta p(u_x + v_y)\, dA + \int_A \delta u[\mu\nabla^2 u - p_x - \rho(uu_x + vu_y)]\, dA + \int_A \delta v[\mu\nabla^2 v - p_y - \rho(uv_x + vv_y)]\, dA = 0. \tag{3}$$

On application of Green's theorem

$$- \int_A [\delta(u_x + v_y)p + \delta p(u_x + v_y)]\, dA + \mu \int_A (\delta u_x u_x + \delta v_x v_x + \delta u_y u_y + \delta v_y v_y)\, dA + \rho \int_A [\delta u(uu_x + vu_y) + \delta v(uv_x + vv_y)]\, dA = \int_b [\delta u(\mu u_n - \alpha p) + \delta v(\mu v_n - \beta p)]\, db, \tag{4}$$

where n is the direction of the outward normal.

RAYLEIGH-RITZ METHOD

For the moment, it is convenient to concentrate on the linear case. This facilitates the use of the Rayleigh-Ritz method.

With omission of the nonlinear terms

$$- \int_A [\delta(u_x + v_y)p + \delta p(u_x + v_y)]\, dA + \mu \int_A (\delta u_x u_x + \delta v_x v_x + \delta u_y u_y + \delta v_y v_y)\, dA = \int_b [\delta u(\mu u_n - \alpha p) + \delta v(\mu v_n - \beta p)]\, db. \tag{5}$$

Under the constraint of solenoidal velocity components

$$\mu \int_A (\delta u_x u_x + \delta v_x v_x + \delta u_y u_y + \delta v_y v_y)\, dA = \int_b [\delta u(\mu u_n - \alpha p) + \delta v(\mu v_n - \beta p)]\, db. \tag{6}$$

Thus, the desired solution is found from the stationary condition of the variational functional

$$I(u,v) = \frac{\mu}{2}\int_A (u_x^2 + v_x^2 + u_y^2 + v_y^2)\, dA - \int_b [u(\mu u_n - \alpha p) + v(\mu v_n - \beta p)]\, db. \quad (7)$$

In fact, it is this expression which forms the basis of each of the subsequent approaches.

An important detail is that there is no equivalent Bubnov-Galerkin method. The explanation stems from this method's cancellation of the arbitrary nodal parameters at the inner product stage, so that substitution of the solenoidal constraint does not result in the suppression of the pressure.

Solenoidal Approach

The solenoidal constraint is a relation between the velocity components and can be imposed at an element level.

The interpolation functions $\underline{N}^{uu}, \underline{N}^{uv}, \underline{N}^{vu}, \underline{N}^{vv}$ define the local behaviour within an element in the form

$$\begin{bmatrix} u \\ v \end{bmatrix} = \begin{bmatrix} \underline{N}^{uu} & \underline{N}^{uv} \\ \underline{N}^{vu} & \underline{N}^{vv} \end{bmatrix} \begin{bmatrix} \underline{\tilde{u}} \\ \underline{\tilde{v}} \end{bmatrix}. \quad (8)$$

Differentiation of the variational functional with respect to the nodal parameters $\underline{\tilde{u}}, \underline{\tilde{v}}$ leads to

$$\begin{bmatrix} \underline{K}^{uu} & \underline{K}^{uv} \\ \underline{K}^{vu} & \underline{K}^{vv} \end{bmatrix} \begin{bmatrix} \underline{\tilde{u}} \\ \underline{\tilde{v}} \end{bmatrix} = \begin{bmatrix} \underline{f}^{u} \\ \underline{f}^{v} \end{bmatrix}, \quad (9)$$

where

$$\underline{K}^{uu} = \mu \sum_e \int_{Ae} (\underline{N}_x^{uu^T} \underline{N}_x^{uu} + \underline{N}_x^{vu^T} \underline{N}_x^{vu} + \underline{N}_y^{uu^T} \underline{N}_y^{uu} + \underline{N}_y^{vu^T} \underline{N}_y^{vu})\, dA$$

$$\underline{K}^{uv} = \mu \sum_e \int_{Ae} (\underline{N}_x^{uu^T} \underline{N}_x^{uv} + \underline{N}_x^{vu^T} \underline{N}_x^{vv} + \underline{N}_y^{uu^T} \underline{N}_y^{uv} + \underline{N}_y^{vu^T} \underline{N}_y^{vv})\, dA$$

$$\underline{K}^{vu} = \mu \sum_e \int_{Ae} (\underline{N}_x^{uv^T} \underline{N}_x^{uu} + \underline{N}_x^{vv^T} \underline{N}_x^{vu} + \underline{N}_y^{uv^T} \underline{N}_y^{uu} + \underline{N}_y^{vv^T} \underline{N}_y^{vu})\, dA \quad (10.1)$$

$$\underline{K}^{vv} = \mu \sum_e \int_{Ae} (\underline{N}_x^{uv^T} \underline{N}_x^{uv} + \underline{N}_x^{vv^T} \underline{N}_x^{vv} + \underline{N}_y^{uv^T} \underline{N}_y^{uv} + \underline{N}_y^{vv^T} \underline{N}_y^{vv})\, dA$$

$$\underline{f}^{u} = \sum_e \int_{be} [\underline{N}^{uu^T} (\mu u_n - \alpha p) + \underline{N}^{vu^T} (\mu v_n - \beta p)]\, db$$

$$\underline{f}^{v} = \sum_e \int_{be} [\underline{N}^{uv^T} (\mu u_n - \alpha p) + \underline{N}^{vv^T} (\mu v_n - \beta p)]\, db \quad (10.2)$$

confirm symmetry. It follows that solution specification requires

$$\begin{aligned} u &= u' \quad \text{or} \quad \mu u_n - \alpha p = \mu u_n' - \alpha p',\ \mu v_n - \beta p = \mu v_n' - \beta p' \\ v &= v' \quad \text{or} \quad \mu u_n - \alpha p = \mu u_n' - \alpha p',\ \mu v_n - \beta p = \mu v_n' - \beta p' \end{aligned} \tag{11}$$

to be the boundary conditions.

Penalty Function Approach

The solenoidal constraint can also be incorporated by the creation of the perturbed functional

$$\begin{aligned} I(u,v) &= \varepsilon \int_A (u_x + v_y)^2 \, dA \\ &+ \frac{\mu}{2} \int_A (u_x{}^2 + v_x{}^2 + u_y{}^2 + v_y{}^2) \, dA - \int_b [u(\mu u_n - \alpha p) + v(\mu v_n - \beta p)] \, db \end{aligned} \tag{12}$$

where ε is a large constant.

The interpolation functions $\underline{N}^{uu}, \underline{N}^{vv}$ define the local behaviour within an element in the form

$$\begin{bmatrix} u \\ v \end{bmatrix} = \begin{bmatrix} \underline{N}^{uu} & \underline{0} \\ \underline{0} & \underline{N}^{vv} \end{bmatrix} \begin{bmatrix} \underline{\tilde{u}} \\ \underline{\tilde{v}} \end{bmatrix}. \tag{13}$$

Differentiation of the perturbed functional with respect to the nodal parameters $\underline{\tilde{u}}, \underline{\tilde{v}}$ leads to

$$\begin{bmatrix} \underline{K}^{uu} & \underline{K}^{uv} \\ \underline{K}^{vu} & \underline{K}^{vv} \end{bmatrix} \begin{bmatrix} \underline{\tilde{u}} \\ \underline{\tilde{v}} \end{bmatrix} = \begin{bmatrix} \underline{f}^{u} \\ \underline{f}^{v} \end{bmatrix}, \tag{14}$$

where

$$\begin{aligned} \underline{K}^{uv} &= 2\varepsilon \sum_e \int_{Ae} \underline{N}_x^{uu^T} \underline{N}_y^{vv} \, dA \\ \underline{K}^{vu} &= 2\varepsilon \sum_e \int_{Ae} \underline{N}_y^{vv^T} \underline{N}_x^{uu} \, dA \\ \underline{K}^{uu} &= 2\varepsilon \sum_e \int_{Ae} \underline{N}_x^{uu^T} \underline{N}_x^{uu} \, dA + \mu \sum_e \int_{Ae} (\underline{N}_x^{uu^T} \underline{N}_x^{uu} + \underline{N}_y^{uu^T} \underline{N}_y^{uu}) \, dA \\ \underline{K}^{vv} &= 2\varepsilon \sum_e \int_{Ae} \underline{N}_y^{vv^T} \underline{N}_y^{vv} \, dA + \mu \sum_e \int_{Ae} (\underline{N}_x^{vv^T} \underline{N}_x^{vv} + \underline{N}_y^{vv^T} \underline{N}_y^{vv}) \, dA \end{aligned} \tag{15.1}$$

$$\begin{aligned} \underline{f}^{u} &= \sum_e \int_{be} \underline{N}^{uu^T} (\mu u_n - \alpha p) \, db \\ \underline{f}^{v} &= \sum_e \int_{be} \underline{N}^{vv^T} (\mu v_n - \beta p) \, db \end{aligned} \tag{15.2}$$

confirm symmetry. It follows that solution specification requires

$$u = u' \quad \text{or} \quad \mu u_n - \alpha p = \mu u_n' - \alpha p'$$
$$v = v' \quad \text{or} \quad \mu v_n - \beta p = \mu v_n' - \beta p' \tag{16}$$

to be the boundary conditions.

An equation for the recovery of the pressure can be extracted from the first variation of the perturbed functional, in that

$$2\varepsilon \int_A \delta(u_x + v_y)(u_x + v_y)\, dA$$
$$+ \mu \int_A (\delta u_x u_x + \delta v_x v_x + \delta u_y u_y + \delta v_y v_y)\, dA \tag{17}$$
$$= \int_b [\delta u(\mu u_n - \alpha p) + \delta v(\mu v_n - \beta p)]\, db.$$

Identification with equation (5) then reveals

$$p = -\varepsilon(u_x + v_y). \tag{18}$$

Moreover, equation (18) is needed for the suppression of the pressure in the nonlinear case.

Lagrange Multiplier Approach

The solenoidal constraint can also be incorporated by the creation of the perturbed functional

$$I(\lambda,u,v) = \int_A \lambda(u_x + v_y)\, dA$$
$$+ \frac{\mu}{2} \int_A (u_x{}^2 + v_x{}^2 + u_y{}^2 + v_y{}^2)\, dA - \int_b [u(\mu u_n - \alpha p) + v(\mu v_n - \beta p)]\, db \tag{19}$$

where λ is an extra variable.

The interpolation functions $\underline{N}^{\lambda\lambda}, \underline{N}^{uu}, \underline{N}^{vv}$ define the local behaviour within an element in the form

$$\begin{bmatrix} \lambda \\ u \\ v \end{bmatrix} = \begin{bmatrix} \underline{N}^{\lambda\lambda} & \underline{0} & \underline{0} \\ \underline{0} & \underline{N}^{uu} & \underline{0} \\ \underline{0} & \underline{0} & \underline{N}^{vv} \end{bmatrix} \begin{bmatrix} \underline{\tilde{\lambda}} \\ \underline{\tilde{u}} \\ \underline{\tilde{v}} \end{bmatrix}. \tag{20}$$

Differentiation of the perturbed functional with respect to the nodal parameters $\underline{\tilde{\lambda}}, \underline{\tilde{u}}, \underline{\tilde{v}}$ leads to

$$\begin{bmatrix} \underline{0} & \underline{K}^{\lambda u} & \underline{K}^{\lambda v} \\ \underline{K}^{u\lambda} & \underline{K}^{uu} & \underline{0} \\ \underline{K}^{v\lambda} & \underline{0} & \underline{K}^{vv} \end{bmatrix} \begin{bmatrix} \underline{\tilde{\lambda}} \\ \underline{\tilde{u}} \\ \underline{\tilde{v}} \end{bmatrix} = \begin{bmatrix} \underline{0} \\ \underline{f}^{u} \\ \underline{f}^{v} \end{bmatrix}, \tag{21}$$

where

$$\underline{K}^{\lambda u} = -\sum_e \int_{Ae} \underline{N}^{\lambda\lambda^T} \underline{N}_x^{uu} \, dA$$

$$\underline{K}^{u\lambda} = -\sum_e \int_{Ae} \underline{N}_x^{uu^T} \underline{N}^{\lambda\lambda} \, dA$$

$$\underline{K}^{\lambda v} = -\sum_e \int_{Ae} \underline{N}^{\lambda\lambda^T} \underline{N}_y^{vv} \, dA \qquad (22.1)$$

$$\underline{K}^{v\lambda} = -\sum_e \int_{Ae} \underline{N}_y^{vv^T} \underline{N}^{\lambda\lambda} \, dA$$

$$\underline{K}^{uu} = \mu \sum_e \int_{Ae} (\underline{N}_x^{uu^T} \underline{N}_x^{uu} + \underline{N}_y^{uu^T} \underline{N}_y^{uu}) \, dA$$

$$\underline{K}^{vv} = \mu \sum_e \int_{Ae} (\underline{N}_x^{vv^T} \underline{N}_x^{vv} + \underline{N}_y^{vv^T} \underline{N}_y^{vv}) \, dA$$

$$\underline{f}^{u} = \sum_e \int_{be} \underline{N}^{uu^T} (\mu u_n - \alpha p) \, db \qquad (22.2)$$

$$\underline{f}^{v} = \sum_e \int_{be} \underline{N}^{vv^T} (\mu v_n - \beta p) \, db$$

confirm symmetry. It follows that solution specification requires

$$\lambda = \lambda' \qquad (23.1)$$

to be a datum value,

$$\begin{aligned} u = u' \quad &\text{or} \quad \mu u_n - \alpha p = \mu u_n' - \alpha p' \\ v = v' \quad &\text{or} \quad \mu v_n - \beta p = \mu v_n' - \beta p' \end{aligned} \qquad (23.2)$$

to be the boundary conditions.

An equation for the recovery of the pressure can be extracted from the first variation of the perturbed functional, in that

$$\begin{aligned} &\int_A [\delta(u_x + v_y)\lambda + \delta\lambda(u_x + v_y)] \, dA \\ &+ \mu \int_A (\delta u_x u_x + \delta v_x v_x + \delta u_y u_y + \delta v_y v_y) \, dA \\ &= \int_b [\delta u(\mu u_n - \alpha p) + \delta v(\mu v_n - \beta p)] \, db. \end{aligned} \qquad (24)$$

Identification with equation (5) then reveals

$$p = -\lambda. \qquad (25)$$

Moreover, equation (25) is needed for the suppression of the pressure in the nonlinear case.

Discussion

In addition to the obvious similarities, a not so apparent characteristic which is common to all the approaches is the relaxation of the solenoidal constraint. This is consistent with the transformation to the weaker statement in the integral formulation. For the solenoidal approach, it is not possible to construct a conforming element which is pointwise solenoidal. A compromise is to impose the constraint on the average over the element. Even so

$$\int_{Ae} (u_x + v_y)\, dA = 0 \tag{26}$$

implies

$$\int_{be} (\alpha u + \beta v)\, db = 0 \tag{27}$$

which contradicts the definition of conforming independent nodal parameters. However, small errors in certain terms are sufficient to overcome this dilemma [7]. There is also a restriction on the order of the formula used to evaluate the penalty function [8]. There is also a restriction on the order of the polynomial used to represent the Lagrange multiplier [9]. These manifestations of the relaxation reinforce the notion that all the approaches are variations on the same theme.

UNIVERSAL ORTHOGONAL METHOD

With the establishment of the origins of the approaches, it is appropriate to elaborate on the solenoidal approach for the nonlinear case. This necessitates the use of a universal orthogonal method which is distinguished by its circumvention of the employment of a variational functional. The other approaches follow a similar line.

Under the constraint of solenoidal velocity components, equation (4) becomes

$$\begin{aligned} &\mu \int_A (\delta u_x u_x + \delta v_x v_x + \delta u_y u_y + \delta v_y v_y)\, dA \\ &+ \rho \int_A [\delta u(uu_x + vu_y) + \delta v(uv_x + vv_y)]\, dA \\ &= \int_b [\delta u(\mu u_n - \alpha p) + \delta v(\mu v_n - \beta p)]\, db. \end{aligned} \tag{28}$$

The nonlinear terms can be handled by the introduction of the complete first order expansions

$$\begin{aligned} uu_x + vu_y + u^*u_x{}^* + v^*u_y{}^* &= u^*u_x + v^*u_y + uu_x{}^* + vu_y{}^* \\ uv_x + vv_y + u^*v_x{}^* + v^*v_y{}^* &= u^*v_x + v^*v_y + uv_x{}^* + vv_y{}^* \end{aligned} \tag{29}$$

where an asterisk signifies the value from the prior iteration of the variable or its derivatives.

Again, the cross-coupled local behaviour is

$$\begin{bmatrix} u \\ v \end{bmatrix} = \begin{bmatrix} \underline{N}^{uu} & \underline{N}^{uv} \\ \underline{N}^{vu} & \underline{N}^{vv} \end{bmatrix} \begin{bmatrix} \underline{\tilde{u}} \\ \underline{\tilde{v}} \end{bmatrix}. \tag{30}$$

But, now, on substitution into the integral formulation

$$\begin{bmatrix} \delta\underline{\tilde{u}}^T & \delta\underline{\tilde{v}}^T \end{bmatrix} \begin{bmatrix} \underline{K}^{uu} & \underline{K}^{uv} \\ \underline{K}^{vu} & \underline{K}^{vv} \end{bmatrix} \begin{bmatrix} \underline{\tilde{u}} \\ \underline{\tilde{v}} \end{bmatrix} = \begin{bmatrix} \delta\underline{\tilde{u}}^T & \delta\underline{\tilde{v}}^T \end{bmatrix} \begin{bmatrix} \underline{f}^u \\ \underline{f}^v \end{bmatrix}, \tag{31}$$

whereupon

$$\begin{bmatrix} \underline{K}^{uu} & \underline{K}^{uv} \\ \underline{K}^{vu} & \underline{K}^{vv} \end{bmatrix} \begin{bmatrix} \underline{\tilde{u}} \\ \underline{\tilde{v}} \end{bmatrix} = \begin{bmatrix} \underline{f}^u \\ \underline{f}^v \end{bmatrix} \tag{32}$$

provides the solution on acceptable convergence of successive iterations. As shown by the definitions

$$\begin{aligned}
\underline{K}^{uu} &= \mu \sum_e \int_{Ae} (\underline{N}_x^{uu\,T} \underline{N}_x^{uu} + \underline{N}_x^{vu\,T} \underline{N}_x^{vu} + \underline{N}_y^{uu\,T} \underline{N}_y^{uu} + \underline{N}_y^{vu\,T} \underline{N}_y^{vu})\, dA \\
&+ \rho \sum_e \int_{Ae} \underline{N}^{uu\,T} (u^*\underline{N}_x^{uu} + u_x^*\underline{N}^{uu} + v^*\underline{N}_y^{uu} + u_y^*\underline{N}^{vu})\, dA \\
&+ \rho \sum_e \int_{Ae} \underline{N}^{vu\,T} (u^*\underline{N}_x^{vu} + v_x^*\underline{N}^{uu} + v^*\underline{N}_y^{vu} + v_y^*\underline{N}^{vu})\, dA \\
\underline{K}^{uv} &= \mu \sum_e \int_{Ae} (\underline{N}_x^{uu\,T} \underline{N}_x^{uv} + \underline{N}_x^{vu\,T} \underline{N}_x^{vv} + \underline{N}_y^{uu\,T} \underline{N}_y^{uv} + \underline{N}_y^{vu\,T} \underline{N}_y^{vv})\, dA \\
&+ \rho \sum_e \int_{Ae} \underline{N}^{uu\,T} (u^*\underline{N}_x^{uv} + u_x^*\underline{N}^{uv} + v^*\underline{N}_y^{uv} + u_y^*\underline{N}^{vv})\, dA \\
&+ \rho \sum_e \int_{Ae} \underline{N}^{vu\,T} (u^*\underline{N}_x^{vv} + v_x^*\underline{N}^{uv} + v^*\underline{N}_y^{vv} + v_y^*\underline{N}^{vv})\, dA \\
\underline{K}^{vu} &= \mu \sum_e \int_{Ae} (\underline{N}_x^{uv\,T} \underline{N}_x^{uu} + \underline{N}_x^{vv\,T} \underline{N}_x^{vu} + \underline{N}_y^{uv\,T} \underline{N}_y^{uu} + \underline{N}_y^{vv\,T} \underline{N}_y^{vu})\, dA \\
&+ \rho \sum_e \int_{Ae} \underline{N}^{uv\,T} (u^*\underline{N}_x^{uu} + u_x^*\underline{N}^{uu} + v^*\underline{N}_y^{uu} + u_y^*\underline{N}^{vu})\, dA \\
&+ \rho \sum_e \int_{Ae} \underline{N}^{vv\,T} (u^*\underline{N}_x^{vu} + v_x^*\underline{N}^{uu} + v^*\underline{N}_y^{vu} + v_y^*\underline{N}^{vu})\, dA \\
\underline{K}^{vv} &= \mu \sum_e \int_{Ae} (\underline{N}_x^{uv\,T} \underline{N}_x^{uv} + \underline{N}_x^{vv\,T} \underline{N}_x^{vv} + \underline{N}_y^{uv\,T} \underline{N}_y^{uv} + \underline{N}_y^{vv\,T} \underline{N}_y^{vv})\, dA \\
&+ \rho \sum_e \int_{Ae} \underline{N}^{uv\,T} (u^*\underline{N}_x^{uv} + u_x^*\underline{N}^{uv} + v^*\underline{N}_y^{uv} + u_y^*\underline{N}^{vv})\, dA \\
&+ \rho \sum_e \int_{Ae} \underline{N}^{vv\,T} (u^*\underline{N}_x^{vv} + v_x^*\underline{N}^{uv} + v^*\underline{N}_y^{vv} + v_y^*\underline{N}^{vv})\, dA
\end{aligned} \tag{33.1}$$

$$\underline{f}^u = \rho \sum_e \int_{Ae} [\underline{N}^{uu^T} (u^*u_x^* + v^*u_y^*) + \underline{N}^{vu^T} (u^*v_x^* + v^*v_y^*)] \, dA$$

$$+ \sum_e \int_{be} [\underline{N}^{uu^T} (\mu u_n - \alpha p) + \underline{N}^{vu^T} (\mu v_n - \beta p)] \, db \qquad (33.2)$$

$$\underline{f}^v = \rho \sum_e \int_{Ae} [\underline{N}^{uv^T} (u^*u_x^* + v^*u_y^*) + \underline{N}^{vv^T} (u^*v_x^* + v^*v_y^*)] \, dA$$

$$+ \sum_e \int_{be} [\underline{N}^{uv^T} (\mu u_n - \alpha p) + \underline{N}^{vv^T} (\mu v_n - \beta p)] \, db$$

symmetry does not prevail. To complete the solution

$$\begin{aligned} u &= u' \quad \text{or} \quad \mu u_n - \alpha p = \mu u_n' - \alpha p', \ \mu v_n - \beta p = \mu v_n' - \beta p' \\ v &= v' \quad \text{or} \quad \mu u_n - \alpha p = \mu u_n' - \alpha p', \ \mu v_n - \beta p = \mu v_n' - \beta p' \end{aligned} \qquad (34.1)$$

are the boundary conditions,

$$\begin{aligned} u &= u^* \\ v &= v^* \end{aligned} \qquad (34.2)$$

are the initial estimates.

Details of a suitable element for use with this approach are contained in [7].

SAMPLE APPLICATION

Figure 1 and Figure 2 furnish some evidence of the merit of this approach. The flow in a duct with an expansion is the object of considerable research and so proves a worthwhile test problem. Such a flow is of value because it exhibits various important features. At the entrance to the duct, a boundary layer forms on each wall. Its thickness increases with the distance from the entrance. With an expansion in the duct, a recirculation region occurs in the corner. Its size depends on the abruptness of the expansion.

The situation chosen here is set out in part (a). The characteristic length is the step height. The characteristic speed is the source velocity. These define the Reynolds number Re. Incrementing this flow parameter in stages, with the null solution as the initial estimate, convergence is achieved after 4 iterations at a Reynolds number of 10 then 5 iterations at a Reynolds number of 100. Individual profiles are displayed in part (b). Vectors for the separate Reynolds numbers are illustrated in part (c) plus part (d). In spite of the courseness of the discretisation, it is possible to discern the expected features.

CONCLUSIONS

The previous sections discuss the origins of several useful approaches. It is demonstrated that each can be interpreted as an alternative means for the imposition of the solenoidal constraint. In the process, the solenoidal approach, notable for its relative absence from the published literature, is revealed to be an attractive proposition. Its basis is its trial functions which are devised so as to have zero divergence. The difficulty, of course, lies in the construction of the element. Unfortunately, space does not permit examination of this particular aspect here.

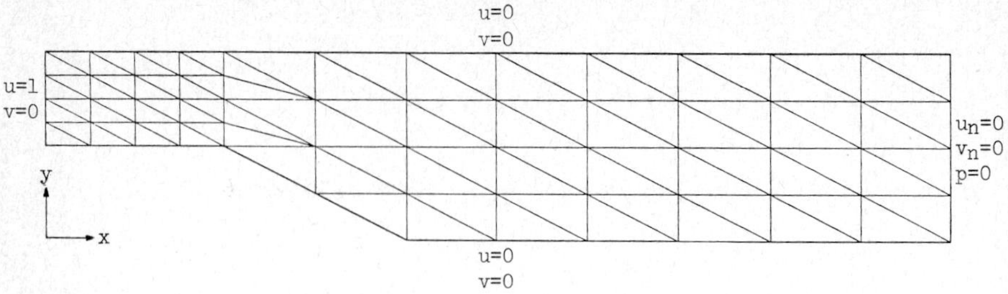

(a) - Problem statement

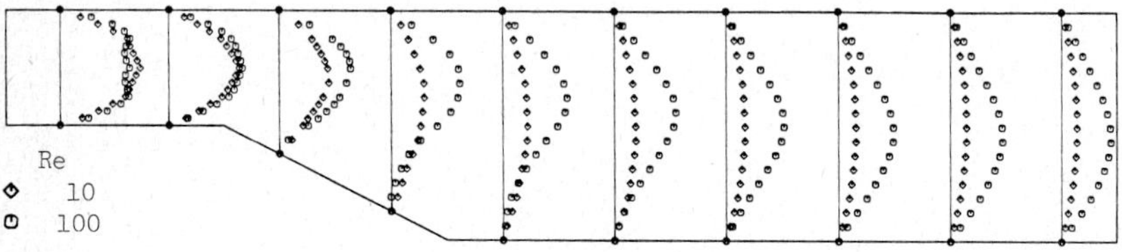

(b) - Velocity profiles

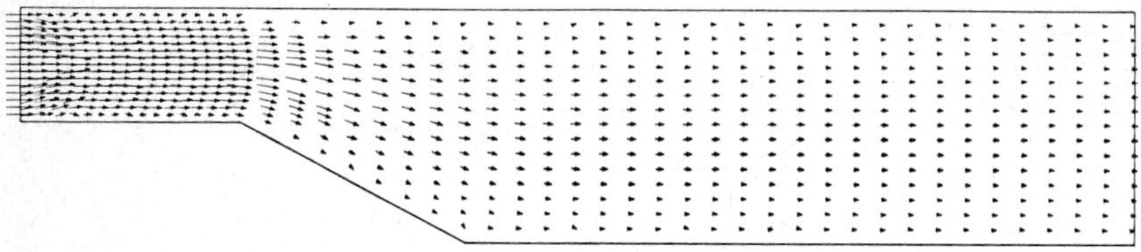

(c) - Velocity vectors at Re=10

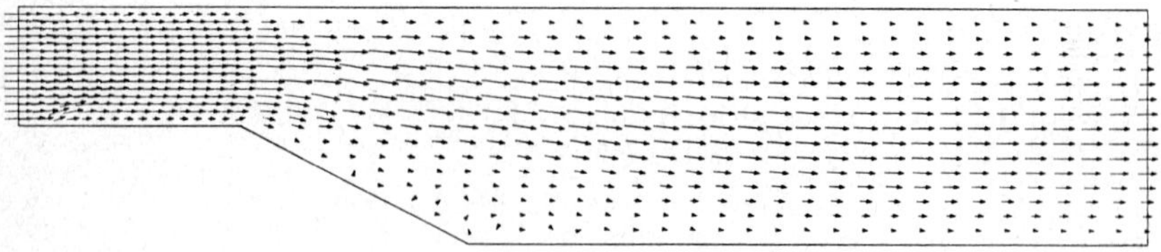

(d) - Velocity vectors at Re=100

Figure 1 - Entrance flow with gentle expansion

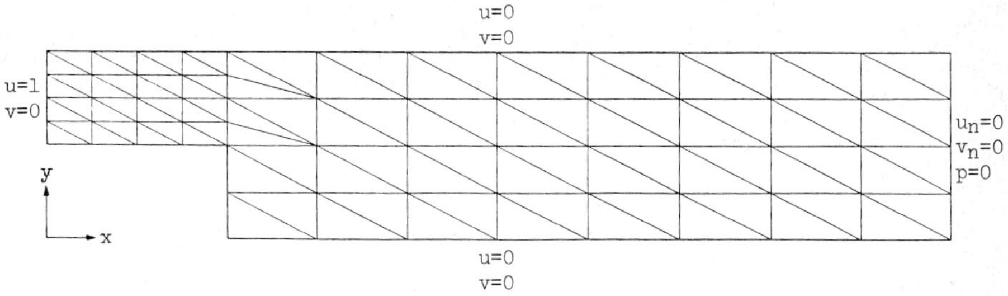

(a) - Problem statement

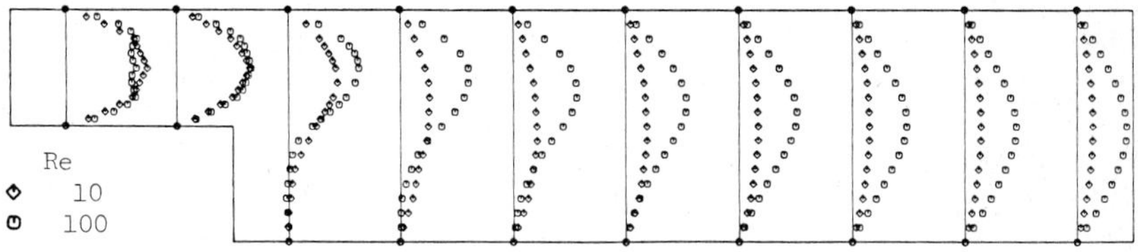

(b) - Velocity profiles

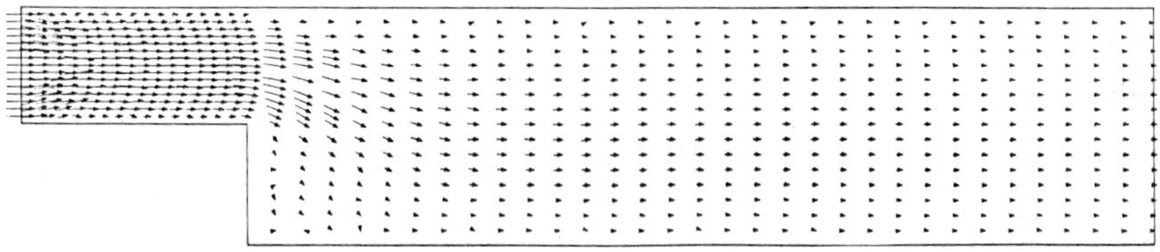

(c) - Velocity vectors at Re=10

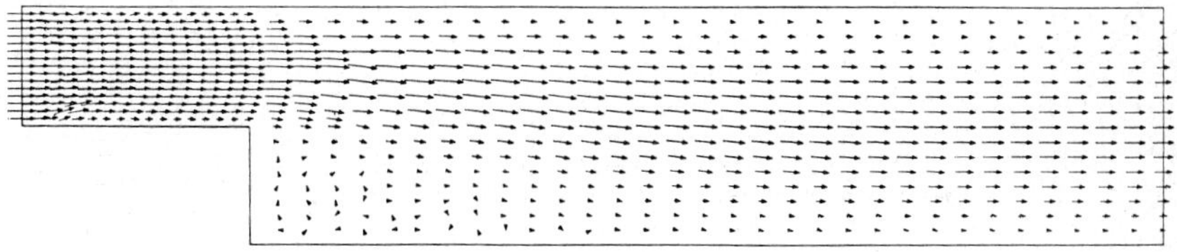

(d) - Velocity vectors at Re=100

Figure 2 - Entrance flow with sudden expansion

REFERENCES

[1] Temam R. - "Some finite element methods in fluid flow" - Proceedings of the Sixth International Conference on Numerical Methods in Fluid Dynamics, Tbilisi, 1978, 34-55

[2] Mack N. - "An uncoupled approach for primitive variable solutions to viscous incompressible flows" - Proceedings of the Third International Conference on Finite Elements in Flow Problems, Banff, 1, 1980, 121-131

[3] Hughes T.J., Taylor R.L., Levy J.F. - "A finite element method for incompressible viscous flows" - Proceedings of the Second International Symposium on Finite Element Methods in Flow Problems, Rapallo, 1976, 3-15

[4] Bercovier M., Engelman M. - "A finite element for the numerical solution of viscous incompressible flows" - Journal of Computational Physics, 30, 1979, 181-201

[5] Hood P., Taylor C. - "Navier-Stokes equations using mixed interpolation" - Proceedings of the International Symposium on Finite Element Methods in Flow Problems, Swansea, 1974, 121-132

[6] Gartling D.K., Nickell R.E., Tanner R.I. - "A finite element convergence study for accelerating flow problems" - International Journal for Numerical Methods in Engineering, 11, 1977, 1155-1174

[7] Mack N. - A Solenoidal Approach to Viscous Flow Simulation, Thesis, Doctor of Philosophy, University of Sydney, to appear in 1983

[8] Zienkiewicz O.C., Godbole P.N. - "Viscous incompressible flow with special reference to non-Newtonian (plastic) fluids" - Finite Elements in Fluids, 1, 1975, 25-55

[9] Tuann S.Y., Olson M.D. - A Study of Various Finite Element Solution Methods for the Navier-Stokes Equations, Structural Research Series, Report 14, Department of Civil Engineering, University of British Columbia, 1976

Computational Techniques & Applications: CTAC-83
J. Noye & C. Fletcher (Editors)

TIME-SPLITTING AND THE GROUP FINITE ELEMENT FORMULATION

C.A.J. Fletcher

University of Sydney
Sydney, N.S.W. 2006
Australia

It is shown that the *group* finite element formulation is substantially more economical and no less accurate than the conventional finite element formulation particularly if higher dimensions or higher-order nonlinearities occur. Solutions to the two-dimensional Burgers' equations illustrate the improvement. It is also shown that higher-order trial functions are particularly uneconomical for three-dimensional problems. The group formulation also permits *directional mass and difference operators* to be extracted when used with Lagrange trial functions. In turn this allows a very efficient three level time-split algorithm to be developed which is used to obtain steady-state solutions for the flow over a backward-facing step.

1. INTRODUCTION

For steady flow problems it is very common to construct an equivalent unsteady formulation which is marched in time until the transient has disappeared leaving the required steady solution. Since the transient solution is of no particular interest time merely provides a convenient iteration path. Alternatively the time-dependent terms may be modified to provide faster convergence; this is the basis of the pseudo-transient technique (Mallinson and de Vahl Davis, 1973).

One of the major attractions of the equivalent unsteady approach is the very large radius of convergence that the method provides; the steady-state solution can be reached from virtually any reasonable and, a lot of extremely unreasonable, starting solutions. By contrast attempts to use a Newton-Raphson technique to solve the steady equations iteratively almost requires a knowledge of the exact solution if the number of unknowns, N, in the problem is large. This is because (Rheinboldt, 1974), for the Newton-Raphson method, the radius of convergence shrinks as N increases.

For the equivalent unsteady formulation, since the transient solution is not required, the execution time spent in reaching the steady state should be minimised. This effectively rules out explicit schemes, since the stability restriction on the time-step requires a large number of time-steps to reach the steady state and consequently a large execution time. Although a fully implicit scheme would avoid the stability restriction, the requirement of solving a non-narrow banded system of equations at every time-step implies a large execution time per time-step and, consequently, a large overall execution time.

However by manipulating the implicit equations into a factored or time-split form (Gourlay, 1977) it is possible to replace the original multi-dimensional system of equations by a sequence of smaller one-dimensional system of equations. Each one-dimensional system of equations is narrow banded and, as a result, can be efficiently solved. Although time-splitting is often used in finite difference formulations of flow problems it has only recently been demonstrated to be effective with the finite element method (Fletcher, 1980, 1981, 1982a).

For the conventional finite element method part of the difficulty is associated with the treatment of the nonlinear convective terms. The requirement that a separate trial solution be introduced in each element for each dependent variable prevents an efficient splitting from being achieved. In addition the product of the individual trial solutions in the convective term produces a large number of terms in the resulting algebraic equations whose manipulation is particularly uneconomic. This problem is aggravated in higher dimensions or if higher-order trial functions are used or if higher-order nonlinearities occur.

Fortunately both problems with the convective terms can be overcome by adopting a *group* finite element formulation (Fletcher 1983a). In the group formulation the convective terms are recast in conservation form and a single trial solution is introduced for the group of terms. This technique is demonstrated using Burgers' equations in Section 2, where it is shown that a significant gain in economy can be obtained while still achieving a small gain in accuracy.

Even with the help of the group formulation the finite element method is still relatively uneconomic in higher dimensions or if higher-order elements are used due to the increased *connectivity* associated with the integration over all spatial dimensions.

For Lagrange elements this increased connectivity can be readily demonstrated by isolating the *directional mass and difference operators* (Fletcher and Srinivas, 1983). The mass and difference operators and the general problem of increased connectivity in higher dimensions are discussed in Section 3.

The explicit extraction of the mass and directional difference operators also permits a *consistent* time-split formulation to be developed. Incorporation of this formulation into a three-level time-marching framework produces a very robust procedure for obtaining the steady-state solution. This algorithm is described in Section 4 and applied to the problem of separating incompressible viscous flow over a step.

The above procedures and algorithms have also been applied to viscous compressible laminar and turbulent flow (Srinivas and Fletcher, 1983a). Both the group formulation and the time-split algorithm can be extended to handle domains of irregular geometry by applying the finite element formulation in the *generalised co-ordinate* plane. This aspect is described in a companion paper (Srinivas and Fletcher, 1983b).

2. GROUP FINITE ELEMENT FORMULATION

The group formulation avoids the products of nodal values that are generated by the conventional finite element method as a result of introducing separate trial solutions for each dependent variable in the convective terms. This can be illustrated by considering the one-dimensional Burgers' equation, i.e.

$$u_t + uu_x - \frac{1}{Re} u_{xx} = 0 \quad . \tag{1}$$

Here Re is a Reynolds number-like parameter. In the conventional finite element method a trial solution is introduced for u, as

$$u = \sum_j N_j(x) u_j(t) \quad , \tag{2}$$

where $N_j(x)$ are the trial functions and $u_j(t)$ are the nodal values of u. Substitution into eq. (1) and application of the Galerkin integral condition generates a system of ordinary differential equations. If linear trial functions are applied on a uniform grid the system of ordinary differential equations has the following form,

$$\frac{1}{6}\dot{u}_{i-1} + \frac{2}{3}\dot{u}_i + \frac{1}{6}\dot{u}_{i+1} + \frac{1}{3}(u_{i-1}+u_i+u_{i+1})\frac{(u_{i+1}-u_{i-1})}{2\Delta x} - \frac{1}{Re}\frac{(u_{i-1}-2u_i+u_{i+1})}{\Delta x^2} = 0 \tag{3}$$

where i denotes a grid point in the x direction and $\dot{u} \equiv u_t$.

The group finite element formulation is applied in two stages. First the convective terms are written in *divergence or conservation* form. Thus eq. (1) becomes

$$u_t + 0.5\,(u^2)_x - \frac{1}{Re}u_{xx} = 0 \quad . \tag{4}$$

The second stage consists of introducing an additional trial solution for the group, u^2. That is

$$u^2 = \sum_j N_j(x)u_j^2(t) \quad . \tag{5}$$

Application of the Galerkin method, with linear trial functions in eqs. (2) and (5) and a uniform grid, produces the result

$$\frac{1}{6}\dot{u}_{i-1} + \frac{2}{3}\dot{u}_i + \frac{1}{6}\dot{u}_{i+1} + \frac{1}{2}(u_{i+1} + u_{i+1})\frac{(u_{i+1} - u_{i-1})}{2\Delta x} - \frac{1}{Re}\frac{(u_{i-1}-2u_i+u_{i+1})}{\Delta x^2} = 0 \tag{6}$$

In one dimension the difference between the group and conventional finite element formulations, as indicated by eqs. (3) and (6), is not very great. Both formulations demonstrate comparable convergence properties, accuracy and economy for trial functions of linear, quadratic and cubic order (Fletcher, 1982b).

For incompressible flow the convective terms are similar to those in eqs. (1) or (4); that is quadratic nonlinearities occur. For compressible flow terms of the form $\rho u v_x$ or $\rho v u_y$ appear. In the conventional finite element method representation of such terms requires the product of individual trial solutions for each of the dependent variables. This implies a large number of coupled nodal values and consequently an uneconomic method. This problem of increased connectivity is aggravated in three dimensions or if higher-order trial functions are used. Here connectivity may be quantified as the number of nodal groups appearing in the algebraic expressions after application of the finite element method. When the pseudo-transient formulation is used with an efficient algorithm for solving the implicit system of equations associated with each grid-line the execution time is dominated by the evaluation of the steady-state residual. The execution time is directly proportional to the operation count. Therefore an operation count comparison for the group and conventional finite element methods will also provide an indirect comparison of the relative economy of the two methods. Such a comparison is shown in Table 1.

Equation system	Convective non-linearity	Conventional F.E.M.		Group F.E.M.		Conventional R.O.C. / Group R.O.C.
		Connectivity (convective nonlinearity)	Residual operation count	Connectivity (convective nonlinearity)	Residual operation count	
2-D Burgers' equations	quadratic	49	828	9	206	4
3-D Burgers' equations	quadratic	343	12603	27	1308	9
2-D viscous compressible flow	cubic	225	6772	9	404	16.8
3-D viscous compressible flow	cubic	3375	217065	27	2349	92.4

Table 1 Connectivity and operation count for linear conventional and group finite element methods.

The residual operation counts shown in Table 1 are the number of equivalent additions (one multiplication or division = three additions or subtractions) to evaluate all the steady-state residuals. The 2-D Burgers' equations require the evaluation of two residuals whereas 3-D viscous compressible flow requires the evaluation of four residuals. The operation counts are based on the use of linear trial functions in rectangular (two dimensions) and brick (three dimensions) elements.

An examination of Table 1 demonstrates that the connectivity of the convective terms for the conventional finite element method *grows rapidly* with an increase in the dimension of the problem or in the order of the nonlinearity. A substantial increase in the connectivity would also be expected if higher-order trial functions were used. In contrast there is a small increase in connectivity with dimension for the group finite element method but *none* with the order of the nonlinearity.

The relative magnitude of the connectivity is also reflected in the relative magnitude of the residual operation counts for the conventional and group finite element methods. The ratio of the operation counts, shown in the last column of Table 1, indicates the ratio of the execution times for the conventional and group finite element methods. It is clear that the conventional finite element method is particularly *uneconomical* as the dimension or the order of the nonlinearity increases.

2.1 Two-dimensional Burgers' Equations

Numerical experiments with the two-dimensional Burgers' equations (Fletcher, 1983b) indicate that the execution time for the conventional finite element method is about two and a half times that of the group finite element method. This is in reasonable agreement with the residual operation count ratio of four shown in Table 1, since the execution time comparison includes the formation and solution of the tridiagonal system of equations, which is common to both the conventional and group methods.

The two-dimensional Burgers' equations are

$$u_t + uu_x + vu_y - \frac{1}{Re}(u_{xx} + u_{yy}) = 0$$

$$\text{and} \quad v_t + uv_x + vv_y - \frac{1}{Re}(v_{xx} + v_{yy}) = 0 \quad . \tag{7}$$

where Re is a Reynolds number-like parameter.

Using the Cole-Hopf transformation exact solutions of eqs. (7) can be constructed without difficulty (Fletcher, 1983c). A typical solution for a moderate internal gradient is shown in Fig. 1.

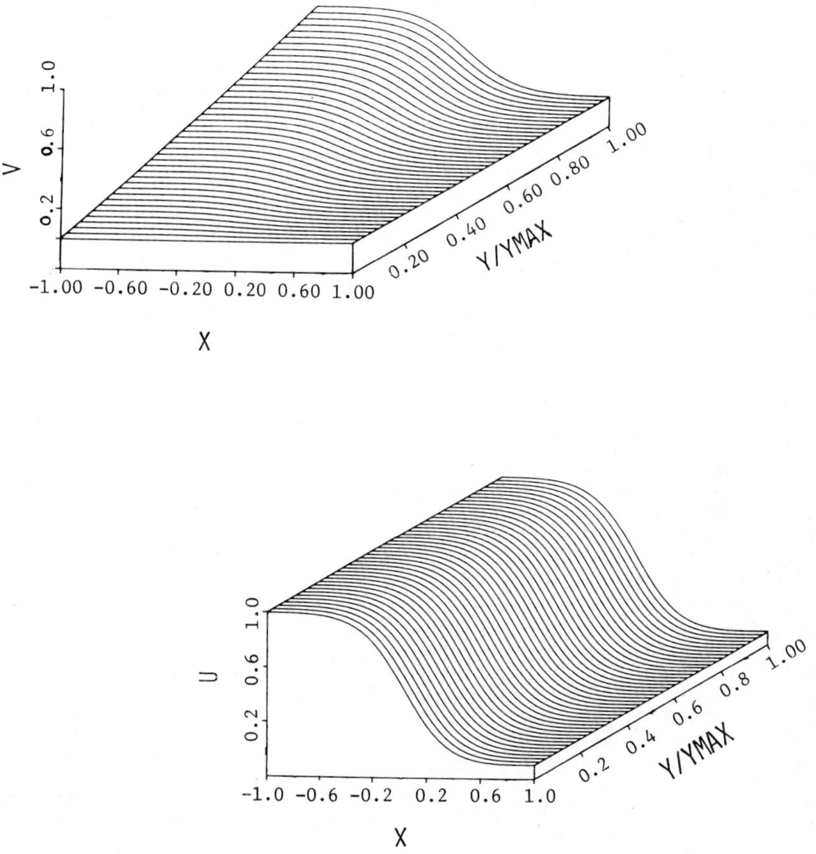

Figure 1
Exact steady-state solutions of the two-dimensional Burgers' equation with a moderate internal gradient.

An advantage of being able to obtain exact solutions of the two-dimensional Burgers' equations is that the accuracy and convergence properties of the group and conventional finite element methods can be compared precisely. The variation of the r.m.s. error in the steady-state solution (u only) with grid size is shown in Fig. 2. Linear trial functions have been used with both the conventional and group finite element formulations.

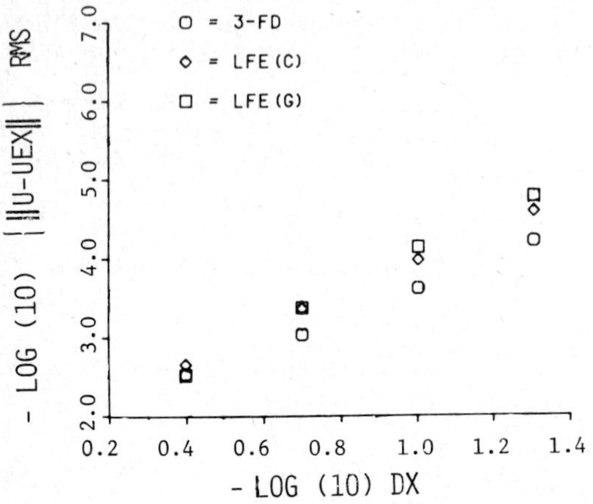

Figure 2
Spatial convergence properties for a moderate internal grid with Re = 10; 3-FD = three-point finite difference method; LFE(C) = conventional linear finite element method; LFE(G) = group linear finite element method.

The results shown in Fig. 2 indicate that both the group and conventional finite element formulations are more accurate than the three-point finite difference scheme. However a more important feature of Fig. 2 is that there is *no loss of accuracy* associated with introducing the group formulation; in fact there is a small increase in accuracy.

Application to a number of problems (Fletcher, 1983a) suggests that the group finite element formulation is considerably more economical than the conventional finite element and is no less accurate.

3. MASS AND DIFFERENCE OPERATORS

As well as being very economical the group formulation facilitates the extraction of the directional mass and difference operators which is an important intermediate step in generating a consistent time-split algorithm. This is demonstrated for the vorticity transport equation.

3.1 Vorticity Transport Equation

For two-dimensional incompressible viscous flow the governing equations are

$$\frac{\partial \zeta}{\partial t} + \frac{\partial}{\partial x}(u\zeta) + \frac{\partial}{\partial y}(v\zeta) - \frac{1}{Re}\left(\frac{\partial^2 \zeta}{\partial x^2} + \frac{\partial^2 \zeta}{\partial y^2}\right) = 0 \qquad (8)$$

$$\frac{\partial^2\psi}{\partial x^2} + \frac{\partial^2\psi}{\partial y^2} = \zeta \tag{9}$$

where $\zeta = \partial u/\partial y - \partial v/\partial x$ and $u = \partial\psi/\partial y$, $v = -\partial\psi/\partial x$.

In eqs. (8) and (9) ζ is the vorticity and ψ is the streamfunction. Equation (8) is the vorticity transport equation which is derived from the x and y momentum equations.

The group finite element formulation is applied with linear Lagrange rectangular elements. Neighbouring elements may vary in size with the restriction that elements remain rectangular as indicated in Fig. 3.

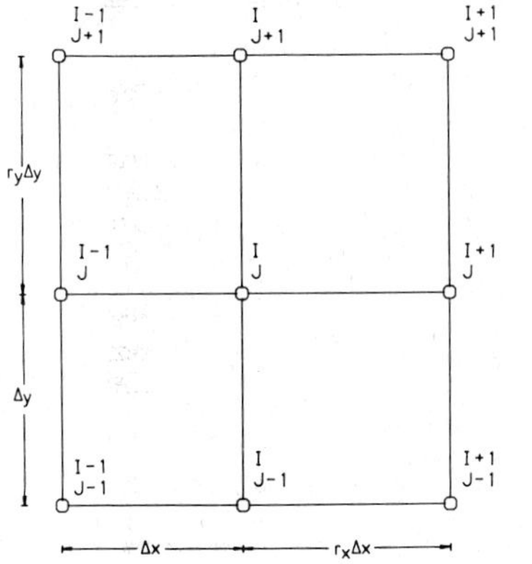

Figure 3
Global grid notation

The resulting system of ordinary differential equations can be written (after dividing by $\Delta x\ \Delta y$)

$$M_x \otimes M_y \dot{\zeta} + M_y \otimes L_x\, u\zeta + M_x \otimes L_y v\zeta - \frac{1}{Re}\left\{M_y \otimes L_{xx} + M_x \otimes L_{yy}\right\} \zeta = 0 \tag{10}$$

where the *directional mass operators* are

$$M_x \equiv \left\{\frac{1}{6}\ , \frac{(1+r_x)}{3}\ , \frac{r_x}{6}\right\}\ , \ M_y^t \equiv \left\{\frac{r_y}{6}\ , \frac{(1+r_y)}{3}\ , \frac{1}{6}\right\} \tag{11}$$

and the *directional difference operators* are

$$L_x \equiv \frac{1}{2\Delta x}\left\{-1,0,1\right\}\ , \quad L_y^t \equiv \frac{1}{2\Delta y}\left\{1,0,-1\right\} \tag{12}$$

$$L_{xx} \equiv \frac{1}{\Delta x^2}\left\{1,\ -(1+1/r_x),\ 1/r_x\right\}\ ,\quad L^t_{yy} \equiv \frac{1}{\Delta y^2}\left\{1/r_y,\ -(1+1/r_y),\ 1\right\} \qquad (13)$$

and $\otimes$ denotes the tensor product.

The difference operators, e.g. L_x, are the same as occur in the finite difference method, at least for a uniform grid. The integral nature of the finite element method is responsible for the appearance of the directional mass operators M_x, M_y. These operators effectively *transform* three-point finite difference formulae into nine-point finite element formulae, for linear elements in two dimensions.

In three dimensions terms like

$$M_y \otimes M_z \otimes L_x F$$

appear, where F is the group of dependent variables that replaces $u\zeta$ in eq. (10). In three dimensions the tensor product involving M_y and M_z effectively converts a three-point finite difference representation into a twenty seven-point finite element representation.

The extraction of explicit directional mass and difference operators is possible due to the *product nature* of Lagrange trial functions i.e.

$$N_j(x,y,z) = N_j^{(x)}(x)\, N_j^{(y)}(y)\, N_j^{(z)}(z)\ . \qquad (14)$$

However such is not the case for Serendipity trial functions of any order higher than linear (Fletcher, 1983d).

In eq. (14) the one-dimensional component, $N_j^{(x)}(x)$, would have the following form for linear elements,

$$N_j^{(x)}(\xi) = 0.5(1 + \xi_j\xi) \qquad (15)$$

where $-1 \leqslant \xi \leqslant 1$ and $\xi_j = \pm 1$. Here ξ is an element-based co-ordinate. The various operators, in eqs. (11) - (13), are obtained from

$$M_x = \sum_e \int_{-1}^{1} N_j^{(x)}(\xi) \quad N_k^{(x)}(\xi)\ d\xi \qquad (16)$$

$$L_x = \sum_e \int_{-1}^{1} \frac{\partial N_j^{(x)}(\xi)}{\partial x} \quad N_k^{(x)}(\xi)\ d\xi \qquad (17)$$

$$\text{and}\quad L_{xx} = \sum_e \int_{-1}^{1} \frac{\partial N_j^{(x)}(\xi)}{\partial x}\ \frac{\partial N_k^{(x)}(\xi)}{\partial x}\ d\xi \qquad (18)$$

where $\sum_e$ denotes a summation over all elements. In eq. (18) there would also be a line integral contribution if node k was on the boundary. The general form of eqs. (16) to (18) is equally applicable to higher-order trial functions, although the operators will then have more components. Thus for quadratic elements M_x is

$$\text{for corner nodes:}\quad M_x \equiv \frac{1}{15}\left\{-1, 2, 8, 2, -1\right\}$$

for midside nodes: $M_x \equiv \frac{2}{15}\left\{0,1,8,1,0\right\}$. (19)

Even for the group formulation the appearance of the directional mass operators implies an increased connectivity in higher dimensions or if higher-order trial functions are used. The increased connectivity implies a reduced economy. Whether the increase in accuracy associated with the appearance of the directional mass operators outweighs the reduction in economy is not known at present.

3.2 Connectivity for Higher-Order Elements or Higher Dimensions

The increase in connectivity for the finite element method that is associated with the directional mass operators can be partially quantified by considering a simple linear equation such as Laplace's equation,

$$\nabla^2 u = 0 . \tag{20}$$

Since eq. (20) is linear the group and the conventional finite element formulations coincide.

Application of the finite difference or finite element method would produce a system of algebraic equations,

$$\underline{K}\,\underline{u} = \underline{B} \tag{21}$$

where $\underline{K}$ contains the algebraic coefficients produced by the discretisation and $\underline{B}$ additionally contains the known values of $\underline{u}$ on the boundary. Each row of $\underline{K}$ contains as many terms as there are connected nodes i.e. the connectivity considered in Section 2.

If eq. (21) is solved iteratively the execution time per iteration would be proportional to the number of non-zero terms in each row. If eq. (21) is solved by a direct bandwidth solver the execution time will be proportional to the square of the number of non-zero terms in each row (Jennings, 1977).

The precise number of non-zero terms in each row will depend on whether the equation is centered at a corner or midside node, for the finite element method. To provide a comparison with the finite difference method, for which each row will contain the same number of terms, an rms average number of non-zero terms per row has been calculated for the finite element method. The results are shown in Table 2.

Dimension	FINITE ELEMENT (LAGRANGE) METHOD		FINITE DIFFERENCE METHOD	
	Order of Shape Function	Average Number of Terms	N-Point Formula	Number of Terms
1	Linear	3	3	3
	Quadratic	4	5	5
	Cubic	5	7	7
2	Linear	9	3	5
	Quadratic	17	5	9
	Cubic	29	7	13
3	Linear	27	3	7
	Quadratic	70	5	13
	Cubic	140	7	19

Table 2 Average Number of Nonzero Terms in $\underline{K}$

For the finite difference formulae in two and three dimensions three-, five- and seven-point formulae have been assumed in each coordinate direction.

It is clear from Table 2 that the connectivity (number of nonzero terms in each row) increases more rapidly with the finite element method if either the dimension is increased or higher-order trial functions are used. Therefore the use of high-order trial functions in three dimensions must produce a significantly higher accuracy if the finite element method is to be computationally more efficient than the finite difference method. Results for the one and two-dimensional Burgers' equations (Fletcher, 1983b) suggest that the use of trial functions higher than linear is not cost-effective in *more than one dimension*.

4. CONSISTENT TIME-SPLIT ALGORITHM

The results and discussion in the previous sections indicate that a group finite element formulation based on linear rectangular elements will be computationally more efficient (accuracy per unit execution time) than either the conventional finite element method or the use of higher-order trial functions.

In this section a consistent time-split algorithm is developed for the system of ordinary differential equations, (10) to (13), obtained from the application of the group Galerkin finite element method to the vorticity transport equation, (8). Subsequently sample results are presented for the flow of a viscous, incompressible fluid past a backward-facing step.

A three-level finite difference expression is introduced for the time-derivative ($\dot{\zeta} \equiv \zeta_t$) in eq. (10) and the result is

$$M_x \otimes M_y \left\{ \alpha \frac{\Delta\zeta^{n+1}}{\Delta t} + (1-\alpha) \frac{\Delta\zeta^n}{\Delta t} \right\} = \beta \, RHS^{n+1} + (1-\beta) RHS^n \tag{22}$$

where $\Delta\zeta^{n+1} = \zeta^{n+1} - \zeta^n$, $\Delta\zeta^n = \zeta^n - \zeta^{n-1}$ and

$$RHS = \frac{1}{Re} (M_y \otimes L_{xx} + M_x \otimes L_{yy})\zeta - M_y \otimes L_x u\zeta - M_x \otimes L_y v\zeta \, . \tag{23}$$

In eq. (22) α and β are parameters introduced to weight the time levels n and n+1 at which $\Delta\zeta$ and RHS are evaluated. In eq. (22) the solutions at time levels n-1 and n are known and the solution at time level n+1 is to be obtained, by constructing an algorithm for $\Delta\zeta^{n+1}$.

To obtain a linear system of equations for $\Delta\zeta^{n+1}$ it is necessary to expand RHS^{n+1} as a Taylor series about time level n. This allows eq. (22) to be rearranged as

$$\alpha(M_x \otimes M_y - \Delta t\beta \, \partial(RHS)/\partial\zeta) \, \Delta\zeta^{n+1} = \Delta t \, RHS^{n,\beta} - (1-\alpha) \, M_x \otimes M_y \Delta\zeta^n \tag{24}$$

where u and v in $RHS^{n,\beta}$ are evaluated at time, $t^n+\beta\Delta t$. All other terms are evaluated at t^n.

Equation (24) is factorised into

$$\alpha\left[M_x - \Delta t \frac{\beta}{\alpha} (\frac{1}{Re} L_{xx} - L_x u)\right] \otimes \left[M_y - \Delta t \frac{\beta}{\alpha} (\frac{1}{Re} L_{yy} - L_y v)\right] \Delta\zeta^{n+1}$$

$$= \Delta t \, RHS^{n,\beta} - (1-\alpha) \, M_x \otimes M_y \, \Delta\zeta^n \, , \tag{25}$$

which introduces an additional term of $O(\Delta t^2)$. The present *consistent* splitting preserves the directional mass operators which are expected to contribute to the accuracy of the transient solution. Previously (Fletcher, 1982) it was found necessary to lump the equivalent of the mass operators in order to effect the splitting.

Equation (25) can be solved as an efficient two-stage algorithm. In the first stage the following equation is solved

$$\left[M_x - \Delta t \frac{\beta}{\alpha}\left(\frac{1}{Re} L_{xx} - L_x u\right)\right] \Delta\zeta^* = \frac{\Delta t}{\alpha} RHS^{n,\beta} - \frac{(1-\alpha)}{\alpha} M_x \otimes M_y \Delta\zeta^n \tag{26}$$

and in the second stage

$$\left[M_y - \Delta t \frac{\beta}{\alpha}\left(\frac{1}{Re} L_{yy} - L_y v\right)\right] \Delta\zeta^{n+1} = \Delta\zeta^* \quad . \tag{27}$$

The implicit terms (left hand side) in eqs. (26) and (27) are associated with a particular direction and permit a decoupling for individual grid-lines. This allows the efficient implementation of a generalised Thomas algorithm for each grid line in turn. The streamfunction equation, (9), can also be solved using an equivalent algorithm (Fletcher and Srinivas, 1983) to eqs. (26) and (27).

Different choices for α and β in eqs. (26) and (27) are possible. The choice $\alpha = 1$, $\beta = 0.5$ produces the trapezoidal (Crank-Nicolson) scheme which has a truncation error of $O(\Delta t^2)$ and only requires two levels of data, n and n+1. Although this scheme has been widely used in the past it demonstrates relatively slow convergence to the steady state when used with a pseudo-transient formulation.

This slow convergence is associated with the choice $\beta = 0.5$. A more robust, but three-level, scheme is provided by the choice $\alpha = 1.5$, $\beta = 1.0$. This scheme is unconditionally stable and has a truncation error of $O(\Delta t^2)$. The convergence of the two schemes is compared in Fig. 4.

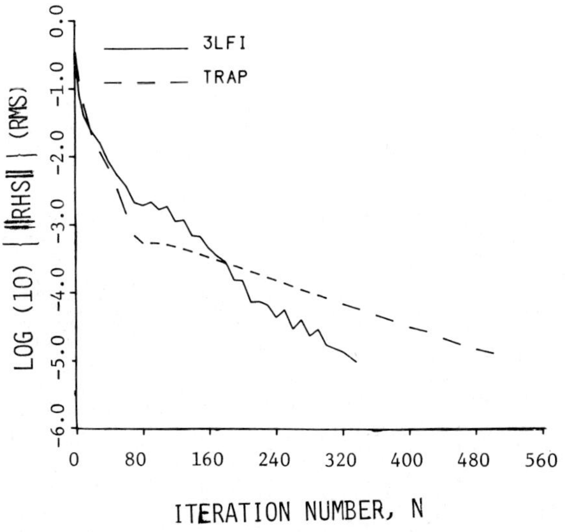

Figure 4
Convergence comparison for time-marching algorithms.
3LFI ≡ three-level fully implicit; TRAP ≡ trapezoidal (Crank-Nicolson)

In Fig. 4 $\|RHS\|_{rms}$ is the rms value of RHS in eq. (23). As the steady state is approached $\|RHS\|_{rms} \to 0$. The results shown in Fig. 4 were based on the flow over a rearward-facing step (Fig. 5) with a step-height Reynolds number, $Re_h = u_\infty h/\nu$ = 50, and a 59 x 59 grid. The results shown in Fig. 4 indicate that the three-level fully implicit scheme converges faster.

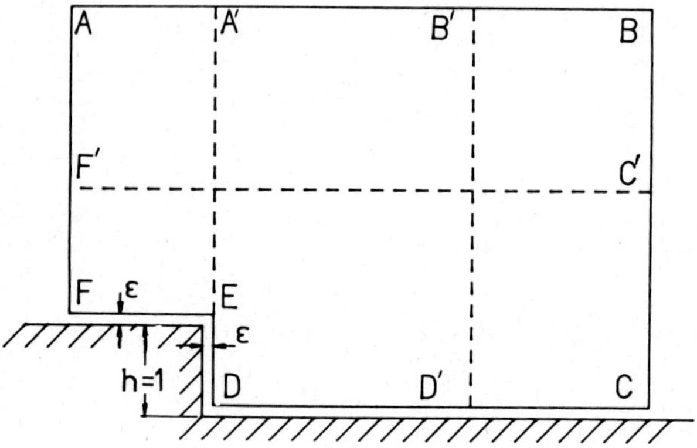

Figure 5
Computational domain for the flow over a rearward-facing step.

4.1 Flow Past a Rearward-Facing Step

The flow over a rearward-facing step is characterised by a bubble of slowly moving fluid downstream of the step (ED in Fig. 5). The bubble length grows with increasing step-height Reynolds number. The length of the separation bubble (reattachment length) has been investigated experimentally and typical results are shown in Fig. 6.

Also shown in Fig. 6 are computed solutions using the present time-split method. The various cases are based on different mesh sizes and overall domain sizes. The details are given in Table 3.

The distribution of vorticity at the edge of the computational domain adjacent to the solid surface is shown in Fig. 7 for two step-height Reynolds numbers. The solutions are similar at both Reynolds numbers. The vorticity rises to a maximum adjacent to E. The vorticity is actually singular at the sharp edge E and this is the reason for separating the computational domain from the solid surface by the surface layer, E shown in Fig. 5 (Fletcher and Srinivas, 1983). After falling to a large negative value in negotiating the sharp edge E, the vorticity rises to zero adjacent to D but is again negative downstream of D associated with the reverse flow in the separation bubble. After reattachment the surface vorticity climbs to a maximum and then decreases as C (Fig. 5) is approached. The decrease corresponds to the development of a new boundary layer after reattachment.

The corresponding pressure distribution adjacent to the solid surface is shown in

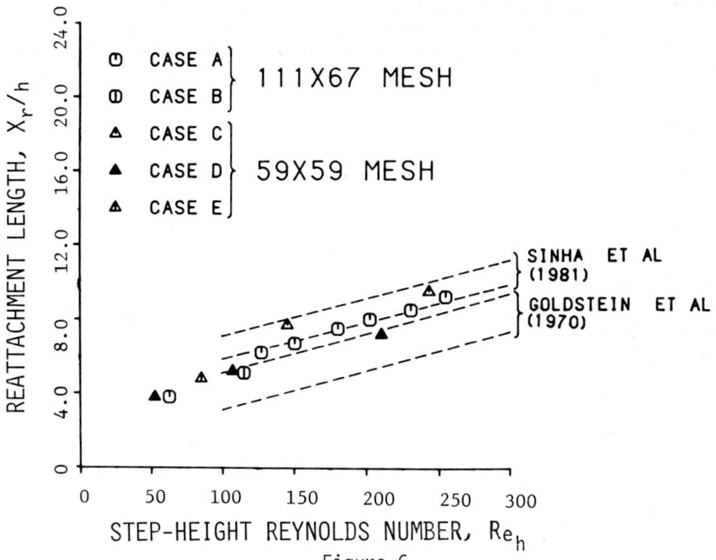

Figure 6
Variation of reattachment length with step-height Reynolds number

CASE	A	B	C	D	E
Grid (ABxBC)	111x67	111x67	59x59	59x59	59x59
Length AB	27.65	44.94	88.75	28.89	34.43
Length BC	8.67	6.69	10.73	8.91	6.43
$\Delta x_{min} = (\Delta y_{min})$	0.063	0.063	0.114	0.114	0.051
Δx_{max}	0.372	0.810	2.830	1.390	3.850
Δy_{max}	0.482	0.273	0.467	0.340	0.630
Bound. layer thick.(FF')	1.266	1.266	2.273	1.136	1.010
Re (in eq. 8)	250-1000	300	600-1000	150-600	300
Stepheight Re	63-253	119	145-242	53-211	89

Table 3 Grid details for results shown in Figs. 6 to 8

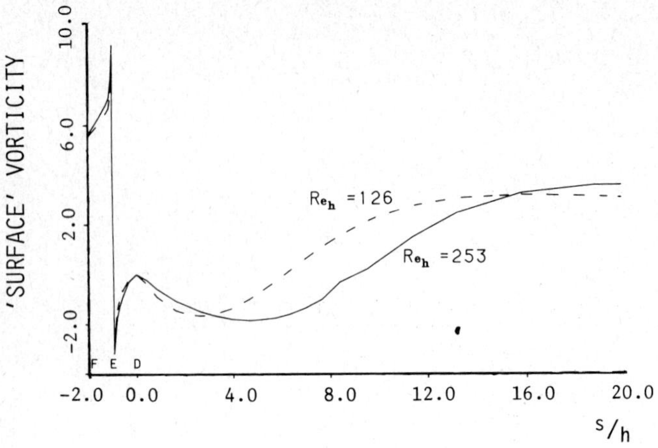

Figure 7
Vorticity distribution for a rearward-facing step

Fig. 8. The expansion of the flow about E causes a slightly sub-freestream pressure to persist in the separation bubble, until reattachment is approached. The pressure rises through reattachment and subsequently subsides to zero as the

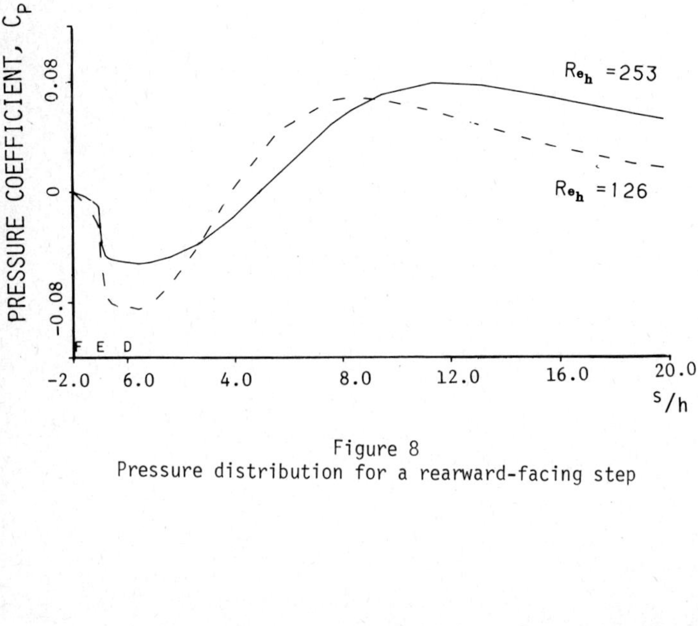

Figure 8
Pressure distribution for a rearward-facing step

new boundary layer, downstream of reattachment, develops. The pressure distribution is consistent with that obtained in experiments. The results shown in Fig. 6 to 8 demonstrate that the present method is capable of producing accurate results for complex flows.

5. CONCLUSION

In this paper it has been shown that the group finite element method is computationally more efficient than the conventional finite element method and that linear trial functions are considerably more economical than higher-order trial functions.

The combination of the group finite element method and linear Lagrange elements permits directional mass and difference operators to be explicitly extracted. As a result a computationally efficient consistent three-level time-split algorithm can be developed. The fully implicit three-level scheme is shown to converge to the steady state faster than the trapezoidal (Crank-Nicolson) scheme. The time-split algorithm has been used to obtain the solution for viscous incompressible flow over a backward-facing step.

ACKNOWLEDGEMENT

The author is grateful to the Australian Research Grants Committee for the assistance in purchasing the Perkin-Elmer 3220 Computer on which most of the current results were obtained.

REFERENCES

C.A.J. Fletcher (1980), "On the Application of Alternating Direction Implicit Finite Element Methods to Flow Problems," *Proc. 3rd Int. Conf. Finite Elements in Flow Problems*, Calgary, Canada, 217-228.

C.A.J. Fletcher (1981), "An Alternating Direction Implicit Finite Element Method for Compressible, Viscous Flows," *Lecture Notes in Physics*, 141, Springer-Verlag, 182-187.

C.A.J. Fletcher (1982a), "On an Alternating Direction Implicit Finite Element Method for Flow Problems," *Comp. Meth. Appl. Mech. Eng.*, 30, 307-322.

C.A.J. Fletcher (1982b), "A Comparison of the Finite Element and Finite Difference Methods for Computational Fluid Dynamics" in *Finite Element Flow Analysis* (ed. T. Kawai) Univ. of Tokyo Press, 1003-1010.

C.A.J. Fletcher (1983a), "The Group Finite Element Formulation," *Comp. Meth. Appl. Mech. Eng.*, 37, 225-243.

C.A.J. Fletcher (1983b), "A Comparison of Finite Element and Finite Difference Solutions of the One- and Two-Dimensional Burgers' Equations," *J. Comp. Phys.* (to appear).

C.A.J. Fletcher (1983c), "Generating Exact Solutions of the Two-Dimensional Burgers' Equations," *Int. J. Num. Meth. Fluids* (to appear).

C.A.J. Fletcher (1983d), "*Computational Galerkin Methods*," Springer-Verlag, New York.

C.A.J. Fletcher and K. Srinivas (1984), "Stream Function Vorticity Revisited," *Comp. Meth. Appl. Mech. Eng.*, (to appear).

R.J. Goldstein, V.L. Ericksen, R.M. Olson and E.R.G. Eckert (1970, "Laminar Separation, Reattachment and Transition of the Flow over a Downstream-Facing Step," *J. Basic Engineering*, 92, 732-741.

A.R. Gourlay (1977), "Splitting Methods for Time Dependent Partial Differential Equations," in *The State of the Art in Numerical Analysis* (ed. D. Jacobs), Academic Press, London, 757-796.

A. Jennings (1977), "Matrix Computation for Engineers and Scientists," Wiley, London, 152.

G.D. Mallinson and G. de Vahl Davis (1973), "The Method of the False Transient for the Solution of Coupled Elliptic Equations," *J. Comp. Phys.* 12, 435-461.

W.C. Rheinboldt (1974), "On the Solution of Large, Sparse Sets of Nonlinear Equations," TR-324, University of Maryland.

S.N. Sinha, A.K. Gupta and M.M. Oberai (1981), "Laminar Separating Flow over Backsteps and Cavities, Part I: Backsteps, *AIAA J.*, 19, 1527-1530.

K. Srinivas and C.A.J. Fletcher (1983a), "Finite Element Solutions for Laminar and Turbulent Compressible Flow," *Int. J. Num. Meth. Fluids* (to appear).

K. Srinivas and C.A.J. Fletcher (1983b), "A Generalised Coordinate Time-Split Finite Element Method for Compressible, Viscous Flow," Proc. CTAC-83 (ed. J. Noye) North-Holland, Amsterdam.

Computational Techniques & Applications: CTAC-83
J. Noye & C. Fletcher (Editors)

NUMERICAL SOLUTION OF THE PARABOLISED THREE DIMENSIONAL STEADY FLOW IN AXIALLY SYMMETRICAL PIPES

D. Yashchin, M. Israeli, M. Wolfshtein[1]
Technion, Israel Institute of Technology
Haifa
ISRAEL

The paper is concerned with the solution of steady three-dimensional laminar flow in axially-symmetrical pipes. The governing equations are parabolised in order to economise in computer time demands. The axial pressure gradient is estimated from integral mass conservation considerations. Mesh staggering is utilised to increase the stability and accuracy. The staggering is also utilised to enhance a novel treatment of the singularity of the equations, on the pipe axis. The solution is obtained by marching, utilizing an ADI scheme. Results for axisymmetrical and non axisymmetrical flows are presented. The axisymmetrical results show good agreement with previous ones.

INTRODUCTION

Three-dimensional viscous flows occur frequently in practical aerodynamics, for example, in flows through inlets, curved ducts, corners, etc. Such flows may be predicted by solution of the parabolized three-dimensional Navier-Stokes equations. The finite difference equations are solved by marching integration in the direction of the main flow.

Parabolization of the Navier-Stokes equations is possible when:

(a) there is a dominant direction for the flow;

(b) the diffusion of momentum and mass is negligible in this direction;

(c) the downstream pressure field has little influence on the upstream field.

Some numerical techniques exist for three-dimensional flow in rectangular ducts, e.g. Patankar and Spalding (1972) or Briley (1974). In the present work Briley's method is used for non-symmetrical flows in round pipes, with constant cross-section. Briley uses a "non staggered grid" mesh, while Patankar and Spalding utilize a "staggered grid" in the lateral xy-plane. We use a staggered grid in all three directions, z, r and θ. The order of the operations on the velocity components is different: first, estimates of the cross-stream velocity components u and v are found, later the main-stream velocity w and the correction for it are calculated and finally the corrections for u and v are computed.

Another novel point is the numerical treatment of the duct centre, which requires a special arrangement of the grid, and special difference approximations. Stability and second order accuracy are maintained by the use of ADI techniques.

[1]Present address: School of Mechanical and Industrial Engineering,
The University of New South Wales.

The parabolized Navier-Stokes equations may be referred to as three-dimensional boundary layer equations. The description of a numerical procedure for their solution consists of two parts:

(a) discussion of the mathematical foundation.

(b) description of the numerical method used to solve the equations.

FORMULATION OF THE EQUATIONS

The Three-Dimensional Navier-Stokes Equation for Incompressible Flow

The differential equations for steady state, incompressible three-dimensional fluid flow in cylindrical coordinates are:

I. Momentum equations:

$$u\frac{\partial u}{\partial r} + \frac{v}{r}\frac{\partial u}{\partial \theta} - \frac{v^2}{r} + w\frac{\partial u}{\partial z} = -\frac{1}{\rho}\frac{\partial P}{\partial r} + \nu\left[\frac{\partial^2 u}{\partial z^2} + \frac{\partial^2 u}{\partial r^2} + \frac{1}{r}\frac{\partial u}{\partial r} - \frac{u}{r^2} + \frac{1}{r^2}\frac{\partial^2 u}{\partial \theta^2} - \frac{2}{r^2}\frac{\partial v}{\partial \theta}\right] \quad (1)$$

$$u\frac{\partial v}{\partial r} + \frac{v}{r}\frac{\partial v}{\partial \theta} + \frac{uv}{r} + w\frac{\partial v}{\partial z} = -\frac{1}{\rho r}\frac{\partial P}{\partial \theta} + \nu\left[\frac{\partial^2 v}{\partial z^2} + \frac{\partial^2 v}{\partial r^2} + \frac{1}{r}\frac{\partial v}{\partial r} - \frac{v}{r^2} + \frac{1}{r^2}\frac{\partial^2 v}{\partial \theta^2} + \frac{2}{r^2}\frac{\partial u}{\partial \theta}\right] \quad (2)$$

$$u\frac{\partial w}{\partial r} + \frac{v}{r}\frac{\partial w}{\partial \theta} + w\frac{\partial w}{\partial z} = -\frac{1}{\rho}\frac{\partial P}{\partial z} + \nu\left[\frac{\partial^2 w}{\partial r^2} + \frac{1}{r}\frac{\partial w}{\partial r} + \frac{1}{r^2}\frac{\partial^2 w}{\partial \theta^2} + \frac{\partial^2 w}{\partial z^2}\right] \quad (3)$$

II. Continuity equation :

$$\frac{\partial u}{\partial r} + \frac{u}{r} + \frac{1}{r}\frac{\partial v}{\partial \theta} + \frac{\partial w}{\partial z} = 0 \quad (4)$$

In these equations u, v, w are velocity components in the r, θ, z directions respectively, ν is the kinematic viscosity, ρ is the density.

Parabolization of the Equations

The geometry of the problem is shown in Figure 1. The primary flow is in the z-direction and the secondary flow is the r-θ plane. In this type of flow it is reasonable to assume

$$\frac{r}{z} << 1$$

and dimensional analysis of equations (1)-(4) gives the following conditions

$$\frac{u}{w} << 1$$

$$\frac{v}{w} << 1$$

$$\frac{\partial P}{\partial r}, \frac{1}{r}\frac{\partial P}{\partial \theta} << \frac{\partial P}{\partial z}$$

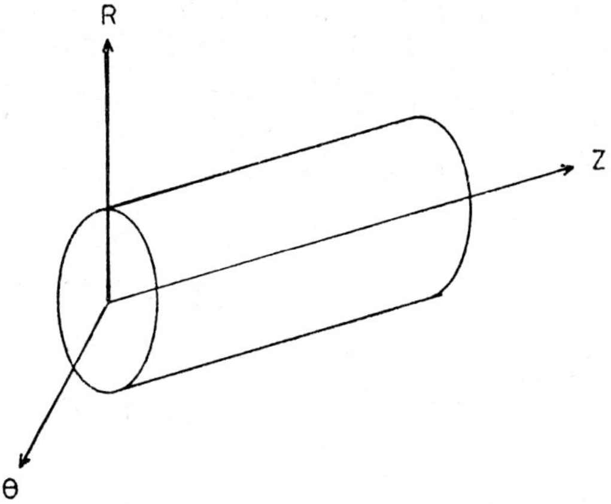

Figure 1
The coordinate system for three-dimensional axisymmetrical duct flow problem.

As a result, it is possible to split the pressure into two parts

$$P = P_o + p(r,\theta,z)$$

where:

P_o is the inviscid contribution to the pressure, and

p is the viscous contribution to the pressure.

such that

$$\frac{p(r,\ ,z)}{P_o(z)} << 1$$

and

$$\frac{\partial P}{\partial z} = \frac{dP_o}{dz}$$

Application of these approximations to the Navier-Stokes equations results in the following set of equations:

$$u\frac{\partial u}{\partial r} + \frac{v}{r}\frac{\partial u}{\partial \theta} - \frac{v^2}{r} + w\frac{\partial u}{\partial z} = -\frac{1}{\rho}\frac{\partial p}{\partial r} + \nu\left[\frac{\partial^2 u}{\partial r^2} + \frac{1}{r}\frac{\partial u}{\partial r} - \frac{u}{r^2} + \frac{1}{r^2}\frac{\partial^2 u}{\partial \theta^2} - \frac{2}{r^2}\frac{\partial v}{\partial \theta}\right] \quad (5)$$

$$u\frac{\partial v}{\partial r} + \frac{v}{r}\frac{\partial v}{\partial \theta} + \frac{uv}{r} + w\frac{\partial v}{\partial z} = -\frac{1}{\rho r}\frac{\partial p}{\partial \theta} + \nu\left[\frac{\partial^2 v}{\partial r^2} + \frac{1}{r}\frac{\partial v}{\partial r} - \frac{v}{r^2} + \frac{1}{r^2}\frac{\partial^2 v}{\partial \theta^2} + \frac{2}{r^2}\frac{\partial u}{\partial \theta}\right] \quad (6)$$

$$u \frac{\partial w}{\partial r} + \frac{v}{r}\frac{\partial w}{\partial \theta} + w \frac{\partial w}{\partial z} = \frac{1}{\rho}\frac{dP_o}{\partial z} + \nu \left[\frac{\partial^2 w}{\partial r^2} + \frac{1}{r}\frac{\partial w}{\partial r} + \frac{1}{r^2}\frac{\partial^2 w}{\partial \theta^2}\right] \quad (7)$$

The equations to be solved then are (4)-(7). As can be easily seen from (5)-(7) the second derivatives of u, v, w with respect to z are neglected altogether and only those with respect to r and θ remain. As such, these equations look like parabolic equations in which z is the direction of marching. In fact, the solution of these equations is achieved by marching and this, of course, is much cheaper than the solution of the full Navier-Stokes equations (1)-(4).

Integration of the continuity equation (4) over the cross-section of the duct yields the integral mass flow equation

$$\int_o^{2\pi} d\theta \int_o^R w(r,\theta)\, r dr = \dot{m}/\rho \quad (8)$$

where $\dot{m}$ is the mass flow rate through the duct.

Finite difference solution of Eqs. (4)-(7) does not guarantee that Eq. (8) is satisfied, and therefore this equation must be checked after every integration step.

Initial and Boundary Conditions

For a complete specification of the problem the initial and boundary conditions for all the variables must be specified. The initial conditions are

$$w(r,\theta)_{z=h_z/2} = w_{in}(r,\theta)$$

$$u(r,\theta)_{z=0} = u_{in}(r,\theta)$$

$$v(r,\theta)_{z=0} = v_{in}(r,\theta)$$

$$p(r,\theta)_{z=0} = p_{in}(r,\theta)$$

The distribution of the velocity components is given in the points where it is required.

The boundary conditions are:

$$w(R,\theta) = u(R,\theta) = v(R,\theta) = 0$$

$$p(R,\theta) = 0$$

THE NUMERICAL METHOD

In this section a finite-difference procedure for solving Eq. (4)-(7) based on the ADI method is discussed. In this method like in Briley (1974) the equations are locally linearized. The velocity and pressure distributions in the secondary flow are calculated simultaneously by solving a Poisson equation as in Briley's method.

We realize that Eqs. (4)-(7) have a singularity at r=0. This singularity is not physical since the flow has no singularity whatsoever at r=0, but is due to the transformation $(x, y) \to (r, \theta)$. This singularity may cause a drop in the order of accuracy of the finite difference equations if a uniform grid in the r-direction is used. One way to overcome this problem is to rewrite the r-derivative terms in the Navier-Stokes equations in a particular form as follows.

Momentum Equations:

$$w\frac{\partial w}{\partial z} = -\frac{1}{\rho}\frac{dP_o}{dz} + \nu\,[\underline{\frac{1}{r}\frac{\partial}{\partial r}(r\frac{\partial w}{\partial r})} + \frac{1}{r^2}\frac{\partial^2 w}{\partial \theta^2}] - \frac{u}{r}\,[\underline{\frac{\partial}{\partial r}(wr) - w}] - \frac{v}{r}\frac{\partial w}{\partial \theta} \tag{9}$$

$$w\frac{\partial v}{\partial z} = -\frac{1}{\rho r}\frac{\partial p}{\partial \theta} + \frac{2\nu}{r^2}\frac{\partial u}{\partial \theta} - \frac{uv}{r} + \nu\,\underline{[\frac{1}{r}\frac{\partial}{\partial r}(r\frac{\partial v}{\partial r}) - \frac{v}{r^2}]} - \underline{\frac{u}{r}\,[\frac{\partial (vr)}{\partial r} - v]}$$

$$+ \nu\,(\frac{1}{r^2}\frac{\partial^2 v}{\partial \theta^2}) - \frac{v}{r}\frac{\partial v}{\partial \theta} \tag{10}$$

$$w\frac{\partial u}{\partial z} = -\frac{1}{\rho}\frac{\partial p}{\partial r} + \frac{v^2}{r} - \frac{2\nu}{r^2}\frac{\partial v}{\partial \theta} - \frac{u}{r}\,[\underline{\frac{\partial (ur)}{\partial r} - u}] + \nu\,\underline{[\frac{1}{r}\frac{\partial^2}{\partial r^2}(ru) - \frac{1}{r^2}\frac{\partial}{\partial r}(ru)]}$$

$$- \frac{v}{r}\frac{\partial u}{\partial \theta} + \frac{\nu}{r^2}\frac{\partial^2 u}{\partial \theta^2} \tag{11}$$

and the continuity equation

$$\frac{1}{r}\frac{\partial}{\partial r}(ru) + \frac{1}{r}\frac{\partial v}{\partial \theta} + \frac{\partial w}{\partial z} = 0 \tag{12}$$

Mesh staggering

The method of solution of this problem utilizes a staggered grid in all three-directions r, θ, z.

Figure 2 shows a general layout of the r-θ plane at a $z = kh_z$ distance from the edge of the duct. A more detailed illustration of a specific flow field element is presented in Figure 3. This is a blow up of the hatched element of Fig. 2. The precise definition of the variables at the various staggered grid points are set here and the reader should refer to these illustrations when deriving the difference equations.

One should note that the momentum equations will be discretized at the grid mid-points α, β and γ, as shown in Fig. 3. The pressure P is given at the main grid points $\bar{p}$, and the pressure derivatives with respect to r, θ and z will be given at the mid-points α, β and γ respectively. One should also note that the first point $\bar{p}$ in the r-direction is taken to be $r = 1/2\ h_r$, as suggested by de Vahl Davis (1979).

The Finite Difference Equations

To implement the procedure, the flow region is discretized by grid points having equal spacings h_r and h_θ, in the r - θ directions, respectively, and a variable axial step size h_z. The subscripts m and n, and the superscript k denote the location of the grid point in r, θ, z directions, respectively. The following shorthand difference operator notation is used for the finite difference formulae.

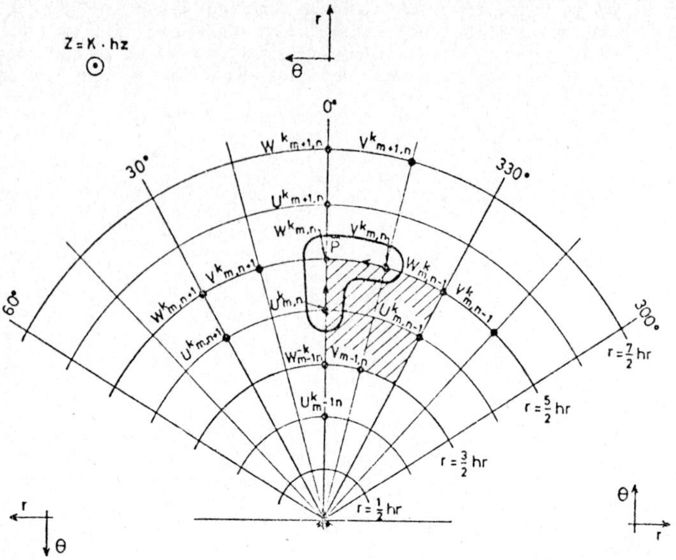

Figure 2
r-θ plane at z^k. Projection of mesh staggering on the plane $z=k.h_z$. u, v are discretized on this plane, whereas w is shifted forward by 1/2 h_z. the z axis forms a right hand triad with r and θ axes.

$$\delta_z f^k_{m,n} = \frac{f^{k+1}_{m,n} - f^k_{m,n}}{h_z} = \left.\frac{\partial f}{\partial z}\right|^k_{m,n} + O(h_z^2) \tag{13}$$

$$\delta_\theta f^k_{m,n} = \frac{f^k_{m,n+1} - f^k_{m,n-1}}{2h_\theta} = \left.\frac{\partial f}{\partial \theta}\right|^k_{m,n} + O(h_\theta^2) \tag{14}$$

$$\delta^2_\theta r^k_{m,n} = \frac{f^k_{m,n+1} - 2f^k_{m,n} + f^k_{m,n-1}}{h^2_\theta} = \left.\frac{\partial^2 f}{\partial \theta^2}\right|^k_{m,n} + O(h_\theta^2) \tag{15}$$

In the above formula, $f^k_{m,n} = f\ (r_m,\ \theta_n,\ z^k)$ is a dummy symbol representing any of the independent variables, defined at grid point r_m, θ_n, z^k. We should note, however that r_m is defined as follows:

$$r_m = \begin{cases} (m-1)\ h_r & \text{in the u-momentum equation} \\ (m-\frac{1}{2})\ h_r & \text{otherwise.} \end{cases} \qquad m=1,2,3...M$$

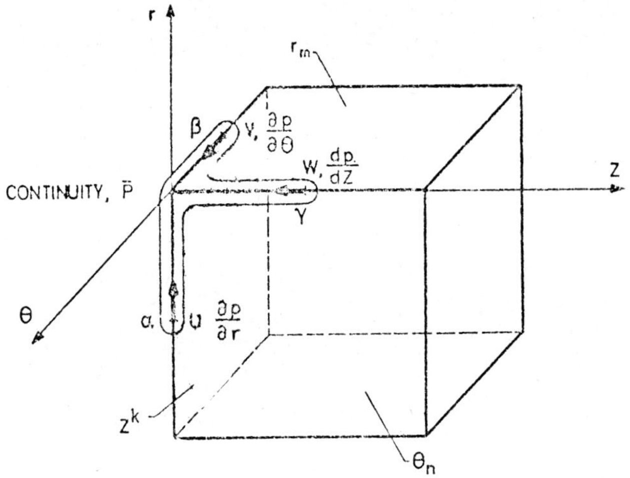

Figure 3

A three-dimensional flow element.

$\alpha - r_m = (m-1)h_r \; ; \; \theta_n = (n-1)h_\theta \; ; \; z^k = (k-1)h_z.$

$\beta - r_m = (m-1/2)h_r \; ; \; \theta_n = (n-3/2)h_\theta \; ; \; z^k = (k-1)h_z.$

$\gamma - r_m = (m-1/2)h_r \; ; \; \theta_n = (n-1)h_\theta \; ; \; z^k = (k-1/2)h_z.$

$P - r_m = (m-1/2)h_r \; ; \; \theta_n = (n-1)h_\theta \; ; \; z^k = (k-1)h_z.$

To obtain the radial difference approximations to second order in the local truncation error we usually have to use central differences. When we approach the origin r = 0, central differences require points which are on the other side of the origin and consequently have different θ coordinates. In an ADI procedure, coupling of such points in the implicit solution step, will destroy the tri-diagonal structure of the matrices and result in complicated and time consuming algorithms. If we difference consistently the forms (9), (10), (11), (12) we find that the schemes that result are of second order and do not require values of the variables at the center or on the other side of the origin. The required forms are as follows:

$$\frac{1}{r_m}\,\delta_r\,(r_m \delta_r f^k_{m,n}) = \frac{1}{r_m}\,\frac{(r\,\delta_r f)^k_{m+1/2,n} - (r\,\delta_r f)^k_{m-1/2,n}}{h_r}$$

$$= \frac{1}{r_m}\,\frac{r_{m+1/2}\dfrac{f^k_{m+1,n} - f^k_{m-1,n}}{h_r} - r_{m-1/2}\dfrac{f^k_{m,n} - f^k_{m-1,n}}{h_r}}{h_r} =$$

$$= \frac{1}{h_r^2}\left[f_{m+1,n}^{k}\frac{r_{m+1/2}}{r_m} - f_{m,n}^{k}\left(\frac{r_{m+1/2}}{r_m} + \frac{r_{m-1/2}}{r_m}\right) + f_{m-1,n}^{k}\frac{r_{m-1/2}}{r_m}\right]$$

$$= \frac{1}{r}\frac{\partial}{\partial r}\left(r\frac{\partial f}{\partial r}\right)_{m,n}^{k} + O(h_r^{\,2}) \tag{16}$$

$$\delta_r(f_{m,n}^{k} r_m) = \frac{f_{m+1/2,n}^{k}\, r_{m+1/2} - f_{m-1/2,n}^{k}\, r_{m-1/2}}{h_r} =$$

$$\frac{1}{h_r}\left(\frac{f_{m+1,n}^{k} + f_{m,n}^{k}}{2} r_{m+1/2} - \frac{f_{m,n}^{k} + f_{m-1,n}^{k}}{2} r_{m-1/2}\right) =$$

$$= \frac{1}{2h_r}\left[f_{m+1,n}^{k} r_{m+1/2} + f_{m,n}^{k}(r_{m+1/2} - r_{m-1/2}) - f_{m-1,n}^{k} r_{m-1/2}\right]$$

$$= \frac{\partial}{\partial r}(rf)_{m,n}^{k} + O(h_r^2) \tag{17}$$

Those derivatives are for $f = \begin{cases} w_{m,n}^{k} \\ v_{m,n}^{k} \end{cases}$ or for any other function at the half staggered grid point. The other derivatives are given by:

$$\frac{1}{r_m}\delta_r^2(r_m f_{m,n}^{k}) = \frac{1}{r_m}\frac{r_{m+1}f_{m+1,n}^{k} - 2r_m f_{m,n}^{k} + r_{m-1}f_{m-1,n}^{k}}{h_r^{\,2}} = \frac{1}{r}\frac{\partial^2}{\partial r^2}(rf)_{m,n}^{k} + O(h_r^2) \tag{18}$$

$$\delta_r(r_m f_{m,n}^{k}) = \frac{r_{m+1}f_{m+1,n}^{k} - r_{m-1}f_{m-1,n}^{k}}{2h_r} = \frac{\partial}{\partial r}(rf)_{m,n}^{k} + O(h_r^2) \tag{19}$$

These difference equations are good for $f_{m,n}^k = u_{m,n}^k$ or any other functions at at fully staggered grid point.

In the continuity equation the central point for the finite difference approximation is (r, θ, z) and the equation becomes:

$$\operatorname{div} \vec{v} = \frac{1}{r_m}\frac{(ru)_{m+1,n}^{k} - (ru)_{m,n}^{k}}{h_r} + \frac{1}{r_m}\frac{v_{m,n+1}^{k} - v_{m,n}^{k}}{h_\theta} + \frac{w_{m,n}^{k+1} - w_{m,n}^{k}}{h_z} +$$

$$+O(h_r^2, h_\theta^2, h_z^2) = 0 \tag{20}$$

If any variable is not defined at a point where it is required, it is computed as the average of the values of this variable at the neighbouring two or four points.

Description of Numerical Solution

Application of Briley's method to the present equations gives the following procedure:

(1) v_p^{k+1} and u_p^{k+1} are computed from the momentum equations (10) and (11);

(2) w^{k+1} is computed from the axial momentum equation (9), with p^{k+1} determined implicitly to ensure that the axial mass flow relation (8) is satisfied;

(3) v_c^{k+1} and u_c^{k+1} are computed from equation (12), using a velocity potential. This is a Neumann problem for the Poisson equation;

(4) $\frac{\partial p}{\partial r}^{k+1}$ and $\frac{\partial p}{\partial \theta}^{k+1}$ are computed from equations (9) and (10). (At this point all the velocity components at k+1 are known).

In this method the velocity components, u, v and w are obtained by using an ADI method marching in the z-direction, and the pressure is calculated from the upstream pressure gradient.

Calculation of Cross-Section Velocities

Equations (10)-(11) can be written in the form:

$$\frac{\partial f}{\partial z} = \frac{1}{w} [L^2 f + S_f) \tag{21}$$

where: L^2 - is a second-order differential operator

S_f - is a source term, given by:

for $f_{m,n}^k = u_{m,n}^k$: $S_f = -\frac{1}{\rho}\frac{\partial p}{\partial r} + \frac{v^2}{r} - \frac{2\nu}{r^2}\frac{\partial v}{\partial \theta}$,

and for $f_{m,n}^k = v_{m,n}^k$: $S_f = -\frac{1}{\rho r}\frac{\partial p}{\partial r} - \frac{uv}{r} + \frac{2\nu}{r^2}\frac{\partial u}{\partial \theta}$,

Now, equation (21) can be written in the following finite difference form

$$f_p = A_N^f f_N + A_S^f f_S + A_E^f f_E + A_W^f f_W + C^f \tag{22}$$

where the A-coefficients originate from the operator L^2 and C^f is the "discretized" source term. The $\frac{\partial p}{\partial r}$ and $\frac{\partial p}{\partial \theta}$ are estimated from the previous pressure. Usually the upstream values of $\frac{\partial p}{\partial r}$ and $\frac{\partial p}{\partial \theta}$ are sufficiently accurate. The equation (22) is solved separately for u and v-velocity components with A_N, A_S, A_E and A_W known from upstream.

The correct values of lateral velocity components are assumed to be given by the following form:

$$u_{m,n}^{k+1} = u_p + u_c$$

$$v_{m,n}^{k+1} = v_p + v_c \tag{23}$$

where

u_p and v_p are the predictions of $u^{i+1}_{m,n}$ and $v^{k+1}_{m,n}$ computed from the secondary flow momentum equations (10)-(11);

u_c and v_c are corrections to u_p and v_p computed so as to satisfy the continuity equation (12). It is assumed that the velocity corrections are irrotational, and a velocity potential ϕ is introduced such that

$$u_c = \frac{\partial\phi}{\partial r}$$

$$v_c = \frac{1}{r}\frac{\partial\phi}{\partial\theta} \qquad (24)$$

The boundary conditions require that the normal velocity vanishes at the walls. The boundary condition for the potential ϕ is:

$$\left(\frac{\partial\phi}{\partial r}\right)_{r=R} = 0 \qquad (25)$$

By substitutiong (23) and (24) in (12) we obtain:

$$\frac{1}{r}\frac{\partial}{\partial r}\left(r\frac{\partial\phi}{\partial r}\right) + \frac{1}{r^2}\frac{\partial^2\phi}{\partial\theta^2} + \text{SOUR} = 0 \qquad (26)$$

where

$$\text{SOUR} = \frac{1}{r}\frac{\partial}{\partial r}(ru_p) + \frac{1}{r}\frac{\partial}{\partial\theta}(v_p) + \frac{\partial w}{\partial z} = 0 \qquad (27)$$

The resulting difference equation is:

$$\frac{1}{h_r^2}\left[\phi_{m+1,n}\frac{r_{m+1/2}}{r_m} - 2\phi_{m,n} + \phi_{m,n}\frac{r_{m-1/2}}{r_m}\right] + \frac{1}{r^2 h_\theta^2}\left[\phi_{m,n+1} - 2\phi_{m,n} + \phi_{m-1,n}\right] +$$

$$+ \frac{r_{m+1}u^k_{m+1,n} - r_m u^k_{m,n}}{r_m h_r} + \frac{1}{r_m}\frac{v^k_{m,n+1} - v^k_{m,n}}{h_\theta} + \frac{w^{k+1}_{m,n} - w^k_{m,n}}{h_z} = 0 \qquad (28)$$

Equation (26) is a Poisson equation with Neumann boundary condition (25). This equation may be solved by an ADI method, or by a Fast Direct Method.

The Cross-Stream Pressure Gradient

For the next step, i.e. calculation of $u^{k+2}_{m,n}$ and $v^{k+2}_{m,n}$ we need the estimated values of $\frac{\partial p}{\partial r}$ and $\frac{\partial p}{\partial\theta}$. We obtain these values from the momentum equations (10-(11) where we used the corrected values of all three-components of the velocities.

$$\frac{1}{\rho}\frac{\partial p}{\partial r} = - w\frac{\partial u}{\partial z} + \frac{v^2}{r} - \frac{2\nu}{r^2}\frac{\partial v}{\partial\theta} - \frac{u}{r}\left[\frac{\partial(ur)}{\partial r} - u\right] + \nu\left[\frac{1}{r}\frac{\partial^2}{\partial r^2}(ru) - \frac{1}{r^2}\frac{\partial}{\partial r}(ru)\right] -$$

$$- \frac{v}{r}\frac{\partial u}{\partial\theta} + \frac{\nu}{r^2}\frac{\partial^2 u}{\partial\theta^2} \qquad (29)$$

$$\frac{1}{\rho}\frac{\partial p}{\partial \theta} = - w\frac{\partial v}{\partial z} + \frac{2\nu}{r^2}\frac{\partial u}{\partial \theta} - \frac{u}{r}\frac{\partial}{\partial r}(vr) + \nu\left[\frac{1}{r}\frac{\partial}{\partial r}\left(r\frac{\partial v}{\partial r}\right) - \frac{v}{r^2}\right] - \frac{v}{r}\frac{\partial v}{\partial \theta} + \frac{\nu}{r^2}\frac{\partial^2 v}{\partial \theta^2} \quad (30)$$

Where $\frac{\partial u}{\partial z}$ and $\frac{\partial v}{\partial z}$ are approximated by

$$\frac{\partial u}{\partial z} = \delta_z u^k_{m,n}$$

$$\frac{\partial v}{\partial z} = \delta_z v^k_{m,n} \quad (31)$$

Calculation of w-Velocity Component

The axial momentum equation (9) can be written in the form:

$$w\frac{\partial w}{\partial z} = M^2 w + Tw \quad (32)$$

where:

M^2 - is a second-order differential operator such that

$$M^2 w = \nu\left[\frac{1}{r}\frac{\partial}{\partial r}\left(r\frac{\partial w}{\partial r}\right) + \frac{1}{r^2}\frac{\partial^2 w}{\partial \theta^2}\right] - \frac{u}{r}\left[\frac{\partial}{\partial r}(wr) - w\right] - \frac{v}{r}\frac{\partial w}{\partial \theta}, \quad (33)$$

Tw - is a source term given by

$$Tw = - \frac{1}{\rho}\frac{dP_o}{dz}. \quad (34)$$

The finite-difference form of Eq. (32) is:

$$w_P = A^w_N w_N + A^w_S w_S + A^w_E w_E + A^w_W w_W + C^w, \quad (35)$$

where

$$C^w = - \frac{1}{w}\frac{dP_o}{dz}.$$

Equation (35) is solved after the A's and C are calculated using upstream values and the velocity components u and v are approximated by the predicted u_p and v_p in (23).

When we solve Eq. (35) the pressure gradient is unknown and we approximate it by the upstream pressure gradient. Therefore, the value of the velocity component w is not exact. Correction for this value can be obtained from the integral mass conservation Equation (8):

$$\int_0^{2\pi}\int_0^R w_i r dr d\theta - \dot{m}/\rho = \Delta m_i, \quad (36)$$

where

w_i - is the i-th iteration for w, based on $\left.\frac{dP_o}{dz}\right|_i$,

Δm_i - the added mass, should be theoretically zero.

Because of the linear character of the difference equations we can find $\frac{dP_o}{dz}$ for zero added mass flow: $\Delta m = 0$, as shown in Figure 4:

$$\left.\frac{dP_o}{dz}\right|_3 = \frac{\Delta m_1 \left.\frac{dP_o}{dz}\right|_2 - \Delta m_2 \left.\frac{dP_o}{dz}\right|_1}{\Delta m_1 - \Delta m_2} \tag{37}$$

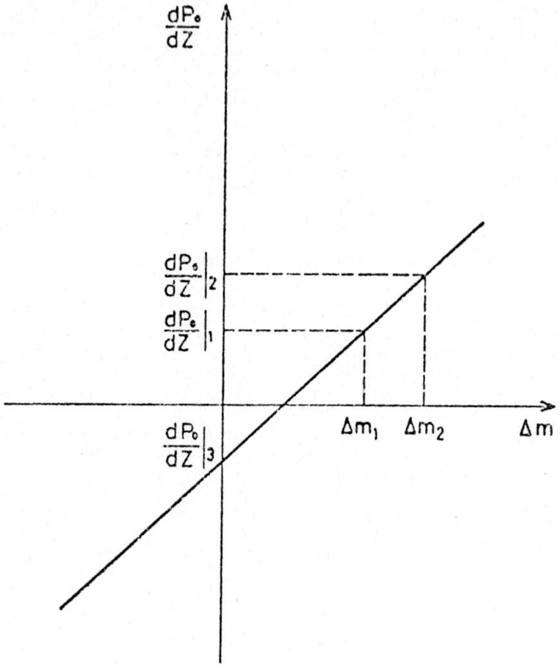

Figure 4
The pressure gradient correction.

Once we have found the pressure gradient $\frac{dP_o}{dz}$, that annihilates, Δm, the velocity component w can be found from Eq. (35).

The solution of the Finite-Difference Equations

Equations (9) - (11) may be written in a general form as:

$$\frac{\partial f}{\partial z} = A^2(f) + S_d + O(h^2) \tag{38}$$

and they can be solved by the ADI method. To describe the method we replace Eq. (21) by its finite-difference analog:

$$\frac{\partial f}{\partial z} = (A_r + A_\theta)\ f + S_d + O(h^2) \tag{39}$$

where

A_r, A_θ - are matrix operators,

f - represents u, v, w

S_d - is a finite-difference form of the source term.

Equation (23) is solved by the Peaceman-Rachford ADI method as described by Isenberg and de-Vahl Davis (1975):

$$(I - \frac{h_z}{2} A_r) f^* = (I + \frac{h_z}{2} A_\theta) f^k + \frac{h_z}{2} S_D^{k+1/2}$$

$$(I - \frac{h_z}{2} A_\theta) f^{k+1} = (I + \frac{h_z}{2} A_r) f^* + \frac{h_z}{2} S_D^{k+1/2} \tag{40}$$

where

I - denotes the unit matrix, and A_r is given by (41)

	$f_{m-1,n}$	$f_{m,n}$	$f_{m+1,n}$
$A_r W^k_{m,n}$	$\frac{1}{W^k_{m,n}} \frac{\nu}{h_r^2} \frac{r_{m-1/2}}{r_m} +$ $+ \frac{u_{m,n}}{2h_r} \frac{r_{m-1/2}}{r_m}$	$- \frac{1}{W^k_{m,n}} \frac{2\nu}{h_r^2} -$ $- \frac{u^k_{m,n}}{2r_m}$	$\frac{1}{W^k_{m,n}} \frac{\nu}{h_r^2} \frac{r_{m+1/2}}{r_m} -$ $- \frac{u_{m,n}}{2h_r} \frac{r_{m+1/2}}{r_m}$
$A_r U^k_{m,n}$	$\frac{1}{W_{m,n}} \frac{u_{m,n}}{2h_r} \frac{r_{m-1}}{r_m} +$ $+ \nu \frac{r_{m-1}}{r_m} \frac{1}{h_r^2} + \frac{0.5}{h_r r_m}$	$\frac{1}{W^k_{m,n}} \frac{u^k_{m,n}}{r_m} - \frac{2\nu}{h_r^2}$	$\frac{1}{W^k_{m,n}} \frac{u_{m,n}}{2h_r} \frac{r_{m+1}}{r_m} +$ $+ \nu \frac{r_{m+1}}{r_m} \frac{1}{h_r^2} - \frac{0.5}{h_r r_m}$
$A_r V^k_{m,n}$	$\frac{1}{W^k_{m,n}} \frac{\nu}{h_r^2} \frac{r_{m-1/2}}{r_m} +$ $\frac{u_m}{2h_r} \frac{r_{m-1/2}}{r_m}$	$- \frac{1}{W^k_{m,n}} \frac{2\nu}{h_r^2} -$ $\frac{u^k_{m,n}}{2r_m} + \frac{\nu}{r_m^2}$	$\frac{1}{W^k_{m,n}} \frac{\nu}{h_r^2} \frac{r_{m+1/2}}{r_m} -$ $\frac{u_{m,n}}{2h_r} \frac{r_{m+1/2}}{r_m}$

(41)

and A_θ is given by:

$$A\,f = \frac{1}{W^k_{m,n}} \left\{ f_{m,n-1} \left[\frac{\nu}{h_\theta^2 r_m^2} + \frac{v^k_{m,n}}{2h_\theta r_m}\right] - f_{m,n} \frac{2\nu}{r_m^2 h_\theta^2} + f_{m,n+1} \left[\frac{\nu}{h_\theta^2 r_m^2} - \frac{v^k_{m,n}}{2h_\theta r_m}\right] \right\} \tag{42}$$

The velocity components u, v and w in euqations (9)-(11) are defined at the points, where the equation (38) is solved.

The solution method consists of the following steps:

1. The solution of equation (41) in the r-direction for each row successively, and for each variable, by a standard tridiagonal algorithm (Thomas - algorithm).

2. The solution of equation (42) in the θ-direction for each row successively, and for each variable, by tridiagonal algorithm, but with periodical boundary conditions i.e. $f^k_{m,1} = f^k_{m,N+1}$. The solution of (42) is the new value $f^{k+1}_{m,n}$.

If the Poisson's equation (26) is solved by ADI method too, Eq. (4)) is solved repeatedly until a stationary value is obtained for $f^k_{m,n}$.

Computational details

(a) Initial Conditions

The following initial values should be specified:

$$u^o_{m,n} = u \text{ in } (r,\ \theta)$$

$$v^o_{m,n} = v \text{ in } (r,\ \theta)$$

$$w^o_{m,n} = w \text{ in } (r,\ \theta)$$

$$p^o_{m,n} = p \text{ in } (r,\ \theta) \tag{43}$$

(b) Solution of The Two Parabolic Equations for the Lateral Velocities

$$\frac{\partial u_p^{k+1}}{\partial z} = \frac{1}{2}\,[A_r(u^k,\ v^k,\ w^k) + A_\theta(u^k,\ v^k,\ w^k)]\ u^k - \frac{1}{w^k}\,[\frac{1}{\rho}\frac{\partial p}{\partial r} + \frac{(v^k)^2}{r} - \frac{2\nu}{r^2}\frac{\partial v^k}{\partial \theta}] \tag{44}$$

$$\frac{\partial v_p^{k+1}}{\partial z} = \frac{1}{2}\,[A_r(u^k,\ v^k,\ w^k) + A_\theta(u^k,\ v^k,\ w^k)]\ v^k - \frac{1}{w^k}\,[\frac{1}{\rho r}\frac{\partial p}{\partial \theta} + \frac{u^k v^k}{r} - \frac{2\nu}{r^2}\frac{\partial v^k}{\partial \theta}] \tag{45}$$

(c) Solution of the parabolic equation for the axial velocity

$$\frac{\partial w^{k+1}}{\partial z} = \frac{1}{2}\,[A_r(u^{k+1},\ v^{k+1},\ w^k) + A_\theta(u^{k+1},\ v^{k+1},\ w^k)]\ w^k - \frac{1}{\rho w^k}\frac{dP_o}{dz} \tag{46}$$

where A_r and A_θ are obtained from Eqs. (41)-(42). The value of the pressure gradient is obtained from Eq. (37).

(d) Remarks on the Solution of Neumann Problem for Poisson's Equation

In order to solve Eq. (26) for θ attention must be given to the integral constraint required when solving the Poisson equations with normal derivative boundary conditions. The two-dimensional Poisson equation for ϕ with source distribution f(r, θ) and Neumann boundary conditions has a solution only if the following condition is satisfied.

$$\int_A f(r, \theta)dA = \int_C \frac{\partial \phi}{\partial r} dS \tag{47}$$

where

A - is the area enclosed by C

r - is the outward normal to C

S - is a distance along C.

In our problem $f(r, \theta)$ is not known exactly but an approximation to it is obtained by discretization. Therefore, we do not expect (47) to be satisfied exactly. Hence we compute the total correction E defined by

$$E = \int_A f(r, \theta)dA - \int_C \frac{\partial \phi}{\partial} dS. \tag{48}$$

The average correction is then defined to be

$$f_{ave} = \frac{E}{A} \tag{49}$$

and the corrected source term is then:

$$f_{correct}(r, \theta) = f(r, \theta) - f_{ave}. \tag{50}$$

It is with this $f_{correct}(r, \theta)$ that Poisson's equation for ϕ in (39) is solved.

(e) Computation of the true value of the lateral velocities

$$u^{k+1}_{m,n} = u^{k+1}_{m,n,_p} + \frac{\partial \phi}{\partial r} \tag{51}$$

$$u^{k+1}_{m,n} = v^{k+1}_{m,n,_p} + \frac{1}{r}\frac{\partial \phi}{\partial \theta} \tag{52}$$

(f) Treatment of w and u-velocities component on the axis of the duct

As we have seen the numerical solution for w has been obtained at the grid points for which $r_m = (m-1/2)\ h_r$. However, we are also interested in finding the value of w at r=0. We find the approximate value w to fourth order by the formula:

$$w^k(o, \theta) = \frac{9}{16}\,[w^k(\frac{1}{2}\,h_r, \theta) + w(\frac{1}{2}\,h_r, \theta + \pi)] - \frac{1}{16}\,[w^k(\frac{3}{2}\,h_r, \theta) + w(\frac{3}{2}\,h_r, \theta + \pi)] \tag{53}$$

Because of the "staggered grid" the equation for the u-component of the velocity at r = 0 has a singular character. The non-physical solution of the equation can be filtered out mathematically. The value of the u-component velocity at r=0 is:

$$u^k(0, \theta), = \frac{2}{3}\,[u^k(h_r, \theta) - u^k(h_r, \theta + \pi)] - \frac{1}{6}\,[u^k(2\,h_r, \theta) - u^k(2\,h_r, \theta + \pi)] \tag{54}$$

RESULTS

The numercial method and the computer program were tested on two problems. The first one was that of an axially-symmetrical plug flow entering a circular pipe with a radius of 1m, at a velocity of 0.01m/s. The kinematic viscosity was 1.5×10^{-5} and the Reynolds number was 1333. The development of the profile of the axial velocity W with η = z/R/Re is shown in Figure 5, and compared with

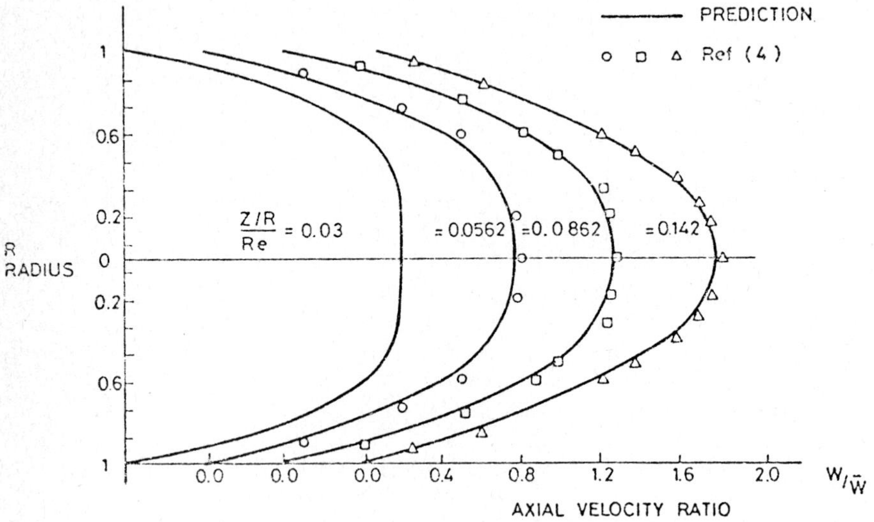

Figure 5
The development of the axial velocity profile,
Re = 1333, 9x8 mesh points

the calculations of Williams (1969). Good agreement is obtained. The growth of the velocity on the axis due to velocity diminution near the wall is shown in Figure 6. Thus we may conclude that the program appears to be correct.

As an example of the capabilities of the program we chose a case with a linear shear at the pipe inlet. All the conditions are identical with the previous example but the axial velocity at the inlet is given by

$$W_{in} = 0.01\ (1 + \frac{Y}{R})\ \text{m/s} \tag{55}$$

The maximal velocity growth is shown in Fig. 6. The maximum velocity starts from a high value obtained due to the linear velocity distribution and falls down to a shallow minimum from which it rises monotonously. The plug flow results shown on the same figure start from a lower initial value and rise until the velocity reaches its fully developed value of 1.8. The two lines are virtually identical beyond η = Z/R/Re = 0.15.

The axial velocity profiles along the y-axis are shown in Fig. 7. The develop-

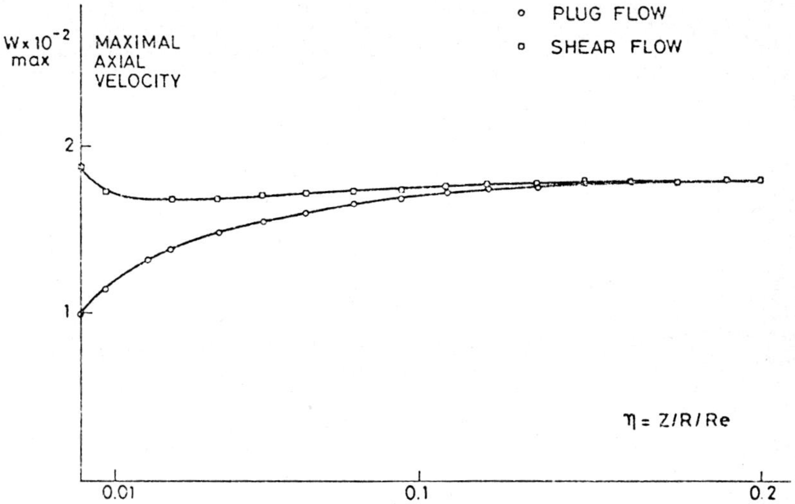

Figure 6
The axial velocity distribution on the axis for a uniform shear inlet velocity Eq. (55), Re = 1333, 9x8 mesh points.

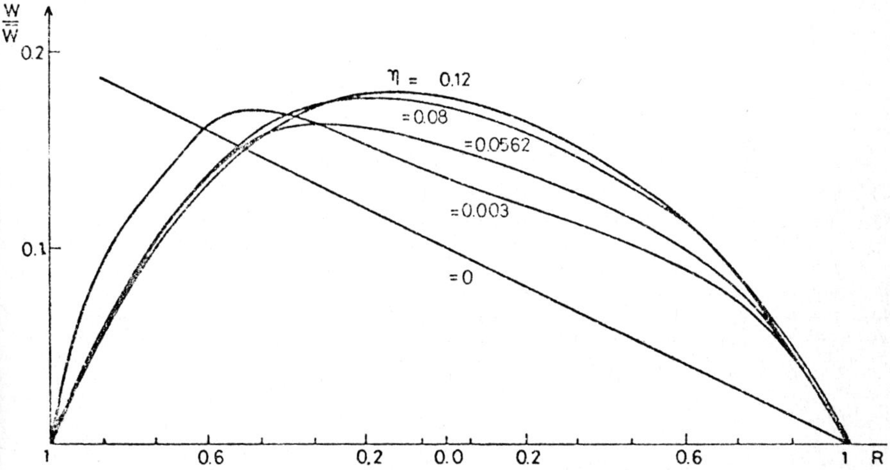

Figure 7
The development of the axial velocity profile, same conditions as in Fig. 6.

ment of the profile from a linear asymmetrical one to a symmetrical parabola is easily seen. It can be seen by comparison to Fig. 5 that the distance required for the development of the profile is not strongly dependent on the initial condition.

The lateral velocity profiles, u (in the r-direction) and v (in the θ-direction) in the y-direction are shown in Figures 8 and 9 respectively.

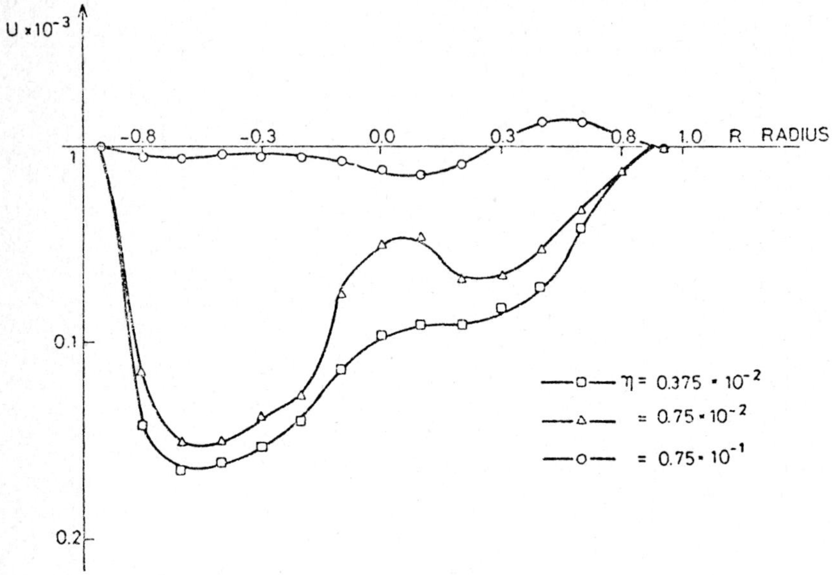

Figure 8
The development of the radial velocity (u) profile, same conditions as in Fig. 6.

The growth of u near the walls, due to the boundary layer growth, is clearly seen at a small distance from the inlet, but near the duct center the radial velocity is smaller showing that the stream lines at this region are not strongly inclined. Further downstream, u vanishes as we expect at the fully developed part of the pipe. The circumferential velocity is plotted at $\theta = \pm 22.5$ and at two axial locations. The symmetry is easily seen, and the fairly large lateral velocities in the nonsymmetrical flow region is apparent.

CONCLUSIONS

This paper is concerned with non-symmetrical flow in symmetrical ducts. The origin of the natural cylindrical coordinate system is a singular point in such a coordinate system and the lateral velocities are multi-valued there. This problem was resolved by suitable formulation of the radial derivatives and by mesh staggering.

The numerical method is similar to Briley's. However, mesh staggering improved the numerical efficiency. Unlike Briley's results, the report is concerned with circular ducts, as discussed above.

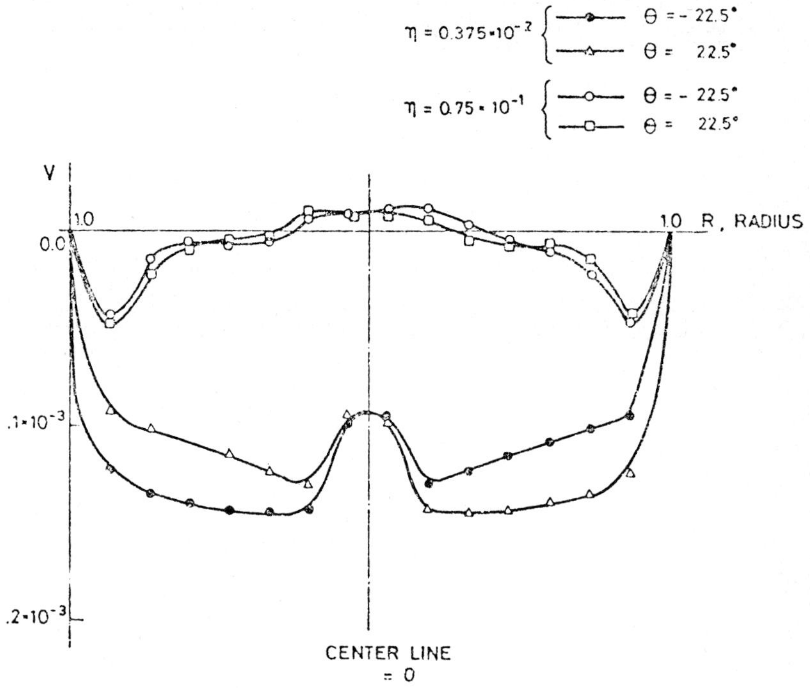

Figure 9
The development of the circumferential velocity (v) profile,
same conditions as in Fig. 6

Sample calculation shows the influence of shear on the velocity profiles. This influence appears to be limited to the entrance region, and the symmetrical parabolic flow is established at about the same distance from the inlet as in symmetrical flow.

The method can be easily extended to axially symmetrical ducts with curved walls, and to curved axes. Turbulence and heat and mass transfer may be easily incorporated.

REFERENCES

[1] Briley, W.R., Numerical method for predicting three-dimensional steady viscous flow in ducts. J. of Computational Physics, 14 (1974) 8-28.

[2] de Vahl Davis, G., A note on a mesh for use with polar coordinates, Numerical Heat Transfer, 2 (1979) 261-266.

[3] Isenberg, J and de Vahl Davis, G., in: Topics in Transport Phenomena, ed. C. Gutfinger, John Wiley & Sons, N.Y. 1975, 457-553.

[4] Levy, R., McDonald, H. and Briley, W.R., Calculation of three-dimensional turbulent subsonic flows in transition ducts. Sixth International Conference on Numerical Methods in Fluid Dynamics, Proceedings, Tbilisi (1978) 361-369.

[5] Patankar, S.V. and Spalding, D.B., Calculation procedure for heat, mass and momentum transfer in three-dimensional parabolic flows, Int. J. of Heat and Mass Transfer, 15 (1972) 1787-1806.

[6] Williams, G.P., Numerical integration of the three-dimensional Navier-Stokes equations for incompressible flow, J. Fluid Mech. 37 (1969) 727-750.

Computational Techniques & Applications: CTAC-83
J. Noye & C. Fletcher (Editors)

THE EFFECT OF A SMALL BLOWING ON VORTEX-BREAKDOWN OF A SWIRLING FLOW

Keiichi Karashima

The Institute of Space and Astronautical Science
Komaba, Meguro-ku, Tokyo, Japan

Shoji Kitama

National Space Agency of Japan
Hamamatsu-cho, Minato-ku, Tokyo, Japan

A numerical study of an interaction between a swirling flow and a coaxial jet is developed to clarify aerodynamic mechanism and feasibility on control of vortex-breakdown by means of a small blowing. Application of the blowing to a broken vortex-flow having a bubble induces an amount of leeward movement and shrinking of the bubble simultaneously, because it enhances the vortex core to depress considerably the axial deceleration of the rotating flow. It is concluded that the blowing can offer a significant improvement in breakdown characteristics of the swirling flow.

INTRODUCTION

From viewpoint of practical applications as well as an academic interest, the development of control techniques of vortex-breakdown seems to become increasingly important in fluid mechanics, because the breakdown can often be observed in many rotating flows of engineering interest such as, for example, the the swirling flow in a duct and the leading-edge vortex system on low aspect-ratio lifting surfaces at low subsonic speeds.

In general, the vortex-breakdown is recognized as occurrence of stagnation in the swirling flow and, if it occurs, the flow field undergoes significant modification that may be characterized by abrupt flow deceleration, streamtube expansion and pressure rise, etc. Many researches have been conducted on the vortex-breakdown in both analytical and experimental methods (refs. 1 to 10), and a considerable amount of comprehensive information is available pertinent to flow visualization, internal structure of the breakdown and effects of flow parameters such as Reynolds number and Rossby number, etc.

In the aircraft engineering there are many theoretical and experimental investigations (refs. 11 to 17) made for low speed flow characteristics of a vortex system induced by flow separation at a swept leading-edge of low aspect-ratio lifting surfaces, and it is well known that the leading-edge vortex can offer significant improvements in lifting efficiency of the wings unless breakdown of the vortex occurs up to high angles of attack.

One of the promising method for enhancing stability of the leading-edge vortex is the so-called spanwise blowing. A number of experimental studies (refs. 18 to 23) have been devoted to clarify feasibility and aerodynamic mechanism of lift augmentation by the blowing, and it has been deduced that the momentum addition due to the blowing aids in formation and strengthening of the vortex core to delay breakdown to higher angles of attack, thus resulting in a considerable lift-increment of the wings. However, it seems that there have been developed few analytical attempts for elucidating the deduction.

It is the purpose of this study to develop a numerical analysis of an interaction between a rotating flow and a small jet in order to present a theoretical account of the aerodynamic mechanism on response of the vortex-breakdown to a blowing. To simplify the analysis the flow is assumed to be incompressible and axisymmetric. Moreover, an initial-boundary value problem is formulated in such a way that the basic swirling flow field is assumed to consist of the simplified model proposed by Grabowski and Berger (1976), to which is applied a small coaxial jet at the windward boundary of the computational domain. The virtual system of the unsteady Navier-Stokes equations having Chorin's (1967) artificial compressibility is solved using the ADI algorithm, and the asymptotic steady-state solutions are obtained.

It must be noted that the present analysis is not a direct simulation to the breakdown control by means of the spanwise blowing because of the unrealistic flow model. However, since the model preserves an essential physical process of a jet-rotating flow interaction, the results may predict some theoretical prospects on controllability of the vortex-breakdown by use of a coaxial blowing.

BASIC FORMULATIONS

With the assumption of the Grabowski-Berger model as the basic swirling flow, the present initial-boundary value problem may be formulated essentially in the same way as that in ref. 9 except the boundary conditions appropriate to a small coaxial jet.

Let the origin of a cylindrical coordinates system (x^*, r^*, θ) be taken at the windward boundary of the computational domain, x^*-axis being aligned with the swirling flow (see Fig. 1), where the asterisk indicates dimensional quantities.

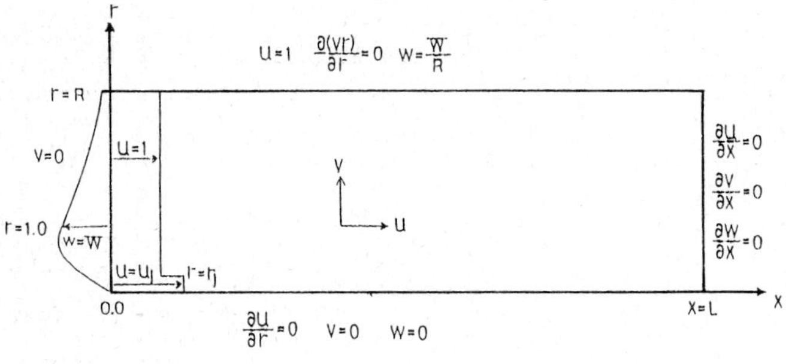

Fig. 1. Coordinates and boundary conditions.

For simplicity the nondimensional variables defined by the relations

$$x = \frac{x^*}{r_c^*}, \qquad r = \frac{r^*}{r_c^*}, \qquad p = \frac{p^* - p_\infty^*}{\rho_\infty^* U_\infty^{*2}} \qquad (1)$$

$$u = \frac{u^*}{U_\infty^*}, \qquad v = \frac{v^*}{U_\infty^*}, \qquad w = \frac{w^*}{U_\infty^*}$$

are introduced, where u,v and w are velocity components, p; the pressure and ρ; the density. U_∞^* and r_c^* denote free stream velocity and radius of vortex core at the windward boundary, respectively.

The virtual system of the unsteady Navier-Stokes equations for incompressible, axisymmetric flow may be written using the Chorin's artificial compressibility technique as

$$\frac{\partial u}{\partial t} + Fv\frac{\partial u}{\partial y} + u\frac{\partial u}{\partial x} = -\frac{\partial p}{\partial x} + \frac{1}{Re}\left[F^2\frac{\partial^2 u}{\partial y^2} + (G + FH)\frac{\partial u}{\partial y} + \frac{\partial^2 u}{\partial x^2} \right] \tag{2.1}$$

$$\frac{\partial v}{\partial t} + Fv\frac{\partial v}{\partial y} + u\frac{\partial v}{\partial x} - Hw^2 = -F\frac{\partial p}{\partial y} + \frac{1}{Re}\left[F^2\frac{\partial^2 v}{\partial y^2} + (G + FH)\frac{\partial v}{\partial y} + \frac{\partial^2 v}{\partial x^2} - H^2 v \right] \tag{2.2}$$

$$\frac{\partial w}{\partial t} + Fv\frac{\partial w}{\partial y} + u\frac{\partial w}{\partial x} + Hvw = \frac{1}{Re}\left[F^2\frac{\partial^2 w}{\partial y^2} + (G + FH)\frac{\partial w}{\partial y} + \frac{\partial^2 w}{\partial x^2} - H^2 w \right] \tag{2.3}$$

$$\gamma\frac{\partial p}{\partial t} + FH\frac{\partial(u/H)}{\partial y} + \frac{\partial u}{\partial x} = 0 \tag{2.4}$$

$$F = \frac{1}{\alpha\beta}\exp(-\alpha y), \qquad G = \frac{1}{\alpha\beta^2}\exp(-2\alpha y)$$
$$H = \frac{1}{\beta}\left[\exp(\alpha y) - 1 \right], \qquad Re = \frac{U_\infty^* r_c^*}{\nu} \tag{2.5}$$

where γ and Re denote the artificial compressibility factor and core Reynolds number, respectively, and the coordinate transformation defined by the equation

$$y = \frac{1}{\alpha}\log\left(1 + \frac{r}{\beta}\right) \tag{3}$$

has been employed. α and β in Eqs.(2.5) and (3) are arbitrary constants to be so determined as to generate favourable grid-spacing in the r-direction.

Since the Grabowski-Berger model has been assumed as the basic swirling flow, there is no essential difference in boundary conditions between the present formulation and ref. 9 except the conditions for axial velocity component at the windward boundary. For simplicity it is assumed in the present approach that the axial velocity distribution at the windward boundary consists of a uniform velocity equal to the free stream one, upon which is superimposed a uniform disturbance velocity due to the injection in the region ($0 \le r \le r_j \ll 1$), where r_j is the radius of the jet. The other boundary conditions may be predicted as follows. The circumferential velocity of the vortex core at the inlet is assumed to be given by a cubic polynomial of r. The symmetric conditions are given along the axis, whereas the asymptotic solutions to a rotating potential flow are assumed at the external radial boundary ($y = R \gg 1$). At the leeward boundary ($x = L \gg 1$), on the other hand, are imposed the conditions of continuous outflow with zero velocity gradient.

With the coordinate transformation given by Eq.(3) these boundary conditions may be summarized as at x = 0

$$u = \begin{cases} u_j & (0 \leq y \leq a) \\ 1.0 & (a \leq y \leq b) \end{cases} \tag{4.1}$$

$$v = 0 \tag{4.2}$$

$$w = \begin{cases} \frac{W}{H}\left(2 - \frac{1}{H^2}\right) & (0 \leq y \leq a) \\ WH & (a \leq y \leq b) \end{cases} \tag{4.3}$$

$$a = \frac{1}{\alpha} \log\left(1 + \frac{r_j}{\beta}\right), \qquad b = \frac{1}{\alpha} \log\left(1 + \frac{R}{\beta}\right) \tag{4.4}$$

at x = L

$$\frac{\partial u}{\partial x} = \frac{\partial v}{\partial x} = \frac{\partial w}{\partial x} = 0 \tag{4.5}$$

at r = 0

$$\frac{\partial u}{\partial r} = v = w = 0 \tag{4.6}$$

at r = R

$$u = 1.0, \qquad \frac{\partial (vr)}{\partial r} = 0, \qquad w = \frac{W}{R} \tag{4.7}$$

where W denotes the inlet circumferential velocity at the edge of the vortex core (r = 1.0).

COMPUTATIONAL SCHEME

In the numerical approach to the unsteady Navier-Stokes equations for incompressible flow, it is generally required to solve a Poisson equation at every time step in order to obtain pressure field. However, the computation will be simplified if the artificial compressibility technique is applied to the continuity equation, because the virtual system of the Navier-Stokes equations thus obtained can be integrated easily using a conventional time-dependent algorithm. Although the instantaneous solutions to the virtual system do not make any sense from the physical point of view, the asymptotic steady-state solutions may be reduced to those to the corresponding steady equations.

The computational conditions used in the present analysis are summarized as follows;

$$\begin{aligned} &L = 9.0, \qquad R = 3.0 \\ &\Delta x = \Delta y = 0.1, \qquad \Delta t = 0.01, \qquad \gamma = 1.0 \\ &\alpha = 1.0051, \qquad \beta = 0.15470 \end{aligned} \tag{5}$$

where the values of α and β are so chosen as to yield 20 radial grids within the vortex core in the physical plane.

On the other hand, the convergence criterion for the present iterative scheme is defined as

$$\text{Max}\ (\ \Delta f_i^n\ ) < 0.5 \times 10^{-4}$$
$$\Delta f_i^n = |\ f_i^n - f_i^{n-100}\ | \qquad (6)$$

where f_i is the i-th component of a four-dimensional column vector defined by the dependent variables (u, v, w, p) and the subscript n indicates the number of time step.

The numerical calculation was carried out using the ADI algorithm, where the computational process consisted of two steps. The first step is a preliminary calculation of the basic swirling flow having a breakdown bubble (Grabowski-Berger model), which is characterized by two parameters. Those are the core Reynolds number Re defined by Eq.(2.5) and the circulation coefficient that is defined as

$$\Omega = \frac{\Gamma^*_{r=1}}{2r_c^* U_\infty^*} = \pi W \qquad (7)$$

where $\Gamma^*_{r=1}$ is the inlet circulation at the edge of the vortex core. Therefore, with given Re and Ω together with the nonblowing condition (u_j =1.0), integration of the basic equations, Eqs.(2.1) to (2.4), is carried out starting from the initial conditions of uniform stream everywhere in the flow field until the convergence criterion is satisfied.

The next step is the calculation of an interacting flow field between the basic flow and a jet, where the injection velocity u_j and the blowing radius r_j are the characteristic parameters. Consequently, giving u_j and r_j a priori, the basic equations are integrated again starting from the solutions to the basic flow until the convergence criterion is satisfied.

However, in computing the cases for increased injection velocities, a stepwise computation, which is made in such a way that the solutions to the last u_j are utilized as initial conditions for the next one, was adopted, because it was efficient in the sense of not only enhancing numerical stability but also saving computational time.

RESULTS AND DISCUSSIONS

In order to understand intuitively the effect of blowing on vortex-breakdown it will be helpful to demonstrate several illustrations on variation of streamline pattern with injection velocity. Fig. 2 shows the streamline pattern for nonblowing and blowing in the case of the basic flow defined by Re = 100 and Ω = 3.14. In this case the circulation coefficient is relatively small, so that the breakdown occurs at a far distance downstream from the inlet to generate a small region of recirculating flow (breakdown bubble), the contour of which is nearly ellipsoidal, as shown in Fig. 2(a). On the other hand, Fig. 2(b) reveals a remarkable result that a small blowing affects the breakdown to induce simultaneously a leeward movement and shrinking of the bubble to a considerable extent. Since the bubble is so small as to be affected

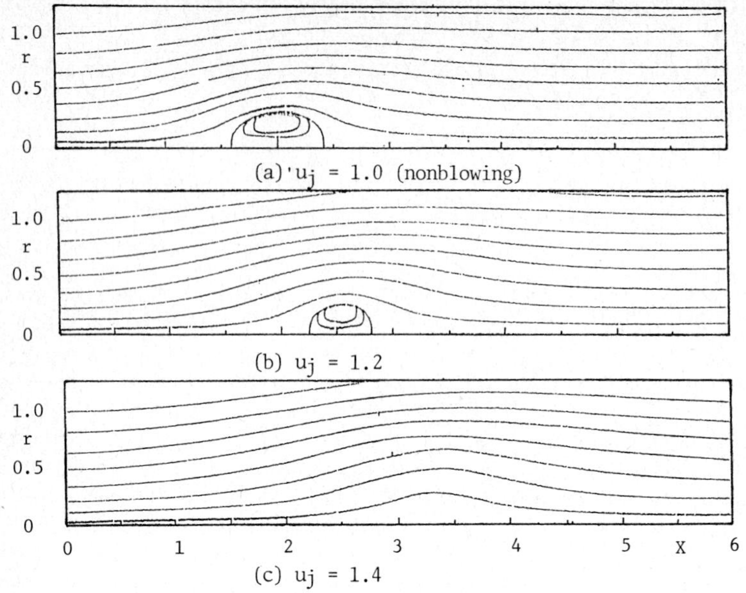

Fig. 2. Strealine patterns. Re = 100, Ω = 3.14, r_j = 0.228.

entirely by the injection, the change in location and size of the bubble seems to be continuous for increasing injection velocity. Thus, the bubble moves farther downstream shrinking but retaining roughly the similar shape, and it disappears finally, as is seen in Fig. 2(c).

The shape of the bubble in the basic flow and its response to the blowing are modified strikingly as the circulation is increased. In Figs. 3(a) to 3(c) are presented the streamline patterns for Re = 100 and Ω = 3.77. In the light of the result shown in Fig. 2(a) it will be easily found that an increase of the circulation results in a windward movement and a sizable growth of the bubble in the basic flow simultaneously, and the most significant modification takes place in the bubble shape. In fact, the rear contour of the boundary streamline, which separates the internal recirculating flow from the external main stream, is tremendously distorted inward near the axis. The reason for this may be laid on the effect of the reversed velocity induced by negative vorticities existing just aft of the bubble.

In the case of a small blowing (u_j = 1.2), the change in flow pattern is characterized mainly by a large scale of leeward distortion in the front contour of the bubble, whereas the windward distortion in the rear one is only appreciable. As the result, the injection turns out a marked shrinking of the bubble with mutual proximity of the forward and rearward stagnations on the axis, as shown in Fig. 3(b).

As the injection increases further, the change in bubble shape continues to proceed in the similar manner, and the forward stagnation encounters the rearward one at a certain critical blowing, beyond which no stagnation exists on the axis, whereas the recirculating flow still continues to exist in a closed region at a certain radial distance. This flow pattern can be obviously seen in Fig. 3(c),

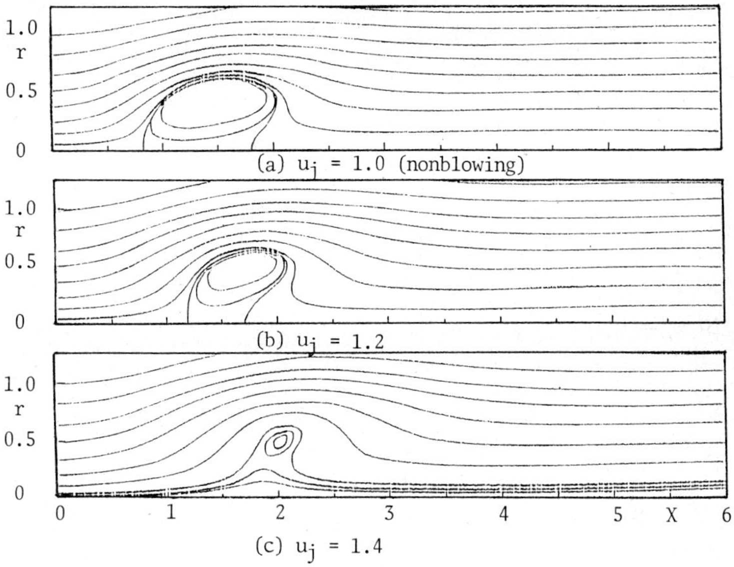

Fig. 3. Streamline patterns. Re = 100, Ω = 3.77, r_j = 0.228.

where the bubble is transformed into an annulus like a doughnut, looking as if it were blown through by the injected stream. For further increase of the injection velocity the annular bubble vanishes finally, although a sizable distortion of the streamline still remains in the flow field.

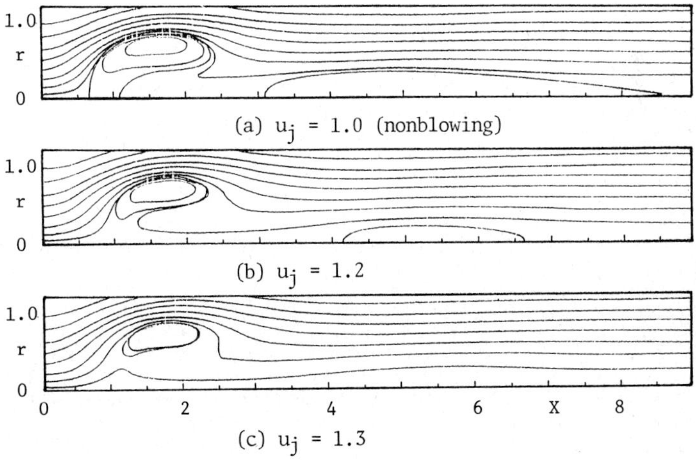

Fig. 4. Streamline patterns. Re = 100, Ω = 4.40, r_j = 0.228.

Figs. 4(a) to 4(c) present the streamline patterns in the case of Re = 100 and Ω = 4.40. Since the negative vorticities in the downstream vicinity of the bubble are intensified as the circulation grows, the velocity recovery of the broken vortex flow near the axis is insufficient to preserve an ordinary flow and this, in turn, generates a long secondary bubble at a certain distance downstream of the primary one, as is seen in Fig. 4(a). In the case of blowing, the flow field seems to become more sensitive to the injection, because the primary bubble has been transformed into an annulus even at u_j = 1.2 (see Fig. 4(b)). However, the secondary bubble shrinks faster to vanish with increase of the injection velocity. The seemingly predominating effect of the blowing on the secondary bubble may be attributed to the fact that the associated recirculating flow is so weak and rather stagnant as to be blown out by only a small amount of momentum addition.

Fig. 5 shows pressure distribution along the axis of symmetry in the case of Re = 100 and Ω = 3.77. The pressure in the basic flow rises abruptly as the breakdown is approached and is roughly constant in the recirculating region. This characteristic seems to be similar qualtitatively to that observed in the process of a boundary layer separation. Downstream of the bubble the pressure first decreases slightly because of the vorticity induced flow and, then, it increases gradually. The same is true qualtitatively also in the case of blowing and no specific change in axial pressure takes place even for the injection that is strong enough to induce an annular bubble except the differences in overall pressure level and location of the peak pressure.

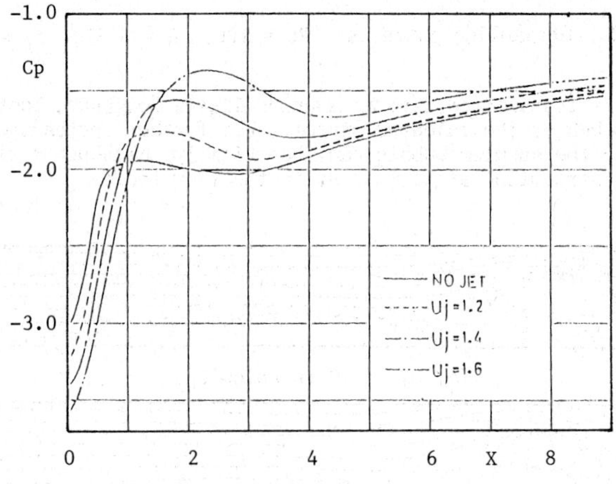

Fig. 5. Pressure distribution on the axis of symmetry. Re = 100, Ω = 3.77, r_j = 0.228.

Detailed examination of local flow structure clearly reveals a noteworthy fact that the momentum addition due to the blowing certainly enhances the core of the swirling flow in such a way that deceleration of both axial and circumferential velocity components is alleviated tremendously near the axis and the breakdown is, therefore, delayed. In case the blowing is strong enough, the flow deceleration is diminished so considerably as to bring about no stagnation on the axis, thus resulting in that the bubble is blown through by the injected stream near the axis to become annular.

The foregoing statement may be recognized from the results shown in Fig. 6, which indicates representative examples on radial distribution of axial and circumferential velocity components of the local flow at various stations in

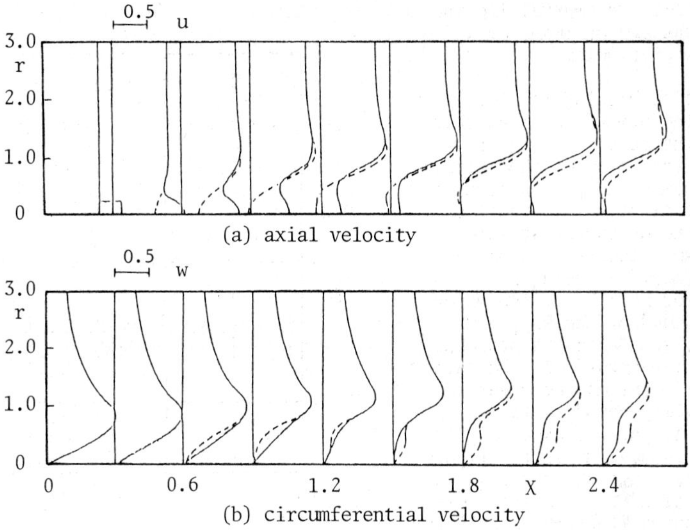

Fig. 6. Radial distribution of axial and circumferential velocity components. Re = 100, Ω = 3.77, r_j = 0.228.

the case of Re = 100, Ω = 3.77 and u_j = 1.4. In the figure the corresponding nonblowing data are also plotted in the broken lines to emphasize the effect of the blowing. The results indicate obviously that the injection contributes mainly to depress deceleration of the axial velocity and partially to augment the circumferential velocity near the axis, resulting in no stagnation on the axis of symmetry. It must be noted that the radial velocity component seems to have little effect on the vortex-breakdown, because it is of smaller order of magnitude than the other two everywhere in the flow field.

Fig. 7 shows the change in location and length of the bubble with blowing velocity in the case of Re = 100 and Ω = 3.77, where ΔX_B is defined as the axial displacement of the forward stagnation point relative to the nonblowing position and l_B is the distance between the two stagnation points on the axis. The qualitatively same trend can be obtained for the circulation coefficients other than 3.77, indicating that the rate of bubble displacement to u_j is decreased for larger Ω, whereas the rate of bubble shrinking increases. Anyway, the results shown in Fig. 7 should be remarked as quantitative information

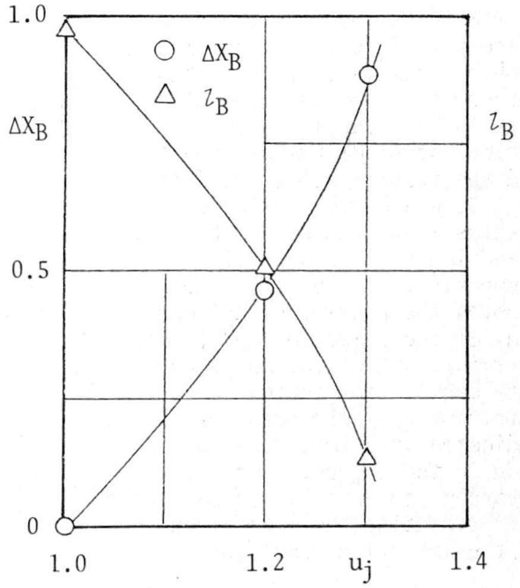

Fig. 7. Displacement and length of of the bubble. Re = 100, Ω = 3.77, r_j = 0.228.

on feasibility of controlling the vortex-breakdown by means of a small coaxial blowing.

The effect of blowing on occurrence of the breakdown is another point of interest. In Fig. 8 is presented variation of the critical Reynolds number Re_{cr} with circulation coefficient. It is defined as the Reynolds number at which the minimum velocity u_{min} on the axis vanishes. Since u_{min} is negative for $Re > Re_{cr}$, the upper region bounded by the Re_{cr} - Ω_{cr} curve corresponds to the presence of the vortex-breakdown. In this sense, the curve may be regarded as a stability boundary for existence of the unbroken swirling flow. It is obvious in the figure that the curve is shifted upward by application of a coaxial blowing, indicating that the stability limit on the unbroken rotating flow is extended to higher Reynolds numbers. This, in turn, implies that the breakdown is delayed by the injection. Therefore, the result seems to offer an analytical support to the experimental deduction on the mechanism of lift augmentation by means of the spanwise blowing mentioned previously.

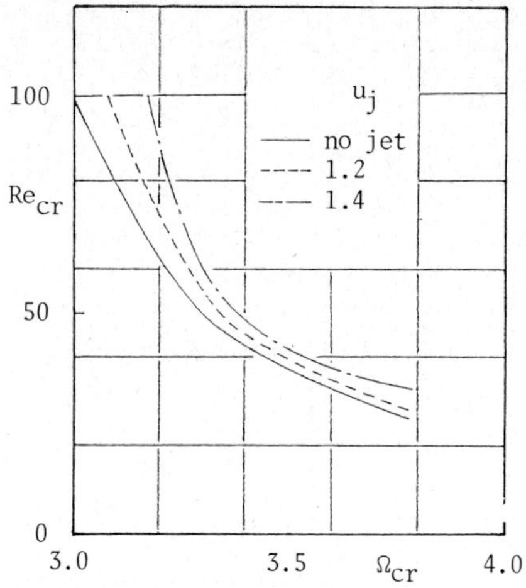

Fig. 8. Critical Reynolds number and circulation for occurrence of vortex-breakdown. $r_j = 0.228$.

It remains to discuss the effectiveness of injection mass flux on bubble displacement. For this purpose the numerical computations were further carried out for various Reynolds numbers and r_j, and the results are plotted in Fig. 9, where the blowing effectiveness λ and the nondimensional injection area S_j are defined, respectively, in Eq.(8), where m_j denotes the nondimensional mass flux of the injection and is equal to $r_j^2 u_j$. Since the computed data can be represented approximately by a single mean curve indicated in a full line, as is seen in the figure, λ may be regarded as a similarity parameter for predicting the effectiveness of the injection mass flux.

$$\lambda = \frac{\Omega}{Re} \left(\frac{dX_B}{dm_j} \right)_{u_j = 1} \qquad (8)$$

$$S_j = r_j^2$$

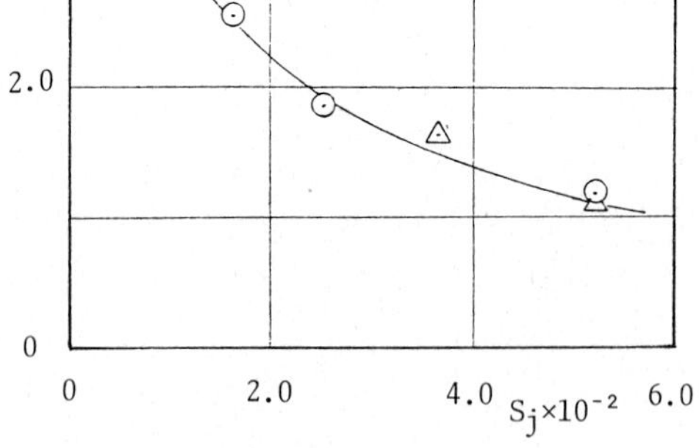

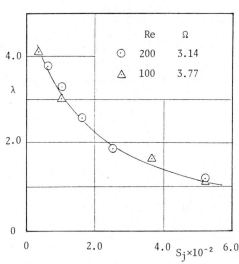

Fig. 9. Blowing effectiveness for displacement of the bubble.

The figure reveals that the smaller injection area is more effective in the sense of mass flux efficiency. Apart from the physical meanings of the blowing effectiveness, the result shown in Fig. 9 seems to be consistent qualitatively with those of the other kind of small blowings such as the local injection and film cooling, etc., although they may be used for different objectives of alleviating the aerodynamic heating.

CONCLUSION

With emphasis on feasibility of controlling the vortex-breakdown, a numerical analysis has been developed on the interaction between an axisymmetric broken vortex flow and a small coaxial jet. With the assumption that the basic swirling flow consists of the Grabowski-Berger model, the virtual system of the unsteady Navier-Stokes equations has been solved using the ADI algorithm and the results may be summarized as follows.

Application of a small blowing to a broken vortex flow having a breakdown bubble induces an amount of leeward movement and shrinking of the bubble simultaneously, because, in the region upstream of the bubble, the momentum transfer from the injected stream diminishes considerably the axial deceleration of the flow near the axis.

As the injection mass flux is increased, the reverse flow near the axis tends to vanish, whereas a closed recirculating region still continues to exist at a certain radial distance. Since the bubble is blown through by the injected stream near the axis, it is transformed into an annulus like a doughnut. For further increase of the injection velocity, the annular bubble disappears finally.

Detailed examination of local flow structure reveals that the blowing aids in augmentation of axial velocity component of the rotating core near the axis to delay breakdown to higher core Reynolds numbers in case the circulation coefficient is fixed, and this trend becomes increasingly conspicuous as the circulation grows. Therefore, the result leads to a statement that the blowing can offer a significant improvement in stability limit of the swirling flow.

Finally, it is suggested that the present results may predict some theoretical prospects on controllability of the vortex-breakdown by means of a small coaxial blowing.

REFERENCES :

[1] Benjamin, T.B., Theory of the Vortex Breakdown Phenomenon, J. Fluid Mech. 14 (1962) 593-629.
[2] Hall, M.G., A Numerical Method for Solving the Equations for a Vortex Core, A.R.C. R&M-3467 (1965).
[3] Lavan, Z., Nielsen, H. and Fejer, A.A., Separation and Flow Reversal in Swirling Flows in Circular Ducts, Phys. Fluids 12 (1969) 1747-1757.
[4] Torrance, K.E. and Kopecky, R.M., Numerical Study of Axisymmetric Vortex Breakdown, NASA CR-1865 (1971).
[5] Sarpkays, T., Vortex Breakdown in Swirling Conical Flows, AIAA J. 9 (1971) 1792-1799.
[6] Chigier, N.A., Gasdynamics of Swirling Flow in Combustion System, Astronautica Acta. 17 (1972) 387-395.
[7] Faler, J.H. and Leibovich, S., Disrupted States of Vortex Flow and Vortex Breakdown, Phys. Fluids 20 (1977) 1385-1400.
[8] Faler, J.H. and Leibovich, S., An Experimental Map of the Internal Structure of a Vortex Breakdown, J. Fluid Mech. 86 (1978) 313-335.

[9] Grabowski, W.J. and Berger, S.A., Solutions of the Navier-Stokes Equations for Vortex Breakdown, J. Fluid Mech. 75 (1976) 525-544.
[10] Nakamura, Y. and Uchida, S., Numerical Solutions of Navier-Stokes Equations for Axisymmetric Weak Swirling Flows in a Pipe, Transactions Japan Soc. Aero. Space Sci. 24 (1982) 222-226.
[11] Ludwieg, H., Zur Erklarung der Instabilitat der uber angestellten Delta-flugeln auftretenden freien Wirbelkerne, Z. Flugwiss. 10 (1962) 242-249.
[12] Earnshaw, P.B., Measurements of Vortex-Breakdown Position at Low Speed on a Series of Sharp-Edged Symmetrical Models, A.R.C. CP-828 (1965).
[13] Polhamus, E.C., A Concept of Vortex Lift of Sharp-Edged Delta Wings Based on a Leading-Edge-Suction Analogy, NASA TN D-3769 (1966).
[14] Bird, J.D., Tuft-Grid Survey at Low Speeds for Delta Wings, NASA TN D-5049 (1969).
[15] Wentz, W.H. and Kohlman, D.L., Vortex Breakdown on Slender Sharp-Edged Wings, J. Aircraft 8 (1971) 156-161.
[16] Polhamus, E.C., Prediction of Vortex-Lift Characteristics by a Leading-Edge-Suction Analogy, J. Aircraft 8 (1971) 193-199.
[17] Luckring, J.M., Aerodynamics of Strake-Wing Interactions, J. Aircraft 16 (1979) 756-762.
[18] Bradley, R.G., Smith, C.W. and Wray, W.O., An Experimental Investigation of Leading-Edge Vortex Augmentation by Blowing, NASA CR-132415 (1974).
[19] Coe, P.L.Jr. and Kulla, D., Effect of Upper-Surface Blowing on Static Longitudinal Stability of a Swept Wing, J. Aircraft 9 (1974) 537-539.
[20] Bradley, R.G. and Wray, W.O., A Conceptual Study of Leading-edge Vortex Enhancement by Blowing, J. Aircraft 11 (1974) 34-38.
[21] Cambell, J.F., Effects of Spanwise Blowing on the Pressure Field and Vortex-Lift Characteristics of a 44° Swept Trapezoidal Wings, NASA TN D-7907 (1975).
[22] Erickson, G.E. and Cambell, J.F., Flow Visualization of Leading-Edge Vortex Enhancement by Spanwise Blowing, NASA TM X-72702 (1975).
[23] Cambell, J.F., Augmentation of Vortex Lift by Spanwise Blowing, J. Aircraft 13 (1976) 727-732.
[24] Chorin, A.J., A Numerical Method for Solving Incompressible Viscous Flow Problems, J. Comp. Phys. 2 (1967) 12-26.

POROUS MEDIA, THERMAL PROBLEMS

Computational Techniques & Applications: CTAC-83
J. Noye & C. Fletcher (Editors)

NUMERICAL TECHNIQUES FOR ESTIMATING THE BEHAVIOUR OF UNSTABLE, IMMISCIBLE FLOW IN POROUS MEDIA

Geoffrey R.G. Woodham, David F. Bagster

Department of Chemical Engineering
The University of Sydney
Sydney, N.S.W. 2006
Australia

W.V. Pinczewski

Department of Chemical Engineering and Industrial Chemistry
The University of New South Wales
Kensington, N.S.W. 2033
Australia

Optimal exploitation of oil reservoirs by fluid injection is usually hampered by the instability of the interface caused by unfavourable viscosity ratios. This paper examines whether the numerical models commonly used to predict the recovery efficiencies in actual reservoirs under stable displacement conditions can be applied when the displacement is unstable. It is shown that unstable displacement cannot be simulated using conventional relative permeabilities coupled with perturbations to the governing differential equations. Empirically derived, pseudo relative permeabilities are demonstrated to be potentially useful in simulating the behaviour of unstable floods.

INTRODUCTION

The problem of viscous instability in multiphase flow through porous media has received considerable attention in the literature (Engelberts and Klinkenberg, 1951; Chuoke et al. 1959; de Haan, 1959; Saffman and Taylor, 1958; Scheidegger, 1974). Since most displacement processes aimed at improving the recovery of oil from petroleum reservoirs involve the displacement of viscous oil by a less viscous phase eg. water, gas etc., such operations are inherently unstable. Under such conditions the displacement may be characterized by a highly irregular frontal advance in which the displacing phase penetrates the resident oil through distinct channels, Fig. 1. Such unstable displacements are said to exhibit "viscous fingering" and are characterized by considerably reduced oil recoveries at the breakthrough of the displacing phase. In contrast, stable displacement is characterized by an essentially flat front yielding relatively high oil recoveries.

Flock et al. (1977) and Peters and Flock (1981) have used this "breakthrough recovery" to quantify the degree of viscous instability in laboratory experiments. Typically, the experimentally determined breakthrough recovery is correlated with a dimensionless rate function, I, representing the ratio of viscous to capillary forces. At intermediate displacement rates, viscous and capillary forces are of similar magnitude resulting in a well defined, stable displacement front with breakthrogh recoveries being largely insensitive to the displacement rate. At high displacement rates viscous forces dominate, destabilizing the displacement front and resulting in a gradual decline in the breakthrough recovery with increasing displacement rate.

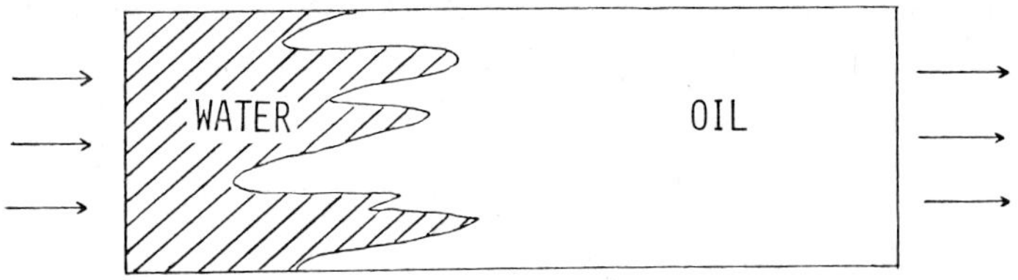

Fig. 1 - Illustration of an unstable displacement front

Although numerical models are widely used in the petroleum recovery industry to predict displacement efficiencies in actual reservoirs under stable displacement conditions, it is not clear whether these models are capable of predicting the progressive decline in recovery with increasing displacement rate which is observed under conditions when the displacement front is unstable. Previous work by Rachford (1964) in which he attempted to use numerical reservoir models to simulate the formation and growth of viscous fingers by introducing random perturbations in rock and fluid properties, suggests that these models may be used to predict recovery efficiencies under unstable conditions. In a more recent study, Flock et al. (1977) were unable to predict the experimentally observed decline in recovery efficiency with displacement rate by numerically solving the appropriate Buckley-Leverett equation.

The present paper presents the results of a preliminary study in the use of numerical models to simulate immiscible, adverse-mobility-ratio floods. The work is presented in two parts. Firstly, it is demonstrated that the introduction of random perturbations into the numerical solution of the governing fluid flow equations, as was done by Rachford (1964), cannot lead to a successful simulation of unstable displacements. Secondly, we outline a modification to conventional numerical models based on the introduction of pseudo relative permeabilities which can lead to a realistic prediction of the decline in recovery efficiency with increasing displacement rate.

CONVENTIONAL MODEL FORMULATION

Typical of the commercially used numerical reservoir simulation models is the formulation presented by Spillette et al. (1973). The equations of continuity for the displacing and displaced phases are combined with expressions for Darcy's law for each phase, based on the conventional concept of relative permeability, to produce the following two equations.

$$\frac{\partial}{\partial x}(\lambda_t \frac{\partial p_n}{\partial x}) - \frac{\partial}{\partial x}(\lambda_w \frac{\partial p_c}{\partial x}) + \frac{\partial}{\partial y}(\lambda_t \frac{\partial p_n}{\partial y}) - \frac{\partial}{\partial y}(\lambda_w \frac{\partial p_c}{\partial y}) + Q_n + Q_w = 0 \qquad (1)$$

$$-\frac{\partial}{\partial x}(f_w v_t) - \frac{\partial}{\partial x}(f_w \lambda_n \frac{\partial p_c}{\partial x}) - \frac{\partial}{\partial y}(f_w v_t) - \frac{\partial}{\partial y}(f_w \lambda_n \frac{\partial p_c}{\partial y}) + Q_w = \phi \frac{\partial S_w}{\partial t} \qquad (2)$$

where

$$\lambda_{nw} = kk_{rnw} / \mu_{nw} \tag{3}$$

$$\lambda_{w} = kk_{rw} / \mu_{w} \tag{4}$$

$$\lambda_{t} = \lambda_{nw} + \lambda_{w} \tag{5}$$

$$f_{w} = \lambda_{w} / \lambda_{t} \tag{6}$$

$$p_{c} = p_{nw} - p_{w} \tag{7}$$

The above system of equations is referred to as the total velocity formulation and has been demonstrated by Peaceman (1977) to be very stable. Although the permeability, k, is an intrinsic property of the porous medium, the relative permeabilities k_{rw} and k_{rnw} must be determined by conventional laboratory core flood experiments. In this way the relative permeabilities are found to be very strong functions of fluid saturation.

The physical situation studied was identical to that described by Rachford (1964). It consisted of an end-to-end displacement in a rectangular region measuring 45.7 cm x 14.7 cm having a thickness of 1.0 cm. Initially the model contains oil at the irreducible water saturation. Water is injected along the entire width of one end of the model and oil is produced along the entire width of the opposite end. The appropriate boundary conditions for the case where the produced oil is at atmospheric pressure are:

$$S_{w}(x,y,\ t=0) = S_{wor},\ 0 \leqslant x \leqslant L,\ 0 \leqslant y \leqslant W$$

$$\frac{\partial p_n}{\partial y} = \frac{\partial p_w}{\partial y} = 0,\ 0 \leqslant x \leqslant L,\ y = 0$$

$$\frac{\partial p_n}{\partial y} = \frac{\partial p_w}{\partial y} = 0,\ 0 \leqslant x \leqslant L,\ y = W \tag{8}$$

$$Q_{INJ} = Q_t,\ \frac{\partial p_n}{\partial x} = \frac{\partial p_w}{\partial x} = 0,\ x = 0,\ 0 \leqslant y \leqslant W$$

$$Q_w = f_w Q_t,\ p_n = 1\ \text{atm},\ \frac{\partial p_w}{\partial x} = 0,\ x = L,\ 0 \leqslant y \leqslant W$$

The system of equations (1)-(8) was solved by the sequential finite-difference method outlined by Peaceman (1977) using a 40 x 20 grid. The elliptic "pressure" equation was solved for p_n by line successive overrelaxation with the one-dimensional additive correction proposed by Watts (1971, 1973).

The estimated values of p_n at the new time step were substituted into the hyperbolic "saturation" equation which was then solved by line successive overrelaxation for S_w at the new time. An acceleration parameter of unity (i.e. Gauss Seidel iteration) was found to be optimal for the hyperbolic equation. In the study using pseudo relative permeability curves (see below), it was not necessary to retain the fine resolution in the y direction and a one dimensional grid, (40x1), was used.

SIMULATION RESULTS USING CONVENTIONAL RELATIVE PERMEABILITY CURVES

In this section we present the results of numerical experiments similar to those reported by Rachford (1964) in which he attempted to simulate viscous fingering by introducing small, random perturbations into the numerical solution for displacements where the mobility ratio is unfavourable i.e. unstable displacements. Rachford (1964) argued that similar perturbations would be present in actual field displacements as a result of the random nature of real porous media and that these perturbations should grow in the numerical model just as they do in the actual reservoir rock.

Following Rachford (1964) we have used the conventional relative permeability and capillary pressure curves shown in Fig. 2. These were determined from measurements for a 38 cp water phase displacing a 165 cp oil phase. The properties of the porous medium used are shown in Table 1.

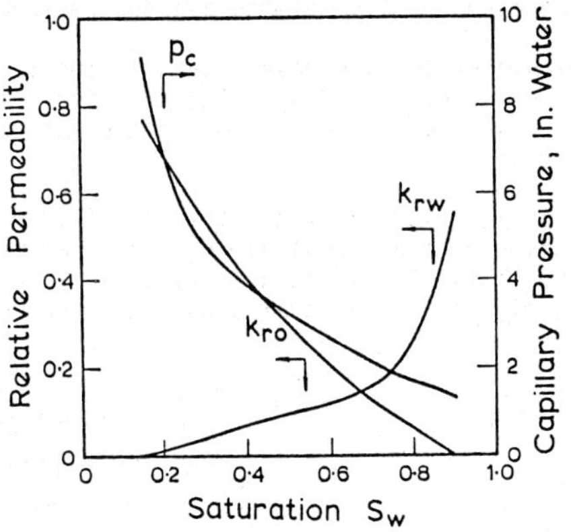

Fig. 2 - Relative Permeability and Capillary Pressure data from Rachford (1964)

TABLE 1

POROUS MEDIUM PROPERTIES

Length (cm)	45.7
Width (cm)	14.7
Height (cm)	1.0
Porosity	0.46
Absolute Permeability (Darcys)	36
Permeability to water @ S_{wi} (Darcys)	20
Permeability to oil @ S_{wor} (Darcys)	28
Irreducible Water Saturation	0.15
Water Saturation @ Residual Oil	0.90
Initial Water Saturation, S_{INIT}	0.15

Figures 3 and 4 (for v = 0.776 cm/sec) show the computed cumulative oil recoveries and produced water oil ratio (WOR) as a function of cumulative pore volume of water injected respectively, for the following cases;

i) Homogeneous case - this represents the base-case with all rock and fluid properties held constant.

ii) Capillary Pressure (Saturation) distribution case - the initial saturation of a small rectangle (4 x 4 grid blocks) was set to 0.5 (elsewhere the grid was at the connate water saturation). This provides an approximation to the classical analysis of Chuoke et al. (1959) who considered a small perturbation on an initially flat interface between two immiscible fluids in a porous medium.

iii) Flat permeability distribution case - permeability is perturbed at each point by selecting values from a pseudo-random set with a uniform distribution and a range of ± 5% of the average permeability.

iv) Normal permeability distribution case - this is identical to the distribution used by Rachford (1964). The selected values were taken from a normal set having a mean equal to the average permeability and a standard deviation of 1.8 darcys.

v) Log-Normal permeability distribution - this is similar to case (iv) except that the distribution is log-normal having a similar mean and a standard deviation of 0.05. This distribution is typical of permeability distributions usually found in actual sandstone reservoirs.

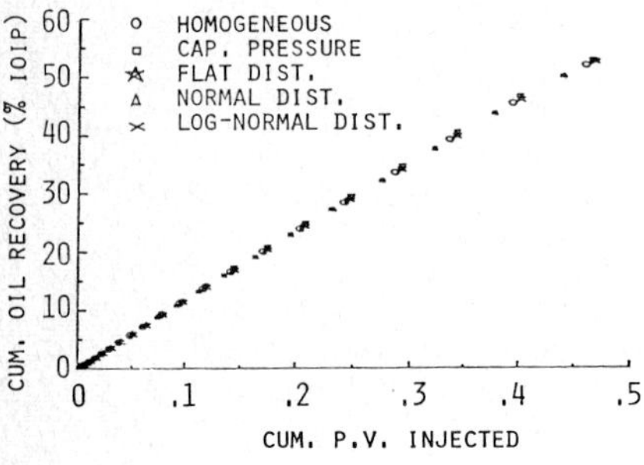

Fig. 3 - Predicted cumulative recoveries using conventional relative permeability curves

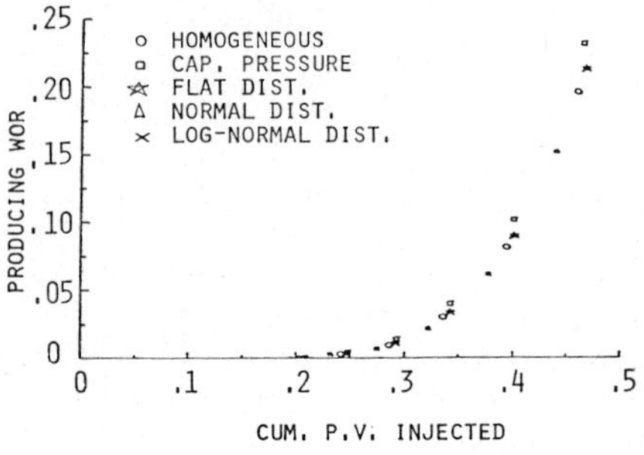

Fig. 4 - Predicted Producing Water/Oil ratios using conventional permeability curves

As is evident from Figures 3 and 4 there is no significant difference between the recovery performance computed for the above cases. The perturbations in absolute permeability in cases iii-v did initiate small fluctuations in the saturation but none of these grew to form viscous fingers. Moreover, in further calculations shown in Fig. 5 for case (i) where the injection rate was varied from v = 0.003 cm/sec to v = 0.776 cm/sec, there is no indication of a decline in recovery with increasing injection rate as is observed in laboratory core floods conducted under similar conditions. Utilizing the criterion proposed by Peters and Flock (1981) (see below) for the onset of instability, we find that the displacement is unstable for all v > 0.0293 cm/sec.

SIMULATION RESULTS USING PSEUDO RELATIVE PERMEABILITY CURVES

Since none of the above perturbation techniques produced the experimentally observed decline in recovery with increasing displacement rate we conclude that some other phenomenon is responsible for the observed displacement behaviour under unstable conditions.

The decline in breakthrough recovery which characterises unstable displacements suggests that the effective mobility of the displacing phase relative to that of the displaced phase is greater for unstable displacement conditions than for stable conditions. This suggests that we must modify the "stable" or conventional relative permeability curves if our numerical model is to predict displacement behaviour under unstable conditions. Further, the experimentally observed decline in breakthrough recovery with increasing displacement rate suggests that the modification of the conventional relative permeability curves should also be rate dependent.

Peters and Flock (1981) have recently proposed a generalised criterion for the onset of instability in two phase displacements in porous media. They correlate the tendency towards instability with a dimensionless number,

$$I_{SR} = \frac{(M-1)(v-v_c)\mu_w W^2 H^2}{C^*\sigma k_{wor}(W^2+H^2)} \tag{9}$$

where

$$v_c = \frac{k_{wor}(\rho_w-\rho_o)g\cos\theta}{\mu_w(M-1)} \tag{10}$$

and C* is a constant related to system wettability.

For displacements in cylindrical sand packs Peters and Flock (1981) show that breakthrough recovery declines gradually over the range 13.56 $< I_{SC} <$ 1000, where 13.56 is the critical value above which the displacement is unstable. Fig. 6 shows a typical result from the experiments reported by these authors.

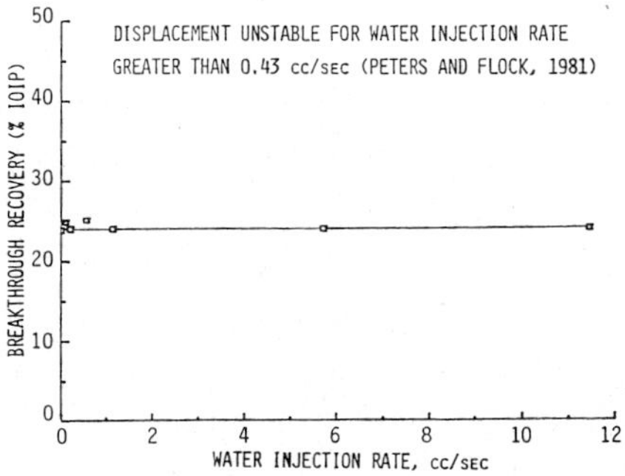

Fig. 5 - Predicted breakthrough recovery for displacement in a homogeneous rectangular sand pack

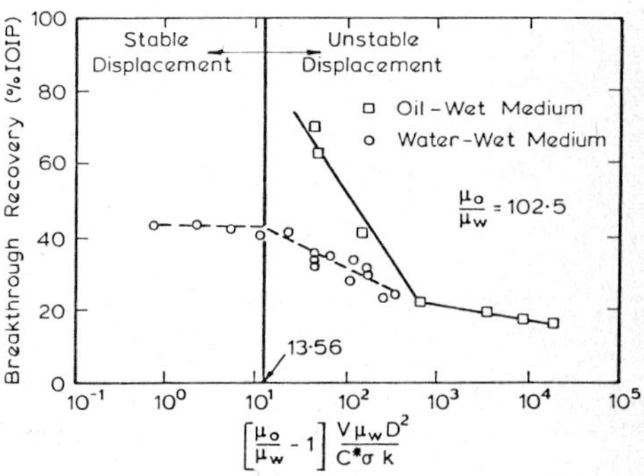

Fig. 6 - Recovery data from Peters and Flock (1981)

Using the dimensionless number I_{SR}, we now define some, as yet undetermined, function $\alpha(I_{SR})$ which modifies the conventional "stable" laboratory relative permeability according to

$$k^*_{rw} = \alpha(I_{SR})k_{rw} \tag{11}$$

where $\alpha(I_{SR})$ will be equal to or greater than unity depending on whether the displacement velocity is below or above that required for stability of the displacement front. When $\alpha(I_{SR})$ is greater than unity the displacement is unstable and k^*_{rw} is a modified or "pseudo" relative permeability.

Just as the conventional relative permeability values must be determined with a laboratory flooding experiment so must the pseudo relative permeabilities or values of $\alpha(I_{SR})$ be determined with laboratory experiments under unstable conditions.

In the absence of sufficiently detailed laboratory flood data we assume that the function $\alpha(I_{SR})$ is a monotonically increasing function of I_{SR}, which has the value of 1 at the onset of instability (I_{SR} = 9.87). For the purpose of illustration we further assume that $\alpha(1000)$ = 1.6 and that $\alpha(I_{SR})$ plots as a straight line function of I_{SR} on semi-log paper,

$$\text{i.e.} \quad \alpha(I_{SR}) = 0.3 \log_{10}(I_{SR}) + 0.70 \tag{12}$$

$$\text{over} \quad 9.87 < I_{SR} < 1000$$

Using equations (11) and (12) to define a pseudo relative permeability and using this to replace the conventional relative permeability in the numerical simulation model, we obtain the result shown in Fig. 7 where we plot the computed breakthrough oil recovery (η_{BT}) as a function of the dimensionless scaling number I_{SR}. For our model conditions we see that the predicted breakthrough recovery (solid line) declines appreciably with increasing displacement rate from the constant value of 24% calculated using conventional relative permeability curves (see Fig. 5). This result is in good agreement with the trends in available experimental data (de Haan, 1959; Flock et al. 1977; Peters and Flock, (1981) (cf. Fig. 6).

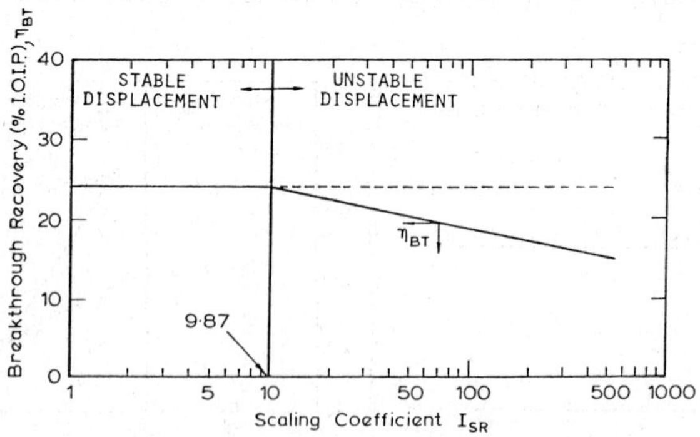

Fig. 7 - Comparison of predicted breakthrough recovery using conventional relative permeability (dotted line) and pseudo relative permeability (solid line).

DISCUSSION

The results of the present study confirm that the usual formulation of numerical models to simulate displacement in porous media using conventional relative permeability curves is not capable of predicting the performance of displacements under conditions when the displacement front is not stable. The results clearly demonstrate that the introduction of perturbations, random or systematic into the numerical solution scheme as proposed by Rachford (1964), does not lead to a realistic simulation of the growth of viscous fingers when the displacement front is unstable. The phenomenon of instability in porous media would therefore appear to be related to the interaction between the two phases on a pore or microscopic scale.

In the formulation of numerical models the only point at which the interaction between the displacing and displaced phases enters the calculations is through the relative permeability and capillary pressure curves. It therefore follows that both relative permeability and capillary pressure will be different for stable and unstable displacement conditions. From experience with numerical simulation models (Peaceman, 1977) it is known that relative permeability has the dominating influence on displacement efficiency with capillary pressure having only a minor effect. This leads us to the concept of pseudo relative permeability to describe fluid flow under unstable displacement conditions. We have demonstrated that this concept, based on a modification of conventional relative permeability curves, allows numerical models to predict performance when the displacement front is unstable which is in qualitative agreement with experimental data.

Although we have suggested a functional form for the relationship between pseudo relative permeability and conventional relative permeability which depends on displacement rate (or I_{SR}), this relationship is only speculative. The establishment of the form of the relationship will depend on a comprehensive comparison between computations of the form outlined in this paper and carefully conducted displacement experiments over a wide range of displacement rates. No such comparison is possible at this time because of the lack of appropriate experimental data.

The usefulness of the proposed concept of pseudo relative permeability in predicting the effect of viscous fingering on recovery efficiency under actual field conditions, will depend both on the development of techniques to determine pseudo relative permeabilities on the basis of laboratory scale tests and the further establishment of the equivalence between the instabilities developed in laboratory scale floods and actual field conditions; the latter was pioneered by Chuoke et al. (1959) and later continued by Peters and Flock (1981). Since we may anticipate viscous fingers in the field having dimensions of the order of the core diameters used in conventional laboratory floods this equivalence is not yet clearly established. It is interesting to note that Todd and Longstaff (1972) have reported some success in simulating the behaviour of miscible floods using a numerical reservoir simulator coupled with an empirical mixing model which accounts for large scale mixing due to viscous fingering. It is not unreasonable to accept a similar level of success with the presently proposed pseudo relative permeability concept for unstable immiscible displacement.

Whereas the concept of pseudo relative permeability is empirical, it is no more empirical than the conventional relative permeability concept. Although absolute permeability is an intrinsic property of reservoir rocks, relative permeability is not. Rather, it must be determined for each rock type from laboratory core flood tests, just as we have suggested that the pseudo relative permeability must be determined.

CONCLUSIONS

On the basis of the present study we conclude that it is not possible to predict the performance of unstable displacement in porous media using numerical models based on the conventional relative permeability concept.

The growth of viscous fingers does not appear to be associated with random perturbations in permeability and saturation, which occur because of the random nature of the porous medium, as has been previously suggested. Rather, it would appear that the instability is more closely related to the interaction between the individual phases in the pore spaces of the porous medium, i.e. relative permeability.

An empirical concept of a modified or pseudo relative permeability has been demonstrated to be a potentially useful tool in extending the ability of conventional numerical reservoir simulation models to predict the performance of displacements where the displacement front is unstable. A form for the pseudo relative permeability function has been suggested which provides computed results which are in qualitative agreement with laboratory flood experiments which show a decline in the breakthrough recovery with increasing displacement rate.

NOMENCLATURE

C = Chuoke's wettability number, dimensionless
C^* = Peters' wettability number, dimensionless
g = gravitational acceleration, m/s^2
f_w = water fractional flow function, dimensionless
I_{SC} = Peters' dimensionless number (scaling coefficient) for a cylindrical system
I_{SR} = Peters' dimensionless number (scaling coefficient) for a rectangular system
k = absolute permeability, darcy
k_{ro} = standard laboratory relative permeability to oil
k^*_{rw} = pseudo relative permeability to water for adverse mobility ratio conditions
k_{rw} = standard laboratory relative permeability to water
k_{wor} = permeability to water at residual oil saturation, darcy
L = length of rectangular system, cm
W = width of rectangular system, cm
H = height of rectangular system, cm
M = end point mobility ratio
p_c = capillary pressure, atm
p_n = non-wetting phase pressure, atm
p_w = wetting phase pressure, atm
Q_t = total fluid injection/production rate per unit volume of porous medium, s^{-1}
Q_w = water injection/production rate per unit of porous medium, s^{-1}
S_w = water saturation
S_{INIT} = initial water saturation
S_{wi} = irreducible water saturation

S_{wor} = water saturation at residual oil conditions
t = real time, seconds
v = constant superficial velocity, cm/s
v_c = critical superficial velocity, cm/s
x,y = rectangular coordinates, cm
α = rate parameter, dimensionless
λ_n = non-wetting phase mobility, darcy/cp
λ_w = wetting phase mobility, darcy/cp
λ_t = total mobility, darcy/cp
μ_o = oil viscosity, cp
μ_w = water viscosity, cp
π = 3.14159..
ρ_o = oil density, g/cm^3
ρ_w = water density, g/cm
σ = interfacial tension, dyne/cm
θ = angle rectangular system makes with the vertical, radians
$\eta_{B.T.}$ = breakthrough oil recovery, % Initial Oil In Place

REFERENCES

[1] Engelberts, W.F. and Klinkenberg, L.J. (1951), Laboratory Experiments on the Displacement of Oil by Water from Packs of Granular Materials, Proc. Third World Petroleum Congress, Sec II, 544.

[2] Chuoke, R.L., van Meurs, P. and van der Poel, C. (1959), The Instability of Slow, Immiscible, Viscous Liquid-Liquid Displacement in Permeable Media, Trans. AIME, 216, 188.

[3] de Haan, H.J. (1959), Effect of Capillary Forces in the Water-Drive Processes, Proc. Fifth World Pet. Cong., Sec II, 1-13.

[4] Saffman, P.G. and Taylor, G.I. (1958), The Penetration of a Fluid into a Porous Medium or Hele-Shaw Cell Containing a More Viscous Liquid, Proc. Roy. Soc., A245, 312-329.

[5] Scheidegger, A.E. (1974), The Physics of Flow Through Porous Media, University of Toronto Press (Can).

[6] Rachford, H.H. Jr. (1964), Instability in Water Flooding Oil from Water-Wet Porous Media Containing Connate Water, Soc. Pet. Eng.J., 133-148.

[7] Flock, D.L., Peters, E.J., Baird, H., Wilborg, R., and Kloepfer, J., The Influence of Frontal Instabilities During Viscous Oil Displacement, The Oil Sands of Canada - Venezuela, CIM, 17, 380-385.

[8] Peters, E.J., Flock, D.L. (1981), The Onset of Instability During Two-Phase Immiscible Displacement in Porous Media, Soc. Pet. Eng.J., 249-258.

[9] Todd, M.R., and Longstaff, W.J. (1972), The Development, Testing, and Application of a Numerical Simulator for Predicting Miscible Flood Performance, JPT, 874.

[10] Spillette, A.G., Hillstead, J.G. and Stone, H.L. (1973), A High-Stability Sequential Solution Approach to Reservoir Simulation, Soc.Pet.Eng. 48th Ann. Meet., Las Vegas, Nev., S.P.E. Paper, no. 4542.

[11] Watts, J.W. (1971), An Iterative Matrix Solution Method Suitable for Anisotropic Problems, Soc. Pet. Eng. J., 47-51.

[12] Watts, J.W. (1973), A Method for Improving Line Successive Overrelaxation in Anisoptropic Problems - A Theoretical Analysis, Soc. Pet. Eng. J., 105-118.

[13] Peaceman, D.W. (1977), Fundamentals of Numerical Reservoir Simulation, Elsevier Scientific Publishing Company.

Computational Techniques & Applications: CTAC-83
J. Noye & C. Fletcher (Editors)

FINITE ELEMENT APPROXIMATIONS OF UNIDIRECTIONAL NON-LINEAR SEEPAGE FLOWS

S.-S. Chow

Centre for Mathematical Analysis
The Australian National University
GPO Box 4 Canberra ACT 2601
Australia

In this paper, we consider a class of fluid flow problems in porous media under a generalized seepage law. Specializing in the case of unidirectional flows, we arrived at a two point boundary value problem with mixed boundary conditions. When the linear finite element method is applied, we obtain a system of nonlinear algebraic equations. The solution of such a system by conventional methods is very expensive. By utilizing a nonlinear version of LU decomposition we show these equations may by solved in a stable and efficient manner. We then discuss the application of this solution procedure to some singular two point boundary value problems and to some Dirichlet problems. We also briefly describe some acceleration methods and give results from numerical experiments.

1. INTRODUCTION

Darcy's law is commonly adopted in the description of fluid percolation in a statistically isotropic and homogeneous porous medium. In many instances the use of this prevailing choice of seepage law yields satisfactory results that are adequate for practical applications. This is especially true when the inertia effect is negligible and the flow is in the prelamina regime. On the other hand, when one needs to take into account the inertia effect arising from rapid fluid motion, one can no longer rely on Darcy's law to provide an accurate picture of the fluid motion. In order to ensure that realistic solutions are obtained, it is necessary to take due consideration of the presence of nonlinear effects and modify the seepage law accordingly. These adjustments are required, for example, in the study of fluid percolation through rockfill dams or in areas adjacent to a pumping well in a coarse grained aquifer [6], [7].

In this paper, we consider a class of fluid flow problems in porous media under a generalized seepage law. Specializing in the case of unidirectional flows, we arrived at a two point boundary value problem with mixed boundary conditions. When the linear finite element method is applied, we obtain a system of nonlinear algebraic equations. The solution of such a system by conventional methods is very expensive. By utilizing a nonlinear version of LU decomposition we show these equations may be solved in a stable and efficient manner. We then discuss the application of this solution procedure to some singular two point boundary value problems and to some Dirichlet problems. We also briefly describe some acceleration methods and give results from numerical experiments.

2. NONLINEAR SEEPAGE FLOWS

Consider a packed column of unit length consisting of a homogeneous, porous material with some fluid passing through it in a longitudinal direction. Suppose the grain size of the particles in the column are such that if one utilizes Darcy's law to represent the relationship between the flux and the hydraulic gradient, one obtains an incorrect or incomplete description of the fluid motion. Under such circumstances, it becomes necessary to employ nonlinear seepage laws that are relevant to the problem. An example of this situation arises in the construction of mathematical models for various chemical processes in reactor towers.

For flows in the nonlinear or high Reynold's number regime, two popular choices of seepage laws are the Forchheimer's law [3]

$$-\frac{dH}{dx} = a\,q + b|q|q\,, \tag{1}$$

and the Missbach's law [6]

$$-\frac{dH}{dx} = c|q|^{\alpha}q \tag{2}$$

where q is the velocity in the flow direction x, H is the hydraulic head and a, b, c, α are non-negative parameters whose values are determined by experimental data. There are of course many other versions of nonlinear seepage laws, see for example [1] and [4]. In the sequel, we shall employ the following generalized seepage law

$$-\frac{dH}{dx} = L(|q|)q\,, \tag{3}$$

where $L : \mathbb{R}^+ \to \mathbb{R}^+$ is a continuous function with $L(s)s$ being a strictly increasing and unbounded function of $s \in \mathbb{R}^+$. This generalized law expresses the experimentally observable relationship between the hydraulic gradient and the fluid velocity. It also represents, in a very general manner, most of the nonlinear seepage laws so far proposed. For example, when $L(s)$ takes the form $a + bs$ and cs^{α}, (3) is reduced to (1) and (2) respectively.

If there is a source or sink represented by a function $f = f(x)$ within the column, one may write the equation of continuity as

$$\frac{dq}{dx} = f(x)\,. \tag{4}$$

Suppose that the inflow velocity at $x=0$ is given by some constant η and that the value of the hydraulic head at $x=1$ is prescribed by some constant H_1, that is,

$$q(0) = \eta \ , \tag{5}$$

$$H(1) = H_1 \ . \tag{6}$$

Using this information, we can immediately integrate (4) and then (3) to obtain

$$q(y) = \int_0^y f(t)\,dt + \eta \ , \quad 0 \le y \le 1 \ , \tag{7}$$

$$H(x) = H_1 + \int_x^1 L(|q(y)|)q(|y|)\,dy \ , \quad 0 \le x \le 1 \ . \tag{8}$$

In general, it is not possible to carry out the integration in (7) and (8) analytically. Thus, if one desires to obtain a good representation of the hydraulic head, some sort of approximation procedure is inevitable. One option is to use (7) and (8) to evaluate H at selected nodal points $x_0, x_1, \ldots, x_N$ with the aid of numerical integration procedures, and then employ some suitable interpolation scheme to obtain an approximation of hydraulic head. However, this strategy is, computationally speaking, not very economical. For each nodal point x_i, the use of numerical integration necessitate the need to evaluate the function q at several points. Each of these function evaluations leads in turn to another invocation of numerical integration routine. When one needs to compute the function value of H at a large number of nodal points, as in the case of some singular problems which we shall discuss later, the cost is high. Thus, one is interested in seeking out alternative strategies that will minimize the computational effort required.

In what follows, we shall describe a simple, efficient and easy to implement procedure for computing a good approximation of the hydraulic head. As the solution process is based on the finite element method, we first set out to formulate (3) - (6) in a form which facilitates the use of the finite element method. Now, because the function $L(s)s$ is strictly increasing for all $s \ge 0$, it possesses a unique inverse defined on $\mathbb{R}^+$ which we shall denote by $k(t)t$, where $k(t)$ is a positive, continuous function of $t > 0$ with the property that $k(t)t = 0$ at $t = 0$. For example, when $L(s) = a + bs$, $k(t) = \left(\frac{a}{2} + \sqrt{(\frac{a}{2})^2 + bt}\right)^{-1}$ and when $L(s) = cs$, $k(t) = (t/c)^{p-2}$, where $p = (\alpha+2)/(\alpha+1)$. Note however that in general it is not possible to obtain an explicit analytic expression for k. Fortunately, as we shall see later, the function k is only used in the derivation of the algorithm and its presence is not required in the actual computational stages.

Taking modulus on both sides of (3) we deduce that

$$q = -k\left(\left|\frac{dH}{dx}\right|\right)\frac{dH}{dx} . \tag{9}$$

Combining this with the fact that q satisfies (4) and making use of the substitution $u(x) = H(x) - H_1$, it is easy to see that (3) - (6) is equivalent to the following two point boundary value problem:

$$-(k(|u'|)u')' = f \quad \text{in} \quad]0,1[, \tag{10}$$

$$-k(|u'|)u' = \eta \quad \text{at} \quad x = 0 , \tag{11}$$

$$u = 0 \quad \text{at} \quad x = 1 . \tag{12}$$

With the view of applying the finite element method let us put this problem in a weak form. Let $\phi = \phi(x)$ be an absolutely continuous function defined on the unit interval and vanishes at $x = 1$. Suppose u is a solution of (10) - (12). Then clearly u also satisfies the equality

$$\int_0^1 k(|u'|)u'\,\phi'\,dx = \int_0^1 f\,\phi\,dx + \eta\,\phi(0) . \tag{13}$$

Conversely, if u is a twice differentiable function in the unit interval and satisfies (13) for every ϕ possessing the properties described above, then, providing that u vanishes at $x=1$, we may conclude that u is a solution of (10) - (12). We remark that it is not difficult to show the existence and uniqueness in a certain Banach space of a function u that satisfies (12) and (13).

3. FINITE ELEMENT APPROXIMATIONS

Let us now proceed to examine the application of the linear finite element method to (12) - (13). The use of linear elements not only simplifies the treatment but, more importantly, enables us to minimize the need to perform numerical integration. We begin by partitioning [0,1] into $N+1$ (not necessarily evenly spaced) subintervals: $0 = x_0 < x_1 < \ldots < x_N < x_{N+1} = 1$. Let $h_i = x_{i+1} - x_i$ for $i = 0(1)N$ and let $h = \max_i\{h_i\}$. Recall that one obtains a linear finite element space from the spanning set of the hat functions $\{\phi_i\}_{i=0}^{N+1}$, where

$$\phi_i(x) = \begin{cases} (x-x_{i-1})/h_{i-1} & \text{if} \quad x_{i-1} \le x \le x_i , \\ (x_{i+1}-x)/h_i & \text{if} \quad x_i \le x \le x_{i+1} , \\ 0 & \text{otherwise.} \end{cases}$$

If $u_h = \sum_{i=0}^{N+1} \alpha_i\phi_i$ is a finite element approximation of (12) - (13) with α_i's being parameters to be determined, one sees immediately that $\alpha_{N+1} = 0$ for u_h to satisfy the essential boundary condition (12). Moreover, because each of the ϕ_j's, $0 \le j \le N$, is absolutely continuous and takes the value zero at $x = 1$, it is clear that $\{\alpha_i\}_{i=0}^{N}$ may be determined by the $N+1$ algebraic equations:

$$\sum_{j=0}^{N} \int_{x_j}^{x_{j+1}} k(|u_h'|)u_h' \phi_h' \, dx = \int_0^1 f \phi_j \, dx + \eta \phi_j(0) \;, \quad j = 0(1)N \,. \tag{14}$$

As $\phi_j'(x) = h_{j-1}^{-1}$ for $x \in]x_{j-1}, x_j[$, $\phi_j' = -h_j^{-1}$ for $x \in]x_j, x_{j+1}[$ and $\phi_j'(x) = 0$ for $x \notin [x_{j-1}, x_{j+1}]$, we see that u_h' is a piecewise constant function. Thus, over each subinterval $]x_j, x_{j+1}[$, $j = 0(1)N$, $k(|u_h'|)$ takes constant value. This observation, together with the fact that ϕ_j's have small supports, enables us to derive

$$\begin{aligned}\int_0^1 k(|u_h'|)u_h' \phi_0' \, dx &= \int_{x_0}^{x_1} k(|\alpha_0\phi_0' + \alpha_1\phi_1'|) \left(\frac{\alpha_0}{h_0^2} - \frac{\alpha_1}{h_0^2}\right) dx \\ &= k\left(\left|\frac{-\alpha_0+\alpha_1}{h_0}\right|\right)\frac{\alpha_0}{h_0} - k\left(\left|\frac{-\alpha_0+\alpha_1}{h_0}\right|\right)\frac{\alpha_1}{h_0} \;,\end{aligned} \tag{15}$$

and, for $j = 1(1)N$,

$$\begin{aligned}&\int_0^1 k(|u_h'|)u_h' \phi_j' \, dx \\ &= k\left(\left|\frac{-\alpha_{j-1}+\alpha_j}{h_{j-1}}\right|\right)\left(\frac{-\alpha_{j-1}}{h_{j-1}}\right) + k\left(\left|\frac{-\alpha_j+\alpha_{j+1}}{h_j}\right|\right)\left(\frac{-\alpha_{j+1}}{h_j}\right) \\ &\quad + k\left(\left|\frac{-\alpha_{j-1}+\alpha_j}{h_{j-1}}\right|\right)\frac{\alpha_j}{h_{j-1}} + k\left(\left|\frac{-\alpha_j+\alpha_{j+1}}{h_j}\right|\right)\frac{\alpha_j}{h_j} \,.\end{aligned} \tag{16}$$

It is now clear that we may apply (15) and (16) to "evaluate" the left hand side of (14) without resorting to numerical integration. Because $\alpha_{N+1} = 0$, $\phi_0(0) = 1$ and $\phi_j(0) = 0$ for $j \neq 0$, we obtain from (14)

$$k\left(\left|\frac{-\alpha_0+\alpha_1}{h_0}\right|\right)\frac{\alpha_0-\alpha_1}{h_0} = \int_{x_0}^{x_1} f \phi_0 \, dx + \eta \,, \tag{17}$$

for $j = 1(1)N - 1$,

$$k\left(\left|\frac{-\alpha_{j-1}+\alpha_j}{h_{j-1}}\right|\right)\frac{-\alpha_{j-1}+\alpha_j}{h_{j-1}} + k\left(\left|\frac{-\alpha_j+\alpha_{j+1}}{h_j}\right|\right)\frac{\alpha_j-\alpha_{j+1}}{h_j} = \int_{x_{j-1}}^{x_{j+1}} f \phi_j \, dx \,, \tag{18}$$

$$k\left(\left|\frac{-\alpha_{N-1}+\alpha_N}{h_{N-1}}\right|\right)\frac{-\alpha_{N-1}+\alpha_N}{h_{N-1}} + k\left(\left|\frac{-\alpha_N}{h_N}\right|\right)\frac{\alpha_N}{h_N} = \int_{x_{N-1}}^{x_{N+1}} f \phi_N \, dx \,. \tag{19}$$

Observe that when $L \equiv 1$ and the subintervals are of equal length, $k \equiv 1$ and so (10) - (12) is just the Poisson equation. In this case, (17) - (19) may be rewritten in terms of the familiar tridiagonal system

$$\begin{bmatrix} 1 & -1 & & & & \\ -1 & 2 & -1 & & & \\ & -1 & 2 & -1 & & \\ & & \ddots & \ddots & \ddots & \\ & & & -1 & 2 & -1 \\ & & & & -1 & 2 \end{bmatrix} \underline{\alpha} = h\,\underline{\tilde{d}}\ , \tag{20}$$

where $\underline{\alpha} = (\alpha_0, \alpha_1, \ldots, \alpha_N)^T$ and $\tilde{d}_j$, the $(j+1)$th element of $\underline{\tilde{d}}$, represents the terms on the right hand side of (17) - (19) for $j = 0(1)N$.

If one employs standard methods for solving nonlinear algebraic equations (for example, Newton's method) to obtain a solution set $\{\alpha_i\}_{i=0}^{N}$ for (17) - (19), the computational cost is very high. Since that in general we cannot express the function k explicitly in analytic form, the application of conventional methods demands a considerable number of evaluations of k (and possibly its derivatives), each of which involves the root finding of a single nonlinear algebraic equation. This is certainly not desirable. By contrast, recall that in the solution strategy using (7) and (8), we only need to evaluate the function L, whose analytic form is explicitly given. Thus, an essential criterion in designing an algorithm that is competitive is the circumvention of any computation involving the function k.

In the linear case, an effective and efficient procedure for solving (20) is to decompose the tridiagonal matrix into LU form and then perform forward and backward substitutions. This is commonly known as the Thomas algorithm. In (17) - (19) one observes that the term $(\alpha_j - \alpha_{j+1})/h_j$ appears in every equation, with j not necessarily representing the same index in each occurrence. This points to the possibility of simplifying the system by making the substitution

$$\beta_j = (\alpha_{j-1} - \alpha_j)/h_{j-1}\ , \quad j = 1(1)N\ , \tag{21}$$

$$\beta_{N+1} = \alpha_N/h_N\ . \tag{22}$$

Indeed, by writing (17) - (19) in terms of β_j's, one has

$$k(|\beta_1|)\beta_1 = \tilde{d}_0\ , \tag{23}$$

$$-\,k(|\beta_j|)\beta_j + k(|\beta_{j+1}|)\beta_{j+1} = \tilde{d}_j\ , \quad j = 1(1)N\ . \tag{24}$$

Note that when $k \equiv 1$ and the nodal points are equally spread, (21) - (22) and (23) - (24) reduce to the same upper and lower triangular systems one obtains while performing a LU decomposition in (20). In this way, one may interpret this substitution procedure as a nonlinear version of the LU decomposition.

Now, using the recurrence relations (23) - (24) and setting

$$d_j = \sum_{i=0}^{j} \tilde{d}_i \ , \qquad j = 0(1)N \ ,$$

we see that (23) - (24) reduces to

$$k(|\beta_{j+1}|)\beta_{j+1} = d_j \ , \ j = 0(1)N \ ,$$

which one can solve easily to obtain

$$\beta_{j+1} = L(|d_j|)d_j \ , \quad j = 0(1)N \ .$$

To compute the parameters α_j's , we simply use (21) - (22) to perform a backward substitution. The solution process outlined above is clearly stable and easy to implement. Numerical experience also indicates that in many cases one may obtain a reasonably accurate approximation by this method.

Before considering some applications and extensions of this procedure for computing a piecewise linear approximation to (3)-(6), we summarize the algorithm here:

Algorithm 1

(i) set $d_0 := \int_{x_0}^{x_1} f \ \phi_0 \ dx + \eta$,

(ii) for $j := 1(1)N$, compute

$$\beta_j := L(|d_{j-1}|)d_{j-1} \ ,$$

$$d_j := d_{j-1} + \int_{x_{j-1}}^{x_{j+1}} f \ \phi_j \ dx \ ,$$

set $\beta_{N+1} := L(|d_N|)d_N$,

(iii) set $\alpha_N := h_N \ \beta_{N+1}$,

for $i := N(-1)1$, compute

$$\alpha_{i-1} := h_{i-1} \ \beta_i + \alpha_i \ .$$

Using this strategy, we see that there is no need to compute the function k and hence calculations may be carried out in an efficient manner. Once the parameters $\{\alpha_i\}_0^N$ are available, one obtains an approximation H_h of the hydraulic head by setting $H_h = H_1 + \sum_{i=0}^{N} \alpha_i \ \phi_i$.

4. APPLICATIONS

Consider the following nonlinear elliptic problem in an n-dimensional unit sphere B with boundary ∂B :

$$- \nabla\cdot(k(|\nabla\tilde{u}|)\nabla\tilde{u}) = f(|\underline{x}|) \ , \quad \underline{x} \in B \ , \tag{25}$$

$$\tilde{u} = 0 \qquad \text{on} \quad \partial B \ , \tag{26}$$

where $k : \mathbb{R}^+ \to \mathbb{R}^+$ is a continuous function with the property that $k(t)t$ is strictly increasing and tends to infinity as $t \to \infty$. This problem may be regarded as a model for the non-Darcy flow of a fluid in a ball of porous material. We shall assume here, however, that the function k is given but an explicit analytic expression for $L(s)s$, the unique inverse function of $k(t)t$, may not be available.

Because of the spherical symmetry of the problem (25) - (26), it is not difficult to deduce that the radial part $u = \tilde{u}(|x|)$ of the solution $\tilde{u}$ satisfies the singular two point boundary value problem

$$- (x^b k(|u'|)u')' = x^b f(x) \ , \quad x \in]0,1[\tag{27}$$

$$u'(0) = u(1) = 0 \ , \tag{28}$$

where $b = n-1 \geq 0$. Obviously, when $b = 0$, the above problem is identical to (10) - (12) with $\eta = 0$. When $b > 0$, we note that instead of (14) we have the algebraic equations:

$$\sum_{j=0}^{N} \int_{x_j}^{x_{j+1}} x^b k(|u_h'|)u_h' \phi_j' \, dx = \int_0^1 x^b f \phi_j \, dx \ , \quad j = 0(1)N \ . \tag{29}$$

Proceeding as before we obtain the equalities (17) - (19) with $\eta = 0$, $x^b f$ replacing f in each of the integrals on the right hand side and

$\gamma_j k\left(\left|\frac{-\alpha_j+\alpha_{j+1}}{h_j}\right|\right)$ replacing $k\left(\left|\frac{-\alpha_j+\alpha_{j+1}}{h_j}\right|\right)$, $j = 0(1)N$, in each of the integrals on the left hand side, where

$$\gamma_j = \frac{1}{h_j}\int_{x_j}^{x_{j+1}} x^b \, dx = \frac{1}{b+1}\left(x_{j+1}^{b+1} - x_j^{b+1}\right)/h_j \ , \quad j = 0(1)N \ .$$

Thus, to compute the parameters α_i's in this case, we only need to make minor modifications to algorithm 1. We summarize the steps as follows

<u>Algorithm 2</u>

(i) set $d_0 := \int_{x_0}^{x_1} x^b f \phi_0 \, dx$,

(ii) for $j := 1(1)N+1$, if the function L is available in closed form,

$$\beta_j := L(|d_{j-1}/\gamma_{j-1}|)\ (d_{j-1}/\gamma_{j-1}) ,$$

otherwise, solve the single nonlinear equation

$$k(|\beta_j|)\beta_j = d_{j-1}/\gamma_{j-1} \tag{30}$$

for β_j . If $j \le N$, set

$$d_j := \int_{x_{j-1}}^{x_{j+1}} x^b f\ \phi_j\ dx + d_{j-1} .$$

(iii) solve for α_i's as in algorithm 1.

Notice that the assumed increasing property of $k(t)t$ implies the uniqueness of solutions of (30) for each j . Thus the algorithm is stable. Also, because no restriction is imposed on the placement of the nodal points, a good strategy is to concentrate the nodes near $x = 0$ where singularity occurs. It should also be pointed out that when the function L is not known explicitly, it is computationally more economical to apply algorithm 2 than to numerically integrate the integrals

$$g(y) = \int_0^y t^b f(t)\,dt , \quad H = H_1 + \int_x^1 L(|y^b g(y)|)\,y^b g(y)\,dy .$$

This is due to the fact that one usually needs to solve a lesser number of nonlinear equations of the form (30) in the former procedure.

5. DIRICHLET CONDITIONS

Until now we have assumed that the given boundary conditions are of mixed type (5) - (6). In this section we consider another physically interesting set of boundary conditions. Suppose that the values of the hydraulic head are prescribed at both ends of a packed column, then, instead of (5) and (6), we have

$$H(0) = H_0 , \quad H(1) = H_1 ,$$

where H_0 and H_1 are constants. Letting $u(x) = H(x) - H_0 - (H_1-H_0)x$, it is not difficult to check that u satisfies the differential equation (10) and the homogeneous Dirichlet conditions $u(0) = u(1) = 0$.

Let us now apply the linear finite element method to this problem. For the finite element approximation $u_h = \sum_{i=0}^{N+1} \alpha_i \phi_i$ to satisfy the essential boundary conditions, we must have $\alpha_0 = \alpha_{N+1} = 0$. If we discretize the problem as before and generate the algebraic equations (16) with the aim of determining the remaining coefficients

$\{\alpha_i\}_{i=1}^{N}$, we soon find the substitution process (21) leads to a nonlinear system whose solution is difficult to compute.

On the other hand, if we employ the integrals (7) and (8) to solve this Dirichlet problem, we find that it is necessary to solve the following equation for the constant η appearing in (5):

$$H_0 - H_1 = \int_0^1 L\left(\left|\int_0^y f(t)\,dt+\eta\right|\right)\left(\int_0^y f(t)\,dt+\eta\right)dy \ . \tag{31}$$

The solution of this equation is not straight forward. However, once η is ascertained, one may proceed to compute H using (7) and (8) or using algorithm 1.

Instead of calculating η from (31), another approach is to use an iterative scheme to determine η and hence the solution of the Dirichlet problem. This method may be described as follows.

Algorithm 3

(i) guess η

(ii) solve the boundary value problem (10) - (12) with the aid of algorithm 1.

(iii) modify η and repeat (ii) until $|\alpha_0|$ is sufficiently small.

Note that this process may be regarded as a form of shooting method. However, it differs from the usual shooting methods in that at each iteration step, we solve a boundary value problem with a method which exploits the structure of the problem instead of solving an initial value problem with some general method. Thus we expect this algorithm would provide a more efficient means in securing an approximate solution to the Dirichlet problem.

6. NUMERICAL EXAMPLES

Apart from the various practical uses of algorithms 1 - 3, these algorithms are also valuable in the theoretical studies of the behaviour of the finite element approximations. To illustrate this let us examine some numerical examples.

Example 1 Here we assume the Forchheimer law (1) holds with $a = 4.36$ and $b = 1.27$. We also assume $f \equiv -2,23$ and let $\eta = 0.95$. For fixed p , the $W^{1,p}$-norm of u is given by $(\int_2^1 |u'|^p dx)^{1/p}$ and the L^p-norm by $(\int_0^1 |u|^p dx)^{1/p}$. Let N+1 be the number of equally spaced subintervals, we estimate the L^∞ norm of u by the maximum of the values of $|u|$ at the points $i/(N+1)$, $i = 0(1)N$. We also use the following relative error measure for the finite element approximamation u_h of the solution u :

$$\frac{1}{N+1} \sum_{\substack{i=0 \\ u(x_i)\neq 0}}^{N} \frac{|u(x_i)-u_n(x_i)|}{|u(x_i)|} \quad ; \quad x_i = i/(N+1) \ .$$

Setting $p = 3/2$ we have the following table:

N+1	$W^{1,p}$	L^p	$W^{1,2}$	L^2	L^∞	relative
8	4.37-1	2.04-2	4.72-1	2.10-2	4.58-3	1.55-3
16	2.18-1	5.05-3	2.36-1	5.22-3	1.16-3	3.78-4
32	1.09-1	1.27-3	1.18-1	1.31-3	2.89-4	9.85-5
64	5.46-2	3.17-4	5.91-2	3.28-4	7.40-5	2.56-5
128	2.79-2	7.92-5	2.95-2	8.19-5	1.83-5	6.38-6
256	1.40-1	1.98-5	1.48-2	2.05-5	4.61-6	1.60-6

Example 2 Suppose the governing seepage law is Missbach's law (2) with $c = 0.67$ and $\alpha = 1.6$. Further, suppose that $f \equiv 1.03$ and $\eta = 2.96$, then upon setting $p = (\alpha+2)/(\alpha+1) = 1.3846154$, we obtain the following results:

N+1	$W^{1,p}$	L^p	$W^{1,2}$	L^2	L^∞	relative
8	4.35-1	2.00-2	4.81-1	2.08-2	4.06-3	1.86-4
16	2.18-1	4.99-3	2.41-1	5.21-3	1.02-3	4.89-5
32	1.09-1	1.25-3	1.20-2	1.30-3	2.54-4	1.25-5
64	5.44-2	3.12-4	6.01-2	3.25-4	6.35-5	3.18-6
128	2.81-2	7.79-5	3.01-2	8.14-5	1.59-5	8.00-7
256	1.31-2	1.94-5	1.50-2	2.03-5	3.97-6	2.01-7

Example 3 Consider the boundary value problem

$$-\left[\left(e^{|u'|}-1\right)\mathrm{sgn}(u')\right]' = f(x)$$

$$u'(0) = u(1) = 0 \ .$$

It is easy to check that $u(x) = \frac{1}{2}(1-x^2)$ is a solution of this problem when $f(x) = e^x$. This equation is of interest as its coefficient does not satisfy a polynomial growth condition. We summarize the result in the following table.

N+1	$W^{1,1}$	L^1	$W^{1,2}$	L^2	L^∞	relative
8	2.97-2	1.01-3	3.61-2	1.15-3	6.51-4	8.39-4
16	1.48-2	2.52-4	1.80-2	2.88-4	1.63-4	2.17-4
32	7.39-3	6.31-4	9.02-3	7.20-5	4.07-5	5.53-5
64	3.69-3	1.58-5	4.51-3	1.80-5	1.02-5	1.40-5
128	1.85-3	3.94-6	2.26-3	4.50-6	2.54-6	3.51-6
256	1.02-3	9.65-7	1.13-3	1.13-6	6.36-7	8.79-7

Example 4 As a final check, we investigate the boundary value problem of example 3 with $f \equiv 0$ and boundary conditions $u'(0) = 1$ and $u(1) = 0$. In this case $\eta = 1 - e = 1.718281828$ and $u(x) = 1 - x$ is a solution of the problem. As expected, the piecewise linear finite element solution is exact up to machine precision.

In all cases considered above, we see that the error of the finite element approximations is of order h and h^2 in the $W^{1,2}$ and L^2 norm respectively. Also, the order of convergence in the L^∞ norm appears to be quadratic. The problem of establishing *a priori* error estimates for the nonlinear examples considered above is largely unexplored. Theoretical studies concerning the order of convergence of finite element approximations relating to examples (1) and (2) may be found in [2].

7. CONCLUDING REMARKS

As the algorithms 1 - 3 impose no special structural or growth conditions on the nonlinear coefficient, one can take advantage of the efficiency and the ease of implementation of these algorithms to study the behaviour of the error of the finite element approximations for various types of nonlinear problems. Moreover, one can also utilize algorithm 2 to study the problem of optimal grading of the mesh points for various singular problems.

In the event that a higher order approximation is desired, we may employ the accelerating methods of Lin [5] and Xie [8]. Suppose we have computed a piecewise linear finite element approximation u_h of (10) - (12). First we refine the partition by introducing additional nodal points $x_{i+\frac{1}{2}} = (x_{i+1}+x_i)/2$, $i = 0(1)N$. Using this new partition we construct a set of quadratic or cubic basis functions $\{\psi_i\}_{i=0}^{1N+3}$ on $[0,1]$. Then a higher order approximation $u_h^* = \sum_{i=0}^{2N+3} \beta_i \psi_i$, where β_i's are parameters to be determined, may be obtained either by solving the linear system

$$\int_0^1 k(|u_h'|)(u_h^*)'\psi_j' \, dx = \int_0^1 f\,\psi_j \, dx + \eta\,\psi_j(0) \quad , \qquad j = 0(1)2N+2 \ ,$$

or by the Newton's method which we now describe.

Let $s_h = \sum_{i=0}^{2N+2} \gamma_i \psi_i$, where γ_i's are parameters to be determined by computing a solution for the linear system of equations

$$\int_0^1 \frac{d}{dt}(k(t)t)\Big|_{t=|u_h'|} s_h' \psi_j' \, dx$$

$$= \int_0^1 k(|u_h'|)u_h' \psi_j' \, dx - \int_0^1 f\,\psi_j \, dx + \eta\,\psi_j(0) \quad , \qquad j = 0(1)2N+2 \ .$$

Once s_h is available, we set $u_h^* = u_h - s_h$. Note that when $k \equiv 1$, that is, the linear case, the above methods reduce to the usual method for computing higher order approximations. Also, if the function k is not available in explicit form, we can simply apply the relations $k(t)\,\ell(s) = 1$ and $\frac{d}{dt}(k(t)t) \cdot \frac{d}{ds}(\ell(s)s) = 1$, where $k(t)t = s$ if $\ell(s)s = t$, to determine the desired functional values.

8. ACKNOWLEDGEMENT

I wish to express my gratitude to Bob Anderssen and Frank de Hoog for their constant encouragement and valuable comments. I would also like to thank Dorothy Nash for typing this paper.

REFERENCES

[1] J. Bear, Dynamics of fluids in porous media, American Elsevier, New York, 1972.

[2] S.-S. Chow, Finite element error estimates for nonlinear problems of monotone type, Ph.D. Thesis, Dept. of Mathematics, Faculty of Science, Australian National University, (April 1983).

[3] P. Forchheimer, Wasserbewegung durch Boden, Z. Ver. Deutsch. Ing. 45, (1901) 1782-1788.

[4] A. Hannoura and F. Barends, Non Darcy Flow - A State of the Art, in: Flow and Transport in Porous Media, (eds) A. Verruigt and F. Barends), A.A. Balkema, Rotterdam, Netherlands, 1981.

[5] Lin Qun, Methods to increase the accuracy of low-degree element solution in nonlinear problems, in: Computing methods in applied sciences and engineering, (eds) R. Glowinski and J.-L. Lions, North Holland, Amsterdam, 1980.

[6] D.H. Norrie and G. de Vries, A survey of the finite element applications in fluid mechanics, in: Finite elements in fluid, Vol. 3, (eds) R.H. Gallagher, O.C. Zinkiewicz and J.T. Oden, Wiley, London, 1978, 363-395.

[7] R.E. Volker, Nonlinear flow in porous media by finite elements, J. Hyd. Div. A.S.C.E. 95, HY-6 (1969) 2093-2114.

[8] Xie Gan-Guen, Li Jian-hao and Guo Gong-hao, A fast convergent finite element method for computation of nonlinear magnetic induction and corresponding generalizations (to appear).

Computational Techniques & Applications: CTAC-83
J. Noye & C. Fletcher (Editors)

LINEAR MODELS FOR MANAGING SOURCES OF GROUNDWATER POLLUTION

Steven M. Gorelick
U.S. Geological Survey
Menlo Park, California 94025
U.S.A.

Sven-Åke Gustafson
NADA, Royal Inst. of Technology
S-10044 Stockholm 70, Sweden
and
Centre for Mathematical Analysis
Australian National University
GPO Box 4, Canberra, ACT 2601, Australia

We discuss mathematical models for the problem of maintaining a specified groundwater quality while permitting solute waste disposal at various facilities distributed over space. The pollutants are assumed to be chemically inert and their concentrations in the groundwater are governed by linear equations for advection and diffusion. Our aim is to determine a disposal policy which maximises the total amount of pollutants released during a fixed time T while meeting the condition that the concentration everywhere is below prescribed levels.

1. INTRODUCTION

In the present paper we treat linear models for the management of the sources of groundwater pollution. Our discussion is a continuation of the presentation given in [3] and papers referenced there. In particular, we extend the mathematical treatment to the case when water quality standards must be met in an entire region. We shall exploit some obvious similarities between the present water pollution management problem and the air pollution control problem examined in [1] and [4]. We shall also show that the theory of semi-infinite programming as given in [2] may be successfully brought to bear on groundwater pollution management models. We will develop a scheme for computational treatment.

2. PRINCIPAL COMPONENTS OF THE MANAGEMENT MODEL

The management problem is to maintain satisfactory water quality in a given groundwater system during a prescribed *management period* T. We consider a single chemically inert pollutant and the *water quality* is defined in terms of the concentration of the pollutant under study. All known sources are listed in a *source inventory* which contains the technical data about each disposal facility, in particular the release intensity $c_r(t)$ for source r, and for $0 \leq t \leq T$. By using a numerical simulation model of transient solute transport we may calculate the concentration contribution caused by a certain source, *the concentration response* R_r, if the release intensity of the source is described by the time-dependent function c_r. This concentration response varies with time and space and hence R_r is a function depending both on a space-variable s, specifying *where* the concentration occurs and a time variable t indicating *when* it occurs. Thus we write $R_r(s,t)$. The advection-diffusion model is linear. Hence concentration contributions from several sources can be added according to the superposition principle. Let the total released amount corresponding to c_r be v_r . Then

$$v_r = \int_0^T c_r(t)\,dt \ .$$

Consider next a source inventory of n sources with release intensities $c_1, c_2, \ldots, c_n$. The corresponding concentration contributions are $R_1, R_2, \ldots, R_n$ and the released amounts are $v_1, v_2, \ldots, v_n$. Then the total released quantity is w where

$$w = \sum_{r=1}^{n} v_r = \int_0^T \left(\sum_{r=1}^{n} c_r(t) \right) dt \ , \tag{1}$$

and the pollutant concentration at the point s and time t is given by

$$R(s,t) = \sum_{r=1}^{n} R_r(s,t) \ . \tag{2}$$

Let a groundwater standard be imposed, defined by means of a function $f(s,t)$ which specifies the maximum allowable concentration at point s and time t . Thus a permissible release policy must be such that

$$\sum_{r=1}^{n} R_r(s,t) \le f(s,t) \ , \ s \in S \ , \ t \in [0,T] \ . \tag{3}$$

Here S is the domain of water quality control. If we are concerned with water quality at N selected inspection wells situated at $s_1, s_2, \ldots, s_N$, then (3) takes the special form

$$\sum_{r=1}^{n} R_r^i(t) \le f^i(t) \ , \ i = 1,2,\ldots,N \ , \tag{4}$$

where we have written $R_r^i(t)$ for $R_r(s_i,t)$ and $f^i(t)$ for $f(s_i,t)$.

The goal is to determine a permissible release policy which maximises a prescribed preference function. In [3] the total disposal capacity is chosen to be the function to optimise. Thus we seek to maximise (1) subject to the constraint (3) and the condition

$$c_r(t) \ge 0 \ , \ r = 1,2,\ldots,n \ , \ t \in [0,T] \ . \tag{5}$$

3. REDUCTION TO A SEMI-INFINITE PROGRAM

As formulated in the preceding section our management problem requires that functions $c_1, c_2, \ldots, c_n$ should be determined such that the integral (1) is maximised subject to the infinitely many constraints (3). However, it seems reasonable only to consider such policy functions c_r which are determined by finitely many parameters only. Therefore we follow the well-established usage of of numerical methods and consider only functions c_r in a pre-selected finite-dimensional space. Then we only need to determine finitely many parameters subject to infinitely many constraints, i.e. solve a semi-infinite program. In the present argument we shall require c_r to be piecewise linear functions with jump discontinuities allowed at prescribed nodes. Thus we select an integer k and a partition $0 = t_0 < t_1 < .. < t_k = T$ of $[0,T]$. Next we define the functions $u_1, u_2, \ldots, u_{3k}$ as follows:

$$u_{3i-2}(t) = 1 \,, \quad u_{3i-1}(t) = \frac{t-t_{i-1}}{t_i-t_{i-1}} \,, \quad u_{3i}(t) = \frac{t_i-t}{t_i-t_{i-1}} \,,$$

$$t_{i-1} \le t < t_i \,, \; i = 1,2,\ldots,k \,, \tag{6a}$$

$$u_{3i-j}(t) = 0 \,, \; t < t_{i-1} \,, \; t \ge t_i \,, \; j = 0,1,2, \quad i = 1,2,\ldots,k \,. \tag{6b}$$

Any piecewise linear function U_r with break-points at $t_0, t_1, \ldots, t_k$ may thus be written, if U_r is also required to be nonnegative

$$U_r(t) = \sum_{j=1}^{3k} x_{rj} u_j(t) \,, \; x_{rj} \ge 0 \,, \; j = 1,2,\ldots,3k \,, \; r = 1,2,\ldots,n \,. \tag{7}$$

The representation (7) will become unique, if we also require

$$x_{r,3i-2} \cdot x_{r,3i-1} \cdot x_{r,3i} = 0 \,, \quad i = 1,\ldots,k \tag{8}$$

To ensure continuity one must impose the constraints

$$\lim_{t \to t_i} \sum_{j=1}^{3k} x_{rj} u_j(t) = \sum_{j=1}^{3k} x_{rj} u_j(t_i) \,,$$

giving the relations

$$x_{r,3i-2} + x_{r,3i-1} = x_{r,3i+1} + x_{r,3i+3} \,, \quad i = 1,2,\ldots,k-1 \,.$$

Hence the disposal policy of a single source is described by means of 3k parameters x_{rj}. Let R_{rj} be the concentration response caused by selecting $c_r = u_j$ where $3i-2 \le j \le 3i$ for an integer i. Then

$$R_{rj}(s,t) = 0 \,, \quad t \le t_{i-1} \,, \tag{9}$$

since $u_j(t) = 0$ for $t < t_{i-1}$. If the times $t_1, t_2, \ldots, t_k$ are equidistant, i.e. $t_i = iT/k$, a major simplification is possible, since

$$u_j(t) = u_{j+3}(t+T/k) \,, \quad j = 1,\ldots,3k-3 \,.$$

We then get the useful recurrence relations

$$\begin{cases} R_{r,1}(s,t) = R_{r,4}(s,t+T/k) = R_{r,7}(s,t+2k/T) \ ,\ldots \\ R_{r,2}(s,t) = R_{r,5}(s,t+T/k) = R_{r,8}(s,t+2k/T) \ ,\ldots \\ R_{r,3}(s,t) = R_{r,6}(s,t+T/k) = R_{r,9}(s,t+2k/T) \ ,\ldots \end{cases} \tag{10}$$

Therefore it is only needed to calculate the response functions corresponding to u_1, u_2, u_3 since the remaining response function values may be found by combining (9) and (10) with the calculated responses for u_1, u_2, u_3. However, this procedure must be carried out for each of the n sources. The quantity v_r released with the disposal policy U_r of (7) may be written

$$v_r = \sum_{j=1}^{3k} x_{rj} d_j \ , \tag{11}$$

where $d_{3i-2} = t_i - t_{i-1}$, $d_{3i-1} = d_{3i} = (t_i - t_{i-1})/2$.

Hence the value of the preference function (1) becomes

$$w = \sum_{j=1}^{3k} d_j \sum_{r=1}^{n} x_{rj} \ , \tag{12}$$

and the constraints (3) take the form

$$\sum_{j=1}^{3k} \sum_{r=1}^{n} x_{rj} R_{rj}(s,t) \le f(s,t) \ , \ s \in S \ , \ 0 \le t \le T \ , \tag{13}$$

$$x_{rj} \ge 0 \ , \quad r = 1,\ldots,n \ , \quad j = 1,2,\ldots,3k \ . \tag{14}$$

The task of maximising (12) subject to the constraints (13), (14) will be called *Problem (P)*. It is a semi-infinite program and the theory and computational methods of [2] are applicable. We note that this problem has $3nk$ variables and that (13) defines a linear inequality constraint for each pair (s,t).

4. DUALITY RESULTS

The dual of the problem defined by (12), (13), (14) will be termed *Problem (D)*. It reads

Determine an integer q , pairs (s_ℓ, t_ℓ) , $\ell = 1,\ldots,q$ $s_\ell \in S$, $0 \le t_\ell \le T$, reals $\rho_1,\ldots\rho_q$ such that the expression

$$\sum_{\ell=1}^{q} \rho_\ell f(s_\ell, t_\ell) \ , \tag{15}$$

is rendered a minimum subject to the constraints

$$\sum_{\ell=1}^{q} \rho_\ell \, R_{rj}(s_\ell, t_\ell) \geq d_j \ , \quad r = 1,\ldots,n \ , \quad j = 1,\ldots,3k \ , \tag{16}$$

$$\rho_\ell \geq 0 \ , \quad \ell = 1,\ldots,q \ . \tag{17}$$

Both problems are consistent and have a joint optimal value which is assumed. The complementary slackness condition gives the following relations which optimal solution satisfy besides (13), (14) and (16), (17)

$$\rho_\ell \left\{ \sum_{j=1}^{3k} \sum_{r=1}^{n} x_{rj} \, R_{rj}(s_\ell, t_\ell) - f(s_\ell, t_\ell) \right\} = 0 \ , \tag{18}$$

$$\ell = 1,\ldots,q \ ,$$

$$x_{rj} \left\{ \sum_{\ell=1}^{q} \rho_\ell R_{rj}(s_\ell, t_\ell) - d_j \right\} = 0 \ , \quad r = 1,\ldots,n \ , \quad j = 1,\ldots,3k \ . \tag{19}$$

Put now

$$\phi(s,t) = \sum_{j=1}^{3k} \sum_{r=1}^{n} x_{rj} \, R_{rj}(s,t)$$

where x_{rj} , $r = 1,\ldots,n$, $j = 1,\ldots,3k$ is an optimal solution to Problem (P), defined by (12), (13), (14). ϕ is the pollutant concentration corresponding to an optimal policy. It is known that the problem (D) has an optimal solution with $q \leq 3kn$, the number of constraints defined by (16). Due to (18) $\phi(s,t)$ reaches the maximum permissible value at the q points s_ℓ at the times t_ℓ . Arguing in a similar way as on p.777 in [3] and in analogy to similar discussions of air pollution abatement problems we show that the infinitely many linear constraints (13) may be replaced by the requirement that (13) is satisfied for the q pairs (s_ℓ, t_ℓ) mentioned above. However, these pairs are in general not known before Problems (P) and (D) are solved.

5. COMPUTATIONAL TREATMENT

An approximate solution to Problems (P) and (D) is calculated by means of *discretisation*, i.e. we replace the infinite set of inequalities (13) by a finite subset obtained by introducing a finite grid, i.e. a finite number of pairs (s_m, t_m) $m = 1,\ldots,M$. The linear programs hereby arising may be solved by means of a suitable computer code. The discussion in [3] is valid for these discretised problems. We point out that if one accepts an optimal solution of these discretised problems the corresponding concentration $\tilde{\phi}$ may violate the standard *between* the gridpoints but not on the grid. Therefore, if we interpolate $\tilde{\phi}$ linearly between two adjacent gridpoints (adjacent in time or space) we will get a function whose values do not exceed those of the standard f . Hence the magnitude of the violation of the standard is bounded by the error resulting when $\tilde{\phi}$ is interpolated linearly between two adjacent gridpoints. It is known that the discretisation error can be made less than any given positive bound by making the grid sufficiently fine. However, it should be enough to make the discretisation error small in comparison to other uncertainties in the model.

REFERENCES

[1] Carbone, R., W.L. Gorr, S.-Å. Gustafson, K.O. Kortanek and J.R. Sweigart, A bargaining resolution of the efficiency versus equity conflict in energy and air pollution regulation, TIMS Studies in the Management Sciences 10 (1978), 95-108.

[2] Glashoff, K. and Gustafson, S.-Å., Linear Optimization and Approximation (Springer-Verlag, New York, Heidelberg, Berlin, 1983).

[3] Gorelick, S.M., A model for managing sources of groundwater pollution, Water Resources Research 18 (4) (1982), 773-781.

[4] Gustafson, S.-Å. and Kortanek, K.O., A comprehensive approach to air quality planning: Abatement, monitoring networks, real time interpolation, Proceedings of 1979 IIASA Workshop on Air Pollution Modelling, Pergamon Press, 1979.

Computational Techniques & Applications: CTAC-83
J. Noye & C. Fletcher (Editors)

NUMERICAL SOLUTION OF A FREE SURFACE DRAINAGE PROBLEM USING A VARIATIONAL INEQUALITY METHOD

John H. Knight

CSIRO Division of Mathematics and Statistics,
GPO Box 1965, Canberra ACT, Australia.

Unconfined groundwater seepage problems are traditionally difficult to solve numerically, since the flow region is unknown. The Baiocchi transformation was developed to give a formulation of this problem in terms of variational inequalities on a known larger domain. Numerical solution methods using successive over relaxation with projection follow naturally from the theory, and are simple to implement. We adapt the Baiocchi method to solve a steady drainage problem with known vertical flow through the free surface, and give numerical approximations for the streamlines and equipotentials of the flow field, as well as the position of the free surface.

INTRODUCTION

Groundwater unconfined seepage problems, which require the position of an unknown free surface, are difficult to solve either analytically or numerically. Analytical solutions use the hodograph transformation and very complicated conformal mappings (e.g. Bear, 1972), and conventional numerical methods using finite differences (or more recently finite elements) iterate to find the flow region, requiring a calculation on a different region during each iteration.

In 1971, Baiocchi introduced a transformation of the dependent variable which allowed a formulation in terms of variational inequalities on a fixed domain. Baiocchi et al. (1973) showed that this formulation led naturally to numerical solution methods, as well as giving existence and uniqueness. Baiocchi and his co-workers at Pavia subsequently studied a large number of seepage problems, as did Bruch and Sloss (1978), Bruch (1979, 1980a, 1980b), and Oden and Kikuchi (1980).

Recently, van der Hoek, Barnes and Knight (1983) adapted the Baiocchi method to investigate drainage problems with known vertical flow through the free surface, and studied existence and uniqueness of solutions. In this paper, an outline and numerical results will be given for a numerical algorithm for solving such drainage problems, using as an example a problem studied by Bear (1972).

THE PHYSICAL PROBLEM

The 'model problem' that we consider involves the steady two-dimensional seepage of uniformly applied irrigation water through a uniform porous soil overlying an impervious subsoil to a series of parallel, regularly spaced drains fully penetrating the porous layer. We assume that the impermeable lower boundary is horizontal, and that there is a well-defined free surface between the fully saturated flow region and the unsaturated region (see Figure 1).

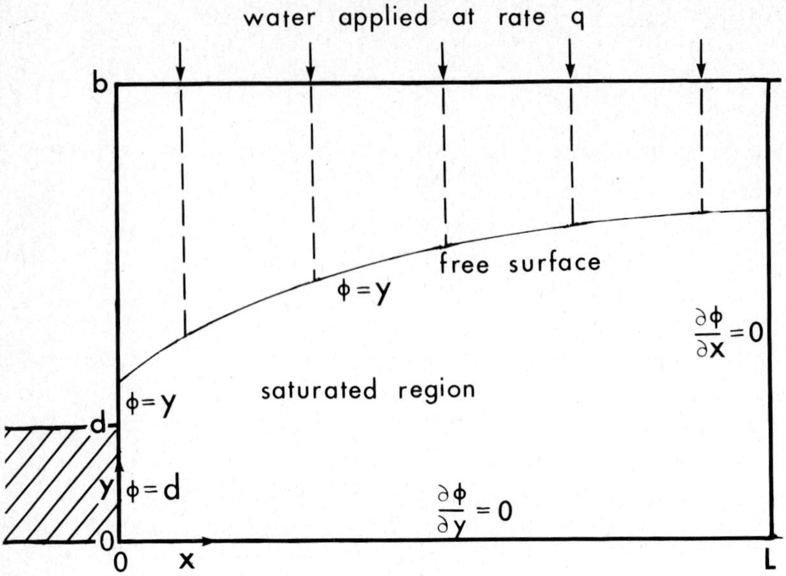

FIGURE 1
Drainage by fully penetrating ditch

We take the height of free water in the ditch to be constant at d, and the rate of water application at the soil surface to be constant at q, which is less than the saturated hydraulic conductivity K. The soil is assumed homogeneous and isotropic, with the flow obeying Darcy's law.

In this case we can define a piezometric head ϕ which is the sum of the pressure head $p/\rho g$ and the height y above the reference level

$$\phi = y + p/\rho g \tag{1}$$

and then by Darcy's law

$$u_1 = k\,\partial\phi/\partial x, \quad u_2 = K\,\partial\phi/\partial y \tag{2}$$

where x and y are the cartesian coordinates, and u_1 and u_2 are the velocity components. By conservation of mass, the piezometric head satisfies Laplace's equation

$$\frac{\partial^2\phi}{\partial x^2} + \frac{\partial^2\phi}{\partial y^2} = 0 \tag{3}$$

If the distance between vertical drains is $2L$, we can by symmetry study one cell of width $2L$ with no horizontal flow at the centre line, or a half-cell of width L. Taking the vertical side of the drain to be at $x = 0$,

$$K\,\partial\phi/\partial x = 0 \quad \text{at} \quad x = L. \tag{4}$$

For simplicity, we will assume here that the unknown free surface does not intersect the soil surface at $y = b$. The more general case with possible intersection is considered by van der Hoek et al. (1983).

BOUNDARY CONDITIONS

Along the horizontal impervious layer at $y = 0$,

$$\partial\phi/\partial y = 0 \ .$$

At the centre line at $x = L$, by symmetry

$$\partial\phi/\partial x = 0 \ .$$

On the vertical boundary at the drain at $x = 0$, below $y = d$, the boundary is in contact with free water at rest in the drain, so the piezometric head is constant and equal to d. Above $y = d$, there is a 'surface of seepage' in contact with air at atmospheric pressure, which can be taken at the reference value of zero. Therefore along the surface of seepage the piezometric head takes the value of $\phi = y$.

The position of the free surface is unknown, so two boundary conditions are necessary to determine it. This surface is also assumed to be at atmospheric pressure, so $\phi = y$ along it. The other boundary condition follows from the assumption of uniform vertical flow through the unsaturated region into the free surface, and will be written as a condition on a stream function.

DIMENSIONLESS COORDINATES

We introduce dimensionless variables

$$x' = x/L,\ y' = y/L,\ d' = d/L,\ \phi' = \phi/L \ ,$$

$$u_1' = u_1/K,\ u_2' = u_2/K,\ q' = q/K,\ 0 < q' < 1 \ ,$$

satisfying

$$(u_1' \ , u_2') = (\partial\phi'/\partial x',\ \partial\phi'/\partial y') \qquad (5)$$

$$\frac{\partial^2\phi'}{\partial x'^2} + \frac{\partial^2\phi'}{\partial y'^2} = 0 \qquad (6)$$

We introduce a dimensionless stream function ψ satisfying

$$\partial\phi'/\partial x' = \partial\psi/\partial y' \ , \quad \partial\phi'/\partial y' = \partial\psi/\partial x' \qquad (7)$$

$$\frac{\partial^2\psi}{\partial x'^2} + \frac{\partial^2\psi}{\partial y'^2} = 0 \qquad (8)$$

In terms of the new variables, the boundary conditions are:

$$\partial\phi'/\partial x' = 0 \quad \text{when} \quad x' = 1,$$

$$\partial\phi'/\partial y' = 0 \quad \text{when} \quad y' = 0,$$

$$\phi'(x',y') = d' \quad \text{when} \quad x = 0,\ 0 < y' < d'$$

$$\phi'(x',y') = y' \quad \text{when} \quad x = 0,\ d' < \psi' < b'$$

On the unknown free surface defined by $y' = h(x')$,

$$\phi'(x',y') = y'$$

and the second boundary condition can be written simply in terms of the stream function as

$$\psi(x',y') = -q'x'.$$

Henceforth only dimensionless variables will be used, and the primes (′) will be dropped.

THE BAIOCCHI TRANSFORMATION

Baiocchi (1971) introduced a new dependent variable w defined on a larger and known domain $D = [0,1] \times [0,b]$, where b is chosen to be above the maximum height of the free surface. We adapt the Baiocchi transformation to the case of nown flow through the free surface, by extending ϕ and ψ continuously to $\bar{\phi}$ and $\bar{\psi}$ defined on the whole domain D. In the saturated region Ω we take

$$\bar{\phi} = \phi, \quad \bar{\psi} = \psi$$

and in the unsaturated region $D - \bar{\Omega}$ above the free surface

$$\bar{\phi} = y \ , \ \bar{\psi} = -qx \ .$$

Note that

$$\frac{\partial\bar{\phi}}{\partial x} - \frac{\partial\bar{\psi}}{\partial y} = 0 \quad \text{in} \quad D \ ,$$

$$\frac{\partial\bar{\phi}}{\partial y} - \frac{\partial\bar{\psi}}{\partial x} = \begin{cases} 0 & \text{in} \quad \Omega \\ 1 - q & \text{in} \quad D - \Omega \ . \end{cases}$$

For (x,y) in D we define the Baiocchi transformation by

$$w(x,y) = \int_y^b [\bar{\phi}(x,\bar{y}) - \bar{y}]d\bar{y} \tag{9}$$

Physically, w is the integral of the pressure head from (x,y) up to the free surface, for (x,y) in the saturated region. Since the pressure is positive below the free surface, and zero above it by construction, w is also positive in the saturated region and zero in the unsaturated region. From the definition (9), it follows that

$$\partial w/\partial y = y - \bar{\phi} \ , \ \partial w/\partial x = -qx - \bar{\psi} \ , \tag{10}$$

and

$$\begin{aligned} \frac{\partial^2 w}{\partial x^2} + \frac{\partial^2 w}{\partial y^2} &= 1 - q - [\frac{\partial\bar{\phi}}{\partial y} + \frac{\partial\bar{\psi}}{\partial x}] \\ &= 1 - q \text{ in } \Omega \\ &= 0 \quad \text{ in } D - \Omega \end{aligned} \tag{11}$$

Along the top boundary, $w = 0$, and on the line of symmetry at $x = 1$, $\partial w/\partial x = 0$. On the side at $x = 0$,

$$w(0,y) = \int_y^d (d - y)dy = (d - y)^2/2 \qquad 0 \leq y \leq d,$$

$$= 0 \qquad d \leq y \leq b.$$

Along the horizontal base $y = 0$

$$\partial w/\partial x = -qx - \bar{\psi} = q(1-x)$$

so

$$w(x,0) = w(0,0) + qx - qx^2/2 = [d^2 + qx(2-x)]/2.$$

VARIATIONAL INEQUALITY FORMULATION

For a theoretical account of variational inequalities and their properties, see Kinderlehrer and Stampacchia (1980) or Oden and Kikuchi (1980). Using the Baiocchi variable w, together with the formulation of the problem as a variational inequality, gives a means of proving existence and uniqueness of the solution (van der Hoek et al, 1983). The complementarity form of the variational inequality leads directly to a discretization of the problem and a numerical solution.

Let K^+ be the closed convex set consisting of all suitably continuous non-negative functions defined on D and satisfying the above boundary conditions. It can be shown (van der Hoek et al, 1983) that w satisfied the variational inequality

$$\int_D [\nabla(v - w).\nabla w]\, dxdy + (1-q)(v - w) \geq 0 \qquad (12)$$

for all v in K^+, and this form can easily be used to establish uniqueness of the solution. For numerical solution it is straightforward to directly discretize the complementarity form derived from Eq. (11),

$$\nabla^2 w = \begin{cases} 1 - q & w > 0, \\ 0 & w \leq 0 \end{cases} \qquad (13)$$

This has replaced the problem of solving Laplace's equation with complicated boundary conditions on an unknown region, by one of solving Eq. (13) which has a more complicated form, but simpler boundary conditions on a known rectangular region.

DISCRETE APPROXIMATION

Following Baiocchi et al (1973), we use a discrete approximation of Eq. (13), and solve it numerically using successive over-relaxation (S.O.R.) with projection. Finite elements can be used to discretize such variational inequalities (Kikuchi, 1977), but we use the usual five-point finite difference approximation to the Laplacian.

Without loss of generality, we can take intervals in each direction and choose b such that $b = N_2/N_1$, where N_1 and N_2 are the numbers of grid spaces in the x and y directions respectively. Setting

$$w_{i,j} = w[i/N_1, j/N_1]$$

and using the superscript 'n' to indicate the nth iterate, we can write the S.O.R. with projection in two steps as:

$$w_{i,j}^{n+1/2} = \frac{1}{4}\left[w_{i-1,j}^{n+1} + w_{i+1,j}^{n} + w_{i,j-1}^{n+1} + w_{i,j+1}^{n} - (1-q)/N_1^2 \right] \tag{14a}$$

for $i = 2,\ldots,N_1$, $j = 2,\ldots,N_2$,

$$w_{i,j}^{n+1/2} = \frac{1}{4}\left[2w_{i-1,j}^{n+1} + w_{i,j-1}^{n+1} + w_{i,j+1}^{n} - (1-q)/N_1^2 \right] \tag{14b}$$

for $i = N_1+1$, $j = 2,\ldots,N_2$;

$$w_{i,j}^{n+1} = \max\left[0,\ w_{i,j}^{n} + \alpha\{w_{i,j}^{n+1/2} - w_{i,j}^{n}\} \right] \tag{15}$$

where α is the relaxation parameter, its range being $1 < \alpha < 2$ for over-relaxation. The value of α giving greatest convergence acceleration depends on the problem parameters and the grid spacing.

Equation (15) also incorporates the 'projection' part of the algorithm, which enforces the non-negativity of w. In the absence of a better guess, w can be taken to be zero throughout D initially. When the S.O.R. with projection has converged, the position of the free surface is given approximately by the boundary between the region where w is positive and the region where it is zero. Approximate values of the piezometric head and stream function can be calculated with a finite difference approximation to Eq.(10).

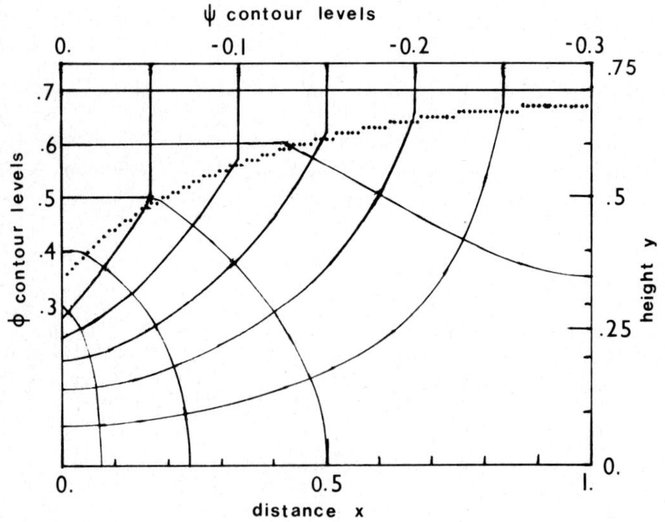

FIGURE 2

Position of free surface and contours of ϕ and ψ

RESULTS

Numerical results were obtained for the parameter values of $q = 0.3$, $d = 0.25$, $b = 1.0$. With 101 grid points in each direction and an initial guess that w was zero in the interior of D, various values of the relaxation parameter α were tried. The value of 1.93 was found to give the best convergence, successive iterates being sufficiently close after 150 iterations. Numerical values of the piezometric head and stream function were calculated from a discretization of Eq. (10), and contours of these quantities were drawn, as shown in Figure 2. The approximate position of the free surface is also shown. Setting up the boundary conditions and initial values of w and programming Eqs. (14) and (15) proved to be very straightforward, and the calculations to produce Figure 2 including finding the contours took only about 4 seconds of processing time on a CDC Cyber 76 computer.

REFERENCES

[1] Baiocchi, C., Sur un problème à frontière libre traduisant le filtrage de liquides à travers des milieux poreux, C.R. Acad. Sci. Paris, Ser. A, 273 (1971) 1215-1217.

[2] Baiocchi, C., Comincioli, V., Guerri, L. and Volpi, G., Free boundary problems in the theory of fluid flow through porous media: a numerical approach, Calcolo 10 (1973) 1-85.

[3] Bear, J., Dynamics of fluids in porous media (American Elsevier, New York, 1972).

[4] Bruch, J.C., A numerical solution of an irrigation flowfield, Intern. J. for Numer. and Anal. Meths. in Geomech. 3 (1979) 23-26.

[5] Bruch, J.C., A survey of free boundary value problems in the theory of fluid flow through porous media: variational inequality approach - Part I, Adv. Water Resour. 3 (1980) 65-80.

[6] Bruch, J.C., A survey of free boundary value problems in the theory of fluid flow through porous media: variational inequality approach - Part II, Adv. Water Resour. 3 (1980) 115-124.

[7] Bruch, J.C. and Sloss, J.M., A variational inequality method applied to free surface seepage from a triangular ditch, Water Resour. Res. 14 (1978) 119-124.

[8] Kikuchi, N., An analysis of the variational inequalities of seepage flow by finite element methods, Quart. Appl. Math. 35 (1977) 149-163.

[9] Kinderlehrer, D. and Stampacchia, G., An introduction to variational inequalities and applications (Academic Press, New York, 1980).

[10] Oden, J.T. and Kikuchi, N., Theory of variational inequalities with applications to problems of flow through porous media, Int. J. Engng. Sci. 18 (1980) 1173-1284.

[11] van der Hoek, J., Barnes, C.J. and Knight, J.H., A free surface problem arising in the drainage of a uniformly irrigated field: existence and uniqueness results, J. Austral. Math. Soc. (Ser.A) 35 (1983) (to appear).

Computational Techniques & Applications: CTAC-83
J. Noye & C. Fletcher (Editors)

THERMAL PROFILES FOR HIGH-LEVEL RADIOACTIVE WASTE BURIAL

Jerard M. Barry
Paul C. Miskelly
John P. Pollard

Australian Atomic Energy Commission
Lucas Heights Research Laboratories
Private Mail Bag, Sutherland NSW 2232
Australia

Several solutions for the long-term storage of high-level radioactive waste have been proposed. Essentially they involve the solidification of the spent nuclear material and its storage in the most stable environment possible. The conventional method is to solidify such wastes, by vitrification in borosilicate glass but, at high temperatures and specific hydrothermal conditions, these wasteforms have shown enhanced leachability. This problem can be overcome by reducing the concentration of the heat source (i.e. the radioactive material) but promising alternatives are available, such as the Australian SYNROC process.

A computational model for the determination of temperature distribution produced by radioactive decay, particularly the temperature within a cylinder of SYNROC material buried deep beneath the Earth's surface, is discussed.

1. INTRODUCTION

Apart from its radiation effects, the decay of radioactive waste material generates significant quantities of heat. To evaluate a suitable means of long-term disposal by burial, it is important to determine the maximum temperature reached in the material and its container. This has an important bearing on the long-term behaviour of the material used to confine the waste. At the Lucas Heights Research Laboratories, studies are being undertaken into the suitability of trapping and storing nuclear high-level waste in a synthetic rock-like material. (The material, SYNROC, was developed at the Australian National University by Professor A.E. Ringwood.) As part of a thermal analysis, it is assumed that a canister containing SYNROC-bound radioactive material is buried 600 metres below the Earth's surface in a non-ventilated rock formation. An investigation of the time-dependent heat build-up at the canister's surface is made for this postulated worst-case scenario.

2. MODEL DEVELOPMENT

The burial conditions are shown in Figure 1. It is presumed that the canisters are buried as a horizontal arrray in direct contact with solid rock to a height of 600 metres above and that the only means of cooling is by conduction through the rock. For the worst-case situation and to simplify the problem, heat generation and removal from a canister well inside the array boundary is studied. Because such a canister is surrounded by similar heat sources, heat rejection occurs only vertically, in both directions. It is further assumed that heat loss is symmetrical about the line of canisters.

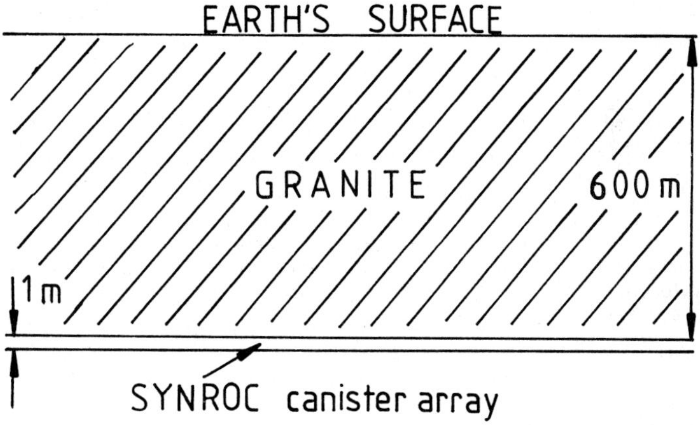

Figure 1
SYNROC burial conditions - schematic

These conditions are readily modelled by considering the heat flow in a single vertical bar of rock, 600 metres long, one end at the Earth's surface and the other in contact with a canister with a known heat generating source. This conceptualisation (rotated to the horizontal for ease of presentation) is shown in Figure 2.

Figure 2
Heat transfer pathway from SYNROC canister to earth's surface

The bar of cross-sectional area A and length L of 600 metres has boundary conditions

$$\frac{\partial T}{\partial x} = 0 \text{ at } x = 0, \text{ for all } t, \text{ and}$$

$$T = 0 \text{ at } x = L, \text{ for all } t \tag{1}$$

and the initial condition

$$q = q_0 \text{ at } x = 0, \quad t = 0 \text{ where } q_0 = 16 \text{ W m}^{-2} \quad , \tag{2}$$

which then decays at the known rate

$$q = q_0 e^{-\lambda t}, \text{ where t is time in years, and} \tag{3}$$

$$\lambda = 0.02333 \text{ (corresponding a 30 year half-life).}$$

The problem is to find the temperature profile at the canister surface

$$T = T(t) \text{ at } x = 0 \quad . \tag{4}$$

This condition is readily modelled by a one-dimensional form of the general heat transfer equation where $\delta(x)$ is the usual delta function:

$$\rho C_p \frac{\partial T}{\partial t} = K \frac{\partial^2 T}{\partial x^2} + q\, \delta(x) \quad . \tag{5}$$

The implementation and solution of this equation on a hybrid computer, and its concordance with the results obtained from a purely digital computation, form the basis of this paper.

3. HYBRID MODEL

3.1 Why Analogue/Hybrid?

Because the general form of this equation is readily solved by analogue techniques (Korn and Korn, 1964) and skilled people and the necessary facilities were available, it seemed likely that an analogue implementation would produce the speediest solution to the problem. As is usual in modelling, however, further problems were found and this novel means of solution had to be supplemented by a more established technique for verification.

3.2 Hybrid Model Derivation

The solution of partial differential equations on an analogue computer requires that they be written as a set of ordinary differential equations. For this approach, one variable is discretised while the other is treated by continuous integration. This is called lumped parameter modelling.

3.2.1 Lumped-parameter model

In this model, the space dimension is discretised into N segments, that is, the rock slab is sectioned. This visualisation is shown by Figure 3. Equation 5 is rewritten in central difference form

$$\frac{\partial T_n}{\partial t} = \frac{K}{\rho C_p \Delta x^2}\,[T_{n-1} - 2T_n + T_{n+1}] \quad , \tag{6}$$

where $1 \leq n \leq N$.

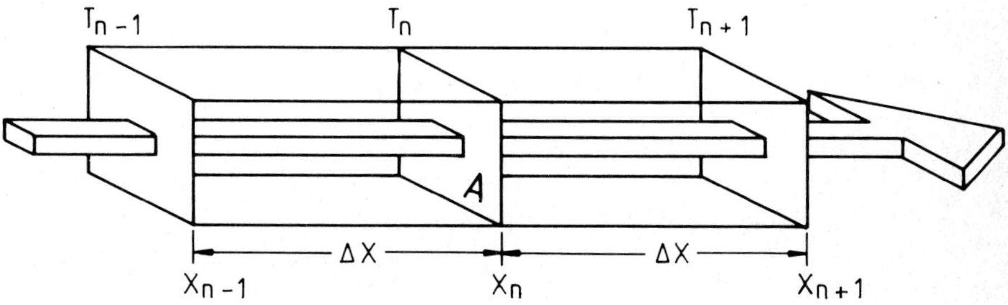

Figure 3
Lumped parameter element of rock column

3.2.2 Boundary and initial conditions

At the Earth's surface, the temperature is T = 0, for all time t. At x = 0 however, the temperature profile is not known a priori. Instead, the profile of the heat source (the SYNROC) is known (Equation 3). It is assumed that

$$\text{at } x = 0, \quad \frac{\partial T}{\partial x} = 0 \quad \text{for all } t \quad . \tag{7}$$

This latter condition is incorporated by assuming that the heat flow is symmetrical at x = 0, (Figure 4); half the heat flows in either direction and, as the material properties are similarly symmetrical, it follows that

$$T_{-1} = T_1 \quad , \tag{8}$$

and the difference approximation for n = 0 may be written as

$$\frac{\partial T_0}{\partial t} = \frac{2K}{\rho C_p \Delta x^2}(T_1 - T_0) + \frac{q}{\rho C_p \Delta x} \quad . \tag{9}$$

Initial conditions are as stated in Section 2.

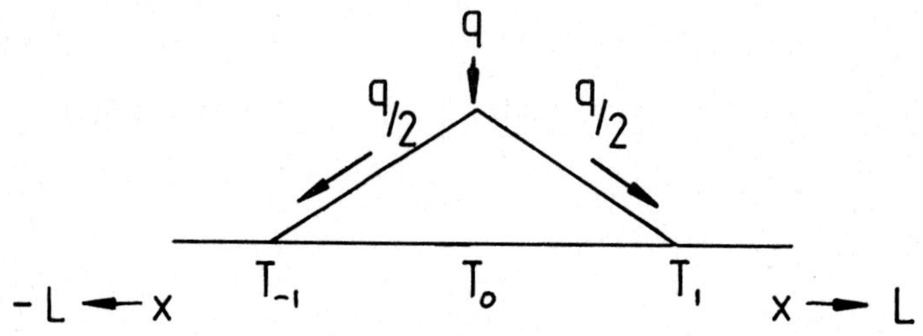

Figure 4
Heat Input Boundary Condition

3.3 Implementation

The analogue solution technique is similar in concept to an explicit digital method, being a lumped-parameter model. The space dimension is divided into segments, and an equation written for each segment, coupling it to adjacent segments, and the initial and boundary conditions are defined. In a first attempt, the x dimension (600 metres) was divided into 15 equal-length segments to give time-dependent temperature profiles 40 metres apart starting from the origin (the hot end). Although coarse, this demonstrates the feasibility of an analogue solution, each run taking 30 sec and representing a 3000 year interval.

In the next analogue attempt, a variable segment length was used. A fine mesh was selected near the origin because the initial results suggested that the spatial temperature gradients were largest in this region, as would be expected, necessitating care with the interpolation.

As a further check, a hybrid technique was used to store and replay one time-temperature profile; this is more fully explained in Section 3.2.2. The aim was to achieve a progressively finer mesh near the x origin on successive model runs by subdividing the first segment after each run. This subdivision of the first segment was continued until the solutions for T_0 converged.

3.3.1 Analogue circuitry

(i) Heat input term - This is the solution to the ordinary differential equation

$$\frac{\partial q}{\partial t} = -\lambda q \quad , \tag{10}$$

$$\text{where } q = q_0 \text{ at } t = 0. \tag{11}$$

It is readily set up as a simple integration circuit as in Figure 5(a).

(ii) the lumped-parameter circuit for T_0, shown in Figure 5(b).

(iii) The lumped parameter circuit for T_n, $n \neq 0$, shown in Figure 5(c).

3.3.2 Sampling Technique

As already suggested, it was thought that the analogue model was not accurately modelling the region near $x = 0$ where the spatial temperature gradient is greatest. A hybrid method was devised to use a finer mesh near the origin on successive solution runs. The intention was to iterate with successively finer meshes until the solutions for T_0 converged. This method is illustrated in Figure 6. The sampling techniqueis due to a colleague, C.P. Gilbert.

The first run models the bar in 15 segments of 40 metres each. The T_1 profile is sampled very accurately in real time and stored digitally as 100 equally spaced (in time) sample points. The digitised profile is called T_s. The next run models the first 40 metres of the bar in ten 4-metre segments with the stored version of T_1 (T_s) from the first run replayed as its end temperature T_9. Again, the time profile of the new T_1 is sampled very accurately and stored. On the next run, a 4-metre segment of the bar is modelled with the stored temperature profile being again replayed.

parameter coefficients

$C_0 = q_0/2$

$C_1 = \lambda$

$C_2 = \frac{1}{\rho C\rho \Delta x}$

$C_3 = \frac{K}{\rho C\rho \Delta x^2}$

$C_4 = 2C_3$

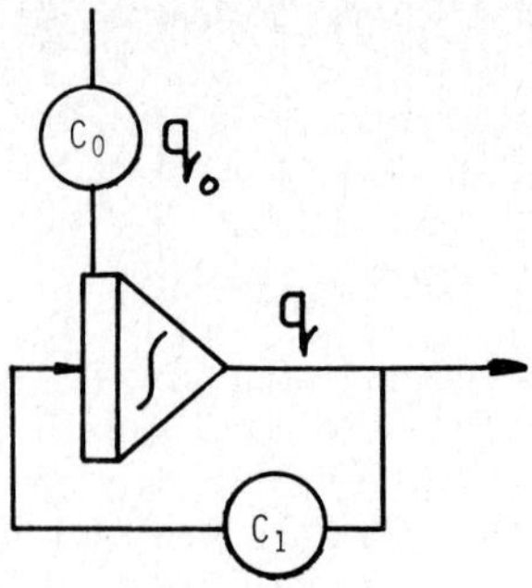

Figure 5(a)
Heat Input term generator

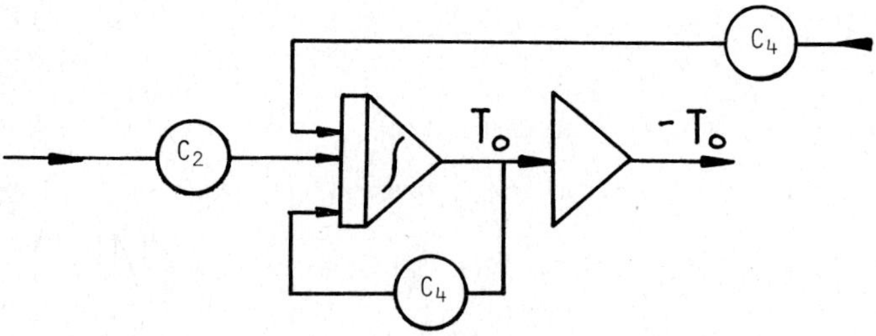

Figure 5(b)
Circuit module for lumped parameter element for T_0

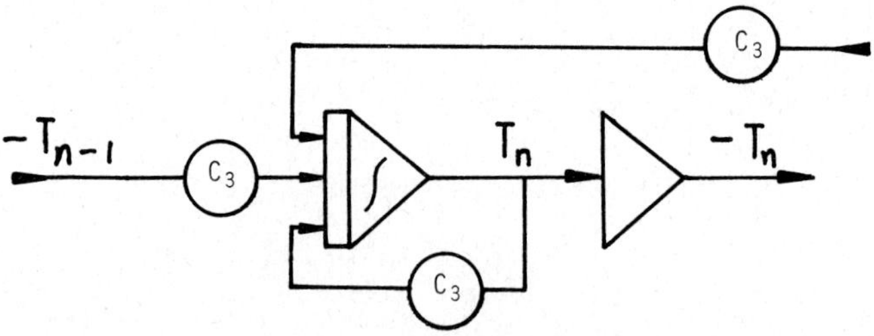

Figure 5(c)
Circuit module for lumped parameter element for T_n, n = 0

Figure 5
Analogue circuit modules for heat diffusion equation

On each run, the time solution for T_1 is stored by sampling into digital storage by monitoring the solution run with a digital program. After resetting the analogue computer coefficients to represent a shorter segment of the bar, the digital program commences the next solution and replays the stored solution as the boundary condition at the cooler end of the shorter section.

3.3.3 Hybrid model performance

Only the three runs shown in Figure 6 were necessary to obtain convergence, requiring a solution time of about two minutes. Further, the hybrid solution vindicated the pure analogue approach initially adopted.

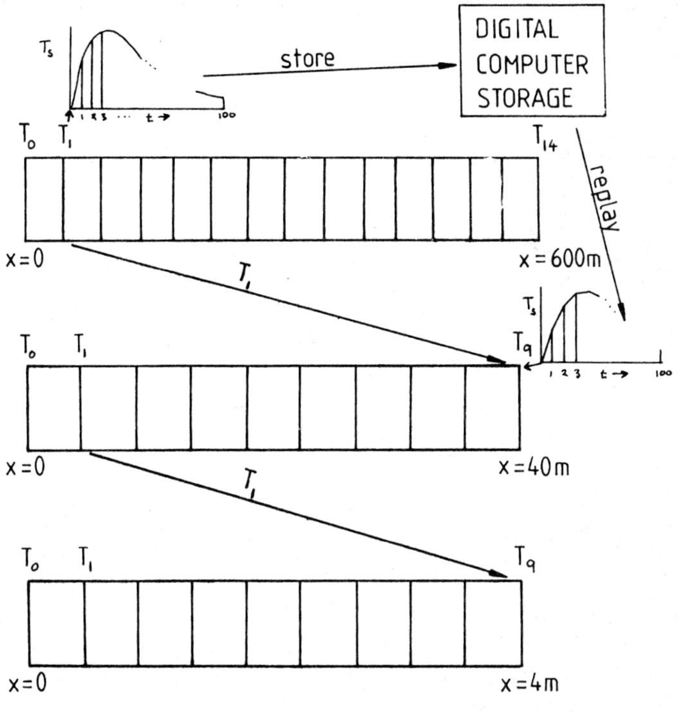

Figure 6
Technique for generating a finer mesh near the origin

4. PURE DIGITAL IMPLEMENTATION

To support the hybrid technique described in Section 3, a digital solution was obtained with a more usual approach. The solution was obtained quickly (i.e. with minimal program development time) by transforming the heat conduction equation into a form suitable for solution with the existing nuclear code POW3D (Barry, Harrington and Pollard, in preparation). POW3D solves neutron diffusion and source problems which conform to a general equation presented here in a simplified form (i.e. neglecting delayed neutron precursors):

$$-\frac{1}{v_g}\frac{\partial \phi_g}{\partial t}(\underline{r},t) = -\nabla . D_{n,g}(\underline{r},t)\,\nabla\phi_g(\underline{r},t) + \sigma_{rem\ g}(\underline{r},t)\phi_g(\underline{r},t)$$

$$- \sum_{g'} \sigma_{sg'\to g}(\underline{r},t)\ \phi_{g'}(\underline{r},t)$$

$$- \chi_g \sum_{g'} \frac{\nu}{k}\ \sigma_{fg'}(\underline{r},t)\ \phi_{g'}(\underline{r},t) - S_g(\underline{r},t) \quad , \qquad (12)$$

where k is the effective steady state multiplication, $S_g(\underline{r},t)$ is the external neutron source for energy group g, χ_g is the fission spectrum, $\sigma_{s\ g'\to g}(\underline{r},t)$ is the scattering matrix from group g' to g, $\sigma_{rem\ g}(\underline{r},t)$ is the removed cross section for group g (consisting of adsorption and outscatters), $\nu\sigma_{fg}$ is the fission emission, $D_{n,g}(\underline{r},t)$ denotes directional diffusion coefficients, v_g the average velocity for group g, $\phi_g(\underline{r},t)$ is the group neutron flux to be calculated, and t is the time. The spatial aspects allow three-dimensional slab, cylinders, spheres or point models to be constructed.

By reducing the number of energy groups to one, eliminating delayed neutron groups, carefully rescaling Equation 12, reducing the fission emission term to a negligible size, defining the constants appropriately, and using the external neutron source as the heat generator, the neutron diffusion and heat transfer equations become equivalent. The code POW3D was designed for reactor physics problems. It consequently handles boundary conditions where the neutron flux is reflective ($\frac{\partial\phi_g}{\partial n} = 0$) or where it is extrapolated to zero beyond the reactor boundaries ($D_{n,g}\frac{\partial\phi_g}{\partial n} + \frac{\phi_g}{3d} = 0$), where d is the extrapolation parameter.

A reflective boundary condition was applied at the centre of the cylinder of nuclear waste and the temperature at the right hand boundary extrapolated to zero after a 600 metre thickness of the Earth's crust was included in the model. Although the transformation may seem inconvenient, this was preferable to developing alternative software. It employs a complex time integration procedure (Pollard, 1973) which reduces to the Crank-Nicholson method in the absence of the delayed neutron groups term.

Control over the integration times is easily specified by the user without the need to write a special-purpose code. A finite-difference approximation based on the edge flux method (Wachspress, 1966) leads to a tridiagonal system of linear equations for the one-dimensional problem. These equations are solved by a direct method. POW3D is a very facile system featuring free-format input, easy adjustment of meshes and convenient displays of all input data. Its use was justified as reasonable solutions were obtained within two hours of commencing the analysis.

In the digital solution, the choice of mesh is not restricted by the lack of integrators as in the analogue approach. The real time for integration, however, may be much slower and numerical instabilities are always possible. Fortunately the latter did not eventuate. Various spatial refinements and time intervals were attempted. The digital solutions converged quickly to the results shown in Figure 7, where the temperature at the surface of the cylinder for a period of 160 years is displayed. This is in agreement with the analogue/hybrid result.

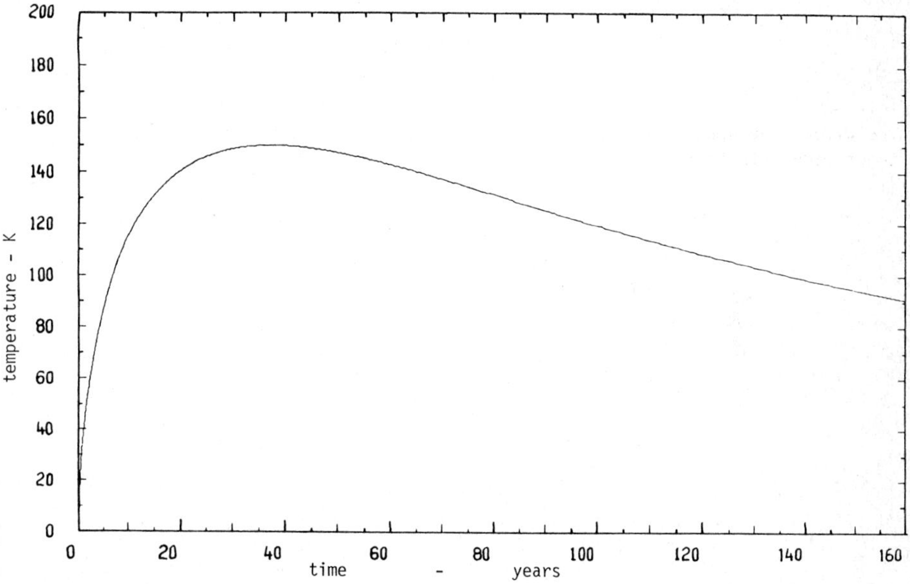

Figure 7
Predicted temperature rise of SYNROC canister above ambient - K

5. CONCLUSION

Two different approaches to the implemention of the equations for a model of thermal profiles yielded the same solution. Further, one of them was an explicit method (analogue/hybrid) and the other an implicit one. Although the solution run times were comparable, the implementation time for the analogue/hybrid method would have been shorter, if a completely new digital code relying on standard integration packages had been created.

6. REFERENCES

[1] Barry, J.M., Harrington, B.V. and Pollard, J.P., POW3D, a Computer Code for the Solution of 3-D Neutron Diffusion Equations, (AAEC/E report, in preparation).

[2] Korn, G. A., Korn, T. M., Electronic Analog and Hybrid Computers (McGraw-Hill, 1964) pp.540-541.

[3] Pollard, J.P., Numerical Methods Used for Neutronics Calculations, Ph. D. thesis, University of New South Wales (1973).

[4] Wachspress, E.L., Iterative Solution of Elliptic Systems (Prentice Hall, New Jersey, 1966).

7. SYMBOL TABLE

C_p - heat capacity of granite ($J\ kg^{-1}\ K^{-1}$)

L - depth of rock covering SYNROC array (m)

λ - overall rate of decay of radioactive waste ($year^{-1}$)

K - thermal conductivity ($W\ m^{-1}\ K^{-1}$)

q - heat flux from SYNROC ($W\ m^{-2}$)

ρ - density of granite ($kg\ m^{-3}$)

t - time (years)

T - relative temperature to that of the Earth's surface (K)

T_n - relative temperature at position n along the slab (K)

Computational Techniques & Applications: CTAC-83
J. Noye & C. Fletcher (Editors)

ON A FINITE DIFFERENCE METHOD OF PLANE THERMOELASTIC PROBLEM IN MULTIPLY-CONNECTED REGION EXHIBITING TEMPERATURE DEPENDENCIES OF MATERIAL PROPERTIES

Yoshihiro Sugano

Department of Mechanical Engineering
University of Osaka Prefecture
Sakai 591, Osaka
Japan

A formulation in terms of Airy's stress function is given for the plane thermoelastic problem in the multiply-connected region exhibiting the temperature dependencies of material properties. New Michell integral conditions necessary for the assurance of single-valuedness of the rotation and displacements are derived taking into account the temperature dependencies of material properties. The fundamental equations of the plane thermoelastic problem formulated are solved by the finite difference method. Numerical calculations are carried out for a rectangular plate with a rectangular hole made of aluminium alloy.

INTRODUCTION

The determination of the thermal stresses in bodies with various holes plays an important role in the design of nuclear reactors, gas and steam turbines and chemical plants.

Takeuti and Sekiya (1968) determined the steady plane thermal stresses in a polygonal cylinder with a circular hole under uniform internal heat generation by the use of the five elementary functions method and a point-matching technique. Hulbert (1970) reported a solution of the thermal stress problem in tube sheets by the boundary point least squares method. Takeuti and Noda (1973) applied their five elementary functions method to a transient thermoelastic problem in a polygonal cylinder with a circular hole by the aid of Laplace transform and the point-matching technique. These solutions to the plane thermal stress problems based on classical thermoelasticity have been obtained assuming that the material properties do not depend on the temperature field. Therefore, they can not truely describe the actual thermal-stress distributions in the structural elements in which severe temperature gradients occur.

As far as the author is aware, no solution has been reported to the plane thermoelastic problem in the multiply-connected region exhibiting temperature-dependent material properties in which the single-valuedness of rotation and displacements must be guaranteed.

In this paper, therefore, we present a formulation of the plane thermoelastic problem in a multiply-connected region having the temperature-dependent material properties in terms of Airy's stress function. New Michell integral conditions necessary for the assurance of single-valuedness of the rotation and displacements are derived taking into account the temperature dependencies of material properties.

The nonlinear heat-conduction equation and the system of fundamental thermoelastic equations with the temperature-dependent coefficients are solved numerically by the use of the finite difference method.

Numerical calculations are carried out for a rectangular plate with a rectangular hole made of aluminium alloy.

FORMULATION OF PROBLEM

We consider a finite rectangular region with a rectangular hole as shown in Fig.1 as an example of the multiply-connected region without loss of generality.

TEMPERATURE FIELD

The finite rectangular region of dimension $2l_x \times 2l_y$ with a rectangular hole of dimension $2l_x' \times 2l_y'$, initially at the same uniform temperature T_0 as the surrounding media, is heated by an abrupt temperature rise of the inner surrounding medium. Taking into account the variation in thermal properties with temperature, we may express the governing equation for the transient temperature field as follows:

$$\frac{\partial}{\partial x}[\lambda(T)\frac{\partial T}{\partial x}]+\frac{\partial}{\partial y}[\lambda(T)\frac{\partial T}{\partial y}]-2\delta_{0l}h(T-T_0)/b=c(T)\gamma(T)\frac{\partial T}{\partial t} \tag{1}$$

where $\lambda(T)$, $c(T)$ and $\gamma(T)$ are the thermal conductivity, specific heat and density depending on the temperature, respectively, δ_{0l} is Kronecker's delta ($l=0$ for the rectangular plate of thickness b, $l=1$ for the infinitely long rectangular cylinder) and h is the heat transfer coefficient on the upper and lower surfaces of the rectangular plate.
From the above description, the initial and boundary conditions may be expressed as

$$T=T_0, \qquad t=0 \tag{2}$$

$$\lambda(T)\frac{\partial T}{\partial x}+h_0(T-T_0)=0, \qquad x=l_x \tag{3}$$

$$\lambda(T)\frac{\partial T}{\partial y}+h_0(T-T_0)=0, \qquad y=l_y \tag{4}$$

$$\lambda(T)\frac{\partial T}{\partial x}-h_1(T-T_1)=0, \qquad x=l_x' \tag{5}$$

$$\lambda(T)\frac{\partial T}{\partial y}-h_1(T-T_1)=0, \qquad y=l_y' \tag{6}$$

In addition, the symmetry of the thermal conditions and the region with respect to the x and y axes yields

$$\frac{\partial T}{\partial y}=0, \quad l_x' \leq x \leq l_x, \quad y=0 \ ; \qquad \frac{\partial T}{\partial x}=0, \quad x=0, \quad l_y' \leq y \leq l_y. \tag{7}$$

PLANE THERMOELASTIC PROBLEM IN MULTIPLY-CONNECTED REGION EXHIBITING TEMPERATURE DEPENDENCIES OF MATERIAL PROPERTIES

Consider a general treatise based on stress function method of the plane thermoelastic problem in the rectangular region (plate or cylinder) with a rectangular hole and having the temperature-dependent material properties.

The stress-strain relations are given by

$$\begin{aligned} \varepsilon_{xx}&=[\sigma_{xx}-\nu_e(T)\sigma_{yy}]/E_e(T)+\Psi_e(T) \\ \varepsilon_{yy}&=[\sigma_{yy}-\nu_e(T)\sigma_{xx}]/E_e(T)+\Psi_e(T) \\ \varepsilon_{xy}&=[1+\nu_e(T)]\sigma_{xy}/E_e(T) \end{aligned} \tag{8}$$

where

$$E_e(T)=E(T), \quad \nu_e(T)=\nu(T), \quad \Psi_e(T)=\int^T\alpha(\rho)d\rho, \qquad \textit{for plane stress}$$

$$E_e(T)=E(T)/[1-\nu^2(T)], \quad \nu_e(T)=\nu(T)/[1-\nu(T)], \quad \Psi_e(T)=[1+\nu(T)]\int^T\alpha(\rho)d\rho \tag{9}$$

for plane strain

and $E(T)$, $\nu(T)$ and $\alpha(T)$ are the Young's modulus, Poisson's ratio and linear thermal

expansion coefficient, respectively.
In the absence of body forces, the equations of equilibrium are

$$\frac{\partial \sigma_{xx}}{\partial x} + \frac{\partial \sigma_{xy}}{\partial y} = 0, \qquad \frac{\partial \sigma_{xy}}{\partial x} + \frac{\partial \sigma_{yy}}{\partial y} = 0. \tag{10}$$

The compatibility equation for strain components is

$$\frac{\partial^2 \varepsilon_{xx}}{\partial y^2} + \frac{\partial^2 \varepsilon_{yy}}{\partial x^2} = 2\frac{\partial^2 \varepsilon_{xy}}{\partial x \partial y}. \tag{11}$$

The equilibrium equations of (10) are identically satisfied by expressing the stress components in terms of Airy's stress function as follows:

$$\sigma_{xx} = \frac{\partial \chi^2}{\partial y^2}, \qquad \sigma_{yy} = \frac{\partial \chi^2}{\partial x^2}, \qquad \sigma_{xy} = -\frac{\partial^2 \chi}{\partial x \partial y}. \tag{12}$$

Substituting eqs.(8) into eq.(11) and then expressing the stress components in terms of stress function by the use of eqs.(12), we may obtain the following fundamental differential equation for the plane thermoelastic problem having the temperature dependencies of material properties.

$$\begin{aligned}
&\nabla^4\chi - \frac{2}{E_e(T)}\Big[\frac{\partial E_e(T)}{\partial x}\frac{\partial}{\partial x}(\nabla^2\chi) + \frac{\partial E_e(T)}{\partial y}\frac{\partial}{\partial y}(\nabla^2\chi)\Big] + \Big\{\frac{1}{E_e(T)}\Big[\frac{2}{E_e(T)}\Big(\frac{\partial E_e(T)}{\partial x}\Big)^2 \\
&- \frac{\partial^2 E_e(T)}{\partial x^2}\Big] - \frac{\nu_e(T)}{E_e(T)}\Big[\frac{2}{E_e(T)}\Big(\frac{\partial E_e(T)}{\partial y}\Big)^2 - \frac{\partial^2 E_e(T)}{\partial y^2}\Big] - \frac{\partial^2 \nu_e(T)}{\partial y^2} + \frac{2}{E_e(T)}\frac{\partial \nu_e(T)}{\partial y} \\
&\times \frac{\partial E_e(T)}{\partial y}\Big\}\frac{\partial^2\chi}{\partial x^2} + \Big\{\frac{1}{E_e(T)}\Big[\frac{2}{E_e(T)}\Big(\frac{\partial E_e(T)}{\partial y}\Big)^2 - \frac{\partial^2 E_e(T)}{\partial y^2}\Big] - \frac{\nu_e(T)}{E_e(T)}\Big[\frac{2}{E_e(T)}\Big(\frac{\partial E_e(T)}{\partial x}\Big)^2 \\
&- \frac{\partial^2 E_e(T)}{\partial x^2}\Big] - \frac{\partial^2 \nu_e(T)}{\partial x^2} + \frac{2}{E_e(T)}\frac{\partial \nu_e(T)}{\partial x}\frac{\partial E_e(T)}{\partial x}\Big\}\frac{\partial^2\chi}{\partial y^2} + 2\Big\{\frac{[1+\nu_e(T)]}{E_e(T)}\Big[\frac{2}{E_e(T)}\frac{\partial E_e(T)}{\partial x} \\
&\times \frac{\partial E_e(T)}{\partial y} - \frac{\partial^2 E_e(T)}{\partial x \partial y}\Big] + \frac{\partial^2 \nu_e(T)}{\partial x \partial y} - \frac{1}{E_e(T)}\frac{\partial \nu_e(T)}{\partial y}\frac{\partial E_e(T)}{\partial x} - \frac{1}{E_e(T)}\frac{\partial \nu_e(T)}{\partial x}\frac{\partial E_e(T)}{\partial y}\Big\}\frac{\partial^2\chi}{\partial x \partial y} \\
&= -E_e(T)\nabla^2\Psi_e(T)
\end{aligned} \tag{13}$$

where $\nabla^2 = \partial^2/\partial x^2 + \partial^2/\partial y^2$.

In the absence of external forces, the mechanical boundary conditions in terms of stress function may be expressed as

$$\chi = 0, \quad \partial\chi/\partial n' = 0 \qquad \text{on outer boundary} \tag{14}$$

$$(\chi)_{p_i} = c_1 x_{p_i} + c_2 y_{p_i} + c_3, \quad (\partial\chi/\partial n')_{p_i} = c_1 \cos(x,n')_{p_i} + c_2 \cos(y,n')_{p_i} \qquad \text{on inner boundary} \tag{15}$$

where n' is an arbitrary direction which does not coincide with the tangential direction to the boundary curve and p_i is an arbitrary point on the boundary curve. The unknown constants c_1, c_2 and c_3 included in the boundary conditions must be determined from physical consideration.

Michell (1899) has pointed out that the physical conditions are obtained from the consideration of single-valuedness of rotation and displacements. However, there seems to be no work on the conditions which must be considered to guarantee the single-valuedness of rotation and displacements in the multiply-connected region

exhibiting the temperature dependencies of material properties.
In this paper, therefore, we derived the conditions in terms of Airy's stress function as follows:

$$\oint_L \frac{\partial}{\partial n}\left[\frac{1}{E_e(T)}\nabla^2\chi + \Psi_e(T)\right]ds + \oint_L \frac{\partial}{\partial y}\left[\frac{1+\nu_e(T)}{E_e(T)}\right] d\left(\frac{\partial\chi}{\partial x}\right)$$

$$- \oint_L \frac{\partial}{\partial x}\left[\frac{1+\nu_e(T)}{E_e(T)}\right] d\left(\frac{\partial\chi}{\partial y}\right) = 0 \qquad (16)$$

$$\oint_L \left(x\frac{\partial}{\partial s} - y\frac{\partial}{\partial n}\right)\left[\frac{1}{E_e(T)}\nabla^2\chi + \Psi_e(T)\right]ds + \oint_L \left[\frac{1+\nu_e(T)}{E_e(T)}\right] d\left(\frac{\partial\chi}{\partial x}\right)$$

$$+ \oint_L y\frac{\partial}{\partial x}\left[\frac{1+\nu_e(T)}{E_e(T)}\right] d\left(\frac{\partial\chi}{\partial y}\right) - \oint_L y\frac{\partial}{\partial y}\left[\frac{1+\nu_e(T)}{E_e(T)}\right] d\left(\frac{\partial\chi}{\partial x}\right) = 0 \qquad (17)$$

$$\oint_L \left(y\frac{\partial}{\partial s} + x\frac{\partial}{\partial n}\right)\left[\frac{1}{E_e(T)}\nabla^2\chi + \Psi_e(T)\right]ds + \oint_L \left[\frac{1+\nu_e(T)}{E_e(T)}\right] d\left(\frac{\partial\chi}{\partial y}\right)$$

$$- \oint_L x\frac{\partial}{\partial x}\left[\frac{1+\nu_e(T)}{E_e(T)}\right] d\left(\frac{\partial\chi}{\partial y}\right) + \oint_L x\frac{\partial}{\partial y}\left[\frac{1+\nu_e(T)}{E_e(T)}\right] d\left(\frac{\partial\chi}{\partial x}\right) = 0 \qquad (18)$$

where L is a closed integral path including the inner boundary curve, n is the normal direction to L and s represents the arc length along the curve L.
Finally, the plane thermoelastic problem in the multiply-connected region exhibiting the temperature dependencies of material properties has been formulated in terms of Airy's stress function by eqs.(12)-(18).
In general, it is difficult to obtain the exact solution to the plane thermoelastic problem in the rectangular region with a rectangular hole even if the material properties are independent of the temperature. Hence we now present how a system of the fundamental equations of the plane thermoelastic problem in the multiply-connected region exhibiting the temperature dependencies of material properties is solved by the use of the finite difference method.

FINITE DIFFERENCE FORMULATION

We take the finite difference grids with spatial intervals Δx and Δy in the x and y directions, respectively and Δt as the time interval, and use the subscripts i and j and the superscript k to denote the quantities of the i-th and j-th discrete points in the x and y directions, respectively and the k-th discrete time.
Then, we may obtain the finite difference forms of eqs.(1)-(6) by replacing the partial derivatives by the central differences in the spatial coordinates and the pure implicit method in time as follows:

$$\left[(\lambda^{*k}_{i+1,j}+2\lambda^{*k}_{i,j}+\lambda^{*k}_{i-1,j})/(\Delta\xi)^2 + (\lambda^{*k}_{i,j+1}+2\lambda^{*k}_{i,j}+\lambda^{*k}_{i,j-1})/(\Delta\eta)^2 + (\rho^{*k}_{i,j}+\rho^{*k-1}_{i,j})/\Delta\tau\right.$$

$$\left.+4\delta_{0l}m/\beta\right]T^k_{i,j} - (\lambda^{*k}_{i+1,j}+\lambda^{*k}_{i,j})T^k_{i+1,j}/(\Delta\xi)^2 - (\lambda^{*k}_{i,j}+\lambda^{*k}_{i-1,j})T^k_{i-1,j}/(\Delta\xi)^2$$

$$- (\lambda^{*k}_{i,j+1}+\lambda^{*k}_{i,j})T^k_{i,j+1}/(\Delta\eta)^2 - (\lambda^{*k}_{i,j}+\lambda^{*k}_{i,j-1})T^k_{i,j-1}/(\Delta\eta)^2$$

$$= (\rho^{*k}_{i,j}+\rho^{*k-1}_{i,j})T^{k-1}_{i,j}/\Delta\tau + 4\delta_{0l}mT_0/\beta \qquad (19)$$

$$T^0_{i,j} = T_0, \qquad \tau=0 \qquad (20)$$

$$(\lambda^{*k}_{i+1,j}+\lambda^{*k}_{i-1,j})(T^k_{i+1,j}-T^k_{i-1,j})/4\Delta\xi + m_0(T^k_{i,j}-T_0)=0, \quad \xi=1 \tag{21}$$

$$(\lambda^{*k}_{i,j+1}+\lambda^{*k}_{i,j-1})(T^k_{i,j+1}-T^k_{i,j-1})/4\Delta\eta + m_0(T^k_{i,j}-T_0)=0, \quad \eta=\bar{l}_y \tag{22}$$

$$(\lambda^{*k}_{i+1,j}+\lambda^{*k}_{i-1,j})(T^k_{i+1,j}-T^k_{i-1,j})/4\Delta\xi - m_1(T^k_{i,j}-T_1)=0, \quad \xi=\bar{l}'_x \tag{23}$$

$$(\lambda^{*k}_{i,j+1}+\lambda^{*k}_{i,j-1})(T^k_{i,j+1}-T^k_{i,j-1})/4\Delta\eta - m_1(T^k_{i,j}-T_1)=0, \quad \eta=\bar{l}'_y \tag{24}$$

where $\lambda(T)$, $c(T)$ and $\gamma(T)$ were disjointed into two factors, the first one dimensional and invariant, denoted by the subscript nought, and the second one dimensionless and a function of temperature, denoted by an asterisk:

$$\lambda(T)=\lambda_0\lambda^*(T), \qquad c(T)=c_0c^*(T), \qquad \gamma(T)=\gamma_0\gamma^*(T) \tag{25}$$

and the following dimensionless quantities were introduced:

$$\begin{aligned}&\xi=x/l_x, \quad \eta=y/l_x, \quad \tau=\kappa_0t/l_x^2, \quad \kappa_0=\lambda_0/(c_0\gamma_0), \quad m=hl_x/\lambda_0, \quad m_0=h_0l_x/\lambda_0\\ &m_1=h_1l_x/\lambda_0, \quad \bar{l}_y=l_y/l_x, \quad \bar{l}'_x=l'_x/l_x, \quad \bar{l}'_y=l'_y/l_x, \quad \beta=b/l_x,\\ &\rho^*(T)=c^*(T)\gamma^*(T)\end{aligned} \tag{26}$$

In eqs.(19)-(24), the superscript k denotes the unknown values at the current dimensionless time $\tau=\tau_k$ and the superscript $k-1$ the known ones at $\tau=\tau_{k-1}$.

The trouble in the nonlinear heat-conduction problem described above is that at the current time $\tau=\tau_k$ the values of the thermal properties $\lambda^{*k}_{i,j}$ and $\rho^{*k}_{i,j}$ are unknown together with the temperature at all grid points. We used an iterative procedure because of the inherent difficulty. At the first step we use $\lambda^{*k-1}_{i,j}$ and $\rho^{*k-1}_{i,j}$ at $\tau=\tau_{k-1}$ as first approximation values of the thermal properties at $\tau=\tau_k$ and solve the simultaneous equations for the grid-point temperatures obtained from the application of eq.(19) to all grid points including the grid points on the boundaries. At the second step we recompute the values of thermal properties at $\tau=\tau_k$ by substituting the temperature founded at the first step into eqs.(25), and then solve the simultaneous equations to obtain the improve values of the temperature at $\tau=\tau_k$. This process is continued until the sufficient convergence to the grid-point temperatures is obtained, and then we proceed to the next time step.

The fundamental differential equation (13) for the plane thermoelastic problem may be expressed in the finite difference form as follows:

$$\begin{aligned}&C_1\tilde{\chi}_{i,j} + C_2\tilde{\chi}_{i+1,j} + C_3\tilde{\chi}_{i-1,j} + C_4\tilde{\chi}_{i,j+1} + C_5\tilde{\chi}_{i,j-1} + C_6\tilde{\chi}_{i+2,j} + C_7\tilde{\chi}_{i-2,j}\\ &+ C_8\tilde{\chi}_{i,j+2} + C_9\tilde{\chi}_{i,j-2} + C_{10}\tilde{\chi}_{i+1,j+1} + C_{11}\tilde{\chi}_{i+1,j-1} + C_{12}\tilde{\chi}_{i-1,j+1}\\ &+ C_{13}\tilde{\chi}_{i-1,j-1} = -F(T_{i,j})\end{aligned} \tag{27}$$

where $\tilde{\chi}=\chi/l_x^2$ and C_1 - C_{13} are typically

$$\begin{aligned}C_1=&\ 2[3(\Delta\xi)^4+3(\Delta\eta)^4+4(\Delta\xi)^2(\Delta\eta)^2]-2[b_3(\Delta\xi)^4+b_2(\Delta\eta)^4+(b_1+b_4)(\Delta\xi)^2(\Delta\eta)^2]\\ C_2=&-4(\Delta\eta)^2[(\Delta\xi)^2+(\Delta\eta)^2]+(\Delta\eta)^2[(b_1-b_5)(\Delta\xi)^2+(b_2-b_5)(\Delta\eta)^2]\\ C_3=&-4(\Delta\eta)^2[(\Delta\xi)^2+(\Delta\eta)^2]+(\Delta\eta)^2[(b_1+b_5)(\Delta\xi)^2+(b_2+b_5)(\Delta\eta)^2]\\ b_1=&-\nu_{ei,j}(E_{ei,j+1}-E_{ei,j-1})^2/2E^2_{ei,j}+(E_{ei,j+1}-E_{ei,j-1})(\nu_{ei,j+1}-\nu_{ei,j-1})\\ &/2E_{ei,j}+\nu_{ei,j}(E_{ei,j+1}+E_{ei,j-1}-2E_{ei,j})/E_{ei,j}-\nu_{ei,j+1}-\nu_{ei,j-1}+2\nu_{ei,j}\end{aligned}$$

$$
\left.
\begin{aligned}
& b_2=(E_{ei+1,j}-E_{ei-1,j})^2/2E^2_{ei,j}-(E_{ei+1,j}+E_{ei-1,j}-2E_{ei,j})/E_{ei,j} \\
& b_3=(E_{ei,j+1}-E_{ei,j-1})^2/2E^2_{ei,j}-(E_{ei,j+1}+E_{ei,j-1}-2E_{ei,j})/E_{ei,j} \\
& F(T_{i,j})=E_{ei,j}(\Delta\xi)^2(\Delta\eta)^2[(\alpha_{i+1,j}-\alpha_{i-1,j})(T_{i+1,j}-T_{i-1,j})(\Delta\eta)^2/4 \\
& \qquad +(\alpha_{i,j+1}-\alpha_{i,j-1})(T_{i,j+1}-T_{i,j-1})(\Delta\xi)^2/4+\alpha_{i,j}(T_{i+1,j}+T_{i-1,j}-2T_{i,j}) \\
& \qquad \times(\Delta\eta)^2+\alpha_{i,j}(T_{i,j+1}+T_{i,j-1}-2T_{i,j})(\Delta\xi)^2] \qquad \textit{for plane stress} \\
& F(T_{i,j})=E_{ei,j}(\Delta\xi)^2(\Delta\eta)^2\{[(\nu_{i+1,j}+\nu_{i-1,j}-2\nu_{i,j})(\Delta\eta)^2+(\nu_{i,j+1}+\nu_{i,j-1}-2\nu_{i,j}) \\
& \qquad \times(\Delta\xi)^2](\textstyle\int^T\alpha(\rho)d\rho)_{i,j}+\alpha_{i,j}[(\nu_{i+1,j}-\nu_{i-1,j})(T_{i+1,j}-T_{i-1,j})(\Delta\eta)^2 \\
& \qquad +(\nu_{i,j+1}-\nu_{i,j-1})(T_{i,j+1}-T_{i,j-1})(\Delta\xi)^2]/2+(1+\nu_{i,j})[(\alpha_{i+1,j}-\alpha_{i-1,j}) \\
& \qquad \times(T_{i+1,j}-T_{i-1,j})(\Delta\eta)^2+(\alpha_{i,j+1}-\alpha_{i,j-1})(T_{i,j+1}-T_{i,j-1})(\Delta\xi)^2]/4 \\
& \qquad +(1+\nu_{i,j})\alpha_{i,j}[(T_{i+1,j}+T_{i-1,j}-2T_{i,j})(\Delta\eta)^2+(T_{i,j+1}+T_{i,j-1}-2T_{i,j}) \\
& \qquad \times(\Delta\xi)^2]\} \qquad \textit{for plane strain}
\end{aligned}
\right\} \qquad (28)
$$

Since the plane thermoelastic problem considered here is symmetric with respect to the x and y axes, the conditions (17) and (18) of the single-valuedness of the displacements u_x and u_y are identically satisfied and the unknown constants c_1 and c_2 in eqs.(15) become zero because of the symmetry of Airy's stress function. If we take the integral path as shown in Fig.2, the remaining condition (16) necessary for the assurance of single-valuedness of the rotation may be expressed as

$$
\begin{aligned}
& \int_{S_1}^{S_2}(W_1\tilde{\chi}_{i,j}+W_2\tilde{\chi}_{i+1,j}+W_3\tilde{\chi}_{i-1,j}+W_4\tilde{\chi}_{i,j+1}+W_5\tilde{\chi}_{i,j-1}+W_6\tilde{\chi}_{i+1,j+1}+W_7\tilde{\chi}_{i+1,j-1} \\
& +W_8\tilde{\chi}_{i-1,j+1}+W_9\tilde{\chi}_{i-1,j-1}+W_{10}\tilde{\chi}_{i+2,j}+W_{11}\tilde{\chi}_{i-2,j}+F_1)d\eta \\
& +\int_{S_3}^{S_2}(V_1\tilde{\chi}_{i,j}+V_2\tilde{\chi}_{i+1,j}+V_3\tilde{\chi}_{i-1,j}+V_4\tilde{\chi}_{i,j+1}+V_5\tilde{\chi}_{i,j-1}+V_6\tilde{\chi}_{i+1,j+1}+V_7\tilde{\chi}_{i+1,j-1} \\
& +V_8\tilde{\chi}_{i-1,j+1}+V_9\tilde{\chi}_{i-1,j-1}+V_{10}\tilde{\chi}_{i,j+2}+V_{11}\tilde{\chi}_{i,j-2}+F_2)d\xi=0
\end{aligned}
\qquad (29)
$$

where

$$
\left.
\begin{aligned}
& W_1=f_1/\Delta\eta+f_3f_4/\Delta\xi,\quad W_2=-[f_3+f_4/2E_{ei,j}(\Delta\xi)^2]/\Delta\xi,\quad W_3=[f_3-f_4/2E_{ei,j}(\Delta\xi)^2]/\Delta\xi \\
& W_4=-(f_1+f_4/E_{ei,j}\Delta\xi\Delta\eta)/2\Delta\eta,\quad W_6=(1/E_{ei,j}\Delta\xi\Delta\eta+f_2/4)/2\Delta\eta,\quad W_5=W_4 \\
& W_7=(1/E_{ei,j}\Delta\xi\Delta\eta-f_2/4)/2\Delta\eta,\quad W_8=-W_6,\quad W_9=-W_7,\quad W_{10}=1/2E_{ei,j}(\Delta\xi)^3,\quad W_{11}=-W_{10} \\
& V_1=f_2/\Delta\xi+f_3f_5/\Delta\eta,\quad V_2=-(f_2+f_5/E_{ei,j}\Delta\xi\Delta\eta)/2\Delta\xi,\quad V_3=V_2, \\
& V_4=-[f_3+f_5/2E_{ei,j}(\Delta\eta)^2]/\Delta\eta,\quad V_5=[f_3-f_5/2E_{ei,j}(\Delta\eta)^2]/\Delta\eta, \\
& V_6=(1/E_{ei,j}\Delta\xi\Delta\eta+f_1/4)/2\Delta\xi,\quad V_7=-V_6,\quad V_8=(1/E_{ei,j}\Delta\xi\Delta\eta-f_1/4)/2\Delta\xi \\
& V_9=-V_8,\quad V_{10}=1/2E_{ei,j}(\Delta\eta)^3,\quad V_{11}=-V_{10} \\
& f_1=[\nu_{ei+1,j}-\nu_{ei-1,j}-(1+\nu_{ei,j})(E_{ei+1,j}-E_{ei-1,j})/E_{ei,j}]/E_{ei,j}\Delta\xi\Delta\eta \\
& f_2=[\nu_{ei,j+1}-\nu_{ei,j-1}-(1+\nu_{ei,j})(E_{ei,j+1}-E_{ei,j-1})/E_{ei,j}]/E_{ei,j}\Delta\xi\Delta\eta \\
& f_3=[(\Delta\xi)^2+(\Delta\eta)^2]/E_{ei,j}(\Delta\xi)^2(\Delta\eta)^2,\quad f_4=(E_{ei+1,j}-E_{ei-1,j})/E_{ei,j} \\
& f_5=(E_{ei,j+1}-E_{ei,j-1})/E_{ei,j},\quad F_1=\alpha_{i,j}(T_{i+1,j}-T_{i-1,j})/2\Delta\xi \\
& F_2=\alpha_{i,j}(T_{i,j+1}-T_{i,j-1})/2\Delta\eta, \qquad \textit{for plane stress} \\
& F_1=\{(\nu_{i+1,j}-\nu_{i-1,j})[\textstyle\int^T\alpha(\rho)d\rho]_{i,j}+\alpha_{i,j}(1+\nu_{i,j})(T_{i+1,j}-T_{i,j-1})\}/2\Delta\xi\Delta\eta
\end{aligned}
\right|
$$

$$F_2=\{(\nu_{i,j+1}-\nu_{i,j-1})[\int^T\alpha(\rho)d\rho]_{i,j}+\alpha_{i,j}(1+\nu_{i,j})(T_{i,j+1}-T_{i,j-1})\}/2\Delta\eta \quad \textit{for plane strain} \tag{30}$$

The integration of eq.(29) is evaluated numerically by means of Simpson's rule.

Finally, to solve the plane thermoelastic problem formulated in terms of Airy's stress function in the rectangular region with a rectangular hole and having the temperature-dependent material properties by the finite difference method is reduced to the determination of the values of the stress function at all grid points excluding those on the boundaries and the unknown constant c_3 by solving the simultaneous equations obtained from the application of eq.(27) to only the grid points in the interior of the region and the numerical integration of eq.(29) along the integral path L. Once the values of the stress function are determined, the desired thermal stress components can be easily calculated by the following finite difference representation of eq.(12).

$$\begin{aligned}(\sigma_{xx})_{i,j}&=(\tilde{\chi}_{i,j+1}+\tilde{\chi}_{i,j-1}-2\tilde{\chi}_{i,j})/(\Delta\eta)^2, \quad (\sigma_{yy})_{i,j}=(\tilde{\chi}_{i+1,j}+\tilde{\chi}_{i-1,j}-2\tilde{\chi}_{i,j})/(\Delta\xi)^2\\ (\sigma_{xy})_{i,j}&=-(\tilde{\chi}_{i+1,j+1}-\tilde{\chi}_{i+1,j-1}-\tilde{\chi}_{i-1,j+1}+\tilde{\chi}_{i-1,j-1})/(4\Delta\xi\Delta\eta)\end{aligned} \tag{31}$$

NUMERICAL CALCULATIONS AND DISCUSSION

Numerical calculations were carried out for the temperature and the plane thermal stresses in the rectangular plate with a rectangular hole made of the aluminium alloy having the following material properties (Morinaga(1970), JSME(1980))

$$\rho(T)=c(T)\gamma(T)=0.235\times10^3+0.157\times10T \ (KJ/Kg.K)$$
$$\lambda(T)=0.247-0.122\times10^{-3}T+0.123\times10^{-6}T^2 \ (KW/m.K)$$
$$E(T)=0.724\times10^2-0.156\times10^{-1}T-0.684\times10^{-4}T^2 \ (GPa)$$
$$\alpha(T)=0.228\times10^{-4}+0.183\times10^{-7}T \ (1/K)$$
$$\nu(T)=0.34 .$$

In addition, the following values were used

$$\bar{l}_y=1, \quad \bar{l}'_x=\bar{l}'_y=0.5, \quad m_0=1, \quad m_1=\infty, \quad m=0, \quad T_0=0^\circ c, \quad T_1=80^\circ c$$

$$\beta=0.05, \quad \Delta\xi=\Delta\eta=0.03125,\ 0.0625, \quad \Delta\tau=0.001\sim0.01.$$

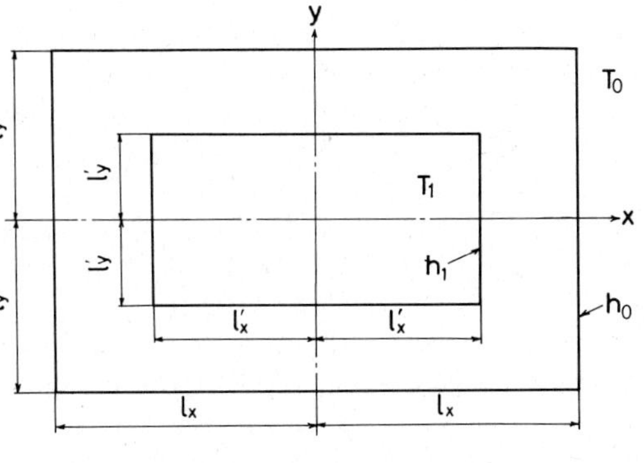

Figure 1. Rectangular region with a rectangular hole.

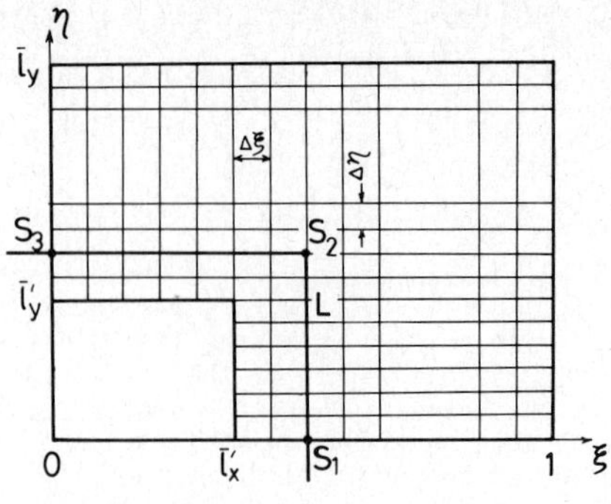

Figure 2. Integral path L including inner boundary.

In the following figures, the solid lines and the dashed lines indicate the distributions of the transient temperature and thermal stress in the y direction calculated by the use of the temperature-dependent material properties and the material constants at the initial temperature, respectively.

First, figure 3 shows the temperature distribution on the x axis. As illustraled in this figure, when the temperature dependencies of the thermal properties are taken into account, the temperature is lower than that calculated by the use of the thermal properties at the initial temperature because the thermal conductivity of aluminium alloy becomes smaller with the increase of the temperature rise; while the temperature gradient is larger because the inner boundary is kept at the constant temperature. However, in general, the effect of the temperature dependencies of the material properties on the temperature distribution is small.

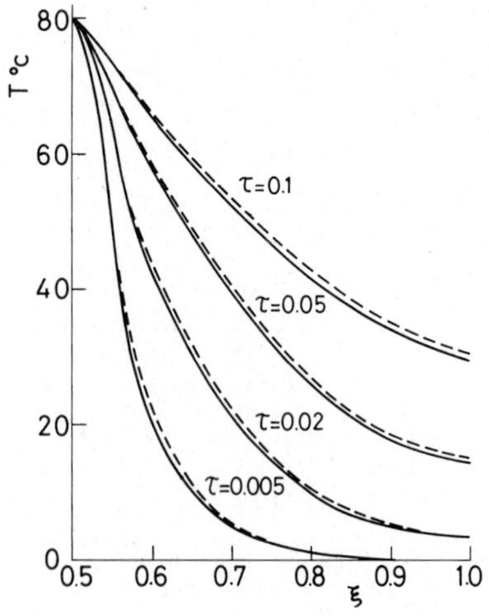

Figure 3. Temperature distribution on x axis.

Figures 4 and 5 show the distributions of σ_{yy} on the x axis and the outer boundary $\xi=1$, respectively. It is found from figure 4 that the absolute value of the compressive thermal stress on the inner boundary calculated by the use of the temperature-dependent material properties is larger than that obtained by using the material properties at the initial temperature. This is due to that the temperature gradient becomes larger by taking into account the temperature dependencies of the thermal properties, and that the value of the product of Young's modulus and the linear thermal expansion coefficient with the temperature rise becomes larger.

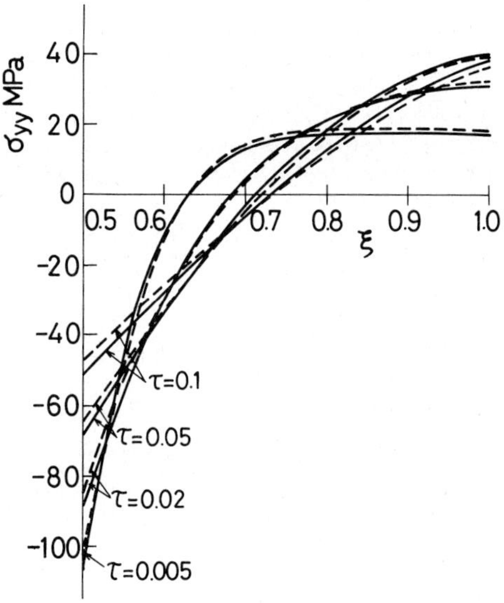

Figure 4. Distribution of σ_{yy} on x axis.

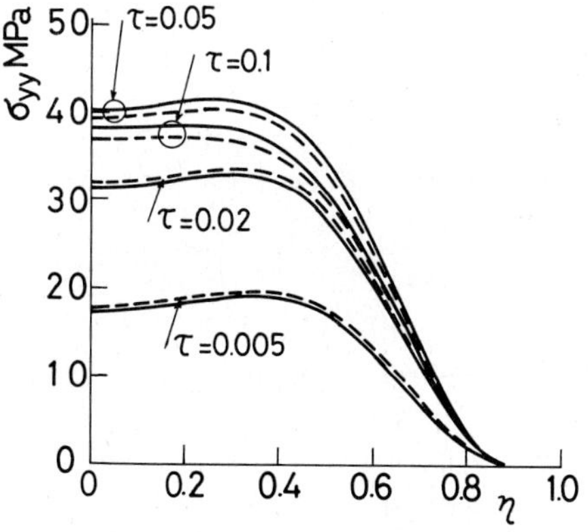

Figure 5. Distribution of σ_{yy} on outer boundary.

REFERENCES:

[1] Takeuti, Y. and Sekiya, T., Thermal stresses in a polygonal cylinder with a circular hole under internal heat generation, Z.A.M.M. 48 (1968) 237-246.

[2] Hulbert, L.E., Solution of thermal stress problems in tube sheets by the boundary point least squares method, J. Eng. for Industry, Trans. ASME 92 (1970) 339-349.

[3] Takeuti, Y. and Noda, N., Transient thermoelastic problem in a polygonal cylinder with a circular hole, J. Appl. Mech. 40 (1973) 935-940.

[4] Michell, J.H., On the direct determination of stress in an elastic solid, with application to the theory of plates, Proc. of London Math. Soc. 31 (1899) 100-124.

[5] Morinaga, T.(ed.), Handbook of Aluminium Working Technology (Nikkankogyo, Tokyo 1970).

[6] The Japan Society of Mechanical Engineers, The Modulus of Elasticity of Metals and Alloys (JSME, Tokyo, 1980).

TURBULENCE, ACOUSTICS, PLASMAS

Computational Techniques & Applications: CTAC-83
J. Noye & C. Fletcher (Editors)

APPLICATION OF THE DORODNITSYN BOUNDARY LAYER FORMULATION TO WALL BLOWING

R.W. Fleet and C.A.J. Fletcher

University of Sydney
Sydney, NSW 2006,
Australia

The Dorodnitsyn boundary layer formulation is extended to include the effects of wall blowing (and suction) in the normal direction. Results are obtained for both the finite element and spectral Dorodnitsyn formulations and compared with a typical finite difference package, STAN5, and experimental results due to McQuaid (1967). For the zero-blowing case all three methods produce accurate results but the finite element formulation is an order-of-magnitude more economical than the other methods. For the blowing case the best agreement is obtained between the experimental results and the spectral Dorodnitsyn formulation.

1. INTRODUCTION

Boundary layer flows are governed by partial differential equations that are parabolic in character. As a result the most successful computational algorithms have been of a marching type. Very often the governing equations have been reduced to a system of ordinary differential equations as the first stage of the solution process. This initial discretisation may be due to the finite difference method (Blottner, 1975), finite element method (Fletcher and Fleet, 1983a), spectral method (Fletcher, 1978) or method of integral relations (Holt, 1977).

Whereas finite difference methods typically treat the velocity components u and v as the dependent variables and the co-ordinates x and y (in two dimensions) as the independent variables there are many advantages in adopting a Dorodnitsyn formulation which uses a non-dimensional normal velocity gradient as the dependent variable and x and u as the independent variables.

An immediate computational advantage is that an infinite domain in the y direction is replaced by a finite domain in u; u is scaled to vary between zero and unity in traversing the boundary layer. The scaling of u means that the grid automatically captures the boundary layer growth in the downstream direction. In (x,y) space periodic readjustment of the boundary layer grid at the downstream stations is computationally expensive.

In the Dorodnitsyn formulation it is convenient to specify a uniform grid in the u direction. For the finite element Dorodnitsyn formulation this permits a higher accuracy to be achieved. In contrast in physical space a non-uniform grid is invariably required which implies, for the finite difference or finite element method, a larger truncation error than if a uniform grid is used.

The use of a uniform grid in u-space in the Dorodnitsyn formulation provides high resolution in physical space adjacent to the wall. This is indicated for a typical laminar velocity profile in Fig. 1a. However the high resolution is even more important for a typical turbulent velocity profile (Fig. 1b).

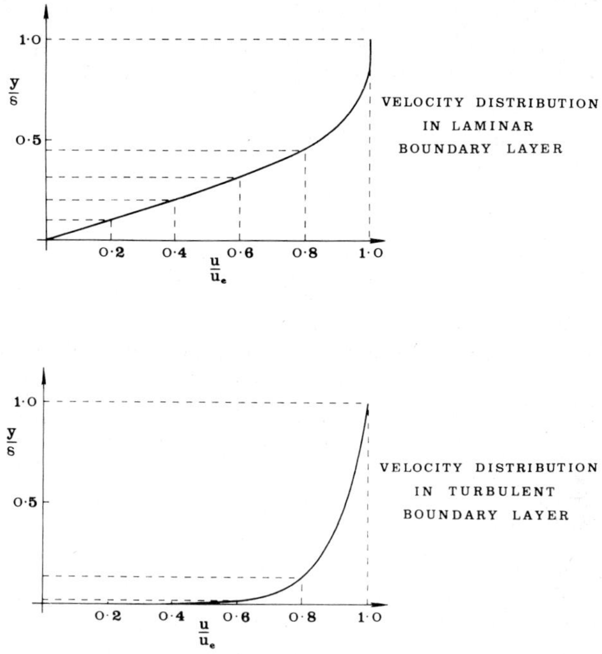

Figure 1
Dorodnitsyn grid for laminar and turbulent velocity profiles

For two-dimensional flows the Dorodnitsyn formulation offers the additional advantage of avoiding the explicit appearance of the normal velocity component, v. Although it can be recovered if required. Consequently only one equation is solved with the Dorodnitsyn formulation.

By choosing the non-dimensional velocity gradient as the dependent variable the shear stress is computed accurately. This is particularly important in determining the skin friction behaviour.

The above advantages have been exploited for both spectral and finite element Dorodnitsyn formulations in laminar (Fletcher and Holt, 1976; Fleet and Fletcher, 1982) and turbulent (Fletcher and Fleet, 1983b ; Fleet and Fletcher, 1983) boundary layer flow. It has also been shown that the Dorodnitsyn boundary layer formulation is effective in analysing the turbulent boundary layers that form in axisymmetric diffusers when swirl is present in the main flow (Fletcher, 1983a).

In this paper the Dorodnitsyn formulation is extended to include flows with blowing or suction at the wall in the normal direction. In Section 2 the Dorodnitsyn formulation is developed for both the spectral and finite element methods. In Section 3 results are presented and discussed for turbulent boundary layer flows in both the blowing and zero-blowing cases. In each case solutions obtained with the spectral and finite element formulations are compared with solutions obtained with a representative finite difference code, STAN5, and with experimental results.

2. DORODNITSYN FORMULATIONS

In this section the Dorodnitsyn boundary layer formulation is described for two-dimensional flow. Subsequently the specific modifications required by the spectral and finite element Dorodnitsyn formulations are introduced.

The equations governing steady two-dimensional turbulent boundary layer flow can be written

$$\partial u/\partial x + \partial v/\partial y = 0 \tag{1}$$

$$u\partial u/\partial x + v\partial u/\partial y = u_e \partial u_e/\partial x + \frac{1}{Re}\,\partial\left\{(1 + \nu_T/\nu)\partial u/\partial y\right\}/\partial y \tag{2}$$

Equation (1) is the continuity equation and eq. (2) is the x-momentum equation. In eq. (2) ν_T is an eddy viscosity. The expression $\rho\nu_T\partial u/\partial y$ replaces the Reynolds stress, $-\rho\overline{u'v'}$. The equation system, (1) and (2) is parabolic and requires initial conditions

$$u(x_o,y) = u_i(y) \quad \text{and} \quad v(x_o y) = v_i(y) \tag{3}$$

and boundary conditions,

$$u(x,0) = \quad , \quad v(x,0) = v_w \quad \text{and} \quad u(x,\infty) = u_e(x) \quad , \tag{4}$$

where v_w is the prescribed normal velocity at the wall.

In eqs. (2) and (4), $u_e(x)$ is the known velocity distribution at the outer edge of the boundary layer. In eqs. (1) to (4) u, u_e and v have been nondimensionalised with a reference velocity u_∞ and x and y have been nondimensionalised with a reference length L. Consequently the Reynolds number, $Re = u_\infty L/\nu$.

Dorodnitsyn (1962) developed the present formulation in order to apply the method of integral relations (MIR) to problems involving a viscous boundary layer. Previously (see Holt, 1977) MIR had been applied to inviscid, compressible flow problems, often involving shock waves.

Dorodnitsyn introduced the following variables

$$\xi = \int_o^x u_e(x')dx' \quad , \quad \eta = Re^{\frac{1}{2}} u_e\, y$$

$$u' = u/u_e \quad , \quad v' = Re^{\frac{1}{2}}\, v/u_e \quad \text{and} \quad w = v' + \eta u'\{\partial u_e/d\xi\}/u_e$$

As a result eqs. (1) to (4) become

$$\frac{\partial u'}{\partial \xi} + \frac{\partial w}{\partial \eta} = 0 \tag{5}$$

$$u'\frac{\partial u'}{\partial \xi} + w\frac{\partial u'}{\partial \eta} = \frac{1}{u_e}\frac{\partial u_e}{\partial \xi} + \frac{\partial}{\partial \eta}\left\{(1+\nu_T/\nu)\,\frac{\partial u'}{\partial \eta}\right\} \tag{6}$$

with auxiliary conditions,

$$u' = 0 \quad , \quad w = Re^{\frac{1}{2}}\, v_w/u_e \text{ at } \eta = 0 \quad \text{and} \quad u' = 1 \text{ at } \eta = \infty \tag{7}$$

A general weight (test) function, $f_k(u')$, is incorporated into the product,

$$f_k \times \text{eq.}(5) + \frac{\partial f_k}{\partial u'} \times \text{eq.}(6) = 0 \quad ,$$

to give (after dropping the superscript ′)

$$\frac{\partial}{\partial \xi}(uf_k) + \frac{\partial}{\partial \eta}(wf_k) = \frac{1}{u_e}\frac{\partial u_e}{\partial \xi}\frac{\partial f_k}{\partial u}(1-u^2) + \frac{\partial f_k}{\partial u}\frac{\partial}{\partial \eta}\left\{(1+\nu_T/\nu)\frac{\partial u}{\partial \eta}\right\} \tag{8}$$

An integration with respect to η is made,

$$\frac{\partial}{\partial \xi}\int_0^\infty uf_k \, d\eta + \left[wf_k\right]_0^\infty = \frac{1}{u_e}\frac{du_e}{d\xi}\int_0^\infty \frac{df_k}{\partial u}(1-u^2)d\eta + \int_0^\infty \frac{df_k}{\partial u}\frac{\partial}{\partial \eta}\left\{(1+\frac{\nu_T}{\nu})\frac{\partial u}{\partial \eta}\right\} d\eta \tag{9}$$

The function, f_k, is chosen so that $f_k(\infty) = 0$. The final stage of the Dorodnitsyn formulation changes the variable of integration from η to u and introduces new dependent variables,

$$T = 1/\theta^* = \partial u/\partial \eta \ . \tag{10}$$

As a result eq. (9) becomes

$$\frac{\partial}{\partial \xi}\int_0^1 uf_k \, \theta^* du - \left[wf_k\right]_{u=0} = \frac{1}{u_e}\frac{du_e}{d\xi}\int_0^1 \frac{\partial f_k}{\partial u}(1-u^2)\,\theta^* du + \int_0^1 \frac{\partial f_k}{\partial u}\frac{\partial}{\partial u}\left\{(1+\nu_T/\nu)T\right\}du \tag{11}$$

Equation (11) is the Dorodnitsyn turbulent boundary layer formulation with a prescribed blowing or suction at the wall acting in the normal direction.

In the original Dorodnitsyn formulation trial solutions of the following form are introduced,

$$\theta^* = \frac{1}{(1-u)}(b_\circ + b_1 u \ldots b_{N-1} u^{N-1}) \tag{12}$$

$$\text{and} \qquad T = (1-u)(c_\circ + c_1 u \ldots c_{N-1} u^{N-1}) \tag{13}$$

The factor (1-u) in eqs. (12) and (13) ensures that $T = 0$ at $u = 1$. Substitution of eqs. (12) and (13) into eq. (11) with

$$f_k(u) = (1-u)^k \tag{14}$$

produces a system of ordinary differential equations for $b_k(\xi)$, etc. The original Dorodnitsyn formulation is effective as long as N is small in eqs. (12) and (13). However if N is large there is not much difference between f_k and f_{k+1} and the system of ordinary differential equations for $b_k(\xi)$ becomes progressively more ill-conditioned as N increases.

The spectral and finite element Dorodnitsyn formulations overcome this problem in different ways; these formulations are described in the next two sections.

2.1 Spectral Dorodnitsyn Formulation

In this formulation the trial functions, u^k, in eqs. (12) and (13) and the weight function, f_k, in eq. (14) are replaced by related *orthonormal* functions, $g_k(u)$. The orthonormal functions are constructed from the Dorodnitsyn functions, $f_k(u)$, as follows

$$g_k(u) = \sum_{r=1}^{k} B_{rk} f_r(u) \tag{15}$$

where the coefficients B_{rk} are evaluated via the Gram-Schmidt orthonormalisation process (Isaacson and Keller, 1966) so that

$$\begin{aligned}(g_j, g_k) = \int_0^1 g_j(u)\, g_k(u)\, u/(1-u)\, du &= 1 \quad \text{if} \quad j = k \\ &= 0 \quad \text{if} \quad j \neq k\end{aligned} \tag{16}$$

The trial solution, eq. (12), is replaced with

$$\theta = \frac{1}{(1-u)}\left\{ b_o + \sum_{j=1}^{N-1} b_j g_j(u) \right\} \tag{17}$$

With g_k replacing f_k in eq. (11), and substitution of eq. (17), the following is obtained

$$\frac{d}{d\xi}\int_o^1 \left[b_o + \sum_{j=1}^{N-1} b_j g_j(u)\right] g_k(u) \left\{u/(1-u)\right\} du = C_k \tag{18}$$

where

$$C_k = \left[w g_k\right]_{u=0} + \frac{1}{u_e}\frac{du_e}{d\xi}\int_o^1 \frac{dg_k}{du}(1-u^2)\,\theta^* du + \int_o^1 \frac{dg_k}{du}\frac{\partial}{\partial u}\left\{\frac{(1+\nu_T/\nu)}{\theta^*}\right\} du \tag{19}$$

However because $g_j(u)$ has been chosen to satisfy eq. (16), it is possible to simplify eq. (18) significantly, to give

$$V_k \frac{db_o}{d\xi} + \frac{db_k}{d\xi} = C_k \quad \text{for} \quad k = 1, \ldots N-1 \tag{20}$$

and

$$V_N \frac{db_o}{d\xi} = C_N \quad \text{for} \quad k = N \tag{21}$$

Using eqs. (20) and (21), the following explicit ordinary differential equation is obtained,

$$\frac{db_k}{d\xi} = C_k - C_N V_k/V_N \quad , \quad k = 1, \ldots N-1 \tag{22}$$

The coefficients, V_k, in eqs. (20) to (22) are given by

$$V_k = \int_0^1 g_k(u)\; u/(1-u)\; du\;. \tag{23}$$

The problem reduces to the solution of eqs. (21) and (22) which can be marched in the ξ direction using the variable step, variable order Gear method (Gear, 1971). At each step in the ξ direction, substitution of the b_k values into eq. (17) gives θ^* and hence T.

The spectral Dorodnitsyn formulation has been applied to incompressible (Fletcher and Holt, 1975) and compressible (Fletcher and Holt, 1976) laminar flows and to incompressible (Yeung and Yang, 1981; Fletcher and Fleet, 1982) and compressible (Fleet and Fletcher, 1983) turbulent flows.

2.2 Finite Element Dorodnitsyn Formulation

This formulation will be described for independent variables x and u, and dependent variables T and $\theta^*(\equiv 1/T)$ as before. Equation (11) then takes the form

$$\frac{\partial}{\partial x}\int_0^1 uf_k\,\theta^* du - Re^{\frac{1}{2}}\, v_w f_k(0) = \frac{1}{u_e}\frac{du_e}{dx}\int_0^1 \frac{df_k}{du}(1-u^2)\,\theta^*\, du$$

$$+\, u_e \int_0^1 \frac{df_k}{du}\frac{\partial}{\partial u}\left\{(1+\nu_T/\nu)T\right\} du \tag{24}$$

where v_w is the prescribed normal velocity at the wall.

Trial solutions, of the following form, are introduced for θ^* and $(1+\nu_T/\nu)T$.

$$\theta^* = \sum_{j=1}^{M} N_j(u)/(1-u)\theta_j(x) \tag{25}$$

and

$$(1+\nu_T/\nu)T = \sum_{j=1}^{M} (1-u)N_j(u)\;(1+\nu_T/\nu)_j\tau_j(x) \tag{26}$$

In eqs. (25) and (26) the factor (1-u) has been introduced to ensure that θ^* and T have the correct behaviour at the edge of the boundary layer. The terms $N_j(u)$ are one-dimensional shape functions, typically linear or quadratic.

It may be noted that in eq. (26) the trial solution has been introduced for groups of terms. This is an example of the *group* finite element formulation (Fletcher, 1983b). In the present problem the eddy viscosity, ν_T, is typically a complicated function of the velocity u and normal co-ordinate, η. The use of the group formulation allows ν_T to be evaluated at the nodes only. This provides a substantial contribution to the economy of the present formulation (see Section 3.1).

The test function, $f_k(u)$ in eq. (24), is represented by

$$f_k(u) = (1-u)N_k(u) \quad . \tag{27}$$

The form of $f_k(u)$ in eq. (27) ensures that $f_k(u) = 0$ at $u = 1.0$. This avoids the explicit appearance of v, evaluated at the outer edge of the boundary layer, in eq. (24).

Equations (25) to (27) are substituted into eq. (24) to produce a *modified Galerkin method* (Fletcher, 1983c). The various integrals can be evaluated to give the following system of ordinary differential equations for the nodal values τ_j and θ_j,

$$\sum_j CC_{kj} \frac{d\theta_j}{dx} = Re^{\frac{1}{2}} v_w \delta_{1k} + \frac{1}{u_e}\frac{du_e}{dx} \sum_j EF_{kj}\theta_j + u_e \sum_j AA_{kj}(1+\nu_T/\nu)_j \tau_j \tag{28}$$

where $\delta_{1k} = 1$ if $k = 1$

$= 0$ if $k \neq 1$.

The coefficients in eq. (28) can be evaluated, once and for all, from

$$CC_{kj} = \int_0^1 N_k N_j \, u \, du \; , \; EF_{kj} = \int_0^1 N_j \left\{\frac{dN_k}{du}(1-u) - N_k\right\}(1+u)\, du$$

$$\text{and } AA_{kj} = \int_0^1 \left(\frac{dN_j}{du}(1-u) - N_j\right)\left(\frac{dN_k}{du}(1-u) - N_k\right) du \quad . \tag{29}$$

The system of equations, (28), has a particularly compact form due to simultaneously prescribing trial solutions for θ_j and τ_j. However this feature prevents eq. (10) being satisfied except at the nodes, where $\theta_j = 1/\tau_j$, or in the limit $M \to \infty$.

To avoid a restriction on the step-size, Δx, it is convenient to construct an implicit algorithm to march eq. (28) downstream. Equation (28) is approximated by

$$\sum_j CC_{kj} \, \Delta\theta_j^{n+1} = \Delta x \left\{wS^{n+1} + (1-w)S^n\right\} \tag{30}$$

where $S = Re^{\frac{1}{2}} v_w \delta_{1k} + \frac{1}{u_e}\frac{du_e}{dx} \sum_j EF_{kj}\theta_j + u_e \sum_j AA_{kj}(1+\nu_T/\nu)_j \tau_j$.

$\Delta\theta_j^{n+1} = \theta_j^{n+1} - \theta_j^n$ and n denotes the downstream location.

The parameter w controls the degree of implicitness. Typically $w \geqslant 0.6$ for stable solutions, without restriction on Δx. To permit a linear system of equations to be developed for $\Delta\theta_j^{n+1}$, the following approximation is introduced,

$$S^{n+1} \approx S^n + \left(\frac{\partial S}{\partial \theta_j}\right)^n \Delta\theta_j^{n+1} \quad . \tag{31}$$

As a result eq. (30) can be put into the following form,

$$\sum_j CCC_{kj} \, \Delta\theta_j^{n+1} = P_k \tag{32}$$

where $CCC_{kj} = CC_{kj} - w \, \Delta x \left[\left(\frac{1}{u_e} \frac{du_e}{dx} \right)^{n+1} EF_{kj} - u_e^{n+1} AA_{kj} \, G_j^n \right]$

$$G_j^n = \left\{ (1+\nu_T/\nu)\tau^2 - \left[\partial(\nu_T/\nu)/\partial\theta \right] \tau \right\}_j^n$$

and

$$P_k = \Delta x \left[Re^{\frac{1}{2}} v_w \delta_{1k} + \left\{ w \left[\frac{1}{u_e} \frac{du_e}{dx} \right]^{n+1} + (1-w) \left[\frac{1}{u_e} \frac{du_e}{dx} \right]^n \right\} \sum_j EF_{kj} \, \theta_j^n \right.$$

$$\left. + \left\{ w \, u_e^{n+1} + (1-w) \, u_e^n \right\} \sum_j AA_{kj} (1+\nu_T/\nu)_j^n \, \tau_j^n \right] .$$

The matrix equation, (32), is tridiagonal for linear elements and pentadiagonal for quadratic elements. The Thomas algorithm can be generalised to take account of the varying bandwidth of CCC associated with the use of quadratic shape functions at midside and corner nodes.

To maintain the inherent economy of the finite element Dorodnitsyn formulation, eq. (32) is marched from $x^{(n)}$ to $x^{(n+1)}$ *without iteration.* Numerical convergence tests (Fletcher and Fleet, 1983a) indicate that the scheme is accurate to $O(\Delta x)$, only. However good agreement is obtained with known exact and experimental results, without having to restrict the size of the marching step, Δx.

The present algorithm actually allows the stepsize, Δx, to vary as the solution develops; details are given by Fletcher and Fleet (1983b).

3. TURBULENT WALL-BLOWING FLOWS

In this section results will be presented for the application of the spectral and finite element Dorodnitsyn formulations to the flow over a flat plate both with and without blowing in the normal direction. The solutions obtained with the Dorodnitsyn formulations will be compared with results obtained with a typical finite difference package, STAN5 (Reynolds, 1976).

Both Dorodnitsyn formulations use an eddy viscosity procedure to represent the Reynolds shear stress. The finite element Dorodnitsyn formulation uses a van Driest modified mixing length representation for the region adjacent to the wall. In the region away from the wall a Clauser eddy formulation is used. The spectral Dorodnitsyn formulation uses the Spalding/Kleinstein model for the eddy viscosity in the region adjacent to the wall. A desirable feature of the Spalding/Kleinstein model is that it expresses the eddy viscosity in terms of u, directly. The details of the turbulence models used are given by Fletcher and Fleet (1983b).

3.1 Zero-Blowing Case

For the flow past a flat plate without blowing detailed experimental measurements are available due to Wieghardt and Tillmann (1951) that were used as a test case at the 1968 Stanford Conference on the computation of turbulent boundary layers (Coles and Hirst, 1968).

Solutions with the finite element Dorodnitsyn formulation (DOROD-FEM) have been obtained with 11 equally-spaced points across the boundary layer. This may be contrasted with STAN5 which requires 33-39 points to obtain an accurate representation of the solution. The spectral Dorodnitsyn formulation (DOROD-SPEC) uses 6 unknown coefficients in eq. (17). The feature of requiring relatively few parameters in the Dorodnitsyn formulations to represent the solution across the boundary layer follows from using u as the dependent variable instead of y. As might be expected there is also a corresponding reduction in execution time. For a range of problems involving different pressure gradients (Fletcher and Fleet, 1983b) it is found that DOROD-FEM is typically *ten times as economical* as STAN5.

The variation of the skin friction coefficient with downstream position is shown in Fig. 2. The skin friction coefficient is the non-dimensionalised shear stress at the wall, i.e.

$$\tau_w = c_f \tfrac{1}{2}\rho u_e^2 \tag{33}$$

and measures the retarding effect of the wall on the flow. It is clear that all three methods are predicting accurately the reduction in skin friction coefficient, c_f, with downstream location.

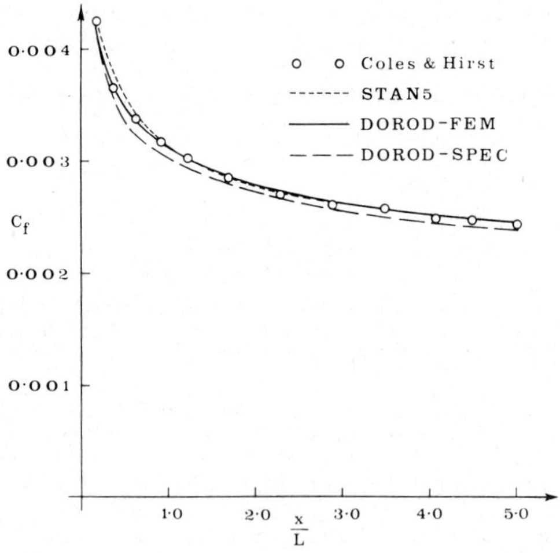

Figure 2
Skin friction variation for zero-blowing case.

Two important length scales, the displacement thickness (δ^*) and the momentum thickness (θ), are related to the boundary layer thickness (δ) as follows,

$$\delta^* = \int_0^\delta (1 - u/u_e)\, dy \tag{34}$$

$$\text{and} \quad \theta = \int_0^\delta \frac{u}{u_e}\left(1 - \frac{u}{u_e}\right) dy \quad . \tag{35}$$

The various parameters in eqs. (33) to (35) are dimensional quantities.

The displacement and momentum thicknesses grow downstream (Fig. 3). All three methods predict the downstream growth with DOROD-FEM and DOROD-SPEC giving better agreement with the experimental results than STAN5.

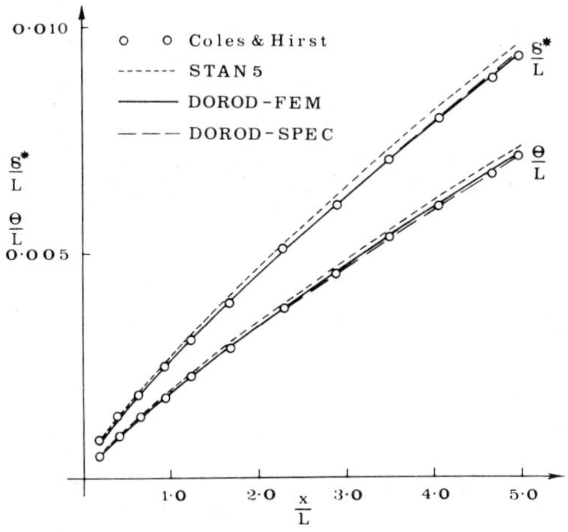

Figure 3
Displacement and momentum thickness variation for zero-blowing case.

3.2 Wall Blowing Case

This case considers the flow over a flat plate with continuous blowing at the plate in the normal direction. The magnitude of the blowing is specified by

$$F = v_w/u_e \ , \tag{36}$$

where v_w is the normal velocity at the wall and u_e is the velocity at the outer edge of the boundary layer. Here u_e = 50 ft/sec in order to match the experimental conditions of McQuaid (1967). In Figs. 4 to 7 results are presented for values of F up to 0.0046.

Computational solutions have been obtained with DOROD-SPEC, DOROD-FEM and STAN5. DOROD-SPEC has used 5 unknown coefficients in eq. (17) to represent the solution across the boundary layer. DOROD-FEM has used 11 equally-spaced points in u and STAN5 has used 28-32 points across the boundary layer.

Solutions for the variation of skin friction coefficient with downstream location, obtained by DOROD-SPEC, are shown in Fig. 4. Also shown in Fig. 4 are values of

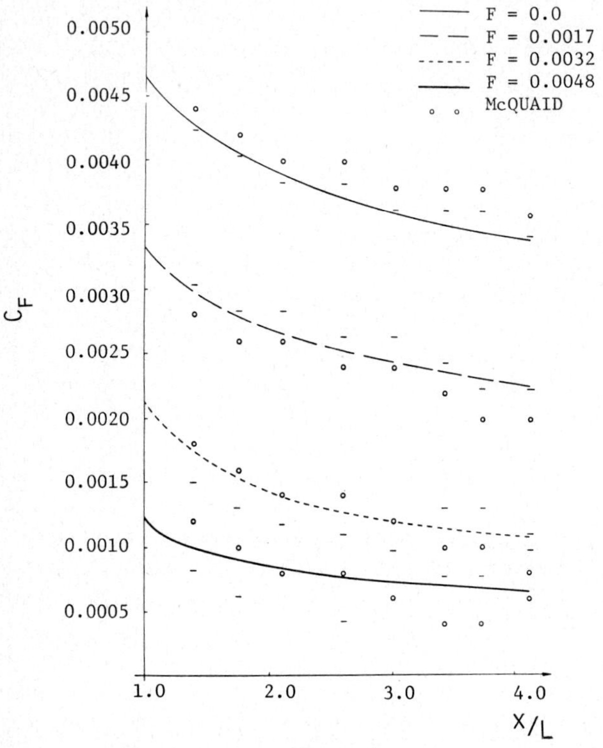

Figure 4
Skin friction variation for continuous wall-blowing; DOROD-SPEC.

skin friction coefficient calculated from the experiments of McQuaid (1967). McQuaid measured the velocity distribution across the boundary layer and calculated the skin friction from the momentum integral equation,

$$\frac{c_f}{2} = \frac{d\delta_m}{dx} - \frac{v_w}{u_e} + (H+2)\,\frac{\delta_m}{u_e}\frac{du_e}{dx} \tag{37}$$

where δ_m ($\equiv \theta$ in eq. (35)) is the momentum thickness and $H = \delta^*/\delta_m$. The horizontal bars shown in Fig. 4 provide an estimate of the possible error in the evaluation of c_f from eq. (37) applied to the experimental data.

It is clear from Fig. 4 that increasing values of F cause a reduction in skin friction and that the skin friction variation with x/L is well-predicted by DOROD-SPEC.

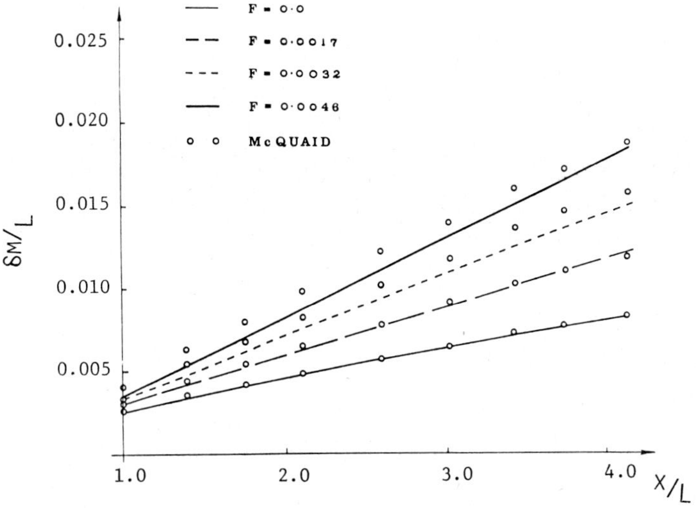

Figure 5
Momentum thickness variation for continuous wall-blowing, DOROD-SPEC.

The variation of the momentum thickness with downstream location is shown in Fig. 5. Increased wall-blowing (increasing F) produces a larger value of the momentum thickness at the same downstream position. The DOROD-SPEC results show close agreement with the experimental results for all the blowing cases considered.

The corresponding momentum thickness variation for DOROD-FEM is shown in Fig. 6. A difficulty was encountered in obtaining a suitable starting velocity distribution, u(y), for DOROD-FEM from the experimental data. Therefore an estimate was used for this profile. As a result good agreement with McQuaid's results is not obtained for small values of x/L. However the agreement is reasonable for large values of x/L.

Solutions, obtained using STAN5, for the momentum thickness variation are shown in Fig. 7.

The STAN5 solutions correctly predict the trend of increasing momentum thickness with blowing but generally slightly underpredict the magnitude.

Based on these and other results, DOROD-SPEC generally produces more accurate solutions than the other methods for non-zero wall blowing.

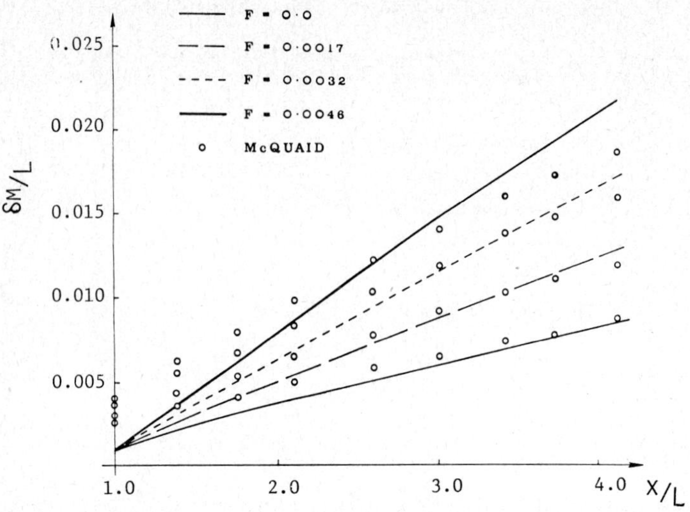

Figure 6
Momentum thickness variation for continuous wall-blowing; DOROD-FEM

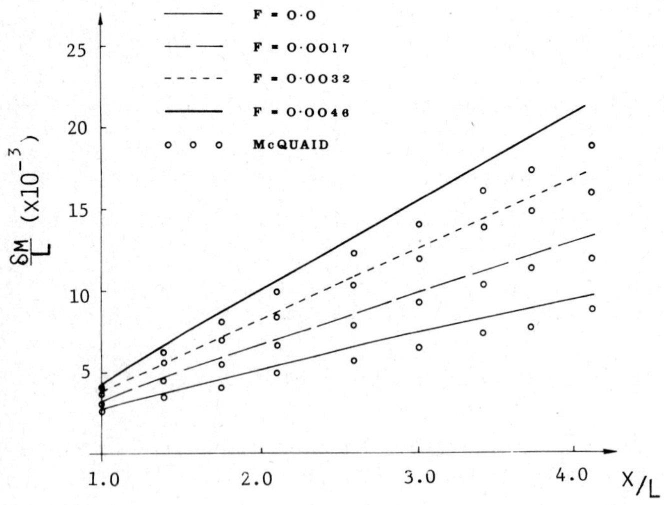

Figure 7
Momentum thickness variation for continuous wall-blowing; STAN5

CONCLUSION

The Dorodnitsyn boundary layer formulation produces both accurate and economical solutions. The finite element Dorodnitsyn formulation is, typically, an order-of-magnitude more economical, for the same accuracy, than a representative finite difference method, STAN5. The Dorodnitsyn formulation is applied to the flow past a flat plate with wall-blowing in the normal direction and it is found that the spectral Dorodnitsyn formulation is more accurate than either the finite element Dorodnitsyn formulation or STAN5.

REFERENCES

F.G. Blottner (1975), "Computational Techniques for Boundary Layers," AGARD Lecture Notes No. 73.

P. Coles and E. Hirst (eds) (1968), "Computation of Turbulent Boundary Layers - 1968", AFOSR-ISP Stanford Conference, Vol. 2.

A.A. Dorodnitsyn (1962), "General Method of Integral Relations and its Application to Boundary Layer Theory," *Advances in Aero. Sci.*, 3, 207-219.

R.W. Fleet and C.A.J. Fletcher (1982), "A Comparison of the Finite Element and Spectral Methods for the Dorodnitsyn Boundary Layer Formulation," in *Finite Elements in Engineering* (ed. P.J. Hoadley), Univ. of Melbourne, 59-63.

R.W. Fleet and C.A.J. Fletcher (1983), "Application of the Dorodnitsyn Boundary Layer Formulation to Turbulent Compressible Flow," Proc. Eighth Australasian Fluid Mechanics Conference, Newcastle, Australia.

C.A.J. Fletcher (1978), "Application of an Improved Method of Integral Relations to the Supersonic Boundary Layer Flow about Cones at large Angles of Attack," in *Numerical Simulation of Fluid Motion* (ed. J. Noye), North-Holland, 537-550.

C.A.J. Fletcher (1983a), "Internal Swirling Flows by the Dorodnitsyn Finite Element Formulation," Proc. 3rd Int. Conf. Num. Meth. Laminar and Turbulent Flow, Seattle, U.S.A.

C.A.J. Fletcher (1983b), "The Group Finite Element Formulation," *Comp. Meth. Appl. Mech. Eng.*, 37, 225-243.

C.A.J. Fletcher (1983c), "*Computational Galerkin Methods*," Springer-Verlag, New York.

C.A.J. Fletcher and M. Holt (1975), "An Improvement to the Method of Integral Relations," *J. Comp. Phys.* 18, 154-164.

C.A.J. Fletcher and M. Holt (1976), "Supersonic Flow about Inclined Cones," *J. Fluid Mechanics*, 74, 561-591.

C.A.J. Fletcher and R.W. Fleet (1982), "A Dorodnitsyn Finite Element Boundary Layer Formulation," Proc. Eighth Int. Conf. Num. Meth. Fluid Dynamics, Aachen, W. Germany.

C.A.J. Fletcher and R.W. Fleet (1983a), "Dorodnitsyn Finite Element Formulation for Laminar Boundary Layer Flow," *Int. J. Num. Meth. Fluids* (to appear).

C.A.J. Fletcher and R.W. Fleet (1983b), "A Dorodnitsyn Finite Element Formulation for Turbulent Boundary Layers," *Comp. and Fluids* (to appear).

C.W. Gear (1971), "The Automatic Integration of Ordinary Differential Equations," *Comm. A.C.M.*, 14, 176-179.

M. Holt (1977), "*Numerical Methods in Fluid Dynamics*, "Springer-Verlag, Heidelberg.

E. Isaacson and H.B. Keller (1966), "*Analysis of Numerical Methods*," Wiley, New York.

J. McQuaid (1967), Experiments on Incompressible Turbulent Boundary Layers with Distributed Injection", ARC R&M No. 3549.

K. Wieghardt and W. Tillmann (1951), "Investigations of the Wall Shearing Stress in Turbulent Boundary Layers," NACA TN 1285.

Computational Techniques & Applications: CTAC-83
J. Noye & C. Fletcher (Editors)

DIRECT SIMULATION OF BURGULENCE

P. Orlandi and M. Briscolini

Dipartimento di Meccanica e Aeronautica
Università degli Studi di Roma "La Sapienza"
Via Eudossiana, 18 - 00184 - Roma
ITALY

Burgers' equation has been solved by a finite difference scheme at very low viscosity. In order to describe very accurately the very thin shock region a coordinate transformation has been used. The small number of grid points in the space together with an implicit non-iterative scheme for the time discretization allow to have the solution by a low computing time. The scheme has been tested by a steady case provided with analytical solution. In the unsteady dissipating case, from the velocity distribution in the physical space, the energy spectra and the energy transfer term have been yielded. The transfer energy term derived from the direct simulation allows to draw some understanding on the two-points closures models.

INTRODUCTION

The increasing of computers capability gave rise, recently, to the direct numerical simulations of three-dimensional turbulent flows [1,2]. These calculations are feasible only by people which can use supercomputers. In our country, at the moment, a supercomputer is not available, therefore direct simulation of two or three-dimensional flows can not be carried out. Even having at disposal a supercomputer, the solutions of flows in two or three dimensions at high Reynolds numbers are impossible. These reasons brought the authors of the present paper to consider the Burgers' equation, which in some aspect is analogous to the Navier Stokes equations and offers insight into some of the properties of turbulence. The Burgers' equation lacks a pressure term and therefore incompressibility of the velocity is not required. However in both cases the advection term produces regions of large velocity gradients, which implies transfer of energy from low to high wavenumbers, where it is dissipated. If the viscosity is very low the dissipation occurs in a very thin region which looks like a shock front. The shock front structure implies a $E(k) \simeq k^{-2}$ [3], which differs from the $k^{-5/3}$ Kolmogoroff law obtained for the Navier-Stokes equations. These similarities and differences make the Burgers' equation very attractive together with the most important aspect that direct numerical simulation at high Reynolds number can be done.

Several numerical schemes can be used for the direct simulation of Burgers' equation. Basdevant [4] used Fourier spectral methods. He analysed the influence of the number of the degree of freedom on the results. Others used, for a different case [5], cubic spline technique. In this paper we have chosen a finite difference scheme which, by a coordinate transformation, gives a higher resolution in the shock region. Our method requires a limited number of collocation points even at very low viscosity. While the Fourier spectral methods, at high Reynolds numbers, to describe accurately the shock structure require a very large number of degree of freedom which make the calculation almost impossible.

Once the velocity field is known in the physical space, all the quantities in the wave number space can be easily derived. Of particular interest is the transfer term T(k) which occurs in the energy balance equation

$$\frac{\partial E(k)}{\partial t} + 2\nu k^2 E(k) = T(k) \qquad (1)$$

This term is the one must be modelled in the two-points closure models. These closure models overcome the limitations of the Reynolds averaged models because they take into consideration the energy transfer between different scales. The most used of these models is the Eddy Damped Quasi-Normal Markovianization (EDQNM), which requires assumptions on the relaxation time. In this paper the relaxation time has been calculated by the values of T(k) obtained by the direct simulation and compared with the distribution usually assumed in the EDQNM closures.

In the discretization of the Burgers' equation by a finite difference scheme different kind of schemes can be used. To find the schemes giving the best results a comparison between numerical and analytical solutions is necessary. Since the analytical solution of the case considered by Basdevant [4] is not valuable at small t, the numerical schemes have been tested in the case of the following equation

$$\frac{\partial U}{\partial t} = U\frac{\partial U}{\partial x} + \nu\frac{\partial^2 U}{\partial x^2} \qquad (2)$$

which admits steady solution for particular boundary conditions. Our aim is devoted to analyse the following aspects:

a) to compare different time discretization schemes in order to use a scheme allowing large time steps in presence of highly non-uniform grids;
b) to analyse how differently behave the conservative and non-conservative schemes for the advective term, particularly when highly non-uniform grids are used;
c) to investigate how different coordinate transformations affect the solution.

Once the test of the numerical scheme has been accomplished for each one of the above mentioned aspects, the results of the direct simulation can be used to draw some insight on closure hypothesis.

NUMERICAL SCHEME

To investigate how the solution is affected by different numerical approximations, the Burgers' equation can be written as follows

$$\frac{\partial U}{\partial t} + C\{(1-\beta)U\frac{\partial U}{\partial x} + \frac{\beta}{2}\frac{\partial U^2}{\partial x}\} = \nu\frac{\partial^2 U}{\partial x^2} \qquad (3)$$

The constant C takes values +1 when we consider the unsteady case. C instead is -1 when we check the numerical methods solving Eq.(2). The convective term in Eq.(3) for $\beta = 1$ is expressed in conservative form, while for $\beta = 0$ is expressed in non-conservative form. The advective term, if C is positive, increases the positive velocity gradients and decreases the negative ones, and viceversa in the case of negative C.

A) Space discretization.

If the viscosity is very small the velocity profile shows a shock structure. In order to describe, accurately, the shock by a finite difference scheme, which emploies a uniform grid, a large number of computational points are required. A coordinate transformation, which places the necessary number of mesh points in the shock region, and which places a limited number of grid points in the large part of the field where the velocity gradients are low, can be used. This coordinate transformation has been used previously [6,7] to solve two dimensional flow fields. In order to have an appraisal of the truncation errors we prefer to give the coordinate transformation by analytical expressions and to calculate the derivatives of the transformation by a finite difference scheme. If the transformation is $x = x_G(\eta)$, where η is the "new" coordinate equally spaced, the second derivative can be expressed as

$$\left(\frac{\partial^2 U}{\partial x^2}\right)_E = \frac{\partial^2 U}{\partial \eta^2} \cdot \left(\frac{d\eta}{dx_G}\right)^2 - \frac{\partial U}{\partial \eta} \cdot \frac{(d^2 x_G/d\eta^2)}{(dx_G/d\eta)^3} \tag{4}$$

or

$$\left(\frac{\partial^2 U}{\partial x^2}\right)_C = \frac{\partial}{\partial \eta}\left(\frac{d\eta}{dx_G}\,\frac{\partial U}{\partial \eta}\right)\frac{d\eta}{dx_G} \tag{5}$$

If we wish to analyse which of the two expressions predict the best results, equation (3) can be written as

$$\frac{\partial U}{\partial t} + C\frac{d\eta}{dx_G}\left\{(1-\beta)\,U\frac{\partial U}{\partial \eta} + \frac{\beta}{2}\frac{\partial U^2}{\partial \eta}\right\} = \nu\left\{\gamma\frac{\partial}{\partial \eta}\left(\frac{d\eta}{dx_G}\,\frac{\partial U}{\partial \eta}\right)\frac{d\eta}{dx_G} + \right.$$
$$\left. + (1-\gamma)\left[\frac{\partial^2 U}{\partial \eta^2}\left(\frac{d\eta}{dx_G}\right)^2 - \frac{\partial U}{\partial \eta}\,\frac{d^2 x_G/d\eta^2}{(dx_G/d\eta)^3}\right]\right\} \tag{6}$$

When the partial differential operators occurring in Eqs.(4) and (5) are expressed by

$$\left(\frac{\partial A}{\partial \eta}\right) = \left(\frac{\delta A}{\delta \eta}\right)_I = \frac{A_{I+1} - A_{I-1}}{2\,\Delta\eta}$$
$$\left(\frac{\partial A}{\partial \eta}\right) = \left(\frac{\delta A}{\delta \eta}\right)_{I+\frac{1}{2}} = \pm\frac{A_{I\pm 1} - A_I}{\Delta\eta} \tag{7}$$
$$\left(\frac{\partial^2 A}{\partial \eta^2}\right) = \left(\frac{\delta^2 A}{\delta \eta^2}\right)_I = \frac{A_{I+1} - 2A_I + A_{I-1}}{\Delta\eta^2}$$

The truncation errors expression can be obtained by using Taylor's series expansion. For Eq.(4) we have

$$\left.\frac{\delta^2 U}{\delta x^2}\right|_I^E = \frac{\partial^2 U}{\partial x^2} + \frac{\Delta\eta^2}{12}\left\{U_{4x}\,x_\eta^2 + 8\,U_{3x}\,x_{2\eta} + 9\,U_{2x}\frac{x_{2\eta}^2}{x_\eta^2}\right\} \tag{8}$$

and for Eq.(5)

$$\left.\frac{\delta^2 U}{\delta x^2}\right|_I^C = \frac{\partial^2 U}{\partial x^2} + \frac{\Delta\eta^2}{12}\left\{U_{4x}\,x_\eta^2 + 4\,U_{3x}\,x_{2\eta} + 2\,U_{2x}\left(\frac{x_{3\eta} - x_{2\eta}}{x_\eta}\right)\right\} \tag{9}$$

The operators on the left hand side of equations (8) and (9) represent equations (4) and (5), where the differentiation symbol have been substituted by the operators in Eq.(7). The general quantity Φ_{ny} represents $\partial^n\Phi/\partial y^n$.

B) Time discretization.

Previous researchers to solve Burgers' equation used predictor corrector algorithms [8,9]. The equation (6) on the general location $x_I = x_G(\eta_I)$ can be written as

$$\left.\frac{\partial U}{\partial t}\right|_I = \left[L_A(\beta,U)_I + L_\nu(\gamma)_I\right]U_I \tag{10}$$

where the operators $L_A(\beta,U)$ and $L_\nu(\gamma)$ are given by

$$L_A(\beta,U) = C\left(\frac{\delta x_G}{\delta \eta}\right)^{-1}\left\{(1-\beta)\,U\frac{\delta}{\delta \eta} + \frac{\beta}{2}\frac{\delta U}{\delta \eta}\right\} \tag{11}$$

$$L_\nu(\gamma) = \nu\left\{\gamma\left(\frac{\delta x_G}{\delta \eta}\right)^{-1}\frac{\delta}{\delta \eta}\left[\left(\frac{\delta x_G}{\delta \eta}\right)^{-1}\frac{\delta}{\delta \eta}\right] + (1-\gamma)\left(\frac{\delta x_G}{\delta \eta}\right)^{-2}\left[\frac{\delta^2}{\delta \eta^2} + \right.\right.$$
$$\left.\left. - \frac{d^2 x_G}{d\eta^3}\cdot\left(\frac{dx_G}{d\eta}\right)^{-1}\frac{\delta}{\delta \eta}\right]\right\} \tag{12}$$

The velocity distribution at the new time step, by the predictor corrector scheme, can be obtained by

$$U_I(t+\Delta t) = U_I(t) + \left[L_A(\beta,U^*)_I + L_\nu(\gamma)_I\right]\left[\alpha\,U_I(t+\Delta t) + (1-\alpha)U_I(t)\right] \tag{13}$$

where the non linear operator $L_A(\beta,U)$ is function of the velocity U^* calculated by the predictor step

$$U_I^* = U_I(t) + \left[L_A(\beta,U(t))_I + L_\nu(\gamma)_I\right](\alpha' U_I^* + (1-\alpha')U_I) \qquad (14)$$

Recently Briley and McDonald [10] proposed an implicit non-iterative scheme, which has been applied from one of the authors of this paper to several kind of parabolic partial differential equations [7,11]. This scheme discretizes the time derivative as

$$\frac{\partial U}{\partial t} = \frac{1}{\Delta t}\left\{\frac{(1+\zeta)\Delta - \zeta\nabla}{1+\theta\Delta}\right\} U^n \qquad (15)$$

The forward operator Δ means $\Delta U^n = U^{n+1} - U^n$ while the backward one means $\nabla U^n = U^n - U^{n-1}$. If the right hand side of equation (10) is substituted into Eq.(15) and the operator Δ is applied to the operator $L_A(\beta,U)$ and $L_\nu(\gamma)$, the following equation is yielded

$$\Delta U_I^n - \frac{\zeta}{1+\zeta}\nabla U_I^n = \frac{\Delta t}{1+\zeta}\{(L_A(\beta,U^n)_I + L_\nu(\gamma)_I)U_I^n + \theta(L_A(\beta,U^n)_I + L_\nu(\gamma)_I)\Delta U_I^n + \\ + L_A(\beta,\Delta U^n)_I U_I^n\} + (\theta - \zeta - \tfrac{1}{2})\,0(\Delta t^2) + 0(\Delta t^3) \qquad (16)$$

The equation (16) as well the system of equations (13) and (14) represent a tridiagonal system of algebraic equations, which can be easily solved by the Thomas' algorithm. The implicit non-iterative algorithm retains a wider range of time discretization schemes, infact it represents in addition to the Euler implicit ($\theta = 1$, $\zeta = 0$) and the Crank Nicolson ($\theta = \frac{1}{2}$, $\zeta = 0$) also the three points backward ($\theta = 1$, $\zeta = 0.5$) schemes.

TEST OF NUMERICAL SCHEMES

A) Time discretization.

The properties of the two time marching schemes, above described, can be investigated by solving Burgers' equations provided with analytical solutions. A steady solution has been chosen in order to save computer time. The Eq.(3) with $C = -1$ and with $U \pm 1$ for $x = \pm 0.5$ as boundary conditions, gives the following analytical solution

$$U = \mathrm{tgh}\; x/2\nu \qquad (17)$$

At very low values of ν, e.g. $\nu = 10^{-3}$, a very accurate numerical solution can be obtained only if a reasonable number of mesh points describes the sharp discontinuity around $x = 0$. Since the discontinuity is sharper as smaller is the viscosity, the coordinate transformation must change with the value of the viscosity. In order to analyse the difference between the implicit non-iterative and the predictor corrector schemes in Eqs. (11) and (12) the assumption of $\beta = 0$ and $\gamma = 1$ has been done.

Initial conditions are necessary to solve Eq.(10) and the steady solution should not depend on them. The predictor corrector scheme by using the coordinate transformation

$$x_G(\eta) = \frac{1}{2}\left[1 + \frac{\mathrm{tgh}\,2S(\eta-1)}{\mathrm{tgh}\,S}\right] \qquad 0.5 < \eta < 1$$
$$x_G(\eta) = \frac{1}{2}\left[-1 + \frac{\mathrm{tgh}\,2S\eta}{\mathrm{tgh}\,S}\right] \qquad 0 < \eta < 0.5 \qquad (18)$$

with $S > 4$ never converged to a steady solution if a discontinuous initial condition was assumed. On the contrary this scheme gave a steady solution with $S = 8$ assuming as initial condition the equation (17) with $\nu = 10^{-2}$. This behavior emphasizes that the predictor corrector scheme is not able to damp the high wave number oscillations which can be generated during the calculation. Moreover the predictor corrector scheme with $\alpha = \frac{1}{2}$ and $\alpha' = 0$ did not converge to a steady solution also with the smooth initial condition. The maximum time step to converge in one hundred steps in the case of $\alpha = \frac{1}{2}$ and $\alpha' = 1$ was $\Delta t = 0.1$.

The implicit non-iterative scheme did not present any of the disadvantages shown by the predictor corrector schemes. The steady solution was obtained with only six time steps with $\Delta t = 0.5$. Both time discretization schemes gave the maximum percent error $E_U = (U_N - U_E)/U_E$ on the velocity prediction $E_U = 1.16\%$, at $x = 3.68 \cdot 10^{-3}$ where $U_E = 0.95099$, and they did not show any oscillation around the value $U = 1.$ as was predicted by the calculation of Ref. [8] which used uniform grids. Since the large superiority of the implicit non-iterative scheme above the predictor corrector schemes, the analysis of the effects of the other factors has been done by the former scheme.

B) <u>Space discretization.</u>

The equation (6) has been solved by the implicit non-iterative scheme to analyse the effect of the coordinate transformation given by Eq.(18). All the calculation have been carried out with 81 grid points in the entire field. The following values of the parameters occurring in Eqs. (16) and (6) have been assumed: $\theta = 1$, $\zeta = 0.5$ (three points back); $\beta = 0$ (non conservative form); $\gamma = 1$ (compact form for the second derivative). Fig. 1a shows the largest errors are obtained by a very weak stretching because the major part of calculation points are located in the region where U is constant. The best distribution of the errors is obtained by a coordinate transformation which gives small grids in the region where there are large velocity gradi-

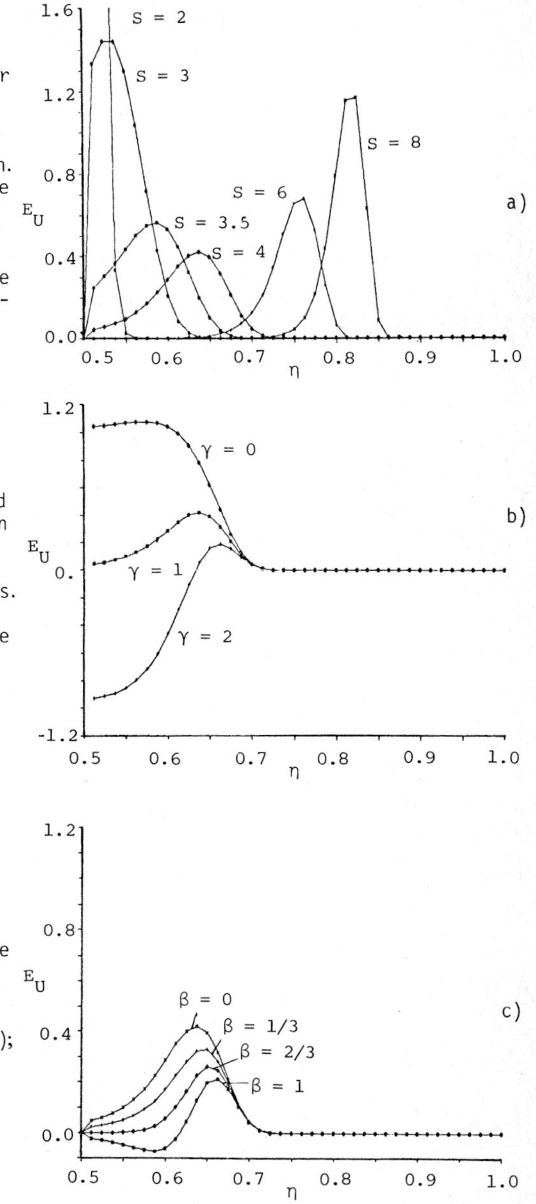

Fig. 1 - Errors distributions versus η .
(a) Coordinate stretching effects.
(b) $\frac{\partial^2}{\partial x^2}$ approximation effects.
(c) Advective term effects.

ents. At higher values of S too many points are located near $x = 0$ giving rise to grids distribution with very small meshes near $x = 0$ and large meshes around the location of maximum $\partial^2U/\partial x^2$.

Fig. 1b shows the error distribution obtained with different approximations for the second derivative. The case of $\gamma = 2$ corresponds to a combination of the two schemes of Eq.(8) and (9), such that the truncation errors do not depend any longer from U_{3x}. Fig. 1b shows the best distribution is obtained by the compact form. This behavior agrees with the behavior of the truncation errors yielded by Eqs. (8) and (9). The velocity and coordinate transformation derivatives which appear in Eqs. (8) and (9) have been calculated analytically from the equations (17) and (18).

C) Advective terms.

The treatment of advective terms by conservative and non-conservative form is the subject of many discussion, particularly in the numerical solution of transonic aerodynamics. It is well known that the conservative formulation gives a much better description of the shock region with coarser meshes. Fig. 1c shows the error distribution obtained with several values of β and with $S = 4$. The errors decrease increasing γ, until to reach the best distribution with $\beta = 1$. However the variations on the errors distributions changing β are smaller than the variations yielded by changing the parameters S and γ.

UNSTEADY CASE

A) Direct simulation.

The physical case considered by Basdevant [4] has been solved by the numerical scheme described in the previous sections. This case has the following initial and boundary conditions

$$U(0,x) = -\sin x \; ; \; U(t,\pm 1) = 0$$

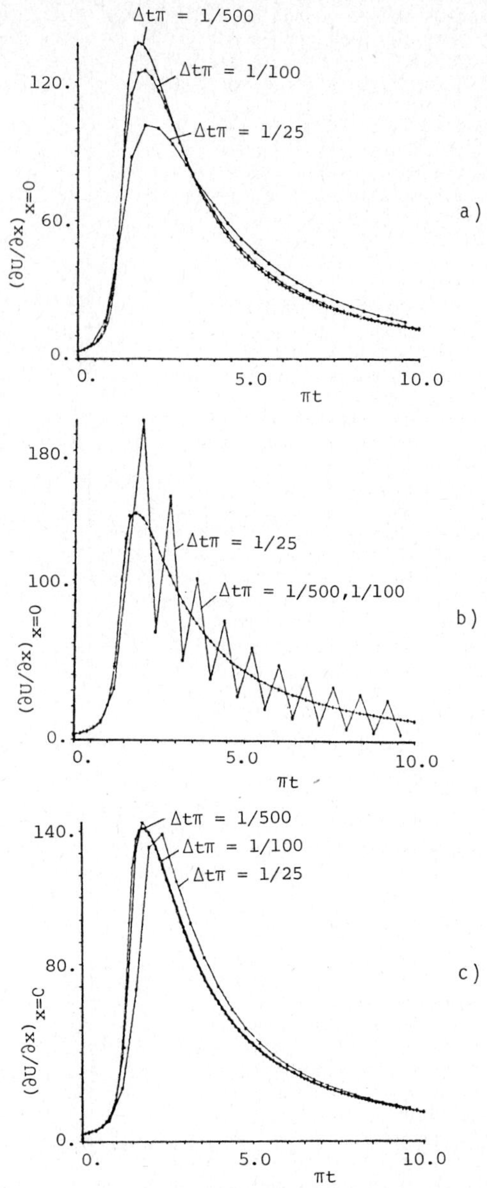

Fig. 2 - Slope distribution versus time with different time steps.
(a) Euler Implicit.
(b) Crank Nicolson.
(C) Three points backward.

A further check of the numerical scheme has been done particularly to analyse how the time discretization scheme affects the solution. The number of grid points was held equal to 81 in all the calculations done. Figs. 2a-c show the distribution of $(\partial U/\partial x)_{x=0}$ versus time for the Euler implicit $[O(\Delta t)]$, the Crank-Nicolson $[O(\Delta t^2)]$ and the three points backward $[O(\Delta t^2)]$ with different time steps sizes. Fig. 2a shows the Euler implicit scheme gives a $(\partial U/\partial x)_{x=0}$ which largely depends on the size of the time step. Fig. 2b shows the Crank-Nicolson behaves very well at intermediate values of the time step, while at large values of the time step the scheme gives rise to very large oscillations around the exact solution. Fig. 2c shows the three points backward scheme is able to give accurate solutions also with large time steps. The maximum time step allowable, for the three points backward scheme, was 80 times bigger than the time step used by Basdevant to obtain the reference solution.

Fig. 3 shows the $\frac{\partial U}{\partial x})_{x=0}$ calculated using for the second derivative discretization both the compact and the expanded forms. The compact form gives result which agrees better with the reference solution of Basdevant. The results of Figs. 2a-c and 3 show that also in the unsteady case the best prediction have been obtained with the three points backward scheme and the second derivative expressed by the compact form. A comparison between the results obtained by the advective term in conservative and non conservative form has been done, without find any appreciable difference.

The Burgers' equation written in the form of Eq.(1) shows that the energy is dissipated by the viscous term at greater wave numbers as smaller becomes the viscosity. Large wave numbers correspond to very small space dimensions. As a consequence if the viscosity decreases the energy is dissipated in very thin regions

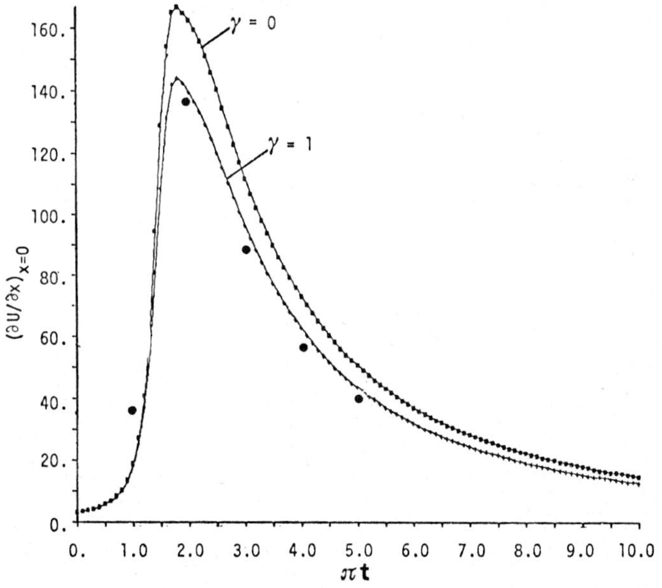

Fig. 3 - Slope distributions at different second derivative approximations. • Ref.[4] data.

where very high velocity gradients occur. Fig. 4 shows the distribution in time of $\log \nu(\partial U/\partial x)_{x=0}$ at four different values of $\pi\nu$, 10^{-1}, 10^{-2}, 10^{-3}, 10^{-4}. The maximum of the gradients occurs at values of t which weakly depend on the viscosity and which tend to the value $t = \{(\partial U/\partial x)^{-1}_{x=0}\}_{t=0} = 1/\pi$ yielded by the inviscid analysis of the Burgers' equation [12]. During the post shock period the gradients at low viscosity decay following a law which does not depend on the viscosity. From the velocity field in the physical space the coefficients of the Fourier series can be derived

$$U_k(t) = \frac{1}{2\pi} \int_{-\pi}^{+\pi} U(x,t)\, e^{-ikx}\, dx \tag{19}$$

Analogously from the numerical values of $U \frac{\partial U}{\partial x}$ the coefficients $T_c(k)$ can be calculated by

$$T_c(k) = \frac{1}{2\pi} \int_{-\pi}^{+\pi} U \frac{\partial U}{\partial x}\, e^{-ikx}\, dx \tag{20}$$

Since the calculation was done by non-uniform meshes, interpolation schemes must be used to have the values in correspondence to the $N_F = 2K_F + 1$ equally spaced X locations necessary to calculate the integrals in Eqs. (19) and (20). Once the values of $U(k)$ and $T_c(k)$ have been calculated the transfer and the dissipation terms of Eq.(1) can be calculated. Fig. 5 shows the profile of the enstrophy versus the wave number calculated with $\pi\nu = 10^{-2}$ and at $\pi t = 1, 2, 6, 10$. The $E(k) \simeq k^{-2}$, characteristic of the shock structure, is well predicted. The amplitude of the region where $E(k) \sim k^{-2}$ decreases in time, in the post shock stage. The region of constant enstrophy at $\pi\nu = 10^{-1}$ is almost negligible, while at $\pi\nu = 10^{-3}$ it is larger than the one at $\pi\nu = 10^{-2}$. Fig. 6 shows the profiles of kT

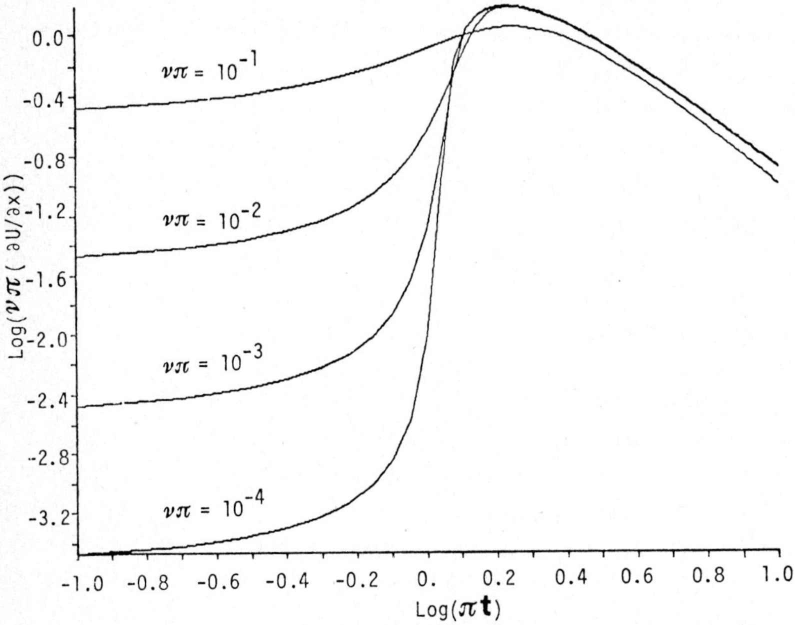

Fig. 4 - Slope distributions versus time at different viscosity.

at $\pi\nu = 10^{-2}$ and at $\pi t = 1, 2, 6, 10$, T has been calculated by $T(k) = 2(T_c(k)\, U(-k) + T_c(-k)\, U(k))$. A very large transfer of energy from very low k to intermediate k is obtained before the shock formation. In the post shock stage energy is transferred from low to high wave numbers where it is dissipated. In the region where $E \sim k^{-2}$ the transfer is much smaller than in the other regions. The same low values of kT occur in the inertial region ($E \sim k^{-5/3}$) when high Reynolds number three-dimensional isotropic turbulence flows are calculated [13].

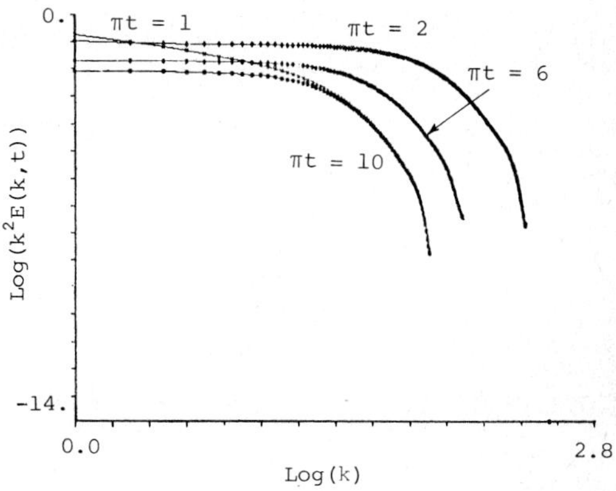

Fig. 5 - Enstrophy distribution versus log k, at different time.

The calculation of the total energy $Q = \sum_1^{k_F} E(k)$ as well the total enstrophy $\Omega = \sum_1^{k_F} k^2 E$ have been done in order to check the global energy conservation properties $\sum_{1k}^{k_F} T(k) = 0$. Fig. 7 shows the energy Q in the post shock stage decreases following the law $Q \sim t^{-n}$ with $n \simeq 5/3$. A comparison with the law given by Ref. [14] $Q(t) \simeq t^{2(\alpha-1)}$ predicts for our calculation a value of $\alpha = 1/6$ which is smaller than the value $\alpha = 1/2$ Tatsumi and Kida evaluated by a theoretical analysis [14]. Tatsumi and Kida said $\alpha = 1/2$ was in good agreement with the numerical values obtained by Crown and Canavan by a Wiener Hermite expansion [5]. The decay law with $\alpha = 1/2$ was obtained in Ref. [5] at low R when strong shocks are not generated. To demonstrate the Wiener Hermite expansion did not represent, at high Reynolds numbers, the energy cascade in Ref. [5] were reported energy decay distributions calculated directly from the Burgers' equation at different R. At values of $R > 5$ the values of the exponent n agree well with the value $n = 5/3$ obtained by our calculation.

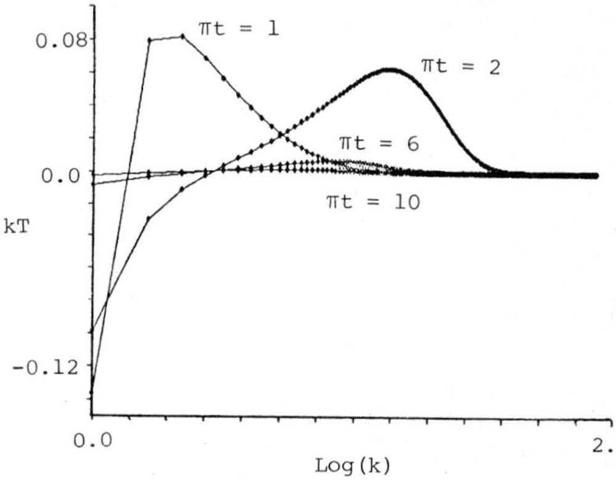

Fig. 6 - Energy transfer term versus log k, at different viscosity.

B) <u>Comparison with two points closure.</u>

Reid [15] obtained the expression of the transfer term by the assumption of a normal probability distribution for the fourth-order correlations terms which appear in the equation for the third order correlation terms. Since is well known this assumption can give negative values of energy for the three-dimensional isotropic turbulence, the Eddy Damped Quasi Normal Markovianization theory has been developed [16]. To derive the transfer term of Eq.(1) the equations for the triple correlation terms must be considered. If in these equations the effect of

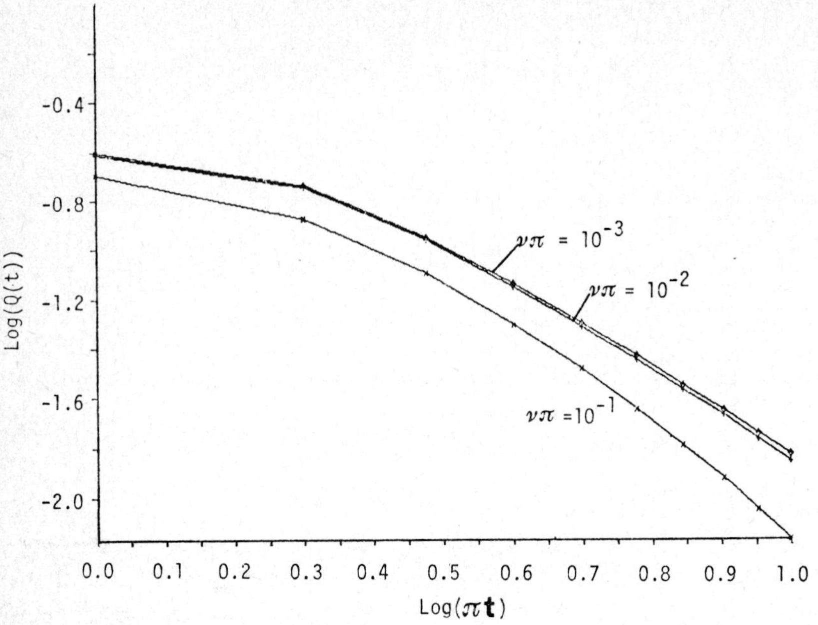

Fig. 7 - Total energy distribution versus time at different viscosity.

the viscosity is neglected we obtain

$$\frac{\partial T(k)}{\partial t} = k^2 \sum_{p=-k_F}^{+k_F} \left[E(k+p)\,E(p) - 2\,E(k)\,E(p)\right] - \mu(k)\,T(k)$$

The Markovianized form of this equation is

$$T(k) = k^2\, \theta_k(t) \sum_{p=-k_F}^{k_F} \left[E(k+p)\,E(p) - 2\,E(k)\,E(p)\right] \tag{21}$$

where $\theta_k(t)$ in the general case is $\theta_k(t) = \frac{1 - e^{-\mu(k)t}}{\mu(k)}$, and at large t with a good approximation $\theta_k \simeq \mu(k)^{-1}$. A more complete expression can be derived without neglecting the viscosity effect, in the equations for the triple correlation terms, it is

$$\frac{\partial T(k)}{\partial t} = \sum_{p=-k_F}^{k_F} \left[k\,E(p)\,E(q) + p\,E(q)\,E(k) + q\,E(k)\,E(p)\right] - \left[\eta(k) + \eta(p) + \eta(q)\right] T(k)$$

where $\eta(k) = \nu k^2 + \mu(k)$ and $k + p + q = 0$

The Markovianized form is

$$T(k) = k \sum_{p=-k_F}^{k_F} \theta_{kpq}(t) \left[k\,E(p)\,E(q) + p\,E(q)\,E(k) + q\,E(k)\,E(p)\right] \tag{22}$$

where in the general case

$$\theta_{kpq}(t) = \frac{1 - e^{-\left[\eta(k)+\eta(p)+\eta(q)\right]t}}{\eta(k) + \eta(p) + \eta(q)} \tag{23}$$

Fig. 8 shows $\theta_k(t)$ versus k at several time locations calculated by Eq.(21). In this paper we are not interested to derive the energy spectra by Eq.(1) with the assumption of the EDQNM theory. This will be the aim of a next paper, where from the profiles of $\theta_k(t)$ a law for $\mu(k)$ will be yielded. This law allows to obtain $\theta_{kpq}(t)$ which substituted in Eq.(23) together to Eqs. (22) and (1) gives the evolution of the spectra without any limitations on the Reynolds number.

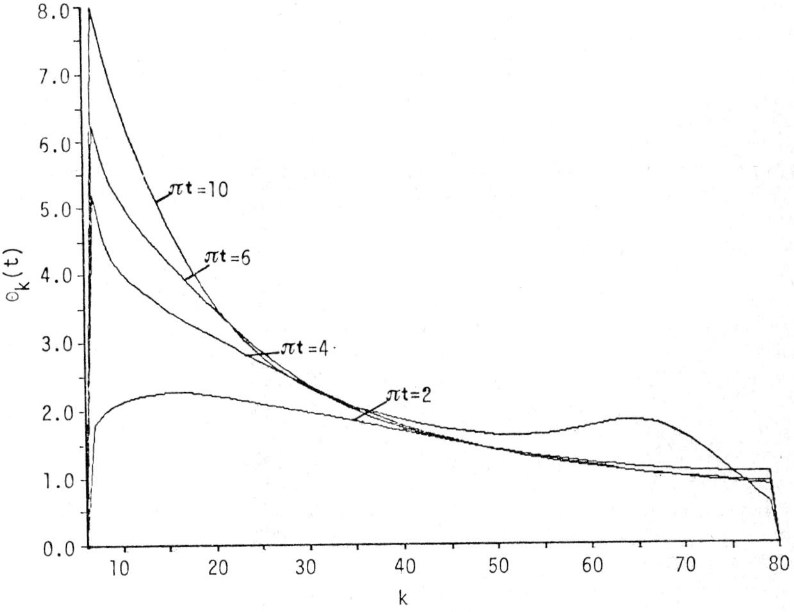

Fig. 8 - Relaxation time distribution versus k at different time locations.

CONCLUSIONS

The solution of Burgers' equation carried out in this paper using different numerical methods showed that a very accurate solution can be obtained at very low viscosity, with a limited number of grid points. An implicit non-iterative scheme has been used which allows very large time steps. The advective term in conservative form did not show a very large improvement with respect to the non conservative one, when high coordinate stretching are used. This numerical method can be easily extended to time-dependent non uniform grids, and to equations with dissipation given by higher order derivatives.

ACKNOWLEDGEMENTS

We would like to thank the 'Centro Scientifico IBM - Roma' which is supporting one of the authors (M.B.) on his thesis. A particular thanks also to C. Basdevant who gives to us his results before to be published. Gratitude is also expressed to Drs. A. Vulpiani and R. Benzi for useful discussions. This work was for the major part supported by C.N.R.

REFERENCES

1. R.S. Rogallo, "Numerical Experiments in Homogeneous Turbulence", NASA TM 81315 (1981).
2. P. Moin, J. Kim, "Numerical Investigation of Turbulent Channel Flow", J. Fluid Mech, 18 (1982).
3. P.G. Saffman, "Lectures on Homogeneous Turbulence", Topics in Nonlinear Physics, Springer (1968).
4. C. Basdevant, "Study of the Burgers equations using Fourier Spectral Methods", Laboratorie de Meteorologie Dynamique Ecole Normale Superieure (1982).
5. S.C. Crow, G.H. Canavan, "Relationship Between a Wiener Hermite Expansion and an Energy Cascade", J. Fluid Mech., 41, 387 (1970).
6. D. Cunsolo, P. Orlandi, "Accuracy in Non-Orthogonal Grid Reference Systems", Proceedings of Numerical Methods in Laminar and Turbulent Flow, Swansea, July 1978.
7. P. Orlandi, J.F. Ferziger, "Implicit Non-Iterative Schemes for Unsteady Boundary Layers", AIAA Journal, Vol. 19, No. 11, 1408 (1981).
8. Qin Meng-Zhao, "An Implicit Scheme for Nonlinear Evolution Equation", J. Comput. Phys., 48, 57 (1982).
9. J. Douglas, B.F. Jones, "On Predictor-Corrector Methods for Nonlinear Parabolic Differential Equations", S.I.A.M., 11, 195 (1963).
10. W.R. Briley, H. McDonald, "Solutions of Three-Dimensional Compressible Navier-Stokes Equations by an Implicit Technique", Proceedings fo the Fourth International Conference on Numerical Methods in Fluid Dynamics, Boulder, Colorado, June 1974, Springer (1975).
11. P. Orlandi, "Implicit Non-Iterative Scheme for Turbulent Unsteady Boundary Layers", Seventh International Conference on Numerical Methods in Fluid Dynamics, Springer (1980).
12. J.M. Burgers, "A Mathematical Model Illustrating the Theory of Turbulence", Adv. Appl. Mech., 1 (1948).
13. L. Crocco, P. Orlandi, "A Transformation of the Triadic Integral for Isotropic Turbulence", submitted to Journal of Fluid Mechanics.
14. T. Tatsumi, S. Kida, "Statistical Mechanics of the Burgers Model of Turbulence", J. Fluid Mech., 55, 659 (1972).
15. W.H. Reid, "On the Transfer of Energy in Burgers' Model of Turbulence", Appl. Sci. Res., 6A, 85 (1956).
16. S.A. Orszag, "Analytical Theories of Turbulence", J. Fluid Mech., 41, 363 (1970).

Computational Techniques & Applications: CTAC-83
J. Noye & C. Fletcher (Editors)

MODELLING TURBULENT RECIRCULATING FLOWS IN COMPLEX GEOMETRIES

Garry D. Tong

COMPUTATIONAL FLUID MECHANICS INTERNATIONAL Pty. Ltd.
Computer Applications Centre
South Australian Institute of Technology
The Levels, S.A., 5098, Australia.

The essential features of modelling turbulent flows containing zones of recirculation are discussed. First, the requirement to adequately represent momentum transfer through a shear layer from a main-stream inducing flow to a recirculating (closed streamline) zone. The k-ε model is introduced as a minimum global length scale model for this complex flow type in which there is flow separation, a dominant internal shear layer, reattachment and redevelopment. The finite element method is then introduced as an appropriate numerical method with utilisation of the natural boundary condition arising from the integral formulation and the potential to control artificial mixing (numerical diffusive and dispersive effects) by local grid refinement.

INTRODUCTION

Recirculating flows are a commonly occurring class of flow in industrial and environmental fluid mechanics. They are characterised by flow separation in the form of a detaching boundary layer, internal mixing layer and regions of re-attachment and redevelopment. As a general class of flow they have received long-standing investigation, for example the well-known early experimental work of Abbott and Kline (1962) through to the more recent analytical and numerical work of, for example, Lean and Weare (1979).

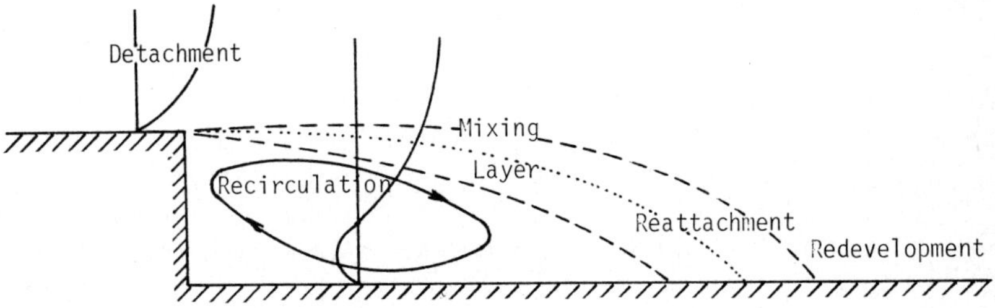

Figure 1
Characteristics of a Recirculating Flow

Figure 1 identifies the complex flow type for possibly the simplest case of a re-circulating flow, that occurring in the lee of a backward facing step. The diagram is to conceptually reinforce the difficulty of arbitrarily assigning a turbulence length scale throughout such a flow field and therefore highlighting the need for solving field equations for turbulence momentum transfer, especially through the mixing layer, to sustain the recirculation zone. The importance of a length scale in determining mixing rates is convincingly demonstrated in the numerical experiments of Birch (1976).

Launder and Spalding (1974) present a turbulence model comprising two general transport equations, one for k, turbulence kinetic energy and one for ε, the dissipation rate of turbulence kinetic energy. From dimensional arguments, ε can be considered as an equivalent length scale variable through the relationship,

$$\varepsilon = c_D \frac{k^{3/2}}{L} \qquad \text{where } c_D \simeq 1. \tag{1}$$

The equations can be derived directly from the Navier-Stokes equations, see Tong (1982)[1] with reference to Hanjalic (1970) and Svensson (1978). Two assumptions, further to a Reynolds decomposition, are the assumption of isotropy and the use of the Boussinesq assumption relating stress to strain-rate in a manner analagous to a viscous fluid in laminar motion. The Prandtl-Kolmogorov law,

$$\nu_t = c'_\mu k^{1/2} L \tag{2}$$

gives the eddy viscosity field and links the k and ε fields.

With (1) it becomes,

$$\nu_t = c_\mu \frac{k^2}{\varepsilon} \tag{3}$$

where $c_\mu = c'_\mu c_D \simeq 0.09$.

The relatively rigorous form of the model together with its length scale (equivalent) equation gives it a unique position within the hierarchy of turbulence models, it being; *the simplest model not requiring an imposed length scale input.*

The main thrust of this paper is the finite element solution of the set of advective/diffusive transport equations summarised in the following section. Before moving to that aspect it is useful to preview the advective/diffusive nature of the transport equation for k and in so doing establish a fundamental limitation of turbulence models in which mixing lengths are arbitrarily imposed.

k equation

$$\frac{\partial k}{\partial t} + u_j \frac{\partial k}{\partial x_j} - \nu_t \left(\frac{\partial u_i}{\partial x_j} + \frac{\partial u_j}{\partial x_i}\right)\frac{\partial u_i}{\partial x_j} + \varepsilon = \frac{\partial}{\partial x_j}\left(\frac{\nu_t}{\sigma_k}\frac{\partial k}{\partial x_j}\right) \tag{4}$$

advection   production   dissipation   diffusion

The assumption is now made that the turbulence is in local equilibrium, i.e. the local rate of change of k, advection of k and diffusion of k are all negligible. Under these conditions the equation reduces to a simple balance between production and dissipation which, for a simple shear layer and invoking (1), is:

$$\nu_t \left(\frac{\partial u}{\partial z}\right)^2 = c_D \frac{k^{3/2}}{L}. \tag{5}$$

On further invoking (2) to replace k, (5) becomes:

$$\nu_t = \left(\frac{c'^3_\mu}{c_D}\right)^{1/2} L^2 \left|\frac{\partial u}{\partial z}\right| \qquad (6)$$

Clearly this is the same functional form of the well-known mixing length expressions such as

$$\ell_m = c\delta \qquad (7)$$

where here δ = L, the local layer or flow width and

$$c = \left(\frac{c'^3_\mu}{c_D}\right)^{1/4} .$$

For $c'_\mu = c_\mu \simeq 0.09$ and $c_D \simeq 1$, c becomes 0.16 which is of the same order of magnitude as the tabulated values given by Reynolds (1974) for various classes of turbulent flow. This general agreement serves as support for the values of c_D and c'_μ adopted by the model developers.

The main point of interest, however, is that, under the assumption of local equilibrium, the k-equation reduces to a mixing length expression for eddy viscosity and hence virtually precludes the use of a formula of this type for more general non-local equilibrium conditions.

Models which adopt a mixing length closure are thus based on a level of parameterisation which requires an input of flow width and factor which reflects the spreading rates of various turbulent flow types. While experimental results show that there is some universality in spreading rate, the questions of where in the domain this empirical input should be applied, and what is a typical flow width at a specific location, still have to be answered.

The answers to these questions supply information about the dynamic equilibrium or "memory" properties of the motion and are difficult to supply a priori in a complex flow such as one containing a zone of recirculation. The same considerations are automatically accounted for in the transport equations of the k-ε model as a result of its derivation from fundamentals.

The final introductory remarks are to focus attention again on the need to numerically solve advective/diffusive equations. A considerable literature has been built up around this demanding task in recent years, see for example, Leonard (1979) and Tong (1980). The question of numerical accuracy by control over numerical diffusion and dispersion was certainly the major factor in the decision to try a finite element discretisation on the equations of the next section and it is the purpose of the paper to report the results obtained for the backward facing step experimental rig of Baker (1977). Although more novel procedures such as the Petrov-Galerkin formulation were kept in mind as potentially useful techniques, some encouraging results were obtained with the commonly used Bubnov-Galerkin formulation together with continual appeal back to the physics of the problem. This strategy indicated an appropriate control for the algorithm and suggested some rather heavy-handed modifications required to reach the results presented.

THE EQUATIONS

The equations for both the mean flow and turbulence models are assembled here for convenience.

Mean Flow Equations

$$u\frac{\partial u}{\partial x} + w\frac{\partial u}{\partial z} - \frac{\lambda}{\rho}\frac{\partial}{\partial x}\left(\frac{\partial u}{\partial x} + \frac{\partial w}{\partial z}\right) = 2\frac{\partial}{\partial x}\left(\nu_t\frac{\partial u}{\partial x}\right) + \frac{\partial}{\partial z}\left(\nu_t\left(\frac{\partial u}{\partial z} + \frac{\partial w}{\partial x}\right)\right), \quad (8a)$$

$$u\frac{\partial w}{\partial x} + w\frac{\partial w}{\partial z} - \frac{\lambda}{\rho}\frac{\partial}{\partial z}\left(\frac{\partial u}{\partial x} + \frac{\partial w}{\partial z}\right) = \frac{\partial}{\partial x}\left(\nu_t\left(\frac{\partial u}{\partial z} + \frac{\partial w}{\partial x}\right)\right) + 2\frac{\partial}{\partial z}\left(\nu_t\frac{\partial w}{\partial z}\right). \quad (8b)$$

Turbulence Equations

$$u\frac{\partial k}{\partial x} + w\frac{\partial k}{\partial z} = \frac{\partial}{\partial x}\left(\frac{\nu_t}{\sigma_k}\frac{\partial k}{\partial x}\right) + \frac{\partial}{\partial z}\left(\frac{\nu_t}{\sigma_k}\frac{\partial k}{\partial z}\right) + P_h - \varepsilon, \quad (9a)$$

$$u\frac{\partial \varepsilon}{\partial x} + w\frac{\partial \varepsilon}{\partial z} = \frac{\partial}{\partial x}\left(\frac{\nu_t}{\sigma_\varepsilon}\frac{\partial \varepsilon}{\partial x}\right) + \frac{\partial}{\partial z}\left(\frac{\nu_t}{\sigma_\varepsilon}\frac{\partial \varepsilon}{\partial z}\right) + C_1\frac{\varepsilon}{k}P_h - C_2\frac{\varepsilon^2}{k}, \quad (9b)$$

where

$$P_h = \nu_t\left\{2\left(\frac{\partial u}{\partial x}\right)^2 + 2\left(\frac{\partial w}{\partial z}\right)^2 + \left(\frac{\partial u}{\partial z} + \frac{\partial w}{\partial x}\right)^2\right\}.$$

The mean flow equations are given in penalty function form, see Zienkiewicz and Heinrich (1979) or Bercovier and Engelman (1979). Essentially this means the incompressibility condition is violated and written as a penalised constraint equation in which incompressibility is satisfied up to a predetermined level as set by the penalty parameter, λ. This penalty form of the incompressibility equation, $p = -\lambda \operatorname{div} \underline{u}$, is substituted into the momentum equations thus reducing the number of mean flow equations by one and eliminating the pressure variable from the primary computation. The pressure field can be accurately recaptured from the computed velocity field via a Poisson equation.

The general transport form of the k and ε equations is noted and also the presence of significant source/sink terms. Such source dominated equations are potentially troublesome in that they can create steep gradients within a solution field thereby triggering the usual problem of numerical diffusion and dispersion. Such is the case in complex geometries where there is an intensification of turbulence (vorticity) associated with a detaching boundary layer at singular points and high levels of both production of k and its dissipation rate through the mixing layer.

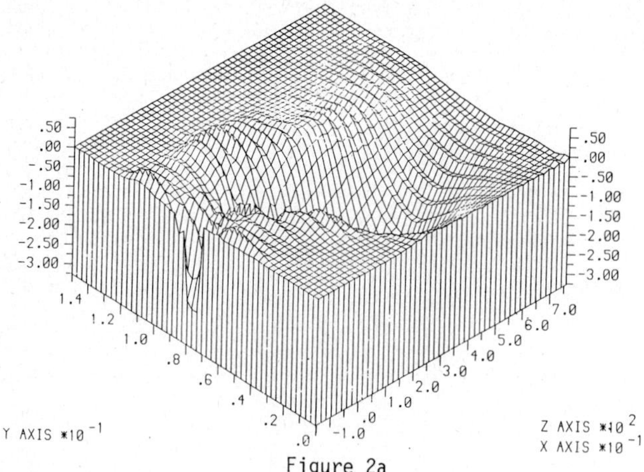

Figure 2a
Production Field (k) - Backward Facing Step Configuration

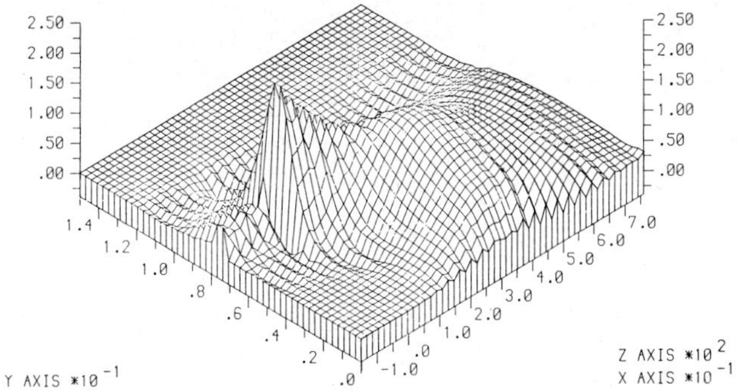

Figure 2b
Dissipation Rate Field (ε) - Backward Facing Step Configuration

THE ALGORITHM

The penalty formulation in the mean flow equations provides a convenient two-variable/ two-equation system and it seemed imminently sensible to structure the code to run the mean flow model and turbulence model in series iteration. This of course represents a decoupling of the full system but one which was pursued in the interests of computational efficiency. The equations dictate the required feed-back mechanism. Mean velocities (u_i) and strain-rate fields ($\partial u_i/\partial x_j$) drive the turbulence model which, in turn, supplies an updated eddy viscosity field via the Prandtl-Kolmogorov law, $\nu_t = c_\mu k^2/\varepsilon$. The other non-linear connection between the two models is in the form of a special boundary condition developed to link the wall shear stress or shear velocity represented in mean flow wall law behaviour with the setting of Dirichlet boundary conditions in the k-ε model along wall boundaries. This work has been reported in detail, Tong (1982)[2], and is introduced as part of the weighted residual formulation in the next section.

THE FINITE ELEMENT DISCRETISATION AND STRESS BOUNDARY CONDITIONS

The method of weighted residuals is the general class of solution technique on which the finite element method is based where no direct variational statement of the problem can be made. The method consists of introducing a "trial" solution which leads to a "residual" error. This error is weighted over the domain and equated to zero, i.e. the integral equation to be solved is formed by driving the residual error to zero over the domain (i.e. it is driven to zero in a global sense). In the case where the weighting function takes the same form as the trial function, the method becomes the well-known Galerkin-(Bubnov) method.

Now complex recirculating flows invariably occur in complex geometries and, as previously stated, althought the finite element method was chosen primarily because of its facility for grid refinement in regions of potentially steep gradients, the method of weighted residuals, as an integral method, proved to be an essential feature in establishing the computational algorithm. Integration (by Greens theorem) of weighted second derivatives yields a natural boundary condition which is an appropriate way of incorporating stresses on wall boundaries and, as a direct result, the setting of k and ε along wall boundaries.

On the reasonable assumption that local equilibrium conditions prevail adjacent to wall boundaries, i.e. production and dissipation are in balance while advection and diffusion of turbulence kinetic energy is small then (5) is rewritten

$$\nu_t \left(\frac{\partial u}{\partial z}\right)^2 = \varepsilon \tag{10}$$

or

$$\nu_t^2 \left(\frac{\partial u}{\partial z}\right)^2 = c_\mu k^2 \tag{11}$$

substituting the Prandtl-Kolmogorov law (3).

On the further assumption that the shear stress in the fluid is approximately equal to the wall shear stress near a wall, $\tau \simeq \tau_w$, then,

$$\rho \nu_t \frac{\partial u}{\partial z} = \rho u_*^2 . \tag{12}$$

Substituting (12) in (11) gives,

$$k = \frac{u_*^2}{c_\mu^{1/2}} \tag{13}$$

while (12) in (10) and invoking the log law, $u/u_* = 1/k \ln(zu_*/\nu \cdot E)$ gives,

$$\varepsilon = \frac{u_*^3}{kz} . \tag{14}$$

This shows the dependence of boundary conditions for k and ε on the wall shear stress or shear velocity and hence the recognition of the naturally arising boundary condition as a statement of total stresses on boundaries is an essential part of the analysis. The log law provides the economy of an analytical representation through the boundary layer and links u to u_*.

The formulation is as follows.

Mean Flow Model (Momentum x)

$$\int_\Omega W\{u\frac{\partial u}{\partial x} + w\frac{\partial u}{\partial z} - \frac{\lambda}{\rho}\frac{\partial}{\partial x}(\frac{\partial u}{\partial x} + \frac{\partial w}{\partial z}) - 2\frac{\partial}{\partial x}(\nu_t \frac{\partial u}{\partial x}) - \frac{\partial}{\partial z}(\nu_t (\frac{\partial u}{\partial z} + \frac{\partial w}{\partial x}))\} d\Omega = 0 \tag{15}$$

On integration by parts,

$$\int_\Omega W(u\frac{\partial u}{\partial x} + w\frac{\partial u}{\partial z}) d\Omega + \int_\Omega ([\frac{\lambda}{\rho}(\frac{\partial u}{\partial x} + \frac{\partial w}{\partial z}) + 2(\nu_t \frac{\partial u}{\partial x})]\frac{\partial W}{\partial x} + \nu_t (\frac{\partial u}{\partial z} + \frac{\partial w}{\partial x})\frac{\partial W}{\partial z}) d\Omega$$

$$- \int_\Gamma W \underbrace{\boxed{(\frac{\lambda}{\rho}(\frac{\partial u}{\partial x} + \frac{\partial w}{\partial z})\hat{n}_x + 2(\nu_t \frac{\partial u}{\partial x})\hat{n}_x + \nu_t (\frac{\partial u}{\partial z} + \frac{\partial w}{\partial z})\hat{n}_z)}}_{\text{Cauchy traction vector, } \sigma_{ij}} d\Gamma = 0 \tag{16}$$

The integration over the domain is carried out in a standard manner using 9-noded quadrilateral elements and the introduction of quadratic weight and shape functions. Again the usual practice was adopted of integrating over a standard "unit" element in the ξ, η computational plane with a derived Jacobian mapping back to the physical plane, see Zienkiewicz (1977).

The integration of the boundary stress integrals was more innovative, see Tong (1982), a key feature being the extraction of velocity gradients over edge elements by sampling at reduced Gauss points and extrapolating bilinearly out to element boundaries.

Turbulence Model

After some manipulation the following linearised forms of the equations were used in the code.

k-equation

$$\int_\Omega W\{UCK\frac{\partial k}{\partial x} + VCK\frac{\partial k}{\partial z} + \frac{\varepsilon}{k}\cdot k\}d\Omega$$

$$+ \frac{\nu_t}{\sigma_k}\int_\Omega(\frac{\partial k}{\partial x}\frac{\partial W}{\partial x} + \frac{\partial k}{\partial z}\frac{\partial W}{\partial z})d\Omega - \frac{\nu_t}{\sigma_k}\int_\Gamma W(\frac{\partial k}{\partial x}\hat{n}_x + \frac{\partial k}{\partial z}\hat{n}_z)d\Gamma$$

$$= \int_\Omega W\{P_h\}d\Omega \qquad (17)$$

ε-equation

$$\int_\Omega W\{UCE\frac{\partial \varepsilon}{\partial x} + VCE\frac{\partial \varepsilon}{\partial z} + C_2\frac{\varepsilon}{k}\cdot \varepsilon\}d\Omega$$

$$+ \frac{\nu_t}{\sigma_\varepsilon}\int_\Omega(\frac{\partial \varepsilon}{\partial x}\frac{\partial W}{\partial x} + \frac{\partial \varepsilon}{\partial z}\frac{\partial W}{\partial z})d\Omega - \frac{\nu_t}{\sigma_\varepsilon}\int_\Gamma W(\frac{\partial \varepsilon}{\partial x}\hat{n}_x + \frac{\partial \varepsilon}{\partial z}\hat{n}_z)d\Gamma$$

$$= \int_\Omega W\{C_1\frac{\varepsilon}{k}\cdot P_h\}d\Omega \qquad (18)$$

where

$$UCK = u - \frac{1}{\sigma_k}\frac{\partial \nu_t}{\partial x}, \quad VCK = w - \frac{1}{\sigma_k}\frac{\partial \nu_t}{\partial z}$$

$$UCE = u - \frac{1}{\sigma_\varepsilon}\frac{\partial \nu_t}{\partial x}, \quad VCE = w - \frac{1}{\sigma_\varepsilon}\frac{\partial \nu_t}{\partial z}$$

and P_h, the production term is as previously defined.

The line integrals for the k and ε equations were taken as zero representing no flux of k or ε, respectively, normal to a boundary as the natural boundary condition.

To illustrate the discrete form of these equations the following weight and shape function expansions are introduced,

$$W = \sum_i N_i$$

$$k = \sum_j N_j \; k_j \; , \; \frac{\partial k}{\partial x} = \sum_j \frac{\partial N_j}{\partial x} k_j, \quad \frac{\partial k}{\partial z} = \sum_j \frac{\partial N_j}{\partial z} k_j$$

$$\varepsilon = \sum_j N_j \; \varepsilon_j \; , \; \frac{\partial \varepsilon}{\partial x} = \sum_j \frac{\partial N_j}{\partial x} \varepsilon_j, \quad \frac{\partial \varepsilon}{\partial z} = \sum_j \frac{\partial N_j}{\partial z} \varepsilon_j$$

to give:

$$\sum \int_\Omega \{N_i \, [UCK \frac{\partial N_j}{\partial x} \cdot k_j + VCK \frac{\partial N_j}{\partial z} \cdot k_j + (\frac{\varepsilon}{k}) N_j \cdot k_j]$$
$$+ \frac{\nu_t}{\sigma_k} \frac{\partial N_i}{\partial x} [\frac{\partial N_j}{\partial x} \cdot k_j] + \frac{\nu_t}{\sigma_k} \frac{\partial N_i}{\partial z} [\frac{\partial N_j}{\partial z} \cdot k_j \,]\} d\Omega$$
$$= \sum \int_\Omega N_i \, \{P_h\} d\Omega \qquad (19)$$

and,

$$\sum \int_\Omega \{N_i \, [UCE \frac{\partial N_j}{\partial x} \cdot \varepsilon_j + VCE \frac{\partial N_j}{\partial z} \cdot \varepsilon_j + C_2 (\frac{\varepsilon}{k}) N_j \, \varepsilon_j \,]$$
$$+ \frac{\nu_t}{\sigma_\varepsilon} \frac{\partial N_i}{\partial x} [\frac{\partial N_j}{\partial x} \cdot \varepsilon_j \,] + \frac{\nu_t}{\sigma_\varepsilon} \frac{\partial N_i}{\partial z} [\frac{\partial N_j}{\partial z} \cdot \varepsilon_j]\} d\Omega$$
$$= \sum \int_\Omega N_i \, \{C_1 \frac{\varepsilon}{k} \, P_h\} d\Omega \qquad (20)$$

THE BACKWARD FACING STEP COMPUTATIONS

The test rig used to demonstrate the computational procedure is the backward facing step configuration of Baker (1977).

The finite element mesh was set up as in Figure 3 and boundary conditions applied as in Figure 4.

Inlet conditions

The approach conditions took the form of a theoretical "flat plate" turbulent boundary layer specified two step heights upstream of the step face with a partial slip velocity ($\frac{1}{2}u_0$) of $5ms^{-1}$ on the approach face. Following Imperial College practice, McGuirk (1980), k was typically set to a value between .1% and 1% of u_0 squared while ε was estimated using $\varepsilon = C_D \, k^{3/2}/L$. These conditions are summarised in Table 1.

Initial Conditions

The mean flow field was initiated with Stokes conditions, $u = w = 0 \; ms^{-1}$, and run up from arbitrary high eddy viscosity values, $\nu_t = 0.15$, 0.015 and 0.003 m^2s^{-1}

before computed values of ν_t from the turbulence model were introduced.

A heavy relaxation factor, 0.9, was kept on the eddy viscosity field between successive series iterations.

On the expectation that production terms for both k and ε are dominant through the mixing layer the following arbitrary small values of k and ε were chosen; $k = 0.1\ m^2s^{-2}$, $\varepsilon = 1.8\ m^2s^{-3}$.

Node No.	Location z(m)	u (ms^{-1})	k (m^2s^{-2})	ε (m^2s^{-3})
1	.076	5.000	.500	50.0
2	.078	6.138	.407	15.0
3	.080	6.838	.300	10.0
4	.090	8.241	.150	4.0
5	.100	8.983	.063	0.5
6	.125	9.872	.063	0.5
7	.150	10.000	.063	0.5
8	.494	10.000	.063	0.5
9	.838	10.000	.063	0.5

Table 1
Inflow conditions at two step heights upstream of step face

Constants in the k-ε model

Prandtl numbers σ_k and σ_ε were run up from 0.1 to 0.5 in steps of 0.1 per series iteration. Since there is evidence that C_μ is not a universal constant there is no serious concern that σ_k and σ_ε are a factor of 2 below the values recommended by Launder and Spalding (1974).

C_μ	C_1	C_2	σ_k	σ_ε
0.09	1.44	1.92	0.5	0.5

Table 2

Constants in the k-ε model for results presented

Obtaining of results

It remains to comment on the difficulty in obtaining a converged solution field, the appeal back to fundamental physics and the computational steps required to achieve the results presented.

Initially the algorithm control was set to achieve non-linear convergence within the k-ε model as established by the mean square error, $\Sigma\|\Delta u_i\|^2/\Sigma\|u_i+\Delta u_i\|^2$, being less than $(1\%)^2$.

Convergence of the k-ε model, with an interim mean flow field, could not be achieved and instability quickly became evident with a rapidly increasing turbulence field for both k and ε. Negative values of both k and ε also appeared.

Appeal was made back to the inertial behaviour of a two-dimensional flow field as suggested by Fjortoft (1953) and Kraichnan (1967) and upon which an explanation of the current phenomenon was based in Tong (1982)[3]. It was contended that, at the scales of the mean motion, the computations were experiencing a type of Fjortoft/Kraichnan backward energy cascade (energy transfer to lower wave numbers) and that certain impositions had to be made on the computations to enforce stable forward cascade characteristics.

The following two changes were made to the computations, both of which were found to be necessary to obtain the converged results (8 compound or series iterations), given below.

(i) Supply of a non-stationary strain-rate field to the turbulence model by successively updating the mean flow field using a partially converged k-ε, hence ν_t, field.

(ii) Enforcement of a forward energy cascade by using the modulus of ε and also of k, negative k values being non-physical and a result of numerical dispersive effects.

CONCLUDING REMARKS

Essential features of modelling turbulent recirculating flows in complex geometries have been presented. Important aspects of physics have been discussed together with their relationship to numerical aspects.

The need to control numerical mixing is of foremost importance when attempting to implement the physics of turbulence momentum transfer. This dual consideration of physics/numerics does not stop there however but is shown to play a useful role in the development and running of the computational algorithm.

With this approach to model development a computational code has been developed which can simulate two-dimensional wind tunnel results down to turbulence quantities and with as much credibility as experimental results, for a complex flow field such as that occurring downstream of a backward facing step.

ACKNOWLEDGEMENTS

The work reported here-in was supported by Hydraulics Research Station, Wallingford, U.K. as an extra mural research contract with the Department of Civil Engineering, University College of Swansea, Wales, U.K. The author acknowledges with gratitude the support of both organisations and permission from HRS, Wallingford to publish the work.

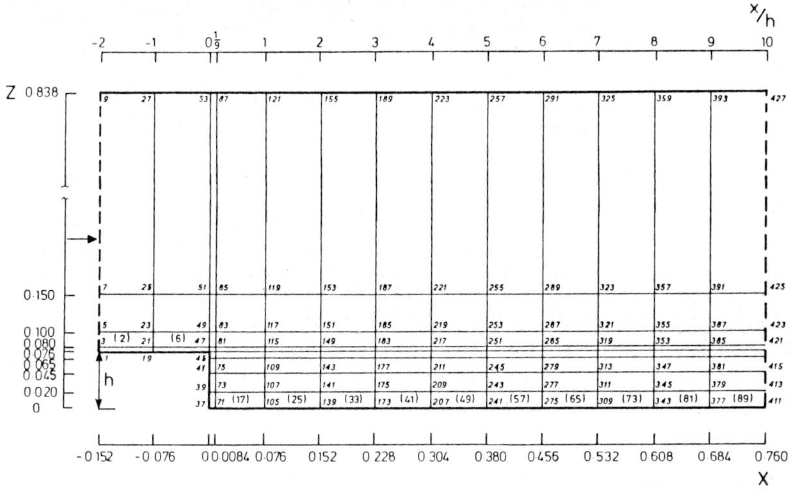

Figure 3 : Mesh - Backward Facing Step Experimental Rig, Baker (1977)

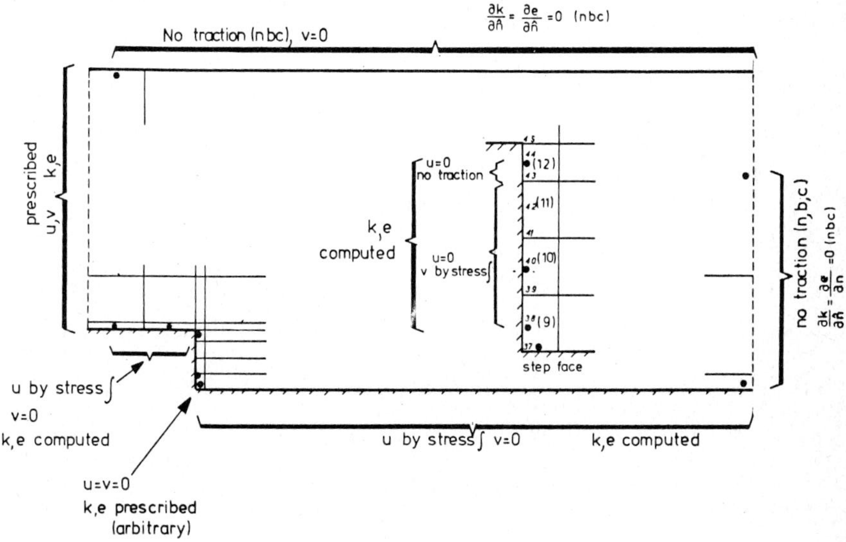

Figure 4 : Boundary Condition Types - Backward Facing Step

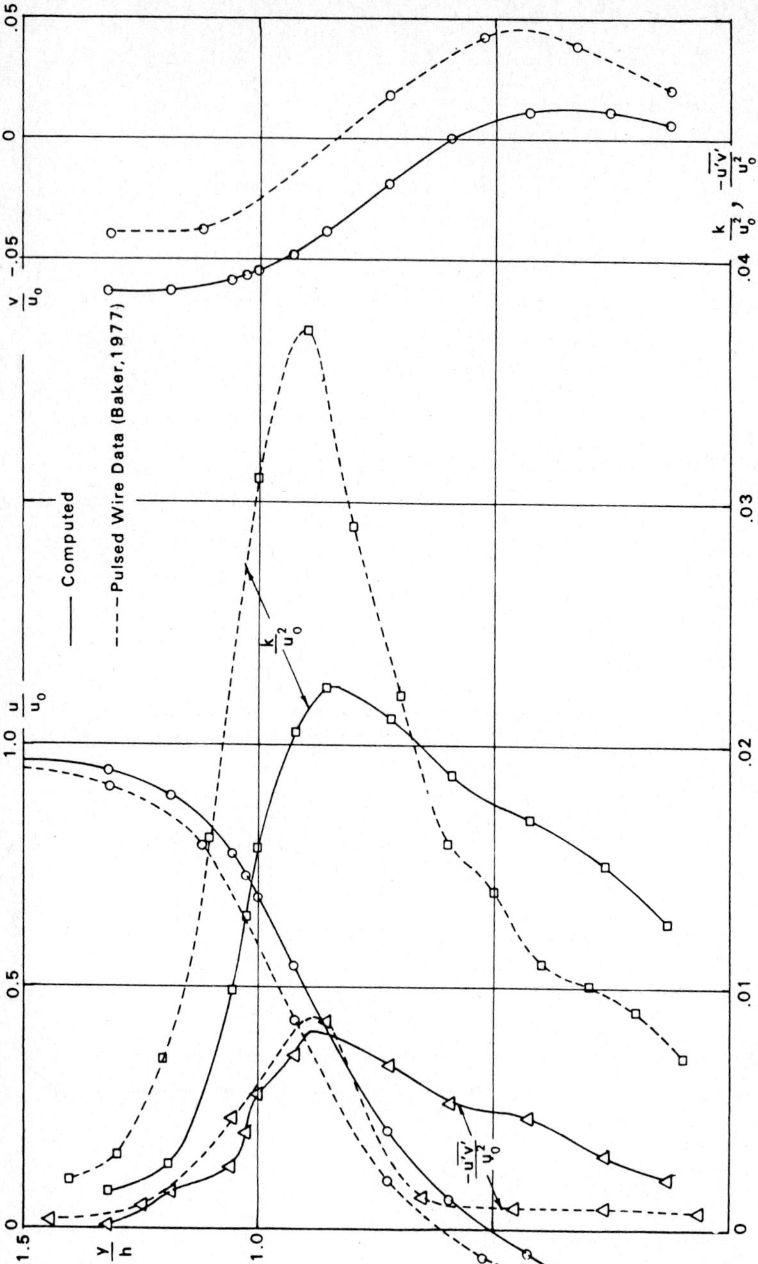

Figure 5A : Mean flow profiles, Turbulence kinetic energy and Reynolds stress at x/h = 2 - Backward Facing Step

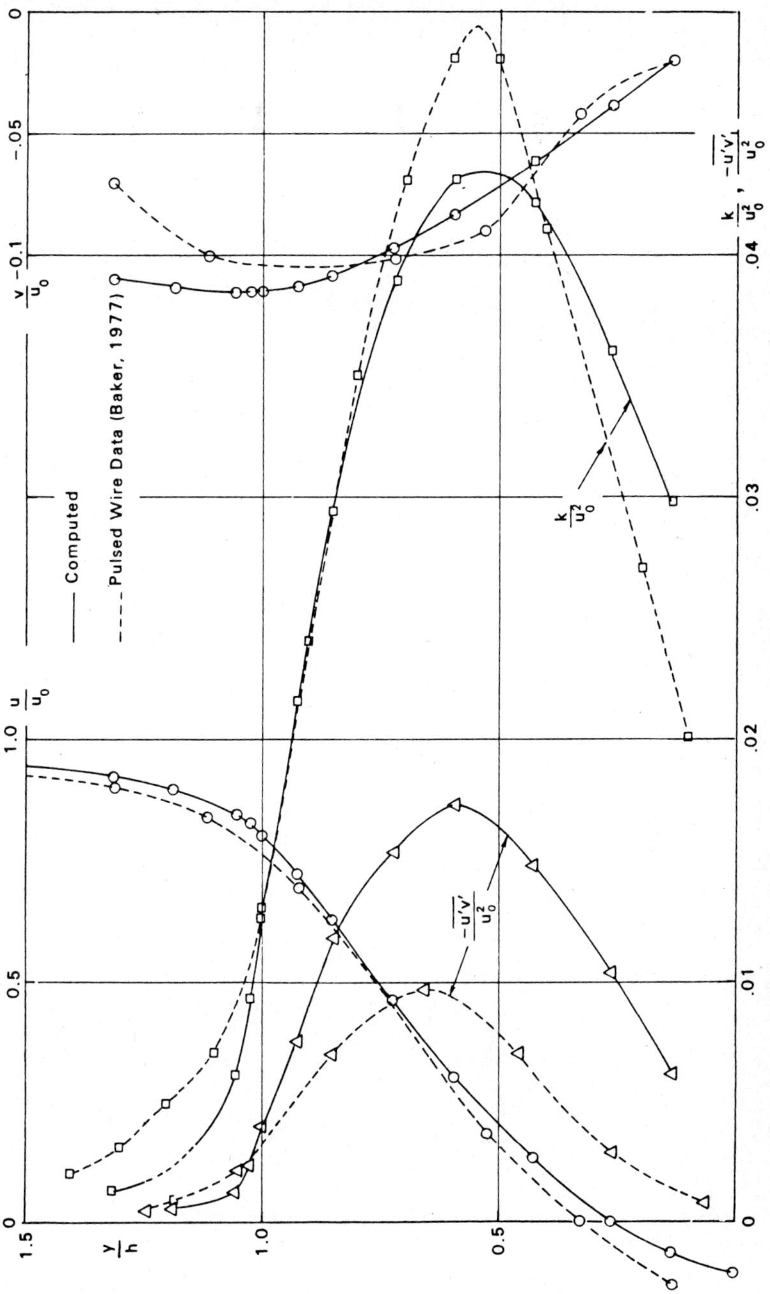

Figure 5B : Mean flow profiles, Turbulence kinetic energy and Reynolds stress at x/h = 4 - Backward Facing Step

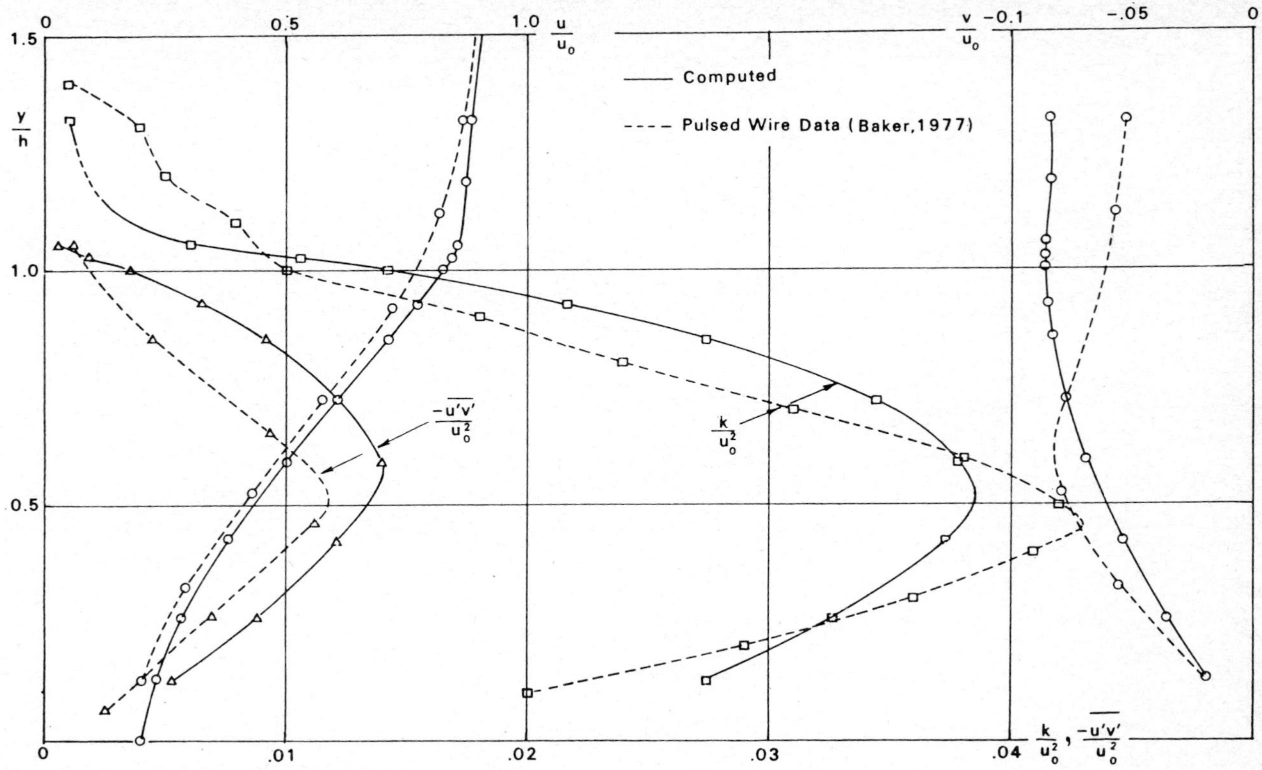

Figure 5C : Mean flow profiles, Turbulence kinetic energy and Reynolds stress at x/h = 6 - Backward Facing Step

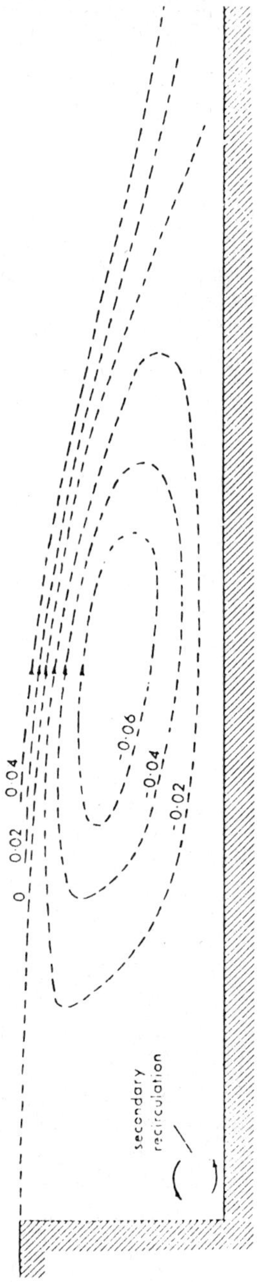

Experimental (From Baker, 1977)

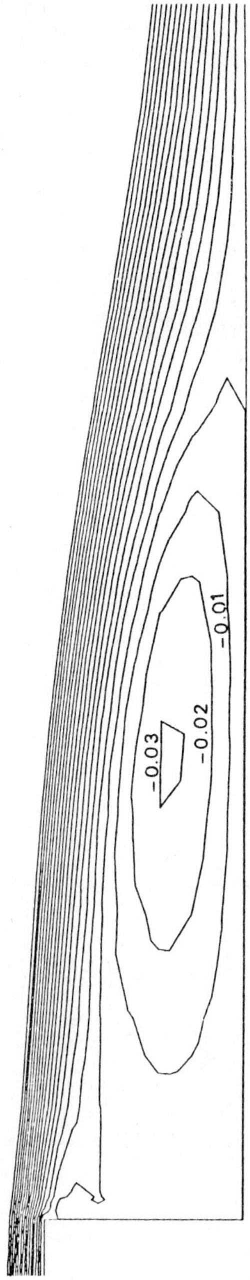

8th Compound Iteration (converged)

Figure 6 : Streamlines (normalised) - Backward Facing Step

REFERENCES

Abbott, D.E. and Kline, S.J., Experimental investigation of subsonic turbulent flow over single and double backward facing steps, ASME Journal of Basic Engineering, 84, Series D. (1962), 317-325.

Baker, S., Regions of Recirculating Flow Associated with Two-Dimenionsal Steps, Ph.D. Thesis, University of Surrey, (1977).

Birch, S.F., A critical Reynolds number hypothesis and its relation to phenomenological turbulence models, Proc. 1976 Heat Transfer and Fluid Mechanics Institute, Stanford University Press (1976), 152-164.

Bercovier, M. and Engelman, M.S., A finite element for the numerical solution of viscous incompressible flows, Journal of Computational Physics, 30, (1979), 181-201.

Fjortoft, R., On the changes in the spectral distribution of kinetic energy for two-dimensional, non-divergent flow. Tellus, Vol.5, 3, August (1953), 225-230.

Hanjalic, K., Two-Dimensional Asymmetric Turbulent Flow in Ducts, Ph.D. Thesis, Department of Mechanical Engineering, Imperial Collete, University of London (1970).

Kraichnan, R.H., Inertial ranges in two-dimensional turbulence. The Physics of Fluids, Vol.10, 7, July (1967), 1417-1423.

Launder, B.E. and Spalding, D.B., The numerical computation of turbulent flows, Computer Methods in Applied Mechanics and Engineering, 3, (1974), 269-289.

Lean, G.H. and Weare, T.J., Modelling two-dimensional circulating flow, ASCE Journal Hyd.Div., 105, Jan.(1979), 17-26.

Leonard, B.P., A survey of finite differences of opinion on numerical muddling of the incomprehensible defective-confusion equation, In : Finite Element Methods for Convection Dominated Flows (ed. Hughes, T.J.R.), ASME Proceedings AMD - Vol.34, (1979).

McGuirk, J.J., Private communication (1980).

Reynolds, A.J., Turbulent Flows in Engineering, Wiley, London (1974), 373.

Svensson, U., A Mathematical Model of the Seasonal Thermocline, Report No. 1002, Department of Water Resources Engineering, Lund Institute of Technology, Lund, Sweden (1978).

Tong, G.D., Environmental hydrodynamic modelling - problem identification and a strategy to model assessment. In : Industrialised Embayments and their Environmental Problems (eds. Collins, M.B. et.al.), Pergamon Press, Oxford (1980).

Tong, G.D.[1], Computation of Turbulent Recirculating Flow, Ph.D. Thesis, Dept. of Civil Engineering, University of Wales, Swansea, (January 1982).

Tong, G.D.[2], A treatment of wall boundaries for (k-ε) turbulence modelling within an integral (finite element) formulation. Proceedings 4th Int.Symposium on Finite Elements in Flow Problems, Tokyo, July (1982).

Tong, G.D.[3], Computation of recirculating flows in two-dimensions using finite elements andthe k-ε model. Proceedings IAHR Symposium on Refined Modelling of Flows, Paris, September (1982), 343-353.

Zienkiewicz, O.C., The Finite Element Method (3rd edition), McGraw-Hill, U.K., (1977), 189-191.

Zienkiewicz, O.C. and Heinrich, J.C., A unified treatment of steady-state shallow water and two-dimensional Navier-Stokes equations - Finite element penalty function approach, Computer Methods in Applied Mechanics and Engineering, 17/18, (1979), 673-698.

Computational Techniques & Applications: CTAC-83
J. Noye & C. Fletcher (Editors)

PRACTICAL SOLUTIONS OF THE PARABOLIC EQUATION MODEL FOR UNDERWATER ACOUSTIC WAVE PROPAGATION

D.J. Kewley, L.T. Sin Fai Lam and G. Gartrell*
Weapons Systems Research Laboratory
Defence Science & Technology Organization
Department of Defence
GPO Box 2151
ADELAIDE SA 5001

This paper reports an investigation on the performance of the split-step Fast Fourier Transform (FFT) and the Implicit Finite Difference (IFD) methods for solving the parabolic equation for underwater acoustic wave propagation. The emphasis is on testing two existing implementations of these methods on two real cases: a deep water problem and a shallow water problem. The effect of the starting fields using the vertical linear array (VLA), the Gaussian spike and the normal mode methods is examined. Two local error norms are used to assess the "correctness" of the solutions. The results show that the VLA starting field is acceptable as a simple and realistic technique. The performance of the FFT and IFD methods is shown to depend upon the example considered. For similar step sizes the FFT method is superior for the deep water case and the IFD method is superior for the shallow water case.

INTRODUCTION

The development of the parabolic equation (PE) approximation [1,2] for the solution of underwater acoustic wave propagation has led to increased sophistication in the types of problems now studied. These include range dependence [3], shallow water [4], and three-dimensional propagation [5]. The first practical method of solving the PE was due to Tappert and Hardin [6] who developed a split-step algorithm employing the Fast Fourier Transform(FFT). Since then it has been found that other solution methods may be applicable, for instance, using ordinary differental equations [7] and finite difference methods [8].

This paper compares performance, by way of examples, of the FFT method and the Implicit Finite Difference (IFD) method and the appropriateness of the initial data and the sea bottom boundary conditions. Several initial data generating methods are compared and the effects of different step sizes investigated. These methods are applied to two real data examples.

THE ACOUSTICAL PROBLEM

One of the basic problems in underwater acoustics is to determine the propagation (or transmission) loss of a given signal through the ocean, possibly reflecting off the sea surface and bottom, out to ranges of hundreds of kilometres. As the

*Now at ARDU, RAAF Edinburgh, South Australia

ocean is in reality a stochastic medium, due to fluctuations in the acoustic propagation properties in space and time, a first approximation is to assume that it can be modelled as having either constant or gradually varying properties. The cylindrically symmetric Helmholtz equation for underwater acoustic propagation is thus given by

$$\frac{\partial^2 P}{\partial r^2} + \frac{1}{r}\frac{\partial P}{\partial r} + \frac{\partial^2 P}{\partial z^2} + \kappa_o^2(n^2 + i\alpha)P = 0, \tag{1}$$

where $\kappa_o = \omega/c_o$ is a reference wavenumber, $n = n(r,z) = c_o/c(r,z)$ is the index of refraction, P is the acoustic pressure, c_o is the reference sound speed, $c(r,z)$ = sound speed at range r and depth z, $\omega = 2\pi f$, f is the frequency and $\alpha(r,z)$ is the volume attenuation coefficient. When the far field approximation, $\kappa_o r \gg 1$, is made then the acoustic pressure is

$$P(r,z) = \psi(r,z)\,(2/i\pi\kappa_o r)^{\frac{1}{2}} e^{i\kappa_o r} \tag{2}$$

Substitution into (1) gives

$$\frac{\partial^2 \psi}{\partial r^2} + 2i\kappa_o\frac{\partial \psi}{\partial r} + \frac{\partial^2 \psi}{\partial z^2} + \kappa_o^2(n^2 - 1 + i\alpha)\psi = 0 \tag{3}$$

A further approximation is made by representing the acoustic field by sound rays which have only small angles of inclination to the horizontal. This is the paraxial approximation

$$\left|\frac{\partial^2 \psi}{\partial r^2}\right| \ll \left|2i\kappa_o\frac{\partial \psi}{\partial r}\right| \tag{4}$$

Equation (3) becomes

$$2i\kappa_o\frac{\partial \psi}{\partial r} + \frac{\partial^2 \psi}{\partial z^2} + \kappa_o^2(n^2 - 1 + i\alpha)\psi = 0, \tag{5}$$

which is the underwater acoustic parabolic wave equation for propagation introduced by Tappert [1] using the above assumptions.

The boundary conditions needed to solve (5) are the sea surface and bottom properties and an initial acoustic pressure profile. The sea surface will be considered here as a perfect reflector causing a 180° phase shift. The sea bottom is considered to have a constant positive gradient of sound speed up to a constant sub-bottom value. This models a layer of sediment above a rock basement. The initial data is generated by three methods: the normal mode method [3] of solving equation (1), a Gaussian spike [1] or a vertical linear array [9].

From the solution, the propagation loss in decibels is calculated as

$$PL = 10\log_{10}\left[|P(z,r)|^2/(P_o/r_o)^2\right], \tag{6}$$

where P_o/r_o is the pressure of the point source at unit distance $r_o = 1$ m.

SPLIT-STEP METHOD (FFT)

The Tappert and Hardin method [6] of solving (5) first uses an operator form

$$\frac{\partial\psi}{\partial r} = ik_o \frac{(n^2-1+i\alpha)}{2} + \frac{i}{2\kappa_o}\frac{\partial^2\psi}{\partial z^2} \tag{7}$$

This is solved using

$$\psi(r+\Delta r,z) \simeq \exp[i\Delta r\,\kappa_o\frac{(n^2-1+i\alpha)}{2}]\;F^{-1}(\exp[-i\frac{\Delta r\,s^2}{2\kappa_o}]\;F(\psi(r,z))), \tag{8}$$

where $F(\psi(r,z)) = \int_{-\infty}^{\infty} \psi(r,z)e^{-izs}dz$ is a Fourier transform and F^{-1} denotes the inverse Fourier transform.

The solution requires ψ and $\frac{\partial\psi}{\partial z} \to 0$ as $z \to \pm\infty$ and in the current implementation [9] to satisfy these conditions an approximation is made by choosing a sufficiently large number of points in the Fast Fourier Transforms. The error (difference in differential equations) incurred by using the split-step method is proportional to $(\kappa_o\Delta r\,(\partial n/\partial z))^2$. A discussion on errors is given in Tappert [1] and McDaniel [2].

IMPLICIT FINITE DIFFERENCE METHOD (IFD)

The method of Lee and Papadakis [7] solves (5) by using the Crank-Nicolson method for variable coefficients. Full details are provided in Lee and Botseas [10].

The local truncation error is of order $(\Delta r^3 + \Delta r\,\Delta z^2)$.

INITIAL FIELD CALCULATION

The simplest method of finding a starting field is to use the Gaussian spike [1]. The field used here [10] is given by

$$\psi(0,z) = A[\exp(-\frac{(z-z_s)^2}{B}) - \exp(-\frac{(-z-z_s)^2}{B})], \tag{9}$$

where $B = 2/\kappa_o$, $A = (B)^{-\frac{1}{2}}$ and z_s is the depth of the signal source.

Another method, being developed here, is to calculate a table of phase and amplitude values at a series of equispaced points across a vertical line a short distance from the location of the point source. For the purposes of generating the initial field, sound speed is assumed to be locally depth-invariant at the source, so that the wavefronts intersecting the line may be considered to be spherical. The table values must be calculated at suitably close depth intervals along the line for a distance subtending the source angle of interest, with table values outside this angle set to zero. The paraxial approximation limitation of the PE model precludes source grazing angles much in excess of 25°, which in turn restricts the vertical coverage of the non-zero data values (referred to as the aperture) and so limits the error introduced by considering that the source region sound speed is depth invariant.

The most complex method involves solving the full elliptic Helmholtz equation. The normal mode solution technique [3] determines the starting field close to the source as the summation of all contributory modes each with the appropriate excitation amplitude and phase. Considerable computation is necessary to define the modes, particularly in deep ocean cases where as many as several hundred distinct trapped modes may exist. If the modes are defined with sufficient precision the logical result of this summation is that the modes reinforce in the near vicinity of the source to recreate the near field wave structure, but cancel everywhere else. The method involves extensive computation and interpretation, the major portion of which may in deep water cases be directed towards establishing what is known already, i.e. the starting field intensity is close to zero everywhere but adjacent to the source.

Direct field methods such as the Gaussian spike and mini vertical linear array are simpler than the normal mode technique since they concentrate on direct calculation of the near field in the vicinity of the source and simply fill the remainder of the starting field array with zeroes. This is a much more computationally efficient procedure and does not require any operator intervention. This paper examines the effects of these starting fields upon the solution generated by the FFT and IFD methods.

RESULTS AND DISCUSSION

We have made a comparison of the FFT and the IFD methods using two cases: 1. a deep water problem, 2. a shallow water problem. We have investigated the following effects on the numerical solution:

(1) the starting field using: (a) a Gaussian spike field, (b) a vertical linear array (VLA) starting field, and (c) a normal mode field;

(2) the stability of the result with step in range Δr and step in depth Δz.

The calculations were made in batch mode using an IBM 3033 computer and the FORTRAN 77 compiler (VS/FORTRAN).

A Deep Water Problem

A deep water sound propagation problem is considered in a region bounded by a pressure release and a flat horizontal bottom. The following set of input parameters is used :

Source depth	=	244 m
Receiver depth	=	241 m
Bottom depth	=	5390 m
Reference sound speed	=	1505.3 m/s

Sound speed gradient g ($\equiv dc/dz$) in sediment is $g = 1\ s^{-1}$

Total horizontal range	=	200 km
Frequency	=	63 Hz
Attenuation	=	0.085 dB/wavelength (for sediment and rock)

The density is held constant everywhere.

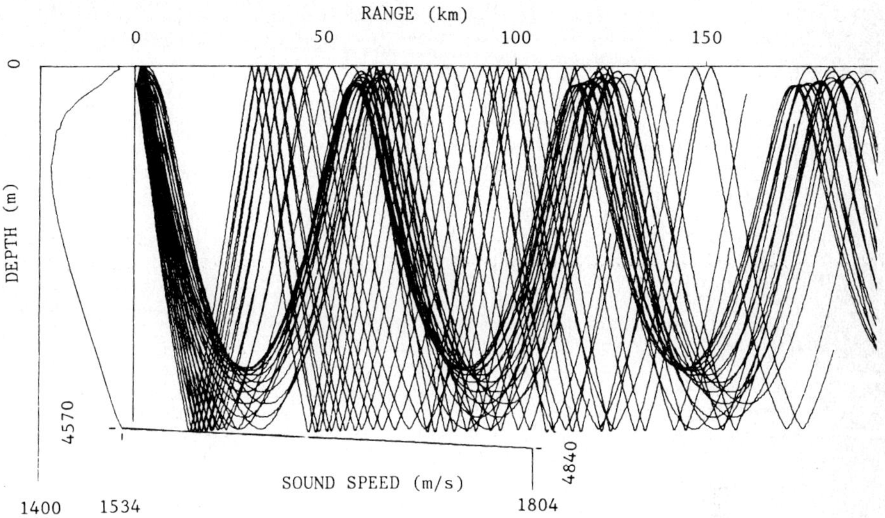

Figure 1. Acoustic ray tracing for the deep water problem. The sound speed profile is shown.

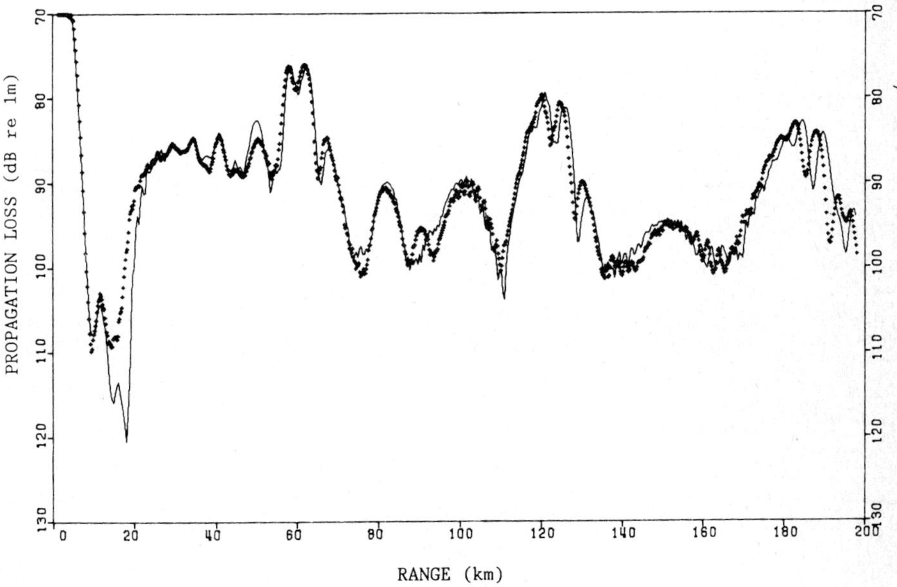

Figure 2. Comparison of FFT and IFD methods using a VLA starting field. + + + FFT Δr=200m, Δz=10m ——— IFD Δr=10m, Δz=5m

Figure 1 shows the acoustic rays from the source depth propagating through the ocean, reflecting from the subsea bottom and sea surface and refracting under the influence of the ocean sound speed profile. The sound speed profile, seen in figure 1, varies with depth but is held constant in range. The ray tracing shows that (for rays having a small inclination to the horizontal (±15°)) there will be (1) a high loss region between 5 to 20 km, (2) a region dependent upon a bottom bounce, (3) a ray convergence region of low loss and (4) further bottom bounce and convergence regions. The higher angled rays will not be modelled well because of the paraxial approximation used to obtain (4).

Our calculations show that with the FFT method we can use the values of $\Delta r = 200$ m and $\Delta z = 10$ m. With the IFD method a much smaller value of step range is necessary, $\Delta r = 10$ m, $\Delta z = 5$ m. The same VLA starting field is used in each case. Figure 2 compares the results using the FFT and the IFD methods averaged over 2 km intervals to assist the comparison. The results show the features predicted by the ray trace plot with a complex range structure. Both methods show the same structure with small differences past 20 km. The very high loss at shorter ranges shown by both methods does not match reality as this region must be dependent on bottom bounces of high angled rays, so their differences are not significant when comparison is made with experiments. The high loss is due to the paraxial approximation.

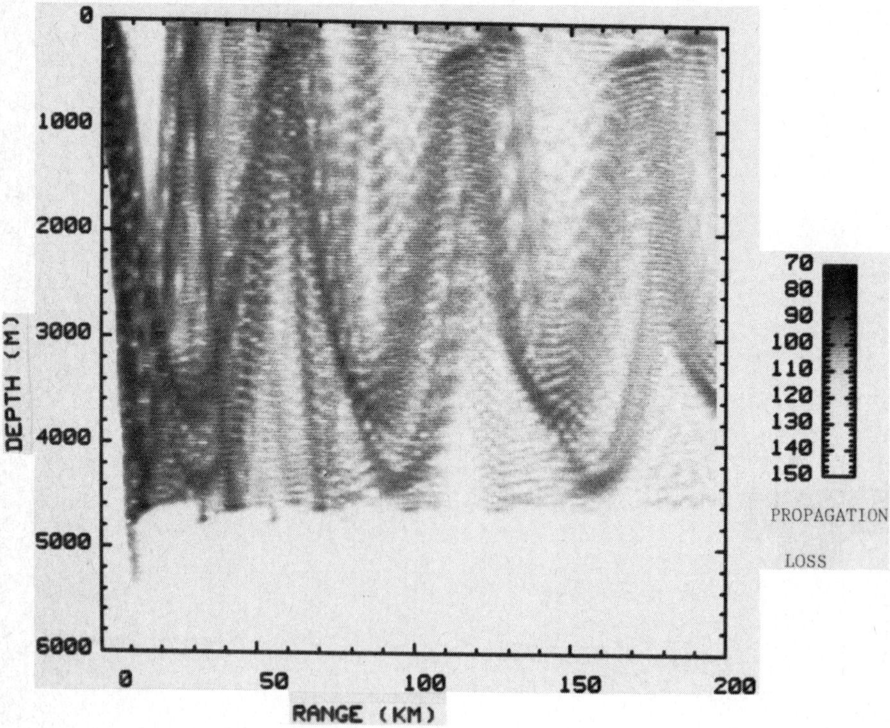

Figure 3. The two dimensional propagation loss for the deep water problem using the IFD method with a VLA starting field and Δr=10m, Δz=5m

The two dimensional acoustic field is seen in figure 3 using the above IFD results. The reflections and refractions predicted by the ray tracing plot, figure 1, are verified. The penetration by waves into the sea bed is shown to occur. Complex interference patterns not predicted by simple ray tracings are also seen.

To check the validity of the parabolic approximation we have also calculated a "norm" of local error $||\varepsilon||$ defined by

$$||\varepsilon|| = \int \varepsilon(z,r)|\psi|^2 dz / \int |\psi|^2 \, dz \, . \tag{10}$$

Here, ε is defined by $\varepsilon = n^2 - 1 + i\alpha = c_o^2/c^2 - 1 + i\alpha$, and it is determined by environmental conditions and by the choice of c_o. The ψ is the particular solution of (5) under consideration at range r.

It is seen from table 1 that $||\varepsilon||$ is small when a satisfactory set of values for Δr and Δz is chosen. The maximum values occur at ranges corresponding to high values of ε and ψ, ie the first bottom bounce. However, the difference between the FFT and IFD solutions when $\Delta r = 200$ m and $\Delta z = 10$ m are used for both, is clearly seen in figure 4 to be substantial although the $||\varepsilon||$ values are comparable.

A further "norm" of μ is defined by

$$||\mu|| = \int \psi^* \mu \psi \, dz / \int |\psi|^2 dz$$

$$= \int \left| \frac{1}{\kappa_o} \frac{\partial \psi}{\partial z} \right|^2 dz / \int |\psi|^2 \, dz \tag{11}$$

where $\mu = (\partial^2/\partial z^2)/\kappa_o^2$. Since $\partial\psi/\partial z$ is the vertical wavenumber and $\kappa_o^{-1} \, \partial\psi/\partial z$ is the corresponding angle to the horizontal, $||\mu||$ gives the mean signal angle of propagation with respect to horizontal [1]. The values of $||\mu||$ are given in table 1. The values which correspond to an acceptable solution are expected to be less than 0.04 based upon Tappert's example [1]. The IFD 200 m solution is seen to have a much higher set of $||\mu||$ values which confirms the solution failure seen in figure 4. The two curves in figure 2 show similar values of $||\mu||$ which are in turn reflected in the agreement between them. Note however that there is an order of magnitude difference in CPU execution times between the two solutions which can become a significant consideration in practical usage.

Comparison between the curves in figure 5 shows that, using a given set of values of Δr and Δz, the VLA starting field gives a deeper first minimum in the propagation loss than the Gaussian spike field. Our calculations indicate the VLA starting field gives a better description of the first bounce zone, caused by the deep reflection of the acoustic wave from the bottom. The values of $||\mu||$ are seen to be more sensitive to the starting field requiring a smaller step size.

TABLE 1. A deep water problem

	Δr(m)	Δz(m)	$\|\|\varepsilon\|\|_{min}$	$\|\|\varepsilon\|\|_{max}$	$\|\|\varepsilon\|\|_{av}$	$\|\|\mu\|\|_{min}$	$\|\|\mu\|\|_{max}$	$\|\|\mu\|\|_{av}$	CPU (sec)
FFT (VLA)	200	10	0.012 (139)*	0.023 (31)	0.017	0.013 (152)	0.086 (6)	0.034	73
	200	20	0.011 (139)	0.024 (31)	0.017	0.010 (153)	0.068 (7)	0.030	37
	200	40	0.011 (1.5)	0.027 (27)	0.019	0.015 (27)	0.033 (8)	0.024	19
	400	10	0.002 (139)	0.023 (31)	0.017	0.013 (151)	0.086 (6)	0.035	38
FFT (GAUSS)	200	10	0.012 (139)	0.037 (10)	0.018	0.015 (15)	0.165 (4)	0.046	72
IFD (VLA)	10	5	0.011 (139)	0.022 (11)	0.017	0.014 (152)	0.092 (6)	0.036	754
	10	10	0.011 (142)	0.023 (14)	0.017	0.014 (155)	0.090 (7)	0.037	226
	50	5	0.011 (140)	0.028 (17)	0.017	0.014 (154)	0.095 (7)	0.038	205
	200	10	0.010 (20)	0.025 (36)	0.018	0.041 (175)	0.092 (11)	0.059	39
IFD (GAUSS)	10	5	0.011 (140)	0.058 (10)	0.018	0.015 (153)	0.258 (4)	0.050	777

* Distance in km where $||\varepsilon||_{min}$ occurs, and similarly for $||\varepsilon||_{max}$, $||\mu||_{min}$ and $||\mu||_{max}$

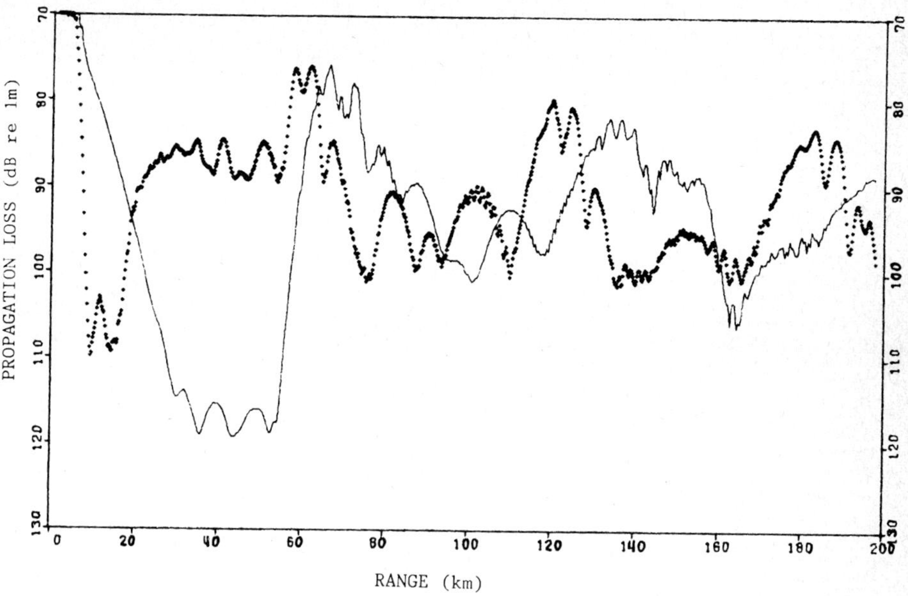

Figure 4. Comparison of FFT and IFD methods using a VLA starting field and same Δr and Δz. + + + FFT Δr=200m Δz=10m , —— IFD

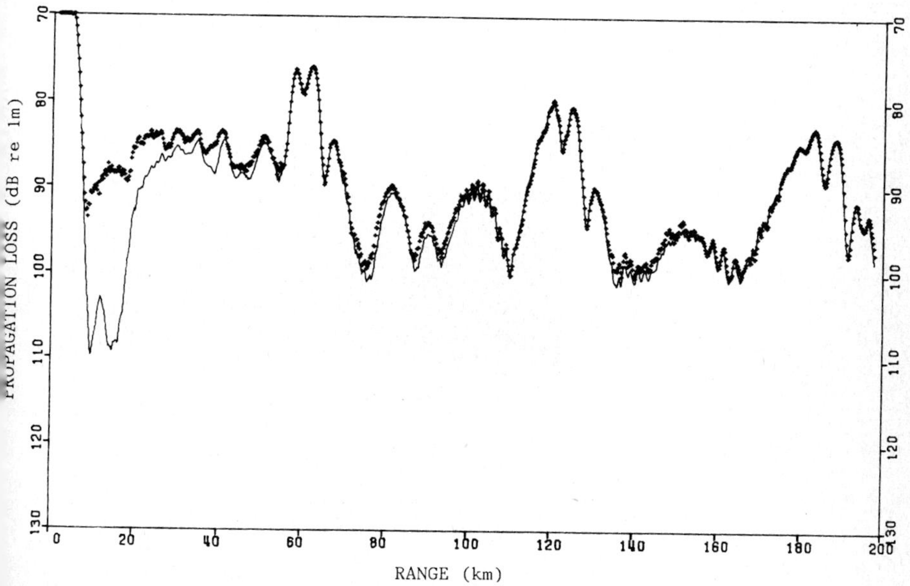

Figure 5. Comparison of FFT method using VLA and Gaussian starting fields. + + + Gaussian Δr=200m Δz=10m, —— VLA

It will be seen from Table 1 with Δr=200 m, the FFT method yields comparable values of $||\varepsilon||$ and $||\mu||$ when the depth step size Δz is increased from 10 to 20 m. A similar result is obtained using the IFD method with Δr=10 m and Δz is increased from 5 to 10 m. However, comparison of the plots of the propagation loss shows that, with the larger step size Δr, the predicted values for the first shadow region up to 20 km are lower in magnitude, and for larger ranges the results are similar but the structures are displaced up to 2 km to the right. This indicates that, although $||\varepsilon||$ and $||\mu||$ are useful for monitoring the convergence of the solutions, care needs to be exercised in choosing the optimum set of values of Δr and Δz.

Calculations using the normal mode starting field are in progress and will be presented at the conference.

A Shallow Water Problem

The region of the shallow water sound propagation considered in this work is bounded by a pressure release and a flat horizontal bottom. The following set of input parameters is used:

Source depth	=	18.3 m
Receiver depth	=	55 m
Bottom depth	=	680 m
Reference sound speed	=	1510.6 m/s
Frequency	=	50 Hz
Sound speed gradient g	=	10 s^{-1}
Attenuation	=	0.085 dB/wavelength (for sediment and rock)
Total horizontal range	=	60 km

The density is held constant everywhere.

The ray tracing of the sound wave propagation is shown in figure 6. The shallow water causes the rays to undergo many reflections and, unlike the deep water problem, the spatial structure is clear.

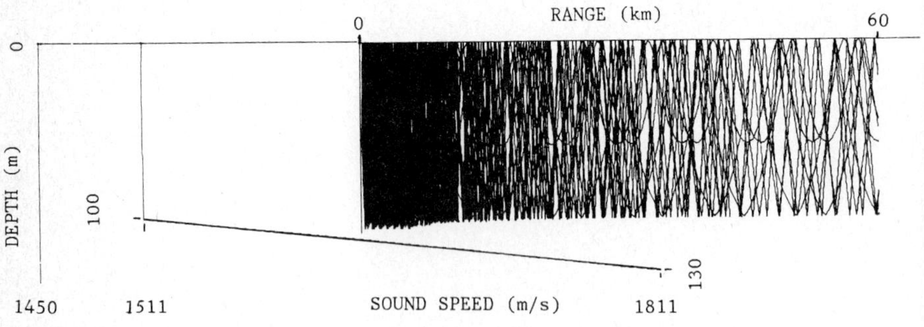

Figure 6. Acoustic ray tracing of shallow water problem.

TABLE 2. A shallow water problem

	Δr(m)	Δz(m)	$\|\|\varepsilon\|\|_{min}$	$\|\|\varepsilon\|\|_{max}$	$\|\|\varepsilon\|\|_{av}$	$\|\|\mu\|\|_{min}$	$\|\|\mu\|\|_{max}$	$\|\|\mu\|\|_{av}$	CPU (sec)
FFT (VLA)	50	2	0.002 (60)*	0.174 (0.5)	0.020	0.036 (60)	0.191 (13)	0.109	44
	100	2	0.006 (52)	0.176 (1.2)	0.029	0.084 (48)	0.352 (3)	0.129	24
	200	2	0.132 (46)	0.256 (11)	0.185	0.190 (1)	3.32 (60)	1.636	14
	25	2	0.0006 (57)	0.165 (0.5)	0.015	0.028 (59)	0.145 (2)	0.059	87
	25	4	0.0005 (47)	0.161 (0.4)	0.014	0.020 (59)	0.131 (20)	0.042	43
FFT (GAUSS)	50	2	0.0005(0.1)	0.126 (0.4)	0.020	0.051 (59)	0.239 (15)	0.150	45
IFD (VLA)	50	2	0.0005(56)	0.134 (1)	0.0137	0.021 (52)	0.121 (1)	0.043	21
	50	4	0.0005 (53)	0.153 (1)	0.0137	0.019 (51)	0.121 (1)	0.040	11
	100	2	0.0007(60)	0.141 (3)	0.030	0.022 (57)	0.123 (2)	0.056	14
	200	2	0.027 (58)	0.162 (27)	0.098	0.048 (60)	0.134 (32)	0.101	10
	25	2	0.0004(57)	0.133 (1)	0.011	0.020 (59)	0.121 (1)	0.039	35
IFD (GAUSS)	50	2	0.0005(53)	0.100 (2)	0.017	0.023 (55)	0.203 (1)	0.074	21

* Distance in km where $||\varepsilon||_{min}$ occurs, and similarly for $||\varepsilon||_{max}$, $||\mu||_{min}$ and $||\mu||_{max}$.

The results of our calculations with the FFT and the IFD methods using the same initial VLA starting field are summarised in table 2. The error norms indicate that the appropriate IFD and FFT step sizes are $\Delta r = 50$ m, $\Delta z = 2$ m and $\Delta r = 25$ m, $\Delta z = 2$ m, respectively. The corresponding CPU execution times show the IFD method being about four times faster, in contrast to the deep water problem.

Figure 7 shows the propagation loss predicted by these two methods using the corresponding appropriate step sizes. Comparison between the ray tracings of the sound wave propagation, shown in figures 1 and 6 indicates the reason for the IFD method being better than the FFT method for treating the shallow water problem, and vice versa for the deep water problem. In the shallow water problem, the acoustic rays undergo many reflections. The properties of the sea bottom are therefore important and these features are taken explicitly into account with a horizontal interface treatment in the IFD method [10].

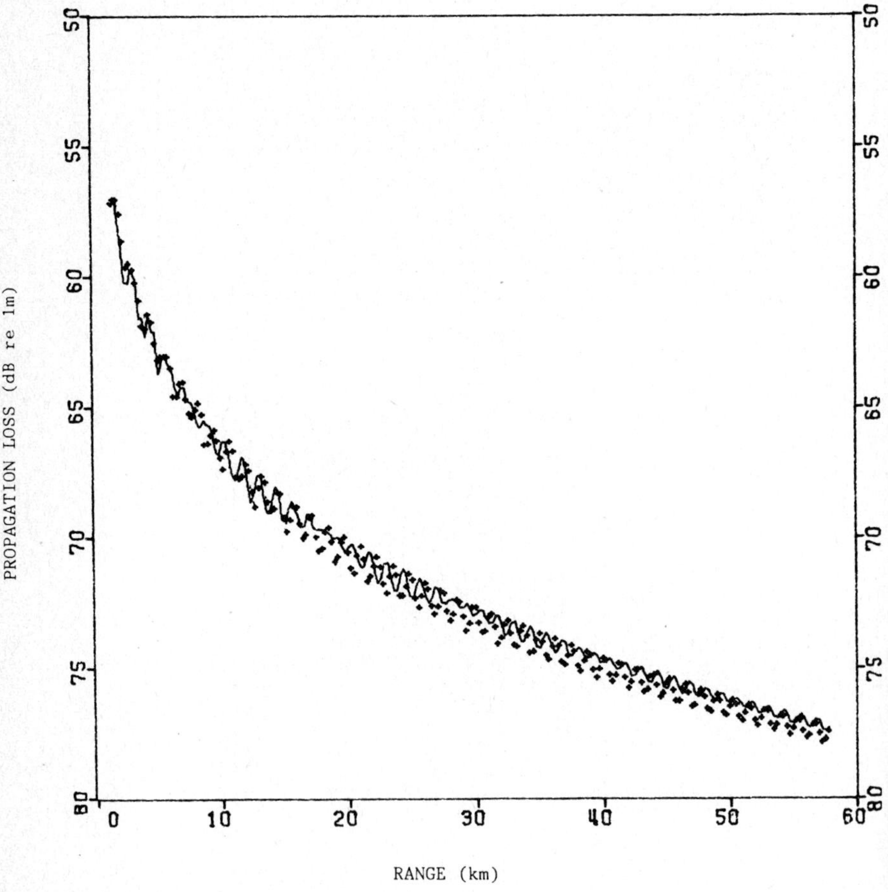

Figure 7. Comparison of FFT and IFD methods using a VLA starting field. + + + FFT Δr=25m Δz=2m , —— IFD Δr=50m Δz=2m.

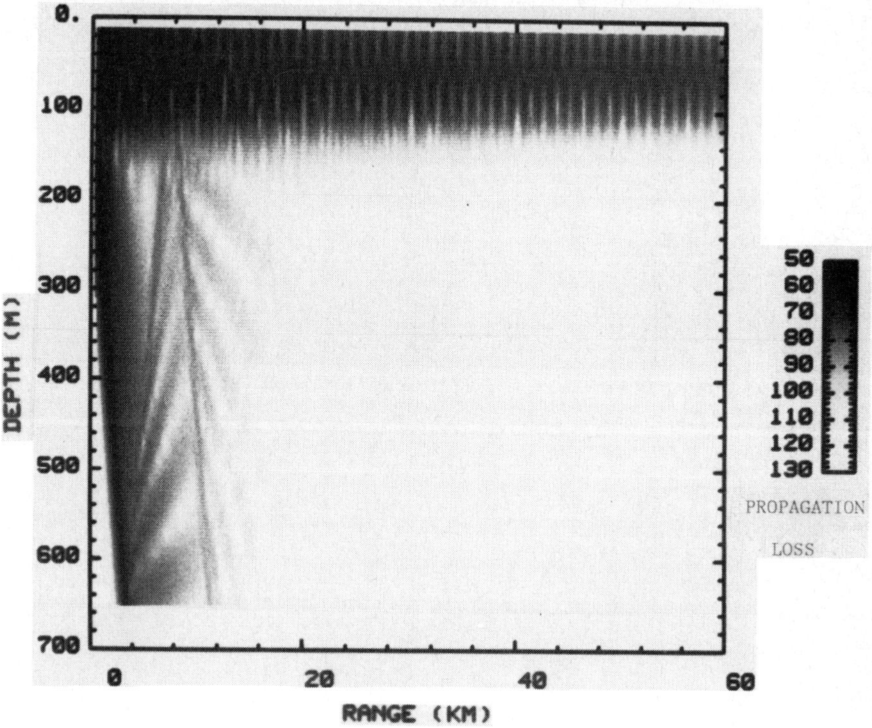

Figure 8. Two dimensional propagation loss for the shallow water problem using the IFD method with a VLA starting field Δr=50m, Δz=2m.

The two dimensional acoustic field is seen in figure 8 using the above IFD results. The large number of reflections off the sea bottom is seen to produce complex interference patterns which are the cause of the oscillations seen in figure 7. At shorter ranges the sea bed wave penetration is clearly very important and was not taken into account by the ray tracing method.

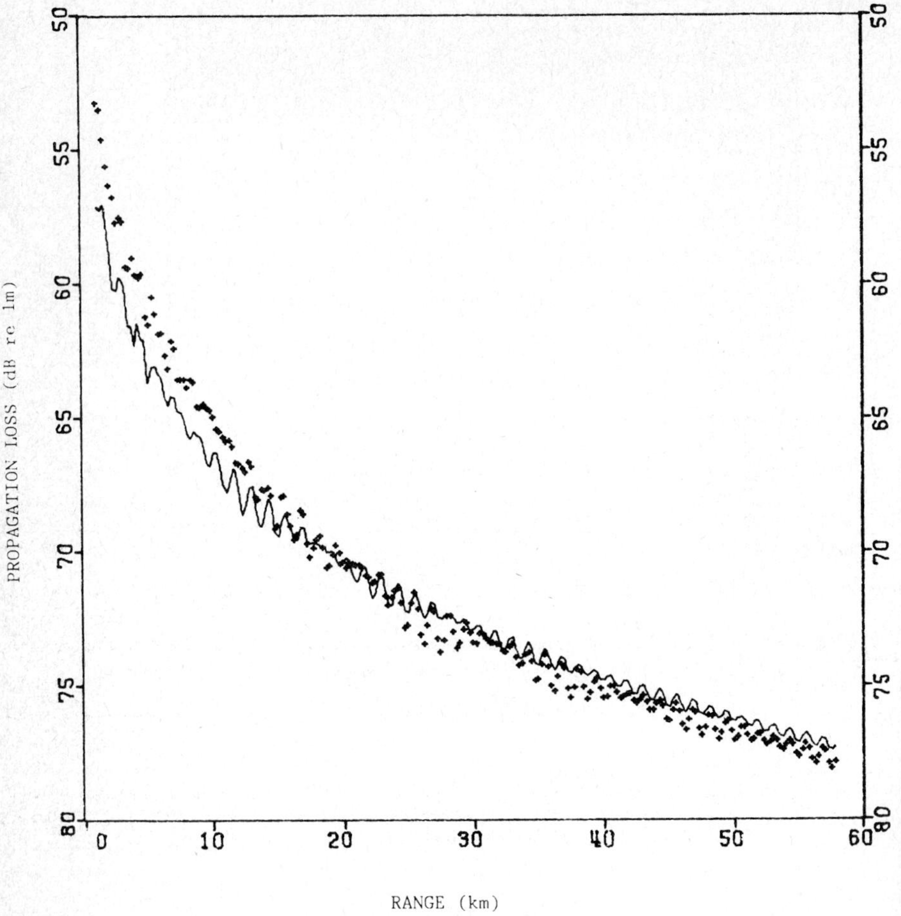

Figure 9. Comparison of IFD method using VLA and Gaussian starting fields.
+ + + Gaussian , —— VLA Δr=50m Δz=2m.

The effect of the starting field is shown in figure 9. The calculations have been done with the IFD method using the VLA and Gaussian spike starting fields with the same step sizes Δr = 50 m and Δz = 2 m. The results obtained using the Gaussian spike show lower loss at ranges less than 15 km and higher loss thereafter, compared with the VLA results. Examination of table 2 shows that both $||\varepsilon||_{av}$ and $||\mu||_{av}$ are significantly larger for the Gaussian case compared to the VLA case. This suggests that the step size needs to be reduced to treat the higher angle components of the Gaussian spike starting field. These high angle components are not included in the VLA technique and cannot, in any case, be accurately modelled because of the paraxial assumption (eq. 4) used to derive the parabolic equation (eq. 5).

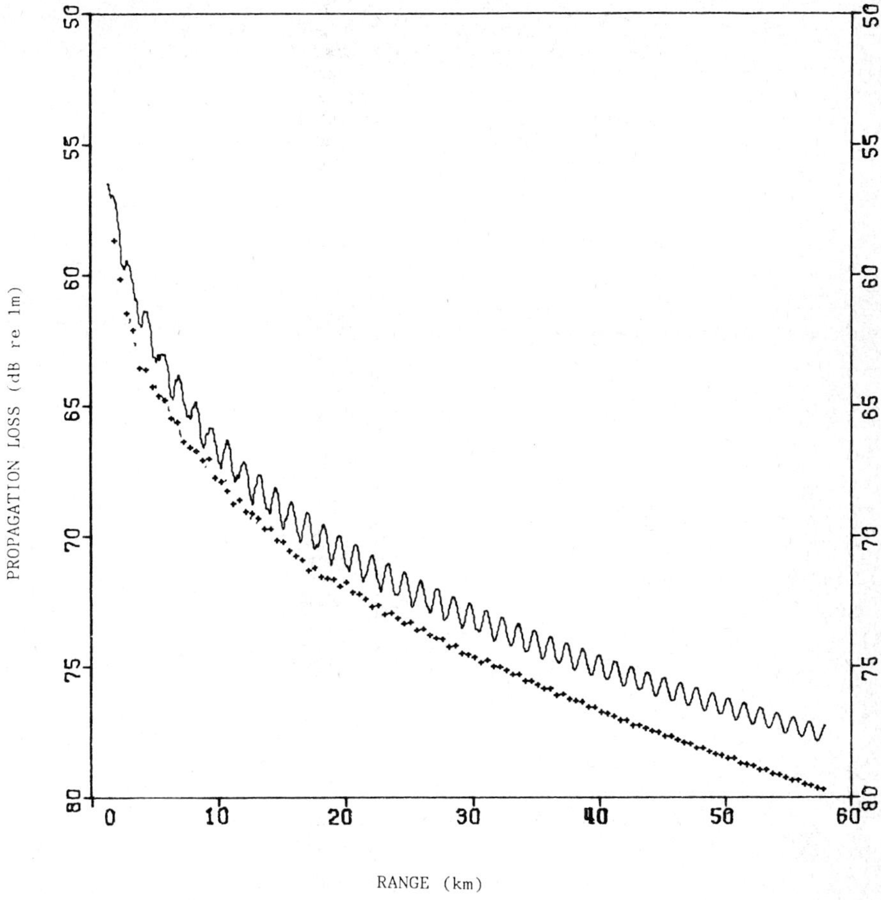

Figure 10. Comparison of FFT method using VLA starting field and different step sizes. + + + Δr=100m Δz=2m, —— Δr=25m Δz=2m.

Finally, figure 10 shows the effect on the predicted propagation loss when the range step size Δr is increased from 25 to 100 m, using a constant Δz = 2 m. The calculations have been done using the FFT method. The predicted loss is seen to be only up 5 dB less with Δr = 100 m than with Δr = 25 m. However, examination of table 2 shows that the error norms $||\varepsilon||_{av}$ and $||\mu||_{av}$ are much larger when Δr = 100 m is used.

SUMMARY

A comparison of two methods of solving the parabolic equation for underwater acoustic propagation shows that the environment will determine which method is the more accurate. The two local error norms of Tappert usually show when to suspect the solution as inaccurate but not always. The effect of the initial starting field was investigated and shows that the VLA method is probably the most appropriate due to its simplicity and appeal to reality.

REFERENCES

[1] Tappert, R.J., The parabolic approximation method, in: Keller,J.B and Papadakis,J.S. (eds.), Lecture Notes in Physics, Vol. 70 (SpringerVerlag, New York, 1977).

[2] McDaniel, S.T., Parabolic approximations for underwater sound propagation, J. Acoust. Soc. Am 58 (1975) 1178-1185.

[3] DiNapoli, F.R. and Deavenport, R.L., Numerical models of underwater acoustic propagation, in: DeSanto, J.A. (ed.), Topics in Current Physics (Springer-Verlag, Heidelberg, 1979).

[4] Jensen, F.B. and Kuperman W.A., Range-dependent bottom-limited propagation modelling with the parabolic equation, in: Kuperman, W.A. and Jensen, F.B. (eds.), Bottom-Interacting Ocean Acoustics (Plenum Press, New York, 1980).

[5] Perkins, J.S. and Baer R.N., An approximation to the three dimensional parabolic-equation method for acoustic propagation, J. Acoust. Soc. Am. 72 (1982) 515-522.

[6] Tappert F.D. and Hardin R.H., Computer simulation of long-range ocean acoustic propagation using the parabolic equation method. Proceedings of the Eighth International Congress on Acoustics, London, Vol 2 p.452 (1974).

[7] Lee, D. and Papadakis J.S., Numerical solutions of the parabolic wave equation: An ordinary-differential equation approach, J. Acoust. Soc. Am. 68 (1980) 1482-1488.

[8] Lee, D. Botseas, G and Papadakis, J.S., Finite-difference solutions to the parabolic wave equation, J. Acoust. Soc. Am. 70 (1981) 795-1488.

[9] Gartrell, G., A parabolic equation propagation loss model, Technical Report WSRL-0034-TR (1978).

[10] Lee, D and Botseas, G., IFD: An Implicit Finite-difference computer model for solving the parabolic equation, NUSC Technical Report 6659 (May 1982).

Computational Techniques & Applications: CTAC-83
J. Noye & C. Fletcher (Editors)

ACOUSTICAL RADIATION IN MOVING FLOWS: A FINITE ELEMENT APPROACH

R. J. Astley

Department of Mechanical Engineering
University of Canterbury
Christchurch
New Zealand

An axysymmetric finite element model is presented for the prediction of radiation patterns generated by stationary acoustical sources in moving flows. The acoustical field is represented in an outer region by wave envelope elements which incorporate some features of ray acoustical behaviour. These are compatibly matched to a conventional finite element mesh in a region close to the generating mechanism. Results are presented for problems involving vibrating cylinders and spheres in subsonic mean flows. Good agreement is established between computed and exact solutions.

1. INTRODUCTION

The simulation of wave radiation into an unbounded region is a problem which presents many practical difficulties when conventional numerical schemes are used. Since the computational domain is of necessity of finite extent the boundary condition imposed at its outer surface must be such as to permit radiation (but not reflection) of the computed solution. The obvious treatment at such boundaries involves the use of a Sommerfeld type of radiation condition but is appropriate only if the boundary is sufficiently distant from the radiating surface. In effect the boundary must be many wavelengths distant from the source. In practice it is often prohibitively expensive to carry a conventional numerical solution into the 'far field' in this way. To do so involves computation of a solution which comprises many spatial wavelength variations within the computational domain. This requires a fine mesh if the local wavelike nature of the solution is to be resolved. When the characteristic wavelength of the problem is relatively small compared with the geometrical length scale (i.e. the dimensions of the generating mechanism or radiating surface) this problem is further exacerbated. In this case the far field boundary condition may only satisfactorily be imposed at distances from the source which are 'large' in comparison both with the characteristic wavelength and the geometrical lengthscale. A problem involving all of these troublesome characteristics is that of fan noise radiation from the inlets of turbofan aircraft engines. The geometrical lengthscale is typically of the order of 1 metre or more, with characteristic acoustical wavelengths (at frequencies of interest) which are considerably smaller. The necessity of using a purely numerical solution to represent the acoustical field in the close vicinity of such inlets appears inevitable given the complex geometry of the boundaries and the equally complex variation of the mean flow variables. The extension of a conventional numerical model into the far field is, however, impracticable given the number of mesh points that would then be required. It is in an attempt to develop numerical techniques for the turbofan inlet problem that much of the work reported in this paper is directed.

In the present paper a conventional finite element (FE) scheme is proposed for the representation of the 'near' field. It is matched compatibly to a 'wave envelope' FE scheme in an intermediate region which extends to a far field boundary. The shape functions of the wave envelope elements are modified to accommodate locally wavelike solutions. These modifications are based on ray acoustical considerations. The method differs, however, from the pure application of ray acoustical theory in that no high frequency approximations are involved and the full acoustical equations are preserved throughout the numerical scheme. At the outer boundary of the wave envelope region a convected form of the Sommerfeld condition is applied as a non-reflecting boundary condition.

This general approach has already been used successfully in the absence of mean flow [1,2] and applied in the flow case to acoustical propagation in ducted regions [3]. The scheme is developed for an axisymmetric model and is tested against exact solutions for the acoustical field in the vicinity of vibrating cylindrical membranes and 'juddering' spheres. In both cases a uniform subsonic mean flow is assumed at large distances from the source. The effect of the radiating body in distorting the mean flow is fully included in the analysis.

2. GEOMETRY AND GOVERNING EQUATIONS

The geometry of the problem is shown in figure 1. The acoustical field in a region R (subdivided into regions R_1 and R_2) is generated by a vibrating surface at C_o. Cylindrical symmetry is assumed about the z axis with r forming a cylindrical radial co-ordinate. Spherical polar co-ordinates r' and θ are also used and are defined as indicated in figure 1. A steady mean flow $\underline{U}(r,z)$ exists at all points external to C_o, and approaches a uniform adverse flow $\underline{U}_\infty$ (see figure 1) at large distances from the radiating surface. The mean flow and the acoustical perturbation are assumed to be irrotational. The acoustical velocity $\underline{u}^*(r,z,t)$ may therefore be written

$$\underline{u}^*(r,z,t) = \underline{\nabla}[\phi^*(r,z,t)],$$

where ϕ^* is an acoustical velocity potential. The linearised equation for the acoustical field is then given by [4]

$$c^{*2}\,\nabla^2\phi^* - 2\underline{U}.\underline{\nabla}\phi^*_t - \tfrac{1}{2}\underline{\nabla}\phi^*.\underline{\nabla}|\underline{U}|^2 - \phi^*_{tt} - (\gamma-1)\phi^*_t\underline{\nabla}.\underline{U}$$
$$- (\gamma-1)(\underline{U}.\underline{\nabla}\phi^*)(\underline{\nabla}.\underline{U}) - \underline{U}.\underline{\nabla}(\underline{U}.\underline{\nabla}\phi^*) = 0, \qquad (1)$$

where c^{*2} is the local sound speed and γ is the ratio of specific heats (the flow is assumed inviscid and non-heat-conducting). It is now assumed that the acoustical disturbance is a superposition of time harmonic components of the form

$$\phi^*(r,z,t) = \mathrm{Re}\left\{\phi(r,z)^{i\omega t}\right\}$$

Equation (1) may be rewritten in terms of ϕ as

$$c^2\nabla^2\phi - 2ik\underline{M}.\underline{\nabla}\phi - \tfrac{1}{2}\underline{\nabla}\phi.\nabla|M|^2 + k^2\phi - (\gamma-1)ik\phi\underline{\nabla}.\underline{M} \qquad (2)$$
$$-(\gamma-1)(\underline{M}.\underline{\nabla}\phi)\underline{\nabla}.\underline{M} - \underline{M}.\underline{\nabla}(\underline{M}.\underline{\nabla}\phi) = 0$$

where $k = \omega/c_\infty$ (c_∞ denotes the sound speed at large distances from C_o). The vector mach number $\underline{M}$ (= $\underline{U}/c_\infty$) has now replaced $\underline{U}$ and the non-dimensional local sound speed, c, is given by

$$c^2 = 1 - \frac{\gamma-1}{2}\left[|\underline{M}|^2 - |\underline{M}_\infty|^2\right], \qquad (3)$$

($\underline{M}_\infty$ is the vector mach number at large distances from C_o, i.e. $\underline{U}_\infty/c_\infty$)..

Before proceeding further we observe that in the absence of the mean flow (i.e. $\underline{U} = \underline{M} = \underline{0}$) equations (1) and (2) reduce to the wave equation and Helmholtz

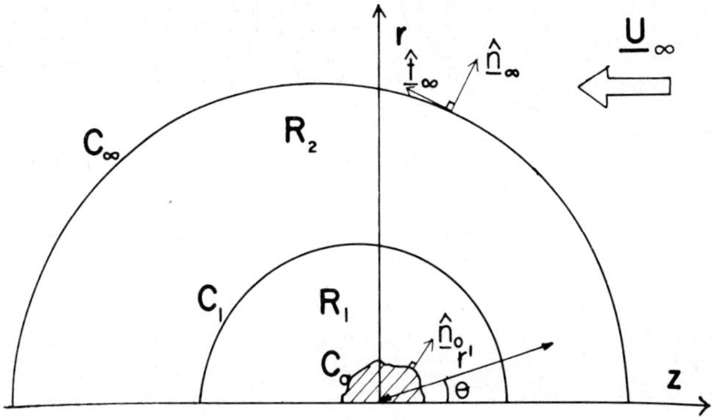

Figure 1. Geometry

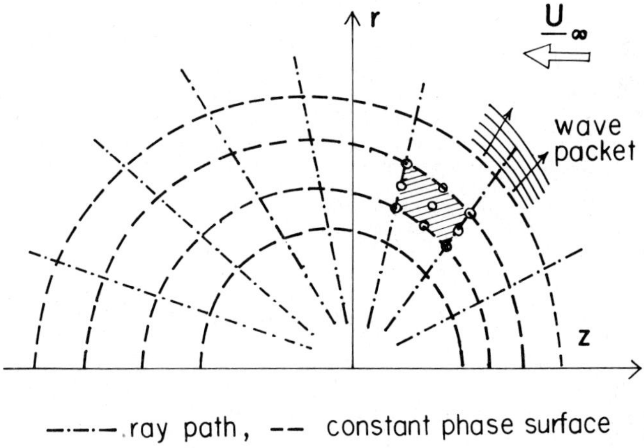

Figure 2. F.E. Discretisation of outer region.

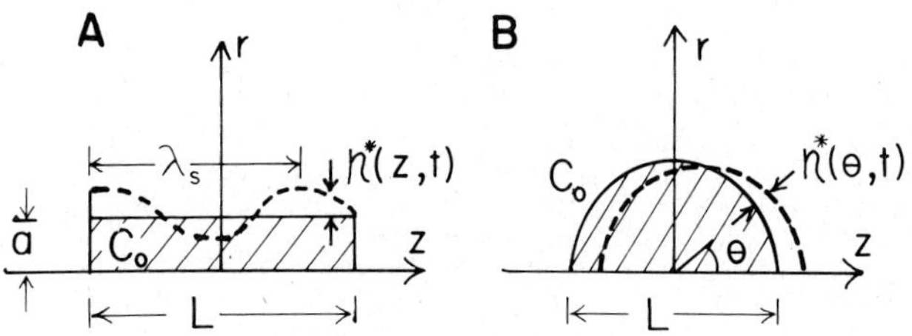

Figure 3. Geometry of the radiating surface;
(A) Vibrating cylindrical membrane
(B) Juddering sphere

equation respectively. In their full form they are equations which describe wave-like disturbances but include convective effects associated with the mean flow. A typical wavelength of the resulting acoustical field, both with and without flow, is given by $\lambda = 2\pi/k$. Typically k will be sufficiently large (i.e. the frequency ω will be sufficiently large) for this wavelength to be of the same order as, or smaller than, the geometrical lengthscale associated with the boundary C_o. The analysis is not therefore restricted to "compact" acoustical sources.

The boundary condition to be imposed at the radiating surface is of the form

$$\underline{\nabla}\phi.\hat{\underline{n}}_o = f(s) \text{ at } C_o, \tag{4}$$

where $\hat{\underline{n}}_o$ is a unit normal vector as indicated in figure 1, and s is a curvilinear co-ordinate along the boundary C_o. The function f(s) may be determined from the amplitude and phase of the vibrating boundary about its mean position. Some care must be taken to correctly include all the convective terms [5] in specifying f(s). The explicit form of f(s) will be presented at a later stage when specific geometries and displacements are considered for the radiating surface.

The boundary condition to be imposed at the far field boundary, denoted by C_∞ in figure 1, demands careful attention. If we assume that this boundary is many wavelengths away from the source and that the source is itself small compared with the curvature of C_∞ then we would expect the solution in the vicinity of C_∞ to behave locally like a 'wave packet' which originated in the vicinity of the origin. The ray paths and constant phase surfaces of such wave packets may be deduced from ray-acoustic theory [6]. For an undisturbed, uniform, adverse flow of magnitude U_∞ the local behaviour of a wave packet in the far field is given by

$$\phi^* \sim e^{i(\omega t - k\psi(z,r))} \tag{5}$$

where $$\psi(z,r) = \left[M_\infty z + (z^2 + (1 - M_\infty^2)r^2)^{\frac{1}{2}} \right] / (1 - M_\infty^2).$$

The surfaces of constant phase, ($\psi(z,r)$ = const), are then a family of offset circles as illustrated in figure 2. The ray paths in this case are straight lines. If we now assume that the same "local" behaviour is to be present in the computed solution near the outer boundary equation (5) may be used to derive an appropriate boundary condition. If C_∞ is chosen as a constant phase surface - i.e. if C_∞ is given by $\psi(z,r)$ = constant - the same assumption that the solution behaves locally like an outward propagating wave packet may be restated as an impedance condition,

$$\underline{\nabla}\phi.\hat{\underline{n}}_\infty = -ik|\underline{\nabla}\psi|\phi \text{ at } C_\infty, \tag{6}$$

where $\hat{\underline{n}}_\infty$ is a unit vector normal to C_∞, as shown in figure 1. In the case with no mean flow this boundary condition reduces to the straightforward Sommerfeld condition,

$$\frac{\partial\phi}{\partial r'} = -ik\phi \quad \text{as } r' \to \infty .$$

Boundary condition (6) completes the specification of the problem. An appropriate numerical solution is now sought using the method of weighted residuals with a finite element discretisation.

3. THE RESIDUAL SCHEME

A solution to equation (2) is now sought which is valid in the region R and which satisfies boundary conditions (4) and (6) at C_o and C_∞. A weighted residual scheme is used with a trial function $\tilde{\phi}$ and a family of weight functions W_i (i = 1, 2,....n). It is first convenient to rewrite equation (2) so that all the second order terms in ϕ come within a single divergence operator. When this is done and the field residual is multiplied by the weight function W_i then integrated over the volume of revolution generated by the region R and finally equated to zero we obtain

$$\int_R W_i\left\{\underline{\nabla}.(c^2\underline{\nabla}\tilde{\phi} - \underline{M}.(\underline{M}.\underline{\nabla}\tilde{\phi})) - 2ik\underline{M}.\underline{\nabla}\tilde{\phi} + k^2\tilde{\phi} + (\tfrac{\gamma}{2}-1)\underline{\nabla}|\underline{M}|^2.\underline{\nabla}\tilde{\phi} - (\gamma-1)ik\underline{\nabla}.\underline{M}\tilde{\phi} - (\gamma-2)(\underline{M}.\underline{\nabla}\tilde{\phi})\underline{\nabla}.\underline{M}\right\} r\,dr\,dz = 0. \qquad (7)$$

The first term in the integral over R - i.e. the term of the form $W_i\underline{\nabla}.(\;)$ - may be transformed to a boundary integral by the use of the Green's theorem. This gives

$$\int_R \left\{- \underline{\nabla}W_i.(c^2\underline{\nabla}\tilde{\phi} - \underline{M}(\underline{M}.\underline{\nabla}\tilde{\phi})) - 2ik\,W_i\,\underline{M}.\underline{\nabla}\tilde{\phi} + k^2W_i\tilde{\phi} + (\tfrac{\gamma}{2}-1)W_i\underline{\nabla}|\underline{M}|^2.\underline{\nabla}\tilde{\phi} - (\gamma-1)ik\,W_i(\underline{\nabla}.\underline{M})\tilde{\phi} - (\gamma-2)W_i(\underline{M}.\underline{\nabla}\tilde{\phi})(\underline{\nabla}.\underline{M})\right\} r\,dr\,dz$$

$$+ \int_{C_\infty} W_i\left\{c^2\underline{\nabla}\tilde{\phi} - \underline{M}(\underline{M}.\underline{\nabla}\tilde{\phi})\right\}.\hat{\underline{n}}_\infty r\,ds - \int_{C_o} W_i\left\{c^2\underline{\nabla}\tilde{\phi} - \underline{M}(\underline{M}.\underline{\nabla}\tilde{\phi})\right\}.\hat{\underline{n}}_o r\,ds = 0, \qquad (8)$$

(ds is used to denote integration along the contours C_o and C_∞ in the r-z plane of revolution). The boundary conditions on C_∞ may now be incorporated in the residual scheme by rewriting the integral over C_∞ as

$$\int_{C_\infty} W_i\left\{(c^2 - M_n^2)\underline{\nabla}\tilde{\phi}.\hat{\underline{n}}_\infty - M_nM_t\underline{\nabla}\tilde{\phi}.\hat{\underline{t}}_\infty\right\} r\,ds$$

where M_n and M_t denote the components of $\underline{M}$ at C_∞ in the normal and tangential directions represented by the unit vector $\hat{\underline{n}}_\infty$ and $\hat{\underline{t}}_\infty$ of figure 1. The term $\underline{\nabla}\tilde{\phi}.\hat{\underline{n}}_\infty$ in the above integral may now be replaced - using boundary condition (6) - by $\{-ik|\nabla\psi|\tilde{\phi}\}$ and the integral rewritten

$$\int_{C_\infty} W_i\left\{(c^2 - M_n^2)(-ik|\underline{\nabla}\psi|\tilde{\phi}) - M_nM_t\underline{\nabla}\tilde{\phi}.\hat{\underline{t}}_\infty\right\} r\,ds$$

This integral then replaces the original integral over C_∞ in equation (8).

The boundary condition on C_o is included in the weighted field residual equation in a similar fashion, i.e. by incorporating boundary condition (4) into equation (8) via the boundary integral over C_o. The last integral on the left hand side of equation (8) may then be replaced by

$$- \int_{C_o} W_i\, c^2 f(s) r\,ds. \qquad (9)$$

At this stage it is convenient to specify the trial function $\tilde{\phi}$ in more precise form as a linear combination of basis functions $\phi_i(r,z)$, (i = 1,......n) multiplied by unknown coefficients a_i (i = 1, ...n). $\tilde{\phi}$ may therefore be written

$$\tilde{\phi} = \sum_{i=1}^{n} a_i\,\phi_i(r,z)$$

The above expression is substituted into equation 8 (with the modifications to the boundary integrals already described). A set of linear equations results which may be written

$$\underset{n\times n}{[K]}\;\underset{n\times 1}{\{a\}} = \underset{n\times 1}{\{F\}} \qquad (10)$$

where the components of [K] and {F} are given by

$$K_{ij} = \int_R \Big\{ - \underline{\nabla} W_i . (c^2 \underline{\nabla}\phi_j - \underline{M}(\underline{M}.\underline{\nabla}\phi_j)) - 2ik\, W_i \underline{M}.\underline{\nabla}\phi_j + k^2 W_i \phi_j$$

$$+ (\tfrac{\gamma}{2}-1) W_i \underline{\nabla}|\underline{M}|^2 . \underline{\nabla}\phi_j - (\gamma-1) ik (\underline{\nabla}.\underline{M}) W_i \phi_j - (\gamma-2) W_i (\underline{M}.\underline{\nabla}\phi_j)(\underline{\nabla}.\underline{M})\, r\, dr\, dz$$

$$+ \int_{C_\infty} \Big\{ - W_i (c^2 - M_n^2) ik |\underline{\nabla}\psi| \phi_j - W_i M_n M_t \underline{\nabla}\phi_j . \hat{\underline{t}}_\infty \Big\}\, r\, ds, \tag{11}$$

and

$$F_i = \int_{C_0} \Big\{ c^2 W_i f(s) \Big\}\, r\, ds. \tag{12}$$

Subject to the choice of suitable basis and weight functions a solution to the problem is now obtained by solving equations (10) for the unknown coefficients a_i.

4. CHOICE OF WEIGHT AND BASIS FUNCTIONS

The solution region is divided into inner and outer subregions R_1 and R_2 which have a common interface at C_1. This interface is chosen to coincide with a surface of constant ray acoustical phase, i.e. $\psi(r,z) = \text{const}$. Within the region R_1, close to the radiating surface, a conventional finite element discretisation is used. The region is subdivided into a large number of nine noded isoparametric rectangles giving a fine mesh capable of resolving the wavelength scale of the solution. The basis functions ϕ_i and weight functions W_i are defined in the usual way by the global shape functions of the finite element mesh, i.e.

$$\phi_i(r,z) = W_i(r,z) = N_i(r,z), \tag{13}$$

where $N_i(r,z)$ is defined implicitly inside each element by the appropriate element shape function. The coefficients a_i therefore become nodal values of $\tilde{\phi}$. The components of [K] and {F} are obtained by evaluating the integrals of expressions (11) and (12) element by element and assembling the equivalent element matrices in the usual way.

In the outer region R_2 the computational domain is again divided into nine noded isoparametric rectangles. This time, however, the mesh is much coarser than in the inner region with elements extending over several wavelengths. Basis functions within this region are now defined which incorporate a wave envelope factor, i.e. ϕ_i is defined by

$$\phi_i(r,z) = N_i(r,z) e^{-ik(\psi(r,z) - \psi_i)}, \tag{14}$$

where N_i denotes the conventional global shape function (for a node within R_2), $\psi(r,t)$ is the ray acoustical phase function and ψ_i is the value of $\psi(r,z)$ evaluated at node i. By defining the basis function ϕ_i in this way it has the usual 'shape' function property of equating to unity at node i and zero at all other nodes. The coefficients a_i are thus preserved as nodal values of $\tilde{\phi}$. The presence, however, of the harmonic term in the definition of ϕ_i permits the existence of outward propagating wave packets within each element, i.e. the shape function $N_i(r,z)$ is only required to resolve the amplitude variation of the computed solution and any divergence of its harmonic behaviour from the asymptotic ray acoustic form. Most of the wavelike detail in the solution is therefore absorbed in the harmonic term and many outwardly propagating wavelengths may be accurately represented within a single element. Some care must be taken to ensure interelement compatibility, i.e. to ensure that $\tilde{\phi}$ is everywhere continuous. This is achieved if the boundaries of the elements are themselves formed by ray paths and lines of constant phase. A typical 'wave envelope' element of this type is illustrated in figure 2.

The weighting functions in the wave envelope region are now chosen as the complex conjugates of the basis functions, i.e.

$$W_i = N_i(r,z)e^{ik(\psi(r,z) - \psi_i)} \tag{15}$$

This specification - rather than the more obvious Galerkin specification of $W_i = \phi_i$ - has the practical advantage of removing the harmonic components from the integrand of equation (11). This results from the cancellation in each term of this integrand of the factor $e^{-ik\psi}$ (associated with W_i) with the factor $e^{+ik\psi}$ (associated with ϕ_i). The element contributions required to assemble K_{ij} may thus be calculated using standard forms of numerical quadrature. In the current analysis nine point Gauss-Legendre integration is used within each element both in the inner and outer regions. This constitutes a considerable practical advantage of the current approach when compared for example with analagous infinite element formulations for which special integration schemes would be required [1,7]. It is in fact somewhat difficult to envisage a suitable integration scheme for an infinite domain which could cope accurately with the convective terms within the integrand in addition to a harmonic variation of the form specified by equation (14).

5. TEST CASES: THE BOUNDARY CONDITION AT C_o

Two test cases were selected for assessment of the numerical scheme. In the first test case the contour C_o is a cylindrical surface of finite length L and radius a. The cylinder wall vibrates so that the radial displacement $\eta^*(z,t)$ (see figure 3(a)) is given by

$$\eta^*(z,t) = \eta_o e^{i\omega t - ik_s z} \tag{16}$$

This displacement represents a cylindrically symmetric structural wave of amplitude η_o and wavelength $\lambda_s = 2\pi/k_s$ travelling along the cylinder in the positive z direction (see figure 3(a)). The characteristics of the acoustical fields generated by such sources in the absence of mean flow have been discussed extensively in the context of radiation from air moving ductwork [8].

The second test case is that of a 'juddering' sphere, i.e. a rigid sphere which vibrates with an amplitude ε in the z direction. The normal displacement of the boundary $\eta^*(\theta,t)$ (see figure 3(b)) is then given by

$$\eta^*(\theta,t) = e^{i\omega t}\varepsilon\cos\theta \tag{17}$$

The general form of the acoustical boundary condition at a vibrating surface in the presence of mean flow is given by (see equation (20) of reference 5)

$$\underline{\nabla}\phi^*.\underline{\nabla}\Gamma/|\nabla\Gamma| = -\frac{1}{|\underline{\nabla}\Gamma|}\left(\frac{\partial}{\partial t} + \underline{U}.\underline{\nabla}\right)\sigma + \sigma\ \underline{\nabla}\Gamma.\underline{\nabla}(U.\underline{\nabla}\Gamma)/|\nabla\Gamma|^3, \tag{18}$$

where the equation of the vibrating surface is

$$\Gamma(r,z) + \sigma(r,z,t) = 0 \tag{19}$$

In the case of the vibrating cylinder Γ and σ are given by $\Gamma = (r-a)$ and $\sigma = \eta^* = \eta_o e^{i\omega t - ik_s z}$. After some manipulation equation (18) then gives

$$\left.\frac{\partial\phi}{\partial r}\right|_{r=a} = i\omega\eta_o(1 + M'k_s/k)e^{-ik_s z}, \tag{20}$$

where M' is the (adverse) mach number along the cylindrical boundary. If a/L is small so that the mean flow is effectively undisturbed equation (20) may be written

$$\left(r\frac{\partial\phi}{\partial r}\right)\Bigg|_{r=a} = i\omega a\eta_o(1 + M_\infty k_s/k)e^{-ik_sz},$$

giving $$r\,f(s) = -\,i\alpha\,(1 + M_\infty k_s/k)\,e^{-ik_sz}\,,\ \text{on}\ C_o \qquad (21)$$

where f(s) is as specified by boundary condition (4) and $\alpha = \omega a\eta_o$. Equation (21) may now be used to give an explicit form for (rf(s)) within the integral for F_i (i.e. equation (9)). Equation (21) may also be used to develop an exact expression for ϕ in the limiting case $\frac{a}{L} \to 0$. The vibrating cylinder may then be regarded as a distribution of monopole sources along a line segment of length L. The equivalent creation of fluid mass per unit length, m(z) say, is given by

$$m(z) = 4\pi i\alpha\rho_\infty(1 + M_\infty k_s/k)e^{-ik_sz}, \qquad (22)$$

where ρ_∞ is the density of the gas at infinity (and at all other points since the flow is undisturbed). The acoustical velocity potential, ϕ^E, generated by such a source distribution is given by

$$\phi^E(r,z) = \frac{i\alpha}{2}(1 + M_\infty k_s/k)\int_{-\frac{L}{2}}^{+\frac{L}{2}}\left\{\frac{1}{R'}e^{-ik_sz' - ik\psi}\right\}dz' \qquad (23)$$

where $$R' = ((z-z')^2 + (1 - M_\infty)r^2)^{\frac{1}{2}},$$

and $$\Psi = (M(z-z') + R')/(1 - M_\infty^2).$$

The integral in equation (23) may be performed numerically to give ϕ^E at any prescribed point in the flow. It may be shown without difficulty that if standard far field approximations are used equation (23) is equivalent to equation (8) of reference (8) in the absence of flow.

The second test case is that of a 'juddering' sphere, i.e. a rigid sphere of diameter L vibrating about the origin with amplitude ε along the z axis. The equation of the displaced boundary is then given by equation (19) with

$$\Gamma = (r' - L/2)\ \text{and}\ \sigma = \eta^*(\theta,t) = \varepsilon\cos\theta e^{i\omega t}. \qquad (24)$$

The mean flow, at low mach numbers, may be approximated by the inviscid incompressible, irrotational flow about a sphere. The tangential component of the vector velocity at the surface of the sphere is then given by

$$U_\theta = \frac{3}{2}(M_\infty c_\infty)\sin\theta. \qquad (25)$$

Substitution of equations (24) - for Γ and σ - and equations (25) - for $\underline{U}$ - into boundary condition (17) then yields

$$\frac{\partial\phi}{\partial r'} = (2\varepsilon c_\infty/L)\left[ik\cos\theta + 3M_\infty\cos^2\theta - \frac{3}{2}M_\infty\sin^2\theta\right] = -f(\theta) \qquad (26)$$

Equation (26) now defines the function f required for the evaluation of the integral in equation (9). An exact solution for the same problem, restricted to low mach numbers and wavenumbers for which kL is of order 1, is given by [5]

$$\phi^E(r',\theta) = \left(\frac{2\varepsilon c_\infty}{L}\right) e^{-ikM_\infty((r' + (a^3/2r'^2))\cos\theta)} \times \frac{1}{k}\left\{-\frac{M_\infty k^2a^2}{2}\frac{h_o^{(2)}(kr)}{h_o^{(2)}(ka)} + ika\frac{h_1^{(2)}(kr)}{h_1^{(2)}(ka)}\cos\theta + \frac{M_\infty(3 - k^2a^2)}{2}\frac{h_2^{(2)}(kr)}{h_2^{(2)}(ka)}(3\cos^2\theta - 1)\right\}, \quad (27)$$

where $a = L/2$ and $h_n^{(2)}(\ )$ is the spherical Henkel function of the second kind and n^{th} order. The above equation is in fact the complex conjugate of that derived in reference (5). This arises from an initial assumption of time harmonic behaviour of the form $e^{+i\omega t}$ instead of $e^{-i\omega t}$.

6. RESULTS

A comparison of computed and exact solutions is now presented for the two test cases described in the previous section. Results are presented in the form of;

(i) contour plots of computed values of $|\phi|$ superimposed on a representation of the FE mesh (the contours are plotted at equal increments on a logarithmic scale),

(ii) directivity plots of computed and exact values of $|\phi|$ at the outer boundary of the computational domain.

The first type of plot serves to illustrate the general nature of the solution and the second provides a concise comparison between computed and exact values in the region of greatest interest, i.e. at large distances from the source.

Computed contours of $|\phi|$ for the first test case are shown in figures 4A-4C. The solutions are calculated for acoustical and structural wavenumbers given by $kL = 2\pi$ and $k_sL = \pi$ with mach numbers at infinity of 0.0, 0.33 and 0.5. The vibrating cylinder is located at the origin. In each case the FE mesh is indicated by broken lines. Also included in figures 4A-4C are a representation of the characteristic acoustical and structural wavelengths. In the coarse outer region of the mesh - where wave envelope elements are used - the characteristic acoustical wavelength is clearly much smaller than the dimensions of the element. It has been found in previous no-flow applications that optimal results are obtained using a gradual rather than abrupt, increase in the size of the wave envelope elements at the boundary C_1 between the conventional and wave-envelope regions. This technique has been adopted in generating the meshes for figures 4A-4C with the result that the interface C_1 is somewhat indistinct and is indicated therefore by a heavy broken line.

The general nature of the solutions represented by figures 4A-4C is much as one would expect with the no flow solution of figure 4A being convected in the direction of the mean flow as the mach number increases. The mesh itself is modified by the mean flow, being generated in the outer region by the ray paths and phase surfaces of a source at the origin. The treatment of the radiation condition at the outer boundary appears to be satisfactory since there are no signs of spurious reflections in the contours of figures 4A-4C (these would manifest themselves as standing wave patterns). The accuracy of the computed solutions is confirmed by a comparison with the exact solution presented in figures 5A-5C for the same parameters. Values of $|\phi|$ and $|\phi^E|$ are plotted against θ around the outer boundary of the computational domain. In both cases $|\phi|$ is normalised by dividing by the factor α - see equation (21) - which includes the amplitude of the structural wave. Correspondence between the exact and

(A)

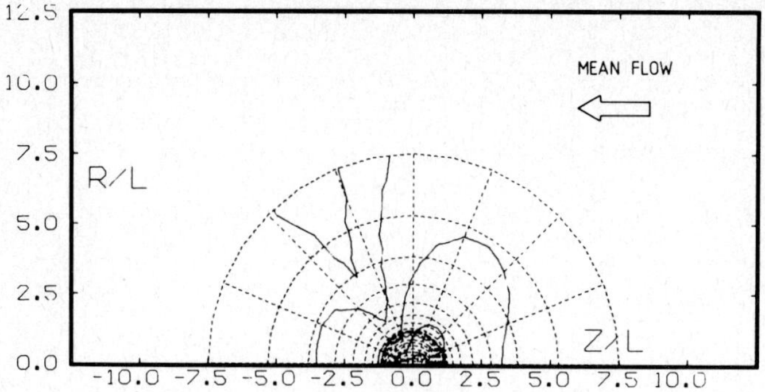

(B)

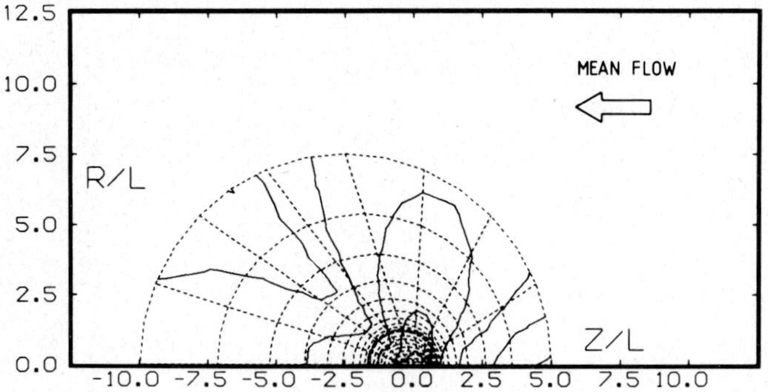

(C)

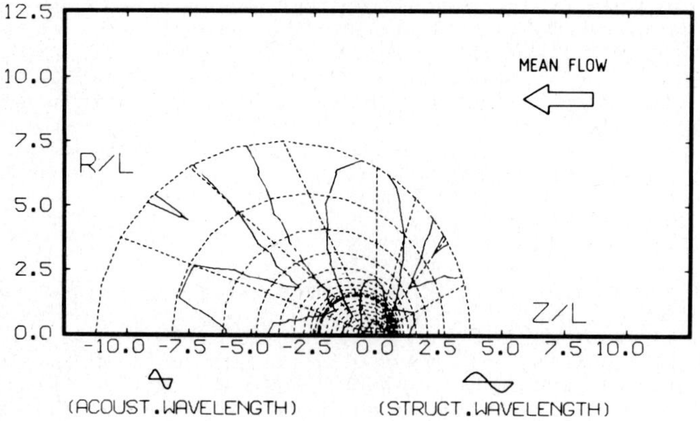

Figure 4. Computed contours of $|\phi|$, first test case; (A) $M_\infty = 0.0$, (B) $M_\infty = 0.33$, (C) $M_\infty = 0.5$

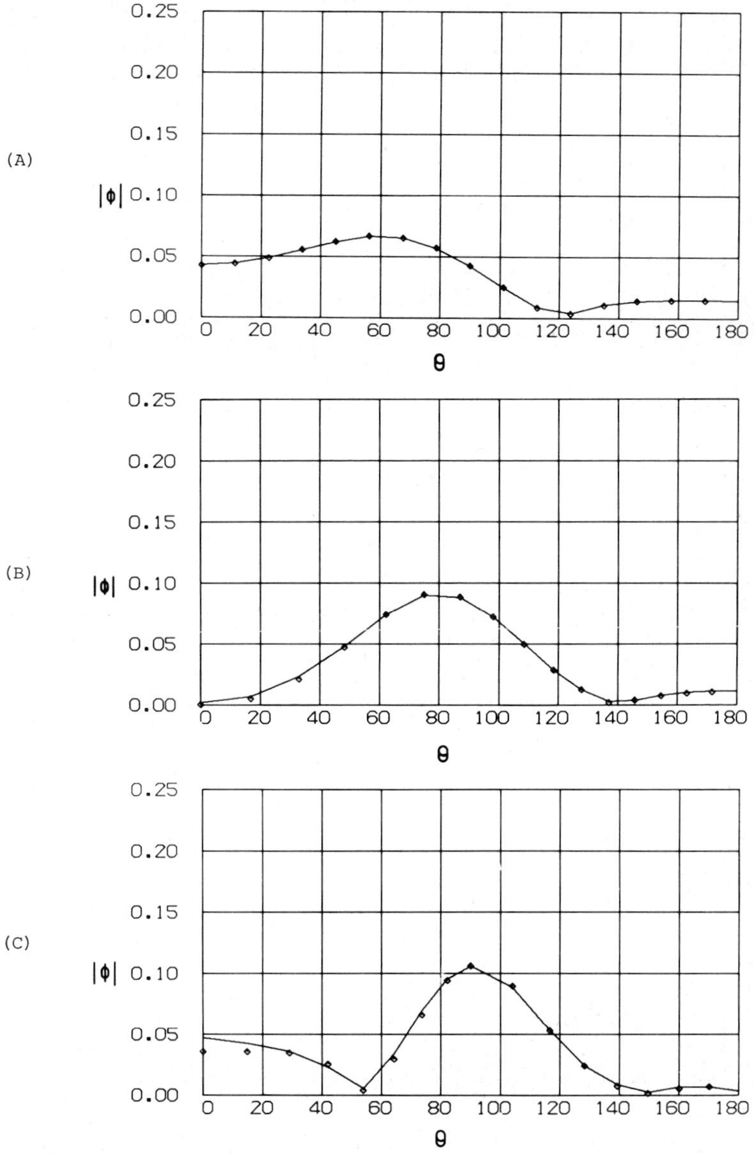

Figure 5. Exact and computed far field directivity patterns, first test case; ◇ exact, —— computed: (A) $M_\infty = 0.0$, (B) $M_\infty = 0.33$, (C) $M_\infty = 0.5$

computed solutions is seen to be extremely good for all three mach numbers.

The mean flow is undisturbed in test case 1 and the results obtained, although they indicate the accuracy of the method in representing the convective radiation condition, do not truly address the problem of predicting the effects of mean flow variations. The second test case, however, does include such effects since the mean flow is now influenced by the presence of the sphere. Computed contours of $|\phi|$ for the second test case are shown in figures 6A and 6B. The mach number at infinity is 0.0 (for figure 6A) and 0.2 (for figure 6B). The frequency in both cases is given by kL = 2. The relatively low values of mach number and frequency in this case are required by the exact solution which becomes invalid unless M is small and kL is of order 1. The acoustical field in the no-flow case - i.e. figure 6A with $M_\infty = 0.0$ - has a symmetrical dipole-like appearance as one might expect. When the mach number is increased an asymmetry develops as shown in figure 6B. A comparison of exact and computed values of $|\phi|$ for these two cases is shown in figures 7A and 7B. Values of $|\phi|$ are plotted against θ after first being normalised by dividing both solutions by the factor $(2\varepsilon c_\infty/L)$. Good correspondence is again evident between the existing and computed values. The small discrepancies in the second case - with $M_\infty = 0.2$ - are probably not in fact indications of errors in the computed solution since the 'exact' solution is approaching the limits of its own validity at these values of the mach number.

7. CONCLUSIONS

The wave envelope scheme which is proposed in this paper for wave radiation in the presence of mean flow appears to give good results for simple but representative test cases. The ability to resolve many wavelengths within a single element and to accurately predict far field characteristics at modest computational cost and without the restrictive assumptions implicit in ray theory offer considerable potential for further development. Similar techniques could be applied, without great modification, to wave propagation in a variety of inhomogeneous media where wavelengths are of the same order as, or smaller than, the lengthscale of inhomogenities (the inhomogenities in the present case being associated with the mean flow). Although the current scheme is applied to the axisymmetric problem, its modification for use in two dimensional or fully three dimensional applications is relatively straightforward.

8. ACKNOWLEDGEMENTS

The work reported in this paper was supported by the NASA Langley Research Centre under Grant NAG-1-198.

REFERENCES

[1] Astley, R.J. and Eversman, W., Finite element formulations for acoustical radiation, Journal of Sound and Vibration, 88(1), (1983), 47-64.

[2] Astley, R.J., Wave envelope and infinite element schemes for acoustical radiation, Int. Journal for Numerical Methods in Fluids, (in publication).

[3] Astley, R.J. and Eversman, W., A note on the utility of a wave envelope approach in finite element duct transmission studies, Journal of Sound and Vibration, 76(4), (1981), 595-601.

[4] Nayfeh, A.H., Shaker, B.S. and Kaiser, J.E., Transmission of sound through non-uniform circular ducts with compressible mean flow, AIAA Journal, 18, (1980), 515-525.

(A)

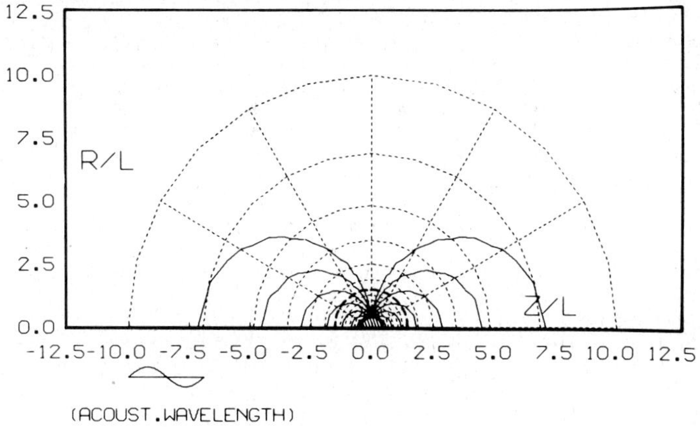

(B)

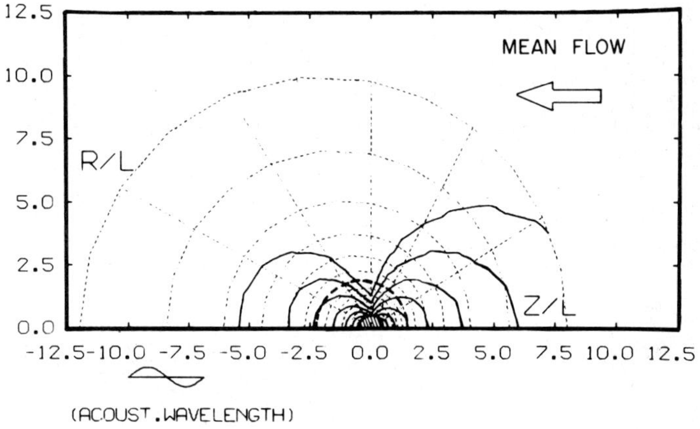

Figure 6: Computed contours of $|\phi|$, juddering sphere;
(A) $M_\infty = 0.0$ (B) $M_\infty = 0.2$

[5] Taylor, K., Acoustic generation by vibrating bodies in homentropic potential flow at low mach number, Journal of Sound and Vibration, 65(1), (1979), 125-136.

[6] Lighthill, M.J., Waves in fluids, (C.U.P. Cambridge 1978).

[7] Bettess, P. and Zienkiewicz, O.C., Diffraction and refraction of surface waves using finite and infinite elements, Int. Journal for Numerical Methods in Engineering, 11, (1977), 1271-1290.

[8] Cummings, A., Low frequency acoustic radiation from duct walls, Journal of Sound and Vibration, 71(2), (1980), 201-226.

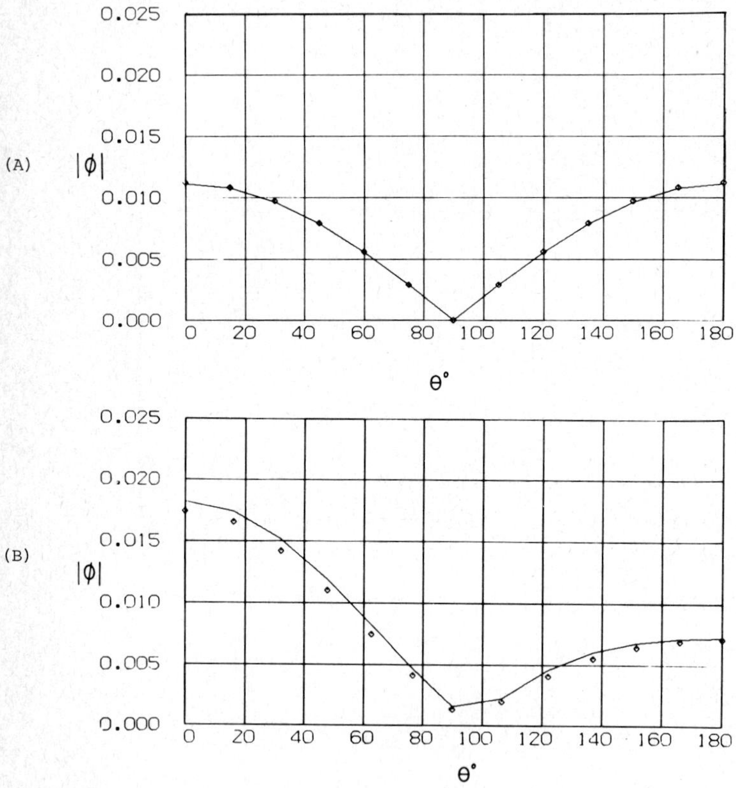

Figure 7. Exact and computed far field directivity patterns; juddering sphere; ◇ exact, —— computed; (A) $M_\infty = 0.0$, (B) $M_\infty = 0.2$

Computational Techniques & Applications: CTAC-83
J. Noye & C. Fletcher (Editors)

NUMERICAL MODEL OF A TWO-FLUID FULLY IONIZED DENSE PLASMA

P. Lalousis*

CSIRO Division of Applied Physics, Sydney, Australia

and

H. Hora

Department of Theoretical Physics, University of NSW, Australia

A fully ionized plasma for inertial confinement fusion may be treated as two coupled fluids, one consisting of electrons and the other of ions. Hydrodynamic equations for each of the fluids describe the conservation of mass, momentum and energy. The fluids are coupled by frictional forces and macroscopic electric fields. The numerical solution of the continuity and momentum equations for a two-fluid adiabatic plasma, coupled by Poisson's equation is discussed here.

INTRODUCTION

Laser-driven fusion involves the heating and compression of a deuterium-tritium pellet by laser radiation to the point of thermonuclear ignition (see e.g. Hora (1981)). For a plasma excited by radiation from a neodymium-glass laser operating at 1.06 μm, the critical density is about $10^{21}cm^{-3}$. In laser fusion we are interested in plasma densities near the critical number density because of the predominance of the nonlinear force over the thermokinetic force (Hora (1981)), and because most of the absorption of the laser light occurs near this density.

Plasmas exhibit fluid-like behaviour due mainly to electrostatic coupling between electrons and ions. Whereas an ordinary fluid is dominated by binary collisions, hot plasmas are controlled by long-range Coulomb interactions which give the plasma a fluid-like behaviour. Although the Coulomb interaction between two single charged particles is small, the collective motion of plasma particles creates electric fields of very large magnitude.

From the macroscopic point of view plasma behaviour may be modelled by two approaches:

1. Two-fluid model: the electrons and ions are treated as separate conducting fluids, coupled by momentum exchange and Coulomb interactions. Separate conservation equations describe the two fluids, and Poisson's equation is used to evaluate the Coulomb interaction.

2. One-fluid model: the plasma is treated as one fluid by assuming charge neutrality, $n_e = Zn_i$, throughout the plasma. One set of equations is used to describe conservation of mass and momentum in the plasma, but separate energy equations may be used. Poisson's equation is not required.

The motivation for developing a truly two-fluid code was to study the nonthermal direct electrodynamic interaction between laser energy and a fully ionized plasma. The particular emphasis is on the action of nonlinear (ponderomotive) forces (Hora (1981)) in which the optical electromagnetic field acts on the plasma electrons, which then transfer their energy to the ions electrostatically. The numericical treatment of two-fluid energy equations, nonlinear force and absorption via inverse bremmstrahlung will be presented in a forthcoming paper. In this paper we describe the numerical methods used to solve the equations describing conservation of mass and momentum, and Poisson's equation for a two-fluid, inhomogeneous plasma under adiabatic conditions.

EQUATIONS OF A TWO-FLUID PLASMA

The Eulerian equations used here have been derived by Lalousis and Hora (1983).

Continuity

electrons:

$$\frac{\partial \rho_e}{\partial t} \quad \frac{\partial(\rho_e u_e)}{\partial x} = 0 \tag{1.1a}$$

ions:

$$\frac{\partial \rho_i}{\partial t} + \frac{\partial(\rho_i u_i)}{\partial x} = 0 \tag{1.1b}$$

Here n_e, n_i are the number densities of the electrons and ions, m_e, m_i are the masses of electrons and ions, u_e, u_i are the velocities of the electrons and ions, and $\rho_e = m_e n_e$, $\rho_i = n_i m_i$ are the mass densities of the electons and ions respectively.

Conservation of momentum

electrons:

$$\frac{\partial(\rho_e u_e)}{\partial t} = - \frac{\partial(\rho_e u_e^2}{\partial x} - \frac{\partial P_e}{\partial x} - n_e\, e\, E + \rho_e \nu (u_i - u_e) \tag{1.2a}$$

ions:

$$\frac{\partial(\rho_i u_i)}{\partial t} = - \frac{\partial(\rho_i u_i^2)}{\partial x} - \frac{\partial P_i}{\partial x} + n_i Z\, e\, E - \rho_e \nu (u_i - u_e) \;. \tag{1.2b}$$

Here P_e, P_i are the electron and ion pressures respectively, E is the electric field due to charge separation, ν is the phenomenological electron-ion collision frequency given by Hora (1981), and e is the electronic charge.

Poisson's equation

The potential Φ is given by

$$\frac{\partial^2 \Phi}{\partial x^2} = - 4\pi e (Z n_i - n_e) \tag{1.3}$$

The electric field is defined as $E = - \frac{\partial \Phi}{\partial x}$. By differentiating Poisson's equation (1.3) with respect to time, substituting the continuity equations for ions and electrons, and integrating over x one obtains the following formulation:

$$\frac{\partial E}{\partial t} = 4\pi e (u_e n_e - Z\, u_i n_i) \tag{1.3a}$$

The (time dependent) integration constant is determined by the initial conditions. Poisson's equation is computed in the form (1.3) using the initial disturbance of the charge density, then at subsequent times equation (1.3a) is used. Poisson's equation in the form of equation (1.3a) has been used in particle simulation codes (Devatit (1981)).

To close the system of partial differential equations we use the thermodynamic equation of state

$$P = c\, n^{\gamma}$$

where c is a constant and γ is the adiabatic constant. For a one dimensional pressure $\gamma = 3$. Hence, for adiabatic compression where heat flow is negligible,

$$\frac{\partial P_e}{\partial x} = \gamma\, k\, T_e \frac{\partial n_e}{\partial x} \quad \text{and} \quad \frac{\partial P_i}{\partial x} = \gamma\, k\, T_i \frac{\partial n_i}{\partial x} \, .$$

Equations (1.1a, 1.1b, 1.2a, 1.2b, 1.3a) are then five equations describing the behaviour of the five variables n_e, n_i, u_e, u_i and E, where the initial conditions at time $t = 0$ have to be specified.

FINITE DIFFERENCE ALGORITHM

Equations (1.1a - 1.3a) are hyperbolic and expressed in conservative form. The two-step Lax-Wendroff method was chosen to solve the continuity and momentum equations. This method is second-order in time and centered in space, it is effective when used to treat an Eulerian formulation (Ritchmyer and Morton 1967), and it is easy to implement. The Lax-Wendroff method involves two steps, the auxiliary and the main step. Poisson's equation is differenced implicitly on each of these steps, using the advanced values of number densities and velocities obtained at each step from the Lax-Wendroff scheme. In the following finite difference equations spatial discretization is annotated using subscript 'j' and temporal discretization is annotated using superscript 't'.

Auxiliary step

Continuity equations:

$$\rho_{j+\frac{1}{2}}^{t+\frac{1}{2}} = \frac{1}{2}(\rho_{j+1}^{t} + \rho_{j}^{t}) - \frac{\Delta t}{2} \frac{(\rho u)_{j+1}^{t} - (\rho u)_{j}^{t}}{\Delta x}$$

Momentum equations:

$$(\rho u)_{j+\frac{1}{2}}^{t+\frac{1}{2}} = \frac{1}{2}((\rho u)_{j+1}^{t} + (\rho u)_{j}^{t}) + \frac{\Delta t}{2} \left(- \frac{(\rho u^2)_{j+1}^{t} - (\rho u^2)_{j}^{t}}{\Delta x} - \frac{P_{j+1}^{t} - P_{j}^{t}}{\Delta x} \right.$$

$$\left. \pm\, n_{j+\frac{1}{2}}^{t} e\, E_{j+\frac{1}{2}}^{t} \pm \rho_{j+\frac{1}{2}}^{t} \nu_{j+\frac{1}{2}} (u_{j+\frac{1}{2}}^{+} - u_{j+\frac{1}{2}}^{-})^{t} \right)$$

where

$$n_{j+\frac{1}{2}}^{t} = \frac{1}{2}(n_{j+1}^{t} + n_{j}^{t}) \, , \quad E_{j+\frac{1}{2}}^{t} = \frac{1}{2}(E_{j+1}^{t} + E_{j}^{t}) \quad \text{etc.,}$$

since the variables at integral values of Δt are computed at the j grid points. Thus the advanced velocities and number densities needed for Poisson's equation are

$$u_{j+\frac{1}{2}}^{t+\frac{1}{2}} = \frac{(\rho u)_{j+\frac{1}{2}}^{t+\frac{1}{2}}}{\rho_{j+\frac{1}{2}}^{t+\frac{1}{2}}} \quad \text{and} \quad n_{j+\frac{1}{2}}^{t+\frac{1}{2}} = \frac{\rho_{j+\frac{1}{2}}^{t+\frac{1}{2}}}{m_{e,i}} \, .$$

Poisson's equation:

$$E_{j+\frac{1}{2}}^{t+\frac{1}{2}} = \frac{1}{2}\left(E_{j+1}^{t} + E_{j}^{t}\right) + \Delta t 2\pi e \left((un)_{j+\frac{1}{2}}^{-} - Z\,(un)_{j+\frac{1}{2}}^{+} \right)^{t+\frac{1}{2}}$$

Main step

Continuity equations:

$$\rho_j^{t+1} = \rho_j^{t} - \Delta t \frac{(\rho u)_{j+\frac{1}{2}}^{t+\frac{1}{2}} - (\rho u)_{j-\frac{1}{2}}^{t+\frac{1}{2}}}{\Delta x}$$

Momentum equations:

$$(\rho u)_j^{t+1} = (\rho u)_j^{t} + \Delta t \left(- \frac{(\rho u^2)_{j+\frac{1}{2}}^{t+\frac{1}{2}} - (\rho u^2)_{j-\frac{1}{2}}^{t+\frac{1}{2}}}{\Delta x} - \frac{P_{j+\frac{1}{2}}^{t+\frac{1}{2}} - P_{j-\frac{1}{2}}^{t+\frac{1}{2}}}{\Delta x} \pm n_j^{t+\frac{1}{2}}\, e\, E_j^{t+\frac{1}{2}} \pm \rho_j^{t+\frac{1}{2}}\, \nu_j (u_j^{-} - u_j^{+})^{t+\frac{1}{2}} \right)$$

where

$$E_j^{t+\frac{1}{2}} = \frac{1}{2}\left(E_{j+\frac{1}{2}}^{t+\frac{1}{2}} + E_{j-\frac{1}{2}}^{t+\frac{1}{2}}\right) \qquad \text{etc.,}$$

since the variables at the auxiliary step were computed at the $j+\frac{1}{2}$ grid points. Thus

$$n_j^{t+1} = \frac{\rho_j^{t+1}}{m_{e,i}} \quad \text{and} \quad u_j^{t+1} = \frac{(\rho u)_j^{t+1}}{\rho_j^{t+1}} .$$

Poisson's equation:

$$E_j^{t+1} = E_j^{t} + \Delta t 4\pi e \left((un)_j^{-} - Z\,(un)_j^{+} \right)^{t+1}$$

The value of $E_j^{t+\frac{1}{2}}$ obtained at the auxiliary step is not used to integrate from $t+\frac{1}{2}$ to $t+1$ by $\frac{\Delta t}{2}$, because the auxiliary values have only first-order accuracy (Richtmyer and Morton 1967).

The numerical procedure described above is similar to the coupled sound and heat flow problem described by Richtmyer and Morton (1967, §10.4). The procedure is in effect explicit because continuity and the momentum equations are first solved explicitly to obtain the values of the number densities and velocities at times $t+\frac{1}{2}$ and $t+1$, which are then used in the finite difference representation of Poisson's equation.

Time step Δt

The stability condition of the Lax-Wendroff method is

$$(|u| + c)\frac{\Delta t}{\Delta x} < 1.$$

where u is the velocity of the fluid, and c is the velocity of sound in the fluid. But the momentum equations of the electron and ion fluids are coupled by Poisson's equation, and in the presence of any perturbation an electric field is created which tries to restore charge neutrality. This electric field will evolve on a time scale of the order of a plasma period

$$t_p = \frac{2\pi}{\omega_p}$$

where ω_p is the plasma frequency. The phase velocity of the electrostatic relaxation is therefore much greater than the velocity of sound in the electron fluid. Hence, to resolve this relaxation Δt must be less than the plasma period, or the 'computational speed' (Roache 1972) $\frac{\Delta x}{\Delta t}$ must be greater than the phase speed V_ϕ of the wave generated by these equations. From computational experimentation it was found that by taking $\Delta t \approx \frac{0.1}{\omega_p}$ results were obtained that were accurate to 5 significant figures.

Boundary Conditions

Second order extrapolation is applied at the boundaries. At the auxiliary step, the conservative variable U is computed at the N+½ and at the ½ grid points, where N is the number of grid points representing the continuum:

$$U_{N+\frac{1}{2}}^{t+\frac{1}{2}} = \frac{1}{2}(3U_N^t - U_{N-1}^t) - \frac{\Delta t}{2}\frac{(F_N^t - F_{N-1}^t)}{\Delta x}$$

$$U_{\frac{1}{2}}^{t+\frac{1}{2}} = \frac{1}{2}(3U_1^t - U_2^t) - \frac{\Delta t}{2}\frac{(F_2^t - F_1^t)}{\Delta x}$$

If extrapolation to the boundaries is performed only at the N+½ grid point in the auxiliary step and at the 1 grid point at the main step (as one is tempted to do at a first glance of the two-step Lax-Wendroff method) a spurious wave is formed after a few time steps. Chu and Sereny (1974) and Roache (1972) discuss a number of different boundary conditions for numerical fluid codes.

RESULTS AND DISCUSSION

The following numerical example starts with an initial linear density distribution for a plasma with temperature $T_e = T_i = 10^7$K. The plasma is stationary, it has no electrostatic field, and has a maximum hydrogen density of 10^{21}cm^{-3}, corresponding to the cut-off for neodymium-glass laser light (figure 1). Initially the electrons move rapidly towards -x while the ions move more slowly because of their mass difference. Electric fields of up to 3×10^6 V/cm are created within the first half of the plasma period (figure 2). The time duration of one plasma period is with respect to the minimum number density, $5 \times 10^{20}\text{cm}^{-3}$ which is 5 fsec.

After ten periods (figure 3) the time dependent field shows a wavelike structure. Later times (40 to 42 plasma periods, figure 4) show a nearly constant electrostatic field profile with values of up to 10^6 V/cm. The velocity of the electron fluid (figure 5) exhibits the same behaviour as the electrostatic field, but the oscillations out of phase by $\frac{\pi}{2}$ radians.

The collision frequency ν increases with decreasing temperature. This influences the damping process of the oscillating electrostatic fields and electron velocities

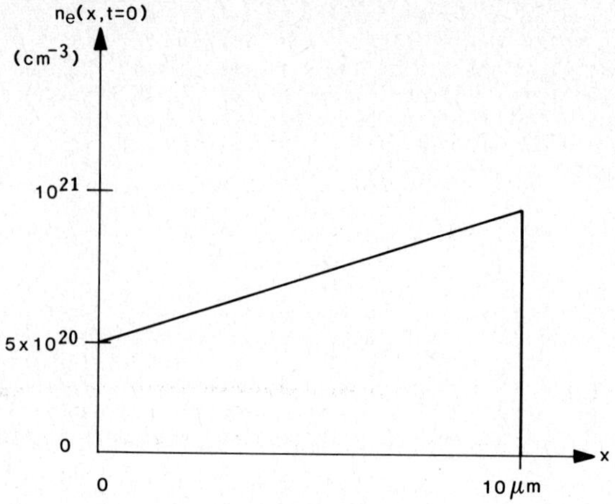

Figure 1
Initial density profile for a 10 μm thick slab hydrogen plasma, Z=1, with $n_i = n_e$. The initial ion and electron velocities are zero, and electron and ion temperatures are $T_e = T_i = 10^7$ or 10^5 K.

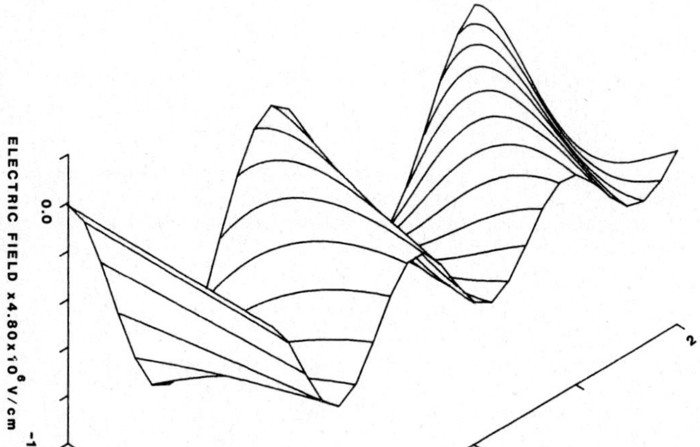

Figure 2
Time dependent development of the electrostatic field along the density ramp of plasma temperature 10^7 K.

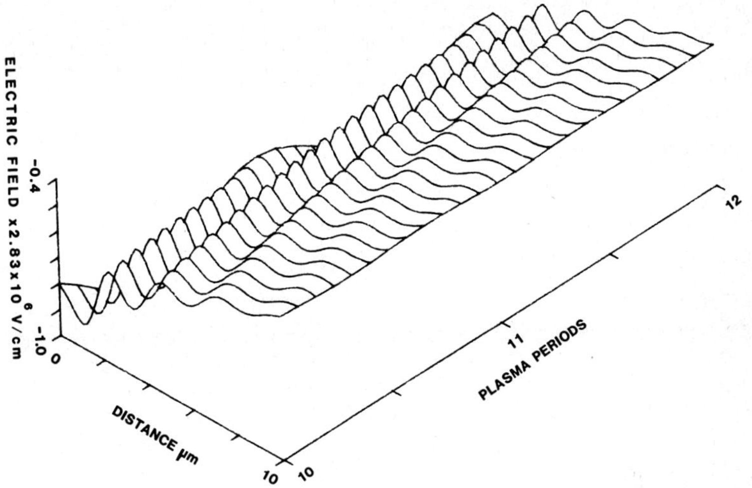

Figure 3
Same as figure 2 but for times 10 to 12 plasma periods.

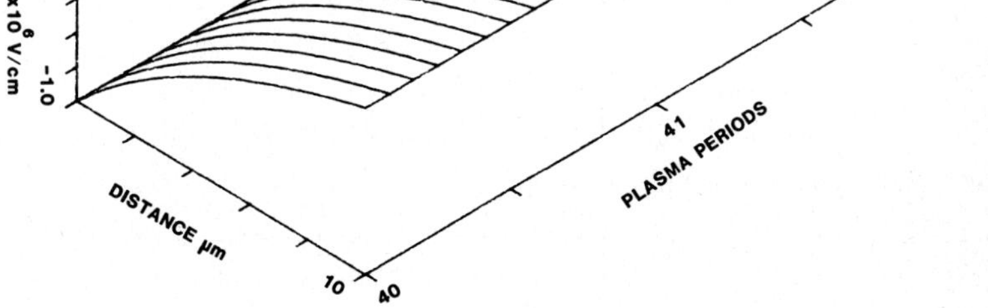

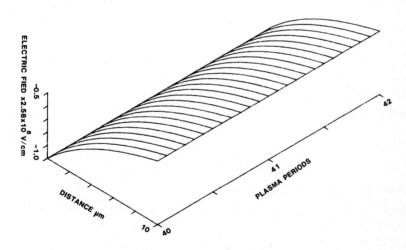

Figure 4
Same as figure 2 but for times 40 to 42 plasma periods.

seen in figures 2 to 5. To illustrate this, a calculation with an initial temperature $T_e = T_i = 10^5$ K in the hydrogen plasma slab as described in figure 1 was made. Figure 6 shows the electrostatic field oscillations for the first two plasma periods. Figure 7 reproduces the same case for 10 to 12 periods showing that the damping of the nearly stationary electric field has occurred for these earlier times compared with the case of figure 4 for higher temperature. For neither temperature does the velocity of the ions exhibit any oscillatory behaviour. Figure 8 shows the velocity of the ions for temperature of 10^7 K at times 4, 12, and 42 plasma periods. Initially the ions expand more slowly than the electrons. However, after 40 plasma periods (for temperature 10^7 K) the ions and electrons move with the same velocity: this flow is therefore ambipolar. At a temperature of 10^5 K this ambipolar flow occurs much earlier.

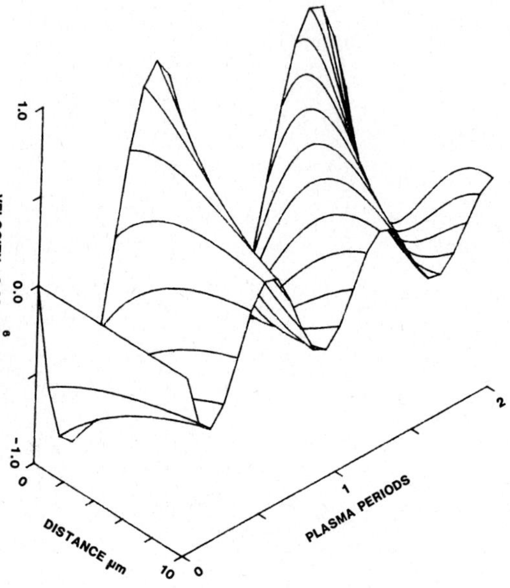

Figure 5
Time dependent development of the electron velocity from figure 1 for temperature 10^7 K.

In the numerical example discussed above we described the way electrostatic fields are set up in the presence of density gradients in dense plasma, based on certain initial conditions. The numerical results are in good agreement with an approximate analytical solution obtained by Lalousis and Hora (1983).

Since a problem involving the explicit solution of Poisson's equation requires time steps below the local plasma period, some researchers have devised techniques (so-called plasma period dilation techniques) which use implicit methods (e.g. Devatit (1981)). These methods can take time steps such that $\Delta t\omega_p \gg 1$. However, for hyperbolic equations the suitability of implicit methods is not clear (Potter (1977)) since information on the dynamics of the electron fluid would be lost.

These computations have been performed on a number of different computers (CDC 72, CDC 7600 and CRAY-1S). It was found that by avoiding constructs that inhibit vectorization, the code ran 9 times faster on a CRAY-1S than on a CDC 7600.

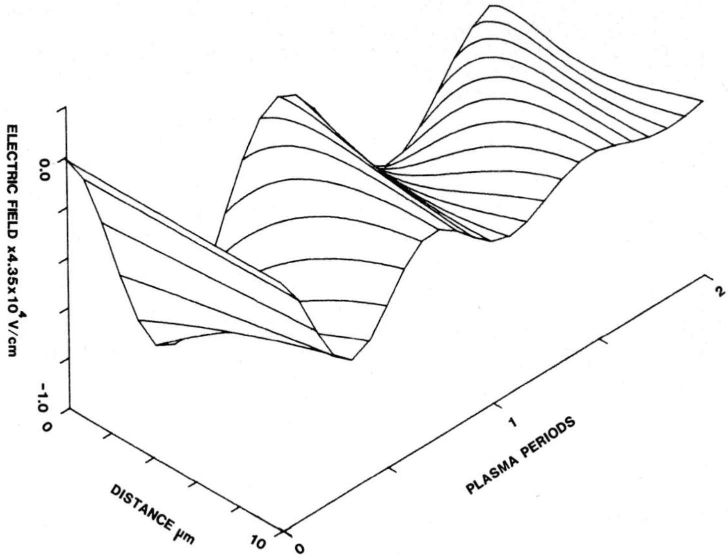

Figure 6
The time dependent development of the electrostatic field E from figure 1 for temperature 10^5 K.

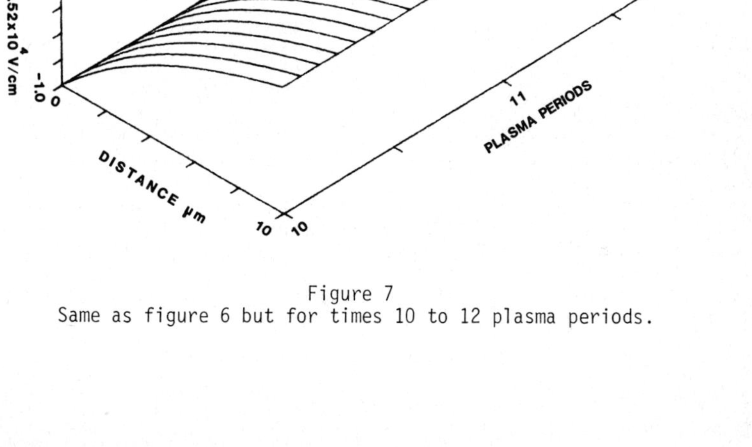

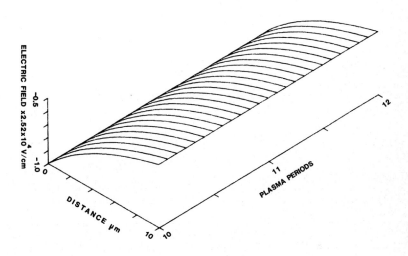

Figure 7
Same as figure 6 but for times 10 to 12 plasma periods.

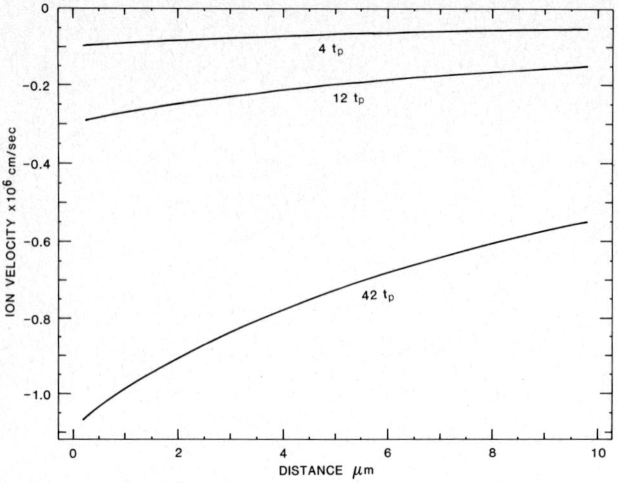

Figure 8
Ion velocity from figure 1 for temperature 10^7 K
at times 4, 12, and 42 plasma periods.

* This work is related to a Ph.D. degree at the University of NSW.

REFERENCES

[1] Chu, C.K. and Sereny, A., Jrnl. Comp. Phys. 15 (1974) 476-491.

[2] Devatit, J., Jrnl. Comp. Phys. 42 (1981) 337-366.

[3] Hora, H., Physics of Laser Driven Plasmas (Wiley, New York, 1981).

[4] Lalousis, P. and Hora, H., Laser and Particle Beams, Vol.1, 3 (1983).

[5] Potter, D., Computational Physics (Wiley-Interscience 1977).

[6] Roache P.J., Computational Fluid Dynamics (Hermosa, Albuquerque 1972).

[7] Richtmyer R.D. and Morton, K.W.D., Difference Methods for Initial Value Problems (Interscience 1967).

Computational Techniques & Applications: CTAC-83
J. Noye & C. Fletcher (Editors)

SOLUTIONS OF ARC PLASMA DECAY WITH SELF-INDUCED RADIAL FLOW USING THE ORTHOGONAL COLLOCATION METHOD

H. Lee

CSIRO Division of Applied Physics
Sydney, Australia 2070

The behaviour of the transient decay of an electric arc discharge is similar to the non-linear diffusion problem, but with the addition of radial convection. This convection is due to the strong change in the mass density gradient in the non-stationary temperature distribution of the arc. The orthogonal collocation method is used to solve the radial temperature and velocity distributions in the one-dimensional cylindrical arc. The numerical implementation of the method is simpler and faster than that of the finite difference method. A case study of the decay of a nitrogen arc at atmospheric pressure is presented.

INTRODUCTION

The thermal behaviour of a transient decay of a wall stabilized arc in a one-dimensional cylindrical geometry (Jones et al. 1971) can be represented by the energy balance equation and the mass continuity equation of the arc plasma.

These equations have been solved previously using an explicit finite difference method by Lowke et al. (1973). In the present investigation, we have solved these equations using the orthogonal collocation method.

In the explicit finite difference method, a small time step is required so that the solutions are stable (Richtmyer and Morton (1967)). It is also necessary for this type of calculations to use more than 20 grid points to represent the radial temperature and velocity distributions because of the non-linear temperature-dependent thermal conductivity of the arc gas. In the orthogonal collocation method, accurate solutions can be obtained using only three collocation points to represent the arc column. In addition, the computing time is greatly reduced.

The present calculations using the orthogonal collocation method have been implemented on a personal computer.

STATEMENT OF THE PROBLEM

The energy balance equation of the wall stabilized arc assuming the plasma is axially homogeneous and rotationally symmetric is given by,

$$\rho c_p \frac{\partial T}{\partial t} = \frac{1}{r}\frac{\partial}{\partial r}\left(rk\frac{\partial T}{\partial r}\right) - \rho c_p V \frac{\partial T}{\partial r} - 4\pi\varepsilon \qquad (1)$$

$$R_W \geqslant r \geqslant 0$$

The mass continuity equation for this problem is,

$$\frac{\partial \rho}{\partial t} = -\frac{1}{r}\frac{\partial}{\partial r}(r\rho V) \tag{2}$$

where T is the temperature, V the radial velocity, t the time, r the arc radius variable and R_w the arc radius. The material properties which are non-linear functions of temperature are ρ the mass density of the gas, C_p the specific heat, k the thermal conductivity, and ε the radiation emission coefficient.

It is more convenient to transform the temperature variable to a heat flux potential variable so that the temperature dependent thermal conductivity can be more easily accounted in equation (1). This is done by integral transform,

$$S = \int_0^T k\,dT' \tag{3}$$

Equation (1), using this transform, is reduced to:

$$\frac{\rho C_p}{k}\frac{\partial S}{\partial t} = \frac{1}{R_w^2}\frac{1}{x}\frac{\partial}{\partial x}\left(x\frac{\partial S}{\partial x}\right) - \frac{\rho C_p}{k}\frac{V}{R_w}\frac{\partial S}{\partial x} - 4\pi\varepsilon \tag{4}$$

after normalising the radius r, by x where

$$x = \frac{r}{R_w} \tag{5}$$

The boundary conditions for the temperature are,

$$\frac{\partial T}{\partial x}(0,t) = 0, \text{ for symmetry}$$

$$T(1,t) = T_w, \text{ fixed wall temperature.}$$

The radial velocity has the following two boundary conditions given by,

$$V(0,t) = 0$$

$$V(1,t) = 0.$$

The initial radial velocity assumes the steady-state condition of no flow or zero velocity and the initial value for the temperature is assumed from the steady-state solution (Lee 1983).

METHOD OF SOLUTION

a) Trial function in orthogonal polynomial

The orthogonal collocation method is applied to equations (2) and (4) and solved for the case of transient decay of a 100 A arc in nitrogen at 1 atmosphere.

The trial function chosen is even order, shifted Legendre polynomial (Villadsen and Stewart (1967)) given by,

$$S^{(n)} = S(1) + (1 - x^2) \sum_{i=0}^{n-1} a_i^{(n)} P_i(x^2) \tag{6}$$

where a_i are the undetermined constants, $S(1)$ a boundary condition and $P_i(x^2)$ the orthogonal polynomial. This trial function can be shown to satisfy the functional integral,

$$\int_0^1 (1 - x^2) P_i(x^2) P_n(x^2)\, xdx = C_j \delta_{jn} \tag{7}$$

$$j = 1,2,\ldots n-1$$

where c_j are constants and δ_{jn} are the Kronecker values. Further, it can be shown that when the trial function is adjusted to satisfy the operator equation, i.e. equation (4), at the zeros of the orthogonal polynomial, the residual function obtained will either vanish everywhere or else it will be another similar polynomial, $P_n(x^2)$. Equation (7) is also the Galerkin functional given that it contains the trial function and the residual function. In the orthogonal collocation method this functional is known by the constant terms on the RHS of the equation and the need to solve the functional explicitly is obviated. The method then reduces to finding the solutions of the residual function at the discrete zeros of the orthogonal polynomial.

b) Solution of radial velocity

By taking advantage of the given initial condition for radial flow velocity, i.e. $V(x,0) = 0$, the mass continuity equation (2) can be coupled to equation (4) using the time derivative of S rather than ρ. Equation (2) can be simplified to,

$$\frac{\partial \rho}{\partial t} = \frac{\partial \rho}{\partial S} \cdot \frac{\partial S}{\partial t} = -\frac{1}{R_w} \frac{1}{x} \frac{\partial}{\partial x} (x \rho V) \tag{8}$$

or

$$\frac{1}{R_w^2} \frac{1}{x} \frac{\partial}{\partial x} \left(x \frac{\partial S}{\partial x}\right) - \frac{\rho C_p}{k} \frac{V}{R_w} \frac{\partial S}{\partial x} - 4\pi\varepsilon = -\frac{\rho C_p}{k} \cdot \frac{\partial S}{\partial \rho} \cdot \frac{1}{x} \frac{\partial}{\partial x} (x \rho V) \tag{9}$$

using equation (4). Note that the new equation for radial velocity is independent of t explicitly, and solutions of V can be obtained for $t>0$ once S is evaluated in equation (4) in the forward time step.

c) Solving by matrix equations

The normal procedure for solving S is to substitute the trial function (6) into the energy equation (4) and determine the a_i constants. However, it has been shown by Villadsen and Stewart (1967) that a more efficient computing procedure is to first obtain solutions at the collocation points in terms of a generalised polynomial equation. In this procedure, the trial function shown in equation (6) can be written as a series polynomial.

$$S(x) = \sum_{i=1}^{n+1} \cdot x^{2i-2}\, b_i \tag{10}$$

after changing the limits of i. The values of S in equation (4) at each collocation point in terms of this polynomial is simply,

$$S(x_j) = \sum_{i=1}^{n+1} x_j^{2i-2} b_i \qquad (11)$$

$$i,j = 1, 2, \ldots n+1$$

where x_j are the zeros of the orthogonal polynomial which satisfy the functional integral in equation (7). The vector b_i can be obtained when all the $S(x_j)$ have been evaluated. This is given by the multiplication of the inverse matrix in equation (11) and $S(x_j)$,

$$b_i = \left[\sum_{i=1}^{n+1} x_j^{2i-2} \right]^{-1} S(x_j) \qquad (12)$$

The first derivative in S obtained using equation (11) is given by,

$$\frac{\partial S}{\partial x}(x_j) = \sum_{i=1}^{n+1} A_{ji} S(x_i) \qquad (13)$$

where

$$A_{ji} = (2i-2) \sum_{i=1}^{n+1} x_j^{2i-3} \left[\sum_{i=1}^{n+1} x_j^{2i-2} \right]^{-1} \qquad (14)$$

Similarly, the second derivative in equation (4) can be shown to be,

$$\frac{1}{x_j} \frac{\partial}{\partial x_j} \left(x_j \frac{\partial S}{\partial x_j} \right) = \sum_{i=1}^{n+1} B_{ji} S(x_i) \qquad (15)$$

where

$$B_{ji} = (2i-2)^2 \sum_{i=1}^{n+1} x_j^{2i-4} \left[\sum_{i=1}^{n+1} x_j^{2i-2} \right]^{-1} \qquad (16)$$

In the present example for convenience the same trial function has been used to describe the radial velocity and this means the A matrix in equation (14) can be applied to the velocity V.

The system of equations for the time-dependent heat flux potential, S, at the collocation points can be obtained using equations (4) and (15). This is given by,

$$\frac{dS(x_j)}{dt} = \frac{k}{\rho C_p}\bigg|_{S(x_j)} \left[\frac{1}{R_w^2} \sum_{i=1}^{n+1} B_{ji} S(x_i) - 4\pi\varepsilon(S(x_j)) \right] - \frac{V(x_j)}{R_w} \sum_{i=1}^{n+1} A_{ji} S(x_i)$$

$$j=1,2\ldots n \qquad (17)$$

The solutions of $S(x_j)$ in the first time step can be easily evaluated because initially $V(x_j) = 0$. For the subsequent time step, the radial velocity at the collocation points can be calculated from the system of equations derived in (9), i.e.,

$$-\frac{1}{R_w}\left[\frac{V(x_j)}{x_j} + \sum_{i=1}^{n+1} A_{ji}\, V(x_i)\right] = \frac{k}{\rho^2 C_p}\cdot\frac{\partial\rho}{\partial S}\bigg|_{S(x_j)} \cdot \left[\frac{1}{R_w^2}\sum_{i=1}^{n+1} B_{ji}\, S(x_i) - 4\pi\varepsilon(S(x_j))\right]$$

$$j=1,2\ldots n \qquad (18)$$

where $V_1 = 0$ is a given boundary condition. Note that equation (18) is a system of simultaneous linear equations, the RHS term is known from the solutions calculated in the previous equation (17).

NUMERICAL PROCEDURE

The matrix sets shown in equations (14) and (16) are first evaluated using the zeros of the shifted Legendre polynomial obtained from tables (Finlayson (1972), Villadsen and Stewart (1967)). The size of these square matrixes can either be n+1 or n+2. This will depend on whether direct solution is required for the centre at x=0, in which case the size is n+2. Although the final calculation can describe the solution at the centre or any other radial position using equations (12) and (10), the present study has shown that more accurate results are obtained when the extra equation is used.

The system of time-dependent differential equations shown in (17) is solved using the standard fourth order Runge-Kutta integration scheme. Initial conditions can be obtained from steady-state results (Lee (1983)) or experimental measurements of d.c. arcs. The radial velocity can be calculated using equation (18). The elements of the LHS matrix which determine the velocity terms are constants, so a direct multiplication of this inverted matrix with a variable column vector leads to the solution of the radial velocity at the collocation points.

RESULTS AND DISCUSSIONS

The study is made of the case of a nitrogen arc confined in a 5 mm cylindrical tube at atmospheric pressure. The arc is initially burning with a current of 100 A at t = 0 before it is switched off at the beginning of the decaying process.

Calculations have been made for the 3 and 6 collocation point cases. This leads to the solutions for both the radial velocity and temperature distributions being approximated by a polynomial of up to eighth and fourteenth order in x respectively. The thermodynamic properties of the arc gas are stored in lookup tables located at intervals of 1000 K. But during the calculations the linear interpolations of these values are done with reference to the heat flux potential rather than the temperature.

Time steps of 1 μs and 2.5 μs have given similar and stable solutions for both the cases of 3 and 6 collocation points. The time step is limited by the time constant of the gas. This is given by the thermal diffusivity and the radius square of the arc as,

$$\zeta = \frac{\rho C_p}{k} R_w^2 \qquad (19)$$

Solutions obtained from the 3 and 6 point cases are compared wherever possible with previous published finite difference solutions and experimental measurements. However, there is an inherent difficulty, in that the radiation emission data used for both the present calculations and the finite difference calculations are different. In the finite difference method Lowke et al. (1973) solved not only the energy and mass continuity equations but the radiation flux equations as well. However, in that case the radiation emission coefficient was evaluated at a time step of 5 μs while the main calculations of temperature and radial velocity were done at smaller time steps. In the present calculations the values of the radiation emission coefficient have been taken from recent accurate measurements by Ernst et al. (1973). The advantages of using this approach are the accuracy of the radiation data and increased computing speed.

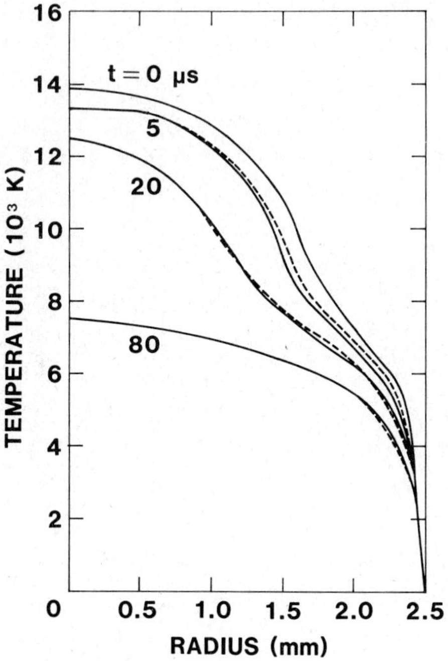

Figure 1
Calculated radial temperature distributions at time = 0, 5, 20, and 80 μs. At time = 0, conditions are that of a 100 A arc in nitrogen at 1 atmosphere.

In Figure 1 the radial temperature distributions of the arc at 5, 20 and 80 μs after the start of the decay are shown for both 3 and 6 collocation point cases. The results are very similar except for the early 5 μs distributions. All the curves show a shoulder near the 6000 to 7000 K region which corresponds to the molecular dissociation temperture of nitrogen. The decay of the centre temperature in Figure 2 is compared with the result from finite difference method (Lowke et al. (1973)). The initial starting conditions are similar for both methods, but the results from the orthogonal collocation method are slightly higher in the early period of the decay. This disparity may be due to the dissimilar radiation emission data used in the two methods of calculation.

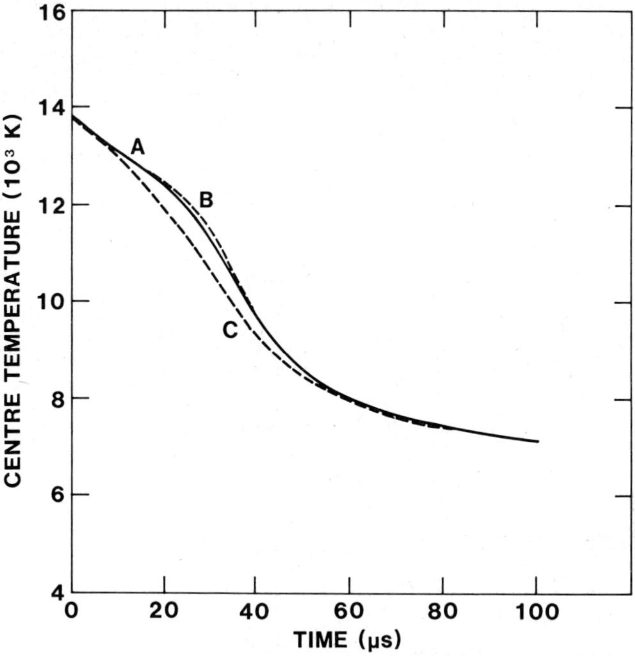

Figure 2
The decay of centre temperature for the case shown in Figure 1. Curves: (A) Calculations using 3 collocation points; (B) Calculations using 6 collocation points, and (C) Finite difference method (Lowke et al. (1973)) with 20 grid spaces.

In Figure 3 the results of radial velocity distributions obtained by the finite difference method and the orthogonal collocation method are shown for 3 instants of time during the arc decay. There are significant differences between the results from the two methods, especially at 5 μs and 20 μs. In the present calculations, the coefficient $\partial\rho/\partial S$ is approximated using the same look-up tables. This problem is apparent in the sensitivity of the results for the radial velocity.

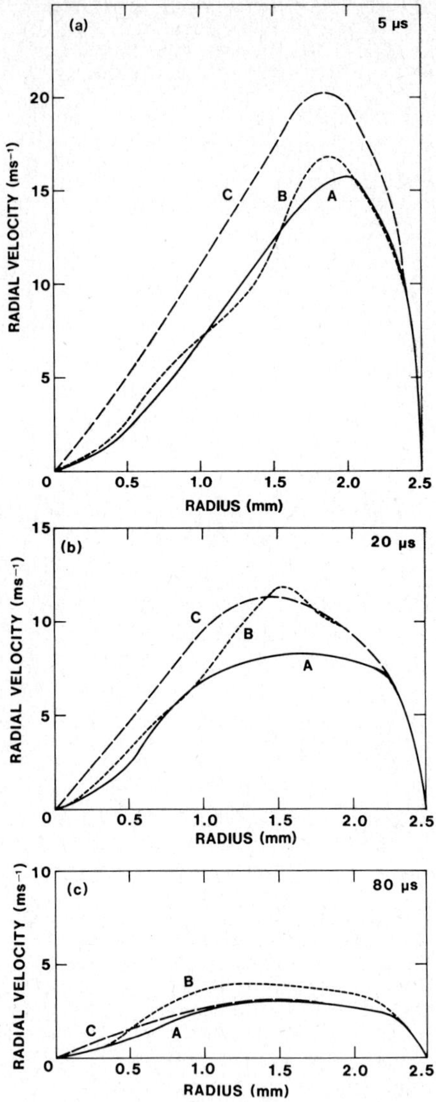

Figure 3
Radial velocity distributions for the case shown in Figure 1. (a) Time = 5 μs; (b) 20 μs and (c) 80 μs. Curves: (A) calculations using 3 collocation points; (B) calculations using 6 collocation points and (C) finite difference method (Lowke et al. (1973)).

It is experimentally difficult to measure both the arc temperature and radial flow velocity during the arc decay. But measurement of the arc conductance decay is possible and this has been made by Hertz (1971) using biased segments of the wall confining the arc. In Figure 4, arc conductance is shown as a function of time for measurements and calculations. There is a fair agreement in the results except that the measured values are slightly higher than predicted by the calculations for time greater than 80 μs. This type of measurement is susceptible to sheath problems when the arc is cold and for this reason the measured arc conductance in the late times could be suspect.

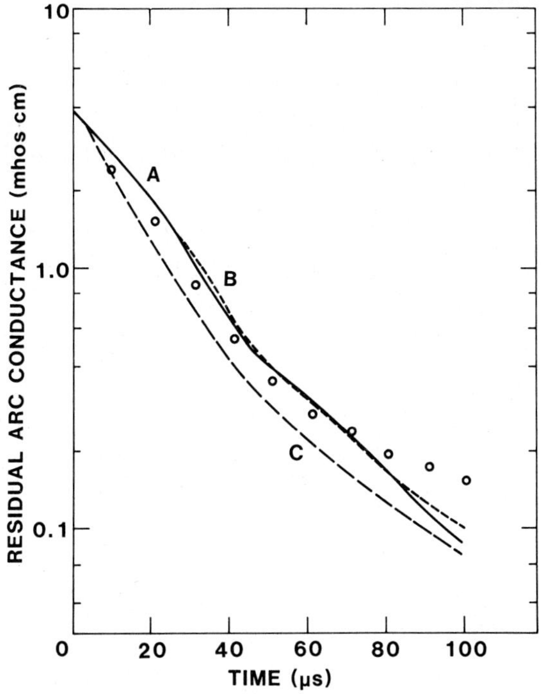

Figure 4
The residual arc conductance decay. Curves:
(A) Calculations using 3 collocation points;
(B) Calculations using 6 collocation points, and
(C) Finite difference method (Lowke et al. (1973)).
Circles: Experimental measurements (Hertz (1971)).

The time-dependent solutions of the temperature and radial velocity distributions have been calculated using a personal computer, which is based on a Z-80 A microprocessor with advanced Microsoft Basic Interpreter. The computing time to calculate the solutions for a 6 collocation point case is approximately 20 minutes. Compared with this figure, solution time on a Burrough 5500 main-frame computer using the explicit finite difference method was reported to be about 14 minutes for a similar set of calculations (Lowke et al. (1973)). The speed advantage of some of the fast mainframe computers compared to the personal computers typically exceeds 1000, especially when considerable floating point arithmetic calculations are performed.

CONCLUSIONS

In high temperature plasma arc research, the transient arc is frequently studied to determine the insulation properties of the gas type at different operating pressures. This is relevant in the design of power apparatus and devices such as plasma switches, electric fuses and power system circuit breakers. Consequently, an efficient and accurate method for predicting the arc properties would be of interest. Such a method would be of value for treating other time-dependent problems of the arc where external circuit considerations are important in the understanding of pulsed phenomena.

This present study of the wall stabilized transient arc shows that the orthogonal collocation method offers an advantage in computing speed over the finite difference method. In addition, the results from using only 3 collocation points are very similar to those from using 6 collocation points, indicating that reasonably accurate calculations may be obtained with only a few points.

ACKNOWLEDGEMENTS

The author wishes to thank his colleagues for the encouragement and valuable assistance given to him during the preparation of this work.

REFERENCES

[1] Ernst, K.A., Kopainsky, J.G. and Maecker, H., The energy transport, including emission and absorption in N_2-arcs of different radii, IEEE Trans. Plasma Sci. PS-1 3 (1973) 3-16.

[2] Finlayson, B.A., The method of weighted residuals and variational principles, (Academic Press, New York, 1972).

[3] Hertz, W., Messung und Deutung des Leitwertabklingens zylindrischer Bogen, Z. Physik 245 (1971) 105-125.

[4] Jones, G.R., Freeman, G.H. and Edels, H., Transient temperature distributions in cylindrical arc columns following abrupt current changes, J. Phys. D. 4 (1971) 236-245.

[5] Lee, H., Solutions of plasma arc equations by the orthogonal collocation method. (Accepted for publication - J. Phys. D. 1983).

[6] Lowke, J.J., Voshall, R.E. and Ludwig, H.C., Decay of electrical conductance and temperature of arc plasmas, J. Appl. Phys. 44 (1973) 3513-3523.

[7] Richtmyer, R.D. and Morton, K.W., Difference methods for initial value problems. (Interscience, New York, 1967).

[8] Villadsen, J.V. and Stewart, W.E., Solution of boundary-value problems by orthogonal collocation, Chem. Eng. Sci. 22 (1967) 1483-1501.

Computational Techniques & Applications: CTAC-83
J. Noye & C. Fletcher (Editors)

FLUX-CORRECTED TRANSPORT ON A NON-UNIFORM MESH IN PLASMA BOUNDARY PROBLEMS

R. Morrow and L.E. Cram

CSIRO Division of Applied Physics,
Sydney, Australia 2070

It is shown how the flux corrected transport (FCT) algorithm of Boris and Book (J. Comput. Phys. 11, (1973) 38-69) can be used to solve the continuity equations for electrons and ions on a non-uniform mesh, to facilitate the study of plasma structure near electrodes. Extreme (even abrupt) mesh size changes can be accommodated giving results comparable to the results for a uniform mesh. The evolution of a plasma cloud adjacent to a conducting electrode is described as a specific application of the method.

INTRODUCTION

Theoretical studies of gaseous discharges require a solution of the equations describing the drift and diffusion of charged particles in an electric field. Often, the electric field varies strongly in space and time, placing stringent requirements on the accuracy and stability of the numerical technique used to solve the drift and diffusion equations. Previous work (Morrow (1981)) has shown that the Flux-Corrected Transport (FCT) algorithm is well-suited to solving the hyperbolic continuity equations of this kind and examples of such an application may be found in Morrow (1982) and Morrow and Lowke (1982).

However, the application of the FCT algorithm to problems in which the plasma is in contact with a conducting electrode has revealed the existence of a boundary layer or sheath which is difficult to resolve with a uniform mesh. Thus, a variable mesh FCT algorithm has been developed. This algorithm, briefly presented by Boris and Book (1976a), is described in part 2 of this paper.

The multi-fluid continuity equations must be solved simultaneously with Poisson's equation for the electric field. The "method of disks" of Davies et al (1964) has been extended to allow a more accurate representation of the charge density distribution in regions where steep changes occur. The new method involves the evaluation of an influence matrix operator which operates on the coarse space-charge distribution to generate an accurate value of the electric field. The operator is generated once only. Part 3 of the paper describes this method.

The variable-mesh FCT algorithm, combined with the accurate Poisson solver, have been applied successfully to describe the development of the cathode fall region of a discharge. The application of these techniques in this context is described in part 4 of this paper.

FCT ON A NON-UNIFORM MESH

a) Solution of Drift Equation

The equation to be solved is

$$\frac{\partial \rho}{\partial t} = \frac{\partial}{\partial x}(\rho w) \tag{1}$$

where ρ is the particle density, and w the velocity. Although the FCT algorithm on a uniform mesh can be derived by a formal procedure, we have found it desirable to return to the original geometrical arguments of Boris and Book (1973) to devise a scheme that has the correct forward, backward, and centered differences.

The transport stage of the algorithm can be understood in terms of figure 1, which shows how adjacent trapezoidal fluid elements are convected and distorted in a single time step. If the spatial mesh is defined as the set $(x_i | i = 1, N)$, two interleaved mesh spacings may be defined:

$$\delta x_{i+\frac{1}{2}} = x_{i+1} - x_i, \tag{2}$$

and

$$\delta x_i = \tfrac{1}{2}(\delta x_{i+\frac{1}{2}} + \delta x_{i-\frac{1}{2}}). \tag{3}$$

In terms of these spacings, the new densities defined in figure 1(b) are

$$\begin{aligned} \rho_- &= \rho^0_{i-1}\delta x_{i-\frac{1}{2}}/[\delta x_{i-\frac{1}{2}} + \delta t(W_i - W_{i-1})], \\ \rho^1_0 &= \rho^0_i \delta x_{i-\frac{1}{2}}/[\delta x_{i-\frac{1}{2}} + \delta t(W_i - W_{i-1})], \\ \rho^2_0 &= \rho^0_i \delta x_{i+\frac{1}{2}}/[\delta x_{i+\frac{1}{2}} + \delta t(W_{i+1} - W_i)], \end{aligned} \tag{4}$$

and

$$\rho_+ = \rho^0_{i+1}\delta x_{i+\frac{1}{2}}/[\delta x_{i+\frac{1}{2}} + \delta t(W_{i+1} - W_i)].$$

The new density values may be interpolated and extrapolated back to the original mesh points to give the following generalization of eq. (7) in Boris and Book (1973):

$$\begin{aligned} \rho_i^{n+1} = {} & \frac{\delta x_{i+\frac{1}{2}}}{\delta x_i}[\rho_i^n Q_+ + \tfrac{1}{2}(\rho^n_{i+1} - \rho^n_i)Q_+^2] \\ & + \frac{\delta x_{i-\frac{1}{2}}}{\delta x_i}[\rho_i^n Q_- + \tfrac{1}{2}(\rho^n_{i-1} - \rho^n_i)Q_-^2], \end{aligned} \tag{5}$$

where

$$Q_+ = (\tfrac{1}{2} - W_i\delta t/\delta x_{i+\frac{1}{2}})/[1 + \frac{\delta t}{\delta x_{i+\frac{1}{2}}}(W_{i+1} - W_i)] \tag{6}$$

and

$$Q_- = (\tfrac{1}{2} + W_i\delta t/\delta x_{i-\frac{1}{2}})/[1 + \frac{\delta t}{\delta x_{i-\frac{1}{2}}}(W_i - W_{i-1})]. \tag{7}$$

For a uniform velocity field these equations become

$$\begin{aligned} \rho_i^{n+1} = {} & \rho_i^n - \tfrac{1}{2}[\frac{\delta x_{i+\frac{1}{2}}}{\delta x_i}\varepsilon_{i+\frac{1}{2}}(\rho^n_{i+1} + \rho^n_i) - \frac{\delta x_{i-\frac{1}{2}}}{\delta x_i}\varepsilon_{i-\frac{1}{2}}(\rho^n_i + \rho^n_{i-1})] \\ & + [\frac{\delta x_{i+\frac{1}{2}}}{\delta x_i}\nu_{i+\frac{1}{2}}(\rho^n_{i+\frac{1}{2}} - \rho^n_i) - \frac{\delta x_{i-\frac{1}{2}}}{\delta x_i}\nu_{i-\frac{1}{2}}(\rho^n_i - \rho^n_{i-1})] \end{aligned} \tag{8}$$

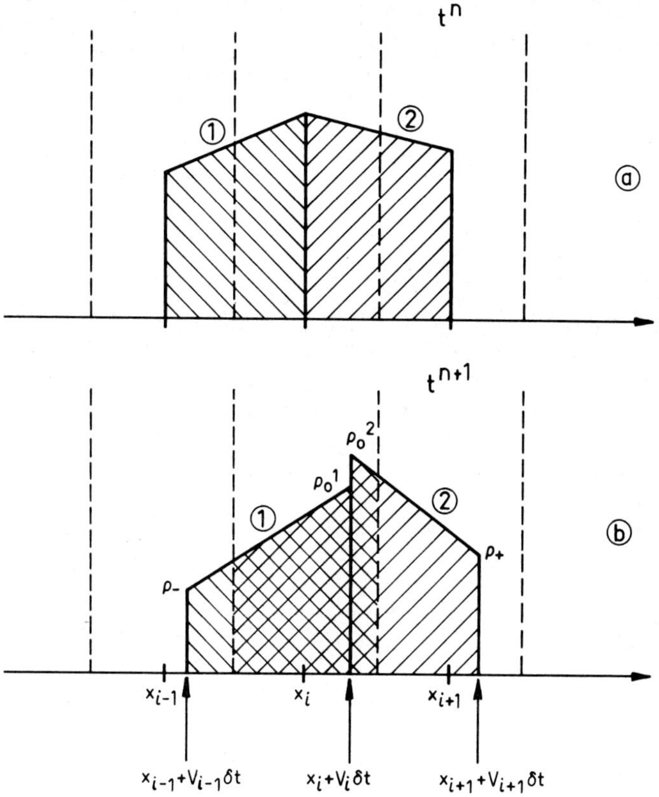

Figure 1
Transport stage of the SHASTA algorithm using a non-uniform mesh.

with

$$\varepsilon_{i+\frac{1}{2}} = W\delta t/\delta x_{i+\frac{1}{2}}, \tag{9}$$

and

$$\nu_{i+\frac{1}{2}} = \frac{1}{8} + \frac{1}{2}\,\varepsilon^2_{i+\frac{1}{2}}. \tag{10}$$

By comparing eq. (8) in the present paper with eq. (8) in Boris and Book (1973), the reader will see how the asymmetry of the non-uniform mesh appears. The comparison also shows how the transportive and diffusive elements of the SHASTA algorithm have been modified, and corresponding changes may be made in other more refined implementations of the transport stage of the FCT algorithm.

To develop eq. (8) further, we follow Boris and Book (1976a) and consider an arbitrary 3-point approximation to the continuity equation:

$$\tilde{\rho}_i = a_i\rho^o_{i-1} + b_i\rho^o_i + c_i\rho^o_{i+1}, \tag{11}$$

where the coefficients (a_i, b_i, c_i) are to be chosen to satisfy as many as possible of the requirements listed by Boris and Book (1976a). Conservation of mass demands

$$a_{i+1} + b_i + c_{i-1} = 1. \tag{12}$$

Criteria that allow other desirable properties to be provided in the algorithm may be identified by equating coefficients in (11) and (8). We find

$$a_i = \frac{\delta x_{i-\frac{1}{2}}}{\delta x_i}(\nu_{i-\frac{1}{2}} + \frac{1}{2}\varepsilon_{i-\frac{1}{2}}),$$

$$b_i = \frac{\delta x_{i-\frac{1}{2}}}{\delta x_i}(\tfrac{1}{2} + \tfrac{1}{2}\varepsilon_{i-\frac{1}{2}} - \nu_{i-\frac{1}{2}}) + \frac{\delta x_{i+\frac{1}{2}}}{\delta x_i}(\tfrac{1}{2} - \tfrac{1}{2}\varepsilon_{i+\frac{1}{2}} - \nu_{i+\frac{1}{2}}), \tag{13}$$

and

$$c_i = \frac{\delta x_{i+\frac{1}{2}}}{\delta x_i}(\nu_{i+\frac{1}{2}} - \tfrac{1}{2}\varepsilon_{i+\frac{1}{2}}).$$

A convenient and conservative form of the parameter ε is given by

$$\varepsilon_{i+\frac{1}{2}} = W_{i+\frac{1}{2}}\delta t/\delta x_{i+\frac{1}{2}}, \tag{14}$$

where

$$W_{i+\frac{1}{2}} = \tfrac{1}{2}(W_i + W_{i+1}). \tag{15}$$

Any value of $\nu_{i+\frac{1}{2}} \geqslant 0.5|\varepsilon_{i+\frac{1}{2}}|$ (for all i) will then ensure the positivity of $\tilde{\rho}_i$, but algorithms of this kind will be subject to strong numerical diffusion. This defect may be corrected by introducing antidiffusive fluxes. On a non-uniform mesh, the antidiffusive fluxes analogous to eq. (18) of Boris and Book (1976a) are

$$\phi_{i+\frac{1}{2}} = \mu_{i+\frac{1}{2}}\frac{\delta x_{i+\frac{1}{2}}}{\delta x_i}(\tilde{\rho}_{i+1} - \tilde{\rho}_i)$$

and

$$\phi_{i-\frac{1}{2}} = \mu_{i-\frac{1}{2}}\frac{\delta x_{i-\frac{1}{2}}}{\delta x_i}(\tilde{\rho}_i - \tilde{\rho}_{i-1}), \tag{16}$$

where $\{\mu_{i+1}\}$ are antidiffusive coefficients. The explicit SHASTA algorithm uses

$$\nu_{i+\frac{1}{2}} = \frac{1}{8} + \frac{1}{2}\varepsilon^2_{i+\frac{1}{2}} \text{ and } \mu_{i+\frac{1}{2}} = \frac{1}{8}, \tag{17}$$

where $\varepsilon_{i+\frac{1}{2}}$ is given by (14). A more accurate antidiffusive scheme is provided by the Phoenical LPE SHASTA algorithm (Boris and Book, 1976b), which has the following form on a non-uniform mesh:

$$\phi_{i+\frac{1}{2}} = \mu_{i+\frac{1}{2}}\frac{\delta x_{i+\frac{1}{2}}}{\delta x_i}[\tilde{\rho}_{i+1} - \tilde{\rho}_i + \frac{1}{6}\{-\frac{\delta x_{i+3/2}}{\delta x_{i+1}}(\rho^o_{i+2} - \rho^o_{i+1}) + \frac{\delta x_{i+\frac{1}{2}}}{\delta x_{i+1}}(\rho^o_{i+1} - \rho^o_i)$$
$$+ \frac{\delta x_{i+\frac{1}{2}}}{\delta x_i}(\rho^o_{i+1} - \rho^o_i) - \frac{\delta x_{i-\frac{1}{2}}}{\delta x_i}(\rho^o_i - \rho^o_{i-1})\}],$$

and (18)

$$\phi_{i-\frac{1}{2}} = \mu_{i-\frac{1}{2}}\frac{\delta x_{i-\frac{1}{2}}}{\delta x_i}[\tilde{\rho}_i - \tilde{\rho}_{i-1} + \frac{1}{6}\{-\frac{\delta x_{i+\frac{1}{2}}}{\delta x_i}(\rho^o_{i+1} - \rho^o_i) + \frac{\delta x_{i-\frac{1}{2}}}{\delta x_i}(\rho^o_i - \rho^o_{i-1})$$
$$+ \frac{\delta x_{i-\frac{1}{2}}}{\delta x_{i-1}}(\rho^o_i - \rho^o_{i-1}) - \frac{\delta x_{i-3/2}}{\delta x_{i-1}}(\rho^o_{i-1} - \rho^o_{i-2})\}].$$

In this algorithm, one chooses

$$\nu_{i+\frac{1}{2}} = \frac{1}{6} + \frac{1}{3}\varepsilon^2_{i+\frac{1}{2}} \text{ and } \mu_{i+\frac{1}{2}} = \frac{1}{6}(1 - \varepsilon^2_{i+\frac{1}{2}}). \tag{19}$$

To prevent the antidiffusive schemes (16) or (18) from introducing spurious maxima or minima in the solution, or accentuating existing maxima, the antidiffusive flux estimates must be limited. The limiting criteria given by Boris and Book (1976a - eq. 20) or Zalesak (1979 - Sec. IV) do not need to be modified when applied on a non-uniform grid. We use Zalesak's flux limiting algorithm for our examples below. The limited antidiffusive fluxes $\{\tilde{\phi}_{i+\frac{1}{2}}\}$ may then be used to compute the final value of the updated density.

$$\rho^1_i = \tilde{\rho}_i - \tilde{\phi}_{i+\frac{1}{2}} + \tilde{\phi}_{i-\frac{1}{2}}. \tag{20}$$

b) <u>Test Cases</u>

Our first test shows the transport, without diffusion, of rectangular and triangular density pulses across the non-uniform mesh defined by

$$x_i = x_{i-1} + \delta x_i \quad (i = 2.....,N) \tag{21}$$

where $x_1 = 0$, $\delta x_i = 1.0 - 0.75 \exp(-((i - M)/k)^2)$, $M = \frac{N-1}{2} + 1$, and $K = \frac{N-1}{5}$. The mesh is adjusted to give the required value of x_N. In all cases we used $N = 101$ and $W = 3.786 \times 10^8 \text{ cms}^{-1}$.

Figure 2(a) shows how the mesh spacing varies with x. Figures 2(b) and 2(c) show the forms of the pulses as they are transported across the mesh, and figure 2(d) and 2(e) compare the final pulse shapes from figure 2(b) and 2(c) with the corresponding shapes at the same time following transport across a uniform N-point mesh. The differences between the pulses after traversing the non-uniform and uniform meshes are due almost entirely to the difficulty of precisely matching the two mesh arrays. It is evident that the pulses traverse the non-uniform grid with no significant distortion, and no parasitic structures emerge.

Encouraged by this success, we considered a second test involving severe and sudden distortions in the mesh:

$$\begin{aligned} x_i &= x_{i-1} + \delta x & (i = 2, 26) \\ x_i &= x_{i-1} + \delta x/4 & (i = 27, 76) \\ x_i &= x_{i-1} + \delta x & (i = 77, 101) \end{aligned} \tag{22}$$

where δx is adjusted to give the required value of x_N, and $x_1 = 0$. Figure 3(a) exhibits the variation of mesh size with x, Figure 3(b) shows a rectangular pulse as it traverses the mesh, and Figure 3(c) compares the final pulse with the result of transport across a uniform mesh. Although there is evidence of slight perturbations induced by transport across this highly distorted mesh, the algorithm appears to be well-behaved even in this extreme test. We would, however, urge the reader to heed the advice of Zalesak (1981) and to use smooth mappings from the index $\{i\}$ to the spatial mesh $\{x_i\}$.

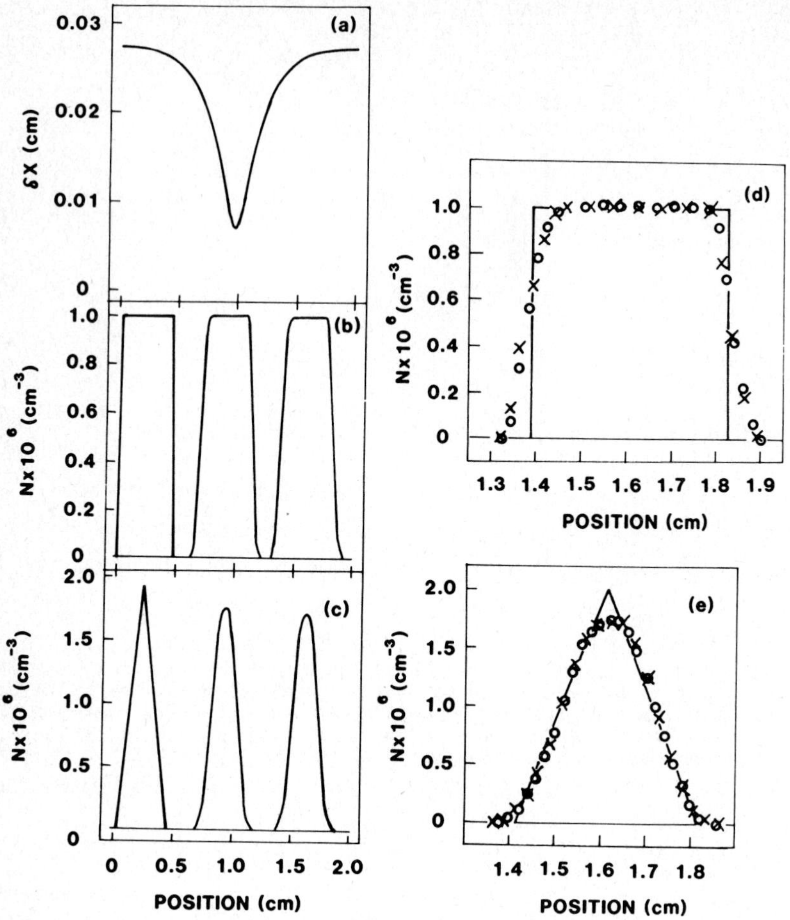

Figure 2
Results of numerical studies of the transport of square and triangular pulses over a non-uniform mesh. (a) mesh variation. (b) square wave propagation from left to right over mesh shown in (a). (c) triangular wave propagation from left to right over mesh shown in (a). (d) position of the ideal square wave solution (———) compared with solutions for a uniform mesh, 0, and the non-uniform mesh, X, after 800 time steps, or 36.11 ns. (e) position of the ideal triangular wave solution (———) compared with solutions for a uniform mesh, 0, and a non-uniform mesh, X, after 800 time steps, or 36.11 ns.

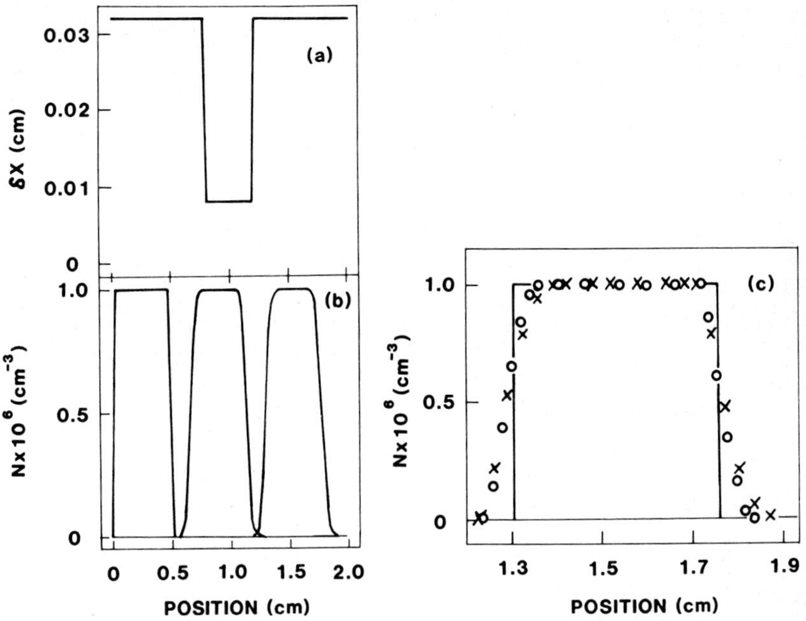

Figure 3
Results of numerical studies of the transport of a square wave over an abrupt 4:1 step in the mesh size. (a) step mesh variation. (b) square wave propagation from left to right over mesh shown in (a). (c) comparison of the final curves, after 640 time steps or t = 33.81 ns, for: the ideal solution, ———; a uniform mesh, O; the step mesh variation, X.

SOLUTION OF POISSON'S EQUATION

In many gas discharge problems of practical interest the electric current flows through a narrow conducting filament or channel between the electrodes. Because the charge flow takes place predominantly in the axial direction, one-dimensional continuity equations provide an acceptably accurate description of this configuration. However, the one-dimensional form of Poisson's equation does not provide an accurate representation of the electric field. A satisfactory estimate of the axial electric field may be evaluated by using the method of disks (Davies et al 1964), which represents the discharge filament by a cylinder with a uniform radial distribution and a non-uniform axial distribution of charge. With such a model the axial component of the electric field at a point x on the axis is

$$E(x) = 2\pi\{\int_{-x}^{0} \rho(x + x')[-1 - \frac{x'}{\sqrt{x'^2 + r^2}}]dx' + \int_{0}^{d-x} \rho(x + x')[1 - \frac{x'}{\sqrt{x'^2 + r^2}}]dx'\} \qquad (23)$$

where $\rho(x)$ is the charge density, r is the cylinder radius, and the gap extends from x=0 to x=d. In applying this equation, the electrode boundary conditions are implemented by including "images" of the charge $\rho(x)$ reflected appropriately at the surface of the perfectly conducting electrodes.

A numerical representation of the integral (23) is provided by the operator equation

$$E_i = \sum_j F_{ij}\, \rho_j \qquad (24)$$

where $E_i = E(x_i)$, $\rho_j = \rho(x_j)$, and F_{ij} is an influence matrix. A first representation of F_{ij} may be found by trapezoidal quadrature on the mesh $(x_i | i=1, N_x)$, but it is found that greater precision is required in regions where the charge density ρ is changing rapidly with x.

A more accurate quadrature is effected by subdividing each interval (x_i, x_{i+1}) into a finer mesh. The charge density distribution on the fine mesh is evaluated by an even-order Lagrange interpolation, using only the central interval for accuracy. The resulting quadrature formula may be again written in the form (24), where F_{ij} now contains the fine interpolation of ρ. Near the endpoints of the interval (o,d), it is found preferable to reduce the order of the Lagrange interpolation, rather than to extrapolate from a one sided charge distribution. It should be emphasised that the matrix operator F_{ij} is determined only by geometry, and is therefore computed only once.

AN APPLICATION

To illustrate the application of the method, we consider the time-dependent development of a cloud of plasma initially located close to the cathode of a uniformly stressed 2 cm gap. In Figure 4 the cathode is at 0 cm and anode at 2 cm, both are perfectly conducting and absorbing. The electron velocity, We, is given by (Morrow and Lowke 1982)

$$We = 2.157 \times 10^{16} \times (E/N)^{0.6064} \text{ cms}^{-1}$$

while the positive ion velocity, Wp, is given by

$$Wp = 5.5 \times 10^{19} \times E/N \text{ cms}^{-1}$$

where E is the local electric field in Vcm^{-1} and N is the neutral number density given by

$$N = p \times 3.2959 \times 10^{16} \text{ cm}^{-3}$$

where p is the gas pressure in torr.

In this case the gas pressure was 50 torr and the initial electric field was -2.5 k V/cm. The initial distribution of electrons and positive ions is chosen to be cylindrical with a diameter of 0.5 cm, over which the densities are constant, while in the axial direction the density varies as

$$Ne(x) = Np(x) = 2 \times 10^{11} \exp\left[-(x/0.667)^2\right] \text{ cm}^{-3}$$

As a boundary condition it is assumed that positive ions incident on the cathode are perfectly absorbed. For the model treated here, no electrons can reach the cathode. Production of ions or electrons by processes in the gap or on the electrode are neglected.

Figures 4(a), (c) and (e) show, respectively, the development of the electron density, ion density, and electric field in the gap. Figure 4(b), (d) and (f) show the corresponding variables in the region near the electrode. The structure of the non-uniform mesh is exponentially expanding as illustrated by dots on figures 4(a), (b).

It can be seen that the variable mesh permits an accurate treatment of the formation of the cathode sheath. Electrons are repelled from the cathode, leaving a positive ion sheath and an intense space-charge layer. Outside the sheath, the electrons and ions separate slowly to produce a region of zero electric field in which the space-charge field cancells the impressed Laplacian field. This phenomenon of field expulsion is a characteristic feature of plasmas in regions remote from boundaries. On the anode side of the field-free region there is an excess of electrons, and the net negative space charge is again associated with a strong electric field. We see in fact the temporal development of an anode sheath. The time scale for this process is slower than that of the cathode sheath, since it is related to the drift velocity of positive ions, rather than electrons. For the problem considered here, a steady state is never attained, because charge is lost at the electrodes, and not renewed in the gap.

The calculations illustrated in figure (4) are made on a non-uniform mesh of N_x = 51 points, and represent 1900 time steps for a total duration of 30 ns. The time step is variable and set at half the C.F.C condition. The influence matrix F_{ij} was calculated using fourth order Lagrange interpolation with 10 finer mesh points between the main grid points, and with third order interpolation being used at the boundary. The computer time required for the calculation was 36 sec on a C.D.C. Cyber 76.

Studies of test problems of this kind show the use of the FCT on a non-uniform mesh provides robust and accurate solutions of the equations of charged particle flow under the influence of space charge fields. The method will provide a valuable tool for theoretical studies of gas discharges.

SUMMARY

Starting from first principles, it has been shown that modifications to the FCT SHASTA algorithm can be developed which allow these algorithms to be applied to non-uniform meshes. These modifications carry over to the more sophisticated algorithms such as Phoenical LPE SHASTA and the results obtained are comparable with those for a uniform mesh. Even for an abrupt step change in mesh size of 4:1, there is no sign of the distortions and reflections often referred to in the literature. In all cases presented, the total number density of particles is conserved to a very high order of accuracy.

The development of such accurate methods of solving hyperbolic equations on a non-uniform mesh will allow self-consistent solutions of problems in discharge physics, including the details at the electrode boundaries as well as in the body of the plasma. In addition, these solutions can be obtained with medium-sized computers.

Further work is continuing to apply these methods to adaptive non-uniform meshes and to consider their extension to two dimensions.

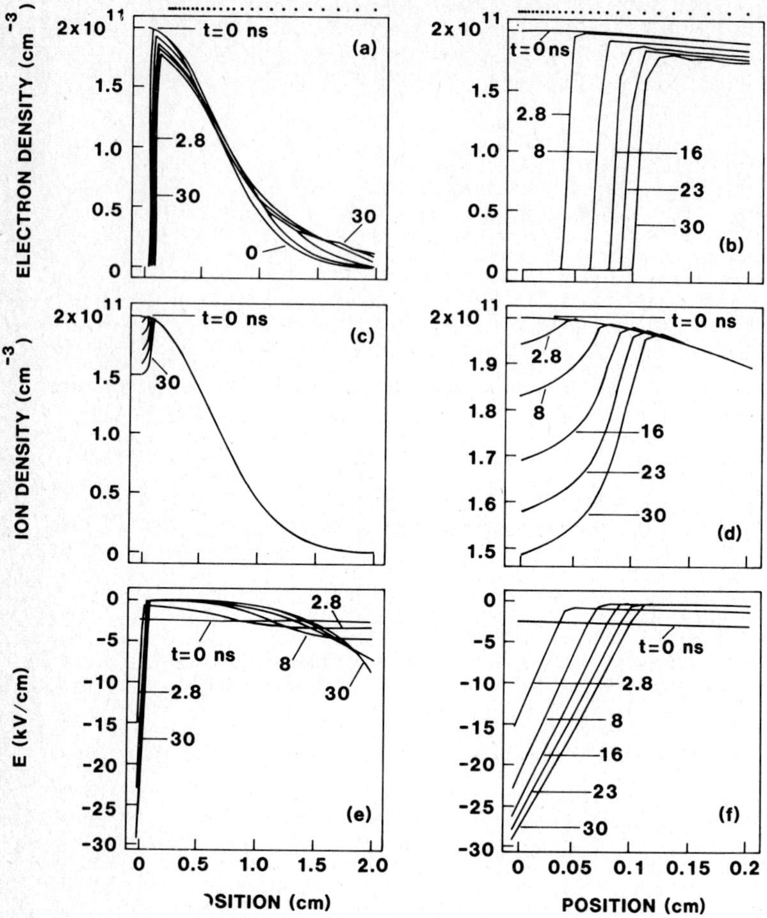

Figure 4

Distributions of electrons, positive ions and electric field for a plasma near a conducting wall (cathode) with an initially uniform electric field. Results are computed on an exponentially expanding mesh, shown as dots above figures a) and b). Curves are presented for 0, 2.8, 8, 16, 23 and 30 ns after the start of the calculations corresponding to 0, 100, 400, 900, 1400 and 1900 computational steps respectively. (times are shown on curves in ns). Graphs (a), (c) and (e) are the electron density, positive ion density and electric field distributions respectively over the entire space between the electrodes, while graphs (b), (d) and (f) resolve these distributions near the cathode.

REFERENCES

(1) Morrow, R., Numerical Solution of Hyperbolic Equations for Electron Drift in Strong Non-Uniform Electric Fields, J.Comp.Phys., 43 (1981) 1-15.

(2) Morrow, R., Space-charge Effects in High-density Plasmas, J.Comp.Phys., 46 (1982) 454-461.

(3) Morrow, R. and Lowke, J.J., Trichel Pulses in Oxygen: A Simple Theory, 7th International Conference on Gas Discharges (London, 1982) 124-127.

(4) Boris, J.P. and Book, D.L., Solution of Continuity Equations by the Method of Flux-corrected Transport in: John Killeen, ed., Methods in Computational Physics, (Academic Press, New York 1976a) 85-130.

(5) Davies, A.J., Evans, C.J. and Llewellyn-Jones, F., Electrical Breakdown of Gases: the Spatio-temporal Growth of Ionization in Fields Distorted by Space-charge, Proc. Roy. Soc., A 281 (1964) 164-183.

(6) Boris, J.P. and Book, D.L., Flux-corrected Transport: I.SHASTA, A Fluid Transport Algorithm that works, J.Comp.Phys. 11 (1973) 38-69.

(7) Boris, J.P. and Book, D.L., Flux-corrected Transport III. Minimal-Error FCT Algorithms, J.Comput.Phys., 20. (1976b) 397-431.

(8) Zalesak, S.T., Fully Multidimensional Flux-corrected Transport Algorithms for Fluids, J.Comput.Phys., 31, (1979), 335-362.

(9) Zalesak, S.T., Advances in Computer Methods for Partial Differential Equations-IV, (R. Vichnevetsky and R.S. Stepleman, eds. IMACS, 1981)

Computational Techniques & Applications: CTAC-83
J. Noye & C. Fletcher (Editors)

SINGULAR FINITE ELEMENT METHODS IN PLASMA STABILITY COMPUTATIONS - A SIMPLE MODEL

Robert L. Dewar

Department of Theoretical Physics
Research School of Physical Sciences
The Australian National University
G.P.O. Box 4
Canberra, A.C.T. 2601
Australia

and

Raymond C. Grimm

Plasma Physics Laboratory
Princeton University
P.O. Box 451
Princeton N.J. 08544
U.S.A.

The implementation of boundary layer asymptotic matching techniques for the computation of the resistive mode stability properties of toroidal magnetic containment devices is a challenging area for the application of singular finite element techniques. In order to expose some of the problems in as simple a manner as possible, we analyze a simple singular ordinary differential equation, soluble in Whittaker functions. We examine two possible choices of singular element. Algebraic manipulation computer programs have been written to calculate the matrix elements and analyze the convergence. Convergence is found to be sensitive to the order of truncation of the Frobenius series, and to the ratio of the node spacing of the regular elements to the width of the singular elements, but not to the choice of type of element.

I. INTRODUCTION

Over the past decade finite element methods have been used extensively for the computation of unstable eigenmodes of axisymmetric magnetic containment devices proposed for fusion experiments.[1] These calculations have been based on the ideal, perfectly conducting hydromagnetic fluid model of a plasma.

Although this model has nice Hermiticity properties, it is known to break down near certain singular magnetic surfaces, where even a small amount of resistivity can play an important role in a narrow boundary layer.[2]

This phenomenon can be treated by calculating a complete set of weak solutions of the ideal equations, without requiring that they lie within the Hilbert space appropriate to the ideal model. These singular solutions can then be matched onto solutions in the boundary layer obtained by other techniques.

In order to compute the full set of weak solutions via the Galerkin method, techniques must be developed for handling highly singular finite elements, which are constructed to have the correct asymptotic behaviour in the neighbourhood of the singular surfaces.

This method has been successfully applied to zero-pressure toroidal plasma problems[3,4] by adapting the Princeton finite element code PEST.[1] Convergence problems have been encountered, however, with finite-pressure plasmas, where the singularities are stronger.

In Sec. II we introduce a simple singular ordinary differential equation with some features in common with the ideal plasma problem, in order to develop some analytical understanding of convergence. In Sec. III we show how Galerkin's method may be applied to calculate generalized function solutions, and how to compute matrix elements involving finite elements not belonging to the Hilbert space. Two possible choices for such elements (designated A and B) are introduced in Sec. IV, and convergence of the Galerkin method is analyzed in Sec. V using a Taylor expansion approach. The various conclusions of this study are summarized in Sec. VI.

II. THE MODEL

The equation we study is

$$Fy = 0 , \tag{1}$$

where the "linearized force operator" F is given by

$$F = \frac{d}{dx} x^2 \frac{d}{dx} + (a+bx+cx^2) , \tag{2}$$

with a, b, and c being real constants. Equation (1) is singular at $x = 0$, and models the equation[5] for the radial component of the ideal quasistatic hydromagnetic fluid displacement in the neighbourhood of a "rational surface", i.e. a magnetic surface where $\underset{\sim}{B}.\underset{\sim}{\nabla}$ is a singular operator ($\underset{\sim}{B}$ being the equilibrium magnetic field).

Corresponding to the assumption that the plasma is stable against ideal interchange instabilities,[5] we assume $a < 1/4$, so that we can write

$$a = 1/4 - \mu^2 , \tag{3}$$

where μ is a real, positive constant. We seek weak solutions with support on the positive real axis, so that the general solution is

$$y = a_b y_b + a_s y_s \ , \qquad (4)$$

where a_b and a_s are arbitrary constants, and y_b and y_s are, respectively, the "big" and "small" solutions given by

$$y_b = x_+^{p_b} \, e^{-z/2} \, F(-\mu-\lambda+\tfrac{1}{2}|-2\mu+1|z) \ , \qquad (5)$$

$$y_s = x_+^{p_s} \, e^{-z/2} \, F(\mu-\lambda+\tfrac{1}{2}|2\mu+1|z) \ , \qquad (6)$$

where $F(\cdot|\cdot|\cdot)$ is a confluent hypergeometric function,[6] $z \equiv (-4c)^{\frac{1}{2}}x$, $\lambda \equiv b(-4c)^{-\frac{1}{2}}$, $p_b \equiv -\frac{1}{2} - \mu$, and $p_s \equiv -\frac{1}{2} + \mu$. The special case $b = 0$ is solvable in modified Bessel functions. We take the generalized function x_+^p to be the canonically regularised fractional power function with positive support defined by Gel'fand and Shilov.[7]

In more general cases, a power series solution corresponding to Eqs. (4-6) can be constructed by the method of Frobenius.[8] As is well known, the Frobenius method must be modified when $p_s - p_b \equiv 2\mu$ is an integer. In our case this can be avoided by redefining the big solution to be proportional to $z^{-1}W_{\lambda,\mu}(z)$, where $W_{\lambda,\mu}(\cdot)$ is the Whittaker function.[6] There are also complications when p_b is a negative integer ($\mu = 1/2, 3/2, 5/2 \ldots$) owing to the poles of $x_+^{p_b}$ at these values.[7] These should be treated in the context of the more physically relevant problem of the solutions with support on the entire real line, in which case we work with the odd and even functions[7] $|x|^p$ and $|x|^p \operatorname{sgn} x$. In the interests of simplicity, however, we here confine our attention to the cases when μ is not an integer multiple of $\frac{1}{2}$, thus unfortunately excluding the zero-pressure plasma cases[4] $\mu = \frac{1}{2}$.

We complete our specification of the model by prescribing the boundary conditions

$$y(x) \sim x_+^{p_b} \quad \text{as} \quad x \to 0 \ , \qquad (7)$$

and

$$y(1) = 0 \ . \qquad (8)$$

Equation (7) implies that $a_b = 1$. The task is to compute

$$a_s = -y_b(1)/y_s(1) \qquad (9)$$

by expanding in finite elements, and using Galerkin's method.[9] The coefficient

a_s is the analogue of the quantity Δ' which determines stability to resistive tearing modes.[4]

III. GALERKIN'S METHOD WITH SINGULAR ELEMENTS

The small solution, and most of the big solution, lie in the Hilbert space $\mathcal{H}_+$ which is the completion of the space of real differentiable functions, vanishing at $x = 1$, and for $x < 0$, with finite norm $\|\cdot\|$ defined by

$$\|\phi\|^2 \equiv \int_{-\infty}^{1} dx[\phi^2+x^2(\phi')^2], \quad \forall\phi\in\mathcal{H}_+ . \tag{10}$$

Functions in $\mathcal{H}_+$ may have arbitrary discontinuities at $x = 0$, provided they grow more slowly than $|x|^{-\frac{1}{2}}$ as $x \to 0$. This allows us to restrict attention to the solutions with support on $[0,1]$.

The part of y_b with support in the neighbourhood of $x = 0$ is excluded from $\mathcal{H}_+$ since $p_b < -\frac{1}{2}$. However, y_b is a weak solution in the sense that

$$\langle\phi,Fy_b\rangle = 0 , \quad \forall\phi\in\mathcal{H}_+ , \tag{11}$$

where the inner product $\langle\phi,\psi\rangle$ is defined by

$$\langle\phi,\psi\rangle \equiv \int_{-\infty}^{1} \phi\psi\, dx . \tag{12}$$

Thus, by expanding y in the form Y + error, where

$$Y \equiv \alpha_b\phi_b + \sum_{i=0}^{N} \alpha_i\phi_i , \tag{13}$$

where $\phi_i\in\mathcal{H}_+$ for $i = 0,1,2...$, and $(\alpha_b\phi_b-y_b)\in\mathcal{H}_+$, we can construct the Galerkin approximation by solving the equations

$$\sum_{j=0}^{N} \langle\phi_i,F\phi_j\rangle\alpha_j = -\alpha_b\langle\phi_i,F\phi_b\rangle . \tag{14}$$

We require that $\langle\phi_i,F\phi_j\rangle$ be negative definite, so that Eq. (14) can always be inverted. This corresponds physically to the assumption that the plasma is ideally stable. The matrix can be calculated by integrating by parts to obtain only first derivatives. However, since ϕ_b does not belong to $\mathcal{H}_+$, we cannot correctly compute $\langle\phi_i,F\phi_b\rangle$ by integrating by parts. Instead we now give a formula for $\langle\phi_i,F\phi_j\rangle$ which does not require integration by parts, and therefore applies also when ϕ_j is replaced by ϕ_b.

We assume that ϕ_j is piecewise differentiable

$$\phi_j = \sum_k \sigma_j^k(x)\, H_k(x) \ , \tag{15}$$

where $\sigma_j^k(x)$ is a smooth function (except at $x = 0$) and $H_k(x)$ is a "box car" function: unity between nodes x_k and x_{k+1} and zero elsewhere. We require ϕ_j to be continuous (to belong to $\mathcal{H}_+$), so that

$$\sigma_j^k(x_k) = \sigma_j^{k-1}(x_k) \ . \tag{16}$$

Then we can show that

$$\langle\phi_i, F\phi_j\rangle = \sum_k \langle\phi_i, \{H_k F + [\delta(x-x_k) - \delta(x-x_{k+1})]x^2 \frac{d}{dx}\}\sigma_j^k\rangle \ , \tag{17}$$

where $\delta(\cdot)$ is the Dirac delta function. Note that Eq. (16) has led to the cancellation of the $\delta'(\cdot)$ terms. Equation (17) is readily amenable to machine computation using an algebraic manipulation code, and a REDUCE[10] program has been written to construct matrix elements with various choices of ϕ_j. By choosing one of the ϕ_i to go like $x_+^{p_s}$ in the neighbourhood of the origin, we directly obtain an approximation to a_s as the appropriate coefficient α_i (to within a normalization factor).

IV. CHOICES OF ELEMENT

Except near the origin, we use "tent functions": piecewise linear elements as depicted below

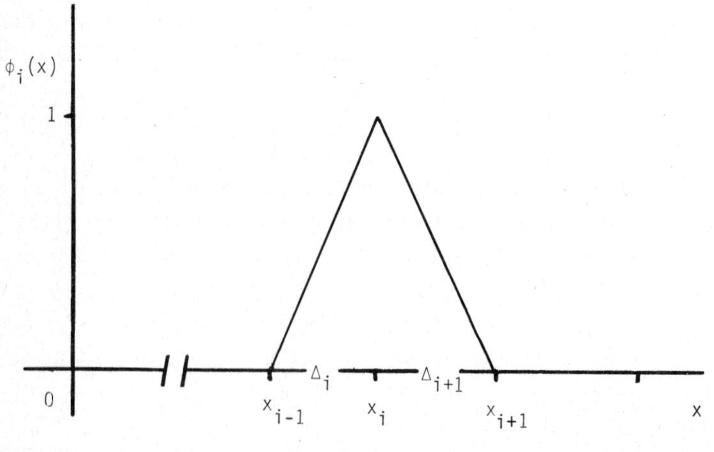

We call these the *regular elements*, and we refine the mesh progressively as $x_i \to 0$ by distributing the mesh so that the spacings $\Delta_i \equiv x_i - x_{i-1}$ and $\Delta_{i+1} \equiv x_{i+1} - x_i$ are given by

$$\Delta_i = \frac{2\varepsilon}{1+\varepsilon} x_i , \quad \Delta_{i+1} = \frac{2\varepsilon}{1-\varepsilon} x_i , \tag{18}$$

where ε is a small parameter, one of the two convergence parameters in the problem. The solution of this difference equation is

$$x_i = x_1 \left[\frac{1+\varepsilon}{1-\varepsilon}\right]^{i-1} . \tag{19}$$

The first node is at x_1, which we denote by δ, the second convergence parameter.

Near the origin we use two *singular* elements to describe the asymptotic x^{p_b} and x^{p_s} behaviours. In previous work[4] we have used what we shall now call Type A elements:

$$\phi_b^A(x) = H(\delta-x) \sum_{r=0}^{R_b} c_r^b \, \delta^r (z_+^{p_b+r} - z_+^k) . \tag{20.b}$$

There is a similar Eq. (20.s) for the small element. Here $z \equiv x/\delta$, c_r^b is the coefficient of x^{p_b+r}/δ^{p_b} in the Frobenius expansion for y_b/δ^{p_b} (so $c_0^b \equiv 1$), R_b is the order to which the Frobenius expansion is taken, and k is a constant. We consider the cases $k = 0$ and $k = 1$, modelling the even and odd elements of the two-sided problem. The function $H(\cdot)$ is the Heaviside step function. With Type A singular elements, the first regular element must be

$$\phi_1^A(x) = H(\delta-x)z_+^k + H_1(x)(x_2-x)/\Delta_2 . \tag{21}$$

Although they minimize the support of the singular trial functions, the Type A elements have the unsatisfying feature that a spurious z^k behaviour is introduced near the origin. The first regular element, if it does its job, must simply subtract out the spurious behaviour. This suggests that we can save an element, and possibly make the convergence better, by combining ϕ_1^A and ϕ_b^A into a single element. Likewise for ϕ_s^A. We call the new singular trial functions Type B elements. The big element is

$$\phi_b^B = H(\delta-x)z_+^{p_b} \sum_{r=0}^{R_b} c_r^b x^r + H_1(x)(x_2-x)\phi_b^B(\delta)/\Delta_2 . \tag{22.b}$$

The first regular element is now just the standard tent function $\phi_2(x)$. Although the support of the Type B singular elements is wider, the absence of the spurious z^k terms means that the matrix elements are in fact simpler.

We illustrate the types of element below

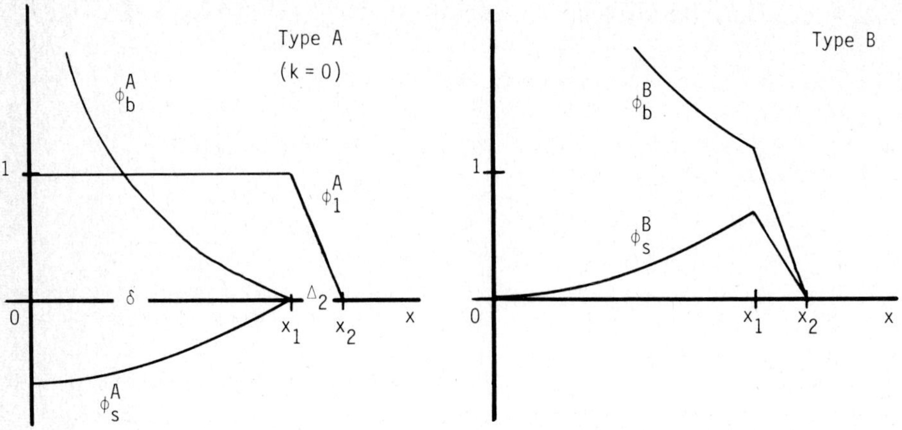

V. CONVERGENCE

We examine convergence by obtaining an asymptotic approximation to the truncation error using Taylor expansion about the nodes. The algebra is handled using REDUCE. For simplicity we take $b = 0$ and only the leading term of the Frobenius expansion, but since $c_1 \equiv 0$ when $b = 0$, we are effectively taking two terms.

For the regular elements, which are of unit height, the coefficient α_i is assumed to be an approximation $Y(x_i)$ to the true solution, $y(x_i)$. The finite element equation, Eq. (14), is for $i > 1$ (or 2 for Type B elements) a homogeneous second order difference equation. We assume the node spacing to be given by Eq. (18), assume the Ansatz

$$Y(x) = y(x) + \varepsilon y_1(x) + \varepsilon^2 y_2(x) + \ldots$$

where y_i are analytic error functions, and Taylor expand the difference equation. Using Eq. (1) to eliminate higher derivatives of y we find $y_1 \equiv 0$, and

$$(x^2 y_2')' + (\tfrac{1}{4} - \mu^2 + cx^2) y_2 = \tfrac{1}{3}\{[\mu^4 - \tfrac{9}{2}\mu^2 + \tfrac{17}{16} + 2cx^2(\tfrac{1}{4} - \mu^2 + \tfrac{c}{2}x^2)]y + 4xy'\} \tag{23}$$

We do not need an explicit solution, but merely write

$$Y = A_b y_b(x)[1+\varepsilon^2 e_b(x)] + A_s y_s(x)[1+\varepsilon^2 e_s(x)] + O(\varepsilon^4) , \tag{24}$$

where, as $x \to 0$ the asymptotic behaviours are as below

$$\begin{aligned} y_b(x) &\sim x^{-\frac{1}{2}-\mu}\left[1 - \frac{cx^2}{4(1-\mu)} + \ldots\right], \quad e_b(x) \sim (\tfrac{2}{3} - \mu_2)\ln x , \\ y_s(x) &\sim x^{-\frac{1}{2}+\mu}\left[1 - \frac{cx^2}{4(1+\mu)} + \ldots\right], \quad e_s(x) \sim (\tfrac{2}{3} + \mu_2)\ln x , \end{aligned} \tag{25}$$

where

$$\mu_2 \equiv (\mu^4 - 9\mu^2/2 - 15/16)/6\mu .$$

Note that ε is related to Δ_2 and δ by $\varepsilon \approx \Delta_2/2\delta$.

The right-hand boundary condition on Y is $Y(1) = 0$. We obtain the left-hand boundary condition by eliminating the unknown coefficient of the small-singular element in terms of $Y(x_1)$, either by using the first of the finite element Eqs. (14), in the case of Type-A elements, or directly from the definitions in the case of Type-B elements. The next finite element equation then relates $Y(x_1)$ and $Y(x_2)$. Taylor expanding around $x_1 = \delta$, up to $O(\delta^2)$ beyond the leading order [assuming $\varepsilon = O(\delta)$], and multiplying the result by the appropriate factor to make the right-hand sides equal, we find that the left-hand sides are also equal. Thus we conclude: *Type* A *and Type* B *elements give identical results* (at least up to $O(\delta^2)$ and neglecting rounding errors).

The boundary condition is

$$(\mu - \tfrac{1}{2})Y - \delta Y' + \frac{\delta^2\rho^2}{6}\left[(\mu^2 - \tfrac{1}{4})Y - 2\delta Y'\right] - \frac{c\delta^2}{4(\mu+1)}\left[(\mu + \tfrac{3}{2})Y - \delta Y'\right] + O(\delta^3) = 2\mu\delta^{-\frac{1}{2}-\mu} \tag{26}$$

where Y and Y' are evaluated at $x = x_1 = \delta$, and $\rho \equiv \Delta_2/\delta^2$. For the time being we assume $\rho = O(1)$. This is a difference approximation to the Wronskian[8] between y and the big solution.

Inserting Eq. (24) in Eq. (26) we find

$$A_b = 1 - \varepsilon^2\left[e_b(\delta) - \delta\frac{e_b'(\delta)}{2\mu}\right] - \frac{\delta^2\rho^2}{12\mu}(\mu^2 + 2\mu + \tfrac{3}{4}) + O(\delta^3) . \tag{27}$$

There is a similar expression for A_s, which converges to a_s given by Eq. (9). Thus Y will always converge to y as Δ_2 and $\delta \to 0$. However, our task is to find under what conditions a_s is approximated by $\alpha_\star \equiv \delta^{-p_s}\alpha_s$, where α_s is the coefficient of the small singular element, Eq. (20.s) or (22.s). By considering the Type B elements we see that $\alpha_s = [Y(\delta) - \delta^{-\frac{1}{2}-\mu}\phi_b(\delta)]/\phi_s(\delta)$. Thus $\alpha_\star$ results from the difference of two large terms. Using Eq. (27) we find

$$\alpha_\star = A_s[1+\varepsilon^2 e_s(\delta) - c\delta^2/4(1+\mu)] + \delta^{-2\mu}\left[\frac{\varepsilon^2}{2\mu}\left(\frac{2}{3} - \mu_2\right) - \frac{\delta^2\rho^2}{12\mu}\left(\mu^2+2\mu+\frac{3}{4}\right) - \frac{c\delta^2}{4(1-\mu)} + O(\delta^3)\right] . \tag{28}$$

It is the term containing $\delta^{-2\mu}$ which gives the trouble. By taking $\varepsilon = o(\delta^\mu)$ we can always make the first two troublesome terms converge to zero as $\delta \to 0$. However, unless

$$\mu < 1 \tag{29}$$

the term proportional to c will diverge as $\delta \to 0$.

VI. CONCLUSIONS

The analysis of Sec. V confirms that our past[4] practice of extrapolating first to $\varepsilon = 0$, then $\delta = 0$ was a valid procedure, which should converge in the zero-pressure case $\mu = \frac{1}{2}$, as observed.

We have also found that if $\mu \geqslant 1$ the solution will diverge as $\delta \to 0$, as has also been observed. Although Type B elements are easier to calculate and analyze, and require one less element, their convergence properties appear to be identical to those of the Type A elements. This shows that the Galerkin method automatically nulls out the spurious z_+^k dependence introduced by the Type A elements.

To obtain convergence for $\mu \geqslant 1$, we would have to include the first three terms of the Frobenius expansion. We anticipate that this would give convergence for $\mu < 3/2$. Intuitively, one must be sceptical that the highly localized information required to compute the Frobenius expansion to this order can really affect global stability, which suggests that an ideal approximation is adequate for surfaces with $\mu \sim 1$. However this remains to be proved. The availability of algebraic manipulation programs makes higher order calculations practicable, even in situations more complicated than our model problem.

Another moral can be drawn from the analysis of Section V, and that is that the calculation of $\alpha_\star$ is sensitive to small errors, such as rounding errors, due to the near cancellation of large terms as $\delta \to 0$. Another source of such errors could be a mismatch between the accuracy of the singular elements and of the regular elements near $x = \delta$. This suggests that *both* singular and regular elements should be computed analytically out to $x = 2\delta$ (say).

In our numerical analysis we have not exploited the variational nature of the Galerkin method.[9] There are probably better ways of analyzing the problem.

ACKNOWLEDGEMENTS

This work was partly supported by U.S. Department of Energy contract DE-AC02-76-CH03073. The REDUCE calculations were performed interactively on Cray-I computers at the U.S. National Magnetic Fusion Energy Computer Center, Livermore, California, using the MIDAS data transmission service of the Overseas Telecommunications Commission (Australia),

REFERENCES

[1] Grimm, R.C., Dewar, R.L. and Manickam, J., Ideal MHD Stability Calculations in Axisymmetric Toroidal Coordinate Systems, J. Comput. Phys. 49 (1983) 94-116.

[2] Glasser, A.H., Greene, J.M. and Johnson, J.L., Resistive Instabilities in General Toroidal Plasma Configurations, Phys. Fluids 18 (1975) 875-888.

[3] Grimm, R.C., Dewar, R.L., Manickam, J., Jardin, S.C., Glasser, A.H. and Chance, M.S., Resistive Instabilities in Tokamak Geometry, in: Proceedings of the 9th International Conference on Plasma Physics and Controlled Nuclear Fusion Research, Baltimore, 1-8 September, 1982 (IAEA, Vienna, 1983).

[4] Manickam, J., Grimm, R.C., and Dewar, R.L., Resistive MHD Stability in Tokamaks, in: Energy Modeling and Simulation, Eds. A.S. Kydes et al., Volume IV of IMACS Transactions on Scientific Computation (North-Holland, Amsterdam, 1983) pp. 355-360.

[5] Newcomb, W.A., Hydromagnetic Stability of a Diffuse Linear Pinch, Ann. Phys. (N.Y.) 10 (1960) 232-267.

[6] Whittaker, E.T. and Watson, G.N., A Course of Modern Analysis (Cambridge U. Press, Cambridge, 1965) 4'th ed. Ch. XVI.

[7] Gel'fand, I.M. and Shilov, G.E., Generalized Functions (Academic, N.Y., 1964) Ch. I.

[8] Ince, E.L., Ordinary Differential Equations (Dover, N.Y., 1956) Chs. VII and XVI.

[9] Strang, G. and Fix, G.J., An Analysis of the Finite Element Method (Prentice-Hall, Englewood Cliffs N.J., 1973).

[10] Hearn, A.C. and Griss, M.L., Reduce 2 User's Manual, 2nd. ed. (NMFECC Cray-I version).

DYNAMICS, STRUCTURES

Computational Techniques & Applications: CTAC-83
J. Noye & C. Fletcher (Editors)

SOLUTIONS OF SOME MIXED BOUNDARY PROBLEMS IN CONSOLIDATION THEORY

J.R. Booker, B.Sc, Ph.D, D.Eng.
and
J.C. Small, B.Sc(Eng), Ph.D.

School of Civil and Mining Engineering
The University of Sydney
Sydney, Australia

In this paper a method is presented which may be used to obtain the time-settlement behaviour of a circular footing resting on a deep clay stratum. The governing equations are solved for several different boundary conditions on the upper surface of the clay. The loading applied may be considered to be completely flexible or completely rigid, while the drainage conditions may be such that the entire upper surface is permeable or impermeable over the loaded region and permeable elsewhere. The solution is found in Laplace transform space and a numerical inversion technique is then used to obtain the time dependent solution.

INTRODUCTION

When a foundation resting on a saturated soil is loaded the soil will tend to compress and the foundation will undergo an initial settlement. During this compression the pore water pressure will rise and excess pore pressures will develop. Subsequently pore water will tend to flow from regions of high excess pore pressure to regions of lower pore water pressure and the excess pore water pressure will dissipate, thus causing the foundation to settle with time.

McNamee and Gibson (1960) have examined the behaviour of a perfectly flexible circular footing on a deep clay layer. This work was later extended to examine the behaviour of a perfectly flexible rectangular footing, Gibson and McNamee (1963); Booker (1974) studied the behaviour of perfectly flexible circular and rectangular footings resting on a layer of finite depth. The behaviour of a rigid circular footing on a deep clay layer was examined by Chiarella and Booker (1975) and the settlement behaviour of a circular raft of finite flexibility was investigated by Small and Booker (1983).

All these investigations have assumed that the upper surface of the soil was completely free to drain. In many practical situations this is not the case, quite often the foundation is relatively impermeable and so drainage beneath the foundation will be severely impeded. In this paper a solution is obtained to the problem of the settlement of a perfectly rigid or perfectly flexible circular footing, which may be either completely permeable or completely impermeable, resting on a deep clay layer.

BASIC EQUATIONS

The problem of a circular footing resting on a deep homogeneous layer of saturated soil is shown schematically in Fig. 1.

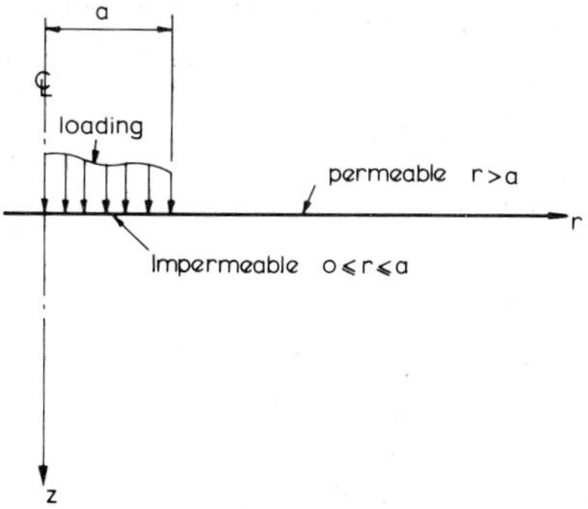

Figure 1

Clearly this problem is axisymmetric, under these circumstances the equations of equilibrium reduce to:

$$\begin{aligned} G(\nabla^2 u - u/r^2) - (\lambda + G)\ \partial\varepsilon_v/\partial r &= \partial p/\partial r \\ G\nabla^2 w - (\lambda + G)\ \partial\varepsilon_v/\partial z &= \partial p/\partial z \end{aligned} \tag{1}$$

where $\varepsilon_v = -(\partial u/\partial r + u/r + \partial w/\partial z)$

and u, w, p denote the radial displacement, the vertical displacement and the excess pore pressure respectively; where λ, G are the Lame and shear modulus.

The equation of volume constraint together with Darcy's Law become

$$\varepsilon_v = -\int_0^t k/\gamma_w \nabla^2 p\ dt \tag{2}$$

where k is the coefficient of permeability and γ_w is the unit weight of water.

Equations (1,2) can be combined to show that

$$\varepsilon_v = \int_0^t c\nabla^2 \varepsilon_v\ dt \tag{3}$$

where $c = k(\lambda + 2G)/\gamma_w$ is the coefficient of one dimensional consolidation.

Equations (1,3) can be simplified by introduction of simultaneous Hankel and Laplace transforms defined as follows:

$$\begin{aligned} (W,P) &= \int_0^\infty r\ J_0(\rho r)(w,p)\ dr \\ U &= \int_0^\infty r\ J_1(\rho r)\ u\ dr \end{aligned} \tag{4}$$

$$(\bar{U},\bar{W},\bar{P}) = \int_0^\infty e^{-st}(U,W,P)\,dt$$

Equations (1-3) then become:

$$G(\partial^2\bar{U}/\partial z^2 - \rho^2\bar{U}) = -(\lambda + G)\rho\bar{E}_v - \rho\bar{P}$$
$$G(\partial^2\bar{W}/\partial z^2 - \rho^2\bar{W}) = (\lambda + G)\partial\bar{E}_v/\partial z + \partial\bar{P}/\partial z \quad (5)$$
$$\partial^2\bar{E}_v/\partial z^2 - \zeta^2\bar{E}_v = 0$$

where $\bar{E}_v = -(\rho\bar{U} + \partial\bar{W}/\partial z)$

and $\zeta^2 = \rho^2 + s/c$

If we assume the consolidating soil occupies the region $z > 0$ then the solution which remains bounded as $z \to \infty$ is:

$$\begin{bmatrix} \rho\bar{U} \\ \rho\bar{W} \\ \bar{P}/2G \end{bmatrix} = \begin{bmatrix} -(1-\rho z)e^{-\rho z} & \frac{\rho^2 c}{s}\, e^{-\zeta z} & -\rho e^{-\rho z} \\ \rho z\, e^{-\rho z} & \frac{\rho\zeta c}{s}\, e^{-\zeta z} & \rho e^{-\rho z} \\ e^{-\rho z} & -\eta e^{-\zeta z} & 0 \end{bmatrix} \begin{bmatrix} \bar{A} \\ \bar{B} \\ \bar{C} \end{bmatrix} \quad (6)$$

where $\zeta = \sqrt{(\rho^2 + s/c)}$ has a positive real part and $\bar{A}$, $\bar{B}$, $\bar{C}$ are constants to be determined. Also $\eta = (\lambda + 2G)/2G$.

Other quantities of interest are the vertical and shear stress and the surface flow:

$$\sigma_{zz} = p - \lambda\,\varepsilon_v - 2G\,\partial w/\partial z \quad (7)$$
$$\sigma_{rz} = -G(\partial w/\partial r + \partial u/\partial z)$$
$$q = -\int_0^t k/\gamma_w \cdot \partial p/\partial z\,dt$$

Now if we define the Hankel transforms

$$(S_{zz}, Q) = \int_0^\infty r\,J_0(\rho r)\,(\sigma_{zz}, q)\,dr$$
$$(8)$$
$$S_{rz} = \int_0^\infty r\,J_1(\rho r)\,\sigma_{rz}\,dr$$

we find that

$$\begin{bmatrix} \bar{S}_{zz}/2G \\ \bar{S}_{rz}/2G \\ \rho\bar{Q} \end{bmatrix} = \begin{bmatrix} z\rho e^{-\rho z} & \frac{\rho^2 c}{s} e^{-\zeta z} & -\rho e^{-\rho z} \\ -(1-\rho z)\rho e^{-\rho z} & \frac{\rho\zeta c}{s} e^{-\zeta z} & -\rho e^{-\rho z} \\ \frac{\rho^2 c}{\eta s} e^{-\rho z} & -\frac{\rho\zeta c}{s} e^{-\zeta z} & 0 \end{bmatrix} \begin{bmatrix} \bar{A} \\ \bar{B} \\ \bar{C} \end{bmatrix} \qquad (9)$$

METHOD OF SOLUTION

We wish to consider the behaviour of circular footings acting on the surface of a half space. It will be assumed in all cases that the footing is smooth and that free drainage is possible outside the footing and also the load applied to the footing is uniform and of intensity p_{AV}. There are thus four basic cases to consider. These will be dealt with in order of increasing complexity.

(a) Perfectly Flexible, Completely Permeable Footing

In this case we see immediately that:

$$P_0 = 0 \qquad (10a)$$

$$S_{rz0} = 0 \qquad (10b)$$

$$\bar{S}_{zz0} = \bar{p}_{AV}\, H(\rho,a) \qquad (10c)$$

where $H(\rho,a) = aJ_1(\rho a)/\rho$ and where the subscript zero indicates the value at the surface. This is sufficient information to determine the constants $\bar{A}$, $\bar{B}$, $\bar{C}$ appearing in equations (6,9) and thus to find the complete solution in transform space. In particular we find:

$$\rho\bar{W}_0 = \bar{M}\,\bar{S}_{zz0}/2G$$

where $\bar{M} = \eta(\zeta + \rho)\left[\rho(1-\eta) - \eta\zeta\right]$

(b) Rigid Permeable Footing

This case is similar to the previous one, and it is found that equations (10a,b) still hold. The contact pressure distribution is now unknown but we find that:

$$\rho\bar{W}_0 = \bar{M}\,\bar{S}_{zz0}/2G \qquad (11)$$

Now let us assume that σ_{zz} can be represented, to sufficient accuracy in the form:

$$\sigma_{zz0} = \sum_{i=1}^{m} a_i\phi_i(r) \qquad (12a)$$

so that:

$$S_{zz0} = \sum_{i=1}^{m} a_i \, \Phi_i(\rho) \tag{12b}$$

where

$$\Phi_i = \int_0^{\infty} r \, J_0(\rho r)\phi_i(r) \, dr$$

We now define generalised deflections:

$$w_i = \int_0^{a} r \, \phi_i(r)w(r)dr \tag{12c}$$

We thus obtain from equation (11) the relationship

$$\bar{\underset{\sim}{w}} = \bar{T} \, \bar{\underset{\sim}{a}} \tag{13a}$$

where $\underset{\sim}{w} = (w_1, \ldots, w_m)^T$

$\underset{\sim}{a} = (a_1, \ldots, a_m)^T$

and T is the m x m matrix with components

$$\bar{T}_{ij} = \int_0^{\infty} \bar{M} \, \Phi_i(\rho)\Phi_j(\rho) \, d\rho .$$

Now for a rigid footing we know that $w(r) = \delta$, $0<r<a$ and thus we may write

$$\underset{\sim}{w} = \delta\underset{\sim}{\lambda} \tag{13b}$$

where $\lambda_i = \int_0^{a} r\Phi_i(r) \, dr$

To complete the analysis we need the equation of equilibrium and so from equation (12a), we have

$$\underset{\sim}{\lambda}^T \underset{\sim}{a} = p_{AV} \, a^2/2 \tag{13c}$$

Equations (13a,b,c) may be written in the compact form

$$\begin{bmatrix} \bar{T} & -\underset{\sim}{\lambda} \\ -\underset{\sim}{\lambda}^T & 0 \end{bmatrix} \begin{bmatrix} \bar{\underset{\sim}{a}} \\ \bar{\delta} \end{bmatrix} = \begin{bmatrix} 0 \\ -\bar{p}_{AV} \, a^2/2 \end{bmatrix} \tag{14}$$

Equation (14) may be used to calculate the coefficients a and hence the value of S_{zz0}. We now have sufficient information to determine the coefficients $\bar{A}$, $\bar{B}$, $\bar{C}$ occurring in equations (6,9) and hence may obtain the solution in transform space.

(c) Perfectly Flexible Impermeable Footing

Suppose now the raft is completely flexible but is impermeable, we can show from equations (6,9) that the surface pore pressure and flow are related by the equation:

$$\rho\bar{P}_0 = -\rho(\bar{L}/\bar{K}).\ \bar{S}_{zz0} + (2G\rho^2/\bar{K}).\bar{Q}_0 \tag{15}$$

where $\bar{L} = -(1+\bar{M})$

and $\bar{K} = 1+\bar{M} - \rho^2 c/\eta s$

Now let us assume that the surface stress is given by an expression of the form (12a), while the flow outside the footing can be approximated by an expression having the form

$$q = \sum_{j=1}^{n} b_j\psi_j(r) \tag{16}$$

so that Q_0 may be represented in the form

$$Q_0 = \sum_{j=1}^{n} b_j\Psi_j(\rho) \tag{16b}$$

where $\Psi_j = \int_a^\infty r\, J_0(\rho r)\ \psi_j(r)dr$

If we now define the generalised pore pressures

$$p_j = \int_a^\infty r\ p(r)\psi_j(r)dr$$

we find that equation (15) reduces to

$$\bar{p} = -\bar{Y}^T\bar{a} + \bar{Z}\bar{b} \tag{17}$$

where $p = (p_1,\ldots,p_n)^T$

$b = (b_1,\ldots,b_n)^T$

and Y, Z are the m x n, n x n matrices with components:

$$\bar{Y}_{ij} = \int_0^\infty (\bar{L}/\bar{K})\Phi_i\Psi_j d\rho$$

$$\bar{Z}_{ij} = \int_0^\infty (2G\rho^2/\bar{K})\Psi_i\Psi_j d\rho$$

The excess pore pressure outside the footing will be zero and so the generalised pore pressures will also vanish. Since the footing is perfectly flexible the contact pressure will be known and so the coefficients a will be prescribed. In fact for uniform loading we see that $m = 1$, $a_1 = p_{AV}$ and $\Phi_1 = H(\rho,a)$.

We thus find that the coefficients $\underset{\sim}{b}$ are given by

$$\bar{\underset{\sim}{b}} = \bar{Z}^{-1} \bar{Y}^T \bar{\underset{\sim}{a}} \tag{18}$$

and so we may regard $\bar{S}_{zz0}$, $\bar{Q}_0$, $\bar{S}_{rz0}$ as known and hence we have through equations (6,9) sufficient to obtain the complete solution in transform space.

(d) Rigid Impermeable Footing

Equations (6,9) can be used to establish the following relationship:

$$\begin{bmatrix} -\rho\bar{W}_0 \\ \rho\bar{P}_0 \end{bmatrix} = \begin{bmatrix} (\bar{L}^2-\bar{K}\bar{M})/2G\bar{K} & -\rho\bar{L}/\bar{K} \\ -\rho\bar{L}/\bar{K} & 2G\rho^2/\bar{K} \end{bmatrix} \begin{bmatrix} \bar{S}_{zz0} \\ \bar{Q}_0 \end{bmatrix} \tag{19}$$

If we now adopt representations of the type (12,16) we find that:

$$\begin{bmatrix} \bar{\underset{\sim}{w}} \\ \bar{\underset{\sim}{p}} \end{bmatrix} = \begin{bmatrix} \bar{X} & \bar{Y} \\ -\bar{Y}^T & \bar{Z} \end{bmatrix} \begin{bmatrix} \bar{\underset{\sim}{a}} \\ \bar{\underset{\sim}{b}} \end{bmatrix} \tag{20}$$

where $\bar{X}$ is an m x m matrix with coefficients

$$\bar{X}_{ij} = \int_0^\infty ((\bar{L}^2-\bar{K}\bar{M})/2G\bar{K})\phi_i\phi_j d\rho$$

and where the other quantities were defined previously.

We now use the condition that the footing is rigid (13b), the condition of equilibrium (13c) and the fact that the pore pressures vanish outside the loaded area to show that

$$\begin{bmatrix} \bar{X} & -\bar{Y} & -\underset{\sim}{\lambda} \\ -\bar{Y}^T & \bar{Z} & \underset{\sim}{0} \\ -\underset{\sim}{\lambda}^T & \underset{\sim}{0}^T & 0 \end{bmatrix} \begin{bmatrix} \bar{\underset{\sim}{a}} \\ \bar{\underset{\sim}{b}} \\ \bar{\delta} \end{bmatrix} = \begin{bmatrix} \underset{\sim}{0} \\ \underset{\sim}{0} \\ -p_{AV}\, a^2/2 \end{bmatrix} \tag{21}$$

The coefficients $\bar{\underset{\sim}{a}}$, $\bar{\underset{\sim}{b}}$ can be found from this equation and hence again the solution can be found in transform space.

NUMERICAL SOLUTION

The solution outlined in the previous section depends on the approximating functions ϕ_i, ψ_i. It was found that most satisfactory results could be obtained by employing the following functions

$$\phi_m(r) = (1-r^2/a^2)^{-\frac{1}{2}} \tag{22a}$$

$$\phi_i(r) = 1 \quad a_i < r < a_{i+1} \qquad i = 1,\ldots, m-1$$

$$= 0 \quad \text{elsewhere}$$

where $a_1 = 0$ and $a_m = a$

and $\psi_j(r) = 1 \quad b_i < r < b_{i+1} \qquad i = 1,\ldots, n$

$$= 0 \quad \text{elsewhere}$$

where $b_1 = a$.

This leads to the transformed functions

$$\Phi_m = a \sin(\rho a)/\rho \tag{22b}$$

$$\Phi_i = H(\rho,a_{i+1}) - H(\rho,a_i) \qquad i = 1,\ldots, m-1$$

and

$$\Psi_i = H(\rho,b_{j+1}) - H(\rho,b_j) \qquad j = 1,\ldots, n.$$

Once these approximating functions have been selected the analysis can go ahead as described and we find the solution in transform space. The Laplace transform may be inverted by a remarkably efficient algorithm developed by Talbot (1979), while the Hankel transform may be inverted by employing the Hankel inversion theorem and numerical quadrature.

EXAMPLES

As an illustration of the foregoing theory, solutions were obtained for circular loadings applied to a clay layer. In all cases the soil was assumed to be a very deep porous elastic medium, while the loaded area $0 \leqslant r \leqslant a$ on the surface (see Fig. 2) was assumed impermeable, with the region outside this, $r>a$ being permeable. Loadings were chosen to be instantaneous loadings, applied at $t = 0^+$ and thereafter held constant, although any type of time dependent loading for which the Laplace transform may be found is easily incorporated. Two extremes for the behaviour of the loaded area were considered; a perfectly flexible or a perfectly rigid behaviour. These would correspond to extremely flexible or extremely stiff rafts respectively.

(a) Uniform Loading (perfectly flexible)

Firstly solutions were obtained for perfectly flexible loadings i.e. the loading p_{AV} is uniform over the region $0 \leqslant r \leqslant a$. The resultant time-deflection behaviour is plotted in Fig. 2 where the dimensionless time τ used in this plot is defined as:

$$\tau = ct/a^2$$

and t = time

$$c = \frac{k}{\gamma_w} E_s \frac{(1-\nu_s)}{(1+\nu_s)(1-2\nu_s)} \quad \text{is the coefficient of consolidation}$$

E_s = Young's modulus of the soil
ν_s = Poisson's ratio of the soil

k/γ_w is the permeability per unit weight of water.

The dimensionless deflection at the centre of the loading is plotted in the figure for the case where $\nu_s = 0$. This solution is compared with two other solutions obtained by McNamee and Gibson (1960). Curve (i) is for an entirely permeable upper surface z = 0, while curve (ii) is for a completely impermeable surface. As can be seen from Fig. 2 the present solution asymptotes between the two curves (i) and (ii), behaving more like the solution for an impermeable surface at small times and more like the solution for a permeable surface at large times. This is not unexpected, as for both a completely impermeable surface and a partially permeable one, initial dissipation of pore water pressures beneath the loading would mainly be due to radial flow of pore water. Figure 3 shows pore water pressure dissipation beneath a uniform loading for the case where $\nu_s = 0.3$. At small times the pore pressure p has the same distribution as the applied load but as time increases it dissipates away to zero.

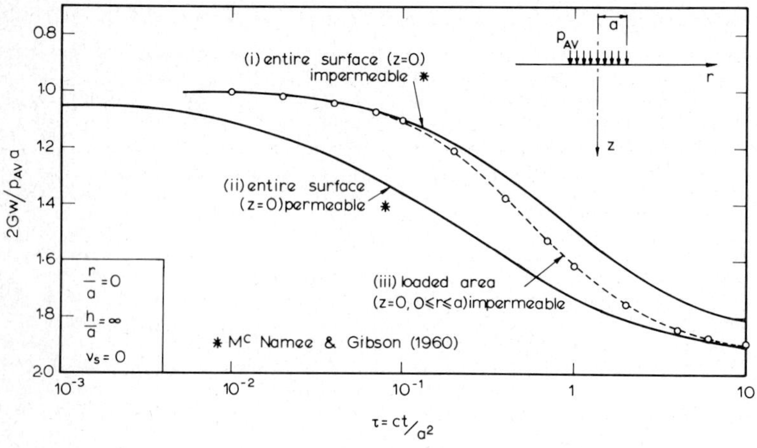

Figure 2
Time-Deflection Behaviour of Circular Loading

(b) Rigid Loading (perfectly rigid)

Results for this type of loading are presented in Fig. 4 which shows pore pressure dissipation beneath the loaded area on the surface z = 0. The theoretical contact stress distribution at time $\tau = 0$ is the same as that for a rigid disc applied to a uniform elastic half space (Muki, 1961). This is shown as a broken line in the figure and it may be seen that the pore pressures calculated at small times (e.g. $\tau = 0.0001$) also have this distribution. As consolidation proceeds the soil no longer behaves as if it is a single phase elastic material and the contact stress varies.

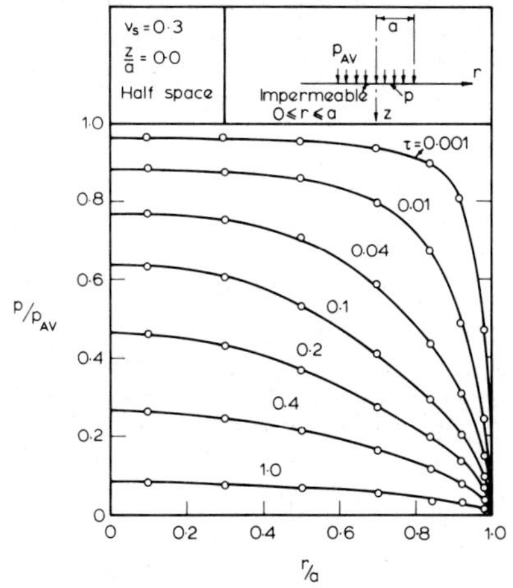

Figure 3
Pore Pressure - Time Behaviour. Uniform Loading on Impermeable Circular Region.

This effect is shown in Fig. 5 where contact stresses are plotted over the loaded region 0⩽r⩽a. For small times the contact stresses increase over the central portion of the loaded area and reduce at the edge. The maximum stress at the centre occurs at a time factor of about τ = 0.1 before reducing at higher times. Contact stress distributions are shown only for time factors in excess of τ = 0.1 for clarity, however it may be noted that the final distribution of contact stress (τ = 100) approaches the theoretical distribution for a rigid loading applied to a uniform elastic half space as would be expected.

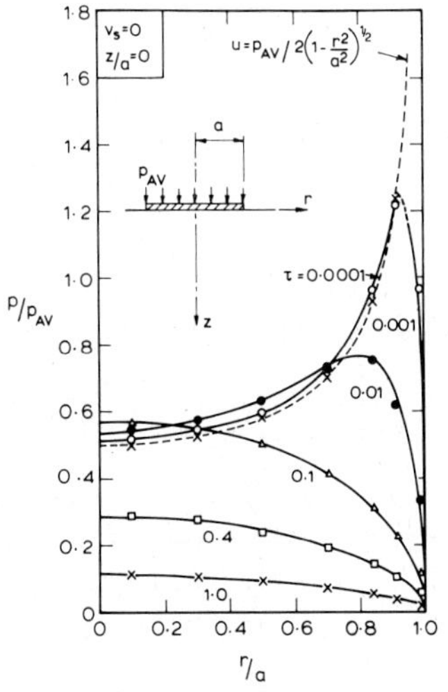

Figure 4
Pore Pressure Variation with Time

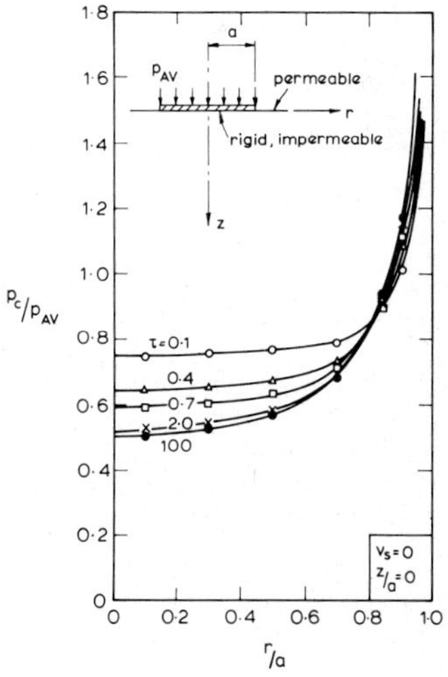

Figure 5
Contact Stress Variation with Time

Finally, curves of the degree of settlement U of a rigid loading were obtained for various values of Poisson's ratio ν_s, of the soil, and these are presented in Fig. 6. The effect of increasing Poisson's ratio of the soil is to increase the degree of consolidation at intermediate time factors τ. Degree of consolidation is defined as

$$U = \frac{\delta(t) - \delta(0)}{\delta(\infty) - \delta(0)} \tag{23}$$

where $\delta(t)$ = deflection of the rigid loaded area at any time t.

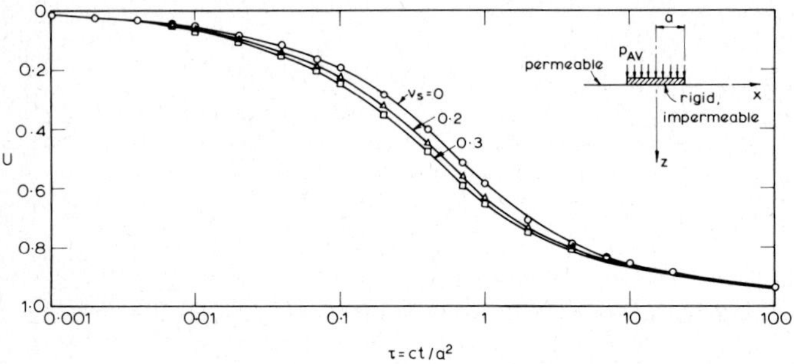

Figure 6
Degree of Consolidation Versus Time Factor

CONCLUSIONS

A method has been presented which enables analytic solutions to be found to the problem of an axisymmetric surface loading on a consolidating clay layer. The method has been shown by means of examples, to be particularly effective in obtaining solutions to problems involving flexible or rigid loadings which are impermeable over their area of application.

REFERENCES

[1] Booker, J.R., The Consolidation of a Finite Layer Subject to a Surface Loading, Int. Jnl. Solids Structs., 10, 1974, 1053-1065.

[2] Chiarella, C. and Booker, J.R., The Time-deflection Behaviour of a Rigid Die Resting on a Deep Clay Layer, Q. Jnl. Mech. Appl. Math., XXVIII, Pt.3, 1975, 317-328.

[3] Gibson, R.E. and McNamee, J., A Three-Dimensional Problem of the Consolidation of a Semi Infinite Clay Stratum, Q. Jnl. Mech. Appl. Math., XVI, Pt.2, 1963, 115-127.

[4] McNamee, J. and Gibson, R.E., Plane Strain and Axially Symmetric Problems of the Consolidation of a Semi Infinite Clay Stratum, Q. Jnl. Mech. Appl. Math., XIII, Pt.2, 1960, 210-227.

[5] Muki, R., Axisymmetric Problems of the Theory of Elasticity for a Semi Infinite Solid and a Thick Plate, Progress in Solid Mechanics, 1, North Holland Publishing Co., Amsterdam, 1961.

[6] Small, J.C. and Booker, J.R., The Time-Deflection Behaviour of a Circular Raft of Finite Flexibility on a Deep Clay Layer, (to be published in Int. Jnl. Num. Anal. Meth. in Geomechs.), 1983.

[7] Talbot, A., The Accurate Numerical Inversion of Laplace Transforms, Jnl. Inst. Math. Applics., 23, 1979, 97-120.

Computational Techniques & Applications: CTAC-83
J. Noye & C. Fletcher (Editors)

THE BEHAVIOUR OF A LINED CIRCULAR TUNNEL IN VISCOELASTIC GROUND

J. P. Carter and J.R. Booker,

School of Civil and Mining Engineering
University of Sydney
Sydney, Australia

The problem of a long tunnel of circular cross-section, passing through viscoelastic ground has been studied. The tunnel cavity is assumed to be lined with a long cylindrical tube of elastic material which makes intimate contact with the surrounding medium. The interactive response of the structural lining and the ground as the cavity is formed has been investigated. Solutions for the time dependent displacements of the lining, the stresses at the interface with the ground, and the hoop force and bending moment induced in the lining have been presented. The importance of the creep properties of the ground and the structural properties of the lining in determining the overall response, has been illustrated.

INTRODUCTION

When a tunnel is driven through a soil or rock mass it may be necessary to place around the walls of the cavity some form of thin lining. This may be necessary to provide structural support for the surrounding ground or merely to act as an impermeable membrane, preventing the ingress of ground water. In both cases the lining, if in intimate contact with the soil or rock, will be subjected to loading due to the presence of in situ stresses. Furthermore, the loading which the lining carries may vary with time, particularly if the ground around the opening creeps in response to changes in its stress regime.

In order to provide a rational design for these linings it is necessary to make realistic estimates of the stresses acting at the interface with the ground and to be able to estimate the thrust and bending moment induced within the lining.

In this paper the problem of a long tunnel of circular cross-section, driven through a pre-stressed, infinite, viscoelastic medium is investigated. The tunnel walls are assumed to be lined with a long thin cylindrical tube of elastic material which makes intimate contact with the surrounding ground.

The behaviour of the lining is time dependent because the material surrounding it creeps under loading. The development of the governing equations for the lining is carried out in Laplace transform space, and simple expressions are obtained for the displacements. These transforms have been inverted numerically and solutions have been obtained for the variations with time of the lining displacement, stresses at the ground interface, and the hoop thrust and bending moments induced in the lining.

A parametric study has been made of the problem in order to determine the importance of the creep properties of the ground upon the behaviour of the lining.

This work is an extension of previous studies of the behaviour of lined tunnels in elastic ground, e.g. Carter, Booker and Moore, 1983 and Hoeg, 1968.

PROBLEM DESCRIPTION

It is assumed that the ground through which the tunnel passes can be modelled as a single phase, linear, isotropic, viscoelastic material. Two different types of model have been investigated and the details for each are set out in a later section. Each model requires the specification of three material parameters. It is also assumed that initially the interface between the tunnel lining and the surrounding material is circular in cross section with radius a, and that the lining has an annular cross section with outer radius a and thickness d, i.e. the lining and the surrounding ground everywhere make perfect radial contact. The axis of the tunnel is assumed to be horizontal and the tunnel itself is assumed to be long so that the problem is one of plane strain. A typical, vertical cross section is shown in Fig.1.

Before the cavity is created the stress state in the ground (the in situ stress state, sometimes called the virgin or field stress) is homogeneous with normal stress σ_V which acts in the vertical direction, and normal stress σ_H, which acts in all horizontal directions. The horizontal stress may be expressed as a proportion of the vertical stress, thus

$$\sigma_H = N\sigma_v \qquad (1)$$

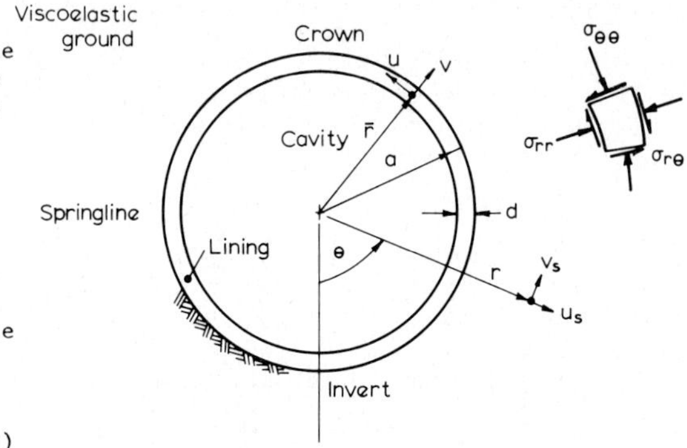

FIG. 1 COORDINATE DESCRIPTION

For a circular opening it is more convenient to express the in situ stress state in terms of polar coordinates, see Fig.1. It follows that initially the stresses acting on a circular boundary are

$$\sigma_{rr}(r,\theta) = \sigma_m + \sigma_d \cos 2\theta \qquad (2a)$$

$$\sigma_{\theta\theta}(r,\theta) = \sigma_m - \sigma_d \cos 2\theta \qquad (2b)$$

$$\sigma_{r\theta}(r,\theta) = - \sigma_d \sin 2\theta \qquad (2c)$$

where $\sigma_m = (\sigma_v + \sigma_H) =$ the mean stress

$\sigma_d = (\sigma_v - \sigma_H) =$ the deviator stress

and θ is the polar angle measured from the direction vertically downwards.

It is assumed that the cavity is created and the lining installed in such a way that the in situ stress state initially is undisturbed. It is only after the lining is in place that the normal and the shear stress at the inside face of the lining are removed. This procedure may be a reasonable model for the case in practice where the lining is jacked horizontally into the ground ahead of the excavation face. In any case, this model of the installation process is likely to give the largest predictions of stress acting on the lining and thus, for design purposes, could be considered conservative.

The problem to be analysed is the structural interaction between the elastic tube lining and the surrounding ground. This interaction will be time dependent because the ground is idealised as a viscoelastic material. In order to simplify the analysis the in situ stress state can be considered as the sum of two parts; and the overall effects will be found from the superposition of the results for each case. The two cases are:

Case I, removal of the mean stress, for which the boundary conditions are

$$\left.\begin{aligned} \Delta\sigma_{rr} &= -\sigma_m \\ \Delta\sigma_{r\theta} &= 0 \end{aligned}\right\} \quad \text{at} \quad r = a - d \tag{3a}$$

Case II, removal of the deviator stress, for which the boundary conditions are

$$\left.\begin{aligned} \Delta\sigma_{rr} &= -\sigma_d \cos 2\theta \\ \Delta\sigma_{r\theta} &= +\sigma_d \sin 2\theta \end{aligned}\right\} \quad \text{at} \quad r = a - d \tag{3b}$$

The symbol Δ is used to represent a change in the appropriate quantity.

The form of the in situ stress state suggests that the anaylsis can be made in terms of the Fourier components of the field quantities, e.g. at any time the stresses for each case may be written as

$$\sigma_{rr} = S_{rr} \cos n\theta \tag{4a}$$

$$\sigma_{\theta\theta} = S_{\theta\theta} \cos n\theta \tag{4b}$$

$$\sigma_{r\theta} = S_{r\theta} \sin n\theta \tag{4c}$$

where S_{rr}, $S_{\theta\theta}$ and $S_{r\theta}$ are the appropriate Fourier coefficients. It is obvious that for Case I, $n = 0$ and for Case II, $n = 2$.

ANALYSIS

It is convenient to begin an analysis of this time dependent, interaction problem by considering separately the behaviour of the viscoelastic ground and the behaviour of the elastic lining. Once the governing equations have been developed for each material then the interaction effects may be investigated. In another paper (Carter, Booker and Moore, 1983) the problem of a lined tunnel in linear elastic ground has been studied. The governing equations have been developed in that paper and they may be adapted, using the viscoelastic analogy, for the current problem.

The solution to the interaction problem will depend on the conditions assumed at the interface between the tunnel lining and the surrounding ground. Two extreme conditions can be identified, viz (a) a perfectly rough interface, implying compatability of radial and circumferential displacement across the interface, and (b) a perfectly smooth interface, implying zero shear stress at the interface. In reality the condition may be somewhere between these two extremes, i.e. there may be some finite limit to the shear stress which can be developed between the ground and the lining. Nevertheless, it is useful to examine these two extremes.

Before the governing equations are given explicitly, it is necessary to define some relevant quantities.

After the material has been excavated from within the tunnel it is assumed that the Fourier components of the radial normal stress and the shear stress acting at the interface have values denoted by R_n and T_n respectively.

Thus the general stresses at the interface are given by

$$\sigma_{rr}(a,\theta) = R_n \cos n\theta \tag{5a}$$

$$\sigma_{r\theta}(a,\theta) = T_n \sin n\theta \tag{5b}$$

Under the imposed loading the ring shaped lining will deform with radial and circumferential components of displacement given generally as

$$u(\bar{r}, \theta) = X_n \cos n\theta \tag{6a}$$

$$v(\bar{r}, \theta) = Y_n \sin n\theta \tag{6b}$$

where $\bar{r} = a - d/2$ (see Fig.1). A moderately thin ring (d a) can be analysed using the bending theory for a thin curved beam, and it will be sufficiently accurated to assume that $\bar{r} \approx a$ and that

$$u(\bar{r}, \theta) \simeq u(a, \theta) \tag{7a}$$

$$v(\bar{r}, \theta) \simeq v(a, \theta) \tag{7b}$$

It is possible to derive closed form expressions for the Laplace transforms of the various quantities. In what follows a bar shall denote a Laplace transform, viz

$$\bar{f}(s) = \int_0^\infty e^{-st} f(t)dt \tag{8}$$

where the symbol t is used to represent time, measured from zero at the instant of tunnel excavation. The transformed shear modulus of the ground $\bar{G}_s$ is related to the deviatoric creep function $J_D(t)$ as follows

$$\bar{G}_s = \frac{1}{s\bar{J}_D} \tag{9}$$

For reasons discussed later only viscoelastic materials with a constant value of Poisson's ratio ν_s will be considered.

For these materials the volumetric creep is coupled to the deviatoric (shear) creep and the transformed bulk modulus $\bar{K}_s$ of the ground is related to the transformed shear modulus $\bar{G}_s$ by

$$\bar{K}_s = \frac{2(1 + \nu_s)}{3(1 - 2\nu_s)} \bar{G}_s \tag{10}$$

For the lining it is also convenient to define the following quantities

$$S_A = E_\ell^* \frac{d}{a^2} = \frac{1}{a^2} \times \text{Membrane stiffness of the ring} \tag{11}$$

$$S_B = E_\ell^* \frac{d^3}{a^4} = \frac{1}{a^4} \times \text{Bending stiffness of the ring} \tag{12}$$

where $$E^*_\ell = \frac{E_\ell}{(1 - \nu_\ell^{\,2})}$$

and E_ℓ, ν_ℓ are Young's modulus and Poisson's ratio of the ring material, respectively.

The solutions for both rough and smooth interface are set out below.

(a) Perfectly rough interface

Case I, n = 0

For the removal of the in situ mean stress the solution is as follows

$$\overline{X}_o = -\frac{\sigma m}{s}\left(\frac{1}{\frac{2\overline{G}_s}{a} + S_A}\right) \tag{13a}$$

$$Y_o = 0 \tag{13b}$$

$$R_o = -S_A X_o \tag{13c}$$

$$T_o = 0 \tag{13d}$$

The force induced in the lining is given by

$$F_o = -S_A a X_o \tag{13e}$$

There are no bending effects associated with the case n = 0, i.e.

$$M_o = 0$$

Case II, n = 2

For the removal of the in situ deviator stress the Laplace tranforms of displacement can be obtained from the solution of the following set of equations

$$\begin{bmatrix} S_{11} & S_{12} \\ S_{12} & S_{22} \end{bmatrix} \begin{bmatrix} \overline{X}_2 - \overline{Y}_2 \\ \overline{X}_2 + \overline{Y}_2 \end{bmatrix} = \begin{bmatrix} 2\sigma_d/s \\ 0 \end{bmatrix} \tag{14a,b}$$

where
$$S_{11} = -\frac{2\bar{G}_s}{a}\left(\frac{1}{3-4\nu_s}\right) - \frac{1}{2}S_A - 2S_B$$
$$S_{12} = +\frac{3}{2}S_A - 6S_B$$
$$S_{22} = -\frac{6\bar{G}_s}{a} - \frac{9}{2}S_A - 18S_B$$

The interface stress are given by the following relations

$$\begin{bmatrix} R_2 - T_2 \\ R_2 + T_2 \end{bmatrix} = -\begin{bmatrix} \frac{1}{2}S_A + 2S_B, & -\frac{3}{2}S_A + 6S_B \\ -\frac{3}{2}S_A + 6S_B, & \frac{9}{2}S_A + 18S_B \end{bmatrix}\begin{bmatrix} X_2 - Y_2 \\ X_2 + Y_2 \end{bmatrix} \tag{14c,d}$$

The hoop force and the bending moment induced in the lining can be obtained from

$$F_2 = -S_A a(X_2 + 2Y_2) \tag{14e}$$

$$M_2 = -S_B a^2(4X_2 + 2Y_2) \tag{14f}$$

(b) Perfectly smooth interface

Case I, $n = 0$

For the case where $n = 0$ the solution for a perfectly smooth interface is the same as that for a perfectly rough interface and is given by equations (13).

Case II, $n = 2$

The solution in this case is obtained from

$$X_2 = -\frac{\sigma d}{s}\left\{\frac{3\left(\dfrac{3-4\nu_s}{5-6\nu_s}\right)}{\dfrac{9S_AS_B}{(S_A+S_B)} + \dfrac{6\bar{G}_s}{a(5-6\nu_s)}}\right\} \tag{15a}$$

$$Y_2 = -\left\{\frac{S_A + 4S_B}{2(S_A+S_B)}\right\}X_2 \tag{15b}$$

$$R_2 = \left\{\frac{9S_AS_B}{(S_A+S_B)}\right\}X_2 \tag{15c}$$

$$T_2 = 0 \tag{15d}$$

$$F_2 = - S_A a(X_2 + 2Y_2) \tag{15e}$$

$$M_2 = - S_B a^2(4X_2 + 2Y_2) \tag{15f}$$

NUMERICAL INVERSION OF TRANSFORMS

Closed form expressions for the Laplace transforms of the lining displacements have been set out in the previous section. In order to recover values for all coefficients at any selected time these transforms must be inverted. This can be achieved by an application of the Complex Inversion Theorem, i.e. for the general function f(t) the inversion is given by

$$f(t) = \frac{1}{2\pi i} \int_C \bar{f}(s)e^{st}\, ds \tag{16}$$

where s is a complex variable and C is any contour in the complex plane such that all singularities of f(s) lie to the left of C. For the results presented in this paper the integration was performed numerically using the contour and the efficient numerical scheme developed by Talbot (1979).

RESULTS

In an earlier study of the creep behaviour of unlined tunnels (Carter and Booker, 1983) several different material models were investigated. It was found that creep had the greatest influence on the time dependent behaviour of the tunel when an ideal material, involving coupled shear and volumetric creep with constant Poisson's ratio, was used to model the ground response. Hence, in this study of lined tunnels, only creep models assuming a constant value of Poisson's ratio will be adopted. Two such models, one involving terminating shear creep, and the other involving a form of non-terminating shear creep have been investigated.

Terminating Shear Creep

In the first instance we consider the ground to be a viscoelastic material which exhibits shear and volumetric creep, which are coupled such that the value of Poisson's ratio is constant. Furthermore, the creep is assumed to terminate at large time, i.e. the rheological model for the shear behaviour is as shown in Fig.2. For this type of model the following quantities are defined.

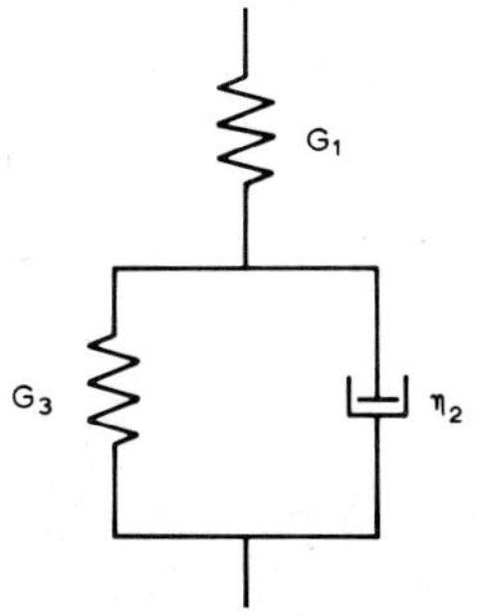

FIG. 2 MODEL FOR TERMINATING SHEAR CREEP

For shear deformations, this type of material has a relaxation function of the form

$$J_D(t) = A - B \exp\{-\alpha t\} \tag{17}$$

If it is subjected to a constant increment in shear stress τ_o then it will exhibit a shear strain γ given by

$$\gamma = \tau_o \left[\frac{1}{G_o} \exp\{-T_D\} + \frac{1}{G}\left(1 - \exp\{-T_D\}\right)\right] \tag{18}$$

where $G_o = G_1 = \dfrac{1}{A - B}$ = Shear 'modulus' at t = 0

$G = \dfrac{G_1 G_3}{G_1 + G_3} = \dfrac{1}{A}$ = Shear 'modulus' at t =

$T_D = \dfrac{G_3 t}{\eta_2} = \alpha t$ = Viscoelastic time factor

and thus will creep from an initial strain $\gamma_o = \tau_o/G_o$ to a final shear strain $\gamma_\infty = \tau_o/G_\infty$.

In order to calculate the time dependent response of this type of material the Laplace transform of the shear 'modulus' is required, i.e.

$$\bar{G}_s = \frac{1}{s\bar{J}_D} = \frac{s + \alpha}{(A - B)s + A\alpha} \tag{19}$$

A parametric study has been carried out using this model to investigate the effects of various parameters on the behaviour of the tunnel lining-ground system. In particular, the effects of the following have been studied:

(a) E^*_ℓ/G_o = the ratio of the effective Young's modulus of the lining material to the initial shear 'modulus' of the ground;
(b) d/a = the ratio of the lining thickness to the radius of the lining - soil interface;
(c) G_o/G_∞ = the ratio of the intitial to final values of the shear 'modulus' of the ground;
(d) The interface condition, either perfectly smooth or perfectly rough.

In all cases of the present study a value of ν_s =0.5 was adopted for the Poisson's ratio of the ground, i.e. the ground material was considered to be incompressible. Thus the model could be used, for example, to represent the creep behaviour of clay under undrained conditions.

(a) Effect of modulus ratio, E^*_ℓ/G_o

Results were calculated for the case where d/a = 0.01, G_o/G = 2 and the ratio E^*_ℓ/G_o had a number of different values. The interface between the lining and the ground was assumed to be perfectly smooth. The results, presented in Figs.3 to 6, show the variation with non-dimensional time T_D of the Fourier coefficients for: the lining displacements (X = radial component, Y = circumferential component); the radial stress transmitted across the interface, R; the hoop force developed in the lining, F; and the bending moment developed in the lining, M. Results for Case I - after removal of the in situ mean stress - are shown in Figs.3 and 4, and for Case II - after removal of the in situ deviator stress - are given in Figs.5 and 6. Where more than one quantity has the same form of variation with time then all such quantities are shown on the same graph (e.g. R, F and M for Case II are shown on Fig.6).

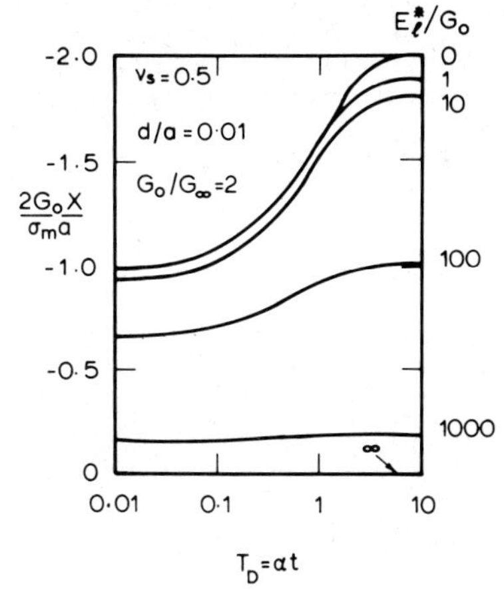

FIG. 3 COEFFICIENT OF RADIAL DISPLACE- VERSUS TIME FOR CASE I - SMOOTH SURFACE

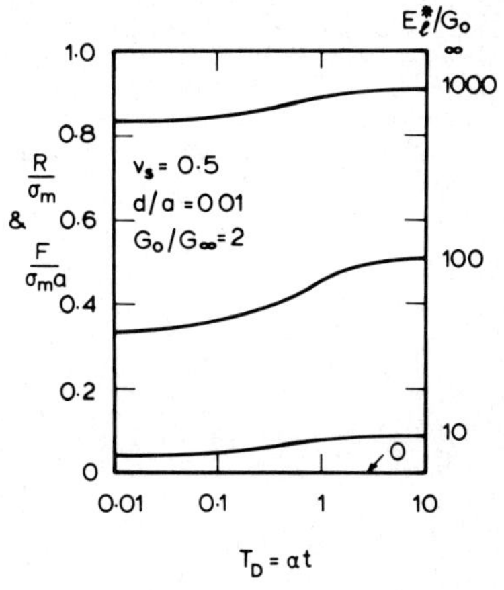

FIG. 4 COEFFICIENTS OF RADIAL STRESS AND HOOP THRUST VERSUS TIME FOR CASE I - SMOOTH INTERFACE

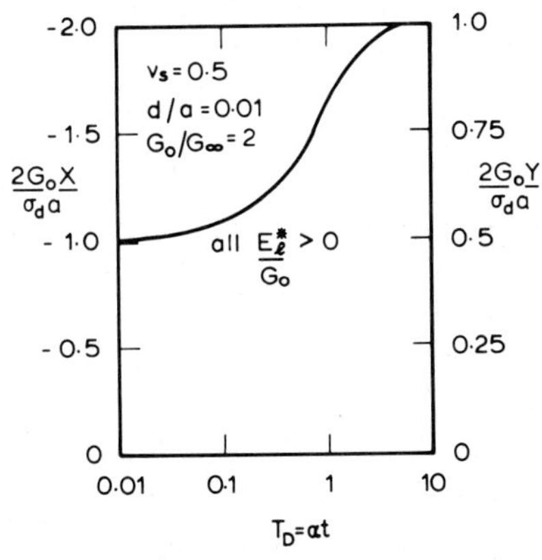

FIG. 5 COEFFICIENTS OF RADIAL AND CIRCUMFERENTIAL DISPLACEMENT VERSUS TIME FOR CASE II - SMOOTH INTERFACE

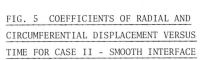

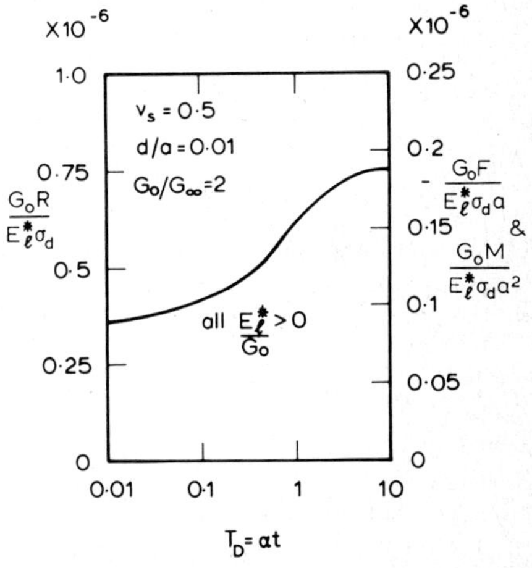

FIG. 6 COEFFICIENTS OF RADIAL STRESS, HOOP THRUST AND BENDING MOMENT VERSUS TIME FOR CASE II - SMOOTH INTERFACE

In case I the circumferential displacement and the bending moment are always zero, i.e. the deformations involve only the membrane action of the lining. The modulus ratio E_ℓ^*/G_o has an effect on the magnitudes of the radial displacement, the contact pressure and the thrust in the lining. As the lining becomes stiff compared with the ground, the displacements become smaller (see Fig.3) while the contact pressure and the hoop thrust approach limiting values of σ_m and $\sigma_m a$ respectively (see Fig.4).

In case II both the radial and the circumferential displacements are independent of the modulus ratio (see Fig.5). This trend is contrary to the general behaviour of elastic bodies (displacements inversely proportional to elastic modulus) and is no doubt due to the nature of the loading for this case. The fact that a single curve is plotted in Fig.6 indicates that the magnitudes of the quantities R, F and M are all proportional to the ratio E_ℓ^*/G_o, i.e. for any given soil and lining geometry (d/a = constant), the stiffer the lining material then the greater is the load which will be attracted to it.

The results shown in Figs.3 to 6 indicate that the quantities X, Y, R, F and M all increase in magnitude as the ground around the tunnel lining creeps. The tunnel displaces more and greater loads are placed upon it as the ground becomes less stiff with time.

(b) Effect of the lining thickness, d/a

Two different tunnel geometries have been studied, viz d/a = 0.01 and 0.1, and in both cases the smooth tunnels were assumed to be in the same type of ground, i.e. E_ℓ^*/G_o = 100 and G_o/G_∞ = 2. The effect of the parameter d/a on the behaviour of the tunnels is shown in Figs.7 to 10.

For Case I (Figs.7 and 8), the thicker the lining then the smaller is the radial displacement and the larger are the forces acting on it and, in addition, the smaller is the proportional variation of these quantities with time.

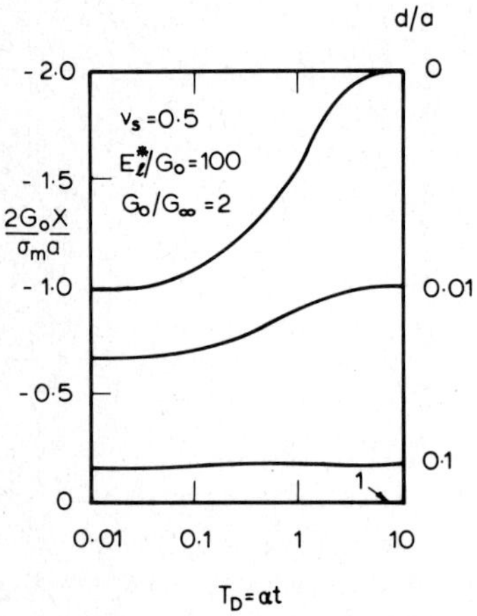

FIG. 7 COEFFICIENT OF RADIAL DISPLACEMENT VERSUS TIME FOR CASE I - SMOOTH INTERFACE

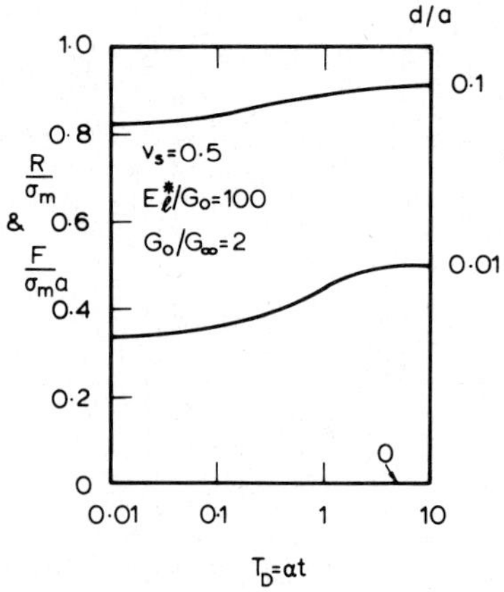

FIG. 8 COEFFICIENTS OF RADIAL STRESS AND HOOP THRUST VERSUS TIME FOR CASE I - SMOOTH INTERFACE

For Case II, the displacements of these two linings are almost the same (Fig.9). The results in Fig.10 show that for these two linings the contact stress, the hoop thrust and the bending moments are functions only of the bending stiffness ($E_\ell^* d^3$) of the tunnel lining and not of its membrane or hoop stiffness ($E_\ell^* d$). This result is only true for this limited choice of lining and ground properties and lining thicknesses. The dependence of the tunnel response on the bending and membrane stiffnesses of the lining has been treated in greater detail elsewhere (Carter, Booker and Moore, 1983).

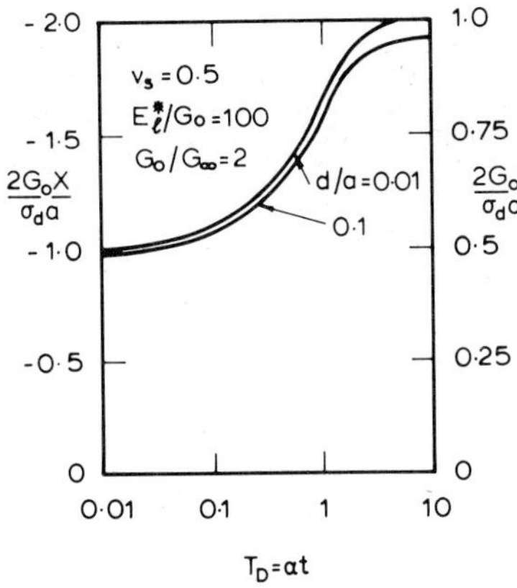

FIG. 9 COEFFICIENTS OF RADIAL AND CIRCUMFERENTIAL DISPLACEMENT VERSUS TIME FOR CASE II - SMOOTH INTERFACE

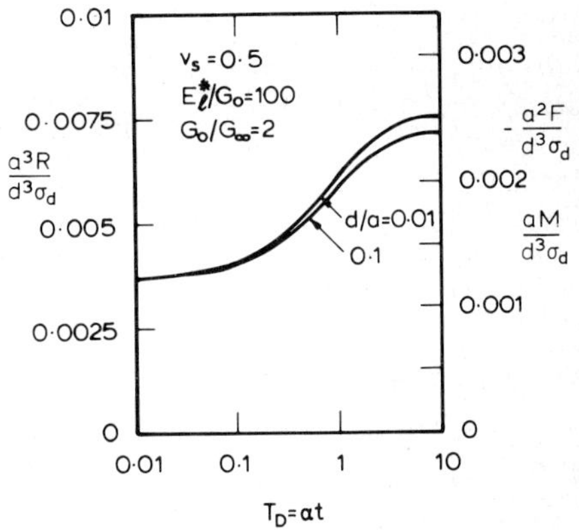

FIG. 10 COEFFICIENTS OF RADIAL STRESS, HOOP THRUST AND BENDING MOMENT VERSUS TIME FOR CASE II - SMOOTH INTERFACE

(c) Effects of the creep properties of the ground, G_o/G_∞

One particular value of the modulus ratio, $E_\ell^*/G_o = 100$, and one tunnel thickness, $d/a = 0.01$, were selected for a study of the effects of the creep properties of the ground on the behaviour of a smooth lining. Three values were selected for the ratio G_o/G_∞, viz 2, 5 and 10, in order to investigate the effects of the amount of ground creep on the behaviour of the lining.

The results of this study are presented in Fig.11 for the Case I loading, and Fig.12 for the Case II loading. Again it has been possible to show the variations with time of more than one quantity on each plot.

For Case I, Fig.11 shows that if the ground material has creep defined by $G_o/G_\infty = 10$ and $\nu_s = 0.5$ then the magnitudes of the quantities X, R and F increase by slightly more than 100% of their initial values.

For Case II, Fig.12 shows that the quantities X, Y, R, F and M undergo even greater proportional changes with time. For a ground material characterised by $G_o/G_\infty = 2$ the displacements and loads ultimately double in magnitude as the ground creeps, and where $G_o/G_\infty = 10$ these quantities increase by a factor of 10. The results indicate that for the Case II loading the ratio of the magnitude of the final response to its initial value is exactly equal to the modulus ratio G_o/G_∞ of the creeping ground.

(d) Effect of the interface condition, smooth or rough

All of the results presented so far have been for the case of a perfectly smooth interface between the tunnel lining and the surrounding ground, i.e. no shear stress can be developed at the interface. It is of some interest to compare a typical set of these results with other results for the case where all of the material and geometric parameters are the same, but the interface is considered to be perfectly rough, i.e. no slip can occur between the lining and the ground, they always maintain perfect contact (or bonding). The case selected is characterised by $d/a = 0.01$, $E^*_\ell/G_o = 100$ and $G_o/G_\infty = 2$.

It can be noted that the interface condition has no effect on the results for the Case I loading condition, i.e. all quantities are the same in both cases and in particular there can be no circumferential displacement and no shear stress developed at the interface.

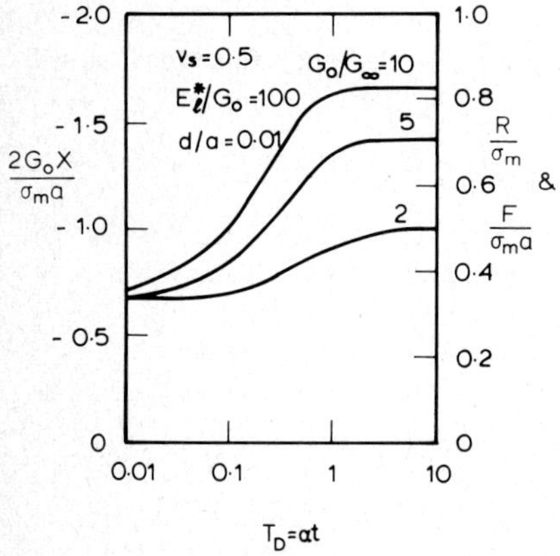

FIG. 11 COEFFICIENT OF RADIAL DISPLACEMENT VERSUS TIME FOR CASE I - SMOOTH INTERFACE

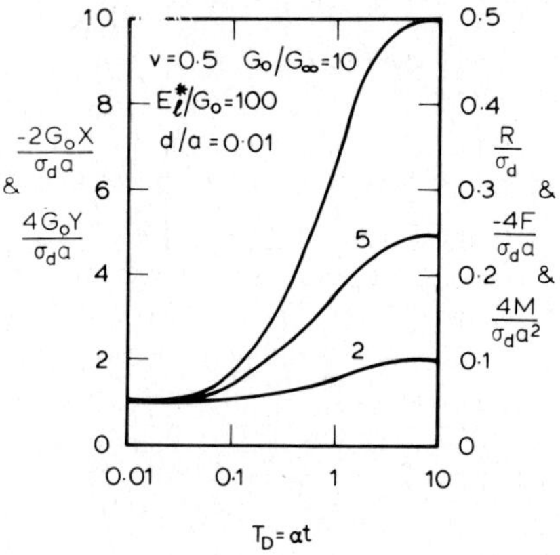

FIG. 12 COEFFICIENT OF RADIAL DISPLACEMENT VERSUS TIME FOR CASE II - SMOOTH INTERFACE

Results for the Case II loading condition for both a smooth and a rough interface are shown in Figs.13 to 18. Fig.13 indicates that the radial displacement of the lining is unaffected by the interface condition but Fig.14 indicates that at all times the circumferential displacement for a rough interface is between about 35% to 50% greater than that for a smooth interface. Shear stress can be developed at a rough interface and the variation with time of this shear is shown in Fig.15 (on the ordinate T = the Fourier coefficient for the interface shear stress).

For Case II the interface condition also has an effect on both the magnitude and the sign of the forces acting on and within the tunnel lining - see Figs.16 to 18. Apart from the shear stress, by far the greatest difference occurs in the case of the radial stress at the interface and the hoop thrust induced in the lining. The variation with time of the radial contact stress is indicated in Fig.16. At the rough interface the magnitude of this stress is always between about 5000 and 7000 times greater than at a smooth interface. In addition, these stresses are of opposite sign. A similar result is true for the hoop thrust, where values for the rough interface are between 5000 and 15000 times those for a smooth interface and are also of opposite sign. In contrast to this both linings exhibit a similar behaviour in bending. Fig.18 shows that the bending moments induced in the rough and smooth linings are of similar orders of magnitude, with those in the smooth lining always just greater than the moment in the rough lining.

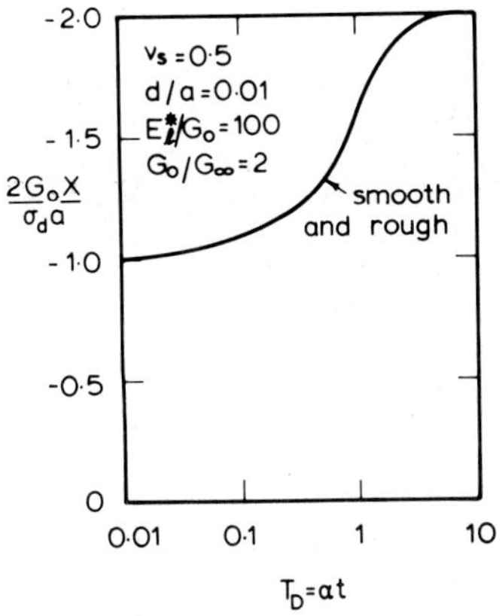

FIG. 13 COEFFICIENT OF RADIAL DISPLACEMENT VERSUS TIME FOR CASE II - SMOOTH AND ROUGH INTERFACES

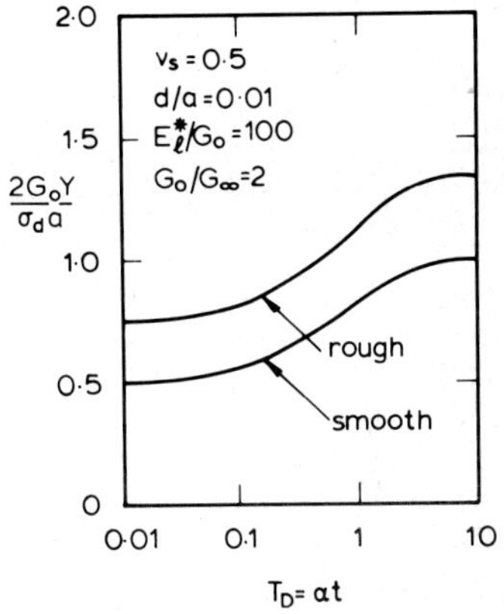

FIG. 14 COEFFICIENT OF CIRCUMFERENTIAL DISPLACEMENT VERSUS TIME FOR CASE II - SMOOTH AND ROUGH INTERFACES

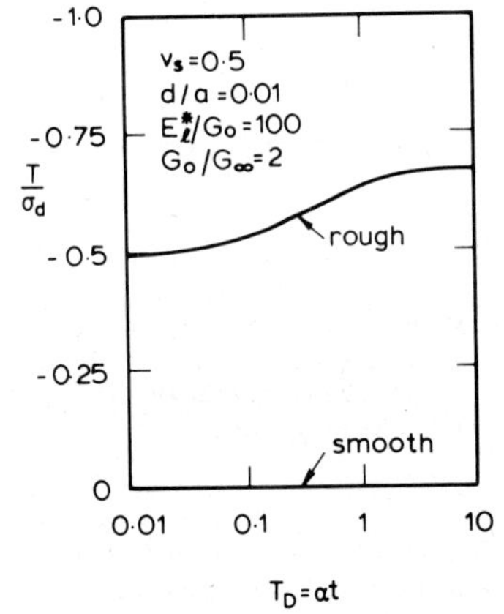

FIG. 15 COEFFICIENT OF SHEAR STRESS VERSUS TIME FOR CASE II - SMOOTH AND ROUGH INTERFACES

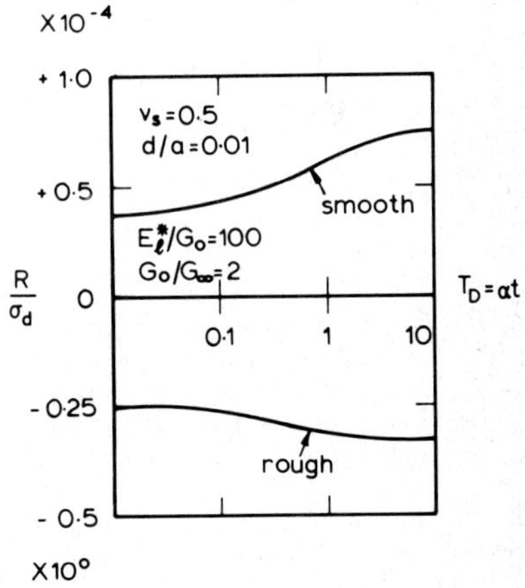

FIG. 16 COEFFICIENT OF RADIAL STRESS VERSUS TIME FOR CASE II - SMOOTH AND ROUGH INTERFACES

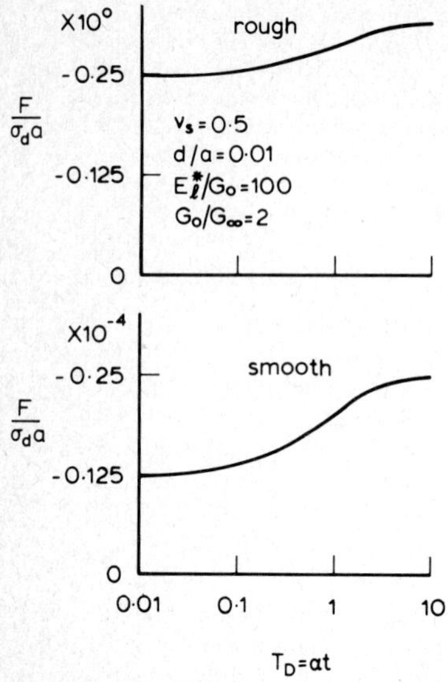

FIG. 17 COEFFICIENT OF HOOP THRUST VERSUS TIME FOR CASE II - SMOOTH AND ROUGH INTERFACES

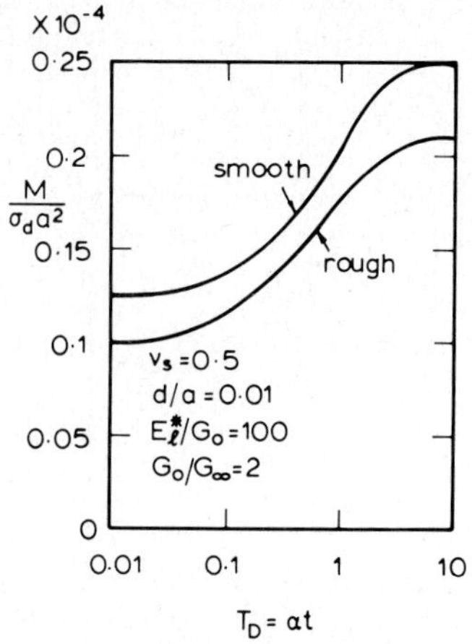

FIG. 18 COEFFICIENT OF BENDING MOMENT VERSUS TIME FOR CASE II - SMOOTH AND ROUGH INTERFACES

Non-terminating Shear Creep

The second model to be adopted to represent the response of the ground surrounding the tunnel assumes a non-terminating deviatoric creep and a constant value of Poisson's ratio. For shear deformations this type of material has a relaxation function of the form

$$J_D(t) = A + B\,\ell n\,\{1 + \alpha t\}$$

If it is subjected to a constant increment in shear stress τ_o then it will exhibit a shear strain γ given by

$$\gamma = \tau_o\,[A + B\,\ell n(1 + \alpha t)]$$

and thus will creep indefinitely from an initial strain of $\gamma_o = A\tau_o$.

In order to calculate the time dependent response of this type of material the Laplace transform of the shear 'modulus' is required. This is

$$\bar{G}_s = \frac{1}{s\bar{J}_D} = \frac{1}{A + B\,\exp\{s/\alpha\}\,E_1(s/\alpha)}$$

where $E_1(z) = \int_z^\infty \frac{e^{-zt}}{t}\,dt \,, \quad |\arg z| < \pi$

is related to the exponential integral.

Computations have been made using this material model with the values B/A = 1, 2 and 10 and $\nu_s = 0.5$. The results of these computations for a smooth tunnel have been plotted on Fig.19 for Case I and on Fig.20 for Case II. For a tunnel in any given soil the Fourier coefficients of the radial displacement (X), the contact stress (R) and the hoop thrust (F) all have the same variation with time in Case I. However, the type of material surrounding the tunnel does have an effect on this time dependence. In general, the larger the value of B/A then the smaller is the 'modulus' of the ground at any given time, and hence the larger is the ratio of the modulus of the lining to the modulus of the ground at that time. At large times the ground has no effective stiffness and thus the entire in situ ground stress must be carried by the lining. This situation will come about sooner in viscoelastic ground with a larger value of B/A. For Case I the ultimate condition is given by

$$X = \left\{ \frac{(1 - \nu_\ell^2)}{E\ (d/a)} \right\} \sigma_m a$$

$$R = \sigma_m$$

$$F = \sigma_m a$$

Fig.19 indicates that the time required for the ultimate condition to be closely approached is large, e.g. for the material with B/A = 10, ninety percent of the asymptote value is reached at $T_D = 10^8$. This may be compared with the results from the previous section where terminating creep has very little additional effect on the tunnel lining at non-dimensional times T_D greater than about 10.

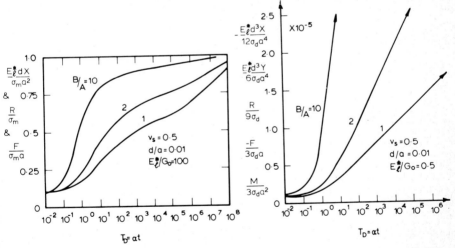

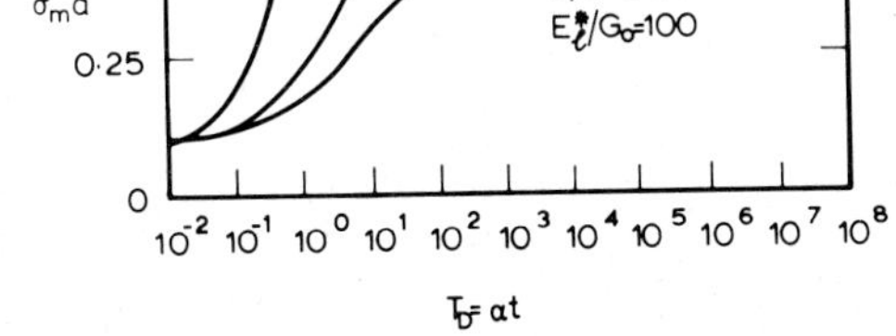
FIG. 19 SOLUTION FOR CASE I FOR NON-TERMINATING CREEP AND SMOOTH SURFACE

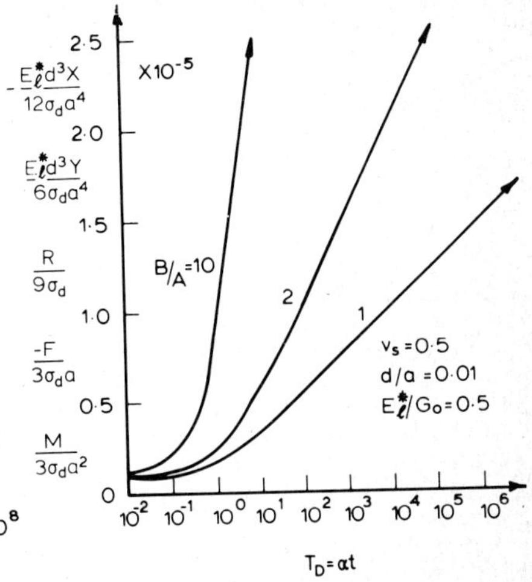
FIG. 20 SOLUTION FOR CASE II FOR NON-TERMINATING CREEP AND SMOOTH INTERFACE

For Case II loading the value of the parameter B/A also has an effect on the time dependence of the behaviour of the lining - see Fig.20. However, at 'early' and 'intermediate' times (T_D less than about 10^8) the variation with time of the displacement (X and Y), the contact stress (R), the hoop thrust (F) and the bending moment (M) are all effectively the same. Expressions for the limiting values (corresponding to $T_D \to \infty$) of the Fourier coefficients of various quantities are

$$X = -\left\{\frac{(1-\nu_\ell^2)}{\frac{1}{12}E_\ell(d/a)^3}\right\}.\ \sigma_d a\ .\left\{1+\frac{1}{6}(d/a)^2\right\}$$

$$Y = \left\{\frac{(1-\nu_\ell^2)}{\frac{1}{12}E_\ell(d/a)^3}\right\}.\ \sigma_d a\ .\left\{\frac{1}{12}+\frac{1}{18}(d/a)^2\right\}$$

$$R = \sigma_d$$

$$F = -\sigma_d a$$

$$M = \tfrac{1}{2}\sigma_d a^2$$

A comparison of Figs.19 and 20 shows that the lining will be required to carry almost all of the in situ mean stress (Case I) much sooner than all of the in situ deviator stress (Case II).

CONCLUSIONS

The problem of a long tunnel of circular cross section piercing viscoelastic ground has been studied. The cavity was assumed to be lined with a long, thin cylindrical tube of elastic material making intimate contact with the surrounding medium. The interaction of the ground and the lining as the cavity is formed and also the subsequent behaviour of the lining has been studied.

Typical solutions have been given for the time dependent displacements of the lining, the stresses acting on it and the thrust and bending moment induced within it. It was found convenient to separate the overall problem into two components: one involving the mean initial stress in the ground; and the other involving the difference between the initial vertical and horizontal normal stress. It was demonstrated that for the mean stress problem, the stiffer is the lining material compared with the soil or rock, the smaller are the effects of ground creep on the changes in the lining behaviour. For the deviator stress problem it is the bending stiffness of the lining which is most important in determining its response. Also for the deviator problem, the condition assumed at the interface between the lining and the surrounding ground has a significant effect on the thrust induced in the lining and on its variation with time. The interface condition has only a moderate effect on the moments induced in the lining.

It has also been demonstrated that if the ground around the tunnel undergoes non-terminating shear creep, then eventually the lining will be called upon to support the entire in situ stress state. Furthermore, it will be required to carry the mean stress component much sooner than the deviatoric component of the initial stress state.

REFERENCES

[1] Carter, J.P., Booker, J.R., Creep and Consolidation around Circular Openings in Infinite Media, Int. J. Solids Structures, 19 (1983) to appear.

[2] Carter, J.P., Booker, J.R. and Moore, I.D., The Analysis of a Lined Circular Opening in Elastic Ground, Univ. Sydney, School of Civil and Mining Engg, Research Report (1983).

[3] Hoeg, K., Stresses Against Underground Structural Cylinders. J. Soil Mechs. Fndns Divn, A.S.C.E., 94 (1968) 833-858.

[4] Talbot, A., The Accurate Numerical Inversion of Laplace Transforms. J. Inst. Maths Applics, 23 (1979) 97-120.

Computational Techniques & Applications: CTAC-83
J. Noye & C. Fletcher (Editors)

COMPUTATIONAL IMPROVEMENTS OF THE DYNAMIC DEFORMATION METHOD

Richard Kohoutek

Department of Civil Engineering
University of Melbourne
Parkville 3052
Australia

In this paper the solution of the differential equation can be formulated without use of the hyperbolic terms. This leads to the elimination of numerical instability for higher frequencies and to a considerable improvement for quasi-static cases. Further enhancement is achieved by the computational analysis of the frequency functions. The commonly recommended and used bisection method for finding zeroes (natural frequencies) can be replaced by higher convergence methods, substantially reducing the computing time required for an analysis.

INTRODUCTION

The dynamic deformation methods has been used for several decades in the dynamic analysis of continuous beams and frames. Its major advantage over some other mathematical models of real structures is the inclusion of continuous mass and stiffness distributions, particularly important for higher modes of vibrations.

The recent paper of Gartner and Olgac [5] improves the numerical stability of the eigenfunctions for a uniform beam. In this paper we will show how similar ideas may be applied to the more general case of the analysis of planar frame. The basic formulation employed is that of Kloušek [9,11]. The new form of the frequency functions is further enhanced by computational analysis.

Reliable and general algorithm for the determination of asymptotes is suggested, utilizing specific properties of the problem, to avoid possible misinterpretation of asymptotes and zeroes.

DIFFERENTIAL EQUATION

The differential equation for an unloaded beam under transverse vibrations is

$$EI \frac{\partial^4 v(x,t)}{\partial x^4} + \mu \frac{\partial^2 v(x,t)}{\partial t^2} = 0 \tag{1}$$

([9], [11], [14] and [8]) where μ is a beam mass per unit length, EI flexural stiffness, $v(x,t)$ is the deflection.

If the motion is harmonic then

$$v(x,t) = v(x)\sin \omega t, \tag{2}$$

and the partial differential equation (1) will become

$$EI \frac{d^4 v(x)}{dx^4} - \mu \omega^2 v(x) = 0. \tag{3}$$

The homogeneous differential equation (3) is valid for a beam of constant mass μ, constant flexural stiffness EI and for an arbitrary boundary condition.

For a structure comprising several beams a system of differential equations (3) has to be solved subject to additional constraints for the intersections of the beams. This leads to the determination of 4n constants, where n is a number of beams in the structure. The dynamic deformation method described elsewhere [9, 11, 8, 14], reduces the problem to the evaluation of a stiffness matrix A for the condition

$$Av = 0 \tag{4}$$

where v is a vector of nodal deformations and the elements of a matrix A are some linear combination of the frequency functions ([8, 11], also Appendix 1, 2).

The original frequency functions formulated by Koloušek [9, 10, 11] have been derived assuming the solution of equation (3) in the form

$$v(x) = A \cos \frac{\lambda}{\ell} x\ B \sin \frac{\lambda}{\ell} x + C \cosh \frac{\lambda}{\ell} x + D \sinh \frac{\lambda}{\ell} x \tag{5}$$

with frequency functions in Appendix 1. This form of frequency functions have two intervals with inherent computational problems. One interval is for higher values of argument λ, which leads to a small difference of large numbers, hence introducing a large error. Furthermore, there is eventually an overflow due to the evaluation of sinh λ and cosh λ. With the second interval there is a problem of numerical stability for small values of $\lambda(0.0 \rightarrow 0.2)$.

IMPROVEMENT FOR LARGE VALUES OF λ

The difficulty is virtually removed by considering the formulation of a solution for the differential equation (3) in the form

$$v(x) = A \cos \frac{\lambda}{\ell} x + B \sin \frac{\lambda}{\ell} x + C\ e^{\frac{-\lambda}{\ell} x} + E\ e^{\frac{-\lambda}{\ell} (\ell - x)}. \tag{6}$$

The solution in the form (6) avoids sinh λ, cosh λ in frequency functions improving the accuracy, numerical stability and eliminating the overflow problems.

IMPROVEMENT FOR SMALL VALUES OF λ

All frequency functions have limits for $\lambda = 0$ which are shown in Appendix 2. In the design of a program for frequency functions it is a very simple task to test the value of λ with a branching and replace the calculated value with a limit value. Small values of λ can occur only for small values of a frequency, that is static ($\omega = 0$) or quasistatic analysis. All examples to follow were calculated using floating point arithematic with a mantissa of ten digits. Without loss of generality the computational analysis is performed on $F_1(\lambda)$.

The series expansion of $F_1(\lambda)$ around zero argument is

$$F_1(\lambda) = 2.0 + 0.007142857\lambda^4 + 0.000015704\lambda^8 + 0.000000032\lambda^{12} + \ldots \tag{7}$$

which for small values of an argument ($\lambda < 1$) will return the correct functional value (10 digit mantissa).

The original formulation by Koloušek is

$$\bar{F}_1(\lambda) = -\lambda \frac{\sinh \lambda - \sin \lambda}{\cosh \lambda \ \cos \lambda - 1} \tag{8}$$

and the form derived by the author is

$$\bar{\bar{F}}_1(\lambda) = -\lambda \; \frac{1-e^{-\lambda}(2\sin\lambda \; + e^{-\lambda})}{\cos\lambda - e^{-\lambda}(2-e^{-\lambda}\cos\lambda)} . \tag{9}$$

The ratio of two relative errors is adopted as a measure of accuracy and therefore numerical performance of formulas (8) and (9). Relative error for the original function is

$$\Delta_1 = \left| \frac{\bar{F}_1(\lambda) - F_1(\lambda)}{F_1(\lambda)} \right| \tag{10}$$

and for the new formulation the relative error is

$$\Delta_1 = \left| \frac{\bar{\bar{F}}_1(\lambda) - F_1(\lambda)}{F_1(\lambda)} \right| . \tag{11}$$

Hence the ratio of (10) to (11) is

$$\Delta = \frac{\Delta_1}{\Delta_2} = \left| \frac{\bar{F}_1(\lambda) - F_1(\lambda)}{\bar{\bar{F}}_1(\lambda) - F_1(\lambda)} \right| . \tag{12}$$

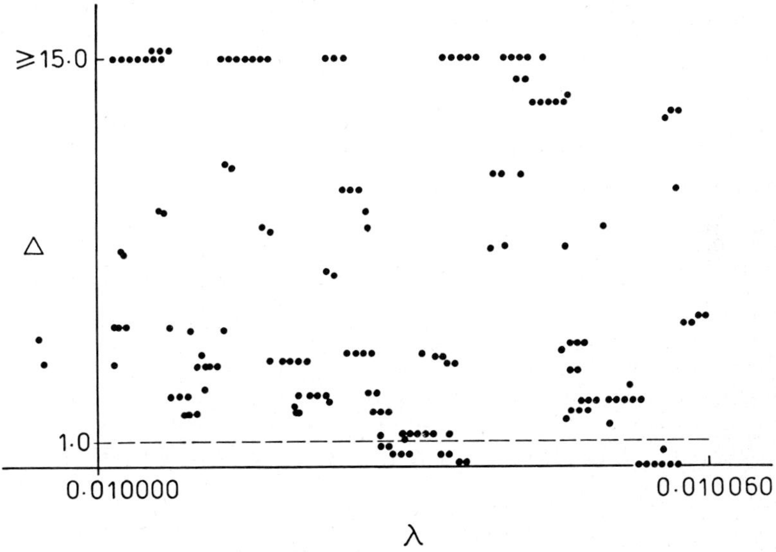

FIG. 1 RATIO OF RELATIVE ERRORS Δ (EQUATION 12) AS FUNCTION OF λ (USING 9)

Clearly, if $\Delta > 1$, the performance of the new formulation is better than the original, whereas for $\Delta < 1$ the original form has a smaller error. The equation (12) has been evaluated for the interval λ between 0.010 and 0.01006 with the increment $\varepsilon = 2E - 7$ and the results are in Fig. 1. Functions $F_1(\lambda)$, $\bar{F}(\lambda)$ and $F_1(\lambda)$ have been calculated as defined by equations (7), (8) and (9) respectively and evaluated from left to the right.

While Fig. 1 demonstrates general improvement, because the majority of ratios are greater than unity, further improvement of the formula (9) is possible. The calculation generates an error due to a finite number of digits available for performing "pseudo" operations on "digital numbers" (Householder, 1974).

Analysis of the expression (9) reveals the error arising from pseudo division, because the divisor has a large relative error. A number of significant digits may be increased if the divisor is evaluated in the form (left to right)

$$(e^{-\lambda} \cos \lambda - 1)e^{-\lambda} - e^{-\lambda} + \cos \lambda \tag{13}$$

as may be seen from Fig. 2.

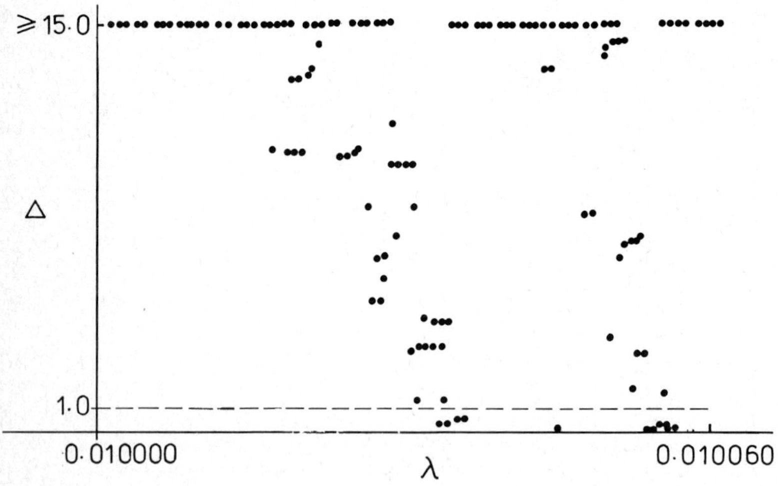

FIG. 2 RATIO OF RELATIVE ERRORS Δ (EQUATION 12) AS FUNCTION OF λ (USING 13)

Further improvement is possible if the divisor is in the form

$$(\cos \lambda - e^{-\lambda}) + (e^{-\lambda}(\cos \lambda - 1)) \quad (14)$$

as shown in Fig. 3.

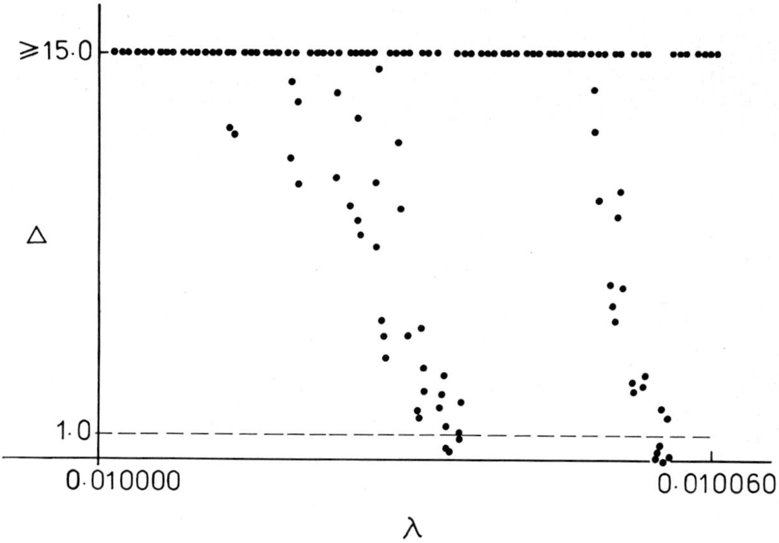

FIG. 3 RATIO OF RELATIVE ERRORS Δ (EQUATION 12) AS FUNCTION OF λ (USING 14)

Considerable improvement as shown in the figures above for $F_1(\lambda)$ without additional computing effort is also possible for other frequency functions.

FRAME ANALYSIS

However, frequency functions are only elements in the evaluation of a determinant |A| (4) for free or forced vibrations of a structure using the dynamic deformation method. The evaluation of a determinant (free vibrations) or the solution of a system (4) (forced vibration) creates its own computational errors related to the algorithm adopted. In examining the improved performance of a new formulation of frequency functions the algorithm for a determinant evaluation will not change, only elements (frequency functions) will be calculated in different ways as for $F_1(\lambda)$ above. The nondimensional analysis of the frame in Fig. 4 is performed to show the improvement for a simple structure.

There are two important sets of frequencies when the equation (4) is valid :

(a) frequencies at which $|A| = 0$ - these are resonant frequencies ;

(b) frequencies at which one or more elements of A become infinite-these are anti-resonant frequencies (asymptotes).

(a) RESONANT (NATURAL) FREQUENCIES

The matrix is generally Hermitian, but only the real symmetric matrix will be considered here. The natural frequencies may be located by finding frequencies at which the determinant changes sign through zero. From the iteration methods available, the bisection method is the safest, but unfortunately it is also very slow. The calculation effort to produce a single functional value of the determinant can be considerable, especially for large matrices. It is possible to improve the situation with the new formulation of frequency functions (improved numerical stability) by the use of higher convergence methods, and thus minimizing the number of evaluations of the determinant.

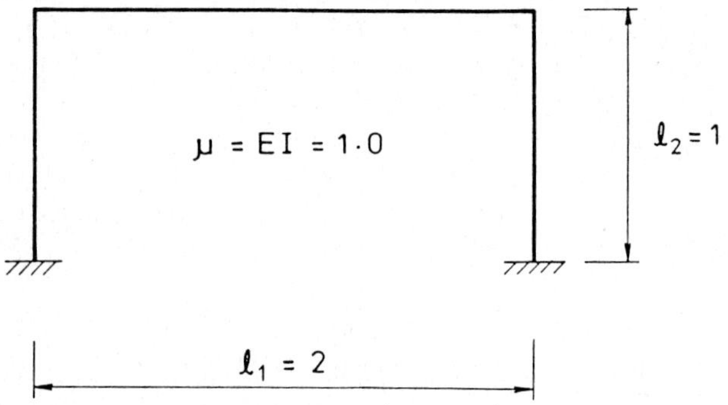

FIG. 4 ONE BAY FRAME

With matrix element given by Appendix 3, three intervals will be examined for a numerical stability of the determinant: $\lambda_1 \doteq 0.1$, $\lambda_2 \doteq 0.5$, $\lambda_3 \doteq 1.0$ by increasing the last significant digit and monitoring the determinant value. The increment for each interval thus created is ε_1 = 1E - 10, ε_2 = 1E - 10 and ε_3 = 1E - 9, respectively.

First the original formulation is compared with that of the author for each interval λ_i, i = 1,2,3. The evaluation of frequency functions is as given in Appendix 1 and Appendix 2 with the determinant in Figs. 5-7.

As may be seen from Figs. 5, 6 and 7 significant improvement has been achieved for $\lambda \doteq 0.1$, $\lambda \doteq 0.5$ with the determinant oscillating in the last three significant digits. For $\lambda \doteq 1.0$, the last two digits are unstable due to determinant evaluation and there is no improvement.

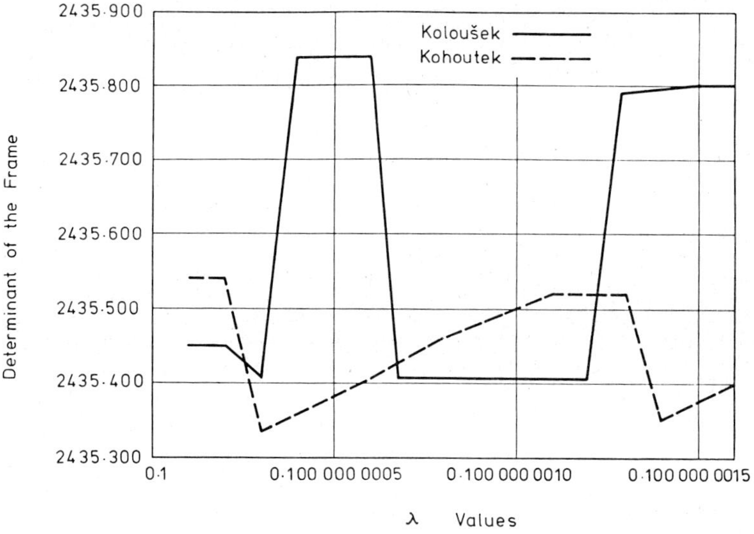

FIG. 5 DETERMINANT OF THE FRAME IN FIG. 4
$\lambda = 0{\cdot}1,\ 0{\cdot}100\,000\,0016,\ \varepsilon_1 = 1\text{E}{-}10$

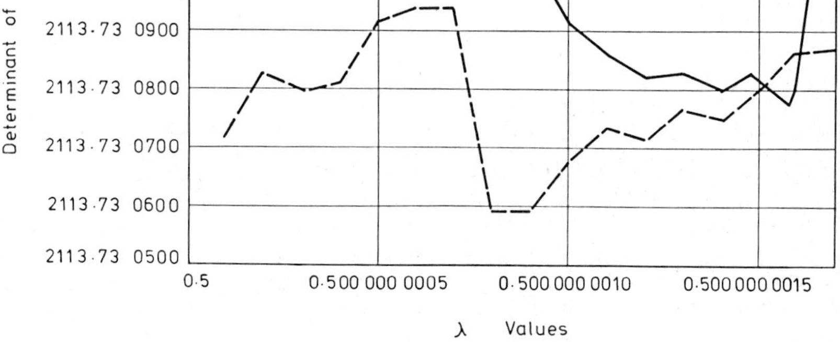

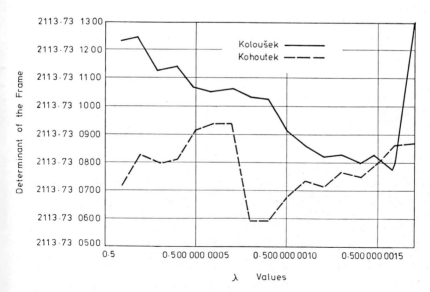

FIG. 6 DETERMINANT OF THE FRAME IN FIG. 4
$\lambda = 0{\cdot}5,\ 0{\cdot}500\,000\,0017,\ \varepsilon_2 = 1\text{E}{-}10$

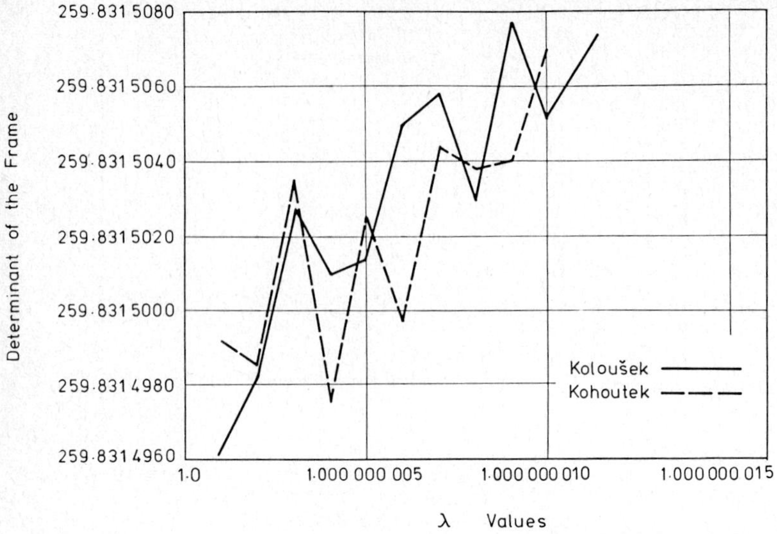

FIG. 7 DETERMINANT OF THE FRAME IN FIG. 4

$\lambda = 1{\cdot}0,\ 1{\cdot}000\,000\,016,\ \varepsilon_3 = 1\text{E}-9$

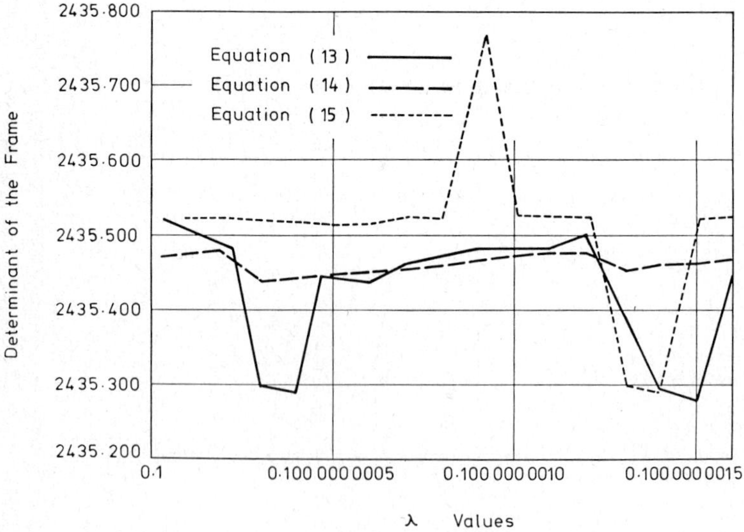

FIG. 8 DETERMINANT OF THE FRAME IN FIG. 4

$\lambda = 0{\cdot}1,\ 0{\cdot}100\,000\,0016,\ \varepsilon_1 = 1\text{E}-10$

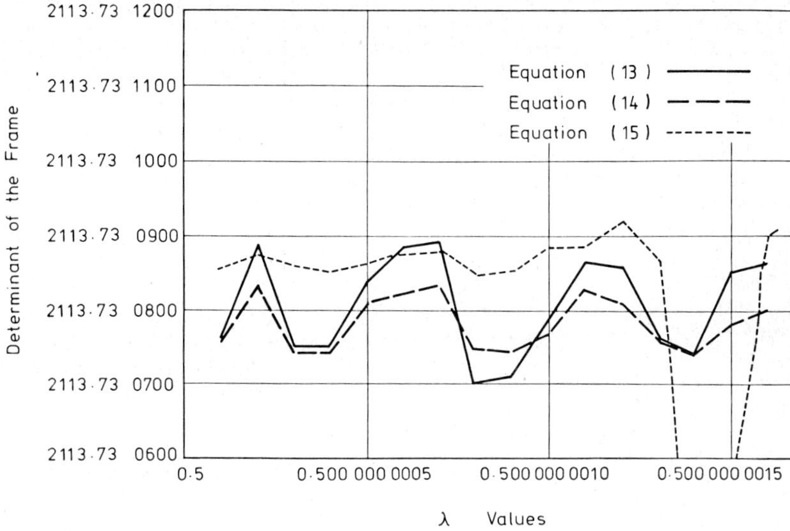

FIG. 9 DETERMINANT OF THE FRAME IN FIG. 4
$\lambda = 0{\cdot}5,\ 0{\cdot}500\,000\,0017,\ \varepsilon_2 = 1\text{E}-10$

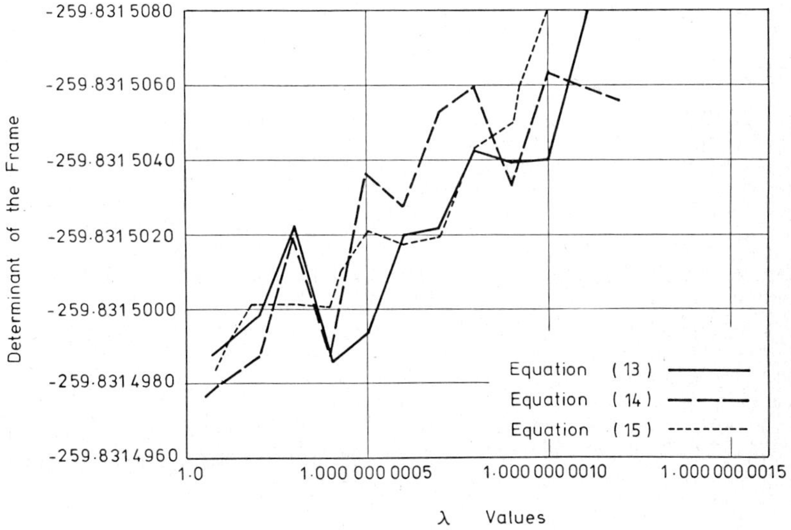

FIG. 10 DETERMINANT OF THE FRAME IN FIG. 4
$\lambda = 1{\cdot}0,\ 1{\cdot}000\,000\,0016,\ \varepsilon_3 = 1\text{E}-9$

Utilising equations (13) and (14) as before and for the same intervals, the determinant values to the same scale are shown in Figs. 8, 9, 10 with improved stability. However, further improvement is possible if the divisor is evaluated in the form

$$\cos \lambda(1 + e^{-2\lambda}) - 2e^{-\lambda} \tag{15}$$

with the determinant shown in Figs. 8, 9 and 10. Sudden "jumps" are due to rounding off for numerator and divisor for different values of the argument. The numerical stability of equation (15) when compared with equation (14) (Fig.10) leads to substantial savings in the evaluation of the determinant, when a root (natural frequency) is calculated.

The first natural frequency has been calculated using the secant method with the following result :

Equation (15) (Fig. 10)

Iteration	Determinant
1	6.484093763
2	1.075973133
3	0.004319721
4	0.000001537
	(0.9182395521) (root)

Equation (14) (Fig. 10)

Iteration	Determinant
1	6.484094570
2	1.075973000
3	0.004317660
4	0.000002853
5	0.000002508
6	-0.000019795
7	-0.000001931
	(0.9182395530) (root)

As may be seen the rate of convergence is almost doubled due to numerical stability using equation (15) instead of equation (14) in frequency functions.

DISCUSSION

Superior performance of equation (15) over equation (14) is primarly due to numerical stability, demonstrated in Figs. 8, 9, 10 where for lower values of an argument in Figs. 8 and 9 marred by "jumps". An interesting question arises of *a priori* comparison of a performance for functions defined by (13), (14) and (15). All functions are evaluated at discrete points with all "oscillations" due to a numerical process unrelated to neither physical nor mathematical formulation of the problem. Natural splines [3] could be used as a measure of how "unsmooth" a particular process is in performing computational evaluation of identical analytical function.

Let us consider a function f(x) satisfying

$$f(x_i) = y_i + \varepsilon_i \tag{16}$$

with ε_i being an error of computation for each discrete point x_i, i =1,2,3 ..., n, we may fit a natural interpolating spline of degree 2k - 1 through x_i, y_i and call it s_k. Consider interval [a,b] containing all x_i, then any $f(x_i)$ satisfying (16) and such that $f^{(k)} \in L_2(a,b)$ satisfies

$$\int_a^b (f^{(k)})^2 \geq \int_a^b (s^{(k)})^2 =: \nu_k \tag{17}$$

where the parameter ν_k in (17) may be calculated form the data, and for various k, will indicate the "smoothness".

(b) ANTI-RESONANT FREQUENCIES (ASYMPTOTES)

At these frequencies one or more of the terms in A will be infinite, due to zero divisions. This will lead to one or more of the components of $\underset{\sim}{u}$ being zero. In order to find these frequencies we need to find the zeroes of the reciprocals $1/a_{ij}$. The matrix A is diagonally dominant, that is off diagonal elements are not increasing faster than the main diagonal elements as a function of a frequency. Hence, in practice, it is found sufficient to consider only the zeroes of the diagonal terms $1/a_{ii}$. The function $1/a_{ii}$ is typically "flat" around a zero and the chances of missing a zero are small. Often a number of different terms $1/a_{ii}$ have common zeroes; all of these will be asymptotes of the determinant $|A|$. There is virtually no additional computing required to establish the positions of asymptotes.

NUMBER OF NATURAL FREQUENCIES BETWEEN ASYMPTOTES

An excellent paper by Williams and Wittrick [15] gives a method of finding a count of natural frequencies for a continuous system without their evaluation. The algorithm is an extension of a theorem by Rayleigh [12] for a discrete system. A simple and more general solution is to apply the theorem directly to the diagonal elements of the upper triangular matrix $\underset{\sim}{\bar{A}}$ numerically.

The first step is to determine the positions of the asymptotes, then following the theorem of William and Wittrick to establish a sign count of the main diagonal $\bar{a}_{ii}$. Let $\bar{\omega}_j$ and $\bar{\omega}_{j+1}$ be two neighbouring asymptotes. Put $_1\omega = \bar{\omega}_j + \Delta\omega$ and $_2\omega = \bar{\omega}_{j+1} - \Delta\omega$, for some $\Delta\omega$ small positive number.
If $\overset{+}{_a\mu_j}$ is the number of negative (positive) elements $\bar{a}_{ii}$ at $_1\omega$, and $\overset{-}{_a\mu_{j+1}}$ is the corresponding count at $_2\omega$, then the number of natural frequencies (between frequencies $_1\omega$, $_2\omega$) is

$$J = \overset{-}{_a\mu_{j+1}} - \overset{+}{_a\mu_j} \tag{18}$$

CONCLUSIONS

An improvement of accuracy and numerical stability using a new form of frequency function for large values of λ is self evident by comparison of Appendix 1 and Appendix 2. For small values of λ the improved new formulation may be further enhanced by paying some attention to computational procedure in the evaluation of frequency functions as has been demonstrated in Figs. 5 - 10. However, very small values of $\lambda(\lambda \cong 0.01)$ are unlikely in practice; since a dynamic analysis will convert to a quasi-static analysis.

The determination of asymptotes is crucial for the evaluation of zeroes between them. This step must be performed first, before the number and the position of zeroes is established. The algorithm presented here covers automatically any variation induced by modelling (see also [8] for some problems due to modelling).

ACKNOWLEDGEMENT

This work has been undertaken while on sabbatical leave from the University of Melbourne, Australia at the Solid Mechanics Division of the University of Waterloo, Canada. Particular thanks are due to :

Dr. G.M.L. Gladwell of the Solid Mechanics Division for pointing out the paper [5], and Dr. J. Kautsky from the Flinders University, South Australia for suggesting an application of splines to the problem.

REFERENCES

[1] Armstrong, I.D. (1969), "The Natural Frequencies of Multi-storey Frames", The Structural Engineer, Vol. 47, No. 8, pp. 290-308.

[2] Baťa, M. and Plachý (1978), "Vyšetřování dynamických učinků na stavební konstrukce", (Analysis of vibrations of engineering structures) SNTL, Praha.

[3] Carl de Boor (1978), "A Practical Guide to Splines", Springer-Verlag, New York Heidelberg Berlin.

[4] Conte, S.D. and C. de Boor (1980), "Elementary Numerical Analysis" McGraw-Hill, International student edition.

[5] Gartner, J.R. and Olgac, N. (1982), "Improved numerical computation of uniform beam characteristic values and characteristic function". Journ. of Sound and Vibrations, 84(4) pp 481-489.

[6] Householder, A.S. (1974), "Principles of Numerical Analysis", Dover, New York.

[7] Isaacson, E. and H.B. Keller (1966), "Analysis of Numerical Methods", J. Wiley & Sons, New York.

[8] Kohoutek, R. (1983), "Analysis and design of foundations for vibrations", Ch. 4, University of Melbourne, May.

[9] Koloušek, V. (1942), "Kmitání Patrových Rámů", (Vibrations of Multi-storey Frames), Technický Obzor, Vol. 50, pp 327-331, 352-355, 366-368, 374-378, Praha.

[10] Koloušek, V. (1947), "Vibrations Amorties Des Portiques", Publication Preliminaries. III Congress AIPC pp 681-688. Liege.

[11] Koloušek, V. (1973), "Dynamics in Engineering Structures", Praha-Academia, London - Butterworth.

[12] Rayleigh, J.W.S. (1945), "The Theory of Sound", Vol. 1, p. 119, Dover, New York.

[13] Tranberg, C.H. (1977), "On the Free and Forced Vibration of Structural Frames", Ph.D. Thesis, University of Queensland.

[14] Veletsos, A.S. and Newmark, N.M. (1957), "Natural Frequencies of Continuous Flexural Members", Trans. ASCE Vol. 122, pp 249-285.

[15] Williams, F.W. and Wittrick, W.H. (1970), "An Automatic Computational Procedure for Calculating Natural Frequencies of Skeletal Structures", Int. J. Mech. Sc., Vol. 12, pp 781-791.

[16] Wittrick, W.H. & Williams, F.W. (1970), "A General Algorithm for Computing Natural Frequencies of Elastic Structures", Quart. Journ. Mech. and Applied Math., Vol. XXIV, pp 263-284.

APPENDIX 1

FREQUENCY FUNCTIONS $F_i(\lambda)$

(After Koloušek)

Vibration without damping; shear and rotary inertia not considered.

$$F_1(\lambda) = -\lambda \frac{\sinh\lambda - \sin\lambda}{\cosh\lambda\cos\lambda - 1}$$

$$F_2(\lambda) = -\lambda \frac{\cosh\lambda\sin\lambda - \sinh\lambda\cos\lambda}{\cosh\lambda\cos\lambda - 1}$$

$$F_3(\lambda) = -\lambda^2 \frac{\cosh\lambda - \cos\lambda}{\cosh\lambda\cos\lambda - 1}$$

$$F_4(\lambda) = \lambda^2 \frac{\sinh\lambda\sin\lambda}{\cosh\lambda\cos\lambda - 1}$$

$$F_5(\lambda) = \lambda^3 \frac{\sinh\lambda + \sin\lambda}{\cosh\lambda\cos\lambda - 1}$$

$$F_6(\lambda) = -\lambda^3 \frac{\cosh\lambda\sin\lambda + \sinh\lambda\cos\lambda}{\cosh\lambda\cos\lambda - 1}$$

$$F_7(\lambda) = \lambda \frac{2\sinh\lambda\sin\lambda}{\cosh\lambda\sin\lambda - \sinh\lambda\cos\lambda}$$

$$F_8(\lambda) = \lambda^2 \frac{\sinh\lambda + \sin\lambda}{\cosh\lambda\sin\lambda - \sinh\lambda\cos\lambda}$$

$$F_9(\lambda) = -\lambda^2 \frac{\cosh\lambda\sin\lambda + \sinh\lambda\cos\lambda}{\cosh\lambda\sin\lambda - \sinh\lambda\cos\lambda}$$

$$F_{10}(\lambda) = -\lambda^3 \frac{\cosh\lambda + \cos\lambda}{\cosh\lambda\sin\lambda - \sinh\lambda\cos\lambda}$$

$$F_{11}(\lambda) = \lambda^3 \frac{2\cosh\lambda\cos\lambda}{\cosh\lambda\sin\lambda - \sinh\lambda\cos\lambda}$$

$$F_{12}(\lambda) = \lambda^3 \frac{\cosh\lambda\cos\lambda + 1}{\cosh\lambda\sin\lambda - \sinh\lambda\cos\lambda}$$

$$F_{13}(\lambda) = -\lambda^3 \frac{\sinh\lambda - \sin\lambda}{2\sinh\lambda\sin\lambda}$$

$$F_{14}(\lambda) = -\lambda^3 \frac{\cosh\lambda\sin\lambda - \sinh\lambda\cos\lambda}{2\sinh\lambda\sin\lambda}$$

$$F_{15}(\lambda) = -\lambda \frac{\cosh\lambda\sin\lambda - \sin\lambda\cos\lambda}{\cosh\lambda\cos\lambda + 1}$$

$$F_{16}(\lambda) = \lambda^2 \frac{\sinh\lambda\sin\lambda}{\cosh\lambda\cos\lambda + 1}$$

$$F_{17}(\lambda) = -\lambda^3 \frac{\cosh\lambda\sin\lambda + \sinh\lambda\cos\lambda}{\cosh\lambda\cos\lambda + 1}$$

APPENDIX 2

FREQUENCY FUNCTIONS $F_i(\lambda)$

Vibration without damping; shear and rotary inertia not considered.

FUNCTION	LIMIT (λ = 0)
$F_1(\lambda) = -\lambda \dfrac{1 - e^{-\lambda}(2 \sin\lambda + e^{-\lambda})}{f_1(\lambda)}$	$F_1(0) = 2.\overline{00}$
$F_2(\lambda) = -\lambda \dfrac{\sin\lambda - \cos\lambda + e^{-2\lambda}(\sin\lambda + \cos\lambda)}{f_1(\lambda)}$	$F_2(0) = 4.\overline{00}$
$F_3(\lambda) = -\lambda^2 \dfrac{1 + e^{-\lambda}(e^{-\lambda} - 2\cos\lambda)}{f_1(\lambda)}$	$F_3(0) = 6.\overline{00}$
$F_4(\lambda) = \lambda^2 \dfrac{\sin\lambda\,(1-e^{-2\lambda})}{f_1(\lambda)}$	$F_4(0) = -6.\overline{00}$
$F_5(\lambda) = \lambda^3 \dfrac{1 + e^{-\lambda}(2\sin\lambda - e^{-\lambda})}{f_1(\lambda)}$	$F_5(0) = -12.\overline{00}$
$F_6(\lambda) = -\lambda^3 \dfrac{\sin\lambda + \cos\lambda + e^{-2\lambda}(\sin\lambda - \cos\lambda)}{f_1(\lambda)}$	$F_6(0) = 12.\overline{00}$

$$f_1(\lambda) = \cos\lambda - e^{-\lambda}(2 - e^{-\lambda}\cos\lambda)$$

$F_7(\lambda) = \lambda \dfrac{2\sin\lambda\,(1-e^{-2\lambda})}{f_2(\lambda)}$	$F_7(0) = 3.\overline{00}$
$F_8(\lambda) = \lambda^2 \dfrac{1 + e^{-\lambda}(2\sin\lambda - e^{-\lambda})}{f_2(\lambda)}$	$F_8(0) = 3.\overline{00}$
$F_9(\lambda) = -\lambda^2 \dfrac{\sin\lambda + \cos\lambda + e^{-2\lambda}(\sin\lambda - \cos\lambda)}{f_2(\lambda)}$	$F_9(0) = -3.\overline{00}$

APPENDIX 2 (CONTD)

FUNCTION	LIMIT ($\lambda = 0$)
$F_{10}(\lambda) = -\lambda^3 \dfrac{1 + e^{-\lambda}(2\cos\lambda + e^{-\lambda})}{f_2(\lambda)}$	$F_{10}(0) = -3.\overline{00}$
$F_{11}(\lambda) = \lambda^3 \dfrac{2\cos\lambda\,(1 + e^{-2\lambda})}{f_2(\lambda)}$	$F_{11}(0) = 3.\overline{00}$
$F_{12}(\lambda) = \lambda^3 \dfrac{\cos\lambda\,(1 + e^{-2\lambda}) + 2e^{-\lambda}}{f_2(\lambda)}$	$F_{12}(0) = 3.\overline{00}$
$f_2(\lambda) = \sin\lambda - \cos\lambda + e^{-2\lambda}(\sin\lambda + \cos\lambda)$	
$F_{13}(\lambda) = -\lambda^3 \dfrac{1 - e^{-\lambda}(2\sin\lambda + e^{-\lambda})}{2\sin\lambda\,(1-e^{-2\lambda})}$	$F_{13}(0) = 0.\overline{00}$
$F_{14}(\lambda) = \lambda^3 \dfrac{\cos\lambda\,(1-e^{-2\lambda}) - \sin\lambda\,(1 + e^{-2\lambda})}{2\sin\lambda\,(1 - e^{-2\lambda})}$	$F_{14}(0) = 0.\overline{00}$
$F_{15}(\lambda) = -\lambda \dfrac{\sin\lambda - \cos\lambda + e^{-2\lambda}(\cos\lambda + \sin\lambda)}{f_3(\lambda)}$	$F_{15}(0) = 0.\overline{00}$
$F_{16}(\lambda) = \lambda^2 \dfrac{\sin\lambda\,(1 - e^{-2\lambda})}{f_3(\lambda)}$	$F_{16}(0) = 0.\overline{00}$
$F_{17}(\lambda) = -\lambda^3 \dfrac{\sin\lambda + \cos\lambda + e^{-2\lambda}(\sin\lambda - \cos\lambda)}{f_3(\lambda)}$	$F_{17}(0) = 0.\overline{00}$
$f_3(\lambda) = \cos\lambda + e^{-\lambda}(2 + e^{-\lambda}\cos\lambda)$	

APPENDIX 3

A vector of deformation is

$$u = \xi_1, u_1, v_1, \xi_2, u_2, v_2$$

and elements of the matrix are

$$a_{11} = \frac{EI_1}{\ell_1} F_2(\lambda_1) + \frac{EI_2}{\ell_2} F_2(\lambda_2) \qquad a_{34} = \frac{EI_2}{\ell_2^2} F_3(\lambda_2)$$

$$a_{12} = \frac{EI_1}{\ell_1^2} F_4(\lambda_1) \qquad a_{36} = \frac{EI_2}{\ell_2^3} F_5(\lambda_2)$$

$$a_{13} = -\frac{EI_2}{\ell_2^2} F_4(\lambda_2) \qquad a_{44} = \frac{EI_2}{\ell_2} F_2(\lambda_2) + \frac{EI_3}{\ell_3} F_2(\lambda_3)$$

$$a_{14} = -\frac{EI_2}{\ell_2} F_1(\lambda_2) \qquad a_{45} = \frac{EI_3}{\ell_3^2} F_4(\lambda_3)$$

$$a_{16} = -\frac{EI_2}{\ell_2^2} F_3(\lambda_2) \qquad a_{46} = \frac{EI_2}{\ell_2^2} F_4(\lambda_2)$$

$$a_{22} = -\frac{EI_1}{\ell_1^3} F_6(\lambda_1) + \frac{EA_2}{\ell_2} \psi_2 \cot \psi_2 \qquad a_{55} = \frac{EA_2}{\ell_2^2} \psi_2 \cot \psi_2 + \frac{EI_3}{\ell_3^3} F_6(\lambda_3)$$

$$a_{25} = -\frac{EA_2}{\ell_2} \psi_2 \operatorname{cosec} \psi_2 \qquad a_{66} = \frac{EI_2}{\ell_2^3} F_6(\lambda_2) + \frac{EA_3}{\ell_3} \psi_3 \cot \psi_3$$

$$a_{33} = \frac{EA_1}{\ell_1} \psi_1 \cot \psi_1 + \frac{EI_2}{\ell_2^3} F_6(\lambda_2)$$

$a_{15} = 0$; $a_{23} = 0$; $a_{24} = 0$; $a_{26} = 0$; $a_{35} = 0$; $a_{56} = 0$.

Computational Techniques & Applications: CTAC-83
J. Noye & C. Fletcher (Editors)

MACRO ELEMENTS FOR THE ANALYSIS OF BEAM-SLAB SYSTEMS

B.W. Golley and J. Petrolito

Dept. of Civil Engineering, Faculty of Military Studies,
University of New South Wales,
Duntroon, ACT
Australia

A variable degree-of-freedom plate bending element is presented. The displacement function within an element satisfies the governing thin plate equation, substantially reducing the number of equations requiring generation and solution for the accurate analysis of beam-plate structures compared to existing finite elements. Large elements corresponding to structural units bounded by beams may be used, requiring a minimum of data preparation.

INTRODUCTION

Many commonly occurring structures are composed of beams, columns and slabs laid out in a rectangular pattern. The analysis of these structures requires discretisation into a large number of rectangular plate finite elements in which both in-plane and out-of-plane displacements are considered in addition to beam elements. A range of rectangular plate bending elements is available, of which the twelve degree-of-freedom non-conforming element (Zienkiewicz and Cheung (1964)) and the sixteen degree-of-freedom conforming element (Bogner et al (1965)) are the best known. A rectangular element incorporating trigonometric basis functions was considered by Chakrabarti (1971).

To reduce the number of elements, several plate bending panel elements have been proposed. French et al (1975) have used the sixteen degree-of-freedom element, modified to eliminate the corner twist degrees of freedom to give corner displacements compatible with the usual beam degrees of freedom. Gutkowski and Wang (1976) have proposed a "finite panel" in which displacements are combinations of trigonometric and hyperbolic functions. Symmetrical loadings only may be considered. Golley (1979) and Yan (1979, 1982) have proposed variable degree-of-freedom panel elements in which normal moments are continuous between elements, and displacements are C_0 continuous. The mixed elements produce accurate results with minimum discretisation, but the unusual degrees-of-freedom are difficult to incorporate into beam-plate structures.

Rectangular elements for in-plane loadings are among the simplest of finite elements and a number of types are discussed by Tottenham and Brebbia (1970).

Despite the available finite elements, it is however uncommon to analyse beam-column-slab structures using finite elements (Hrabok and Hrudley (1983)). Most structures in this class are analysed using approximate methods. It is common practice to analyse buildings as equivalent frames, with the slab contributing some approximately determined stiffness to the beams. The equivalent frame is then analysed by discretisation into a minimum number of elements, each corresponding to a single beam or column.

The term macro element has been used to describe a finite element suitable for modelling a domain with minimum discretisation (Parsons and GangaRao (1979)). In this paper, a variable degree-of-freedom rectangular plate bending element and an associated beam element are described. In-plane loads are not considered but are the subject of continuing study. In general, the plate bending elements correspond to structural units bounded by beams, permitting discretisation into a minimum number of elements.

Degrees of freedom at plate corners enable conventional elements to be used for columns. Degrees of freedom associated with various edges of an element may be varied from edge to edge enabling elements with large aspect ratios to be treated economically. The elements have variable degrees of freedom enabling convergence to be readily assessed. The governing differential equation is satisfied within each plate element, greatly reducing the equations requiring solution for engineering accuracy to be obtained.

ASSUMPTIONS

It is assumed that

(1) Within each element, the plate thickness is constant and the element behaves according to isotropic thin plate theory.

(2) Edge beams are rigidly connected to the plate with the plate and beam neutral axes coincident.

If edge beams have their neutral axes offset from the plate neutral axis, in-plane stresses are introduced into the plate. The analysis of this case requires that the present plate bending element be coupled with an analogous plane stress element. This problem will be discussed by Petrolito (1983).

THIN PLATE THEORY

The usual assumptions of beam and thin plate theory apply (Timoshenko and Woinowsky-Krieger (1959)). Accordingly, the middle surface deflection, w, of the plate should satisfy the equation

$$w,_{xxxx} + 2w,_{xxyy} + w,_{yyyy} = \nabla^4 w = \frac{p}{D} \tag{1}$$

in which p is the intensity of the transverse load and D is the flexural rigidity of the plate per unit length. A comma denotes partial differentiation.

PLATE DISPLACEMENT FUNCTIONS

Displacement functions within each element satisfy the following conditions.

(1) The governing differential equation is satisfied within the element.

(2) Corner degrees of freedom are the displacement w and the derivatives $w,_x$ and $w,_y$. Hence at element corners, displacements are C_1 continuous.

(3) On interelement boundaries other than at corners, displacements are C_0 continuous.

(4) On any edge, displacements are approximated by a quintic polynomial plus a truncated sine series.

Interelement normal rotations are not continuous other than at the corners. However, using a modified potential energy functional to obtain element stiffness matrices, C_1 continuity is approximated in a weighted integral sense. For convergence to exact values with minimum discretisation, it can be shown that the quintic polynomials in condition (4) are required, rather than lower order polynomials.

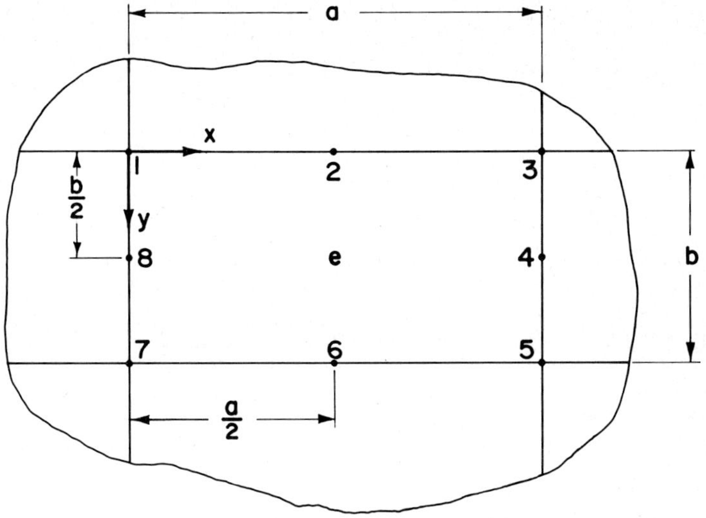

Figure 1
Typical Element

In a typical element, e, shown in Figure 1 the displacement, w^e, is approximated by

$$w^e = [N_p^e(x,y)]\{\phi_p^e\} + \sum_{k=1}^{K} [N_{Wk}^e(x,y)]\{\phi_{Wk}^e\} + \sum_{k=1}^{K^*} [N_{Mk}^e(x,y)]\{\phi_{Mk}^e\} + w_o^e \tag{2}$$

The four terms in Equation (2) are obtained by superposing the cases shown in Figure 2.

The first term is derived as follows. The displacement within the element, $w_1^e(x,y)$, is taken as the sum of the biharmonic polynomials

$$w_1^e(x,y) = [P(x,y)]\{A\} \tag{3}$$

where $[P(x,y)]$ are the biharmonic polynomials given in Appendix 1 and $\{A\}$ are parameters.

The element degrees of freedom $\{\phi_p^e\} = [w_1^e \;\; w_{,x1}^e \;\; w_{,y1}^e \;\; w_2^e \;\; w_{,x2}^e \;\; w_3^e \;\; w_{,x3}^e \;\; w_{,y3}^e \;\; w_4^e \;\; w_{,y4}^e \;\; w_5^e \;\; w_{,x5}^e \;\; w_{,y5}^e \;\; w_6^e \;\; w_{,x6}^e \;\; w_7^e \;\; w_{,x7}^e \;\; w_{,y7}^e \;\; w_8^e \;\; w_{,y8}^e]^T$ are nodal displacements and rotations at corner and midside nodes. Note that two rotations are specified at corner nodes, but only a tangential rotation is specified at midside nodes. $\{\phi_p^e\}$ is expressed in terms of $\{A\}$ as

$$\{\phi_p^e\} = [C]\{A\} \tag{4}$$

Combining equations (3) and (4) gives

$$w_1^e(x,y) = [P(x,y)][C]^{-1}\{\phi_p^e\} = [N_p^e(x,y)]\{\phi_p^e\} \tag{5}$$

The terms in $[N_p^e(x,y)]$ have been evaluated explicitly and are given by Petrolito (1983).

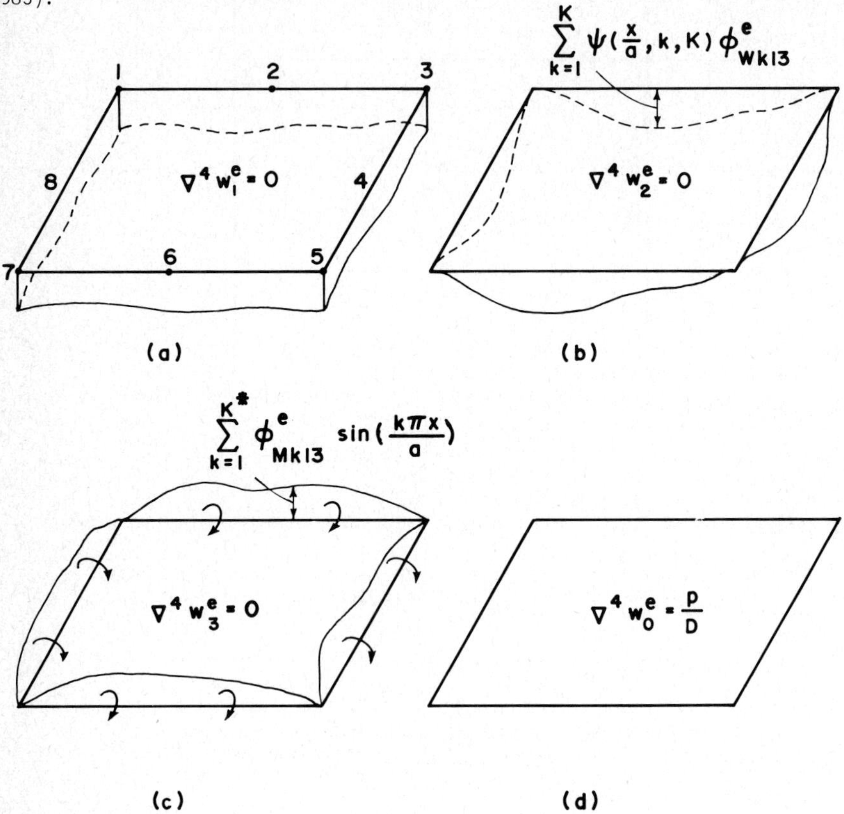

Figure 2
Displacement Approximation within Element

The second term in Equation (2) results from boundary displacements of the form shown in Figure 2(b), with boundary normal moments zero. The boundary displacement on the edges are approximated by truncated sine series, constrained such that derivatives are zero at the corners. Hence on say the edge y = 0 the displacement is approximated by

$$w^e_{2|y=0} = \sum_{k=1}^{K} \psi(\frac{x}{a}, k, K)\ \phi^e_{Wk13} \tag{6}$$

where if K is even

$$\psi(\frac{x}{a}, k, K) = \sin\frac{k\pi x}{a} - (\frac{k}{K+1}) \sin\frac{(K+1)\pi x}{a}, \quad k = 1,3 \ldots K-1$$

$$\psi(\frac{x}{a}, k, K) = \sin\frac{k\pi x}{a} - (\frac{k}{K+2}) \sin\frac{(K+2)\pi x}{a}, \quad k = 2,4 \ldots K \tag{7a}$$

and if K is odd

$$\psi(\tfrac{x}{a}, k, K) = \sin\frac{k\pi x}{a} - (\frac{k}{K+2}) \sin\frac{(K+2)\pi x}{a}, \quad k = 1,3 \ldots K$$

$$\psi(\tfrac{x}{a}, k, K) = \sin\frac{k\pi x}{a} - (\frac{k}{K+1}) \sin\frac{(K+1)\pi x}{a}, \quad k = 2, 4 \ldots K - 1 \ (K > 1)$$

(7b)

Displacements on the other boundaries are approximated by similar expressions.

With boundary displacements specified by Equation (6) and with normal moments zero on the boundaries, Levy series solutions (Timoshenko and Woinowsky-Krieger (1959)) to the homogeneous form of Equation (2) are obtained in terms of the coefficients ϕ^e_{Wk13} where the subscript 13 refers to edge 13. Hence for the kth term in the series

$$w^e_{2k} = [N^e_{Wk}]\{\phi^e_{Wk}\} \tag{8}$$

where $\{\phi^e_{Wk}\} = [\phi^e_{Wk13} \ \ \phi^e_{Wk35} \ \ \phi^e_{Wk57} \ \ \phi^e_{Wk17}]^T$. The terms in $[N^e_{Wk}]$ are given by Petrolito (1983). Summing up the solutions in Equation (8) gives the second term of Equation (2).

In the above it has been assumed for simplicity that the same number of terms are taken on each side of the element. This is not necessary in general and the use of different numbers of terms on each side leads to computational efficiency in some cases.

The third term in Equation (2) is the displacement due to the application of normal boundary moments in the form of truncated sine series with zero boundary displacements, as shown in Figure 2(c). On say y = 0, the normal moment is approximated by

$$M^e_{n3|y=0} = \sum_{k=1}^{K^*} \phi^e_{Mk13} \sin\frac{k\pi x}{a} \tag{9}$$

With moments on the other boundaries similarly approximated, Levy series solutions to the homogeneous form of Equation (2) are obtained in terms of the coefficients, ϕ^e_{Mk13} etc.

Hence for the kth term in the series

$$w^e_{3k} = [N^e_{Mk}]\{\phi^e_{Mk}\} \tag{10}$$

where $[\phi^e_{Mk} = \phi^e_{Mk13} \ \phi^e_{Mk35} \ \ \phi^e_{Mk57} \ \ \phi^e_{Mk17}]^T$. The terms in $[N^e_{Mk}]$ are given by Petrolito (1983). Summing up the solutions in Equation (10) gives the third term of Equation (2).

The last term in Equation (2) is a particular solution to the governing equation with all sides of the element simply supported (see Figure 2(d)). For the most general loading, Navier's double series solution is required, while Levy's single series solutions may be appropriate in more restricted cases of loadings (Timoshenko and Woinowsky-Krieger (1959)).

PLATE ELEMENT FORMULATION

The element shape functions chosen are such that normal rotations are only continuous at corners. To enforce continuity of normal rotations on the sides requires the use of a modified total potential energy principle (Pian and Tong (1969)) in which additional Lagrange multipliers are introduced. This results in a mixed set of equations which are computationally undesirable. However the mixed formulation may be avoided by following the procedure outlined below, in which the final variables are displacements and rotation parameters only.

The total potential energy in element e, Π_p^e, is

$$\Pi_p^e = \int_{A^e} (u^e - pw^e)\, dA \tag{11}$$

where u^e is the strain energy per unit area and A^e is the area of the element. Substituting for w^e from Equation (2) and performing the integration gives

$$\Pi_p^e = \frac{1}{2}[\{\phi_P^e\}^T \quad \{\phi_W^e\}^T \quad \{\phi_M^e\}^T] \begin{bmatrix} [K_{PP}^e] & [K_{PW}^e] & [K_{PM}^e] \\ [K_{PW}^e]^T & [K_{WW}^e] & [\ 0\ ] \\ [K_{PM}^e]^T & [\ 0\ ] & [K_{MM}^e] \end{bmatrix} \begin{Bmatrix} \{\phi_P^e\} \\ \{\phi_W^e\} \\ \{\phi_M^e\} \end{Bmatrix}$$

$$- [\{\phi_P^e\}^T \{\phi_W^e\}^T \{\phi_M^e\}^T] \begin{Bmatrix} \{L_P^e\} \\ \{L_W^e\} \\ \{L_M^e\} \end{Bmatrix} \tag{12a}$$

or

$$\Pi_p^e = \frac{1}{2}\{\phi^e\}^T [K^e]\{\phi^e\} - \{\phi^e\}^T\{L^e\} \tag{12b}$$

The terms in $[K^e]$ and $\{L^e\}$ have been evaluated explicitly by Petrolito (1983). The submatrix $\{L_p^e\}$ contains infinite series terms arising from the particular solution, w_o^e.

Weighted integrals of normal rotations on each side of the plate are now introduced as follows. On the side y = 0, weighted integrals of normal rotation, $\{\phi_{13}^e\}$, are defined as

$$\{\theta_{13}^e\} = \int_0^a (\{\lambda\}\ w^e_{,y}|_{y=0})\ dx \tag{13}$$

where $\{\lambda\} = [\sin\frac{\pi x}{a} \quad \sin\frac{2\pi x}{a} \ \ldots \ \sin\frac{K^*\pi x}{a}]^T$. Similar weighted integrals are defined on the other three sides. Collecting the weighted integral terms gives

$$\{\theta^e\} = [C_p^e]\{\phi_p^e\} + [C_W^e]\{\phi_W^e\} + [K_{MM}^e]\ \{\phi_M^e\} + \{\theta_o^e\} \tag{14}$$

where $\{\theta^e\} = [\{\theta_{13}^e\}^T \ \{\theta_{35}^e\}^T \ \{\theta_{57}^e\}^T \ \{\theta_{17}^e\}^T]^T$ and the term $\{\theta_o^e\}$ arises from the particular solution, w_o^e. Expressions for the terms in $[C_p^e]$, $[C_W^e]$ and $\{\theta_o^e\}$ are given by Petrolito (1983).

Solving Equation (14) for $\{\phi_M^e\}$ gives

$$\{\phi_M^e\} = [K_{MM}]^{-1}\{\{\theta^e\} - [C_P^e]\{\phi_P^e\} - [C_W^e]\{\phi_W^e\} - \{\theta_o^e\}\} \tag{15}$$

Substituting for $\{\phi_M^e\}$ in Equation (12) gives the total potential energy of the plate as

$$\Pi_P^e = \frac{1}{2}\{\bar{\phi}^e\}^T[\bar{K}^e]\{\bar{\phi}^e\} - \{\bar{\phi}^e\}^T\{\bar{L}^e\} \tag{16}$$

where $\{\bar{\phi}^e\} = [\{\phi_P^e\}^T\{\phi_W^e\}^T\{\theta^e\}^T]^T$. $[\bar{K}^e]$ and $\{\bar{L}^e\}$ are the stiffness matrix and load vector for the plate. Hence the total potential energy of the plate is now expressed in terms of displacement parameters. Assembly and solution of the element equations now follows standard finite element procedures with some modifications required to account for the variable degrees of freedom.

BEAM ELEMENT

A variable degree of freedom element suitable to be coupled to the plate element is derived as follows. On say the edge y = 0, the vertical displacement of the beam is taken as being compatible with the plate displacement on that edge. Hence the displacement is expressed as a quintic polynomial in terms of the six polynomial parameters on that edge plus a truncated sine series in terms of the series displacement parameter on that edge, constrained in the same way as the plate element. The rotation of the beam about its longitudinal axis is expressed as a function of the two corner rotations of the edge and the weighted integrals of normal rotations of the edge.

The stiffness matrix for the beam is calculated from the strain energy of the beam under the assumption of beam bending theory and St. Venant's torsion. Expressions for the stiffness matrix are given by Petrolito (1983). The beam stiffness matrices are added to the plate stiffness matrix to give the complete stiffness matrix for the structure.

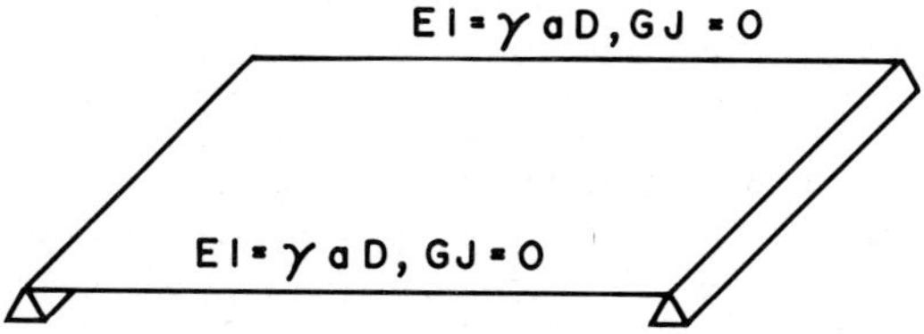

Figure 3
Example 2. Beam supported square plate (a x a)

EXAMPLES

Two examples were analysed using the proposed element. In the first example a uniformly loaded square plate simply supported on all sides was analysed. Due to symmetry only a quarter of the plate was considered. The particular solution was taken as a Levy series solution (Timoshenko and Woinowsky-Krieger (1959)). Infinite series arise in the terms of $\{L_P^e\}$. These series were truncated at N = 5, 10 and 25 terms. Poisson's ratio was taken as 0.3. Table 1 shows the results for the central deflection, w_1, central moment, M_1, quarter point deflection, w_2 and quarter point moment, M_2 for varying number of series terms, K, K^* and N.

Table 1. Factors for Simply Supported Plate

N	$K = K^*$	DOF	α_1	α_2	β_1	β_2
5	1	28	4.335	2.179	3.084	2.953
	2	36	4.130	2.141	3.159	2.938
	3	44	4.092	2.136	3.606	2.944
	4	52	4.078	2.134	3.940	2.944
	10	100	4.065	2.132	4.812	2.944
10	1	28	4.335	2.179	3.075	2.953
	2	36	4.131	2.141	3.147	2.938
	3	44	4.091	2.136	3.579	2.944
	4	52	4.078	2.134	3.910	2.944
	10	100	4.064	2.132	4.678	2.944
25	1	28	4.335	2.179	3.074	2.953
	2	36	4.131	2.141	3.146	2.938
	3	44	4.091	2.136	3.575	2.944
	4	52	4.078	2.134	3.906	2.944
	10	100	4.064	2.132	4.659	2.944
Exact			4.062	2.132	4.789	2.944

$$w_1 = \frac{\alpha_1 pa^4}{1000D} \qquad w_2 = \frac{\alpha_2 pa^4}{1000D} \qquad M_1 = \frac{\beta_1 pa^2}{100} \qquad M_2 = \frac{\beta_2 pa^2}{100}$$

Table 2. Factors for Beam Supported Plate

$\gamma = \frac{EI}{aD}$	$K = K^*$	DOF	α	β
0	1	28	13.37	10.49
	3	44	13.14	11.29
	5	60	13.11	11.78
	10	100	13.10	12.25
Exact			13.09	12.25
1	1	28	6.505	4.939
	3	44	6.279	5.460
	5	60	6.260	5.986
	10	100	6.253	6.491
Exact			6.240	6.592
10	1	28	4.612	3.312
	3	44	4.371	3.816
	5	60	4.351	4.362
	10	100	4.344	4.894
Exact			4.341	5.019

$$w = \frac{\alpha pa^4}{1000D} \qquad M_x = \frac{\beta pa^2}{100}$$

In the second example, the uniformly loaded square plate shown in Figure 3 was analysed using a quarter of the plate. Table 2 shows the results for the central deflection, w, and the central moment, M_x, for various ratios of beam flexural stiffness, EI, to plate stiffness, aD, and for varying number of series terms K and K^*. The particular solution was taken as a Levy series solution truncated at N = 25. The beam torsional stiffness,GJ, was taken as zero and Poisson's ratio was taken as 0.3.

DISCUSSION OF EXAMPLES

It is seen from Table 1 that deflections and moments converge rapidly with increasing K and N. Displacements and moments converge most slowly on the boundary of the element. Away from the boundary, convergence is very rapid. The infinite series terms in $\{L_p^e\}$ converge very rapidly with an order of convergence of approximately $1/N^4$. It should be noted that treating the full plate as a single element leads to the exact solution (i.e. deflection = $w_o^e(x,y)$).

Table 2 again shows the rapid convergence of both deflections and moments. Deflections converge to one percent or better with as few as 3 series terms. Moment convergence is not as rapid with more series terms being required to achieve comparable accuracy.

CONCLUSIONS

The proposed element leads to the accurate determination of displacements and bending moments in beam-plate structures with few degrees of freedom and minimum discretisation. The formulation presented is in terms of displacement parameters only, which enables the element to be coupled with beam and column elements for the analysis of floor systems using standard finite element procedures.

REFERENCES

[1] Bogner, F.K., Fox, R.L., and Schmit, L.A., The Generation of Interelement-Compatible Stiffness and Mass Matrices by the Use of Interpolation Formulae, Proc., Conf. on Matrix Methods in Structural Mechanics, Wright Patterson Air Force Base, Ohio, U.S.A. (1965) 397-443.

[2] Chakrabarti, S., Trigonometric Function Representation for Rectangular Plate Bending Elements, Int. Jrnl. for Numerical Methods in Eng. 3 (1971) 261-273.

[3] French, S., Kabaila, A.P., and Pulmano, V., Single Element Panel for Flat Plates, Proc., American Society of Civil Engineers 101, ST9 (1975) 1801-1812.

[4] Golley, B.W., A Rectangular Variable Degree of Freedom Plate Bending Panel Element, Proc., 3rd Int. Conf. in Australia on Finite Element Methods, Sydney, Australia (1979) 37-47.

[5] Gutkowski, R.M., and Wang, C.K., Continuous Plate Analysis by Finite Panel Method, Proc., American Society of Civil Engineers 102, ST3 (1976) 629-643.

[6] Hrabok, M.M., and Hrudey, T.M., Finite Element Analysis in Design of Floor Systems, Proc., American Society of Civil Engineers 109, ST4 (1983) 909-925.

[7] Parsons, T.J., and GangaRao, H.V.S., Macro Element Analysis of a Skew Plate in Bending, Proc., 3rd American Society of Civil Engineers Engineering Mechanics Specialty Conf., University of Texas, Austin (1979) 27-32.

[8] Petrolito, J., Variable Degree-of-Freedom Elements in Stress Analysis, Ph.D. Thesis, Dept. of Civil Eng., Faculty of Military Studies, Univ. N.S.W. (1983, in preparation).

[9] Pian, T.H.H., and Tong, P., Basis of Finite Element Methods for Solid Continua, Int. Jrnl. for Numerical Methods in Eng. 1 (1969) 3-28.

[10] Timoshenko, S., and Woinowsky-Krieger, S., Theory of Plates and Shells (McGraw-Hill, New York, 1959).

[11] Tottenham, H., and Brebbia, C.A., Finite Element Techniques in Structural Mechanics (Southampton Univ. Press, Southampton, 1970).

[12] Yan, Z., Solution of Plate Structures by Finite Panel Method, presented at National Conf. on Finite Element Methods, Nanning, China (1979).

[13] Yan, Z., A Mixed Method of Finite Panel and its Application, Proc., Int., Conf. on Finite Element Methods, Shanghai, China (1982) 821-823.

[14] Zienkiewicz, O.C., and Cheung, Y.K., The Finite Element Method for the Analysis of Elastic Isotropic and Orthotopic Slabs, Proc., Institution of Civil Engineers 28 (1964) 471-488.

APPENDIX 1

$$[P(x,y)] = [1 \quad x \quad y \quad x^2 \quad xy \quad y^2 \quad x^3 \quad x^2y \quad xy^2 \quad y^3 \quad x^3y \quad xy^3$$

$$x^4 - 3x^2y^2 \quad y^4 - 3x^2y^2 \quad x^4y - x^2y^3 \quad xy^4 - x^3y^2 \quad x^5 - 5x^3y^2 \quad y^5 - 5x^2y^3$$

$$x^5y - \frac{5}{3}x^3y^3 \quad xy^5 - \frac{5}{3}x^3y^3]$$

Computational Techniques & Applications: CTAC-83
J. Noye & C. Fletcher (Editors)

VIBRATION ANALYSIS OF A ROTATING BLADE USING DYNAMIC DISCRETIZATION

Chris Norwood

Department of Mechanical Engineering
Footscray Institute of Technology
Footscray, Victoria
Australia

An analysis of the frequencies and modes of vibration of a rotating blade is presented using dynamic discretization, as developed by Downs (1980). The method involves the discretization of the stiffness and mass properties of blade segments and the solution of the ensuing eigen value problem.

INTRODUCTION

The determination of the vibration characteristics of rotating cantilever beams is of importance in the design of blades for turbomachinery, aircraft propellors and helicopter blades. Both frequencies and mode shapes need to be accurately evaluated.

There is a need for precise eigen value determination in order to obtain acceptable accuracy in the corresponding eigen vector. Levy (1971) notes that usually for frequencies in error by more than 3-5% the mode shapes predicted are no better than poor. Bending stresses are elated to the second derivative of the eiger vector and so lack of precision in frequency determination precludes the possibility of accurate stress distribution prediction.

Propellors and blades are generally of complex geometry so that numerical methods must be used. Additional factors introduced by considering rotation of the blade include hub radius, rotational speed, setting angle and shroud loading and/or coupling.

For pretwisted blades of asymmetric cross-section the bending, torsional and axial modes of vibration are coupled together both by pretwist and section asymmetry. Rotation induces further coupling between the modes due to the centrifugal forces.

The work presented in this paper provides an analysis of the frequencies and modes of vibration of a rotating cantilever beam. The analysis is performed for a non-twisted doubley symmetric blade and includes the effects of rotary inertia and shear deformation. These simplifications limit the mode coupling to that due to rotation. The beam is however free to have any form of taper or mass distribution.

EQUATIONS OF MOTION

The equations of motion for the general case of an undamped Timoshenko beam of doubley symmetric viable section, subjected to a distributed axial force are derived by Norwood et al (1980).

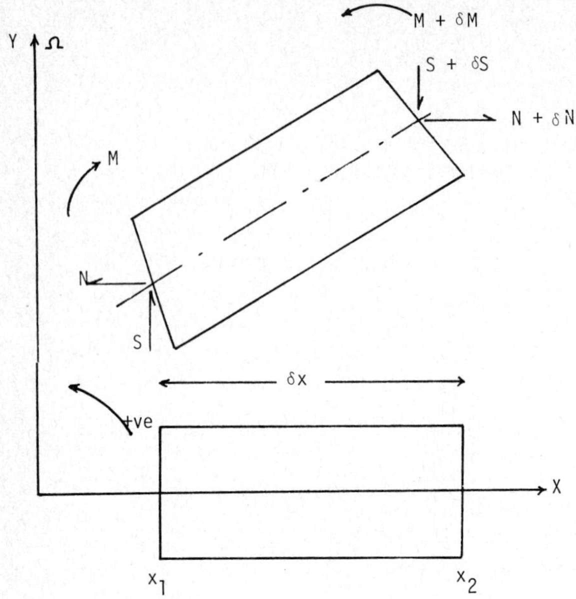

Figure 1
Deflections, Forces and Moments for an Element

Figure 1 shows an element dx; initially in the undeformed state coincident with the OX axis; in the deflected state. Allocating deflections, forces and moments according to the convention given then the general equations given by Norwood et al become:

$$N = N_R + \int_x^{x_2} \rho A \Omega^2 x \, dx \tag{1}$$

$$S = R_L + \int_{x_1}^{x} \rho A . \omega^2 y \, dx \tag{2}$$

$$M = -M_L + N_R \int_{x_1}^{x} \frac{dy}{dx} dx + \int_{x_1}^{x} \frac{dy}{dx} \int_x^{x_2} \rho A \Omega^2 x \, dx \, dx$$
$$+ R_L (x - x_1) + \int_{x_1}^{x} \int_{x_1}^{x} \rho A \omega^2 y \, dx \, dx - \int_{x_1}^{x} A k^2 \rho \frac{d^3 y_b}{dx \, dt^2} \tag{3}$$

and $$M = EI \frac{d^2 y_b}{dx^2} \tag{4}$$

$$S = -k' A G \frac{dy_3}{dx} \tag{5}$$

$$y = y_s + y_b \tag{6}$$

By introducing the non-dimensional length $x_* = x/\ell$, where ℓ is the segment length, moving the origin to the left hand end of the segment and writing the parameters which influence the beam deformation in the form $\rho A = \rho Am_0 f(Am)$ etc., where ρAm_0 is the mass per unit length at the left hand end and $f(Am)$ is a function of X_* which describes the variation of ρA along the segment, equations (1) to (3) become:

$$N = N_R + \rho Am_o \, \Omega^2 \, \ell^2 \int_{x_*}^{1} (x_{1*} + x_*) \, f(Am) \, dx_* \tag{7}$$

$$S = R_L + \rho Am_o \, \omega^2 \, \ell \int_0^{x_*} f(Am) \, dx_* \tag{8}$$

$$\frac{M}{\ell} = \frac{-M_L}{\ell} + N_R \int_0^{x_*} \ell \, \emptyset \, dx_*$$

$$+ \rho Am_o \, \Omega^2 \, \ell \int_0^{x_*} \ell\emptyset \int_{x_*}^{1} (x_{1*} + x_*) \, f(Am) \, dx_* \, dx_*$$

$$+ R_L \, x_* + \rho Am_o \, \omega^2 \, \ell \int_0^{x_*}\int_0^{x_*} y \, f(Am) \, dx_* \, dx_*$$

$$- \rho \frac{Im_o \omega^2}{\ell} \int_0^{x_*} \ell\emptyset_b \, f(Im) \, dx_* \tag{9}$$

where $\emptyset_b = \frac{dy_b}{dx}$, $\emptyset_s = \frac{dy_s}{dx}$ and $\emptyset = \frac{dy}{dx}$

By using separate notation to distinguish between "mass area" and "shear area" and moment of inertia and bending stiffness in this way, the segment may consist of different layers of materials, the sections of which vary along its length.

Using equations (4) and (9), integrating and rearranging gives:

$$\frac{EI_o}{\ell^3} (\ell \, \emptyset_b - \ell \, \emptyset_{bL}) = \frac{-M_L}{\ell} \int_0^{x_*} \frac{1}{f(I)} \, dx_*$$

$$+ \frac{N_R}{\ell} \int_0^{x_*} \frac{1}{f(I)} \int_0^{x_*} \ell \, \emptyset \, dx_* \, dx_*$$

$$+ \rho Am_o \, \Omega^2 \, \ell \int_0^{x_*} \frac{1}{f(I)} \int_0^{x_*} \ell\emptyset \int_{x_*}^{1} (x_{1*} + x_*) \, f(Am) \, dx_* \, dx_* \, dx_*$$

$$+ R_L \int_0^{x_*} \frac{x_*}{f(I)} \, dx_* + \rho Am_o \omega^2 \ell \int_0^{x_*} \frac{1}{f(I)} \int_0^{x_*}\int_0^{x_*} y f(Am) \, dx_* \, dx_* \, dx_*$$

$$- \rho \frac{Im_o{}^2}{\ell} \int_0^{x_*} \frac{1}{f(I)} \int_0^{x_*} \ell\emptyset_b \, f(Im) \, dx_* \, dx_* \tag{10}$$

and integrating again gives

$$\frac{EI_o}{\ell^3} (y_b - y_{bL} - \emptyset_{bL} \, x_*) = \frac{-M_L}{\ell} \int_0^{x_*}\int_0^{x_*} \frac{1}{f(I)} \, dx_* \, dx_*$$

$$+ \rho \, Am_o \Omega^2 \ell \int_0^{x_*}\int_0^{x_*} \frac{1}{f(I)} \int_0^{x_*} \ell\emptyset \int_{x_*}^{1} (x_{1*} + x_*) \, f(Am) \, dx_* \, dx_* \, dx_* \, dx_*$$

$$+ R_L \int_0^{x_*}\int_0^{x_*} \frac{x_*}{f(I)} \, dx_* \, dx_* + \rho Am_o \omega^2 \ell \int_0^{x_*}\int_0^{x_*} \frac{1}{f(I)} \int_0^{x_*}\int_0^{x_*} y f(Am) dx_* dx_* dx_* dx_*$$

$$- \rho \frac{Im_o \omega^2}{\ell} \int_0^{x_*}\int_0^{x_*} \frac{1}{f(I)} \int_0^{x_*} \ell\emptyset_b \, f(Im) \, dx_* dx_* dx_* \tag{11}$$

Substituting into equation (5) and integrating

$$\ell\emptyset_s = \frac{-R_L \ell}{k'A_oG}\frac{1}{f(A)} - \frac{\rho Am_o \omega^2 \ell^2}{k'A_oG}\frac{1}{f(A)}\int_0^{x_*} f(Am)\, y\, dx_*$$

and
$$y_s = \frac{-R_L \ell}{k'A_oG}\int_0^{x_*}\frac{1}{f(A)}\, dx_* - \frac{\rho Am_o \omega^2 \ell^2}{k'A_oG}\int_0^{x_*}\frac{1}{f(A)}\int_0^{x_*} f(Am) y\, dx_*\, dx_* \qquad (12)$$

Substituting equations (11) and (12) into equation (6)

$$\frac{EI_o}{\ell^3}(y - y_L - x_*\emptyset_{bL}) = \frac{-M_L}{\ell}\int_0^{x_*}\int_0^{x_*}\frac{1}{f(I)}\, dx_*\, dx_*$$
$$+ \frac{N_R}{\ell}\int_0^{x_*}\int_0^{x_*}\frac{1}{f(I)}\int_0^{x_*}\ell\emptyset\, dx_*\, dx_*\, dx_*$$
$$+ \rho Am_o \Omega^2 \ell \int_0^{x_*}\int_0^{x_*}\frac{1}{f(I)}\int_0^{x_*}\ell\emptyset\int_{x_*}^{1}(x_{1*} + x_*)\, f(Am)\, dx_*\, dx_*\, dx_*\, dx_*$$
$$+ R_L\left\{\int_0^{x_*}\int_0^{x_*}\frac{x}{f(I)}\, dx_*\, dx_* - \frac{\ell}{k'A_oG}\frac{EI_o}{\ell^3}\int_0^{x_*}\frac{1}{f(A)}\, dx_*\right\}$$
$$+ \rho Am_o \omega^2 \ell\left\{\int_0^{x_*}\int_0^{x_*}\frac{1}{f(I)}\int_0^{x_*}\int_0^{x_*} y\, f(Am)\, dx_*\, dx_*\, dx_*\, dx_*\right.$$
$$\left. - \frac{\ell}{k'A_oG}\int_0^{x_*}\frac{1}{f(A)}\int_0^{x_*} y\, f(Am)\, dx_*\, dx_*\right\}$$
$$-\rho\frac{Im_o \omega^2}{\ell}\int_0^{x_*}\int_0^{x_*}\frac{1}{f(I)}\int_0^{x_*}\ell\emptyset_b\, f(Im)\, dx_*\, dx_*\, dx_* \qquad (13)$$

By applying the dynamic discretization technique (Downs (1979)) to equations (7) to (10) and (13) the stiffness and equivalent mass matrices for the segment can be formed. For a beam of more than one segment the overall stiffness and mass matrices are formed by superposition. Frequencies and mode shapes are then obtained by the solution of the higher order eigen-value problem.

VALIDATION

For a cantilever, length ℓ, the frequencies of vibration can be expressed in dimensionless form

$$\lambda = \omega\left(\rho\frac{Am_o \ell^4}{EI_o}\right)^{\frac{1}{2}}$$

where ρ is the material density, ω the natural frequency (rad/s), A_o the area at the encastre point and EI_o the bending stiffness at the encastre point.

Similarly for a rotating beam the dimensionless frequency of rotation can be expressed as

$$\alpha = \Omega\left(\rho\frac{Am_o \ell^4}{EI_o}\right)^{\frac{1}{2}}$$

where Ω is the frequency of rotation (rad/s).

Table 1 shows the dimensionless frequencies for a non-rotating uniform cantilever using both Euler and Timoshenko theory, calculated using the dynamic discretization programme. The entries in brackets are exact analytical solutions.

Table 2 shows the variation in natural frequency with rotational speed for both Timoshenko and Euler theory for a uniform cantilever. Comparison is made for modes 1 and 2 with results given by Boye et al (1954). A comparison of the displacement profiles for a rotating and non-rotating beam is shown in figure 2 for the first four nodes, for Timoshenko Theory.

Mode	Euler	Timoshenko
1	3.516 (3.516)	3.323 (3.324)
2	22.035 (22.035)	16.289 (16.289)
3	61.697 (61.697)	36.708 (36.708)
4	120.902 (120.902)	58.288 (58.288)
5	199.860 (199.860)	80.213 (80.213)
6	284.556 (284.556)	94.452 (94.452)

Table 1
Dimensionless Frequencies for a Uniform Cantilever

Mode \ α^2	0	5	20	50
1 Euler	3.516 (3.516)	4.278 (4.167-4.278)	5.985 (5.684-5.987)	8.365 (7.897-8.375)
Timoshenko	3.324	4.024	5.525	7.427
2 Euler	22.035 (22.034-22.725)	22.757 (22.683-23.483)	24.800 (24.536-25.425)	28.447 (27.887-29.024)
Timoshenko	16.289	16.864	18.458	21.200
3 Euler	61.697	62.416	64.519	68.511
Timoshenko	36.708	37.178	38.523	40.946
4 Euler	120.902	121.644	123.843	128.108
Timoshenko	58.288	58.719	59.984	62.292

Table 2
Dimensionless Flap Frequencies for a Rotating Uniform Cantilever

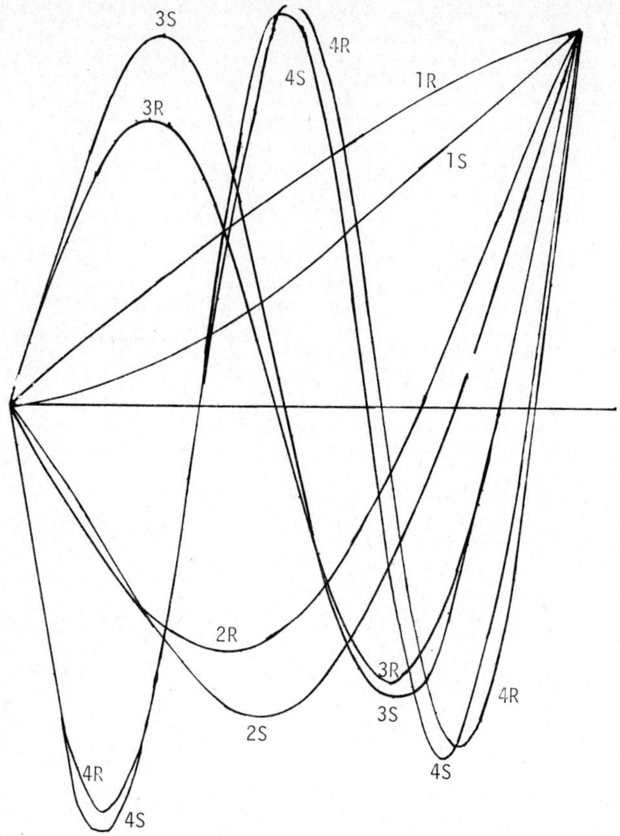

Figure 2
Comparison of Displacement Profiles of a Stationary and a Rotating Uniform Cantilever Calculated by Dynamic Discretization Using Timoshenko Theory

RESULTS FOR MODEL ROTOR

A diagram of the model helicopter rotor blade is shown in figure 3. It was subdivided into 20 segments for the discretization process. The prime consideration in the subdivision was the ease with which the area and stiffness functions along each segment could be described mathematically.

The natural frequencies of vibration for the model rotor are given in table 3.

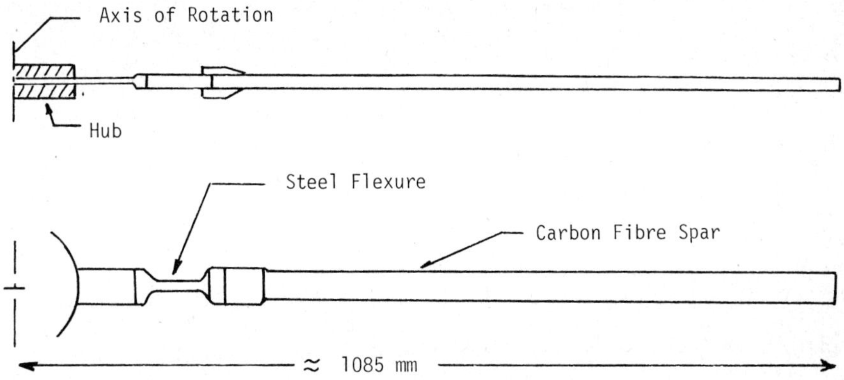

Figure 3
Diagram of Model Rotor

Mode \ Rotational Speed (Hz)	0	5	10
1st Flap	5.7 (5.8)	7.6 (7.3)	11.4 (11.0)
2nd Flap	32.9 (33.1)	34.9 (34.7)	40.5 (39.4)
3rd Flap	87.2 (87.8)	89.7	95
4th Flap	168.2 (166.1)	170.3	176.6
1st Lag	11.9 (12.0)	12.2 (12.25)	12.4 (13.1)
2nd Lag	106.2 (107.4)	106.9 (107.9)	109.1 (109.9)
3rd Lag	294	294.7	296.7
4th Lag	543.7	544.3	546.1

Table 3
Vibration Frequencies for Model Rotor Blade

The entries in brackets are experimental results supplied by RAE Farnborough. There is excellent agreement shown between the two sets of results. Even closer agreement would result from a more accurate value being used for the rotational flexibility at the encastre point (due mainly to the flexibility of the hub mounting). Although a finite stiffness, (determined by experiment), was used for the encastre point the programme model is sensitive to changes in this stiffness and hence any inaccuracy in the value will cause an error in the frequencies.

The computer time used in the analysis is quite modest with a total of approximately 500 mil units required for the discretization process and determination of the first four modes.

CONCLUSION

Dynamic discretization may be used to predict the vibration frequencies of rotating beams. It offers a viable alternative to conventional Finite Element analysis. Care, however, must be taken in the modelling of the physical system as the predicted frequencies are dependent on the end conditions assumed.

REFERENCES

[1] Downs, B., Vibration Analysis of Continuous Systems by Dynamic Discretization, J. Mech. Design, Trans ASME, 102 (1980) 391-398.

[2] Levy, R., Guyan Reduction Solutions Recycled for Improved Accuracy, NASTRAN Users Experiences, NASA (1971), 201-220.

[3] Norwood, C.J., Cubitt, N.J. and Downs, B., A Note on the Equations of Motion for the Transverse Vibration of a Timoshenko Beam Subjected to an Axial Force, Jrnl. of Sound and Vibration, 70 (1980), 475-479.

[4] Boyce, W.E., Di Prima, R.C. and Handelman, G.H., Vibrations of Rotating Beams of Constant Section, Proc. 2nd U.S. Nat. Congress of App. Mech., Michigan (1954).

Computational Techniques & Applications: CTAC-83
J. Noye & C. Fletcher (Editors)

THE TIME SETTLEMENT BEHAVIOUR OF A RIGID RAFT SUBJECT TO MOMENT

J.C. Small, BSc(Eng),PhD.

School of Civil and Mining Engineering
The University of Sydney
Sydney, Australia

Solutions are found to the time-settlement behaviour of a rigid circular raft foundation which is subject to an applied moment. The solution method involves approximating the contact stress (which exists between the raft and the soil on which it is resting) by a series. The unknown coefficients of the series are determined by invoking displacement compatibility between raft and soil and moment equilibrium. Once they have been determined the contact stress and hence the complete solution may be obtained.

INTRODUCTION

The problem of the consolidation of soil layers is of great interest to foundation engineers. When a structure is placed on a layer of clay the loading applied to the foundation causes an increase in pressure in the water which exists in the pores of the soil. Flow of water then occurs from areas of high water pressure to areas of low water pressure, and as this occurs the applied load is transferred to the soil fabric or skeleton, resulting in consolidation of the soil. The resulting settlements may be large enough to cause damage to the structure and it is therefore desirable to be able to predict what these deflections may be.

Most of the analytic solutions to the problem of settlement of foundations resting on a layer of saturated clay have been obtained for an infinitely flexible foundation. (McNamee and Gibson (1960), Gibson, Schiffman and Pu (1970), Gibson and McNamee (1963), Booker (1974)). Exceptions are Chiarella and Booker (1975) who considered the settlement of a rigid circular disc subject to symmetric loading, and Booker and Small (1983) who have considered the settlement of a circular raft of finite flexibility acted on by a uniform normal stress. The problem of non symmetric loading of foundations seems to have received very little attention.

In this paper the settlement of a rigid circular disc subjected to pure moment is considered. This result, together with the solution for the settlement of a circular disc subject to a central load (Chiarella and Booker (1975) or Booker and Small (1983)) may then be used to examine the settlement behaviour of a rigid circular disc subject to general loading conditions.

THEORY

It is well known (Gibson and McNamee (1963)) that the solution for the vertical settlement w of the surface $z = 0$ of a deep layer of clay which is idealized as a porous elastic medium may be written as

$$\bar{w} = \int_{-\infty}^{+\infty} \int_{-\infty}^{+\infty} Q(\alpha,\beta)\ F(\alpha,\beta,s)\ e^{i(\alpha x+\beta y)}\ d\alpha\ d\beta \tag{1a}$$

where

$$Q(\alpha,\beta) = \frac{1}{4\pi^2} \int_{-\infty}^{+\infty} \int_{-\infty}^{+\infty} q(x,y)\ e^{-i(\alpha x+\beta y)}\ dxdy \tag{1b}$$

is the Fourier transform of the general surface loading $q(x,y)$. Equation 1a has been simplified by the application of a Laplace transform where the superior bar denotes that a transform has been applied, e.g.

$$\bar{f} = \int_0^{\infty} f(t)\ e^{-st} dt$$

$F(\alpha,\beta,s)$ is a function which depends upon the Laplace transform parameter s and is therefore the time dependent portion of the solution. The expression for the function is given in Appendix I for the particular case of a deep clay layer with a permeable, shear free upper surface.

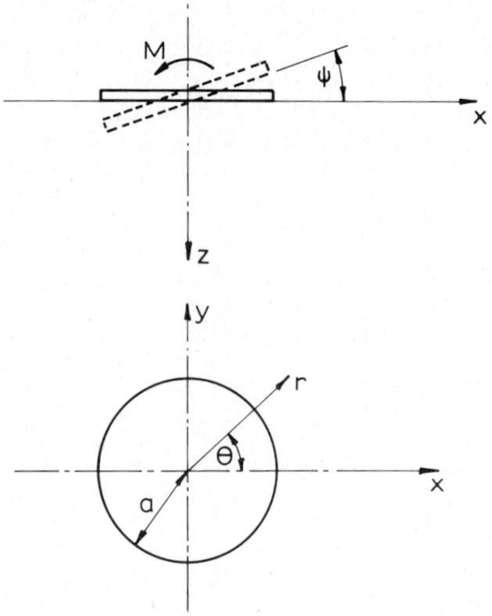

Figure 1
Circular Raft on Deep Clay Layer

If we now consider the problem of a moment M which is applied to a circular raft of radius a (see Fig. 1) we may notice that the deflection $w(r,\theta)$ along a radius at angle θ is given by

$$w(r,\theta) = w(r,0)\cos\theta \tag{2}$$

where $w(r,0)$ is the vertical deflection at $\theta = 0$ (the direction perpendicular to the axis of rotation)

Recomposition of the solution into Fourier components reveals that under these circumstances, the contact stress may be written in the form

$$q_c(r,\theta) = q_c(r,0)\cos\theta \tag{3}$$

If we now make the substitution into Eq.(1b)

$$\begin{aligned} x &= r\cos\theta \\ y &= r\sin\theta \\ \alpha &= \rho\cos\varepsilon \\ \beta &= \rho\sin\varepsilon \\ q(x,y) &= q_c(r,0)\cos\theta \end{aligned}$$

then

$$Q(\alpha,\beta) = \frac{1}{4\pi^2} \int_0^{\infty} \int_0^{2\pi} q_c(r,0)\cos\theta \; e^{-i\rho r\cos(\theta-\varepsilon)} \; r\,dr\,d\theta \tag{4}$$

Making use of the well known integral representation for Bessel functions

$$J_1(z) = \frac{1}{i\pi} \int_0^{\pi} e^{iz\cos\theta} \cos\theta \, d\theta \tag{5}$$

we find that

$$Q(\alpha,\beta) = \frac{-i}{2\pi} \int_0^{\infty} q\,(r,0)\, J_1(\rho r)\, \cos\varepsilon \; r\,dr \tag{6}$$

where $\rho = \sqrt{\alpha^2 + \beta^2}$

As the form of the contact stress $q_c(r,0)$ is not known, we can approximate it with a series i.e.

$$q_c(r,0) = \sum_{n=0}^{N} c_n \phi_n \tag{7}$$

where $\phi_n = r(1-r^2/a^2)^n \qquad n=0,1,2,\ldots N-1$

$\phi_N = r(1-r^2/a^2)^{-\frac{1}{2}}$

and where c_n are the unknown coefficients in the series.

The terms of the above series are carefully chosen, as it is known that the solution for the contact stress under a rigid circular plate subject to a moment is proportional to $r(1-r^2/a^2)^{-\frac{1}{2}}$ if the plate is on an elastic half space. Other terms are added to the series in order to approximate the changes in contact stress which occur as consolidation of the soil proceeds.

Substituting the approximation of Eq. 7 into Eq. 4 leads to

$$Q(\alpha,\beta) = \frac{-i\cos\varepsilon}{2} \sum_{n=0}^{N} c_n \Phi_n \tag{8a}$$

$$\text{where} \quad \Phi_n = \int_0^\infty r\, \phi_n\, J_1(\rho r)\, dr \tag{8b}$$

The integral occurring in Eq. 8b may be evaluated analytically and the result is given in Appendix II.

We may now evaluate the vertical deflection $\bar{w}$ by substitution of the value of $Q(\alpha,\beta)$ into Eq. 1a.

$$\bar{w} = \frac{-i}{2\pi} \int_0^\infty \int_0^{2\pi} \sum_{n=0}^{N} c_n\, \Phi_n\, F(\alpha,\beta,s) \cos\varepsilon\, e^{i\rho r\cos(\theta-\varepsilon)}\, \rho\, d\rho\, d\varepsilon \tag{9}$$

Again using the integral representation of the Bessel function of Eq. 5 we find

$$\bar{w} = -\int_0^\infty \rho\, F(\alpha,\beta,s) \sum_{n=0}^{N} c_n\, \Phi_n\, J_1(\rho r) \cos\theta\, d\rho \tag{10}$$

The above result confirms that the vertical deflection w is proportional to cos θ and is therefore consistent with the assumption of Eq. 2.

It is now convenient to define the generalised deflection δ where

$$\bar{\delta}_m = \int_0^a r\phi_m\, \bar{w}\, dr \tag{11}$$

Hence

$$\bar{\delta}_m = \sum_{n=0}^{N} \bar{c}_n \int_0^a F(\alpha,\beta,s)\, \Phi_n\, \Phi_m \cos\theta\, d\rho \tag{12}$$

If the raft is to be rigid then we have the condition that

$$w(r,\theta) = \psi r \cos\theta \tag{13}$$

where ψ is the rotation of the raft.

If $\bar{\delta}_m$ is now computed using the definition of w given in Eq. 13 we may equate the resulting value with that of Eq. 12. This must follow if the raft is to behave like a rigid disc. This leads to the following set of equations:

$$\Lambda\bar{c} - \alpha\,\bar{\psi} = 0 \tag{14}$$

$$\text{where} \quad \Lambda_{mn} = \int_0^a \rho F(\alpha,\beta,s)\, \Phi_n\, \Phi_m\, d\rho$$

$$\alpha_m = \int_0^a r^2\, \phi_m\, dr$$

As the integral defining Λ is quite complex it is best evaluated using numerical integration such as Gaussian quadrature. Moment equilibrium must also be satisfied for the rigid raft and so

$$\int_0^a \int_0^{2\pi} q(r,0) \cos\theta \,.\, r\cos\theta \,.\, r\,dr\,d\theta = M \tag{15}$$

Making the substitution of Eq. 7 for $q_c(r,0)$ the above equation becomes

$$\sum_{n=0}^{N} \bar{c}_n \, \alpha_n = \bar{M}/\pi \tag{16}$$

Combining equations (14, 16) leads to the final set of equations to be solved

$$\begin{bmatrix} \Lambda & -\underset{\sim}{\alpha} \\ -\underset{\sim}{\alpha}^T & 0 \end{bmatrix} \begin{bmatrix} \underset{\sim}{\bar{c}} \\ \bar{\psi} \end{bmatrix} = \begin{bmatrix} \underset{\sim}{0} \\ -\bar{M}/\pi \end{bmatrix} \tag{17}$$

Values for the coefficients $\bar{c}_n$, and the rotation of the footing $\bar{\psi}$ need to have an inverse Laplace transform applied in order to find the solution at any time t. This may be done using numerical means such as Talbot's algorithm (Talbot 1979). Once the c_n values are known the contact stress may be evaluated by summing the series of Eq. 7.

EXAMPLES

As an illustration of the foregoing theory the problem of a rigid circular raft foundation of radius a, subject to a moment M (which is applied at time $t = 0^+$ and thereafter held constant) was analysed. The raft rests upon a deep clay layer and it is assumed that the surface of the layer z=0 is shear free and free to drain. It is also assumed that the Poisson's ratio of the soil is $\nu_s = 0$.

Values of the angle of rotation of the raft versus time are presented in the non-dimensional plot of Fig. 2. The time factor used in the plot is defined as

$$\tau = \frac{ct}{a^2} \tag{18}$$

and c is defined in Appendix I. Rotation of the raft can be seen to increase with time from an initial value to a final value, and it is possible to provide a check on these two limiting values by calculating what they should be using elastic theory.

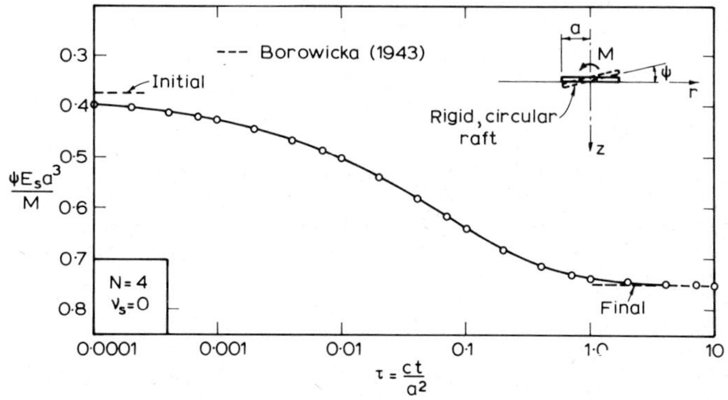

Figure 2
Rotation of Rigid Circular Raft Versus Time. Deep Clay Layer.

Initially, since the moment is applied rapidly to the raft, the pore water does not have time to flow out of the soil and so the soil behaves as if incompressible (i.e. the Poisson's ratio of the soil is effectively $\nu_u = \frac{1}{2}$ and the modulus of the soil $E = 2(1+\nu_u)G$). Finally, after all pore pressures have dissipated, the soil behaves like an elastic material with Poisson's ratio ν_s' and modulus E_s. (G is the shear modulus of the soil = $E_s/(1+\nu_s)2$).

By using the appropriate values of E, solutions for initial and final rotations were evaluated using results presented by Borowicka (1943) and these are shown on Fig. 2. The calculated time-rotation curve may be seen to lie between the calculated values.

It is also possible to compare initial and final contact stress $q_c(r,0)$ distributions with the elastic solutions of Borowicka (1943). These comparisons are made in Fig. 3, and as the stress distributions do not depend on the value of Poisson's ratio the contact stresses should be the same initially and finally. It may be seen from the figure that this is so, with the stresses calculated for a very small time $\tau = 0.00001$ and a large time $\tau = 1000$ showing close agreement with Borowicka's results. At intermediate times (e.g. $\tau = 0.01$) the contact stress tends to increase slightly over the central portion of the raft.

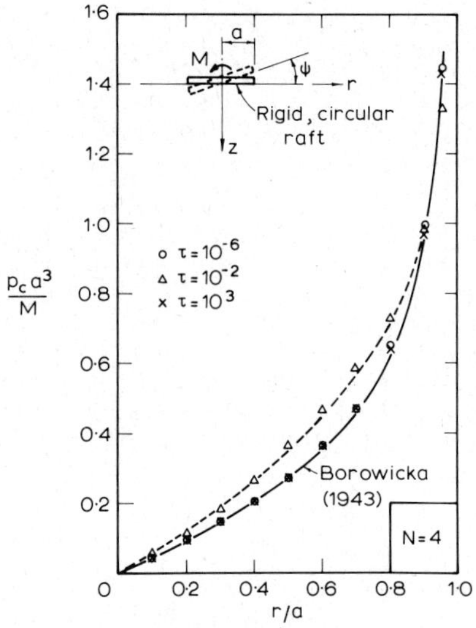

Figure 3
Contact Stress Beneath Rigid Circular Raft Subjected to a Moment

It is of interest to note that very few terms are needed to obtain reasonably accurate results. Only four terms in the series of Eq. 7 were used in order to obtain the results of Figs. 2 and 3.

The effect of Poisson's ratio on the rate of rotation may be seen in the plot of Fig. 4 where the degree of rotation U is plotted for various values of Poisson's ratio ν_s against the time factor τ where U is defined as:

$$U = \frac{\psi(t) - \psi(0)}{\psi(\infty) - \psi(0)} \tag{19}$$

and ψ(t) is the rotation at any time t.

As may be seen from Fig. 4, the effect of an increase in Poisson's ratio is to cause a larger degree of consolidation at any given time factor τ.

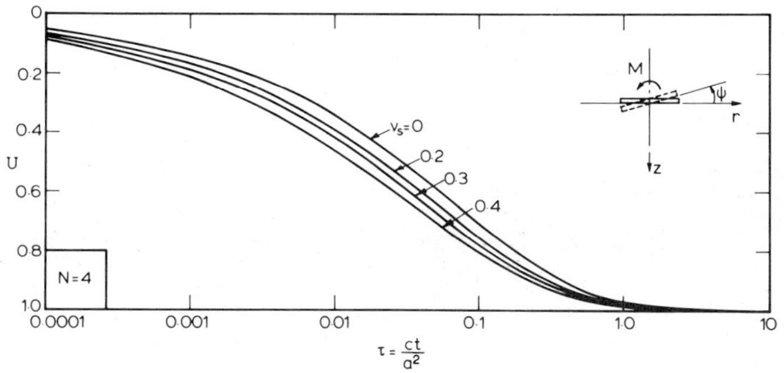

Figure 4
Degree of Rotation Versus Time for Various Values of Poisson's Ratio

In order to enforce the condition of a rigid raft, the generalised deflections δ_m (see Eq. 11) of the soil were made equal to those which would result from a rigid rotation. This eliminates the need for matching deflections at various points across the base of the raft.

A check can be made on the accuracy of the method by comparing the actual deflections as obtained from Eq. 10 with those calculated by Eq. 13. This was done and the results are presented in Fig. 5. In this figure, the full lines show the deflection of the raft w(r) (at various times τ) as calculated from the rotation (i.e. using Eq. 13). The points plotted on the figure show the values of deflections calculated at various radii using Eq. 10. It can be seen that very close agreement results, indicating that by matching generalised deflections, rigid body behaviour is accurately modelled.

CONCLUSIONS

A method has been presented which enables the time-settlement behaviour of a rigid raft subjected to a pure moment to be computed. Numerical methods are used to evaluate the more complex integrals and to invert the Laplace transformations, and it is shown by way of example, that fairly accurate solutions result.

One aspect of the solution method is that the unknown contact stress between the raft and the soil is approximated by a series. Results presented show that very few terms (approximately four) are required to obtain a reasonably accurate solution to the problem, and this makes the method extremely efficient.

The solutions obtained in this paper may be combined with solutions for a rigid raft subjected to vertical loading to obtain the solution for the more general load case which often exists in practice.

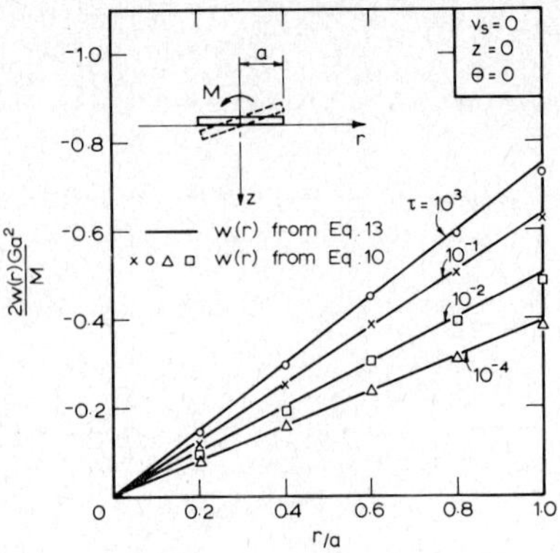

Figure 5
Comparison of Calculated Deflections

REFERENCES

[1] Booker, J.R., The Consolidation of a Finite Layer Subject to a Surface Loading, Int. Jnl. Solids Structs., 10, 1974, 1053-1065.

[2] Borowicka, H., Uber ausmittig belaste starre Platten auf elastich-isotropem Untergrund, Ingenieur-Archiv., Berlin, 1, 1943, 1-8.

[3] Chiarella, C. and Booker, J.R., The Time-Settlement Behaviour of a Rigid Die Resting on a Deep Clay Layer, Q. Jnl. Mech. Appl. Math., Vol.XXVIII, Pt.3, 1975, 317-328.

[4] Gibson, R.E. and McNamee, J., A Three-Dimensional Problem of the Consolidation of a Semi Infinite Clay Stratum, Q. Jnl. Mech. Appl. Math., Vol.XVI, Pt.2, 1963, 115-127.

[5] Gibson, R.E., Schiffman, R.L. and Pu, S.L., Plane Strain and Axially Symmetric Consolidation of a Clay Layer on a Smooth Impervious Base, Q. Jnl. Mech. Appl. Math., XXIII, Pt.2, 1970, 210-227.

[6] McNamee, J. and Gibson, R.E., Plane Strain and Axially Symmetric Problems of the Consolidation of a Semi Infinite Clay Stratum, Q. Jnl. Mech. Appl. Math., XIII, Pt.2, 1960, 210-227.

[7] Talbot, A., The Accurate Numerical Inversion of Laplace Transforms, Jnl. Inst. Math. Applics., 23, 1979, 97-120.

APPENDIX I

$$F(\alpha,\beta,s) = F(\rho,s) = -\frac{(\xi+\rho)}{\xi + \frac{\rho\nu_s}{(1-\nu_s)}} \cdot \frac{1}{\rho}$$

where ν_s is the Poisson's ratio of the soil

$$\xi = \sqrt{\rho^2 + s/c}$$

c is the coefficient of consolidation of the soil

$$c = \frac{k}{\gamma_w} \frac{E_s(1-\nu_s)}{(1+\nu_s)(1-2\nu_s)}$$

k/γ_w is the permeability of the soil per unit wt. of water

E_s is the Young's modulus of the soil.

The above expression for $F(\rho,s)$ only applies to a deep clay layer with a permeable, shear free surface at $z = 0$.

APPENDIX II

$$\Phi_n = \int_0^\infty r\,\phi_n\,J_1(\rho r)\,dr$$

$$= \int_0^a r\,(1-r^2/a^2)^n\,J_1(\rho r)\,dr$$

$$\Phi_n = (2a/\rho)^n \frac{\Gamma(n+1)}{\rho} J_{n+2}(\rho a)$$

Computational Techniques & Applications: CTAC-83
J. Noye & C. Fletcher (Editors)

A RELIABILITY-BASED OPTIMUM DESIGN COMPUTATIONAL TECHNIQUE FOR PLASTIC STRUCTURES

Dan M. Frangopol

Department of Civil Engineering
University of Colorado at Boulder
Boulder, Colorado
U.S.A.

The purpose of this paper is to present a reliability-based computational technique to design optimal plastic structures. In this computation, the correlation between loads and the correlation between plastic moments if accounted for by using a technique which incorporates the effect of the statistical dependence between any two collapse mechanisms. Based on this technique, a computer program was developed which automatically designs optimal plastic structures with up to two hundred collapse mechanisms. Some of the investigation in this paper is concerned also with the sensitivity of the plastic optimal solutions to load and to resistance correlations.

INTRODUCTION

The aim of structural design is to achieve a structural system in an optimal fashion in accordance with a set of predefined needs and safety requirements. Based on the viewpoint that safety only has meaning in a probabilistic sense, the rational approach for sizing members to achieve an optimum solution is the reliability-based design. During the past twenty years considerable effort has been done to show that optimization techniques have potential for practical applications in reliability-based structural design. In the reliability-based approach, the loads and resistances are treated as random variables, and failure is defined as the event that the load effect exceeds the structural resistance. The risk measure that corresponds to this event is the probability of failure which is obtained from a systematic analysis of the uncertainties in all variables. These uncertainties are associated with the inherent randomness as well as modelling and prediction errors included in the formulation.

Structural systems need a large number of variables to describe the complete structure and the load conditions (Moses (1983)). Under any given load condition these systems may undergo various degree of damage and present, generally, numerous possible failure modes. One of the most difficult questions in structural system reliability is how to deal with the statistical dependence between mode failure events in both the analysis and optimum design formulations which consider all the load and resistance uncertainties. Such dependence may arise due to common material and/or loading sources and/or similar fabrication, inspection and control methods. Any two failure modes are independent if they do not have any common random plastic moments and loads, and if all the basic variables implicated in the two individual failure mode equations are mutually independent. In general, the failure modes of a structural system are inter-dependent and the assumption of independence between them may be very innapropriate.

This paper presents a reliability-based computational technique to design optimal plastic structures. In this computation, the evaluation of the system failure probability P_f is based on upper and lower bounds (reliability P_s is defined herein as $1-P_f$, where P_f is the probability of failure). These bounds on system failure probability incorporate the effect of the statistical dependence among any pair of failure modes through the coefficient of correlation between their linear safety margins (i.e., collapse mode expressions). Using Ditleven's method (1982) of conditional bounding and Vanmarcke's concept (1971) of failure mode decomposition, very close upper and lower bounds on the true probability of system failure are obtained. For each bound, the method only requires a numerical calculation of an integral with respect to a single parameter.

Optimizing plastic structures in a reliability context raises questions as to the meaning of the optimum. Several alternatives for reliability-based optimization have been suggested (Moses (1973), Frangopol (1976), Moses (1977), Frangopol and Rondal (1977), Parimi and Cohn (1978), Frangopol (1983)). These include: (a) minimization of the total expected cost of the structure (including the initial cost and the penalty associated with failure); (b) minimization of the total structural weight (or cost) for a specified allowable failure probability; and (c) minimization of the total probability of failure for a fixed weight (or cost). In the present paper, the approach adopted is to minimize the total structural weight subject to an allowable probability of failure of the entire structure. Minimization of weight for a specified reliability ensures an optimum distribution of safety amongst different (inter-dependent or independent) failure modes. The optimization problem is a nonlinear constrained minimization. A computer program was developed for solving the problem of distributing the overall preassigned probability of failure among collapse modes such that the total structural weight is a minimum. The basic idea in this program, which automatically designs optimal plastic frames with up to two hundred collapse mechanisms, is to use upper and lower bounds on optimum weight to develop iterative search strategies. The main feature is that at each stage of the search procedure better upper and lower bounds on the optimum weight function are established. One may, at any step, decide to stop the procedure if the bounds are sufficiently close. The optimization technique used in this program falls into the class of direct methods of nonlinear programming and can be considered a form of the feasible directions method, as the objective function is improved in each step from one feasible design to a better feasible design.

In all reliability-based optimization problems it is necessary to subject the results to sensitivity analysis in order to determine the influence of the input statistical parameters (including distribution functions, coefficients of correlation and coefficients of variation) on optimum solutions. Only this type of sensitivity studies will provide a means for evaluating the results obtained by optimization in order to achieve a practical optimum solution (Moses (1970), Frangopol (1976), Parimi and Cohn (1978)). Obviously, the practical optimum structure is not necessarily the lightest one, since in order to achieve it one may have to use standard available shape sizes (Frangopol (1977)), or simply because of other engineering reasons (e.g., the dominant mode of failure for the practical optimum solution is less dangerous for the loss of human life than that of the lightest one). In this final stage of the optimization process, where engineering judgement and experience are required, the interaction designer-computer should function in a complementary manner in finding the practical optimum solution (Somekh and Kirsch (1981), Balling, Pister and Pollak (1983), Frangopol (1984)). The computer should perform those aspects of the optimization process which are repetitious, taking into account the modification of input according to the changed set of design variables generated from sensitivity analysis, while the structural engineer should perform those aspects which require judgement and experience. So, until the stage arrives when the designer will be able to put his judgement and experience into a computer language, the dynamic interaction with the computer is useful to execute those decisions which cannot be automated. However, considerable time and effort is spent in generating the data for sensitivity analysis and in interpreting the results. A description of the central computing facility at the University of Colorado is given in Reference [9]. The graphical interaction with the computer provides the designer with the means to define potential reliability-based optimum solutions, perform and evaluation of these solutions by sensitivity analysis, choose the solution which will produce the best performance and display the results. The ease and speed with which this process can be performed by graphic illustration of the computational results makes such an interaction particularly attractive for the evaluation of alternative solutions.

The proposed computational technique is illustrated on a portal frame example with random loads and plastic moments. The numerical results obtained and the accompanying sensitivity studies show that this technique provides a powerful tool to obtain a practical reliability-based optimum solution for plastic structures.

ASSUMPTIONS

The basic assumptions adopted herein for the structural reliability optimization of plastic structures are: (a) the members are prismatic and straight; (b) shear deformation is neglected; (c) the load and plastic hinge positions are deterministic; (d) the loads and plastic moments are assumed to be statistical independent random variables; (e) the statistical dependence among loads and the statistical dependence among plastic moments is accounted for through the coefficients of correlation between loads and between plastic moments; (f) the necessary statistical information is assumed to be available; (g) the failure of a structure by collapse is defined as in the simple bending plastic theory (an indeterminate plastic structure fails by collapse when enough plastic hinges make possible the occurrence of a plastic mechanism); (h) local fracture, instability and other possible causes of failure are avoided; (i) the topology and geometry of the structure are given in a deterministic (non-statistical) fashion; and (j) the design variables are the cross-sectional dimensions of the structural members.

BOUNDING PROBABILITY OF PLASTIC COLLAPSE FAILURE

In general, structural system topology may be classified as series (or weakest-link) in which any component failure causes system collapse, and parallel (or fail-safe) in which alternate load paths exist (Moses (1974)). In plastic collapse analysis both series and parallel systems are present.

For a structure with a number n of possible modes of failure, the possibility that failure may occur by any of these modes must be considered. The probability of system failure, P_f, can be expressed in terms of the mode failure events F_i, i=1,2,...,n, as follows:

$$P_f = P[F] = P[\bigcup_{i=1}^{n} F_i] \tag{1}$$

where the symbol U signifies the union of the mode failure events (e.g., $P[F_\ell \cup F_m]$ means the probability that either event F_ℓ or F_m will occur). Consequently, the operation of combining mode failure events is governed by the series system concept because, according to equation (1), the failure of the entire structure by collapse is defined as the occurrence of any individual failure mode. This is analogous to a weakest-link chain which fails if any link reaches its capacity.

Let $P[F_i]$ be the probability of occurrence of the i-th individual collapse mode, and the safety margin

$$Z_i = U_i - E_i = \sum_{j=1}^{s} \theta_{ji}M_j - \sum_{\ell=1}^{t} \delta_{\ell i}Q_\ell \tag{2}$$

the difference between the internal and external work associated with the i-th mechanism. In equation (2) M_j is the random plastic moment capacity at the j critical section, θ_{ji} is the corresponding relative rotation, Q_ℓ is the ℓ-th random concentrated load causing the occurrence of the i-th collapse mode and $\delta_{\ell i}$ is the corresponding displacement. In equation (2) the resistance term is given by the internal work U_i and the load term is given by the external work E_i. Both U_i and E_i depend on the sum of several random resistance and load variables, respectively. Due to parallel action of resistances (the s plastic moments in the i-th collapse mode) an individual collapse mode event

$$F_i = (Z_i < 0) \tag{3}$$

occurs only after all the s critical sections reach their maximum (plastic) moment capacity. Thus, an individual collapse mode event is governed by the parallel system concept. This parallel action permits the redistribution of moments from an overstressed section (plastic hinge) to remaining critical sections, following the first plastic hinge occurrence (no such redistribution is possible in statically determinate structures because the occurrence of only one plastic hinge will result in formation of a mechanism with consequent collapse).

Therefore, it is emphasized that the reliability analysis of a plastic structure is complicated by the fact that it combines both topologies: parallel, for each individual collapse event F_i, and series, for the overall collapse event $F = \bigcup_{i=1}^{n} F_i$. The numerical difficulties in pursuing an exact evaluation of the probability of failure of such a structure are complex because of the correlation between various failure modes caused by common loads or plastic moments, and inherent statistical dependence among the loading characteristics and among the strength characteristics.

The exact evaluation of the overall probability (1) can be based on the following general result (Moses and Kinser (1967)):

$$P_f = P[F_1] + \sum_{i=2}^{n} P[F_i]P[S_1 \cap S_2 \cap \ldots \cap S_{i-1}|F_i] \tag{4}$$

in which S_i is the event of survival of the failure mode i, $S_i = (Z_i > 0)$. The probability of survival of the mode i, is the complement of the probability of failure via this mode with respect to unity, i.e. $P[S_i] = 1 - P[F_i]$. In general the numerical effort for the evaluation of the conditional probabilities in equation (4) is exorbitant even for small systems. These difficulties can be overcome by limiting the scope of analysis to bounds on P_f.

Commonly used approximations of overall failure probability (Cornell (1967)) are based either on the assumption of the perfect statistical dependence of the mode failure events (lower bound), or on that of their statistical independence (upper bound):

$$\max_{i=1}^{n} P[F_i] \le P_f \le 1 - \prod_{i=1}^{n} P[S_i] \tag{5}$$

Another upper bound on P_f can be obtained from equation (4) by observing that the following sum of probabilities exceeds P_f:

$$P[F_1] + P[F_2]P[S_1|F_2] + P[F_3]P[(S_1 \cap S_2)|F_3] + \ldots \geq P_f \qquad (6)$$

This upper bound gives a value of P_f between the values obtained by the assumption of independent failure modes and the true probability of failure. The left side of the preceding inequality has many attractive features. Its value depends on the ordering of failure modes. Consequently, this bound can be optimized, in order to obtain the closest upper bound, according to the ordering rule given by Moses and Kinser (1967) and Vanmarcke (1971).

The upper and lower bounds in the inequalities (5) may be widely different because the correlation between failure modes is not included in the formulation. Recently, there have been considerable improvements in bounding by conditioning the system failure probability with inter-dependent failure modes (Ditlevsen (1982)). These narrow bounds, which incorporate the effect of the statistical dependence between any two failure modes, are as follows:

$$P[F_1] + \sum_{i=2}^{n} \max\{0; P[F_i] - \sum_{j=1}^{i-1} P[F_i \cap F_j]\} \leq P_f \qquad (7)$$

$$\sum_{i=1}^{n} P[F_i] - \sum_{i=2}^{n} \max_{j<i} P[F_i \cap F_j] \geq P_f \qquad (8)$$

Note that these bounds only involve computation of the probability of occurrence of individual modes, $P[F_i]$, and of the probability of occurrence of pairs of modes, $P[F_i \cap F_j]$. The bounds (7) and (8) almost coincide for all of these correlation coefficients between failure modes not larger than about 0.6 provided the total failure probability P_f is smaller than about 0.001. Without restriction on the magnitude of the failure mode correlation coefficients, very narrow bounds can be obtained on basis on inequalities (7) and (8), using Ditlevsen's method (1982) of conditional bounding. For problems with many failure modes, this method is very efficient if the Vanmarcke's concept (1971) of failure mode decomposition is also considered. The main feature of this concept is that at each stage of the design process a subset of failure modes (basic modes) is isolated and a relatively simple auxiliary problem is solved to determine upper and lower bounds on the true optimum value of the objective function. These bounds suggest also that the original complex system with n failure paths (failure modes) can be approximated for reliability analysis by a much simpler system involving fewer failure paths.

RELIABILITY-BASED OPTIMIZATION TECHNIQUE

The system reliability bounds developed above can be incorporated in an optimization procedure. The reliability-based optimization problem in the automated design of plastic structures consists in the optimization of sizes of members for a specified overall failure probability of the structure. This problem can be stated as follows (Moses (1973), Frangopol (1976), Parimi and Cohn (1978)):

minimize $\quad W = W(\bar{M}) \qquad (9)$

subject to $\quad P_f = P_f(\bar{M}) \leq P_f^* \qquad (10)$

The design is defined by the vector

$$\bar{M} = \{\bar{M}_1, \bar{M}_2, \ldots, \bar{M}_n\}^T \qquad (11)$$

of the mean values of the plastic moments capacities (member strengths) at all critical sections, the objective function W is the total structural weight (or cost), as a function of $\bar{M}$, $P_f = P_f(\bar{M})$ is the probability of system failure and P_f^* is the specified allowable failure probability of the structure.

An alternative approach minimizes the probability of collapse for a fixed weight (or cost). However, the approach generally adopted in the design process is to minimize the weight (or cost) for a specified probability of failure of the structure.

In mathematical terminology, the optimization problem given by equations (9) and (10) is a constrained minimization. The objective function $W(\bar{M})$ is a linear (deterministic) function of the mean values of plastic moment capacities $\bar{M}$ and the reliability constraint (10) is nonlinear. A good minimization procedure is needed because no explicit function for the overall failure probability P_f exists without evaluating numerical integrations.

In developing the computer optimization program care has been taken to make the input as simple as possible and to keep the storage requirements at a minimum. The engineering modeling, which means identification, description and enumeration of the n various failure modes of a structure is generated by the computer (a failure mode is defined as the minimum set of platic hinges to form a collapse mechanism in the structure). Probabilistic calculations to determine correlation coefficients between any two failure modes $\rho(Z_i,Z_j)$ and individual mode failure probabilities $P(F_i)$, $i=1,2,...n$, are then automatically computed for the following probability distributions: normal, lognormal, gamma, extreme value Type I (Gumbel), Type II (Frechet) and Type III (Weibull). Additional distribution functions may be added if desired.

Using Ditlevsen's method (1982) of conditional bounding and Vanmarcke's concept (1971) of failure mode decomposition, which incorporate the effect of statistical dependence between any two failure modes, upper and lower bounds of the true probability of overall plastic collapse P_f and of the objective function W can thus be generated relatively inexpensively. The difference between lower and upper bounds is successively reduced to obtain optimum solution. The main feature of the computational technique used is that it does not rely on one-dimensional search to compute a step size at any iteration. The principles of this technique and methods for step size calculation are similar to those proposed by Belsare and Arora (1983).

Using as input an acceptable design and interaction, the computer is first used to determine the optimization algorithm which will be the most appropriate in function of the size and complexity of the problem. Three optimization algorithms are available: the penalty functions algorithm, the feasible directions algorithm and the proposed algorithm.

For plastic structures with up to about 3 design variables and about 30 failure modes the penalty functions algorithm is very efficient. The feasible directions algorithm is generally recommended for more complex problems. In determining the direction vector $\{D\}$ which changes the initial design in the minimum weight direction until the reliability constraint (10) is met, both conditions of feasibility (a move in that direction does not cause constraint violation) and usability (a move in that direction results in a reduction of the objective function W) must be satisfied (Kirsch (1981)). Then, a move is made in a direction which continues to reduce the objective function without leaving the feasible domain. A graphical interaction program is used to observe the convergence of the step-by-step optimization technique of the feasible directions method with the purpose of choosing the steepest descent direction and eliminating unnecessary cycles near the optimum.

The proposed algorithm can be considered a form of the feasible direction method, as it is characterized by the property that when a design is unsafe (infeasible) successive designs are monotonically decreasing in the maximum amount of safety constraint violation, and when a design is feasible successive designs are also feasible and monotonically decreasing in objective weight function. Therefore, this algorithm is desirable for interactive computer aided design because the user is guaranteed a better design with each iteration. The basic idea is to first compute bounds for the optimum weight structure. The design space between these bounds is systematically searched to reach the optimum solution. There are three basic steps in the algorithm similar to those applied by Belsare and Arora (1983): (a) a weight function reduction step; (b) a constraint correction step which is executed only when the safety constraint (10) is violated; and (c) a constraint weight step which is executed from one unsafe design that is not too far from the feasible region.

With the proposed procedure it is possible to plot on the screen the objective function against the number of iterations. The procedure is continued until the optimum is reached or when the movement in the design domain produces no change in the value of the objective function. Since this optimum point may be a local minimum of the objective function, the design procedure should be started from a number of different initial points. A graphical structure display program can be used to interactively build the starting structural beam or frame and to modify this starting structure at any later stage in the optimization pocess. A high degree of confidence in the optimum solution is achieved when the same design point is determined from several starting points.

A perturbation procedure is also available for call from optimization algorithm at steps for which sensitivity analysis is required. This procedure perturbs each variable of interest and re-evaluates the optimum solution. The sensitivity of the theoretical optimum structure to: (a) the type of load distribution; (b) the type of plastic moment distribution; (c) the coefficient of variation of load; (d) the coefficient of variation of plastic moment; (e) the coefficient of correlation between loads; (f) the coefficient of correlation between plastic moment capacities; and (g) the specified allowable failure probability, is interactively examined. In this final step of the optimization process a graphical interaction program can be used for comparing alternative izosafety and/or izoweight optimized structures in order to obtain a practical optimum design. Within this decision context, the use of interactive computer graphics significantly alleviates the problems associated with input data generation, and output result interpretation.

DESIGN APPLICATION

Limited numerical experience has been gained with the computational optimization technique proposed in this paper. It involves reliability-based optimum design of plastic beams and frames with up to two-hundred inter-dependent failure modes. The minimum weight reliability-based design of the rectangular steel portal frame shown in Figure 1, with two random plastic moments (M_B for the beam and M_C for the columns) is considered for illustration.

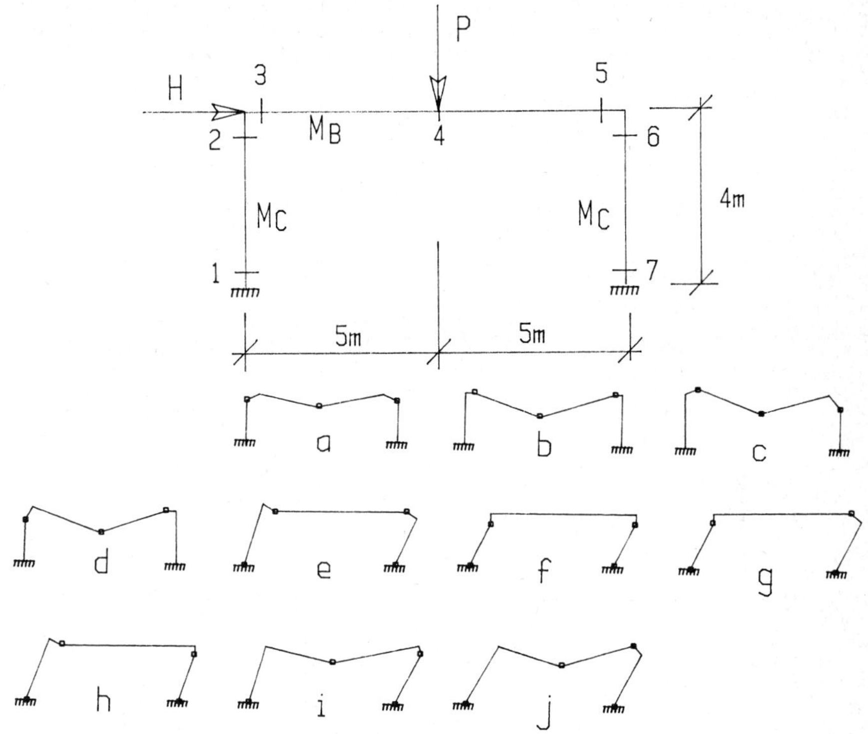

Figure 1
Rectangular Portal Frame

The example frame has seven critical sections and ten possible failure modes (denoted a, b,...,j in Figure 1). It is assumed that for the same critical section the random plastic moment is equal in positive or negative bending. The acting loading is described by two random concentrated forces: H and P. The following input data have been assumed: (a) the loads, H and P, and the plastic moments, M_B and M_C, are normally distributed with the coefficients of variation $V(H) = 20\%$, $V(P) = 15\%$, $V(M_B) = V(M_C) = 10\%$; (b) the mean values of the loads are $\bar{P} = 53.80$ kN and $\bar{H} = 10.10$ kN: (c) the specified overall failure probability of the frame is $P_f = P_f^* = 10^{-5}$; (d) the statistical dependence between the loads P and H, or between the plastic moments M_B and M_C, is accounted for by using the correlation coefficient $\rho(P,H)$ or $\rho(M_B,M_C)$, respectively (the correlation coefficient is a measure of the linear dependence between two random variables). In the following we assume: independent loads $\rho(P,H)=0$ and completely dependent plastic moments $\rho(M_B,M_C) = 1$(this last assumption implies that the moment capacities along a member and between columns are also perfectly correlated). The optimum solution is given in Figure 2. The dominant mode of failure which corresponds to this solution is the mode a in Figure 1; its contribution to the overall probability P_f is 75.5%.

Figures 3 and 4 summarize typical reliability-based optimization sensitivity results of the frame shown in Figure 1.

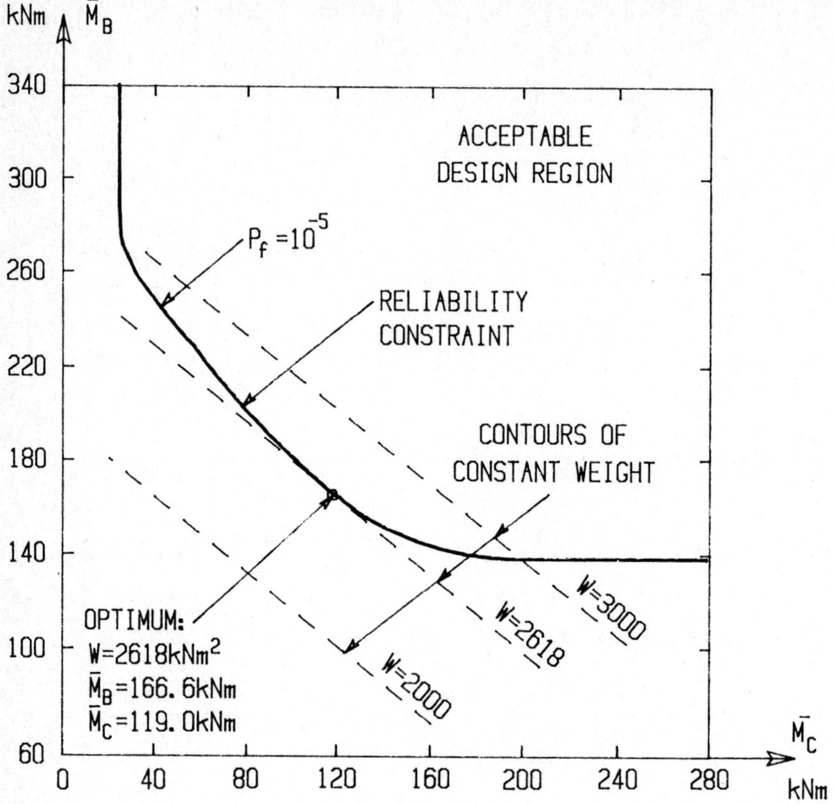

Figure 2
Design Space with Reliability Constraint and Optimum Solution

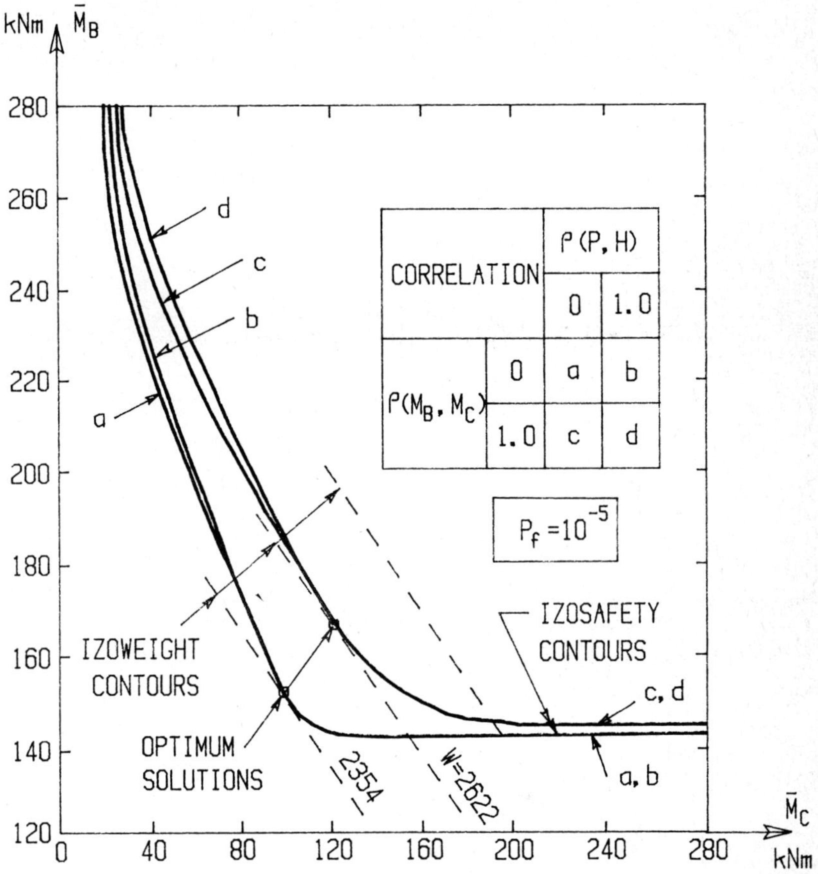

Figure 3
Influence of Correlations on the Optimum Solution

Figure 3 shows that the feasible design region increases with decreasing correlation between beam and column moment capacities and between vertical and horizontal loads. So, an assumption of independent beam-column moment capacities may be very inappropriate and on the unconservative side when statistical dependence between these variables arises. Another conclusion for this particular example is that the optimum solution is insensitive to the correlation between lateral and gravity loads. The reason is that the dominant failure mode for the optimum solution is not the combined mechanism i or j (see Figure 1). A further conclusion is that the same collapse mode is not found to dominate at the optimum when the beam-column correlation is taken or is not taken into account. As a matter of fact, for independent beam-column moment capacities the dominant failure mode is b (with 36.5% contribution to the overall failure probability) and for perfect dependent beam-column moment capacities the dominant failure mode is a (with 75.5% contribution). Note also that the contribution of the failure mode b to the overall P_f is 11.3% in the case of perfect dependence between beam-column moment capacities, and the contribution of the failure mode a is 34.3% in the case of statistical independence between beam-column moment capacities.

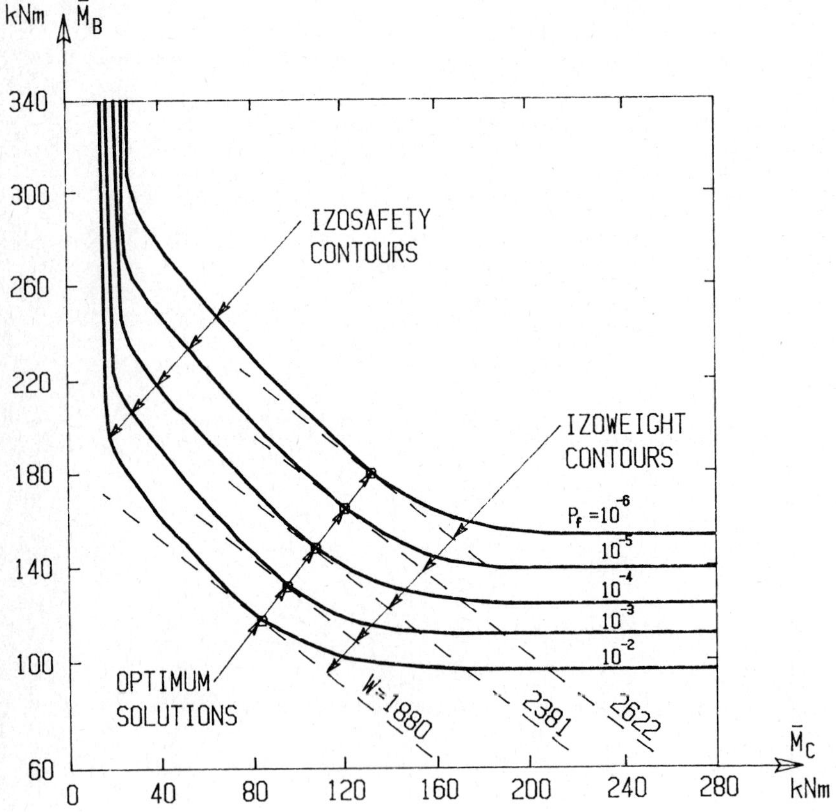

Figure 4
Influence of the Overall Probability of Failure on the Optimum Solution

For completely dependent plastic moment capacities, figure 4 shows the sensitivity of the optimum solution to the overall probability of failure. It should be noted that the objective function increases with increasing the reliability of the frame. In each case considered, the same failure mode (mechanism a) is found to dominate at the optimum. Another conclusion is that the contribution of the dominant failure mode to the overall P_f is quite insensitive to the prescribed value of the probability of collapse P_f (the contribution of mode a is about 75% for all values of P_f in the interval 10^{-2} and 10^{-6}).

CLOSURE

In this paper a computational technique for reliability-based plastic optimum design is presented which incorporate the statistical dependence between any two failure modes. The procedure to obtain the best design is based on the minimization of the total expected structural weight for a specified overall failure probability. The basic idea of the optimization technique is to compute bounds on the overall probability of collapse and on the optimum solution using Ditlevsen's method (1982) of conditional bounding and Vanmarcke's concept (1981) of failure mode decomposition. In order to reach the optimum solution, the design space between these bounds is systematically searched using an algorithm similar to that proposed for deterministic optimization by Belsare and Arora (1983).

Based on this algorithm a computer program was developed which automatically designs optimal plastic structures with up to two hundred failure modes. This program can be linked to a computer-aided design system, based on combining automated optimization algorithms, graphical display programs and interaction. A perturbation procedure is also available for sensitivity analysis. Limited numerical experience with the example frame reported here and additional examples given in Reference [9] has been gained with the computational technique proposed. However, this technique provides a powerful tool to obtain a practical optimum solution offering the potential for application within a reliability-based code context, which is an important goal in structural design.

REFERENCES

[1] Balling, R.J., Pister, K.S. and Polak, E., Delight-Struct: an optimization-based computer-aided design environment for structural engineering, Computer Methods in Applied Mechanics and Engineering 38 (1983) 237-251.

[2] Belsare, S.V. and Arora, J.S., An algorithm for engineering design optimization, Internl. Jrnl. for Numerical Methods in Engineering 19 (1983) 841-858.

[3] Cornell, C.A., Bounds on the reliability of structural systems, Jrnl. of the Structural Division 93 (1967) 171-200.

[4] Ditlevsen, O., System reliability bounding by conditioning, Jrnl. Engng. Mech. Div. 108 (1982) 708-718.

[5] Frangopol, D.M., Probabilistic Study of Structural Reliability, Ph.D. Thesis (in French), Dept. of Strength of Materials and Stability of Structures, Univ. of Liège (Nov. 1976).

[6] Frangopol, D.M., Discrete-optimum probability solution of plastic structures, in: Advances in Civil Engineering through Engineering Mechanics (Proc. 2nd Annual ASCE Eng. Mech. Div. Speciality Conference, New YorK, 1977).

[7] Frangopol, D.M. and Rondal, J., Optimum probability-based design of plastic structures, Engineering Optimization 3 (1977), 17-25.

[8] Frangopol, D.M., Reliability analysis and optimization of plastic structures, in: Augusti, G., Borri, A. and Vannucchi, G., (eds.) Proc. 4th Intern. Conf. on Applications of Statistics and Probability in Soil and Structural Engineering (Pitagora Ed., Bologna, 1983).

[9] Frangopol, D.M., Interactive reliability-based optimum CAD of frames structures, submitted for consideration as a contribution paper to the 6th Intern. Conf. on Computers in Design Engineering (Butterworth Sc. Ltd., Guildford, 1984).

[10] Kirsch, U., Optimum Structural Design (McGraw-Hill, New York, 1981).

[11] Moses, F. and Kinser, D.E., Analysis of structural reliability, Jrnl. of the Structural Division 93 (1967) 147-164.

[12] Moses, F., Sensitivity studies in structural reliability, in: N.C. Lind (ed.), Structural Reliability and Codified Design (Solid Mechanics Division, Study No. 3, University of Waterloo, 1970).

[13] Moses, F., Design for reliability-concepts and applications, in: R.H. Gallagher and O.C. Zienkiewicz (eds.), Optimum Structural Design (John Wiley & Sons, New York, 1973).

[14] Moses, F., Reliability of structural systems, Jrnl. of the Structural Division 100 (1974) 1813-1820.

[15] Moses, F., Structural system reliability and optimization, Computers & Structures 7 (1977) 283-290.

[16] Moses, F., System reliability developments in structural engineering, Structural Safety 1 (1983) 3-13.

[17] Parimi, S.R. and Cohn, M.Z., Optimal solutions in probabilistic structural design, Jrnl. Méc. Appliquée 2 (1978) 47-90.

[18] Somekh, E. and Kirsch, U., Interactive optimal design of truss structures, Computer-Aided Design 13 (1981) 253-259.

[19] Vanmarcke, E.H., Matrix formulation of reliability analysis and reliability based design, Computers & Structures 3 (1971) 757-770.

MISCELLANEOUS

Computational Techniques & Applications: CTAC-83
J. Noye & C. Fletcher (Editors)

AIRBLAST SIMULATIONS USING FLUX-CORRECTED TRANSPORT CODES

David L. Book

Naval Research Laboratory
Washington DC
Mark A. Fry, Science Applications, Inc.,
Mclean, Va.

Flow fields resulting from the passage of a blast wave through a mixture of high-explosive detonation products and air can be significantly altered as compared to a single-phase medium. The description of the two-phase flow has been numerically simulated in one-(1-D) and two-dimensional (2-D) calculations using multi-dimensional flux-corrected transport (FCT) techniques. Calculated and experimentally determined flow fields have been compared.

FCT is a finite difference technique for solving fluid equations in problems where sharp discontinuities arise. Shocks, slip surfaces, and contact surfaces are accurately modelled through the application of diffusion and anti-diffusion at each timestep. This diffusion is just large enough to prevent dispersive ripples and guarantees that all conserved quantities remain positive. Simulation of two-phase flow is a natural outgrowth of the FCT technique. Each species is treated as a separate fluid with the source terms in the energy and momentum equations providing the coupling. Moreover, the timestep-split feature of the most recent version of FCT allows direct extension from 1-D to 2-D simulation. Results are compared with shock tube and field test data.

INTRODUCTION

The method of Flux-Corrected Transport (FCT) was developed about ten years ago (Boris and Book (1973), Book, et al (1975) and Boris and Book (1976)) as a means for solving systems of hyperbolic equations in such a way that physically positive quantities remain positive. Its principal application has been to compressible fluids, i.e., fluids in which some of the flow speeds are comparable with or greater than the local speed of propagation of waves in the medium. Examples include plasmas, combusting systems, and gas-dynamic flows.

Subsequently the method has been extended and generalized in a number of important respects. These include the construction of general-purpose algorithms for solving fluid equations on moving nonuniform grids in curvilinear coordinate systems (Boris 1976), strictly multidimensional FCT algorithms (Zalesak 1979) (as opposed to timestep-splitting of one-dimensional algorithms), and variants of these. In addition, numerous techniques have been developed by other workers which incorporate the positivity-preserving property in other ways (Harten (1978) Van Leer (1979) and Collela (1983)). The weight of opinion among computational scientists who deal with problems of compressible flow now strongly affirms the superiority of positivity-preserving methods over conventional ones. In a word, the advantagesare improved robustness, flexibility, resolution, and freedom from nonphysical numerical artifacts.

In this paper we describe some applications of FCT to airblast problems, i.e., to calculations of the blast wave and fireball produced by an explosion in the atmosphere near the ground. As this is one of the applications for which FCT was originally devised, it is not surprising that the method yields satisfactory results in a wide range of cases with very little need for computational development. Nevertheless, some of the refinements we have employed in this connection are of interest both in their own right and for other users who may need to carry out similar calculations.

In most of the airblast calculations we have carried out, four distinct stages are discernible: (i) initialization, usually employing a Chapman-Jouguet model (Kuhl et al (1981)) to describe conditions at the time the detonation wave reaches the periphery of the explosive charge; (ii) free-field (one-dimensional) evolution, before the shock reaches the ground; (iii) subsequent two-dimensional evolution using axisymmetric (r-z) geometry; and (iv) postprocessing of the resulting solution to analyze and display the results.

The advances described in this paper relate to all but the first of these stages. They may be cataloged as follows: refinements in the free-field solution; development of a regridding algorithm for the early-time (high-over pressure) two-dimensional solution; improvements in boundary conditions in the two-dimensional stage; and exploitation of tracer particles as a diagnostic, particularly in the late stages of fireball evolution, after shock breakaway. In the body of the paper one section is devoted to each of these topics.

REFINEMENTS IN THE FREE-FIELD SOLUTION

Figure 1 shows one-dimensional (spherical) profiles calculated using a form of the general-purpose FCT algorithm ETBFCT (Boris 1976). Although these profiles are accurate and physically correct on the average, they exhibit numerous flat shelves and accompanying sharp drops, especially in the region of the rarefaction wave. This phenomenon is known as "terracing". In this instance we attribute its presence to the nonlinear action of the flux limiter stage of FCT on dispersive errors from the transport stage.

To explain what this means, let us describe a class of FCT algorithms which includes the two variants referred to above. To advance the one-dimensional continuity equation

$$\frac{\partial \rho}{\partial t} = - \frac{\partial}{\partial x} (\rho v) \tag{1}$$

one time step on a uniform mesh with a given flow velocity v, we suppose the old cell-centered value ρ_j^0 and v_j are known. A general three-point conservative finite-difference operator acting on ρ_j^0 can be written in the form

$$\begin{aligned} \rho_j^T = \rho_j^0 &- \varepsilon_{j+\frac{1}{2}} \rho_{j+\frac{1}{2}} + \varepsilon_{j-\frac{1}{2}} \rho_{j-\frac{1}{2}} \\ &+ \gamma_{j+\frac{1}{2}} (\rho_{j+1}^0 - \rho_j^0) - \gamma_{j-\frac{1}{2}} (\rho_j^0 - \rho_{j-1}^0), \end{aligned} \tag{2}$$

where $\rho_{j+\frac{1}{2}} = \frac{1}{2}(\rho_j + \rho_{j+1})$. To obtain a first-order approximation to Eq. (1), we must define $\varepsilon_{j+\frac{1}{2}}$ to be the Courant number $v_{j+\frac{1}{2}} \delta t/\delta x$, where $v_{j+\frac{1}{2}} = \frac{1}{2}(v_j + v_{j+1})$; likewise, for a conventional scheme with second-order (spatial) accuracy, we must let $\gamma_{j+\frac{1}{2}} = \frac{1}{2} \varepsilon_{j+\frac{1}{2}}^2$. In FCT, however, a numerical diffusion is applied to

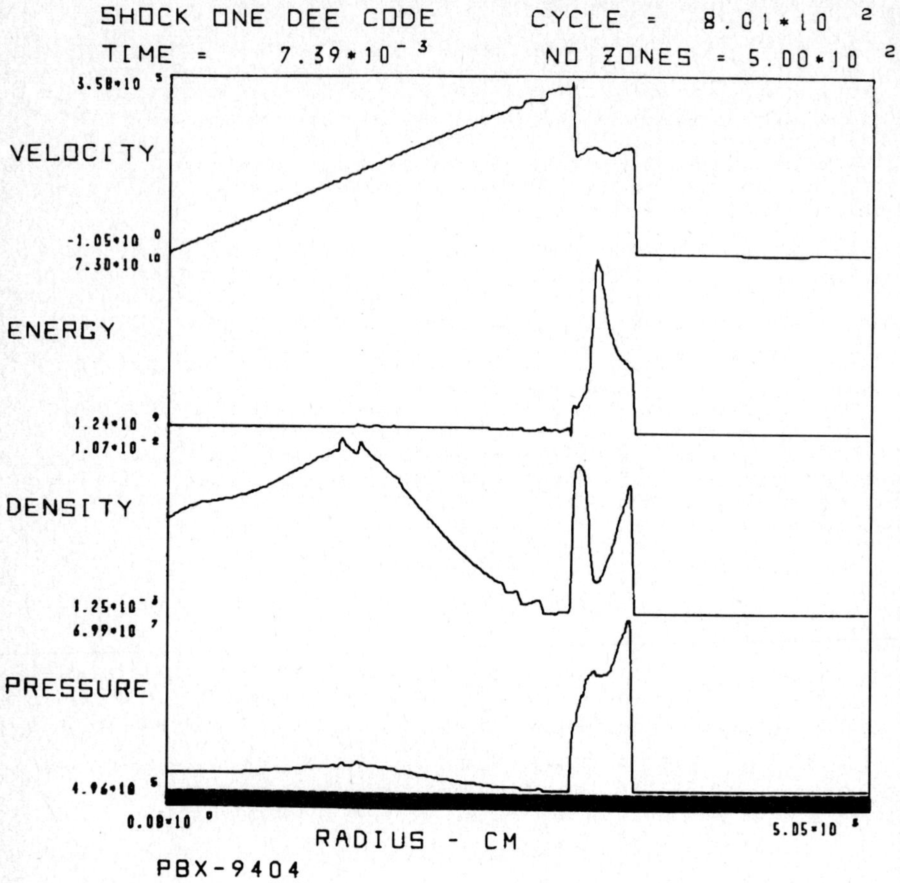
SHOCK ONE DEE CODE CYCLE = 8.01*10 2
TIME = 7.39*10-3 NO ZONES = 5.00*10 2
3.58*10 5
VELOCITY
-1.05*10 0
7.30*10 10
ENERGY
1.24*10 9
1.07*10-2
DENSITY
1.25*10-3
6.99*10 7
PRESSURE
4.96*10 5
0.00*10 0
RADIUS - CM
5.05*10 3
PBX-9404

Fig. 1

ρ_j^T according to

$$\rho_j^{TD} = \rho_j^T + \nu_{j+\frac{1}{2}}(\rho_{j+1}^o - \rho_j^o) - \nu_{j-\frac{1}{2}}(\rho_j^o - \rho_{j-1}^o). \tag{3}$$

Evidently the γ's and the ν's combine in the same way in ρ_j^{TD}, so we leave both unspecified for the moment.

If we now applied another diffusion operation, this time with the opposite sign, it would just tend to cancel the diffusion terms already present in ρ_j^{TD}. Instead we write for the new densities ρ_j^n

$$\rho_j^n = \rho_j^{TD} - \phi_{j+\frac{1}{2}}^c + \phi_{j-\frac{1}{2}}^c, \tag{4}$$

where the "corrected" fluxes $\phi_{j+\frac{1}{2}}^c$ differ from the "raw" fluxes $\phi_{j+\frac{1}{2}} = \mu_{j+\frac{1}{2}} (\rho_{j+1}^o - \rho_j^o)$ in two ways: we use $\{\rho_j^T\}$ instead of $\{\rho_j^o\}$ in the definition of the raw fluxes (this allows us to define γ so as to optimize some property in the difference scheme, as will be seen shortly); and we <u>correct</u> the fluxes, so that the anti-diffusion process (which evidently tends to make all gradients steeper) can enhance no extrema already present in $\{\rho_j^{TD}\}$, nor introduce any new ones. The simplest formula for achieving this in all possible situations is that used in "strong flux limiting":

$$\phi_{j+\frac{1}{2}}^c = S \cdot \{\max[0, \min(|\phi_{j+\frac{1}{2}}|, \Delta_{j-\frac{1}{2}}, \Delta_{j+3/2})]\}, \tag{5}$$

where $S = \text{sign}\ \phi_{j+\frac{1}{2}}$ and $\Delta_{j+\frac{1}{2}} = \rho_{j+1}^{TD} - \rho_j^{TD}$.

At most points in a gently varying profile no correction is required and $\phi_{j+\frac{1}{2}}^c = \phi_{j+\frac{1}{2}}$. In that case we can perform a Von Neumann analysis, calculating the complex propagator or amplification factor $A = A_r + iA_i = \rho_j^n/\rho_j^o$ for a sinusoidal density profile $\rho_j^o = \exp(ijk\,\delta x)$, where k is the wave number. Writing $\beta = k\delta x$ and $\varepsilon = v\delta t/\delta x$, we expand in powers of β the amplification

$$|A| = 1 + A_2\,\beta^2 + A_4\,\beta^4 + \ldots\ldots\ldots \tag{6}$$

and the relative phase error

$$R = \frac{1}{\beta\varepsilon}\tan^{-1}(A_i/A_r) - 1 = R_2\,\beta^2 + R_4\,\beta^4 + \ldots\ldots\ldots\ldots\ldots \tag{7}$$

It is easy to show that the condition that A_2 vanish is

$$\gamma + \nu - \mu = \tfrac{1}{2}\,\varepsilon^2. \tag{8}$$

The condition that R_2 vanish is

$$\mu = \frac{1}{6} - \frac{1}{6}\,\varepsilon^2. \tag{9}$$

There remains one free parameter which can be chosen so as to build some desirable property into the algorithm (this is the reason for introducing γ). The choice $\gamma = \frac{1}{2}\varepsilon^2$ confers no particular benefit. Letting $\gamma = 0$ as in ETBFCT (Boris (1976)) reduces the operation count and yields a useful general-purpose algorithm, which is, however, subject to the terracing problems mentioned previously. There are two obvious candidates: $\gamma = \frac{1}{4}\varepsilon^2$, which implies $A_4 = 0$, and $\gamma = \frac{1}{5} + \frac{1}{5}\varepsilon^2$,

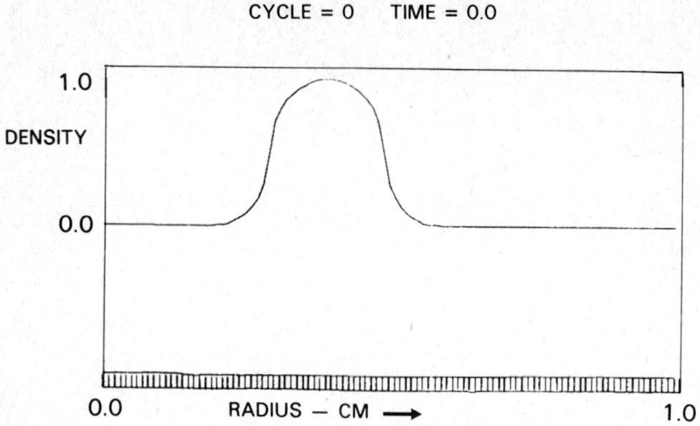

Fig. 2a

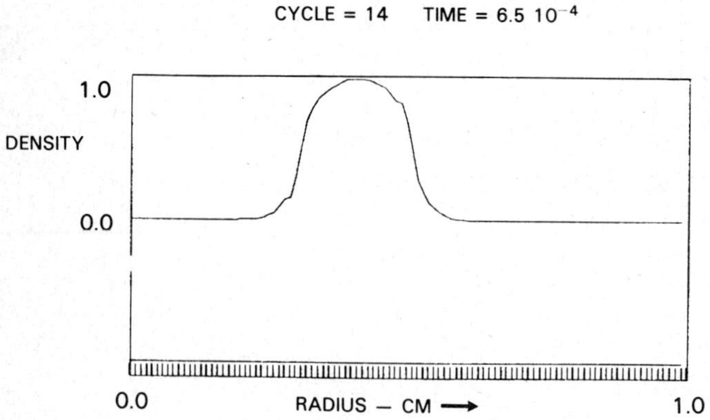

Fig. 2b

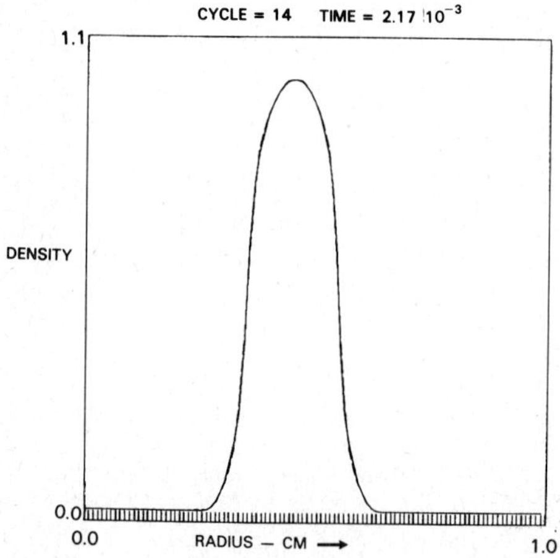

Fig. 2c

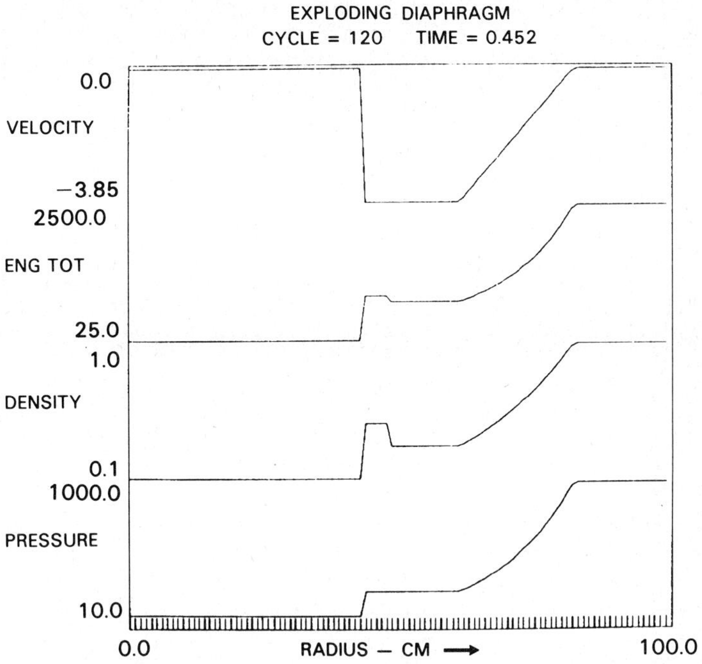
EXPLODING DIAPHRAGM
CYCLE = 120 TIME = 0.452
0.0
VELOCITY
−3.85
2500.0
ENG TOT
25.0
1.0
DENSITY
0.1
1000.0
PRESSURE
10.0
0.0
RADIUS — CM →
100.0

Fig. 3a

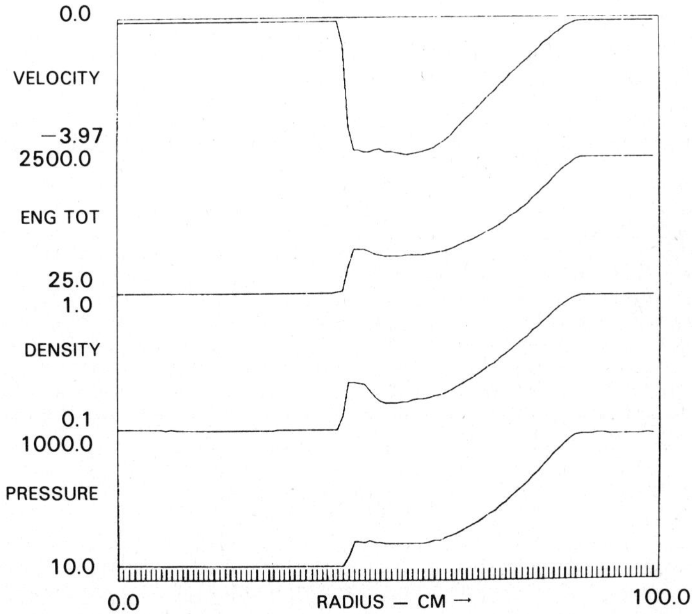
0.0
VELOCITY
−3.97
2500.0
ENG TOT
25.0
1.0
DENSITY
0.1
1000.0
PRESSURE
10.0
0.0
RADIUS — CM →
100.0

Fig. 3b

which implies $R_4 = 0$.

Figure 2 shows this for a test involving passive advection with a uniform velocity. A test profile, initially consisting (Fig. 2a) of a half ellipse, connected by two quarter ellipses to a straight line of zero slope, is advected with a Courant number $\varepsilon = 0.2$. The results obtained using the general-purpose algorithm (Fig. 2b) and the new multicoefficient version (Fig. 2c) are shown for comparison. It is clear that the latter represents a distinct improvement, a conclusion that continues to hold for other test profiles.

The results of a different test are shown in Fig. 3. In Fig. 3a, the profile obtained from an analytic solution of the bursting-diaphragm (Riemann) problem with 10-to-1 density ratio and 10^3-to-1 pressure are plotted, while in Fig. 3b those calculated with the new FCT algorithm are shown for comparison. Here again the results agree fairly closely. The agreement is worst near the contact surface, which, unlike the shock, has no physical self-steepening mechanism.

In chemical explosions a characteristic feature is the contact surface between the detonation products and the air compressed ahead of the blast (Fig. 4). In the absence of diffusion (usually a good approximation in actual explosion, if not numerically), the contact surface remains sharp, and the two media differ in density across this discontinuity by $\sim$ 30%.

In an airblast, a layer of air is compressed beneath the burst, so that the detonation products never reach the ground. The shock wave propagates down to the ground, reflects upward, and propagates back through the layer of compressed air. When it encounters the contact surface, it is partly transmitted and partly reflected downward again. This process can repeat several times. At finite ground range the shocks reflect obliquely and soon propagate away from the fireball, but directly beneath the burst the process is only terminated when the rising fireball recedes from the ground sufficiently far. (Experimentally and computationally the multiply reflected shocks usually become indistinguishable from noise after the second or third reflection.)

Evidently it is important for computational purpose to model the contact surface accurately if these shock reverberations are to be described properly. If the contact surface in the simulation is diffuse, the shock reflected from it will be weak or undetectible. (If physical diffusion processes are present experimentally but absent in the simulation, however, the latter will predict excessively strong reflection.) The results shown in Fig. 4a, calculated using 200 zones, while close to the physically correct profiles, exhibit considerable diffusion at the contact surface. This is in contrast with the results shown in Fig. 4b, obtained using 400 zones, in which the contact surface is much more distinct. A consequence of improving the resolution is shown in Fig. 5, which depicts the pressure history that a sensor directly beneath the bursts shown in Fig. 4 would measure. The peaks are due to the initial shock and the shock re-reflected from the contact surface, respectively. Note that both maxima increase when the resolution improves, but the second peak rises more. Although in this case further improvement in the resolution makes little difference, it is clear that the second peak might in some cases be raised above the first. Thus a qualitative change in the solution can result from improving the resolution. This is of course also applicable to the accuracy of the two-dimensional results calculated after the one-dimensional solution is laid down on the 2D mesh.

REGRIDDING ALGORITHM FOR TWO-DIMENSIONAL SOLUTION

To study the transition to Mach reflection extremely high resolution is required, as may be seen from the following discussion. The Mach stem at transition has vanishing length. If we are restricted to resolution of features several (say,

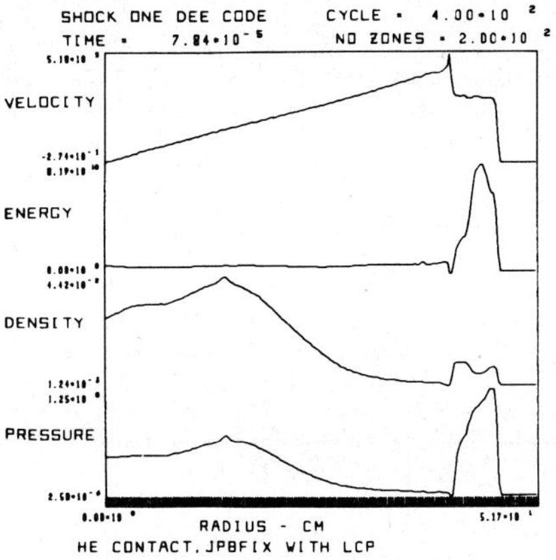

Fig. 4a

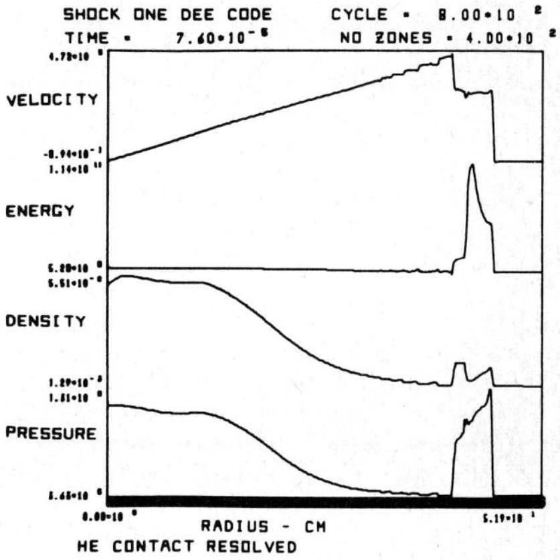

Fig. 4b

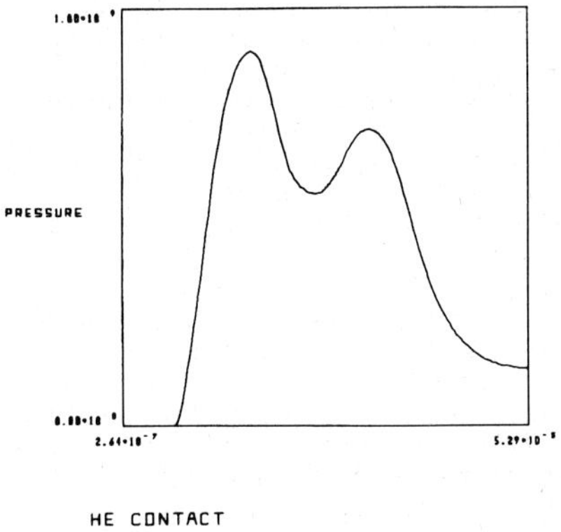

Fig. 5a

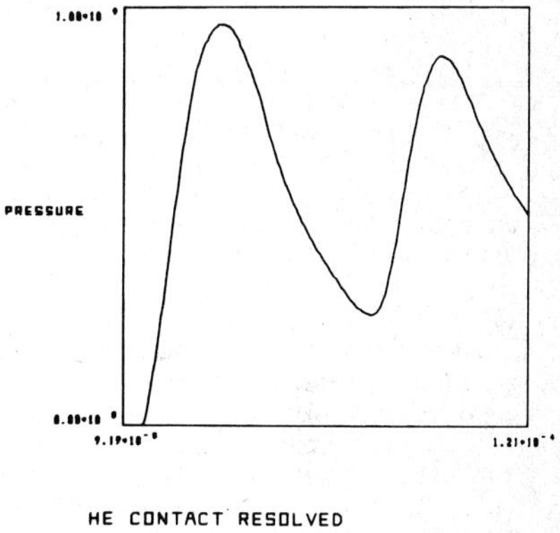

Fig. 5b

a factor of three) times the mesh size $\delta x \sim \delta y$, then the ground range at which transition occurs will be overestimated by an amount $\delta \ell$ equal to this length divided by the tangent of the opening angle χ, which is typically a small fraction of a radian, so that $\delta \ell \sim 30\ \delta x$. Since core limitations restrict us to meshes with a total of no more than 10^4 - 10^5 zones and since transition typically taken place at a ground range comparable with the height of burst (HOB), if we use a uniform mesh the error $\delta \ell$ would be as much as ten or twenty percent of the HOB.

Fortunately, this is unnecessary. We can grid up near where transition takes place, using cells smaller by one or two orders of magnitude in each direction, with a proportionate reduction in $\delta \ell$. Since the mesh is rectilinear, this implies the presence of extremely elongated (non-square) zones on portions of the grid, with consequent errors in the solution. This creates no difficulties if the details of the flow there are unimportant and if those regions do not generate errors which propagate into the regions of chief interest.

However, a shock wave propagating from coarsely gridded region abruptly into a finely gridded one sheds sound waves as it steepens up, seriously degrading the solution at and behind the front (Boris and Book (1976)). Hence it is necessary to ensure that the shock at the front of the blast wave be allowed to propagate some distance through a finely gridded region prior to reaching the point of reflection, that the transition be gradual, and most significantly, that this be true not just at reflection but at all times up until then. The only way to avoid using an inordinate number of fine zones is to move the highly resolved region with the reflection point, starting at the instant the blast wave reaches the ground and continuing as long as the Mach reflection process is of interest (Fig. 6). This method exploits the "sliding rezone" (continuous regridding) capability of the ETBFCT algorithm.

Typically we start with a mesh ~ 150 x 150 zones, having ~ 100 fine zones of dimension $\delta x = \delta y$ and ~ 50 coarse zones of dimension $\Delta x = \Delta y$, with $\Delta x/\delta x = 10$, in each coordinate direction. The fine zones are those closest to the axes. Let the cell boundaries be specified by $\{x_i\}$ and $\{y_j\}$. We introduce a smooth transition region by diffusing these interfaces repeatedly according to

$$x_i' = x_i + \eta(x_{i+1} - 2x_i + x_{i-1}) \tag{10}$$

$$y_j' = y_j + \eta(y_{j+1} - 2y_j + y_{j-1}), \tag{11}$$

where the diffusion coefficient η is of order 0.1. The diffusion process is repeated N times (N ~ 30) to generate a transition region N cells wide. The initial reflection necessarily occurs in the fine-zoned region.

As the reflection point moves outward, the grid is redefined each cycle. We search on density (or pressure) to find the smallest radial displacement such that conditions deviate from ambient by a significant amount (3%). Using this as a center, we define new cell boundaries, smooth $\{x_i\}$ and $\{y_i\}$ as before, and deposit the transported fluid quantities into the cells of this new grid at the conclusion of the transport calculation. As the fine-zoned region marches outward, coarse cells disappear ahead of it and reappear at small radii (Fig.6).

This method has proved successful on a number of HOB calculations. A certain amount of tuning is necessary to ensure that enough zones are present in both the highly resolved and transition regions, but there is no difficulty in tracking the reflection point, nor is any significant detail lost through diffusion.

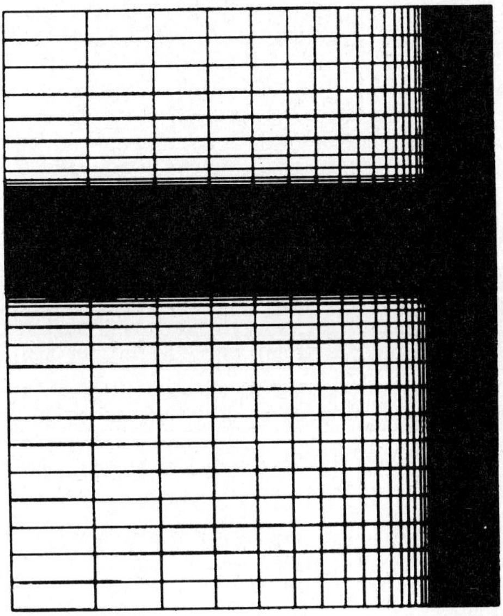

Fig. 6b

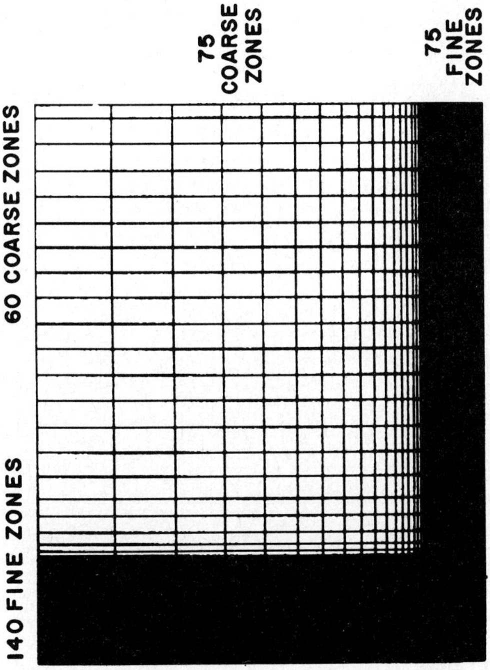

Fig. 6a

BOUNDARY CONDITIONS

The conditions for reflection at a planar solid boundary in a perfectly invisuid flow are

$$\frac{\partial \rho}{\partial n} = \frac{\partial p}{\partial n} = \frac{\partial vt}{\partial n} = v_n = 0, \tag{12}$$

where $\frac{\partial}{\partial n}$ represents a normal derivative and the subscripts n and t label the normal and tangential components of velocity. These conditions are imposed at the interface between the last cell of the computational mesh and a guard cell juxtaposed to it. To ensure that the normal derivative of a quantity f vanish, it is sufficient to set the value in the guard cell equal to that in the last internal cell, i.e., $f_{N+1} = f_N$. To ensure that the quantity itself vanish, we simply set $f_{N+1} = -f_N$. (Special attention must be paid to the equation describing the component of momentum normal to the boundary in order that no momentum is diffused through the interface.) Equation (12) applies to any summetry plane or axis, and so to the axis of symmetry in the HOB problem at r=0. At large values of r and z (i.e., at the peripheral boundary — the outside and top of the mesh), we impose the ambient conditions $\rho = \rho_a$, $p = p_a$, $\underset{\sim}{v} = 0$.

This is satisfactory until shocks propagate to the boundary. If we are interested in following the shocks further, it is necessary at this point to begin moving the grid outward (as described in the previous section), or to lay down the solution on a larger, coarser mesh. If we are interested in characteristic late-time phenomena such as vortex formation and fireball rise, however, shocks are irrelevant and it is appropriate to use simple outflow (or transmission) boundary conditions, namely

$$\frac{\partial \rho}{\partial n} = \frac{\partial p}{\partial n} = \frac{\partial \underset{\sim}{v}}{\partial n} = 0 \tag{13}$$

Higher-order extrapolation have been recommended in the literature (see e.g., Roache1972), but have not been found particularly useful in connection with FCT, apparently because of the way they interact with FCT's tendency to clip off the tops of peaked profiles. At very late times a small but significant modification, suggested by Boris (personal communication, 1983) has been employed. This involves introducing a relaxation term at the peripheral boundary in the form

$$f_{N+1} = f_N + \alpha(f_a - f_N) \tag{14}$$

Here the relaxation coefficient satisfies $\alpha \approx c_a \delta t/L$, where c_a is the ambient speed of sound and L is a typical scale size (e.g., fireball cloud diameter) in the problem. Typically α is small $\sim 10^{-2} - 10^{-3}$. Its effect is to cause the peripheral values of the fluid variables to relax toward their ambient values, simulating the influence of inflow from the surrounding atmosphere.

Also at very late times (typically tens of seconds), gravity must be included in the vertical momentum equation to describe the effect of buoyancy. The equation we advance has the form

$$\frac{\partial(\rho v)}{\partial t} + \frac{\partial}{r\partial r}(r\rho uv) + \frac{\partial}{\partial z}(\rho v^2) + \frac{\partial}{\partial z}(p - p_a) \tag{15}$$

$$+ (\rho - \rho_a)g = 0.$$

The ambient quantities, which are taken from tables, are the same as those used in the boundary conditions. This guarantees that the ambient quantities represent a hydrostatic balance, which truncation errors will not destabilize.

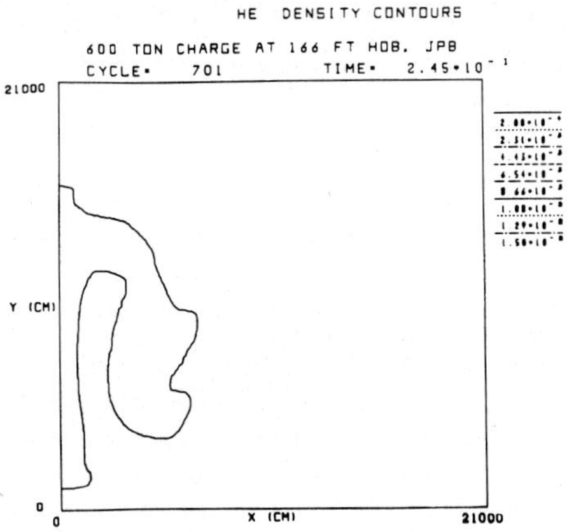

Fig. 7a

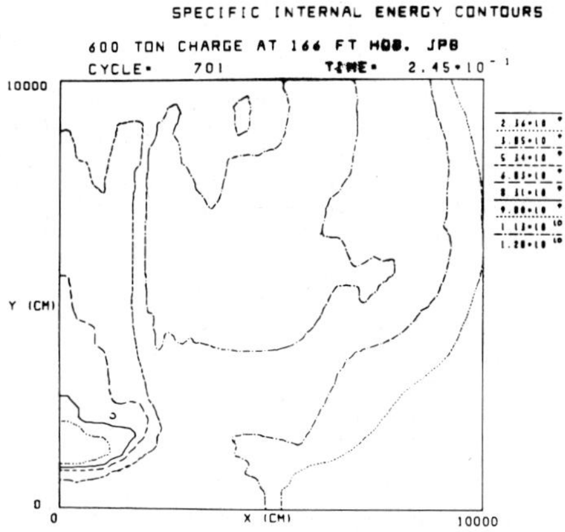

Fig. 7b

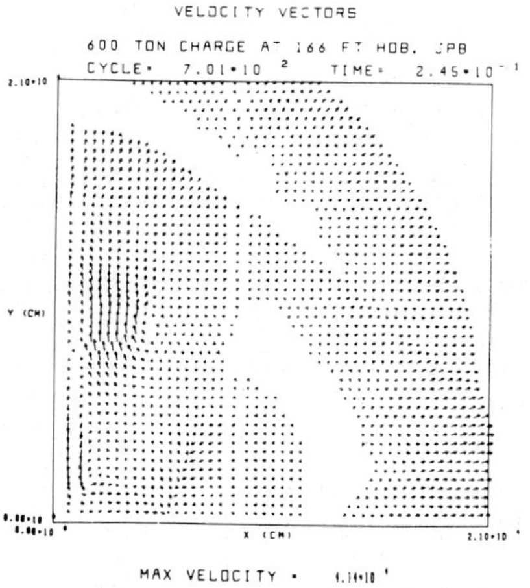

Fig. 7c

The energy equation also acquires an additional term due to gravity; this, however, plays only a minor role.

Figure 7 shows the late time evolution of a fireball calculated for a 600-ton burst of high explosive (HE) set off at a height of 50.6 m. Note the elongation of the cloud and the characteristic vortices which appear especially clearly in the HE density contours.

GRAPHICS DEVELOPMENTS

Approximately one quarter of our computational costs are accumulated in generating pictures showing the results of the calculation. These take the form of pressure histories; plots of surface pressure and density vs range; velocity vector (arrow) plots; contours of constant total and HE product mass density, total and internal energy density, pressure, and vorticity; instantaneous locations or trajectories of passively advected tracer particles; and movies of the time variation of these displays. Both the latter and the use of distinguishing colors are particularly revealing, although for obvious reasons examples can not be presented here. In three-dimensional work we have carried out (Fry and Book (1983), unpublished), the problems of displaying results become signficantly greater.

The approach we have taken is to perform the calculation in its entirety first, often with several restarts, producing only enough pictures to make sure everything is going well, and dumping out the results (essentially the two-dimensional ρ, u, v, and p arrays) at intervals onto magnetic tape. This means that if we require *a posteriori* a quantity which cannot be deduced from the data in the dump files (such as the total impulse on the ground sensors or the pressure history at a location where no sensor existed), the calculation has to be rerun. This is true also if the dump interval chosen is too long, but in many cases intermediate values can be obtained satisfactorily by interpolation. For example, a movie of some 700 frames was made using linear interpolation between twenty-seven dumps. (Cubic interpolation would have required generating a new data tape containing only the arrays being processed, since the contents of four dumps cannot be fitted into core simultaneously.) The graphics calculations, like the hydrodynamics calculation themselves, were performed on a Texas Instruments ASC, a two-pipe vector machine with approximately half a million words of core. The plots were output on a Tektronix 4020 scope; the color movie was generated using a Dicomed graphics system.

This approach, in which we generate graphics by postprocessing a limited number of dump files, is a compromise between generating all the graphics in the course of the calculation and saving the results from every timestep. (This would entail prohibitive disk or I/O requirements.) In general it has worked well, particularly as we have learned from experience what quantities should be saved and how often they should be dumped.

In concluding, we remark that passive advection of tracer particles is among the most useful of the graphical displayswe have used, particularly in movies. Figure 8 shows tracer particle histories over two time intervals for the calculation described previously. Note that four distinct vortices are revealed in these pictures, two in the "forward" direction and two reversed. In previous work (both experimental and in simulations), only one of each had been detected. Moreover, we are not restricted to passive advection, which is appropriate to the motion of infinitesimal particles in the blast wind fields. We have also modeled finite-sized particles (dust and debris) by balancing inertial forces against gravity and the Stokes drag according to

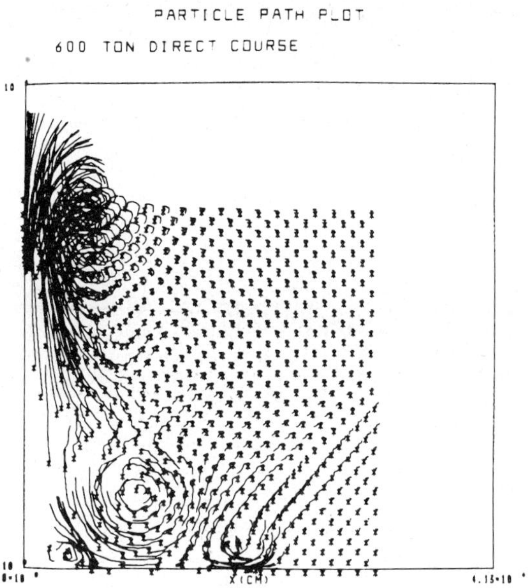

Fig. 8a

Time Interval
1.8 sec → 3.97 sec

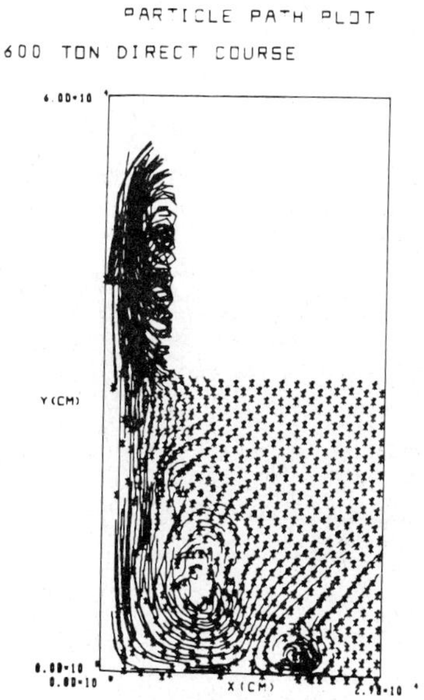

Fig. 8b

Time Interval
3.97 → 7.3 sec

$$m\frac{dV}{dt} = -mg + D(v - V), \tag{16}$$

where m and V are the particle mass and velocity, respectively, and D is the drag coefficient. By using (16) with a spectrum of particle masses $\{m_i\}$ and appropriate initial conditions obtained from an ejecta model, we can generate realistic maps showing how much material is entrained by the blast and where it is located at any given time. This is invaluable for comparison with observations of field tests, where the obscuring debris clouds are among the most conspicuous features.

ACKNOWLEDGEMENT

We would like to thank Mrs Debra Miller for preparing this manuscript. This work was supported by the U.S. Defense Nuclear Agency under Subtask N99QAXAH, Work Unit #00061, Work Unit "Cloud Calculations", and Under Subtask Y99QAXSG, Work Unit #00057, Work Unit Title "Airblast Calculations".

REFERENCES

(1) Book, D.L., Boris, J.P., and Hain, K.H., Flux-Corrected Transport: II. Generalizations of the Method, J.Comput.Phys., 18, (1975) 248.

(2) Boris, J.P., Flux-Corrected Transport Modules for Solving Generalized Continuity Equations, NRL Memo Report No. 3237 (1976).

(3) Boris, J.P., and Book, D.L., Flux-Corrected Transport: I. SHASTA, A Fluid Transport Algorithm that Works, J.Comput.Phys., 11 (1973) 38-69.

(4) Boris, J.P., and Book, D.L., Flux-Corrected Transport: III Minimal-Error FCT Algorithm, J.Comput.Phys., 20 (1976) 397-431.

(5) Collela, P., J.Comput.Phys. (1983, in press); see also Woodward, P.R., and Collela, P., High Resolution Difference Schemes for Compressible Gas Dynamics in Proc. Seventh Inter. Conf. Num. Methods in Fluid Dynamics, Springer-Verlay, New York (1981).

(6) Harten, A., Math.Comput. 32 (1978) 363.

(7) Kuhl, A.L., Fry, M.A., Picone, J.M., Book, D.L., and Boris, J.P., FCT Simulation of HOB Airblast Phenomena, NRL Memo Rept. No. 4613 (1981).

(8) Roache, P.J., Computational Fluid Dynamics, Hermosa, Albuquerque (1972).

(9) Van Leer, B., J.Comput.Phys., 32, (1979) 101.

(10) Zalesak, S.T., Fully Multidimensional Flux-Corrected Transport Algorithms for Fluids, J.Comput.Phys., 31 (1979) 335-362.

Computational Techniques & Applications: CTAC-83
J. Noye & C. Fletcher (Editors)

NUMERICAL SOLUTION OF EIGENVALUE PROBLEMS FOR ORDINARY DIFFERENTIAL EQUATIONS

Alan L. Andrew

Mathematics Department, La Trobe University,
Bundoora, Victoria 3083,
Australia

Some recent developments in the numerical solution of eigenvalue problems for ordinary differential equations are surveyed. The main emphasis is on the Sturm-Liouville problem, especially the computation of higher eigenvalues and the corresponding eigenfunctions. The discussion ranges from available software to the computation of bounds and a recent conjecture concerning a certain extrapolation technique is settled. Applicability of the methods discussed to more general eigenvalue problems is indicated, and certain problems in which the eigenvalue occurs nonlinearly are considered.

1. INTRODUCTION

The main aim of this paper is to survey some recent developments in the numerical solution of eigenvalue problems for ordinary differential equations and to provide users with advice on appropriate methods for specific problems and with a guide to the relevant literature. Also in Section 4 a reason is given for some previously unexplained numerical results noted in [10].

The term "eigenvalue problem" is taken to refer only to linear homogeneous equations, and coefficient functions are assumed to be real-valued. The term "nonlinear eigenvalue problem", used by some writers to describe certain nonlinear parameter-dependent equations, is used in Section 5 of this paper to mean linear equations with nonlinear dependence on the eigenparameter - that is equations of the form

$$A(\lambda)x = 0 \qquad (1)$$

where the linear differential operator $A(\lambda)$ depends nonlinearly on the eigenvalue, λ [8].

Sections 2,3 and 4 deal mainly with the regular Sturm-Liouville problem

$$(ry')' + (\lambda p-q)y = 0 \quad \text{on } [a,b] \qquad (2a)$$

$$\alpha(ry')(a) + \beta y(a) = \gamma(ry')(b) + \delta y(b) = 0 \qquad (2b)$$

where $p,q > 0$, $|\alpha|+|\beta| > 0 < |\gamma|+|\delta|$ and $a,b \in R$. Particular attention is given to the relatively difficult problem of computing higher eigenvalues and the corresponding eigenfunctions. However, much of the cited literature deals with more general problems (nonseparated boundary conditions, higher order equations, non-selfadjoint problems etc.) and some of the more important of these extensions are mentioned in the text. Singular Sturm-Liouville problems are also considered, including the bound-state problem for the radial Schrödinger equation [3,79]. Good additional references for problems on infinite intervals are [53,54].

A recurring theme is the advantage of making an appropriate change of variables before beginning numerical computation. For example, higher eigenvalues of (2) may be obtained much more accurately for a given amount of effort if (2a) is first transformed to the Liouville normal form

$$y'' + (\lambda - q)y = 0. \qquad (3)$$

This may be done by the Liouville transformation and is generally advantageous even if a numerical approximation is required for the integrals occurring in the transformation [62].

The following abbreviations are used:

BVP -	boundary value problem,	DE -	differential equation,
EVP -	eigenvalue problem,	ODE -	ordinary differential equation ,
IVP -	initial value problem,	PDE -	partial differential equation.

2. FINITE DIFFERENCE AND VARIATIONAL METHODS

Two important methods for the numerical solution of EVPs for both ODEs and PDEs are finite difference methods [42-44] and variational methods [24,55,81] (or more generally projection methods [72]) which approximate the solution of the DE by the solution of a matrix EVP for which excellent software is available [29,89]. Variational methods, as described for quite general self adjoint EVPs in [24] and for Sturm-Liouville problems in [64], are hard to beat for accuracy for the fundamental eigenvalue, the error in an eigenvalue of such problems being proportional to the square of the error in the corresponding eigenfunction.

However, since the accuracy of both variational and finite difference methods falls off rapidly with the order of the eigenvalue, they are recommended only for the first few eigenvalues, unless modified. Moreover, devices such as Richardson extrapolation [42] or deferred correction [23] are useful mainly for low eigenvalues. The simple centred difference formula with (uniform) mesh length h gives an error of order k^4h^2 in the kth eigenvalue of (2). With h^2- extrapolation this becomes k^6h^4 - a significant improvement only if kh is small. Numerov's method [21], which has been highly recommended by physicists for solution of the Schrödinger equation, gives an error of order k^6h^4 without the need for extrapolation. For example it approximates the kth eigenvalue, $k^2\pi^2$, of

$$y'' + \lambda y = 0, \quad y(0) = y(1) = 0$$

by

$$\frac{12[1-\cos(kh\pi)]}{h^2[5+\cos(kh\pi)]} = k^2\pi^2 - \frac{k^6h^4\pi^6}{240} + O(k^8h^6).$$

The rapid growth of error with k is due to the fact that standard finite difference and finite element methods approximate the eigenfunctions by piecewise low degree polynomials which are not good approximations for highly oscillatory eigenfunctions. Raptis and others [71] have developed a finite difference scheme for the Schrödinger equation which involves approximating the eigenfunctions by more appropriate functions and which can achieve a great improvement in accuracy, but unfortunately the coefficients in this scheme depend on k. A much simpler and highly effective method for improving the accuracy of computed higher eigenvalues of regular Sturm-Liouville problems in Liouville normal form is proposed in [63]. At negligible extra cost, a correction described there reduces the error in eigenvalues computed by simple centred differences from $O(k^4h^2)$ to $O(kh^2)$. It is shown in [58] that this correction is also very useful in the numerical solution of the important inverse Sturm-Liouville problem of calculating coefficients from the eigenvalues, a typical application being the determination from seismic data of the variation with depth of the density of the earth [60]. The possible use of similar corrections for other difference schemes is an obvious field for further investigation.

An important bonus of variational methods is that they can provide upper bounds for eigenvalues. The usefulness of these bounds is reduced by the difficulty of computing lower bounds of comparable accuracy. There is a vast theoretical literature [28,35,86,87] on the computation of these complementary bounds, but they require much more effort for a given accuracy than the variational bounds and have not gained great popularity with users. Wendroff [88] has suggested a simple method of using the upper bounds for Sturm-Liouville and certain other EVPs given by the Ritz method to obtain (much less sharp) lower bounds at negligible extra cost, and a slight refinement of this is given in [74]. Similar ideas may be used to obtain two-sided bounds by finite difference methods [38,47]. Another limitation of these bounds is that they are rigorous only if all numerical quadratures required are performed exactly and the resulting matrix equations are solved without roundoff error. Although appropriate implementation of variational methods can normally ensure that the result of such errors is very small [64], the accuracy obtainable by variational methods for the fundamental eigenvalue is so great that these extra errors can become dominant, making the computed upper bounds sometimes (slightly) lower than the true value.

Although most classical theory for variational and finite difference methods makes important smoothness assumptions on the coefficients, problems whose coefficients have jump discontinuities often arise in the analysis of composite structures in engineering [17] and the study of free oscillations of models of the earth [60]. Consequently there has been much recent interest in the use of variational methods for such problems [12,17]. Two sided bounds for eigenvalues of such problems are also readily obtained by the method of Section 4 [10].

3. SHOOTING METHODS

Probably the two best currently available general purpose software packages for numerical solution of Sturm-Liouville problems are SLEIGN [14,16], developed by Bailey and others at Sandia Laboratories, and the NAG library package DO2KDF, DO2KAF, DO2KEF [31,32,69]. The similarity of these two packages, which use a shooting method, is indicative of considerable agreement on the best implementation of such methods for Sturm-Liouville problems. An excellent critical comparison of these two packages, with special attention to their user interface and their methods of error control, is given in [69] which also includes a useful discussion of the theory, the results of systematic numerical tests and suggestions for an implementation of the best features of both codes.

Because of the relatively highly developed state of software for IVPs [85], shooting methods, which replace an EVP by a sequence of IVPs for the same DE, have proved very effective for solving a wide class of EVPs. The basic theory is given in [44].

The classical shooting method for solving Sturm-Liouville problems is described in [42]. With this approach, standard IVP routines approximate y and ry' in (2a) by piecewise polynomials. This causes problems for high eigenvalues similar to those noted in Section 2, though the method is still very satisfactory for low eigenvalues. For high eigenvalues, performance can be improved by using the Prüfer transformation [68],

$$y = \rho \sin\theta, \quad ry' = \rho \cos\theta ,$$

as the functions ρ and θ vary more slowly than y and ry'. This was used in an early version of SLEIGN which is described in detail in [16]. However, for high eigenvalues, the higher derivatives of θ, which are critical factors in the accuracy of IVP routines, are still too large [76] and better accuracy may be obtained with the modified Prüfer transformation, which has

$$\tan\theta = [r(\lambda p-q)]^{\frac{1}{2}} y/ry' .$$

Unfortunately this causes problems for automatic routines unless $\lambda p-q > 0$. However Bailey [13] noted that this problem could be avoided and much of the advantage of the modified Prüfer transformation retained by using the "slightly modified" or scaled Prüfer transformation

$$y = \rho \sin \circ \theta, \; ry' = z\rho \cos \circ \theta$$

where z is a suitable constant. This converts (2a) into two first order ODEs, the first of which,

$$\theta' = (z/r)\cos^2 \circ \theta + (\lambda p-q)z^{-1}\sin^2 \circ \theta \; ,$$

is independent of ρ. The constant z is chosen so that, "on average", the coefficients of $\cos^2 \circ \theta$ and $\sin^2 \circ \theta$ are of about the same magnitude. This is the strategy used in SLEIGN [14], while DO2KDF is very similar but allows a variable z and makes some other minor modifications [69].

A thorough analysis of such methods needs to take into account the method used to solve the IVPs. A complete analysis of all three Prüfer methods for the regular problem (2) has been given by Paine [56] for the cases when the IVPs are solved by the Heun method or the classical fourth order Runge-Kutta method. His results show the advantage of first transforming (2a) to the Liouville normal form (3). (This can be done by the user before using SLEIGN or the NAG library.) His results for (2) with the Heun method and for (3) with the fourth order Runge-Kutta method are stated without proof in [60]. Several results proved for the Heun method in [56] are stronger than those mentioned in [60]. For the scaled transformation, Paine proved sharper results when $pr=1$ and for the modified transformation he proved sharper results for the general case and still sharper results when $(pr)' = p'' = 0$. For problems in Liouville normal form with $q \in C^3$, he showed that if either the modified or scaled Prüfer transformation is used then there exist positive constants H and c such that the estimate $\tilde{\lambda}_k$ obtained for the kth eigenvalue, λ_k, with step length h, satisfies

$$|\lambda_k - \tilde{\lambda}_k| \leqslant ch^2 \quad \text{whenever} \quad 4\lambda_k h^2 \leqslant \pi^2 \text{ and } h < H \; .$$

The software packages described in this section have been designed to cope efficiently with singular problems such as the Schrödinger equation [3]. It is noted in [69] that SLEIGN is especially robust for singular problems, though at the cost of being slightly less efficient than the NAG routines for "easy" problems. The solution by shooting methods of EVPs for other singular ODEs can also be facilitated by a suitable change of variables [19].

An extension of SLEIGN, called SLEIGN2, is available for solving multiparameter analogues of Sturm-Liouville problems which sometimes arise when PDEs are solved by the method of separation of variables [15]. The possibility of solving such problems by Prüfer methods was earlier suggested in [37].

4. APPROXIMATING THE COEFFICIENTS

Another method which has proved very effective for computing higher eigenvalues and the corresponding eigenfunctions of Sturm-Liouville problems involves approximating the coefficients, generally by step functions or other piecewise (low order) polynomials [9,10,26,50,57,61,62,66,67]. That is, a partition

$$\pi := \{a = x_0 < x_1 < \ldots < x_n = b\}$$

is chosen and the coefficients p,q,r are replaced by approximations $\tilde{p},\tilde{q},\tilde{r}$ satisfying $\tilde{r} \in PC_\pi^1[a,b]$, $\tilde{p},\tilde{q} \in PC_\pi[a,b]$, $\tilde{r} \geqslant r_0$ (const.) > 0, $\tilde{p} \geqslant p_0$ (const.) > 0, where $PC_\pi^m[a,b]$ denotes the set of functions $u:[a,b] \to R$ such that

$$u \in C^m[x_0,x_1) \cap C^m(x_1,x_2) \cap \ldots \cap C^m(x_{n-1},x_n]$$

and $u^{(m)}$ has left and right hand limits at $x_1,\ldots,x_{n-1}$, and $PC_\pi := PC_\pi^0$. Approximations $\tilde{\lambda}$ and $\tilde{y}$ to eigenvalues and eigenfunctions of (2) are then obtained by solving a system of equations on the various subintervals

$$(\tilde{r}\tilde{y}')' + (\tilde{\lambda}\tilde{p}-\tilde{q})\tilde{y} = 0 \quad \text{on } (x_{i-1},x_i),\ i=1,\ldots,n \tag{4a}$$

coupled by the continuity conditions

$$\tilde{y} \in C[a,b], \quad \tilde{r}\tilde{y}' \in C\,[\tilde{r};a,b] \tag{4b}$$

where, as in [10], the notation $u \in C\,[\tilde{r};a,b]$ means there exists $u_1 \in C\,[a,b]$ such that $u(x) = u_1(x)$ for all $x \in [a,b]$ except perhaps the points of discontinuity of $\tilde{r}$. The boundary conditions (2b) remain, giving

$$\alpha(\tilde{r}\tilde{y}')(a) + \beta\tilde{y}(a) = \gamma(\tilde{r}\tilde{y}')(b) + \delta\tilde{y}(b) = 0. \tag{4c}$$

This method has been called the "piecewise analytic method" because the approximate eigenfunction is piecewise analytic but, as this is also true of approximations obtained by shooting or finite element methods, the name seems less suitable than "approximate DE method" [9] which is used here.

With this method, jump discontinuities in the coefficients present no difficulty provided all points of discontinuity are used as grid points, x_i. For regular problems in Liouville normal form with $q \in C^2[a,b]$, Paine and de Hoog [62] showed that, if $\tilde{q} \in PC_\pi[a,b]$ is the step function satisfying $\tilde{q}((x_{i-1} + x_i)/2) = q((x_{i-1} + x_i)/2)$, $i=1,\ldots,n$, then there are numbers n_0 and c such that, for all $n > n_0$,

$$|\lambda_k - \tilde{\lambda}_k| \leqslant cn^{-2} \text{ for } k=1,2,\ldots,[n/2]$$

when n equal subintervals are used. Paine [57] showed that uniformly accurate second order estimates may also be obtained for $k > [n/2]$ by means of a simple correction. The approximate DE method is readily adapted to singular problems and has been widely used for the solution of Schrödinger's equation [2,27,34,40, 78,79]. The success of the method for higher eigenvalues is essentially due to the fact that the coefficients are generally less oscillatory than the higher eigenfunctions.

A survey of approximate DE methods for regular Sturm-Liouville problems was given in [9] but some recent related work should be noted. Much of this concerns the solution of the radial Schrödinger equation using a perturbative correction technique suggested by Gordon [34] and further developed by Ixaru and others [2,40]. Smooke has given a detailed analysis of these perturbative corrections for a system of coupled radial Schrödinger equations [79] as well as for a single such equation [78] and has also investigated their use for certain PDEs [77]. Similar methods have also been used for stiff systems of IVPs [18]. Use of approximate DE methods without perturbative correction has also been examined for Hill's equation [27] and for EVPs for certain higher order ODEs [26].

Three suggestions for further investigation were made in [9]. The first was the need for a systematic comparison of the various methods that have been used for the numerical solution of the approximating equations (4). The optimum choice may depend on the particular problem and on whether eigenfunctions are required as well as eigenvalues. To the author's knowledge no further work has been done on this in the subsequent two years. Another suggestion concerned the generalization of the approximate DE method for the solution of other problems, a question addressed in the previous paragraph. However more remains to be done. For example, following an idea of Leighton [50] and Pruess [66], it was shown in [10], using comparison theorems, that approximate DE methods could be used to obtain rigorous two-sided bounds for Sturm-Liouville eigenvalues rather than merely estimates. Since comparison theorems are known for much more general problems [73], it seems worth investigating the computation of eigenvalue bounds for these more general problems by similar methods.

The remaining question asked in [9] has however now been answered. This concerns the generality of convergence rates observed in some numerical examples. It was proved in [10] for a special case and in [9] in general that, as $h := \max(x_i - x_{i-1}) \to 0$, the convergence to the true eigenvalues of the bounds given by the algorithm of [10] is of the first order in h. However it was noted that all published numerical results indicated that the mean of the upper and lower bounds showed $O(h^2)$ convergence. Moreover, for the example used by Leighton [50], h^2 - extrapolation of these means was observed in [10] to give $O(h^4)$ convergence. It was asked in [9] whether this high order convergence was sufficiently general to give us "something for nothing" by using results calculated as bounds to obtain, at negligible extra cost, higher order estimates. At that time the results in [10] were the only published results sufficiently accurate to test the extrapolation hypothesis. Other results gave only 4 or 5 significant figures and the extrapolated means were so close to the true eigenvalues that their rate of convergence could not be deduced.

The second order convergence of the means has now been proved under very general conditions [61]. (Actually the proof shows only that the convergence is at least $O(h^2)$, but numerical results confirm that it is not generally faster than this.) However recent numerical results of Paine [59] show that, while h^2- extrapolation usually (but not always) increases the accuracy significantly, it certainly does not produce $O(h^4)$ convergence in general. After learning of these results, the author at last realised that the $O(h^4)$ convergence of the extrapolated means for Leighton's atypical example could be explained by the following simple argument, in which the terms $P_i^{\pm}$, $R_i^{\pm}$, c_i have the same meaning as in [10].

In Leighton's example, $rp = K$ (constant), $q = 0$ and $p' \geqslant 0$. (A similar argument holds when $p' \leqslant 0$.) Hence $R_i^+ = r(x_{i-1})$, $R_i^- = r(x_i)$, $P_i^+ = p(x_i)$, $P_i^- = p(x_{i-1})$, $c_i = |\lambda K|^{\frac{1}{2}}$ and the reasoning of Theorem 3 of [61] shows that the upper bounds for positive eigenvalues computed by equation (3.2) of [10] satisfy

$$\tan(\theta_1-\theta_0) = \tan[(\lambda/K)^{\frac{1}{2}} \Sigma_{i=1}^{n} (x_i-x_{i-1})p(x_{i-1})]$$

where

$$\theta_0 := \frac{\pi}{2} \text{ if } \beta = 0 \quad\text{and}\quad \theta_0 := -\tan^{-1}[\alpha(\lambda K)^{\frac{1}{2}}/\beta] \quad \text{otherwise,}$$

and

$$\theta_1 := \frac{\pi}{2} \text{ if } \delta = 0 \quad\text{and}\quad \theta_1 := -\tan^{-1}[\gamma(\lambda K)^{\frac{1}{2}}/\delta] \quad \text{otherwise.}$$

Similarly the lower bounds and the exact eigenvalues satisfy

$$\tan(\theta_1-\theta_0) = \tan[(\lambda/K)^{\frac{1}{2}} \Sigma_{i=1}^{n} (x_i-x_{i-1})p(x_i)]$$

and

$$\tan(\theta_1-\theta_0) = \tan[(\lambda/K)^{\frac{1}{2}} \int_{x_0}^{x_n} p(t)dt]$$

respectively. (As in [61], a similar argument holds for negative eigenvalues with tan replaced by tanh, but this is not needed for Leighton's example which has all eigenvalues positive.) It follows that the upper and lower bounds, $\lambda_k^+(n)$ and $\lambda_k^-(n)$, obtained for the eigenvalue λ_k using n equal subintervals will be $\lambda_k/(1+\varepsilon_{n-})^2$ and $\lambda_k/(1+\varepsilon_{n+})^2$ respectively, where

$$1 + \varepsilon_{n+} = \Sigma_{i=1}^{n} (x_i-x_{i-1})p(x_i)/\int_{x_0}^{x_n} p(t)dt$$

and

$$1 + \varepsilon_{n-} = \Sigma_{i=1}^{n}(x_i - x_{i-1})p(x_{i-1}) \Big/ \int_{x_0}^{x_n} p(t)dt.$$

Extrapolation of the means gives

$$[4\lambda_k^+(2n) + 4\lambda_k^-(2n) - \lambda_k^+(n) - \lambda_k^-(n)]/6$$

$$= \lambda_k[4(1+\varepsilon_{2n-})^{-2} + 4(1+\varepsilon_{2n+})^{-2} - (1+\varepsilon_{n-})^{-2} - (1+\varepsilon_{n+})^{-2}]/6$$

$$= \lambda_k[6-4(2\varepsilon_{2n-} - 3\varepsilon_{2n-}^2 + 4\varepsilon_{2n-}^3 + 2\varepsilon_{2n+} - 3\varepsilon_{2n+}^2 + 4\varepsilon_{2n+}^3) + 2\varepsilon_{n-} - 3\varepsilon_{n-}^2 + 4\varepsilon_{n-}^3$$
$$+ 2\varepsilon_{n+} - 3\varepsilon_{n+}^2 + 4\varepsilon_{n+}^3 + O(\varepsilon^4)]/6 \qquad (5)$$

by the binomial theorem, where $O(\varepsilon^4)$ denotes terms of fourth and higher order in the small quantities $\varepsilon_{n\pm}$, $\varepsilon_{2n\pm}$. Now in Leighton's example p is linear (i.e. $p'' = 0$). This implies that

$$\varepsilon_{n-} = -\varepsilon_{n+} \quad (\forall n)$$

so that all the odd terms in (5) cancel, and moreover that

$$|\varepsilon_{2n\pm}| = \tfrac{1}{2}|\varepsilon_{n\pm}| \qquad (\forall n)$$

so that the second order terms also cancel, leaving only the fourth and higher order terms. By the standard error formula for the rectangle (quadrature) rule, $\varepsilon_{n\pm} = O(n^{-1})$ so that the fourth order convergence of the extrapolated means in this very special case now follows from (5). These results illustrate the danger of concluding too much from a single numerical example.

5. NONLINEAR EIGENVALUE PROBLEMS

Unlike SLEIGN, the NAG library routines discussed in Section 2 deal with problems which are more general than the classical Sturm-Liouville problem in three ways. (i) The coefficient function $(\lambda p-q)$ of (2a) is replaced by a more general function $Q(\lambda)$ which may depend nonlinearly on λ. (ii) For each fixed λ, the functions $Q(\lambda)$ and r, or their derivatives, may contain jump discontinuities. (iii) The boundary conditions may depend (possibly nonlinearly) on λ. The user supplies a single routine to compute r, Q and $\partial Q/\partial\lambda$.

An example of this special type of nonlinear dependence on the eigenvalue parameter, which is considered in [69], is the Taylor-Goldstein equation from fluid mechanics, with the wave speed taken as eigenparameter, λ. When the upper boundary is a free surface, then the boundary condition there also involves λ nonlinearly. It is reported in [69] that DO2KDF worked well for this problem.

More general problems with nonlinear dependence on the eigenvalue parameter often occur in applications [8] and most of the theory in [44] is valid for problems with quite general nonlinear dependence on the eigenvalue. Usually shooting methods have proved suitable for ODEs of this type. Invariant imbedding methods [75] and finite difference methods [11] have also proved useful. Of course some problems are harder than others. A comparison of several finite difference, shooting and invariant imbedding techniques for a difficult nonlinear EVP from fluid mechanics is given in [30].

Numerical solution of very general linear operator equations, which include ODEs as a special case, with quite general holomorphic (nonlinear) dependence on λ in (1), has also received considerable recent attention [41,48]. Numerical examples applying some of those methods to ODEs are included for example in [51,52,83].

Application of the method of separation of variables to PDEs of the form

$$\Sigma_{i=0}^{n} \left(\frac{\partial}{\partial t}\right)^{i} A_i \, u = 0 ,$$

where the A_i are linear differential operators involving only the space variables, leads to the polynomial EVP

$$\Sigma_{i=0}^{n} \lambda^{i} A_i \, v = 0, \tag{6}$$

and completeness theorems needed to justify this technique have been established under very general conditions [70]. Other nonlinearities in λ are sometimes introduced by numerical devices used to cope with infinite intervals [54] or by changes of variables designed to overcome other difficulties. Conversely a change of variables may also be used to "linearize" nonlinear EVPs, making them linear in λ. This often (but not always) simplifies numerical solution. For the finite dimensional problem where the A_i are matrices, the recommended method for computing the complete eigensystem of (6), especially when (as is usual) n is not too large, is based upon a linearization [65] obtained by introducing the vector $(v^T, \lambda v^T, \ldots, \lambda^{n-1} v^T)^T$. The same linearization may be used when (6) is a DE. Various other linearizations have also been used. A rather general approach to linearization is described in [33], but not in the context of numerical solution.

A linearization with which the author became involved some years ago is one due to Chandrasekhar [20] for the DEs governing small adiabatic nonradial oscillations of stars. For the derivation of these equations and a review of the early literature see [49] and for more recent references see [25,82]. The equations have been solved successfully for various stellar models by shooting methods [25,82], and a comparison between shooting and Chebyshev expansion methods is given in [39]. The DEs are a singular fourth order system with the eigenvalue occurring quadratically. In theoretical work the perturbation of gravitational potential is often neglected giving a simpler second order system [49]

$$\left(\frac{f_1 w'}{\lambda + f_2}\right)' = \left(\frac{f_3}{\lambda} + f_4\right) w$$

where λ is the eigenvalue and the functions $f_1, \ldots, f_4$ depend on the stellar model. Both the full (fourth order) and the simplified (second order) systems have a sequence of normal modes ("p-modes") with eigenvalues tending to ∞ and another ("g-modes") with eigenvalues tending to 0.

Chandrasekhar suggested computing solutions by variational methods using his formulation, $Tu = \lambda u$ where, for the full system, T is a Hermitian integro-differential operator and, for the simplified system, T and T^{-1} are both Hermitian differential operators [6]. The author found that this gave excellent results for the lower p-modes (as with Sturm-Liouville problems) but failed for the g-modes except in some cases where the trial functions were very close to the exact eigenfunctions. (See [4,5] for details.) Sobouti [80] confirmed that the results depend critically on the choice of trial functions.

The classical theory of variational methods requires the underlying operator either to be compact (and hence bounded) or else to have a compact inverse, a condition not satisfied in this case where eigenvalues cluster at both 0 and ∞. For one class of nonlinear EVPs, variational principles similar to those for compact operators have been proved [1,36]. The basic idea behind this approach is that separate (nonquadratic) "Rayleigh functionals" are associated with separate (generally infinite) subsets of the eigenvalues. Chandrasekhar's variational principle is not covered by this theory or the slightly more general theory of [84]. However, there has been a great deal of work on factorising polynomial operator pencils following ideas of [46], so that if $(\lambda I-X)$ is a factor of the left hand side of (6) then the eigenvalues of X are eigenvalues of (6). Perhaps such a factorisation might yield an effective method of computing g-modes.

As a first step towards a general theory of variational methods for such non-standard problems, the author [5,6,7] proved a number of related results but the general theory is still lacking. A proper understanding of these methods might also improve understanding of another recently proposed method [45] which yields p-modes but not g-modes.

REFERENCES

[1] Abramov, Ju.Š., On the theory of nonlinear eigenvalue problems, Soviet Math. Dokl. 14 (1973), 1271-1275.

[2] Adam, Gh., Adam , S. and Corciovei, A., Local and accumulated truncation errors in a class of perturbative numerical methods, Rev. Roumaine Phys. 25 (1980), 39-53.

[3] Adamová, D., Eigenvalues of the Schrödinger operator via the Prüfer transformation for a general class of central potentials, Czechoslovak J. Phys. B31 (1981), 1225-1237.

[4] Andrew, A.L., Non-radial oscillations of massive stars by variational methods, Acad. Roy. Belg. Bull. Cl. Sci. 54 (1968), 1046-1055.

[5] Andrew, A.L., Application of the Ritz method to non-standard eigenvalue problems, J. Austral. Math. Soc. 10 (1969), 367-384.

[6] Andrew, A.L., Variational solution of some nonlinear eigenvalue problems. I, J. Math. Anal. Appl. 32 (1970), 400-413.

[7] Andrew, A.L., Variational solution of some nonlinear eigenvalue problems. II, J. Math. Anal. Appl. 33 (1971), 425-432.

[8] Andrew, A.L., Eigenvalue Problems with Nonlinear Dependence on the Eigenvalue Parameter. A Bibliography, Tech. Rep., Math. Dept., La Trobe University (1974).

[9] Andrew, A.L., Computation of higher Sturm-Liouville eigenvalues, in: Meek, D.S. and van Rees, G.H.J. (eds.) Congressus Numerantium 34 (1982), 3-16.

[10] Andrew, A.L., de Hoog, F.R. and Robb, P.J., Leighton's bounds for Sturm-Liouville eigenvalues, J. Math. Anal. Appl. 83 (1981), 11-19.

[11] Antia, H.M., Finite-difference method for generalized eigenvalue problem in ordinary differential equations, J. Comput. Phys. 30 (1979), 283-295.

[12] Babuška, I. and Osborn , J.E., Numerical treatment of eigenvalue problems for differential equations with discontinuous coefficients, Math. Comp. 32 (1978), 991-1023.

[13] Bailey, P.B., A slightly modified Prüfer transformation useful for calculating Sturm-Liouville eigenvalues, J. Comput. Phys. 29 (1978), 306-310.

[14] Bailey, P.B., SLEIGN, An Eigenvalue-Eigenfunction Code for Sturm-Liouville Problems, Rep. SAND77-2044, Sandia Laboratories (1978).

[15] Bailey, P.B., The automatic solution of two-parameter Sturm-Liouville eigenvalue problems in ordinary differential equations, Appl. Math. Comput. 8 (1981), 251-259.

[16] Bailey, P.B., Gordon, M.K. and Shampine, L.F., Automatic solution of the Sturm-Liouville problem. ACM Trans. Math. Software 4 (1978), 193-208.

[17] Bielak, J., Some remarks on bounds to eigenvalues of Sturm-Liouville problems with discontinuous coefficients, Z. Angew. Math. Phys. 32 (1981), 647-657.

[18] Bienstock, S. and Gordon, R.G., Piecewise analytic integration of chemical rate equations. I. The algorithm, II. Error analysis, implementation and applications, J. Chem. Phys. 77 (1982), 2902-2911, 2912-2927.

[19] Bramley, J.S. and Sloan, D.M., Note on the Riccati method for differential eigenvalue problems showing algebraic growth, J. Math. Anal. Appl. 83 (1981), 351-356.

[20] Chandrasekhar, S., A general variational principle governing the radial and the non-radial oscillations of gaseous masses, Astrophys. J. 139 (1964), 664-674.

[21] Chawla, M.M. and Katti, C.P., On Noumerov's method for computing eigenvalues, BIT 20 (1980), 107-109.

[22] Childs, B., Scott, M., Daniel, J.W., Denman, E. and Nelson, P. (eds.), Codes for Boundary-Value Problems in Ordinary Differential Equations (Springer, Berlin, 1979).

[23] Chu, K.-W.E., Deferred correction for the ordinary differential equation eigenvalue problem, Bull. Austral. Math. Soc. 26 (1982), 445-454.

[24] Ciarlet, P.G., Schultz, M.H. and Varga, R.S., Numerical methods for higher-order accuracy for nonlinear boundary value problems III. Eigenvalue problems, Numer. Math. 12 (1968), 120-133.

[25] Cox, J.P., Theory of Stellar Pulsation (Princeton University Press, Princeton, 1980).

[26] Dähnn, J. Ein direktes Verfahren zur numerischen Behandlung einer Klasse von selbstadjungierten, positiv definiten Eigenwertaufgaben, Preprint (Neue Folge) Humboldt-Univ. Berlin, Sekt. Math. 2 (1980).

[27] Dähnn, J., Kastelewicz, A. and Schalm, G., Zur numerishen Berechnung der Eigenwerte und Eigenfunktionen der Hillschen Differentialgleichung mit periodischen Randbedingungen, Wiss. Z. Humboldt - Univ. Berlin Math. Natur. Reihe 28 (1979), 495-504.

[28] Fichera, G. Numerical and Quantitative Analysis (Pitman, London, 1978).

[29] Garbow, B.S., Boyle, J.M., Dongarra, J.J., and Moler, G.B., Matrix Eigensystem Routines - EISPACK Guide Extension (Springer, Berlin , 1977).

[30] Gladwell, I., On the numerical solution of a differential nonlinear eigenvalue problem on an infinite range, Appl. Math. Comput. 4 (1978), 121-137.

[31] Gladwell, I., The development of the boundary-value codes in the ordinary differential equations chapter of the NAG library, in: Ref. [22], 122-143.

[32] Gladwell, I. and Sayers, D.K. (eds.) Computational Techniques for Ordinary Differential Equations (Academic Press, London, 1980).

[33] Gohberg, I.C., Kaashoek, M.A. and Lay, D.C., Equivalence, linearization and decomposition of holomorphic operator functions. J. Functional Analysis 28 (1978), 102-144.

[34] Gordon, R.G., New method for constructing wave functions for bound states and scattering, J. Chem. Phys. 51 (1969), 14-25.

[35] Gould, J.H., Variational Methods for Eigenvalue Problems (Univ. Toronto Press, Toronto, 1978).

[36] Hadeler, K.P., Nonlinear eigenvalue problems, in: Numerische Behandlung von Differentialgleichungen, ISNM 27 (Birkhäuser, Basel, 1975), 111-129.

[37] Hargrave, B.A., Numerical approximation of eigenvalues of Sturm-Liouville systems, J. Comput. Phys. 20 (1976), 381-396.

[38] Hubbard, B.E., Bounds for eigenvalues of the Sturm-Liouville problem by finite difference methods, Arch. Rational Mech. Anal. 10 (1962), 171-179.

[39] Hurley, M., Roberts, P.H. and Wright, K., The oscillations of gas spheres, Astrophys. J. 143 (1966), 535-551.
[40] Ixaru, L. Gr., Cristu, M.I. and Popa, M.S., Choosing step sizes for perturbative methods to solve the Schrödinger equation, J. Comput. Phys. 36 (1980), 170-181.
[41] Jeggle, H. and Wendland, W., On the discrete approximation of eigenvalue problems with holomorphic parameter dependence, Proc. Roy. Soc. Edinburgh Sect. A 78 (1977), 1-29.
[42] Keller, H.B., Numerical Methods for Two-Point Boundary Value Problems (Ginn-Blaisdell, Waltham, Mass., 1968).
[43] Keller, H.B., Numerical solution of boundary value problems for ordinary differential equations: Survey and some recent results on difference methods, in: Aziz, A.K. (ed.), Numerical Solutions of Boundary Value Problems for Ordinary Differential Equations (Academic Press, New York, 1975), 27-88.
[44] Keller, H.B., Numerical Solution of Two Point Boundary Value Problems (SIAM, Philadelphia, 1976).
[45] Knölker, M. and Stix, M., A convenient method to obtain stellar eigenfrequencies, Solar Phys. 82 (1983), 331-341.
[46] Krein, M.G. and Langer, H., On some mathematical principles in the linear theory of damped oscillations of continua, Integral Equations Operator Theory 1 (1978), 364-399 and 539-566 (first published in Russian, 1965).
[47] Kuttler, J.R., Upper and lower bounds for eigenvalues of torsion and bending problems by finite difference methods, Z. Ang ew. Math. Phys. 21 (1970), 326-342.
[48] Lancaster, P., A review of numerical methods for eigenvalue problems nonlinear in the parameter, in: Numerik und Andwendungen von Eigenwertaufgaben und Verzweigungsproblemen, ISNM 38 (Birkhäuser, Basel, 1977), 43-67.
[49] Ledoux, P. and Walraven, Th., Variable stars, in: Flügge, S. (ed.) Handbuch der Physik 51 (Springer, Berlin , 1958), 353-604.
[50] Leighton, W., Upper and lower bounds for eigenvalues, J. Math. Anal. Appl. 35 (1971), 381-388.
[51] Linden, H., Upper bounds for eigenvalues of a nonlinear eigenvalue problem, J. Math. Anal. 63 (1978), 753-761.
[52] Linden, H., The method of orthogonal invariants for quadratic eigenvalue problems, Applicable Anal. 9 (1979), 53-61.
[53] Markowich, P.A., Eigenvalue problems on infinite intervals, Math. Comp. 39 (1982), 421-441.
[54] Markowich, P.A. and Weiss, R., Nonlinear eigenvalue problems on infinite intervals, SIAM J. Math. Anal. 14 (1983), 431-449.
[55] Mikhlin, S.G., The Numerical Performance of Variational Methods (Wolters - Noordhoff, Groningen, The Netherlands, 1971).
[56] Paine, J.W.,Numerical Approximation of Sturm-Liouville Eigenvalues, Ph.D. Thesis, Australian National Univ. (1979).
[57] Paine, J.,Correction of Sturm-Liouville eigenvalue estimates, Math. Comp. 39 (1982), 415-420.
[58] Paine, J., A numerical method for the inverse Sturm-Liouville problem, SIAM J. Sci. Statist. Comput. (To appear).
[59] Paine, J.W., Unpublished numerical results, 1982.
[60] Paine, J.W. and Anderssen, R.S., Uniformly valid approximation of eigenvalues of Sturm-Liouville problems in geophysics, Geophys. J. Roy. Astronom. Soc. 63 (1980), 441-465.
[61] Paine, J.W. and Andrew, A.L., Bounds and higher order estimates for Sturm-Liouville eigenvalues, J. Math. Anal. Appl. (To appear).
[62] Paine, J. and de Hoog, F., Uniform estimation of the eigenvalues of Sturm-Liouville problems, J. Austral. Math. Soc. Ser. B 21 (1980), 365-383.
[63] Paine, J., de Hoog, F.R. and Anderssen, R.S., On the correction of finite difference eigenvalue approximations for Sturm-Liouville problems, Computing 26 (1981), 123-139.
[64] Parlett, B.N. and Johnston, O.G., Numerical implementation of variational methods for eigenvalue problems, in: Rice, J.R. (ed.) Mathematical Software (Academic Press, New York, 1971), 357-368.

[65] Peters, G. and Wilkinson, J.H., $Ax = \lambda Bx$ and the generalized eigenproblem, SIAM J. Numer. Anal. 7 (1970), 479-492.
[66] Pruess, S.A., Estimating the Eigenvalues of Sturm-Liouville Problems by Approximating the Differential Equation, Ph.D. Thesis, Purdue Univ. (1970).
[67] Pruess, S., Estimating the eigenvalues of Sturm-Liouville problems by approximating the differential equation, SIAM J. Numer. Anal. 10 (1973), 55-68.
[68] Prüfer, H., Neue Her leitung der Sturm-Liouvilleschen Reihenentwicklung stetiger Funktionen, Math. Ann. 95 (1926), 499-518.
[69] Pryce, J.D., Two Codes for Sturm-Liouville Problems, Rep. CS-81-01, School of Mathematics, Univ. Bristol (1981).
[70] Radzievskii, G.V, The problem of the completeness of root vectors in the spectral theory of operator-valued functions, Russian Math. Surveys 37 (1982), 91-164.
[71] Raptis, A.D., Two-step methods for the numerical solution of the Schrödinger equation, Computing 28 (1982), 373-378.
[72] Reddien, G.W., Some projection methods for the eigenvalue problem, Applicable Anal. 6 (1976), 61-73.
[73] Reid, W.T., Sturmian Theory for Ordinary Differential Equations (Springer, New York, 1980).
[74] Robb, P.J., Computation of Two-sided Bounds for Eigenvalues of Sturm-Liouville Problems, M.Sc. Thesis, La Trobe Univ. (1979).
[75] Scott, M.R., Invariant Imbedding and its Application to Ordinary Differential Equations : an introduction (Addison - Wesley, Reading, Mass., 1973).
[76] Shampine, L.F., Efficiency of phase function methods for Sturm-Liouville eigenvalues, J. Inst. Math. Appl. 23 (1979), 413-420.
[77] Smooke, M.D., Error Estimates for Piecewise Perturbation Series Solutions of Parabolic and Hyperbolic Equations, Rep. SAND80-8627 (Sandia Laboratories, 1980).
[78] Smooke, M.D., Piecewise analytical perturbation series solutions of the radial Schrödinger equation: one dimensional case, SIAM J. Sci. Statist. Comput. 3 (1982), 195-222.
[79] Smooke, M.D., Error estimates for piecewise perturbation series solutions of the radial Schrödinger equation, SIAM J. Numer. Anal. 20 (1983), 279-295.
[80] Sobouti, Y., A definition of the g- and p-modes of self-gravitating fluids, Astronom. and Astrophys. 55 (1977), 327-337.
[81] Strang, G. and Fix, G.J., An Analysis of the Finite Element Method (Prentice-Hall, Englewood Cliffs, N.J., 1973).
[82] Tassoul, J.-L., Theory of Rotating Stars (Princeton Univ. Press, Princeton, 1978).
[83] Terray, J., and Lancaster, P., On the numerical calculation of eigenvalues and eigenvectors of operator polynomials, J. Math. Anal. Appl. 60 (1977), 370-378.
[84] Voss, H. and Werner, B., A minimax principle for nonlinear eigenvalue problems with applications to nonoverdamped systems, Math. Methods Appl. Sci. 4 (1982), 415-424.
[85] Watts, H.A., Initial value integrators in BVP codes, in: Ref. [22], 19-39.
[86] Weinberger, H.F., Variational Methods for Eigenvalue Approximation (SIAM, Philadelphia, 1974).
[87] Weinstein, A. and Stenger, W., Methods of Intermediate Problems for Eigenvalues. Theory and ramifications (Academic Press, New York, 1972).
[88] Wendroff, B., Bounds for eigenvalues of some differential operators by the Rayleigh-Ritz method, Math. Comp. 19 (1965), 218-224.
[89] Wilkinson, J.H. and Reinsch, C., Handbook for Automatic Computation: Linear Algebra (Springer, Berlin, 1971).

Computational Techniques & Applications: CTAC-83
J. Noye & C. Fletcher (Editors)

LINEAR PROGRAMMING METHODS FOR THE INVERSION OF DATA

R.S. Anderssen
Division of Mathematics and Statistics
CSIRO, PO Box 1965
Canberra City, ACT 2601
Australia

Sven-Åke Gustafson
NADA, Royal Inst. of Technology
S-10044 Stockholm 70, Sweden
and
Centre for Mathematical Analysis
Australian National University
GPO Box 4, Canberra, ACT 2601, Australia

Since indirect measurement problems are invariably improperly posed, it is necessary to take this fact into account when constructing computational procedures for their approximate solution. Usually, some form of "stabilization" is used to remove from consideration physically unrealistic solutions. In recent papers proposing the use of linear programming procedures, the need for such stabilization has been overlooked. In this paper, singular value decomposition is used for stabilizing such procedures.

1. INTRODUCTION

In many practical situations, the required information can only be obtained from indirect measurements in the sense that some effect of the phenomenon of interest, not the phenomenon itself, is all that can be observed. For example, in seismic exploration, information about the velocity of seismic waves as a function of depth below the Earth's surface must be obtained from indirect measurements such as the travel times of seismic waves from an explosion or earthquake to seismometers placed at various locations around this source. The price that must be paid for using such indirect measurements is the added scientific and mathematical sophistication required to ensure that they yield conclusions consistent with the information they contain - this is the essence of (geophysical) data inversion.

For example, if the underlying mathematical formalism reflects the fact that the data being used to derive the required information only measures some consequence of the phenomenon of interest, then it usually fails to satisfy the second and third of Hadamard's conditions for a problem to be *properly posed*. They are: (i) the problem has a solution; (ii) the solution is unique; and (iii) the solution depends "continuously" on the data. For this reason, (geophysical) data inversion is invariably *improperly posed*.

Data inversion problems for which (ii) fails arise naturally as an essential characteristic of mathematical formalisms in geophysics. They are symptomatic of situations where non-uniqueness and bifurcation occur naturally. For example, Backus and Gilbert (1967) have examined quite general conditions under which a finite set of observations (data) fails to uniquely resolve continuous structure, unless the continuous structure is parametrized such that the number of parameters is less than the number of independent observations.

By its very nature, a computational procedure is usually limited to approximating problems with uniquely defined solutions. As a direct consequence, (i) and (ii) represent the minimum requirements which any mathematical formalism must satisfy before standard computational procedures can be applied. Thus, when (ii) fails to hold, it is necessary to either introduce additional constraints which guarantee a uniquely defined solution, or resort to methods which allow the family of solutions supported by the given data to be analysed to see whether or not they consistently determine properties (i.e. possess common properties) which are of significance or interest.

Much of the discussion about data inversion in the geophysical literature centers on which of these two alternatives is the preferred option and on how they should be implemented.

In this paper, we examine the use of linear programming methods for exploring the family of solutions supported by given data.

The failure of (iii) does not of itself represent a bar to the application of standard computational procedures, but the results thereby generated do. In fact, even a slight failure of (iii), in the sense that the uniquely defined solution is only mildly sensitive to small perturbations in the data, can pose difficulties numerically. For example, even the numerical differentiation of observational data can exhibit acute sensitivity to the actual form of the observational data. Thus, ways of coping with such sensitivity numerically represent an important consideration in any discussion of computational procedures for the inversion of data.

In a nutshell, it is necessary to introduce some form of "stabilization" to remove from consideration geophysically unrealistic solutions. In papers published on the use of linear programming methods for geophysical data inversion, the necessity to introduce such stabilization has been ignored. It is the central consideration of the present paper.

2. THE GENERAL METHODOLOGY FOR IMPROPERLY POSED PROBLEMS

Starting with the improperly posed mathematical formalism of some geophysical data inversion problem, there are two independent ways to proceed to obtain a well-posed computational process. In the classical approach, "stabilization" is introduced into the mathematical formalism before discretization. This strategy includes the mathematical approach to regularization (de Hoog (1980) and Lukas (1980)). One of its advantages is that it can sometimes yield an explicit characterization of the resulting regularization solution (e.g. the regularization interpretation of smoothing splines) (Wahba (1977)). However, it is not this approach which will be pursued here, but the alternative algebraic approach.

In the algebraic approach, the mathematical formalism is initially discretized to obtain a corresponding algebraic formalism before the "stabilization" is applied. Its advantage is that numerous ways exist for stabilizing algebraic systems for which there is potentially readily available software.

We assume that the exact discretized algebraic system takes the form

$$A \underset{\sim}{x} = \underset{\sim}{b} \tag{1}$$

where A is an $m \times n$ matrix, $\underset{\sim}{x}$ is a vector of discretized values of the phenomenon of interest and $\underset{\sim}{b}$ is a vector of discretized values of the observed effect of the phenomenon, and that the only information available about the right hand side vector $\underset{\sim}{b}$ is observational; i.e. one does not observe $\underset{\sim}{b}$ but

$$\underset{\sim}{d} = \underset{\sim}{b} + \underset{\sim}{\varepsilon} \tag{2}$$

where ε corresponds to a vector of observational errors.

Thus, the problem is not simply to solve (1), but to solve (1) given that the only information available about $\underset{\sim}{b}$ is $\underset{\sim}{d}$. Computationally, the various ways in which this can be done can be classified as either uniqueness modelling or non-uniqueness modelling.

Uniqueness Modelling

In uniqueness modelling, additional assumptions about the form of the solution vector $\underset{\sim}{x}$, the relative size of m and n and the structure of the error vector $\underset{\sim}{\varepsilon}$ are invoked so that (1) and (2) together with these assumptions yield a unique estimate of $\underset{\sim}{x}$. Most of the methods used in applications can be classified as either statistical modelling or algebraic regularization.

Statistical Modelling

In the statistical approach, it is assumed that

(i) $m > n$ (the problem is overdetermined with (considerably) more observations than the number of components defining the solution $\underset{\sim}{x}$);

(ii) the error vector $\underset{\sim}{\varepsilon}$ has a known probabilistic structure.

The solution $\underset{\sim}{x}$ can either correspond to parameters in some model of the required phenomenon of interest or to point estimates of it. The form will depend on how the discretization of the mathematical formalism to yield (1) has been accomplished.

For example, if $m > n$ and it is assumed that the components of $\underset{\sim}{\varepsilon}$ are normally distributed errors with zero mean and the same variance σ^2, then the unique estimate of $\underset{\sim}{x}$ becomes the least squares solution (assuming the rank of A equals n)

$$\hat{\underset{\sim}{x}} = (A^T A)^{-1} A^T \underset{\sim}{d} . \qquad (3)$$

The disadvantage of this direct statistical approach is that it does not allow for the possibility that (1) has been obtained as the discretization of an improperly rather than a properly posed problem. When (1) results from the discretization of an improperly posed problem, A will be poorly conditioned in the sense that the linear independence of the columns of A will be weak. As a direct consequence, $\hat{\underset{\sim}{x}}$ will be poorly determined in that it will contain high frequency contributions which have no meaning (geophysically).

Algebraic Regularization

In order to allow for the fact that A has been obtained as the discretization of an improperly posed problem, conditions are invoked which aim to remove from any unique estimate of $\underset{\sim}{x}$ the meaningless high frequency contributions which are symptomatic of the statistical approach. The aim is to damp out of any unique estimate of $\underset{\sim}{x}$ contributions from $\underset{\sim}{\varepsilon}$ which would have arisen if more standard methods had been used to solve (1) and (2). Mathematically, this can be achieved by seeking an estimate of $\underset{\sim}{x}$ which not only minimizes some measure of the residual

$$\underset{\sim}{r} = A \underset{\sim}{x} - \underset{\sim}{d} , \qquad (4)$$

but also some measure of $\underset{\sim}{x}$ or a linear transformation of $\underset{\sim}{x}$, $B\underset{\sim}{x}$, which discourages the use of high frequency contributions in the construction of this estimate, and guarantees its uniqueness.

There are various ways in which this can be done. The most common is to construct a trade-off between the minimization of the chosen measure on $\underset{\sim}{r}$, $\rho(\underset{\sim}{r})$, and the chosen measure on $B\underset{\sim}{x}$, $\sigma(B\underset{\sim}{x})$; viz.

$$\min_{\underset{\sim}{x}} \rho(\underset{\sim}{r}) + \alpha\, \sigma(B\underset{\sim}{x}) , \quad 0 < \alpha < \infty . \qquad (5)$$

When ρ and σ correspond to the Euclidean norms of the vectors $\underset{\sim}{r}$ and $B\underset{\sim}{x}$, the standard form for *algebraic regularization* is obtained; viz.

$$\min_{\underset{\sim}{x}} \{\|A \underset{\sim}{x} - \underset{\sim}{d}\| + \alpha \|B\underset{\sim}{x}\|\} , \quad 0 < \alpha < \infty ; \qquad (6)$$

with the corresponding Euler-Lagrange equations for the unique minimizer $\bar{\underset{\sim}{x}}$ defined by the matrix equation

$$A^T A\underset{\sim}{x} + \alpha B^T B\underset{\sim}{x} = A^T \underset{\sim}{d} . \qquad (7)$$

There is no optimal choice for the linear transformation $B\underset{\sim}{x}$. Where possible it should be based on the problem context. A popular choice is to take $B\underset{\sim}{x}$ to be the matrix of 2nd order divided differences defined on the components of $\underset{\sim}{x}$; viz.

$$B\underset{\sim}{x} = \begin{bmatrix} 0 & & & & & \\ 1 & -2 & 1 & & & \\ & 1 & -2 & 1 & & \\ & & \ddots & \ddots & \ddots & \\ & & & 1 & -2 & 1 \\ & & & & & 0 \end{bmatrix} \begin{bmatrix} x_1 \\ x_2 \\ x_3 \\ \vdots \\ x_{n-1} \\ x_n \end{bmatrix} . \tag{8}$$

When $B = I$, the Euler-Lagrange equations become

$$(A^T A+\alpha I)\bar{\underset{\sim}{x}} = A^T \underset{\sim}{d} \tag{9}$$

which corresponds to the matrix equation representation for stabilized least squares and ridge regression. It shows immediately the role played by the stabilization term $\alpha\|B\underset{\sim}{x}\|$ in (6). For suitably large α , it gives effective computational linear independence to the matrix $A^T A + \alpha I$.

Computationally, the singular value decomposition of A (cf. Strang (1976)) can be used to construct the stabilized least squares solution $\bar{\underset{\sim}{x}}$ defined by (9).

Non-Uniqueness Modelling

In non-uniqueness modelling, one aims to characterize the family of geophysically realistic solutions which, with respect to the mathematical formalism being used to model the problem, are determined by given data. The nature of such problems (where the mathematical formalism is continuous while the number of data is finite) is such that, without invoking additional assumptions, they will support infinitely many solutions if the conditions defining the mathematical formalism and data are not inconsistent. This situation has been analysed in some detail by Backus and Gilbert (1967). Non-uniqueness modelling exploits the fact that, though there are infinitely many solutions, it does not follow that they are infinitely varied; and therefore seeks the common structural features (important geophysically) which they exhibit. The advantage of such information will be that it has been obtained under less restrictive assumptions than those invoked to yield a single solution.

Structurally, non-uniquness modelling represents one way of implementing the maximum entropy criterion of determining solutions which are consistent with the given data but are maximally non-committal with regard to unavailable data (cf. Ables (1974)).

The two distinct types of methods which have been used to implement non-uniqueness modelling are Monte Carlo Inversion and Linear Programming Inversion.

Monte Carlo Inversion

Starting with an *a priori* defined set of realistic admissible solutions for the problem, the essence of Monte Carlo Inversion is to randomly sample this set in search of members which satisfy the given problem with a suitable accuracy (Anderssen and Seneta (1971)). The common structural features which this set of "successful" solutions displays reflects the information contained in the data and can be used to answer questions about the problem on the basis of the data being used.

This methodology has been applied successfully to various problems in geophysics (e.g. Press (1970), and Anderssen, Worthington and Cleary (1972)). Its major disadvantage is its expense computationally. For most problems, it will be necessary to sample a huge number of admissible solutions before even a small set of successful ones are found.

Linear Programming Inversion

If it is assumed that the only information known about the errors $\underset{\sim}{\varepsilon}$ in (2) are lower and upper bounds $\underset{\sim}{\ell}$ and $\underset{\sim}{u}$; viz.

$$\underset{\sim}{\ell} \leq \underset{\sim}{\varepsilon} \leq \underset{\sim}{u} \quad \text{(component-wise)} \; ; \tag{10}$$

non-uniqueness modelling reduces to finding the family of solutions which satisfy (1), (2) and (10).

Johnson (1972) proposed the following linear programming method for the solution of this problem: for $i=1,2,\ldots,n$, determine

$$\underline{x}_i = \min_{\underset{\sim}{x}} e_i^T \underset{\sim}{x} \quad \text{and} \quad \bar{x}_i = \max_{\underset{\sim}{x}} e_i^T \underset{\sim}{x} \; , \tag{11}$$

subject to the linear constraints

$$\underset{\sim}{\ell} \leq A \underset{\sim}{x} - \underset{\sim}{d} \leq \underset{\sim}{u} \; , \quad \underset{\sim}{x} \geq 0 \; , \tag{12}$$

where e_i denotes the unit vector with zeros in all positions but the i-th. It is often referred to as *linear programming inversion;* or *LP-inversion*. There are two ways to interpret this linear programming procedure. On the one hand, it determines the box (with faces (sides) parallel to the coordinate planes which define the positive "quadrant" in $\mathbb{R}^n$) which contains the simplex of feasible solutions defined by the constraints (12). On the other, it defines in solution-space an envelope within which all geophysically realistic solutions, consistent with the constraints (12), will lie.

It follows naturally from the first interpretation that, because the box contains the simplex, a larger set of solutions than just the feasible solutions are determined. The non-feasible solutions in this set will correspond to geophysically unrealistic solutions (such as highly oscillatory solutions) which lie within the envelope of the second interpretation. In fact, one of the defects of LP-inversion is its failure to characterize more explicitly what defines a geophysically realistic solution.

Two disadvantages associated with LP-inversion are: (i) the need to repeatedly solve the same basic linear programming problem; (ii) the fact that the objective functions defined by (11) do not stabilize the information contained in (12). As it stands, the only stabilization is determined by the choice of $\underset{\sim}{\ell}$ and $\underset{\sim}{u}$.

In the form (11) and (12), LP-inversion has been used to solve the travel-time inversion problem in geophysics (Garmany *et al* (1979)).

3. STABILIZATION OF LP-INVERSION

We base the stabilization of LP-inversion on the two independent ways in which the singular value decomposition of the matrix A can be stabilized so that spurious oscillations in the solution of

$$A \underset{\sim}{x} = \underset{\sim}{b} , \quad \underset{\sim}{d} = \underset{\sim}{b} + \underset{\sim}{\varepsilon} ,$$

are damped out. Since the improperly posed nature of a formulation is a direct consequence of the structure of A , this would seem the most natural way to approach the stabilization of LP-inversion.

If rank(A) = r $\le$ min(m,n) , then *the singular value decomposition* of A takes the form (Golub and Reinsch (1970))

$$A = UDV^T , \tag{13}$$

where U is the orthogonal matrix of the unit eigenvectors of AA^T , V is the orthogonal matrix of the unit eigenvectors of A^TA , and D is the diagonal matrix of non-negative square roots, σ_i , of the eigenvalues of A^TA (called *the singular values* of A) ordered so that

$$\sigma_1 \ge \sigma_2 \ge \dots \ge \sigma_{r-1} \ge \sigma_r > \sigma_{r+1} = \dots = \sigma_r = 0 . \tag{14}$$

The two methods of stabilization we consider are:

Truncated LP-Inversion For i=1,2,...,n , determine

$$\underset{\sim}{x}_i = \min_{\underset{\sim}{x}} \underset{\sim}{e}_i^T \underset{\sim}{x} \quad \text{and} \quad \bar{x}_i = \max_{\underset{\sim}{x}} \underset{\sim}{e}_i^T \underset{\sim}{x} , \tag{15}$$

subject to the linear *truncated* constraints

$$\underset{\sim}{\ell} \le A_p \underset{\sim}{x} - \underset{\sim}{d} \le \underset{\sim}{u} , \qquad \underset{\sim}{x} \ge 0 , \tag{16}$$

where A_p denotes the truncated singular value decomposition of A of order p ; viz.

$$A_p = UD_pV^T , \quad D_p = \text{diag}(\sigma_1,\sigma_2,\ldots,\sigma_p,0,\ldots,0) , \tag{17}$$

with $p \le r$.

Regularized LP-Inversion For $i=1,2,\ldots,n$, determine

$$\underline{x}_i = \min \underset{\sim}{e}_i^T \underset{\sim}{x} \quad \text{and} \quad \bar{x}_i = \max \underset{\sim}{e}_i^T \underset{\sim}{x} \tag{18}$$

subject to the linear regularized constraints

$$\underset{\sim}{\ell} \le \bar{A}_\alpha \underset{\sim}{x} - \underset{\sim}{d} \le \underset{\sim}{u} , \qquad \underset{\sim}{x} \ge 0 , \tag{19}$$

where $\bar{A}_\alpha$ denotes the regularized singular value decomposition of A ; viz.

$$\bar{A}_\alpha = U(D+\alpha I)V^T , \quad \alpha > 0 . \tag{20}$$

The first step is to show that both these procedures do in fact stabilize LP-inversion. Clearly, if some modification is to stabilize LP-inversion, it must produce some form of contraction in the shape and size of the simplex of feasible solutions. We therefore introduce the following definition.

DEFINITION 1 A modification to an LP-inversion is called *stabilizing* if it yields a contraction in the magnitudes of the coordinates of the feasible solutions; and *strongly stabilizing* if it yields a contraction in the shape and size of the simplex of feasible solutions defined by the original LP-inversion formulation.

It follows immediately from this definition that, for a modification of LP-inversion which is strongly stabilizing, the size of the box which contains the simplex of feasible solutions is reduced, and the envelope which contains all geophysically realistic solutions becomes narrower (though the change is not necessarily uniform).

Clearly, the simplest modifications to the LP-inversion (11) and (12) which are strongly stabilizing are ones based on increasing the lower bounds $\underset{\sim}{\ell}$ and/or decreasing the upper bounds $\underset{\sim}{u}$. For the two modifications of LP-inversion introduced in this section, we can prove

THEOREM 1 *Truncated LP-inversion is strongly stabilizing.*

Proof It is only necessary to examine the effect of replacing A with A_r on the constraints (12) which can be rewritten as

$$\underset{\sim}{\ell} \le UD\underset{\sim}{p} - \underset{\sim}{d} \le \underset{\sim}{u} , \quad \underset{\sim}{p} = V^T \underset{\sim}{x} , \quad \underset{\sim}{x} \ge 0 , \tag{21}$$

where the coordinate transformation $V^T \underset{\sim}{x} = \underset{\sim}{p}$ simply redefines the identification of the set of feasible solutions. On replacing A by A_r in (12) the corresponding form of (21) becomes

$$\underset{\sim}{\ell} \le U^T D_r \underset{\sim}{p} - \underset{\sim}{d} \le \underset{\sim}{u} \; , \quad \underset{\sim}{p} = V^T \underset{\sim}{x} \; , \quad \underset{\sim}{x} \ge 0 \; . \tag{22}$$

Thus, the effect of replacing A by A_r is to delete from consideration the (r+1)-th to n-th components of $\underset{\sim}{p}$; the least important geophysically of the components of $\underset{\sim}{p}$. In this way, we see that the truncation of LP-inversion yields an LP-inversion over a reduced number of variables. Clearly, it stabilizes LP-inversion strongly by contracting the set of feasible solutions to those which lie in the intersection of the hyperplane defined by $\underset{\sim}{p}_r = (p_1 p_2, \ldots, p_r)$ and the original simplex of feasible solutions. #

This proof also indicates that the truncation of LP-inversion is a geophysically realistic way to stabilize LP-inversion since the effect of replacing $D\underset{\sim}{p}$ by $D_r\underset{\sim}{p}$ is to delete from consideration the projections of $\underset{\sim}{x}$ associated with the more oscillatory components of $A^T A$.

<u>THEOREM 2</u> *Regularized LP-inversion is stabilizing, and conditionally strongly stabilizing.*

<u>Proof</u> We proceed in the same manner as in the proof of Theorem 1. We first observe that (21) can be rewritten as

$$\underset{\sim}{\ell} \le U\underset{\sim}{\phi} - \underset{\sim}{d} \le \underset{\sim}{u} \; , \quad \underset{\sim}{\phi} = D\underset{\sim}{p} = DV^T \underset{\sim}{x} \; , \quad \underset{\sim}{x} \ge 0 \; . \tag{23}$$

Next, on replacing A by $\bar{A}_\alpha$ in (12), the corresponding form of (23) becomes

$$\underset{\sim}{\ell} \le U\bar{\underset{\sim}{\phi}} - \underset{\sim}{d} \le \underset{\sim}{u} \; , \quad \bar{\underset{\sim}{\phi}} = (D+\alpha I)\bar{\underset{\sim}{p}} = (D+\alpha I)V^T\bar{\underset{\sim}{x}} \; , \quad \bar{\underset{\sim}{x}} \ge 0 \; . \tag{24}$$

Let F denote the set of feasible solutions defined by the simplex

$$\underset{\sim}{\ell} \le U\underset{\sim}{\phi} - \underset{\sim}{d} \le \underset{\sim}{u} \; .$$

Since $V^T\underset{\sim}{x} = \underset{\sim}{p}$ simply defines a change of coordinates, properties of the set of feasible solutions F when changing from the $\underset{\sim}{x}$'s to the $\bar{\underset{\sim}{x}}$'s can be proved by examining their properties when changing from the $\underset{\sim}{p}$'s to the $\bar{\underset{\sim}{p}}$'s. Since it follows from (23) and (24) that

$$\underset{\sim}{p} = D^{-1}\underset{\sim}{\phi} \quad \text{and} \quad \bar{\underset{\sim}{p}} = (D+\alpha I)^{-1}\underset{\sim}{\phi} \; , \quad \underset{\sim}{\phi} \in F \; ,$$

we obtain that a change from the $\underset{\sim}{p}$'s to the $\bar{\underset{\sim}{p}}$'s corresponds to a decrease in the magnitudes of the coordinates of the $\bar{\underset{\sim}{p}}$'s which proves that regularized LP-inversion is stabilizing. As an immediate consequence, it follows that regularized LP-inversion defines a contraction in the set of feasible solutions F , and hence is strongly stabilizing, if the origin of the $\underset{\sim}{\phi}$'s (and hence the $\underset{\sim}{p}$'s) lies within or on the feasible set F . #

It is clear from the proofs of both these theorems that the stabilization is accomplished by modifying the most the more oscillatory components in $\underset{\sim}{x}$.

REFERENCES

[1] Ables, J.G., Maximum entropy spectral analysis, Astron. Astrophys. Suppl. 15 (1974), 383-393.

[2] Anderssen, R.S. and Seneta, E., A simple statistical estimation procedure for Monte Carlo inversion in geophysics, Pure Appl. Geophysics 91 (1971), 5-13.

[3] Anderssen, R.S., Worthington, M.H. and Cleary, J.R., Density modelling by Monte Carlo inversion - I methodology, Geophys, J.R. Astronom. Soc. 29 (1972), 433-444.

[4] Backus, G.E. and Gilbert, F.J., Numerical application of a formalism for geophysical inverse problems, Geophys. J.R. Astronom. Soc. 13 (1967), 247-276.

[5] Garmany, J., Orcutt, J.A. and Parker, R.L., Travel time inversion: a geometric approach, J.G.R. 84 (1979), 3615-3622.

[6] Glashoff, K. and Gustafson, S.-Å., Linear Optimization and Approximation (Springer-Verlag, New York, Heidelberg, Berlin, 1983).

[7] Golub, G.H. and Reinsch, C., Singular value decomposition and least squares solutions, Numer. Math. 14 (1970), 403-420.

[8] de Hoog, F.R., Review of Fredholm equations of the first kind, in: Anderssen, R.S., de Hoog, F.R. and Lukas, M.A. (eds), The Application and Numerical Solution of Integral Equations (Sijthoff and Noordhoff, Alphen aan den Rijn, The Netherlands, 1980).

[9] Johnson, C.E., Regional Earth Models from Linear Programming Methods, M.Sc. Thesis, MIT (Sept. 1972).

[10] Lukas, M.A., Regularization, in: Anderssen, R.S., de Hoog, F.R. and Lukas, M.A. (eds), The Application and Numerical Solution of Integral Equations (Sijthoff and Noordhoff, Alphen aan den Rijn, The Netherlands, 1980).

[11] Press, F., Earth models consistent with geophysical data, Phys. Earth Planet. Interiors, 3 (1970), 3-22.

[12] Strang, G., Linear Algebra and Its Applications (Academic Press, New York, 1976).

[13] Wahba, G., Practical approximate solutions to linear operator equations when the data are noisy, SIAM J. Numer. Anal. 14 (1977), 651-667.

Computational Techniques & Applications: CTAC-83
J. Noye & C. Fletcher (Editors)

COMPUTING THE FOLIAGE ANGLE DISTRIBUTION FROM CONTACT FREQUENCY DATA

D.R. Jackett and R.S. Anderssen

CSIRO Division of Mathematics and Statistics
Box 1965 GPO, Canberra ACT 2601.

Abstract

In 1965, Miller derived an inversion formula for the integral equation which defines the known contact frequency in terms of the unknown foliage angle distribution of leaves in a canopy. Subequently, in 1967, he showed that the average foliage density could be computed directly from the contact frequency. This formalised mathematically Warren Wilson's experimental verification that the leaf area index could be estimated as a linear combination of the measured contact frequency. Recently, it has been shown that Miller's result is a special case of a general transformation which allows linear functionals defined on the foliage angle distribution to be evaluated as linear functionals defined on the contact frequency. In this paper, we explore the consequences of this result for the underlying computational problem.

1. Introduction In order to economize on available resources and time, the structure and status of plant canopies are often assessed from indirect measurements,such as hemispherical photographs from the ground and satellite monitoring from above. Though direct measurements could be made they would be tedious and logistically difficult when the plants (e.g. trees) are tall. Even destructive methods such as tree felling are of little assistance because the orientation and relative position of the leaves are key factors.

For these and other reasons, considerable attention has been devoted to the use of indirect measurements to study the structure and status of plant canopies. The key process under examination is the interaction of sunlight with leaves. Clearly, the amount of sunlight absorbed by the leaves will depend on their orientation (defined by their normals) and area, and the direction of the sunlight.

Since the normals of the leaves do not follow a clearly defined deterministic pattern, their angle distribution must be stochastically modelled. The probability density function model in popular use is the azimuthally averaged foliage angle density function $g(\alpha)$, where $g(\alpha)\, d\alpha$ defines the contribution to the foliage density of leaves inclined to the horizontal at angles between α and $\alpha + d\alpha$. This is based on the assumption that the azimuth angle distribution of the foliage normals is uniform. Clearly, this type of model replaces the locational variation with a grouped variation and involves an equivalencing based on a trade-off between leaf area and leaf number.

The indirect measurements can take various forms, but we limit attention to those measurements which can be related back to the azimuthally averaged mean canopy projection $f(\beta)$ in the elevation direction β. This projection is the concept of common reference between the indirect measurements and the foliage angle density $g(\alpha)$. On the one hand, a number of authors (Reeve in Warren Wilson [1960], Philip [1965], and Smith, Oliver and Berry [1977]) have derived the integral equation which defines $f(\beta)$ in terms of $g(\alpha)$

$$f(\beta) = \int_0^{\pi/2} k(\alpha,\beta)\, g(\alpha)\, d\alpha, \qquad 0 \leqslant \alpha \leqslant \pi/2, \tag{1}$$

where

$$k(\alpha,\beta) = \begin{cases} \cos\alpha \sin\beta & \alpha \leqslant \beta, \\ \cos\alpha \sin\beta \left[1+\frac{2}{\pi}(\tan\theta - \theta)\right] & \alpha \geqslant \beta, \end{cases} \tag{2}$$

with

$$\theta = \cos^{-1}(\tan\beta/\tan\alpha), \qquad 0 \leqslant \theta \leqslant \pi/2. \tag{3}$$

On the other, $f(\beta)$ can be determined from various types of indirect measurements.

For example, if the indirect measurements correspond to the proportion of azimuthally averaged gap $P_o(\beta)$ (as a function of the elevation angle β), as obtained from hemispherical photographs of the canopy (Anderson [1964,1971]), then, using an appropriate stochastic model for the leaves (e.g. leaves of equal area with centroids randomally dispersed by a spatially stationary angular (Poisson) distribution), it follows that

$$P_o(\beta) = \exp(-f(\beta)/\sin\beta), \tag{4}$$

or equivalently

$$f(\beta) = -\sin\beta \ln(P_o(\beta)). \tag{5}$$

Together, (1)-(5) define the relationship between the indirect measurements $P_o(\beta)$ and the phenomenon of interest $g(\alpha)$. However, this does not solve the problem, but only yields a foundation on which the study of the structure and status of a plant canopy can be built. In reality the practitioner not only requires a recipe for solving (1)-(3), but wants the value of some key indicator such as the leaf area index

$$F = \int_0^{\pi/2} g(\alpha)\, d\alpha \tag{6}$$

given that the only information about $f(\beta)$ is

$$d_i = f(\beta_i) + \varepsilon_i, \quad 0 \leqslant \beta_1 < \beta_2 < \ldots < \beta_n \leqslant \pi/2. \tag{7}$$

where the ε_i denote observational (random) errors, and given that n is small and the variance of ε_i is large.

Prior to 1960, this was a virtually insurmountable problem. Warren Wilson's [1960, 1963] way around it was to seek the angle β_1 and the weight w_1 such that, with only one measurement of $f(\beta)$ at the elevation angle β_1,

$$w_1 f(\beta_1) \tag{8}$$

yields the best estimate of F from among all such estimates. This obviously generalizes to seeking the angles $\beta_1, \beta_2, \ldots, \beta_m$ and the weights $w_1, w_2, \ldots, w_m$ such that, with only m measurements of $f(\beta)$,

$$w_1 f(\beta_1) + w_2 f(\beta_2) + \ldots + w_m f(\beta_m) \tag{9}$$

yields the best estimate of F from among all such estimates.

Various attempts were made to determine the angles and weights in (9) (Philip [1965]). However, in 1967, Miller [1967] obtained the full solution by showing that they must correspond to the angles and weights which define any one of the possible quadrature formulae for the integral

$$2 \int_0^{\pi/2} \cos\beta \, f(\beta) \, d\beta \quad . \tag{10}$$

Using the inversion formula

$$g(\alpha) = -\tan\alpha \sec^3\alpha \quad x$$

$$\int_\alpha^{\pi/2} \frac{d}{d\beta}\{\cos^3\beta[f(\beta) + f''(\beta)]\} (\tan^2\beta - \tan^2\alpha)^{-\frac{1}{2}} d\beta \tag{11}$$

he had previously derived (Miller [1963]), Miller simply substituted it for $g(\alpha)$ in (6) and showed that

$$\int_0^{\pi/2} g(\alpha) \, d\alpha = 2 \int_0^{\pi/2} \cos\beta \, f(\beta) \, d\beta. \tag{12}$$

That (1)-(3) is improperly posed is an automatic consequence of the differentiation in the inversion formula (11). On the basis of the above result, this difficulty can be circumvented if one only works with the linear functionals defined on $f(\beta)$; e.g.

$$\int_0^{\pi/2} \phi(\beta) \, f(\beta) \, d\beta, \qquad \phi(\beta) \text{ known.} \tag{13}$$

But the functionals of interest are defined on $g(\alpha)$ (Anderssen, Jackett and Jupp [1983])

$$\int_0^{\pi/2} \theta(\alpha)\, g(\alpha)\, d\alpha, \qquad \theta(\alpha) \text{ known.} \tag{14}$$

Thus, it is necessary to generalize Miller's result (12): for a given $\theta(\alpha)$, determine the corresponding linear functional defined on $f(\beta)$ (it will often take the form (13), but possibly in a generalized sense because it may involve point functionals at $\alpha = 0$ and $\alpha = \pi/2$). This has been done by Anderssen and Jackett [1983].

In the present paper, we examine some of the computational aspects which arise in replacing the evaluation of a known functional defined on $g(\alpha)$ by the evaluation of its counterpart defined on $f(\beta)$.

2. Computation of the Linear Functionals

To aid the discussion of the computation of the linear functionals which arise in the study of foliage angle density, it is desirable to have some examples of the practically significant functionals. We give four examples:

a. Leaf Area Index,

$$\int_0^{\pi/2} g(\alpha)\, d\alpha, \quad \theta(\alpha) = 1; \tag{15}$$

b. Vertical and Horizontal Projectors of the Leaf Area Index (c.f. Suits [1972]),

$$L_{\sin\alpha}(g) = \int_0^{\pi/2} \sin\alpha\, g(\alpha)\, d\alpha, \qquad \theta(\alpha) = \sin\alpha, \tag{16}$$

$$L_{\cos\alpha}(g) = \int_0^{\pi/2} \cos\alpha\, g(\alpha)\, d\alpha, \qquad \theta(\alpha) = \cos\alpha; \tag{17}$$

c. The Segmented Foliage Density (c.f. Ross [1981]),

$$L_{\chi[a,b]}(g) = \int_0^{\pi/2} \chi[a,b]\, g(\alpha)\, d\alpha, \quad 0 \leqslant a < b \leqslant \pi/2 \tag{18}$$

where

$$\theta(\alpha) = \chi[a,b] = \begin{cases} 1 & \text{if } \alpha \in [a,b], \\ 0 & \text{if } \alpha \notin [a,b]; \end{cases} \tag{19}$$

d. The Moments of $g(\alpha)$,

$$L_{\alpha^p}(g) = \int_0^{\pi/2} \alpha^p\, g(\alpha)\, d\alpha, \quad \theta(\alpha) = \alpha^p, \ p > 0. \tag{20}$$

Since the leaf area index measures by how many times the area of leaves exceeds the basal area of the vertical cylinder within which the leaves are located, it is obvious that it is an important quantity for anyone concerned with studying the properties of vegetation. The vertical and horizontal projectors of leaf area index have also been used extensively. For example, forest researchers use these quantities (as well as the leaf area index) as the fundamental variables in photosynthesis and radiative transfer models of canopies.

The segmented foliage density is extremely useful when the data is noisy and sparse, which is usually the practical situation. In this case it is unrealistic to expect the inversion techniques to yield a reliable estimate of $g(\alpha)$, particularly considering the improperly posed nature of the original integral equation and the inversion formula. In any case, what is often required is only a rough qualitative picture of $g(\alpha)$, such as a piecewise constant approximation with only 3 or 4 steps on $[0,\pi/2]$. This can be adequately supplied by the segmented density.

Since $g(\alpha)$ represents a probability density function, the moments of $g(\alpha)$ are obviously useful for determining the statistical properties of $g(\alpha)$.

A theoretical transformation from functionals on $g(\alpha)$ to functionals on $f(\beta)$ is given by the following.

Proposition 2.1. Let f be such that f'' exists and is absolutely continuous on $[0,\pi/2]$ and $f'(0) = f'(\pi/2) = 0$, and assume that $\theta(\alpha)$ is a given function on $[0,\pi/2]$ such that $\theta'(\alpha)$ exists and is continuous. Then the linear functionals

$$L_\theta(g) = \int_0^{\pi/2} \theta(\alpha)\, g(\alpha)\, d\alpha \tag{21}$$

can be transformed to

$$L_\theta(g) = \theta(0)f(\pi/2) + \int_0^{\pi/2} \psi(\beta)\, \sin\beta\, \left[f(\beta) + f''(\beta)\right] d\beta, \tag{22}$$

where

$$\psi(\beta) = \int_0^{\beta} \frac{d}{d\alpha}\left(\sec\alpha\, \theta(\alpha)\right) \frac{\cos\alpha\, \cos\beta}{\sqrt{\sin^2\beta - \sin^2\alpha}}\, d\alpha \tag{23}$$

Proof: See Anderssen and Jackett [1983].

This result demonstrates a significant difference between the mathematical and computational solutions of a problem. Either $\theta(\alpha)$ above yields simple formula for the associated functional, or, when Proposition (2.1) can be explicitly applied, $\theta(\alpha)$ produces functionals which must be evaluated from a formula very difficult to implement. For example, the functionals a) and b) reduce easily to $2\int_0^{\pi/2} \cos(\beta)f(\beta)\, d\beta$, and $\pi f(0)/2$ and $f(\pi/2)$ respectively, whereas functional c) cannot be transformed by (2.1), and the functionals d) lead to nasty computational formulae.

For a general $\theta(\alpha)$ for which Proposition (2.1) does not simplify, two major difficulties arise. Firstly, the functional involves a second differentiation on the contact frequency f, and unless this differentiation can be shifted to the ψ (which is usually not explicitly the case), large errors will be incurred. Secondly, the evaluation of ψ involves a (square root) singularity, which for most choices of $\theta(\alpha)$ produce elliptic integrals. These integrals in general can only be evaluated by numerical means, thereby introducing further error, which is magnified if the differentiation is moved from f to ψ.

What is required therefore is an alternative to Proposition (2.1) which is more amenable to computation. Fortunately this can be achieved using a generalization of the leaf area index. Thus, consider the functional on the contact frequency defined by

$$I(f;\alpha) = \int_{\alpha}^{\pi/2} \frac{\partial\sqrt{\sin^2\beta - \sin^2\alpha}}{\partial\beta} f(\beta)\, d\beta \quad , \tag{24}$$

where the leaf area index is clearly $2I(f;0)$.

Despite the elliptic nature of $I(f;\alpha)$, it can be computed very accurately and efficiently. Indeed using an even grid of (n+1) points on the interval $[0,\pi/2]$, the product mid-point integration scheme yields the following approximation

$$I(f;\alpha) = \sum_{i=0}^{n-1} f(\bar{\beta}_i) \sqrt{\sin^2 \beta_{i+1} - \sin^2\alpha}$$

$$- \sum_{i=0}^{n-1} f(\bar{\beta}_i) \sqrt{\sin^2 \beta_i - \sin^2\alpha} \tag{25}$$

where $\bar{\beta}_i = \dfrac{\beta_i + \beta_{i+1}}{2}$, $\beta_i = ih$, $i = 0, 1, \ldots, n$, and $h = \dfrac{\pi}{2n}$. Notice this algorithm does not involve any differentiation on the contact frequency.

The segmented foliage density function has a simple mathematical representation in terms of $I(f;\alpha)$.

Proposition 2.2

$$\int_{\alpha}^{\pi/2} g(\alpha)\, d\alpha = 2I(f;\alpha) - \cot\alpha\, I'(f;\alpha) + I''(f;\alpha) \tag{26}$$

where the differentiation is with respect to α.

Proof: See Appendix 1.

What has been achieved is that the differentiation on f in the inversion formula has been transferred to differentiation (of lower order) on $I(f;\alpha)$. Since $I(f;\alpha)$ is also a smoothing of the contact frequency data, this has obvious computational advantages.

Far more important than the evaluation of the segmented density is that Proposition (2.2) yields a stable algorithm for the evaluation of linear functionals on the leaf angle distribution which is independent of any differentiation on the contact frequency. We have

Proposition 2.3 If $\theta(\alpha)$ is such that $\theta'(\alpha)$ exists and is absolutely continuous, then

$$\int_0^{\pi/2} \theta(\alpha)\, g(\alpha)\, d\alpha = -\int_0^{\pi/2} \{2\theta(\alpha) + \cot\alpha\, \theta'(\alpha) + \theta''(\alpha)\}\, I'(f;\alpha)\, d\alpha$$

$$-\theta'(\pi/2)\, f(\pi/2) \qquad (27)$$

Proof: See Appendix 2.

As before, product integration techniques can be used to efficiently evaluate integrals of the form

$$\int_0^{\pi/2} \Phi(\alpha)\, I'(f;\alpha)\, d\alpha.$$

In fact, the discretization used above yields a formula similar to (25). The evaluation of such functionals is now free from differentiation of the contact frequency data, at the expense however of differentiation of $\theta(\alpha)$. Since most of the functionals used in practice have well behaved $\theta(\alpha)$, this presents no computational problems.

Numerical experimentation with synthetic data has been performed, the details of which can be found in Anderssen, Jackett and Jupp [1983]. The conclusions reached there indicate that when equation (27) is used for the leaf area index and the first three moments (i.e., $\theta(\alpha) = 1$, $\theta(\alpha) = \alpha^p$, $p = 1,2,3$), it yields accurate estimates of these quantities even for quite noisy data (e.g. 10%). Also, when equation (26) is used to construct the segmented foliage density, the resulting function, although not as accurate as the functionals above, does preserve the qualitative properties of the underlying $g(\alpha)$. The results also confirm the advantages of working with the contact frequency functionals rather than the corresponding foliage density functionals.

Appendix 1

In this appendix we prove

$$\int_{\alpha}^{\pi/2} g(\alpha)\, d\alpha = 2I(f;\alpha) - \cot\alpha\, I'(f;\alpha) + I''(f;\alpha), \text{ where}$$

$$I(f;\alpha) = \int_{\alpha}^{\pi/2} \frac{\partial\sqrt{\sin^2\beta - \sin^2\alpha}}{\partial\beta} f(\beta)\, d\beta$$

Proof: From Anderssen and Jackett [1983] we know that

$$g(\alpha) = -\sec\alpha \frac{d}{d\alpha} \left(\cos\alpha \int_{\alpha}^{\pi/2} \frac{\partial\tau(\alpha,\beta)}{\partial\beta} \cos\beta \left[f(\beta) + f''(\beta)\right] d\beta\right)$$

and

$$\int_{\alpha}^{\pi/2} g(\alpha)\, d\alpha = \int_{\alpha}^{\pi/2} \left[\tau(\alpha,\beta) \sin\beta + \frac{\partial\tau(\alpha,\beta)}{\partial\beta} \cos\beta\right] \left[f(\beta) + f''(\beta)\right] d\beta$$

where

$$\tau(\alpha,\beta) = \arctan\left[\frac{\sqrt{\sin^2\beta - \sin^2\alpha}}{\cos\beta}\right],$$

which implies

$$\frac{\partial\tau(\alpha,\beta)}{\partial\beta} = \frac{\sin\beta}{\sqrt{\sin^2\beta - \sin^2\alpha}}.$$

Thus

$$-\cos\alpha\, g(\alpha) = \frac{d}{d\alpha}\left[\cos\alpha \left\{\int_{\alpha}^{\pi/2} g(\alpha')\, d\alpha' - \int_{\alpha}^{\pi/2} \tau(\alpha,\beta) \sin\beta \left[f(\beta) + f''(\beta)\right] d\beta\right\}\right],$$

which yields

$$\int_{\alpha}^{\pi/2} g(\alpha) = I_1(f;\alpha) - \cot(\alpha) \frac{d}{d\alpha} I_1(f;\alpha), \tag{1.1}$$

where it can be shown

$$I_1(f;\alpha) = \int_{\alpha}^{\pi/2} \tau(\alpha,\beta) \sin\beta \left[f(\beta) + f''(\beta)\right] d\beta$$

$$= \int_{\alpha}^{\pi/2} \frac{\partial\tau(\alpha,\beta)}{\partial\beta} \cos\beta\, f(\beta)\, d\beta - \int_{\alpha}^{\pi/2} \frac{\partial\tau(\alpha,\beta)}{\partial\beta} \sin\beta\, f'(\beta)\, d\beta. \tag{1.2}$$

However

$$\int_{\alpha}^{\pi/2} \frac{\partial\tau(\alpha,\beta)}{\partial\beta} \cos\beta\, f(\beta)\, d\beta = I(f;\alpha) \tag{1.3}$$

and

$$\int_{\alpha}^{\pi/2} \frac{\partial\tau(\alpha,\beta)}{\partial\beta} \sin\beta\, f'(\beta)\, d\beta$$

$$= \int_{\alpha}^{\pi/2} \sqrt{\sin^2\beta - \sin^2\alpha}\, f'(\beta)\, d\beta + \int_{\alpha}^{\pi/2} \frac{\sin^2\alpha}{\sqrt{\sin^2\beta - \sin^2\alpha}} f'(\beta)\, d\beta$$

$$= \sec\alpha\, f(\pi/2) - I(f;\alpha) + \tan\alpha\, I'(f;\alpha) \tag{1.4}$$

The result now follows on combining (1.1), (1.2), (1.3), and (1.4).

Appendix 2

In this appendix we show that if $\theta(\alpha)$ is such that $\theta'(\alpha)$ exists and is absolutely continuous, then

$$\int_0^{\pi/2} \theta(\alpha)\, g(\alpha)\, d\alpha = -\int_0^{\pi/2} \{2\theta(\alpha) + \cot\alpha\, \theta'(\alpha) + \theta''(\alpha)\}\, I'(f;\alpha)\, d\alpha - \theta'(\pi/2)\, f(\pi/2)$$

Proof: Clearly

$$\int_0^{\pi/2} \theta(\alpha)\, g(\alpha)\, d\alpha$$

$$= -\int_0^{\pi/2} \theta(\alpha) \frac{d}{d\alpha} \{\int_{\alpha}^{\pi/2} g(\alpha')\, d\alpha'\}\, d\alpha$$

$$= -\int_0^{\pi/2} \theta(\alpha) \frac{d}{d\alpha} \{2I(f;\alpha) - \cot\alpha\, I'(f;\alpha) + I''(f;\alpha)\}\, d\alpha.$$

Observing $I'(f;0) = I''(f;\pi/2) = 0$ and $I'(f;\pi/2) = -f(\pi/2)$, the result follows using several integrations by parts.

References

M.C. Anderson [1964] Studies of the woodland climate I, The photographic computation of light conditions, **J.Ecol.** 52, 27-41.

M.C. Anderson [1966] Radiation and crop structure, In.Z.Sestak, J. Catsky and P.G. Jarvis (Editors), Plant Photosynthetic Production/Manual of Methods, Junk, The Hague, pp.412-466.

R.S. Anderssen and D.R. Jackett [1983] Linear functionals of foliage angle density, **J.Aust.Math.Soc,** Series B (accepted).

R.S. Anderssen, D.R. Jackett and D.L.B. Jupp [1983] Linear functionals of the foliage angle distribution as tools to study the structure of plant canopies, **Aust.J.Bot** (Submitted).

J.B. Miller [1963] An integral equation from phytology, **J.Aust.Math.Soc.** 4, 397-402.

J.B. Miller [1967] A formula for average foliage density. **Aust.J. Bot.** 15, 141-144.

J.R. Philip [1965] The distribution of foliage density with foliage angle estimated from inclined point quadrat observations, **Aust.J.Bot.** 13, 357-366.

J. Ross [1981] The Radiation Regime and the Architecture of Plant Stands (W. Junk, The Hague).

J.A. Smith, R.,E. Oliver and J.K. Berry [1977] A comparison of two photographic techniques for estimating foliage angle distribution. **Aust.J.Bot** 25, 545-553.

G.H. Suits [1972] The calculation of the directional reflectance of a vegetative canopy. **Rem.Sens.Environ.** 2, 117-125.

J. Warren Wilson [1960] Inclined point quadrats, with Appendix by J.E. Reeve, **The New Phytologist** 59, 1-8.

J. Warren Wilson [1963] Estimation of foliage denseness and foliage angle by inclined point quadrats, **Aust.J.Bot.** 11, 95-105.

Computational Techniques & Applications: CTAC-83
J. Noye & C. Fletcher (Editors)

THE MARKOV CHAIN TECHNIQUE APPLIED TO THE SURVIVAL OF VEHICLES IN A MINEFIELD

I.S. Williams

Directorate of Operational Analysis-Army
Department of Defence
Russell Offices
Canberra, ACT, 2600
Australia

The Markov Chain technique has been applied to the problem of determining the effectiveness of various anti-tank minefields. The methodology allows incorporation of both scattered and conventional patterned minefields containing mixtures of mines with a range of fuze types. Columns of tanks fitted with a variety of breaching devices can be investigated using this methodology.

INTRODUCTION

Minefields have long been used in warfare as a means of delaying and disrupting an enemy advance. Modern minefields fall into either of two categories: patterned or scattered. Patterned minefields are those in which individual mines are buried in approximately parallel rows. Within a row, mines are regularly positioned with inter-mine spacing s_o. Scattered minefields are those in which mines have been randomly distributed over an area. The distribution of mines in a scattered minefield will depend on the method of mine delivery, though it will normally correspond to a series of overlapping clusters of mines.

The mines themselves consist of a fuze and an explosive charge. In the case of anti-tank mines, the fuze will detonate the explosive charge in response to the presence of some distinctive feature of a tank's signature. Different fuzes are designed to respond to different aspects of the tank's signature such as: the pressure of the tank track, the magnetic signature of the tank, the acoustic signature of the tank, or movement in the mines orientation due to any physical disturbance. A single minefield could consist of groups of mines armed with different fuzes.

When forced to breach a minefield, a column of tanks can have several methods at their disposal. The simplest method would be to drive through in a well defined column, accepting any casualties. A better approach is to fit track width ploughs, or heavy track width rollers to the lead tank (or tanks) in the column. When available, specialised mine-destroying vehicles such as flail tanks, remotely controlled tanks, or vehicles simulating a tank's magnetic signature can be used to lead the column. Another widely used method is to attempt to destroy or detonate mines using some form of explosives, prior to the column of tanks entering the minefield. Each method suffers from some limitation, and specialised mine fuzes have been devised to counter most breaching methods. For example, a plough will disturb the orientation of a mine in its path, causing destruction of the plough if the mine has an 'anti-disturbance' fuze. Fuzes which require two pressure impulses before detonation are not detonated by a roller, however will destroy the tank pushing the roller.

The interaction of tanks with minefields is a topic widely studied by Defence analysts using a range of Operations Research techniques. The most common method is to use a Monte Carlo approach with computer simulation. The main disadvantage of this approach is the requirement for exceedingly long computation time. A simple mathematical approach has been proposed by Zacks (1967) though this does not permit incorporation of the complexities of mine warfare such as ploughs, rollers and double-impulse fuzed mines. Analysts in the United Kingdom (Davies, 1976) have developed a minefield model based on the Markov Chain method, which though efficient in computer time and more realistic than the simple mathematical approach, is still deficient in many respects. A major disadvantage is that the approach is limited to patterned minefields. This paper describes work carried out in the Australian Department of Defence (Williams, 1982). The Markov Chain model has been extended to more realistically incorporate mine-tank interactions and also to allow study of scattered minefields.

BREACHING PATHS

Consider a straight path of width w passing completely through a minefield (Figure 1). Such a path will contain some number of mines M (where a mine is considered in the path if its centre lies in the path). A column of tanks passing along, and within such a path, can be expected to 'encounter' at most M mines. Assuming no prior knowledge, the choice of 'best' path for crossing a minefield will be uncertain and so the choice of breaching path is likely to be a somewhat arbitrary decision. Assuming, therefore, that a breaching path is equally likely to be chosen anywhere across the minefield front, it is no longer possible to talk about a single path with M mines. Instead, a minefield can be described by a probability distribution P(M,w) which is the probability of a path of width w through the minefield containing M mines.

The function P(M,w) can be evaluated for some specific minefields. Consider a patterned minefield consisting of R independent parallel rows of mines (Figure 1a). If a path is chosen such that it makes an angle θ with the mine row direction, then the 'effective' separation of mines within a row will be

$$S = s_o \sin\theta$$

If the path is chosen to have a width w = S, then the path will contain exactly R mine centres. The position of mines across the path width will be random and independent (due to the independence of the rows). For such a patterned minefield, the distribution of numbers of mine centres in a path will be

$$P(R,S) = 1.0$$

$$P(M,S) = 0 \text{ for } M \neq R$$

A requirement of any path is that the width w is greater than the width of a column of tanks. Normally, s_o is chosen to be slightly larger than the width of a column of tanks and minefields are oriented such that $\theta \simeq 90^o$. In practical situations, therefore, the width of a column of tanks will satisfy the requirement of being narrower than the effective mine separation.

In a scattered minefield in which mines have been randomly and independently laid (Figure 1b) the number of mine centres in any path across the minefield will be given by a Poisson distribution

$$P(M,w) = \frac{[E(M)]^M}{M!} e^{-E(M)}$$

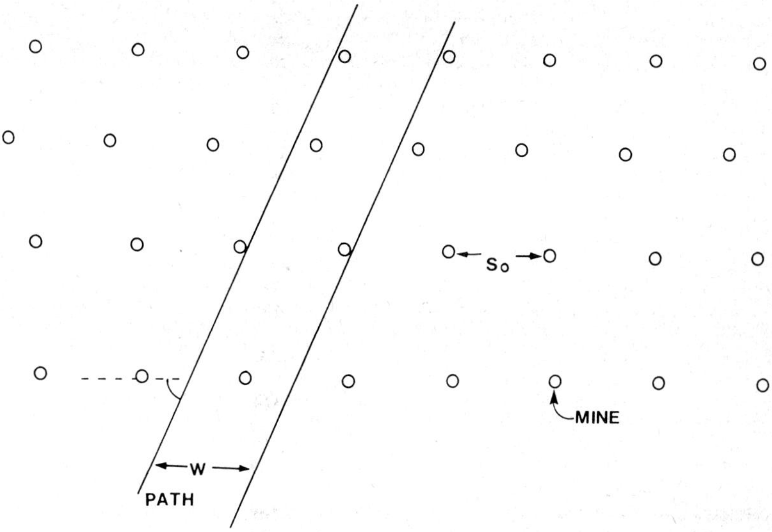

(a) A Patterned Minefield

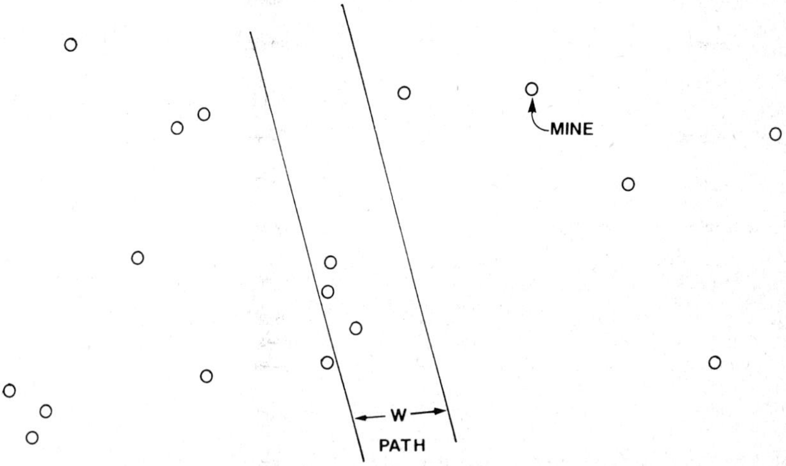

(b) A Scattered Minefield

Figure 1
Breach Path through Minefield

where

$$E(M) = dwl$$

is the average number of mines expected to be found in a path of width w and length l through a minefield of mine density d.

Scattered minefields often correspond to many slightly overlapping clusters of mines. Such a distribution if of low density may not be well represented by the Poisson function. As cluster arrangement and density can vary significantly, it is not possible to derive a simple function to describe P(M,w). Simple statistical simulation[1] must be used to provide an estimate of the distribution for any 'clustered' scattered minefield. Unlike the patterned or 'Poisson' scattered minefield, the clustered minefield will not have mines independently and randomly spread across the width of the path. Simulation has shown, however, that for any realistic clustered minefield, the mine locations across the path width can be considered effectively independent and random.

BREACHING STATES

When tanks are required to breach a minefield, they can be considered to be in one of two states: operational or non-operational. Though the varying degrees of damage possible has a significant long term effect on tank repair, the only short term consideration is whether damage is sufficient to make the tank non-operational for the immediate conflict. A column of N tanks can exist therefore in any of N + 1 states, numbered 0 to N and corresponding to the number of still operational tanks. A specialised breaching tank, such as a plough tank can exist in more than two states (e.g. operating with one or two ploughs damaged). If a breaching vehicle, or group of breaching vehicles, can exist in a total of L states (defined as appropriate for the particular breaching system), then N conventional tanks led by the breaching system can exist in a total of L(N+1) states. When a column of tanks encounters a mine in their path, there is a possibility of the column transitioning from one state to another. Ignoring any accumulation of damage due to the encounter of successive mines, the passage of a column of tanks through a minefield can be considered to be a succession of independent interactions with mines. Such a situation is simply described as a Markov process and is amenable to analysis using the Markov Chain method.

For a column of tanks which can exist in any one of L(N+1) states, it is possible to calculate the probability a(i,j) of the column transitioning from a state i to a state j on encountering a mine in its path. These transition probabilities can be represented by a L(N+1) square transition matrix A. The column of tanks exists initially in a state corresponding to all tanks and breaching systems fully operational. This can be represented by an initial state row vector I. The final state row vector after encountering a mine will be

$$F = IA$$

If the column encounters M mines successively in the path (assuming that two or more mines are never simultaneously encountered), then the final state vector becomes

$$F = IA^M$$

If a column of tanks passes through a minefield along a path type with a mine probability function P(M,w), then the final state of the column of tanks will be

$$F = \sum_{M=1}^{\infty} P(M,w)IA^M$$

DEMOBILIZATION PROBABILITIES

The problem then reduces to one of determining the transition probabilities which make up the matrix A. If a mine is known to exist in a path of width w and is equally likely to be anywhere across the path, then the probability of the ith tank or breaching device being demobilized by the mine when the column is in state Q will be

$$h_Q(i) = \frac{L_Q(i)}{(w - \sum_{k=1}^{i-1} L_Q(k))} \qquad i > 1$$

$$h_Q(1) = L_Q(1)/w$$

where i corresponds to the systems (tank or breaching device) position in the column (the leading system being numbered i=1) $L_Q(i)$ is the lethal width of the ith system when the column is in state Q. The lethal width is simply the width of system which is vulnerable[2] to a particular mine multiplied by the probability of that mine type detonating and demobilizing the system.

A minefield may contain more than one mine type and so, in general, the effective lethal width will be

$$L_Q(i) = \sum_{k=1}^{\infty} \rho(k) L_Q(i,k)$$

where $\rho(k)$ is the proportion of mines of type k present in the minefield and $L_Q(i,k)$ is the lethal width of the ith system against type k mines for a column in state Q. For any single mine type k, the lethal width is, in its most general form,

$$L_Q(1,k) = g_1(k,0:)V_1(k,0:)$$

$$L_Q(2,k) = g_2(k,0:)V_2(k,0:) + g_2(k,1:1)V_2(k,1:1)$$

and for n > 2

$$L_Q(n,k) = g_n(k,0:)V_n(k,0:) + \sum_{x_1=1}^{n-1} g_n(k,1:x_1)V_n(k,1:x_1)$$

$$+ \sum_{j=2}^{n-1} \left[\sum_{x_1=1}^{n-j} \cdots \sum_{x_j\; x_{j-1}+1}^{n-1} g_n(k,j:x_1,\ldots,x_j) V_n(k,j:x_1,\ldots,x_j) \right]$$

where $g_n(k,j:x_1,\ldots,x_j)$ is the probability of mine type k not being detonated by j of the systems (or breaching devices) ahead identified by the numbers $x_1,x_2,\ldots,x_j$ and finally detonating and demobilizing the nth system in the column.

$V_n(k,j:x_1...x_j)$ is the vulnerable width of the nth system which covers the same segment of path as j of the preceding systems numbered $x_1, x_2, ..., x_j$. Stated fairly simply, the lethal width of the nth system is calculated by dividing its vulnerable width into various segments defined by the degrees of 'shadowing' of all systems ahead. The width of these regions is multiplied by the probability of a mine demobilizing the nth system, having not detonated under any previous systems. When demobilization probabilities differ for various regions of a tank or breaching device (e.g. track and tank body), the system must be subdivided into regions with common demobilization probabilities.

To illustrate with a simple example, consider a column of N identical tanks crossing a segment of path containing a track attack (i.e. pressure fuzed) mine. Assume that these mines have a probability f of demobilising the first tank whose vulnerable region passes over the mine. Assume further that a mine which is not activated by one tank in its vulnerable region, will not be activated by subsequent tanks. The lethal widths for state 1 (all tanks operating) become

$$L_1(1,TA) = 2(T+2r)f$$

$$L_1(n,TA) = 2C(n)f$$

Where TA stands for track attack mine, T is the width of one track, r is the mine activation radius[3] and C(n) is the misalignment (assumed small compared with the tank width) with respect to the column of tanks ahead.

Calculations of lethal widths, and hence probabilities, reduce to simple geometric constructions. In general, it is assumed that

$$g_n(k,j:x_1,...,x_j) = 0 \text{ for } j > 1$$

Some adjustment must be made to lethal widths when plough tanks are used, as the ploughs have the effect of concentrating mines just outside their clearance zone. When breaching is preceded by explosive clearance techniques, lethal widths are amended by reducing the g functions as appropriate to account for the probability of some mines being destroyed by the explosives. Equations representing lethal widths for a large number of breaching situations have been calculated by Williams (1982).

TRANSITION PROBABILITIES

When one tank is demobilized, any following tanks must either divert around on a new path or wait until the damaged tank is removed. Previous work (Davies, 1976; Williams, 1982) has considered the effect of diverting around demobilized tanks in a patterned minefield. In a realistic Poisson scattered minefield, the choice of bypassing policy will have little effect on expected losses. In the following calculations, it is assumed that demobilized tanks are removed and that following tanks remain on the same path.

Consider a column of N standard tanks with no breaching devices. The probability of the nth tank being demobilized when the column is in state Q will be $h_Q(n)$. The column of tanks will be defined by states 0 to N representing the number of tanks operational. The transition probabilities between states (numbered 0 to N) when encountering a mine will be:

$$a(i,j) = 0 \text{ for } i < J$$

as there cannot be an increase in the number of tanks;

$$a(i,j) = 0 \text{ for } i > j + 1$$

as it is assumed that it is not possible for one mine to damage more than one tank;

$$a(i,i) = \prod_{j=1}^{i} \left[1-h_i(j) \right]$$

and

$$a(i,i-1) = \sum_{x=1}^{i} \left[h_i(x) \prod_{j=1}^{x-1} \left\{ 1-h_i(j) \right\} \right]$$

To allow simple generalization of formulae, it is defined that

$$\prod_{i=1}^{0} \text{(any function)} = 1$$

If a column of N tanks is preceded by a breaching system which can occupy L different states, then a state can be defined by the integer

$$i = (N+1)i_1 + i_2$$

where i_1 is the state of the breaching system alone (between 0 and L) and i_2 is the state of the N following tanks. A state j is defined similarly for breaching device state j_1 and following tanks state j_2. It is assumed that a single mine can destroy at most only one tank or breaching device. The transition probabilities become

$$a(i,j) = 0 \text{ for } i_2 < j_2$$

$$a(i,j) = 0 \text{ for } i_2 > j_2 + 1$$

$$a(i,j) = H(i_1,j_1) \prod_{n=1}^{i_2} \left[1-h_1(x+n) \right]$$

$$\text{for } i_1 = j_1 \text{ and } i_2 = j_2$$

$$a(i,j) = H(i_1,j_1) \sum_{k=1}^{i_2} \left\{ h_i(x+k) \prod_{n=1}^{k-1} \left[1-h_i(x+n) \right] \right\}$$

$$\text{for } i_1 = j_1 \text{ and } i_2 = j_2 + 1$$

$$a(i,j) = H(i_1,j_1) \text{ for } i_1 \neq j_1$$

where $H(i_1,j_1)$ is the probability of the breaching device transitioning from state i_1 to state j_1. The probability $H_i(x+n)$ is the probability of losing system x + n in state i if the first following tank has system number x + 1 (i.e. $h_i(1),\ldots,h_i(x)$ are the loss probabilities for the x components of the breaching device in state i).

CONCLUSION

The outcome of tanks crossing a minefield is defined by the final state vector F. This is calculated by first calculating lethal widths from the geometry of the breaching column; converting lethal widths to demobilization probabilities and then transition probabilities; and finally carrying out the matrix multiplication weighted by the probability distribution for mines in the path. The whole process can be carried out fairly quickly on a computer even for complex breaching systems and long columns of tanks. A FORTRAN program has been written to carry out the computation (Williams, 1982).

This model allows any minefield type to be studied with a fairly high degree of reliability. The effects of varying mine densities, compositions and fuze types can be studied. Factors such as driving misalignment and 'channelling' of plough-cleared mines can be included giving the model a high degree of realism. Lethal widths for a large number of breaching methods have been derived and these can be extended or added to whenever considered necessary. This model approaches the realism of the most complex simulation models, however with the advantage of dramatically lower computation times. The parametric nature of the model allows sensitivity analysis to be carried out and key parameters identified.

REFERENCES

[1] Davies, F.V., UK Defence Publication (1976).

[2] Williams, I.S., DOA-A Report 4, Australian Department of Defence (1982).

[3] Zacks, S., Naval Research Logistics Quarterly (September 1967) 329.

FOOTNOTES

1. A given minefield need be simulated only once in order to determine P(M,w), which can then be incorporated easily with any breaching arrangement. Full minefield simulation requires a complete Monte Carlo simulation for each minefield-breaching arrangement combination.

2. The vulnerable width of a system is the width of breaching path which, if it contains a mine centre, can cause system demobilization. This is obtained by combining the vulnerable dimensions of the system with the activation radius of a mine.

3. A requirement of the model is that the width of the column of tanks is less than w. For realistic mines with a non-zero activation radius, r, the column of tanks must have a width less than (w-r). In other words, the total vulnerable width V_{tot} of the column must lie within the path of width w.

Computational Techniques & Applications: CTAC-83
J. Noye & C. Fletcher (Editors)

THE NONLINEAR COOLING OF A SEMI-INFINITE SOLID - PADE APPROXIMATION METHODS

A.J. O'Connor

School of Science
Griffith University, Nathan, Qld., 4111
Australia

When a body loses heat by radiation the heat flux at the surface is proportional to the fourth power of the temperature. If the geometry is essentially one dimensional the surface temperature u(t) obeys a nonlinear Volterra integral equation

$$u(t) = f(t)-\int_0^t k(t-s)\, u^4(s)\, ds$$

Typically u drops rapidly near t=0 and then evens out to a relatively slow decay. There are standard numerical ways to solve this equation but they require quite short time steps. When the solid is a semi-infinite and has a uniform initial temperature we are able to use Pade techniques in appropriate variables to obtain an excellent analytic approximation to u. Practical error bounds can also be proven. Similar approximations could be found for any initial temperature distribution.

INTRODUCTION

Non linear Volterra integral equations occur in several parts of applied mathematics - for instance in heat transfer by radiation or convection, chemical reactions in laminar boundary layer flows and gas absorption in liquids [1,9,12]. The general mathematical properties of their solutions have been well studied [10, 11] and there are reliable schemes for their numerical solutions [2].

In this paper we look at one particular equation in detail. It is

$$u(t) = 1 - \pi^{-\frac{1}{2}} \int_0^t u^4(s)\, (t-s)^{-\frac{1}{2}}\, ds \tag{1}$$

The solution u(t) is the surface temperature of a semi-infinite slab of material which loses heat by black body radiation (the precise details are given later). When a direct numerical method is applied to (1) very small step sizes are needed near t=0 to get accurate solutions (eg steps of 2.5×10^{-4} out to t=0.005 and then steps of 5×10^{-3} out to t=1 were needed to obtain 6 significant figures).

It is possible though to obtain such accuracy over a much wider range by studying the analytic properties of u and by choosing Pade approximants which mimic the most important singularities of u in the complex $t^{\frac{1}{2}}$ plane.

Very similar techniques would work for any Volterra equation with a weakly singular algebraic kernel., eg.

$$x(t) = f(t) + \int_0^t (t^a-s^a)^{-b}\, p[x(s)]\, ds \tag{2}$$

where, for instance, f is an analytic function of t near t=0 and p is an entire function.

The most important properties of u(t) in the present case are

(i) that u is an analytic function of $t^{\frac{1}{2}}$ at every point on the positive real axis [10].

(ii) near t=0, u can be expanded as a power series in $t^{\frac{1}{2}}$. This power series converges when $|(\alpha t)^{\frac{1}{2}}|<1$ where $\alpha=38.748670$.

(iii) u has a singularity at $(\alpha t)^{\frac{1}{2}} = -1$ and near this singularity $u \sim A(1+(\alpha t)^{\frac{1}{2}})^{-1/6}$.

(iv) as $t\to\infty$, $u(t) \sim (\pi t)^{-1/8}$.

There are 6 sections to this paper.

1. The physical problem.
2. Analytic properties of u.
3. Asymptotic behaviour of u.
4. Pade techniques and numerical results.
5. Error estimation and improvement.
6. Possible applications to finite solids.

The reader interested in numerical techniques could skim sections 2 and 3 and concentrate on 4 and 5.

1. The Physical Problem

Equation (1) comes from the following problem -

A semi-infinite slab of material fills the region z<0. This material has uniform thermal properties and at t=0 its temperature is θ_0 everywhere. It can only lose heat by black body radiation from the surface z=0.

So the temperature θ depends only on the depth z and time t. $\theta(z,t)$ obeys

$$D\frac{\partial^2\theta}{\partial z^2} = \frac{\partial\theta}{\partial t} \qquad z<0,\ t>0 \qquad (3)$$

$$k\frac{\partial\theta}{\partial z} = -\varepsilon\sigma\theta^4 \qquad z=0,\ t>0 \qquad (4)$$

$$\theta\ (z,t)\to\theta_0 \text{ as} \qquad z\to\infty,\ t>0 \qquad (5)$$

$$\theta(z,0) = \theta_0 \qquad \text{all } z \qquad (6)$$

An easy Green's function arguement relates $\theta(z,t)$ to the heat lost from the surface at all times s prior to t viz

$$\theta(z,t) = \theta_0 - \int_0^t (D/\pi(t-s))^{-\frac{1}{2}}\, e^{-z^2/4D(t-s)}\ \varepsilon\sigma\theta^4(0,s)\ ds \qquad (7)$$

There are natural time and length scales for this problem. The natural "radiation length" and "radiation time" are

$$\ell_{rad} = k/\varepsilon\sigma\theta_0^3 \text{ and } Dt_{rad} = \ell_{rad}^2.$$

Over a few "radiation times" the temperature of the solid only changes significantly within a few "radiation lengths" of the surface. Taking graphite or carbon steel for instance we have [15]

$D = 0.1\ cm^2/sec$, $k = 0.5$ watts/cm/°k,

$\varepsilon = 0.8$ and $\sigma = 5.7$ watts/cm/°k^4

(ε is the emissivity of the material and σ the universal Stefan-Boltzmann constant)

Typical scales are then

Table 1.

θ_0 (°k)	length scale (cm)	time scale (sec)
100	10^5	10^{11} (3000 yrs)
1,000	10^2	10^5 (1 day)
10,000	0.1	0.1

In one "radiation time" the surface temperature drops by one third.

If we replace θ by $\theta_0 u$ and use z for z/ℓ_{rad} and t for t/t_{rad} we get the non dimensional equation (1) from (7) when z=0;

$$u(t) = 1 - \int_0^t u^4(s)\,[\pi(t-s)]^{-\frac{1}{2}}\,ds \qquad (1)$$

2. Analytic Properties of u

Mann and Wolf [10] have already proven that (1) has a unique continuous solution which decreases steadily from 1 to 0 as t tends to infinity. A naive way to try to solve (1) is by iteration. This generates a sequence of "approximate solutions" $u_0, u_1, \ldots\ u_n \ldots$

$$u_{n+1}(t) = 1 - \pi^{-\frac{1}{2}} \int_0^t u_n^4(s)\,(t-s)^{-\frac{1}{2}}\,ds \qquad (8)$$

Taking $u_0=1$ gives $u_1 = 1 - 2\pi^{-\frac{1}{2}}t^{\frac{1}{2}}$ etc. In general u_n is a polynomial in $t^{\frac{1}{2}}$ and it is possible to show that when $t^{\frac{1}{2}}$ is small this sequence of polynomials converges to an analytic function of $t^{\frac{1}{2}}$ which solves (1). This is a simple application of the contraction mapping theorem. We have then

$$u(t) = \sum_{n=0}^{\infty} (-1)^n\, u_n\, t^{n/2} \qquad (9)$$

with every $u_n>0$. In fact it is easy to calculate the u_n's (in double precision). To do this accurately it is best to compute a scaled version of (9) viz

$$u(t) = \sum_{n=0}^{\infty} (-1)^n\, w_n(\lambda t)^{n/2} \qquad (10)$$

When λ is about 40, the w_n values decrease steadily.

Approximate values are

n	0	1	5	10	20	30
u_n	1	1.128	450	$2.4\text{x}10^6$	$1.1\text{x}10^{14}$	$7\text{x}10^{21}$

The rapid increase in u_n and the alternating coefficients suggest that this series has a finite radius of convergence and has a singularity on the negative $t^{\frac{1}{2}}$ axis.

We can use numerical methods to estimate the radius of convergence. Such problems have been studied in statistical mechanics by Domb et al [3,8] and in hydrodynamics by van Dyke [14]. The usual assumption is that the dominant singularity is a power law and that u can be written as

$$u(t) = A\,(1 + (\alpha t)^{\frac{1}{2}})^{-p} + \text{less singular terms} \qquad (11)$$

If we assume that the less singular term is analytic inside a region larger than $|(\alpha t)^{\frac{1}{2}}|<1$ we have

$$\alpha = \lim_{n\to\infty} u_{n+2}/u_n \tag{12}$$

and

$$p-1 = \lim_{n\to\infty} n^2\left(\frac{u^2_{n+1}}{u_n\, u_{n+2}} - 1\right) \tag{13}$$

These sequences converge slowly but the convergence can be accelerated by extrapolating against $1/n$. Eventually small errors in the u_n's accumulate to make further extrapolation useless. (so it is very important to calculate the u's in double precision). Hunter and Guerrieri's method (which is probably the best) gives

$$\alpha = 38.748670$$

$$p = 0.1666 \tag{14}$$

(it is hard to estimate the accuracy of these results but the calculated values settled down to within 3 in the last quoted figure). This suggests that the dominant singularity in u is a $-1/6$th power at $(\alpha t)^{\frac{1}{2}}=-1$.

We can also justify this analytically. It is convenient to define a function v(t) by

$$v(t) = \sum_{n=0}^{\infty} u_n\, t^{n/2} \tag{15}$$

v obeys the integral equation

$$v(t) = 1 + \pi^{-\frac{1}{2}} \int_0^t v^4(s)\, (t-s)^{-\frac{1}{2}}\, ds \tag{16}$$

$$= Mv(t)$$

Suppose that v has a power law singularity at $(\alpha t)^{\frac{1}{2}}=1$ of the type

$$v(t) \sim d(1-(\alpha t)^{\frac{1}{2}})^{-p} \sim c(\beta-t)^{-p} \tag{17}$$

(where $\beta = \alpha^{-1}$ and $c = d(2\beta)^p$)

The right hand side of (16) then behaves like

$$\pi^{-\frac{1}{2}}\, c^4 \int_0^t (\beta-s)^{-4p}\, (t-s)^{-\frac{1}{2}}\, ds$$

$$\sim \pi^{-\frac{1}{2}}\, c^4\, (\beta-t)^{\frac{1}{2}-4p}\, B(4p-\tfrac{1}{2},\tfrac{1}{2}). \tag{18}$$

Here B is the Beta function. Matching the singularities on both sides of (16) gives

$$-p = \tfrac{1}{2}-4p$$

$$c = \pi^{-\frac{1}{2}}\, c^4\, B(4p-\tfrac{1}{2}, \tfrac{1}{2}) \tag{19}$$

Hence $p = 1/6$ and $c = 0.624263$.

In the original form we have

$$v(t) \sim A\, (1-(\alpha t)^{\frac{1}{2}})^{-1/6}$$

$$u(t) \sim A\, (1+(\alpha t)^{\frac{1}{2}})^{-1/6} \tag{20}$$

where A = 1.023059 near their respective singularities.

Of course this is not a rigorous proof but we can make some rigorous statements. We can prove two theorems which give bounds on α and show that the singularity on the real axis cannot be worse than a $-\frac{1}{6}$th power. There is also good numerical evidence that this is the only singularity on the circle of convergence.

Theorem 1 If the power series for u converges within the circle $|(\alpha t)^{\frac{1}{2}}|<1$ then α obeys the inequalities

$$\alpha_1 = 33.792 < \alpha < \alpha_2 = 45.836$$

Proof This is based on a comparison arguement. The coefficient u_{n+1} is determined by $u_0, u_1, \ldots, u_n$. In fact if we define

$$[v(t)]_n = \sum_{r=0}^{n} u_r t^{r/2} \tag{21}$$

we have, using the notation of (16),

$$v_{n+1} = \text{coefficient of } t^{(n+1)/2} \text{ in } M([v]_n)$$

Suppose we define $w(x) = (1-x^{\frac{1}{2}})^{-1/6}$ and we assume that for $0<i<n$

$$v_i > \text{the ith coefficient of } w(\alpha_1 t)$$

Then if we choose α, so that

$$\text{the (n+1)st coefficient in } w(\alpha_1 t) >$$
$$\text{the (n+1)st coefficient in } M([w(\alpha_1 t)]_n)$$

we will be able to conclude that $v(t) > w(\alpha_1 t)$. A little algebra shows that this is the same as $\alpha_1 > \chi_n^2$ where

$$\chi_n = \binom{-2/3}{n} B(\tfrac{1}{2}n+1, \tfrac{1}{2})/\pi^{-\frac{1}{2}} \binom{-1/6}{n+1} \tag{22}$$

Equally if we choose $\alpha_2<\chi_n^2$ for all n, we get $v(t)<w(\alpha_2 t)$. So we should take $\alpha_2 = \min \chi_n^2 = 144\,\pi^{-1}$ and $\alpha_1 = \max \chi_n^2 = 2(\Gamma(1/6)/\Gamma(2/3))^2$.

Theorem 2 The singularity in v(t) (and hence in u(t)) on the real axis can be no worse than a $-\frac{1}{6}$th power since we can get a constant K so that

$$|(1-(\alpha t)^{\frac{1}{2}})^{1/6}\, v(t)| < K$$

for all t between 0 and $\alpha^{-1} = \beta$

Proof Since the expansion v(t) has positive coefficients v(t) is monotone increasing between 0 and β. Thus

$$v(t) > \pi^{-\frac{1}{2}} \int_0^t v^4(s)\,(t-s)^{-\frac{1}{2}}\, ds$$

and

$$v^n(t) > \pi^{-n/2} \int_0^t ds_1 \ldots \int_0^t ds_n\, v^4(s_1) \ldots v^4(s_n)\,(t-s_1)^{-\frac{1}{2}} \ldots (t-s_n)^{-\frac{1}{2}}$$

$$= n!\,\pi^{-n/2} \int_0^t ds_1 \int_0^{s_1} ds_2 \ldots \int_0^{s_{n-1}} ds_n\,(t-s_1)^{-\frac{1}{2}} \ldots (t-s_n)^{-\frac{1}{2}}\, v^{4n}(s_n)$$

$$= \pi^{-n/2}\, 2^n \int_0^t (t-s)^{\frac{1}{2}n-1}\, v^{4n}(s)\, ds \qquad (23)$$

If $\gamma<\beta$ and $\phi(t) = (\gamma-t)^{1/6} v(t)$ we get

$$\int_0^\gamma \phi^n(t)\, dt > 2^n\, \pi^{-n/2}\, B(\tfrac{1}{6}n+1, \tfrac{1}{2}) \int_0^\gamma \phi^{4n}(t)\, dt$$

$$> a^n \int_0^\gamma \phi^{4n}(t)\, dt \qquad (24)$$

We can use this to show that ϕ cannot be very large. If $M = \max \phi(t)$ and w is the width of the largest interval on which $\phi(t) > \frac{1}{2}M$ we have

$$\gamma M^n > a^n\, w(\tfrac{1}{2}M)^{4n}$$

Taking the nth root and letting $n\to\infty$ gives $a^{-1/3}>M$. This is independent of γ and so the theorem is true.

Is the singularity at $|(\alpha t)^{\frac{1}{2}}| = 1$ the only singularity on the circle of convergence? This is a delicate question. Suppose we write

$$v(t) = \sum_{n=0}^{\infty} c_n (\alpha t)^{n/2} \qquad (25)$$

The behaviour of v on its circle of convergence is the same as that of $f(z) = \sum_{n=0}^{\infty} c_n z^n$ on the unit circle. We can use Dirichlet's test [6,p250] here.

Suppose that $f(z)$ has the unit circle as its circle of convergence and that $c_n\to 0$ and $\sum_{n=0} |c_n - c_{n-1}| < \infty$. Then the only singularity of f on the unit circle is at $z=1$.

If we assume that $c_n c_{n+2} > c_{n+1}^2$ is always true we can use the Dirichlet condition as follows.

c_{n+1}/c_n is an increasing sequence which tends to 1. Hence c_n is a decreasing sequence which will aproach a limit c.

Then $v(t) > c(1-(\alpha t)^{\frac{1}{2}})^{-1}$

So if c is positive we contradict theorem 2 and so c_n is a decreasing sequence which approaches 0. So by the Dirichlet test u and v have only one singularity on their circle of convergence.

Although I have not been able to prove this condition on the coefficients c_n it is easy to check it up to n=80 since it is the same as the condition $u_n u_{n+2} > u_{n+1}^2$. I think this is good evidence.

Finally we can try to determine the other singular terms in the expansion. I have not been able to prove that

$$\lim_{t\to\beta} (1-(\alpha t)^{\frac{1}{2}})^{1/6} v(t) = A \qquad (26)$$

but it is, I think, plausible. To find the lower terms it is convenient to change variables in (16). Put $t = \beta(1-e^{-x})$ and $f(x) = (\beta-t)^{1/6} v(t)$.
Then

$$f(x) = \beta^{1/6}\, e^{-x/6} + \int_0^x R(x-y)\, f^4(y)\, dy$$

$$R(u) = \pi^{-\frac{1}{2}}\, e^{-u/6}\, (1-e^{-u})^{-\frac{1}{2}} \qquad (27)$$

Assuming that f tends to a limit c as $x\to\infty$ we can iterate (27) taking $f_0=c$ to get

$$f_1 = c + e^{-x/6}(c_1) + \dots$$
$$f_2 = c + e^{-x/6}(c_1+c_2x) + \dots$$

or in terms of u

$$u(t) = A\,(1+(\alpha t)^{\frac{1}{2}})^{-1/6} + \text{a power series in } \log\,(1+(\alpha t)^{\frac{1}{2}}). \quad (28)$$

It is quite easy to re-express u in terms of this log variable. If $x=\log(1+(\alpha t)^{\frac{1}{2}})$ and we put

$$u(t) = A(1+(\alpha t)^{\frac{1}{2}})^{-1/6} + \sum_{n=0}^{\infty} b_n x^n \quad (29)$$

we find that the b_n's decrease quickly until n reaches 50 when they start to increase slowly. (eg $b_1=0.02306$, $b_5=0.53\times10^{-4}$, $b_{10}=0.55\times10^{-18}$).

3. The Asymptotic Properties of u

It is easiest to rewrite (1) as

$$u^4(t) = \pi^{-\frac{1}{2}}\frac{d}{dt}\int_0^t (1-u(s))\,(t-s)^{-\frac{1}{2}}\,ds \quad (30)$$

Taking $u_0\equiv0$ as a starting point we get $u_1(t) = (\pi t)^{-1/8}$ and by iteration

$$u_9(t) = \sum_{j=1}^{8} c_j t^{-j/8} + d\,t^{-9/8}\ln t + \dots \quad (31)$$

The coefficients are derived most easily by putting (31) into (29), expanding and equating coefficients.
The expansions

$$\int_1^t s^{-a}(t-s)^{-\frac{1}{2}}\,ds \sim \pi^{-\frac{1}{2}}\left\{\frac{\Gamma(\frac{1}{2}-a)}{\Gamma(\frac{3}{2}-a)}t^{\frac{1}{2}-a} + \frac{1}{a-1}t^{-\frac{1}{2}} + \frac{\pi^{-\frac{1}{2}}}{2(a-2)}t^{-3/2} + \dots\right\} \qquad a \neq 1,2\,.. \quad (32)$$

and

$$\int_1^t s^{-1}(t-s)^{-\frac{1}{2}}\,ds \sim \log t.\ t^{-\frac{1}{2}} + 2t^{-3/2} + \dots \quad (33)$$

are useful. They can be quickly derived using Mellin transforms. This method is due to Trivedi et al [13] and Handlesman et al [5] .

We get the following numbers

Table 2

j	c_j	j	c_j
1	0.866 675	5	-0.001 486
2	-0.152 998	6	-0.000 693
3	-0.020 655	7	-0.000 488
4	-0.004 661	8	-0.000 598

d = -0.000 65

It is not hard to prove that $\lim_{t\to\infty} (\pi t)^{1/8} u(t) = 1$ using an arguement based on the Laplace transform and the Tauberian theorem but I have not been able to justify this expansion any further.

4. Analytic Approximations to u

Our strategy is to obtain compact yet accurate approximations to a power series by using Pade approximants (ie. ratios of polynomials). We can try to incorporate the different analytic properties of u viz.

(i) $u(t) \sim t^{-1/8}$ at infinity

(ii) $u(t) \sim (1 + (\alpha t)^{\frac{1}{2}})^{1/6}$ near the origin and especially on the negative $t^{\frac{1}{2}}$ axis.

Since u is really a function of $t^{\frac{1}{2}}$ we introduce a variable $\tau = (\alpha t)^{\frac{1}{2}}$.

Various schemes are possible.

(a) We can only use the asymptotic behaviour of u. Then $u'/u \sim -1/4\tau$ and so we try to approximate u'/u by a ratio $P(\tau)/Q(\tau)$. P and Q are polynomials of degree N and N+1 and $P/Q \sim -1/4\tau$ at infinity.

Cross multiplying and equating to zero the leading coefficients of $u'Q-uP$ gives a set of linear equations for the coefficients of P and Q. This set becomes very ill conditioned when N exceeds 5 so we use N=4.

$$\text{If } P(\tau)/Q(\tau) = \sum_{j=1}^{N} \frac{r_j}{\tau+p_j} \qquad (34)$$

We approximate u by

$$\log u(\tau) = \sum_{j=1}^{N} r_j \{\log p_j - \log(p_j+\tau)\} \qquad (35)$$

(b) We can do better than this by including our knowledge of the singularity of the singularity near 0. The function

$$w(\tau) = (1+\tau)\, u^6(\tau) \qquad (36)$$

will be better behaved near $\tau=0$ and decays as $\tau^{-\frac{1}{2}}$ at infinity. So we can look for a constrained [N,N+1] Pade approximant to w'/w. Again N=4 is a good value to use. In fact beyond this the denominator and numerator have nearly coincident factors and so are really of lower order than they might at first appear.

(c) Finally we use (28) and subtract out the dominant singularity at $\tau=-1$. We re-express the remaining power series in τ as one in $x=\log(1+\tau)$. Since this "tail" has no special behaviour at infinity we just approximate it by an [N,N] Pade approximant.

The end result is then

$$u(t) \sim A(1+\tau)^{-1/6} - P(x)/Q(x) \qquad (37)$$

We can now let N go up to 6 and this approximation is definitely superior to (a) or (b).

Some typical results are:-

Table 3

t	"Exact"	Pade (a)	Pade (b)	Pade (c)	Asymptotic
0.01	0.915 147	0.915 147	0.915 147	0.915 147	-
0.1	0.816 887	0.816 887	0.816 887	0.816 887	-
1.0	0.686 570	0.686 6	0.686 6	0.686 570	0.685 1
5.0	-	0.592 1	0.592 1	0.592 300	0.592 2
10.0	-	0.552 8	0.552 8	0.553 207	0.553 1
100.0	-	0.433	0.433	0.434 73	0.434 7
1000.0	-	0.333	0.332	0.336 8	0.336 5

Notes

(i) The "exact" values are obtained by the product integral method of de Hoog and Weiss. To obtain these steps of 2.5×10^{-4} were taken out to 0.005 and then larger steps of 5×10^{-3} out to 1.

(ii) (a) is calculated from the constrained [4,5] Pade fit to u'/u.

(b) is calculated from the constrained [4,5] Pade fit to w'/w

(c) is calculated from the [6,6] fit explained in (c) above. Its coefficients are given in the appendix.

(iii) Fit (c) gives remarkable agreement to u over a very wide range - especially since the power series from which it was derived only converged for $t < 0.025$

5. Error Estimation and Reduction

One advantage of having an analytic approximation for u is that we can accurately calculate the residual $r(t)$. If we call the exact solution of (1), u and an approximate solution w the residual is defined by

$$r(t) = \pi^{-1/2} \int_0^t w^4(s)\,(t-s)^{-1/2}\,ds \;\; - w(t) - 1 \qquad (38)$$

Using approximation (c) we get

t	0.1	1.0	5.0	20.0	50
$\lvert r(t)\rvert$	10^{-8}	10^{-7}	10^{-7}	2×10^{-6}	8×10^{-6}

We can even estimate the error $|u(t)-w(t)|$ from $r(t)$. If we write $u=w+e$ we get

$$r(t) = e(t) + \int_0^t \pi^{-1/2}\,(t-s)^{-1/2}\,\{4u^3e + 6u^2e^2 + 4ue^3 + e^4\}(s)\,ds \qquad (39)$$

If $\sigma(t) = \max_{o<s<t} |e(s)|$ we get

$$\sigma < |r| + (1-u)\,\{4(\sigma/u)+6(\sigma/u)^2 + 4(\sigma/u)^3 + (\sigma/u)^4\} \qquad (40)$$

Here $\sigma=\sigma(t)$ and we estimate integrals like

$$\int_0^t \pi^{-1/2}\,(t-s)^{-1/2}\,u^3(s)\,ds < u(t)^{-1} \int_0^t \pi^{-1/2}\,(t-s)^{-1/2}\,u^4(s)\,ds$$

$$= (1-u)/u \qquad (41)$$

If $4(1-u)/u < 0.95$ we get

$$\sigma < 20|r| + 35\sigma^2 + 29\sigma^3 + 46\sigma^4 \quad (42)$$

and so for $t<0.1$ (where $w=0.81$) we get $\sigma < 20 \times 10^{-8}$ + terms of order 10^{-15}.

This simple argument breaks down when t gets larger but we can still estimate the error. We rewrite (38) as an "almost linear equation".

$$e(t) + \pi^{-\frac{1}{2}} \int_0^t 4u^3(s)\, e(s)\, (t-s)^{-\frac{1}{2}}\, ds = r_1(t) \quad (43)$$

We can solve this using a resolvent kernel

$$e(t) = r_1(t) - \int_0^t R(t,s) r_1(s)\, ds \quad (44)$$

Miller has studied these kernels and has proven that [11 , Thm. 6.3]

$$0 < R(t,s) < 4\pi^{-\frac{1}{2}} \{\max_{0<s<t} u^3(s)\} (t-s)^{-\frac{1}{2}}$$

$$= 4\pi^{-\frac{1}{2}} (t-s)^{-\frac{1}{2}} \quad (45)$$

So substituting for r_1 we get

$$\sigma(t) < |r(t)| + [4t + 2(t/\pi)^{\frac{1}{2}}]\, [6\sigma^2 + 4\sigma^3 + \sigma^4]\, (t) \quad (46)$$

We get the following results

t	10	100	1000
$\|r(t)\|$	2×10^{-6}	6×10^{-4}	8×10^{-4}
$\sigma(t)$	2×10^{-6}	15×10^{-4}	16×10^{-4}

This confirms table 3 - our approximation to u is really very accurate.

Should we wish to improve it we could use this Pade formula in conjunction with the product integral method of de Hoog and Weiss. The numerical problems in solving (1) directly are:-

(a) the rapid decrease in u near 0 (because of the $t^{\frac{1}{2}}$ powers in its expansion)

(b) the weak singularity in the kernel of (1).

The Pade approximant gives a very accurate representation of u near t=0. So we could take a cutoff point a and below a accept the Pade approximation for u. Above a we could use the Pade approximation as a starting value to obtain a better value.

At $t=a+nh$ we can replace (1) by

$$u_n = 1 - d_n - \sum_{j=0}^{n} u^4_{n-j}\, w_j \quad (47)$$

Here

$$d_n = \int_0^a \pi^{-\frac{1}{2}} (nh-s)^{-\frac{1}{2}} u^4(s)\, ds \quad (48)$$

and the w_j's are the weights for an accurate product integral representation, ie.

$$\pi^{-\frac{1}{2}} \int_{t-nh}^{t} f(s)\ (t-s)^{-\frac{1}{2}}\ ds = \sum_{j=0}^{n} w_j\ f(t-jh) \qquad (49)$$

(at least for piecewise quadratics).

In essence we are using the second step in the method of de Hoog and Weiss to improve our approximate solution. Typical results are summarized below.

Table 4 Improved results for u

t	a	h	Pade value for u(t)	Improved value for u(t)
5.0	0.9	0.01	0.592 299 80	0.592 299 89
	2.0	0.01		0.592 299 91
10.0	2.0	0.02	0.533 206 68	0.553 206 92
		0.04		0.553 206 94
	4.0	0.02		0.553 206 95
		0.05		0.553 206 98

Actually this improvement is hardly needed but it might perhaps be useful in other cases.

6. Possible Application to Finite Solids

It is not a simple matter to extend this approach to finite solids. The integral kernel is more complicated and the analytic approach may be too involved. However it is certainly possible to use the present results to obtain starting solutions. For instance consider a finite slab which loses heat by radiation from only one face. The differential equation is then:

$$D\frac{\partial^2\theta}{\partial z^2} = \frac{\partial\theta}{\partial t} \qquad -L\leqslant z\leqslant 0,\ t\geqslant 0$$

$$k\frac{\partial\theta}{\partial z} = -\varepsilon\sigma\theta^4 \qquad z=0,\ t\geqslant 0$$

$$\frac{\partial\theta}{\partial z} = 0 \qquad z=-L$$

$$\theta(z,0) = \theta_0 \qquad \text{all } z \qquad (50)$$

There are now two time scales - the "radiation time" already defined and a "diffusion time", t_{diff}

$$t_{diff} = L^2/D \qquad (51)$$

Using the rescaled temperature one gets

$$u(t) = 1-\pi^{-\frac{1}{2}} \int_0^t (t-s)^{-\frac{1}{2}}\ a(t-s)\ u^4(s)\ ds \qquad (52)$$

$$\text{where } a(t) = \sum_{n=-\infty}^{\infty} e^{-n^2 r/t} \qquad (53)$$

$$r = t_{diff}/t_{rad} \qquad (54)$$

Since a is close to 1 when t is of order r/20 it is possible to replace the exact solution of (52) by the solution for the semi-infinite solid for times up to r/20.

It is then possible to use the Green's function to obtain the internal temperatures and to use a standard method such as the Crank-Nicholson or the Galerkin method to solve the problem.

This has the advantage of giving a smoother starting solution. However the method outlined here is probably more appropriate to problems with a relatively simple kernel in the intergral equation.

REFERENCES

[1] P.L. Chambre and A. Acrivos, On chemical surface reactions in laminar boundary layer flows, Jnl. Appl. Physics 27 (1956), 1322-1328.

[2] F. de Hoog and R. Weiss, High order methods for class of Volterra integral equations with weakly singular kernels, SIAM J. Num. Analysis, 11(1974), 1166-1180.

[3] D.S. Gaunt and A.J. Guttmann, Asymptotic analysis of coefficients, in: C. Domb (ed), Phase transitions and critical phenomena, vol. 3, (Academic Press, New York, 1974).

[4] G. Gripenberg, Asymptotic solutions of some nonlinear Volterra integral equations, SIAM J. Math. Anal. 12 (1981), 595-602.

[5] R.A. Handlesman and W.E. Olmstead, Asymptotic solutions of a class of non-linear Volterra integral equations, SIAM J. Appl. Math, 22 (1972), 373-384.

[6] G.H. Hardy, A course of Pure Mathematics (Cambridge Univ. Press, Cambridge, 1908).

[7] C. Hunter and B. Guerrieri, Deducing the properties of singularities of functions from their Taylor series coefficients, SIAM Jnl. Appl. Maths, 39 (1980), 248-263.

[8] D.L. Hunter and G.A. Baker Jr., Methods of series analysis I and II, Phys. Rev. B 7 (1973), 3346-3376 and 3377-3392.

[9] M.J. Lighthill, Contributions to the theory of heat transfer through a laminar boundary layer, Proc. R. Soc. Lond (A) 202 (1950), 359-377.

[10] W.R. Mann and H.R. Wolf, Heat transfer between solids and gases under non-linear boundary conditions, Quarterly of Appl. Math, 9 (1951), 163-184.

[11] R.K. Miller, Non linear Volterra integral equations (W.A. Benjamin, Menlo Park, California 1971).

[12] W.E. Olmstead, A nonlinear integral equation associated with gas absorption in a liquid, Zeit. f. Angewandte Mat und Physik, 28 (1977), 512-523.

[13] V.K. Trivedi and I,J. Kumar, On a Mellin transform technique for the asymptotic solution of a nonlinear Volterra equation, Proc. R. Soc. Lond. (A), 352 (1977), 339-343.

[14] M. van Dyke, Extended Stokes series: laminar flow through a loosely coiled pipe, J. Fluid Mechanics 86 (1978), 129-145.

[15] Y.S. Touloukian et al, Thermophysical properties of matter, vols 1-13 (Plenum, New York, 1970).

Computational Techniques & Applications: CTAC-83
J. Noye & C. Fletcher (Editors)

COMPUTATIONAL ASPECTS ASSOCIATED WITH THE DIRECT USE OF INDIRECT MEASUREMENTS: REFRACTIVE INDEX OF BIOLOGICAL LENSES

R.S. Anderssen,
CSIRO Division of Mathematics and Statistics,
GPO Box 1965, Canberra, ACT 2601, Australia
and
M.C.W. Campbell,
Physiology, JCSMR, and Applied Mathematics, R.S. Phy.S.S.,
GPO Box 4, Canberra, ACT 2601, Australia.

In many practical situations, available experimental data corresponds to indirect measurements. Some effect of the phenomenon of interest is measured, not the phenomenon itself. How these data are manipulated to answer relevant questions about the phenomenon depends crucially on the nature of the data and the form of the question. In some situations, a direct analysis of the data can bypass computational difficulties associated with the underlying improperly posed nature of the relationship between the phenomenon and the indirect measurements. In this paper, we explore this facet of the analysis of indirect measurements and illustrate the relevant points with reference to a number of applications. In particular, we discuss in some detail the non-destructive determination of the refractive index structure of biological lenses.

1. INTRODUCTION

When the available data correspond to indirect observations of some phenomenon of interest, the way in which they are manipulated in order to answer relevant questions about the phenomenon depends crucially on the nature of the data and the form of the question. In fact, three distinct possibilities can be identified which reflect some of the limitations and difficulties associated with using indirect measurements.

They are:

1. The questions are such that they can be answered directly from the data without having to solve explicitly some mathematical formulation which identifies the interrelationship between the indirect measurements and the phenomenon. As illustrated in Section 2, this does not imply that implicit use has not been made of the interrelationship.

2. When the questions are sufficiently sophisticated, they can only be answered by solving an appropriate form of the interrelationship between the indirect measurements and the phenomenon of interest. The improperly posed nature of the interrelationship now becomes a key issue. This aspect is not pursued in this paper. The interested reader is referred to de Hoog (1980) and Lukas (1980) and the references cited there.

3. The information contained in indirect measurements about the phenomenon of interest is to a certain extent determined by the nature of the measurements (e.g. the information content of photographs). For example, in the size distribution problem discussed in Section 2, measurements of the size distribution of spherical carbon particles on random plane sections can be used to determine information about the size distribution of the carbon spheres themselves, but not their number density or any information about their relative positions in the steel. For such information, additional indirect measurements must be made. Thus, the type of data collected must be matched to the question being asked about the phenomenon of interest.

In this paper, the aim is to examine 1. and its relationship to the non-destructive determination of the refractive index structure of biological lenses.

2. DIRECT USE OF INDIRECT MEASUREMENTS

Within the framework of the direct use of indirect measurements discussed in 1. of Section 1, we can identify two different situations:

1(a). In many situations, the question being asked is such that it can be answered from a direct (statistical) analysis of the data after it (the question) has been transformed, through the manipulation of the interrelationship between the indirect measurements and the phenomenon, to an appropriate form. Clearly, this situation can only occur when the question being asked is such that such manipulations are possible. In fact, one actually "solves" the interrelationship between the indirect measurements and the phenomenon, but in an implicit manner. The best known examples of this situation arise when the question being asked can be formulated as a linear functional defined on the phenomenon. This linear functional can then be transformed analytically to corresponding linear functionals defined on the data.

Some examples are:

Linear Functionals of the Size Distribution of Spherical Carbon Particles in Steel

In the study of the impact strength of steel, the carbon particles are modelled as spheres the size distribution $u(x)$ of which is determined from measurements d_i, $i=1,2,\ldots,n$, of the size distribution $s(x)$ of their circular cross-sections on random planes taken through the steel. The relationship between the indirect measurements d_i, $i=1,2,\ldots,n$, and the phenomenon of interest $u(x)$ is given by (Anderssen (1980), Section 2)

$$s(y) = \frac{y}{m} \int_y^a u(x)(x^2-y^2)^{-\frac{1}{2}}dx \quad , \tag{1}$$

$$m = \int_0^a xu(x)dx = \frac{\pi}{2}\left(\int_0^a \frac{s(x)}{x}\, dx \right)^{-1} \quad , \tag{2}$$

$$d_i = s(x_i) + \varepsilon_i \;,\; 0 \leqslant x_1 < x_2 < \ldots < x_n \leqslant a \;,\; \varepsilon_i \equiv \text{random errors}, \tag{3}$$

where $[o,a]$ defines the support of $u(x)$. Reliable numerical procedures for the construction of approximations to $u(x)$ which are consistent with the information contained in the data d_i, $i=1,2,\ldots,n$, are available [Anderssen and Jakeman (1975)].

However, the decision to accept a batch of steel is not based on the explicit structure of $u(x)$, but on the values of certain key linear functionals

$$\int_0^a \theta(x)u(x)dx \quad , \qquad \theta(x) \text{ known}, \tag{4}$$

defined as $u(x)$. The functionals commonly used by metallurgists for this purpose are [Hyam and Hutting (1956)]:

a. average sphere radius m as defined by (2) (which must be estimated before (1)-(3) can be solved);

b. average particle surface area

$$A_u = 4\pi \int_0^a x^2 u(x)dx \; ; \tag{5}$$

c. average particle volume

$$V_u = \frac{4\pi}{3} \int_0^a x^3 u(x)dx \; . \tag{6}$$

If the known inversion formula for (1) is substituted for $u(x)$ in (5) and (6) and the alternative form for m given in (2) is used, then the evaluation of these functionals is reduced to corresponding functionals defined on $s(x)$:

$$m = \frac{\pi}{2}\left(\int_0^a \frac{s(x)}{x} dx\right)^{-1} \; , \tag{7}$$

$$A_u = 16m \int_0^a xs(x)dx \; , \tag{8}$$

$$V_u = 2\pi m \int_0^a x^2 s(x)dx \; . \tag{9}$$

In this way, the need to first solve (1)-(3) for $u(x)$ before computing the functionals (5) and (6) is bypassed. #

Linear Functionals of the Foliage Angle Distribution

Miller (1967) has shown how average foliage density of a plant canopy can be computed from projection measurements made at different angles on the foliage in the canopy; i.e. from the contact frequency of the canopy which can be derived, for example, from hemispherical photographs taken from the ground. This formalized mathematically Warren Willson's (1963) experimental verification that leaf area index could be estimated as a linear combination of measured contact frequency data. Recently, Anderssen and Jackett (1983) and Jackett and Anderssen (1983) have shown that Miller's result is a special case of a general transformation which allows linear functionals defined on the foliage angle distribution to be evaluated as linear functionals defined on the contact frequency. This circumvents the need to first solve an integral equation which involves a $2\frac{1}{2}$ -differentiation of the measured contact frequency. #

1(b). In some situations, the data themselves can actually be used to answer the question being asked in that the interrelationship between indirect measurements and phenomenon can be ignored. One simply exploits the fact that in some situations, the nature of the question is such that its answer is contained in the data. The process of doing this will be easier than solving some interrelationship between the indirect meaurements and the phenomenon. Three examples are:

The Growth and Development of Components in a Biological System

Lower dimensional (stereological) observations of the different components which constitute some biological system such as the cell cycle of Chlorella (Atkinson et al (1974)) are used to identify and quantify the different stages of the components' contribution to the system's growth and development. Since the same measure is made on each of the different components, the data themselves yield a direct comparative measure of their growth. Such techniques have been used to show that, among yeasts and Chlorella, the enzymes are usually accumulated discontinuously during the cell division cycle (Atkinson *et al* (1974)). #

Point Counts and Area Fraction Estimation of Volume Fraction

When, as discussed in the last example, stereological measures are used to infer comparative three-dimensional information and structure about the system under examination, it is tacitly assumed that any relative variation in the stereological measures consistently measures a corresponding variation in the three-dimensional system. Underlying justification for this is the basic stereological fact that area fractions on random plane sections taken through the system yield (unbiased) estimates of the corresponding volume fractions; and further that comparative point counts made using a uniform grid on a plane section yield (unbiased) estimates of the corresponding area fractions. A detailed discussion of the use and interpretation of stereological estimators can be found in Underwood (1970). #

Axial Symmetry in Cylindrical and Spherical Structures

In an examination of the refractive index of circular structures such as optical fibres and biological lenses, it is not necessary to first determine the three-dimensional refractive index structure before concluding that it is axial symmetric. In fact, it is only necessary to compare measurements of some property made on radial traverses across the circular face of the structure to decide whether it has an axial symmetric structure. #

It is this last aspect which we pursue in the remainder of this paper.

3. NON-DESTRUCTIVE MEASUREMENT OF REFRACTIVE INDEX IN CRYSTALLINE LENSES

The gradient of refractive index within the crystalline lens contributes much of the optical power of the eye. In the past, nonanatomical models of the lens have been used to predict the optical properties of eyes. A knowledge of the refractive index distribution within the crystalline lens will enable the image on the retina to be defined accurately.

Previous method of measuring the refractive index either directly, using a Schlieren interferometric method (Nakao et al (1968)) or indirectly, by measuring protein concentrations by microradiography (Philipson (1969)) or by microspectrophotometry (Bando et al (1976)), have all involved taking sections of frozen crystalline lenses. Cutting undistorted sections of frozen crystalline lenses is very difficult and, because of changes in the stress distribution in the lens, freezing may affect the refractive index distribution. In the rat, formation of a cold cataract near the lens centre is not reversible when the section is thawed.

The way around such difficulties is to use a non-destructive method for determining the refractive index.

The effect that a lens has on light incident upon it is predetermined by its refractive index. The changes of phase and direction of the incident beam are determined by the path the incident beam traverses through the lens which in turn is determined by its refractive index. In fact, under appropriate assumptions, the extent of the change can be defined in terms of the relationship which connects it with the refractive index along the path transversed.

Different relationships are obtained depending on the type of measurement made as well as the shape and refractive index structure assumed for the lens. Explicit relationships are known for the situation where the lens is assumed to be rotationally symmetric about its optic axis and to have a radially symmetric structure with radius ρ and refractive index $n(r)$ in the planes containing the optic axis.

When the measurements correspond to the deflection angle $\psi(w)$ shown in Figure 1, the relationship is given by (Pask (1983))

$$\psi(w) = \begin{cases} -2\cos^{-1}\left(\frac{w}{\rho}\right) + 2wn(\rho)\int_{r_o}^{\rho}\{r[r^2n^2(r) - w^2n^2(\rho)]^{\frac{1}{2}}\}^{-1}dr\ , & w \leq \rho \\ 0\ , & w \geq \rho\ , \end{cases} \tag{10}$$

where r_o is the solution of the equation

$$r_o^2n^2(r_o) - w^2n^2(\rho) = 0\ . \tag{11}$$

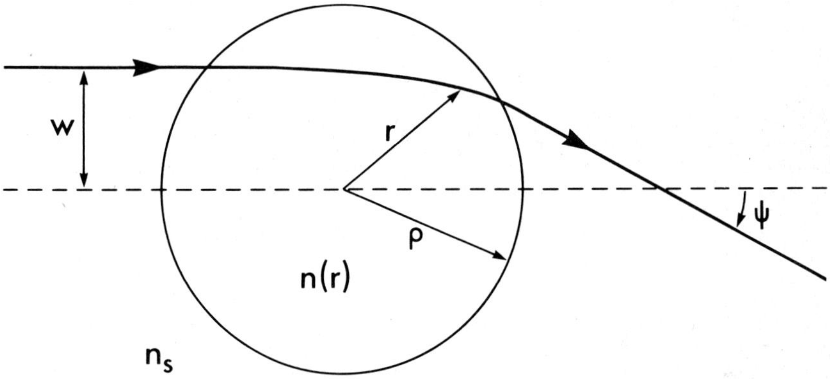

FIGURE 1

When the measurements correspond to the optical pathlength $\Phi(p)$ of rays which traverse the lens, the relationship becomes (Vest (1975))

$$\Phi(p) = 2\int_{p}^{\bar{p}} n^2(dr/d\eta)\{r(\eta^2-p^2)^{\frac{1}{2}}\}^{-1}d\eta\ , \tag{12}$$

where $\eta=rn(r)$, $p = \rho n(\rho)$, and p denotes the ray parameter which is constant along any path and is defined by (cf. Figure 1 in Vest (1975))

$$p = rn(r)\sin i\ , \tag{13}$$

with $\tan i = p/\{[rn(r)]^2 - p^2\}^{\frac{1}{2}}$. The inversion formulas for the standard Abel equation (Anderssen (1976)) can be used to solve both (10) (Pask (1983)) and (12) (Vest (1975)).

For more complex lens geometrics and refractive index structures, much more complex relationships are needed to define the connection between the measurements, the paths traversed and the refractive index (Sweeney and Vest (1973) and Chu and Whitebread (1979)). For example, if the above assumption that the lens is spherical is only changed to assuming that it is rotationally symmetric about its optic axis, then it is necessary to determine the two-dimensional refractive index structure which is rotated. Now, however, the refractive index structure cannot be recovered from a single set of either deflection angle or optical pathlength measurements. It is necessary to repeat them as the lens s rotated about an axis through its center and perpendicular to its optic axis (Francois, Sasaki and Adams (1982)).

Because the situation is even more complex if the assumption of rotational symmetry does not hold, it is crucial to avoid dropping it unless necessary. For example, it would be grossly inefficient to first determine the three-dimensional refractive index structure of a lens and then find that it was infact rotationally symmetric about its optic axis, if it were possible to conclude that it was rotationally symmetric by some simpler procedure.

This is certainly possible. In fact, it is only necessary to compare measurements made on radial traverses across the face of the lens (perpendicular to the optic axis). The use of deflection angle measurements for this purpose are discussed in Section 4.

4. USE OF DEFLECTION ANGLE MEASUREMENTS TO TEST FOR ROTATIONAL SYMMETRY IN LENSES

Deflection angle measurements made on 4 radial traverses across the face of a lens from an eye of a cat are shown in Figure 2. It is immediately clear from these data that each individual radial traverse is seeing the same structure, which yields the tentative hypothesis that

"lenses from the eyes of cats are rotationally symmetric".

Though the use of the observational data in this manner has circumvented the need to solve the underlying improperly posed relationship (such as (10)), it has not removed the need

(i) to analyse the data in some appropriate manner to confirm rigorously the above tentative hypothesis, or to make additional measurements of this or another kind on this or other lenses from the eyes of cats;

(ii) to solve the underlying improperly posed relationship in order to answer other questions about the refractive index structure of the cat eye lens (e.g. the two-dimensional structure in the planes containing the optic axis when rotational symmetry can be assumed).

We do not pursue **(ii)** here. For **(i)**, we try two independent approaches: an examination of the singular values of the data matrix generated by the radial traverses; and the modelling of the radial traverses as a two-way table with one observation per cell to which a standard two-way analysis of variance is applied.

The results are shown in Table 1. It is clear from the relative sizes of the singular values of the (4×20)-data matrix D , constructed so that the radial traverses define its rows, that D can be interpreted as a slightly perturbed rank-one matrix; and hence, on the basis of the data contained in Figure 2, can be taken as confirmaion that, except for "random" errors, the radial traverses generating D are the same. The disadvantage of this test is its lack of a formal basis of comparison for one data matrix D with another; and the fact that, without appealing to additional information, this test only proves that the rows of D are constant multiples of each other.

This dificulty is circumvented if statistical testing of hypotheses is used (cf. Rohatgi (1976), Chapter 9).

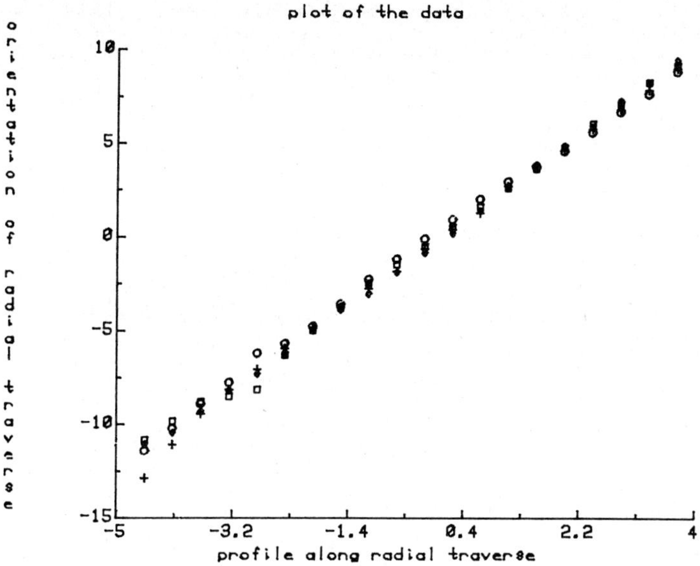

FIGURE 2

TABLE 1

a. THE SINGULAR VALUES OF THE (4×20)-DATA MATRIX GENERATED WITH THE RADIAL TRAVERSES DEFINING ITS ROWS:

56.28, 2.12, 1.34, 0.93

b. TWO-WAY ANALYSIS OF VARIANCE APPLIED TO THE (4×20)-DATA MATRIX GENERATED WITH THE RADIAL TRAVERSES DEFINING ITS ROWS, WHERE THE BLOCKS CORRESPOND TO THE DISTANCES ALONG THE RADIAL TRAVERSES AND THE TREATMENTS TO THE DIFFERENT ORIENTATIONS OF THE RADIAL TRAVERSES.

BLOCK, TREATMENT AND ERROR SUM OF SQUARES
3115.76, 1.52, 7.31

DEGREES OF FREEDOM
19, 3, 57

BLOCK AND TREATMENT F.RATIOS
1276.79, 3.96

By interpreting the (4,20)-data matrix D as a two-way table with one observation per cell, where the blocks correspond to the distances along the radial traverses and the treatments to the different orientations of the radial traverses, we can apply a standard two-way analysis of variance (IMSL Subroutine ARCBAN) to obtain the results in part b. of Table 1. Since from a table of the F-distribution, we find that

$$F_{19,57,0.01} = 2.24$$

and

$$F_{3,57,0.01} = 4.15 ,$$

we reject the hypothesis that the block effects are the same (since BLOCK F-RATIO >> $F_{19,57,001}$) and accept the hypothesis that the treatments (the radial traverses) are the same (since TREATMENT F-RATIO < $F_{3,57,0.01}$).

This approach corresponds to assuming that the radial traverses are parallel and then testing whether they are in fact the same curve. The 0.01 level of significance is used to minimize the possibility of rejecting the conclusion that they are the same curve, because of the obvious disadvantages of working with a three-dimensional refractive index structure when in fact it is only two-dimensional.

The use of the two-way table interpretation for the (4×20)-data matrix D is not the only way in which statistical testing of hypotheses can be introduced into the analysis of the structure of the radial traverses. For example, multivariate statistical tests could be applied. One possibility is the following. If X denotes the matrix obtained from D by subtracting the fourth row from the first three rows, and X is interpreted as a (3×20)-data matrix whose columns are assumed to have been drawn <u>independently</u> from a 3-variate normal distribution with mean vector $\underset{\sim}{\mu}$ and covariance Σ, Hotelling's T^2-statistic can be used to test the hypothesis. In fact, it is only necessary to observe that the statistic

$$\gamma = \frac{20\times 17}{3} \, \underset{\sim}{\bar{x}}^T \, (XX^T)^{-1} \underset{\sim}{\bar{x}} ,$$

where $\underset{\sim}{\bar{x}}$ denotes the sample mean vector (average of the columns of X), has an $F_{3,17,\alpha}$-distribution.

For the (4,20)-data matrix used in the calculations of Table 1, we find that

$$\gamma = 2.565 .$$

Since $F_{3,17,0.01} = 5.185$, we accept the hypothesis that $\underset{\sim}{\mu}=0$ and conclude that the radial traverses have the same structure.

The use of the above multivaraite test is based on the assumption that the columns of X are independently sampled. This implies that it would have been more appropriate if the radial traverses had been made at random on a number of randomly chosen lenses. This and other aspects associated with the application and interpretation of statistical tests will be discussed elsewhere.

Acknowledgement

The authors wish to thank the following people for their assistance and advice: David Jackett for assistance with the computations; Jeff Wood for comments about the two-way analysis of variance; and Bill Davis for advice about multivariate tests.

REFERENCES

R.S. Anderssen and A.J. Jakeman (1975) Abel-type integral equations in stereology. II. Computational methods of solution and the random spheres approximation, J.Micros. 105(1975), 135-153.

R.S. Anderssen (1980) On the use of linear functionals for Abel-type integral equations in applications, in The Application and Numerical Solution of Integral Equations (Eds. R.S. Anderssen, F.R. de Hoog and M.A. Lucas), Sijhoff and Noordhoff, Alphen aan den Rijn, The Netherlands.

R.S. Anderssen and D.R. Jackett (1983) Linear functionals of foliage angle density, J.Aust.Math. Soc, Series B.(in press)

A.W. Atkinson, Jr., P.C.L. John and B.E.S. Gunning (1974) The growth and division of the single mitachondrion and other organelles during the cell cycle of Chlorella, studied by quantitative stereology and three dimensional reconstruction, Protoplasma 81 (1974), 77-109.

M. Bando, A. Nakajima, M. Nakagawa and T. Hiraoka (1976) Measurement of protein distribution in human lens by microspectrophotometry, Exp. Eye Res 22 (1976), 389-392.

P.L. Chu and T. Whitebread (1979) Non destructive determination of refractive index profile of an optical fiber: fast Fourier transform method, Appl. Optics 18 (1979), 1117-1122.

F.R. de Hoog (1980) Review of Fredholm equations of the first kind, in The Application and Numerical Solution of Integral Equations (Eds R.S. Anderssen, F.R. de Hoog and M.A. Lukas), Sijthoff and Noordhoff, Alphen aan den Rijn, The Netherlands.

P.-L. Francois, I. Sasaki and M.J. Adams (1982) Practical three-dimensional profiling of optical fiber preforms, IEEE J. Quantum Electronics QE-18 (1982), 524-535.

E.D. Hyam and J. Nutting (1956) The tempering of plain carbon steels, J. Iron and Steel Inst. 148 (1956), 148-165.

D.R. Jackett and R.S. Anderssen (1983) Computing the foliage angle distribution from contact frequency data, Proceedings of Computational Techniques and Applications Conference 1983, Sydney, August 29-31, 1983.

M.A. Lukas (1980) Regularization, in The Application and Numerical Solution of Integral Equations (Eds R.S. Anderssen, F.R. de Hoog and M.A. Lukas), Sijthoff and Noordhoff, Alphen aan den Rijn, The Netherlands.

J.B. Miller (1967) A formula for average foliage density, Aust. J. Bot. 15 (1967), 141-144.

S. Nakao, S. Fujimoto, R. Nagata and K. Iwata (1968) Model of refractive-index distribution in the rabbit crystalline lens, J.Opt.Soc.Ann. 58 (1968), 1125-1130.

C. Pask (1983) The theory of non-destructive lens index distribution measurement, in Modelling the Eye with Gradient Index Optics (Ed. A. Hughes), Cambridge University Press, Cambridge.

B. Philipson (1969) Distribution of protein within the normal rat lens, Invest. Ophthalmol. 8 (1969), 258-270.

V. Rohatgi (1976) An Introduction to Probability Theory and Mathematical Statistics, Wiley-Interscience, New York, 1976.

D.W. Sweeney and C.M. Vest (1973) Reconstruction of three-dimensional refractive index fields from multidimensional interferometric data, Appl. Optics 12 (1973), 2649-2664.

E.E. Underwood (1970) Quantitative Stereology, Addison and Wesley, Reading, Mass., 1970.

C.M. Vest (1975) Interferometry of strongly refracting axisymmetric phase objects, Appl. Optics 14 (1975), 1601-1606.

J. Warren Wilson (1963) Estimation of foliage denseness and foliage angle by inclined point quadrats, Aust. J. Bot. 11 (1963), 95-105.

Computational Techniques & Applications: CTAC-83
J. Noye & C. Fletcher (Editors)

PROBLEMS WITH DERIVED VARIABLE METHODS FOR THE NUMERICAL SOLUTION OF THREE-DIMENSIONAL FLOWS

J.A. Reizes, E. Leonardi and G. de Vahl Davis

School of Mechanical and Industrial Engineering,
The University of New South Wales, Kensington,
N.S.W., Australia, 2033.

In this paper we discuss problems associated with ensuring the satisfaction of global continuity when the vector potential/vorticity and the velocity/vorticity methods are used for the numerical solution of three-dimensional fluid flow problems. It is shown that the vector potential/vorticity formulation can be satisfactorly used for simply-connected regions and a new method is presented which can be applied to multiply connected regions.

INTRODUCTION

Although the vector potential, the three-dimensional equivalent of the stream function, has been known for more than a century [14] it has not been extensively used in three-dimensional numerical calculations. One reason for this neglect has been the confusion resulting from the difficulty in specifying the appropriate boundary conditions. The first usable boundary conditions were found by Aziz and Hellums [2], but could only be employed for enclosed flows in simply-connected regions. The boundary conditions for through-flow problems, also in simply-connected regions, were derived by Hirasaki and Hellums [4] but were far too complex to be exploited. In a later paper [5] they introduced a scalar potential which was utilized to represent the through-flow velocity at the boundaries, and which led to the same boundary conditions on the vector potential for through-flow problems as for the enclosed flow case. Once again these boundary conditions apply only to simply-connected regions.

A number of researchers have used the Hirasaki and Hellums boundary conditions to solve three-dimensional problems in enclosures (e.g. [8, 9]) and with through flow [1, 3]. Richardson and Cornish [12] have shown that these boundary conditions do not apply to multiply-connected regions, and have derived a set supposedly applicable to those regions. Ozoe , Okomoto and Churchill [10] and Ozoe, Shibata and Churchill [11] avoided the use of the new boundary conditions by making assumptions from experimental results regarding flow symmetry in natural convection in cylindrical annuli. They were thus able to change a multiply-connected region to a simply-connected region and were therefore able to use the Hirasaki and Hellums boundary conditions.

In an earlier paper [7] we had difficulties in using the Richardson and Cornish boundary conditions in the investigation of natural convection in rotating annuli. As the result we examined the suitability of the various boundary conditions which have been proposed. It is shown in this paper that the Hirasaki and Hellums boundary conditions are applicable to simply-connected regions, whereas the Richardson and Cornish boundary conditions are not adequate for multiply-connected regions. The reason for the inadequacy is that the normal velocity at the boundaries is not satisfactorily represented. In effect the continuity equation has been used implicitly but not, as would be required, explicitly enforced. Since this is also true for the velocity/vorticity scheme, this formulation is also shown to be inadequate.

It is also demonstrated that it is possible to modify the Richardson and Cornish boundary conditions so that fluid flows in multiply-connected regions can be calculated; however, this requires the introduction of the scalar potential which increases the cost of the solution in terms of the computer time and storage required.

MATHEMATICAL BACKGROUND

Two methods of numerical solution of fluid flow problems will be discussed in this section. The vector potential/vorticity scheme is presented first because of its apparent advantage of satisfying the continuity equation automatically. The simpler velocity/vorticity formulation is then presented.

VECTOR POTENTIAL/VORTICITY

It can be shown quite generally [12] that the velocity field in a fluid may be represented by

$$\underline{V} = \nabla \times \underline{\Psi} + \nabla\Phi, \qquad (1)$$

in which $\underline{V}$ is the velocity, $\underline{\Psi}$ is the vector potential and Φ is the scalar potential. In solving fluid flow problems with this formulation for the velocity field it is usual to transform the Navier-Stokes equation into the vorticity transport equation, with the vorticity $\underline{\zeta}$, defined as

$$\underline{\zeta} = \nabla \times \underline{V}. \qquad (2)$$

The relationship between the vorticity and the vector potential is established by substituting equation (1) into equation (2) which yields

$$\underline{\zeta} = -\nabla^2\underline{\Psi} + \nabla(\nabla.\underline{\Psi}). \qquad (3)$$

The vector potential Ψ is not single valued since a new vector potential,

$$\underline{\Psi}' = \underline{\Psi} + \nabla B, \qquad (4)$$

in which B is any continuous and differentiable scalar field, will yield the same result as $\underline{\Psi}$ when substituted in equation (1). To simplify equation (3) a number of authors [4, 5, 12, 14] have proposed that $\underline{\Psi}$ be solenoidal viz.,

$$\nabla.\underline{\Psi} = 0, \qquad (5)$$

so that equation (3) becomes

$$\nabla^2\underline{\Psi} = -\underline{\zeta}, \qquad (6)$$

which is the simplest form of the relationship between the vorticity and the vector potential.

It should be noted that the introduction of equation (5) does still not make $\underline{\Psi}$ unique. Suppose that $\underline{\Psi}$ yields a certain velocity field and is solenoidal. If B in equation (4) is a harmonic funtion, that is,

$$\nabla^2 B = 0, \qquad (7)$$

the same velocity field is obtained from equation (1) if $\underline{\Psi}'$ is used, whilst $\underline{\Psi}'$ is itself solenoidal. Because equation (6) is linear, for any given vorticity field the $\underline{\Psi}$ field is uniquely determined from equation (6) for a particular set of boundary conditions. Since it is possible to generate any number of vector potential fields, which differ from each other by the gradient of one or more harmonic functions, to yield the same velocity field, a unique set of boundary conditions does not exist for the solution of $\underline{\Psi}$. In fact, it is possible to have an infinite number of boundary conditions for the solution of $\underline{\Psi}$ which will yield the same solution for the velocity field with $\underline{\Psi}$ solenoidal. This is the reason for the different boundary conditions on the vector potential which have been mentioned in the literature [4, 5, 12, 15].

The equation of continuity for an incompressible fluid, the vorticity transport equation in which the Boussinesq approximation has been used and the energy transport equation can be written in non-dimensional form as [13]

$$\nabla.\underline{V} = 0, \tag{8}$$

$$\frac{\partial\underline{\zeta}}{\partial t} + \nabla \times (\underline{\zeta} \times \underline{V}) = \text{RaPr}\nabla \times (\theta\hat{\underline{g}}) + \text{Pr}\nabla^2\underline{\zeta} \tag{9}$$

and

$$\frac{\partial\theta}{\partial t} + \nabla.(\underline{V}\theta) = \nabla^2\theta, \tag{10}$$

in which Ra is the Rayleigh number, Pr is the Prandtl number, θ is the non-dimensional temperature and g is the gravitational acceleration. The same remarks can be applied to compressible fluids but the relations become more complex [6].

The substitution of equation (1) into equation (8) leads to

$$\nabla^2\Phi = 0. \tag{11}$$

The solution of equations (6), (9), (10) and (11) together with equation (1) and a set of appropriate boundary conditions should lead to the solution of any incompressible fluid flow problem. Note that the solution of the above equations may not lead to a vector potential which is solenoidal since this condition is not forced, so that it may be better to use equation (3) rather than equation (6). Further, it has been shown by Wong and Reizes [15] that the introduction of the scalar potential leads to numerical errors which are not immediately obvious so that the use of the scalar potential should be avoided. Also, as discussed by Richardson and Cornish [12] the simplest boundary conditions are different for simply and multiply-connected regions, although, of course, the boundary conditions for multiply-connected regions perforce apply to simply-connected regions. The derivations for the boundary conditions which have been found in the literature [4, 5, 12] are extremely complex, and simple derivations are presented below which also make apparent the reason for the boundary conditions developed for simply-connected regions not being applicable in multiply-connected regions.

VELOCITY/VORTICITY

If the curl of equation (2) is taken, a Poisson equation which can be easily handled numerically results viz.,

$$\nabla \times \underline{\zeta} = -\nabla^2\underline{V} + \nabla(\nabla.\underline{V}). \tag{12}$$

For an incompressible fluid equation (8) applies so that equation (12) becomes

$$\nabla^2\underline{V} = -\nabla \times \underline{\zeta}. \tag{13}$$

Once again, in principle at least, the solution of equations (9), (10) and (13) with appropriate boundary conditions leads to the solution of any incompressible fluid flow problem. This is a much simpler approach than the vector potential/vorticity formulation; however, it must be noted that equation (8) is assumed to be satisfied without being forced, with the result that there is no guarantee that the calculated velocity field will be solenoidal.

BOUNDARY CONDITIONS ON THE VECTOR POTENTIAL

Following Hirasaki and Hellums [5] it is assumed that the scalar potential is related to the normal velocity at the boundaries by

$$(V_n)_b = (\partial\Phi/\partial n)_b, \tag{14}$$

in which n refers to the direction normal to the boundary and b to the boundary. Richardson and Cornish [12] obtained this relation as part of the general derivation of their boundary conditions, whereas Hirasaki and Hellums assumed it.

SIMPLY-CONNECTED REGIONS

The equation of continuity for an incompressible fluid can be written for the entire region as

$$\iint_A \underline{V} \cdot d\underline{A} = 0 \tag{15}$$

in which A is the surface area. Upon the substitution of equation (1) into equation (15) this relationship becomes,

$$\iint_A (\nabla \times \underline{\Psi} + \nabla\Phi) \cdot d\underline{A} = 0. \tag{16}$$

The integration of equation (14) over the whole surface of the domain yields

$$\iint_A (\nabla\Phi) \cdot d\underline{A} = \iint_A V_n dA = 0, \tag{17}$$

so that equation (16) simplifies to

$$\iint (\nabla \times \underline{\Psi}) \cdot d\underline{A} = 0. \tag{18}$$

But by Stokes' theorem,

$$\iint_A (\nabla \times \underline{\Psi}) \cdot d\underline{A} = \int_s \underline{\Psi} \cdot d\underline{s}, \tag{19}$$

in which s is the contour enclosing the surface. One solution of equation (19) is that the vector potential, $\underline{\Psi}$, is everywhere perpendicular to the boundary. Thus a possible set of boundary conditions would be

$$(\Psi_{t_1})_b = (\Psi_{t_2})_b = 0, \tag{20}$$

in which the subscripts t_1 and t_2 refer to the two tangential directions at the boundary.

Since the Stokes theorem can only be used in simply-connected regions, it is apparent that a different approach has to be adopted when dealing with multiply-connected regions.

Because at this stage, it has been assumed that the vector potential is solenoidal, it follows that

$$\frac{\partial\Psi_n}{\partial n} = -\frac{\partial\Psi_{t_1}}{\partial s_{t_1}} - \frac{\partial\Psi_{t_2}}{\partial s_{t_2}} \tag{21}$$

which, when combined with equation (20) leads to the final boundary condition for a simply-connected region, viz.,

$$\left[\frac{\partial \Psi_n}{\partial n}\right]_b = 0. \qquad (22)$$

It should be noted that these boundary conditions ensure that the correct normal velocity is maintained at the boundary of the region. For example, if there is no through flow (i.e. the flow is confined to an enclosure), it follows from equations (11) and (14) that Φ=0, which together with equation (20) ensures that the normal velocity is zero at the boundaries.

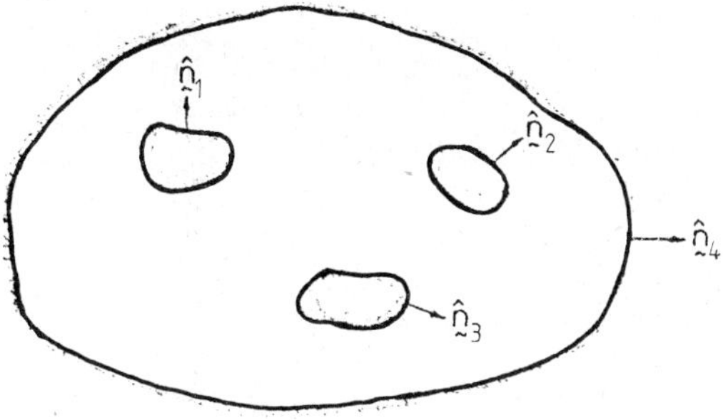

Figure 1. Schematic diagram of a multiply connected region with m bodies

MULTIPLY CONNECTED REGIONS

As mentioned above, it is not possible here to use the continuity equation to set up the boundary conditions in multiply-connected regions. Instead the fact that the vector potential is solenoidal is employed. Suppose we have a multiply-connected region with m bodies as shown in figure 1. It follows from Gauss' divergence theorem that

$$\iiint_v (\nabla.\underline{\Psi})dv = \iint_{A_1} \underline{\Psi}.d\underline{A} + \iint_{A_2} \underline{\Psi}.d\underline{A} + \quad . \quad . \quad . \quad + \iint_{A_m} \underline{\Psi}.d\underline{A}, \qquad (23)$$

in which v is volume. Because $\underline{\Psi}$ is solenoidal it follows that the sum of the integrals on the right hand side is zero so that a possible solution of equation (23) is

$$(\Psi_n)_{b_1} = (\Psi_n)_{b_2} = (\Psi_n)_{b_3} = \ldots = (\Psi_n)_{b_m} = 0, \qquad (24)$$

and the remaining boundary conditions are obtained from the tangential velocities viz,

$$\frac{1}{F_{t_2} F_n} \frac{\partial(F_{t_2}\Psi_{t_2})}{\partial s_n} = -V_{t_1} + \frac{1}{F_{t_1}} \frac{\partial \Phi}{\partial s_{t_1}}$$

and

$$\frac{1}{F_{t_1} F_n} \frac{\partial(F_{t_1}\Psi_{t_1})}{\partial s_n} = V_{t_2} - \frac{1}{F_{t_2}} \frac{\partial \Phi}{\partial s_{t_2}}, \qquad (25)$$

in which F refers to scale factors.

It is important to note that if these boundary conditions are used, the normal velocity at the boundaries is not correctly represented. From equation (1)

$$V_n = \frac{1}{F_{t_1} F_{t_2}} \left[\frac{\partial F_{t_1} \Psi_{t_1}}{\partial s_{t_2}} - \frac{\partial F_{t_2} \Psi_{t_2}}{\partial s_{t_1}} \right]_b + \frac{1}{F_n} \frac{\partial \Phi}{\partial n}, \qquad (26)$$

which together with equation (14) lead to

$$\left[\frac{\partial (F_{t_2} \Psi_{t_2})}{\partial s_{t_1}} \right]_b = \left[\frac{\partial (F_{t_1} \Psi_{t_1})}{\partial s_{t_2}} \right]_b , \qquad (27)$$

for the normal velocity to be satisfactorily described at the boundaries. Equation (27) is implicitly assumed by Richardson and Cornish [12] , but is not forced. As the result it is possible to have "leaks" in impermeable boundaries if these boundary conditions are used. That is, the boundary conditions for multiply-connected regions, as given by Richardson and Cornish [12], are the necessary, but not sufficient conditions. As a consequence the results we had published earlier [7] are incorrect.

Since the tangential velocity is used in the vorticity boundary conditions, one obvious way of overcoming the above difficulty is to replace one of the conditions in equation (25) with equation (27). However, both terms in equation (27) are tangential derivatives which makes their implementation in a numerical scheme extremely difficult.

In order to demonstrate that the Richardson and Cornish boundary conditions and the velocity/vorticity formulation can lead to serious errors, three-dimensional natural convection between concentric horizontal cylinders was studied using the vector potential/vorticity scheme with the Hirasaki and Hellums and the Richardson and Cornish boundary conditions as well as the velocity/vorticity method.

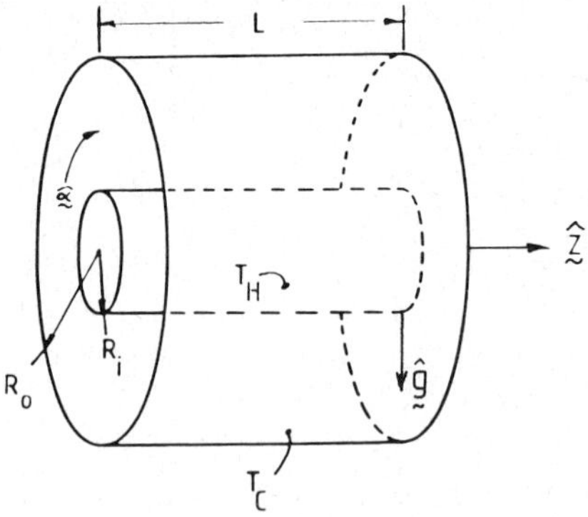

Figure 2. Arrangement of the annulus

PROBLEM DEFINITION

The geometry of the horizontal annulus is illustrated in figure 2. It is assumed that all the surfaces are stationary and impermeable, the outer cylinder is at a uniform cold temperature T_c, the inner cylinder is at a uniform hot temperature T_h and the end plates are adiabatic. Equations (9) and (10) have been non-dimensionalised with $R_o - R_i$ for distance, $\nu/(R_o-R_i)$, (in which ν is the kinematic viscosity) for velocity, $\nu/(R_o-R_i)^2$ for vorticity and the non-dimensional temperature θ is defined as $(T-T_c)/(T_h-T_c)$, in which T is the temperature at any point.

The boundary conditions for the vorticity are obtained from the definition of vorticity (equation (2)) and are given by:

$$\zeta_r = -\partial v_\alpha/\partial z \ , \ \zeta_\alpha = \partial v_r/\partial z \quad \text{and} \quad \zeta_z = 0 \tag{28}$$

on the end walls, and

$$\zeta_r = 0, \ \zeta_\alpha = -\partial v_z/\partial r \quad \text{and} \quad \zeta_z = 1/r \ \partial(rV_\alpha)/\partial r \tag{29}$$

on the cylinder walls.

The steady state solution of the appropriate equations together with the relevant boundary conditions was obtained with the use of the method of the false transient [8].

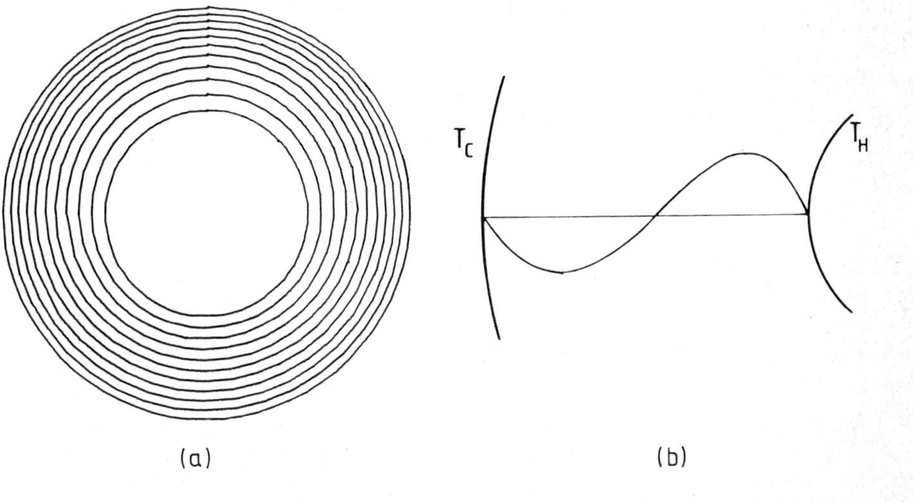

Figure 3. (a) Isotherms and (b) azimuthal velocity distribution on the horizontal diameter at the mid plane of the annulus. Hirasaki and Helllums boundary conditions

RESULTS AND DISCUSSION

The difficulties encountered in using the Richardson and Cornish boundary conditions on the vector potential or the velocity/vorticity formulation clearly occur at all values of radius ratio (R_o/R_i), aspect ratio (ratio of the length of the annulus to the radial gap), Rayleigh and Prandtl numbers. Only one annulus, with a radius ratio of two, aspect ratio of two, Ra=500 and Pr=0.72 has therefore been considered. The grid used in all cases is 9x20x9 (r, α, z).

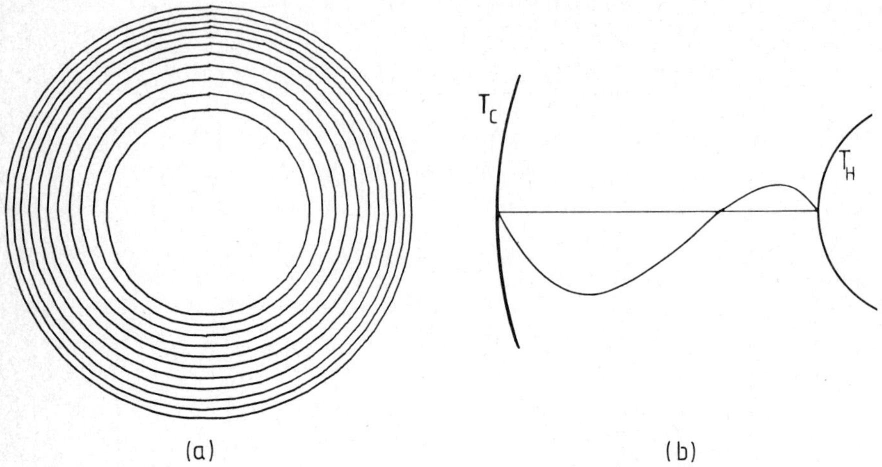

Figure 4. (a) Isotherms and (b) azimuthal velocity distribution on the horizontal diameter at the mid plane of the annulus. Richardson and Cornish boundary conditions.

Because of the symmetry of the flow about the vertical plane through the axis of the cylinders, the problem can be made simply-connected by making this plane a boundary of the solution region; thus permitting the use of the Hirasaki and Hellums boundary conditions. The azimuthal velocity distribution on the horizontal diameter and the temperature field at the mid-plane of the annulus when the Hirasaki and Hellums boundary conditions are used are given in figure 3.

The results for the same variables obtained from the solution of the same problem using the Richardson and Cornish boundary conditions are given in figure 4. It is apparent that the two results are not the same. Whereas the downward flow is balanced by the upward flow in figure 3 (as must be the case since $V_z = 0$ on this plane), the downward flow is much larger than the upward flow in figure 4. In other words the radial velocity on the inner cylinder is not "seen" as zero when the Richardson and Cornish boundary conditions are used. This is perhaps best illustrated by calculating $\nabla.V$ with the assumption that the velocity vector on the boundary is zero. When the Hirasaki and Hellums boundary conditions are used $\nabla.V$ is of the order of the round off error (10^{-15} on the Cyber 171 used for these calculations) throughout the entire region of interest. However, when the Richardson and Cornish boundary conditions are used, the dilatation one mesh point from the boundaries is of the order of 4. It is therefore readily apparent that the normal velocity at the boundaries is not "seen" by the program as zero.

The final demonstration of this effect is given in figure 5 in which the radial velocities one mesh point from the inner cylinder at the centre plane of the annulus are compared for the two cases. The radial velocity at this radius is always substantially larger in the case of the Richardson and Cornish boundary conditions than when the Hirasaki and Hellums boundary conditions are used. This can be interpreted as indicating that there is an apparent flow through the impermeable inner cylinder. Similar effects occur at the outer cylinder but are less severe. The errors are a function of the curvature which follows directly from the form of the continuity equation in cylindrical coordinates. In any case it is readily apparent that the Richardson and Cornish boundary conditions have not yielded the correct solution.

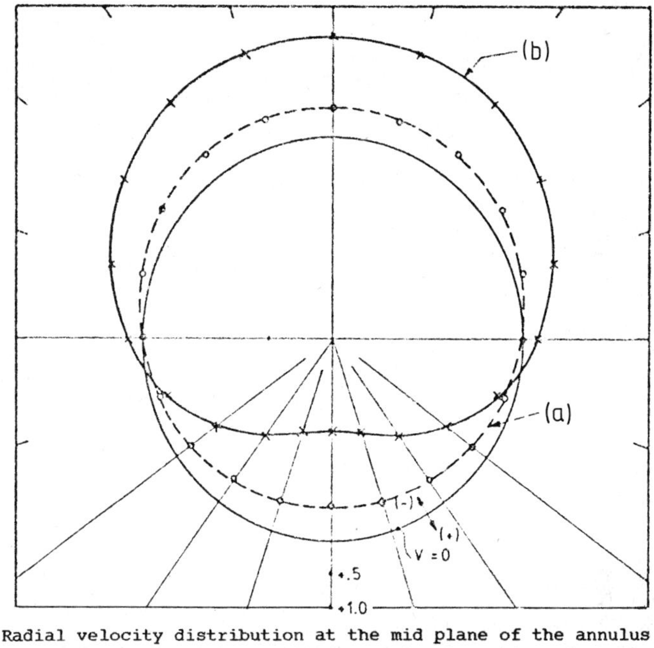

Figure 5. Radial velocity distribution at the mid plane of the annulus one mesh point from the inner cylinder (a) Hirasaki and Hellums boundary conditions (b) Richardson and Cornish boundary conditions

It should be noted that the isotherms in figures 3 and 4 do not differ greatly. This is because at a Rayleigh number of 500 the heat transfer is essentially by conduction so that large errors in the flow fields do not affect the temperature distribution significantly.

The solution of the same problem using the velocity/vorticity formulation is shown in figure 6. This solution is similar to that obtained using the Richardson and Cornish boundary conditions (figure 4). The leaks on the inner cylinder are once again greater than those on the outer cylinder. This means that the error is a function of the curvature. The error can possibly be reduced by inceasing the number of mesh points; however, it should be noted that a satisfactory solution was obtained with the same number of mesh points when the vector potential/vorticity formulation with the Hirasaki and Hellums boundary conditions was used. It follows that great caution has to be exercised when the velocity/vorticity method is employed.

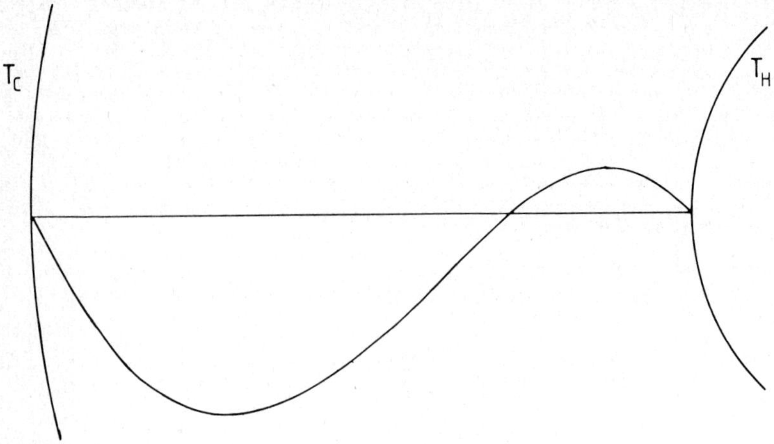

Figure 6. Velocity distribution on the horizontal diameter at the mid plane of the annulus. Velocity/vorticity formulation.

MODIFICATION OF THE RICHARDSON AND CORNISH BOUNDARY CONDITIONS

Since the boundaries appear to be leaking when the Richardson and Cornish boundary conditions are used and the scalar potential was used by Hirasaki and Hellums [5] to represent the normal velocity it is proposed from equation (26) that Φ be defined by

$$\left[\frac{\partial\Phi}{\partial n}\right]_b = \frac{-F_n}{F_{t_1}F_{t_2}}\left[\frac{\partial F_{t_1}\Psi_{t_1}}{\partial s_{t_2}} - \frac{\partial F_{t_2}\Psi_{t_2}}{\partial s_{t_1}}\right]_b + F_n(V_n)_b \qquad (30)$$

When equation (30) was used instead of equation (14), results almost identical to those in figure 3 were obtained. However, the program took many more iterations to converge and the time per iteration was longer than that for the Hirasaki and Hellums boundary conditions because Φ now needs to be calculated.

The above procedure means that even if a multiply-connected region is enclosed by impermeable walls Φ has to be included in the calculation procedure and the continuity equation is not automatically satisfied [15]. As a result a finer mesh may have to be used if the same accuracy is required as would have been achieved with the use of the vector potential alone in simply-connected regions. However, the introduction of the scalar potential permits the vector potential/vorticity formulation to be used in multiply-connected regions.

REFERENCES

[1] Aregbesola, Y.A.S. and Burley, D.M. - The Vector and Scalar Potential Method for the Numerical Solution of Two- and Three-Dimensional Navier-Stokes Equations, Jrnl. Computational Physics 24 (1977) 389-415.

[2] Aziz, K. and Hellums, J.D. - Numerical Solution of the Three-Dimensional Equations of Motion for Laminar Convection, Physics of Fluids 10 (1967) 314-324.

[3] de Vahl Davis, G. and Wolfshtein, M. - A Study of a Three Dimemensional Free Jet Using the Vorticity/Vector Potential Method, Fourth Int. Conf. on Numerical Methods in Fluid Dynamics (Colorado, 1974).

[4] Hirasaki, G.J. and Hellums, J.D. - A General Formulation of the Boundary Conditions on the Vector Potential in Three-Dimensional Hydrodynamics, Quarterly of Applied Maths. 26 (1968) 331-342.

[5] Hirasaki, G.J. and Hellums, J.D. - Boundary Conditions on the Vector Potential in Viscous Three-Dimensional Hydrodynamics, Quarterly of Applied Maths. 28 (1970) 293-296.

[6] Leonardi, E. and Reizes, J.A. - Chapter 18 in Numerical Methods in Heat Transfer R.W. Lewis, K. Morgan and O. Zienkiewicz (eds) (John Wiley and Sons, Chichester, U.K., 1981).

[7] Leonardi, E. Reizes, J.A. and de Vahl Davis, G. - Natural Convection in a Rotating Annulus, Proc of the 2nd Conference on Numerical Methods in Laminar and Turbulent Flow C. Taylor and B.A. Schrefler (eds), (Pineridge Press, Swansea, U.K., 1981) 995-1006.

[8] Mallinson, G. and de Vahl Davis, G. - Three-Dimensional Natural Convection in a Box: A Numerical Study, Jrnl. Fluid Mechanics 83 (1977) 1-31.

[9] Mallinson, G., Graham, A.D. and de Vahl Davis, G. - Three-Dimensional Flow in a Closed Thermosyphon, Jrnl. Fluid Mechanics 109 (1981) 259-276. 1981

[10] Ozoe, H., Okomoto, T. and Churchill, S.W. - Natural Convection in a Vertical Annular Space Heated from Bellow, Heat Transfer - Japanese Research 8 (1980) 82-93.

[11] Ozoe, H., Shibata, T. and Churchill, S.W. - Natural Convection in an Inclined Circular Cylindrical Annulus Heated and Cooled on its end Plates, International Jrnl. Heat and Mass Transfer 24 (1980) 727-737.

[12] Richardson, S.M. and Cornish, A.R.H. - Solution of Three-Dimensional Incompressible Flow Problems, Jrnl. Fluid Mechanics 82 (1977) 309-319.

[13] Roache, P. Computational Fluid Mechanics (Hermosa Publishers, Albuquerque, New Mexico, 1976).

[14] Truesdall, C The Kinematics of Vorticity (Indiana University Press, Bloomington, Indiana, 1954).

[15] Wong, A.K. and Reizes, J.A. - An Effective Vorticity-Vector Potential Formulation for the Numerical Solution of Three-Dimensional Duct Flow Problems, to appear in Jrnl. Comptutational Physics.

Computational Techniques & Applications: CTAC-83
J. Noye & C. Fletcher (Editors)

A HYBRID METHOD FOR SOLVING THE BOUNDARY VALUE PROBLEMS FOR HELMHOLTZ'S EQUATION IN TWO-DIMENSIONAL DOMAINS

A.P. Raiche and Z.K. Tarlowski

CSIRO Division of Mineral Physics
P.O. Box 136
North Ryde, NSW 2113
Australia

A new hybrid method (SRHM) combining the method of summary representation and either finite differences or finite elements is described and used to solve the Helmholtz equation on a two-dimensional rectangular domain. The method offers significant computational advantages when local variations in physical properties can be confined to one rectangular subdomain for which either a finite difference or finite element solution is used. The rest of the domain is divided into other rectangular subdomains in each of which the physical properties are constant. The solution in each of these domains is found using the method of summary representation.

INTRODUCTION

There are a number of numerical methods for solving the partial differential equations of mathematical physics. The most frequently used methods are those based on finite differences, finite elements and integral equations.

In many problems, such as those encountered in geophysics, accurate solutions require approximating meshes extending over a large domain even though the local variations in physical properties ($\gamma(x,y)$) may be restricted to a relatively small subdomain. Use of the finite difference or finite element method over the whole domain can lead to numerically ill-posed problems as well as being very consumptive of computer resources.

For the problem of a small central subdomain with locally varying γ, surrounded by larger subdomains in each of which γ is a constant (see Figure 1) one would like to use a local finite difference or finite element method combined with a more efficient, perhaps analytic, method over the rest of the whole domain. Lee et al (1981) used the finite element method in conjunction with the integral equation method. However the use of the integral equation method places severe restrictions on the structure of the domain exterior to the region of local variation.

In this paper we define a new hybrid method (SRHM) using the method of summary representation in combination with the finite difference or finite element method to solve the Helmholtz equation on two-dimensional domains. Since the method of summary representation is not widely known, it is described briefly in the next section.

SOLUTION OF THE BOUNDARY-VALUE PROBLEM FOR HELMHOLTZ'S EQUATION ON A RECTANGULAR DOMAIN BY THE METHOD OF SUMMARY REPRESENTATION

For a limited class of discretization procedures, the method of summary representation can be used to find analytical solutions to finite difference

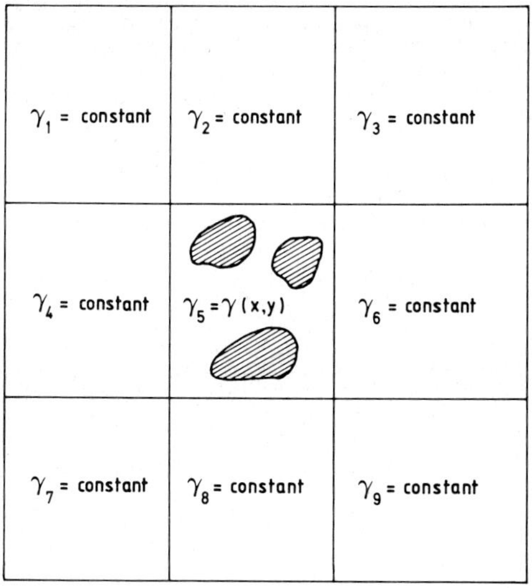

Figure 1
Division of Large Domain into Appropriate Subdomains

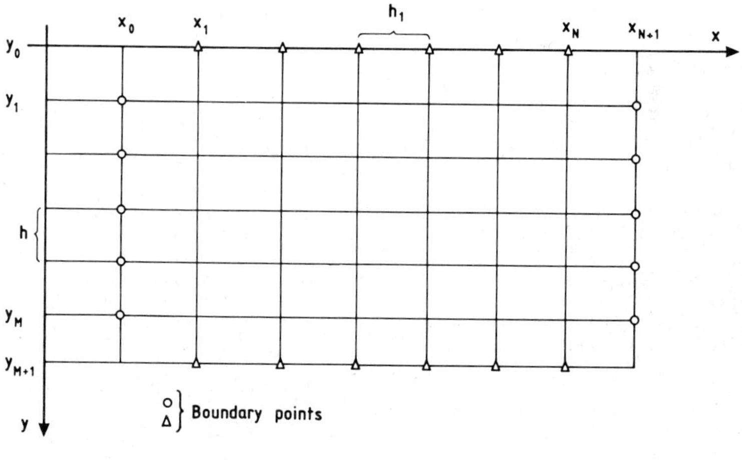

Figure 2
Rectangular Domain

analogs of many of the differential equations of mathematical physics. In general, the method consists, first, of finding an analytical solution to the one-dimensional finite difference eigenvalue problem. For self-adjoint finite difference boundary value problems expressed over N points, a basis of N orthogonal eigenvectors, corresponding to N distinct real eigenvalues, can be found. These functions, expressed as a matrix of N eigenfunctions at N points, form the basis of the so-called P transform (Polozhii (1965)).

The two-dimensional problem is solved by using a five point difference scheme to discretize the equation over a mesh such as that shown in Figure 2 with N interior points in the x-direction and M interior points in the y-direction. This gives a system of MN equations in which the x and y variations are coupled. It is assumed that the boundary values at $x = x_o$, $x = x_{N+1}$, $y = y_o$ and $y = y_{M+1}$ are known.

Applying the abovementioned P transform to this system of equations decouples the x and y variations into M systems of N equations in one variable. But since each of these resulting one-dimensional equations can be solved analytically, the solution at any given point in the entire region can be expressed in analytic form in terms of the one-dimensional eigenfunctions and the boundary values.

We begin a more explicit description of the method by writing the Helmholtz equation in two dimensions.

$$\frac{\partial^2 u(x,y)}{\partial x^2} + \frac{\partial^2 u(x,y)}{\partial y^2} + \gamma^2(x,y) = f(x,y) \qquad (1)$$

where f(x,y) is a known source function and $\gamma(x,y)$ contains the physical property description. (In electromagnetic geophysics $\gamma^2 = i\omega\mu\sigma(x,y)$ where ω = the angular frequency of the electromagnetic source, $\mu = 4\pi \times 10^{-7}$, and $\sigma(x,y)$ is the electrical conductivity.)

We discretize this equation using a basic five point difference scheme. However, as stated earlier, the method of summary representation works only for a limited class of discretization procedures because of the difficulty of finding analytic eigenfunctions for the one-dimensional problem. At present, we have only been able to find these for constant mesh spacing in the x and y directions and for $\gamma(x,y)$ constant on the rectangular subdomain. Thus, assuming constant mesh spacings h_1 and h in the x and y directions respectively, the discretized version of equation (1) at an interior point (x_k, y_j) becomes

$$u(x_k,y_{j+1})+u(x_k,y_{j-1})+\alpha^2[u(x_{k+1},y_j)+u(x_{k-1},y_j] -2b\, u(x_k,y_j) = h^2 f(x_k,y_j) \qquad (2)$$

for k = 1,2,...N and j = 1,2...M.

$\alpha = h/h_1$ = the ratio of internodal spacings

$x_{k+1} = x_k + h_1$

$y_{j-1} = y_j - h$

$b = 1 + \alpha^2 + \gamma^2 h^2/2$

The MN equations (one for each interior point) represented by equation (2) can be rewritten in terms of M (N-dimensional) vector equations as

$$\underline{u}(y_{j+1}) + \underline{u}(y_{j-1}) - 2T\, u(y_j) = \underline{S}(y_j) \qquad (3)$$

for j = 1,2....M and where

$$\underline{u}(y_j) = \begin{bmatrix} u(x_1,y_j) \\ u(x_2,y_j) \\ \cdot \\ \cdot \\ \cdot \\ u(x_{N-1},y_j) \\ u(x_N,y_j) \end{bmatrix}, \quad \underline{S}(y_j) \begin{bmatrix} h^2 f(x_1,y_j)-\alpha^2 u(x_0,y_j) \\ h^2 f(x_2,y_j) \\ \cdot \\ \cdot \\ \cdot \\ h^2 f(x_{N-1}, y_j) \\ h^2 f(x_N,y_j)-\alpha^2 u(x_{N+1}, y_j) \end{bmatrix}$$

$\underline{u}(y_j)$ is a vector consisting of the unknown function values along a given row of nodes $y = y_j$. $\underline{S}(y_j)$ is a generalized source vector whose first and N^{th} components are augmented by the known boundary values, $u(x_0,y_j)$ and $u(x_{N+1},y_j)$ respectively, i.e. the boundary values at the left and right hand sides of the mesh.

T is a tridiagonal matrix which can be expressed as

$$T = \frac{\alpha^2}{2} R + b\, I_N \tag{4}$$

where I_N is the N-dimensional identity matrix and

$$R = \begin{bmatrix} 0 & 1 & 0 & & & & \\ 1 & 0 & 1 & & & & \\ & 1 & 0 & 1 & 0 & & \\ & & \cdot & \cdot & \cdot & & \\ & 0 & & \cdot & \cdot & \cdot & \\ & & & & 1 & 0 & 1 \\ & & & & & 1 & 0 \end{bmatrix} \tag{5}$$

In equation (3), the x and y variations are coupled because T is not a diagonal matrix. The choice of constant mesh spacing has given R a simple form. The restriction of constant $\gamma(x,y)$ has meant that any transform which diagonalizes R will also diagonalize T.

For all self-adjoint representations of the T matrix, a P transform exists such that

$$P\,T\,P^{-1} = \Lambda = \begin{bmatrix} \lambda_1 & & & \\ & \lambda_2 & & 0 \\ & & \ddots & \\ 0 & & & \lambda_n \end{bmatrix} \tag{6}$$

Thus equation (3) can be transformed as

$$P\underline{u}(y_{j+1}) + P\underline{u}(y_{j-1}) - 2PTP^{-1}P\underline{u}(y_j) = P\underline{S}(y_j) \tag{7}$$

Defining $\underline{u}' = P\,\underline{u}$, we see that for each value of j = 1,2, ... M, equation (7) represents N independent equations of the form

$$u'_k(y_{j+1}) + u'_k(y_{j-1}) - 2\,\lambda_k u'_k(y_j) = S'_k(y_j) \tag{8}$$

For each value of k = 1, 2, ... N, equation (8) represents a one-dimensional finite difference equation defined over M points with known boundary conditions of y_0 and y_{M+1} (top and bottom of mesh). Because of the simple form of equation (8), i.e. constant coefficients, we will be able to find an analytic solution for it. If the requirement of constant mesh spacing in the y-direction had not been made, equation (8) would have had variable coefficients and thus would have been more difficult to solve. Specifically, the solution to equation (8) can be expressed (Polozhii (1965))

$$u_k'(y_j) = A_k r_k^{\,j} + B_k r_k^{\,-j} + W_k'(y_j) \tag{9}$$

where $r_k^{\,j} = (\lambda_k + (\lambda_k^2 - 1)^{1/2})^j$ and

$$W_k'(y_j) = \sum_{i=1}^{j-1} \frac{r_k^{\,j-i} - r_k^{\,-(j-i)}}{r_k - r_k^{\,-1}} S_k'(y_i)$$

A_k and B_k are found by satisfying equation (9) at the boundaries $y=y_o$ and $y=y_{M+1}$. When this is done, equation (9) can be expressed in vector form as

$$\underline{u}'(y_j) = F'_T(j)\,\underline{u}'(y_o) + F'_B(j)\,\underline{u}'(y_{M+1}) + \underline{K}'(j) \tag{10}$$

$\underline{u}'(y_o)$ is the transformed vector containing the boundary values at the top of the mesh.

$\underline{u}'(y_{M+1})$ is the transformed vector containing the boundary values at the bottom of the mesh.

$\underline{K}'(j)$ is an N-component vector which contains the source and left and right hand boundary contributions to the value of $\underline{u}$ at any interior point. Its components $K_k'(j)$ are given by

$$K_k'(j) = \frac{1}{(r_k - r_k^{\,-1})(1-r_k^{\,2M+2})} \left[\sum_{m=1}^{j-1} r_k^{\,j-m}(1-r_k^{\,2M})(1-r_k^{\,2M+2-2j})\, S_{km}' + \sum_{m=j}^{M} r^{M-j}(1-r_k^{\,2j})(1-r_k^{\,2M+2-2m})\, S_{km}' \right] \tag{11}$$

$F_T'(j)$ and $F_B'(j)$ are N-dimensional diagonal matrices which relate the top and bottom boundary values to the interior solution. In component form

$$\left[F_T'(j)\right]_{kk} = \frac{r_k^{\,j}\,(1-r_k^{\,2M+2-2j})}{1-r_k^{\,2M+2}} \tag{12}$$

and

$$\left[F_B'(j)\right]_{kk} = \frac{r_k^{\,M+1-j}(1-r_k^{\,2j})}{1-r_k^{\,2M+2}} \tag{13}$$

Using $\underline{u} = P^{-1}\underline{u}'$, $F_T = P^{-1}F'_T P$, $F_B = P^{-1}F'_B P$, and $\underline{K} = P^{-1}\underline{K}'$ and premultiplying equation (10) by the inverse P transform yields

$$\underline{u}(y_j) = F_T(j)\,\underline{u}(y_o) + F_B(j)\,\underline{u}(y_{M+1}) + \underline{K}(j) \tag{14}$$

Thus, under the restrictions of uniform mesh spacing and constant $\gamma(x,y)$, we can write the solution directly at any point in the mesh in terms of the boundary value vectors and generalized source vectors.

It will be of interest to note that since F'_T and F'_B are diagonal matrices, we know the inverse of the matrices F_T and F_B, namely

$$F_T^{-1} = P^{-1} \ (F_T^{'})^{-1} P \tag{15a}$$

$$F_B^{-1} = P^{-1} (F_B^{'})^{-1} P \tag{15b}$$

The vector $\underline{K}(j)$ contains the source contributions as well as the contributions from the right and left hand boundaries. We will find it convenient to decompose $\underline{K}(j)$ in order to make all of the boundary combinations explicit. Thus

$$\underline{K}\,(j) = F_L\, \underline{v}\,(x_o) + F_R\, \underline{v}\,(x_{N+1}) + \underline{H}(j) \tag{16}$$

where $\underline{v}(x_o)$ and $\underline{v}\,(x_{N+1})$ are M-dimensional vectors containing the boundary values at the left and right hand sides of the mesh, and F_L and F_R are N x M dimensional matrices. $\underline{H}(j)$ is the residue from $\underline{K}(j)$. Equation (14) then becomes

$$\underline{u}\,(y_j) = F_T(j)\, \underline{u}(y_o) + F_B(j)\, \underline{u}(y_{M+1}) + F_L\, \underline{v}(x_o) + F_R\, \underline{v}(x_{N+1}) + H(j) \tag{17}$$

The preceding development has assumed the existence of a P transform which diagonalized the T matrix. To find this transform, we work with the following one-dimensional finite difference eigenvalue problem.

$$u_{n+1} - 2u_n + u_{n-1} = 2\lambda' u_n \qquad n=1,2,\ldots.N \tag{18}$$

subject to the boundary conditions: $u_o = u_{N+1} = 0.$

We can find a solution of the form

$$u_n = C_1 \cos n\,\Theta + C_2 \sin n\,\Theta \tag{19}$$

where $\Theta = \cos^{-1}(1 + \lambda')$. $u_o = 0$ implies that $C_1 = 0$. $u_{N+1} = 0$ implies that $\sin\,(N+1)\,\Theta = 0$ which has the roots $\Theta_k = \frac{k\pi}{N+1}$. (20)

Thus we can find N distinct eigenvalues

$$\lambda'_k = \cos\left(\frac{k\pi}{N+1}\right) - 1 \tag{21}$$

The normalized eigenfunction solutions to equation (15) are

$$L_k\,(x_n) = \left(\frac{2}{N+1}\right)^{1/2} \sin\left(\frac{kn\pi}{N+1}\right) \tag{22}$$

We next define an (N x N) P matrix whose columns consist of the eigenvectors

$$P_{kn} = L_k\,(x_n) = \left(\frac{2}{N+1}\right)^{1/2} \sin\left(\frac{kn\pi}{N+1}\right) \tag{23}$$

This matrix is idempotent, i.e. $P^{-1} = P$. It remains to show that this is the matrix which diagonalizes the T matrix. Note that equation (18) could be expressed in matrix form as

$$RL_k - 2I_N L_k = 2\lambda_k' L_k \tag{24}$$

where the R matrix has been defined in equation (5) and I_N is the N-dimensional identity matrix. Since $P^2 = I_N$, the L_k are eigenvectors for the R matrix or

$$RL_k = 2\bar{\lambda}_k L_k \tag{25}$$

Since $T = \frac{\alpha^2}{2} R + bI$, PTP = Λ where Λ is the diagonal matrix required by equation (6). That is

$$\Lambda_{kk} = \lambda_k = \alpha^2 \cos\left(\frac{k\pi}{N+1}\right) + b = \left(\frac{h}{h_1}\right)^2 \left[1 + \cos\left(\frac{k\pi}{N+1}\right)\right] + 1 + \frac{\gamma^2 h^2}{2} \tag{26}$$

We have now derived a fast, efficient way to calculate solutions for the Helmholtz equation using equation (14). This was a consequence of some restrictions. By requiring uniform discretization in the y direction and constant γ(y), equation (8) had constant coefficients and hence was easy to solve analytically. However, in analogy with ordinary differential equations, there exists a class of difference equations with variable coefficients, which can be solved in terms of finite difference special functions. Thus, one could presumably define a non-uniform discretization procedure which would correspond to one of these difference equations.

Similarly, the restriction of uniform x discretization and constant γ(x) allowed us to find the P transform analytically. Once again, in principle one could find other discretization procedures and γ(x) variations which would yield an equation with variable coefficients but with an analytic solution.

Before concluding this section, one other possibility should be considered. Suppose we still require uniform y discretization and constant γ(y) but allow variable γ(x) and non-uniform x discretization. Although we might not be able to find an analytic form for P, we could find the eigenvalues and eigenvectors of the T matrix using numerical procedures. In particular, if the x discretization procedure and γ(x) were symmetric about the x midpoint we could, at least numerically, always find orthogonal eigenfunctions for the T matrix. We would certainly lose a lot of our former computational advantage but if we had only the one region to consider, this method should be significantly faster than the usual finite difference procedure of solving the MN equations.

In practice, we are seldom interested in problems where γ(x,y) is constant or under severe restriction over the whole domain. Thus the major advantage of the method of summary representation is to use it for regions of varying γ(x,y). In the next section, we show how the hybrid method is constructed.

THE SUMMARY REPRESENTATION HYBRID METHOD (SRHM)

We begin the description of SRHM by referring to Figure 3. For the sake of illustration, the domain is divided into nine regions. In region 5, $\gamma_5(x,y)$ varies as a function of x and y but in each of the other regions γ must be a constant. The horizontal discretization is uniform in each region as follows.

In regions 1, 4, and 7, there are N_1 internal nodes with separation h_1. In regions 2, 5 and 8 there are N_2 internal nodes in the x direction with separation h_2. In regions 3, 6 and 9 there are N_3 internal nodes with separation h_3. The vertical discretization is done in similar fashion. Regions 1, 2 and 3 have M_1 internal nodes along the y axis with separation g_1. Regions 4, 5 and 6 have M_2 internal nodes with separation g_2. Regions 7, 8 and 9 have M_3 internal nodes with separation g_3. This gives a total of N nodes ($N = N_1+N_2+N_3+4$) in the x direction and M nodes ($M = M_1+M_2+M_3+4$) in the y direction.

$\underline{Y}_1$ and $\underline{Y}_4$ are N-dimensional boundary value vectors containing the known boundary values at the top and bottom of the mesh respectively. $\underline{Y}_2$ and $\underline{Y}_3$ are unknown N-dimensional internal boundary value vectors.

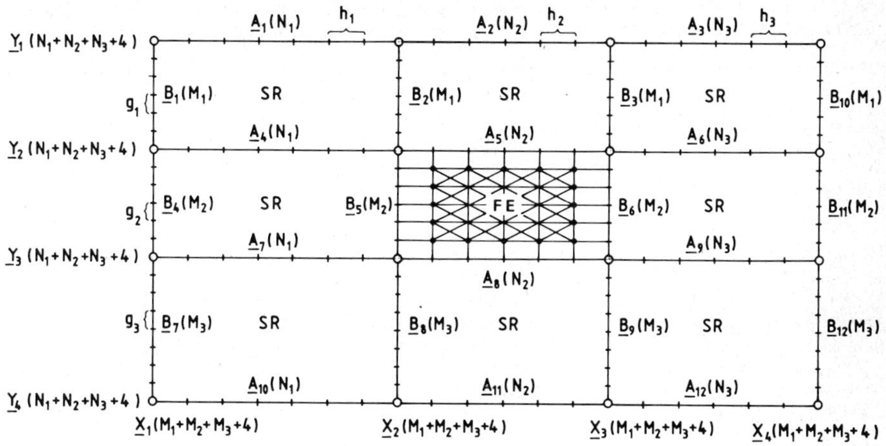

Figure 3
Approximating Mesh for Large Domain
SR: summary representation method; Fe: finite element method

Similarly, $\underline{X}_1$ and $\underline{X}_4$ are M-dimensional vectors containing the known boundary values at the nodes on the left and right hand sides of the mesh respectively. $\underline{X}_2$ and $\underline{X}_3$ are the unknown internal boundary value vectors.

We next subdivide the vectors to form the boundaries for each region. For example $\underline{Y}_1$ will be subdivided into $\underline{A}_1$, $\underline{A}_2$ and $\underline{A}_3$ of dimensionality N_1, N_2, and N_3 respectively. $\underline{A}_1$, $\underline{A}_2$, and $\underline{A}_3$ form the top boundaries of regions 1, 2 and 3 respectively. Similarly we subdivide $\underline{Y}_2$ into $\underline{A}_4$, $\underline{A}_5$, and $\underline{A}_6$; $\underline{Y}_3$ into $\underline{A}_7$, $\underline{A}_8$, $\underline{A}_9$, and $\underline{Y}_4$ into $\underline{A}_{10}$, $\underline{A}_{11}$ and $\underline{A}_{12}$. A similar subdivision occurs for the X vector; i.e. $\underline{X}_1$: $\underline{B}_1$, $\underline{B}_4$, $\underline{B}_7$; $\underline{X}_2$: $\underline{B}_2$, $\underline{B}_5$, $\underline{B}_8$; $\underline{X}_3$: $\underline{B}_3$, $\underline{B}_6$, $\underline{B}_9$; and $\underline{X}_4$: $\underline{B}_{10}$, $\underline{B}_{11}$, $\underline{B}_{12}$.

For all regions except region 5, the solution will be expressed using summary representation. Since $\gamma(x,y)$ is variable in region 5, one possible method of solution would be finite differences. In this case, the internodal distances h_2 and g_2 would have to be sufficiently small to cope with the variations in $\gamma(x,y)$. A consequence of this is that regions 2,8,4 and 6 could be considerably over-discretized. A more efficient method would involve using a finite element solution in region 5, subject to the constraint that an internal perimeter of nodes with the g_2, h_2 spacing be placed just inside the region 5 boundary as shown in Figure 3.

Using equation (17) we can express the summary representation solution in any of the eight regions of constant γ in the analytic form in terms of the known and unknown boundary value vectors. For example we could write the solution in region 6 as

$$\underline{u}^{(6)}(y_j) = F_T^{(6)}(j)\,\underline{A}_6 + F_B^{(6)}(j)\,\underline{A}_9 + F_L^{(6)}\underline{B}_6 + F_R^{(6)}\underline{B}_{11} + \underline{H}^{(6)}(y_j) \qquad (27)$$

where $F_T^{(6)}$ and $F_B^{(6)}$ are known $N_3 \times N_3$ matrices, and $F_L^{(6)}$ and $F_R^{(6)}$ are known $N_3 \times M_2$ matrices of the form described in the previous section. $\underline{u}^{(6)}(y_j)$ is an N_3-dimensional, as yet unknown, solution along the horizontal row of nodes at y =

y_j. $\underline{A}_6$, $\underline{A}_9$ and $\underline{B}_6$ are unknown boundary value vectors. $\underline{B}_{11}$ is a known exterior vector and $\underline{H}^{(6)}(y_j)$ contains the source function.

It will also be useful to write the solution in each region in terms of the vectors $\underline{v}(x_k)$, which correspond to columns of nodes rather than rows. For example in region 6, $\underline{v}^{(6)}(x_k)$ is the M_2-dimensional vector containing the solution along a vertical column of nodes at $x = x_k$. In analogy with equation (27) we can express this type of solution as

$$\underline{v}^{(6)}(x_k) = G_L^{(6)}(k)\,\underline{B}_6 + G_R^{(6)}(k)\,\underline{B}_{11} + G_T^{(6)}\underline{A}_6 + G_B^{(6)}\underline{A}_9 + \underline{K}^{(6)}(x_k) \tag{28}$$

where G_L and G_R are known $M_2 \times M_2$ matrices, G_T and G_B are known $M_2 \times N_3$ matrices, and $\underline{K}$ is an M_2-dimensional source vector.

The next step is to find a solution for region (5). In fact, the two types of solution for region 5, $\underline{u}_5$ and $\underline{v}_5$ can be expressed in exactly the same way as for the other regions

$$\underline{u}^{(5)}(y_j) = F_T^{(5)}(j)\,\underline{A}_5 + F_B^{(5)}(j)\,\underline{A}_8 + F_L^{(5)}\,\underline{B}_5 + F_R^{(5)}\,\underline{B}_6 + \underline{H}^{(5)}(j) \tag{29}$$

$$\underline{v}^{(5)}(x_k) = G_L^{(5)}(k)\,\underline{B}_5 + G_R^{(5)}(k)\,\underline{B}_6 + G_T^{(5)}\underline{A}_5 + G_B^{(5)}\underline{A}_8 + \underline{K}^{(5)}(k) \tag{30}$$

In this case, the $F^{(5)}$ and $G^{(5)}$ matrices cannot be expressed in analytic form. They are the result of multiplying known matrices by the numerical inverse of the $N_5 \times N_5$ global finite element matrix where N_5 is the total number of nodes in region 5.

In order to solve the Helmholtz equation at any internal point, we need to find the twelve internal boundary value vectors $\underline{A}_4$, $\underline{A}_5$, $\underline{A}_6$, $\underline{A}_7$, $\underline{A}_8$, $\underline{A}_9$, $\underline{B}_2$, $\underline{B}_3$, $\underline{B}_5$, $\underline{B}_6$, $\underline{B}_8$, and $\underline{B}_9$. To do this we need to have equations which link these vectors with the interior solutions on either side.

Consider the junction between four regions of different but constant values of γ as shown in Figure 4. We can integrate equation (1) over the shaded area bounded by $x_1 < x < x_2$ and $y_1 < y < y_2$ with the result

$$\int_{y_1}^{y_2}\left[\left.\frac{\partial u}{\partial x}\right|_{x_2} - \left.\frac{\partial u}{\partial x}\right|_{x_1}\right] dy + \int_{x_1}^{x_2}\left.\frac{\partial u}{\partial y}\right|_{y_2} - \left.\frac{\partial u}{\partial y}\right|_{y_1} dx = \int_{y_1}^{y_2}\int_{x_1}^{x_2}(f-\gamma^2 u)\,dx\,dy \tag{31}$$

The next step is to discretize this equation using a five point difference scheme

$$\left[\frac{(u_{k+1,j} - u_{k,j})}{h_3} - \frac{(u_{kj} - u_{k-1,j})}{h_1}\right]\frac{h_2 + h_4}{2} + \left[\frac{u_{k,j+1} - u_{kj}}{h_4} - \frac{u_{kj} - u_{k,j-1}}{h_2}\right]\frac{h_1 + h_3}{2}$$
$$= f_{jk}\,\frac{(h_1 + h_3)(h_2 + h_4)}{2} - \frac{1}{4}\left[\gamma_1 h_1 h_2 + \gamma_2 h_2 h_3 + \gamma_3 h_3 h_4 + \gamma_4 h_4 h_1\right] \tag{32}$$

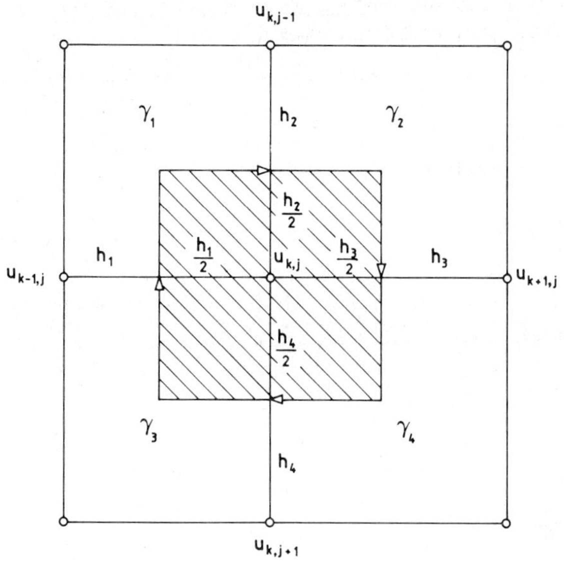

Figure 4
Five Point Difference Scheme

If we were to write this for a row of N_i nodes, the vector form would be

$$a_1 \underline{u} (y_{j-1}) + a_2 \underline{u}(y_{j+1}) + Q_j \underline{u} (y_j) = a_3 \underline{f}(y_j) \tag{33}$$

Alternatively, equation (32) applied along a column of M_i nodes would be

$$b_1 \underline{v}(x_{k-1}) + b_2 v(x_{k+1}) + Z_k v(x_k) = b_3 \underline{f}(x_k) \tag{34}$$

Q_j and Z_k are $N_i \times N_i$ and $M_i \times M_i$ tridiagonal matrices respectively.

Using equation (33) across the six internal horizontal boundaries and equation (34) across the six internal vertical boundaries, along with equations similar to equations (27) and (28) for each of the nine regions, the twelve unknown internal boundary value vectors can be expressed in terms of the twelve known external boundary value vectors.

Once the six internal A vectors have been found, the solution at any point (x_k, y_j) in region i can be expressed directly as

$$\underline{u}^{(i)}(y_j) = F_T^{(i)}(j)\, \underline{A}^{(i)}(y_j) + F_B^{(i)}(j)\, \underline{A}^{(i+3)}(y_j) + \underline{K}^{(i)}(j) \tag{35}$$

At the moment, this version of SRHM exists only in theoretical form. Computer code is currently being written.

In earlier work (Tarlowski, Raiche and Nabighian (1983)), we used a simpler hybrid algorithm. The total domain was divided into horizontal layers as shown in Figure 5. The layers of constant γ were solved using summary representation. The layer containing variable $\gamma(x,y)$ was solved using finite differences.

The formal solution to this layered hybrid scheme was much less complex than that described above for the block hybrid scheme. This was because the only solutions required were always of the form of equation (35), i.e. one had only to match solutions across horizontal boundaries.

However, the layered finite difference solution would require the numerical inversion of a much larger matrix than would be the case for the block finite element solution; simply because there are a lot more nodes in an entire layer than there are in a limited region of that layer. We do not yet know which scheme is computationally more efficient. In the next section we present some results from our layered version of SRHM.

GEOPHYSICAL APPLICATION OF SRHM

A common problem in electromagnetic geophysics is to model the effect of an electromagnetic plane wave vertically incident upon an earth with a two-dimensional conductivity structure, $\sigma(x,y)$. The plane wave has an exp $(-i\omega t)$ time dependance where ω is the angular frequency. An idealized geological cross section is shown in Figure 6.

For the situation described above, Maxwell's equations can be reduced to a scalar Helmholtz equation for E, the electric field parallel to the strike direction.

$$\nabla^2 E(x,y) + \gamma^2(x,y)\, E(x,y) = 0 \tag{36}$$

where $\gamma^2 = i\omega\mu\sigma(x,y)$ and $\mu = 4\pi \times 10^{-7}$. In regions of constant σ, say σ_j, we can define a skin depth $\delta_j = |\gamma_j|^{-1}$. The skin depth is the distance over which a plane electromagnetic wave will have its amplitude attenuated by a factor of e.

We applied the layered version of SRHM to calculate a model consisting of an air layer (layer 0), an overburden layer (layer 1), a heterogeneous layer (layer 2), and a basement layer (layer 3). The solutions for layers 0, 1, and 3, were expressed using summary representation since these layers had constant conductivities of 0, 0.1, and 0.002 S/m respectively. The method of finite differences was used for layer 2 which had a target of σ = 1 S/m in an otherwise uniform host of σ = 0.002 S/m.

In the absence of the anomalous block in layer 2, the model has an analytic solution which was invoked to give the boundary values on all four sides of the total domain.

The summary representation layers were given vertical discretizations of 0.1 δ_j with the exception of layer 0 which was discretized at 0.1 δ_1. The finite difference layer was discretized at 0.2 δ_T where δ_T was the target skin depth.

The horizontal discretization was constant. Because of the type of boundary conditions, the accuracy of the model was much less sensitive to horizontal discretization than it was to vertical discretization.

In order to compare the speed of SRHM with that of the finite difference method used for the entire domain, we ran four different models for N = 21, 41, 61, and 81, where N is the number of nodes in the horizontal direction. The results are shown in Table 1 for a VAX 11/780.

Although the ratio of nodes in the finite difference layer to the total number of nodes in all four layers of the domain remained constant, the solution time should

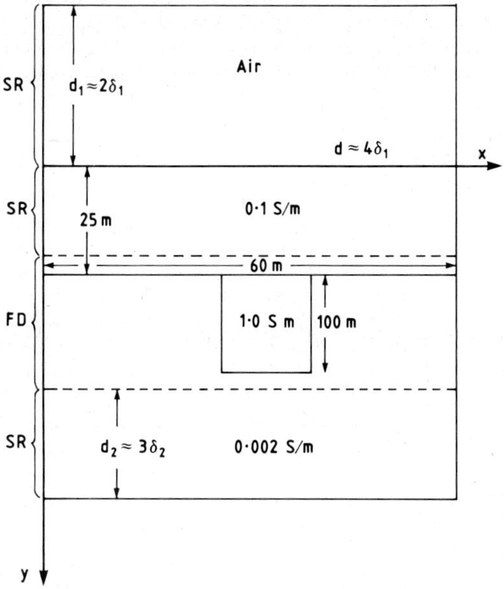

Figure 5
Two-dimensional Model
δ_1 = skin depth in overburden
δ_2 = skin depth in basement
FD = finite difference method

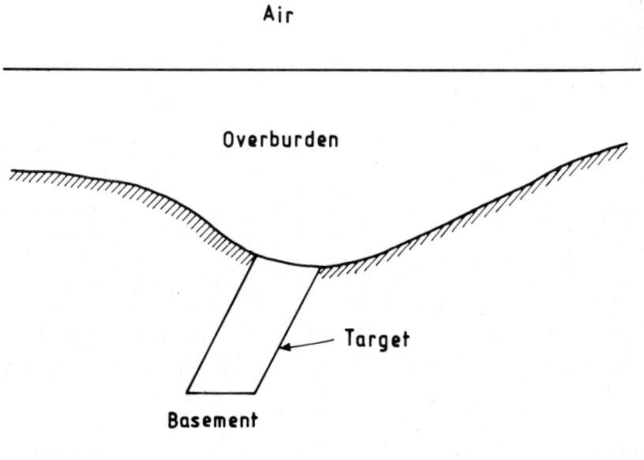

Figure 6
Two-dimensional Conductive Structure

TABLE 1. Comparison of Computation Times for Hybrid and Finite Difference Methods

Total number of internal nodes	Number of internal nodes in the layer with an inhomogeneity	Time in seconds for the hybrid method	Time in seconds for the finite difference method
1323	231	44.51	73.51
2583	451	101.40	294.70
3843	671	270.50	853.8
5103	891	529.04	1934.00

increase as the cube of the number of finite difference nodes and as the square of the number of summary representation nodes. Thus, as N increases, SRHM will have an increasing speed advantage over the finite difference method when used alone. As seen from Table 1, SRHM had a speed advantage factor of between 3 and 4.

It had been argued by Polozhii (1965) that the summary representation method should have better precision that the finite difference method. We tested this for several layered models for which analytic solutions existed. The summary representation solution exhibited marginally more accurate behaviour than the finite difference solution.

CONCLUSIONS

The use of summary representation in conjunction with finite difference/finite element methods can offer a significant speed advantage over the sole use of finite differences or finite elements. SRHM offers more geometric flexibility than hybrid methods combining integral equation solutions with finite elements. The layered version of SRHM has in practice demonstrated a speed advantage. In principle, the block version of SRHM should be even more effective but this has yet to be shown.

REFERENCES

[1] Lee, K.H., Pridmore, D.F. and Morrison, H.F. (1981) A hybrid three-dimensional electromodelling scheme, Geophysics, 46, 796-805.

[2] Polozhii, G.N. (1965) The Method of Summary Representation for Numerical Solution of Problems of Mathematical Physics (Pergamon).

[3] Tarlowski, C.Z., Raiche, A.P. Nabighian, M. The use of summary representation for EM modelling, submitted to Geophysics.

Computational Techniques & Applications: CTAC-83
J. Noye & C. Fletcher (Editors)

THE COMPUTATION OF AXISYMMETRICAL VORTEX SHEETS

W. K. Soh

Department of Mechanical Engineering
The University of Wollongong
New South Wales
AUSTRALIA

The technique for the simulation of axisymmetrical jet flows from a nozzle using vortex sheets is extended to include the treatment for no-flow boundaries. A matching process for the evaluation of the flows near to the nozzle edge is presented. This enables the Kutta condition to be applied correctly. It is found that the rate for the shedding of vorticity is significantly smaller than that predicted under the assumption of steady state and uniform flow.

1. INTRODUCTION

The simulation of time dependent flows using vortex sheets and vortices to represent free shear layers has found many applications in the studies of fluid flow problems. For example, Delcourt and Brown [1] have produced many features of turbulent mixing layer using cluster of point vortices. Smith [2] modelled the flow separations from leading-edges of slender delta-wings and good agreements with measured data were obtained. Recently, Hoeijmakers and Vaatstra [3] have improved the accuracy of the computation. The accuracy of their technique was demonstrated in their application to a variety of aerodynamic problems.

The computation of vortex sheet in axisymmetric flow has been applied to the modelling of jet flows. Acton [4] carried out numerical studies of large eddies generated by axisymmetric jets and the effects when harmonic excitations were applied. The boundary condition for the nozzle wall was not imposed and the shedding of vorticity was evaluated using the assumption of steady state and uniform flow. However, reasonable agreements were obtained and conclusions were made that a substantial part of the large-scale structure in a jet was essentially axisymmetric. It must be noted that in many practical situations, a large part of the jet flow is turbulent. However, in the studies of vortex pairing in jet noise generation, the shadow graphs which were taken by Kibens [5] showed that the flow in the immediate region of the nozzle were essentially laminar. Well defined axisymmetric vortex sheets, attached to the nozzle edge, were observed and extended to a distance more than one diameter of the nozzle. This suggests that the shedding of vorticity has to be evaluated from the modelling of a vortex sheet shed from the nozzle. As the shed vorticity depends on the ambient fluid velocity as much as the velocity of the jet just outside the nozzle the movements of the ambient fluid plays an important part in this modelling process. The velocity of the ambient fluid near to the nozzle can only be evaluated if the boundary condition on the nozzle wall is imposed properly. Moreover, a well defined wall boundary in the flow model is a requirement for the application of the Kutta condition at the nozzle edge. The aim of this work is to extend the technique for the computation of a vortex sheet into axisymmetric flow so that boundary conditions on all walls are satisfied.

There are basically two approaches for treating no-flow boundary condition. One is to place a finite number of singularity on the rapid boundaries. The strengths of these singularities are such that the boundary conditions can be satisfied. As the distribution of these singularities is arbitrary, walls of irregular shapes can easily be represented. However, in the evaluation of the velocity at a point on the vortex sheet, the velocities induced by these singularities have to be computed. This involves the evaluations of transcendtal functions and the computational process is quite labourious.

The second approach is to split the Stokes stream function into two parts: ψ_{vs} represents a vortex sheet in infinite fluid and its complementary ψ_b. It is such that the sum of these two parts satisfies all the boundary conditions. The discretization process for these stream functions are different. For ψ_{vs}, it is carried out on the vortex sheet and is Lagrangian. For ψ_b, it is solved for a matrix of field points covers the whole computational region and these field points are Eulerian. It turns out that the number of iteration required to update the values of ψ_b at every time step is very small. The evaluation of the velocity due to ψ_b on the pivotal point only involves interpolation from four neighbouring field points. Thus the overall computational effort is small compared with the first approach.

The second approach is adopted for this work and the details of the technique are presented here.

2. SCHEME FOR DISCRETIZATION AND COMPUTATIONS

2.1 The Stokes Stream Functions

The Stokes stream function used for axisymmetrical flow is divided into two components: ψ_{vs} and ψ_b. Let x and ω be the axial and radial coordinates. The component ψ_{vs} is the stream function for free vortex sheet in infinite fluid. Thus it is expressed as an integration along the vortex sheet:

$$\psi_{vs} = \int \frac{\gamma}{\pi}(r_1 + r_2).(F_1(\lambda) - E_1(\lambda))ds \tag{1}$$

$$r_1 = [(x - x(s))^2 + (\omega - \omega(s))^2]^{\frac{1}{2}}$$

$$r_2 = [(x - x(s))^2 + (\omega - \omega(s))^2]^{\frac{1}{2}}$$

F_1 and E_1 are the complete elliptical integrals of the first and second kinds respectively and the modulus λ is given by:

$$\lambda = \frac{r_2 - r_1}{r_2 + r_1}$$

The discretization of the integral is to replace the vortex sheet into N segments of size Δs. The circulation in the kth segment is given by $\Gamma_k = \gamma(s_k)\Delta s$. The velocity induced by the vortex sheet on a point s_j on the sheet itself can be evaluated from (1) by considering the Principle value of the integral. In terms of the above discretization, this amounts to neglecting the contribution by the jth segment onto itself. Denoting $(u_{vs})_j$ and $(w_{vs})_j$ as the axial and radial velocity components due to the vortex sheet, their discrete formulae are given by:

$$(u_{vs})_j = \frac{1}{2\pi\omega_j}\sum_{k=1,k\neq j}^{N} \Gamma_k \frac{(r_1)_{jk} + (r_2)_{jk}}{(r_1)_{jk}(r_2)_{jk}}$$

$$[(\omega_j - \lambda_{jk}\omega_k)(F_1(\lambda_{jk}) - E_1(\lambda_{jk})] + (\omega_k(1+\lambda_{jk}^2) - 2\lambda_{jk}\omega_j)[F_1(\lambda_{jk}) - E_1(\lambda_{jk}))]$$

$$\tag{2}$$

$$(w_{vs})_j = \frac{-1}{2\pi\omega} \sum_{k=1,k\neq 1}^{N} \Gamma_k (x_j - x_k) \frac{(r_1)_{jk} + (r_2)_{jk}}{(r_1)_{jk}(r_2)_{jk}}$$

$$[(F_1(\lambda_{jk}) - E_1(\lambda_{jk})] + 2\lambda_{jk}[(F_1'(\lambda_{jk} - E_1'(\lambda_{jk}))] \quad (3)$$

where

$$(r_1)_{jk} = [(x_j - x_k)^2 + (\omega_j - \omega_k)^2]^{\frac{1}{2}}$$

$$(r_2)_{jk} = [(x_j - x_k)^2 + (\omega_j + \omega_k)^2]^{\frac{1}{2}}$$

the modulus,

$$\lambda_{jk} = \frac{(r_2)_{jk} - (r_1)_{jk}}{(r_2)_{jk} + (r_1)_{jk}}$$

F_1' and E_1' are the derivatives of F_1 and E_1 with respect to the modulus λ. Hence,

$$F_1' = \frac{E_1 - (1-\lambda^2)F_1}{\lambda(1-\lambda^2)}$$

and

$$E_1' = \frac{E_1 - F_1}{\lambda}$$

For potential flow the complementary stream function ψ_b satisfies the partial differential equation:

$$\frac{\partial^2\psi_b}{\partial x^2} + \frac{\partial^2\psi_b}{\partial\omega^2} - \frac{1}{\omega}\frac{\partial\psi_b}{\partial\omega} = 0 \quad (4)$$

Thus, if the value of ψ_b is specified on the boundary, it can be solved on a matrix of field points in the computation domain. Consider that a grid of axial spacing Δx and radial spacing $\Delta\omega$ covers the domain. The iteration process which solves for ψ_b in the mth column and the nth row is given by :

$$(\psi_b)_{m,n} \leftarrow (\psi_b)_{m,n} + W[-(\psi_b)_{m,n} + \frac{\Delta\omega^2}{2D}((\psi_b)_{m+1,n} + (\psi_b)_{m-1,n})$$

$$+ \frac{\Delta x^2}{2D}((\psi_b)_{m,n+1} + (\psi_b)_{m,n-1}) - \frac{\Delta\omega.\Delta x^2}{4D}((\psi_b)_{m,n+1} - (\psi_b)_{m,n-1})]$$

(5)

where W is the relaxation factor and $D = \Delta x^2 + \Delta\omega^2$. The velocity at the jth pivotal point due to the complementary stream function ψ_b is evaluated from the mesh points surrounding this point. For instance, when the jth pivotal point lies inside the rectangle with corners at (m,n),(m+1,n),(m,n+1) and (m+1,n+1), then,

$$(u_b)_j = \frac{1}{2\omega_j\Delta\omega}[A_1((\psi_b)_{m+2,n+1} - (\psi_b)_{m,n+1}) + A_2((\psi_b)_{m+1,n+1} - (\psi_b)_{m-1,n+1})$$

$$+ A_4((\psi_b)_{m+2,n} - (\psi_b)_{m,n}) + A_3((\psi_b)_{m+1,n} - (\psi_b)_{m-1,n})] \quad (6)$$

$$(w_b)_j = \frac{-1}{2\omega_j\Delta x}\,[A_1((\psi_b)_{m+1,n+2} - (\psi_b)_{m+1,n}) + A_4((\psi_b)_{m+1,n+1} - (\psi_b)_{m+1,n-1})$$

$$- A_2((\psi_b)_{m,n+2} - (\psi_b)_{m,n}) + A_3((\psi_b)_{m,n+1} - (\psi_b)_{m,n-1})] \qquad (7)$$

where A's are the weighted areas

$$A_1 = \frac{(x_j - x_m)(\omega_j - \omega_n)}{\Delta x \Delta\omega}$$

$$A_2 = \frac{(x_{m+1} - x_j)(\omega_j - \omega_n)}{\Delta x \Delta\omega}$$

$$A_3 = \frac{(x_{m+1} - x_j)(\omega_{n+1} - \omega_j)}{\Delta x \Delta\omega}$$

and $A_4 = 1 - A_1 - A_2 - A_3$.

For the case where the pivotal point falls on the center line (n=1), the limit for ω towards zero has to be considered and the resulting finite different scheme is given by:

$$(w_b)_j = \frac{-1}{2\Delta\omega\Delta x}\,[A_1((\psi_b)_{m+1,2} - (\psi_b)_{m,2}) + A_2((\psi_b)_{m+1,2} - (\psi_b)_{m-1,2})] \qquad (8)$$

Thus the velocity component for a jth pivotal point on the vortex sheet are the sums of the respective components given in equation (2),(3),(6) and (7):

$$u_j = (u_{vs})_j + (u_b)_j$$

$$w_j = (w_{vs})_j + (w_b)_j \qquad (9)$$

2.2 The Boundary Condition For ψ_b

The boundary conditions for ψ_b are evaluated after the values for ψ_{vs} are known. They are in the form $\psi_b = G(x,\omega) - \psi_{vs}$ where G implements the boundary conditions. Consider the control area for a jet flow in pipe as shown in Figure 1. Here DE is the center line and the nozzle wall is at CF where F is the nozzle edge. The line AB represents the pipe wall. The jet flows in from CD with uniform velocity and the inflow to the pipe is also uniform and passes through BC. The outflow AE is far enough from the nozzle so that the velocity distribution has become uniform.

Let R and P be the radii for the nozzle and the pipe respectively. Assume that the velocity of the jet at CD is V_j and the inflow to the pipe at BC is V_f. The boundary condition for ψ_b is given by:

$$\psi_b = \begin{cases} 0 & \text{along DE} \\ \frac{1}{2}\omega^2\,[V_f + (\frac{R}{P})^2(V_j - V_f)] & \text{along AE} \\ \frac{1}{2}\,[V_f P^2 + R^2(V_j - V_f)] - (\psi_{vs})_{AB} & \text{along AB} \\ \frac{1}{2}\,[V_f\,\omega^2 + R^2(V_j - V_f)] - (\psi_{vs})_{BC} & \text{along BC} \end{cases}$$

$\frac{1}{2} R^2 V_j - (\psi_{vs})_{CF}$ along CF

$\frac{1}{2} \omega^2 V_j - (\psi_{vs})_{CD}$ along CD (10)

It is evident that the above conditions imply one dimensional uniform inflows and outflows.

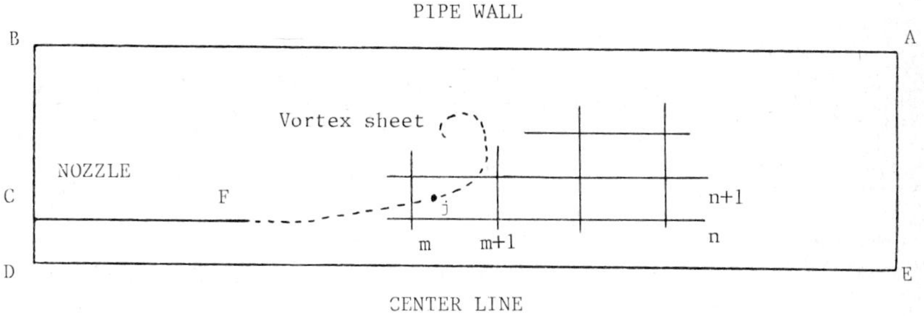

Figure 1. Control area for the computation of jet flow in pipes.

2.3 The Shedding of Vorticity

The component of the flow represented by the complimentary stream function ψ_b has a singularity at the nozzle edge. The shedding of a vortex sheet at this edge will require that the Kutta condition be satisfied at this point. The process of introducing the Kutta condition is a straight forward process if the flow field can be expressed explicitly. For example, in the flow over a semi-infinite flat plate the Kutta condition simply requires that the velocity at a point in the transformed plane which corresponds to the point at the nozzle edge, be made zero. This is sufficient to ensure finite velocity at the edge in the physical plane. For the present axisymmetric flow, the stream function ψ_b is computed at discrete points. The flow field in the small region very near to the nozzle edge can be approximated by the flow around a semi-infinite plate in two dimensional Cartesian plane. It is found that the error in this approximation is of the order $\frac{\Delta x}{R}$ where Δx is the spacing of the grid and R is the radius of the nozzle.

Consider the flow around a semi-infinite plate extended from the origin to the negative horizontal axis, by introducing conformal transformation:

$$z = -\zeta^2 \tag{11}$$

which maps the flow field into the upper half of the ζ-plane. The velocity is given by

$$U - iV = \frac{U_c i}{2.\sqrt{z}} \tag{12}$$

Here z and ζ are complex variables and the value of U_c has to be established by matching with the computed flow field. Let (H,R) be the coordinates for the nozzle edge, and let V_w be the computed velocity at $(H+\Delta x,R)$ which exclude the contribution of those vorticity found within the circle radius Δx center at the nozzle edge (H,R). From equation (12), by setting z equal to Δx, the constant U_c is given by

$$U_c = -2.\sqrt{(\Delta x)}.V_w \tag{13}$$

Suppose the $(n-m)^{th}$ to the nth pivotal points are found within this circle, then the Kutta condition is given by

$$U_c + \frac{1}{2\pi i} \sum_{k=n-m}^{n-1} \left[-\frac{1}{\zeta_k} + \frac{1}{\overline{\zeta_k}} \right] + \frac{\gamma_n}{2\pi i} \cdot [\zeta_\varepsilon - \overline{\zeta_\varepsilon}] = 0 \qquad (14)$$

where ζ_k are the image of the pivotal points at z_k; γ_n is the vorticity density of the nth segment which is attached to the shedding edge and ζ_ε is the image of the other end point of this attached segment. Thus the strength of this n^{th} segment is given by

$$\Gamma_n = \gamma_n . \Delta s \qquad (15)$$

where Δs is the assigned segment length. This nascent segment is then replaced by a pivotal point at the mid-point of the segment, that is $(H + \frac{1}{2}\Delta s, R)$ in the cylindrical coordinate and be convected accordingly.

The time step Δt of this numerical process has to be determined in order that the flow model is a consistent one. The size of this time step comes from the condition that the pressure be continuous at the nozzle edge. When this condition is applied to the shedding edge where the vortex sheet is attached we have,

$$\frac{d\Gamma}{dt} = \gamma \mid q_{vs} \mid \qquad (16)$$

where $\frac{d\Gamma}{dt}$ is the rate at which vorticity is being shed, γ is the vorticity density at the edge and q_{vs} is the fluid velocity at the shedding edge.

In computation the rate of shedding vorticity is replaced by $\frac{\gamma_n \Delta s}{\Delta t}$. The value of $\gamma \mid q_{vs} \mid$ is the flux of vorticity being shed out and is represented by $\gamma_n \mid q_n \mid$. This representation is exact in steady state as the flux shed out of the edge (H,R) should be the same as the flux flow passes the nth point, i.e. $(H + \frac{1}{2}\Delta s, R)$. However, in unsteady flow, owing to the fact that Δs is small and $\gamma_n \mid q_n \mid$ is the averaged flux over the nth segment, this representation of $\gamma \mid q_{vs} \mid$ by $\gamma_n \mid q_n \mid$ is a justifiable approximation. Equation (16) thus becomes:

$$\Delta t = \frac{\Delta s}{\mid q_n \mid} \qquad (17)$$

2.4 Relocation of Pivotal Points

The stability of this numerical process depends on the relative distance between neighbouring pivotal points. When a pair of pivotal points is closer together than their adjacent points, the error incurred in the discrete formula tends to be amplified as computation progresses and eventually leads to numerical instability. A method for stabilizing the computation is to relocate the pivotal points so that the distance between adjacent points are equal. Naturally, the circulation in each segment has to be adjusted so that the vorticity distribution along the vortex sheet remains unchanged through this adjustment.

The algorithm for this process is to choose a reference point on the vortex sheet and construct the function for the circulation F(s) between this reference point and any points on the sheet using the length of the sheet s between these pair of points as the independent variable. The coordinates of the original set of pivotal points is expressed in terms of s. When equal spacing of pivotal points is required, the coordinates of these points can be interpolated. The values of the function for circulation F(s)* for these points are also derived in the same way. The circulations in these new segments are obtained by inverting the function F(s)*, so that the circulation in a segment is the difference of the values of the function F(s)* at the extremities of the segment. Details of this relocation process are reported by Fink and Soh [6].

2.5 Outline of Computations

The computation begins with a specified geometry of the domain, the mesh size required for the field points and the size of the vortex sheet segment Δs. Initially $\psi_{vs} = 0$ and ψ_b is solved for all mesh points by the iteration scheme given in (5) subjected to the constraints imposed by the boundary conditions. The nascent pivotal point is then placed at a distance $\frac{1}{2}\Delta s$ outside the nozzle edge and on the tangent extends from the nozzle inner wall. The circulation of the segment associated with this nascent pivotal point is evaluated using the Kutta condition which involves equations (12), (14) and (15). This is then followed by the evaluation of ψ_b. The time step Δt is derived from (17). The velocities for the pivotal points are evaluated using (9), (2), (3), (6) and (7). The pivotal points are then displayed by integrating equation (9) with respect to t using the first order Euler method for integration. This process repeats from the placement of the nascent pivotal point to the application of the Kutta Condition until the specified time interval has elapsed.

3. COMPUTATIONS AND RESULTS FOR JET FLOW IN PIPE

The numerical scheme described in the last section was applied to the computations for the flow of a jet in a pipe. The ratio of the nozzle diameter, $\frac{R}{P}$ was 0.2. The rates of the inlet velocity to the jet velocity, $\frac{V_f}{V_j}$ ranged from 0.0 to 0.2. Convergence tests were carried out for the case where $\frac{V_f}{V_j} = 0.08$. It is found that the variations of the stream functions is less than 1.0% between the results using a segment size Δs equal to 0.1 and that equal to 0.05. Hence $\Delta s = 0.1$ were used to compute the results presented here. The mesh size for the field points were $\Delta x = 0.2$, and $\Delta\omega = 0.125$. Figure 2 shows the streamline for $\frac{V_f}{V_j} = 0.02$ at two time frames. A recirculation region was formed from shortly after the initial time. The stagnation point on the pipe wall upstream of this recirculation region is found to be quite steady as it remains within one quarter of the pipe diameter downstream of the nozzle exit. The other stagnation point which is downstream of the recirculation region continues to travel with the flow downstream indicating that the potential flow in this area is always unsteady. At the later time frame where $\frac{tV_j}{R} = 42$, small eddies beginning to appear on the boundary of the jet. These eddies are unstable and continue to be convected downstream.

For $\frac{V_f}{V_j} = 0.06$ the recirculation region travels downstream as shown in Figure 3. In fact there are no stagnation points found on the pipe wall. This will be the situation for $\frac{V_f}{V_j}$ greater than 0.06.

The comparison with the measurements for laminar jet is not possible due to the unavailability of data. Some degrees of similarity with turbulent jet can be found when compared with the time average measurements of flow by Barchilon & Curtet [7]. However, the recirculation regions of the present results have

a.
b.
c.
d.

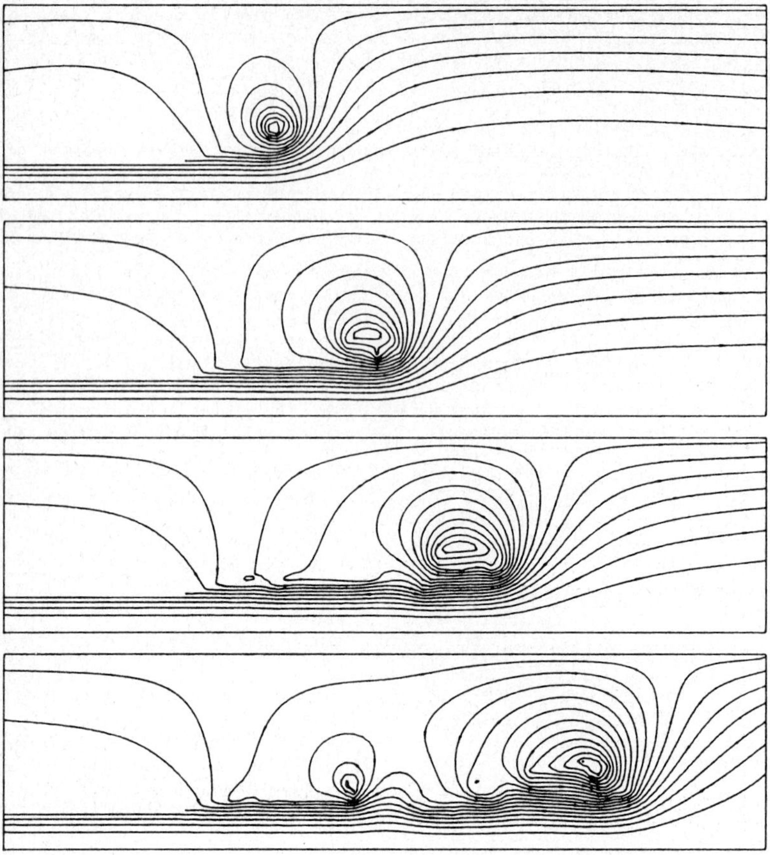

Figure 2. Streamline pattern for

$\frac{V_f}{V_j} = 0.02. \quad \frac{T V_j}{R} =$: (a) 10.1;

(b) 20.5; (c) 31.1 and (d) 42.1.

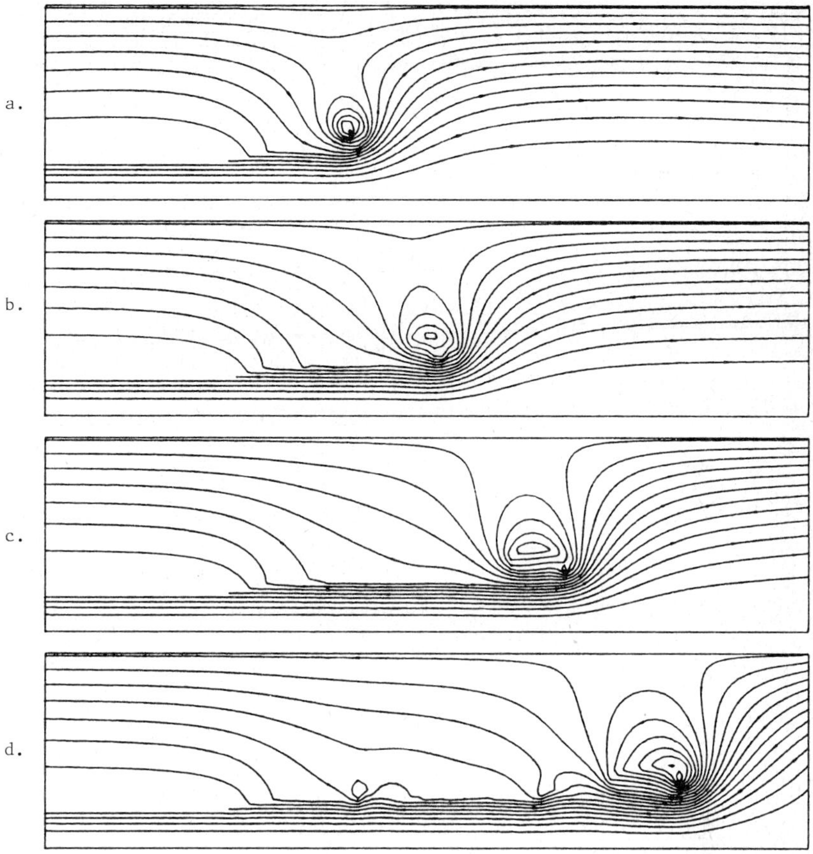

Figure 3. Streamline patterns for $\frac{V_f}{V_j} = 0.06$.

$\frac{T\,V_j}{R} =$: (a) 10.9; (b) 20.2; (c) 31.3

and (d) 40.8.

extended closer to the center line of the pipe than were founded in Barchilon & Curlets' measurements.

The rate for the shedding of vorticity $\frac{d\Gamma}{dt}$ for various values of V_f are plotted in Figure 4. The mean value is about $0.3V_j^2$ and seems to be independent of the values of V_f. The shedding rate is significantly different from that predicted under the assumptions of steady and uniform flow which gives the value of $0.5V_j^2$ when V_f is zero. One will expect that the pressure in the flow will differ by the same order.

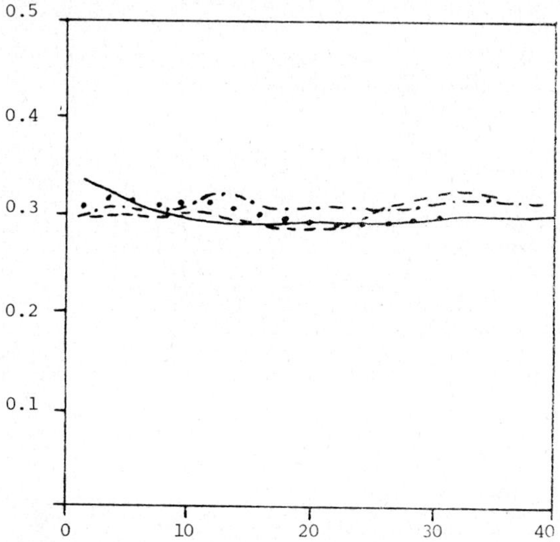

Figure 4. Rate of shedding vorticity

$V_j^{-2} \frac{d\Gamma}{dt}$ plotted against time $\frac{t V_j}{R}$.

$\frac{V_f}{V_j}$ = 0.0, ----- ; 0.02, -,-,-,- ;

0.06, ... and 0.2, ———— .

ACKNOWLEDGEMENT

The support by the Australian Research Grants Scheme for this project is gratefully acknowledged.

REFERENCES

[1] Delcourt, B. & Brown, G.L. Modelling Plane Turbulent Flows with Point-Vortices: A Numerical Experiment, The Sixth Australasian Hydraulics and Fluid Mechanics Conference, The Institution of Engineers, Australia. (1979) 321-325.

[2] Smith, J.H.B. Improved Calculations of Leading-edge Separation from Slender-thin Delta-wings, Proc. R. Soc., Lond. A 306 (1968) 67-90.

[3] Hoeijmakers H.W.M. & Vaatstra, W. Higher Order Panel Method Applied to Vortex Sheet Roll-up, AIAA. Jrnl. 21 4 (1983) 516-523.

[4] Acton, L. A Modelling of Large Eddies in an Axisymmetric Jet. Jrnl. Fluid Mechanics. 98 (1980) 1-31.

[5] Kibens, V. On the Role of Vortex Pairing in Jet Noise Generation. The International Symposium on the Mechanics of Sound Generation in Flows, Gottingen, Germany, (1979).

[6] Fink, P.T. & Soh, W.K. A New Approach to Roll-up Calculations of Vortex Sheets, Proc. R. Soc. Lond. 362 (1978) 195-209.

[7] Barchilon, M. & Curtet, R. Some Details of the Structure of an Axisymmetric Confined Jet with Backflow. J. of Basic Engineering, ASME Trans. 86, Dec. (1964) 777-787.

Computational Techniques & Applications: CTAC-83
J. Noye & C. Fletcher (Editors)

EFFICIENT COMPUTATION OF THE MAGNETIC FORCES ACTING ON A CURRENT CARRYING COIL

P.M. HART

Department of Electrical Engineering,
Monash University,
Clayton, Victoria, Australia

This paper presents an efficient method for computing the force per unit length of conductor acting on a current carrying coil due to its magnetic field. The method retains the simplicity inherent in modelling the coil by a filamentary conductor but includes a novel modification which accounts for the actual current distribution across the conductor cross section and which thereby ensures convergence.

1. INTRODUCTION

Consider the force acting on an elemental length of a filament which carries current I_1 and which is subject to an externally generated magnetic field $\underline{B}$. The force per unit length acting on the filament is

$$\underline{\Gamma} = I_1 \hat{\underline{\ell}}_1 \times \underline{B} \tag{1}$$

where $\hat{\underline{\ell}}_1$ specifies the orientation of the filament. If $\underline{B}$ is generated by a closed filamentary coil C2 of which the element of interest is not a part (Figure 1) then

$$\underline{\Gamma} = \frac{\mu_o}{4\pi} I_1 I_2 \hat{\underline{\ell}}_1 \times \oint_{C2} \frac{d\underline{\ell}_2 \times \hat{\underline{u}}_r}{r^2} \tag{2}$$

where $\hat{\underline{u}}_r$ is directed from the element associated with $d\underline{\ell}_2$ towards that associated with $d\underline{\ell}_1$. If the element of interest is part of a filamentary coil, say C1 in Figure 1, then the total force acting on C1 due to the influence of C2 is

$$\underline{F}_{12} = \frac{\mu_o}{4\pi} I_1 I_2 \oint_{C1} \oint_{C2} \frac{d\underline{\ell}_1 \times (d\underline{\ell}_2 \times \hat{\underline{u}}_r)}{r^2} \tag{3}$$

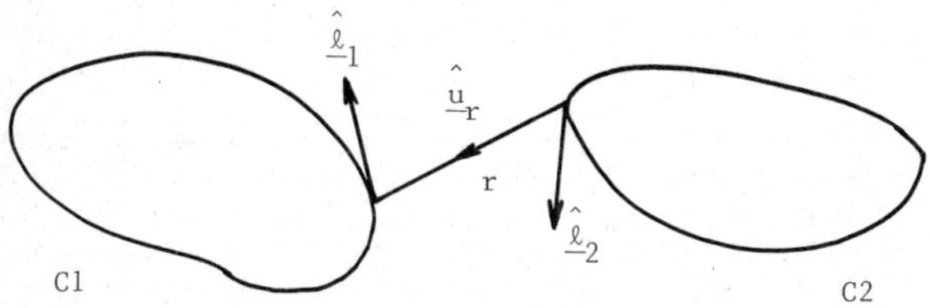

Figure 1.

The filamentary coils.

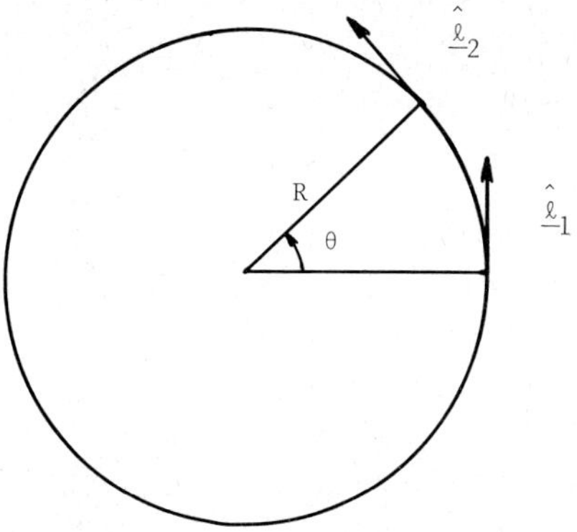

Figure 2.

A filamentary ring coil.

Consider now the force exerted on an elemental filament of interest by other elemental filaments constituting the coil of which it is a member. It is tempting to apply (2) in this case with all references to C2 replaced by corresponding references to C1. However, (2) is singular when $d\underline{\ell}_1$ and $d\underline{\ell}_2$ relate to the same element. This particular difficulty can be overcome by realizing that an elemental filament does not exert a net force on itself and hence we hope to use (2) in the form

$$\underline{\Gamma}_1 = \frac{\mu_o}{4\pi} I_1^{\ 2} \, \hat{\underline{\ell}}_1 \times \left\{ \text{Cauchy Principal Value} \oint_{C1} \frac{d\underline{\ell}_2 \times \hat{\underline{u}}_r}{r^2} \right\} \tag{4}$$

As we will now illustrate for the case of a circular filamentary ring coil, equation (4) is not convergent.

Considering a filamentary ring coil as shown in Figure 2, the force per unit length acting on the first element is, according to (4),

$$\begin{aligned} \Gamma &= \frac{\mu_o}{4\pi} \frac{I^2}{4R} \left\{ \text{C.P.V.} \int_o^{2\pi} \frac{d\theta}{\sin \theta/2} \right\} \\ &= -\frac{\mu_o}{4\pi} \frac{I^2}{4R} \lim_{\delta \to 0} \ln \left[\tan \delta/4 \right] \end{aligned} \tag{5}$$

which is unbounded. This result indicates that the force per unit length acting on a filamentary coil due to its own magnetic field is undefined. We note that the self-inductance of a filamentary coil is also undefined (Graneau (1961), Corson (1962) 233).

In consequence of this result the force per unit length acting on a real coil cannot be computed by modelling the coil as though it were filamentary. This is disappointing from the computational viewpoint as the application of (4) is quite simple, requiring only the centre line of the conductor to be specified. As discussed in Section 2, the nature of the current distribution within the actual conductor cross-section is not involved in the filamentary model and this is the root cause of the singular behaviour of (4). A method of computation which allows a modified form of (4) to be applied to coils having finite conductor cross section is presented in section 3 of the paper.

2. SINGULARITIES IN FORCE CALCULATIONS

The non convergence of equation (4) may be understood by considering two filamentary segments of a closed coil. The coil is envisaged to consist of N straight line segments as shown in Figure 3. The force per-unit length acting on the first segment is according to (4)

$$\underline{\Gamma}_1 = \frac{\mu_o}{4\pi} I^2 \, \hat{\underline{\ell}}_1 \times \{ \lim_{N \to \infty} \sum_{j=2}^{N} \frac{\delta \underline{\ell}_j \times \hat{\underline{u}}_{r1j}}{r_{1j}^{\;2}} \} \tag{7}$$

The contribution to (7) from segments adjacent to the first segment in particular, the second segment, is

$$\delta \underline{\Gamma}_{12} = \frac{\mu_o}{4\pi} I^2 \, \hat{\underline{\ell}}_1 \times \frac{\delta \underline{\ell}_2 \times \hat{\underline{u}}_{r12}}{r_{12}^{\;2}} \tag{8}$$

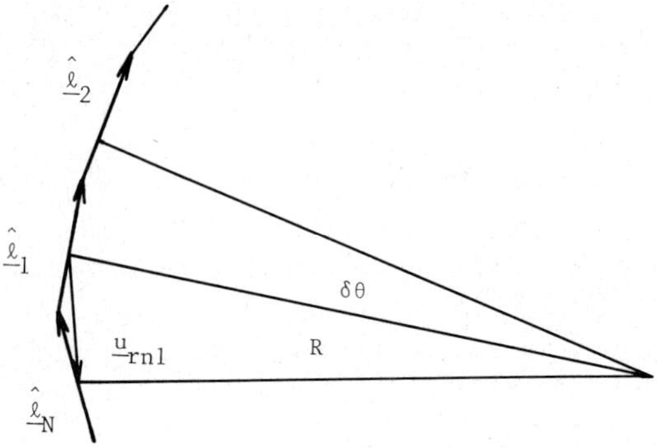

Figure 3.

Adjacent filamentary segments of a coil.

If the coil is described by a local radius of curvature R as shown in Figure 3, then from (8) we obtain

$$\delta \Gamma_{12} = \frac{\mu_o}{4\pi} I^2 \frac{2R\pi \; \mathrm{Sin}\; \pi/N}{N(2R \; \mathrm{Sin}\; \pi/N)^2} \tag{9}$$

In the limit as $N \to \infty$ the contribution to the force per unit length is

$$\delta \Gamma_{12} = \frac{\mu_o}{4\pi} \frac{I^2}{R} \tag{10}$$

That is, if the local radius of curvature of the coil is finite then the contribution to the force per unit length from adjacent segments remains finite even though the segment lengths are taken to zero. This is the source of the non convergence of equation (4). In the special case when the wire is locally straight so that $R \to \infty$, the contribution $\delta\Gamma_{12}$ approaches zero and equation (4) may be applied directly.

Instead of dealing with adjacent filamentary segments consider the interaction between two elemental lengths of actual conductor called "slices", as illustrated in Figure 4. The slices are of circular cross section and carry a uniform current density. The contribution to the force per unit length acting on slice 1 due to slice 2 is

$$\delta\underline{\Gamma}_{12} = \frac{\mu_o}{4\pi^2 a^2} I^2 \, \hat{\underline{\ell}}_1 \times \int_o^{2\pi} \int_o^{a} \int_o^{2\pi} \int_o^{a} \delta\underline{\ell}_2 \times \frac{\hat{\underline{u}}_r(\rho_1, \rho_2, \phi_1, \phi_2)}{r^2} \times \rho_1 \rho_2 d\rho_2 d\phi_2 d\rho_1 d\phi_1 \qquad (11)$$

where ρ_1, ϕ_1 and ρ_2, ϕ_2 are local cylindrical co-ordinates that describe each point on the respective slice cross sections. Although the expression is too complex to allow a closed form solution or approximation, the author has shown by computation that $\delta\Gamma_{12}$ given by (11) approaches zero as the angular separation of the slices approaches zero. This behaviour is fundamentally different to that associated with elemental filaments. The difference arises because the current distribution across the elemental slices has been specified.

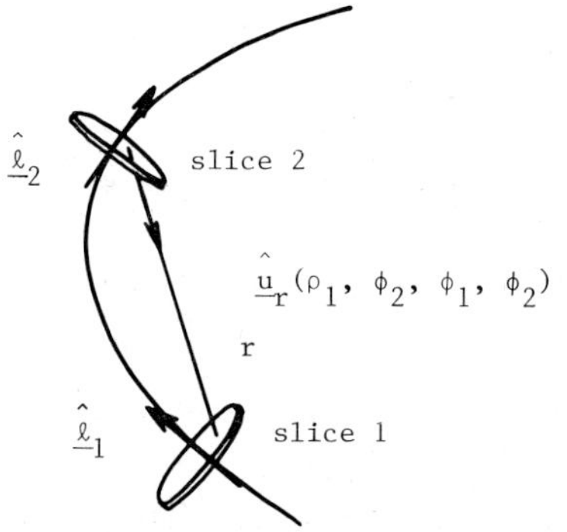

Figure 4.

Two slices of conductor having circular cross section which form part of a coil made of non-filamentary conductor.

Whilst a uniform current distribution was assumed above, this is not necessary. It has been found that $\delta\Gamma_{12}$ associated with a general current distribution approaches zero if the specified current distribution has a two dimensional extent and is not filamentary in nature. Furthermore the slices may have a noncircular section.

3. DEFINITION AND USE OF THE FORCE FACTOR

In order to obtain a method based on equation (4) for computing the force per unit length acting at a point on a non-filamentary coil, we introduce the force factor α which is defined as

$$\alpha = \frac{\text{force magnitude acting on an elemental (observation) slice due to a second elemental (source) slice}}{\text{force magnitude acting on an elemental (observation) filament due to a second elemental (source) filament}} \tag{12}$$

The corresponding elemental slices and filaments are centred on the same point (as shown in Figure 4), and have the same orientation and total current. The force factor is a function of the current distribution within the slices and of the relative position and orientation of the elemental segments. In terms of the length quantities shown in Figure 4,

$$\alpha_{12} = \frac{\left| \iint\limits_{\substack{\text{slice 1}\\ \text{cross section}}} \hat{\underline{\ell}}_1 J_1(\rho_1,\phi_1) dS_1 \times \hat{\underline{\ell}}_2 \times \iint\limits_{\substack{\text{slice 2}\\ \text{cross section}}} J_2(\rho_2,\phi_2) \hat{\underline{r}}\, dS_2 \right|}{I^2 \left| \dfrac{\hat{\underline{\ell}}_1 \times (\hat{\underline{\ell}}_2 \times \hat{\underline{u}}_r)}{r^2} \right|} \tag{13}$$

It is also necessary to compare the directions of the force for the case of elemental slices with that for elemental filaments. The author has shown by computation that the respective forces are virtually colinear for quite arbitrary position and orientation of the elements. Consequently the force exerted on the elemental observation slice is

$$\delta^2 \underline{F}_{\text{slice}} = \alpha\, \delta^2 \underline{F}_{\text{filament}} \tag{14}$$

The force per unit length acting at a point (associated with the i'th segment) on a coil which is imagined to consist of N slice elements and which has a specific current distribution across the disk section is then

$$\underline{\Gamma}_i = \frac{\mu_o}{4\pi} I^2 \hat{\underline{\ell}}_i \times \left\{ \lim_{N \to \infty} \sum_{\substack{j=1\\ j\neq i}}^{N} \frac{\alpha_{ij}\, \delta\underline{\ell}_j \times \hat{\underline{u}}_{rij}}{r_{ij}^{\,2}} \right\} \tag{15}$$

Equation (15) is the basis of the method of computation and it has been found in applications to be convergent.

To facilitate application of the method, a closed form approximation for α_{ij} for quite general position and orientation of the elements is required. Application of (15) would then simply require specification of the coil centre line position and would constitute an efficient method for computing the force per unit length. Such an approximation, applicable to coils that can be described by a local radius of curvature is presented in the following section.

It is known (Bleaney (1965) 127) that the force exerted on one elemental filament by a second such filament does not in general equal the force exerted on the second by the first. This result suggests that a different value for the force factor would be obtained if the descriptions "observation" and "source" were interchanged in the defining equation (12). However, it has been found by computation that this is not so. The force factor depends only on the relative position and orientation of the elements (that is $\alpha_{ij} = \alpha_{ji}$).

The force-factor definition (12) breaks down when the force exerted on the elemental field filament is zero. This occurs when the source filament is colinear with $\hat{\underline{u}}_r$ or otherwise when

$$\underline{d\ell}_{obs} \times (\underline{d\ell}_{source} \times \hat{\underline{u}}_r) = 0 \qquad (16)$$

It has been found by computation that the force exerted on the elemental observation slice in this case may not be zero but is in general relatively small. Hence although the application of (14) will not correctly give $\delta^2 \underline{F}_{slice}$, the resulting error in (15) is generally negligible.

4. APPLICATION OF THE METHOD TO THE RING COIL

A convenient coil shape with which to illustrate the method of computation is the ring coil shown in Figure 5. This coil shape experiences a radially acting force per unit length which may be calculated by applying the principle of virtual work to the known approximate expression for coil inductance. It is noted that the principle of virtual work is useful for calculating forces acting on coils only when the coil shape exhibits a high degree of symmetry.

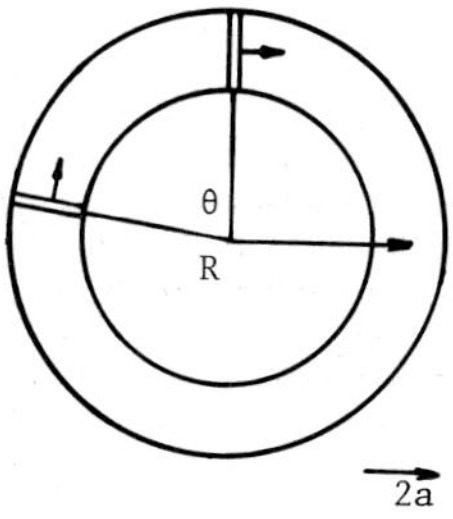

Figure 5.

The ring coil showing two elemental slices.

Figure 5 shows the dimensions of the coil and the positions of two elemental disk segments for which the force factor has been computed by evaluating the integral (13). Figure 6 gives the computed force factor for various values of ring radius and inter-segment angles for uniform current density across a circular cross section. It is seen that the force factor is zero if the elemental disks are adjacent ($\theta = 0$) and approaches unity as the segments recede from one another excepting that if R/a is small (<3) the force factor approaches a value significantly greater than unity. The shapes of the curves suggest that the force factor may be approximately described by an exponential expression. It has been found that

$$\alpha\,(R/a,\ \theta) \approx 1 - \exp\left(\frac{-R}{a}\,\frac{2\theta}{\pi}\right) \qquad (\theta \text{ in radians}) \qquad (17)$$

gives a close approximation to the computed curves for $R/a > 3$.

Figure 7 shows the normalized distance d/a between elemental slice centres as a function of R/a such that the force factor equals .99. Values of d/a lying above the curve are associated with force factor values which are for practical purposes unity.

Figure 8 shows the force factor for various choices of ring radius and inter-segment angle for tubular cross section and uniform current density. Figure 9 shows the force factor for a square cross section and uniform current density. Significant differences are observed between the force factor curves for the three cross sections, particularly for low values of R/a. Figure 10 gives a direct comparison of the force factor curves for the three cross sections with R/a = 3.

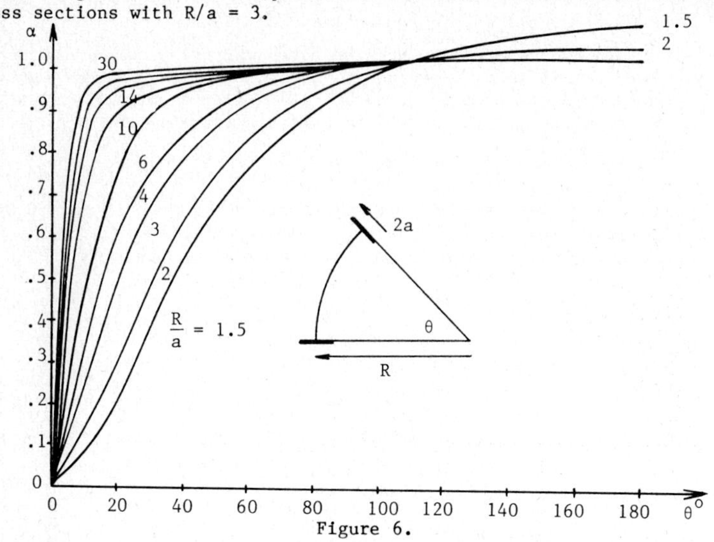

Figure 6.

Force Factor vs. inter-segment angle with R/a as parameter for the circular cross section.

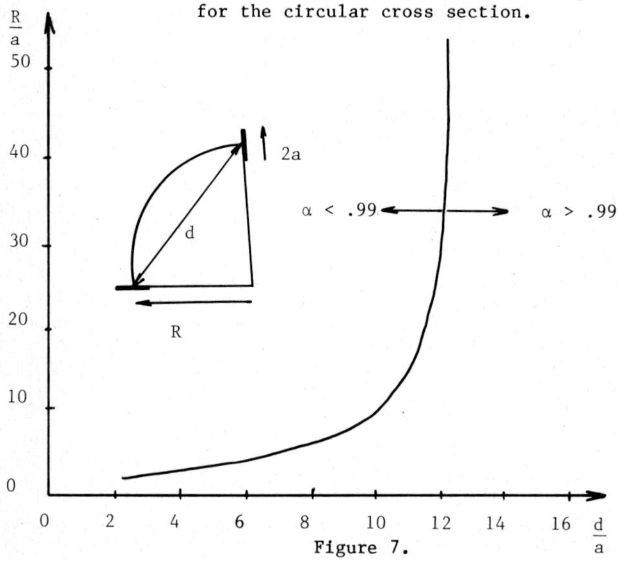

Figure 7.

Normalized distance between segment centres such that α = .99.

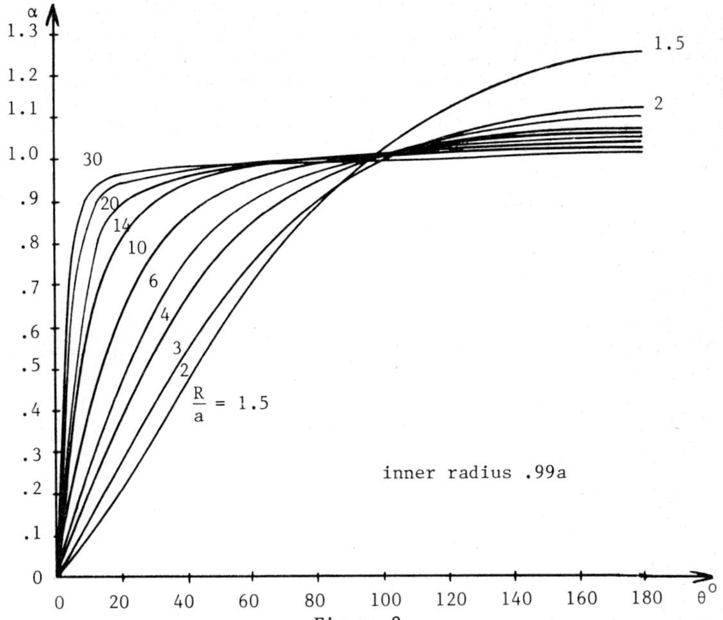

Figure 8.

Force Factor vs. inter-segment angle for the tubular cross section.

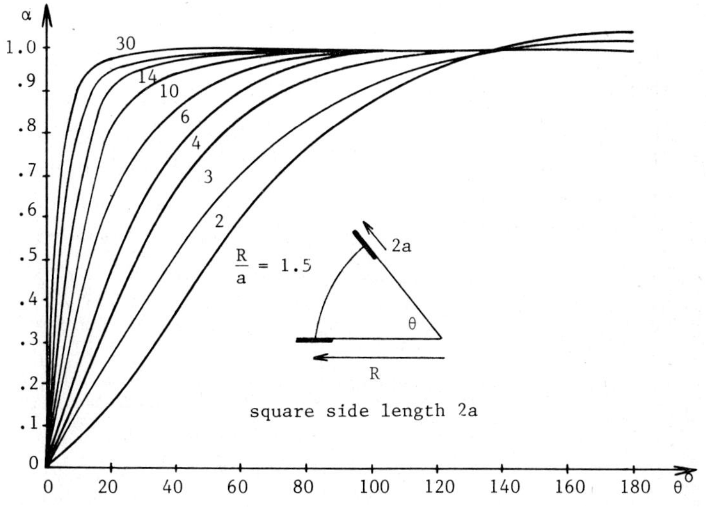

Figure 9.

Force Factor vs. inter-segment angle for the square cross section.

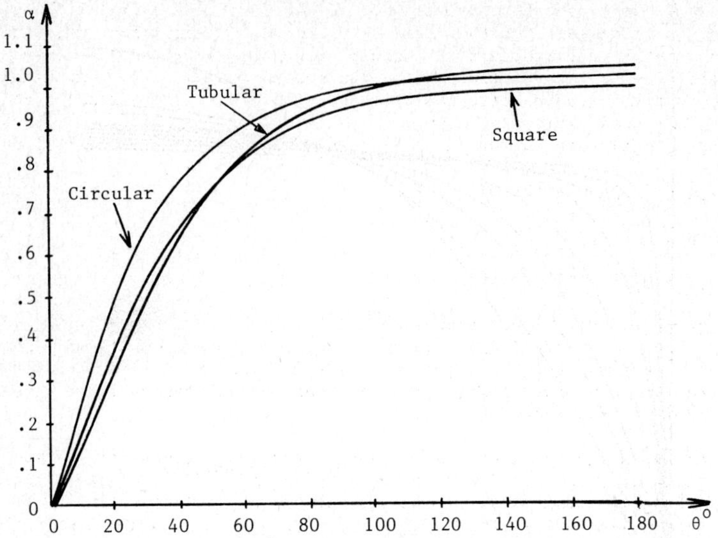

Figure 10.

A comparison of the force factor values for the three cross sections with R/a = 3.

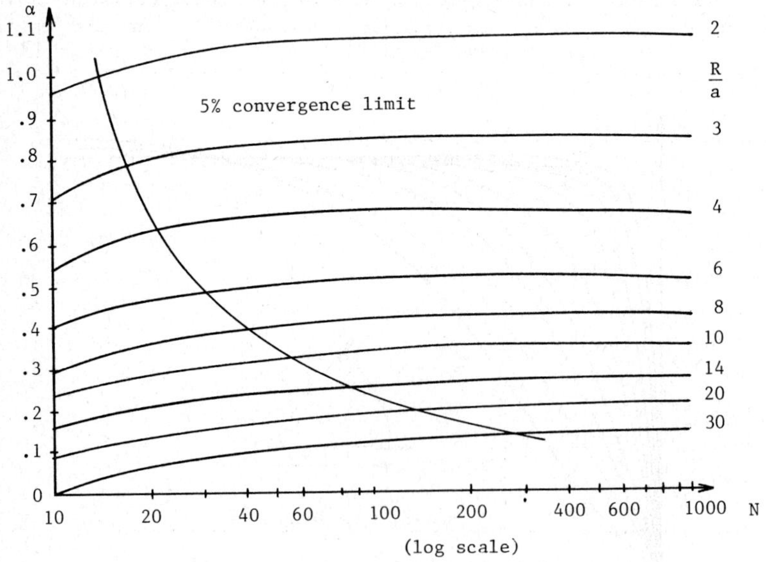

Figure 11.

Convergence behaviour vs. N for the circular cross section.

Table 1 gives a comparison of the normalized force per unit length acting to expand the ring coil for the three conductor cross sections under study. The values have been computed by the application of (15) with N = 1000. Also given for comparison in Table 1 are the normalized force per unit length values applying to the circular cross section with uniform current density which have been derived by applying the principle of virtual work to the approximate expression for the coil inductance (Landau (1960) 139), namely

$$L \;=\; \mu_o R \; \{\ell n \; (8R/a) \; - \; 1.75\} \qquad (18)$$

and hence

$$\Gamma \simeq \frac{\mu_o}{4\pi} I^2 \cdot \frac{1}{R} \{\ell n \; (8R/a) \; - \; 0.75\}. \qquad (19)$$

As (18) is an approximate expression, (19) cannot be assumed to be accurate (particularly for small R/a values). However, comparison of the values given in Table 1 shows that the values calculated using the method based on (15) for the ring coil having circular cross section are generally in agreement with those calculated using (19).

Also shown in Table 1 are values of force per unit length for the ring coil with circular cross section calculated by using the force factor values given by the approximation (17). It is seen that these values are in agreement with those associated with the accurate force factor values given in Figure 6.

Cross-section Description	Circular	Circular	Circular	Tubular	Square
α	as given on Fig. 6	approximation given by (17)	not applicable	as given on Fig. 8	as given on Fig. 9
a	1	1	1	outer = 1 inner = .99	$\sqrt{\frac{\pi}{4}}$
N	1000	1000	not applicable	1000	1000
R/a			(using (19))		
1.5	1.325	1.254	1.157	1.439	1.158
2.0	1.091	1.089	1.011	1.071	.9843
3.0	.8384	.8634	.8094	.7864	.7702
4.0	.6934	.7199	.6789	.6442	.6417
6.0	.5260	.5471	.5202	.4875	.4938
10.0	.3657	.3784	.3632	.3499	.3471
14.0	.2853	.2935	.2835	.2659	.2764
20.0	.2177	.2224	.2163	.2033	.2081
30.0	.1587	.1608	.1578	.1436	.1547

Table 1.

Values of $\Gamma \cdot \frac{4\pi}{I^2 \mu_o}$ for the three cross-sections under consideration

Also given in Table 1 are the normalized force per unit length values for the tubular and the square cross sections respectively. These values are presented for comparison with the circular cross section values as no prior calculations of the forces associated with these sections are known to the author.

The convergence of the method of computation has been investigated. Figure 11 shows the variation of $\Gamma \cdot \frac{4\pi}{I^2 \mu_o}$ with N for the circular cross section with the range of R/a values being considered. Also shown is the 5% convergence boundary based on values for N = 1000. It is seen that convergence is complete for N = 1000 and that the number of segments needed to ensure convergence rises quite rapidly as R/a rises as might be expected.

5. CALCULATION OF THE FORCES ACTING ON ARBITRARY COIL SHAPES

The force factor values given in Figures 6 to 9 apply to elemental slice segments which lie on a circular centre line. Coil shapes which can be described by a local radius of curvature as illustrated in Figure 12, can be approximately treated by using the appropriate force factor curve.

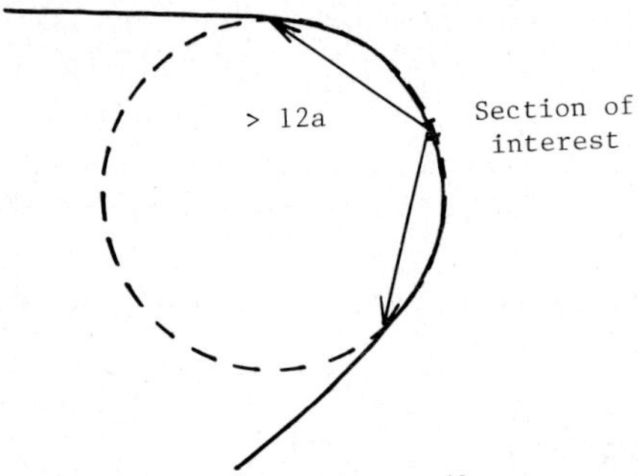

Figure 12.

A coil shape described by a local radius of curvature.

To investigate the use of the method of computation for a more general coil shape the method was applied to a single turn helically wound toroidal coil as shown in Figure 13, for which the local radius of curvature was taken as R_i. The force factor approximation (17) was used in the computation. Figure 13 shows the normalized force per unit length acting at selected points of the coil for a coil with R_i/a = 10 and 10 twists around the circumference. Although no previous force calculations for this coil shape are known to the author, the results confirm convergence of the method in this case.

The author is proceeding with work to obtain a simple approximate expression for the force factor for quite arbitrary positions of the coil elements.

$$\max \Gamma \times \frac{10^7}{I^2} = 14.0$$

$$\min \Gamma \times \frac{10^7}{I^2} = 13.2$$

$$\frac{R}{R_i} = 4$$

20
Normalized
Γ units

$2R_i$ →   R →

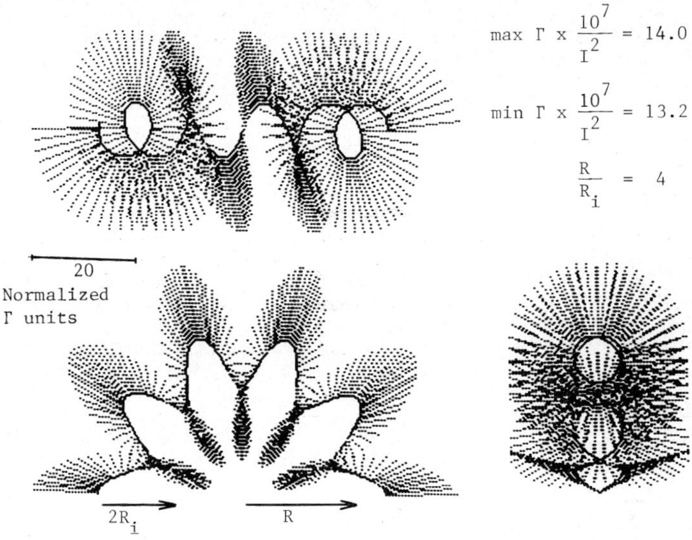

Figure 13.

Orthographic views of the single turn helically wound toroidal
coil with normalized force per unit vectors shown acting
at selected points ($\frac{1}{2}$ view shown for clarity).

6. CONCLUSIONS

The force per unit length of conductor acting on a filamentary coil due to
its own magnetic field is shown to be underdefined. This results from the
lack of information concerning the current distribution across the conductor
cross section that exists in the filamentary model. Consequently the forces
acting on a real coil cannot be calculated by treating the coil as though it
were filamentary. A method of compution is presented which retains the form
and simplicity of the filamentary coil model but ensures convergence by
incorporating the current distribution information via the force factor
(defined in section 3). The force factor can be calculated a priori for a
specified current distribution and conductor cross section.

A study of the forces acting on ring coils with circular, tubular and square
cross sections is presented to illustrate the method of computation.
Computed results are in agreement with the known approximate values of force
per unit length acting on the ring coil of circular cross section.

A closed form approximation for the force factor applicable to ring coils
with circular cross section and uniform current distribution is given which
greatly facilitates application of the method of computation. This
expression may be used to apply the method to any coil of circular cross
section which can be described by a local radius of curvature.

7. ACKNOWLEDGEMENT

The author is grateful to G.K. Cambrell and D.R. Sadedin for fruitful
discussions concerning this work.

LIST OF SYMBOLS

a	wire radius
d	inter-segment distance
Θ	inter-segment angle
R	radius of curvature
$\underline{\hat{\ell}}$	unit tangent vector
$\underline{r}$	radial vector
$\underline{\hat{u}}_r$	unit radial vector
$\underline{\hat{u}}_x, \underline{\hat{u}}_y, \underline{\hat{u}}_z$	unit vectors in a rectangular co-ordinate system
$\underline{dS}$	elemental area vector
$\underline{B}$	magnetic flux density vector
$\underline{F}$	force vector
$\underline{\Gamma}$	force per unit length vector
J	current density magnitude
I	current
μ_o	free space permeability
α	force factor
N	number of coil segments

REFERENCES

[1] Graneau, P., A re-examination of the relationship between self- and mutual inductance, Int. Journal of Electronics and Control, Vol. 12 (1962) 125-132.

[2] Bleaney, B.I. and Bleaney, B., Electricity and magnetism, (Second Edition, Oxford Clarendon Press, 1965).

[3] Landau, L.D. and Lifshitz, E.M., Electrodynamics of continuous media (Pergamon Press, Oxford, 1960).

[4] Corson D. and Lorrain, P., Introduction to electromagnetic fields and waves (Freeman, San Francisco, 1962)

Computational Techniques & Applications: CTAC-83
J. Noye & C. Fletcher (Editors)

COMPUTER SIMULATION OF COLD WORKING OF METALS

W. Thompson, K. Denmeade and S. Barton

Swinburne Institute of Technology

MELBOURNE, AUSTRALIA

The simulation of metal working processes using an EAI 500 Hybrid Computer is described. Differential Equations describing the metal forming processes have been developed and converted into a form suitable for analogue simulation. Machine equations are developed and scaled to suit the range of values likely to be encountered for all relevant parameters in the industrial situation. The simulation allows graphical output on a video terminal of the solution of metal working problems and allows any parameter or parameters to be varied so that a comparison of the effects can be visualised or studied with accuracy and speed.

INTRODUCTION

In industrial metal working, tooling design engineers are faced with many factors having an affect on the forming process; eg. die angle, reduction ratio, decoiler stress (back pull), die pressure (therefore die life), power available and material properties. Some of these factors have counteracting effects on each other making optimisation of the process difficult in a short time frame.

Solution methods available for solving metal forming problems have been tradition--ally:-

(i) trial and error
(ii) semi empirical or "work solutions"
(iii) mathematical model (differential equation)
(iv) slipline field
(v) load bounding

All of these methods have their own particular advantages and disadvantages. Some can not be used for optimisation, and some are unable to describe the effects of varying operating parameters at all. They all (except for "semi empirical") suffer the problem of being very time consuming for both the student of metal forming and the practitioner.

Digital Computers can speed up the solution of problems, but the output from the computer is not usually in a form easily visualised without the use of graphics terminals. Further the solution of differential equations using digital computers requires the use of numerical methods with associated approximations involved. The analogue computer, being able to work with continually changing variables, producing a continuously changing output on a video terminal is suited to solution of the types of problems described above, providing that those problems can be presented to the computer in the correct format, that is the form of a suitable differential equation. Since the basic component of analogue machines are integrating elements, this type of machine should provide a convenient solution route.

The analogue machine relies on an electrical analogy of the physical problem being produced using resistors, capacitors, potentiometers and amplifiers. Physical parameters are then expressed as voltages which are processed by the machine. In simple terms the basic elements of the analogue (and Hybrid) computer are:-

(i) The potentiometer and the digitally controlled amplifier (D.C.A.). The D.C.A. being more accurate than the potentiometer.
(ii) The operational amplifier - these enable the input signal to be multiplied by a negative factor > 1.
(iii) Invertor - this device changes the sign of the signal.
(iv) Integrator - essentially an op. amp. having feed back resistors replaced by capacitors so that the output signal can be altered at a rate dependant on the capacitance. Integration can thus be carried out with respect to the operating time of the computer if an initial condition (I.C.) is provided.
(v) Summer - adds or subtracts signal voltages.

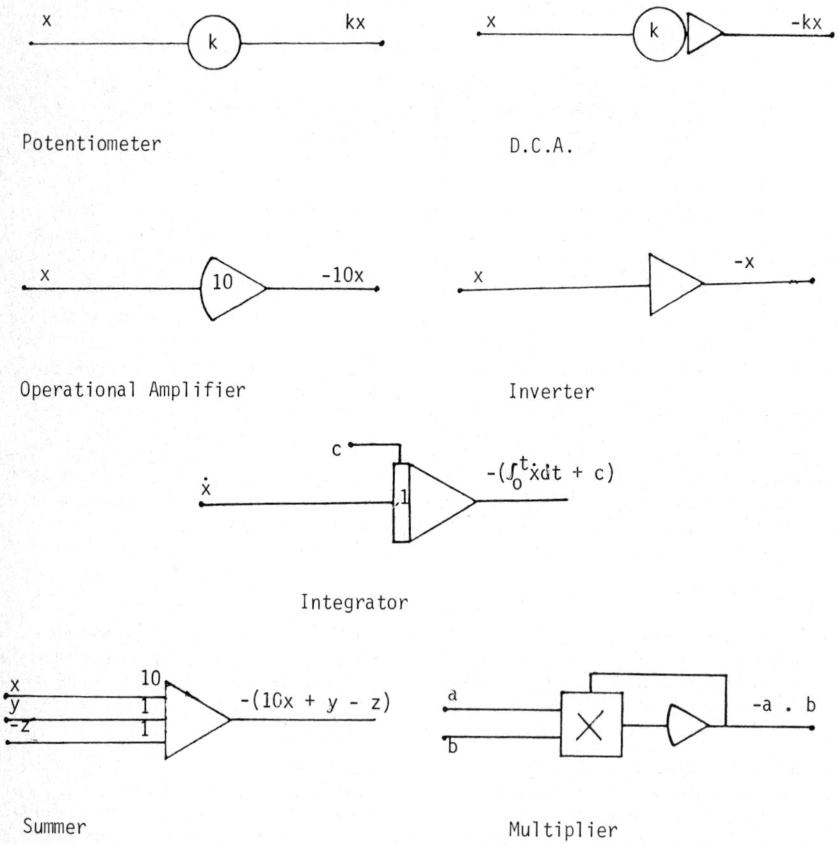

Fig. 1 Basic elements of Analogue (and Hybrid) Computers

Since these machines integrate at a rate proportional to their operating time, all variables to be integrated must be expressed in terms of time. So for metal forming problems it was necessary to represent natural strain (ε) as time.

The method used to develop the analogue circuits for the metal forming problems was to produce a differential equation for that process and then to write that d.e. with the highest order term on the L.H.S. of the equation. The equation then states that if the R.H.S. is integrated the integral of the L.H.S. is produced as follows:

$$\dot{x} = ax + b \quad \text{i.c. being } x = o \text{ at } t = o$$

$$\text{So } -\int_o^t \dot{x}\,dt = -\int_o^t (ax + b)\,dt.$$

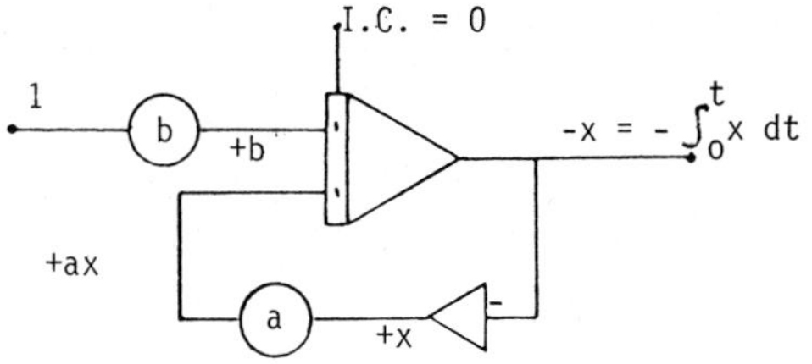

This method is the standard form of writing analogue equations (1).

Fig. 2 Example of simple d.e. Solution

When an analogue computer is controlled by a digital computer, i.e. for data entry, setting of electrical components, etc., it is known as a hybrid computer. This paper describes the application of such a computer, an EAI 500 hybrid computer, to the solution of metal working problems.

RELEVANT WORK

There have been a great number of research articles published on the application of hybrid or analogue computers to the solution of problems in control systems, vibrations and mathematics. None of these were of any help in overcoming the peculiar problems associated with metal working. Rowe (2) attempted solution of simplified metal working problems, and as such has laid the foundation for this study. With regard to the formulation of the process equations, the method known as: - "the slab technique", "the equilibrium method", or the " Mathematical method" has been well known for many years and reference to any text on engineering plasticity is sufficient for a background understanding to be obtained.

DEVELOPMENT OF THE D.E. AND THE ANALOGUE EQUATION

In order to solve the metal working problems chosen using analogue methods it was first necessary to be able to simulate the true stress/true strain characteristics of the material being formed and to develop differential equations for each forming process. These d.e.'s were then converted in the manner described earlier to a form suitable for "patching" onto the hybrid computer.

The processes which have been successfully simulated in this programme were wire drawing, strip drawing, tube drawing on a mandrel, tube drawing on a plug, tube sinking, forging of circular discs and plane strain forging. In all cases, para- -meters which were considered to be of value for industrial optimisation purposes or academic study have been arranged in the analogue equations so that they can be varied individually in discrete increments or "stepped" over a range of values.

The development of the process equations, their conversion to suitable analogue form and time and amplitude scaling have all been carried out using the same fundamental principles, so only one process; ie. wire drawing (being mathematically the simplest) is described in this paper to demonstrate the techniques involved. Results from other processes will be discussed in later papers.

WIRE AND STRIP DRAWING

The assumptions made in the development of this model are well known and have been described by many authors (3) - (9). For this reason they have not been detailed here. Fig. 3 shows the model used for development of the differential equation. This model postulates a volume element which is representative of all elements flowing through the die. This element is then considered as a free body diagram for stress analysis (equilibrium) purposes (Fig. 4) from which the d.e. is developed as follows:-

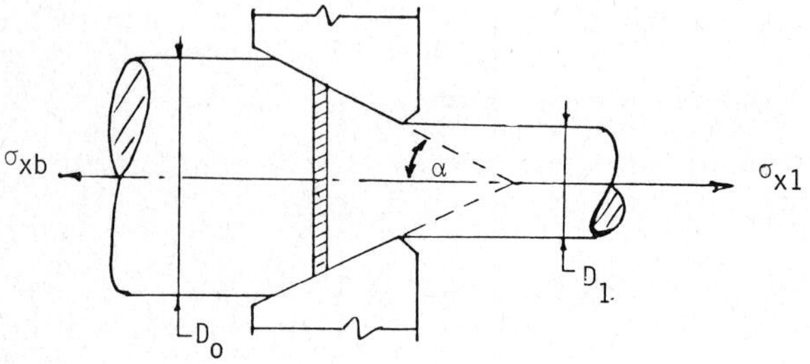

Fig. 3 Wire Drawing Model

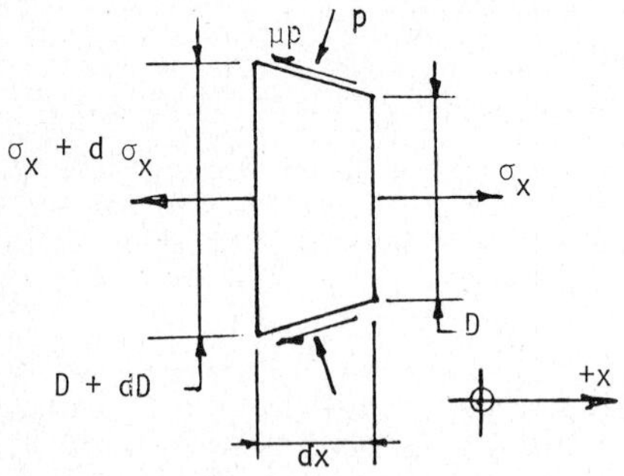

D = Diameter
p = Die pressure
μ = Coef. of friction
σ_x = axial stress
α = semi-die angle

Fig. 4 Free body Diagram

Considering equilibrium of forces in the axial direction.

$$(\sigma x + d\sigma x)\frac{\pi}{4}(D + dD)^2 - (\sigma x)\frac{\pi}{4}D^2 + \int_0^{2\pi} p\frac{dx}{\cos\alpha}\,\text{Sin}\alpha .\ r\,d\theta + \int_0^{2\pi}\mu p\frac{dx}{\cos\alpha}\cos\alpha\ .\ r d\theta = 0 \quad \text{.......................(1)}$$

which simplifies to

$$D d\sigma x + 2\sigma x dD + 2p\,(1+B)\,dD = 0 \quad \ldots\ldots (2)$$

where $B = \mu \mathrm{Cot}\alpha$

This equation has deliberately ignored non-homogenious deformation (redundant work) which has been added as a separate analogue circuit in order that the effects of redundant work (11) (12) can be investigated separately. The properties of the material being formed are introduced in the form of the yield criteria. The Von Mises Criterion usually gives the most accurate modelling and is given by:-

$$(\sigma_1 - \sigma_2)^2 + (\sigma_2 - \sigma_3)^2 + (\sigma_3 - \sigma_1)^2 = 2\bar{\sigma}^2 \quad \ldots\ldots (3)$$

For the hybrid equations the mean effective flow stress symbol $\bar{\sigma}$ is renamed Y or S depending on the process in question (wire or strip). When equation (3) is substituted into equation (2) and simplified the following is obtained:-

$$\frac{d\sigma x}{B\sigma x - (1+B)\,Y} = \frac{2dD}{D} \quad \ldots\ldots (4)$$

where Y is found from $Y = \frac{1}{\varepsilon}\int_0^{\varepsilon} \bar{\sigma}\, de$ ......(5)

Now the strain at any point in the wire drawing operation is given by:-

$$\varepsilon_D = \ell n \left(\frac{Do}{D}\right)^2 \quad \ldots\ldots (6)$$

and so $\frac{d\varepsilon_D}{dD} = \frac{d}{dD}\,(2\ell n Do - 2\ell n\, D)$

ie. $d\varepsilon_D = -\frac{2dD}{D}$ ......(7)

Equation (4) can therefore be rewritten as follows:-

$$\frac{d\sigma x}{B\sigma x - (1+B)\,Y} = \frac{2dD}{D} = -d\varepsilon_D \quad \ldots\ldots (8)$$

or $-\frac{d\sigma x}{d\varepsilon} = B\sigma x - (1+B)\,Y$ ......(9)

Which is now in a form suitable for hybrid Simulation.

Redundant work has been referred to earlier. This is allowed for by calculating the work involved in shearing at inlet and exit to the die. Equations of the form

$$\phi R = A^1 + B^1 \left(\frac{1-r}{r}\right) \mathrm{Sin}\alpha \quad \ldots\ldots (10)$$

result where 'r' is the percentage reduction in area α is the die angle and A^1 & B^1 are constants which can be evaluated from the flow path of the material.

TIME AND AMPLITUDE SCALING

The purpose of amplitude scaling is to establish a relationship between the real world values of the problem variables and constants and the dynamic range of the analogue computer's internal variables representing these variables and constants, ie. computing element input and output voltages. Similiarly a relationship is also required between the problem independant variable and computer time used to rep--resent that variable. This applies regardless of whether the variable is real time or some other quantity as is the case for metal forming and in particular wire drawing.

Before proceeding to the actual task of time and amplitude scaling it is desirable to ascertain how the problem equation (or equations) are to be implemented, ie. how the various analogue computing elements are to be interconnected.

The wire drawing equation to be mechanised as noted earlier is:-

$$-\frac{d\sigma x}{d\varepsilon} = B\sigma x - Y(1+B) \quad \ldots\ldots (9)$$

In order to mechanise this equation it was necessary to establish computer time as a parameter since the independant variable was not time or a time related quantity. This method is known as the "PARAMETRIC TECHNIQUE" and is a simple algebraic trans- -formation. The following transformation definitions were used to achieve this.

Let $\frac{d\sigma x}{d\tau} = \dot{\sigma x}$ and $\frac{d\varepsilon}{d\tau} = \dot{\varepsilon}$

Where τ is computer time

Equation (9) now becomes

$$-\frac{\dot{\sigma x}}{\dot{\varepsilon}} = B\sigma x - Y(1+B)$$

and rearranging yields

$$-\dot{\sigma x} = \{B\sigma x - Y(1+B)\}\,\dot{\varepsilon} \quad \ldots\ldots (11)$$

To implement this equation it was convienient and consistant to let:-

$$\dot{\varepsilon} = k \quad \text{where} \quad o < k \leqslant 1.0 \quad \ldots\ldots (12)$$

Equation (11) is the first of our "MACHINE" equations and was the means by which 'ε' was generated and hence also the variable "Y" in the other machine equation which will be determined later. In this way we were able to establish a time scaling factor which permitted us to choose a suitable rate at which to generate the solutions. The actual value chosen for k depends on the method of displaying the results but more importantly on problem related considerations which should become clear a little later.

Substituting for $\dot{\varepsilon}$ in equation (11) using k of equation (12) leads to the second machine equation which simulates the actual wire drawing process.

$$-\dot{\sigma x} = k\,\{B\sigma x - Y(1+B)\} \quad \ldots\ldots (13)$$

Using equations (12) and (13) the unscaled patch diagrams for the generation of ε and the wire drawing simulation were drawn.

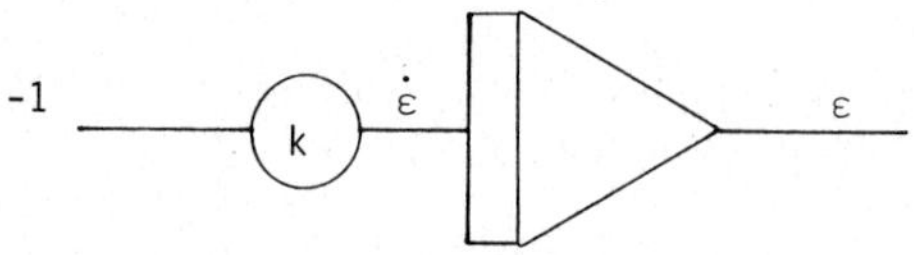

Fig. 5 Mechanisation of $\dot{\varepsilon}$= k to generate ε

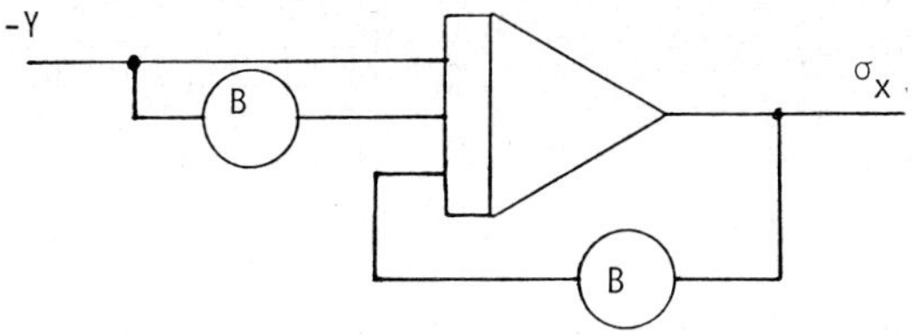

Fig. 6 Unscaled implementation of $\dot{\sigma}x = B\sigma x - Y(1+B)$

It will be observed that we cannot directly connect these two diagrams together since we do not know how to obtain Y from ε. We now proceed to obtain this relationship and hence complete the unscaled patch diagram.

An examination of the stress strain curves for suitable wire drawing materials showed that the relationship between Y and ε was a non-linear one. Plotting the generally used stress strain curves used for determining the drawing stress on log-log graph paper revealed that a simple fractional power law adequately rep--resents this relationship. This law turned out to be one in which the exponent was always less than 1.0 and so we could proceed to derive the required machine equation.

So, we assume that the functional relationship is given by the equation

$$y = Ax^m \quad \text{where m is always less than 1.0 and A is a constant} \quad \ldots\ldots (14)$$

Since such functions cannot be directly implemented we must first obtain a differential equation which can be implemented to enable us to mechanise the required function.

Thus differentiating (14) w.r.t. 'τ' (computer time) yields the d.e.

$$\dot{y} = mAx^{m-1}$$

and by suitable rearrangement and substitution from (14) the following equation (which is easily implemented using a multiplier and integrator) is obtained.

$$\dot{y}x = my \quad \ldots\ldots (15)$$

The following patch diagram implements this equation. Notice that this implement--ation results in the exponent being set on a single potentiometer thus allowing the stress strain curves for different materials to be readily set up and simul--ated.

Replacing x with ε, y with Y (or S), directly transfers our problem variables on to this patch diagram. In deriving equation (15) the constant of proportionality was cancelled out and must be restored. So in implementing eq'n (15) the proportionality constant is restored by inserting a potentiometer at the output of the function block. In this way the final form of the relationship between ε and Y that was mechanised is the form of equation (14).

$$\text{vis} \quad Y = Yo\ \varepsilon^m \quad \ldots\ldots (16)$$

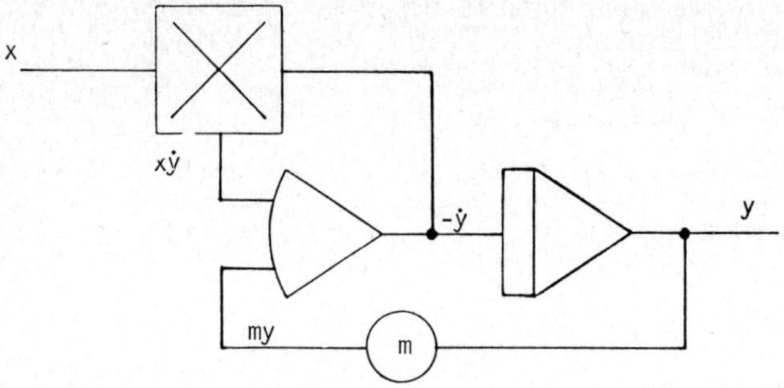

Fig. 7 Unscaled mechanisation of $y = Ax^m$

We were now able to draw the complete wire drawing simulation patch diagram noting that it is a basic simulation of the wire drawing process without any of the optional extras such as back pull or redundant work as described earlier.

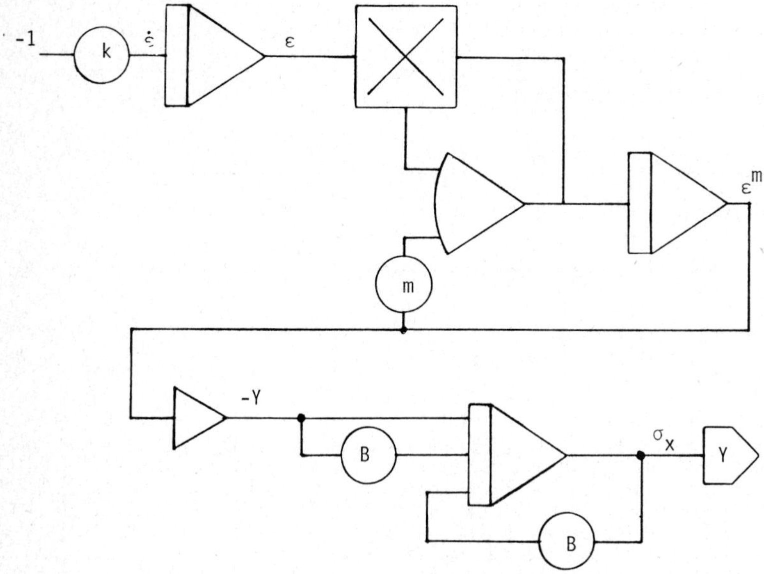

Fig. 8 Unscaled patch diagram of wire drawing process

We could now proceed with the actual task of amplitude and time scaling for the wire drawing simulation. In order to amplitude scale a problem we must first determine by appropriate means the likely maximum values that problem variables will attain during the course of a solution generation. In the case of time scaling we must simply establish a suitable rate at which to generate the solution so that the relationship between the independant variable and computer time is simple and meaningful. We will first time scale the simulation since this has a bearing on the maximum values that the dependant variables are likely to reach.

On the basis of the contraint of maximum stress that a material can withstand during wire drawing we chose a maximum strain of 2. That is ε will be limited to the range 0 to 2.0 Now ε is a function of computer time and reaches a value of 2.0 after 1 of computer time has elapsed. Expresses mathematically this is:

$$\frac{\varepsilon}{\varepsilon_{max}} = \tau \text{ ie. } \frac{\varepsilon}{2} = \tau \quad \ldots\ldots(17)$$

and differentiating yields

$$\frac{d(\varepsilon/2)}{d\tau} = 1 \quad \text{ie. from eqn (12), } k = 1.0.$$

This also leads us to the amplitude scaling of ε which from (16) above is clearly

$$1 \text{ m.u.} = \frac{\varepsilon}{\varepsilon_{max}} = \frac{\varepsilon}{2} \quad \text{i.e. } \varepsilon \text{ is scaled so that } \varepsilon = 2 \text{ at 1 m.u.}$$

The rest of the problem variables which occur in equation (13) required only amplitude scaling as did the proportionality constant of equation (16). The following table sets out the estimated maxima for all of the variables together with the chosen scale factor and final scaled variable. These maximum values are either obtained by calculation or by means of suitable lookup tables or graphs.

VARIABLES	EST. MAX.	SCALING MAX.	SCALED VARIABLE
ε	2	2	$\varepsilon/2$
Y	1200 MPa	1200 MPa	Y/2000
Yo	1200 MPa	1200 MPa	Yo/2000
B	10	10	B/10

TABLE 1 Scaled Variables

Having determined our scaled variables these were substituted into the machine equations to obtain the final scaled equations which establish integrator and summer gains and any static potentiometer settings. These scaling factors also established the relationship between real problem variables and the settings of any potentiometers which are solution specific parameters. An example of such a scaled variable which occurred on a potentiometer (in this case two) is "B" which is a function of the friction coefficient and the semi-die angle. We now proceed with the scaled variable substitution.

Substituting into equation (13) yields $\frac{\sigma_x}{2000} = 1.0 \left\{\frac{B}{10} \cdot \frac{10}{1} \cdot \frac{\sigma_x}{2000} - \frac{Y}{2000}\left(1 + \frac{B}{10} \cdot \frac{10}{1}\right)\right\}$

and into equation (16) gives $\frac{\dot{\varepsilon}^m}{2} \cdot \frac{\varepsilon}{2} = \frac{1}{2} \cdot \frac{m}{1} \cdot \frac{\varepsilon^m}{2}$

Using equations (16), (18) and (19) the complete scaled patch diagram for the wire drawing simulation as shown in Figure 9 was drawn and patched onto the circuit board.

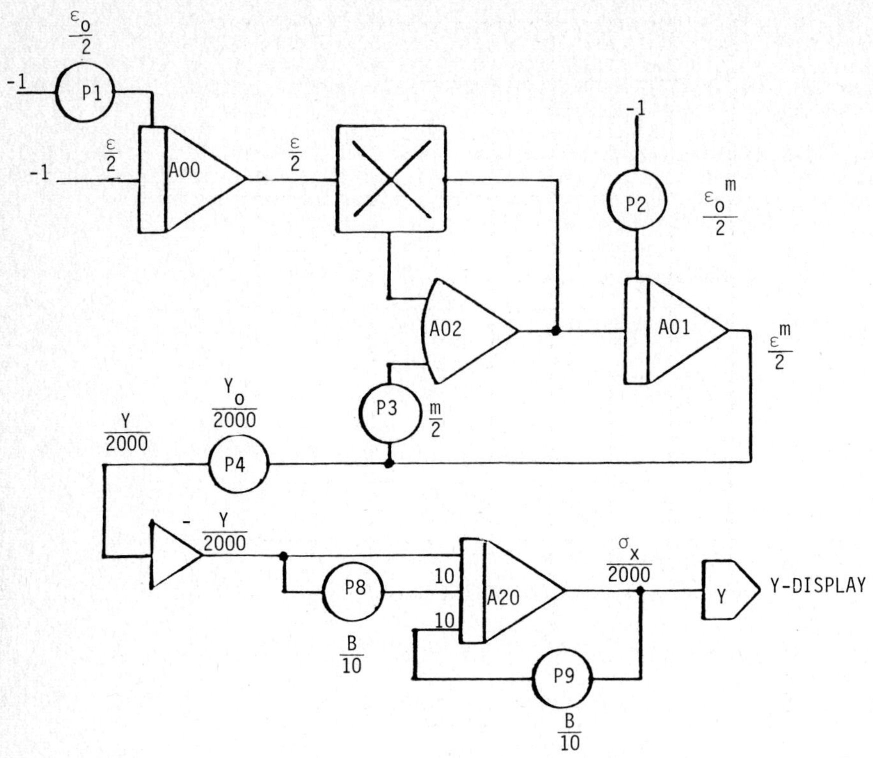

Figure 9 - Scaled Implementation of Wire Drawing Process

HYBRID OUTPUT

The analogue circuits shown earlier were "patched" on to a board, each parameter needing to be varied identified with a potentiometer reference number. Figure 10 shows a typical output for Strip and Wire drawing with no backpull and no redundant work; simulating working in a die of semi-angle $\alpha = 10^{\circ}$ with a coefficient of friction $\mu = 0.1$.

The maximum reduction obtainable in either process is seen to be the same. The curves also show the results of manual calculation to check the computer output. It can be seen that good correlation between hybrid and manual calculation is obtained. As the output from the hard copy unit is too large the following figures have been re-drawn in order to show the types of output and therefore solutions quickly obtainable in an interactive manner from this system. Figure 11 shows the effect of the backpull stress stepwise on the wire drawing process; i.e. the higher the backpull σ_{xb} lower the reduction 'r'% possible.

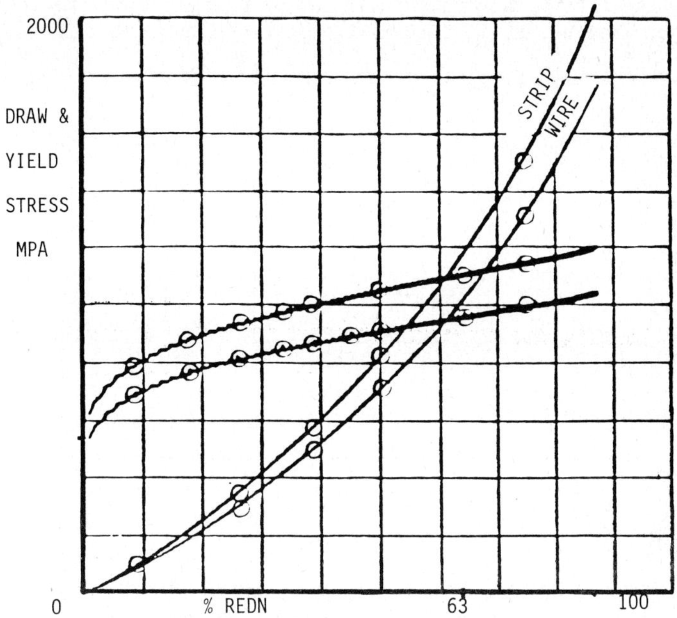

FIGURE 10 Stress/Strain and draw stress curves for Wire & Strip

S	=	Plane strain yield stress	N/mm^2
Y	=	Uniaxial yield stress	N/mm^2
WIRE	=	Draw stress for wire	N/mm^2
STRIP	=	Draw stress for strip	N/mm^2
0	=	Manually calculated values	N/mm^2

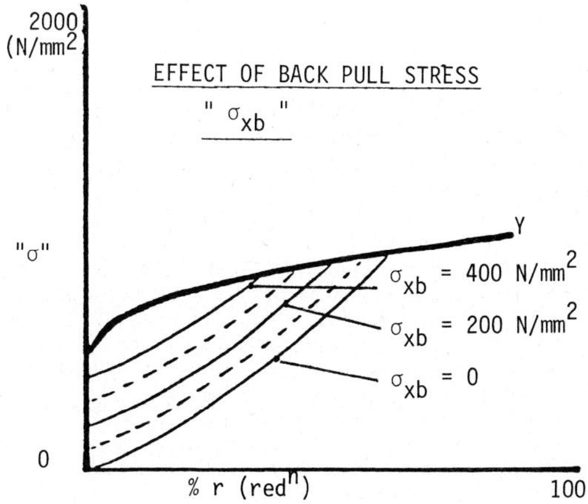

FIGURE 11 Effect of Backpull on Draw Stress in Wire Drawing

Figure 12 shows the draw stress σ_x with and without redundant work, and the die pressure "P" at all points in the die with and without redundant work.

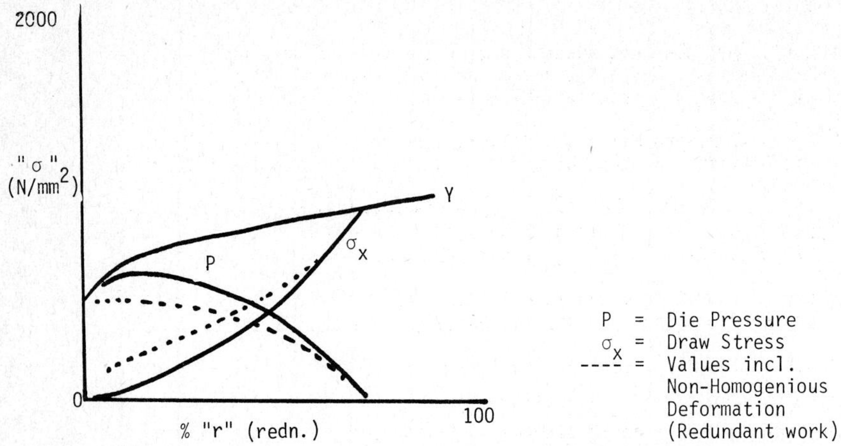

Figure 12 Wire Drawing With and Without Redundant Work

CONCLUSIONS

It has been shown that differential equations of the type found in theoretical plasticity of deforming solids can be solved in an interactive manner by simulating the process on a hybrid computer. Visualisation of the effects of any or all process parameters is easily obtained and the tedium of repeated arithmetic work is eliminated. Many of the approximations usually made in this type of analysis, in particular: - neglecting non-homogenious deformation, the use of the Tresca rather than Von Mises Yield Criteria and the use of straight line approximations to the true stress true strain curve no longer need to be made using this type of solution. More accurate optimisation of industrial processes is therefore possible.

Users and students of metal working theory find the technique easy to follow and now find it easier to conceptualise many of the interacting factors involved in the science of metal working technology.

REFERENCES

(1) Charlesworth, A.S. and Fletcher, J.R., Systematic Analogue Computer Programming. 2nd edn. Pitman, London, 1974.

(2) Rowe, G.W., An Introduction to the Principles of Metalworking. London, Arnold, 1971.

(3) Wistreich, J.G., Investigation of the Mechanics of Wire Drawing. Proc. of the Inst. of Mech. Engrs. London, V169, 1955.

(4) Hill, R. and Tupper, S.J., A New Theory of the Plastic Deformation in Wire Drawing. J. Iron and Steel Inst. V159, 1948.

REFERENCES continued

(5) Davis, E.A. and Dokos, S.J., Theory of Wire Drawing. J. Appl.Mech. V 11, 1944.

(6) Lancaster, P.R. and Rowe, G.W., Experimental Study of the Influence of Lubrication Upon Cold Drawing Under Approximately Plane-Strain Conditions at Low Speeds. Proc.Inst.Mech.Engrs., V178, 1963.

(7) Green, A.P. and Hill, R., Calculations on the Influence of Friction and Die Geometry in Sheet Drawing. J.Mechanics and Physics of Solids, V1, 1953.

(8) Baron, H.G. and Thompson, F.C., Friction in Wire Drawing. J. Inst.Metals, V78, 1950/51.

(9) Sachs, G., Beitrag zur Theorie des Ziehvorganges. Zeitschrift Fuer Angewandte Mathematik und Mechanik, V7, 1927.

(10) Wistreich, J.G., The Fundamentals of Wire Drawing. Metallurgical Reviews. V3, 1958.

(11) Green, A.P., Plane Strain Theories of Drawing. Proc. Inst. Mech. Eng., V174, 1960.

(12) Green, A.P., A Theoretical Investigation of a Ductile Material Between Smooth Flat Dies. Philosophical Mag. Series 7, V42, 1951.

Computational Techniques & Applications: CTAC-83
J. Noye & C. Fletcher (Editors)

AN EXACT NUMERICAL ALGORITHM TO INVERT RATIONAL FRACTION LAPLACE TRANSFORMS

Leon P. Travis

Lecturer, Swinburne Institute of Technology
Hawthorn, Victoria
Australia

An algorithm is presented to calculate the inverse of a Laplace transform in the form of a rational fraction. This short computer algorithm is exact within the limits of the finite representation of the numbers. The computer code is presented along with a simple example.

INTRODUCTION

It has long been known that problems in the physical sciences can be represented easily in the Laplace Domain. While classical analysis in systems and controls makes extensive use of Laplace transforms, transient electrical circuits, discrete models of physical processes such as diffusion also abound. The most commonly occuring type of transform which arises is that of a rational fraction. This paper presents an exact numerical inversion of Laplace transforms in the form of rational fractions.

The literature of numerical inversion is extensive (Davies and Martin, 1979). Examples of the most common approaches to the general inversion problem are numerical quadrature (Bellman, Kalaba, and Lockett, 1966), (Piessens, 1973) and polynomial approximation (Piessens, 1971). Numerical quadrature is reported as being inherently unstable and requires increases of the precision of the inversion parameters with increasing number of integration points in the integration interval. Furthermore, functions (in the time domain) which are unbounded (such as t^2) cannot be successfully inverted at all. Polynomial approximation methods generally involve approximating a Laplace transform with polynomial with a known inverse but again do not work for all functions (Davies and Martin, 1979), and it is recommended that verification by different methods are needed to gain confidence in any results achieved.

The method presented in this paper is accurate, reliable, and simple (a BASIC version takes less than two pages of programming).

METHOD OUTLINE

A stationary, ordinary differential equation with appropriate initial conditions can be found that has the same Laplace transform as the given one. The inversion process is then reduced to one of numerically integrating this equation. A further refinement is obtained by formulating this differential equation in state space form and then determining its transition matrix. The solution follows easily by repeated matrix multiplications.

EQUIVALENT DIFFERENTIAL EQUATION

A Laplace transform of order, n, which is to be inverted is given in equation (1). Clearly, the Laplace transform of differential equation (2) will have the same denominator as (1). Thus the task remaining is to find the initial conditions giving the same numerator as (1). Note that $\bar{Y}(s)$ is the transform of Y and superscript (n) means the nth derivative.

$$\bar{Y}(s) = \frac{a_{n-1}s^{n-1} + \ldots + a_1 s^1 + a_0}{s^n + b_{n-1}s^{n-1} + \ldots + b_1 s + b_0} \tag{1}$$

$$Y^{(n)} + b_{n-1}Y^{(n-1)} + \ldots + b_0 Y = 0 \tag{2}$$

Define the initial conditions of (2) as the set of equations (3).

$$\begin{aligned} Y(0) &= C_0 \\ Y^{(1)}(0) &= C_1 \\ &\ldots \\ Y^{(n-1)}(0) &= C_{n-1} \end{aligned} \tag{3}$$

Now take the Laplace transform of (2) with conditions (3) and compare the resulting numerator with (1) term-by-term. The initial conditions are then given by equations (4).

$$\begin{aligned} C_0 &= a_{n-1} \\ C_1 &= a_{n-2} - (b_{n-1}C_0) \\ C_2 &= a_{n-3} - (b_{n-2}C_0 + b_{n-1}C_1) \\ &\ldots \end{aligned} \tag{4}$$

STATE SPACE FORMULATION

Define a state vector, $\mathbf{X}(t)$, as a column vector with elements as shown in (5). Note the superscript, t, denotes a transpose.

$$\mathbf{X} = (Y \quad Y^{(1)} \quad Y^{(2)} \quad \ldots \quad Y^{(n-1)})^t \tag{5}$$

Equation (2) can now be expressed in state space form (6) where the coefficient matrix **A** has the bottom row as:

$$(-b_0 \quad -b_1 \quad \ldots \quad -b_{n-1})$$

and the diagonal just above the main diagonal, unity and all other elements zero (Ogata, 1967).

$$\dot{\mathbf{X}} = \mathbf{A}\mathbf{X} \tag{6}$$

The transition matrix, $\Phi(h)$ for the time interval, h, is most easily determined by means of a Taylor series expansion (7). For large order inversions, an alternate method of calculating the transition matrix may be necessary. Being determined only once, this transition matrix is used to determine the time domain solution in time intervals of h (8). While this method may not be ideal for large order matrices, it is the simplest to program.

$$\Phi(h) = \mathbf{I} + \mathbf{A}h/1! + \mathbf{A}^2h^2/2! + \ldots \tag{7}$$

$$\mathbf{X}(t+h) = \Phi(h)\mathbf{X}(t) \tag{8}$$

COMPUTER PROGRAM

The program for the method outlined above is shown in figure 1. After the input (lines 230 to 450), the equivalent state space matrix is developed (lines 3000 to 3170), and the required initial conditions are calculated (lines 4000 to 4170). A Taylor series for the transition matrix is built up term-by-term (lines 1000 to 1410). While this program terminates the series arbitrarily at twenty terms, the reader may want to include a more elegant method of estimating the required number of terms. Having the transition matrix, the time domain solution is calculated at interval, h, by a matrix multiplication (lines 590 to 690). A very simple example of inverting the function, $(2/s^3)$ to the time domain is shown in figure 2. Note that the derivatives of the time domain are also calculated as a bonus.

```
10 REM ***********************************************
20 REM *                                             *
30 REM *    LAPLACE INVERSION ver. 1.0   15/1/83     *
40 REM *    EXACT MATRIX PROGRAM  by L.P.Travis      *
50 REM *                                             *
60 REM ***********************************************
70 :
100 REM **********************************************
110 REM *                                            *
120 REM *    INPUT ROUTINE                           *
130 REM *                                            *
140 REM **********************************************
160 :
170 PRINT CHR$(12);
180 PRINT "LINVCRT.BAS   ver. 1.0  by  L.P.Travis 15/1/83"
190 PRINT: INPUT "LAPLACE TRANSFORM ORDER = "; N
200 DIM A(10,10), T1(10,10), T2(10,10), T3(10,10), SUM(10), X(10)
210 X(0) = 0
220 TIME = 0
230 INPUT "TIME INCREMENT = "; DTIME
240 GOSUB 2000
250 PRINT
260 PRINT "CALCULATING THE TRANSITION MATRIX ......"
270 GOSUB 1000: REM ... TO CALC TRANSITION MATRIX
290 PRINT: PRINT
300 PRINT "TRANSITION MATRIX IS BELOW": PRINT
310    FOR I = 1 TO N
320       FOR J = 1 TO N
330          PRINT A(I,J),
340       NEXT J
350       PRINT
360    NEXT I
370 PRINT
380 PRINT "SOLUTION IS LISTED BELOW ............"
390 PRINT: INPUT "PRESS <CR> TO CONTINUE"; Q$
400 PRINT CHR$(12);
410 PRINT "TIME";
420    FOR I = 1 TO N
430       PRINT TAB(14*I); "X(";I;")";
440    NEXT I
450 PRINT
```

FIGURE 1, PROGRAM LISTING (con't next page)

```
480 REM ********************************************************
490 REM *                                                      *
500 REM *    MAIN COMPUTATIONS AND PRINT ANSWER LOOP           *
510 REM *                                                      *
520 REM ********************************************************
530 :
540 FOR ISCREEN = 1 TO 20
550    FOR I = 0 TO N
560       PRINT TAB(14*I); X(I);
570    NEXT I
580    PRINT
590    FOR I = 1 TO N
600       SUM(I) = 0
610       FOR J = 1 TO N
620          SUM(I) = T3(I,J)*X(J) + SUM(I)
630       NEXT J
640    NEXT I
650    TIME = TIME + DTIME
660    X(0) = TIME
670    FOR I = 1 TO N
680       X(I) = SUM(I)
690    NEXT I
700 :
710 NEXT ISCREEN
720 INPUT "MORE RESULTS (Y/N)";Q$
730 IF Q$ = "Y" THEN 520 ELSE END
740 :
1000 REM |------------------------------------------------------------|
1010 REM |                                                            |
1020 REM |    TO CALCULATE TRANSITION MATRIX BY TAYLOR SERIES         |
1030 REM |                                                            |
1040 REM |------------------------------------------------------------|
1050 :
1060    FOR I = 1 TO N
1070       FOR J = 1 TO N
1080          T1(I,J) = 0
1090          T3(I,J) = 0
1100       NEXT J
1110       T1(I,I) = 1
1120       T3(I,I) = 1
1130    NEXT I
1150 REM ..... LOOP "NTERM" TIMES TO APPROX TAYLOR SERIES
1160 NTERM = 20 :REM SETS NUMBER OF TERM IN TAYLOR SERIES
1170    FOR NA = 1 TO NTERM
1180    REM .... FORM NEW TERM FOR TAYLOR SERIES
1190       FOR I = 1 TO N
1200          FOR J = 1 TO N
1210             SS = 0
1220             FOR K = 1 TO N
1230                SS = SS + A(I,K)*T1(K,J)
1240             NEXT K
1250             T2(I,J) = SS*DTIME/NA
1260          NEXT J
1270       NEXT I
1280       FOR I = 1 TO N
1290          FOR J = 1 TO N
1300             T1(I,J) = T2(I,J)
1310          NEXT J
1320       NEXT I
```

FIGURE 1, con't

```
1330        REM .... ADD NEW TERM TO UPDATE "TAYLOR" SUM
1340        FOR I = 1 TO N
1350           FOR J = 1 TO N
1360              T3(I,J) = T3(I,J) + T1(I,J)
1370           NEXT J
1380        NEXT I
1390    NEXT NA
1410 RETURN
1420 :
2000 REM |------------------------------------------------|
2010 REM |                                                |
2020 REM |    INPUT LAPLACE TRANSFORM                     |
2030 REM |                                                |
2040 REM |------------------------------------------------|
2050 :
2060 PRINT: PRINT "INPUT DENOMINATOR IN DESCENDING ORDER"
2070 PRINT "        -----------": PRINT
2080 PRINT "COEF OF S^";N; :INPUT A1
2090    FOR J = N TO 1 STEP -1
2100       PRINT "COEF OF S^";J-1; :INPUT A(N,J)
2110    NEXT J
2120 GOSUB 3000
2130 PRINT: PRINT
2140 PRINT "A matrix ="
2150    FOR I = 1 TO N
2160       FOR J = 1 TO N
2170          PRINT A(I,J);
2180       NEXT J
2190       PRINT
2200    NEXT I
2210 PRINT: PRINT
2220 PRINT "INPUT THE NUMERATOR IN DESCENDING ORDER"
2230 PRINT "              ---------": PRINT
2240    FOR I = N TO 1 STEP -1
2250       PRINT "COEF OF S^"; I-1; :INPUT SUM(I)
2260    NEXT I
2270 GOSUB 4000
2290 PRINT:PRINT "IC VECTOR ="
2300    FOR I = 1 TO N
2310       PRINT X(I),
2320    NEXT I
2330 PRINT
2340 RETURN
2350 :
3000 REM |-----------------------------------------------------|
3010 REM |                                                     |
3020 REM |    CONVERTS LAPLACE TRANSFORM INTO THE "A" matrix   |
3030 REM |                                                     |
3040 REM |-----------------------------------------------------|
3050 :
3060    FOR I = 1 TO N-1
3070       FOR J = 1 TO N
3080          A(I,J) = 0
3090       NEXT J
3100    NEXT I
3110    FOR I = 1 TO N-1
3120       A(I,I+1) = 1
3130    NEXT I
3140    FOR J = 1 TO N
```

FIGURE 1, con't

```
3150         A(N,J) = -A(N,J)/A1
3160      NEXT J
3170 RETURN
3180 :
4000 REM |---------------------------------------------------------|
4010 REM |                                                         |
4020 REM |     CONVERTS LAPLACE TRANSFORM TO I.C. VECTOR           |
4030 REM |                                                         |
4040 REM |---------------------------------------------------------|
4050 :
4060      FOR I = 1 TO N
4070         SUM(I) = SUM(I)/A1
4080      NEXT I
4090 X(1) = SUM(N)
4100      FOR I = 2 TO N
4110         C1 = 0
4120         FOR I9 = 1 TO I-1
4130            C1 = C1 + A(N,I9-I+N+1)*X(I9)
4140         NEXT I9
4150         X(I) = SUM(N-I+1) + C1
4160      NEXT I
4170 RETURN
```

FIGURE 1, con't

X(Ø)	X(1)	X(2)	X(3)
Ø	Ø	Ø	2
1	1	2	2
2	4	4	2
3	9	6	2
4	16	8	2
5	25	1Ø	2
6	36	12	2
7	49	14	2
8	64	16	2
9	81	18	2
1Ø	1ØØ	2Ø	2

FIGURE 2, Inversion of $2/s^3$

X(Ø)	X(1)	X(2)	X(3)	X(4)	X(5)	X(6)
Ø	4Ø	Ø	5.79E-Ø4	-2.28E-Ø4	-1.41E-Ø3	34.3695
1	4Ø.1326	.565Ø45	1.75794	3.26Ø94	.9Ø4363	-6.882Ø8
2	42.11Ø7	3.87966	4.63615	1.96715	-2.33999	-.252538
3	48.54Ø9	9.12282	5.5Ø4Ø7	-.Ø8692Ø	-1.53714	1.22Ø43
4	6Ø.3471	14.3762	4.84113	-1.Ø6362	-.49282	.787572
5	76.9518	18.6317	3.64Ø18	-1.24975	.Ø421581	.319454
6	97.1993	21.6647	2.452Ø9	-1.Ø9674	.223653	.Ø764537
7	119.917	23.6Ø78	1.47463	-.855Ø87	.243954	-.Ø19Ø428
8	144.13	24.6944	.736672	-.626924	.2Ø8Ø17	-.Ø458527
9	169.Ø96	25.15Ø3	.2Ø5915	-.44249	.16Ø867	-.Ø46Ø4Ø3
1Ø	194.282	25.16	-.163528	-.3Ø346	.118541	-.Ø38Ø591
11	219.314	24.8629	-.413678	-.2Ø242	.Ø85Ø479	-.Ø29Ø328
12	243.94	24.3611	-.578Ø7	-.13Ø527	.Ø6ØØ323	-.Ø2128Ø2
13	267.993	23.7269	-.681855	-.Ø8ØØ6Ø3	.Ø419Ø28	-.Ø15263
14	291.367	23.Ø114	-.7433Ø4	-.Ø449879	.Ø28983	-.Ø1Ø8116

FIGURE 3, Sample Problem

AN EXAMPLE

A concrete wall 0.3m thick, insulated one one side and originally at a uniform temperature of 40C, is suddenly exposed on the other side to a hot gas at 860C. It is required that the time for the insulated surface to reach 260C be determined. The thermal properties are given as: thermal conductivity = 0.92W/mK, thermal heat capacity = 837 J/Kg K, density = 2310 Kg/m^3, and the heat transfer coefficient of the hot gasses = 28.4 W/m^2 K.

Dividing the slab into five equal divisions, placing a temperature node at the center of each division, five ordinary differential equations can be derived from an energy balance. Next the Laplace transform of all five differential equations is taken and the resulting four algebraic equations are solved simultaneously. The equation for the insulated side (the time unit is hours) is:

$$\bar{T}_4(s) = \frac{40s^5 + 194.89s^4 + 326.43s^3 + 216.89s^2 + 48.84s + 36.04}{s^6 + 4.872s^5 + 8.161s^4 + 5.422s^3 + 1.221s^2 + 0.04189}$$

The numerical inverse is shown in figure 3 which completes the solution.

CONCLUSIONS

A method, both simple and accurate has been developed that inverts Laplace transforms of the form of a rational fraction. Considerable use could be made of this method in diverse areas of engineering analysis.

REFERENCES

[1] Davies, B. and Martin, B., Numerical Inversion of the Laplace Transform: a Survey and Comparison of Methods, Jrnl. of Computational Physics (1979).

[2] Bellman, R., Kalaba, R., and Lockett, J., Numerical Inversion of the Laplace Transform: Applications to Biology, Economics, Engineering, and Physics (American Elsevier, New York, 1966).

[3] Piessens, R., Gaussian Quadrature Formulas for Bromwich's Integral, Comm. ACM **16** (1973).

[4] Piessens, R., and Branders, M., Numerical Inversion of the Laplace Transform Using Generalized Laguerre Polynomials, Proc. IEE **118** (1971).

[5] Ogata, K., State Space Analysis of Control Systems (Prentice-Hall, N.J., 1967).

Computational Techniques & Applications: CTAC-83
J. Noye & C. Fletcher (Editors)

ON THE PROBLEM OF A THROWN STRING

A. M. Watts

Department of Mathematics
University of Queensland

The motion of a perfectly flexible, inextensible string is simulated using a finite element method to represent the dynamic equations. An implicit scheme is used to solve the ordinary differential equations which are obtained and an iteration is used to satisfy the constraint of inextensibility. The average distance between the ends when the string is in free motion is examined and found to differ from the experimental results which have been reported. In particular, it is found that the results depend on the angular momentum.

INTRODUCTION

It has been reported by Synge [4] that, according to his experiments with a well worn pyjama cord and experiments by other people, when a string is thrown in the air and observed after it has fallen on a horizontal surface, the average distance between the ends of the string is about one-third of its length. The problem raises several questions of statistics and statistical mechanics, in particular the statistical mechanics of a system with infinite degrees of freedom.

Many of the statistical treatments do not consider the dynamics of the problem, at least not explicitly (see, for example, Clarke [2], Bass and Bracken [1]). Kingman [3] after an extensive theoretical treatment of the statistical mechanics finds that for a chain without stiffness, in the limit as the number of links tends to infinity, the mean distance between the ends is zero. With angular momentum, Kingman's results are modified but the average distance is not given.

This problem is also interesting as an exercise in solving a non-linear differential equation, even though the system is described here in terms of Hamilton's principle and the differential equation is not stated. The use of the finite elements is similar to the approximation of the string by a chain with a finite number of links.

We assume that there is an ergodic property holding so that the mean distance between the ends over many experiments is the same as a time average following a particular realization of the motion. One of the questions that we wanted to investigate was whether the angular momentum would have an effect on the result. This was thought to be most likely since it is clearly an invariant of the motion. The energy E is also an invariant of the motion and the quantity that should be considered is the ratio L^2/E. Any scaling of the initial velocities, for example, which would leave this ratio unchanged, does not affect the results except to change the time scale and the scale of the tension variations.

The formulation of the problem and the computer program are designed to cope with three dimensional motion, although the examples calculated so far have only been for plane motion. The effect of tangling of the string cannot be catered for, i.e. it is not possible to put in the constraint that the string cannot cross itself.

FORMULATION

The string is perfectly flexible, uniform and inextensible and the mass and length are equal to unity. Let $\underset{\sim}{X}(s,t)$ be the position of a point in the string at a curvilinear distance s from one end at time t. The kinetic energy is

$$E = \int_0^1 \frac{1}{2}\underset{\sim}{X}_t^{\,2}\,ds .$$

There is no potential energy. Since the string is inextensible, there is the constraint

$$\underset{\sim}{X}_s^{\,2} = 1 .$$

From Hamilton's principle, the integral

$$\int_{t_1}^{t_2} L\,dt = \int_{t_1}^{t_2}\int_0^1 \left[\frac{1}{2}\underset{\sim}{X}_t^{\,2} - \frac{1}{2}T\,(\underset{\sim}{X}_s^{\,2} - 1)\right] ds\,dt$$

is stationary. Here, T is the Lagrange multiplier associated with the constraint but is also the tension in the string. With $\underset{\sim}{X}$ and T as generalized coordinates, the Lagrangian is degenerate in that the time derivative of T does not appear. The corresponding Hamiltonian is similarly degenerate.

NUMERICAL APPROXIMATION

We use a finite element approximation where $\underset{\sim}{X}$ is piecewise linear and T piecewise constant with nodal points equally spaced:

$$\begin{aligned} \underset{\sim}{X}(s,t) &= \sum_{i=0}^{n} \underset{\sim}{X}_i(t)\phi_i(s) , \\ T(s,t) &= \sum_{i=0}^{n-1} T_i(t)\psi_i(s) , \end{aligned} \qquad (1)$$

where

$$\phi_i(s) = \begin{cases} 0 & : \ , |s - s_i| > h \\ 1 - \dfrac{s-s_i}{h} , & s_i \le s \le s_{i+1} \\ 1 - \dfrac{s_i-s}{h} , & s_{i-1} \le s \le s_i \end{cases} \qquad (2)$$

$$\psi_i(s) = \begin{cases} 0 & , s < s_i ,\ s \ge s_{i+1} \\ 1 & , s_i \le s < s_{i+1} \end{cases}$$

$$h = 1/n .$$

The Lagrangian becomes

$$\begin{aligned} L = &\frac{1}{2} \sum_{i,j} \dot{\underset{\sim}{X}}_i \cdot \dot{\underset{\sim}{X}}_j \int_0^1 \phi_i(s)\phi_j(s)ds \\ &- \frac{1}{2} \sum_{i,j,k} T_i \underset{\sim}{X}_j \cdot \underset{\sim}{X}_k \int_0^1 \psi_i \phi_j' \phi_k' ds \\ &+ \frac{1}{2} \sum_i T_i \int_0^1 \psi_i ds , \end{aligned}$$

and after calculation of the integrals

$$L = \frac{1}{2}\sum_{i,j=0}^{n} a_{ij}\dot{\underset{\sim}{X}}_i\dot{\underset{\sim}{X}}_j - \frac{1}{2h}\sum_{i=0}^{n-1} T_i\left[\underset{\sim}{X}_i^2 - 2\underset{\sim}{X}_i \cdot \underset{\sim}{X}_{i+1} + \underset{\sim}{X}_{i+1}^2 - h^2\right] , \qquad (3)$$

where

$$|a_{ij}| = \frac{h}{6}\begin{bmatrix} 2 & 1 & & & & & \\ 1 & 4 & 1 & & & & \\ & 1 & 4 & 1 & & & \\ & & \cdot & \cdot & \cdot & & \\ & & & \cdot & \cdot & \cdot & \\ & & & & \cdot & \cdot & \cdot \\ & & & & 1 & 4 & 1 \\ & & & & & 1 & 2 \end{bmatrix} \tag{4}$$

The Euler-Lagrange equations derived from this Lagrangian are

$$\dot{\underset{\sim}{X}}_i = \underset{\sim}{V}_i \ , \tag{5}$$

$$\sum_{j=0}^{n} a_{ij}\underset{\sim}{V}_j = \begin{cases} \frac{1}{h}\left[T_i(\underset{\sim}{X}_j - \underset{\sim}{X}_{i+1}) - T_{i-1}(\underset{\sim}{X}_{i-1} - \underset{\sim}{X}_i)\right] & i = 1,\ldots,n-1 \\ \frac{1}{h}\ T_i(\underset{\sim}{X}_i - \underset{\sim}{X}_{i+1}) & i = 0 \\ -\frac{1}{h}\ T_{i-1}(\underset{\sim}{X}_{i-1} - \underset{\sim}{X}_i) & i = n \end{cases} \tag{6}$$

$$(\underset{\sim}{X}_{i+1} - \underset{\sim}{X}_i)^2 = h^2 \qquad i = 0,1,\ldots,n-1 \tag{7}$$

The ordinary differential equations (5), (6) are replaced by difference equations in the following way using an implicit scheme. Let $\underset{\sim}{X}_{ij}$, $\underset{\sim}{V}_{ij}$ be the values of $\underset{\sim}{X}_i$, $\underset{\sim}{V}_i$ respectively at $t = kj$, where k is the length of the time step. Equation (5) is approximated by

$$\underset{\sim}{X}_{ij+1} = \underset{\sim}{X}_{ij} + \frac{1}{2}k(\underset{\sim}{V}_{ij+1} + \underset{\sim}{V}_{ij}) \ . \tag{8}$$

Let T_{ij} be the mean value of the tension T_i during the j-th time step. Equation (6) is then approximated by

$$\begin{aligned} \underset{\sim}{V}_{ij+1} + 4\underset{\sim}{V}_{ij+1} + \underset{\sim}{V}_{i-1\ j+1} &= \underset{\sim}{V}_{ij+1} + 4\underset{\sim}{V}_{ij} + \underset{\sim}{V}_{i-1\ j} \\ &+ \frac{3k}{2h^2}\Big[T_{ij+1}(\underset{\sim}{X}_{i+1\ j+1} + \underset{\sim}{X}_{i+1\ j} - \underset{\sim}{X}_{i\ j+1} - \underset{\sim}{X}_{ij}) \\ &- T_{i-1\ j+1}(\underset{\sim}{X}_{ij+1} + \underset{\sim}{X}_{ij} - \underset{\sim}{X}_{i-1\ j+1} - \underset{\sim}{X}_{i-1\ j})\Big] \end{aligned} \tag{9}$$

for $i = 1,\ldots,n-1$, and similarly for $i = 0$, n.

At each time step $\underset{\sim}{V}_{ij+1}$, $\underset{\sim}{X}_{ij+1}$ are calculated using estimate values for T_{ij+1}. These values are then adjusted using a Newton-Raphson method until the equations of constraint (7) are satisfied.

IMPLEMENTATION OF THE NEWTON-RAPHSON METHOD

Let R_ℓ be the residual in the constraint equation (7) for the ℓ-th interval of the string,

i.e. $$f_\ell \equiv \left[(\underset{\sim}{X}_{\ell+1} - \underset{\sim}{X}_\ell)^2 - h^2\right]/h^2 = R_\ell .$$

Then

$$\frac{\partial f_\ell}{\partial T_i} = \frac{\partial f_\ell}{\partial \underset{\sim}{X}_j} \cdot \frac{\partial \underset{\sim}{X}_j}{\partial T_i}$$

$$= 2(\underset{\sim}{X}_{\ell+1} - \underset{\sim}{X}_\ell) \cdot \left(\frac{\partial \underset{\sim}{X}_{\ell+1}}{\partial T_i} - \frac{\partial \underset{\sim}{X}_\ell}{\partial T_i}\right)/h^2 \ .$$

Then the correction to the tension is given by the equation

$$\sum_{i=0}^{n-1} \frac{\partial f_\ell}{\partial T_i} \delta T_i = - R_\ell \ .$$

The partial derivatives $\partial \underset{\sim}{X}_j / \partial T_i$ are given by

$$\partial \underset{\sim}{X}_{\ell+1\ j+1}/\partial T_\ell + 4\partial \underset{\sim}{X}_{\ell\ j+1}/\partial T_\ell + \partial \underset{\sim}{X}_{\ell-1\ j+1}/\partial T_\ell$$

$$= \frac{3\lambda^2}{2} \left(\underset{\sim}{X}_{\ell+1\ j+1} - \underset{\sim}{X}_{\ell\ j+1}\right)$$

To simplify the calculation, these equations are further approximated by lumped mass system, i.e. they are replaced by

$$6 \frac{\partial \underset{\sim}{X}_{\ell\ j+1}}{\partial T_\ell} = 3\lambda^2/2 \ (\underset{\sim}{X}_{\ell\ j+1} - \underset{\sim}{X}_{\ell-1\ j+1}) \ .$$

RESULTS

Complete results have not yet been obtained. However, the results so far show a behaviour that is different from the experimental results. Firstly, without angular momentum, the mean difference between the ends does not seem to have a lower bound above zero, which is in agreement with the results of Kingman. It should be noted though that the running average does not decrease quickly. For a maximum initial velocity of 5, which gives characteristic time of 0.2, the average over time interval from 3 to 4 was approximately 0.2 where the average over the whole time from 0 to 4 was about 0.32. The disagreement with the experimental results could have something to do with the idealizations about the string that have been used here, e.g. that it is perfectly flexible, but it could also be due to the string not having a long enough flight in the experiments.

For the cases with angular momentum, the results were quite different, and depended strongly on the amount of angular momentum as might be expected. Two calculations were done with angular momentum with different values of the ratio of angular momentum squared to kinetic energy. In all calculations, the string was initially straight and the velocity was in a plane in which the string lay, so that the motion was plane. The two examples that were done had initial transverse velocities given by

1.
$$v = \begin{cases} 20\ s & 0 < s < 1/4 \\ 10 - 20\ s & 1/4 < s < 3/4 \\ -20(1-s) & 3/4 < s < 1 \end{cases}$$

2.
$$v = \begin{cases} 5 - 20\ s & 0 < s < 1/4 \\ 0 & 1/4 < s < 3/4 \\ 15 - 20\ s & 3/4 < s < 1 \end{cases}$$

In the first example the distance between the ends reached a mean value of 0.75 while in the second example the mean distance was 0.87, each of these values being reached in a time of approximately 1. The values of L^2/E are respectively 0.094 and 0.130.

REFERENCES:

[1] Bass, L. and Bracken, A.J., The problem of the thrown string, Nature 275 (1978) 205 - 206.

[2] Clarke, R.E., How long is a piece of string? Math. Gaz. 55 (1971) 404 - 407.

[3] Kingman, J.F.C., The thrown string, J.R. Statist. Soc. B 44 (1982) 109 - 138.

[4] Synge, J.L. The problem of the thrown string, Math. Gaz. 54 (1970) 250 - 260.

CONFERENCE REGISTRANTS

CTAC-83: Computational Techniques and Applications Conference, 1983, University of Sydney, New South Wales, Australia.

ANDERSSEN, R.S., *Commonwealth Scientific and Industrial Research Organisation, Division of Mathematics and Statistics, P.O. Box 1965, Canberra City, Australian Capital Territory 2601.*

ANDREW, A.L., *Department of Mathematics, LaTrobe University, Bundoora, Victoria 3083.*

ARCHER, R.D., *School of Mechanical Engineering, University of New South Wales, P.O. Box 1, Kensington, New South Wales 2033.*

ARENICZ, R.M., *Department of Civil Engineering, University of Wollongong, P.O. Box 1144, Wollongong, New South Wales 2500.*

ARNOLD, R.J., *Department of Applied Mathematics, University of Adelaide, Adelaide, South Australia 5001.*

ASTLEY, R.J., *Department of Mechanical Engineering, University of Canterbury, Christchurch 1, New Zealand.*

ATKINSON, J.D., *Department of Mechanical Engineering, University of Sydney, Sydney, New South Wales 2006.*

BAGSTER, D.F., *Department of Chemical Engineering, University of Sydney, Sydney, New South Wales 2006.*

BARRINGTON, F.R., *Department of Mathematics, University of Melbourne, Parkville, Victoria 3052.*

BARRY, J.M., *Lucas Heights Research Laboratories, Private Mail Bag, Sutherland, New South Wales 2232.*

BARTON, N.G., *Commonwealth Scientific and Industrial Research Organisation, Division of Mathematics and Statistics, P.O. Box 218, Lindfield, New South Wales 2070.*

BASU, A., *Department of Mechanical Engineering, University of Wollongong, P.O. Box 1144, Wollongong, New South Wales 2500.*

BEARD, D.A., *12 Shearer Crescent, Salisbury North, South Australia 5108.*

BENJAMIN, B.R., *South Australian Institute of Technology, P.O. Box 1, Ingle Farm, South Australia 5098.*

BENNETT, J.M., *Department of Computer Science, University of Sydney, Sydney, New South Wales 2006.*

BILLS, P.J., *32 Walkerville Terrace, Gilberton, South Australia 5081.*

BLACKMAN, D.R., *Department of Mechanical Engineering, Monash University, Clayton, Victoria 3168.*

BLAIR, D., *Commonwealth Scientific and Industrial Research Organisation, Division of Mineral Physics, P.O. Box 136, North Ryde, New South Wales 2113.*

BOOK, D.L., *Naval Research Laboratory, Code 4040, Washington, District Columbia 20375, United States of America.*

BOURKE, W., *Australian Numerical Meteorological Research Centre, GPO Box 5889AA, Melbourne, Victoria 3001.*

BRADY, B.H.G., *Commonwealth Scientific and Industrial Research Organisation, Division of Geomechanics, P.O. Box 54, Mount Waverly, Victoria 3149.*

BREBBIA, C.A., *Department of Civil Engineering, University of Southampton, Southampton SO9 5NH, United Kingdom.*

BUSH, M.B., *Department of Mechanical Engineering, University of Sydney, Sydney, New South Wales 2006.*

CAMPBELL, J., *Computer Engineering Applications, 56 Berry Street, North Sydney, New South Wales 2060.*

CARTER, J.P., *Department of Civil Engineering, University of Sydney, Sydney, New South Wales 2006.*

CENEK, P.D., *3 Levy Street, Mount Victoria, Wellington 1, New Zealand.*

CHANDLER, G.A., *Department of Mathematics, University of Queensland, St. Lucia, Queensland 4067.*

CHOW, S.S., *Department of Mathematics, Australian National University, GPO Box 4, Canberra, Australian Capital Territory 2600.*

CLEMENTS, D.L., *Department of Applied Mathematics, University of Adelaide, Adelaide, South Australia 5001.*

COULTHARD, M.A., *Commonwealth Scientific and Industrial Research Organisation, Division of Geomechanics, P.O. Box 54, Mount Waverly, Victoria 3149.*

CRAM, L.E., *Commonwealth Scientific and Industrial Research Organisation, Division of Applied Physics, P.O. Box 218, Lindfield, New South Wales 2070.*

CRAVEN, B.D., *Department of Mathematics, University of Melbourne, Parkville, Victoria 3052.*

CRAWFORD, F., *Australian Atomic Energy Commission, Private Mail Bag, Sutherland, New South Wales 2232.*

CURRIE, A., *Compumod Pty. Ltd., GPO Box 4853, Sydney, New South Wales 2001.*

DAVIDSON, M., *Commonwealth Scientific and Industrial Research Organisation, Division of Mineral Physics, Lucas Heights Research Laboratories, Private Mail Bag, Sutherland, New South Wales 2232.*

DAVIS, R.W., *National Bureau of Standards, Washington, District Columbia 20234, United States of America.*

de HOOG, F.R., *Commonwealth Scientific and Industrial Research Organisation, Division of Mathematics and Statistics, P.O. Box 1965, Canberra City, Australian Capital Territory 2601.*

deVAHL DAVIS, G., *School of Mechanical Engineering, University of New South Wales, P.O. Box 1, Kensington, New South Wales 2033.*

DEWAR, R.L., *Department of Theoretical Physics, Australian National University, GPO Box 4, Canberra, Australian Capital Territory 2600.*

DOHERTY, G., *Department of Mathematics, University of Wollongong, P.O. Box 1144, Wollongong, New South Wales 2500.*

DRUMM, M.J. *Department of Mathematics, Capricornia Institute of Advanced Education, Rockhampton, Queensland 4700.*

DUDLEY, D., *Computer Engineering Applications, 56 Berry Street, North Sydney, New South Wales 2060.*

EASTON, A.K., *Department of Mathematics, Swinburne Institute of Technology, Hawthorn, Victoria 3122.*

ELLEM, B.A., *Commonwealth Scientific and Industrial Research Organisation, Division of Mathematics and Statistics, P.O. Box 1965, Canberra City, Australian Capital Territory 2601.*

FLEET, R.W., *Department of Mechanical Engineering, University of Sydney, Sydney, New South Wales 2006.*

FLETCHER, C.A.J., *Department of Mechanical Engineering, University of Sydney, Sydney, New South Wales 2006.*

FORD, D., *Aero Research Laboratories, GPO Box 4331, Melbourne, Victoria 3001.*

FULFORD, G.R., *Department of Mathematics, University of Wollongong, P.O. Box 1144, Wollongong, New South Wales 2500.*

GARDNER, H.J., *Department of Theoretical Physics, Australian National University, GPO Box 4, Canberra, Australian Capital Territory 2600.*

GOLLEY, B.W., *Department of Civil Engineering, Royal Military College, Duntroon, Australian Capital Territory 2600.*

GRAHAM, I.G., *Department of Mathematics, University of Melbourne, Parkville, Victoria 3052.*

GRUNDY, I.H., *45 Brougham Place, North Adelaide, South Australia 5006.*

GUSTAFSON, S.A., *Centre for Mathematical Analysis, Australian National University, GPO Box 4, Australian Capital Territory 2601.*

HART, P.M., *Department of Electrical Engineering, Monash University, Clayton, Victoria 3168.*

HAUSLER, E.P., *Department of Mathematics, Swinburne Institute of Technology, Hawthorn, Victoria 3122.*

HELLIER, A.K., *School of Metallurgy, University of New South Wales, P.O. Box 1, Kensington, New South Wales 2033.*

HOLLAND, P.G., *Lucas Heights Research Laboratories, Private Mail Bag, Sutherland, New South Wales 2232.*

HOLT, M., *Department of Mechanical Engineering, University of California, Berkeley, California 94720, United States of America.*

HOOPER, A.P., *Department of Mathematics, University of Melbourne, Parkville, Victoria 3052.*

HUTCHINSON, M.F., *Commonwealth Scientific and Industrial Research Organisation, Division of Water and Land Research, P.O. Box 1666, Canberra, Australian Capital Territory 2601.*

JACKETT, D.R., *Commonwealth Scientific and Industrial Research Organisation, Division of Mathematics and Statistics, P.O. Box 1965, Canberra City, Australian Capital Territory 2601.*

JENKINSON, J.H., *Department of Mathematics, Australian National University, GPO Box 4, Canberra, Australian Capital Territory 2601.*

JEPPS, G., *Defence Research Centre, Box 2151, G.P.O., Adelaide, South Australia 5001.*

JOE, S., *School of Mathematics, University of New South Wales, P.O. Box 1, Kensington, New South Wales 2033.*

JUNG, K.T., *Kathleen Lumley College, 51 Finniss Street, North Adelaide, South Australia 5006.*

KABAILA, A., *School of Civil Engineering, University of New South Wales, P.O. Box 1, Kensington, New South Wales 2033.*

KACHOYAN, B.J., *Department of Mathematics, University of New South Wales, P.O. Box 1, Kensington, New South Wales 2033.*

KARASHIMA, K., *National Space Agency of Japan, Hamamatsu-cho, Minato-ku, Tokyo, Japan.*

KARIHALOO, B.L., *Department of Civil Engineering, University of Newcastle, Newcastle, New South Wales 2308.*

KELLY, D.W., *School of Mechanical Engineering, University of New South Wales, P.O. Box 1, Kensington, New South Wales 2033.*

KEWLEY, D.J., *Underwater Detection Group, Defence Research Centre, P.O. Box 2151, Adelaide, South Australia 5001.*

KITAMA, S., *National Space Agency of Japan, Hamamatsu-cho, Minato-ku, Tokyo, Japan.*

KNIGHT, J.H., *Commonwealth Scientific and Industrial Research Organisation, Division of Mathematics and Statistics, P.O. Box 1965, Canberra City, Australian Capital Territory 2601.*

KOHOUTEK, R., *Department of Civil Engineering, University of Melbourne, Parkville, Victoria 3052.*

KWOK, S.K., *12/153 Fairway, Nedlands, Western Australia 6009.*

LALOUSIS, P., *Commonwealth Scientific and Industrial Research Organisation, Division of Applied Physics, P.O. Box 218, Lindfield, New South Wales 2070.*

LAYTON, N., *Electricity Commission of New South Wales, P.O.. Box 5257, Sydney, New South Wales 2001.*

LEE, H., *Commonwealth Scientific and Industrial Research Organisation, Division of Applied Physics, P.O. Box 218, Lindfield, New South Wales 2070.*

LEONARD, B.P., *Department of Engineering, City University (Staten Island), New York, New York State, United States of America.*

MACK, N., *Department of Mechanical Engineering, University of Sydney, Sydney, New South Wales 2006.*

McCOWAN, A.D., *Department of Mechanical Engineering, Monash University, Clayton, Victoria 3168.*

McGIRR, M.B., *School of Metallurgy, University of New South Wales, P.O. Box 1, Kensington, New South Wales 2033.*

McINTOSH, P.C., *Department of Mathematics, Monash University, Clayton, Victoria 3168.*

McLEAN, W., *Department of Mathematics, Australian National University, GPO Box 4, Canberra, Australian Capital Territory 2600.*

MANN, K.J., *Department of Mathematics, Chisholm Institute of Technology, Caulfield East, Victoria 3145.*

MAY, R.L., *Department of Mathematics, Royal Melbourne Institute of Technology, LaTrobe Street, Melbourne, Victoria 3000.*

MILLER, A.D., *Centre for Mathematical Analysis, Australian National University, P.O. Box 4, Canberra, Australian Capital Territory 2600.*

MISKELLY, P.D., *Commonwealth Scientific and Industrial Research Organisation, Division of Applied Physics, Building 3, Australian Atomic Energy Commission, Private Mail Bag, Sutherland, New South Wales 2232.*

MITCHELL, A.R., *Department of Mathematics, University of Dundee, Dundee DD1 4HN, Scotland, United Kingdom.*

MITCHELL, W.M., *Department of Applied Mathematics, University of Adelaide, Adelaide, South Australia 5001.*

MOHR, G.A., *School of Engineering, University of Auckland, Auckland, New Zealand.*

MORROW, R., *Commonwealth Scientific and Industrial Research Organisation, Division of Applied Physics, P.O. Box 218, Lindfield, New South Wales 2070.*

MUNRO, S.D., *Department of Chemical Engineering, University of New South Wales, P.O. Box 1, Kensington, New South Wales 2033.*

NORTON, A.H., *7 Dulwich Road, Chatswood, New South Wales 2067.*

NORWOOD, C., *Department of Mechanical Engineering, Footscray Institute of Technology, Ballarat Road, Footscray, Victoria 3011.*

NOYE, B.J., *Department of Applied Mathematics, University of Adelaide, Adelaide, South Australia 5001.*

O'CONNOR, A.J., *School of Science, Griffiths University, Nathan, Queensland 4111.*

O'NEILL, M.J., *Department of Mathematics, Mitchell College of Advanced Education, Bathurst, New South Wales 2795.*

ORLANDI, P., *Aero Institute, Rome University, Via Eudossiana, 16 Roma, Italy.*

PARKER, R.D., *Department of Theoretical Physics, Australian National University, P.O. Box 4, Canberra, Australian Capital Territory 2600.*

PEACOCK, T.E., *Department of Chemistry, University of Queensland, St. Lucia, Queensland 4067.*

PETROLITO, J., *Department of Civil Engineering, Royal Military College, Duntroon, Australian Capital Territory 2600.*

PLATFOOT, R., *Electricity Commission of New South Wales, P.O. Box 5257, Sydney, New South Wales 2001.*

RAICHE, A.P., *Commonwealth Scientific and Industrial Research Organisation, Division of Mineral Physics, P.O. Box 136, North Ryde, New South Wales 2113.*

REIZES, J., *School of Mechanical Engineering, University of New South Wales, P.O. Box 1, Kensington, New South Wales 2033.*

SIN AILAM, L., *Underwater Detection Group, Defence Research Centre, P.O. Box 2151, Adelaide, South Australia 5001.*

SLOAN, I.H., *School of Mathematics, University of New South Wales, P.O. Box 1, Kensington, New South Wales 2033.*

SMALL, J.C., *Department of Civil Engineering, University of Sydney, Sydney, New South Wales 2006.*

SMART, M.G., *Solarch, GSBE, University of New South Wales, P.O. Box 1, Kensington, New South Wales 2033.*

SOH, W.K., *Department of Mechanical Engineering, University of Wollongong, P.O. Box 1144, Wollongong, New South Wales 2500.*

SRINIVAS, K., *Department of Mechanical Engineering, University of Sydney, Sydney, New South Wales 2006.*

STEVENS, M.W., *23 Jamaica Avenue, Fulham Gardens, South Australia 5024.*

SUGANO, Y., *Department of Mechanical Engineering, University of Osaka, Prefecture, Sakai 591, Osaka, Japan.*

SUMMERFIELD, W., *Department of Mathematics, University of Newcastle, Newcastle, New South Wales 2308.*

TAGGART, I.J., *Department of Chemical Engineering, University of New South Wales, P.O. Box 1, Kensington, New South Wales 2033.*

TAIB, B., *Department of Mathematics, University of Wollongong, P.O. Box 1144, Wollongong, New South Wales 2500.*

TARLOWSKI, C.J., *Commonwealth Scientific and Industrial Research Organisation, Division of Mineral Physics, P.O. Box 136, North Ryde, New South Wales 2113.*

THALASSOUDIS, K., *61 Overland Road, Croydon Park, South Australia 5008.*

THOMPSON, W., *Department of Mechanical Engineering, Swinburne Institute of Technology, Hawthorn, Victoria 3122.*

TONG, G.D., *Computer Applications Centre, South Australian Institute of Technology, P.O. Box 1, Ingle Farm, South Australia 5098.*

TRAVIS, L.P., *Department of Mechanical Engineering, Swinburne Institute of Technology, Hawthorn, Victoria 3122.*

VERMA, S., *Commonwealth Scientific and Industrial Research Organisation, Division of Mineral Physics, P.O. Box 136, North Ryde, New South Wales 2113.*

WALSH, B., *Department of Mechanical Engineering, University of Sydney, Sydney, New South Wales 2006.*

WATTS, A.M., *Department of Mathematics, University of Queensland, St. Lucia, Queensland 4067.*

WESSON, V.C., *Phillip Institute of Technology, Plenty Road, Bundoora, Victoria 3083.*

WHATHAM, J.F., *90 Lansdowne Parade, Oatley, New South Wales 2223. (Australian Atomic Energy Commission).*

WILLIAMS, I.S., *Directorate of Operational Analysis - Army, Department of Defence, Russell Offices, Canberra, Australian Capital Territory 2600.*

WOLFSHTEIN, M., *Department of Aero Engineering, The Technion, Haifa, Israel*

WOODHAM, G.R.G., *Department of Chemical Engineering, University of Sydney, Sydney, New South Wales 2006.*

WOODWARD, R.L., *Materials Research Laboratory, P.O. Box 58, Ascot Vale, Victoria 3032.*

ZAPLETAL, E., *Department of Mechanical Engineering, University of Sydney, Sydney, New South Wales 2006.*

LATE REGISTRANT:

FRANGOPOL, D.M., *Department of Civil Engineering, University of Colorado at Boulder, Boulder, Colorado, U.S.A.*